Lecture Notes in Physics

The Lecture Notes in Physics

The series Lecture Notes in Physics (LNP), founded in 1969, reports new developments in physics research and teaching – quickly and informally, but with a high quality and the explicit aim to summarize and communicate current knowledge in an accessible way. Books published in this series are conceived as bridging material between advanced graduate textbooks and the forefront of research and to serve three purposes:

- to be a compact and modern up-to-date source of reference on a well-defined topic
- to serve as an accessible introduction to the field to postgraduate students and nonspecialist researchers from related areas
- to be a source of advanced teaching material for specialized seminars, courses and schools

Both monographs and multi-author volumes will be considered for publication. Edited volumes should, however, consist of a very limited number of contributions only. Proceedings will not be considered for LNP.

Volumes published in LNP are disseminated both in print and in electronic formats, the electronic archive being available at springerlink.com. The series content is indexed, abstracted and referenced by many abstracting and information services, bibliographic networks, subscription agencies, library networks, and consortia.

Proposals should be sent to a member of the Editorial Board, or directly to the managing editor at Springer:

Christian Caron
Springer Heidelberg
Physics Editorial Department I
Tiergartenstrasse 17
69121 Heidelberg / Germany
christian.caron@springer.com

M. Gasperini
J. Maharana (Eds.)

String Theory and Fundamental Interactions

Gabriele Veneziano and Theoretical Physics:
Historical and Contemporary Perspectives

Springer

Editors

Maurizio Gasperini
Università di Bari
Dipartimento di Fisica
Via G. Amendola,173
70126 Bari, Italy
gasperini@ba.infn.it

Jnan Maharana
Institute of Physics
Sachivalaya Marg
Bhubaneswar - 751 005
Orissa, India
maharana@iopb.res.in

M. Gasperini and J. Maharana (Eds.), *String Theory and Fundamental Interactions*, Lect. Notes Phys. 737 (Springer, Berlin Heidelberg 2008), DOI 10.1007/978-3-540-74233-3

Library of Congress Control Number: 2007934340

ISSN 0075-8450
ISBN 978-3-540-74232-6 Springer Berlin Heidelberg New York

Springer is a part of Springer Science+Business Media
springer.com

Typesetting: by the authors and Integra using a Springer LaTeX macro package
Cover design: eStudio Calamar S.L., F. Steinen-Broo, Pau/Girona, Spain

Printed on acid-free paper SPIN: 12070305 5 4 3 2 1 0

To Gabriele
from his friends, with best wishes

Preface

This book has been prepared to celebrate the 65th birthday of Gabriele Veneziano and his retirement from CERN in September 2007. This retirement certainly will not mark the end of his extraordinary scientific career (in particular, he will remain on the permanent staff of the Collège de France in Paris), but we believe that this important step deserves a special celebration, and an appropriate recognition of his monumental contribution to physics.

Our initial idea of preparing a volume of *Selected papers of Professor Gabriele Veneziano*, possibly with some added commentary, was dismissed when we realized that this format of book, very popular in former times, has become redundant today because of the full "digitalization" of all important physical journals, and their availability online in the electronic archives. We have thus preferred an alternative (and unconventional, but probably more effective) form of celebrating Gabriele's birthday: a collection of new papers written by his main collaborators and friends on the various aspects of theoretical physics that have been the object of his research work, during his long and fruitful career.

Selecting a reasonable number of invited contributors and contributed topics has proved to be a very difficult task, given the impressive number of distinguished collaborators (see the full list in the first chapter of this book), and the exceptionally wide spectrum of research interests. After a careful analysis of four decades of work, we have finally decided to invite only a few representative contributions, trying to provide a survey of most of the many faces of Gabriele's activity, and to avoid, at the same time, too many overlaps and too large gaps. We have been assisted in this process by Gabriele himself, but we are responsible for any important omission, of course. We hope, however, that the reader will appreciate the time (and space) limitations of this book, since making a complete and detailed survey of all of Gabriele's activities is surely impossible.

The contributors have been invited to prepare high-level (but not too much specialized) lectures on the assigned themes, with some introductory part and, possibly, some historical perspective concerning their work with Gabriele.

We are very grateful to our colleagues and friends for having accepted our invitation, and for their excellent scientific and pedagogic work:

Daniele Amati
Adi Armoni
Ram Brustein
Alessandra Buonanno
Marcello Ciafaloni
Thibault Damour
Paolo Di Vecchia
Sergio Ferrara
Alberto Giovannini
Massimo Giovannini
Kenichi Konishi
Giuseppe Marchesini
Krzysztof Meissner
Roberto Petronzio
Eliezer Rabinovici
Giancarlo Rossi
Hector Rubinstein
Adam Schwimmer
Mikhail Shifman
Graham Shore
Tomasz Taylor
Luca Trentadue
Henry Tye
Carlo Ungarelli
Gregory Vilkovisky
Miguel Virasoro

Should this book have any form of success and appreciation, the merit will rest on their dedicated and enthusiastic work, and on the many hours of their valuable time spent on the materialization of this project. We would also like to thank Christian Caron, Senior Editor of Physics at Springer, for his kind encouragement, advice, and for many important suggestions.

This book is divided into various parts. The introductory part is fully devoted to Gabriele Veneziano, and contains a short biography summarizing his main successes and achievements (to date), a full updated list of his collaborators and of his publications, and a short interview concerning his personal point of view about the present and future of fundamental physics. We have also included the Latex version of an old, unpublished (and handwritten) note, dating back to 1973, that Gabriele discovered after a long search in his office at CERN. Apart from the genuine historical value of such a document (see, for instance, the comments added by the author for the edition of this book), parts of the original draft are still of interest, and potentially relevant for modern applications.

The rest of the book is divided into the following seven parts:

Part 1 – Dual resonance models and string theory
Part 2 – Perturbative QCD
Part 3 – Non-perturbative QCD
Part 4 – Supersymmetric gauge theories
Part 5 – String dualities and symmetries
Part 6 – String/quantum gravity, black holes, and entropy
Part 7 – String cosmology

Each of these parts contains from a minimum of two to a maximum of five articles (organized in historical or pedagogical order), illustrating different aspects of these fields with special emphasis on the contribution of Gabriele and of his collaborators.

The result is a rather unconventional, "unique" book where old and new scientific results are mixed with personal memories and feelings of the authors, spanning over four decades of research on fundamental interactions and an impressive spectrum of interests, ranging from subnuclear physics to cosmology. We think that it should be easy for the specialized reader to find out his/her preferred topic, and to jump directly to his/her field of interest. We also hope, however, that he/she will be tempted to deviate from this preferred path for enjoying the exploration of other branches of theoretical physics, and learning about their historical development, following the excellent introductions written by leading experts in the field.

To conclude this short introduction we would like to present our warmest thanks to Gabriele Veneziano, teacher and friend, for so many years of enjoyable and rewarding collaboration. We wish him, also on behalf of the other contributors to this volume and of all his friends, a happy 65th Birthday, and many future years full of exciting research projects and outstanding achievements.

Bari and Bhubaneswar,
March 2007

Maurizio Gasperini
Jnan Maharana

Contents

Part VI String dualities and symmetries

Part I

Introduction

Gabriele Veneziano: A Concise Scientific Biography and an Interview

M. Gasperini[1] and J. Maharana[2]

[1] Dipartimento di Fisica, Università di Bari, Via G. Amendola 173, 70126 Bari, Italy and Istituto Nazionale di Fisica Nucleare, Sezione di Bari, Bari, Italy
gasperini@ba.infn.it
[2] Institute of Physics, Bhubaneswar University, 751005 India
maharana@iopb.res.in

Abstract. The aim of these notes is to present a broad brush profile of the scientific activity of Gabriele Veneziano, whose wide spectrum of interests and variety of contributions to fundamental theoretical physics is also reflected by the articles of his collaborators and friends in this book. We thank Gabriele for his kind help in preparing these notes, and for disclosing to us some aspects of his life that we were not aware of. The responsibility of any omission and imprecision will rest on the authors, of course, and we apologize in advance for the (unavoidable) incompleteness of Sect. 1, warning the reader that a full survey of all of Gabriele's activities is outside the scope of this introduction. Finally, we thank Gabriele for his patience in answering our questions that made possible the interview reported in Sect. 3 where, starting from the evocation of his past experience, he illustrates his personal point of view on the present status of fundamental physics, and his expectations for the future.

1 Biographical Notes

Gabriele Veneziano was born on September 7, 1942 in Florence (Italy). After completing his high-school studies (at the Liceo Scientifico "Leonardo da Vinci," Florence) he entered the University of Florence in 1960, where he started studying physics. He took his degrees (Laurea in Fisica) in 1965 defending a thesis on the applications of group theory to strong interactions, under the supervision of Professor Raoul Gatto. A short paper extracted from his thesis became his first scientific publication [1] (here, and in what follows, the quoted numbers refer to the list of publications of Gabriele Veneziano reported at the end of this chapter).

After graduating he won a scholarship of Angelo della Riccia to carry out research in the group directed by Raoul Gatto, who had gathered in Florence a number of brilliant young theorists (like Guido Altarelli, Franco Buccella, Giovanni Gallavotti, Luciano Maiani, and Giuliano Preparata, to

M. Gasperini and J. Maharana: *Gabriele Veneziano: A Concise Scientific Biography and an Interview*, Lect. Notes Phys. **737**, 3–27 (2008)
DOI 10.1007/978-3-540-74233-6_1

mention a few). During that period he wrote a paper on saturation of current algebra sum rules [5] that attracted the attention of Professor Sergio Fubini. Meanwhile (after a conversation with Professor Giulio Racah) he had decided to continue his studies toward a PhD, choosing to apply to the Weizmann Institute of Science in Rehovot (Israel). In July 1966 he got married to Edy Pacifici and, after their honeymoon in Venice, they moved together to Israel.

His official advisor at the Weizmann Institute was Professor Harry J. Lipkin; however, his research activity was mainly carried out under the supervision of Professor Hector Rubinstein (see the contribution of Hector Rubinstein to this volume). In Israel he quickly completed his PhD studies, getting the degree at the end of 1967 (see Fig. 1). The PhD thesis was largely based on his research with Rubinstein and on work done in collaboration with Marco Ademollo (professor in Florence and visiting Harvard at that time) and Miguel Virasoro (who had joined the Weizmann group in the spring of 1967). That work developed important ideas initiated by Sergio Fubini and collaborators on a bootstrap approach to strong interactions based on "superconvergence" and "duality" (see, e.g., [13, 18]).

At the beginning of 1968 he was offered several post-doctoral positions in the United States, and he decided to accept the invitation of MIT (Boston) to join the newly formed Center for Theoretical Physics to which Sergio Fubini and Steven Weinberg had recently moved. Before starting the MIT appointment he spent the whole summer at the TH Division of CERN, where he completed the celebrated paper "Construction of a crossing-symmetric, Regge

Fig. 1. Gabriele Veneziano (left) receiving his PhD diploma at the Weizmann Institute of Science (Rehovot, 1967)

behaved amplitude for linearly-rising trajectories" [20], in which he proposed the scattering amplitude that bears his name, and that is usually regarded as marking the birth of string theory. The model presented in his seminal paper incorporated most of the desired ingredients of an S-matrix theory of strong interactions, and it was largely quoted at the Vienna Conference on High Energy Physics, at the end of that summer.

At MIT he mainly worked with Sergio Fubini to develop generalizations of his earlier work that became known as "dual resonance models." Their work paved the way to the re-interpretation of such models as a theory of strings. In fact, some of the crucial features of string theory, such as the exponential degeneracy of the states [24, 25], the concept of "Fubini–Veneziano" vertex operator [28], and the algebraic structure underlying the Virasoro operators [31], were introduced by them in that period (see the contributions of Paolo di Vecchia, Adam Schwimmer, and Miguel Virasoro to this volume). In that period he also spent a summer at the Lawrence Berkeley Laboratory (California), where he contributed to an influential paper on the "twist" operator [26].

After the birth of his son Ariel (September 1970), and a one-term visit at the Institute for Advanced Studies in Princeton, he undertook a program invoking topological ideas in order to implement unitarity in the context of dual resonance models [30]. This led, in particular, to a model for the "Pomeron" [49], later developed by other researchers into the so-called dual parton model.

In 1972 he came back to the Weizemann Institute as a full professor. In the subsequent 4 years he also spent extended periods at CERN, pursuing the development of the topological unitarization ideas, meanwhile interpreted by Gerard 't Hooft as a $1/N$ expansion [53].

In 1976 he joined the TH Division of CERN, first as a scientific associate, then as a junior staff member (1977–1978), and, finally, as a senior staff member. Later he became Head of the TH Division (1994–1997). The beginning of this period was marked by the birth of his daughter Erika (July 1976), and by a change of direction of his research interests.

He started to work, in particular, on large N expansions in quantum chronodynamics (QCD) [62], its applications to baryon dynamics [64], and Bose–Einstein effects in jet physics [67] (see the contribution of Alberto Giovannini to this volume). Together with Daniele Amati and Roberto Petronzio, he proved the factorization theorem on collinear singularities in perturbative QCD, which forms the basis of the QCD parton model [69, 70] (see the contribution of Roberto Petronzio to this volume). This brought him naturally to devote his activity to the physics of QCD jets, writing some seminal papers with Kenichi Konishi and Akiwa Ukawa [71, 72] (the KUV jet calculus), and with Daniele Amati [74] (pre-confinement) (see the contributions of Marcello Ciafaloni and Giuseppe Marchesini to this volume).

Turning his attention to non-perturbative aspects of QCD, he tackled the $U(1)$ axial problem for a $1/N$ perspective, arriving at the celebrated (and even recently confirmed) Witten–Veneziano formula [77]. Related studies led to an estimate of the electric dipole moment of the neutron induced by a non-vanishing QCD θ-angle [78]. These results were encoded into an effective Lagrangian formalism developed with Paolo di Vecchia [79].

The effective Lagrangian formalism was later applied to supersymmetry (SUSY) Yang–Mills theories [93] and SUSY QCD [95], where the non-perturbative breakdown of non-renormalization theorems was first suggested. The superpotentials derived in those papers, in collaboration with Thomasz Taylor and Shimon Yankielowicz, are still being widely used and cited (often under some other names) in many contexts. The indications of those papers were confirmed by explicit calculations that he later performed with Giancarlo Rossi and collaborators, and that are summarized in [119] (see the contribution of Giancarlo Rossi to this volume). In that period he also pointed out the possible formulation of SUSY Yang–Mills theories in the lattice, and suggested an implementation [115] that is still being attempted.

When string theory was recognized as a promising candidate to unify gravity and gauge interactions (i.e., after the so-called Green–Schwarz revolution in 1984) he came back to the theory that he had to abandon (not without regret) when it appeared inappropriate as a theory of strong interactions. His studies (with various collaborators, including Amit Giveon, Jnan Maharana, and Eliezer Rabinovici) first concentrated on the following directions: the physical consequences of a fundamental length [111], the emergence of new field-theoretic and "stringy" symmetries [113] (see the contributions of Jnan Maharana and Eliezer Rabinovici to this volume), the possible phenomenological consequences of a light dilaton [125], and a background field approach to the study of the T-duality symmetry [133].

A more substantial activity in that period concerned the study of *gedanken* experiments on trans-Planckian string collisions, in collaboration with Daniele Amati and Marcello Ciafaloni [120]. The main purpose of such studies was the understanding of how string theory may reproduce general-relativistic results at large distances, while providing important corrections at string-size distances. The works possibly have applications to an effectively modified uncertainty principle [138] and to the problem of "information loss" in black-hole physics (see the contribution of Daniele Amati to this volume).

While working on string theory he kept alive his interest in the subject of strong interaction phenomenology, producing works on the "spin of the proton" puzzle [132] (see the contribution of Graham Shore to this volume), and on semi-inclusive hard processes [167] (see the contribution of Luca Trentadue to this volume).

Triggered by his wish to find novel applications of string theory (and new possible ways to test it), he then turned his interest toward primordial cosmology and its theoretical and observational challenges. Starting from the study of duality symmetries in cosmological backgrounds [148, 149, 151] (see the contribution of Krzysztof Meissner to this volume) he proposed, in collaboration with Maurizio Gasperini, the so-called pre-big bang scenario [161], which attracted considerable interest in the astrophysical community, stimulating the studies of new mechanisms of inflation (see the contribution of Maurizio Gasperini to this volume). The multiple implications of this scenario were the object of many subsequent studies with various collaborators (see the contributions of Alessandra Buonanno, Thibault Damour, Massimo Giovannini, and Carlo Ungarelli to this volume). Of particular relevance were the phenomenological predictions concerning the generation of magnetic seeds [175], the enhanced production of primordial gravitational waves [178], and the possible axionic origin of the cosmic microwave background (CMB) anisotropy [225], opening a unique observational window on string/Planck-scale physics.

Encouraged by the possibility of concrete experimental verifications of such a string cosmology scenario, he and Maurizio Gasperini also tackled the problem of understanding (or re-interpreting), in such a context, the big bang singularity, by applying either quantum cosmology techniques (in collaboration with Jnan Maharana [185]), or higher-order string corrections (in collaboration with Michele Maggiore [190]), or non-local effects of the quantum back reaction (in collaboration with Massimo Giovannini [235]). The study of the high-curvature, strong-coupling regime (also appropriate to brane inflation, see the contribution of Henry Tye to this volume) led him to obtain, as a by-product, unexpected results on entropy in collaboration with Ram Brustein [207, 213] (see the contribution of Ram Brustein to this volume), and unexpected connections with black-hole physics [211, 240], in collaboration with Thibault Damour. Later developments of the pre-big bang scenario also led to interesting (and testable, in principle) interpretations of the presently observed cosmic acceleration [219].

His most recent interests are mainly focused again on the $1/N$ expansion, with two different ramifications. The first concerns a new version of such expansion, capable of connecting QCD to supersymmetric theories [236], developed in collaboration with Adi Armoni and Mikhail Shifman. The obtained predictions for one-flavor QCD, in particular, have been confirmed by subsequent (phenomenological or lattice) computations (see the contribution of Adi Armoni and Mikhail Shifman to this volume). The second, developed in collaboration with Enrico Onofri and Jacek Wosiek, deals with a Hamiltonian approach to large N dynamics, which, while still limited to quantum mechanics, has already produced interesting results in different branches of mathematical physics, like combinatorics and statistical mechanics [246].

Since 2004 he holds the prestigious Chair of Elementary Particles, Gravitation and Cosmology at the Collège de France, in Paris (see Fig. 2).

We give below a schematic summary of his professional career, his administrative appointments at CERN, his positions and associations, and his prizes and honors.

Fig. 2. Gabriele Veneziano giving the Inaugural Lecture at the Collège de France. Paris, February 17, 2005 (Photo Suzy Vascotto)

1.1 Professional Career

- Research Associate at MIT, Cambridge (USA), 1968–1969
- Visiting Assistant Professor at MIT, 1969–1970
- Visiting Associate Professor at MIT, 1970–1972
- Full Professor at Weizmann Institute of Science, Rehovot (Israel), 1971–1975
- Amos-de Shalit Professor of Physics at Weizmann Institute of Science, 1975–1977
- Junior Staff Member at CERN, TH Division, Geneva, 1977–1978
- Senior Staff Member, CERN, Geneva, 1978–2207
- Head of Theory Division, CERN, Geneva, 1994–1997
- Professor at Collège de France, Paris, since 2004

1.2 Administrative Appointments at CERN

- Member of the SPS (Super Proton Synchrotron) Committee, 1983–1986
- CERN representative to Plenary ECFA (European Committee for Future Accelerators), 1987–1990
- Chairman of the Academic Training Committee, 1990–1994
- Division Leader of the Theory Division, 1994–1997

- Member of the Scientific Information Policy Board, 1997–2000
- Member of the Archives Committee, 2001–2004
- Chairman of the Pauli Committee, 2003–2007

1.3 Positions and Associations

- Recipient of a Chaire Condorcet at LPTENS (Laboratoire de Physique Théorique de l'Ecole Normale Superieure), Paris, 1994
- Co-director (with Gerard 't Hooft and Antonino Zichichi) of the International School on Subnuclear Physics in Erice, Sicily, 1996–2001
- Recipient of a Chaire Blaise Pascal at LPT (Laboratoire de Physique Théorique), Université Paris Sud, Orsay, and IHES (Institut des Hautes Etudes Scientifiques), Bures-sur-Yvette (France), 2000–2002
- Academic Staff Member at Kavli Institute of Theoretical Physics, University of California, Santa Barbara, 2003
- Chairman of the Advisory Committee of the Galileo Galilei Institute in Arcetri (Italy), since 2005
- Member of Accademia delle Scienze di Torino (Italy), since 1994
- Member of Accademia Nazionale dei Lincei, Roma, since 1996
- Member of Académie des Sciences of the Institut de France, Paris, since 2002

1.4 Prizes and Honors

- I. Ya. Pomeranchuk Prize, ITEP, Moscow (May 1999).

  Motivation: "*For his outstanding contributions to quantum field theory and theory of strings.*"

- Gold Medal of the Italian Republic (Diploma di prima classe riservati ai Benemeriti della Scienza e della Cultura), Rome (June 2000).
- Dannie Heineman Prize of the American Physical Society (May 2004).

  Motivation: "*For his pioneering discoveries in dual resonance models which, partly through his own efforts, have developed into string theory and a basis for the quantum theory of gravity.*"

- Enrico Fermi Prize of the Italian Physical Society (September 2005).
- Einstein Medal of the Albert Einstein Gesellschaft, Berne (June 2006).

  Motivation: "*The laureate has made significant contributions to the understanding of string theory.*"

- Commendatore dell'Ordine al Merito della Repubblica Italiana (February 2007).
- Oskar Klein Medal of the Swedish Royal Academy of Sciences, Stockholm (June 2007).

2 List of Collaborators of Gabriele Veneziano (Updated to 2006)

Here we report, in alphabetical order (and to the best of our knowledge), all authors who have published a paper in collaboration with Gabriele Veneziano. Their number is impressive (for a theoretical physicist), and we apologize in advance for any possible omission.

M. Ademollo	D. Amati	A. Armoni
R. Barbieri	A. Bassetto	M. Bishari
V. Bozza	V. Branchina	R. Brustein
F. Buccella	A. Buonanno	L. Caneschi
M. Ciafaloni	R. Crewther	G. Curci
T. Damour	A. C. Davis	V. De Alfaro
D. De Florian	E. Del Giudice	C. DeTar
A. Di Giacomo	P. Di Vecchia	M. J. Duff
R. Durrer	S. Elitzur	M. Fabbrichesi
K. Fabricius	S. Ferrara	V. Ferrari
S. Foffa	D. Freedman	J. Freeman
S. Fubini	G. Furlan	E. Gabathuler
M. Gasperini	R. Gatto	A. Giovannini
M. Giovannini	L. Giusti	A. Giveon
D. Gordon	A. Ghosh	D. Graudenz
M. Grazzini	M. B. Green	N. S. Han
R. Iengo	C. E. Jones	F. Karsch
E. Kohlprath	K. Konishi	T. Kubota
G. Longhi	F. E. Low	R. Madden
M. Maggiore	N. Magnoli	J. Maharana
G. Marchesini	A. Masiero	K. A. Meissner
A. Melchiorri	Y. Meurice	V. F. Mukhanov
S. Narison	F. Nicodemi	S. Okubo
L. B. Okun	E. Onofri	P. Pavlopoulos
P. Pendenza	R. Petronzio	R. Pettorino
F. Piazza	G. Pollifrone	E. Predazzi
E. Rabinovoci	R. Ricci	M. Roncadelli
C. Rosenzweig	G. C. Rossi	H. Rubinstein
M. Sakellariadou	N. Sanchez	M. M. Schaap
A. Schwimmer	L. Sertorio	M. Shifman
G. Shore	T. Taylor	M. Testa
L. Trentadue	H. Tye	A. Ukawa
C. Ungarelli	G. Vilkowisky	F. Vernizzi
M. Virasoro	S. Weinberg	E. Witten
J. Wosiek	S. Yankielowicz	J. E. Young
Y. Zarmi		

3 An Interview with Gabriele Veneziano

MG & JM: Hi Gabriele, and thank you very much for sparing us your valuable time, accepting to answer our questions. We had the privilege of preparing this collection of papers written by your close collaborators, and we would like to ask you a few questions concerning your personal experience with physics during over four decades of successful work. We are interested in your feelings and perspectives about the present status of the research activity in fundamental physics, and your hopes and expectations for the future. But let us start with the past. When (and why) did you decide to devote your professional activity to physics?

GV: *In senior high school in Florence I had a very good teacher of maths and physics, Tebaldo Liverani. He clearly loved those subjects (more maths than physics, he once admitted) and enjoyed teaching them. Probably under his influence, in 1960, myself and two other students in my class decided to enroll at the local university for a degree in either maths or physics. During the summer we had long debates on what to choose, and, eventually, we all opted for physics. I believe that none of us has regretted the choice. This little story tells us how important good-quality teaching is, and not only at the academic level, quite the contrary.*

MG & JM: Do you remember any professor who played a crucial role in influencing your career, both during your studies and at the beginning of your research activity? What should be, in your opinion, the main objectives of undergraduate and graduate courses in physics?

GV: *Besides the high school teacher I have just mentioned, I remember some very good courses at the university, in particular by Professors Mand and Toraldo di Francia. Then, while I was entering my third year, Professor Raoul Gatto arrived to Florence, together with a group of brilliant young theorists, mainly from Rome. His teaching and his presence made me turn in the direction of theoretical particle physics. Without him around I would have probably yielded to some gentle pressure to become a high-energy experimentalist. Later on, at the Weizmann Institute, Hector Rubinstein had a very positive influence on my research. And, finally, at MIT, I learned a lot from working with/under Sergio Fubini. Gatto, Rubinstein, and Fubini had rather different styles in doing theoretical physics and I tried to pick up what I appreciated most from each one of them. Whether I succeeded or not, I certainly owe a lot to all three. What I have appreciated most in all my teachers and mentors has been their passion in doing research together with their professionality. Both are very important attitudes to communicate to the new generations, more important than just giving them a long series of notions. Particularly important is to inject into students a critical, yet constructive, attitude in doing research. Nothing should be taken for granted until it is understood at the deepest possible level.*

MG & JM: Among the many scientific institutions you have visited, where did you find the most pleasant atmosphere and facility of work? What do you think should be of primary care for a laboratory, an institute, or a department of physics in order to encourage the creativity and productivity of its researchers?

GV: *The group in Florence under professor Gatto was a fantastic one. The atmosphere at the Weizmann Institute, particularly in 1966–1968, was also extremely congenial for doing research. Work at the Center for Theoretical Physics at MIT was also carried out under optimal conditions, and the same has always been true for the TH division at CERN. All these places shared the virtue of giving the physicists the time and the means to carry out their research in complete freedom, without administrative burdens and without any demand of short-term results. For instance, at MIT, Fubini and I were working on a program (dual resonance models) which was far from fashionable at the time, but no one tried to push us out of it. I have always been very lucky with the places where I have been working, but also, I must say, with the historical period in which I embarked in theoretical particle physics. A posteriori we can say that the years 1965–1975 were a "golden decade" in theoretical particle physics. We still live, to a large extent, on the great heritage of that period: the standard model, its possible extensions, and string theory.*

MG & JM: You have deeply influenced, in many ways, the past development of fundamental theoretical physics. From your perspective, are you satisfied with the present approach to the physics of fundamental interactions? In particular, what is your attitude toward the main contemporary theoretical "paradigmas"?

GV: *You ask me to stick my neck out. Well, in my opinion, theorists, on the basis of their recent successes with the standard model, have grown a little too arrogant. Some of the ideas around are very well motivated and even beautiful, but it is very hard to find the right way without the input of new data (it is even hard with the data, to be sure, see, e.g., the case of neutrino masses and mixing!). For this reason I am not too excited about the huge activity that is going on in building models for data ... that are not there yet. Perhaps it would be better to wait until those become available and, meanwhile, to put more effort on some of the outstanding theoretical and phenomenological problems that are already in the data, both in particle physics and in cosmology. Just to mention a few: confinement and dynamical symmetry breaking in QCD, and the origin of primordial—as well as of the present—cosmological acceleration. As an example, I don't think that enough effort has been devoted to trying to solve the first two problems I mentioned above at least in the large-N limit. I am pretty convinced that both analytic and numerical large-N techniques can and should be improved. A similar criticism could apply to present mathematical–physics research, mainly concentrated these days on string theory. It looks to me as if we forgot that the main "raison d' être" of modern string theory is the con-*

struction of a fully consistent quantum theory of gravity. Most of the present activity deals with very special (static, supersymmetric) solutions that fail to address the issue of what happens to generic solutions that approach those of General Relativity in some limit, but should look very different near the ubiquitous singularities of the classical solutions. What happens, in string theory, to the big bang singularity? Or to the one inside a black-hole horizon? These are tough problems, of course, but it looks to me that our community tends to ignore these issues in favor of tackling some easier problems (more for sociological than for scientific reasons, I guess). Let me repeat a motto I have voiced a few times: "Let's find tools for our problems, rather than problems for our tools!" In the case of the singularities, for instance, new techniques should be searched for studying string theory in geometries whose curvature radius is much below the string scale. I have the feeling that, by some appropriate duality, this problem should not be too different from that of large curvature radii, which we are already able to deal with. Would it not be wonderful to know about the fate of the big bang singularity—and thus of the beginning of time—in string theory?

MG & JM: A frequently voiced criticism of string theory (see, e.g., L. Smolin's recent book) is that string is not science since it cannot make predictions and, therefore, cannot be falsified. What is your opinion on this?

GV: *I completely disagree. It is fair to say that, at present, we are unable to extract reliable testable predictions from string theory, but this is only due to our present incomplete understanding of such a complicated theory. After all, how many decades had to pass before we could go from Yang–Mills theory to a theory of the weak interactions? For instance, it is often said that, in order to test string theory, one would need such high energies that no (human-built) accelerator will ever be able to produce. But (besides the fact that the Universe itself has provided such enormous energies right after the big bang, and may well have kept some imprint of string theory since then) it is not true that the predictions of string theory are just in the high-energy domain. String theory contains—at the lowest level of approximation—many massless scalar fields that could deeply influence low-energy physics by inducing violations of the equivalence principle, deviations from Newton's law, or space and/or time variations of fundamental constants. The problem is that we are presently unable to understand whether (some of) those massless particles stay massless after the theory is completely solved. If the answer is yes, then superstring theory will be falsified for the same main reason that the old hadronic string was abandoned: strong interactions are short range, but the old string insisted on having massless particles! Another generic prediction of string theory is the existence of extra dimensions of space. If those are not too small they could be revealed at accelerator experiments. But even if they are tiny they could have affected very early cosmology, leaving an imprint of today's cosmological observables.*

MG & JM: What would you like the LHS to discover? And what do you think the LHC will actually discover?

GV: *The best gift the LHC could deliver is ... surprises. The worst would be just a confirmation of the Standard Model by the discovery of a light Higgs boson and nothing else. Unfortunately, given the striking phenomenological successes of the Standard Model, the latter possibility is not easy to exclude. It would amount to some fine-tuning of the Standard Model's parameters, true, but the cosmological constant problem has accustomed us to much worse than that. Another item in any theorist's wish list is the discovery, by the LHC, of a good dark-matter candidate, even better if this will have to do with discovering supersymmetry. Personally, I am quite convinced that supersymmetry will play a role in particle physics, sooner or later. The problem, if supersymmetry lies at too high an energy scale to be reached at the LHC, is that we may never find the motivations (and resources) to push toward the next energy frontier. If I should bet my own money on something, I would say that the LHC will find more than the standard Higgs but not quite what we theorists are expecting or hoping for (like extra dimensions or strong gravity). For instance, I am not fully convinced that the ideas of a dynamical symmetry breaking (of "technicolor" type), or of some compositeness of leptons and quarks explaining the origin of the three families, can already be put to rest. We have not yet understood the non-perturbative dynamics of QCD: how can we be sure that a different gauge theory cannot solve one or both of those questions?*

MG & JM: What are your main suggestions and recommendations to young people at the beginning of their research activity in the field of fundamental theoretical physics?

GV: *To think with their own heads rather than follow the fashion. They should learn of course what has been done by the previous generation, but to follow the latter's prejudices will not help bring out the new ideas we badly need in order to solve the outstanding problems still facing us.*

MG & JM: Do you remember any amusing episode or anecdote concerning your scientific life that you would like to share with us and with the readers?

GV: *An amusing one is the drink I had with Feynman in Caltech after his talk at a conference on QCD. It must have been around 1979–1980. I had been invited to present some results obtained at CERN about how quark and gluon jets evolve and lead, eventually, to a state that looks almost ready to convert into low-mass hadrons. I gave the talk, which was well received. Feynman was in the audience, but I do not remember any question by him, either at the end of my talk or in private afterward. The next day Feynman gave his talk. Apparently he had rewritten it overnight and, consequently, was not very well prepared; but it was brilliant, as usual. His talk was largely inspired by mine and Feynman kept mentioning my results over and over again. I remember he was even mispelling the name of Petronzio by quoting "Veneziano and*

Petronziano," surely joking Mr. Feynman? After the end of the session, I told Feynman I had enjoyed his talk. He must not have been very satisfied with it, since he answered: well, that's because I quoted you all the time, isn't it? And then he added: come, let's have a drink, I want to understand better what exactly you have done. So we went to a nearby pub, had coffee (or was it beer?) and I started to tell him about my work. At some point, before I had finished, he interrupted me and said: "But then you have been cheating me! I thought you had done much more! This is nothing but the Altarelli–Parisi stuff!" I had to sweat a lot to convince him that, indeed, I had done more. Did he get convinced? I am not sure. But at some point I stumbled on his English (too good for me, I guess). I asked him what he meant by a "freying jet," an expression he had used many times. To explain, he pointed at my shirt and said: well your poor-Italian-physicist's shirt is freying...I got it. He also said: you know, he should get together and fix them, referring to some colleagues in Caltech who had also been doing jet physics. I thought that "fixing" them would mean to attack them badly, so I asked "Why be nasty?" But then he reassured me: no, I mean we should just correct what they are doing incorrectly... This was indeed my first and last substantial encounter with Feynman, a person I admired very much for his tremendous talent as a physicist but also for being so straight, so simple, and yet so deep, as a man.

MG & JM: To which subject(s), in particular, would you like to dedicate your future scientific activity?

GV: *Probably the wisest thing for me to do would be to retire from active research and give more time to teaching and to writing. However, for me doing research is a little bit like being addicted to a drug (I'm not sure since I've never been!). It will be difficult to stop abruptly. I would really like to know, for instance, what happens to spacetime singularities in string theory, to understand the origin of cosmic acceleration, and to solve QCD in some suitable large-N limit. But all this sounds like wishful thinking doesn't it?*

MG & JM: Finally, how do you imagine the path that fundamental physics and cosmology will follow in the future? What do you expect, in particular, from string theory and/or M-theory? Is, in your opinion, a successful "theory of everything" really within our reach in a foreseeable future?

GV: *Who said that it is difficult to make predictions, particularly for the future? But, if I have to make some guess, or a bet, I would say that, probably, the new accelerator data will not confirm our simplest theoretical ideas and, in particular, will suggest that there is more structure in today's "elementary particles" than we presently assume. In other words, the desert will blossom. The difficulties we are experiencing with getting the right model from string theory could mean that, like the old strings did not succeed in describing hadrons, the new ones will fail to describe quarks and leptons. Also, about the hierarchy problem, we could be on the wrong track with low-energy SUSY. Possibly, the solutions of the hierarchy and cosmological constant problem are not unrelated.*

Will we arrive one day at a "final theory" and to the end of theoretical physics? I do not think we will ever arrive at a "final theory" (I have given many talks about "Dreams of a Finite Theory" instead) but we may very well come to the end of some branch of physics because of "practical" reasons. I think that Feynman said once that a certain branch of physics may terminate the day the effort to make a tiny step forward (experimentally or theoretically) will be too large to be able to afford it. We may be (slowly!) approaching that limit in high-energy accelerator physics, but I am old enough for not being afraid of it.

References

List of publications of Gabriele Veneziano (updated to 2006)

1. R. Gatto and G. Veneziano, Mass of N_{33} from N/D calculation with $SU(6)_W$ vertices, *Phys. Lett.* **19** (1965) 512.
2. R. Gatto and G. Veneziano, Strong interactions dynamics with vertices invariant under the collinear group, *Phys. Lett.* **20** (1966) 439.
3. F. Buccella, R. Gatto and G. Veneziano, Analysis of sum rules following from local commutation relations of currents, *Nuovo Cimento* **42** (1966) 1019.
4. G. Veneziano, Remarks on the saturation of the sum rules of the chiral algebra, *Nuovo Cimento* **43** (1966) 529.
5. G. Veneziano, On the approximate saturation of the algebra of moments, *Nuovo Cimento* **44** (1966) 295.
6. M. Ademollo, R. Gatto, G. Longhi and G. Veneziano, The $SU(6)_W$ algebra at infinite momentum, its tensor charges, and electric dipoles, *Phys. Lett.* **22** (1966) 521.
7. F. Buccella, G. Veneziano, R. Gatto and S. Okubo, Necessity of additional unitary-antisymmetric q-number terms in the commutator of spatial current components, *Phys. Rev.* **149** (1966) 1268.
8. M. Ademollo, R. Gatto, G. Longhi and G. Veneziano, The $SU(6)_W$ algebra and the commutators of electric dipoles at infinite momentum, *Phys. Rev.* **153** (1967) 1623.
9. M. Ademollo, R. Gatto, G. Longhi and G. Veneziano, Mixing schemes for chiral and collinear algebras, *Nuovo Cimento* **47A** (1967) 334.
10. H.R. Rubinstein and G. Veneziano, Application of current algebra to pion emission, *Phys. Rev. Lett.* **18** (1967) 411.
11. H.R. Rubinstein and G. Veneziano, Connection between Regge pole parameters and local commutation relations, *Phys. Rev.* **160** (1967) 1286.
12. M. Ademollo, H.R. Rubinstein, G. Veneziano and M.A. Virasoro, Saturation of superconvergent sum rules at non-zero momentum transfer, *Nuovo Cimento* **51** (1967) 227.
13. M. Ademollo, H.R. Rubinstein, G. Veneziano and M.A. Virasoro, Bootstrap-like conditions from superconvergence, *Phys. Rev. Lett.* **19** (1967) 1402.
14. H.R. Rubinstein, G. Veneziano and M.A. Virasoro, Fixed poles and compositeness, *Phys. Rev.* **167** (1968) 1441.
15. M. Ademollo, H.R. Rubinstein, G. Veneziano and M.A. Virasoro, Reciprocal bootstrap of the vector and tensor trajectories from superconvergence, *Phys. Lett.* **B27** (1968) 99.

16. D. Amati, R. Jengo, H.R. Rubinstein, G. Veneziano and M.A. Virasoro, Compositeness as a clue for the understanding of the asymptotic behaviour of form factors, *Phys. Lett.* **B27** (1968) 38.
17. H.R. Rubinstein, A. Schwimmer, G. Veneziano and M.A. Virasoro, Generation of parallel daughters from superconvergence, *Phys. Rev. Lett.* **21** (1968) 491.
18. M. Ademollo, H.R. Rubinstein, G. Veneziano and M.A. Virasoro, Bootstrap of meson trajectories from superconvergence, *Phys. Rev.* **176** (1968) 1904.
19. M. Bishari, H.R. Rubinstein, A. Schwimmer and G. Veneziano, Meson bootstraps for unnatural-parity states, *Phys. Rev.* **176** (1968) 1926.
20. G. Veneziano, Construction of a crossing-symmetric, Regge behaved amplitude for linearly-rising trajectories, *Nuovo Cimento* **57A** (1968) 190.
21. M. Ademollo, G. Longhi and G. Veneziano, Spectral function sum rules for tensor currents, *Nuovo Cimento* **58A** (1968) 540.
22. M. Ademollo, G. Veneziano and S. Weinberg, Quantization conditions for Regge intercepts and hadron masses, *Phys. Rev. Lett.* **22** (1969) 83.
23. G. Veneziano, Crossing symmetry Regge behaviour and the idea of duality, *Proc. 6th Coral Gables Conference on "Fundamental Interactions at High Energy"*, Coral Gables, FL, 1969 (Gordon and Breach, New York, 1969), p. 113.
24. S. Fubini and G. Veneziano, Level structure of dual resonance models, *Nuovo Cimento* **64A** (1969) 811.
25. S. Fubini, D. Gordon and G. Veneziano, A general treatment of factorization in dual resonance models, *Phys. Lett.* **B29** (1969) 679.
26. L. Caneschi, A. Schwimmer and G. Veneziano, Twisted propagator in the operatorial duality formalism, *Phys. Lett.* **B30** (1969) 351.
27. G. Veneziano, Elementary particles, *Physics Today* **22** (1969) 31.
28. S. Fubini and G. Veneziano, Duality in operator formalism, *Nuovo Cimento* **67A** (1970) 29.
29. E. Del Giudice and G. Veneziano, Dual models, Pomeranchuk term and crossing symmetry, *Nuovo Cimento Lett.* **3** (1970) 363.
30. A. Di Giacomo, S. Fubini, L. Sertorio and G. Veneziano, Unitarity in dual resonance models, *Phys. Lett.* **B33** (1970) 171.
31. S. Fubini and G. Veneziano, Algebraic treatment of subsidiary conditions in dual resonance models, *Ann. Phys., Amos de Shalit Memorial Volume* **63** (1971) 12.
32. G. Veneziano, Narrow resonance models compatible with duality and their developments, in *Proc. 8th Int. School of Subnuclear Physics "Ettore Majorana"*, Erice, Sicily, 1970 (Academic Press, New York, 1971), p. 94.
33. G. Veneziano, Duality and dual models, in *Proc. 15th Int. Conference on High-Energy Physics*, Kiev, 1970 (Naukova Dumka, Kiev, 1972), p. 437.
34. G. Veneziano, Duality and the bootstrap, *Phys. Lett.* **B34** (1971) 59.
35. D. Gordon and G. Veneziano, Inclusive reactions and dual models, *Phys. Rev.* **D3** (1971) 2116.
36. G. Veneziano, General features of inclusive reactions from duality, *Nuovo Cimento Lett.* **1** (1971) 681.
37. C. E. DeTar, D. Z. Freedman and G. Veneziano, Sum rules for inclusive cross-sections, *Phys. Rev.* **D4** (1971) 906.
38. G. Veneziano, Sum rules for inclusive reactions and discontinuity formulae, *Phys. Lett.* **B36** (1971) 397.
39. M. B. Green and G. Veneziano, Average properties of dual resonances, *Phys. Lett.* **B36** (1971) 477.

40. E. Predazzi and G. Veneziano, A general formulation of inclusive sum rules, *Nuovo Cimento Lett.* **15** (1971) 749.
41. G. Veneziano, Conservation laws in inclusive reactions, in *Rendiconti del 53 Corso Scuola Internazionale di Fisica "Enrico Fermi"*, Varenna, 1971 (Academic Press, New York, 1973), p. 117.
42. S.-H. H. Tye and G. Veneziano, Exotic channels and approach to scaling in inclusive reactions, *Phys. Lett.* **B38** (1972) 30.
43. G. Veneziano, Inclusive approach to unitarity, *Phys. Rev. Lett.* **28** (1972) 578.
44. C. E. Jones, F .E. Low, S.-H. H. Tye, G. Veneziano and J. E. Young, Some general consequences of Regge theory for Pomeranchukon-pole couplings, *Phys. Rev.* **D6** (1972) 1033.
45. G. Veneziano, Trilinear coupling of scalar bosons in the small mass limit, *Nucl. Phys.* **B44** (1972) 142.
46. C. Rosenzweig and G. Veneziano, Unitarity sum rules and soft-pion amplitudes, *Nuovo Cimento* **12A** (1972) 409.
47. S.-H. H. Tye and G. Veneziano, Properties of inclusive reactions in a unitarized dual model of production amplitudes, *Nuovo Cimento* **14A** (1973) 711.
48. G. Veneziano, Duality and multiparticle production, in *Proc. 4th Int. Symposium on Multiparticle Hadrodynamics*, Pavia, 1973 (Istituto Nazionale di Fisica Nucleare, Italy), p. 325.
49. G. Veneziano, Origin and intercept of the Pomeranchuk singularity, *Phys. Lett.* **B43** (1973) 413.
50. G. Veneziano, An introduction to dual models of strong interactions and their physical motivations, *Phys. Rep.* **9C** (1973) 4.
51. G. Veneziano, Unitarity sum rules and the two-Reggeon cut, *Nucl. Phys.* **B69** (1974) 317.
52. G. Veneziano, Regge intercepts and unitarity in planar dual models, *Nucl. Phys* **B74** (1974) 365.
53. G. Veneziano, Large N expansion in dual models, *Phys. Lett.* **B52** (1974) 220.
54. C. Rosenzweig and G. Veneziano, Regge couplings and intercepts from the planar dual bootstrap, *Phys. Lett.* **B52** (1974) 335.
55. A. Schwimmer and G. Veneziano, Saturation of unitarity bounds in planar and non-planar models of multiparticle rescattering, *Nucl. Phys.* **B81** (1974) 445.
56. M. M. Schaap and G. Veneziano, Self-consistent $\rho - p'$ trajectory from the planar dual bootstrap, *Nuovo Cimento Lett.* **12** (1975) 204.
57. G. Marchesini and G. Veneziano, Non-vanishing of the bare triple-Pomeron coupling from s-channel unitarity, *Phys. Lett.* **B36** (1975) 271.
58. M. Ciafaloni, G. Marchesini and G. Veneziano, A topological expansion for high-energy hadronic collisions: I. General properties and connection with the Reggeon calculus, *Nucl. Phys.* **B98** (1975) 472.
59. M. Ciafaloni, G. Marchesini and G. Veneziano, A topological expansion for high-energy hadronic collisions: II. s-channel discontinuities and multiparticle content, *Nucl. Phys.* **B98** (1975) 493.
60. M. Bishari and G. Veneziano, Cut cancellation in the planar integral equation for the Reggeon, *Phys. Lett.* **B58** (1975) 445.
61. G. Veneziano, Harari-Freund and other schemes for the Pomeron in the topological expansion, *Nucl. Phys.* **B108** (1976) 285.
62. G. Veneziano, Some aspects of a unified approach to gauge, dual, and Gribov theories, *Nucl. Phys.* **B117** (1976) 519.

63. J. R. Freeman, G. Veneziano and Y. Zarmi, Constraints on Reggeon amplitudes from analyticity and planar unitarity, *Nucl. Phys.* **B120** (1977) 477.
64. G. C. Rossi and G. Veneziano, A possible description of baryon dynamics in dual and gauge theories, *Nucl. Phys.* **B123** (1977) 507.
65. G. Veneziano, The colour and flavour 1/N expansions, in *Proc. 12th Rencontre de Moriond*, Flaine (1977), ed. J. Tran Thanh Van, Vol. 3, p. 113.
66. G. C. Rossi and G. Veneziano, Electromagnetic mixing of narrow baryonium states, *Phys. Lett.* **B70** (1977) 255.
67. A. Giovannini and G. Veneziano, The Bose–Einstein effect and the jet structure of hadronic final states, *Nucl Phys.* **B130** (1977) 61.
68. G. Veneziano, A topological approach to the dynamics of quarks and hadrons, in *Proc. 9th Ecole d'Et de Physique des Particules*, Gif-sur-Yvette (1977), Vol. 2, p. 23.
69. D. Amati, R. Petronzio and G. Veneziano, Relating hard QCD processes through universality of mass singularities, *Nucl. Phys.* **B140** (1978) 54.
70. D. Amati, R. Petronzio and G. Veneziano, Relating hard QCD processes through universality of mass singularities (II), *Nucl. Phys.* **B146** (1978) 29.
71. K. Konishi, A. Ukawa and G. Veneziano, A simple algorithm for QCD jets, *Phys. Lett.* **B78** (1978) 243.
72. K. Konishi, A. Ukawa and G. Veneziano, On the transverse spread of QCD jets, *Phys. Lett.* **B80** (1979) 259.
73. G. Veneziano, Dynamics of hadronic reactions, in *Proc. XIXth Int. Conference on High-Energy Physics*, Tokyo (1978).
74. D. Amati and G. Veneziano, Preconfinement as a property of perturbative QCD, *Phys. Lett.* **B83** (1979) 87.
75. K. Konishi, A. Ukawa and G. Veneziano, Jet calculus: a simple algorithm for resolving QCD jets, *Nucl. Phys.* **B157** (1979) 45.
76. G. Veneziano, Momentum and colour structure of jets in QCD, in *Proc. 3rd Workshop on Current Problems in High Energy Particle Theory*, Florence (May–June 1979).
77. G. Veneziano, U(1) without instantons, *Nucl. Phys.* **B159** (1979) 213.
78. R. J. Crewther, P. Di Vecchia, G. Veneziano and E. Witten, Chiral estimate of the electric dipole moment of the neutron in QCD, *Phys. Lett.* **B88** (1979) 123.
79. P. Di Vecchia and G. Veneziano, Chiral dynamics in the large N limit, *Nucl. Phys.* **B171** (1980) 253.
80. G.C. Rossi and G. Veneziano, Baryonium physics, *Phys. Rep.* **63** (1980) 153.
81. D. Amati, A. Bassetto, M. Ciafaloni, G. Marchesini and G. Veneziano, A treatment of hard processes sensitive to the infra-red structure of QCD, *Nucl. Phys.* **B173** (1980) 429.
82. P. Di Vecchia and G. Veneziano, Minimal composite Higgs systems, *Phys. Lett.* **B95** (1980) 247.
83. G. Veneziano, Goldstone mechanism from gluon dynamics, *Phys. Lett.* **B95** (1980) 90.
84. G. Veneziano, Quantum chromodynamics, in *From nuclei to particles, Proc. International School of Physics Enrico Fermi*, Varenna (June 1980).
85. G. Marchesini, L. Trentadue and G. Veneziano, Space-time description of colour screening via jet calculus techniques, *Nucl. Phys.* **B181** (1981) 335.
86. P. Di Vecchia, F. Nicodemi, R. Pettorino and G. Veneziano, Large N, chiral approach to pseudoscalar masses, mixings and decays, *Nucl. Phys.* **B181** (1981) 318.

87. G. Veneziano, Tumbling and the strong anomaly, *Phys. Lett.* **B102** (1981) 139.
88. D. Amati, R. Barbieri, A.C. Davis and G. Veneziano, Dynamical gauge bosons from fundamental fermions, *Phys. Lett.* **B102** (1981) 408.
89. P. Di Vecchia, K. Fabricius, G.C. Rossi and G. Veneziano, Preliminary evidence for $U(1)_A$ breaking in QCD from lattice calculations, *Nucl. Phys.* **B192** (1981) 392.
90. P. Di Vecchia, K. Fabricius, G.C. Rossi and G. Veneziano, Numerical check of the lattice definition independence of topological charge fluctuations, *Phys. Lett.* **B108** (1982) 323.
91. D. Amati and G. Veneziano, Metric from matter, *Phys. Lett.* **B105** (1981) 358.
92. D. Amati and G. Veneziano, A unified gauge and gravity theory with only matter fields, *Nucl. Phys.* **B204** (1982) 451.
93. G. Veneziano and S. Yankielowicz, An effective Lagrangian for the pure N = 1 supersymmetric Yang–Mills theory, *Phys. Lett.* **B113** (1982) 231.
94. E. Gabathuler, G. Veneziano and P. Pavlopoulos, Axions, ghosts and pseudoscalars at LEAR, *Phys. Lett.* **B114** (1982) 58.
95. T.R. Taylor, G. Veneziano and S. Yankielowicz, Supersymmetric QCD and its massless limit: an effective Lagrangian analysis, *Nucl. Phys.* **B218** (1983) 493.
96. G. Veneziano, Chiral properties of supersymmetric vacua, *Phys. Lett.* **B124** (1983) 357.
97. G. Veneziano, A supersymmetric variant of Dashen's formula, *Phys. Lett.* **B128** (1983) 199.
98. R. Barbieri, A. Masiero and G. Veneziano, Hierarchy of fermion masses in supersymmetric composite models, *Phys. Lett.* **B128** (1983) 179.
99. F. Karsch, E. Rabinovici, G. Shore and G. Veneziano, The spectrum of a class of supersymmetric theories with false vacua, *Nucl. Phys.* **B242** (1984) 503.
100. G.C. Rossi and G. Veneziano, Non-perturbative breakdown of the non-renormalization theorem in supersymmetric QCD, *Phys. Lett.* **B138** (1984), 195.
101. Y. Meurice and G. Veneziano, SUSY vacua versus chiral fermions, *Phys. Lett.* **B141** (1984) 69.
102. V. De Alfaro, S. Fubini, G. Furlan and G. Veneziano, Stochastic identities in supersymmetric theories, *Phys. Lett.* **B142** (1984) 399.
103. A. Masiero and G. Veneziano, Split light composite supermultiplets, *Nucl. Phys.* **B249** (1985) 593.
104. D. Amati, G.C. Rossi and G. Veneziano, Instanton effects in supersymmetric gauge theories, *Nucl. Phys.* **B249** (1985) 1.
105. V. De Alfaro, S. Fubini, G. Furlan and G. Veneziano, Stochastic identities in quantum theory, *Nucl. Phys.* **B255** (1985) 1.
106. D. Amati and G. Veneziano, Gauge dependence of the Nicolai map in super Yang-Mills theory, *Phys. Lett.* **B157** (1985) 32.
107. A. Masiero, R. Pettorino, M. Roncadelli and G. Veneziano, An attempt at realistic supercompositeness, *Nucl. Phys.* **B261** (1985) 633.
108. D. Amati, Y. Meurice, G.C. Rossi and G. Veneziano, Massive SQCD and the consistency of instanton calculations, *Nucl. Phys.* **B263** (1986) 591.
109. G. Veneziano, Ward identities in dual string theories, *Phys. Lett.* **B167** (1986) 388.
110. J. Maharana and G. Veneziano, Gauge Ward identities of the compactified bosonic string, *Phys. Lett.* **B169** (1986) 177.

111. G. Veneziano, A stringy nature needs just two constants, *Europhys. Lett.* **2** (1986) 199.
112. G.M. Shore and G. Veneziano, Current algebra and supersymmetry, *Int. J. Mod. Phys.* **1** (1986) 499.
113. J. Maharana and G. Veneziano, Strings in a background: a BRS Hamiltonian approach, *Nucl. Phys.* **B283** (1987) 126.
114. K. Konishi and G. Veneziano, Effective action for dynamical supersymmetry breaking, *Phys. Lett.* **B187** (1987) 106.
115. G. Curci and G. Veneziano, Supersymmetry and the lattice: a reconciliation?, *Nucl. Phys.* **B292** (1987) 555.
116. D. Amati, M. Ciafaloni and G. Veneziano, Superstring collisions at Planckian energies, *Phys. Lett.* **B197** (1987) 81.
117. R. Petronzio and G. Veneziano, Constraints from string unification, *Mod. Phys. Lett.* **A2** (1987) 707.
118. G. Veneziano, Mutual focusing of graviton beams, *Mod. Phys. Lett.* **A2** (1987) 899.
119. D. Amati, K. Konishi, Y. Meurice, G.C. Rossi and G. Veneziano, Non-perturbative aspects in supersymmetric gauge theories, *Phys. Rep.* **162** (1988) 169.
120. D. Amati, M. Ciafaloni and G. Veneziano, Classical and quantum gravity effects from Planckian energy superstring collisions, *Int. J. Mod. Phys.* **A7** (1988) 1615.
121. T. Kubota and G. Veneziano, Off-shell effective actions in string theory, *Phys. Lett.* **B207** (1988) 419.
122. V. Ferrari, P. Pendenza and G. Veneziano, Beam-like Gravitational waves and their geodesics, *Gen. Rel. Grav.* **20** (1988) 1185.
123. G. Veneziano, Topics in string theory, in *Proc. DST Workshop in Particle Physics—Superstring Theory* (Kanpur, December 1987), eds. H.S. Mani and R. Ramachandran (World Scientific, Singapore, 1988) p. 1.
124. T.R. Taylor and G. Veneziano, Strings and D = 4, *Phys. Lett.* **B212** (1988) 147.
125. T.R. Taylor and G. Veneziano, Dilaton couplings at large distances, *Phys. Lett.* **B213** (1988) 450.
126. S. Narison and G. Veneziano, QCD tests of G(1.6) = glueball, *Int. J. Mod. Phys.* **A14** (1989) 2751.
127. S. Fubini, J. Maharana, M. Roncadelli and G. Veneziano, Quantum constraints for an interacting superstring, *Nucl. Phys.* **B316** (1989) 36.
128. D. Amati, M. Ciafaloni and G. Veneziano, Can space-time be probed below the string size?, *Phys. Lett.* **B216** (1989) 41.
129. M. Fabbrichesi and G. Veneziano, Thinning out of relevant degrees of freedom in scattering of strings, *Phys. Lett.* **B233** (1989) 135.
130. G. Veneziano, Wormholes, non-local actions and a new mechanism for suppressing the cosmological constant, *Mod. Phys. Lett.* **A4** (1989) 695.
131. T.R. Taylor and G. Veneziano, Quenching the cosmological constant, *Phys. Lett.* **B228** (1989) 210.
132. G. Veneziano, Is there a QCD "spin crisis"?, *Mod. Phys. Lett.* **A4** (1989) 1605.
133. A. Giveon, E. Rabinovici and G. Veneziano, Duality in string background space, *Nucl. Phys.* **B322** (1989) 167.
134. G. Veneziano, Quantum strings and the constants of Nature, in *Proc. 27th Course of the International School of Subnuclear Physics*, Erice, July 1989, ed. A. Zichichi (Plenum Press, 1990) p. 199.

135. T.R. Taylor and G. Veneziano, Quantum Gravity at large distances and the cosmological constant, *Nucl. Phys.* **B345** (1990) 210.
136. G.M. Shore and G. Veneziano, The U(1) Goldberger–Treiman relation and the two components of the proton "spin", *Phys. Lett.* **B244** (1990) 75.
137. G. Veneziano, The spin of the proton and the OZI limit of QCD, in *From Symmetries to Strings: Forty Years of Rochester Conferences* (Okubofest), ed. Ashok Das (World Scientific, Singapore, 1990) p. 13.
138. G. Veneziano, An enlarged uncertainty principle from gedanken string collisions?, in *Proc. Strings '89*, Texas A&M University, March 1989, eds. R. Arnowitt et al. (World Scientific, Singapore, 1990) p. 86.
139. D. Amati, M. Ciafaloni and G. Veneziano, Higher-order gravitational deflection and soft bremsstrahlung in Planckian energy superstring collisions, *Nucl. Phys.* **B347** (1990) 550.
140. G. Veneziano, Quantum string gravity near the Planck scale, in *Proc. 1st Symposium on Particles, Strings and Cosmology*, Northeastern University, March 1990, eds. P. Nath and S. Reucroft (World Scientific, Singapore, 1991) p. 486.
141. S. Ferrara, N. Magnoli, T.R. Taylor and G. Veneziano, Duality and supersymmetry breaking in string theory, *Phys. Lett.* **B245** (1990) 409.
142. N. Sanchez and G. Veneziano, Jeans-like instabilities for strings in cosmological backgrounds, *Nucl. Phys.* **B333** (1990) 253.
143. M. Gasperini, N. Sanchez and G. Veneziano, Highly unstable fundamental strings in inflationary cosmologies, *Int. J. Mod. Phys.* **A6** (1991) 3853.
144. M. Gasperini, N. Sanchez and G. Veneziano, Self-sustained inflation and dimensional reduction from fundamental strings, *Nucl. Phys.* **B364** (1991) 365.
145. Nguyen Suan Han and G. Veneziano, Inflation-driven string instabilities: towards a systematic Large-R expansion, *Mod. Phys. Lett.* **A6** (1991) 1993.
146. G. Veneziano, Inflation-driven string instabilities... and the other way around, (Gatto-Ruegg birthday Conference, Geneva, Nov. 1990), *Helv. Phys. Acta* **64** (1991) 877.
147. G. Veneziano, Strings and Gravity, in *Proc. Texas/ESO-CERN Symposium on Relativistic Astrophysics, Cosmology, and Fundamental Physics*, Brighton, Dec. 1990, eds. J. D. Barrow, L. Mestel and P.A. Thomas (The New York Academy of Sciences, NY, 1991) p. 180.
148. G. Veneziano, Scale factor duality for classical and quantum strings, *Phys. Lett.* **B265** (1991) 287.
149. K.A. Meissner and G. Veneziano, Symmetries of cosmological superstring vacua, *Phys. Lett.* **B267** (1991) 33.
150. K.A. Meissner and G. Veneziano, Manifestly $O(d,d)$ invariant approach to space-time dependent string vacua, *Mod. Phys. Lett.* **A6** (1991) 3397.
151. M. Gasperini, J. Maharana and G. Veneziano, From trivial to non-trivial conformal string backgrounds via $O(d,d)$ transformations, *Phys. Lett.* **B272** (1991) 277.
152. G. Veneziano, Strings in/and inflation, in *Proc. 2nd Symposium on Particles, Strings and Cosmology*, NorthEastern University, March 1991, eds. P. Nath and S. Reucroft (World Scientific, Singapore, 1992) p. 425.
153. M. Gasperini and G. Veneziano, $O(d,d)$-covariant string cosmology, *Phys. Lett.* **B277** (1992) 256.
154. G. Veneziano, Bound on reliable one-instanton cross-sections, *Mod. Phys. Lett.* **A7** (1992) 1661.

155. D. Amati, M. Ciafaloni and G. Veneziano, Planckian Scattering beyond the semi-classical approximation, *Phys. Lett.* **B289** (1992) 87.
156. G.M. Shore and G. Veneziano, The U(1) Goldberger–Treiman relation and the proton "spin": a renormalisation group analysis, *Nucl. Phys.* **B381** (1992) 23.
157. G.M. Shore and G. Veneziano, Renormalization group aspects of $\eta \to \gamma\gamma$, *Nucl. Phys.* **B381** (1992) 3.
158. S. Narison, G.M. Shore and G. Veneziano, A sum rule for the polarized photon structure function g_γ^1, *Nucl. Phys.* **B391** (1993) 69.
159. G.M. Shore and G. Veneziano, The polarized photon structure function g_γ^1 as a probe of chiral symmetry realizations, *Mod. Phys. Lett.* **A8** (1993) 373.
160. M. Gasperini, J. Maharana and G. Veneziano, Boosting away singularities from conformal string background, *Phys. Lett.* **B296** (1992) 51.
161. M. Gasperini and G. Veneziano, Pre Big-Bang in string cosmology, *Astropart. Phys.* **1** (1993) 317.
162. M. Gasperini, M. Giovannini and G. Veneziano, Squeezed thermal vacuum and the maximum scale for inflation, *Phys. Rev.* **D48** (1993) 707.
163. D. Amati, M. Ciafaloni and G. Veneziano, Effective action and all-order gravitational eikonal at Planckian energies, *Nucl. Phys.* **B403** (1993) 707.
164. M. Fabbrichesi, R. Pettorino, G. Veneziano and G.A. Vilkovisky, Planckian energy scattering and surface terms in the gravitational action, *Nucl. Phys.* **B419** (1994) 147.
165. M. Gasperini and G. Veneziano, Inflation, deflation, and frame independence in string cosmology, *Mod. Phys. Lett.* **A8** (1993) 3701.
166. M. Gasperini, R. Ricci and G. Veneziano, A problem with non-Abelian duality?, *Phys. Lett.* **B319** (1993) 438.
167. L. Trentadue and G. Veneziano, Fracture functions: an improved description of inclusive hard processes in QCD, *Phys. Lett.* **B323** (1994) 201.
168. M. Gasperini and G. Veneziano, Dilaton production in string cosmology, *Phys. Rev.* **D50** (1994) 2519.
169. R. Brustein and G. Veneziano, The graceful exit problem in string cosmology, *Phys. Lett.* **B329** (1994) 429.
170. G. Veneziano, Strings, cosmology,... and a particle, in *Proc. PASCOS 1994*, Syracuse, NY, May 1994 (QCD 161:I69:1994), p. 453.
171. G. Veneziano, A new approach to semiclassical gravitational scattering, in *Proc. of the Second Paris Cosmology Colloquium* (Observatoire de Paris, June 1994), eds. H. De Vega and N. Sanchez (World Scientific, Singapore, 1995) p. 322.
172. R. Brustein, M. Gasperini, M. Giovannini, V.F. Mukhanov and G. Veneziano, Metric perturbations in dilaton driven inflation, *Phys. Rev.* **D51** (1995) 6744.
173. S. Narison, G.M. Shore and G. Veneziano, Target independence of the EMC-SMC effect, *Nucl. Phys.* **B433** (1995) 209.
174. M. Gasperini, M. Giovannini, K.A. Meissner and G. Veneziano, Evolution of a string network in backgrounds with rolling horizons, in *String theory in Curved Space Times* (Observatoire de Paris, June 1995), ed. N. Sanchez (World Scientific, Singapore, 1998), p. 49.
175. M. Gasperini, M. Giovannini and G. Veneziano, Primordial magnetic fields from string cosmology, *Phys. Rev. Lett.* **75** (1995) 3796.
176. M. Gasperini, M. Giovannini and G. Veneziano, Electromagnetic origin of the cosmic microwave backgrounds anisotropy, *Phys. Rev.* **D52** (1995) 6651.
177. S. Elitzur, A. Giveon, E. Rabinovici, A. Schwimmer and G. Veneziano, Remarks on nonabelian duality, *Nucl. Phys.* **B435** (1995) 147.

178. R. Brustein, M. Gasperini, M. Giovannini and G. Veneziano, Relic gravitational waves from string cosmology, *Phys. Lett.* **B361** (1995) 45.
179. D. Graudenz and G. Veneziano, Estimating diffractive Higgs boson production at LHC from HERA data, *Phys. Lett.* **B365** (1996) 302.
180. G. Veneziano, String cosmology: basic ideas and general results, in *Proc. of the Third Paris Cosmology Colloquium* (Observatoire de Paris, June 1995), eds. H. De Vega and N. Sanchez (World Scientific, Singapore, 1996).
181. R. Brustein, M. Gasperini, M. Giovannini and G. Veneziano, Gravitational radiation from string cosmology, in *Proc. Int. Europhysics Conference on High Energy Physics* (HEP 95, Brussels, July 1995), eds. J. Lemonne et al. (World Scientific, Singapore, 1996) p. 408.
182. G. Veneziano, String cosmology: concepts and consequences, in *Proc. 4th Course of the International School of Astrophysics D. Chalonge* (Erice, September 1995), eds. N. Sanchez and A. Zichichi (Kluwer Academic Publishers, Dordrecht, The Netherland, 1996).
183. G. Veneziano, Summary of SUSY-95, in *Supersymmetry and Unification of Fundamental Interactions* (SUSY 95), Palaiseau, France, May 1995, eds. I. Antoniadis and H. Videau (Editions Frontieres, Paris, 1996).
184. M. Gasperini and G. Veneziano, Birth of the Universe as quantum scattering in string cosmology, *Gen. Rel. Grav.* **28** (1996) 1301.
185. M. Gasperini, J. Maharana and G. Veneziano, Graceful exit in quantum string cosmology, *Nucl. Phys.* **B472** (1996) 349.
186. G. Veneziano, Summary, in *Proc. 28th Int. Conference on High Energy Physics*, Warsaw, July 1996, eds. Z. Ajduk and A. K. Wroblewski (World Scientific, Singapore, 1997), p. 449.
187. R. Brustein, M. Gasperini and G. Veneziano, Peak and endpoint of the relic graviton background in string cosmology, *Phys. Rev.* **D55** (1997) 3882.
188. G. Veneziano, String cosmology and relic gravitational radiation, in *Proc. Int. Conference on Gravitational Waves: Sources and Detectors*, Pisa, Italy, March 1996.
189. M. Gasperini and G. Veneziano, Singularity and exit problems in two-dimensional string cosmology, *Phys. Lett.* **B387** (1996) 715.
190. M. Gasperini, M. Maggiore and G. Veneziano, Towards a non-singular pre-big bang cosmology, *Nucl. Phys.* **B494** (1997) 315.
191. G. Veneziano, Inhomogeneous pre-big bang string cosmology, *Phys. Lett.* **B406** (1997) 297.
192. G. M. Shore and G. Veneziano, Testing target independence of the Rproton spinS effect in semi-inclusive deep inelastic scattering, *Nucl. Phys.* **B516** (1998) 333-353.
193. D. de Florian, G.M. Shore and G. Veneziano, Target fragmentation at polarized HERA: a test of universal topological-charge screening in QCD, in *Proc. 1997 Workshop on Physics with polarized protonds at HERA*, DESY-Zeuthen and CERN, March–September 1997.
194. G. Veneziano, Theoretical Outlook, in *Proc. Int. EPS Conference on High Energy Physics*, Jerusalem 1997, eds. D. Lellouch, G. Mikenberg and Eliezer Rabinovici (Springer-Verlag, Berlin, 1999).
195. M. Grazzini, L. Trentadue and G. Veneziano, Fracture functions from cut vertices, *Nucl. Phys.* **B519** (1998) 394-404.
196. A. Buonanno, K.A. Meissner, C. Ungarelli and G. Veneziano, Classical inhomogeneities in string cosmology, *Phys. Rev.* **D57** (1998) 2543.

197. J. Maharana, E. Onofri and G. Veneziano, A numerical simulation of pre-big bang cosmology, *JHEP* **4** (1998) 4.
198. A. Buonanno, K. A. Meissner, C. Ungarelli and G. Veneziano, Quantum inhomogeneities in string cosmology, *JHEP* **1** (1998) 4.
199. R. Brustein, M. Gasperini and G. Veneziano, Duality in cosmological perturbation theory, *Phys. Lett.* **B431** (1998) 277.
200. R. Durrer, M. Gasperini, M. Sakellariadou and G. Veneziano, Seeds of large-scale anisotropy in string cosmology, *Phys. Rev.* **D59** (1999) 043511.
201. R. Durrer, M. Gasperini, M. Sakellariadou and G. Veneziano, Massless (pseudo-)scalar seeds of CMB anisotropy, *Phys. Lett.* **B436** (1998) 66.
202. G. Veneziano, Quantum geometric origin of all forces in string theory, in *The Geometric Universe* (Oxford University Press, Oxford, 1998) p. 235.
203. A. Buonanno, T. Damour and G. Veneziano, Pre-big bang bubbles from the gravitational instability of generic string vacua, *Nucl. Phys.* **B543** (1999) 275.
204. M. Gasperini and G. Veneziano, Constraints on pre-big bang models for seeding large-scale anisotropy by massive Kalb-Ramond axions, *Phys. Rev.* **D59** (1999) 043503.
205. A. Ghosh, G. Pollifrone and G. Veneziano, Quantum fluctuations in open pre-big bang cosmology, *Phys. Lett.* **B440** (1998) 20.
206. G. Veneziano, Physics and Mathematics: a happily evolving marriage?, in *Les relations entre les Mathmatiques et la physique theorique*, Festschrift for the 40th anniversary of the IHES (IHES Publications, Bures-sur-yvette 1998), p. 183.
207. G. Veneziano, Pre bangian origin of our entropy and time arrow, *Phys. Lett.* **B454** (1999) 22.
208. G. Veneziano, Entropy bounds and string cosmology, in *Fundamental Interactions: from Symmetries to Black Holes* (Proceedings of conference in honour of F. Englert) (ULB, Bruxelles, March 1999) p. 273.
209. A. Melchiorri, F. Vernizzi, R. Durrer and G. Veneziano, CMB anisotropies and extra dimensions in string cosmology, *Phys. Rev. Lett.* **83** (1999) 4464.
210. A. Ghosh, R. Madden and G. Veneziano, Back reaction to dilaton-driven inflation, *Nucl.Phys.* **B570** (2000) 207.
211. T. Damour and G. Veneziano, Self-gravitating fundamental strings and black holes, *Nucl. Phys.* **B568** (2000) 93.
212. G. Veneziano, Testing string theory by probing the pre-bangian Universe, in *Proc. COSMO-98 Conference*, Asilomar, CA, 1998, ed. D.O. Caldwell (AIP Conference Proceedings, 1999), p. 97.
213. R. Brustein and G. Veneziano, A causal entropy bound, *Phys. Rev. Lett.* **84** (2000) 5695.
214. G. Veneziano, String Cosmology: the pre-big bang scenario, in *Proc. Les Houches Summer School, on The Primordial Universe*, Les Houches, 1999, eds. O. Binetruy et al. (Springer-Verlag, Heidelberg, 2000), p. 581.
215. Valerio Bozza, Gabriele Veneziano, O(d,d)-invariant collapse/inflation from colliding superstring waves, *JHEP* **0010** (2000) 035.
216. R. Brustein, S. Foffa and G. Veneziano, CFT, holography, and causal entropy bound, *Phys. Lett.* **B507** (2001) 270–276.
217. V. Bozza, M. Gasperini and G. Veneziano, Localization of scalar fluctuations in a dilatonic brane-world scenario, *Nucl. Phys.* **B619** (2001) 191.
218. L. Giusti, G.C. Rossi, M. Testa and G. Veneziano, The $U_A(1)$ Problem on the lattice with Ginsparg–Wilson Fermions, *Nucl. Phys.* **B628** (2002) 234–252.

219. M. Gasperini, F. Piazza and G. Veneziano, Quintessence as a run-away dilaton, *Phys. Rev.* **D65** (2002) 023508.
220. M. J. Duff, L. B. Okun and G. Veneziano, Trialogue on the number of fundamental constants, *JHEP* **03** (2002) 023.
221. G. Veneziano, Large-N bounds on, and compositeness limit of, gauge and gravitational interactions, *JHEP* **0206** (2002) 051.
222. E. Kohlprath and G. Veneziano, Black holes from high-energy beam–beam collisions, *JHEP* **0206** (2002) 057.
223. T. Damour, F. Piazza and G. Veneziano, Runaway dilaton and equivalence principle violations, *Phys. Rev. Lett.* **89** (2002) 081601.
224. T. Damour, F. Piazza and G. Veneziano, Violations of the equivalence principle in a dilaton-runaway scenario, *Phys. Rev.* **D66** (2002) 046007.
225. V. Bozza, M. Gasperini, M. Giovannini and G. Veneziano, Assisting pre-big bang phenomenology through short-lived axions, *Phys. Lett.* **B543** (2002) 14.
226. M. Gasperini, and G. Veneziano, The pre-big bang scenario in string cosmology, *Phys. Reports* **373** (2003) 1.
227. V. Bozza, M. Gasperini, M. Giovannini and G. Veneziano, Constraints on pre-big bang parameter space from CMBR anisotropies, *Phys. Rev.* **D67** (2003) 063514.
228. A. Armoni, M. Shifman and G. Veneziano, Exact results in nonsupersymmetric large N orientifold field theories, *Nucl. Phys.* **B667** (2003) 170.
229. V. Bozza, M. Giovannini and G. Veneziano, Cosmological perturbations from a new physics hypersurface, *JCAP* **0305** (2003) 001.
230. M. Gasperini, M. Giovannini and G. Veneziano, Perturbations in a nonsingular bouncing universe, *Phys. Lett.* **B569** (2003) 113.
231. A. Armoni, M. Shifman and G. Veneziano, SUSY relics in one flavor QCD from a new 1/N expansion, *Phys. Rev. Lett.* **91** (2003) 191601.
232. V. Branchina, K. A. Meissner and G. Veneziano, The price of an exact, gauge invariant RG flow equation, *Phys. Lett.* **B574** (2003) 319.
233. A. Armoni, M. Shifman and G. Veneziano, QCD quark condensate from SUSY and the orientifold large N expansion, *Phys. Lett.* **B579** (2004) 384.
234. G. Veneziano, A model for the big bounce, *JCAP* **0403** (2004) 004.
235. M. Gasperini, M. Giovannini and G. Veneziano, Cosmological perturbations across a curvature bounce, *Nucl. Phys.* **B694** (2004) 206.
236. A. Armoni, M. Shifman and G. Veneziano, From Super Yang–Mills theory to QCD: planar equivalence and its implications, in *From Fields to Strings*, eds. M. Shifman et al. (World Scientific, Singapore, 2004) Vol. 1, p. 353.
237. G.C. Rossi and G. Veneziano, Isospin mixing of narrow pentaquark states, *Phys. Lett.* **B597** (2004) 338.
238. A. Armoni, M. Shifman and G. Veneziano, Exact results in a non supersymmetric gauge theory, *Fortsch. Phys.* **52** (2004) 453.
239. G. Veneziano, The myth of the beginning of time, *Sci. Am.* **290**, N5 (2004) 30.
240. G. Veneziano, String-theoretic unitary S-matrix at the threshold of black-hole production, *JHEP* **0411** (2004) 001.
241. A. Armoni and M. Shifman, G. Veneziano, Refining the proof of planar equivalence, *Phys. Rev.* **D71** (2005) 045015.
242. V. Bozza and G. Veneziano, Scalar perturbations in regular two-component bouncing cosmologies, *Phys.Lett.* **B625** (2005) 177.
243. V. Bozza and G. Veneziano, Regular two-component bouncing cosmologies and perturbations therein, *JCAP* **0509** (2005) 007.

244. G. Veneziano, Unconventional scenarios and perturbations therein, *Phys. Scr.* **T117** (2005) 51.
245. A. Armoni, G. Shore and G. Veneziano, Quark condensate in massless QCD from planar equivalence, *Nucl. Phys.* **B740** (2006) 23.
246. G. Veneziano and J. Wosiek, Planar quantum mechanics: an intriguing supersymmetric example, *JHEP* **0601** (2006) 156.
247. G. Veneziano, Cosmology (including neutrino mass limits): a particle theorist's viewpoint (contribution to HEP-EPS 2005), Lisbon, Portugal, July 2005, *PoS HEP 2005* (2006) 403.
248. G. Veneziano and J. Wosiek, A supersymmetric matrix model. II. Exploring higher-fermion-number sectors, *JHEP* **0610** (2006) 033.
249. G. Veneziano and J. Wosiek, A supersymmetric matrix model. III. Hidden SUSY in statistical systems, *JHEP* **0611** (2006) 030.
250. G. Veneziano, Towards a unitary S-matrix description of black-hole formation and decay in string theory, *AIP Conf. Proc.* **861** (2006) 39.

An Unpublished Draft by Gabriele Veneziano (1973): "Non-local Field Theory Suggested by Dual Models"

G. Veneziano

CERN, Theory Unit, Physics Department, CH-1211 Geneva 23, Switzerland, and College de France, 11 Place M. Berthelot, 75005 Paris, France
gabriele.veneziano@cern.ch

Abstract. This article reports an old and incomplete note (written in 1973, mostly at the Weizmann Institute, Rehovot, Israel) about a non-local field theory suggested by dual resonance models, and largely inspired by Yukawa's late work on bilocal fields. It has definite relations to the study of strings in a background (discussed by Ademollo et al.), and to Polyakov's action for a string moving in a tachyonic background. It also suggests, for the first time, a modification of the uncertainty principle coming from the extended nature of strings. The original note is reported in this article using the *slanted* typographical style, for an immediate "visive" separation between the old, original text and the modern comments added by the author in the notes and in the final appendix.

1 Introduction and Content of the Paper

The success of quantum electrodynamics [1] (QED), as well as the recent breakthroughs in weak interactions [2], are a clear confirmation of the soundness of local field theory (LFT) in describing leptonic interactions.

The situation appears much more dubious at the hadronic level. LFT fails to explain the spectrum of hadrons, in particular its amazingly rich structure, and to account for the simple systematics of the SLAC data on electron/nucleon high energy collisions, known as Bjorken scaling. It is also hard to construct field theories which provide strong damping of transverse momenta at high energy, this failure being probably related to the previous ones.

What seems to emerge is the fact that, for strong interactions, already in the several GeV region, local field theory is too singular in position space or, if we prefer, too spread in momentum space. High frequencies are not damped enough to provide a sharp transverse momentum cut off, infinite renormalization constants are needed, and the resulting cutoff brings in scaling violations. Nature, on the other hand, seems to be as naïve as a free field

G. Veneziano: *An Unpublished Draft by Gabriele Veneziano (1973): "Non-local Field Theory Suggested by Dual Models"*, Lect. Notes Phys. **737**, 29–44 (2008)
DOI 10.1007/978-3-540-74233-6_2

theory or, better, as damped as a super-renormalizable LFT. Unfortunately, there are no sound super-renormalizable LFTs in four dimensions.

In order to explain the simple SLAC data crude models have been proposed which get away without an underlying field theory. These ("parton") models are based on a composite picture of the hadron with a large number of constituents giving it a structure. As a consequence, the hadron becomes all but a pointlike object.

Such a composite system can be made such as to enjoy Bjorken scaling. At the same time, a composite hadron can possibly lie on a Regge trajectory and, for an infinitely composite object, a trajectory rising from $-\infty$ to $+\infty$ is quite conceivable. Also, a composite structure will have a spread in position space and can therefore lead to enough damping of large momenta to encompass the above-mentioned difficulties.

It is very difficult to put parton models on a more than descriptive, intuitive level. On the other hand, a much more refined and detailed model has been developed over the past five years which has several attractive features and seems to depart in many respects from any LFT approach. This is the dual resonance model which, started as a simple mathematical realization of the duality idea of Dolen, Horn and Schmit, was developed as far as to represent now, for many, the only possible candidate for a complete theory of hadrons.

Although this theory has not yet produced a completely satisfactory first-order solution to the strong interaction S-matrix, its theoretical consistence and the number of constraints it fulfills can be hardly considered accidental. One of the most amazing properties of this model is the fact that is has a universal length (or mass) scale in it. Calling this length λ we can list here the various quantities related to λ in this dual theory:

1) the slope $dJ/dM^2 \approx \lambda^2 \equiv \alpha'$
2) the size of total cross-sections $\sigma_{\text{tot}} \approx \lambda^2$
3) the cut-off in transverse momenta $\langle p_\perp^2 \rangle \approx 1/\lambda^2$
- ...[1]

The properties of dual models which are related to this length are indeed very suggestive of λ being related to the "size" of the hadron, an intrinsic size which we do not see (yet) in the leptonic world.

This simple observation suggests that we may look at dual models as at some approximation of a non-local (rather than of a local) field theory characterized by this new microscopic constant λ, and which goes into a local theory in the limit $\lambda \to 0$.

Of course this idea of introducing a fundamental length λ in quantum relativistic theories is quite old. In particular, Yukawa has advocated a particular

[1] The original text has a vertical series of dots indicating that several other quantities related to λ were known: an obvious one is the limiting Hagedorn temperature of dual resonance models.

modification of LFT where the introduction of λ can be made quite naturally. Using some ideas of Born, Yukawa managed to constrain this theory further. In spite of some appealing features, however, such attempt of Yukawa has not progressed too far and has encountered problems of higher order corrections.

The aim of this investigation is to point out that dual models can be reformulated as a sort of Yukawa-type non-local field theory, with a lot more of structure in it. We hope that further study along these lines may clarify the physical meaning of duality and of hadronic compositeness. On the other hand, a better physical understanding of the dual formalism could provide new hints for the solution of the remaining problems afflicting dual models such as fermions and currents. On the other hand, this more sophisticated non-local FT could solve some of the problems met by the original Yukawa proposal.

We are thus trying to develop a Non-local-Quantum-Relativistic theory, characterized by the three fundamental constants:

Relativistic – $c \approx 3.010^{10}$ *cm.sec*$^{-1}$
Quantum – $h \approx 6.610^{-27}$ *erg. sec*
Non-Local – $\lambda \approx 2.010^{-14}$ *cm*

It would be of course interesting to analyze other limits besides that of a LFT ($\lambda = 0$). An interesting one, on which we shall have some comments here, is that of a non-local classical field theory.

The plan of this paper is as follows:
In Sect. 2 we review briefly Yukawa's non-local field theory and its local limit. In Sect. 3 we reconsider the zero slope (local) limit of dual models and we argue that there exists an alternative to the results of the type given by Scherk. In Sect. 4 we establish our correspondence principle between a local field and a non-local dual field and discuss its physical meaning in terms of quantum measurements. In Sect. 5 we consider the case of a classical ($h = 0$) non-local field theory as it would emerge from the string picture of the dual model. In Sect. 6 we derive a few simple quantities which could be relevant in the development of the theory. Finally, Sect. 7 contains a few more speculative remarks and our outlook.

2 Yukawa's Non-local Field Theory

Let us consider, at the beginning, a theory of first quantization in which we have introduced as usual operator q_i and p_i such that

$$[q_i, p_j] = ih\delta_{ij} \qquad i = 1, 2, 3, ...(D-1)\,. \tag{1}$$

D is the dimensionality of space-time. For the moment, take $D = 4$. A local field is introduced by Yukawa as a "first-quantized" Hermitian operator U which commutes with q_i, the position operators

$$[q_i, U] = 0\,. \tag{2}$$

Hence position and field can be measured simultaneously, i.e., given a test body, we can measure its position (which implies a point-like body) at a given time as well as the field acting on it at that point and at that time. In other words, we can define the meaning of a field at a point $x = (\boldsymbol{x}, ct)$. If we work in the coordinate representation we shall have, by definition,

$$q_i|\boldsymbol{x}\rangle = x_i|\boldsymbol{x}\rangle\,,$$
$$\langle \boldsymbol{x}'|\boldsymbol{x}\rangle = \delta^{(3)}(\boldsymbol{x} - \boldsymbol{x}')\,, \tag{3}$$

and, because of (2),

$$\langle \boldsymbol{x}|U|\boldsymbol{x}'\rangle = \delta^{(3)}(\boldsymbol{x} - \boldsymbol{x}')\ \ \phi(\boldsymbol{x})\,. \tag{4}$$

Having in mind relativistic invariance, we shall write instead:

$$q_\mu|x\rangle = x_\mu|x\rangle\,, \tag{5}$$

$$\langle x'|x\rangle = \delta^{(4)}(x - x')\,, \tag{6}$$

$$\langle x|U|x'\rangle = \delta^{(4)}(x - x')\phi(x)\,, \tag{7}$$

$$\phi(x) = \frac{\langle x|U|x'\rangle}{\langle x|x'\rangle}\,. \tag{8}$$

Yukawa identifies thus $\phi(x)$ as the local c-number field which then undergoes the usual second quantization procedure. $\phi(x)$ satisfies a wave equation, e.g. a Klein–Gordon equation

$$\left(\frac{\partial}{\partial x_\mu}\ \frac{\partial}{\partial x^\mu} - m^2\right)\ \ \phi(x) = 0\,. \tag{9}$$

This follows from the equation of motion at the U-operator level

$$[p_\mu, [p^\mu, U]] = m^2c^2U\,. \tag{10}$$

Equations (2) and (10) hence characterize, in Yukawa's scheme, a local field theory of a spinless particle of mass m and zero size.

Non-local field theories are then introduced by Yukawa through a modification of (2) to read

$$[q, U] \neq 0\,. \tag{11}$$

As a consequence of (11) we can no longer extract a $\delta^{(4)}(x - x')$ from (2) and we shall have

$$\langle x'|U|x\rangle = U(x', x)\,. \tag{12}$$

Similarly, if we start from eigenstates of p

$$p|k\rangle = k_\mu|k\rangle\,,$$
$$\langle k'|k\rangle = \delta^{(4)}(k - k') \qquad \left(|k\rangle = \frac{1}{(2\pi)}\int d^4x\ e^{ikx}|x\rangle\right)\,, \tag{13}$$

we have, for a local FT,

$$\langle k'|U|k\rangle = \underline{\phi}(k-k')\,, \tag{14}$$

and for a NLFT

$$\langle k'|U|k\rangle = \underline{U}(k,k') = \frac{1}{(2\pi)^4}\int d^4x\; d^4x'\; e^{ikx}e^{-ik'x'}U(x,x')\,. \tag{15}$$

It is convenient to introduce the coordinates

$$\begin{aligned} X &= \frac{x+x'}{2}\,, \qquad r = x-x'\,,\\ K &= \frac{k+k'}{2}\,, \qquad \varDelta = k-k'\,, \end{aligned} \tag{16}$$

and, using $kx - k'x' = Kr + \varDelta\cdot X$, *we have*

$$\begin{aligned} \langle x'|U|x\rangle &= U(x,r)\ \to \text{(in LFT limit)}\ \delta^{(4)}(\mathrm{r})\ \phi(\mathrm{x})\\ \langle k'|U|k\rangle &= \underline{U}(k,\varDelta)\ \to \text{(in LFT limit)}\ \underline{\phi}(\varDelta)\,, \end{aligned} \tag{17}$$

with

$$\begin{aligned} \underline{U}(k,\varDelta) &= \frac{1}{(2\pi)^4}\int d^4x\; d^4r\ \exp(iKr+\varDelta x)U(x,r)\,,\\ \underline{\phi}(\varDelta) &= \frac{1}{(2\pi)^4}\int d^4x\ \exp(i\varDelta x)\ \phi(x)\,. \end{aligned} \tag{18}$$

At this point, in order to restrict the possible choices of NLFT, Yukawa took inspiration from Born reciprocity principle and specified (11) to read:

$$[q_\mu,[q^\mu,U]\,] = \lambda^2 U\,, \tag{19}$$

where λ *has obviously dimensions of a length. Notice the close similarity with (10). Notice that, as a consequence of (10) and (19),*

$$[q,[q,U]\,] = \frac{\lambda^2}{m^2c^2}\,[p,[p,U]] = \frac{(\bar\lambda)^4}{h^2}\,[p,[p,U]]\,, \tag{20}$$

where $\bar\lambda = (\lambda^2\hbar^2\,m^{-2}c^{-2})^{1/4}$ *has also dimensions of a length. Hence,* $[q,[q,U]])$ *and* $[p,[p,U]]$ *are proportional with an assigned constant of proportionality. An immediate consequence of (19) is*

$$(r^2-\lambda^2)U(x,r) = 0 \Rightarrow U(x,r) = \delta(r^2-\lambda^2)\ \phi(x,r)\,, \tag{21}$$

and, as usual, from (10)

$$\underline{U}(k,\varDelta) = \delta(\varDelta^2-m^2)\ \underline{\phi}(k,\varDelta)\,. \tag{22}$$

(21) and (22) are the starting point of Yukawa's approach to a field theory describing particles of mass m *and radius* λ. *LFT is recovered by letting* $\lambda\to 0$.

Yukawa himself pointed out the difficulties inherent in constructing a dynamical system of equations of motions for U. In particular he stressed the fact that a differential formalism (Schrödinger equation) can be made very uneffective because the intial conditions cannot be specified on a spacelike surface as they involve some average over different times as well.

Further developments : field defined on a domain[2].

3 The Zero Slope (Local) Limit of Dual Models

If we want to understand in which sense the dual model can be seen as a non-local extension of ordinary field theory, we have to consider first its own local limit, i.e. the limit $\lambda^2 = \alpha' \to 0$.

This problem was first investigated by Scherk and then further examined by Scherk and others. The result of Scherk, for the generalized Beta function model (GBM), is quite simple. The dual n-point function is given by

$$A_n = \gamma^{n-2} (\alpha')^{(n-4)/2} \sum_{\{P\}} B_n^{\{P\}} , \tag{23}$$

where γ is the dimensionless dual coupling constant $\alpha' = \lambda^2$ and $B_n^{\{}P\}$ is the particular generalized B-function corresponding to the permutation $\{P\}$ of the external legs. B_n is dimensionless and thus the dimensionality of A_n (which comes from the dimensionality of the eigenstates of p, $|p_i\rangle$) is taken care of entirely by the factor $(\lambda)^{n-4}$.

Scherk's limit is defined as the limit of A_n for $\lambda^2 = \alpha' \to 0^+$ with $\gamma/\lambda \equiv g$ fixed (hence $\gamma \to 0$). The limit is taken while keeping $\alpha(m^2) = 0$ with m^2 also kept fixed. m is the mass of the external particles and is also the mass of the lowest state lying on the leading trajectory (assumed here to have $\alpha(0) < 0$). One can see immediately that, in such a limit, the coupling in front of $\sum B_n$ becomes

$$\gamma^{n-2} \, (\alpha')^{\frac{n-4}{2}} \; \longrightarrow \text{(Scherk limit)} \; g^{n-2} \, (\lambda^2)^{(n-3)} \; \longrightarrow 0 \text{ for } \lambda \to 0 \, , g \text{ fixed} \, . \tag{24}$$

The limit is thus $0(\lambda^{2n-6})$ unless B_n can be singular for $\lambda \to 0$. In fact, B_n can be exactly as singular as $(\lambda^2)^{-2n+6}$ if and only if one is sitting near a set of compatible lowest poles such as those of Fig. 1[3].

Hence for finite s_i this term survives as $\lambda \to 0$. An excited pole would not survive because

$$\frac{\Gamma(-\alpha_s)\Gamma(-\alpha_t)}{\Gamma(-\alpha_s - \alpha_t)} \to \frac{(-\alpha_t - 1)}{-\alpha_s + 1} \quad \to (\lambda \to 0) \; \frac{1}{\alpha_s - 1} = \frac{1}{\lambda^2 (s - \frac{1}{\lambda^2})} \to 1 \, . \tag{25}$$

[2] I probably meant to add some mention of a further development of Yukawa's work.

[3] A sketch of a multiperipheral tree-diagram with seven external legs and four internal propagators appears in the original version.

Hence this term is not of order $1/\lambda^2$ if s is finite. Of course, it becomes $O(1/\lambda^2)$ if s becomes $0(1/\lambda^2)$. We have to understand therefore that Scherk's limit also keeps all s_i finite in the limit. In other words, all momenta are supposed to be small compared to the scale $1/\lambda$. This is actually the only meaning we can give to a local limit.

In general, Scherk proved that A_n goes in the limit to the sum of all tree diagrams of a $(g/3!)\phi^3$ theory. A similar result could be proven for loops.

The result of Scherk is certainly correct. On the other hand, there can be other ways to take the limit $\lambda \to 0$. Take for instance the Lovelace–Shapiro model for $\pi\pi$ scattering

$$B_4 = \frac{\Gamma(1-\alpha_s)\,\Gamma(1-\alpha_t)}{\Gamma(1-\alpha_s-\alpha_t)}\,. \tag{26}$$

For s, t finite and $\alpha' \to 0$ with $\alpha(0) = 1/2$ and kept fixed,

$$B_4 \to \frac{\Gamma(1/2)\,\Gamma)1/2)}{\Gamma(0)} = 0(\lambda^2 s, \lambda^2 t) \to 0\,. \tag{27}$$

If it was not for the Adler zero, say $\alpha(0) = 1/3$ or in my original proposal, we would have found

$$B_4 \quad \to (\lambda \to 0) \;\; \text{constant}\,. \tag{28}$$

In general, $B_n \to$ const. for $\lambda \to 0$ unless the region of the small (finite) external momenta happens to take us near a pole (or several poles) of B_n. Hence $A_n \to \gamma^{n-2}(\lambda^2)^{n-4}$ and the limit depends on what we do with γ as $\lambda \to 0$. But, in any case, B_n has no structure on it, in the sense that no singularity appears. It is crucial to have in this limit $\alpha(0)$ fixed and not an integer. There is a little problem, however. Dual models can only be constructed, so far, for on-shell external particles at $p^2 = m^2$. If $m^2 = -(\alpha(0)/\alpha')$, $m^2 \to \infty$ as $\alpha' \to 0$ and therefore p_μ cannot be kept finite. If we let $p_\mu \to 0(1/\lambda)$ then we are back on top of the poles and we get again results à la Scherk. We notice, however, that

1) In a world of pion amplitudes with massless pions we can take p_μ finite. Then the only singularities come from pion poles but, because of the zero slope limit, their contributions are down if there is an Adler condition

$$B_6 \to B_4\,\frac{1}{\alpha' s}\;B_4 \to \lambda^2\,\frac{1}{\lambda^2}\;\lambda^2 \approx \lambda^2 \to 0\,. \tag{29}$$

2) One may hope that in a future formulation of the theory off shell amplitudes can be defined so that one can take the external momenta to be fixed as $\lambda \to 0$. In that limit, $B_n \to$ const.

3) *In a theory with external quarks of zero mass, not appearing as poles, the limit $\lambda \to 0$ (p_μ finite) is again conceivable.*

We note that, in dual models, keeping $\alpha(0)$ fixed is more natural than keeping $\alpha(m^2)$ fixed since many properties do depend on the value of $\alpha(0)$ (or $\alpha' m^2$) and not just on m^2.

We now want to argue that our $\lambda \to 0$ limit may be the correct one physically. This comes from the expression of B_n in the operator formalism. There, B_n is written as a vacuum-expectation value of a product of fields:

$$B_n = \int_0^{2\pi} d\tau_1 \dots d\tau_n \theta(\tau_i - \tau_{i-1}) \langle 0| V(k_1, \tau_1) \quad \dots \quad V(k_n, \tau_n) |0\rangle \,, \tag{30}$$

where

$$\begin{aligned} V(k,\tau) &= : \exp(ik \cdot Q(\tau)) : , \\ Q_\mu(\tau) &= q_\mu + 2\lambda^2 p_\mu \tau + \lambda\sqrt{2} \sum_n \left(\frac{a_{n,\mu}}{\sqrt{n}} \; e^{-in\tau} + \frac{a^\dagger_{n,\mu}}{\sqrt{n}} \; e^{in\tau} \right) , \\ [q_\mu, p_\nu] &= i g_{\mu\nu} \,, \\ [a_{n,\mu}, a^\dagger_{m,\nu}] &= \delta_{n,m} g_{\mu\nu} \,. \end{aligned} \tag{31}$$

For $\lambda \to 0$, $Q(\tau) \to q$ and the vertex $V(k,\tau)$ reduces to the usual $\exp(i\ k\ q)$ of an ordinary local theory and gives, up to a number,

$$B_n = \delta^{(4)}(k_1 + k_2 + \dots + k_n) \,, \tag{32}$$

hence the same as a local interaction ϕ^n to lowest order.

This limit can also be seen in the formalism of Ademollo et al. (strings in an external field) and in the expression of B_n given by Fubini and Veneziano:

$$B_n = \int d\tau_1 \dots d\tau_n \theta(\tau_i - \tau_{i-1}) \langle 0|\phi(\tau_1) \quad \dots \quad \phi(\tau_n)|0\rangle \,. \tag{33}$$

In conclusion the type of correspondence principle that we shall use in this paper will not be that dual amplitudes in the zero slope limit go into the trees of $g\phi^3$ but rather that each B_n goes to the first-order approximation of the highly non-linear local Lagrangian $\gamma^{n-2}\lambda^{n-4}\phi^n$. In the zero-slope limit, the whole dual model would then collapse into the first iteration of a non-polynomial Lagrangian of the type:

$$\mathcal{L}_{\text{int}}(\phi) \approx \frac{1}{\lambda^2} \phi^2 \;\; F(\gamma\, \lambda\, \phi) \,, \tag{34}$$

F being a function of the dimensionless quantity ($\gamma\ \lambda\ \phi$) which can be computed. Of course, for the sum over n the concept of leading order in λ is somehow lost.

4 The Correspondence Principle

We have seen that

$$Q_\mu(\tau) \quad \to (\lambda \to 0) \;\; q_\mu \,. \tag{35}$$

If we consider a field $\phi(q_\mu)$ *this is, in the sense of Yukawa, a local field whereas* $\phi(Q_\mu)$ *in general is not. Hence*

$$\phi(Q_\mu) \quad \to (\lambda \to 0) \;\; \phi(q_\mu) = \text{local field}\,. \tag{36}$$

Also we notice that

$$\overline{Q_\mu(\tau)} \equiv \frac{1}{2\pi} \int_{-\pi}^{+\pi} d\tau \; Q_\mu(\tau) = q_\mu \,. \tag{37}$$

Hence

$$\phi(\overline{Q_\mu(\tau)}) = \phi(q_\mu) = \text{local field}\,. \tag{38}$$

Our correspondence principle will be such that, in the non-local theory, the field at the average position goes into the average of field, i.e.

$$\begin{aligned} \phi(q_\mu) = \phi(\overline{Q_\mu(\tau)}) \quad &\to (\lambda \neq 0) \;\; \overline{\phi(Q_\mu(\tau))}\,, \\ \equiv \frac{1}{2\pi} \int_{-\pi}^{+\pi} \phi(Q_\mu(\tau)) \; d\tau \quad &\to (\lambda \to 0) \;\; \phi(q_\mu)\,, \end{aligned} \tag{39}$$

or, in terms of matrices:

$$\delta(x - x') \; \phi(x) = \langle x | \phi(\overline{Q}) | x' \rangle \to \langle x | \overline{\phi(Q)} | x' \rangle \,. \tag{40}$$

We clearly see that the process of averaging has introduced in a quite essential way a dependence of ϕ *or both* q_μ *and* p_μ *thus making the theory non-local in the sense of Yukawa. This has to be contrasted with recent attempts at constructing a field theory (of the* ∞*-component type) for dual model by introducing a field* $\varphi[X(\sigma, 0)]$ *i.e. a functional of* X *evaluated at* one *value of* τ*. Since* $[X(\sigma, 0),\; X(\sigma', 0)] = 0$*, this still keeps the theory local (multilocal to be more precise), i.e. diagonal in position-space:*

$$\langle x_1, x_2, \ldots x_n | \varphi | x_1', x_2' \ldots x_n' \rangle = \delta(x_1 - x_1') \; \delta(x_2 - x_2') \ldots \phi(x_1, x_2 \ldots x_n) \,. \tag{41}$$

We see that this field depends on half as many variables as our field.

In other words we insist on the physical idea that the extended nature of the dual hadron makes it impossible not only to define a field at a point in space, but also at a point in time. Namely the field one probes with a dual hadronic test body is an average field over a period of time and a region of space related by $\frac{\Delta x}{\Delta t} = c$*. This is even more transparent in the Shapiro-Virasoro model where the average is done over both* σ *and* τ*. The generalized Beta-function model is less symmetric because it corresponds to the case in which the test body is only active at the ends of the string. Yet it is not the* *same as having a pointlike test body since the motion of the ends results from that of the string as a whole.*

The introduction of non-locality is thus made necessary by the sinple fact that, if we average the field over a period of time, that average depends on the trajectory described by the test body. This depends on both the original position and velocity and hence classically ($h = 0$) it is a function of both x and p. Quantum-mechanically, x and p cannot be measured simultaneously and one gets therefore only a matrix representation in x (or p) space.

The above is actually the crucial point at which one is definitively departing from conventional theories. We do not claim that our interpretation is a necessary one for the dual model, but suggest that it is a possible one. Within such an interpretation we now show that dual amplitudes arise as a non-local extension of an ordinary, local Lagrangian (or S-matrix in lowest order).

For a single scalar field theory the only interaction one has is

$$L_I = \phi^n(x)\,. \tag{42}$$

To lowest order the S-matrix for scattering of n particles of n particles of momenta $k_1, k_2, \dots k_n$ is

$$\langle k_1, \dots k_n| \int d^4x\, \phi^n(x)|0, \dots, 0\rangle = \int d^4x e^{ik_1x} \dots e^{ik_nx} = \delta^{(4)}(k_1 + \dots + k_n)\,. \tag{43}$$

Let us see now how to use our correspondence principle to give a non-local extension of such a scattering amplitude. We write:

$$S^{(1)} = \int d^4x_1\, \phi^n(x_1) =$$
$$\int d^4x_1 \dots d^4x_n\, \phi(x_1)\delta(x_1 - x_2) \dots \phi(x_{n-1})\, \delta(x_{n-1} - x_n)\phi(x_n)\,. \tag{44}$$

Using our correspondence principle

$$S^{(1)} = \int d^4x_i \langle x_1|\overline{\phi}|x_2\rangle\, \langle x_2|\overline{\phi}|x_3\rangle\, \langle x_3|\overline{\phi}|x_{n-1}\rangle\, \langle x_{n-1}|\overline{\phi}|x_n\rangle\,. \tag{45}$$

Now the integrations over x_1 and x_n give $\langle 0|$ and $|0\rangle$ respectively (eigenstates of p_μ with zero eigenvalue) and the sum over intermediate x_i, $i = 2, 3, \dots, n-1$ can be replaced by completeness sums in the vector space of Q, $|x\rangle\langle x| = |p\rangle\langle p|$, as well as in the harmonic oscillator basis. Extracting finally the Fourier components with momenta $k_1 \dots k_n$ one finds

$$\langle S^{(1)}\rangle = \int d\tau_1 \, \dots d\tau_n \;\; \langle 0|\exp(ik_1Q(\tau_1)\exp(ik_2\, Q(\tau_2)\dots\exp(ik_n\, Q(\tau_n))|0\rangle\,. \tag{46}$$

This is exactly the n-point dual model provided we add an ordering constraint $\tau_1 \le \tau_2 \le \tau_3 \dots \le \tau_n$.

A hint for how to get the ordering comes from:

$$\int_0^{2\pi} d\tau_i \exp(i\sum k_i Q(\tau_i)) \\ \rightarrow \sum_{\text{orderings}} T_\tau \int_0^{2\pi} d\tau_i \langle 0|\exp(ik_1Q(\tau_{i1})\ldots\exp(ik_nQ(\tau_{in})|0\rangle\,. \tag{47}$$

One should get the l.h.s. of this equation as a first step. Hence this model is capable of producing the dual model interaction in a very natural way. Indeed, if we consider a closed string interacting at all values of σ we get the Shapiro-Virasoro model.

Actually the expression we have obtained is the n-point function only up to an infinite constant since (with $z_i = e^{i\tau_i}$),

$$\left(\frac{1}{2\pi}\right)^n \int_{\tau_i<\tau_{i+1}} d\tau_i \langle 0|V(k_i,\tau_i)|0\rangle = \int \frac{d\tau_a d\tau_b d\tau_c}{|z_a-z_b||z_b-z_c||z_c-z_a|}\cdot B_n = \infty\cdot B_n. \tag{48}$$

In other words, the local interaction giving rise to $A_n = g^{n-2}(\lambda)^{n-4} = B_n$ is

$$\mathcal{L}_{\text{int}} = \sum_{n=3} \mathcal{L}_{\text{int}}^{(n)} \quad , \quad \mathcal{L}_{\text{int}}^{(n)} = \lambda^{-4}\,\frac{G}{g^2}\,\frac{(g\phi\lambda)^n}{n!n}\,, \tag{49}$$

with

$$G = \left(\int \frac{d\tau_a d\tau_b d\tau_c}{|z_a-z_b||z_b-z_c||z_c-z_a|}\right)^{-1}.$$

$g\lambda$ and G would thus play the role of the so-called minor and major coupling constants of a non-polynomial Lagrangian.

One may ask where the infinity has been produced from since, after all, the $\alpha' \to 0$ limit should be finite. We see that the infinity is still there in the $\alpha' \to 0$ limit because our external masses have been fixed to $\alpha' k^2 = \alpha'\mu^2 = -1$; hence, $\mu^2 \to -\infty$ as $\alpha' \to 0$.

If our external masses would not be fixed at values of order $\sim 1/\sqrt{\alpha'}$, but at a finite value as $\alpha' \to 0$, we would not have produced an infinity. On the other hand, the model thus obtained would not have been dual (projective invariant) using with the volume element $\int\prod_i d\tau_i$. In order to get duality, we would have had to use a more complicated volume element such as

$$\int \frac{d\tau_i}{|z_i - z_{i+1}|\ldots}\,.$$

The ideal situation would be one in which the model is dual for external massless particles and, at the same time, it is free of infrared divergences for $\sqrt{\alpha'}k_i \to 0$. This could be possible in a chiral-invariant pion world with a non-integer ρ intercept.

5 Non-Local, Classical Field Theory

We discuss briefly here the case $\hbar = 0$, $\lambda \neq 0$, *i.e., the case of a non-local, non-quantized field theory.*

Having the dual model in mind, consider the classical motion of the free string. The end points of the string describe the classical trajectory (say for $\sigma = 0$*)*

$$x_\mu(\tau) = x_\mu + 2\lambda^2\, p_\mu \tau + i\sqrt{2}\lambda \sum_{n\neq 0} \frac{a_{n,\mu}}{\sqrt{n}}\, e^{-in\tau}\,. \tag{50}$$

This motion is "almost periodic", i.e., periodic with period $\tau_0 = 2\pi$ *up to a linear term* $2\lambda^2 p_\mu \tau$. *During such period of proper time the end of the string shifts its position by the amount* $2(\lambda p_\mu)\cdot(2\pi\lambda)$. *If* p_μ *is, say, in the* z *direction we have*

$$\Delta x = \Delta y = 0\,,\ \ \Delta z_0 = 4\pi\,\lambda\, p_z \lambda\,,\ \ \Delta\, t_0 = 4\pi\,\lambda\, p_0\,\lambda\,,\ \ \frac{\Delta z_0}{\Delta t_0} = \frac{p_z}{p_0} = v\,. \tag{51}$$

Hence $\mathbf{p}/p_0$ *is the* <u>*average*</u> *velocity of the end point.*

Suppose now that we want to define a field $\phi(x)$ *which interacts only with the end point of the string. Since in the classical case we know exactly the motion of the end point we can think of being able to specify the field* $\phi(x)$ *at all the points* $\phi(x_\mu(\tau))$ *namely along the trajectory described by the end point.*

On the other hand, even classically, we may think of having a measuring apparatus incapable of measuring the reactions of the string to the field in a time $\Delta t \to 0$ *and we may demand instead to measure its average reaction during a characteristic interval* $\Delta t_0 = 4\pi\,\lambda^2\, p_0$. *There is a further advantage to that. After a time* Δt_0 *we know exctly where the end of the string ought to be in the absence of ineractions if we just measure its total momentum. For a* $\Delta t \neq \Delta t_0$ *(or a multiple of it), the full knowledge of the internal motion is needed before we can disentangle the free motion from the one produced by the field. Of course we can take* $\Delta t = \Delta t_0/n$ *and we shall only need the first* n *harmonics.*

When the system is quantized this will be even harder. In this case we shall measure, instead of $\phi(x_\mu(\tau))$

$$\bar{\phi}(x_\mu) = \frac{1}{2\pi} \int_{-\pi}^{+\pi} \varphi(x_\mu + 2\lambda^2 p_\mu \tau + \sum_n \ldots) d\tau \ \ \to (\lambda \to 0)\ \ \phi(x_\mu)\,. \tag{52}$$

The field thus defined has become a functional of $x_\mu(\tau)$ *in a sense it is not only a function of what the field is but also depends on the state of the measuring apparatus. This seems to be the lesson to learn. For strong interactions the only way to measure them is to scatter strongly interacting probes. If these have a composite, extended structure then the field measured is a function of the internal motion of the probe as well as a function of actual sources. This may be the clue to duality.*

How should we generalize an interaction of the type $\lambda \int \phi^3(x)dx$? We can try to write

$$
\begin{aligned}
&\lambda \int dx \int_{-\pi}^{+\pi} d\tau_1 d\tau_2 d\tau_3 \; \phi(x+p_1\tau_1)\; \phi(x+p_2\tau_2)\phi(x+p_3\tau_3) \\
&= \lambda \int dx \int dx' dy dz \; \phi(x')\; \phi(y)\phi(x) \int_{-\pi}^{+\pi} d\tau_1 \; d\tau_2 \; d\tau_3 \; \delta(x'-x-p_1\tau_1) \\
&\quad \delta(y-x-p_2\tau)\; \delta(z-x-p_3\tau_1) \\
&= \lambda \int dx\; dy\; dz\; \phi(x)\; \phi(y)\; \phi(z) \int dw\; \theta_{p_1}(w-x)\; \theta_{p_2}(w-y)\; \theta_{p_3}(w-z) \\
&\equiv \lambda \int dx\; dy\; dz\; \phi(x)\; \phi(y)\; \phi(z)\; G(x,y,z;p_1,p_2,p_3)\,. \qquad (53)
\end{aligned}
$$

We get a smoothed interaction, with a smoothing function which depends on the momenta of the three interacting objects. If we let $\lambda \to 0$ at fixed p_i, or $p_i \to 0$ at fixed λ, we recover a local interaction: $G \to \delta(\;\;)\delta(\;\;)$.

6 Smeared Fields

6.1 ?

Consider $\bar{\phi}$ between non-excited states of coordinates x, x'. We have:

$$
\begin{aligned}
\langle x'|\bar{\phi}|x\rangle &= \frac{1}{2\pi}\int d\tau \langle x'|\phi(x_\mu(\tau))|x\rangle = \frac{1}{2\pi}\int d\tau \langle x'|\phi(x_\mu + 2\lambda^2 p_\mu \tau)|x\rangle \\
&= \frac{1}{2\pi}\int d\tau dy_\tau \langle x'|y_\tau\rangle \; \langle y_\tau|x\rangle \phi(y)\,. \qquad (54)
\end{aligned}
$$

Since

$$
\langle x'|y_\tau\rangle = \int d^4p e^{ipx'} e^{-ipy+i\lambda^2p^2\tau} = \int d^4p e^{ip(x'-y)} e^{+i\lambda^2p^2\tau}\,, \qquad (55)
$$

we find:

$$
\begin{aligned}
\langle x'|\bar{\phi}|x\rangle &= \frac{1}{2\pi}\int d\tau \int dy\phi(y) \int d^4p \int d^4p' e^{ip(x'-y)} e^{ip'(y-x)} \\
&= \frac{1}{2\pi}\int dy\phi(y) \int d^4p \int d^4p' \; \frac{sin\pi\lambda^2(p^2-p'^2)}{\pi\lambda^2(p^2-p'^2)} e^{ip(x'-y)} e^{ip'(y-x)} \\
&= \int d\tau \int dy\phi(y) \int d^4P \int d^4k e^{iP(x'-x)} e^{ik(x+x'-2y)} e^{i\lambda^2\tau P\cdot k} \\
&= \int dy\phi(y) \int d\tau \frac{1}{(\lambda^2\tau)^4} \exp\left(i\frac{(x-x')\cdot(x+x'-2y)}{\lambda^2\tau}\right) \\
&\equiv \int dy\phi(y) G(x,x',y)\,. \qquad (56)
\end{aligned}
$$

We can also write:

$$\langle x'|\bar{\phi}|x\rangle = \phi(x,x') = \phi(X,r) = \int dy\phi(y)G(X,r,y)$$
$$= \int dy\phi(y)\int d\tau\frac{1}{(\lambda^2\tau)^4}\exp\left(i\frac{r\cdot(X-2y)}{\lambda^2\tau}\right). \tag{57}$$

We can also get an expression for other quantities:

$$\phi(k,X) = \int d^4re^{ikr}\phi(X,r) = \int dy\phi(y)\frac{1}{2\pi}\int d\tau\frac{1}{(\lambda^2\tau)^4}\delta(\frac{X-2y}{\lambda^2\tau}-k)$$
$$= \frac{1}{2\pi}\int d\tau\phi(X-\lambda^2\tau k)\,, \tag{58}$$

or, in terms of the Fourier-transform of $\phi(y)$,

$$\phi(k,X) = \int d^4q\frac{sin\pi\lambda^2k\cdot q}{\pi\lambda^2k\cdot q}\phi(q)e^{(iqX)}\,, \tag{59}$$

$$\phi(k,\Delta) = \int d^4Xd^4r\exp\left(i(kr+\Delta X)\right)\phi(X,r)\,. \tag{60}$$

Also:

$$\phi(\Delta,r) = \int d^4Xe^{i\Delta X}\phi(X,r) = \frac{1}{2\pi}\int d\tau\delta^{(4)}(r-\lambda^2\tau\Delta)\phi(\Delta)$$
$$= \phi(\Delta)\Theta(-\pi\lambda^2\Delta_\mu < r_\mu < \pi\lambda^2\Delta_\mu)\,, \tag{61}$$

and

$$\phi(\Delta,p) = \int d^4re^{ipr}\phi(\Delta,r) = \int d\tau e^{i\lambda^2\tau p\cdot\Delta}\phi(\Delta)$$
$$= \phi(\Delta)\frac{sin\pi\lambda^2p\cdot\Delta}{\pi\lambda^2p\cdot\Delta}\,. \tag{62}$$

6.2 Various Types of $\phi(y)$ and $\frac{\Delta x}{\Delta p} = \lambda^2$

- $\phi(y) = \text{const}$

$$\langle x|\phi|x'\rangle = \text{const }\delta(r) \quad , \quad \langle r^2\rangle = 0\,. \tag{63}$$

- $\phi(y) = e^{iqy}$

$$\int dy\int d\tau\delta(q-\frac{r}{\lambda^2\tau})e^{iqX} \quad , \quad \sqrt{\langle r^2\rangle}\sim\sqrt{q^2}\lambda^2 \ , \quad \frac{\langle|r|\rangle}{\langle q\rangle} = \lambda^2\,. \tag{64}$$

- $\phi(y) = e^{iky}exp(-\frac{y^2}{\eta^2})$

$$\phi(r,X) = \int dy\int d\tau\frac{1}{(\lambda^2\tau)^4}\exp\left(i\frac{r\cdot X}{\lambda^2\tau}\right)\exp\left(-2i\frac{r\cdot y}{\lambda^2\tau}+iky\right)e^{-y^2/\eta^2}$$
$$= \int d\tau\frac{1}{(\lambda^2\tau)^4}\exp\left(i\frac{r\cdot X}{\lambda^2\tau}\right)\exp\left(-\eta^2(k-\frac{r}{\lambda^2\tau})\right). \tag{65}$$

Thus:

$$r \approx \lambda^2 \tau k \quad , \quad \langle \Delta r \rangle \approx \frac{\lambda^2}{\eta}$$
$$\langle \Delta y \rangle \approx \eta \quad , \quad \langle \Delta p \rangle \approx \frac{1}{\eta} r \, , \tag{66}$$

implying:

$$\frac{\Delta r}{\Delta p} \approx \lambda^2 \, . \tag{67}$$

Conclusion

If the local field is a wave packet of average momentum k and spread $1/\eta$, average position y_0 and spread η, the non-local version has a non-locality parameter $\Delta r \sim \frac{\lambda^2}{\eta}$. Hence $\frac{\Delta r}{\Delta p} \approx \lambda^2$, which is the new indetermination principle.

References

1. [4]
2. [5]

Appendix – Comments by the Author (March 2007)

According to Sect. 1, a seventh (and last) section should have contained some speculative remarks and an outlook, but apparently has never been written. Also, no bibliography has been found with the manuscript.

The following comments on this unpublished manuscript may be of interest and/or of help to the reader:

This draft was probably written at the beginning of 1973, i.e. around the time that QCD was introduced as a candidate theory of strong interactions, but before it was accepted as such. The discovery of asymptotic freedom, the idea of confinement, and the reinterpretation of dual resonance models and string theory as a large-N limit of QCD, have all probably contributed to convince me not to pursue any further the line of thought exposed in this manuscript and to keep it in a drawer.

However, many of the ideas presented there do acquire a definite interest in the context of the reinterpretation of string theory (some 11 years later and after rescaling the length parameter λ by some 20 orders of magnitude) as a unified quantum theory of all interactions, including gravity, Indeed, in my 1986 paper "A stringy Nature needs just two constants" (Europhys.

[4] Presumably a reference to QED precision tests.

[5] Presumably a reference to the proof of renormalizability of the GSW theory.

Lett. 2 (1986) 199), many of the themes presented in this draft, consciously or not, were taken up again. In particular, Born's reciprocity idea – and its implementation in Yukawa's approach – are among the issues common to both works.

Two points got clarified during the 13-year interval between the two papers:

- That α' and λ^2 are conceptually distinct: the first is the inverse of a classical tension, the second is a fundamental length appearing as a result of quantization;
- That Born's reciprocity works, in string theory, as a symmetry between $X' \equiv \partial_\sigma X(\sigma, \tau)$ and P, rather than between x and p, as in Born's or Yukawa's approaches. Precisely, this $X' \leftrightarrow P$ reciprocity gives rise to the famous T-duality of closed strings, or to the connection between Neumann and Dirichlet open strings.

A second point of the manuscript is its reinterpretation of the zero-slope limit of string theory as a low-energy limit in which it reduces to a QFT with a non-polynomial Lagrangian (unlike Scherk's limit of an ordinary QFT). This can be understood today as the result of "integrating out" the massive string modes when the external particles are light and soft. The non-polynomial nature of the Einstein–Hilbert action does indeed come this way in string theory. What was missing in the draft is the idea of defining a one-particle-irreducible functional (the effective action) to avoid the problems of singularities due to the exchange of massless quanta. This makes Sect. 3 somewhat hard to read.

Last, but not least, the manuscript contains (and by far!) the first claim that string theory should lead to a modified uncertainty principle whereby, besides the usual $\Delta x \Delta p > 2\pi\hbar$, the new constraint $\Delta x / \Delta p \sim \alpha'$ should also be imposed. There are statement in this direction in the above-mentioned 1986 paper of mine but not as clearly stated as in the draft (this makes me believe that, by 1986, I had lost track of the draft). It was not until 1989–1990, with the results coming from studying transplanckian string collisions, that the modified uncertainty principle was formulated (independently by D. Gross and by Amati, Ciafaloni and myself) in the form presented at the very end of the manuscript!

Part II

Dual Resonance Models and String Theory

The Birth of the Veneziano Model and String Theory

H. Rubinstein

Albanova Center, Fysikum, Stockholm, Sweden
rub@physto.se

Abstract. In this article I describe the work at the Weizmann Institute just before and when Gabriele arrived.

1 The Weizmann Institute in January 1966 and the Work Leading to the Veneziano Model

1.1 Preliminaries

After two years as a postdoc student at Orsay, France, I went to the Weizmann Institute in 1966. It lasted about 20 years. At Orsay, I had worked with several students: Bernard Diu, Jean Loup Gervais, and also with Jean Basdevant and the late Roger van Royen. Our interest then, one common to many physicists, was the theory of strong interactions. Rehovot was a sleepy town, with a large number of still unpaved streets. I had gone to Weizmann invited by Amos de Shalit to work with Harry Lipkin. The atmosphere at the Institute was very relaxed and friendly. The Department was small, just a few professors and very few students. The research was concentrated in nuclear and atomic physics, and some experimental particle physics led by people educated in cosmic rays experiments.

The Weizmann Institute had taken the lead in the reconciliation with Germany and several young German physicists came to Rehovot for long visits. Germany had instituted a scientific exchange programme, called Minerva, that was a key factor in rapid scientific development. We had a very active time, full of distinguished short time visitors and several postdocs.

The symmetry approach to particle physics $SU(3)$, $SU(6)$ and later $SU(6)_W$ was popular everywhere and Israel had been a leader in the subject thanks to the work of Racah in Jerusalem in atomic and nuclear physics. Amos De Shalit and Igal Talmi continued the tradition at Weizmann in nuclear physics. Harry Lipkin, originally a nuclear physicist, had turned his efforts to the recently

H. Rubinstein: *The Birth of the Veneziano Model and String Theory*, Lect. Notes Phys. **737**, 47–58 (2008)
DOI 10.1007/978-3-540-74233-6_3

established field of particle physics. Haim Harari and Moshe Kugler had recently finished their theses and had gone as postdocs to USA.

Lipkin has been inspired by the late Yuval Ne'eman who had returned from London after having proposed the octet model based on $SU(3)$ [1]. The model had been recently spectacularly confirmed by the discovery of the Ω.

Matters developed rapidly. Murray Gell-Mann, George Zweig and Ne'eman proposed the quarks and we started work in the subject with Lipkin and several graduate students: Moshe Elitzur and Hannah Stern amongst others. The quark idea was not popular in USA. There was great reluctance to accept theories that did not have observable asymptotic states.

At Weizmann and at CERN, and in other places, interest in quarks was intense. A great stream of visitors and postdocs like Florian Sheck, now at Mainz, worked with us in the subject [2].

We did work on the tensor mesons [3], and with Hannah Stern [4] in nucleon–antinuclean annihilation, explaining simply the large mesons multiplicity against phase space intuition. E. Teller wrote to me that we had invented quark chemistry! Also, hadronic mass relations were clarified in work together with P. Federman and I. Talmi [5]. We did show that the octet formula is not always correct, and related masses of the octet and decuplet of baryons without assumptions on the forces, except that these are two body forces.

1.2 The Players Arrive

After my stay at Orsay (1964–1965) my wife and I went to Argentina for 3 months before going to Weizmann. In Buenos Aires, I tought a course on $SU(3)$, invited by J.J. Giambiaggi (of dimensional regularization). One of the students was Miguel Virasoro. He immediately impressed me as an outstanding mind and soon we wrote a paper on $SU(3) \times SU(3)$. We did not published it.

As soon as I came to Weizmann I asked the Head of the Department, Igal Talmi, if we could bring Miguel to the Weizmann Institute. Talmi, without any information but my word, generously agreed to bring him as a postdoc. This was in March 1966. Miguel came in early 1967.

Gabriele Veneziano arrived to Weizmann to complete a Ph.D. in 1966. He came from Florence where he had worked with Raoul Gatto. He came as a student, since Italy did not have an equivalent to a Ph.D. degree.

Lipkin asked me to take him along as a student and I did. My natural inclination to do some dynamical calculations and not only symmetries coincided with Gabriele's interests, and we wrote two papers on commutation relations and Regge poles [6].

It became immediately obvious that Gabriele, as Miguel, was in a class of his own. We started to look to a variety of problems and in particular to analyticity sum rules (see below).

Fig. 1. Picture at the 1966 Rehovot Conference. From left: H. Dahmen, the author, Sergio Fubini, M. Virasoro (standing), G. Veneziano

Soon summer came and we all went abroad. I went to Texas, and afterwards to the Bohr Institute and Gothenburg, where my wife's family had a summer house in southern Sweden. Miguel came with me to Copenhagen and Gabriele went to Italy.

In Copenhagen I taught and I met Professsor Ziro Koba. Two students who will appear in other articles got interested in the subject. These were Holger Nielsen in Copenhagen and Lars Brink in Gothenburg (Fig. 1).

2 The Dominant Problems from 1950 to 1970

Strong interaction physics was the topic that attracted most attention. Driven by a large number of experiments on cross sections, discovery of new particles and resonances, a large classification effort was taking place.

The symmetry approach described above was very successful in classifying particles but it did not have a dynamical principle. It merely related particles and cross sections to other particles and cross sections.

Until the beginning of this period the dynamics has been based on trying to emulate quantum electrodynamics. These perturbative calculations proved unsuccessful.

Under the leadership of Goeffrey Chew at Berkeley the concept of bootstrap emerged. “There are no fundamental particles: any one is as important as any other” was the slogan. Field theory, that could then only be handled perturbatively, was unable to make sensible predictions.

S-Matrix theory became the dogma, and Lagrangian physics was declared obsolete. The analytic structure of scattering amplitudes would – so people thought – constrain the physics, and predictions would ensue.

Other ideas were pursued, like current algebra, which proved to be wanting in depth. Some interesting results like soft pion theorems and the Adler–Weissberger sum rule for the weak axial coupling were important, but the excitement, in my opinion, was not justified. Another field theoretic result that turned out to be most important was the triangle anomaly calculation that showed a very important difference between classical and quantum mechanics. It became a key consideration when field theory returned to the forum.

Field theory was dormant, but important work on spontaneous symmetry breaking and short distance expansions was advancing. It became essential 10 years later. It had to wait for the seminal contributions of Curtis Callan, Kurt Symanzik and Ken Wilson via the renormalization group. Ytzak Frishman and his students did mainly this type of work in the early period at Weizmann [7]. High-energy physics was dominated by the discovery of large number of resonances and new particles. Tulio Regge discovered that the Schrodinger equation allowed continuation in angular momentum for complex values, and linked resonances with different spin. Phenomenologists discovered that all invariant functions in a scattering processes had the form

$$A(s,t) = \beta(t) s^{\alpha(t)}, \tag{1}$$

where s is the direct channel energy and t the momentum transfer. The invariant functions when particles carried spin had to be properly chosen to ensure that only physical singularities were present. This became a sophisticated industry.

These Regge trajectories started to be filled with particles, and it was soon noticed that they were linear in J, and that different families had identical slope. This was evidence that a simple potential would not do. Linearity required physics beyond the potential model. What was nice was the connection of energy and momentum transfer, and the fact that high-spin particles could be fit to a straight line, and that the continuation to negative t corresponds to the scattering angle in the cross channel and could fit the angular distribution.

The steps that led to understanding the particle spectrum and couplings started with the work of Sergio Fubini and collaborators. The thick book by De Alfaro et al. [8] contains a detailed picture of the period. It is a remarkable fact that the book contains little field theory and is based mostly on S-matrix analyticity. The only exciting topics related to field theory were the infinite momentum frame and chiral symmetry.

The first success of the analyticity ideas was the exploration of superconvergence sum rules. Extracting an amplitude with the correct analyticity properties and the appropriate number of helicity flips to ensure a rapid t decrease reads

$$\int^{\infty} ImA(\nu, t)d\nu = 0. \qquad (2)$$

As such this equation is almost tautological, unless a dynamical idea is introduced.

In the first period, Fubini [9] saturated the equation with a few resonances and noticed that the relations between masses and couplings were reasonable. Because of the rapid convergence these sum rules became known as superconvergence relations. The position and strength of the resonances related one to the others.

The real watershed was a paper by R. Dolen, D. Horn and C. Schmid on duality in the pion–nucleon amplitude, written in the fall of 1967 [10]. This paper deviated from the dogma established by quantum electrodynamics.

Though electrodynamics had taught us that singularities in all channels are additive (this was called the interference model), this paper showed that information on the crossed channels was contained in the direct channel. In a pictorial way: the resonances average the symptotic behaviour which is dominated by the Regge trajectories in the t channel. The strong interactions have a different structure that cannot be described by perturbative mechanisms. In modern words, low-energy hadronic physics is a strong interaction realm, as QCD now shows quite clearly, due to infrared slavery.

This duality hypothesis was received with skepticism. When Gabriele proposed the model the reaction was quite negative (see Gabriele's letter, Figs. 2 and 3).

Several groups, including ours, added the Regge behaviour to the sum rules, generalizing the previous equation to

$$\int A(\nu, t)d\nu = \sum_r \beta_r(t) \frac{{\nu_m}^{\alpha_r+n+1}}{\alpha_r + n + 1}. \qquad (3)$$

Physically, one divides the integral in two parts: the low-energy part that is resonance dominated, and the high-energy part that is controlled by Regge trajectories. The next assumption is how to perform the saturation. Notice that it is not adding resonances in the other channel.

Already in the superconvergence case a few resonances saturated the sum rule quite well [11]. The solution gave relations between masses and couplings of resonances. In this scheme, the idea was to relate Regge parameters to resonances. By displacing ν_m we could see the contribution to the sum rule of the individual resonances! The more resonances one includes, the t dependence on both sides agreed better and better.

2.1 A Simple Theoretical Model: $\pi + \pi \to \pi + \omega$

Back in Rehovot more students joined me in the following academic year. Yoram Avni who died prematurely, Mordechai Bishari, Mordechai Milgrom,

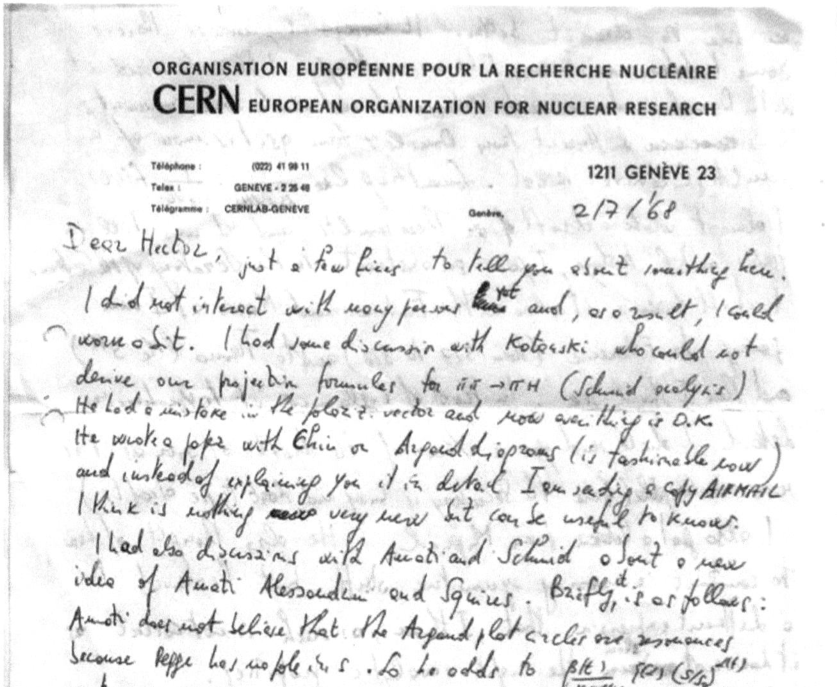
ORGANISATION EUROPÉENNE POUR LA RECHERCHE NUCLÉAIRE
CERN EUROPEAN ORGANIZATION FOR NUCLEAR RESEARCH

Téléphone : (022) 41 98 11
Telex : GENEVE - 2 25 48
Télégramme : CERNLAB-GENÈVE

1211 GENÈVE 23

Genève, 2/7/68

Dear Hector, just a few lines to tell you about something here.
I did not interact with many people yet and, as a result, I could work a bit. I had some discussion with Kotanski who could not derive our projection formulas for $\pi\pi \to \pi H$ (Schmid analysis). He had a mistake in the polariz. vector and now everything is O.K.
He wrote a paper with Chiu on Argand diagrams (is fashionable now) and instead of explaining you it in detail I am sending a copy AIRMAIL. I think is nothing very new but can be useful to know.
I had also discussions with Amati and Schmid about a new idea of Amati Alessandrini and Squires. Briefly, it is as follows: Amati does not believe that the Argand plot circles are resonances because Regge has no poles in s. So he adds to $\frac{\beta(t)}{\sin \pi\alpha(t)} \ldots (s/s_0)^{\alpha(t)}$ a term $\sum_\ell \frac{C_\ell}{s-s_\ell-i\Gamma} P_\ell(\cos\theta_t)$ calling this last poles s-channel resonances.
It is interference model, but, he claims, without double counting. This is because the C_ℓ are not the ones usually considered. In particular even for elastic scattering they are not positive definite because unitarity applies only to the whole amplitude. Furthermore we know from Schmid that the phase of Regge is similar to that of the resonances so experimentally it is hard to separate the two things but can be done if one has a theory of the background like here (Back = Regge exch.)
They seem to find good results by I don't yet agree with their definition of the resonances contribution. Also Schmid was quite sceptical especially about factorization in the s channel.
I have also been thinking about a possible symmetrical Regge form. On the ship I have got an idea and during these days at CERN I have

Fig. 2. Letter from Gabriele to H. R., August 1968, first part

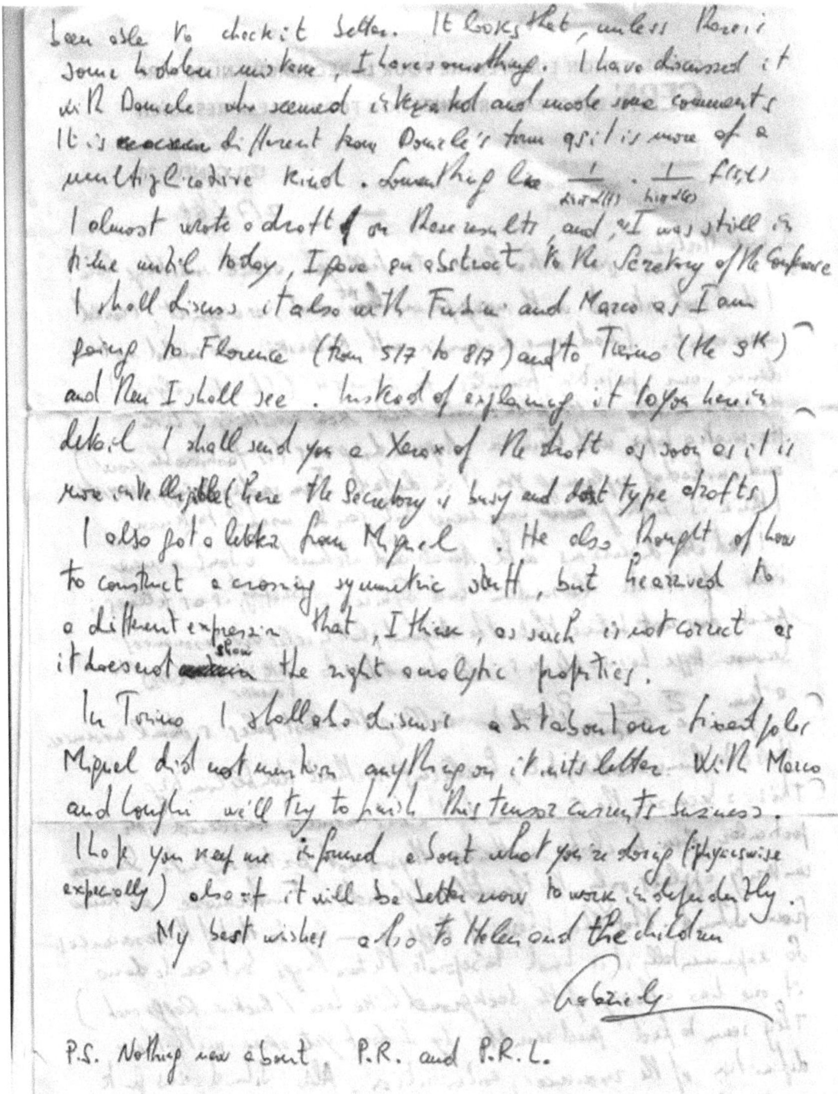
been able to check it better. It looks that, unless there is some hidden mistake, it gives something. I have discussed it with Daniele who seemed interested and made some comments. It is different from Daniele's form q.s. it is more of a multiplicative kind. Something like $\frac{1}{\sin\pi\alpha(s)} \cdot \frac{1}{\sin\pi\alpha(t)} f(s,t)$

I almost wrote a draft on these results, and, as I was still in time until today, I gave an abstract to the Secretary of the Conference.

I shall discuss it also with Fubini and Marco as I am going to Florence (from 5/7 to 8/7) and to Torino (the 9th) and then I shall see. Instead of explaining it to you here in detail I shall send you a Xerox of the draft as soon as it is more intelligible (here the Secretary is busy and don't type drafts).

I also got a letter from Miguel. He also thought of how to construct a crossing symmetric stuff, but arrived to a different expression that, I think, as such is not correct as it does not show the right analytic properties.

In Torino I shall also discuss a bit about our fixed poles. Miguel did not mention anything on it in its letter. With Marco and Longhi we'll try to finish this tensor currents business.

I hope you keep me informed about what you're doing (physicswise especially) also if it will be better now to work independently.

My best wishes also to Helen and the children

Gabriele

P.S. Nothing new about P.R. and P.R.L.

Fig. 3. Letter from Gabriele to H. R., August 1968, cont

Adam Schwimmer and Masud Chaichian amongst others. Also, the first generation of young people trained in particle physics in Israel by Ne'eman and Lipkin was returning.

The return process was not completed until 1968. These scientists included, the late Joe Dothan, Haim Harari, David Horn, Moshe Kugler and Shmuel Nussinov. Harari came from SLAC and had worked with Fred Gilman on dispersion relations and symmetries, somewhat related to our work. He and his students also became involved in sum rule work that led to the relation between the background and diffraction scattering, called the Harari–Freund hypothesis [12].

A small disgression is needed to understand the situation of theoretical and experimental particle physics almost everywhere at the time.

The quark model was still looked at with suspicion in most places, because light quarks were not produced as asymptotic states. Several difficult problems existed. The most puzzling was that the proton was lighter than the neutron. Many people tried to solve the problem but it remained. The other problem was the annoying behaviour of hadronic form factors. Infinite compositeness, as the bootstrap model required, predicted exponential supression with momentum transfer, as first pointed out by S. Mandelstam. The quark model gave the natural answer to both problems. The d quark being heavier than the u quark solves the first problem. We worked with Gabriele, Miguel and Daniele Amati collaborators on form factors. This work proved that proton compositeness required a q^{-4} form factor [13].

However, it was only after the deep inelastic experiments at SLAC that quarks became fashionable in USA.

We continued our work on sum rules and found something that turned out to be important. Together with Marco Ademollo and Adam Schwimmer we developed finite energy sum rules for hadronic amplitudes. We were inspired by Sergio Fubini's work, as already mentioned [14].

The showcase was the fully crossing symmetric $\pi + \pi \to \pi + \omega$ that would soon become the Veneziano model. The solution to these equations was simple and remarkable: they required linearly rising Regge trajectories, in agreement with the hadronic evidence. Moreover, even the couplings looked reasonable. Also, parallel trajectories at lower intercept spaced by 1 unit were unravelled [15]. Here the agreement was spectacular, and the coupling phenomenology worked very accurately. Other reactions like $\pi + \pi \to \pi + A_2$, the A_2 being a spin 2 meson, gave further information [16]. We soon studied all mesonic reactions and results were very good. Meson–nucleon reactions were not working that well. The model was not good for fermions.

3 The Breakthrough

We separated that summer and we all went to Europe planning to continue to USA in the fall. Gabriele made the seemingly trivial but decisive step. I received the letter shown in August.

From

$$A(s,t) = \beta(t) s^{\alpha(t)} \tag{4}$$

he wrote

$$A(s,t) = \frac{\Gamma(1-\alpha(s)\Gamma(1-\alpha(t)}{\Gamma((1-\alpha(s)-\alpha(t))}. \tag{5}$$

This equation has full symmetry between s and t, both at low energies and asymptotically, and obviously multiplies instead of adding different channel resonances [17].

In his paper he realized that the amplitude looked "almost right", but had inherent difficulties. It was difficult to see that it was the starting point of almost 40 years of research that had incredible physical and mathematical developments but has not yet achieved a credible form. The theoretical developments took place rapidly, Fubini and Veneziano [18] and Yohishiro Numbu [19] realized that the equation has a large degeneracy, the Hagedorn spectrum, Miguel with K. Kikkawa and B. Sakita [20] showed that it was a true theory by discovering the loop expansion. Further developments that led towards string theory include the work of many, and this will be covered by others contributors in this book. Lovelace [21] discovered that Lorentz invariance requires 26 space–time dimensions. But even then problems persisted.

The crucial step was Scherk's realization that the model must include gravity!

4 The Early Phenomenology

The appearance of Gabriele's paper, and its phenomenal impact at the International High Energy Vienna conference in September 1968, led to a deluge of papers. I was then starting to edit Nuclear Physics B, and we received in less than 3 months 200 papers on the subject.

The phenomenology of pion–nucleon scattering and five-point functions looked qualitatively promising but not really correct. I will concentrate here on the paper of C. Lovelace [22], later expanded by G. Altarelli and myself [23] on proton–antiproton annihilation at low energy. This paper is perhaps the most intriguing confirmation of the Veneziano formula besides what was already known from the FESR.

The reaction at rest can be thought as the disintegration of a pseudoscalar heavy meson composed by $\bar{p}n$ into three charged pions. By crossing symmetry it is, in a rough approximation, the scattering of a heavy pion on a pion giving two pions. So it is a Veneziano formula and, if the decay is to charged pions, by exoticity there is only one term!

The Dalitz plot was known from Anninos et al. [24] and it is quite remarkable (see Fig. 4). First, as seen in the figure, it has a hole at the centre, and second, doing the conventional fitting with the interference model, it predicted a low-energy resonance that has never been seen. The duality explanation, embodied in the Veneziano formula, explains these features naturally: the hole was a zero caused by the denominator (see (5)) when $\alpha(s) = \alpha(t) = 1/2$. The resonance in the direct channel is a reflection of a resonance in the t channel.

Improvements to the formula make the result plausible, although we know that the theory is inconsistent, and the agreement may all be an accident.

However, it is possible that a consistent string model of QCD will conserve the relevant features of the Veneziano tree-level amplitude curing its shortcomings. At this stage the jury is out.

5 Conclusion

The period 1967–1970 was indeed very productive at Weizmann. The work that culminated in the Veneziano amplitude and the loop expansion of Kikkawa, Sakita and Virasoro made it a respectable theory. But the jewel of the crown was the construction of the Veneziano model.

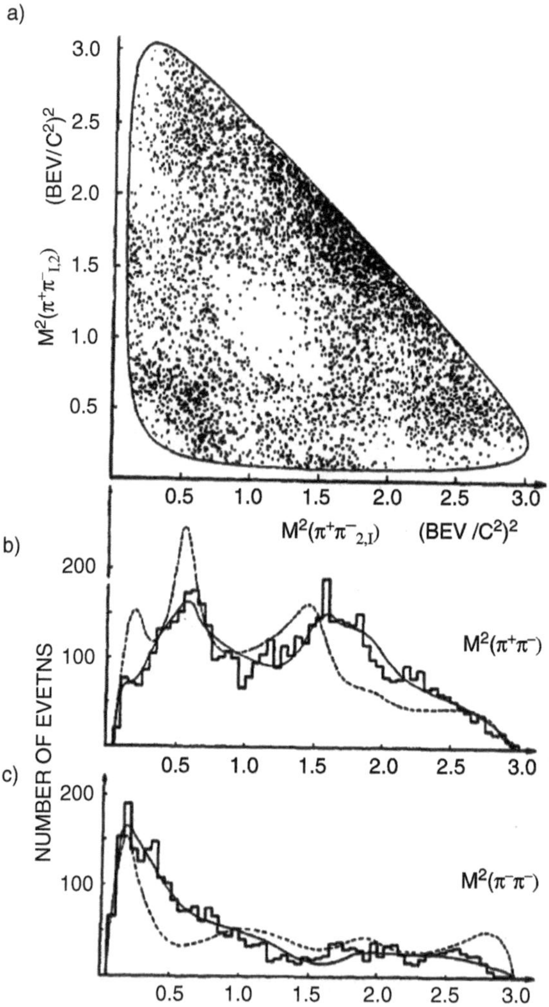

Fig. 4. Dalitz plot of the reaction pseudoscalar to three pions

Gabriele returned to Rehovot after an absence, but soon moved to CERN. Lately he has become interested in cosmology. Miguel moved also to other research topics, and myself also moved to cosmology. Adam Schwimmer is the only of us that has kept working mainly on string theory. Nathan Seiberg, David Kutasov, Offer Aharony and Michael Berkooz became the new Rehovot forces in particle physics.

The developments that led to string theory, in particular the work of Gabriele and Miguel, have not been fully rewarded. It is unquestionable that the little formula above opened a new chapter in theoretical physics that led to important developments with repercussions also in mathematics.

Gabriele's creativity has left an indeleble mark in high-energy physics and cosmology.

References

1. Yuval Ne'eman: Nucl. Phys. **26**, 222 (1961)
2. H. R. Rubinstein, F. Scheck, R. Socolow: Phys. Rev. **154**, 1608 (1967)
3. M. Elitzur, H. J. Lipkin, H. R. Rubinstein, H. Stern: Phys. Rev. Lett. **17**, 420 (1966)
4. H. R. Rubinstein, H. Stern: Phys. Lett. B **21**, 447 (1966)
5. P. Federman, H. R. Rubinstein, I. Talmi: Phys. Lett. B **22**, 208 (1966); H. R. Rubinstein, Phys. Rev. Lett. **17**, 31 (1966)
6. H. R. Rubinstein, G. Veneziano: Phys. Rev. Lett. **18**, 411 (1967); H.R. Rubinstein, G. Veneziano: Phys. Rev. **160**, 5 (1967)
7. Y. Frishman: Phys. Rev. Lett. **25**, 966 (1970)
8. V. De Alfaro, S. Fubini, G. Furlan, C. Rossetti: *Currents in Hadron Physics* (North-Holland, Amsterdam 1973)
9. V. De Alfaro, S. Fubini, C. Rossetti: Nuovo Cimento. Suppl. **6**, 575, 1968
10. R. Dolen, D. Horn, C. Schmid: Phys. Rev. Lett. **19**, 402 (1967)
11. S. Fubini, G. Furlan, C. Rossetti: Nuovo Cimento. **43** 1611. (1966)
12. H. Harari: Proc. Roy. Soc. Lond. A **318**, 355 (1970); P. Freund: Lett. Nuovo Cimento. **4**,147 (1970)
13. D. Amati, R. Jengo, H. R. Rubinstein, G. Veneziano, M. Virasoro: Phys. Lett. B **27**, 38 (1968)
14. M. Ademollo, H. R. Rubinstein, G. Veneziano, M. Virasoro: Phys. Rev. Lett. **19**, 1402 (1967); and also M. Ademollo, H. R. Rubinstein, G. Veneziano, M. Virasoro: Phys. Rev. **176** 1904, 1968
15. H. R. Rubinstein, A. Schwimmer, G. Veneziano, M. Virasoro: Phys. Rev Lett. **21**, 491 (1968)
16. M. Bishari, H. R. Rubinstein , A. Schwimmer, G. Veneziano: Phys. Rev. **176**, 1926 (1968)
17. G. Veneziano: Nuovo Cimento. A **57**, 190 (1968)
18. S. Fubini, G. Veneziano: Nuovo Cimento. A **64**, 811 (1969); Y. Nambu: lecture to be delivered at Copenhagen. The lecture was never delivered because of an accident.
19. Y. Nambu, Copenhagen undelivered lecture

20. S. Kikkawa, B. Sakita, M. Virasoro: Phys. Rev. D **1**, 3258 (1970)
21. C. Lovelace: Proc. Roy. Soc. Lond. A **318** 321 (1970)
22. C. Lovelace: Phys. Lett. B **28**, 269 (1968)
23. G. Altarelli, H. R. Rubinstein: Phys. Rev. **185**, 1469 (1969)
24. P. Anninos et al.: Phys. Rev. Lett. **20**, 402 (1968)

The Birth of String Theory

P. Di Vecchia

Nordita, Blegdamsvej 17, 2100 Copenhagen Ø, Denmark
divecchi@nbi.dk

Abstract. In this contribution we go through the developments that in the years from 1968 to about 1974 led from the Veneziano model to the bosonic string theory. They include the construction of the N-point amplitude for scalar particles, its factorization through the introduction of an infinite number of oscillators and the proof that the physical subspace was a positive-definite Hilbert space. We also discuss the zero slope limit and the calculation of loop diagrams. Lastly, we describe how it finally was recognized that a quantum-relativistic string theory was the theory underlying the Veneziano model.

1 Introduction

The 1960s was a period in which strong interacting processes were studied in detail using the newly constructed accelerators at CERN and other places. Many new hadronic states were found that appeared as resonant peaks in various cross sections, and hadronic cross sections were measured with increasing accuracy. In general, the experimental data for strongly interacting processes were rather well understood in terms of resonance exchanges in the direct channel at low energy, and by the exchange of Regge poles in the transverse channel at higher energy. Field theory that had been very successful in describing QED seemed useless for strong interactions, given the big number of hadrons to accommodate in a Lagrangian and the strength of the pion–nucleon coupling constant that did not allow perturbative calculations. The only domain in which field theoretical techniques were successfully used was current algebra. Here, assuming that strong interactions were described by an almost chiral invariant Lagrangian, that chiral symmetry was spontaneously broken and that the pion was the corresponding Goldstone boson, field theoretical methods gave rather good predictions for scattering amplitudes involving pions at very low energy. Going to higher energy was, however, not possible with these methods.

Because of this, many people started to think that field theory was useless to describe strong interactions, and tried to describe strong interacting

P. Di Vecchia: *The Birth of String Theory*, Lect. Notes Phys. **737**, 59–118 (2008)
DOI 10.1007/978-3-540-74233-6_4

processes with alternative and more phenomenological methods. The basic ingredients for describing the experimental data were at low energy the exchange of resonances in the direct channel, and at higher energy the exchange of Regge poles in the transverse channel. Sum rules for strongly interacting processes were saturated in this way, and one found good agreement with the experimental data that came from the newly constructed accelerators. Because of these successes, and of the problems that field theory encountered to describe the data, it was proposed to construct directly the S matrix without passing through a Lagrangian. The S matrix was supposed to be constructed from the properties that it should satisfy, but there was no clear procedure on how to implement this construction.[1] The word "bootstrap" was often used as the way to construct the S matrix, but it did not help very much to get an S matrix for the strongly interacting processes.

One of the basic ideas that led to the construction of an S matrix was that it should include resonances at low energy and at the same time give Regge behaviour at high energy. But the two contributions of the resonances and of the Regge poles should not be added because this would imply double counting. This was called Dolen, Horn and Schmidt duality [2]. Another idea that helped in the construction of an S matrix was planar duality [3] that was visualized by associating to a certain process a duality diagram, shown in Fig. 1, where each meson was described by two lines representing the quark and the antiquark. Finally, also the requirement of crossing symmetry played a very important role.

Starting from these ideas Veneziano [4] was able to construct an S matrix for the scattering of four mesons that, at the same time, had an infinite number of zero width resonances lying on linearly rising Regge trajectories and Regge behaviour at high energy. Veneziano originally constructed the model for the process $\pi\pi \rightarrow \pi\omega$, but it was immediately extended to the scattering of four scalar particles.

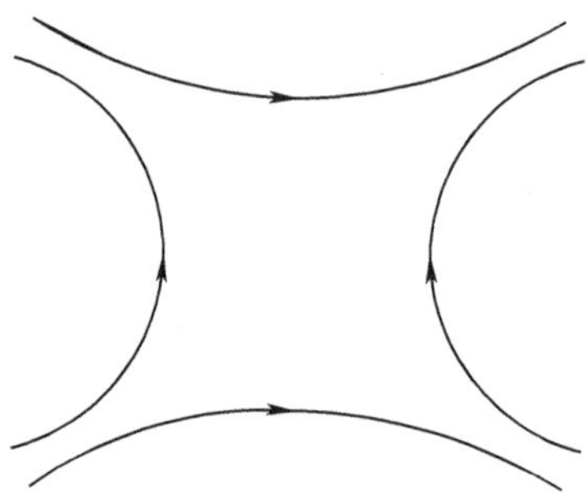

Fig. 1. Duality diagram for the scattering of four mesons

[1] For a discussion of S matrix theory see [1].

In the case of four identical scalar particles, the crossing symmetric scattering amplitude found by Veneziano consists of a sum of three terms:

$$A(s,t,u) = A(s,t) + A(s,u) + A(t,u) \tag{1}$$

where

$$A(s,t) = \frac{\Gamma(-\alpha(s))\Gamma(-\alpha(t))}{\Gamma(-\alpha(s)-\alpha(t))} = \int_0^1 dx x^{-\alpha(s)-1}(1-x)^{-\alpha(t)-1} \tag{2}$$

with linearly rising Regge trajectories

$$\alpha(s) = \alpha_0 + \alpha' s \tag{3}$$

This was a very important property to implement in a model because it was in agreement with the experimental data in a wide range of energies. s, t and u are the Mandelstam variables:

$$s = -(p_1+p_2)^2 \quad , \quad t = -(p_3+p_2)^2 \quad , \quad u = -(p_1+p_3)^2 \tag{4}$$

The three terms in (1) correspond to the three orderings of the four particles that are not related by a cyclic or anticyclic[2] permutation of the external legs. They correspond, respectively, to the three permutations: (1234), (1243) and (1324) of the four external legs. They have only simple pole singularities. The first one has only poles in the s and t channels, the second only in the s and u channels and the third only in the t and u channels. This property follows directly from the duality diagram that is associated to each inequivalent permutation of the external legs. In fact, at that time one used to associate to each of the three inequivalent permutations a duality diagram where each particle was drawn as consisting of two lines that represented the quark and antiquark making up a meson. Furthermore, the diagram was supposed to have only poles singularities in the planar channels which are those involving adjacent external lines. This means that, for instance, the duality diagram corresponding to the permutation (1234) has only poles in the s and t channels as one can see by deforming the diagram in the plane in the two possible ways shown in Fig. 2.

This was a very important property of the duality diagram that makes it qualitatively different from a Feynman diagram in field theory where each diagram has only a pole in one of the three s, t and u channels and not simultaneously in two of them. If we accept the idea that each term of the sum in (1) is described by a duality diagram, then it is clear that we do not need to add terms corresponding to equivalent diagrams because the corresponding duality diagram is the same and has the same singularities. It is now clear

[2] An anticyclic permutation corresponding, for instance, to the ordering (1234) is obtained by taking the reverse of the original ordering (4321) and then performing a cyclic permutation.

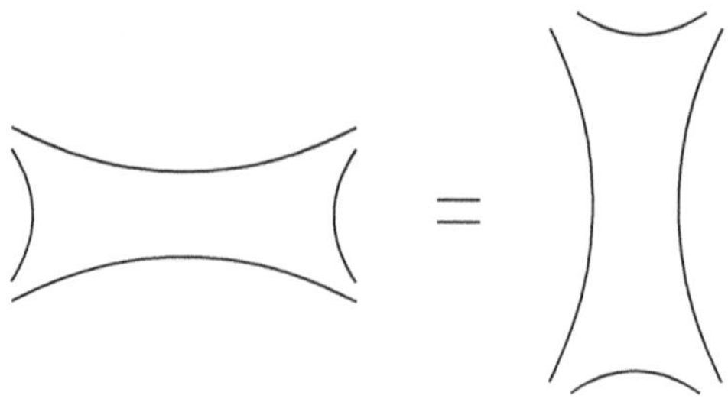

Fig. 2. The duality diagram contains both s and t channel poles

that it was in some way implicit in this picture the fact that the Veneziano model corresponds to the scattering of relativistic strings. But at that time the connection was not obvious at all. The only S matrix property that the Veneziano model failed to satisfy was the unitarity of the S matrix. because it contained only zero width resonances, and did not have the various cuts required by unitarity. We will see how this property will be implemented.

Immediately after the formulation of the Veneziano model, Virasoro [5] proposed another crossing symmetric four-point amplitude for scalar particles that consisted of a unique piece given by

$$A(s,t,u) \sim \frac{\Gamma(-\frac{\alpha(u)}{2})\Gamma(-\frac{\alpha(s)}{2})\Gamma(-\frac{\alpha(t)}{2})}{\Gamma(1+\frac{\alpha(u)}{2})\Gamma(1+\frac{\alpha(s)}{2})\Gamma(1+\frac{\alpha(t)}{2})} \tag{5}$$

where

$$\alpha(s) = \alpha_0 + \alpha' s \tag{6}$$

The model had poles in all three s, t and u channels and could not be written as sum of three terms having poles only in planar diagrams. In conclusion, the Veneziano model satisfies the principle of planar duality being a crossing symmetric combination of three contributions each having poles only in the planar channels. On the other hand, the Virasoro model consists of a unique crossing symmetric term having poles in both planar and non-planar channels.

The attempts to construct consistent models that were in good agreement with the strong interaction phenomenology of the 1960s boosted enormously the activity in this research field. The generalization of the Veneziano model to the scattering of N scalar particles was built, an operator formalism consisting of an infinite number of harmonic oscillators was constructed and the complete spectrum of mesons was determined. It turned out that the degeneracy of states grew up exponentially with the mass. It was also found that the N-point amplitude had states with negative norm (ghosts) unless the intercept of the Regge trajectory was $\alpha_0 = 1$ [6]. In this case it turned out that the model was free of ghosts but the lowest state was a tachyon. The model was called in the literature the “dual resonance model”.

The model was not unitary because all the states were zero width resonances and the various cuts required by unitarity were absent. The unitarity was implemented in a perturbative way by adding loop diagrams obtained by sewing some of the external legs together after the insertion of a propagator. The multiloop amplitudes showed a structure of Riemann surfaces. This became obvious only later when the dual resonance model was recognized to correspond to scattering of strings.

But the main problem was that the model had a tachyon if $\alpha_0 = 1$ or had ghosts for other values of α_0 and was not in agreement with the experimental data: α_0 was not equal to about $\frac{1}{2}$ as required by experiments for the ρ Regge trajectory and the external scalar particles did not behave as pions satisfying the current algebra requirements. Many attempts were made to construct more realistic dual resonance models, but the main result of these attempts was the construction of the Neveu–Schwarz [7] and the Ramond [8] models, respectively, for mesons and fermions. They were constructed as two independent models and only later were recognized to be two sectors of the same model. The Neveu–Schwarz model still contained a tachyon that only in 1976 through the GSO projection was eliminated from the physical spectrum. Furthermore, it was not properly describing the properties of the physical pions.

Actually a model describing $\pi\pi$ scattering in a rather satisfactory way was proposed by Lovelace and Shapiro [9].[3] According to this model the three isospin amplitudes for pion–pion scattering are given by

$$A^0 = \frac{3}{2}\left[A(s,t) + A(s,u)\right] - \frac{1}{2}A(t,u)$$

$$A^1 = A(s,t) - A(s,u) \qquad\qquad A^2 = A(t,u) \tag{7}$$

where

$$A(s,t) = \beta \frac{\Gamma(1-\alpha(s))\Gamma(1-\alpha(t))}{\Gamma(1-\alpha(t)-\alpha(s))} \quad ; \quad \alpha(s) = \alpha_0 + \alpha' s \tag{8}$$

The amplitudes in (7) provide a model for $\pi\pi$ scattering with linearly rising Regge trajectories containing three parameters: the intercept of the ρ Regge trajectory α_0, the Regge slope α' and β. The first two can be determined by imposing the Adler's self-consistency condition, that requires the vanishing of the amplitude when $s = t = u = m_\pi^2$ and one of the pions is massless, and the fact that the Regge trajectory must give the spin of the ρ meson that is equal to 1 when $\sqrt{s}$ is equal to the mass of the ρ meson m_ρ. These two conditions determine the Regge trajectory to be

$$\alpha(s) = \frac{1}{2}\left[1 + \frac{s - m_\pi^2}{m_\rho^2 - m_{\pi^2}}\right] = 0.48 + 0.885s \tag{9}$$

[3] See also [10].

Having fixed the parameters of the Regge trajectory the model predicts the masses and the couplings of the resonances that decay in $\pi\pi$ in terms of a unique parameter β. The values obtained are in reasonable agreement with the experiments. Moreover, one can compute the $\pi\pi$ scattering lenghts:

$$a_0 = 0.395\beta \qquad\qquad a_2 = -0.103\beta \tag{10}$$

and one finds that their ratio is within 10% of the current algebra ratio given by $a_0/a_2 = -7/2$. The amplitude in (8) has exactly the same form as that for four tachyons of the Neveu–Schwarz model with the only apparently minor difference that $\alpha_0 = 1/2$ (for $m_\pi = 0$) instead of 1 as in the Neveu–Schwarz model. This difference, however, implies that the critical space–time dimension of this model is $d = 4$[4] and not $d = 10$ as in the Neveu–Schwarz model. In conclusion, this model seems to be a perfectly reasonable model for describing low-energy $\pi\pi$ scattering. The problem is, however, that nobody has been able to generalize it to the multipion scattering and therefore to get the complete meson spectrum.

As we have seen the S matrix of the dual resonance model was constructed using ideas and tools of hadron phenomenology of the end of the 1960s. Although it did not seem possible to write a realistic dual resonance model describing the pions , it was nevertheless such a source of fascination for those who actively worked in this field at that time for its beautiful internal structure and consistency that a lot of energy was used to investigate its properties and for understanding its basic structure. It turned out with great surprise that the underlying structure was that of a quantum-relativistic string.

The aim of this contribution is to explain the logic of the work that was done in the years from 1968 to 1974[5] in order to uncover the deep properties of this model that appeared from the beginning to be so beautiful and consistent to deserve an intensive study.

This seems to me a very good way of celebrating the 65th anniversary of Gabriele who is the person who started and also contributed to develop the whole thing with his deep physical intuition.

2 Construction of the N-point Amplitude

We have seen that the construction of the four-point amplitude is not sufficient to get information on the full hadronic spectrum because it contains only those hadrons that couple to two ground state mesons and does not see those intermediate states which only couple to three or to a higher number of ground state mesons [12]. Therefore, it was very important to construct the N-point amplitude involving identical scalar particles. The construction of the N-point

[4] This can be checked by computing the coupling of the spinless particle at the level $\alpha(s) = 2$ and seeing that it vanishes for $d = 4$.

[5] Reviews from this period can be found in [11].

amplitude was done in [13] (extending the work of [14]) by requiring the same principles that have led to the construction of the Veneziano model, namely the fact that the axioms of S-matrix theory be satisfied by an infinite number of zero width resonances lying on linearly rising Regge trajectories and planar duality.

The fully crossing symmetric scattering amplitude of N identical scalar particles is given by a sum of terms corresponding to the inequivalent permutations of the external legs:

$$A = \sum_{n=1}^{N_p} A_n \tag{11}$$

Also in this case two permutations of the external legs are inequivalent if they are not related by a cyclic or anticyclic permutation. N_p is the number of inequivalent permutations of the external legs and is equal to $N_p = \frac{(N-1)!}{2}$ and each term has only simple pole singularities in the planar channels. Each planar channel is described by two indices (i, j), to mean that it includes the legs $i, i+1, i+2 \dots j-1, j$, by the Mandelstam variable

$$s_{ij} = -(p_i + p_{i+1} + \cdots + p_j)^2 \tag{12}$$

and by an additional variable u_{ij} whose role will become clear soon. It is clear that the channels (ij) and $(j+1, i-1)$[6] are identical and they should be counted only once. In the case of N identical scalar particles the number of planar channels is equal to $\frac{N(N-3)}{2}$. This can be obtained as follows. The independent planar diagrams involving the particle 1 are of the type $(1, i)$ where $i = 2 \dots N-2$. Their number is $N-3$. This is also the number of planar diagrams involving the particle 2 and not the 1. The number of planar diagrams involving the particle 3 and not the particles 1 and 2 is equal to $N-4$. In general the number of planar diagrams involving the particle i and not the previous ones from 1 to $i-1$ is equal to $N-1-i$. This means that the total number of planar diagram is equal to

$$2(N-3) + \sum_{i=3}^{N-2} (N-1-i) = 2(N-3) + \sum_{i=1}^{N-4} i$$

$$= 2(N-3) + \frac{(N-4)(N-3)}{2} = \frac{N(N-3)}{2} \tag{13}$$

If one writes down the duality diagram corresponding to a certain planar ordering of the external particles, it is easy to see that the diagram can have simultaneous pole singularities only in $N-3$ channels. The channels that allow simultaneous pole singularities are called compatible channels, the others are

[6] This channel includes the particles $(j+1, \dots, N, 1, \dots i-1)$.

called incompatible. Two channels $(i,\, j)$ and $(h,\, k)$ are incompatible if the following inequalities are satisfied:

$$i \leq h \leq j \quad ; \quad j+1 \leq k \leq i-1 \tag{14}$$

The aim is to construct the scattering amplitude for each inequivalent permutation of the external legs that has only pole singularities in the $\frac{N(N-3)}{2}$ planar channels. We have also to impose that the amplitude has simultaneous poles only in $N-3$ compatible channels. In order to gain intuition on how to proceed, we rewrite the four-point amplitude in (2) as follows:

$$A(s,t) = \int_0^1 du_{12} \int_0^1 du_{23}\; u_{12}^{-\alpha(s_{12})-1} u_{23}^{-\alpha(s_{23})-1} \delta(u_{12}+u_{23}-1) \tag{15}$$

where u_{12} and u_{23} are the variables corresponding to the two planar channels (12) and (23) and the cancellation of simultaneous poles in incompatible channels is provided by the δ-function which forbids u_{12} and u_{23} to vanish simultaneously.

We will now extend this procedure to the N-point amplitude. But for the sake of clarity let us start with the case of $N = 5$ [14]. In this case we have five planar channels described by $u_{12}, u_{13}, u_{23}, u_{24}$ and u_{34}. Since we have only two compatible channels only two of the previous five variables are independent. We can choose them to be u_{12} and u_{13}. In order to determine the dependence of the other three variables on the two independent ones, we exclude simultaneous poles in incompatible channels. This can be done by imposing relations that prevent variables corresponding to incompatible channels to vanish simultaneously. A sufficient condition for excluding simultaneous poles in incompatible channels is to impose the conditions:

$$u_P = 1 - \prod_{\bar{P}} u_{\bar{P}} \tag{16}$$

where the product is over the variables $\bar{P}$ corresponding to channels that are incompatible with P. In the case of the five-point amplitude, we get the following relations:

$$u_{23} = 1 - u_{34}u_{12} \quad ; \quad u_{24} = 1 - u_{13}u_{12}$$

$$u_{13} = 1 - u_{34}u_{24} \quad ; \quad u_{34} = 1 - u_{23}u_{13} \quad ; \quad u_{12} = 1 - u_{24}u_{23} \tag{17}$$

Solving them in terms of the two independent ones we get

$$u_{23} = \frac{1-u_{12}}{1-u_{12}u_{13}} \quad ; \quad u_{34} = \frac{1-u_{13}}{1-u_{12}u_{13}} \quad ; \quad u_{24} = 1 - u_{12}u_{13} \tag{18}$$

In analogy with what we have done for the four-point amplitude in (15) we write the five-point amplitude as follows:

$$\int_0^1 du_{12} \int_0^1 du_{13} \int_0^1 du_{23} \int_0^1 du_{24} \int_0^1 du_{34} u_{12}^{-\alpha(s_{12})-1} u_{13}^{-\alpha(s_{13})-1}$$

$$\times u_{24}^{-\alpha(s_{24})-1} u_{23}^{-\alpha(s_{23})-1} u_{34}^{-\alpha(s_{34})-1}$$

$$\times \delta(u_{23} + u_{12}u_{34} - 1)\delta(u_{24} + u_{12}u_{13} - 1)\delta(u_{34} + u_{13}u_{23} - 1) \qquad (19)$$

Performing the integral over the variables u_{23}, u_{24} and u_{34} we get

$$\int_0^1 du_{12} \int_0^1 du_{13} u_{12}^{-\alpha(s_{12})-1} u_{13}^{-\alpha(s_{13})-1}$$

$$\times (1 - u_{12})^{-\alpha(s_{23})-1} (1 - u_{13})^{-\alpha(s_{13})-1} (1 - u_{12}u_{13})^{-\alpha(s_{24})+\alpha(s_{23})+\alpha(s_{34})} \qquad (20)$$

We have implicitly assumed that the Regge trajectory is the same in all channels and that the external scalar particles have the same common mass m and are the lowest lying states on the Regge trajectory. This means that their mass is given by

$$\alpha_0 - \alpha' p_i^2 = 0 \;\; ; \;\; p_i^2 \equiv -m^2 \qquad (21)$$

Using then the relation

$$\alpha(s_{23}) + \alpha(s_{34}) - \alpha(s_{24}) = 2\alpha' p_2 \cdot p_4 \qquad (22)$$

we can rewrite (20) as follows:

$$B_5 = \int_0^1 du_2 \int_0^1 du_3 u_2^{-\alpha(s_2)-1} u_3^{-\alpha(s_3)-1} (1 - u_2)^{-\alpha(s_{23})-1}$$

$$\times (1 - u_3)^{-\alpha(s_{34})-1} \prod_{i=2}^{2} \prod_{j=4}^{4} (1 - x_{ij})^{2\alpha' p_i \cdot p_j} \qquad (23)$$

where

$$s_i \equiv s_{1i} \;\; , \;\; u_i \equiv u_{1i} \; ; \; i = 2, 3 \;\; ; \;\; x_{ij} = u_i u_{i+1} \ldots u_{j-1}. \qquad (24)$$

We are now ready to construct the N-point function [13]. In analogy with what has been done for the four- and five-point amplitudes, we can write the N-point amplitude as follows:

$$B_N = \int_0^1 \ldots \int_0^1 \prod_P [u_P^{-\alpha(s_P)-1}] \prod_Q \delta(u_Q - 1 + \prod_{\bar{Q}} u_{\bar{Q}}) \qquad (25)$$

where the first product is over the $\frac{N(N-3)}{2}$ variables corresponding to all planar channels, while the second one is over the $\frac{(N-3)(N-2)}{2}$ independent δ-functions. The product in the δ-function is defined in (16).

The solution of all the non-independent linear relations imposed by the δ-functions is given by

$$u_{ij} = \frac{(1 - x_{ij})(1 - x_{i-1,j+1})}{(1 - x_{i-1,j})(1 - x_{i,j+1})} \tag{26}$$

where the variables x_{ij} are given in (24). Eliminating the δ-function from Eq. (25) one gets

$$B_N = \prod_{i=2}^{N-2} \left[\int_0^1 du_i u_i^{-\alpha(s_i)-1} (1-u_i)^{-\alpha(s_{i,i+1})-1} \right] \prod_{i=2}^{N-3} \prod_{j=i+2}^{N-1} (1 - x_{ij})^{-\gamma_{ij}} \tag{27}$$

where

$$\gamma_{ij} = \alpha(s_{ij}) + \alpha(s_{i+1;j-1}) - \alpha(s_{i;j-1}) - \alpha(s_{i+1;j}) \quad ; \quad j \geq i+2 \tag{28}$$

It is easy to see that

$$\alpha(s_{i,i+1}) = -\alpha_0 - 2\alpha' p_i \cdot p_{i+1} \quad ; \quad \gamma_{ij} = -2\alpha' p_i \cdot p_j \quad ; \quad j \geq i+2 \tag{29}$$

Inserting them in (27) we get

$$B_N = \prod_{i=2}^{N-2} \left[\int_0^1 du_i u_i^{-\alpha(s_i)-1} (1-u_i)^{\alpha_0-1} \right] \prod_{i=2}^{N-2} \prod_{j=i+1}^{N-1} (1 - x_{ij})^{2\alpha' p_i \cdot p_j} \tag{30}$$

This is the form of the N-point amplitude that was originally constructed. Then Koba and Nielsen [15] put it in the form that is more known nowadays. They constructed it using the following rules. They associated a real variable z_i to each leg i. Then they associated to each channel (i,j) an anharmonic ratio constructed from the variables $z_i, z_{i-1}, z_j, z_{j+1}$ in the following way:

$$(z_i, z_{i+1}, z_j, z_{j+1})^{-\alpha(s_{ij})-1} = \left[\frac{(z_i - z_j)(z_{i-1} - z_{j+1})}{(z_{i-1} - z_j)(z_i - z_{j+1})} \right]^{-\alpha(s_{ij})-1} \tag{31}$$

and finally they gave the following expression for the N-point amplitude:

$$B_N = \int_{-\infty}^{\infty} dV(z) \prod_{(i,j)} (z_i, z_{i+1}, z_j, z_{j+1})^{-\alpha(s_{ij})-1} \tag{32}$$

where

$$dV(z) = \frac{\prod_1^N [\theta(z_i - z_{i+1}) dz_i]}{\prod_{i=1}^N (z_i - z_{i+2}) dV_{abc}} \quad ; \quad dV_{abc} = \frac{dz_a dz_b dz_c}{(z_b - z_a)(z_c - z_b)(z_a - z_c)} \tag{33}$$

and the variables z_i are integrated along the real axis in a cyclically ordered way: $z_1 \geq z_2 \cdots \geq z_N$ with a, b and c arbitrarily chosen.

The integrand of the N-point amplitude is invariant under projective transformations acting on the leg variables z_i:

$$z_i \to \frac{\alpha z_i + \beta}{\gamma z_i + \delta} \quad ; \quad i = 1 \dots N \quad ; \quad \alpha\delta - \beta\gamma = 1 \tag{34}$$

This is because both the anharmonic ratio in (31) and the measure dV_{abc} are invariant under a projective transformation. Since a projective transformation depends on three real parameters, then the integrand of the N-point amplitude depends only on $N-3$ variables z_i. In order to avoid infinities, one has then to divide the integration volume with the factor dV_{abc} that is also invariant under the projective transformations. The fact that the integrand depends only on $N-3$ variables is in agreement with the fact that $N-3$ is also the maximal number of simultaneous poles allowed in the amplitude.

It is convenient to write the N-point amplitude in a form that involves the scalar product of the external momenta rather than the Regge trajectories. We distinguish three kinds of channels. The first one is when the particles i and j of the channel (i,j) are separated by at least two particles. In this case the channels that contribute to the exponent of the factor $(z_i - z_j)$ are the channels (i,j) with exponent equal to $-\alpha(s_{ij}) - 1$, $(i+1, j-1)$ with exponent $-\alpha(s_{i+1,j-1}) - 1$, $(i+1, j)$ with exponent $\alpha(s_{i+1,j}) + 1$ and $(i, j-1)$ with exponent $\alpha(s_{i,j-1}) + 1$. Adding these four contributions, one gets for the channels where i and j are separated by at least two particles

$$-\alpha(s_{ij}) - \alpha(s_{i+1,j-1}) + \alpha(s_{i+1,j}) + \alpha(s_{i,j-1}) = 2\alpha' p_i \cdot p_j \tag{35}$$

The second one comes from the channels that are separated by only one particle. In this case only three of the previous four channels contribute. For instance, if $j = i+2$ the channel $(i+1, j-1)$ consists of only one particle and therefore should not be included. This means that we would get

$$-\alpha(s_{i;i+2}) - 1 + \alpha(s_{1+1;i+2}) + 1 + \alpha(s_{i;i+1} + 1) = 1 + 2\alpha' p_i \cdot p_{i+2} \tag{36}$$

Finally, the third one that comes from the channels whose particles are adjacent, gets only contribution from

$$-\alpha(s_{i;i+1}) - 1 = \alpha_0 - 1 + 2\alpha' p_i \cdot p_{i+1} \tag{37}$$

Putting all these three terms together in (32) and remembering the factor in the denominator in the first equation of (33) we get

$$B_N = \int_{-\infty}^{\infty} \frac{\prod_1^N dz_i \theta(z_i - z_{i+1})}{dV_{abc}} \prod_{i=1}^{N} \left[(z_i - z_{i+1})^{\alpha_0 - 1} \right] \prod_{j>i} (z_i - z_j)^{2\alpha' p_i \cdot p_j} \tag{38}$$

A convenient choice for the three variables to keep fixed is

$$z_a = z_1 = \infty \ ; \ \ z_b = z_2 = 1 \ ; \ \ z_c = z_N = 0 \tag{39}$$

With this choice the previous equation becomes

$$B_N = \prod_{i=3}^{N-1} \left[\int_0^1 dz_i \theta(z_i - z_{i+1}) \right] \prod_{i=2}^{N-1} (z_i - z_{i+1})^{\alpha_0 - 1}$$

$$\times \prod_{i=2}^{N-1} \prod_{j=i+1}^{N} (z_i - z_j)^{2\alpha' p_i \cdot p_j} \tag{40}$$

We now want to show that this amplitude is identical to the one given in (30). This can be done by performing the following change of variables:

$$u_i = \frac{z_{i+1}}{z_i} \ ; \ \ i = 2, 3 \ldots N-2 \tag{41}$$

that implies

$$z_i = u_2 u_3 \ldots u_{i-1} \ ; \ \ i = 3, 4 \ldots N-1 \tag{42}$$

Taking into account that the Jacobian is equal to

$$\det \frac{\partial z}{\partial u} = \prod_{i=3}^{N-2} z_i = \prod_{i=2}^{N-3} u_i^{N-2-i} \tag{43}$$

using the following two relations:

$$\det \frac{\partial z}{\partial u} \prod_{i=2}^{N-1} (z_i - z_{i+1})^{\alpha_0 - 1} = \prod_{i=2}^{N-2} \left[u_i^{(N-1-i)\alpha_0 - 1} \right] \prod_{i=2}^{N-2} (1 - u_i)^{\alpha_0 - 1} \tag{44}$$

and

$$\prod_{i=2}^{N-1} \prod_{j=i+1}^{N} (z_j - z_i)^{2\alpha' p_i \cdot p_j}$$

$$= \prod_{i=2}^{N-2} \prod_{j=i+1}^{N-1} (1 - x_{ij})^{2\alpha' p_i \cdot p_j} \prod_{i=2}^{N-2} u_i^{-\alpha(s_i) - (N-i-1)\alpha_0} \tag{45}$$

and the conservation of momentum

$$\sum_{i=1}^{N} p_i = 0 \tag{46}$$

together with (21), one can easily see that (30) and (40) are equal.

The N-point amplitude that we have constructed in this section corresponds to the scattering of N spinless particles with no internal degrees of freedom. On the other hand, it was known that the mesons were classified according to multiplets of an $SU(3)$ flavour symmetry. This was implemented by Chan and Paton [16] by multiplying the N-point amplitude with a factor, called Chan–Paton factor, given by

$$Tr(\lambda^{a_1}\lambda^{a_2}\dots\lambda^{a_N}) \tag{47}$$

where the λ's are matrices of a unitary group in the fundamental representation. Including the Chan–Paton factors the total scattering amplitude is given by

$$\sum_P Tr(\lambda^{a_1}\lambda^{a_2}\dots\lambda^{a_N})B_N(p_1,p_2,\dots p_N) \tag{48}$$

where the sum is extended to the $(N-1)!$ permutations of the external legs, that are not related by a cyclic permutations. Originally when the dual resonance model was supposed to describe strongly interacting mesons, this factor was introduced to represent their flavour degrees of freedom. Nowadays, the interpretation is different and the Chan–Paton factor represents the colour degrees of freedom of the gauge bosons and the other massive particles of the spectrum.

The N-point amplitude B_N that we have constructed in this section contains only simple pole singularities in all possible planar channels. They correspond to zero width resonances located at non-negative integer values n of the Regge trajectory $\alpha(M^2) = n$. The lowest state located at $\alpha(m^2) = 0$ corresponds to the particles on the external legs of B_N. The spectrum of excited particles can be obtained by factorizing the N-point amplitude in the most general channel with any number of particles. This was done in [17] and [18] finding a spectrum of states rising exponentially with the mass M. Being the model relativistic invariant it was found that many states obtained by factorizing the N-point amplitude were "ghosts", namely, states with negative norm as one finds in QED when one quantizes the electromagnetic field in a covariant gauge. The consistency of the model requires the existence of relations satisfied by the scattering amplitudes that are similar to those obtained through gauge invariance in QED. If the model is consistent they must decouple the negative norm states leaving us with a physical spectrum of positive norm states. In order to study in a simple way these issues, we discuss in the next section the operator formalism introduced already in 1969 [19, 20, 21].

Before concluding this section let us go back to the non-planar four-point amplitude in (5) and discuss its generalization to an N-point amplitude. Using the technique of the electrostatic analogue on the sphere instead of on the disk Shapiro [22] was able to obtain a N-point amplitude that reduces to the four-point amplitude in (5) with intercept $\alpha_0 = 2$. The N-point amplitude found in [22] is

$$\int \frac{\prod_{i=1}^{N} d^2 z_i}{dV_{abc}} \prod_{i<j} |z_i - z_j|^{\alpha' p_i \cdot p_j} \tag{49}$$

where

$$dV_{abc} = \frac{d^2 z_a d^2 z_b d^2 z_c}{|z_a - z_b|^2 |z_a - z_c|^2 |z_b - z_c|^2} \tag{50}$$

The integral in (49) is performed in the entire complex plane.

3 Operator Formalism and Factorization

The factorization properties of the dual resonance model were first studied by factorizing by brute force the N-point amplitude at the various poles [17, 18]. The number of terms that factorize the residue of the pole at $\alpha(s) = n$, increases rapidly with the value of n. In order to find their degeneracy, it turned out to be convenient to first rewrite the N-point amplitude in an operator formalism. In this section we introduce the operator formalism and we rewrite the N-point amplitude derived in the previous section in this formalism.

The key idea [19, 20, 21] is to introduce an infinite set of harmonic oscillators and a position and momentum operators,[7] which satisfy the following commutation relations:

$$[a_{n\mu}, a^{\dagger}_{m\nu}] = \eta_{\mu\nu}\delta_{nm} \quad ; \quad [\hat{q}_{\mu}, \hat{p}_{\nu}] = i\eta_{\mu\nu} \tag{51}$$

where $\eta_{\mu\nu}$ is the flat Minkowski metric that we take to be $\eta_{\mu\nu} = (-1, 1, \dots 1)$. A state with momentum p is constructed in terms of a state with zero momentum as follows:

$$\hat{p}|p\rangle \equiv \hat{p} e^{ip\cdot\hat{q}}|0\rangle = p|p\rangle \quad ; \quad \hat{p}\,|0\rangle = 0 \tag{52}$$

normalized as[8]

$$\langle p|p'\rangle = (2\pi)^d \delta^{(d)}(p + p') \tag{53}$$

In order to avoid minus signs we use the convention that

$$\langle p| = \langle 0| e^{ip\cdot\hat{q}} \tag{54}$$

A complete and orthonormal basis of vectors in the harmonic oscillator space is given by

$$|\lambda_1, \lambda_2, \dots \lambda_i; p\rangle = \prod_{n} \frac{(a^{\dagger}_{\mu_n;n})^{\lambda_{n;\mu_n}}}{\sqrt{\lambda_{n,\mu_n}!}} e^{ip\hat{q}}|0, 0\rangle \tag{55}$$

[7] Actually the position and momentum operators were introduced in [23].

[8] Although we now use an arbitrary d we want to remind you that all original calculations were done for $d = 4$.

where the first $|0\rangle$ corresponds to the one annihilated by all annihilation operators and the second one to the state of zero momentum

$$a_{\mu_n;n}|0,0\rangle = \hat{p}|0,0\rangle = 0 \tag{56}$$

Notice that Lorentz invariance forces to introduce also oscillators that create states with negative norm due to the minus sign in the flat Minkowski metric. This implies that the space spanned by the states in (55) is not positive definite. This is, however, not allowed in a quantum theory and therefore if the dual resonance model is a consistent quantum-relavistic theory we expect the presence of relations of the kind of those provided by gauge invariance in QED.

Let us introduce the Fubini–Veneziano [23] operator

$$Q_\mu(z) = Q_\mu^{(+)}(z) + Q_\mu^{(0)}(z) + Q_\mu^{(-)}(z) \tag{57}$$

where

$$Q^{(+)} = i\sqrt{2\alpha'}\sum_{n=1}^{\infty}\frac{a_n}{\sqrt{n}}z^{-n} \quad ; \quad Q^{(-)} = -i\sqrt{2\alpha'}\sum_{n=1}^{\infty}\frac{a_n^\dagger}{\sqrt{n}}z^{n}$$

$$Q^{(0)} = \hat{q} - 2i\alpha'\hat{p}\log z \tag{58}$$

In terms of Q we introduce the vertex operator corresponding to the external leg with momentum p:

$$V(z;p) =: e^{ip\cdot Q(z)} :\equiv e^{ip\cdot Q^{(-)}(z)} e^{ip\hat{q}} e^{+2\alpha'\hat{p}\cdot p\log z} e^{ip\cdot Q^{(+)}(z)} \tag{59}$$

and compute the following vacuum expectation value:

$$\langle 0,0|\prod_{i=1}^{N} V(z_i,p_i)|0,0\rangle \tag{60}$$

It can be easily computed using the Baker–Haussdorf relation

$$e^A e^B = e^B e^A e^{[A,B]} \tag{61}$$

that is valid if the commutator, as in our case, $[A,B]$ is a c-number. In our case the commutation relations to be used are

$$[Q^{(+)}(z), Q^{(-)}(w)] = -2\alpha'\log\left(1-\frac{w}{z}\right) \tag{62}$$

and the second one in (51). Using them one gets

$$V(z;p)V(w;k) =: V(z;p)V(w;k) : (z-w)^{2\alpha' p\cdot k} \tag{63}$$

and

$$\langle 0,0| \prod_{i=1}^{N} V(z_i,p_i)|0,0\rangle = \prod_{i>j}(z_i-z_j)^{2\alpha' p_i\cdot p_j}(2\pi)^d\delta^{(d)}(\sum_{i=1}^{N} p_i) \tag{64}$$

where the normal ordering requires that all creation operators be put on the left of the annihilation one and the momentum operator $\hat{p}$ be put on the right of the position operator $\hat{q}$. This means that

$$(2\pi)^d\delta^{(d)}(\sum_{i=1}^{N} p_i)B_N = \int_{-\infty}^{\infty}\frac{\prod_1^N dz_i\theta(z_i-z_{i+1})}{dV_{abc}}\prod_{i=1}^{N}\left[(z_i-z_{i+1})^{\alpha_0-1}\right]$$

$$\times\langle 0,0|\prod_{i=1}^{N} V(z_i,p_i)|0,0\rangle \tag{65}$$

By choosing the three variables z_a, z_b and z_c as in (39) we can rewrite the previous equation as follows:

$$(2\pi)^d\delta^{(d)}(\sum_{i=1}^{N} p_i)B_N = \int_0^1 \prod_{i=3}^{N-1} dz_i \prod_{i=2}^{N-1}\theta(z_i-z_{i+1})$$

$$\times\prod_{i=2}^{N-1}\left[(z_i-z_{i+1})^{\alpha_0-1}\right]\langle 0,p_1|\prod_{i=2}^{N-1} V(z_i;p_i)|0,p_N\rangle \tag{66}$$

where we have taken $z_2 = 1$ and we have defined $(\alpha_0 \equiv \alpha' p_i^2; i = 1\ldots N)$:

$$\lim_{z_N\to 0} V(z_N;p_N)|0,0\rangle \equiv |0;p_N\rangle \ \ ; \ \ \langle 0;0|\lim_{z_1\to\infty} z_1^{2\alpha_0}V(z_1;p_1) = \langle 0,p_1| \tag{67}$$

Before proceeding to factorize the N-point amplitude, let us study the properties under the projective group of the operators that we have introduced. We have already seen that the projective group leaves the integrand of the Koba–Nielsen representation of the N-point amplitude invariant. The projective group has three generators L_0, L_1 and L_{-1} corresponding respectively to dilatations, inversions and translations. Assuming that the Fubini–Veneziano fields $Q(z)$ transforms as a field with weight 0 (as a scalar) we can immediately write the commutation relations that $Q(z)$ must satisfy. This means in fact that, under a projective transformation, $Q(z)$ transforms as follows:

$$Q(z) \to Q^T(z) = Q\left(\frac{\alpha z+\beta}{\gamma z+\delta}\right) \ \ ; \ \ \alpha\delta-\beta\gamma = 1 \tag{68}$$

Expanding for small values of the parameters we get

$$Q^T(z) = Q(z) + (\epsilon_1+\epsilon_2 z+\epsilon_3 z^2)\frac{dQ(z)}{dz} + o(\epsilon^2) \tag{69}$$

This means that the three generators of the projective group must satisfy the following commutation relations with $Q(z)$:

$$[L_0, Q(z)] = z\frac{dQ}{dz} \;\; ; \;\; [L_{-1}, Q(z)] = \frac{dQ}{dz} \;\; ; \;\; [L_1, Q(z)] = z^2\frac{dQ}{dz} \tag{70}$$

They are given by the following expressions in terms of the harmonic oscillators:

$$L_0 = \alpha'\hat{p}^2 + \sum_{n=1}^{\infty} n a_n^{\dagger} \cdot a_n \; ; \; L_1 = \sqrt{2\alpha'}\hat{p} \cdot a_1 + \sum_{n=1}^{\infty} \sqrt{n(n+1)} a_{n+1} \cdot a_n^{\dagger} \tag{71}$$

and

$$L_{-1} = L_1^{\dagger} = \sqrt{2\alpha'}\hat{p} \cdot a_1^{\dagger} + \sum_{n=1}^{\infty} \sqrt{n(n+1)} a_{n+1}^{\dagger} \cdot a_n \tag{72}$$

They annihilate the vacuum

$$L_0|0,0\rangle = L_1|0,0\rangle = L_{-1}|0,0\rangle = 0 \tag{73}$$

that is therefore called the projective invariant vacuum, and satisfy the algebra that is called Gliozzi algebra [24][9]:

$$[L_0, L_1] = -L_1 \;\; ; \;\; [L_0, L_{-1}] = L_{-1} \;\; ; \;\; [L_1, L_{-1}] = 2L_0 \tag{74}$$

The vertex operator with momentum p is a projective field with weight equal to $\alpha_0 = \alpha' p^2$. It transforms in fact as follows under the projective group:

$$[L_n, V(z,p)] = z^{n+1}\frac{dV(z,p)}{dz} + \alpha_0(n+1)z^n V(z,p) \;\; ; \;\; n = 0, \pm 1 \tag{75}$$

or in finite form as follows:

$$UV(z,p)U^{-1} = \frac{1}{(\gamma z + \delta)^{2\alpha_0}} V\left(\frac{\alpha z + \beta}{\gamma z + \delta}, p\right) \tag{76}$$

where U is the generator of an arbitrary finite projective transformation.

Since U leaves the vacuum invariant, by using (76) it is easy to show that

$$\langle 0,0| \prod_{i=1}^{N} V(z_i', p)|0,0\rangle = \prod_{i=1}^{N} (\gamma z_i + \delta)^{2\alpha_0} \langle 0,0| \prod_{i=1}^{N} V(z_i, p)|0,0\rangle \tag{77}$$

that together with the following equation:

$$\prod_{i=1}^{N} dz_i' \prod_{i=1}^{N} (z_i' - z_{i+1}')^{\alpha_0 - 1} = \prod_{i=1}^{N} dz_i \prod_{i=1}^{N-1} (z_i - z_{i+1})^{\alpha_0 - 1} \prod_{i=1}^{N} (\gamma z_i + \delta)^{-2\alpha_0} \tag{78}$$

[9] See also [25].

implies that the integrand of the N-point amplitude in (65) is invariant under projective transformations.

We are now ready to factorize the N-point amplitude and find the spectrum of mesons.

From (75) and (76) it is easy to derive the transformation of the vertex operator under a finite dilatation

$$z^{L_0} V(1,p) z^{-L_0} = V(z,p) z^{\alpha_0} \tag{79}$$

Changing the integration variables as follows:

$$x_i = \frac{z_{i+1}}{z_i} \quad ; \quad i = 2,3 \ldots N-2 \quad ; \quad \det \frac{\partial z_i}{\partial x_j} = z_3 z_4 \ldots z_{N-2} \tag{80}$$

where the last term is the jacobian of the trasformation from z_i to x_i, we get from (66) the following expression:

$$A_N \equiv \langle 0, p_1 | V(1,p_2) D V(1,p_3) \ldots D V(1, p_{N-1}) | 0, p_N \rangle \tag{81}$$

where the propagator D is equal to

$$D = \int_0^1 dx x^{L_0 - 1 - \alpha_0} (1-x)^{\alpha_0 - 1} = \frac{\Gamma(L_0 - \alpha_0)\Gamma(\alpha_0)}{\Gamma(L_0)} \tag{82}$$

and

$$A_N = (2\pi)^d \delta^{(d)} \left(\sum_{i=1}^{N} p_i \right) B_N \tag{83}$$

The factorization properties of the amplitude can be studied by inserting in the channel $(1, M)$ or equivalently in the channel $(M+1, N)$ described by the Mandelstam variable

$$s = -(p_1 + p_2 + \ldots p_M)^2 = -(p_{M+1} + p_{M+2} \cdots + p_N)^2 \equiv -P^2 \tag{84}$$

the complete set of states given in (55):

$$A_N = \sum_{\lambda,\mu} \langle p_{(1,M)} | \lambda, P \rangle \langle \lambda, P | D | \mu, P \rangle \langle \mu, P | p_{(M+1,N)} \rangle \tag{85}$$

where

$$\langle p_{(1,M)} | = \langle 0, p_1 | V(1,p_2) D V(1,p_3) \ldots V(1, p_M) \tag{86}$$

and

$$| p_{(M+1,N)} \rangle = V(1, p_{M+1}) D \ldots V(1, p_{N-1}) | p_N, 0 \rangle \tag{87}$$

Introducing the quantity

$$R = \sum_{n=1}^{\infty} n a_n^\dagger \cdot a_n \tag{88}$$

it is possible to rewrite

$$\langle \lambda, P|D|\mu, P\rangle = \sum_{m=0}^{\infty} \langle \lambda, P| \frac{(-1)^m \binom{\alpha_0 - 1}{m}}{R + m - \alpha(s)} |\mu, P\rangle \tag{89}$$

where s is the variable defined in (84). Using this equation we can rewrite (85) as follows:

$$A_N = \sum_{\lambda,\mu} \langle p_{(1,M)}|\lambda, P\rangle \sum_{m=0}^{\infty} \langle \lambda, P| \frac{(-1)^m \binom{\alpha_0 - 1}{m}}{R + m - \alpha(s)} |\mu, P\rangle \langle \mu, P|p_{(M+1,N)}\rangle \tag{90}$$

This expression shows that amplitude A_N has a pole in the channel $(1, M)$ when $\alpha(s)$ is equal to an integer $n \geq 0$ and the states $|\lambda\rangle$ that contribute to its residue are those satisfying the relation

$$R|\lambda\rangle = (n - m)|\lambda\rangle \quad ; \quad m = 0, 1 \ldots n \tag{91}$$

The number of independent states $|\lambda\rangle$ contributing to the residue gives the degeneracy of states for each level n.

Because of manifest relativistic invariance the space spanned by the complete system of states in (55) contains states with negative norm corresponding to those states having an odd number of oscillators with time-like directions (see (51)). This is not consistent in a quantum theory where the states of a system must span a positive-definite Hilbert space. This means that there must exist a number of relations satisfied by the external states that decouple a number of states leaving with a positive-definite Hilbert space. In order to find these relations we rewrite the state in (87) going back to the Koba–Nielsen variables

$$|p_{(1,M)}\rangle = \prod_{i=2}^{M-1} [\int dz_i \theta(z_i - z_{i+1})] \prod_{i=1}^{M-1} (z_i - z_{i+1})^{\alpha_0 - 1}$$

$$\times V(1, p_1) V(z_2, p_2) \ldots V(z_{M-1}, p_{M-1})|0, p_M\rangle \tag{92}$$

Let us consider the operator $U(\alpha)$ that generates the projective transformation that leaves the points $z = 0, 1$ invariant:

$$z' = \frac{z}{1 - \alpha(z - 1)} = z + \alpha(z^2 - z) + o(\alpha^2) \tag{93}$$

From the transformation properties of the vertex operators in (76), it is easy to see that the previous transformation leaves the state in (92) invariant:

$$U(\alpha)|p_{(1,M)}\rangle = |p_{(1,M)}\rangle \tag{94}$$

This means that the generator of the previous transformation annihilates the state in (92):

$$W_1|p_{(1,M)}\rangle = 0 \quad ; \quad W_1 = L_1 - L_0 \tag{95}$$

The explicit form of W_1 follows from the infinitesimal form of the transformation in (93). This condition that is of the same kind of the relations that on-shell amplitudes with the emission of photons satisfy as a consequence of gauge invariance, implies that the residue at the pole in (90) can be factorized with a smaller number of states. It turns out, however, that a detailed analysis of the spectrum shows that negative norm states are still present. This can be qualitatively understood as follows. Due to the Lorentz metric, we have a negative norm component for each oscillator. In order to be able to decouple all negative norm states, we need to have a gauge condition of the type as in (95) for each oscillator. But the number of oscillators is infinite and, therefore, we need an infinite number of conditions of the type as in (95). It was found in [6] that, if we take $\alpha_0 = 1$, then one can easily construct an infinite number of operators that leave the state in (92) invariant. In the next section we will concentrate on this case.

4 The Case $\alpha_0 = 1$

If we take $\alpha_0 = 1$ many of the formulae given in the previous section simplify. The N-point amplitude in (38) becomes

$$B_N = \int_{-\infty}^{\infty} \frac{\prod_1^N dz_i \theta(z_i - z_{i+1})}{dV_{abc}} \prod_{j>i} (z_i - z_j)^{2\alpha' p_i \cdot p_j} \tag{96}$$

that can be rewritten in the operator formalism as follows:

$$(2\pi)^4 \delta(\sum_{i=1}^N p_i) B_N = \int_{-\infty}^{\infty} \frac{\prod_1^N dz_i \theta(z_i - z_{i+1})}{dV_{abc}} \langle 0, 0| \prod_{i=1}^N V(z_i, p_i)|0, 0\rangle \tag{97}$$

By choosing $z_1 = \infty$, $z_2 = 1$ and $z_N = 0$ it becomes

$$(2\pi)^4 \delta(\sum_{i=1}^N p_i) B_N$$

$$= \int_0^1 \prod_{i=3}^{N-1} dz_i \prod_{i=2}^{N-1} \theta(z_i - z_{i+1}) \langle 0, p_1 | \prod_{i=2}^{N-1} V(z_i; p_i) | 0, p_N \rangle \tag{98}$$

where

$$\lim_{z_N \to 0} V(z_N; p_N)|0,0\rangle \equiv |0; p_N\rangle \quad ; \quad \langle 0; 0| \lim_{z_1 \to \infty} z_1^2 V(z_1; p_1) = \langle 0, p_1| \tag{99}$$

Equation (81) is as before, but now the propagator becomes

$$D = \int dx x^{L_0 - 2} = \frac{1}{L_0 - 1} \tag{100}$$

This means that (89) becomes

$$\langle \lambda, P | D | \mu, P \rangle = \langle \lambda, P | \frac{1}{L_0 - 1} | \mu, P \rangle \tag{101}$$

and (90) has the simpler form

$$B_N = \sum_\lambda \langle p_{(1,M)} | \lambda, P \rangle \langle \lambda, P | \frac{1}{R - \alpha(s)} | \lambda, P \rangle \langle \lambda, P | p_{(M+1,N)} \rangle \tag{102}$$

B_N has a pole in the channel $(1, M)$ when $\alpha(s)$ is equal to an integer $n \geq 0$ and the states $|\lambda\rangle$ that contribute to its residue are those satisfying the relation

$$R|\lambda\rangle = n|\lambda\rangle \tag{103}$$

Their number gives the degeneracy of the states contributing to the pole at $\alpha(s) = n$. The N-point amplitude can be written as

$$B_N = \langle p_{(1,M)} | D | p_{(M+1,N)} \rangle \tag{104}$$

where

$$|p_{(1,M)}\rangle = \int \prod_{i=2}^{M-1} [dz_i \theta(z_i - z_{i+1})]$$

$$\times V(1, p_1) V(z_2, p_2) \ldots V(z_{M-1}, p_{M-1} | 0, p_M \rangle \tag{105}$$

Using (79) and changing variables from $z_i, i = 2 \ldots M - 1$ to $x_i = \frac{z_{i+1}}{z_i}$, $i = 1 \ldots M - 2$ with $z_1 = 1$ we can rewrite the previous equation as follows:

$$|p_{(1,M)}\rangle = V(1, p_1) D V(1, p_2) \ldots D V(1, p_{M-1}) | 0, p_M \rangle \tag{106}$$

where the propagator D is defined in (100).

We want now to show that the state in (105) and (106) is not only annihilated by the operator in (95), but, if $\alpha_0 = 1$ [6], by an infinite set of operators whose lowest one is the one in (95). We will derive this by using the formalism developed in [26] and we will follow closely their derivation.

Starting from (70) Fubini and Veneziano realized that the generators of the projective group acting on a function of z are given by

$$L_0 = -z\frac{d}{dz} \;\; ; \;\; L_{-1} = -\frac{d}{dz} \;\; ; \;\; L_1 = -z^2\frac{d}{dz} \tag{107}$$

They generalized the previous generators to an arbitrary conformal transformation by introducing the following operators, called Virasoro operators:

$$L_n = -z^{n+1}\frac{d}{dz} \tag{108}$$

that satisfy the algebra

$$[L_n, L_m] = (n-m)L_{n+m} \tag{109}$$

that does not contain the term with the central charge! They also showed that the Virasoro operators satisfy the following commutation relations with the vertex operator:

$$[L_n, V(z,p)] = \frac{d}{dz}\left(z^{n+1}V(z,p)\right) \tag{110}$$

More in general actually they define an operator L_f corresponding to an arbitrary function $f(\xi)$ and $L_f = L_n$ if we choose $f(\xi) = \xi^n$. In this case the commutation relation in (110) becomes

$$[L_f, V(z,p)] = \frac{d}{dz}\left(zf(z)V(z,p)\right) \tag{111}$$

By introducing the variable

$$y = \int_A^z \frac{d\xi}{\xi f(\xi)} \tag{112}$$

where A is an arbitrary constant, one can rewrite (111) in the following form:

$$[L_f, zf(z)V(z,p)] = \frac{d}{dy}\left(zf(z)V(z,p)\right) \tag{113}$$

This implies that, under an arbitrary conformal transformation $z \to f(z)$, generated by $U = e^{\alpha L_f}$, the vertex operator transforms as

$$e^{\alpha L_f}V(z,p)\, zf(z)\, e^{-\alpha L_f} = V(z',p)z'f(z') \tag{114}$$

where the parameter α is given by

$$\alpha = \int_z^{z'} \frac{d\xi}{\xi f(\xi)} \tag{115}$$

On the other hand, this equation implies

$$\frac{dz}{zf(z)} = \frac{dz'}{z'f(z')} \tag{116}$$

that, inserted in (114), implies that the quantity $V(z,p)\,dz$ is left invariant by the transformation $z \to f(z)$:

$$e^{\alpha L_f} V(z,p) dz e^{-\alpha L_f} = V(z',p)dz' \tag{117}$$

Let us now act with the previous conformal transformation on the state in (105). We get

$$e^{\alpha L_f} |p_{(1,M)}\rangle = \int_0^1 \prod_{i=2}^{M-1} [dz_i \theta(z_i - z_{i+1})] \, e^{\alpha L_f} V(1,p_1) e^{-\alpha L_f}$$

$$\times e^{\alpha L_f} V(z_2,p_2) e^{-\alpha L_f} \ldots\ldots e^{\alpha L_f} V(z_{M-1},p_{M-1}) e^{-\alpha L_f} e^{\alpha L_f} |0,p_M\rangle$$

$$= \int_0^1 \prod_{i=2}^{M-1} \theta(z_i - z_{i+1}) \times e^{\alpha L_f} V(1,p_1) e^{-\alpha L_f}$$

$$\times V(z'_2,p_2) dz'_2 \ldots V(z'_{M-1},p_{M-1}) dz'_{M-1} e^{\alpha L_f} |0,p_M\rangle \tag{118}$$

where we have used (117). The previous transformation leaves the state invariant if both $z=0$ and $z=1$ are fixed points of the conformal transformation. This happens if the denominator in (115) vanishes when $\xi = 0,1$. This requires the following conditions:

$$f(1) = 0 \quad ; \quad \lim_{\xi \to 0} \xi f(\xi) = 0 \tag{119}$$

Expanding ξ near the point $\xi = 1$, we can determine the relation between z and z' near $z = z' = 1$. We get

$$z' = \frac{z e^{-\alpha f'(1)}}{1 - z + z e^{-\alpha f'(1)}} \tag{120}$$

and from it we can determine the conformal factor

$$\frac{dz'}{dz} = \frac{e^{-\alpha f'(1)}}{(1 - z + z e^{-\alpha f'(1)})^2} \to e^{\alpha f'(1)} \tag{121}$$

in the limit $z \to 1$. Proceeding in the same way near the point $z = z' = 0$ we get

$$z' = \frac{z f(0) e^{\alpha f(0)}}{f(0) + z f'(0)(1 - e^{\alpha f(0)}} \to z e^{\alpha f(0)} \tag{122}$$

in the limit $z \to 0$. This means that (118) becomes

$$e^{\alpha\left(L_f - f'(1) - f(0)\right)}|p_{(1,M)}\rangle = |p_{(1,M)}\rangle \tag{123}$$

A choice of f that satisfies (119) is the following:

$$f(\xi) = \xi^n - 1 \tag{124}$$

that gives the following gauge operator:

$$W_n = L_n - L_0 - (n-1) \tag{125}$$

that annihilates the state in (105):

$$W_n|p_{1\ldots M}\rangle = 0 \quad ; \quad n = 1\ldots\infty \tag{126}$$

These are the Virasoro conditions found in [6]. There is one condition for each negative norm oscillator and, therefore, in this case there is the possibility that the physical subspace is positive definite. An alternative more direct derivation of (126) can be obtained by acting with W_n on the state in (106) and using the following identities:

$$W_n V(1,p) = V(1,p)(W_n + n) \quad ; \quad (W_n + n)D = [L_0 + n - 1]^{-1} W_n \tag{127}$$

The second equation is a consequence of the following equation:

$$L_n x^{L_0} = x^{L_0 + n} L_n \tag{128}$$

Equations (127) imply

$$W_n V(1,p) D = V(1,p)[L_0 + n - 1]^{-1} W_n \tag{129}$$

This shows that the operator W_n goes unchanged through all the product of terms VD until it arrives in front of the term $V(1, p_{M-1})|0, p_M\rangle$. Going through the vertex operator it becomes $L_n - L_0 + 1$ that then annihilate the state

$$(L_n - L_0 + 1)|p_M, 0\rangle = 0 \tag{130}$$

This proves (126).

Using the representation of the Virasoro operators given in (108), Fubini and Veneziano showed that they satisfy the algebra given in (109) without the central charge. The presence of the central charge was recognized by Joe Weis[10] in 1970 and never published. Unlike Fubini and Veneziano [26], he used the expression of the L_n operators in terms of the harmonic oscillators

$$L_n = \sqrt{2\alpha' n}\hat{p} \cdot a_n + \sum_{m=1}^{\infty} \sqrt{m(n+m)} a_{n+m} \cdot a_m$$

[10] See noted added in proof in [26].

$$+\frac{1}{2}\sum_{m=1}^{n-1}\sqrt{m(n-m)}a_{m-n}\cdot a_m \quad ; \quad n\geq 0 \quad L_n = L_n^\dagger \tag{131}$$

He got the following algebra:

$$[L_n, L_m] = (n-m)L_{n+m} + \frac{d}{24}n(n^2-1)\delta_{n+m;0} \tag{132}$$

where d is the dimension of the Minkowski space–time. We write here d for the dimension of the Minkowski space, but we want to remind you that almost everybody working in a model for mesons at that time took for granted that the dimension of the space–time was $d = 4$. As far as I remember, the first paper where a dimension $d \neq 4$ was introduced was [27], where it was shown that the unitarity violating cuts in the non-planar loop become poles that were consistent with unitarity if $d = 26$.

In the last part of this section we will generalize the factorization procedure to the Shapiro–Virasoro model whose N-point amplitude is given in (49). In this case we must introduce two sets of harmonic oscillators commuting with each other and only one set of zero modes satisfying the algebra [28]

$$[a_{n\mu}, a^\dagger_{m\nu}] = [\tilde{a}_{n\mu}, \tilde{a}^\dagger_{m\nu}] = \eta_{\mu\nu}\delta_{nm} \quad ; \quad [\hat{q}_\mu, \hat{p}_\nu] = i\eta_{\mu\nu} \tag{133}$$

In terms of them we can introduce the Fubini–Veneziano operator

$$Q(z,\bar{z}) = \hat{q} - 2\alpha'\hat{p}\log(z\bar{z}) + i\frac{\sqrt{2\alpha'}}{2}\sum_{n=1}^{\infty}\frac{1}{\sqrt{n}}\left[a_n z^{-n} - a_n^\dagger z^n\right]$$

$$+i\frac{\sqrt{2\alpha'}}{2}\sum_{n=1}^{\infty}\frac{1}{\sqrt{n}}\left[\tilde{a}_n \bar{z}^{-n} - \tilde{a}_n^\dagger \bar{z}^n\right] \tag{134}$$

We can then introduce the vertex operator

$$V(z,\bar{z};p) =: e^{ip\cdot Q(z,\bar{z})} : \tag{135}$$

and write the N-point amplitude in (95) in the following factorized form:

$$\int \frac{\prod_{i=1}^N d^2 z_i}{dV_{abc}} \langle 0|R\left[\prod_{i=1}^N V(z_i,\bar{z}_i,p_i))\right]|0\rangle$$

$$= (2\pi)^4\delta^{(4)}(\sum_{i=1}^N p_i)\int \frac{\prod_{i=1}^N d^2 z_i}{dV_{abc}} \prod_{i<j}|z_i - z_j|^{\alpha' p_i\cdot p_j} \tag{136}$$

where the radial ordered product is given by

$$R\left[\prod_{i=1}^{N} V(z_i, \bar{z}_i, p_i))\right] = \prod_{i=1}^{N} V(z_i, \bar{z}_i, p_i)) \prod_{i=1}^{N-1} \theta(|z_i| - |z_{i+1}|) + \ldots \quad (137)$$

and the dots indicate a sum over all permutations of the vertex operators.

By fixing $z_1 = \infty, z_2 = 1$ and $z_N = 0$, we can rewrite the previous expression as follows:

$$\int \prod_{i=3}^{N-1} d^2 z_i \langle 0, p_1 | R\left[\prod_{i=2}^{N-1} V(z_i, \bar{z}_i, p_i))\right] |0, p_N\rangle \quad (138)$$

For the sake of simplicity, let us consider the term corresponding to the permutation $1, 2, \ldots N$. In this case the Koba–Nielsen variables are ordered in such a way that $|z_i| \geq |z_{i+1}|$ for $i = 1, \ldots N-1$. We can then use the formula

$$V(z_i, \bar{z}_i, p_i)) = z_i^{L_0 - 1} \bar{z}_i^{\tilde{L}_0 - 1} V(1, 1, p_i) z_i^{-L_0} \bar{z}_i^{-\tilde{L}_0} \quad (139)$$

and change variables

$$w_i = \frac{z_{i+1}}{z_i} \quad ; \quad |w_i| \leq 1 \quad (140)$$

to rewrite (138) as follows:

$$\langle 0, p_1 | V(1, 1, p_i 1) D V(1, 1, p_2) D \ldots V(1, 1, p_{N-1}) |0, p_N\rangle \quad (141)$$

where

$$D = \int \frac{d^2 w}{|w|^2} w^{L_0 - 1} \bar{w}^{\tilde{L}_0 - 1} = \frac{2}{L_0 + \tilde{L}_0 - 2} \cdot \frac{\sin \pi (L_0 - \tilde{L}_0)}{L_0 - \tilde{L}_0} \quad (142)$$

We can now follow the same procedure for all permutations arriving at the following expression:

$$\langle 0, p_1 | P[V(1, 1, p_2) D V(1, 1, p_3) D \ldots V(1, 1, p_{N-1})] |0, p_N\rangle \quad (143)$$

where P means a sum of all permutations of the particles.

If we want to consider the factorization of the amplitude on the pole at $s = -(p_1 + \ldots p_M)^2$ we get only the following contribution:

$$\langle p_{(1 \ldots M)} | D | p_{(M+1 \ldots N)} \rangle \quad (144)$$

where

$$|p_{(M+1 \ldots N)}\rangle = P[V(1, 1, p_{M+1}) D \ldots V(1, 1, p_{N-1}]] |0, p_N\rangle \quad (145)$$

and

$$\langle p_{(1 \ldots M)}| = \langle 0, p_1 | P \left[V(1, 1, p_2) D \ldots V(1, 1, p_M) \right] \quad (146)$$

The amplitude is factorized by introducing a complete set of states and rewriting (141) as follows:

$$\sum_{\lambda,\tilde{\lambda}} \langle p_{1\ldots M}|\lambda,\tilde{\lambda}\rangle \frac{2\pi\langle\lambda,\tilde{\lambda}|\delta_{L_0,\tilde{L}_0}|\lambda,\tilde{\lambda}\rangle}{L_0+\tilde{L}_0-2}\langle\lambda,\tilde{\lambda}|p_{(M+1,\ldots N)}\rangle \tag{147}$$

By writing

$$L_0 = \frac{\alpha'}{4}\hat{p}^2 + R \;\; ; \;\; \tilde{L}_0 = \frac{\alpha'}{4}\hat{p}^2 + \tilde{R} \tag{148}$$

with

$$R = \sum_{n=1}^{\infty} n a_n^{\dagger}\cdot a_n \;\; ; \;\; \tilde{R} = \sum_{n=1}^{\infty} n\tilde{a}_n^{\dagger}\cdot\tilde{a}_n \tag{149}$$

we can rewrite (147) as follows:

$$\sum_{\lambda,\tilde{\lambda}} \langle p_{1\ldots M}|\lambda,\tilde{\lambda}\rangle \frac{2\pi\langle\lambda,\tilde{\lambda}|\delta_{R,\tilde{R}}|\lambda,\tilde{\lambda}\rangle}{R+\tilde{R}-\alpha(s)}\langle\lambda,\tilde{\lambda}|p_{(M+1,\ldots N)}\rangle \tag{150}$$

We see that the amplitude for the Shapiro–Virasoro model has simple poles only for even integer values of $\alpha_{SV}(s) = 2 + \frac{\alpha'}{2}s = 2n \geq 0$ and the residue at the poles factorizes in a sum with a finite number of terms. Notice that the Regge trajectory of the Shapiro–Virasoro model has double intercept and half slope of that of the generalized Veneziano model.

5 Physical States and Their Vertex Operators

In the previous section, we have seen that the residue at the poles of the N-point amplitudes factorizes in a sum of a finite number of terms. We have also seen that some of these terms, due to the Lorentz metric, correspond to states with negative norm. We have also derived a number of "Ward identities" given in (126) that imply that some of the terms of the residue decouple. The question to be answered now is: Is the space spanned by the physical states a positive norm Hilbert space? In order to answer this question, we need first to find the conditions that characterize the on-shell physical states $|\lambda, P\rangle$ and then to determine which are the states that contribute to the residue of the pole at $\alpha(s = -P^2) = n$. In other words, we have to find a way of characterizing the physical states and of eliminating the spurious states that decouple in (102) as a consequence of (126). A state $|\lambda.P\rangle$ contributes at the residue of the pole in (102) for $\alpha(s = -P^2) = n$ if it is on-shell, namely if it satisfies the following equations:

$$R|\lambda, P\rangle = n|\lambda, P\rangle \;\; ; \;\; \alpha(-P^2) = 1 - \alpha' P^2 = n \tag{151}$$

that can be written in a unique equation

$$(L_0 - 1)|\lambda, P\rangle = 0 \tag{152}$$

Because of (126) we also know that a state of the type

$$|s, P\rangle = W_m^\dagger|\mu, P\rangle \tag{153}$$

is not going to contribute to the residue of the pole. We call it a spurious or unphysical state. We start constructing the subspace of spurious states that are on-shell at the level n. Let us consider the set of orthogonal states $|\mu, P\rangle$ such that

$$R|\mu, P\rangle = n_\mu|\mu, P\rangle \;\; ; \;\; L_0|\mu, P\rangle = (1 - m)|\mu, P\rangle \;\; ; \;\; 1 - \alpha' P^2 = n \tag{154}$$

where

$$m = n + n_\mu \tag{155}$$

In terms of these states we can construct the most general spurious state that is on-shell at the level n. It is given by

$$|s, P\rangle = W_m^\dagger|\mu, P\rangle \;\; ; \;\; (L_0 - 1)|s, P\rangle = 0 \tag{156}$$

per any positive integer m. Using (154), (156) becomes

$$|s, P\rangle = L_m^\dagger|\mu, P\rangle \tag{157}$$

where $|\mu, P\rangle$ is an arbitrary state satisfying (154).

A physical state $|\lambda, P\rangle$ is defined as the one that is orthogonal to all spurious states appearing at a certain level n. This means that it must satisfy the following equation:

$$\langle \lambda . P|L_\ell^\dagger|\mu, P\rangle = 0 \tag{158}$$

for any state $|\mu, P\rangle$ satisfying (154). In conclusion, the on-shell physical states at the level n are characterized by the fact that they satisfy the following conditions:

$$L_m|\lambda, P\rangle = (L_0 - 1)|\lambda, P\rangle = 0 \;\; ; \;\; 1 - \alpha' P^2 = n \tag{159}$$

These conditions characterizing the physical subspace were first found by Del Giudice and Di Vecchia [28] where the analysis described here was done.

In order to find the physical subspace, one starts writing the most general on-shell state contributing to the residue of the pole at level n in (154). Then one imposes (159) and determines the states that span the physical subspace. Actually, among these states one finds also a set of zero norm states that are physical and spurious at the same time. Those states are of the form given in

(157), but also satisfy (159). It is easy to see that they are not really physical because they are not contributing to the residue of the pole at the level n. This follows from the form of the unit operator given in the space of the physical states by

$$1 = \sum_{norm \neq 0} |\lambda, P\rangle\langle\lambda, P| + \sum_{zero} [|\lambda_0, P\rangle\langle\mu_0, P| + |\mu_0, P\rangle\langle\lambda_0, P|] \quad (160)$$

where $|\lambda_0, P\rangle$ is a zero norm physical and spurious state and $|\mu_0, P\rangle$ its conjugate state. A conjugate state of a zero norm state is obtained by changing the sign of the oscillators with time-like direction. Since $|\lambda_0, P\rangle$ is a spurious state when we insert the unit operator, given in (160), in (102) we see that the zero norm states never contribute to the residue because their contribution is annihilated either from the state $\langle p_{(1,M)}|$ or from the state $|p_{(M+1,N)}\rangle$. In conclusion, the physical subspace contains only the states in the first term in the r.h.s. of (160).

Let us analyse the first two excited levels. The first excited level corresponds to a massless gauge field. It is spanned by the states $\epsilon^\mu a^\dagger_{1\mu}|0, P\rangle$. In this case the only condition that we must impose is

$$L_1 \epsilon^\mu a^\dagger_{1\mu}|0, P\rangle = 0 \Longrightarrow P \cdot \epsilon = 0 \quad (161)$$

Choosing a frame of reference where the momentum of the photon is given by $P^\mu \equiv (P, 0....0, P)$, (161) implies that the only physical states are

$$\epsilon^i a^{+\dagger}_{1i}|0, P\rangle + \epsilon(a^\dagger_{1;0} - a^\dagger_{1;d-1})|0, P\rangle \;\; ; \;\; i = 1 \ldots d-2 \quad (162)$$

where ϵ^i and ϵ are arbitrary parameters. The state in (162) is the most general state of the level $N = 1$ satisfying the conditions in (159). The first state in (162) has positive norm, while the second one has zero norm that is orthogonal to all other physical states since it can be written as follows:

$$(a^\dagger_{1;0} - a^\dagger_{1;D-1})|0, P\rangle = L^\dagger_1|0, P\rangle \quad (163)$$

in the frame of reference where $P^\mu \equiv (P, ...0, P)$. Because of the previous property it is decoupled from the physical states together with its conjugate

$$(a^\dagger_{1,0} + a^\dagger_{1,d-1})|0, P\rangle \quad (164)$$

In conclusion, we are left only with the transverse $d-2$ states corresponding to the physical degrees of freedom of a massless spin 1 state. At the next level $n = 2$, the most general state is given by

$$[\alpha^{\mu\nu} a^\dagger_{1,\mu} a^\dagger_{1,\nu} + \beta^\mu a^\dagger_{2,\mu}]|0, P\rangle \quad (165)$$

If we work in the centre of mass frame where $P^\mu = (M, \mathbf{0})$ we get the following most general physical state:

$$|Phys>= \alpha^{ij}[a_{1,i}^\dagger a_{1,j}^\dagger - \frac{1}{(d-1)}\delta_{ij}\sum_{k=1}^{d-1} a_{1,k}^\dagger a_{1,k}^\dagger]|0,P\rangle$$

$$+\beta^i[a_{2,i}^\dagger + a_{1,0}^\dagger a_{1,i}^\dagger]|0,P>\rangle$$

$$+\sum_{i=1}^{d-1}\alpha^{ii}\left[\sum_{i=1}^{d-1} a_{1,i}^\dagger a_{1,i}^\dagger + \frac{d-1}{5}(a_{1,0}^{\dagger 2} - 2a_{2,0}^\dagger)\right]|0,P\rangle \qquad (166)$$

where the indices i, j run over the $d-1$ space components. The first term in (166) corresponds to a spin 2 in $(d-1)$-dimensional space and has a positive norm being made with space indices. The second term has zero norm and is orthogonal to the other physical states since it can be written as $L_1^+ a_{1,i}^+|0,P\rangle$. Therefore, it must be eliminated from the physical spectrum together with its conjugate, as explained above. Finally, the last state in (166) is spinless and has a norm given by

$$2(d-1)(26-d) \qquad (167)$$

If $d < 26$ it corresponds to a physical spin zero particle with positive norm. If $d > 26$ it is a ghost. Finally, if $d = 26$ it has a zero norm and is also orthogonal to the other physical states since it can be written in the form

$$(2L_2^\dagger + 3L_1^{\dagger 2})|0> \qquad (168)$$

It does not belong, therefore, to the physical spectrum. The analysis of this level was done [29] with $d = 4$. This did not allow the authors of [29] to see that there was a critical dimension.

The analysis of the physical states can be easily extended [28] to the Shapiro–Virasoro model. In this case the physical conditions given in (159) for the open string, become [28]

$$L_m|\lambda,\tilde{\lambda}\rangle = \tilde{L}_m|\lambda,\tilde{\lambda}\rangle = (L_0-1)|\lambda,\tilde{\lambda}\rangle = (\tilde{L}_0-1)|\lambda,\tilde{\lambda}\rangle = 0 \qquad (169)$$

for any positive integer m. It can be easily seen from the previous equations that the lowest state of the Shapiro–Virasoro model is the vacuum $|0_a, 0_{\tilde{a}}, p\rangle$ corresponding to a tachyon with mass $\alpha' p^2 = 4$, while the next level described by the state $a_{1\mu}^\dagger \tilde{a}_{1\nu}^\dagger|0_a, 0_{\tilde{a}}, p\rangle$ contains massless states corresponding to the graviton, a dilaton and a two-index antisymmetric tensor $B_{\mu\nu}$.

Having characterized the physical subspace one can go on and construct a N-point scattering amplitude involving arbitrary physical states. This was done by Campagna, Fubini, Napolitano and Sciuto [30] where the vertex operator for an arbitrary physical state was constructed in analogy with what has been done for the ground tachyonic state. They associated to each physical state $|\alpha, P\rangle$ a vertex operator $V_\alpha(z, P)$ that is a conformal field with conformal dimension equal to 1:

$$[L_n, V_\alpha(z,p)] = \frac{d}{dz}\left(z^{n+1}V_\alpha(z,p)\right) \qquad (170)$$

and reproduces the corresponding state acting on the vacuum as follows:

$$\lim_{z\to 0} V_\alpha(z;p)|0,0\rangle \equiv |\alpha;p\rangle \;\; ; \;\; \langle 0;0| \lim_{z\to\infty} z^2 V_\alpha(z;p) = \langle \alpha, p| \tag{171}$$

It satisfies, in addition, the hermiticity relation

$$V_\alpha^\dagger(z,P) = V_\alpha(\frac{1}{z}, -P)(-1)^{\alpha(-P^2)} \tag{172}$$

An excited vertex that will play an important role in the next section is the one associated to the massless gauge field. It is given by

$$V_\epsilon(z,k) \equiv \epsilon \cdot \frac{dQ(z)}{dz} e^{ik\cdot Q(z)} \;\; ; \;\; k\cdot\epsilon = k^2 = 0 \tag{173}$$

Because of the last two conditions in (173) the normal order is not necessary. It is convenient to give the expression of $\frac{dQ(z)}{dz}$ in terms of the harmonic oscillators

$$P(z) \equiv \frac{dQ(z)}{dz} = -i\sqrt{2\alpha'} \sum_{n=-\infty}^{\infty} \alpha_n z^{-n-1} \tag{174}$$

It is a conformal field with conformal dimension equal to 1. The rescaled oscillators α_n are given by

$$\alpha_n = \sqrt{n} a_n \;\; ; \;\; \alpha_{-n} = \sqrt{n} a_n^\dagger \;\; ; \;\; n > 0 \;\; ; \;\; \alpha_0 = \sqrt{2\alpha'}\hat{p} \tag{175}$$

In terms of the vertex operators previously introduced the most general amplitude involving arbitrary physical states is given by [30]

$$(2\pi)^4 \delta(\sum_{i=1}^N p_i) B_N^{ex} = \int_{-\infty}^{\infty} \frac{\prod_1^N dz_i \theta(z_i - z_{i+1})}{dV_{abc}} \langle 0,0| \prod_{i=1}^N V_{\alpha_i}(z_i,p_i)|0,0\rangle \tag{176}$$

In the case of the Shapiro–Virasoro model the tachyon vertex operator is given in (135). By rewriting (134) as follows:

$$Q(z,\bar{z}) = Q(z) + \tilde{Q}(\bar{z}) \tag{177}$$

where

$$Q(z) = \frac{1}{2}\left[\hat{q} - 2\alpha'\hat{p}\log(z) + i\sqrt{2\alpha'}\sum_{n=1}^{\infty} \frac{1}{\sqrt{n}}\left[a_n z^{-n} - a_n^\dagger z^n\right]\right] \tag{178}$$

and

$$\tilde{Q}(\bar{z}) = \frac{1}{2}\left[\hat{q} - 2\alpha'\hat{p}\log(\bar{z}) + i\sqrt{2\alpha'}\sum_{n=1}^{\infty} \frac{1}{\sqrt{n}}\left[\tilde{a}_n \bar{z}^{-n} - \tilde{a}_n^\dagger \bar{z}^n\right]\right] \tag{179}$$

we can write the tachyon vertex operator in the following way:

$$V(z, \bar{z}, p) =: e^{ip \cdot Q(z)} e^{ip \cdot \tilde{Q}(\bar{z})} : \tag{180}$$

This shows that the vertex operator corresponding to the tachyon of the Shapiro–Virasoro model can be written as the product of two vertex operators corresponding each to the tachyon of the generalized Veneziano model.

Analogously the vertex operator corresponding to an arbitrary physical state of the Shapiro–Virasoro model can always be written as a product of two vertex operators of the generalized Veneziano model:

$$V_{\alpha,\beta}(z, \bar{z}, p) = V_{\alpha}(z, \frac{p}{2}) V_{\beta}(\bar{z}, \frac{p}{2}) \tag{181}$$

The first one contains only the oscillators α_n, while the second one only the oscillators $\tilde{\alpha}_n$. They both contain only half of the total momentum p and the same zero modes $\hat{p}$ and $\hat{q}$. The two vertex operators of the generalized Veneziano model are both conformal fields with conformal dimension equal to 1. If they correspond to physical states at the level $2n$, they satisfy the following relation ($n = \tilde{n}$):

$$\alpha' \frac{p^2}{4} + n = 1 \tag{182}$$

They lie on the following Regge trajectory:

$$2 - \frac{\alpha'}{2} p^2 \equiv \alpha_{SV}(-p^2) = 2n \tag{183}$$

as we have already seen by factorizing the amplitude in (150).

6 The DDF States and Absence of Ghosts

In the previous section we have derived the equations that characterize the physical states and their corresponding vertex operators. In this section we will explicitly construct an infinite number of orthonormal physical states with positive norm.

The starting point is the DDF operator introduced by Del Giudice, Di Vecchia and Fubini [31] and defined in terms of the vertex operator corresponding to the massless gauge field introduced in (173)

$$A_{i,n} = \frac{i}{\sqrt{2\alpha'}} \oint_0 dz \epsilon_i^{\mu} P_{\mu}(z) e^{ik \cdot Q(z)} \tag{184}$$

where the index i runs over the $d-2$ transverse directions, that are orthogonal to the momentum k. We have also taken $\oint_0 \frac{dz}{z} = 1$. Because of the $\log z$ term appearing in the zero mode part of the exponential, the integral in (184), that is performed around the origin $z = 0$, is well defined only if we constrain the momentum of the state, on which $A_{i,n}$ acts, to satisfy the relation

$$2\alpha' p \cdot k = n \tag{185}$$

where n is a non-vanishing integer.

The operator in (184) will generate physical states because it commutes with the gauge operators L_m:

$$[L_m, A_{n;i}] = 0 \tag{186}$$

since the vertex operator transforms as a primary field with conformal dimension equal to 1 as it follows from (170).

On the other hand it also satisfies the algebra of the harmonic oscillator as we are now going to show. From (184) we get

$$[A_{n,i}, A_{m,j}] = -\frac{1}{2\alpha'} \oint_0 d\zeta \oint_\zeta dz \epsilon_i \cdot P(z) e^{ik\cdot Q(\zeta)} \epsilon_j \cdot P(\zeta) e^{ik'\cdot Q(\zeta)} \tag{187}$$

where

$$2\alpha' p \cdot k = n \quad ; \quad 2\alpha' p \cdot k' = m \tag{188}$$

and k and k' are supposed to be in the same direction, namely,

$$k_\mu = n\hat{k}_\mu \quad ; \quad k'_\mu = m\hat{k}_\mu \tag{189}$$

with

$$2\alpha' p \cdot \hat{k} = 1 \tag{190}$$

Finally, the polarizations are normalized as

$$\epsilon_i \cdot \epsilon_j = \delta_{ij} \tag{191}$$

Since $\hat{k} \cdot \epsilon_i = \hat{k} \cdot \epsilon_j = \hat{k}^2 = 0$ a singularity for $z = \zeta$ can appear only from the contraction of the two terms $P(\zeta)$ and $P((z)$ that is given by

$$\langle 0, 0|\epsilon_i \cdot P(z)\epsilon_j \cdot P(\zeta)|0, 0\rangle = -\frac{2\alpha'\delta_{ij}}{(z-\zeta)^2} \tag{192}$$

Inserting it in (187), we get

$$[A_{n,i}, A_{m,j}] = \delta_{ij} in \oint_0 d\zeta \hat{k} \cdot P(\zeta) e^{-i(n+m))\hat{k}\cdot Q(\zeta)}$$

$$= in\delta_{ij}\delta_{n+m;0} \oint_0 d\zeta \hat{k} \cdot P(\zeta) \tag{193}$$

where we have used the fact that the integrand is a total derivative and therefore one gets a vanishing contribution unless $n + m = 0$. If $n + m = 0$ from (174) and (190) we get

$$[A_{n,i}, A_{m,j}] = n\delta_{ij}\delta_{n+m;0} \quad ; \quad i,j = 1\dots d-2 \tag{194}$$

Equation (194) shows that the DDF operators satisfy the harmonic oscillator algebra.

In terms of this infinite set of transverse oscillators we can construct an orthonormal set of states

$$|i_1, N_1; i_2, N_2; \dots i_m, N_m\rangle = \prod_h \frac{1}{\sqrt{\lambda_h!}} \prod_{k=1}^{m} \frac{A_{i_k,-N_k}}{\sqrt{N_k}} |0,p\rangle \tag{195}$$

where λ_h is the multiplicity of the operator $A_{i_h,-N_h}$ in the product in (195) and the momentum of the state in (195) is given by

$$P = p + \sum_{i=1}^{m} \hat{k} N_i \tag{196}$$

They were constructed in four dimensions where they were not a complete system of states[11] and it took some time to realize that in fact they were a complete system of states if $d = 26$ [32, 33].[12] Brower [32] and Goddard and Thorn [33] showed also that the dual resonance model was ghost free for any dimension $d \leq 26$. In $d = 26$ this follows from the fact that the DDF operators obviously span a positive-definite Hilbert space (see (194)). For $d < 26$ there are extra states called Brower states [32]. The first of these states is the last state in (166) that becomes a zero norm state for $d = 26$. But also for $d < 26$ there is no negative norm state among the physical states. The proof of the no-ghost theorem in the case $\alpha_0 = 1$ is a very important step because it shows that the dual resonance model constructed generalizing the four-point Veneziano formula, is a fully consistent quantum-relativistic theory! This is not quite true because, when the intercept $\alpha_0 = 1$, the lowest state of the spectrum corresponding to the pole in the N-point amplitude for $\alpha(s) = 0$, is a tachyon with mass $m^2 = -\frac{1}{\alpha'}$. A lot of effort was then made to construct a model without tachyon and with a meson spectrum consistent with the experimental data. The only reasonably consistent models that came out from these attempts, were the Neveu–Schwarz [7] for mesons and the Ramond model [8] for fermions that only later were recognized to be part of a unique model that nowadays is called the Neveu–Schwarz-Ramond model.

---

[11] Because of this Fubini did not want to publish our result, but then he went to a meeting in Israel in spring 1971 giving a talk on our work where he found that the audience was very interested in our result and when he came back to MIT we decided to publish our result.

[12] I still remember Charles Thorn coming into my office at CERN and telling me: Paolo, do you know that your DDF states are complete if $d = 26$? I quickly redid the analysis done in [29] with an arbitrary value of the space–time dimension obtaining (166) and (167) that show that the spinless state at the level $\alpha(s) = 2$ is decoupled if $d = 26$. I strongly regretted not to have used an arbitrary space–time dimension d in the analysis of [29].

But this model was not really more consistent than the original dual resonance model because it still had a tachyon with mass $m^2 = -\frac{1}{2\alpha'}$. The tachyon was eliminated from the spectrum only in 1976 through the GSO projection proposed by Gliozzi, Scherk and Olive [34].

Having realized that, at least for the critical value of the space–time dimension $d = 26$, the physical states are described by the DDF states having only $d - 2 = 24$ independent components, open the way to Brink and Nielsen [35] to compute the value $\alpha_0 = 1$ of the Regge trajectory with a very physical argument. They related the intercept of the Regge trajectory to the zero point energy of a system with an infinite number of oscillators having only $d - 2$ independent components

$$\alpha_0 = -\frac{d-2}{2}\sum_{n=1}^{\infty} n \tag{197}$$

This quantity is obviously infinite and, in order to make sense of it, they introduced a cut-off on the frequencies of the harmonic oscillators obtaining an infinite term that they eliminated by renormalizing the speed of light and a finite universal constant term that gave the intercept of the Regge trajectory. Instead of following their original approach we discuss here an alternative approach due to Gliozzi [36] that uses the ζ-function regularization. He rewrites (197) as follows:

$$\alpha_0 = -\frac{d-2}{2}\sum_{n=1}^{\infty} n = -\frac{d-2}{2}\lim_{s\to -1}\sum_{n=1}^{\infty} n^{-s} = -\frac{d-2}{2}\zeta_R(-1) = 1 \tag{198}$$

where in the last equation we have used the identity $\zeta_R(-1) = -\frac{1}{12}$ and we have put $d = 26$. Since the Shapiro–Virasoro model has two sets of transverse harmonic oscillators it is obvious that its intercept is twice that of the generalized Veneziano model.

Using the rules discussed in the previous section we can construct the vertex operator corresponding to the state in (195). It is given by

$$V_{(i;N_i)}(z,P) = \prod_{i=1}^{m} \oint_z dz_i \epsilon_i \cdot P(z_i) e^{iN_i \hat{k}\cdot Q(z_i)} : e^{ip\cdot Q(z)} : \tag{199}$$

where the integral on the variable z_i is evaluated along a curve of the complex plane z_i containing the point z. The singularity of the integrand for $z_i = z$ is a pole provided that the following condition is satisfied.

$$2\alpha' p \cdot \hat{k} = 1 \tag{200}$$

The last vertex in (199) is the vertex operator corresponding to the ground tachyonic state given in (59) with $\alpha' p^2 = 1$.

Using the general form of the vertex one can compute the three-point amplitude involving three arbitrary DDF vertex operators. This calculation has been performed in [37] and since the vertex operators are conformal fields with dimension equal to 1 one gets

$$\langle 0,0|V_{(i^{(1)}_{k_1};N^{(1)}_{k_1})}(z_1,P_1)V_{(i^{(2)}_{k_2};N^{(2)}_{k(2)})}(z_2,P_2)V_{(i^{(3)}_{k_3};N^{(3)}_{k(3)})}(z_3,P_3)|0,0\rangle$$

$$= \frac{C_{123}}{(z_1-z_2)(z_1-z_3)(z_2-z_3)} \tag{201}$$

where the explicit form of the coefficient C_{123} is given by

$$C_{123} = {}_1\langle 0,0|_2\langle 0,0|_3\langle 0,0|e^{\frac{1}{2}\sum_{r,s=1}^{3}\sum_{n,m=1}^{\infty}A^{(r)}_{-n;i}N^{rs}_{nm}A^{(s)}_{-m;i}+\sum_{i=1}^{3}P_i\cdot\sum_{n=1}^{\infty}A^{(r)}_{-n;i}}$$

$$\times\, e^{\tau_0\sum_{r=1}^{3}(\alpha'\Pi_r^2-1)}|N^{(1)}_{k_1},i^{(1)}_{k_1}\rangle_1|N^{(2)}_{k_2},i^{(2)}_{k_2}\rangle_2|N^{(3)}_{k_3},i^{(3)}_{k_3}\rangle_3 \tag{202}$$

where

$$N^{rs}_{nm} = -N^r_n N^s_m \frac{nm\alpha_1\alpha_2\alpha_3}{n\alpha_s + m\alpha_r} \quad ; \quad N^r_n = \frac{\Gamma(-n\frac{\alpha_{r+1}}{\alpha_r})}{\alpha_r n!\Gamma(1-n\frac{\alpha_{r+1}}{\alpha_r}-n)} \tag{203}$$

with

$$\Pi = P_{r+1}\alpha_r - P_r\alpha_{r+1} \quad ; \quad r=1,2,3 \tag{204}$$

Π is independent on the value of r chosen as a consequence of the equations

$$\sum_{r=1}^{3}\alpha_r = \sum_{r=1}^{3}P_r = 0 \tag{205}$$

7 The Zero Slope Limit

In the introduction we have seen that the dual resonance model has been constructed using rules that are different from those used in field theory. For instance, we have seen that planar duality implies that the amplitude corresponding to a certain duality diagram, contains poles in both s and t channels, while the amplitude corresponding to a Feynman diagram in field theory contains only a pole in one of the two channels. Furthermore, the scattering amplitude in the dual resonance model contains an infinite number of resonant states that, at high energy, average out to give Regge behaviour. Also this property is not observed in field theory. The question that was natural to ask, was then: is there any relation between the dual resonance model and field theory? It turned out, to the surprise of many, that the dual resonance model was not in contradiction with field theory, but was instead an extension of a certain number of field theories. We will see that the limit in which a field theory is obtained from the dual resonance model corresponds to taking the slope of the Regge trajectory α' to zero.

Let us consider the scattering amplitude of four ground state particles in (1) that we rewrite here with the correct normalization factor

$$A(s,t,u) = C_0 N_0^4 \left(A(s,t) + A(s,u) + A(t,u) \right) \tag{206}$$

where

$$N_0 = \sqrt{2} g (2\alpha')^{\frac{d-2}{4}} \tag{207}$$

is the correct normalization factor for each external leg, g is the dimensionless open string coupling constant that we have constantly ignored in the previous sections and C_0 is determined by the following relation:

$$C_0 N_0^2 \alpha' = 1 \tag{208}$$

that is obtained by requiring the factorization of the amplitude at the pole corresponding to the ground state particle whose mass is given in (21). Using (21) in order to rewrite the intercept of the Regge trajectory in terms of the mass of the ground state particle m^2 and the following relation satisfied by the Γ-function:

$$\Gamma(1+z) = z\Gamma(z) \tag{209}$$

we can easily perform the limit for $\alpha' \rightarrow 0$ of $A(s,t)$ obtaining

$$\lim_{\alpha' \rightarrow 0} A(s,t) = \frac{1}{\alpha'} \left[\frac{1}{m^2 - s} + \frac{1}{m^2 - s} \right] \tag{210}$$

Performing the same limit on the other two planar amplitudes, we get the following expression for the total amplitude in (206):

$$\lim_{\alpha' \rightarrow 0} A(s,t,u) = \left[\sqrt{2} g (2\alpha')^{\frac{d-2}{4}} \right]^2 \frac{2}{(\alpha')^2} \left[\frac{1}{m^2 - s} + \frac{1}{m^2 - s} + \frac{1}{m^2 - u} \right] \tag{211}$$

By introducing the coupling constant

$$g_3 = 4g(2\alpha')^{\frac{d-6}{4}} \tag{212}$$

Equation (211) becomes

$$\lim_{\alpha' \rightarrow 0} A(s,t,u) = g_3^2 \left[\frac{1}{m^2 - s} + \frac{1}{m^2 - s} + \frac{1}{m^2 - u} \right] \tag{213}$$

that is equal to the sum of the tree diagrams for the scattering of four particles with mass m of Φ^3 theory with coupling constant equal to g_3. We have shown that, by keeping g_3 fixed in the limit $\alpha' \to 0$, the scattering amplitude of four ground state particles of the dual resonance model is equal to the tree diagrams of Φ^3 theory. This proof can be extended to the scattering of N ground state particles recovering also in this case the tree diagrams of Φ^3 theory. It is also valid for loop diagrams that we will discuss in the next section. In conclusion, the dual resonance model reduces in the zero slope limit to Φ^3 theory. The proof that we have presented here is due to Scherk [38].[13]

A more interesting case to study is the one with intercept $\alpha_0 = 1$. We will see that, in this case, one will obtain the tree diagrams of Yang–Mills theory, as shown by Neveu and Scherk [40].[14]

Let us consider the three-point amplitude involving three massless gauge particles described by the vertex operator in (173). It is given by the sum of two planar diagrams. The first one corresponding to the ordering (123) is given by

$$C_0 N_0^3 i^3 Tr\left(\lambda^{a_1}\lambda^{a_2}\lambda^{a_3}\right) \frac{\langle 0,0|V_{\epsilon_1}(z_1,p_1)V_{\epsilon_2}(z_2,p_2)V_{\epsilon_3}(z_3,p_3)|0,0\rangle}{\left[(z_1-z_2)(z_2-z_3)(z_1-z_3)\right]^{-1}} \tag{214}$$

Using momentum conservation $p_1+p_2+p_3=0$ and the mass shell conditions $p_i^2 = p_i \cdot \epsilon_i = 0$, one can rewrite the previous equation as follows:

$$C_0 N_0^3 Tr(\lambda^{a_1}\lambda^{a_2}\lambda^{a_3})\sqrt{2\alpha'}$$

$$\times \left[(\epsilon_1\cdot\epsilon_2)(p_1\cdot\epsilon_3) + (\epsilon_1\cdot\epsilon_3)(p_3\cdot\epsilon_2) + (\epsilon_2\cdot\epsilon_3)(p_2\cdot\epsilon_1)\right] \tag{215}$$

The second contribution comes from the ordering 132 that can be obtained from the previous one by the substitution

$$Tr(\lambda^{a_1}\lambda^{a_2}\lambda^{a_3}) \to -Tr(\lambda^{a_1}\lambda^{a_3}\lambda^{a_2}) \tag{216}$$

Summing the two contributions one gets

$$C_0 N_o^3 Tr(\lambda^{a_1}[\lambda^{a_2},\lambda^{a_3}])\sqrt{2\alpha'}$$

$$\times \left[(\epsilon_1\cdot\epsilon_2)(p_1\cdot\epsilon_3) + (\epsilon_1\cdot\epsilon_3)(p_3\cdot\epsilon_2) + (\epsilon_2\cdot\epsilon_3)(p_2\cdot\epsilon_1)\right] \tag{217}$$

The factor

$$N_0 = 2g(2\alpha')^{(d-2)/4} \tag{218}$$

is the correct normalization factor for each vertex operator if we normalize the generators of the Chan–Paton group as follows:

$$Tr\left(\lambda^i\lambda^j\right) = \frac{1}{2}\delta^{ij} \tag{219}$$

[13] See also [39].

[14] See also [41].

It is related to C_0 through the relation[15]

$$C_0 N_o^2 \alpha' = 2 \tag{220}$$

g is the dimensionless open string coupling constant. Notice that (218) and (220) differ from (207) and (208) because of the presence of the Chan–Paton factors that we did not include in the case of Φ^3 theory.

By using the commutation relations

$$[\lambda^a, \lambda^b] = i f^{abc} \lambda^c \tag{221}$$

and the previous normalization factors we get for the three-gluon amplitude

$$i g_{YM} f^{a_1 a_2 a_3} \left[(\epsilon_1 \cdot \epsilon_2)((p_1 - p_2) \cdot \epsilon_3 \right.$$

$$\left. + (\epsilon_1 \cdot \epsilon_3)((p_3 - p_1) \cdot \epsilon_2) + (\epsilon_2 \cdot \epsilon_3)((p_2 - p_3) \cdot \epsilon_1) \right] \tag{222}$$

that is equal to the three-gluon vertex that one obtains from the Yang–Mills action

$$L_{YM} = -\frac{1}{4} F^a_{\alpha\beta} F^{\alpha\beta}_a \quad , \quad F^a_{\alpha\beta} = \partial_\alpha A^a_\beta - \partial_\beta A^a_\alpha + g_{YM} f^{abc} A^b_\alpha A^c_\beta \tag{223}$$

where

$$g_{YM} = 2g(2\alpha')^{\frac{d-4}{4}} \tag{224}$$

The previous procedure can be extended to the scattering of N gluons finding the same result that one gets from the tree diagrams of Yang–Mills theory. In the next section, we will discuss the loop diagrams. Also, in this case one finds that the h-loop diagrams involving N external gluons reproduces in the zero slope limit the sum of the h-loop diagrams with N external gluons of Yang–Mills theory.

We conclude this section mentioning that one can also take the zero slope limit of a scattering amplitude involving three and four gravitons obtaining agreement with what one gets from the Einstein Lagrangian of general relativity. This has been shown by Yoneya [43].

8 Loop Diagrams

The N-point amplitude previously constructed satisfies all the axioms of S-matrix theory except unitarity because its only singularities are simple poles corresponding to zero width resonances lying on the real axis of the Mandelstam variables and does not contain the various cuts required by unitarity [1].

15 The determination of the previous normalization factors can be found in the appendix of [42].

In order to eliminate this problem, it was proposed already in the early days of dual theories to assume, in analogy with what happens for instance in perturbative field theory, that the N-point amplitude was only the lowest order (the tree diagram) of a perturbative expansion and, in order to implement unitarity, it was necessary to include loop diagrams. Then, the one-loop diagrams were constructed from the propagator and vertices that we have introduced in the previous sections [44]. The planar one-loop amplitude with M external particles was computed by starting from a $(M+2)$-point tree amplitude and then by sewing two external legs together after the insertion of a propagator D given in (100). In this way one gets

$$\int \frac{d^d P}{(2\alpha')^{d/2}(2\pi)^d} \sum_{\lambda} \langle P,\lambda|V(1,p_1)DV(1,p_2)\dots V(1,p_N)D|P,\lambda\rangle \quad (225)$$

where the sum over λ corresponds to the trace in the space of the harmonic oscillators and the integral in $d^d P$ corresponds to integrate over the momentum circulating in the loop. The previous expression for the one-loop amplitude cannot be quite correct because all states of the space generated by the oscillators in (51) are circulating in the loop, while we know that we should include only the physical ones. This was achieved first by cancelling by hand the time and one of the space components of the harmonic oscillators reducing the degrees of freedom of each oscillator from d to $d-2$ as suggested by the DDF operators at least for $d=26$. This procedure was then shown to be correct by Brink and Olive [45]. They constructed the operator that projects over the physical states and, by inserting it in the loop, showed that the reduction of the degrees of freedom of the oscillators from d to $d-2$ was indeed correct. This was, at that time, the only procedure available to let only the physical states circulate in the loop because the BRST procedure was discovered a bit later also in the framework of the gauge field theories!

To be more explicit let us compute the trace in (225) adding also the Chan–Paton factor. We get

$$(2\pi)^d \delta^{(d)}\left(\sum_{i=1}^{M} p_i\right) \frac{NTr(\lambda^{a_1}\dots\lambda^{a_M})}{(8\pi^2\alpha')^{d/2}} N_0^M \int_0^\infty \frac{d\tau}{\tau^{d/2+1}} [f_1(k)]^{2-d} k^{\frac{d-26}{12}} (2\pi)^M$$

$$\times \int_0^1 d\nu_M \int_0^{\nu_M} d\nu_{M-1} \dots \int_0^{\nu_3} d\nu_2 \; \tau^M \prod_{i<j} \left[e^{G(\nu_{ji})}\right]^{2\alpha' p_i\cdot p_j} \; ; k \equiv e^{-\pi\tau} \quad (226)$$

where $\nu_{ji} \equiv \nu_j - \nu_i$,

$$G(\nu) = \log\left(ie^{-\pi\nu^2\tau} \frac{\Theta_1(i\nu\tau|i\tau)}{f_1^3(k)}\right) \quad ; \quad f_1(k) = k^{1/12} \prod_{n=1}^{\infty}(1-k^{2n}) \quad (227)$$

and

$$\Theta_1(\nu|i\tau) = -2k^{1/4}\sin\pi\nu \prod_{n=1}^{\infty}\left(1 - e^{2i\pi\nu}k^{2n}\right)\left(1 - e^{-2i\pi\nu}k^{2n}\right)(1 - k^{2n}) \tag{228}$$

Finally, the normalization factor N_0 is given in (218). We have performed the calculation for an arbitrary value of the space–time dimension d. However, in this way one gets also the extra factor of $k^{\frac{d-26}{12}}$ appearing in the first line of (226) that implies that our calculation is actually only consistent if $d = 26$. In fact, the presence of this factor does not allow one to rewrite the amplitude, originally obtained in the Reggeon sector, in the Pomeron sector as explained below. In the following we neglect this extra factor, implicitly assuming that $d = 26$, but, on the other hand, still keeping an arbitrary d.

Using the relations

$$f_1(k) = \sqrt{t} f_1(q) \quad ; \quad \Theta_1(i\nu\tau|i\tau) = i\Theta_1(\nu|it)t^{1/2}e^{\pi\nu^2/t} \tag{229}$$

where $t = \frac{1}{\tau}$ and $q \equiv e^{-\pi t}$, we can rewrite the one-loop planar diagram in the Pomeron channel. We get

$$(2\pi)^d \delta^{(d)}\left(\sum_{i=1}^{M} p_i\right) \frac{NTr(\lambda^{a_1}\ldots\lambda^{a_M})}{(8\pi^2\alpha')^{d/2}} N_0^M \int_0^{\infty} dt [f_1(q)]^{2-d}(2\pi)^M$$

$$\times \int_0^1 d\nu_M \int_0^{\nu_M} d\nu_{M-1}\ldots\int_0^{\nu_3} d\nu_2 \prod_{i<j}\left[-\frac{\Theta_1(\nu_{ji}|it)}{f_1^3(q)}\right]^{2\alpha' p_i\cdot p_j} \tag{230}$$

Notice that, by factorizing the planar loop in the Pomeron channel, one constructed for the first time what we now call the boundary state [46].[16] This can be easily seen in the way that we are now going to describe. First of all, notice that the last quantity in (230) can be written as follows:

$$\prod_{i<j}\left[-\frac{\Theta_1(\nu_{ji}|it)}{f_1^3(q)}\right]^{2\alpha' p_i\cdot p_j}$$

$$= \prod_{i<j}\left[-2\sin(\pi\nu_{ji})\prod_{n=1}^{\infty}\frac{\left(1 - q^{2n}e^{2\pi i\nu_{ji}}\right)\left(1 - q^{2n}e^{-2\pi i\nu_{ji}}\right)}{(1 - q^{2n})^2}\right]^{2\alpha' p_i\cdot p_j} \tag{231}$$

This equation can be rewritten as follows:

$$\frac{Tr\left(\langle p = 0|q^{2R}\prod_{i=1}^{M} : e^{ip_i\cdot Q(e^{2i\pi\nu_i})} : |p = 0\rangle\right) i^M}{Tr\left(\langle p = 0|q^{2N}|p = 0\rangle\right)} \quad ; \quad R = \sum_{n=1}^{\infty} n a_n^{\dagger}\cdot a_n \tag{232}$$

---

[16] See also the first paper in [47].

where the trace is taken only over the non-zero modes and momentum conservation has been used. It must also be stressed that the normal ordering of the vertex operators in the previous equation is such that the zero modes are taken to be both in the same exponential instead of being ordered as in (59). By bringing all annihilation operators on the left of the creation ones, from the expression in (232), one gets ($z_i \equiv e^{2\pi i \nu_i}$)

$$(2\pi)^d \delta^{(d)} \left(\sum_{i=1}^{\infty} p_i \right) \prod_{i<j} (-2 \sin \pi \nu_{ji})^{2\alpha' p_i \cdot p_j}$$

$$\times \frac{\prod_{i.j} \prod_{n=1}^{\infty} Tr \left(q^{2n a_n^\dagger \cdot a_n} e^{\sqrt{2\alpha'} p_j \cdot \frac{a_n^\dagger}{\sqrt{n}} z_j^n} e^{-\sqrt{2\alpha'} p_i \cdot \frac{a_n}{\sqrt{n}} z_i^{-n}} \right)}{Tr \left(\langle p=0 | q^{2N} | p=0 \rangle \right)} \tag{233}$$

The trace can be computed by using the completeness relation involving coherent states $|f\rangle = e^{f a^\dagger} |0\rangle$:

$$\int \frac{d^2 f}{\pi} e^{-|f|^2} |f\rangle \langle f| = 1 \tag{234}$$

Inserting the previous identity operator in (233), one gets after some calculation

$$(2\pi)^d \delta^{(d)} \left(\sum_{i=1}^{\infty} p_i \right) \prod_{i<j} (-2 \sin \pi \nu_{ji})^{2\alpha' p_i \cdot p_j}$$

$$\times \prod_{i.j=1}^{M} \prod_{n=1}^{\infty} e^{-2\alpha' p_i \cdot p_j e^{2\pi i n \nu_{ji}} \frac{q^{2n}}{n(1-q^{2n})}} \tag{235}$$

Expanding the denominator in the last exponent and performing the sum over n one gets

$$(2\pi)^d \delta^{(d)} \left(\sum_{i=1}^{\infty} p_i \right) \prod_{i<j} (-2 \sin \pi \nu_{ji})^{2\alpha' p_i \cdot p_j}$$

$$\times \prod_{i.j} e^{2\alpha' p_i \cdot p_j \sum_{m=0}^{\infty} \log\left(1 - e^{2\pi i \nu_{ji}} q^{2(m+1)}\right)} \tag{236}$$

that is equal to the last line of (231) apart from the δ-function for momentum conservation. In conclusion, we have shown that (231) and (232) are equal.

Using (231) we can rewrite (230) as follows:

$$\frac{N N_0^M Tr(\lambda^{a_1} \ldots \lambda^{a_M})}{(8\pi^2 \alpha')^{d/2}} \int_0^\infty dt [f_1(q)]^{2-d} (2\pi i)^M \int_0^1 d\nu_M \int_0^{\nu_M} d\nu_{M-1} \ldots$$

$$\ldots \int_0^{\nu_3} d\nu_2 \frac{\sum_\lambda \langle p=0,\lambda|q^{2R}\prod_{i=1}^M : e^{ip_i\cdot Q(e^{2i\pi\nu_i})} : |p=0,\lambda\rangle}{\sum_\lambda \langle p=0,\lambda|q^{2N}|p=0,\lambda\rangle} \tag{237}$$

where the sum over any state $|\lambda\rangle$ corresponds to taking the trace over the non-zero modes. If $d = 26$ we can rewrite (237) in a simpler form

$$\frac{NN_0^M Tr(\lambda^{a_1}\ldots\lambda^{a_M})}{(8\pi^2\alpha')^{d/2}} \int_0^\infty dt\,(2\pi i)^M \int_0^1 d\nu_M \int_0^{\nu_M} d\nu_{M-1}\cdots\int_0^{\nu_3} d\nu_2$$

$$\times \sum_\lambda \langle p=0,\lambda|q^{2R-2}\prod_{i=1}^M : e^{ip_i\cdot Q(e^{2i\pi\nu_i})} : |p=0,\lambda\rangle \tag{238}$$

The previous equation contains the factor $\int dt q^{2R-2}$ that is like the propagator of the Shapiro–Virasoro model, but with only one set of oscillators as in the generalized Veneziano model. In the following we will rewrite it completely with the formalism of the Shapiro–Virasoro model. This can be done by introducing the Pomeron propagator

$$\int_0^\infty dt\; q^{2N-2} = \frac{2}{\pi\alpha'}\hat{D} \;\; ; \quad \hat{D} \equiv \frac{\alpha'}{4\pi}\int \frac{d^2z}{|z|^2} z^{L_0-1}\bar{z}^{\tilde{L}_0-1}; |z| \equiv q = e^{-\pi t} \tag{239}$$

and rewriting the planar loop in the following compact form:

$$\langle B_0|\hat{D}|B_M\rangle \;\; ; \;\; |B_0\rangle \equiv \frac{T_{d-1}}{2}N\prod_{n=1}^\infty e^{a_n^\dagger\cdot\tilde{a}_n^\dagger}|p=0,0_a,0_{\tilde{a}}\rangle \tag{240}$$

where $|B_0\rangle$ is the boundary state without any Reggeon on it,

$$T_{d-1} = \frac{\sqrt{\pi}}{2^{(d-10)/4}}(2\pi\sqrt{\alpha'})^{-d/2-1} \tag{241}$$

and $|B_M\rangle$ is instead the one with M Reggeons given by

$$|B_M\rangle = N_0^M Tr(\lambda^{a_1}\ldots\lambda^{a_M})(2\pi i)^M \int_0^1 d\nu_M \int_0^{\nu_M} d\nu_{M-1}\cdots\int_0^{\nu_3} d\nu_2$$

$$\times\prod_{i=1}^M : e^{ip_i\cdot Q(e^{2i\pi\nu_i})} : |B_0\rangle \tag{242}$$

We want to stress once more that the normal ordering in the previous equation is defined by taking the zero modes in the same exponential. Both the boundary states and the propagator are now states of the Shapiro–Virasoro model. This means that we have rewritten the one-loop planar diagram, where the states of the generalized Veneziano model circulate in the loop, as a tree diagram of the Shapiro–Virasoro model involving two boundary states and a propagator. This is what nowadays is called open/closed string duality.

Besides the one-loop planar diagram in (225), that is nowadays called the annulus diagram, also the non-planar and the non-orientable diagrams were constructed and studied. In particular the non-planar one, that is obtained as the planar one in (225) but with two propagators multiplied with the twist operator

$$\Omega = e^{L_{-1}}(-1)^R \; , \tag{243}$$

had unitarity violating cuts that disappeared [27] if the dimension of the space–time $d = 26$, leaving behind additional pole singularities. The explicit form of the non-planar loop can be obtained following the same steps done for the planar loop. One gets for the non-planar loop the following amplitude:

$$\langle B_R | \hat{D} | B_M \rangle \tag{244}$$

where now both boundary states contain, respectively, R and M Reggeon states. The additional poles found in the non-planar loop were called Pomerons because they occur in the Pomeron sector, that today is called the closed string channel, to distinguish them from the Reggeons that instead occur in the Reggeon sector, that today is called the open string sector of the planar and non-planar loop diagrams. At that time in fact, the states of the generalized Veneziano models were called Reggeons, while the additional ones appearing in the non-planar loop were called Pomerons. The Reggeons correspond nowadays to open string states, while the Pomerons to closed string states. These things are obvious now, but at that time it took a while to show that the additional states appearing in the Pomeron sector have to be identified with those of the Shapiro–Virasoro model. The proof that the spectrum was the same came rather early. This was obtained by factorizing the non-planar diagram in the Pomeron channel [46] as we have done in (244). It was found that the states of the Pomeron channel lie on a linear Regge trajectory that has double intercept and half slope of the one of the Reggeons. This follows immediately from the propagator $\hat{D}$ in Eq. (239) that has poles for values of the momentum of the Pomeron exchanged given by

$$2 - \frac{\alpha'}{2} p^2 = 2n \tag{245}$$

that are exactly the values of the masses of the states of the Shapiro–Virasoro model [48], while the Reggeon propagator in (100) has poles for values of momentum equal to

$$1 - \alpha' p^2 = n \tag{246}$$

However, it was still not clear that the Pomeron states interact among themselves as the states of the Shapiro–Virasoro model. To show this it was first necessary to construct tree amplitudes containing both states of the generalized Veneziano model and of the Shapiro–Virasoro model [49]. They reduced to the amplitudes of the generalized Veneziano (Shapiro–Virasoro) model if we have only external states of the generalized Veneziano (Shapiro–Virasoro) model. Those amplitudes are called today disk amplitudes containing both open and closed string states. They were constructed [49] by using for the Reggeon states the vertex operators that we have discussed in Sect. 5 involving one set of harmonic oscillators and for the Pomeron states the vertex operators given in (181) that we rewrite here

$$V_{\alpha,\beta}(z,\bar{z},p) = V_{\alpha}(z,\frac{p}{2})V_{\beta}(\bar{z},\frac{p}{2}) \tag{247}$$

because now both component vertices contain the same set of harmonic oscillators as in the generalized Veneziano model. Furthermore, each of the two vertices is separately normal ordered, but their product is nor normal ordered. The amplitude involving both kinds of states is then constructed by taking the product of all vertices between the projective invariant vacuum and integrating the Reggeons on the real axis in an ordered way and the Pomerons in the upper half plane, as one does for a disk amplitude.

We have mentioned above that the two vertices are separately normal ordered, but their product is not normal ordered. When we normal order them we get, for instance for the tachyon of the Pomeron sector, a factor $(z-\bar{z})^{\alpha' p^2/2}$ that describes the Reggeon–Pomeron transition. This implies a direct coupling [51] between the $U(1)$ part of gauge field and the two-index antisymmetric field $B_{\mu\nu}$, called Kalb–Ramond field [50], of the Pomeron sector, that makes the gauge field massive [51].

It was then shown that, by factorizing the non-planal loop in the Pomeron channel, one reproduced the scattering amplitude containing one state of the Shapiro–Virasoro and a number of states of the generalized Veneziano model [52]. If we have also external states belonging to the generalized Shapiro–Virasoro model, then by factorizing the non-planar one-loop amplitude in the pure Pomeron channel, one would obtain the tree amplitudes of the Shapiro–Virasoro model [52].

All this implies that the generalized Veneziano model and the Shapiro–Virasoro model are not two independent models, but they are part of the same and unique model. In fact, if one started with the generalized Veneziano model and added loop diagrams to implement unitarity, one found the appearence in the non-planar loop of additional states that had the same mass and interaction of those of the Shapiro–Virasoro model.

The planar diagram, written in (230) in the closed string channel, is divergent for large values of t. This divergence was recognized to be due to exchange, in the Pomeron channel, of the tachyon of the Shapiro–Virasoro model and of the dilaton [47]. They correspond, respectively, to the first two terms of the expansion

$$[f_1(q)]^{-24} = e^{2\pi t} + 24 + O\left(e^{-2\pi t}\right) \tag{248}$$

The first one could be cancelled by an analytic continuation, while the second one could be eliminated through a renormalization of the slope of the Regge trajectory α' [47].

We conclude the discussion of the one-loop diagrams by mentioning that the one-loop diagram for the Shapiro–Virasoro model was computed by Shapiro [53] who also found that the integrand was modular invariant.

The computation of multiloop diagrams requires a more advanced technology that was also developed in the early days of the dual resonance model few years before the discovery of its connection to string theory. In order to compute multiloop diagrams, one needs first to construct an object that was called the N-Reggeon vertex and that has the properties of containing N sets of harmonic oscillators, one for each external leg, and is such that, when we saturate it with N physical states, we get the corresponding N-point amplitude. In the following we will discuss how to determine the N-Reggeon vertex.

The first step toward the N-Reggeon vertex is the Sciuto–Della Selva-Saito [54] vertex that includes two sets of harmonic oscillators that we denote with the indices 1 and 2. It is equal to

$$V_{SDS} = {}_2\langle x=0,0| : \exp\left(-\frac{1}{2\alpha'}\oint_0 dz X_2'(z)\cdot X_1(1-z)\right) : \tag{249}$$

where X is the quantity that we have called Q in (57) and the prime denotes a derivative with respect to z. It satisfies the important property of giving the vertex operator $V_\alpha(z=1)$ of an arbitrary state $|\alpha\rangle$ when we saturate it with the corresponding state

$$V_{SDS}|\alpha\rangle_2 = V_\alpha(z=1) \tag{250}$$

A shortcoming of this vertex is that it is not invariant under a cyclic permutation of the three legs. A cyclic symmetric vertex has been constructed by Caneschi, Schwimmer and Veneziano [55] by inserting the twist operator in (243). But the three-Reggeon vertex is not enough if we want to compute an arbitrary multiloop amplitude. We must generalize it to an arbitrary number of external legs. Such a vertex, that can be obtained from the one in (249) with a very direct procedure, or that can also be obtained by sewing together three-Reggeon vertices, has been written in its final form by Lovelace [56].[17] Here we do not derive it, but we give directly its expression written in [56]:

$$V_{N,0} = \int \frac{\prod_{i=1}^N dz_i}{dV_{abc}\prod_{i=1}^N [V_i'(0)]} \prod_{i=1}^N [{}_i\!<x=0, O_a|]\,\delta(\sum_{i=1}^N p_i)$$
$$\prod_{\substack{i,j=1\\ i\neq j}}^N \exp\left[-\frac{1}{2}\sum_{n,m=0}^{\infty} a_n^{(i)} D_{nm}(\Gamma V_i^{-1}V_j)\, a_m^{(j)}\right] \tag{251}$$

[17] See also [57]. Earlier papers on the N-Reggeon can be found in [58].

where $a_0^{(i)} \equiv \alpha_0^i = \sqrt{2\alpha'}\hat{p}_i$ is the momentum of particle i and the infinite matrix

$$D_{nm}(\gamma) = \frac{1}{m!}\sqrt{\frac{m}{n}}\partial_z^m [\gamma(z)]^n|_{z=0} \ ; \ n,m = 1.. \ ; \ D_{00}(\gamma) = -\log|\frac{D}{\sqrt{AD-BC}}|$$

$$D_{n0} = \frac{1}{\sqrt{n}}(\frac{B}{D})^n \ ; \ D_{0n} = \frac{1}{\sqrt{n}}(-\frac{C}{D})^n \ ; \ \gamma(z) = \frac{Az+B}{Cz+D} \quad (252)$$

is a "representation" of the projective group corresponding to the conformal weight $\Delta = 0$, that satisfies the equations

$$D_{nm}(\gamma_1\gamma_2) = \sum_{l=1}^{\infty} D_{nl}(\gamma_1)D_{lm}(\gamma_2) + D_{n0}(\gamma_1)\delta_{0m} + D_{0m}(\gamma_2)\delta_{n0} \quad (253)$$

and

$$D_{nm}(\gamma) = D_{mn}(\Gamma\gamma^{-1}\Gamma) \qquad \Gamma(z) = \frac{1}{z} \quad (254)$$

Finally, V_i is a projective transformation that maps $0, 1$ and ∞ into z_{i-1}, z_i and z_{i+1}.

The previous vertex can be written in a more elegant form as follows:

$$V_{N,0} = \int \frac{\prod_{i=1}^N dz_i}{dV_{abc}\prod_{i=1}^N [V_i'(0)]} \prod_{i=1}^N [{}_i\langle x = 0, O_a|]\, \delta(\sum_{i=1}^N p_i)$$

$$\times \exp\left\{\frac{i}{4\alpha'}\oint dz \partial X^{(i)}(z)\hat{p}_i \log V_i'(z)\right\}$$

$$\times \exp\left\{-\frac{1}{2}\sum_{\substack{i,j=1\\ i\neq j}}^N \oint dz \oint dy \partial X^{(i)}(z)\log[V_i(z) - V_j(y)]\partial X^{(j)}(y)\right\} \quad (255)$$

where the quantities $X^{(i)}$ are what we called Q, namely the Fubini–Veneziano field, in the previous sections. The N-Reggeon vertex that satisfies the important property of giving the scattering amplitude of N physical particles when we saturate it with their corresponding states, is the fundamental object for computing the multiloop amplitudes. In fact, if we want to compute a M-loop amplitude with N external states, we need to start from the $(N+2M)$-Reggeon vertex and then we have to sew the M pairs together after having inserted a propagator D. In this way we obtain an amplitude that is not only integrated over the punctures z_i $(i = 1 \ldots N)$ of the N external states, but also over the additional $3h - 3$ moduli corresponding to the punctures variables of the states that we sew together and the integration variable of the M propagators. h is the number of loops. The multiloop amplitudes have been obtained in this way already in 1970 [59, 60, 61] and, through the sewing procedure, one obtained functions, as the period matrix, the abelian differentials, the prime form, etc., that are well defined on the Riemann surface! The only thing that was missing, was the correct measure of integrations over the $3h-3$ variables because it was technically not possible to let only the physical states to circulate in the loops. This problem was solved only much later [62, 63] when a BRST invariant formulation of string theory and the light-cone functional integral could be used for computing multiloops. They are two very different approaches that, however, gave the same result. For the sake of completeness, we write here the planar h-loop amplitude involving M tachyons

$$A_M^{(h)}(p_1,\ldots,p_M) = N^h \,\mathrm{Tr}(\lambda^{a_1}\cdots\lambda^{a_M})\, C_h \left[2g_s\,(2\alpha')^{(d-2)/4}\right]^M \times \int [dm]_h^M \prod_{i<j}\left[\frac{\exp\left(\mathcal{G}^{(h)}(z_i,z_j)\right)}{\sqrt{V_i'(0)\,V_j'(0)}}\right]^{2\alpha' p_i\cdot p_j} , \quad (256)$$

where $N^h\,\mathrm{Tr}(\lambda^{a_1}\cdots\lambda^{a_M})$ is the appropriate $U(N)$ Chan–Paton factor, g is the dimensionless open string coupling constant, $\mathcal{C}_h$ is a normalization factor given by

$$\mathcal{C}_h = \frac{1}{(2\pi)^{dh}}\, g_s^{2h-2}\frac{1}{(2\alpha')^{d/2}} , \quad (257)$$

and $\mathcal{G}^{(h)}$ is the h-loop bosonic Green function

$$\mathcal{G}^{(h)}(z_i,z_j) = \log E^{(h)}(z_i,z_j) - \frac{1}{2}\int_{z_i}^{z_j} \omega^\mu\,(2\pi \mathrm{Im}\tau_{\mu\nu})^{-1}\int_{z_i}^{z_j}\omega^\nu , \quad (258)$$

with $E^{(h)}(z_i,z_j)$ being the prime form, ω^μ ($\mu=1,\ldots,h$) the abelian differentials and $\tau_{\mu\nu}$ the period matrix. All these objects, as well as the measure on moduli space $[dm]_h^M$, can be explicitly written in the Schottky parametrization of the Riemann surface, and their expressions for arbitrary h can be found for example in [64]. In particular, the measure on the moduli space is given by

$$[dm]_h^M = \frac{1}{dV_{abc}}\prod_{i=1}^{M}\frac{dz_i}{V_i'(0)}\prod_{\mu=1}^{h}\left[\frac{dk_\mu\,d\xi_\mu\,d\eta_\mu}{k_\mu^2\,(\xi_\mu-\eta_\mu)^2}(1-k_\mu)^2\right] \times [\det(-i\tau_{\mu\nu})]^{-d/2}\prod_\alpha{}'\left[\prod_{n=1}^{\infty}(1-k_\alpha^n)^{-d}\prod_{n=2}^{\infty}(1-k_\alpha^n)^2\right] \quad (259)$$

where k_μ are the multipliers, ξ_μ and η_μ are the fixed points of the generators of the Schottky group.

9 From Dual Models to String Theory

The approach presented in the previous sections is a real bottom-up approach. The experimental data were the driving force in the construction of the Veneziano model and of its generalization to N external legs. The rest of the work that we have described above consisted in deriving its properties. The result is, except for a tachyon, a fully consistent quantum-relativistic model that was a source of fascination for those who worked in the field. Although the model grew out of S-matrix theory where the scattering amplitude is the only observable object, while the action or the Lagrangian have not a central role, some people nevertheless started to investigate what was the underlying microscopic structure that gave rise to such a consistent and beautiful model. It turned out, as we know today, that this underlying structure is that of a quantum-relativistic string. However, the process of connecting the dual resonance model (actually two of them the generalized Veneziano and the Shapiro–Virasoro model) to string theory took several years from the original idea to a complete and convincing proof of the conjecture. The original conjecture was independently formulated by Nambu [20, 65], Nielsen [66] and Susskind [21].[18] If we look at it in retrospective, it was at that time a fantastic idea that shows the enormous physical intuition of those who formulated it. On the other hand, it took several years to digest it before one was able to derive from it all the deep features of the dual resonance model. Because of this, the idea that the underlying structure was that of a relativistic string, did not really influence most of the research in the field up to 1973. Let me try to explain why.

A common feature of the work of [20, 66, 21] is the suggestion that the infinite number of oscillators, that one got through the factorization of the dual resonance model, naturally comes out from a two-dimensional free Lagrangian for the coordinate $X^\mu(\tau, \sigma)$ of a one-dimensional string, that is an obvious generalization of the Lagrangian that one writes for the coordinate $X^\mu(\tau)$ of a point-like object in the proper time gauge

$$L \sim \frac{1}{2}\frac{dX}{d\tau} \cdot \frac{dX}{d\tau} \Longrightarrow L \sim \frac{1}{2}\left[\frac{dX}{d\tau} \cdot \frac{dX}{d\tau} - \frac{dX}{d\sigma} \cdot \frac{dX}{d\sigma}\right] \tag{260}$$

Being this theory conformal invariant the Virasoro operators were also constructed together with their algebra. In this very first formulation, however, the Virasoro generators L_n were just the generators associated to the conformal symmetry of the string world-sheet Lagrangian given in (260) as in any conformal field theory. It was not clear at all why they should imply the gauge conditions found by Virasoro or, in modern terms, why they should be zero classically. The basic ingredient to solve this problem was provided by

[18] See also [67].

Nambu [65] and Goto [68], who wrote the non-linear Lagrangian proportional to the area spanned by the string in the external target space. They proceeded in analogy with the point particle and wrote the following action:

$$S \sim \int \sqrt{-d\sigma_{\mu\nu} d\sigma^{\mu\nu}} \tag{261}$$

where

$$d\sigma_{\mu\nu} = \frac{\partial X_\mu}{\partial \zeta^\alpha} \frac{\partial X_\nu}{\partial \zeta^\beta} d\zeta^\alpha \wedge d\zeta^\beta = \frac{\partial X_\mu}{\partial \zeta^\alpha} \frac{\partial X_\nu}{\partial \zeta^\beta} \epsilon^{\alpha\beta} d\sigma d\tau \tag{262}$$

$X_\mu(\sigma, \tau)$ is the string coordinate and $\zeta^0 = \tau$ and $\zeta^1 = \sigma$ are the coordinates of the string world sheet. $\epsilon^{\alpha\beta}$ is an antisymmetric tensor with $\epsilon^{01} = 1$. Inserting (262) into (261) and fixing the proportionality constant, one gets the Nambu–Goto action [65, 68]

$$S = -cT \int_{\tau_i}^{\tau_f} d\tau \int_0^\pi d\sigma \sqrt{(\dot{X} \cdot X')^2 - \dot{X}^2 X'^2} \tag{263}$$

where

$$\dot{X}^\mu \equiv \frac{\partial X^\mu}{\partial \tau} \qquad X'^\mu \equiv \frac{\partial X^\mu}{\partial \sigma} \tag{264}$$

and $T \equiv \frac{1}{2\pi\alpha'}$ is the string tension, that replaces the mass appearing in the case of a point particle. In this formulation, the string Lagrangian is invariant under any reparametrization of the world-sheet coordinates σ and τ and not only under the conformal transformations. This, in fact, implies that the two-dimensional world-sheet energy–momentum tensor of the string is actually zero as we will show later on. But it took still a few years to connect the Nambu–Goto action to the properties of the dual resonance model. In the meantime, an analogue model was formulated [69] that reproduced the tree and loop amplitudes of the generalized Veneziano model. This approach anticipated by several years the path integral derivation of dual amplitudes. It was very closely related to the functional integral formulation of [70].

However, one needed to wait until 1973 with the paper of Goddard, Goldstone, Rebbi and Thorn [71], where the Nambu–Goto action was correctly treated, all its consequences were derived and it became completely clear that the structure underlying the dual resonance model was that of a quantum-relativistic string. The equation of motion for the string were derived from the action in (263) by imposing $\delta S = 0$ for variations such that $\delta X^\mu(\tau_i) = \delta X^\mu(\tau_f) = 0$. One gets

$$\delta S = \int_{\tau_i}^{\tau_f} \left[\int_0^\pi d\sigma \left(-\frac{\partial}{\partial \tau} \frac{\partial L}{\partial \dot{X}^\mu} - \frac{\partial}{\partial \sigma} \frac{\partial L}{\partial X'^\mu} \right) \delta X^\mu + \frac{\partial L}{\partial X'^\mu} \delta X^\mu |_{\sigma=0}^{\sigma=\pi} \right] = 0 \tag{265}$$

where L is the Lagrangian in (263). Since δX^μ is arbitrary, from (265) one gets the Euler–Lagrange equation of motion

$$\frac{\partial}{\partial\tau}\frac{\partial L}{\partial\dot{X}^{\mu}}+\frac{\partial}{\partial\sigma}\frac{\partial L}{\partial X'^{\mu}}\equiv\frac{\partial}{\partial\zeta^{\alpha}}\left(\frac{\partial L}{\partial(\frac{\partial X^{\mu}}{\partial\zeta^{\alpha}})}\right)=0 \tag{266}$$

and the boundary conditions

$$\frac{\partial L}{\partial X'^{\mu}}=0 \qquad or \qquad \delta X_{\mu}=0 \qquad at \qquad \sigma=0,\pi \tag{267}$$

for an open string and

$$X^{\mu}(\tau,0)=X^{\mu}(\tau,\pi) \tag{268}$$

for a closed string. In the case of an open string, the first kind of boundary condition in (267) corresponds to Neumann boundary conditions, while the second one to Dirichlet boundary conditions. Only the Neumann boundary conditions preserve the translation invariance of the theory and, therefore, they were mostly used in the early days of string theory. It must be stressed, however, that Dirichlet boundary conditions were already discussed and used in the early days of string theory for constructing models with off-shell states [72].

From (263), one can compute the momentum density along the string

$$\frac{\partial L}{\partial\dot{X}^{\mu}}\equiv P_{\mu}=cT\frac{\dot{X}_{\mu}X'^{2}-X'_{\mu}(\dot{X}\cdot X')}{\sqrt{(\dot{X}\cdot X')^{2}-\dot{X}^{2}X'^{2}}} \tag{269}$$

and obtain the following constraints between the dynamical variables X^{μ} and P^{μ}:

$$c^{2}T^{2}x'^{2}+P^{2}=x'\cdot P=0 \tag{270}$$

They are a consequence of the reparametrization invariance of the string Lagrangian. Because of this one can choose the orthonormal gauge specified by the conditions

$$\dot{X}^{2}+X'^{2}=\dot{X}\cdot X'=0 \tag{271}$$

that nowadays is called conformal gauge. In this gauge (269) becomes

$$P_{\mu}=cT\dot{X}_{\mu} \qquad \frac{\partial L}{\partial X'^{\mu}}=-cTX'_{\mu} \tag{272}$$

and therefore the equation of motion in (266) becomes

$$\ddot{X}_{\mu}-X''_{\mu}=0 \tag{273}$$

while the boundary condition in (267) becomes

$$X'_{\mu}(\sigma=0,\pi)=0 \tag{274}$$

The most general solution of the equation of motion and of the boundary conditions can be written as follows:

$$X^{\mu}(\tau,\sigma) = q^{\mu} + 2\alpha' p^{\mu}\tau + i\sqrt{2\alpha'}\sum_{n=1}^{\infty}[a_n^{\mu}e^{-in\tau} - a_n^{+\mu}e^{in\tau}]\frac{cosn\sigma}{\sqrt{n}} \quad (275)$$

for an open string and

$$X^{\mu}(\tau,\sigma) = q^{\mu} + 2\alpha' p^{\mu}\tau + \frac{i}{2}\sqrt{2\alpha'}\sum_{n=1}^{\infty}[\tilde{a}_n^{\mu}e^{-2in(\tau+\sigma)} - \tilde{a}_n^{+\mu}e^{2in(\tau+\sigma)}]\frac{1}{\sqrt{n}}$$

$$+\frac{i}{2}\sqrt{2\alpha'}\sum_{n=1}^{\infty}[a_n^{\mu}e^{-2in(\tau-\sigma)} - a_n^{+\mu}e^{2in(\tau-\sigma)}]\frac{1}{\sqrt{n}} \quad (276)$$

for a closed string. This procedure really shows that, starting from the Nambu–Goto action, one can choose a gauge (the orthonormal or conformal gauge) where the equation of motion of the string becomes the two-dimensional D'Alembert equation in (273). Furthermore, the invariance under reparametrization of the Nambu–Goto action implies that the two-dimensional energy–momentum tensor is identically zero at the classical level (see (271)).

As the Lorentz gauge in QED the orthonormal gauge does not fix completely the gauge. We can still perform reparametrizations that leave in the conformal gauge: they are conformal transformations. Introducing the variable $z = e^{i\tau}$ the generators of the conformal transformations for the open string can be written as follows:

$$L_n = \frac{1}{2\pi i}\oint dz z^{n+1}\left[-\frac{1}{4\alpha'}\left(\frac{\partial X^{\mu}}{\partial z}\right)^2\right] = \frac{1}{2}\sum_{m=-\infty}^{\infty}\alpha_{n-m}\cdot\alpha_m = 0 \quad (277)$$

where

$$\alpha_n^{\mu} = \begin{cases} \sqrt{n}a_n^{\mu} & if\ n>0 \\ \sqrt{2\alpha'}p^{\mu} & if\ n=0 \\ \sqrt{n}a_n^{\dagger\mu} & if\ n<0 \end{cases} \quad (278)$$

They are zero as a consequence of (270) that in the conformal gauge become (271). In the case of a closed string we get instead

$$\tilde{L}_n = \frac{1}{2\pi i}\oint dz z^{n+1}\left[-\frac{1}{\alpha'}\left(\frac{\partial X^{\mu}}{\partial z}\right)^2\right] = 0 \quad (279)$$

$$L_n = \frac{1}{2\pi i}\oint d\bar{z}\bar{z}^{n+1}\left[-\frac{1}{\alpha'}\left(\frac{\partial X^{\mu}}{\partial \bar{z}}\right)^2\right] = 0 \quad (280)$$

In terms of the harmonic oscillators introduced in (276) we get

$$L_n = \frac{1}{2} \sum_{m=-\infty}^{\infty} \alpha_m \cdot \alpha_{n-m} = 0 \quad ; \quad \tilde{L}_n = \frac{1}{2} \sum_{m=-\infty}^{\infty} \tilde{\alpha}_m \cdot \tilde{\alpha}_{n-m} = 0 \quad (281)$$

where for the non-zero modes we have used the convention in (278), while the zero mode is given by

$$\alpha_0^\mu = \tilde{\alpha}_0^\mu = \sqrt{2\alpha'} \frac{p^\mu}{2} \quad (282)$$

In conclusion, the fact that we have reparametrization invariance implies that the Virasoro generators are classically identically zero. When we quantize the theory one cannot and also does not need to impose that they are vanishing at the operator level. They are imposed as conditions characterizing the physical states.

$$\langle Phys'|L_n|Phys\rangle = \langle Phys'|(L_0 - 1)|Phys\rangle = 0 \;\; ; \;\; n \neq 0 \quad (283)$$

These equations are satisfied if we require

$$L_n|Phys >= (L_0 - 1)|Phys >= 0 \quad (284)$$

The extra factor -1 in the previous equations comes from the normal ordering as explained in (198).

The authors of [71] further specified the gauge by fixing it completely. They introduced the light-cone gauge specified by imposing the condition

$$X^+ = 2\alpha' p^+ \tau \quad (285)$$

where

$$X^\pm = \frac{X^0 \pm X^{d-1}}{\sqrt{2}}, \qquad X_\pm = \frac{X_0 \pm X_{d-1}}{\sqrt{2}}. \quad (286)$$

In this gauge the only physical degrees of freedom are the transverse ones. In fact, the components along the directions 0 and $d-1$ can be expressed in terms of the transverse ones by inserting (285) into the constraints in (271) and getting

$$\dot{X}^- = \frac{1}{4\alpha' p^+}(\dot{X}_i^2 + {X'}_i^2), \qquad X'^- = \frac{1}{2\alpha' p^+}\dot{X}_i \cdot {X'}_i \quad (287)$$

that up to a constant of integration determine completely X^- as a function of X^i. In terms of oscillators we get

$$\alpha_n^+ = 0 \;\; ; \;\; \sqrt{2\alpha'}\alpha_n^- = \frac{1}{2p^+} \sum_{m=-\infty}^{\infty} \alpha_{n-m}^i \alpha_m^i; \qquad n \neq 0 \quad (288)$$

for an open string and

$$\alpha_n^+ = \tilde{\alpha}_n^+ = 0 \qquad\qquad n \neq 0 \tag{289}$$

together with

$$\sqrt{2\alpha'}\alpha_n^- = \frac{1}{2p^+} \sum_{m=-\infty}^{\infty} \alpha_{n-m}^i \alpha_m^i$$

$$\sqrt{2\alpha'}\tilde{\alpha}_n^- = \frac{1}{2p^+} \sum_{m=-\infty}^{\infty} \tilde{\alpha}_{n-m}^i \tilde{\alpha}_m^i \tag{290}$$

in the case of a closed string.

This shows that the physical states are described only by the transverse oscillators having only $d-2$ components. Those transverse oscillators correspond to the transverse DDF operators that we have discussed in Sect. 6. The authors of [71] also constructed the Lorentz generators only in terms of the transverse oscillators and they showed that they satisfy the correct Lorentz algebra only if the space–time dimension is $d = 26$. In this way the spectrum of the dual resonance model was completely reproduced starting from the Nambu–Goto action if $d = 26$! On the other hand, the choice of $d = 26$ is a necessity if we want to keep the Lorentz invariance!

Immediately after this, the interaction was also included either by adding a term describing the interaction of the string with an external gauge field [73] or by using a functional formalism [74, 75].

In the following we will give some detail only of the first approach for the case of an open string. A way to describe the string interaction is by adding to the free string action an additional term that describes the interaction of the string with an external field.

$$S_{INT} = \int d^D y \Phi_L(y) J_L(y) \tag{291}$$

where $\Phi_L(y)$ is the external field and J_L is the current generated by the string. The index L stands for possible Lorentz indices that are saturated in order to have a Lorentz invariant action.

In the case of a point particle, such an interaction term will not give any information on the self-interaction of a particle.

In the case of a string, instead, we will see that S_{INT} will describe the interaction among strings because the external fields that can consistently interact with a string are only those that correspond to the various states of the string, as it will become clear in the discussion below.

This is a consequence of the fact that, for the sake of consistency, we must put the following restrictions on S_{INT}:

- It must be a well-defined operator in the space spanned by the string oscillators.

- It must preserve the invariances of the free string theory. In particular, in the "conformal gauge" it must be conformal invariant.
- In the case of an open string, the interaction occurs at the end point of a string (say at $\sigma = 0$). This follows from the fact that two open strings interact attaching to each other at the end points.

The simplest scalar current generated by the motion of a string can be written as follows:

$$J(y) = \int d\tau \int d\sigma \delta(\sigma) \delta^{(d)} [y^\mu - x^\mu(\tau, \sigma)] \tag{292}$$

where $\delta(\sigma)$ has been introduced because the interaction occurs at the end of the string. For the sake of simplicity, we omit to write a coupling constant g in (292).

Inserting (292) into (291) and using for the scalar external field $\Phi(y) = e^{ik\cdot y}$ a plane wave, we get the following interaction:

$$S_{INT} = \int d\tau : e^{ik\cdot X(\tau,0)} : \tag{293}$$

where the normal ordering has been introduced in order to have a well defined operator. The invariance of (293) under a conformal transformation $\tau \to w(\tau)$ requires the following identity:

$$S_{INT} = \int d\tau : e^{ik\cdot X(\tau,0)} : = \int dw : e^{ik\cdot X(w,0)} : \tag{294}$$

or, in other words, that

$$: e^{ik\cdot X(\tau,0)} : \Longrightarrow w'(\tau) : e^{ik\cdot X(w,0)} : \tag{295}$$

This means that the integrand in (294) must be a conformal field with conformal dimension equal to one and this happens only if $\alpha' k^2 = 1$. The external field corresponds then to the tachyonic lowest state of the open string. Another simple current generated by the string is given by

$$J_\mu(y) = \int d\tau \int d\sigma \delta(\sigma) \dot{X}_\mu(\tau, \sigma) \delta^{(d)}(y - X(\tau, \sigma)) \tag{296}$$

Inserting (296) into (291) we get

$$S_{INT} = \int d\tau \dot{X}_\mu(\tau, 0) \epsilon^\mu e^{ik\cdot X(\tau,0)} \tag{297}$$

if we use a plane wave for $\Phi_\mu(y) = \epsilon_\mu e^{ik\cdot y}$. The vertex operator in (297) is conformal invariant only if

$$k^2 = \epsilon \cdot k = 0 \tag{298}$$

and, therefore, the external vector must be the massless photon state of the string. We can generalize this procedure to an arbitrary external field and the result is that we can only use external fields that correspond to on-shell physical states of the string.

This procedure has been extended in [73] to the case of external gravitons by introducing in the Nambu–Goto action a target space metric and obtaining the vertex operator for the graviton that is a massless state in the closed string theory. Remember that, at that time, this could have been done only with the Nambu–Goto action because the σ-model action was introduced only in 1976 first for the point particle [76] and then for the string [77]. As in the case of the photon, it turned out that the external field corresponding to the graviton was required to be on-shell. This condition is the precursor of the equations of motion that one obtains from the σ-model action requiring the vanishing of the β-function [78].

One can then compute the probability amplitude for the emission of a number of string states corresponding to the various external fields, from an initial string state to a final one. This amplitude gives precisely the N-point amplitude that we discussed in the previous sections [73]. In particular, one learns that, in the case of the open string, the Fubini–Veneziano field is just the string coordinate computed at $\sigma = 0$:

$$Q^\mu(z) \equiv X^\mu(z, \sigma = 0) \quad ; \quad z = e^{i\tau} \tag{299}$$

In the case of a closed string we get instead

$$Q^\mu(z, \bar{z}) \equiv X^\mu(z, \bar{z}) \quad ; \quad z = e^{2i(\tau - \sigma)} \;,\; \bar{z} = e^{2i(\tau+\sigma)} \tag{300}$$

Finally, let me mention that with the functional approach Mandelstam [74] and Cremmer and Gervais [79] computed the interaction between three arbitrary physical string states and reproduced in this way the coupling of three DDF states given in (202) and obtained in [37] by using the operator formalism. At this point it was completely clear that the structure underlying the generalized Veneziano model was that of an open relativistic string, while that underlying the Shapiro–Virasoro model was that of a closed relativistic string. Furthermore, these two theories are not independent because, if one starts from an open string theory, one gets automatically closed strings by loop corrections.

10 Conclusions

In this contribution, we have gone through the developments that led from the construction of the dual resonance model to the bosonic string theory trying as much as possible to include all the necessary technical details. This is because we believe that they are not only important from an historical point of view, but are also still part of the formalism that one uses today in many string calculations. We have tried to be as complete and objective as possible, but it could very well be that some of those who participated in the research of these years, will not agree with some or even many of the statements we made. We apologize to those we have forgotten to mention or we have not mentioned as they would have liked.

Finally, after having gone through the developments of these years, my thoughts go to Sergio Fubini who shared with me and Gabriele many of the ideas described here and who is deeply missed, and to my friends from Florence, Naples and Turin for a pleasant collaboration in many papers discussed here.

Acknowledgements

I thank R. Marotta and I. Pesando for a critical reading of the manuscript.

References

1. G.F. Chew: *The Analytic S Matrix* (W.A. Benjamin, Inc., New York, 1966); R.J. Eden, P.V. Landshoff, D.I. Olive, J.C. Polkinghorne: *The Analytic S Matrix* (Cambridge University Press, Cambridge, 1966)
2. R. Dolen, D. Horn, C. Schmid: Phys. Rev. **166**, 1768 (1968); C. Schmid: Phys. Rev. Lett. **20**, 689 (1968)
3. H. Harari: Phys. Rev. Lett. **22**, 562 (1969); J.L. Rosner: Phys. Rev. Lett. **22**, 689 (1969)
4. G. Veneziano: Nuovo Cimento A **57**, 190 (1968)
5. M.A. Virasoro: Phys. Rev. **177**, 2309 (1969)
6. M.A. Virasoro: Phys. Rev. D **1**, 2933 (1970)
7. A. Neveu, J.H. Schwarz: Nucl. Phys. B **31**, 86 (1971); Phys. Rev. D **4**, 1109 (1971)
8. P. Ramond: Phys. Rev. D **3**, 2415 (1971)
9. C. Lovelace: Phys. Lett. B **28**, 265 (1968); J. Shapiro: Phys. Rev. **179**, 1345 (1969)
10. P.H. Frampton: Phys. Lett. B **41**, 364 (1972)
11. V. Alessandrini, D. Amati, M. Le Bellac, D. Olive: Phys. Rep. C **1**, 269 (1971); G. Veneziano: Phys. Rep. C **9**, 199 (1974); S. Mandelstam: Phys. Rep. C **13**, 259 (1974); C. Rebbi: Phys. Rep. C **12**, 1 (1974); J. Scherk: Rev. Mod. Phys. **47**, 123 (1975)
12. F. Gliozzi: Lett. Nuovo Cimento **2**, 1160 (1970)
13. K. Bardakçi, H. Ruegg: Phys. Rev. **181**, 1884 (1969); C.G. Goebel, B. Sakita: Phys. Rev. Lett. **22**, 257 (1969); Chan Hong-Mo, T.S. Tsun: Phys. Lett. B **28**, 485 (1969); Z. Koba, H.B. Nielsen: Nucl. Phys. B **10**, 633 (1969)
14. K. Bardakçi, H. Ruegg: Phys. Lett.B **28**, 671 (1969); M.A. Virasoro: Phys. Rev. Lett. **22**, 37 (1969)

15. Z. Koba, H.B. Nielsen: Nucl. Phys. B **12**, 517 (1969)
16. H.M. Chan, J.E. Paton: Nucl. Phys. B **10**, 516 (1969)
17. S. Fubini, G. Veneziano: Nuovo Cimento A **64**, 811 (1969)
18. Bardakçi, S. Mandelstam: Phys. Rev. **184**, 1640 (1969)
19. S. Fubini, D. Gordon, G. Veneziano: Phys. Lett. B **29**, 679 (1969)
20. Y. Nambu: *Proc. Int. Conf. on Symmetries and Quark Models, Wayne State University 1969* (Gordon and Breach, New York, 1970), p. 269
21. L. Susskind: Nuovo Cimento A **69**, 457 (1970); Phys. Rev. Lett. **23**, 545 (1969)
22. J. Shapiro: Phys. Lett. B **33**, 361 (1970)
23. S. Fubini, G. Veneziano: Nuovo Cimento A **67**, 29 (1970)
24. F. Gliozzi: Lettere al Nuovo Cimento **2**, 846 (1969)
25. C.B. Chiu, S. Matsuda, C. Rebbi: Phys. Rev. Lett. **23**, 1526 (1969): C.B. Thorn: Phys. Rev. D **1**, 1963 (1970)
26. S. Fubini, G. Veneziano: Ann. Phys. **63**, 12 (1971)
27. C. Lovelace: Phys. Lett. B **34**, 500 (1971)
28. E. Del Giudice, P. Di Vecchia: Nuovo Cimento A **5**, 90 (1971); M. Yoshimura: Phys. Lett. B **34**, 79 (1971)
29. E. Del Giudice, P. Di Vecchia: Nuovo Cimento A **70**, 579 (1970)
30. P. Campagna, S. Fubini, E Napolitano, S. Sciuto: Nuovo Cimento A **2**, 911 (1971)
31. E. Del Giudice, P. Di Vecchia, S. Fubini: Ann. Phys. **70**, 378 (1972)
32. R.C. Brower: Phys. Rev. D **6**, 1655 (1972)
33. P. Goddard, C.B. Thorn: Phys. Lett. B **40**, 235 (1972)
34. F. Gliozzi, J. Scherk, D. Olive: Phys. Lett. B **65**, 282 (1976); Nucl. Phys. B **122**, 253 (1977)
35. L. Brink, H.B. Nielsen: Phys. Lett. B **45**, 332 (1973)
36. F. Gliozzi: unpublished; see also P. Di Vecchia: in *Many Degrees of Freedom in Particle Physics*, ed. by H. Satz (Plenum Publishing Corporation, New York, 1978), p. 493
37. M. Ademollo, E. Del Giudice, P. Di Vecchia, S. Fubini: Nuovo Cimento A **19**, 181 (1974)
38. J. Scherk: Nucl. Phys. B **31**, 222 (1971)
39. N. Nakanishi: Prog. Theor. Phys. **48**, 355 (1972); P.H. Frampton, K.C. Wali: Phys. Rev. D **8**, 1879 (1973)
40. A. Neveu, J. Scherk: Nucl. Phys. B **36**, 155 (1973)
41. A. Neveu, J.L. Gervais: Nucl. Phys. B **46**, 381 (1972)
42. P. Di Vecchia, A. Lerda, L. Magnea, R. Marotta, R. Russo: Nucl. Phys. B **469**, 235 (1996)
43. T. Yoneya: Prog. Theor. Phys. **51**, 1907 (1974)
44. K. Kikkawa, B. Sakita, M. Virasoro: Phys. Rev. **184**, 1701 (1969); K. Bardakçi, M.B. Halpern, J. Shapiro: Phys. Rev. **185**, 1910 (1969); D. Amati, C. Bouchiat, J.L. Gervais: Lett. al Nuovo Cimento **2**, 399 (1969); A. Neveu, J. Scherk: Phys. Rev. D **1**, 2355 (1970); G. Frye, L. Susskind: Phys. Lett. B **31**, 537 (1970); D.J. Gross, A. Neveu, J. Scherk, J.H. Schwarz: Phys. Rev. D **2**, 697 (1970)
45. L. Brink, D. Olive: Nucl. Phys. B **56**, 253 (1973); Nucl. Phys. B **58**, 237 (1973)
46. E. Cremmer, J. Scherk: Nucl. Phys. B **50**, 222 (1972); L. Clavelli, J. Shapiro: Nucl. Phys. B **57**, 490 (1973); L. Brink, D.I. Olive, J. Scherk: Nucl. Phys. B **61**, 173 (1973)

47. M. Ademollo, A. D'Adda, R. D'Auria, F. Gliozzi, E. Napolitano, S. Sciuto, P. Di Vecchia: Nucl. Phys. B **94**, 221 (1975);
J. Shapiro: Phys. Rev. D **11**, 2937 (1975)
48. D.I.Olive, J. Scherk: Phys. Lett. B **44**, 296 (1973)
49. M. Ademollo, A. D'Adda, R. D'Auria, E. Napolitano, P. Di Vecchia, F. Gliozzi, S. Sciuto: Nucl. Phys. B **77**, 189 (1974)
50. M. Kalb, P. Ramond: Phys. Rev. D **9**, 2273 (1974)
51. E, Cremmer, J. Scherk: Nucl. Phys. B **72**, 117 (1974)
52. A. D'Adda, R. D'Auria, E. Napolitano, P. Di Vecchia, F. Gliozzi, S. Sciuto: Phys. Lett. B **68**, 81 (1977)
53. J. Shapiro: Phys. Rev. D **5**, 1945 (1975)
54. S. Sciuto: Lett. al Nuovo Cimento **2**, 411 (1969);
A. Della Selva, S. Saito: Lett. al Nuovo Cimento **4**, 689 (1970)
55. L. Caneschi, A. Schwimmer, G. Veneziano: Phys. Lett.B **30**, 356 (1969);
L. Caneschi, A. Schwimmer: Lett. al Nuovo Cimento **3**, 213 (1970)
56. C. Lovelace: Phys. Lett. B **32**, 490 (1970)
57. D.I. Olive: Nuovo Cimento A **3**, 399 (1971)
58. I. Drummond: Nuovo Cimento A **67**, 71 (1970);
G. Carbone, S. Sciuto: Lett. al Nuovo Cimento **3**, 246 (1970);
L. Kosterlitz, D. Wray: Lett. al Nuovo Cimento **3**, 491 (1970);
D. Collop: Nuovo Cimento A **1**, 217 (1971);
L.P. Yu: Phys. Rev. D **2**, 1010 (1970); Phys. Rev. D **2**, 2256 (!970);
E. Corrigan, C. Montonen: Nucl. Phys. B **36**, 58 (1972);
J.L. Gervais, B. Sakita: Phys. Rev. D **4**, 2291 (1971)
59. C. Lovelace: Phys. Lett. B **32**, 703 (1970)
60. V. Alessandrini: Nuovo Cimento A **2**, 321 (1971)
61. D. Amati, V. Alessandrini: Nuovo Cimento A **4**, 793 (1971)
62. P. Di Vecchia, M. Frau, A. Lerda, S. Sciuto: Phys. Lett. B **199**, 49 (1987)
J.L. Petersen and J. Sidenius, Nucl. Phys. B **301**, 247 (1988)
63. S. Mandelstam: in *Unified String Theories*, ed. by M. Green, D. Gross (World Scientific, Singapore), p. 46
64. P. Di Vecchia, F. Pezzella, M. Frau, K. Hornfeck, A. Lerda, S. Sciuto: Nucl. Phys. B **322**, 317 (1989)
65. Y. Nambu: Lectures at the Copenhagen Symposium, 1970 (unpublished)
66. H.B. Nielsen: Paper submitted to the *15th Int. Conf. on High Energy Physics*, Kiev, 1970; Nordita preprint (1969)
67. T. Takabayasi: Progr. Theor. Phys. **44** (1970) 1117;
O. Hara: Progr. Theor. Phys. **46**, 1549 (1971);
L.N. Chang, J. Mansouri: Phys. Rev. D **5**, 2535 (1972);
J. Mansouri, Y. Nambu: Phys. Lett. B **39**, 357 (1972);
M. Minami: Prog. Theor. Phys. **48**, 1308 (1972)
68. T. Goto: Progr. Theor. Phys. **46** 1560 (1971)
69. D. Fairlie, H.B. Nielsen: Nucl. Phys. B **20**, 637 (1970) and **22**, 525 (1970)
70. C.S. Hsue, B. Sakita, M.A. Virasoro: Phys. Rev. **2**, 2857 (1970);
J.L. Gervais, B. Sakita: Phys. Rev. D **4**, 2291 (1971)
71. P. Goddard, J. Goldstone, C. Rebbi, C. Thorn: Nucl. Phys. B **56**, 109 (1973)
72. E.F. Corrigan, D.B. Fairlie: Nucl. Phys. B **91**, 527 (1975)
73. M. Ademollo, A. D'Adda, R. D'Auria, P. Di Vecchia, F. Gliozzi, R. Musto, E. Napolitano, F. Nicodemi, S. Sciuto: Nuovo Cimento A **21**, 77 (1974)

74. S. Mandelstam: Nucl. Phys. B **64**, 205 (1973)
75. J.L. Gervais, B. Sakita: Phys. Rev. Lett. **30**, 716 (1973)
76. L. Brink, P. Di Vecchia, P. Howe, S. Deser, B. Zumino: Phys. Lett. B **64**, 435 (1976)
77. L. Brink, P. Di Vecchia, P. Howe: Phys. Lett. B **65**, 471 (1976);
S. Deser, B. Zumino: Phys. Lett. B **65**, 369 (1976)
78. C. Lovelace: Phys. Lett. B **136**, 75 (1984);
C.G. Callan, E.J. Martinec, M.J. Perry, D. Friedan: Nucl. Phys. B **262**, 593 (1985)
79. E. Cremmer, J.L. Gervais: Nucl. Phys. **76**, 209 (1974)

The Beginning of String Theory: A Historical Sketch

P. Di Vecchia[1] and A. Schwimmer[2]

[1] Nordita, Blegdamsvej 17, 2100 Copenhagen Ø, Denmark
divecchi@nbi.dk
[2] Weizmann Institute, Rehovot 76100, Israel
adam.schwimmer@weizmann.ac.il

Abstract. In this note we follow the historical development of the ideas that led to the formulation of String Theory. We start from the inspired guess of Veneziano and its extension to the scattering of N scalar particles, then we describe how the study of its factorization properties allowed to identify the physical spectrum, and finally we discuss how the critical values of the intercept of the Regge trajectory and of the critical dimension were fixed to be $\alpha_0 = 1$ and $d = 26$.

1 Introduction

The purpose of this note is to follow the historical development of the ideas that led to the formulation of String Theory. As we will discuss, the story consists of a remarkable succession of inspired insights first by Veneziano who guessed the form of the four-point function [1], followed by its extension to an arbitrary number of external legs. At this point the dual resonance model was constructed, and it took some time to analyse its properties and check its consistency through its factorization properties that allowed one to identify the full target Hilbert space of physical states and its critical dimension by the use of various consistency conditions. The natural interpretation of the structure uncovered was that of a string propagating in Minkovski space–time.

We want to stress that all this was achieved without the use of a Lagrangian formulation, but by implementing the basic principles of S-matrix directly on a scattering amplitude in a model containing an infinite number of zero width resonances, where the sum of resonances in one channel represents correctly the resonances in the other channel. As a result, the basic framework of Perturbative String Theory at the operational level was well understood by 1971. Further progress was achieved through the discovery of the Superstring and Space–time Supersymmetry, which led to tachyon free theories. Later some basic concepts used before at a heuristic level, like the origin of the first class constraints necessary for making the spectrum unitary

P. Di Vecchia and A. Schwimmer: *The Beginning of String Theory: A Historical Sketch*, Lect. Notes Phys. **737**, 119–136 (2008)
DOI 10.1007/978-3-540-74233-6_5

and Lorentz invariant, were put on a firm ground starting from the action used in [2].

Further conceptual developments, like the connection between world sheet conformal invariance and target space equations of motion, were only partially understood, and had to wait for the first String Revolution to get a more complete formulation. Finally, the relation between different String Theories through dualities was the result of the second String Revolution.

In this note we will concentrate on the developments during the period 1969–1972.

As we mentioned above three components entering the basic structure of perturbative string theory, i.e.:

- the string world sheet
- the physical spectrum and vertex operators
- the critical dimension

were all correctly identified by the end of 1972, and in this short note we will limit ourselves to the description of the evolution of their understanding. We will not cover other very important developments during the same period, like, e.g. fermionic degrees of freedom on the world sheet (the Neveu–Schwarz–Ramond formalism [3, 4]), compact degrees of freedom on the world sheet leading to internal symmetries [5] and String Field Theory in its light-cone formulation [6].

We will follow the evolution of the ideas, which led to the understanding of the three basic concepts above, outlining the most important conceptual jumps. Just the essential formulae will be given, referring for the detailed derivations to the accompanying paper [7]. We will try to put in perspective the evolution of the ideas by translating the guesses and insights in today's language and understanding, as presented in the standard modern textbooks [8]. We start with a brief reminder of the developments on which the three breakthroughs mentioned above were based.

2 Prehistory: the Discovery of the Dual Scattering Amplitudes

The first step which started the evolution of String Theory was the Veneziano Formula [1]. By a historical accident Veneziano's formula refers to what is today Open String Theory. The analogous formula for Closed String Theory guessed by Virasoro [9] was generalized [10] and analysed later [11] when the basic structure of the open string was already understood. We will follow the historical path and discuss only Open String Theory.

The formula guessed by Veneziano corresponds to what we call today the 2 to 2 scattering amplitude of the bosonic open string tachyons:

$$A(s,t,u) = A(s,t) + A(s,u) + A(t,u), \tag{1}$$

where

$$A(s,t) = \frac{\Gamma(-\alpha(s))\Gamma(-\alpha(t))}{\Gamma(-\alpha(s)-\alpha(t))} = \int_0^1 dx x^{-\alpha(s)-1}(1-x)^{-\alpha(t)-1}, \tag{2}$$

and

$$\alpha(s) = \alpha_0 + \alpha' s \tag{3}$$

is a linearly rising Regge trajectory.

The appearance of the free parameter α_0 instead of the usual value 1 will be discussed below. Moreover, in the Veneziano amplitude, as written above, there is no requirement that the external particles are the spin 0 particles on the leading trajectory $\alpha(s)$. Nevertheless, we will continue to call the external particles "tachyons" because they have negative mass squared if we require them to be on the leading trajectory for $\alpha_0 = 1$.

In Veneziano's original approach the amplitude was supposed to describe scattering of mesons due to strong interactions. The physical principles guiding Veneziano in his guess were the usual analyticity and crossing symmetry requirements of the scattering amplitudes and a new principle, the Dolen–Horn–Schmid (DHS) duality [12].

DHS duality was abstracted from a phenomenological study of hadronic reactions and stated that the scattering amplitude could be decomposed alternatively into a set of s-channel or t-channel poles, each decomposition being complete, and containing, by analytic continuation, the other. This was expressed by the pictorial identity [13, 14] presented in Fig. 1.

In today's language it is qualitatively clear that the DHS requirement is fulfilled if the amplitude is related to the correlator of four vertex operators in a conformal field theory. The two different decompositions which make explicit the pole structure can be represented graphically by two "duality diagrams" related by a continuous deformation, and correspond to the two possible decompositions in conformal blocks of the conformal correlator. This happens if the conformal block is translated into poles in Lorentz invariants constructed from the space–time momenta. This basic feature of String Theory to which DHS duality led, is very far from its phenomenological origin.

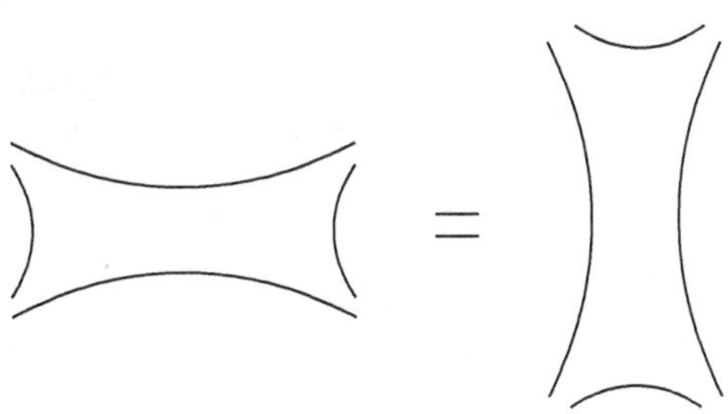

Fig. 1. The duality diagram contains both s- and t-channel poles

Ironically, it seems that present hadron scattering data [15] are not anymore in agreement with DHS duality, which was a feature related to the energy range available at the time.

For the N-point function the DHS duality is generalized by requiring that, for a fixed ordering of the external particles, the amplitude can be represented by any one of the deformations of the respective N-point duality diagram. As described in [7], one way to understand the mechanism by which $A(s,t)$ satisfies the DHS duality is to study its integral representation, and identify the two mutually exclusive integration domains, which produce the poles in the s- and t-channel, respectively. This is generalized for the N-point function by writing it as a sum of terms, each one corresponding to a given ordering of the external legs. Each term has a $(N-3)$-dimensional integral representation. The different deformations of the duality diagram are obtained from the singular contributions to the integral representation of mutually exclusive $(N-3)$- dimensional integration regions.

Based on this idea the unique N-point function was constructed in [16]:

$$B_N = \prod_{i=2}^{N-2} \left[\int_0^1 du_i u_i^{-\alpha(s_i)-1} (1-u_i)^{\alpha_0 - 1} \right] \prod_{i=2}^{N-2} \prod_{j=i+1}^{N-1} (1-x_{ij})^{2\alpha' p_i \cdot p_j}, \quad (4)$$

where

$$s_i \equiv s_{1i} \quad ; \quad x_{ij} = u_i u_{i+1} \ldots u_{j-1}, \quad (5)$$

$$s_{ij} = -(p_i + p_{i+1} + \cdots + p_j)^2, \quad (6)$$

and $p_i, i = 1, 2, .., N$, are the external momenta. We require that the external scalar lies on the leading trajectory as explained in [7]. Starting from this expression Koba and Nielsen [17] put it in the more symmetric $SL(2,R)$ invariant form (see [7] for details)

$$B_N = \int_{-\infty}^{\infty} dV(z) \prod_{(i,j)} (z_i, z_{i+1}, z_j, z_{j+1})^{-\alpha(s_{ij})-1}, \quad (7)$$

where

$$dV(z) = \frac{\prod_1^N \left[\theta(z_i - z_{i+1}) dz_i\right]}{\prod_{i=1}^N (z_i - z_{i+2}) dV_{abc}} \; ; \; dV_{abc} = \frac{dz_a dz_b dz_c}{(z_b - z_a)(z_c - z_b)(z_a - z_c)}, \quad (8)$$

and the variables z_i are integrated along the real axis in a cyclically ordered way: $z_1 \geq z_2 \cdots \geq z_N$ with a, b and c arbitrarily chosen.

The $SL(2,R)$ group mentioned above acts on the integration variables z_i as a Möbius transformation:

$$z_i \to \frac{\alpha z_i + \beta}{\gamma z_i + \delta} \; ; \; i = 1 \ldots N \; ; \; \alpha\delta - \beta\gamma = 1. \quad (9)$$

Using the transformation in (9) for a fixed ordering, one can relate amplitudes corresponding to circularly permuted kinematical invariants and then, adding terms for different orderings, one can show that all the requirements of crossing symmetry are fulfilled. As we understand it today, the Möbius transformations are related to globally defined reparametrizations of the disk which leave invariant the metric up to a conformal factor. This was the first manifestation of the conformal symmetry underlying the world sheet action of String Theory, which played an essential role in the understanding of the theory.

The expression in (7) which was guessed as following from the principles mentioned above, coincides (for $\alpha_0 = 1$) with the tree-level scattering amplitude of N open string tachyons, obtained from calculating the open string path integral on a disk with the insertion of N-tachyon vertex operators after mapping the disk to the upper half plane.

The Koba–Nielsen form of the N-point function was the starting point for the crucial developments which started in 1969. There was a general feeling among the workers in the field that the set of N-point functions represent the result of a unique and consistent underlying theory. While attempts to use the functions to fit hadronic data continued, the search for this theory became the major theoretical challenge. One aspect which became immediately obvious was the necessity to "unitarize" the theory: the presence of zero width poles in the N-point functions showed that the amplitudes should be considered, at best, as "tree diagrams" of an underlying, unknown theory and "loop" diagrams should be added to them. A first attempt [18] to write loop diagrams was by using again a generalized form of the DHS principle, requiring a singularity structure of the amplitudes consistent with deformations of duality diagrams involving loops. The existence of rather involved integrals, found in [18], which fulfil the constraints, reinforced the belief in the existence of an underlying theory. On the other hand, the ambiguities in the amplitudes constructed originating in what we call today "the measure factors", and the impossibility to verify the unitarity, reinforced the necessity of understanding the basic underlying theory.

The approaches used were conditioned by the development of the theoretical tools at the time. Though the path integral formulation of Quantum Field Theory existed, it was not well developed as a calculational tool. This was the case especially for gauge theories where the correct treatment of gauge symmetries achieved a few years later by Faddeev–Popov did not exist. As a consequence, Lagrangian methods based on an action were not very precise, and involved some guess work at different stages. On the other hand, operatorial methods were well developed, and through the Gupta–Bleuler treatment of QED as a prototype even the correct impositions of constraints corresponding to a gauge fixing (at least for the case when the ghosts are decoupled in today's language) were understood. We can roughly divide the search for the underlying theory as the "Lagrangian approach" and the "operatorial approach".

Since we will discuss later in more detail the operatorial approach we start with a description of the evolution of the "Lagrangian" ideas. Researchers following this path tried to guess the underlying Lagrangian which would lead to the N-point functions. This line was open by Nambu, Nielsen and Susskind. Nambu [19] and Susskind [20] proposed that the underlying dynamics of the dual N-point functions corresponds to a generalization of the Schwinger proper time formalism where a relativistic string is propagating in proper time. The equation of motion satisfied by the string coordinates was the two-dimensional D'Alembert equation following from a linearized Lagrangian. Using plausible arguments they obtained expressions similar to the N-point (tree) amplitudes.

Then Nielsen [21] and immediately after Fairlie and Nielsen [22] used this linearized Lagrangian for constructing the "analogue model". The basic observation was that the momentum dependence of the integrands in the Koba–Nielsen amplitudes, and their loop generalizations is related to the energy of two-dimensional electrostatic problems where the momenta are "charges" located on the boundary. Then the electrostatic problem is solved on a disk for the tree amplitude, or on a higher genus two-dimensional surface described by the duality diagram corresponding to the respective loop amplitude. We understand this result today as a simple consequence of the fact that the $ikX(\sigma)$ factor in the exponential of the vertex operator acts as a source for the string coordinates whose propagator is the two-dimensional Coulomb kernel. Though the measure was not correctly reproduced, the "analogue model" is important since it is the first appearance of the two-dimensional world sheet in a mathematical role, rather than just as a picture in the duality diagram. This model is the precursor of the path integral formulation of string theory that was understood completely only later. Furthermore, the "analogue model" motivated the generalization [10] of the Virasoro amplitude [9], and therefore the formulation of the Closed String Theory by simply putting electrostatic sources on a sphere instead that on the boundary of a disk.

A non-linear action, proportional to the area spanned by the string, generalizing the non-linear one for the point-like particle, was also proposed by Nambu and Goto in [23, 24]. But the consequences of its non-linear structure, implying the invariance under an arbitrary reparametrization of the world sheet coordinates, were only clarified few years later with the treatment of [25] that provides a rigorous derivation of the properties of the generalized Veneziano model, though our present understanding of string theory is mostly based on the action used in [2].

The second approach that we will describe in detail in the next section, is based instead on the construction of an operator formalism that made transparent the most important properties of the model as the spectrum of physical states and their scattering amplitudes, and that historically has been essential for relating it to string theory in a completely satisfactory way.

3 The String World Sheet Through Factorization of the N-point amplitudes

The basic observation used in order to uncover the underlying theory in the operatorial approach was that, having a set of N-point functions satisfying DHS duality, crossing symmetry and tree-level analyticity, does not define a consistent set of S-matrix elements, unless the different poles in the various channels can be shown to come from the same set of physical states, the residues being factorized. This means that one should find a set of states, and a set of three-point couplings between these states, such that any expansion of a given ordering contribution to any of the N-point functions is reproduced by the same set of states and couplings.

During 1969 there was an intensive activity in this programme of finding the universal set of states and couplings leading to factorization. We will describe in words the main steps in historical succession, and then describe the complete solution as formulated in [26] at the end of 1969. Through an explicit analysis of the residues of a given pole in [27, 28], it was shown that factorization can be achieved by having an infinite number of intermediate states. An essential step was made in [29], where it was proven that the spectrum is the Fock space of an infinite number of harmonic oscillators. The authors of [29] gave general formulae for the masses of the states in terms of occupation numbers, and for the couplings of the external tachyons to arbitrary pairs of states in terms of matrix elements of vertex operators depending on the harmonic oscillator degrees of freedom. An important result of [29] was the discovery of the existence of the Hagedorn temperature in the theory, a basic feature characterizing String Theories.

We describe now the solution of the factorization problem following [26]. One starts defining the operator $Q_\mu(z)$ by

$$Q_\mu(z) = Q^{(+)}(z) + Q^{(0)}(z) + Q^{(-)}(z), \tag{10}$$

where

$$Q^{(+)} = i\sqrt{2\alpha'}\sum_{n=1}^{\infty}\frac{a_n}{\sqrt{n}}z^{-n} \quad ; \quad Q^{(-)} = -i\sqrt{2\alpha'}\sum_{n=1}^{\infty}\frac{a_n^\dagger}{\sqrt{n}}z^{n};$$

$$Q^{(0)} = \hat{q} - 2i\alpha'\hat{p}\log z, \tag{11}$$

and the vertex operators by

$$V(z;p) =: e^{ip\cdot Q(z)} :\equiv e^{ip\cdot Q^{(-)}(z)}e^{ip\hat{q}}e^{+2\alpha'\hat{p}\log z}e^{ip\cdot Q^{(+)}(z)}. \tag{12}$$

Then it was shown [26] that the integrand of the Koba–Nielsen N-point function is related to the Fock space vacuum matrix element of the product of vertex operators

$$\langle 0,0|\prod_{i=1}^{N} V(z_i,p_i)|0,0\rangle = \prod_{i>j}(z_i - z_j)^{2\alpha' p_i \cdot p_j}(2\pi)^4\delta^{(4)}(\sum_{i=1}^{N} p_i). \tag{13}$$

In order to obtain exactly the Koba–Nielsen expression one has to deal carefully with the fixing of three of the z variables. This is done by extracting the z dependence of the vertex operators using the identity

$$z^{L_0}V(1,p)z^{-L_0} = V(z,p)z^{\alpha_0}, \tag{14}$$

where L_0 is the operator

$$L_0 = \alpha'\hat{p}^2 + \sum_{n=1}^{\infty} na_n^\dagger \cdot a_n. \tag{15}$$

Choosing three consecutive values of z_i to be fixed:

$$z_a = z_1 = \infty \ ; \ z_b = z_2 = 1 \ ; \ z_c = z_N = 0, \tag{16}$$

the Koba–Nielsen amplitude can be rewritten in the operator language as

$$A_N \equiv \langle 0,p_1|V(1,p_2)DV(1,p_3)\dots DV(1,p_{N-1})|0,p_N\rangle, \tag{17}$$

where the "propagator" D is equal to

$$D = \int_0^1 dx x^{L_0-1-\alpha_0}(1-x)^{\alpha_0-1} = \frac{\Gamma(L_0-\alpha_0)\Gamma(\alpha_0)}{\Gamma(L_0)}, \tag{18}$$

and the states (using what we understand today as "operator-state correspondence") are defined as

$$\lim_{z\to 0} V(z;p)|0,0\rangle \equiv |0;p\rangle \ ; \ \langle 0;0|\lim_{z\to\infty} z^{2\alpha_0}V(z;p) = \langle 0,p|. \tag{19}$$

The z_i integrations of the Koba–Nielsen formula which were absorbed in the definition in (18) are translated into integrations over the "proper times" x_i appearing in the propagators.

This provides an explicit solution to the factorization. In fact, one can insert between each V and D a complete set of states of the space spanned by the harmonic oscillators (Fock space) appearing in $Q(z)$. Since D is diagonal in the basis of occupation numbers, poles will appear at $\alpha(s) = 0, 1, 2, ...,$ with factorized residues related in a universal fashion to the matrix elements of the vertex operators.

This solution to the factorization problem was the crucial step in the development of String Theory since, from now on, the N-point functions were clearly related to a theory in which the set of space–time fields is labelled by the states in the Fock space on which the Q_μ fields are realized. The Q_μ fields are, of course, the open string coordinate fields $X^\mu(\sigma,\tau)$ in d space–time dimensions for $\mu = 0, 1, 2, .., d-1$, computed at the endpoint of the string coordinate $\sigma = 0$, where z is related to the other string coordinate τ by $z = e^{i\tau}$. They are Heisenberg operators, their dependence on the world sheet coordinates σ and τ follows from the fact that they are solutions of an equation of motion following from a free linearized Lagrangian. However, as it is described above the Lagrangian was not used in the derivation, the various expressions being obtained by a rewriting of the N-point amplitudes. While the linear spacing between the poles of the Veneziano formula was suggestive of some underlying harmonic oscillator-type structure, only the solution of the factorization problem unveiled the true structure of the theory, i.e. an infinite number of oscillators assembled into a set of fields Q_μ living on a two-dimensional world sheet.

The vertex operators for the emission of tachyons represent insertions on the boundary (for open string theories) of the two-dimensional world sheet. Of course the relation in (17) is the way in which scattering amplitudes are obtained in String Theory, starting from the matrix element of products of vertex operators. The historical way was exactly the opposite, i.e. given the Koba–Nielsen formula, the operators whose matrix elements reproduce the formula were correctly guessed identifying the Hilbert space. Now the fulfillment of the DHS requirements became natural: the Q_μ are massless two-dimensional fields defining a two-dimensional conformal theory and the N-point functions are related to integrals of correlators of the vertex operators in the $SL(2,R)$ invariant vacuum. The integration over the z variables required by the Koba–Nielsen formula, in order to produce the poles in $\alpha(s_{ij})$, is related to the integration over the "proper times" after the mapping of the disk into the half upper plane. The fact that this particular expression is special to a particular gauge (at that time called the orthonormal gauge) was already understood during the first period of String Theory, but it became more transparent and rigorous after Polyakov's seminal paper [2].

Having the decomposition of the amplitudes in "vertices" and "propagators" allows the calculation of loop diagrams by gluing them and taking traces for the loops. The loop diagrams are necessary for producing an S-matrix consistent with unitarity. In this way, one obtained already in 1970 the correct expression in the Schottky parametrization of quantities defined on a Riemann surface as the period matrix, the abelian differentials and the Green's functions [30, 31, 32]. However, the correct measure of integration in the multiloops was not known at the time, since it requires the understanding of ghost contributions. It is clear now that these operatorial expressions in the covariant gauge are the same as those obtained by performing the path integral of the string Lagrangian over the appropriate world sheet.

We know today that the restrictions on the operators V and D which can be used follow from a correct gauge fixing of the string Lagrangian. In the absence of a Lagrangian, again the correct restrictions on V and D were found by a rather tortuous path (from today point of view), which we are going now to describe.

The expressions used above differ from the ones used in the modern formulation in two respects:

i) The vertex operators used were defined for a conformal weight $\alpha' k^2$. This value, related to the mass squared of the open string tachyon, is given in terms of the arbitrary parameter α_0: $\alpha_0 = \alpha' k^2$.

ii) The dimension d of space–time, i.e. the number of string coordinates, was left free.

4 The Virasoro Conditions

We start this section by reminding the reader how the two points mentioned at the end of the previous section are understood today. The starting point today for the bosonic string theory is the σ-model action (the action used in [2]) that, at the classical level, couples the string coordinates to the two-dimensional world sheet metric in a diffeomorphism-invariant and Weyl-invariant manner. Then the requirement that these two "gauge symmetries" (diffeomorphism and Weyl) are not anomalous in the quantum theory fixes the space–time dimension to the value $d = 26$ for the bosonic string.

Once the two "gauge symmetries" are respected at the quantum level, the standard Faddeev–Popov procedure can be applied, in principle in an arbitrary gauge, and a consistent quantization can be performed giving the physical states/operators in the gauge chosen. The states/operators in different gauges are isomorphic leading to the same results when gauge-invariant correlators are calculated. In particular, by choosing a covariant gauge, the Lorentz invariance of the theory follows automatically, while the unitarity of the theory is not obvious. On the other hand, by choosing an explicitly unitary gauge (the light-cone gauge) the unitarity of the theory is completely manifest, while the Lorentz invariance has to be checked. In the covariant gauge the physical states correspond to operators with dimension 1 for the open string and $(1, 1)$ for the closed string. This fixes the leading Regge trajectories to have intercept $\alpha_0 = 1$ or $\alpha_0 = 2$ for the open and closed strings, respectively. In a "physical" gauge, as the light-cone gauge, the states which are now "transverse" correspond to cohomologically equivalent families in the covariant gauge.

We have described above the present procedure for quantizing the bosonic string. However it must also be said that, in practice, one can invert the logic outlined above and fix the Regge intercept and the space–time dimension in the light-cone gauge by requiring that the Lorentz algebra be obeyed at the quantum level. This is, in fact, the way followed in the early days of string theory when the procedure described above was not yet known and this, of course, has led to the above values of the critical dimension and intercepts. Actually, to be more precise, the point of view expressed above has been essential, when we quantize the bosonic string in a covariant gauge, only in order to compute the correct integration measure for multiloop amplitudes. It has not played, in practice, any significant role in the light-cone gauge where the Regge slope and the space–time dimension have been correctly determined by imposing the closure of the Lorentz algebra.

We want to stress here, once more, that none of the ideas based on the Becchi–Rouet–Stora–Tyutin (BRST)-invariant approach (including the σ-model action) were known in the early days of string theory. The Nambu–Goto action was known, but it was not really known how to use it for deriving all the properties obtained using the operator formalism. One had to use alternative methods which amazingly enough led to the correct results. This is what we are going to explain below.

But before we proceed, let us notice that, from the present point of view, the description done in the previous section involved just a conformal theory of d massless fields. Of course, in such a theory any vertex operator is legal, and the correlators of vertex operators on the $SL(2,R)$-invariant vacuum have the block decomposition properties even after integrating over their "proper time" coordinates. Interestingly, even without the understanding that a consistent String Theory should be the gauge fixed version of a Weyl anomaly-free theory, the way to make the theory consistent by restricting i) and ii) was correctly guessed. This was done by looking for some "gauge" conditions that could help in decoupling the negative norm states, required by manifest Lorentz covariance, from the spectrum of the physical states, pretty much in analogy with what was known to happen in QED. We start discussing the way in which the correct gauge conditions were discovered.

In [27] it was pointed out that the residues of the poles on which the amplitude is factorized are not positive definite simply due to the presence of the time components of the oscillators, which in the operator formulation lead to a negative contribution to the scalar product. As a possible way out from this inconsistency of the theory, linear relations between the residues were uncovered leading to the decoupling of some Fock space states from the amplitude. The basic driving idea was that the situation here was analogous to the Gupta–Bleuler quantization of QED. As in QED the Lorentz condition was imposed to characterize the subspace of the physical states, here also some "gauge" conditions, that later on were understood to be due to some first class constraints, were imposed on the spectrum which would eliminate the negative norm states.

In this way, one managed to get the correct result without having to fix the gauge of the diffeomorphisms and Weyl invariance and to introduce the b, c ghost system. This has been possible because the ghosts are decoupled from the string coordinates. As a consequence, the non-trivial BRST cohomology can be realized in terms of the string coordinates only, the ghost ground state not being excited and, for tree diagrams at least, one can calculate consistently using the string coordinates restricted by the first class constraints.

The correct final answer was reached following a rather tortuous, but physical and at that time intuitive path.

We start describing the linear relations [33] mentioned above. In the operatorial formalism there is a realization [34, 35] of the Möbius transformations in (9) in terms of the infinite set of harmonic oscillators. This $SL(2,R)$ algebra has a simple action on the vertex operators and annihilates the vacuum. Its generators L_1, L_0, L_{-1} are

$$L_0 = \alpha' \hat{p}^2 + \sum_{n=1}^{\infty} n a_n^\dagger \cdot a_n \ ; \quad L_1 = \sqrt{2\alpha'}\hat{p} \cdot a_1 + \sum_{n=1}^{\infty} \sqrt{n(n+1)} a_{n+1} \cdot a_n^\dagger \tag{20}$$

and

$$L_{-1} = L_1^\dagger = \sqrt{2\alpha'}\hat{p} \cdot a_1^\dagger + \sum_{n=1}^{\infty} \sqrt{n(n+1)} a_{n+1}^\dagger \cdot a_n. \tag{21}$$

We recognize, of course, the central extension free $SL(2,R)$ subalgebra of the Virasoro algebra, which acts as a symmetry on an arbitrary (conformal field theory) (CFT) correlator, provided it is evaluated on the $SL(2,R)$-invariant vacuum. We remind the reader, however, that the algebra of the Virasoro operators and, more generally, two-dimensional conformal field theories, were not known at the time. Their understanding was a result of the developments we are describing. The $SL(2,R)$ subalgebra generates the Möbius group of the finite transformations of z:

$$z' = \frac{\alpha z + \beta}{\gamma z + \delta}, \tag{22}$$

where $\alpha\delta - \beta\gamma = 1$. The vertex operators have the standard transformation properties under the Möbius group corresponding to the weight $L_0 = \alpha' p^2$. In the expectation value in (17) the information that z_a is fixed appears only through the "bra" vector on the l.h.s. of the matrix element. Therefore, the r.h.s. has a residual symmetry, the subgroup of the Möbius group, which leaves the fixed $z_b = 1, z_c = 0$ unchanged :

$$z' = \frac{z}{1 - \alpha(z-1)} = z + \alpha(z^2 - z) + o(\alpha^2). \tag{23}$$

This subgroup is generated by

$$W_1 = L_1 - L_0. \tag{24}$$

Since the "ket" on the r.h.s. is left invariant by the subgroup in (23) we obtain

$$W_1 |p_{(1,M)}\rangle = 0, \tag{25}$$

where

$$|p_{1,M)}\rangle = V(1, p_M) D \ldots V(1, p_2) |p_1, 0\rangle, \tag{26}$$

independently on the number of VD insertions. Clearly, one gauge condition W_1 is not enough to project out all the negative norm states and additional conditions were searched for. We remark that (25) is not a consequence of any gauge symmetry being valid in any CFT for vertex operators of arbitrary dimensions, provided the vertex operators are inserted at the value $z = 1$. Nevertheless, following the pattern that led to (24), Virasoro [36] realized that, if $\alpha_0 = 1$, the state in (26) is annihilated by an infinite set of "gauge" operators

$$W_n |p_{1,M)}\rangle = 0 \quad ; \quad n = 1, 2, 3, \ldots \tag{27}$$

where

$$W_n = L_n - L_0 - (n-1) \tag{28}$$

with

$$L_n = \sqrt{2\alpha' n}\hat{p} \cdot a_n + \sum_{m=1}^{\infty} \sqrt{m(n+m)} a_{n+m} \cdot a_m$$

$$+\frac{1}{2} \sum_{m=1}^{n} \sqrt{m(n-m)} a_{m-n} \cdot a_m; \quad n \geq 0 \; ; \quad L_{-n} = L_n^{\dagger}. \tag{29}$$

The "gauge" conditions in (27) imply the following equations for the on-shell physical states of the generalized Veneziano model [37]:

$$(L_0 - 1)|Phys\rangle = L_n |Phys\rangle = 0 \quad ; \quad n = 1, 2, \ldots \tag{30}$$

These are exactly the constraints following from the diffeomorphism and Weyl symmetry of the action in presence of a two-dimensional metric, after the gauge fixing that eliminates completely the metric. These constraints annihilate the intermediate states in (17), that are not physical, as we know from the now standard gauge fixing–BRST procedure [8]. We postpone the discussion of the exact conditions under which the constraints eliminate the negative norm states to the next section, since it is closely tied to the recognition of the critical dimension. In conclusion, the correct results were obtained at the tree level without needing to know the underlying Lagrangian and to introduce the ghost degrees of freedom. What is more amazing is that also the correct one-loop measure was correctly obtained by using the Brink–Olive operator, that projected in the subspace of physical states [38]. The correct measure for the multiloop amplitudes was instead determined much later, although it would have been possible, in principle, to determine it by extending the procedure of Brink and Olive to multiloops.

Once the intercept α_0 got fixed to 1, it became clear that the first state on the leading trajectory is a tachyon; its consistent removal was achieved only with the discovery of the superstring and the GSO projection [39]. Imposing the infinite set of Virasoro constraints on the vertex operators corresponds, in today's language that was already used in [40], to the requirement that vertex operators should be primary fields with dimension 1 [40]. Projecting from the Fock space the states which are annihilated by all the Virasoro constraints, and eliminating the zero norm states following the procedure explained in [37], defines the physical Hilbert space which should have positive norm.

Shortly after Virasoro found the constraints (28) it was realized that the L_n operators are the generators of the conformal group in $d = 2$ [33]. The full algebra of the group including the central extension present in the commutator of L_n with L_{-n} was correctly worked out only somehow later [41][1]. In this way the algebra of the Virasoro operators was established and became the basic algebraic structure underlying two-dimensional CFT and String Theory. The central extension discovered by Weis [41], which is understood today as a manifestation of the conformal anomaly [2], has far reaching consequences which we are going to discuss now.

5 The Critical Dimension

The discovery of the critical dimension with its various manifestations shows the serendipity characteristic of this first period of String Theory. Since, as we know it today, the existence of the critical dimension is a consequence of the conformal anomaly cancellation between the string coordinates fields and the b, c ghost system, it is clear that in the absence of the understanding of the coupling to two-dimensional metrics and its gauge fixing which leads to the ghosts, the critical dimension could manifest itself only through its "side effects", i.e. various consistency conditions of the theory. The first calculation pointing to the existence of the critical dimension was done by Lovelace [42]. He calculated the non-planar loop with a number of tachyons as external particles, represented in Fig. 2.

This diagram was proposed earlier [43] as a model for the "Pomeron" which dominates the high-energy elastic scattering amplitude of hadrons and therefore, according to the lore of the time, was described as the Regge pole in the t-channel with the highest intercept. In the calculation the dimension of space–time d and the effective number of dimensions going around in the loop d', were left as free parameters. It was understood at the time that only the physical degrees of freedom which obey the Virasoro gauge conditions circulate

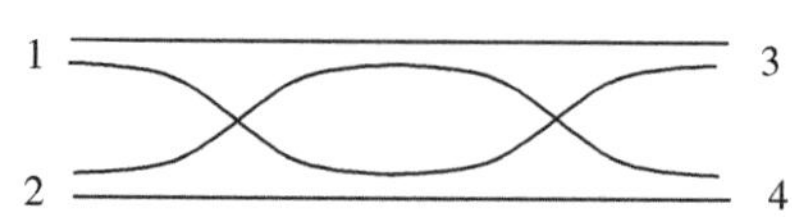

Fig. 2. The doubly twisted open string diagram

[1] See note added in proof of [33].

in the loops but the exact way to implement this fact was not understood[2]. The result of the calculation showed that the singularity in the t-channel became a pole only when $d = 26$ and $d' = 24$ and in this case the intercept of the "Pomeron" Regge trajectory is 2. We understand this result today as a consequence of the conformal invariance of the theory: by a continuous deformation of the world sheet, the diagram in Fig. 2 can be brought to the form in Fig. 3.

Now it is clear that one has a tree diagram, in the t-channel a closed string (the cylinder) being exchanged with the open string tachyons being coupled to the upper and lower disks. However, the conformal deformation of the world sheet on which the above expectation is based is valid only when conformal transformations act as expected classically, i.e. no anomaly is present implying $d = 26$. In addition, we know today that the b, c ghosts circulating in the loop cancel the contribution of two of the space–time string coordinates leading to $d' = 24$. Finally, the intercept 2 is the one required by the correct gauge fixing for the closed string. We identify nowadays the trajectory in the t-channel with the graviton and not the Pomeron, though the connection may come back to haunt us [44]. In the critical case the couplings of the open strings can be factorized and a consistent open–closed theory can be constructed [45, 46].

Further evidence for the existence of the critical dimension came from a close examination of the physical spectrum, i.e. the Hilbert space left after the infinite set of Virasoro conditions are imposed on the Fock space. In [47, 48] it was shown that the physical spectrum, i.e. the ensemble of Fock space states, which satisfy the conditions in (30), has a positive-definite scalar product (it is "ghost free") only when $d \leq 26$. Of course, if the spectrum is ghost free for $d = 26$, it is a fortiori so also for $d < 26$. In order to prove the "no ghost theorem" for $d = 26$ the manipulations used in [47] are very similar to the modern ones based on the BRST formalism, and which are valid provided that the BRST operator Q obeys at the quantum level $Q^2 = 0$. As a corollary

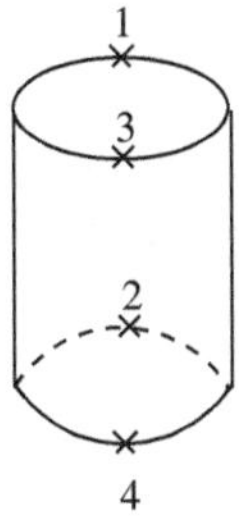

Fig. 3. The diagram of Fig. 2 in the closed string channel

[2] This was clarified few years later by Brink and Olive [38] inserting in the loop the operator that projected into the space of physical states.

of their proof Goddard and Thorn showed that the DDF [49] states form a basis for the physical Hilbert space.

This leads to a third manifestation [25] of the critical dimension which is already very close to our modern understanding. Though the starting point in [25] is the Nambu–Goto action the final results correspond to a correct quantization in light-cone [8] and in covariant gauge [8] of the σ-model action. The DDF states are isomorphic to the states in the light-cone gauge which live in a Hilbert space which has an explicitly positive-definite scalar product. The light-cone gauge is, therefore, unitary; however, Lorentz invariance is not explicit. On the other hand, in the covariant gauge Lorentz invariance is explicit but unitarity is valid only on the physical Hilbert space after the imposition of the conditions of (30). In our modern understanding, the two gauges being equivalent at the critical dimension insures, without further proof, that the spectrum is both unitary and Lorentz invariant. However, at the time one had to prove explicitly that on the spectrum in the light-cone gauge the Lorentz algebra is fully realized. By constructing all the Lorentz generators in [25], it was shown that the algebra correctly closes only if $d = 26$.

We mention finally an interesting interpretation of the central extension (and implicitly of the critical dimension) given by Brink and Nielsen [50]. They related the central extension to the Casimir energy of the string. In our present understanding this is simply the fact that, transforming L_0 to the strip (or cylinder for the closed string) coordinates, an additional term proportional to the central extension appears. This argument was later generalized to an arbitrary CFT in [51], giving a relation between the central extension and energies on finite geometries.

6 Conclusions

In this history-oriented note we briefly reviewed some of the developments that led to what we call today "String Theory". At the end of 1972, a complete theory existed (as summarized in [25]) which, except for the existence of the tachyon, was consistent. Its perturbative spectrum and the precise rules for calculating perturbatively scattering amplitudes were completely understood in the operator formalism. The theory is unitary and Lorentz invariant for $\alpha_0 = 1$ and $d = 26$. All this was obtained starting from a rather strange physical motivation, and involved a long chain of beautiful conceptual insights and guesses. The impressive theoretical structure created in the years 1969–1972, and further intensively developed during the last 25 years, continues to be at the forefront of Theoretical Physics. We dedicate this contribution to Gabriele Veneziano who played a leading role in the developments we described.

References

1. G. Veneziano: Nuovo Cimento A **57**, 190 (1968)
2. A. M. Polyakov: Phys. Lett. B **103**, 207 (1981)
3. A. Neveu, J. H. Schwarz: Nucl. Phys. B **31**, 86 (1971)
4. P. Ramond: Phys. Rev. D **3**, 2415 (1971)
5. K. Bardakci, M. Halpern: Phys. Rev. **D3**, 2493 (1971)
6. S. Mandelstam: Nucl. Phys. B **64**, 205 (1973)
7. P. Di Vecchia: *The birth of string theory*, article in this volume
8. M. B. Green, J.H. Schwarz, E. Witten: *Superstring Theory*, Vol. I (Cambridge University Press, Cambridge 1987);
   J. Polchinski : *String Theory*, Vol. I (Cambridge University Press, Cambridge 1998);
   B. Zwiebach: *A First Course in String Theory* (Cambridge University Press, Cambridge 2004)
9. M. A. Virasoro: Phys. Rev. **177**, 2309 (1969)
10. J. Shapiro: Phys. Lett. B **33**, 361 (1970)
11. E. Del Giudice, P. Di Vecchia: Nuovo Cimento A **5**, 90 (1971);
    M. Yoshimura: Phys. Lett. B **34**, 79 (1971)
12. R. Dolen, D. Horn, C. Schmid: Phys. Rev. **166**, 1768 (1968);
    C. Schmid: Phys. Rev. Letters **20**, 689 (1968)
13. H. Harari, Phys. Rev. Lett. **22**, 562 (1969)
14. J. L. Rosner, Phys. Rev. Lett. **22**, 689 (1969)
15. A. Donnachie, P.V. Landshoff: Phys. Lett. B **296**, 227 (1992)
16. K. Bardakçi, H. Ruegg: Phys. Rev. **181**, 485 (1969);
    C.G. Goebel, B. Sakita: Phys. Rev. Lett. **22**, 256 (1969);
    Chan Hong-Mo, T.S. Tsun: Phys. Lett. B **28**, 485 (1969)
17. Z. Koba, H. B.Nielsen: Nucl. Phys. B **10**, 633 (1969)
18. K. Kikkawa, B. Sakita, M. Virasoro: Phys. Rev. **184**, 1701 (1969)
19. Y. Nambu: Proc. Int. Conf. on *Symmetries and Quark Models*, Wayne State University 1969 (Gordon and Breach, New York 1970) p. 269
20. L. Susskind: Phys. Rev. D **1**, 1182 (1970)
21. H. B. Nielsen: Paper submitted to the *15th Int. Conf. on High Energy Physics* (Kiev, 1970) and Nordita preprint (1969)
22. D. B. Fairlie, H. B. Nielsen: Nucl. Phys. B **20**, 637 (1969)
23. Y. Nambu: Lectures at the Copenhagen Symposium (1970), unpublished
24. T. Goto: Progr. Theor. Phys. **46** (1971) 1560
25. P. Goddard, J. Goldstone, C. Rebbi, C. Thorn: Nucl. Phys. B **56**, 109 (1973)
26. S. Fubini, G. Veneziano: Nuovo Cimento A **67**, 29 (1970)
27. S. Fubini, G. Veneziano: Nuovo Cimento A **64**, 811 (1969)
28. K. Bardakçi, S. Mandelstam: Phys. Rev. **184**, 1640 (1969)
29. S. Fubini, D. Gordon, G. Veneziano: Phys. Lett. **B29**, 679 (1969)
30. C. Lovelace: Phys. Lett. B **32**, 490 (1970)
31. V. Alessandrini: Nuovo Cimento A **2**, 321 (1971)
32. D. Amati, V. Alessandrini: Nuovo Cimento A **4**, 793 (1971)
33. S. Fubini, G. Veneziano: Ann. Phys. **63**, 12 (1971)
34. F. Gliozzi: Lett. al Nuovo Cimento **2**, 846 (1969)
35. C. B. Chiu, S. Matsuda, C. Rebbi: Phys. Rev. Lett. **23**, 1526 (1969);
    C. B. Thorn: Phys. Rev. D **1**, 1963 (1970)

36. M. A. Virasoro: Phys. Rev. D **1**, 2933 (1970)
37. E. Del Giudice, P. Di Vecchia: Nuovo Cimento A **70**, 579 (1970)
38. L. Brink, D. Olive: Nucl. Phys. B **56**, 253 (1973) and Nucl. Phys. B **58**, 237 (1973)
39. F. Gliozzi, J. Scherk, D. Olive: Phys. Lett. B **65** 282 (1976) ; Nucl. Phys. B **122** 253 (1977)
40. P. Campagna, S. Fubini, E Napolitano, S. Sciuto: Nuovo Cimento A **2**, 911 (1971)
41. J. Weis: Unpublished work (1970)
42. C. Lovelace: Phys. Lett. B **34**, 500 (1971)
43. P. G. O. Freund, R. J. Rivers: Phys. Lett. B **29**, 510 (1969)
44. R. C. Brower, J. Polchinski, M. J. Strassler, Chung-I Tan: hep-th/0603115
45. E. Cremmer, J. Scherk: Nucl. Phys. B **50** , 222 (1972);
L. Clavelli, J. Shapiro: Nucl. Phys. B **57**, 490 (1973);
L. Brink, D.I. Olive, J. Scherk: Nucl. Phys. B **61**, 173 (1973)
46. M. Ademollo, A. D'Adda, R. D'Auria, E. Napolitano, P. Di Vecchia, F. Gliozzi, S. Sciuto: Nucl. Phys. B **77**, 189 (1974)
47. P. Goddard, C. B. Thorn: Phys. Lett. B **40**, 235 (1972)
48. R. C. Brower: Phys. Rev. D **6**, 1655 (1972)
49. E. Del Giudice, P. Di Vecchia, S. Fubini: Ann. Phys. **70**, 378 (1972)
50. L. Brink, H. B. Nielsen: Phys. Lett. B **45**, 332 (1973)
51. H. W. J. Bloete, J. L.Cardy, M. P. Nightingale: Phys. Rev. Lett. **56**, 742 (1986)

The Little Story of an Algebra

M. A. Virasoro

Dipartimento di Fisica, Universita di Roma1, Roma, Italy
virasoro@roma1.infn.it

Abstract. The historical path leading to the so-called Virasoro algebra is recalled, and the associated physical context is briefly discussed.

1 Introduction

When I heard about this project to honor our friend Gabriele Veneziano I could not be happier. Gabriele is one of those persons that once you encounter and interact with him you know he will be your friend for ever. If I try to make a balance of my life I realize how lucky I was of encountering him and the rest of the Italian–Argentinian–Israeli mafia on my first post-doc.

Unfortunately, this happiness dissolved soon when I realized that I would have to contribute an article about the algebra. It is not a mystery that I have not invested too much on it. This is not because of any deep reason or because I do not "believe" in it – just the opposite, I am sure contributions to it will remain in the textbooks long after us. But I prefer diversity and perhaps, also I share a diffuse feeling that as we get wiser we should risk working on subjects that are still shapeless. In any case, in a similar occasion celebrating Sakita (another person I truly cherished) I did choose to talk about "Models of the Brain". This time I chose heroically to submerge myself in the past and try to recover some old impressions. I hope Gabriele will appreciate at least my effort.

2 The Context

Let me put the story in context. The place was Madison, Wisconsin, a Midwestern midsize town with a good University and a large student body. The year was 1968, one of those moments in history when everything seemed possible, when the obstacles lay around the corner, hidden to the young and

M. A. Virasoro: *The Little Story of an Algebra*, Lect. Notes Phys. **737**, 137–144 (2008)
DOI 10.1007/978-3-540-74233-6_6

optimist that we were. I was arriving after 4 months spent in Argentina doing important things, like getting married, but no physics.

In Buenos Aires I had suffered a big frustration. Having left Israel in mid-April (leaving my collaborators to finish and write several pending papers), I had noticed, while preparing a CERN seminar, that the "miracles" that we were encountering while saturating the finite energy sum rules with the leading trajectory plus daughters were pointing to a simple mathematical fact: in the imaginary amplitude we were building a beta function:

$$\frac{\Gamma(\alpha(s)+\alpha(t))}{\Gamma(\alpha(s))\Gamma(\alpha(t))} = \sum_k c_k \frac{\alpha(s)^{\alpha(t)-k}}{\Gamma(\alpha(t))}. \qquad (1)$$

For us, interested in obtaining $s-t$ duality, this formula was telling something. It was the Imaginary part of the amplitude but it had to correspond to a full dual amplitude. I tried to use dispersion relations to build the full amplitude but the result was messy, and so I wrote a letter to Gabriele who replied to me immediately saying that he was playing with the same idea but that he has figured it out. He suggested that we wrote our results separately. I looked at my calculations and ... gave up. Gabriele had added a key element when he assumed all resonances to be infinitely narrow. That was not our philosophy in Israel and although he presented it as a mathematical convenient way to deal with an average amplitude, it was the first hint that dual amplitudes represented something different from the total amplitude, more like a Born approximation.

Endowed with this healthy frustration I anticipated my travel to Madison and arrived there with a generous dose of adrenaline. Obviously, Gabriele's paper had opened a Pandora's box and was, to use an expression popular in those times, mind-blowing. But curiously while in Europe the impact was immediate, the reaction in the US was more subdued. The bootstrap approach was identified as a West Coast ideology antagonist to the field theory framework that was instead the main playground for East Coast physicists. As a consequence, only a few (but important) American physicists jumped on the bandwagon. Bunji Sakita was of a different kind. He had a broad background and an extreme curiosity. Putting together his wide perspective, Charlie GoebelUs sheer power of analysis, and a young bright Keiji Kikkawa made Madison the perfect continuation to Rehovoth.

The first year in Wisconsin was a year of adaptation to a new environment. The pace was hectic. Nothing to do with the relaxing Rehovoh atmosphere. We had to learn to expect new results almost everyday and many of them by more than one group simultaneously. The telephone was a key instrument. I was communicating regularly with Gabriele at MIT and Hector in New York.

During the Summer 1969 I visited Europe. I spent a pleasant month discussing Loop Diagrams at Orsay where I met two young graduate students (Joel Scherk and Andre Neveu) who advised by Daniele Amati were interested in Dual Models. I visited CERN (I vividly remember being there when the

Apollo mission was landing on the Moon) and the Niels Bohr Institute where I learned from Holger Nielsen about his analog model, and for the first time I realized the crucial role played by Conformal Symmetry.

When I went back to Wisconsin I began to look carefully at the low lying resonances of the model by calculating by brute force their couplings to n-ground states and checking whether there were cancellations. My luck then was a direct consequence of my laziness. I knew that calculations were much simpler if $\alpha(0) = 1$ (this as a by-product of an earlier work on an alternative dual, crossing-symmetric amplitude), so I was routinely working at that value. Thus when I found that at the first level the ghost decoupled I could continue to the second level and there find that there were some additional decouplings. At that exact moment I heard from Gabriele the unwelcome news that at least two groups [1, 2] had derived the first level results. Fortunately, by then I was used to this kind of frustration and did not rush to publish but continued trying to simplify the calculation. The paper still shows how messy the original calculation was, but how it simplifies considerably once the Fubini–Veneziano [3] creation–annihilation operators $a_n^{\mu,\dagger}, a_n^\mu$ were used. Furthermore, written in that way, the generalization to the mth level became trivial: the operator

$$\begin{aligned} O_m &= \sum_{n=1}^{\infty} \sqrt{n(n+m)} a_{n+m}^\dagger a_n - i\sqrt{2m} P a_m^\dagger \\ &\quad - \sum_{n=1}^{m} \sqrt{n(m-n)} \frac{a_n^\dagger a_m^\dagger}{2} + m - H, \\ H &= \sum_{n=1}^{\infty} n a_n^\dagger a_n - P^2 - 1, \end{aligned} \tag{2}$$

turns out to create a resonance uncoupled to any number of ground states [4].

Thus there are as many uncoupled resonances as there are ghosts and therefore I assumed that all the ghosts had been killed. I was worried that I was trading ghosts for a tachyon. On the other hand, I happily dismissed the possibility that I could be killing good resonances and leaving ghosts alive. I have checked that this was not the case for the first two levels but it was Thorn, Brower and Goddard that took this issue seriously.

More or less at that time (end of the 1969–1970 Winter) Nambu came to Madison and in a seminar he boldly exposed his idea that the Lagrangian for the string was the one of a relativistic massless string. Convinced as I was that the Lagrangian was the conformal invariant one $\partial_\sigma \phi^\mu \partial_\sigma \phi^\mu + \partial_\tau \phi^\mu \partial_\tau \phi^\mu$, I went to him and kindly explained that there were too many resonances for his picture to be correct: the classical string would have two oscillating modes per level and not three. He stared at me, did not answer but was not seriously affected. Of course he was right [5]. One year later Goto [6] in Japan and a year and a half later Chang and Mansouri [7] (Nambu's collaborators) in Chicago proved that through Dirac quantization, a gauge fixing and for a

26-dimensional space–time the two Lagrangians were equivalent. Nambu was an expert in Dirac quantization but I still do not understand what hinted him about this idea. Because of this boldness (or foresight) he is considered the father of the string idea.

During the next year (1970) we watched a sustained effort to build new alternative dual models endowed with similar ghost-killing mechanism. The proposals of Neveu–Schwarz [10] and Ramond [11] models turned out to be richer. The operators O_m were completed with new anti-commuting operators in what turned out to be the first appearance of supersymmetry in physics. In Berkeley, Bardakci and Halpern [13] tried an algebraic approach working directly on the Fock space with creation–destruction operators. The calculations became very complicated. In Wisconsin, our work on loop diagrams led us to believe that the locality of the String degrees of freedom had to be manifested. For this we implemented the functional integral formalism rather than the operator one. With Sakita and his new student Hsue, we tried to fill several gaps. We investigated the anomalous dimension of the vertex operator $\exp(k\phi(z))$ and grasped the role of the full conformal algebra [14]. In a contribution to the Tel Aviv Conference held in April 1971, I presented the programme of studying systematically Conformal invariant Lagrangians to build new models [15]. All the models known at that moment were explained and several proposals could be discussed. I had one unfinished proposal with M. Kaku and M. Yoshimura but in September 1971 I decided to go back to Argentina and my personal interaction with the algebra finished.

The algebra by then had its own life nurtured by physicists like R. Brower, P. Goddard, D. Olive and C.B. Thorn. Any of them could write a much better account of its development. I can only attempt to give an incomplete, personal and probably biased version.

The operators O_m, whose commutators I have computed, generate a subgroup of the full conformal algebra: the transformations that leave the lines $\mathrm{Im}z = 0$ and $\mathrm{Im}z = \pi$, $z = 0$ and $z = -\infty$ invariant. The operators $O^\dagger_m$ leave the same lines, $z = 0$ and $z = \infty$ invariant. Fubini and Veneziano [8] reshuffled these operators with the Hamiltonian, defining

$$\begin{aligned} L_m &= O_m - H, \qquad && m > 0, \\ L_m &= O^\dagger_m - H, \qquad && m < 0, \\ L_0 &= H, \end{aligned} \tag{3}$$

and calculated the remaining commutators to write:

$$[L_n, L_m] = (n - m)L_{n+m} + \delta_{n,-m}\text{Central Charge}. \tag{4}$$

Finally J.H. Weiss noticed and calculated the Central Charge to be added to complete the so-called Virasoro algebra

$$\text{Central Charge} = \frac{m^3 - m}{12}c, \tag{5}$$

with c equal to the number of fields. R. Brower, C.B. Thorn and P. Goddard [9] proved the conjecture that all ghosts had been killed. In their work they found the special role played by the choice of $d = 26$. J. Shapiro constructed the full theory of closed strings proving that in this case there were two commuting Algebras.

In September 1972 I visited the Fermilab to organize a parallel session on Dual Models for the XVI International Conference on High Energy Physics. There I learned about the GGRT paper on the massless string dynamics and the light cone quantization [12]. I found that paper extremely interesting and began to work on the interaction among strings in the light cone gauge, but when I went back to Argentina too many things were happening: a military regime was reaching its end and a new era full of hopes was announced. A whole generation became deeply involved in the process with tragic consequences for many of them, because after a short promising periods events turned for the worst and dragged us into a political eyestorm. In August 1975 Tullio Regge invited me to Princeton and I decided to leave Argentina at least temporarily. At that moment I was interested in Geophysical Fluid Dynamics and was still planning to go back to Argentina, but then the military coup of 1976 definitely convinced me that I had to change plans.

The rest of the decade was kind of quiet also in the front of String Theory and Conformal Invariance. From time to time I was browsing articles and attending Seminars perhaps just to hear, as Hector Rubinstein used to say, the "music". Around 1980 I received a preprint by I.B. Frenkel and V. Kac entitled "Basic Representations of affine Lie algebras and Dual Resonance Models" [16]. I only read the nice introduction and understood that there was an ongoing effort to study the representations of Infinite Dimensional Lie algebras and in this context the developments in dual resonance models of the 1970s were a source of inspiration. Not only had we have found unitary, positive energy representations of an infinite dimensional algebra, but in addition these authors discovered that they can use the Fubini–Veneziano vertex operator to build the representations of the affine Lie (Kac–Moody) algebras.

In mathematics these algebras represent a natural generalization of finite dimensional Lie algebras. An excellent review specially written to introduce this subject to physicists can be found in [17]. Suppose g is defined by

$$[T^a, T^b] = i f^{abc} T^c, \tag{6}$$

where a, b, c run from 1 to dim g and f^{abc} are the structure constants of g, then an affine Kac–Moody algebra is defined by the commutation relations

$$[T^a_m, T^b_n] = i f^{abc} T^c_{m+n} + k m \delta^{ab} \delta_{m,-n} \tag{7}$$

with m any integer and k a central charge. In physics they are known objects. In fact, if we define

$$T^a(\theta) = \sum_{n=-\infty}^{\infty} T^a_n e^{2i\pi n\theta}, \tag{8}$$

we obtain

$$[T^a(\theta), T^b(\phi)] = if^{abc}T^c(\theta)\delta(\theta - \phi) + i\frac{1}{2\pi}k\delta^{ab}\delta'(\theta - \phi). \tag{9}$$

These are the equations of current algebra. Sugawara [18] has shown in 1968 that one could construct an energy–momentum tensor directly from the currents. We also know that from the energy–momentum tensor we can obtain the generators for coordinate transformations. Thus, generically, from representations of the Kac–Moody algebra one can construct representations of the Virasoro algebra. More specifically, the normal ordered bilinear $:\sum_a T^a(\theta)T^a(\theta):$, conveniently normalized and Fourier expanded, gives a set of L_n operators that obey

$$[L_m, T^a_n] = -nT^a_{n+m}. \tag{10}$$

Even my original construction can be written in this new way though obviously in the original Dual Model the g group is abelian.

This interrelation between the Kac–Moody and Virasoro algebras was crucial in the next development. From the point of view of physics, one is interested in representations that have a vacuum state: L_0 should have a spectrum bounded below and (at least in quantum physics) we want the representations to be unitary. Then in a remarkable paper Friedan, Qiu and Shenker [19] proved that the possible values of the central charge c and the ground state energy h had to be

$$\begin{aligned} &\text{either} \quad c \geq 1 \quad \text{and} \quad h \geq 0, \\ &\text{or} \quad c = 1 - \frac{6}{(m+2)(m+3)} \quad \text{and } h = \frac{[(m+3)p - (m+2)q]^2 - 1}{4(m+2)(m+3)}, \end{aligned} \tag{11}$$

with $m = 0, 1, 2, \ldots; p = 1, 2, \ldots, m; q = 1, 2, \ldots, p$. Goddard, Kent and Olive [20] then were able to build all the corresponding representations.

The next chapter in this little story concerns statistical physics systems in two dimensions at a critical point. The conformal group in two dimensions can be seen as rotations plus dilatations that depend on the coordinates. It is not too surprising that local systems that become scale invariant become simultaneously invariant under the full conformal group. This has been stressed before, but became a full programme with the work of Belavin, Polyakov and Zamolodchikov [21]. Their point was that Conformal Invariance imposes constraints on the correlation functions of the different fields defined in the system. If $\phi(z, \overline{z})$ is one of these fields then there are two families of L_n operators acting on them in the following way:

$$[L_n, \phi] = z^{n+1}\frac{\partial}{\partial z}\phi + h(n+1)z^n\phi,$$

$$[\overline{L}_n, \phi] = \overline{z}^{n+1} \frac{\partial}{\partial \overline{z}} \phi + \overline{h}(n+1)\overline{z}^n \phi. \tag{12}$$

In these equations $h, \overline{h}$ are two, possible anomalous, dimensions. They are restricted by the conditions stated above on the lowest eigenvalue of the L_0 operator. Therefore the correlation functions $< 0|\phi(z, \overline{z})\phi(z', \overline{z'})|0 >$ will scale with $h + \overline{h}$. One can the identify which representation is acting on the different systems at their critical point by looking at critical exponents. For instance, $c = 1/2$ corresponds to the Ising Model, $c = 4/5$ to the three-state Potts model and so on.

This is as much as I have been able to follow this subject. There are many new developments about which I am even more ignorant. However, even to a layman this little story shows clearly the advantages of a multi-disciplinary approach. The so-called Virasoro algebra, was known to mathematicians even with its central charge. However, when we discovered it in Physics, the amount of excitement that it produced had the positive effect of a sustained effort to understand it and generalize it. Thus the no-ghost theorem and the Neveu–Schwarz–Ramond generalization. When mathematicians rediscovered our work they had understood other aspects, and in particular the connection with the Kac–Moody algebras, but were happily surprised with the vertex operators that Sergio and Gabriele had introduced. The Kac determinant, a key ingredient for the Friedan et al. classification, could hardly have been discovered by physicists. Furthermore, the excitement on our side had decreased a lot by 1980. The latest discoveries, including the classification of all unitary representations, the restrictions on c and h, were the direct consequence of a fertile dialogue between the two communities. In short this is an edifying story.

References

1. F. Gliozzi: Nuovo Cimento Lett. **22**, 846 (1969)
2. C. Chiu, S. Matsuda, C. Rebbi: Phys. Rev. Lett. **23**, 1526 (1969)
3. S. Fubini, G. Veneziano: Nuovo Cimento A **64**, 811 (1969)
4. M. A. Virasoro: Phys. Rev D **1**, 933 (1970)
5. Y. Nambu: in *Proc. Int. Conf. on Symmetries and Quark Models*, ed. by R. Chand, Wayne State University, 1969 (Gordon and Breach, NY, 1970), p. 269
6. T. Goto: Prog. Theor. Phys. **46**, 1560 (1971)
7. L. N. Chang, F. Mansouri: Phys. Rev. **5**, 2535 (1972)
8. S. Fubini, G. Veneziano: Ann. Phys. **63**, 12 (1971)
9. P. Goddard, C. B. Thorn: Phys. Lett. **4**, 235 (1972)
10. A. Neveu, J.H. Schwarz: Nucl. Phys. B **21**, 86 (1971)
11. P. Ramond: Phys. Rev. D **3**, 2415 (1971)
12. P. Goddard, J. Goldstone, C. Rebbi, C. B. Thorn: Nucl. Phys. B **56**, 109 (1973)
13. K. Bardakci, M. Halpern: Phys. Rev. D **3**, 2493 (1971)
14. C. Hsue, B. Sakita, M. A. Virasoro: Phys. Rev. D **2**, 2857 (1970)

15. M. A. Virasoro: in *Proc. Int. Conf. on Duality and Symmetries in Hadron Physics*, ed. by E. Gotsman, Tel Aviv University, 1971 (The Weizmann Science Press of Israel, Jerusalem), p. 224
16. I. B. Frenkel, V.G. Kac: Invent. Math. **62**, 23 (1980)
17. P. Goddard, D. Olive: Int. J. Mod. Phys. A **1**, 303 (1986)
18. H. Sugawara: Phys. Rev. **170**, 1659 (1968)
19. D. Friedan, Z. Qiu, S. Shenker: Phys. Rev. Lett. **52**, 1575 (1984)
20. P. Goddard, A. Kent, D. Olive: Commun. Math. Phys. **103**, 105 (1986)
21. A.A. Belavin, A. Polyakov, A.B. Zamolodchikov: Nucl. Phys. B **241**, 333 (1984)

Part III

Perturbative QCD

Parton Densities: A Personal Retrospective

R. Petronzio

University of Rome "Tor Vergata" and INFN, Sezione di Roma "Tor Vergata", Roma, Italy
roberto.petronzio@roma2.infn.it

Abstract. The beginning of perturbative QCD and the generalisation of parton evolution probabilities beyond leading order are briefly recalled, together with my personal experience of collaboration and friendships with Gabriele.

My collaboration with Gabriele started at CERN. I was there as a fellow and I got involved with Daniele Amati into a discussion about the general validity of factorisation of mass singularities beyond leading order [1, 2]. The subject started from a stimulating argument by D.J. Politzer [3], who argued that the result about the universality of the leading log result of mass singularities among different processes involving partons in the initial states could be generalised, and lead to universal parton distributions for the normalisation of parton initiated hard processes. The discussion made an extensive use of the Lee–Nauenberg–Kinoshita [4, 5] theorem, and led to arguments in favour of the validity of what is known as the factorisation theorem.

Interacting with Gabriele was very stimulating, easy and rewarding. He was able to stimulate a genuine discussion, in spite of his greater experience in physics, and I could feel I could bring my personal contribution. Later we had many more discussion on several topics of the by that time emerging "perturbative QCD", on subjects like jet definition and pre-confinement. Not always they led to joint publications, but that was not the main aim.

The search on parton densities brought me to a deeper study of an explicit framework by which the study of mass singularities could lead to the generalisation beyond leading order of the probability evolutions, now known as DGLAP probabilities [6, 7, 8, 9]. En passant, it may be worth noticing that a first expression for a part of the leading evolution probabilities appeared in a work [10] with Giorgio Parisi, and was obtained by making the inverse of the Mellin transform of the well-known operator product expansion (OPE) result.

The generalisation of probabilities beyond the leading order was achieved by choosing a suitable calculation scheme. Together with W. Furmanski and G. Curci, that I take this occasion to remember a year after his premature

R. Petronzio: *Parton Densities: A Personal Retrospective*, Lect. Notes Phys. **737**, 147–150 (2008)
DOI 10.1007/978-3-540-74233-6_7

disappearance, we embarked into an explicit factorisation of mass singularities in a light-like gauge, and with dimensional regularisation of both ultraviolet and collinear singularities [11]. We could perform the first two-loop calculations of evolution probabilities directly in the "x" space, both for the flavour non singlet and for the flavour singlet sector. The method could deal with the probabilities in the time-like region as well as in the space-like: the only alternative was the use of the generalisation of the operator product expansion to time-like processes due to A. Mueller [12], and known as the cut vertices.

Many new points became clear to us in the new language: I remember in particular the clarification of the mixing between quark and anti-quark parton densities, occurring in the *non*-singlet case through a peculiar two-loop evolution kernel, and its connection with the distinction at two-loop level between even and odd moments of the probability evolutions. Our kinematical interpretation of the breaking of the relation between space-like and time-like processes has been confirmed by recent three-loop results.

The singlet calculation was more complicated by the larger number of diagrams, and took about 6 months of intense work to Furmanski and myself [13]. I remember we had a discrepancy with the classic [14] result obtained in the moment space through OPE: only a test of gauge invariance and of the supersymmetric relations among probabilities, at the end of additional lengthy checks, did convince us about the validity of our result. Later, the OPE result was corrected to ours.

Shortly after, Furmanski and myself [15] proposed a new method to analyse data that would allow incorporating our two-loop result and yet improve the efficiency of data analysis. The method was based on the use of Laguerre polynomials as a basis for the expansion of the experimental parton densities: the choice was mainly motivated by the easy composition rule of these polynomials under convolution, the standard mathematical operation by which probabilities and parton densities were tied together. The advantage was to avoid specific parametrisation of the experimental parton densities, avoiding then bias in their determination. The real goal was the determination of Λ_{QCD}, in a specific renormalisation scheme.

We first applied the method to NA4 data, with the help of Ruediger Voss and Marc Virchaux, another premature loss in our community. The first results were surprising: Λ_{QCD} in the minimal scheme was of the order of 200 MeV, instead of the usually quoted values around 130 MeV. We also applied the same method to the analysis of the Charm data (although less precise than the NA4 data), with similar findings. The higher value become the standard one some time later. I regret not having pursued the application of the method to more recent data, but both Wojtek and myself moved away from this subject. I kept working on structure functions with Ellis and Furmanski [16], and in particular on higher twists effects [17], a subject that the higher momentum transfers reached by the experiments quickly made not so relevant for the determination of the parton densities.

Parton distributions have been a subject of phenomenological studies by themselves, not only because of their scaling violations. One of my first papers [18] was about a two-stage model of parton distribution in a nucleon, together with Cabibbo, Altarelli and Maiani: a recent paper with Ricco and Simula [19] takes back some of those old ideas with very precise new data: always a "constituent quark" picture seems to emerge from the data, without yet a solid field theoretic description of its nature.

The value of parton densities were also investigated through lattice simulations of QCD in the approximation of neglecting fermion loops [20, 21, 22, 23, 24, 25]. Only the first couple of moments could be evaluated. Today, we are only a few years away from accurate predictions of moments, but delicate issues, like the gluon content at low x, need an approach different from the one based on the operator product expansion, and will not be addressed in a short time.

I would like to end this brief commentary on my activity related to structure functions with the sketch of an idea on which I am currently working on: looking for signatures of the quark–gluon plasma phase transition from a sudden modification of the parton densities. The effect comes from a power-suppressed contribution that occurs also in absence of a new phase of nuclear matter. Structure functions may get a contribution coming from the merging, through a higher twist process, of parton densities belonging to different nucleons that adds up to the yield of ordinary parton densities. In absence of a dense state of matter, the effect is strongly suppressed by the inverse power of the square momentum transfer, times the square of the typical distance of nucleons inside a nucleus. The phase transition brings a nuclear density much higher, by a factor 10, than ordinary matter, and enhances by a factor of about 10 such an effect. The effect remains small, but undergoes a jump, a signature of nuclear matter at unusual density values. Only explicit phenomenological calculations can decide the feasibility of this idea.

My collaboration with Gabriele was not confined to hard processes in QCD, but also on the study of low values of coupling constants running through the renormalisation group equations, from the energy scales of unification schemes and of the onset of string physics [26]. I wish I will be able to keep discussing with Gabriele at CERN or elsewhere in the next years, always learning from him how complex problem can be approached without prejudices, in a simple and physically intuitive manner.

References

1. D. Amati, R. Petronzio, G. Veneziano: Nucl. Phys. B **140**, 54 (1978)
2. D. Amati, R. Petronzio, G. Veneziano: Nucl. Phys. B **146**, 29 (1978)
3. H. D. Politzer: Nucl. Phys. B **129**, 301 (1977)
4. T. Kinoshita: J. Math. Phys. **3, 650** (1962)
5. T. D. Lee, M. Nauenberg: Phys. Rev. **133**, B1549 (1964)

6. G. Altarelli, G. Parisi: Nucl. Phys. B **126**, 298 (1977)
7. L. N. Lipatov: Sov. J. Nucl. Phys. **20**, 94 (1975) [Yad. Fiz. **20** (1974) 181]
8. V. N. Gribov, L. N. Lipatov: Sov. J. Nucl. Phys. **15**, 438 (1972) [Yad. Fiz. **15** (1972) 781]
9. Y. L. Dokshitzer: Sov. Phys. JETP **46**. 641 (1977) [Zh. Eksp. Teor. Fiz. **73** (1977) 1216]
10. G. Parisi, R. Petronzio: Phys. Lett. B **62**, 331 (1976)
11. G. Curci, W. Furmanski, R. Petronzio: Nucl. Phys. B **175**, 27 (1980)
12. A. H. Mueller: Phys. Rev. D **18**, 3705 (1978)
13. W. Furmanski, R. Petronzio: Phys. Lett. B **97**, 437 (1980)
14. E. G. Floratos, D. A. Ross, C. T. Sachrajda: Nucl. Phys. B **152**, 493 (1979)
15. W. Furmanski, R. Petronzio: Nucl. Phys. B **195**, 237 (1982)
16. R. K. Ellis, W. Furmanski, R. Petronzio: Nucl. Phys. B **207**, 1 (1982)
17. R. K. Ellis, W. Furmanski, R. Petronzio: Nucl. Phys. B **212**, 29 (1983)
18. G. Altarelli, N. Cabibbo, L. Maiani, R. Petronzio: Nucl. Phys. B **92**, 413 (1975)
19. R. Petronzio, S. Simula, G. Ricco: Phys. Rev. D **67**, 094004 (2003) [Erratum-ibid. D **68**, 099901 (2003)]
20. M. Guagnelli, K. Jansen, F. Palombi, R. Petronzio, A. Shindler, I. Wetzorke [Zeuthen-Rome (ZeRo) Collaboration]: Eur. Phys. J. C **40**, 69 (2005)
21. M. Guagnelli, K. Jansen, F. Palombi, R. Petronzio, A. Shindler, I. Wetzorke [Zeuthen-Rome (ZeRo) Collaboration]: Phys. Lett. B **597**, 216 (2004)
22. M. Guagnelli, F. Palombi, R. Petronzio, K. Jansen, A. Shindler, I. Wetzorke [Ze-Ro Zeuthen-Roma Collaboration]: Eur. Phys. J. A **17**, 365 (2003).
23. M. Guagnelli, K. Jansen, F. Palombi, R. Petronzio, A. Shindler, I. Wetzorke [Zeuthen-Rome/ZeRo Collaboration]: Nucl. Phys. B **664**, 276 (2003)
24. F. Palombi, R. Petronzio, A. Shindler: Nucl. Phys. B **637**, 243 (2002)
25. A. Bucarelli, F. Palombi, R. Petronzio, A. Shindler: Nucl. Phys. B **552**, 379 (1999)
26. R. Petronzio, G. Veneziano: Mod. Phys. Lett. A **2**, 707 (1987)

Infrared-sensitive Physics in QCD and in Electroweak Theory

M. Ciafaloni

Dipartimento di Fisica, Università di Firenze, Italy and INFN,
Sezione di Firenze, Italy
ciafaloni@fi.infn.it

Abstract. I recall the main ideas about the treatment of QCD infrared physics, as developed in the late 1970s, and I outline some novel applications of those ideas to Electroweak Theory.

1 Infrared-sensitive Observables

The high-energy physics of elementary particles, as described by the Standard Model, gives particular emphasis to states constructed out of *massless* partons or leptons, because of either the original gauge symmetry, or of the QCD chiral symmetry. This in principle introduces a number of problems because of the existence of mass singularities in gauge theories – that is, of infrared and collinear divergences due to the initial or final states being massless. Of course, physical states yield finite cross sections because of QCD confinement, or of electroweak symmetry breaking, or of QED coherent states. However, a remnant of the mass singularities of the problem is that the cross section, besides being dependent on energy and momentum transfers of the process at hand, may also depend on energy through large logarithmic variables, involving some infrared-sensitive mass parameters.

In QCD, avoiding large parameters is vital for the perturbative description of hard processes, characterized by probe(s) with large momentum transfer(s) Q and by a supposedly small coupling. Therefore, the cross section must be infrared safe, that is, sufficiently inclusive in order to cancel the mass singularities according to the KLN and/or Bloch–Nordsieck (BN) theorems [1, 2]. As a consequence, fully inclusive processes are truly perturbative, while the inclusive processes in which some partons of virtuality Q_0 are looked at (in the initial or final state) show anomalous dimensions [3]. However, observables in which soft emission is suppressed (e.g., at the boundary of the phase space) or emphasized (e.g., of multiplicity type) are infrared sensitive [4], and still contain parametrically large logarithms of infrared origin, because of an incomplete cancellation of virtual corrections with real emission.

M. Ciafaloni: *Infrared-sensitive Physics in QCD and in Electroweak Theory*, Lect. Notes Phys. **737**, 151–158 (2008)
DOI 10.1007/978-3-540-74233-6_8

The above observation raises a problem for quite interesting observables (like p_T-form factors and jet multiplicity distributions), but indicates also how to solve it because we know that the infrared behaviour is largely universal due to the QED factorization theorem [1] and generalizations thereof. This fact triggered, in the late 1970s, a number of seminal papers dealing with factorization of the collinear behaviour [5], form factor resummation [6], preconfinement [7], jet evolution [8] and multiplicities [9]. It also appeared that one could describe in full the final state [10] at the level of partons with off-shellness Q_0 much smaller than Q but still large with respect to Λ, the QCD scale, thus providing a ground for event generators [11].

All the above papers are largely based on factorization theorems for various hard processes, and gradually introduce generalized renormalization group techniques in order to predict the logarithmic dependence on the infrared- sensitive parameters at leading-logarithm anomalous dimension level, extended, by further analysis [12], to the subleading ones. The factorization properties are in turn dependent on the cancellation of truly infrared divergent contributions for all such processes, which requires a generalized Bloch-Nordsieck theorem to be valid in QCD, as better established in the 1980s [13]. In fact, the BN theorem states that a cross section which is inclusive over soft *final* states is also infrared safe, irrespective of the fixed, possibly degenerate initial state. In this form, the theorem is not automatically valid, because the nonabelian nature of QCD allows degenerate initial states in a multiplet, which have different charges and thus in general different cross sections for the *same* momentum configuration. This spoils the cancellation of virtual corrections with real emission when summing over final soft states, unless an average over initial colour is performed in order to restore the BN theorem. Fortunately, this averaging is authomatic because of QCD confinement, which allows only colour singlet asymptotic states.

The ideas above have been refined over the years in QCD, leading to an approximate treatment of coherence effects by angular ordering in jet evolution [14], and to a more general treatment of subleading logarithms in form factor calculations [15]. Recently, they have also led to a new interesting development in electroweak theory. Naïvely, one would say that in the latter case the infrared structure is irrelevant because of the spontaneously broken gauge symmetry, which provides a mass for weak bosons and for fermions. However, with the advent of teravolt scale accelerators, we shall soon have access to energies which are much larger than the symmetry breaking scale (say, the W mass) which may act as *infrared* cut-off and thus give rise to parametrically large infrared logarithms in the energy dependence, in addition to the ones of collinear origin. That this is indeed the case was first remarked in the late nineties [17] and soon applied to inclusive observables [18]. The failure of the BN theorem is due again to the nonabelian nature of electroweak theory, where now *no averaging* over flavour is possible, because the initial state consists of electrons, protons, and so on, each of them having a nontrivial weak isospin charge. This also means that double logarithms depending on the electroweak scale affect most cross sections which are apparently infrared safe, so that electroweak radiative corrections are enhanced, sometimes comparable to QCD ones, and to be carefully evaluated in a unified way.

My purpose in this note is to outline, in a few examples, how the novel ideas of the 1970s allow to understand the physics of large logarithms for both QCD and electroweak theory, thus turning a potential problem into a powerful tool. They also lead to a precise calculational framework for the logarithmic energy dependence, for which I refer to the reviews already mentioned [4, 14], and to further dedicated papers [15, 16].

2 QCD Form Factors, Multiplicities, Preconfinement

2.1 Form Factors

An early consequence of the understanding of infrared and collinear behaviours in QCD was the remark [6, 7, 8, 9, 10] that observables where real emission is suppressed are sensitive to the (square of) the partons' Sudakov form factor. The latter is evaluated, at leading logarithmic level, by an evolution equation in μ^2 (the parton virtuality) which is derived by a dispersive argument [4, 6], or by applying [19] Gribov's generalization of the Low theorem [20] as follows:

$$\frac{d\log F_a(Q^2,\mu^2)}{d\log\mu^2} = C_a\frac{\alpha_s(\mu^2)}{2\pi}\log(\frac{Q^2}{\mu^2}) \,, \tag{1}$$

where $C_a = C_F, C_A$ is the Casimir charge of parton $a = q, g$. Note that $\mu^2 > Q_0^2$ plays the role of cut-off for an infrared divergent anomalous dimension, so that F_a shows an exponential suppression which, in the frozen α_s limit, involves two logarithms per power of α_s, one of collinear type and the other of infrared origin. In the case of physical observables, the cut-off on μ^2 should be replaced by a parameter which regulates real emission, like Q^2/N for the parton probability density functions (PDFs) at large moment index N, or $1/B^2$ for impact parameter distributions. The outcome is the characteristic large-N dependence of PDFs for deep inelastic scattering (DIS) and for the Drell–Yan processes and the corresponding p_T-distributions.

For instance, the DIS structure function $F_N(Q^2)$ allows real emission up to gluon momentum fraction $z < 1/N$, and this regulates the anomalous dimension of (1) in the form

$$F_N(Q^2) \simeq \exp[-\frac{C_F}{\pi}\int_{Q_0^2}^{Q^2}\frac{d\mu^2}{\mu^2}\alpha_s(\mu^2)\log Min(\frac{Q^2}{\mu^2},N)] \,. \tag{2}$$

We can see that the anomalous dimension becomes finite and of $logN$ type for $\mu^2 < Q^2/N$, while the "exclusive" limit is reached for $N = Q^2/Q_0^2$, in which case (2) reduces to $F_q^2(Q^2, Q_0^2)$, where Q_0 is the minimal quark virtuality.

2.2 Multiplicities

Actually, the idea underlying [7, 10] is to describe outgoing hadronic jets in semi-inclusive form, at the level of partons of virtuality $Q_0 > \Lambda$, the decay products of the latter being summed over. Here a problem of consistency arises, because Q_0 is a somewhat arbitrary scale, and hadronic distributions should be independent of it. Fortunately, two important properties help. Firstly, multiplicity distributions show a *factorized* Q-dependence with respect to the Q_0 dependence and, secondly, *preconfinement* holds, namely the average mass of "minimal" colour singlets connected to a $q - \bar{q}$ pair is of order Q_0, much smaller than Q. This means that jet evolution can be viewed in two steps, a perturbative QCD evolution from Q down to Q_0 (of order Λ) and a hadronization process at scale Q_0. Thus, the virtue of factorization and preconfinement is that the conversion into hadrons does not affect the Q-dependence, and occurs at a much lower scale.

Of course, the infrared analysis is essential in order to derive the above properties. Factorization of multiplicity distributions is argued for by resumming the double-log Feynman-x dependence of jet distribution functions in the soft region, which eventually leads to a *finite* anomalous dimension with a singular α_s-dependence [9, 10] of type $\gamma_0 \simeq \sqrt{\frac{N_c\alpha_s}{2\pi}}$ [19, 21]. Correspondingly, the average hadronic jet multiplicity has the behaviour

$$\bar{n}(Q^2) \sim exp \int_0^t dt \gamma_0(\alpha_s(t)) \simeq exp \sqrt{\frac{2N_c}{\pi b} \log \frac{Q^2}{\Lambda^2}} \, , \tag{3}$$

and thus grows more rapidly than any power of $\log(Q^2/\Lambda^2) = t$.

The behaviour (3) is remarkably different from the one of QED radiation, essentially because of the gluon charge, implying that the QCD jet evolution is a branching process, leading to a cascade, rather than a bremsstrahlung process off one leg, as in QED. Correspondingly, strong correlations of the final soft partons are present, leading to an approximate KNO scaling of "exclusive" n-parton emission probabilities, which for a gluon jet have the form [4]

$$\frac{\sigma_n}{\sigma_{jet}} \simeq \frac{1}{\bar{n}} \exp[-\frac{1}{2}(\log \frac{n}{\bar{n}})^2], \quad (n \ll \bar{n}) \, . \tag{4}$$

This result shows that the the approximate proportionality of the σ_ns in a gluon jet to the corresponding form factor (1) still holds, at double-log level, as for the electron in QED, but their relationship to the average multiplicity (3) – in the frozen α_s limit – is quite different from QED because of the QCD cascade.

2.3 Preconfinement

On the other hand, preconfinement [7] follows from a veto on the possible final states which are allowed in the minimal colour singlets in which, by definition, a $U(3)$ colour line connects a quark of offshellness Q_0 to the corresponding antiquark. Because of factorization, and of the veto, the inclusive mass distribution of minimal singlets being produced in a jet of mass up to Q is independent of Q and is instead sensitive to the quark form factor, as follows [7, 10]:

$$\frac{M^2 d\sigma}{\sigma_{jet} dM^2} \sim F_q^2(M^2, Q_0^2) \, , \tag{5}$$

so that its average mass is of order Q_0. Therefore, the conversion of partons into hadrons can occur by an interaction of partons which are close in phase space, leading to the so-called *local* parton–hadron duality [22], and to the possibility of building event generators with relatively simple hadronization models [11, 23].

3 Inclusive Electroweak Double Logarithms

The infrared physics outlined above relies on the BN cancellation of virtual and real emission singularities, which in QCD occurs because of the colour averaging in the initial state, as remarked above. Therefore, the form factor behaviour of type (1) shows up only if some veto uncovers the "exclusive" limit of the given hard process. On the other hand, in electroweak (EW) theory the BN theorem *fails* because of the flavour charges of the accelerator beams. For instance, the total cross section for $e_+ e_-$ annihilation into hadrons is an infrared safe observable from the QCD standpoint, but carries nevertheless EW double logarithms, embodied into an enhanced effective coupling

$$\alpha_{eff}(s) = \frac{\alpha_W}{4} (\log \frac{s}{M_W^2})^2 \, , \tag{6}$$

which is of order 0.2 in the teravolt energy range and leads, therefore, to sizeable corrections, of the same order as QCD ones. Besides the expected collinear logarithm, the expression (6) carries an additional one, of infrared origin, due to the violation of the BN theorem.

The analysis of such inclusive double logarithms [18] involves form factors of type (1), where now μ^2 is cut-off by the EW scale $M_W^2 \simeq M_Z^2 = M^2$ and the Casimir C_a refers to the isospin I representation $a = I = 0, 1, ...$ in the t-channel of the lepton–antilepton overlap matrix. For instance, the combinations $\sigma_{e_- \nu} \pm \sigma_{e_- e_+}$ correspond to $I = 0$ ($I = 1$), so that

$$\sigma_{e_+ e_-}(s, M^2) \simeq \frac{1}{2}(\sigma_0 - \sigma_1 F_1(s, M^2)) \simeq \frac{1}{2}(\sigma_0 - \sigma_1 \exp(-2\frac{\alpha_{eff}(s)}{\pi})) \, , \tag{7}$$

where σ_0 corresponds to the isospin averaged cross section and has therefore no double logarithms, while the antisymmetric combination σ_1 is damped by the $I = 1$ form factor, with $C_1 = 2$. We note that, because of the optical theorem, the inclusive form factor is not squared, though referring to a physical cross section in the crossed channel. Note also that in this example $\sigma_1 > 0$, because the neutrino cross section is larger, and therefore the $\sigma_{e_+e_-}/\sigma_0$ ratio *increases* in the teravolt energy range towards its high-energy limit, which is provided by the flavour average.

The above description can be generalized, by collinear factorization, to single logarithmic level and to a generic overlap matrix involving leptons and partons in the initial states, thus coupling the EW and QCD sectors of the Standard Model. The result of this procedure is a set of evolution equations in μ^2 which are similar to the DGLAP equations [24], except that evolution kernels exist in the channels with $I \neq 0$ also, and are infrared singular or, in other words, depend on a logarithmic cut-off, much as in (1). For instance, in the evolution of lepton densities f_l and boson densities f_b, the $I = 0$ evolution kernels coincide with the customary DGLAP splitting functions P_{ba}, while the $I = 1$ ones involve the cut-off-dependent virtual kernels

$$P_f^V = \delta(1-z)(-\log\frac{Q^2}{\mu^2} + \frac{3}{2}),\; P_b^V = \delta(1-z)(-\log\frac{Q^2}{\mu^2} + \frac{11}{6} - \frac{n_f}{6})\,. \quad (8)$$

The corresponding evolution equations have the form

$$-\frac{df_a^1}{d\log\mu^2} = \frac{\alpha_W}{2\pi} f_a^1 P_a^V + \text{regular terms}\,, \quad (9)$$

and have been described in fully coupled form in [25]. Here I just notice that (9) shows a Sudakov behaviour similar to (1) and is consistent with (7) after taking into account the antilepton evolution, which doubles the virtual kernel.

The presence of inclusive double logarithms in spontaneously broken gauge theories remains an intriguing subject. It is mostly an initial state effect and, as such, it is present for any final states of the same class (e.g., flavour blind) and strongly depends on the accelerator beams. Leptonic accelerators maximize it, while hadronic ones (like LHC) provide some partial average on the initial partonic flavours, thus decreasing it. But the effect appears also if the flavour charges are looked at in the final state instead of the initial state, for instance, in gluon fusion processes in which some W's are observed [26]. Furthermore, the effect occurs whenever the soft boson emission mixes several degenerate states having different hard cross sections. Nonabelian theories have it because of the nontrivial multiplets, but also a broken abelian theory shows it whenever the mass eigenstates are not charge eigenstates [27]. An example of the latter type is the mixing of the Higgs boson with the longitudinal gauge boson occurring in a $U(1)$ theory. The Standard Model shows both kinds of effects and, given their magnitude in (6), I think that the coupled evolution equations of parton–lepton distribution functions [25] deserve by now a quantitative study at the teravolt scale.

Perhaps, the most important lesson to be learned from several decades of investigation of infrared-sensitive high-energy physics is that, even at the level of hard processes, the fundamental interactions look much more intertwined, due to the large time nature of asymptotic states which possibly increases their effective couplings. By the same token, because of the large times involved, factorization theorems are at work and allow a good understanding of the infrared dynamics. It remains true, however, that a unified treatment of all degrees of freedom is needed already at Standard Model level – that is, even before discovery of a possible short-distance unification.

Acknowledgements

It is a pleasure to thank Gabriele, as a collaborator and as a friend, for sharing over many years and subjects the excitement of long discussions and, sometimes, of real understanding. I also warmly thank old and new teams on this subject, in particular Stefano, and Paolo and Denis, for various updates of the picture presented here. Finally, I am grateful to the CERN Theory Division for hospitality while this work was being completed, and to the Italian Ministry of University and Research for a PRIN grant.

References

1. F. Bloch, A. Nordsieck: Phys. Rev. **52**, 54 (1937);
   V. V. Sudakov: Sov. Phys. JETP **3**, 65 (1956);
   D. R. Yennie, S. C. Frautschi, H. Suura: Ann. Phys. **13**, 379 (1961)
2. T. Kinoshita: J. Math. Phys.**3**, 650 (1962);
   T. D. Lee, M. Nauenberg: Phys. Rev. **133**, 1549 (1964)
3. Y. L. Dokshitzer, D. Dyakonov, S. I. Troyan: Phys. Rep. **58**, 269 (1980);
   A. H. Mueller: Phys. Rep. **73**, 237 (1981);
   G. Altarelli: Phys. Rep. **81**, 1 (1982)
4. A. Bassetto, M. Ciafaloni, G. Marchesini: Phys. Rep. **100**, 201 (1983)
5. D. Amati, R. Petronzio, G. Veneziano: Nucl. Phys. B **140**, 54 (1978) and B **146**, 29 (1978);
   R. K. Ellis, H. Georgi, M. Machacek, H. D. Politzer, G. C. Ross: Phys. Lett. B **78**, 281 (1978); Nucl. Phys. B **152**, 285 (1979)
6. D. Amati, A. Bassetto, M. Ciafaloni, G. Marchesini, G. Veneziano: Nucl. Phys. B **173**, 429 (1980);
   G. Parisi, R. Petronzio: Nucl. Phys. B **154**, 427 (1979);
   G. Parisi: Phys. Lett. B **90**, 295, (1980);
   G. Curci, M. Greco: Phys. Lett. B **92**, 175 (1980)
7. D. Amati, G. Veneziano: Phys. Lett. B **83**, 87 (1979)
8. K. Konishi, A. Ukawa, G. Veneziano: Nucl. Phys. B **157**, 45 (1979);
   A. Bassetto, M. Ciafaloni, G. Marchesini: Phys. Lett. B **86**, 366 (1979)
9. A. Bassetto, M. Ciafaloni, G. Marchesini: Phys. Lett. B **83**, 207 (1979);
   W. Furmanski, R. Petronzio, S. Pokorski: Nucl. Phys. B **155**, 253 (1979)
10. A. Bassetto, M. Ciafaloni, G. Marchesini: Nucl. Phys. B **163**, 477 (1980)

11. G. C. Fox, S. Wolfram: Nucl. Phys. B **168**, 285 (1980);
R. Odorico: Nucl. Phys. B **172**, 157 (1980); Phys. Lett. B **102**, 341 (1981);
G. Marchesini, B. R. Webber: Nucl. Phys. B **238**, 1 (1984)
12. I. C. Collins, D. E. Soper: Nucl. Phys. B **193**, 381 (1981), B **194**, 445 (1982); Nucl. Phys. B **197**, 446 (1982);
A. Sen: Phys. Rev. D **24**, 3281 (1981);
S. Mukhi, G. Sterman: Nucl. Phys. B **206**, 221 (1982);
J. Kodaira, L. Trentadue: Phys. Lett. B **112**, 66 (1982)
13. R. Doria, J. Frenkel, J. C. Taylor: Nucl. Phys. B **168**, 93 (1980);
G. T. Bodwin, S. J. Brodsky, G. P. Lepage: Phys. Rev. Lett. **47**, 1799 (1981);
A. H. Mueller: Phys. Lett. B **108**, 355 (1982);
W. W. Lindsay, D. A. Ross, C. T. Sachrajda: Nucl. Phys. B **214**, 61 (1983);
P. H. Sörensen, J. C. Taylor: Nucl. Phys. B **238**, 284 (1984);
S. Catani, M. Ciafaloni, G. Marchesini: Phys. Lett. B **168**, 284 (1986); Nucl. Phys. B **264**, 588 (1986)
14. See Yu. L. Dokshitzer, V. A. Khoze, S. I. Troyan, A. H. Mueller: Rev. Mod. Phys. **60**, 373 (1988) and references therein
15. G. Sterman: Nucl. Phys. B **281**, 310 (1987);
S. Catani, L. Trentadue: Nucl. Phys. B **327**, 323 (1989)
16. For an updated list of references, see, e.g., S. Catani et al.: *Proceedings of the CERN Workshop on Standard Model Physics (and More) at the LHC*, ed. by G. Altarelli, M. L. Mangano (CERN, Geneva 2000), Sect. 5
17. P. Ciafaloni, D. Comelli: Phys. Lett. B **446**, 278 (1999); Phys. Lett. B **476**, 49 (2000);
V. S. Fadin, L. N. Lipatov, A. D. Martin, M. Melles: Phys. Rev. D **61**, 094002 (2000);
M. Hori, H. Kawamura, J. Kodaira: Phys. Lett. B **491**, 275 (2000);
J. H. Kuhn, S. Moch, A. A. Penin, V. A. Smirnov: Nucl. Phys. B **616**, 286 (2001)
18. M. Ciafaloni, P. Ciafaloni, D. Comelli: Phys. Rev. Lett. **84**, 4810 (2000); Nucl. Phys. B **589**, 359 (2000)
19. B. I. Ermolaev, V. S. Fadin: JETP Lett. **33**, 269 (1981);
V. S. Fadin: Yad. Fiz. **37**, 408 (1983)
20. F. Low: Phys. Rev. **110**, 974 (1958);
V. N Gribov: Sov. J. Nucl. Phys. **5**, 399 (1967)
21. A. H. Mueller: Phys. Lett. B **104**, 161 (1981)
22. See Ya. I. Azimov, Y. L. Dokshitzer, V. A. Khoze, S. I. Troyan: Z. Phys. C **27**, 65, (1989) and references therein.
23. For an overview see, e.g., G. Marchesini: *From QCD Lagrangian to Monte Carlo Simulation*, this volume
24. V. N. Gribov, L. N. Lipatov: Sov. J. Nucl. Phys. **15**, 438, (1972);
G. Altarelli, G. Parisi: Nucl. Phys. B **126**, 298 (1977);
Y. L. Dokshitzer: Sov. Phys. JETP **46**, 641 (1977)
25. M. Ciafaloni, P. Ciafaloni, D. Comelli: Phys. Rev. Lett. **88**, 102001 (2002);
P. Ciafaloni, D. Comelli: JHEP **0511**, 039 (2005)
26. P. Ciafaloni, D. Comelli: JHEP **0609**, 055 (2006)
27. M. Ciafaloni, P. Ciafaloni, D. Comelli: Phys. Rev. Lett. **87**, 211802 (2001)

From QCD Lagrangian to Monte Carlo Simulation

G. Marchesini

Dipartimento di Fisica, Università di Milano-Bicocca, Milano, Italy
and INFN, Sezione di Milano-Bicocca, Milano, Italy
Giuseppe.Marchesini@mib.infn.it

Abstract. I discuss old and recent aspects of quantum chronodynamics (QCD) jet emission and describe how hard QCD results are used to construct Monte Carlo programs for generating hadron emission in hard collisions. I focus on the program HERWIG at Large Hadron Collider (LHC).

1 The Status

LHC is a discovery machine, it is expected to tell us how to complete the unified theory of elementary interactions. New (heavy) particles are searched to indicate/confirm new symmetries. Events with heavy particles are expected to be accompanied by an intense emission of hadrons at short distances, and this is the domain of perturbative QCD. Therefore, to identify and understand non-standard events a quantitative knowledge of the characteristics of the hard radiation is strongly needed. In 1973 QCD was at the frontier of particle physics (discovery of asymptotic freedom [1] and beginning of quantitative QCD studies), now in 2007 QCD is at the centre of particle studies. The Monte Carlo programs for jet emissions [2, 3, 4] are important instruments for analysing standard and non-standard short distance events. They are the *Summa* of most QCD theoretical results and many present studies aim to improve their quantitative predictions. Thanks to the QCD factorization structure [5], Monte Carlo programs can be interfaced with hard cross sections involving also non-QCD processes (electroweak, supersymmetric, extra dimension, black holes, ...). In this way, Monte Carlo generators can describe both QCD and non-QCD events at short distances.

In this paper I describe the main QCD results which enter the construction of a Monte Carlo generator. They are so many that most of the key points will be recalled in a schematic way, but I hope that this short description could provide an idea of the reliability range of the Monte Carlo generators. For a more detailed description see [6]. Here, aiming to be simple and synthetic, I follow a personal point of view and the focus will be on the Monte Carlo event

G. Marchesini: *From QCD Lagrangian to Monte Carlo Simulation*, Lect. Notes Phys. **737**, 159–180 (2008)
DOI 10.1007/978-3-540-74233-6_9

generator HERWIG [2]. Its general structure is similar to other important Monte Carlo generators [3, 4]. In Sect. 2, I present the scheme of the operations performed by Monte Carlo codes for LHC. The fact that the generation of events can be subdivided into successive stages is physically based on QCD factorization properties. The theoretical basis are discussed/recalled in Sect. 3. In Sect. 4, I discuss the multi-gluon soft distributions and in Sect. 5, I describe in detail a Monte Carlo code for soft emissions. Although important non-soft contributions included in a realistic Monte Carlo are here missed, it provides a simple example containing many important physical effects. In Sect. 6, I discuss non-perturbative effects which enter the Monte Carlo generators. The last section contains final considerations.

2 Structure of Monte Carlo generator

I start describing schematically the way a Monte Carlo code is organized in order to generate hard QCD and non-QCD events at LHC. As a specific illustration I consider the emission of two jets with high E_T. This process is factorized into the elementary hard distribution, the parton densities (structure functions as in DIS (deep inelastic scattering)) and the fragmentation functions (as in e^+e^-):

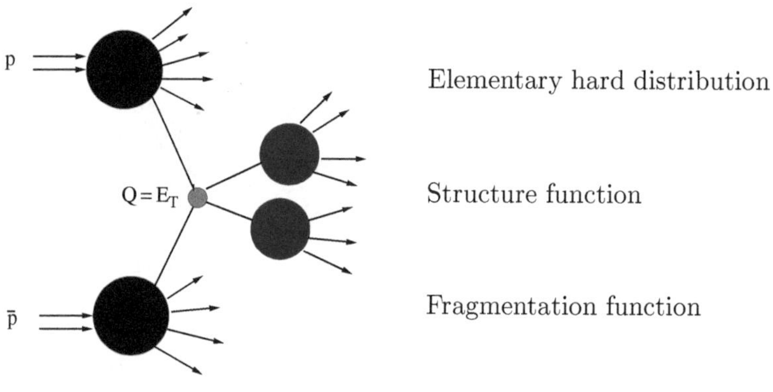

Here are the necessary factorized steps:

- start form the hard elementary distribution $\hat{\sigma}_{ab\to cd}$ with ab the incoming and cd the two outgoing partons. This hard distribution corresponds to QCD jet emission at high E_T. Here one can substitute distributions for other QCD or non-QCD processes. There are many studies of hard distribution for processes relevant for LHC, see [7].
- generate the momenta of the hard incoming (ab) and outgoing (cd) partons (and possible non-QCD particles). Given the hard scales E_T (and possible heavy masses), the momenta are generated (via important sampling)

in computing the total cross section as convolutions of the elementary distribution and the parton densities (structure functions);

- use the initial state space-like evolution (which at the inclusive level gives the structure functions) to generate the "bremsstrahlung" of outgoing initial state partons $k'_1, k'_2 \cdots$. This requires imposing a minimal transverse momentum w.r.t. the collision direction;
- given the outgoing hard QCD partons cd and $k'_1, k'_2 \cdots$, start the QCD shower (parton multiplication). First, from the set of these partons, identify their colour connections and reconstruct the set of the various primary $q\bar{q}$ dipoles. Here one works in the large N_c approximation so that a gluon, from the colour point of view, can be represented as a pair of quark–antiquark lines, a gluon is then associated to two dipoles;
- generate, for each primary dipole, the multi-parton emission according to the coherent branching structure that will be illustrated in the following. This requires imposing a lower bound on the *relative* transverse momenta of final state partons (inside sub-jets);
- match with the exact high-order calculation, if available. It consists in weighting the generated event by comparing [8] the Monte Carlo distribution and the exact square matrix element computed to higher order [7];
- given the system of all emitted partons, generate the final hadrons by using a hadronization model making hadrons out of partons. Using hadronization models based on colour connections and preconfinement [9], such a process should not substantially modify [10] the structure of the hadronic radiation with respect to the partonic one which has been obtained in the previous steps.

In the next sections I describe the QCD basis of these steps.

3 The Long Way to Monte Carlo

QCD has a dimensionless coupling but, even at large-scale Q, when all masses can be neglected, the cross sections do not scale simply as powers of Q^2. This is due to the presence of ultraviolet, collinear and infrared divergences. Ultraviolet divergences are responsible for the presence of the fundamental QCD scale Λ_{QCD} entering the running coupling. Collinear and infrared divergences are well know from QED [11]. Parton distributions can be computed only by fixing a resolution Q_0 (technically, a subtraction point) in the parton transverse momentum. Collinear and infrared divergences are responsible for large enhancements in these distributions which need to be resummed. Monte Carlo generators do actually perform these resummations as I discuss in the following.

The possibility to resum these enhanced terms is based on specific properties of the collinear and infrared singularities: they factories [5, 12, 13].

In this way one can formulate recurrence relations that lead to evolution equations. The fundamental one is the DGLAP evolution equation [14] resumming collinear singularities in parton densities and fragmentation functions. These are single-inclusive quantities, but to reach a complete description of an event one needs many-particle distributions so that the fully exclusive picture can be reconstructed (with given resolutions). The way to this is the jet calculus formulated and constructed by Ken Konishi, Akira Ukawa and Gabriele Veneziano [15] as generalization of the DGLAP evolution equation. Therefore, their work can be considered as the basis of the Monte Carlo parton multiplication. Jet calculus leads the way to the evolution equation for the generating functional [12, 13] of the multi-parton distributions and then to the branching probabilities for parton splitting in a way that could be implemented into Monte Carlo codes. The pioneering Monte Carlo codes [16, 17, 18] were resumming collinear singularities but only after the discovery of coherence of soft gluon radiation, both collinear and infrared enhanced logarithms where correctly resummed. The present Monte Carlo generators [2, 3, 4] fully resum not only the leading collinear and infrared singularities, but also relevant subleading contributions.

In the following I describe the main theoretical points corresponding to the Monte Carlo steps recalled in the previous section.

3.1 Asymptotic Freedom and Physical Coupling

At a short distance the theory becomes free [1] and here the use of perturbation theory is justified. At the two loops one has

$$\alpha_s(Q) \simeq \frac{4\pi}{\beta_0 L}\left(1 - \frac{2\beta_1 \ln L}{\beta_0^2 L} + \dots\right), \qquad L = \ln \frac{Q^2}{\Lambda^2_{\text{QCD}}} \gg 1\,, \tag{1}$$

with $\beta_0 = 11 - \frac{2}{3}n_f$, $\beta_1 = 51 - \frac{19}{3}n_f$ and n_f the number of light flavours.

To account for high-order effects one needs to start from the scheme for the definition of the running coupling. A physical definition [19] is given by the strength of the distribution for the emission of a soft gluon k off a colour singlet pair of a massless quark and antiquark of momenta $p, \bar{p}$. It is given by

$$dw_{p\bar{p}}(k) = C_F \frac{\alpha_s(k_t)}{\pi\, k_t^2} \frac{d^3k}{2\pi|\boldsymbol{k}|}\,, \qquad k_t^2 = 2\frac{(pk)(k\bar{p})}{(p\bar{p})}\,, \tag{2}$$

and corresponds to the coupling associated to the Wilson loop cusp anomalous dimension [20]. The relation to the $\overline{\text{MS}}$ coupling is known at three loops [21]. The argument of the coupling, the transverse momentum k_t relative to the emitting dipole, is obtained by using dispersive methods [12, 22] or, directly, by two-loop calculations [23]. In order to accurately describe soft emissions, the physical coupling with the argument in (2) is used in the Monte Carlo generators.

3.2 Coherence of Soft Gluons and Colour Connection

Successive soft gluon emission takes place into angular ordered regions with intensities related to the colour charges. In the large N_c limit these regions are identified by the parton colour connections. To explain this, one starts from the emission of a soft gluon k off a colour singlet $q\bar{q}$ pair, the dipole (2). This distribution has collinear singularities for $\theta_{pk} = 0$ or $\theta_{k\bar{p}} = 0$. Introducing the angular variable $\xi_{ij} = 1 - \cos\theta_{ij}$, one can isolate the two singular pieces and write

$$w_{p\bar{p}}(k) = \frac{(p\bar{p})}{(pk)(k\bar{p})} = \frac{1}{\boldsymbol{k}^2}\left(\frac{\Psi^{p}_{p\bar{p}}(k)}{\xi_{pk}} + \frac{\Psi^{\bar{p}}_{p\bar{p}}(k)}{\xi_{k\bar{p}}}\right), \; \Psi^{p}_{p\bar{p}}(k) = \tfrac{1}{2}\left(1 + \frac{\xi_{p\bar{p}} - \xi_{pk}}{\xi_{k\bar{p}}}\right) \tag{3}$$

and similarly for the function $\Psi^{\bar{p}}_{p\bar{p}}(k)$ associated to the singularity for $\xi_{k\bar{p}} = 0$. Performing the integration of $\Psi^{a}_{p\bar{p}}(k)$ over the azimuthal angle around a one has

$$\int \frac{d\phi_{ak}}{2\pi}\, \Psi^{a}_{p\bar{p}}(k) = \Theta(\xi_{p\bar{p}} - \xi_{ak})\,, \qquad a = p, \bar{p}\,. \tag{4}$$

This shows that the soft dipole distribution is made up of two collinear pieces, the one singular for k collinear to a ($\xi_{ak} = 0$) is (upon azimuthal averaging) bounded to a cone around a with opening half-angle $\theta_{p\bar{p}}$. Since the $q\bar{q}$ dipole is a colour singlet system, the p and $\bar{p}$ colour lines are "connected".

This coherent structure can be generalized to the soft emission of a gluon k off a colour singlet system made of any number of partons. Consider a $q\bar{q}\,g$ colour singlet of momenta $p, \bar{p}$ and q, respectively. The distribution is given by (for simplicity, we take also the gluon q to be soft)

$$w_{p\bar{p}g}(k) = w_{p\bar{p}}(q) \cdot \left(w_{pq}(k) + w_{q\bar{p}}(k) - \frac{1}{N_c^2} w_{p\bar{p}}(k)\right). \tag{5}$$

Splitting all dipole distributions as in (3), one can classify all collinear singularities in successive emissions within corresponding angular regions. One finds that the piece which is singular for k collinear to a (with $a = p, \bar{p}$ or q) is bounded to a cone around a with opening half-angle θ_{ab} with b the parton colour connected to a (recall that in the planar limit the gluon is equivalent to a quark–antiquark pair).

This angular ordered structure associated to colour connections at large N_c has been extended [24] to the $2 \to 2$ QCD hard processes needed for LHC and used in [2]. Beyond large N_c, the structure of soft radiation off the $2 \to 2$ hard QCD is quite more complex; it involves [25] rotation in the colour space for the hard matrix elements and includes Coulomb phase contributions. This is a very interesting contribution and would be nice if it could be included in a future Monte Carlo generator.

The distribution of a soft gluon k emitted off a colour singlet pair of massive quark and antiquark P and $\bar{P}$ is given by

$$W_{P\bar{P}}(k) = -\tfrac{1}{2}\left(\frac{P}{(Pk)} - \frac{\bar{P}}{(\bar{P}k)}\right)^2 = \frac{(P\bar{P})}{(Pk)(k\bar{P})} - \tfrac{1}{2}\frac{P^2}{(Pk)^2} - \tfrac{1}{2}\frac{\bar{P}^2}{(\bar{P}k)^2}, \quad (6)$$

with $(ij) = E_iE_j(1 - v_iv_j \cos\theta_{ij})$ and $v_i = \sqrt{1 - m_i^2/E_i^2}$. While in the massless case (3) the distribution is collinear singular for k parallel to the emitting charges, in the heavy quark case the collinear singularities are screened: distribution vanishes for k parallel to the heavy quark (or antiquark) P_a and the radiation is suppressed [26, 27] in the cone $\cos\theta_{ak} > v_a$.

The heavy quark screening is included into the Monte Carlo generators. One needs to avoid sharp cut-off around the heavy quark which, taken together with the angular limitations, would leave a *dead cone*, a phase space region without radiation.

3.3 Sudakov Form Factor and Jets

An important element in Monte Carlo generator is the probability that, in a hard process, a parton is not radiating within a given resolution, the Sudakov form factor. To introduce this quantity, consider the *inclusive distributions* (no particle momenta are measured but only energy flows) which are free from collinear and infrared singularities. Classical examples in e^+e^- are the jet-shape distributions $\Sigma(Q, V)$ with

$$V = \sum_i v(k_i)\,. \quad (7)$$

Here the sum runs over all particles in the final state (hadrons in the measurements and partons in the calculations). For $v(k)$ *linear* in the particle momentum, such jet-shape observables are collinear and infrared safe. Actually, individual Feynman diagrams for real emitted partons and virtual corrections are divergent but they are summed in such a way that, order by order, the infinities cancel [11] leaving finite results.

Collinear and infrared safe jet-shape distributions $\Sigma(Q, V)$ have a perturbative expansion with finite coefficients

$$\Sigma(Q,V) = \Sigma_0(Q,V)(1+\alpha_s(Q)\,c_1(V)+\alpha_s^2(Q)\,c_2(V)+\cdots)\,, \quad Q \gg \Lambda_{\rm QCD} \quad (8)$$

with $\Sigma_0(Q, V)$ the Born distribution and $c_i(V)$ finite functions of V expressed in terms of the quark, C_F, or gluon, C_A, colour charges. Actually, by inhibiting the radiation by taking $V \ll 1$, these coefficients are enhanced by powers of $\ln V$. A clever reshuffling of PT (perturbative) series, based on universal nature of soft and collinear radiation (factorization) results [12, 13] in the *exponentiated* answer of the Sudakov form factor $S(Q, V)$

$$\begin{aligned} &\Sigma(Q,V) = \Sigma_0(Q,V)\cdot S(Q,V)\,, \qquad S(Q,V) = e^{-\mathcal{R}(Q,V)}\,, \\ &\mathcal{R}(Q,V) = \sum_{n=1}^{\infty} \alpha_s^n(Q^2)\left(d_n \ln^{n+1} V + s_n \ln^n V + \cdots\right). \end{aligned} \quad (9)$$

The d_n series is referred to as double logarithmic (DL) and s_n as single logarithmic (SL). Reliable predictions for these distributions require the matching [28] of the exact finite order calculation (8) for finite V and the Sudakov resummation (9) for small V.

It is instructive to discuss the emergence of the powers of $\ln V$ in the Sudakov form factor $S(Q,V)$. They result from the incomplete cancellation of real and virtual effects. For $V \ll 1$ the *real* parton production is inhibited, one has $v(k) < V \ll 1$. Since the *virtual* PT radiative contributions remain unrestricted, the *divergences* do cancel in the region $v(k) < V$ leaving only virtual contributions for $v(k) > V$ which produce finite but logarithmically enhanced leftovers. The DL contributions originate from the fact that each gluon emission brings in at most two logarithms (one of collinear, another of infrared origin). This explains the first term $d_1 \ln^2 V$ while the rest of the DL series is generated simply by the presence of the running coupling (1). The SL contributions, are necessary to set the scale of the logarithms ($\ln^n cV = \ln^n V + n \ln c \ln^{n-1} V + \cdots$).

In conclusion, the Sudakov form factor $S(Q,V)$ corresponds to the probability that in e^+e^- the primary quark–antiquark pair remains without accompanying radiation up to resolution $Q_0 = VQ$ for small V.

To obtain the result (9) one uses the fact that the collinear and/or infrared enhanced contributions factories and are resummed by *linear* evolution equations of the DGLAP type. Therefore, after factorization of collinear and infrared singularities (including soft gluon coherence) QCD radiation appears as produced by "independent" gluon emission (bremsstrahlung). Gluon branching (into two gluons or quark–antiquark pair) enters only in reconstructing the running coupling (1) as function of transverse momentum. The fact that here the branching component does not contribute (within SL accuracy) can be understood as a result of real–virtual cancellations of singularities. Indeed, in the collinear limit, the transverse momentum of an emitted gluon is equal to the sum of transverse momenta of its decay products. Therefore, if one measures the total emitted transverse momentum, as in broadening for instance, it is enough to consider the contributions of primary bremsstrahlung gluons. Further branching does not contribute due to unitarity (real–virtual cancellation).

3.4 Structure and Fragmentation Functions

Moving to *less inclusive* measurements one faces infinities. The simple case involves fixing (measuring) momentum of a hadron, e.g. that of the initial proton in DIS (structure function) or of a final hadron (fragmentation function), they are functions of the Bjorken and Feynman variables, respectively

$$x_B = \frac{-q^2}{2(Pq)}, \qquad x_F = \frac{2(Pq)}{q^2}. \tag{10}$$

In DIS q is the large space-like momentum transferred from the incident lepton to the target nucleon P. In e^+e^- annihilation q is the time-like total incoming momentum and P the momentum of the final observed hadron.

In perturbative calculation, replacing the hadron with a parton, one has infinities, real and virtual contributions do not cancel. Soft divergences still cancel but collinear ones do not, making such observables not calculable at the parton level. These effects, however, turn out to be universal and, given a proper technical treatment, can be *factored out* [5] as non-perturbative inputs. What remains under control then is only the Q^2-dependence (scaling violation pattern). This fact is realized in the DGLAP evolution equation which needs, in order to be solved, an *initial condition* at a low virtuality Q_0. This corresponds to a parton resolution (or a factorized subtraction point), which absorbs all large distance divergences. Such "initial condition" cannot be computed by perturbative means and has to be provided by low-scale experimental data.

3.5 DGLAP Evolution Equation for DIS and e^+e^-

To derive the DGLAP evolution equation [14], one needs to study the phase space region leading to collinear singularities. The same Feynman diagrams are involved in the case of structure function (space-like) and fragmentation function (time-like). Therefore they can be studied simultaneously. First note that the Bjorken and Feynman variables (10) are mutually reciprocal: after the crossing operation $P \to -P$, one x becomes the inverse of the other (although in both channels $0 \leq x \leq 1$ thus requiring the analytical continuation).

Such a reciprocity property can be extended to the Feynman diagrams for the two processes and, in particular, to the contributions from mass-singularities. Consider, for DIS (S-case) and e^+e^- annihilation (T-case), the skeleton structure of Feynman graphs in axial gauge and the kinematical relation leading to the mass singularities

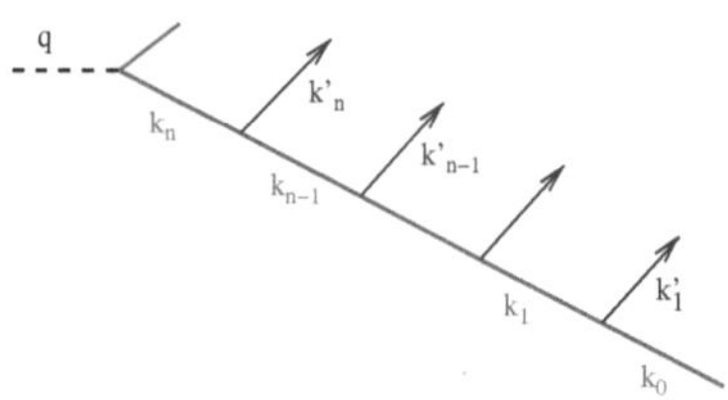

DIS or e^+e^- *skeleton graphs*

$$\frac{|k_i^2|}{k_{i,+}} = \frac{|k_{i-1}^2|}{k_{i-1,+}} + \frac{k'^2_i}{k'_{i,+}} + \frac{k_{i,+}k'_{i,+}}{k_{i-1,+}}\left(\frac{\boldsymbol{k}_{it}}{k_{i,+}} - \frac{\boldsymbol{k}'_{it}}{k'_{i,+}}\right)^2$$

Here $k'_1, \cdots k'_n$ are the outgoing parton systems (sub-jets). For space-like (S: $q^2 < 0$, k_0 entering) and time-like (T: $q^2 > 0$, k_0 outgoing) one has

$$S: \quad \frac{k_{i,+}}{k_{i-1,+}} \equiv z_i \qquad \text{and} \qquad T: \quad \frac{k_{i,+}}{k_{i-1,+}} \equiv z_i^{-1}. \tag{11}$$

The virtuality k_i^2 enters the denominators of the Feynman diagrams. In order for the transverse momentum integration to produce a logarithmic enhancement, the following conditions must be satisfied:

$$\frac{|k_{i-1}^2|}{k_{i-1,+}} \ll \frac{|k_i^2|}{k_{i,+}} \Rightarrow k_{i-1}^2 \ll |k_i^2|\, z_i^{\sigma}, \tag{12}$$

with $\sigma = -1$ for DIS and $\sigma = 1$ for e^+e^-. The same Feynman graphs are contributing and, going from S- to T-channel, the mass singularities are obtained by reciprocity: change z into $1/z$ and the momentum k from space-like to time-like. This fact is at the origin of the Drell–Levy–Yan relation [29] and Gribov–Lipatov [30] reciprocity, which has been largely used in order to obtain the time-like anomalous dimensions from the space-like ones [31, 32]. The ordering (12) in the inverse fluctuation time k^2/k_+ is well known, see for instance [33].

To make the Gribov–Lipatov reciprocity more clear, use the ordering (12) in the computation of the probability $D_\sigma(x, Q^2)$ to find a parton with longitudinal momentum fraction x and virtuality $|k^2|$ up to Q^2 with $\sigma = -1$ for the S-case and $\sigma = 1$ for the T-case. This ordering gives rise to the following *reciprocity respecting equation* [34]:

$$Q^2 \partial_{Q^2} D_\sigma(x, Q^2) = \int_0^1 \frac{dz}{z}\, P(z, \alpha_s)\, D_\sigma\left(\frac{x}{z}, Q^2 z^{\sigma}\right), \quad \sigma = \pm 1, \tag{13}$$

with the same parton splitting kernel $P(z, \alpha_s)$ in the S- or T-channel. This equation, derived simply from kinematical considerations, has been (partially) tested at two [31] and three loops [21, 34, 35].

The reciprocity respecting equation (13) is *non-local* since the derivative of $D_\sigma(x, Q^2)$ in the l.h.s. involves the distribution in the r.h.s. with all virtualities larger or smaller than Q for $\sigma = -1$ or $\sigma = +1$, respectively. For the use in a Monte Carlo generator, one needs to formulate (13) in terms of a *local* evolution equation, a Markov process. Formally this is easy to do: as a hard scale for the parton densities replace Q^2 with $\bar{Q}_+^2 = x\, Q^2$ in the T-case and, by reciprocity, with $\bar{Q}_-^2 = x^{-1}\, Q^2$ in the S-case. The physical meaning of these two different hard scales is well known from the studies of soft gluon coherence [12, 13, 33, 36]: in the T-case is related to the branching angle and in the S-case to the transverse momentum.

It is interesting to illustrate this. The fact that, in the T-case, the ordering variable is not the inverse fluctuation time k^2/k_+ (12) but rather the angle $k^2/k_+^2 \simeq k_t^2/k_+^2 \simeq \theta_k^2$, originates from cancellations [36] due to destructive interference in the region

$$T\text{-case:} \qquad z_i^2 k_i^2 < k_{i-1}^2 < z_i k_i^2\,, \tag{14}$$

thus leaving the angular ordered region $k_{i-1}^2 < z_i^2 k_i^2$. Using reciprocity ($z_i \to z_i^{-1}$) one has that in the S-case the cancelling region (14) becomes

$$S\text{-case:} \qquad |k_i^2| < |k_{i-1}^2| < z_i^{-1} |k_i^2|, \tag{15}$$

thus leaving the transverse momentum ordering $k_{t,i-1}^2 < k_{t,i}^2$. This agrees also, at small x, with the BFKL [37] leading order multi-parton kinematical region.

The cancellation in the region (15) has a well-known physical basis for small x. Consider (see the skeleton graph) the successive emissions $k_{i-2} \to k_{i-1} + k'_{i-1}$ and $k_{i-1} \to k_i + k'_i$ in the region $k_{i,+} \ll k_{i-1,+} \ll k_{i-2,+}$ giving the leading contribution for small x. These cancellations result from taking into account the emission of k_i off the partons k_{i-2} and k'_{i-1} in the region (15). Physically, the process can be viewed upon as an *inelastic* diffraction of the incident particle k_{i-2} in the external gluon field of transverse size of order k_{it}. In the kinematical region (15), the transverse size of the parton fluctuation $k_{i-2} \to k_{i-1} + k'_{i-1}$ is *smaller* than the resolution power of the probe, k_{it}^2. In these circumstances, the destructive interference between k_i interacting with the initial (k_{i-2}) and with the final state ($k_{i-1} + k'_{i-1}$) comes onto the stage. The cancellation under discussion is then equivalent to the general physical observation, due to V.N. Gribov, that inelastic diffraction vanishes in the forward direction.

To deal with very small x, one needs to resum at least all terms $\alpha_s^n \ln^n x$ as given by the BFKL equation [37], which cannot be accounted for by the collinear singularities resummation performed in the Monte Carlo codes. However, the evolution equation in [38] resums leading collinear and $\ln x$ terms (by enlarging the phase space and adding a non-Sudakov form factor) and allows Monte Carlo simulations [39] with the cost of generating events which need to be weighted.

4 Multi-gluon Soft Distributions

Collinear and infrared pieces of the multi-parton QCD distributions factories and can be reproduced by recurrence relations which can be formulated as a Markov branching process. This can be implemented into a Monte Carlo code and the simulation provides a "complete" description of the multi-parton emission in hard process.

I illustrate in detail the case in the *leading soft approximation.* Although important non-soft contributions that are included in a realistic Monte Carlos are here neglected, many important physical effects are well described, in particular, large angle soft emission (without collinear approximation). Moreover, in this approximation the path from multi-gluon soft amplitudes to Monte Carlo is simple to explain. The scheme of the presentation involves the following steps:

- Multi-gluon soft distributions. They are computed in the leading soft approximation and in the planar approximation.

- Recurrence relation for the multi-gluon soft distributions. This is obtained by introducing the *generating functional* for all multi-gluon distributions [12] and deriving the evolution equation. From the generating functional, one computes observables as it will be discussed in Sect. 4.2. For collinear and infrared safe observables such as jet-shape distributions, the cut-off contributes only with power corrections.
- Markov process and Monte Carlo implementation. Here one needs to include proper cutoff for collinear and infrared singularities. This will be discussed in the next section.
- from parton to hadron emission. This will be discussed in Sect. 6.

The starting point is the amplitude for the emission of n soft gluons $q_1, \cdots, q_n$ off a primary colour singlet $q\bar{q}$ pair of momentum $p, \bar{p}$. It is represented as a sum of Chan–Paton factors with the coefficients given by *colour-ordered amplitudes.* We consider the contribution with a single Chan–Paton factor (topological expansion [40])

$$\mathcal{M}_n(p\bar{p}q_1 \cdots q_n) = \sum_{\pi_n} \{\lambda^{a_{i_1}} \cdots \lambda^{a_{i_n}}\}_{\beta\bar{\beta}}\, M_n(pq_{i_1} \cdots q_{i_n}\bar{p})\,, \tag{16}$$

the sum is over the permutation π_n of colour indices, λ^a are the $SU(N_c)$ matrices in the fundamental representation. The softest emitted gluon q_m factorizes and one has [12, 42]

$$M_n(\cdots \ell\, m\, \ell' \cdots) = g_s\, M_{n-1}(\cdots \ell\ell' \cdots) \cdot \left(\frac{q_\ell^\mu}{(q_\ell q_m)} - \frac{q_{\ell'}^\mu}{(q_{\ell'} q_m)}\right). \tag{17}$$

The softest gluon is emitted by the two partons neighbouring in colour space. This approximation is accurate in the soft limit without any collinear approximation. From this factorized structure, one deduces a recurrence relation and computes all colour amplitudes in the soft limit. Summing over the polarization indices, the squared averaged colour amplitude is given, for the fundamental colour permutation, by

$$\begin{aligned} |M_n(pq_1 \cdots q_n\bar{p})|^2 &= |M_0|^2 (2g_s^2)^n\, W_{p\bar{p}}(q_1 \cdots q_n), \\ W_{p\bar{p}}(q_1 \cdots q_n) &= \frac{(p\bar{p})}{(pq_1) \cdots (q_n\bar{p})}\,. \end{aligned} \tag{18}$$

This very simple result for the square amplitude is valid for any energy ordering and depends only on the colour ordering. Note that here one takes the square of the same colour-ordered amplitude. Indeed $M_n(\pi'_n)M_n^*(\pi_n)$ with π_n and π'_n two different colour permutations cannot be expressed in a closed form for any n. On the other hand, contributions from different permutations enter the calculation of the averaged squared amplitude $|\mathcal{M}_n|^2$. A close expression for this distribution for any n is obtained only in the planar approximation [41]. To see this observe that

$$\mathrm{Tr}(\lambda_{\pi_n}\,\lambda_{\pi_n^T}) = 2C_F\left(\frac{N_c}{2}\right)^n\left(1-\frac{1}{N_c}\right)^{n-1}, \tag{19}$$

with $\lambda_{\pi_n} = \{\lambda^{a_1}\cdots\lambda^{a_n}\}$ and $\lambda_{\pi_n^T} = \{\lambda^{a_n}\cdots\lambda^{a_1}\}$. Taking instead two different colour permutations one has that $\mathrm{Tr}(\lambda_{\pi'_n}\,\lambda_{\pi_n^T})$ is suppressed at least by $1/N_c^2$. Therefore, only in the planar approximation one can use the simple result in (18) and obtains [12]

$$|\mathcal{M}_n|^2 = \frac{\sigma_0}{n!}\,(N_c g_s^2)^n \sum_{\pi_n} W_{p\bar{p}}(q_{i_1}\cdots q_{i_n})\,, \tag{20}$$

where $\sigma_0 = 2C_F|M_0|^2$ and symmetrization has been taken into account.

The distributions (18) contain the leading infrared singularities: for any colour permutation one has $W_{p\bar{p}} \sim (\omega_1\cdots\omega_n)^{-2}$ with ω_i the energy of gluon q_i. They contain also the leading collinear singularities for $\theta_{ij}=0$ with ij two partons neighbouring in colour (thus there are up to n collinear singularities).

An alternative way to obtain the the multi-gluon colour amplitude is based on the helicity techniques [43]. For $q\bar{q}$ with $+$ and $-$ polarization, the leading soft contribution is obtained when all gluons have $+$ helicities and the recurrence relation (17) reads (for opposite helicities, the result is the complex conjugate one)

$$\begin{aligned} M_n(\cdots\ell\, m\, \ell'\cdots) &= g_s\, M_{n-1}(\cdots\ell\ell'\cdots)\cdot\frac{\langle q_\ell q_{\ell'}\rangle}{\langle q_\ell q_m\rangle\,\langle q_m q_{\ell'}\rangle}\,,\\ \langle qq'\rangle &= \sqrt{2qq'}\cdot e^{i\phi_{qq'}}\,, \end{aligned} \tag{21}$$

with q_m the softest gluon, z the longitudinal direction and the phase

$$e^{i\phi_{qq'}} = \sqrt{\frac{q_+q'_+}{2qq'}}\left(\frac{\mathbf{q}_t}{q_+}-\frac{\mathbf{q}'_t}{q'_+}\right), \qquad \mathbf{q}_t = q_x + iq_y\,. \tag{22}$$

The solution of this recurrence for the amplitude is very simple; it is the same for any energy ordering and depends only on the colour ordering. For the fundamental permutation, one has

$$M_n(pq_1\cdots q_n\bar{p}) = g_s^n\, M_0\frac{\langle p\bar{p}\rangle}{\langle pq_1\rangle\cdots\langle q_n\bar{p}\rangle}\,, \tag{23}$$

with squared amplitude given by (18). This shows the well-known result that non-planar contributions, obtained from $M_n(\pi_n)\cdot M_N^*(\pi'_n)$ for two different colour orderings, have the same soft singularities but reduced number of collinear singularities.

4.1 Virtual Correction, Generating Functional and Evolution

To compute observables one needs to supplement the multi-gluon soft distributions (20) with the related virtual corrections. For infrared and collinear safe observables, such as jet-shape distributions, the infrared and collinear singularities in (18) has to be cancelled by corresponding singularities in virtual corrections. One way to compute the virtual corrections, at the same level of accuracy in the soft limit as for real emission contribution, consists of performing the integration over the virtual gluon energy by the Cauchy method and then taking the soft limit for the virtual gluon. This way one also regularizes the ultraviolet divergences by neglecting the divergent contribution from the contour at the infinity of the complex energy plane. By properly choosing a constant, this regularization corresponds to the physical scheme in (1). The virtual corrections so computed can be included into the generating functional for the multi-gluon soft distributions. The result of this study not only gives the relevant virtual corrections but, due to the simple structure of (20) in the planar approximation, gives the branching structure of multi-gluon soft emission leading to the Monte Carlo generator.

Consider the soft distribution $d\sigma_{ab}^{(n)}$ for the emission of n gluons off a colour singlet dipole ab (thus one generalizes the primary dipole $p\bar{p}$ to a general dipole with a and b in arbitrary directions). For each emitted soft gluon q_i one introduces a source function $u(q_i)$ and defines the *generating functional* as

$$G_{ab}[E,u] = \sum_n \frac{1}{n!} \int \frac{d\sigma_{ab}^{(n)}}{\sigma_{ab}^{\rm tot}} \prod_i u(q_i)\,, \tag{24}$$

with $E=Q/2$ the hard scale. This functional depends on the directions a and b of the primary dipole. By setting all $u(q_i) = 1$ one has $G_{ab}[E,1] = 1$. Using (20), one has the *real emission* contribution for the generating functional

$$G_{ab}^{\rm real}[E,u] = \sum_n \int \prod_i \left\{ \bar{\alpha}_s u(q_i)\, \frac{d\Omega_{q_i}}{4\pi}\, \omega_i d\omega_i \Theta(E-\omega_i) \right\} \cdot W_{ab}(q_1 \cdots q_n), \tag{25}$$

with $\bar{\alpha}_s = N_c\alpha_s/\pi$. Here one neglects $1/N_c^2$ corrections (planar limit) and uses the soft approximation for the phase space $\omega_i \ll E$. Symmetry of the phase space is used. The condition $G_{ab}[E,1] = 1$ must be satisfied only after including the virtual corrections. To include them, we construct the evolution equation for the generating functional. To this end, we use the fact that the very simple expression (18) has the following factorization property:

$$W_{ab}(q_1 \cdots q_n) = w_{ab}(q_\ell) \cdot W_{a\ell}(q_1 \cdots q_{\ell-1}) \cdot W_{\ell b}(q_{\ell+1} \cdots q_n)\,, \tag{26}$$

with q_ℓ one of the soft gluons and $w_{ab}(q)$ the dipole distribution (3). Taking q_ℓ as the hardest (soft) gluon and differentiating (25) with respect to E, thus setting $\omega_\ell = E$, one obtains [44]

$$E\partial_E G_{ab}[E,u] = \int \frac{d\Omega_q}{4\pi}\, \frac{\bar{\alpha}_s\, \xi_{ab}}{\xi_{aq}\xi_{qb}} \Big\{ u(q)\, G_{aq}[E,u] \cdot G_{qb}[E,u] - G_{ab}[E,u] \Big\}, \tag{27}$$

with $\xi_{ij} = 1-\cos\theta_{ij}$. The negative term in the integrand originates from the virtual corrections obtained via Cauchy integration as mentioned before.

Since they are evaluated within the same soft approximation used for the real contributions, at the inclusive level they cancel against the real contributions giving the correct constraint $G_{ab}[E,1]=1$. Both the real emission (first term in the integrand) and the virtual correction (second term) are collinear and infrared singular. For inclusive observables, (i.e. for suitable sources $u(q)$) these singularities cancel. This evolution equation accounts for coherence of soft gluon radiation [12, 13].

4.2 Observables in the Soft Limit

Using $G_{ab}[E,u]$ one obtains all inclusive distributions in the soft limit. No collinear approximations are involved in (20); therefore, the functional $G_{ab}[E,u]$ gives quantities that involves also large angle soft emission. Let me first recall some observables which are collinear singular around the primary partons a and b.

Collinear Observables

The simplest one is the multiplicity of soft gluons with resolution Q_0. Taking $u(q)=u$ this observable is defined as, see (24),

$$n_{ab}(E) = \partial_u G_{ab}(E,u)\Big|_{u=1} = \sum_n n \frac{\sigma_{ab}^{(n)}}{\sigma_{ab}^{\rm tot}}\,. \tag{28}$$

It is easy to derive from (27) the well-known result [36] for the multiplicity

$$n_{ab}(E) \simeq n_{ab}^{(0)} \exp\left\{\frac{4\pi}{\beta_0}\sqrt{\frac{2N_c}{\pi\alpha_s(E)}}\right\}, \tag{29}$$

with $n_{ab}^{(0)}$ the non-perturbative initial condition. Similarly, one derives the fragmentation function $D_{ab}(x,E)$ by taking the source $u(q)=u(x)$ with x the soft gluon energy fraction

$$D_{ab}(x,E) = \frac{\delta}{\delta u(x)} G_{ab}[E,u]\Big|_{u(x)=1}\,. \tag{30}$$

Soft gluon coherence here is shown by a depletion of radiation [12, 13] at small x.

Observables at Large Angle

The simplest case is the distribution discussed in [45] of heavy systems of mass $\mathcal{M}$ emitted in e^+e^- at large angle $\rho = \frac{1}{2}(1-\cos\theta)$ and small velocity. The heavy system (typically a heavy $q\bar{q}$ system) originates from a gluon in the cascade. The collinear singularities are screened by $\mathcal{M}$ so this distribution is finite and given by a function of the SL quantity

$$\tau = \int_{\mathcal{M}}^{E} \frac{dq_t}{q_t} \bar{\alpha}_s(q_t)\,. \tag{31}$$

It is interesting that this distribution $I(\rho, \tau)$ satisfies an equation with a structure similar to the BFKL equation [45] and then its asymptotic behaviour in τ involves the BFKL characteristic function. One has

$$I(\rho, \tau) \sim \frac{e^{4 \ln 2\, \tau}}{\tau^{3/2}} \cdot \frac{\ln \rho_0/\rho}{\sqrt{\rho}} e^{-\frac{\ln^2 \rho_0/\rho}{2D\tau}}\,, \qquad D = 28\,\zeta(3)\,. \tag{32}$$

The functional $G_{ab}[E, u]$ is suited to give the distributions in the energy emitted away from jets. Such distributions do not have collinear singularities, but only infrared ones. An example in e^+e^- is the distribution in energy recorded *outside* a cone θ_{in} around the thrust (this is a typical "non-global" jet observable [46]):

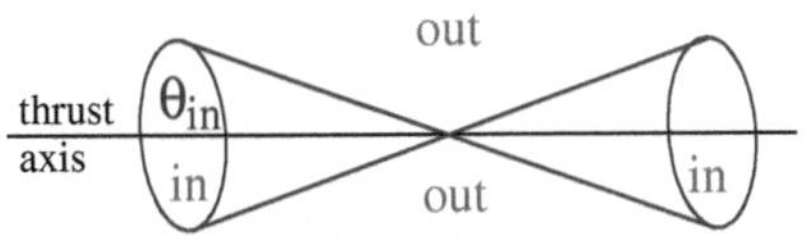

$$\Sigma(E, E_{\text{out}}) = \\ = \sum_n \int \frac{d\sigma_n(E)}{\sigma_{\text{tot}}}\, \Theta\left(E_{\text{out}} - \sum_{\text{out}} k_{ti}\right).$$

Since the jet region is excluded, there are no collinear singularities to SL accuracy and the resummed PT contributions come from large angle soft emission. Here resummation is complex but informative. It brings information on the QCD radiation between jets, a region interesting for understanding colour neutralization among jets.

It is interesting to discuss this quantity in some detail since it illustrates the structure of (27). First observe that the distribution depends on E and E_{out} through the SL function τ given by (31) with $\mathcal{M} \to E_{\text{out}}$. To obtain $\Sigma(\tau)$ from $G_{ab}[E, u]$, one takes $u(q) = 0$ away from jets and $u(q) = 1$ inside the jet region. From (27), one derives the evolution equation [44]

$$\partial_\tau \Sigma_{ab}(\tau) = -s_{ab} \Sigma_{ab}(\tau) + \int_{\text{in}} \frac{d\Omega_q}{4\pi} \frac{\bar{\alpha}_s\, \xi_{ab}}{\xi_{aq}\xi_{qb}} \Big\{ \Sigma_{aq}(\tau) \cdot \Sigma_{qb}(\tau) - \Sigma_{ab}(\tau) \Big\}, \tag{33}$$

with s_{ab} related to the Sudakov form factor

$$S(\tau) = e^{-\tau\, s_{ab}}\,, \qquad s_{ab} = \int_{\text{out}} \frac{d\Omega_q}{4\pi} \frac{\xi_{ab}}{\xi_{aq}\xi_{qb}} \sim \ln \theta_{\text{in}}^{-1}\,. \tag{34}$$

Equation (33) has a bremsstrahlung (first) and branching (second term) components:

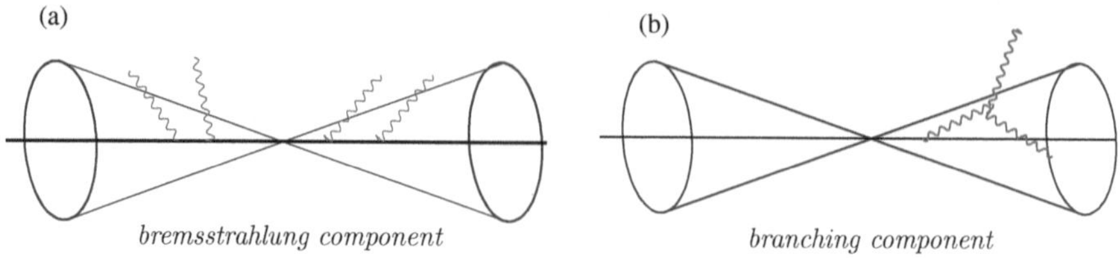

The *bremsstrahlung component* resums contributions from gluons emitted in the recorded region outside the cone. These contributions are the only ones present for the global jet observables considered in the previous subsection. Here, since the collinear singularities are screened by the cone $\theta_{\rm in}$, the Sudakov form factor is a SL function.

The *branching component* resums contributions from gluons emitted inside the jet region. These gluons need to branch in order to generate decay products entering the recorded region. Here real–virtual cancellation is incomplete and virtual enhanced contributions are dominating thus leading to a strong suppression of the distribution which asymptotically turns out to be Gaussian in τ.

The Monte Carlo generator [2] resums only collinear singularities; therefore, it does not fully resum soft emissions at large angles although phenomenologically, it turns out [47] that the most important pieces are correctly reproduced due to soft gluon coherence.

5 Monte Carlo Simulation for Soft Emission

The evolution equation (27) can be formulated as a Markov process and then numerically solved. This Monte Carlo procedure has been introduced in [46] to study non-global distributions. A similar procedure based on dipole branching is used in the Monte Carlo generator [4].

To construct a Monte Carlo generator from (27) one splits the real and virtual corrections. To do so, it is necessary to introduce a cut-off Q_0 in transverse momentum (the argument of $\alpha_{\rm s}$) giving the Sudakov form factor

$$\ln S_{ab}(E) = -\int_{Q_0}^{E} \frac{d\omega_q}{\omega_q} \int \frac{d\Omega_q}{4\pi} \frac{\bar{\alpha}_s\,\xi_{ab}}{\xi_{aq}\xi_{qb}} \cdot \theta(q_{tab} - Q_0), \qquad q_{tab}^2 = 2\,\omega_q^2 \,\frac{\xi_{aq}\xi_{qb}}{\xi_{ab}}, \quad (35)$$

which is the solution of (27) with the real emission piece neglected. Here q_{tab} is the transverse momentum of q with respect to the ab-dipole. Then the evolution equation (27) can be integrated to give (the cut-off Q_0 dependence is implicit)

$$G_{ab}[E] = S_{ab}(E, Q_0) + \int d\mathcal{P}_{ab}(E, \omega_q, \Omega_q)\, u(q)\, G_{aq}[\omega_q, u] \cdot G_{qb}[\omega_q, u]\,, \quad (36)$$

where one has introduced the probability for dipole branching: $(ab) \to (aq)\,(qb)$

$$d\mathcal{P}_{ab}(E,\omega_q,\Omega_q) = \left\{\frac{d\omega_q}{\omega_q}\,\frac{S_{ab}(E)}{S_{ab}(\omega_q)}\right\}\left\{\frac{d\Omega_q}{4\pi}\,\frac{\bar{\alpha}_s\,\xi_{ab}}{\xi_{aq}\xi_{qb}}\right\}\cdot\theta(q_{tab}-Q_0)\,. \tag{37}$$

To see how this could be used in a Monte Carlo simulation one writes $d\mathcal{P}_{ab}(E,\omega,\Omega)$ in the equivalent form (the bound $q_{tab}>Q_0$ is implicit)

$$d\mathcal{P}_{ab}(E,\omega,\Omega) = dr_{ab}(E,\omega)\cdot dR_{ab}(\Omega) \tag{38}$$

with

$$\begin{aligned} r_{ab}(E,\omega_q) &= \frac{S_{ab}(E)}{S_{ab}(\omega_q)}\,, & \int dr_{ab}(E,\omega_q) &= 1 - S_{ab}(E) \\ dR_{ab}(\Omega_q) &= \mathcal{N}_{ab}\,\frac{d\Omega_q}{4\pi}\,\frac{\bar{\alpha}_s\,\xi_{ab}}{\xi_{aq}\xi_{qb}}\,, & \int dR_{ab}(\Omega_q) &= 1\,. \end{aligned} \tag{39}$$

The integral of the branching probability gives

$$\int d\mathcal{P}_{ab}(E,\omega,\Omega) = 1 - S_{ab}(E)\,, \tag{40}$$

and this shows that the Sudakov factor $S_{ab}(E)$ gives the probability for not emitting a gluon within the resolution Q_0 in q_{tab}.

The probability distribution $d\mathcal{P}_{ab}(E,\omega,\Omega)$ can be used to generate Monte Carlo events distributed according to QCD in the soft and planar approximation. Using sets of random numbers $0<\rho<1$, the procedure is as follows:

1. take the ab-dipole with the energy scale E and compare the Sudakov factor $S_{ab}(E)$ with ρ. If $\rho < S_{ab}(E)$ then the ab-dipole does not emit any soft gluon within the resolution. In the opposite case, the dipole is emitting a soft gluon with energy ω_q given by solving the equation $\rho = r_{ab}(E,\omega_q)$;
2. obtain the direction Ω_q by sampling the distribution $dR_{ab}(\Omega_q)$. At this point, from the ab-dipole one has generated two dipoles: aq and qb, both at the new energy scale ω_q;
3. repeat the procedure for each new generated dipole till no dipole emits any more within the resolution.

At the end of this procedure, one is left with a Monte Carlo event: a collection of emitted soft gluons $q_1\cdots q_n$ together with the primary partons a,b. These events are distributed with the QCD probability so they can be used to compute any soft distribution as discussed in Sect. 4.2.

Such a Monte Carlo simulation, based on *evolution equation in energy*, is then a successive emission of softer and softer gluons. Angles are given by the dipole distribution (3) so they are ordered (upon azimuthal average) and coherence is automatically implemented.

In order to obtain a realistic simulation one needs to overcome the soft approximation, that is, to take into account the recoil in the emission and the non-soft pieces of the gluon splitting function (only the singular pieces are present in (27))

$$P_{g\to gg}(z) = N_c \left(\frac{1}{z} + \frac{1}{1-z} + z(1-z) - 2 \right). \tag{41}$$

Similarly, one needs to account also for the quark branching channels. All these points are accounted in the present realistic Monte Carlo generators. Their basis is an *evolution equation in angle* rather than in energy (as (27)). However this implies that one considers collinear approximations in the emission thus soft radiation at large angles are not fully accounted for.

6 From Partons to Hadrons

The above description of the Monte Carlo code refers to the generation of events with emission of partons (possibly together with non-QCD particles) which, due to the presence of collinear and infrared singularities, requires a cut-off Q_0. The main questions are then: how to go from partons to hadrons and how much a phenomenological hadronization model affects and distorts the QCD radiation generated perturbatively. A suggestion on hadronization models which do not substantially modify the peturbative radiation is provided by preconfinement [9].

6.1 Preconfinement

The basis is again the Sudakov function, which suppress the probability of "non-emitting". Consider, in the planar approximation, two colour connected partons emitted in a hard collision at scale Q and with resolution Q_0. Colour connection means that the quark colour line of one parton ends into the antiquark colour line of the other parton (in the planar approximation a gluon could be, from the colour point of view, described as a pair of $q\bar{q}$ colour lines). Thus no gluons are emitted within the resolution Q_0 by this colour line and a Sudakov form factor arises which forces the two colour connected partons to form a system of mass of order Q_0 (even for very large Q). The system of the quark and antiquark in question forms a colour singlet of small mass. Although this is not yet an indication of confinement (the colour system should be localized in space), such a preconfinement property suggests that any hadronization models that associates hadrons to colour connected partons would not distort the perturbative structure of the QCD radiation: parton and hadron flows are similar within the resolution Q_0. Preconfinement is then related to the property of *local hadron–parton duality* [10], which has been phenomenologically well tested.

6.2 Power corrections

Other non-perturbative effects are the power corrections to the observables. They result from the non-convergence the PT expansions even if the coefficients are finite as in (8) and (9). As a consequence, all PT predictions are affected by corrections in powers of $\Lambda_{\rm QCD}/Q$ with coefficients determined by NP (non-perturbative) effects. An important NP effect, present in short distance quantities, is that the running coupling is involved at *any* scale smaller than Q. For example, the average value of V in (7) is given by an integral of the type

$$\langle V \rangle = \int_0^Q \frac{dk_t}{k_t} \alpha_{\rm s}(k_t) \cdot \mathcal{V}(k_t/Q) = v_1\, \alpha_{\rm s}(Q) + v_2\, \alpha_{\rm s}^2(Q) + \cdots , \qquad (42)$$

where the virtual momentum k_t in the Feynman diagrams runs into the large distance region (although the observable is dominated by short distance physics). Since the observable is collinear and infrared finite, for $k_t \to 0$ the Feynman integrand is regular ($\mathcal{V}(k_t/Q) \sim k_t/Q$) so that the integral is finite, apart from the presence of $\alpha_{\rm s}(k_t)$ that enters the confinement region. Mathematically, this is reflected into the fact that, although all PT coefficients in $\alpha_{\rm s}(Q)$ are finite, the expansion is non-convergent [48] (renormalon singularity).

The fact that the running coupling entering the NP region is at the origin of the leading power correction can be checked phenomenologically. From the study of jet-shape observable one finds [49] that, within 10 – 20%, the power corrections are described by the same parameter accounting for the running coupling in the NP region. In the Monte Carlo generators, one sets a cut-off Q_0 in the argument of the coupling and this does bring in these physically relevant power corrections at the perturbative – parton – stage. Instead, power behaving contributions to jet shapes arise at the hadronization level [51].

6.3 Underlying event

Another important NP component in the Monte Carlo for LHC is the presence of radiation besides the one emitted in the hard event. This is typically around the beams as for the peripheral interactions (events at low E_T). Perturbative QCD does not provide indication for this component. Thus there are various models which needs to be studied [50] at the Tevatron together with the extrapolation at LHC.

7 Conclusion

What I have discussed shows that the Monte Carlo generators involves the entire *Summa* of hard QCD results and provide a framework for many future QCD and non-QCD studies. The general attempts to improve the Monte

Carlo generators go in the directions of making the quantitative predictions both more reliable (by adding new theoretical QCD results and phenomenological studies) and more general (by including also electroweak and beyond the standard model physics). As far as the first direction, I have mentioned the works made to include in the Monte Carlo generator the known exact higher-order distributions [8]. As also mentioned, it is interesting to include into the present generators reliable predictions on large angle soft emission (see Sect. 4.2). This would require also the need to account for non-planar corrections by studying colour rotations involved in the colour structure of ensembles of more than three hard partons (see [25]).

The three key elements in a Monte Carlo generator for jet emissions are the QCD factorization properties, the branching algorithm and the procedure for converting partons into hadrons. As I have mentioned, Gabriele Veneziano has either contributed to or started each of these three key developments: The Monte Carlo generators are based on factorization of QCD collinear singularities [5]. Jet calculus [15] leads to the evolution equation for the generating functional for multi-parton distributions which can be formulated as a Markov process. Moreover, the preconfinement property [9] is at the basis of hadronization models that do not destroy the QCD radiation structure.

Acknowledgements

In addition to Gabriele, I am grateful to the many colleagues who shared with me the beauty of QCD and in particular to Bryan Webber, we undertook the risk of conveying incomplete theoretical concepts and results into an event generator, and to Marcello Ciafaloni, Yuri Dokshitzer and Al Mueller, for many discussions during the construction of the original Monte Carlo generator.

References

1. D.J. Gross, F. Wilczek: Phys. Rev. Lett. **30**, 1343 (1973);
   H.D. Politzer: Phys. Rev. Lett. **30**, 1346 (1973)
2. G. Marchesini, B.R. Webber: Nucl. Phys. B **238**, 1 (1984); Nucl. Phys. B **310**, 461 (1988)
   G. Marchesini, B.R. Webber, G. Abbiendi, I.G. Knowles, M.H. Seymour, L. Stanco: Comput. Phys. Commun. **67**, 465 (1992);
   G. Corcella, I.G. Knowles, G. Marchesini, S. Moretti, K. Odagiri, P. Richardson, M.H. Seymour, B.R. Webber: JHEP **0101**, 010 (2001)
3. T. Sjöstrand: Comput. Phys. Commun. **82**, 74 (1994)
4. L. Lönnblad, Comput. Phys. Commun. **71**, 15 (1992)
5. D.Amati, R.Petronzio, G. Veneziano: Nucl. Phys. B **140**, 54 (1978); Nucl. Phys. B **146**, 29 (1978);
   R.K. Ellis, H. Georgi, M. Machacek, H.D. Politzer, G.G. Ross: Nucl. Phys. B **152**, 285 (1979);

S. Libby, G. Sterman: Phys. Rev. D **18**, 3252 (1978);
A.H. Mueller: Phys. Rev. D **18**, 3705 (1978);
C.T. Sachrajda: Phys. Lett. B **73**, 185 (1978); Phys. Lett. B **76**, 100 (1978)
6. R.K. Ellis, W.J. Stirling, B.R. Webber: Camb. Monogr. Part. Phys. Nucl. Phys. Cosmol. **8**, 1 (1996)
7. C. Buttar et al.: *Les Houches physics at Tev colliders 2005*, hep-ph 0604120
8. S. Frixione, B.R. Webber: JHEP **0206**, 029 (2002); S. Frixone, P. Nason, B. R. Webber: JHEP **0308**, 007 (2003); P. Nason, G. Ridolfi: JHEP **0608**, 077 (2006)
9. D. Amati, G. Veneziano: Phys. Lett. B **83**, 87 (1979)
10. Ya.I. Azimov, Y.L. Dokshitzer, V.A. Khoze, S.I. Troian: Z. Phys. C **2**, 65 (1985)
11. T.D. Lee, M. Nauenberg: Phys. Rev. **133**, B1549 (1964);
T. Kinoshita: J. Math. Phys. **3**, 650 (1962)
12. A. Bassetto, M. Ciafaloni, G. Marchesini: Phys. Rep. **100**, 201 (1983)
13. Y.L. Dokshitzer, V.A. Khoze, S.I. Troian, A. H. Mueller: Rev. Mod. Phys. **60**, 373 (1988); *Basics of Perturbative QCD* (Ed. Frontieres, Gif-sur-Yvette, France, 1991)
14. V.N. Gribov, L.N. Lipatov: Sov. J. Nucl. Phys. **15**, 438(1972);
G. Altarelli, G. Parisi: Nucl. Phys. B **126**, 298 (1977);
Y. L. Dokshitzer: Sov. Phys. JETP **46**, 641 (1977)
15. K. Konishi, A. Ukawa, G. Veneziano: Nucl. Phys. B **157**, 45 (1979); Phys. Lett. B **78**, 243 (1978)
16. G.C. Fox , S.Wolfram: Nucl. Phys. B **168**, 285 (1980)
17. R. Odorico: Nucl. Phys. B **172**, 157 (1980)
18. F. Paige, S. Protopopescu: *Supercollider Physics*, ed. by D. Soper (World Scientific, Singapore, 1986)
19. S. Catani, B.R. Webber, G. Marchesini: Nucl. Phys. B **349**, 635 (1991)
Yu.L. Dokshitzer, V.A. Khoze, S.I. Troian: Phys. Rev. D **53** 89 (1996)
20. G.P. Korchemsky: Mod. Phys. Lett. A **4**, 1257 (1989);
G.P. Korchemsky, G. Marchesini: Nucl. Phys. B **406**, 225 (1993)
21. A. Vogt, S. Moch, J.A.M. Vermaseren: Nucl. Phys. B **691**, 129 (2004); Nucl. Phys. B **688**, 101 (2004)
22. D. Amati, A. Bassetto, M. Ciafaloni, G. Marchesini, G. Veneziano: Nucl. Phys. B **173**, 429 (1980)
23. Y. L. Dokshitzer, G. Marchesini, G. Oriani: Nucl. Phys. B **387**, 675 (1992);
Y. L. Dokshitzer, A. Lucenti, G. Marchesini, G.P. Salam: Nucl. Phys. B **511**, 396 (1998), Erratum-ibid. B **593**, 729 (2001)
24. R.K. Ellis, G. Marchesini, B.R. Webber: Nucl. Phys. B **286**, 643 (1987), Erratum-ibid. B **294**, 1180 (1987)
25. N. Kidonakis, G.Sterman: Phys. Lett. B **387**, 867 (1996); Nucl. Phys. B **505**, 321 (1997);
N. Kidonakis, G. Oderda, G. Sterman: Nucl. Phys. B **531**, 365 (1998);
G. Oderda: Phys. Rev. D **61**, 014004 (2000);
R. Bonciani, S. Catani, M. Mangano, P. Nason: Phys. Lett. B **575**, 268 (2003);
A. Banfi, G.P. Salam, G. Zanderighi: Phys. Lett. B **584**, 298 (2004);
Yu.L. Dokshitzer, G. Marchesini: Phys. Lett. B **631**, 118 (2005); JHEP **0601**, 007 (2006)
26. G. Marchesini, B.R. Webber: Nucl. Phys. B **330**, 261 (1990)
27. Yu.L. Dokshitzer, V.A. Khoze S.I. Troian: Phys. Rev. D **53** 89 (1996)
28. S. Catani, G. Turnock, B.R. Webber, L. Trentadue: Phys. Lett. B **263**, 491 (1991)

29. S.D. Drell, D.J. Levy, T.-M. Yan: Phys. Rev. D , 1035 (1970); Phys. Rev. D **1**, 1617(1970)
30. V.N. Gribov, L.N. Lipatov: Sov. J. Nucl. Phys. **1**, 438 (1972)
31. G. Curci, W. Furmanski, R. Petronzio: Nucl. Phys. B **175**, 27 (1980)
32. M. Stratmann, W. Vogelsang: Nucl. Phys. B **496**, 41 (1997)
33. S. Catani, M. Ciafaloni: Phys. Lett. B **150**, 379 (1985);
S. Catani, M. Ciafaloni, G. Marchesini: Nucl. Phys. B **264**, 588 (1986); Phys. Lett. B **168**, 284 (1986)
34. Yu.L. Dokshitzer, G. Marchesini, G.P. Salam: Phys. Lett. B **634**, 504 (2006)
35. A. Mitov, S. Moch, A. Vogt: Phys. Lett. B **638**, 61 (2006)
36. A.H. Mueller: Phys. Lett. B **104**, 161 (1981);
B.I. Ermolayev, V.S. Fadin: JETP Lett. **33**, 285 (1981);
A. Bassetto, M. Ciafaloni, G. Marchesini, A.H. Mueller: Nucl. Phys. B **207**, 189 (1982);
Yu.L. Dokshitzer, V.S. Fadin, V.A. Khoze: Z. Phys. C **15**, 325 (1983); Z. Phys. C **18**, 37 (1983)
37. I.I. Balitsky, L.N. Lipatov: Sov. J. Nucl. Phys. **28**, 822 (1978);
E.A. Kuraev, L.N. Lipatov, V.S. Fadin: Sov. Phys. JETP **45**, 199 (1977)
38. M. Ciafaloni: Nucl. Phys. B **296**, 49 (1988);
S. Catani, F. Fiorani, G. Marchesini: Phys. Lett. B **234**, 339 (1990); Nucl. Phys. B **336**, 18 (1990)
39. G. Marchesini, B.R. Webber: Nucl. Phys. B **349**, 617 (1991); Nucl. Phys. B **386**, 215 (1992);
H. Jung, G.P. Salam: Eur. Phys. J. C **19**, 351 (2001)
40. G. Veneziano: Nucl. Phys. B **117**, 519 1976
41. G.'t Hooft: Nucl. Phys. B **72**, 461 (1974)
42. F. Fiorani, G. Marchesini, L. Reina: Nucl. Phys. B **309**, 439 (1988)
43. S. J. Parke, T. R. Taylor: Phys. Rev. Lett. **56**, 2459 (1986);
M.L. Mangano, S.J. Parke: Phys. Rep. **200**, 301 (1991)
44. A. Banfi, G. Marchesini, G. Smye: JHEP **0208**, 006 (2002)
45. G. Marchesini, A.H. Mueller: Phys. Lett. B **575**, 37 (2003);
see also G. Marchesini, E. Onofri: JHEP **0407**, 031 (2004)
46. M. Dasgupta, G.P. Salam: Phys. Lett. B **512**, 323 (2001); JHEP **0203**, 017 (2002)
47. A. Banfi, G. Corcella, M. Dasgupta: hep-ph 0612282
48. M. Beneke: Phys. Rep. **317**, 1 (1999)
49. For a review see M. Dasgupta, G.P. Salam: J. Phys. G **30** R143 (2004);
R.W. Jones, M. Ford, G.P. Salam, H. Stenzel, D. Wicke: JHEP **0312**, 007 (2003)
50. D. Acosta et al.: Phys. Rev. D **70** 072002 (2004);
see also G. Marchesini, B.R. Webber: Phys. Rev. D **38**, 3419 (1988)
51. B.R. Webber: Phys. Lett .B **339**, 148 (1994);
Y.L. Dokshitzer, B.R. Webber: Phys. Lett. B **352**, 451 (1995)

Fracture Functions

L. Trentadue

Dipartimento di Fisica, Università di Parma, and INFN, Gruppo Collegato di Parma, Parma, Italy
luca.trentadue@cern.ch

Abstract. We present a review of the fracture functions idea. Starting from the original motivations we examine the theoretical developments intervened and some of the phenomenological outputs. Further future applications are also envisaged.

1 Introduction and Motivations

Deep inelastic scattering (DIS) has played a crucial role in the hast four decades for the comprehension of the inner structure of the hadronic interactions. Already from the starting, from the parton model [1] interpretation of the SLAC experiments, it has represented an inavoidable test of the continuously growing inspection of the high-energy experiments and a benchmark for the theoretical description of the most intimate features of the strong interactions dynamics. Quantum chromodynamics (QCD) [2], as the theoretical framework for strong interactions, and the discovery of the asymptotic freedom [3] have given rise to the QCD-improved parton model. A series of new ideas, theoretical tools and hypotheses have then opened a rich and successful phenomenological approach giving rise to a novel interpretation of the experimental results. The separate and complementary role played by the "current" and "target" fragmentation was considered already in the parton model approach to high-energy processes. An heuristic discussion can be found, for example, in Richard Feynman's "Lecture 55" on the "Final Hadronic States in Deep Inelastic Scattering" in his "Photon Hadron Interactions" book [4]. In the framework of perturbative QCD, the physics request of describing semi-inclusively hadronic initial states and the dynamics of target fragmentation was not addressed at the beginning.

The idea of fracture functions originates from the need to extend the description of the semi-inclusive hadronic processes in deep inelastic scattering to include the initial state target fragmentation region. It could seem a natural task, in fact, the one of a complete description of the final state entirely

L. Trentadue: *Fracture Functions*, Lect. Notes Phys. **737**, 181–220 (2008)
DOI 10.1007/978-3-540-74233-6_10

in terms of the collinear and infrared logarithmic structure of QCD in its perturbative phase. The formulation of the initial state dynamics to include the rich complexity of the QCD-improved parton model with his quark and gluon degrees of freedom was not considered. This fact appared even more needed at the time when the new HERA lepton–proton collider was beginning to operate in DESY.

The dynamics of the target fragmentation naturally extends the perturbative region of applicability of the QCD theory. It involves the description of quantitively important processes which are softer than the hard current fragmentation. It, therefore, deals with physics scales which are smaller and at the limits of the perturbative region and, also for this reason, it constitutes a complementary dynamics with respect to the current fragmentation. Both target and current fragmentation have to be taken into account in order to reproduce the entire final state without imposing unnatural cuts to separate them.

Let us at this point recall the idea and the motivations for the fracture functions with the same words we used as taken from [5]: "When one or two hadrons are present in the initial state, collinear singularities cannot be avoided. Asymptotic freedom, however, is still of much importance. Together with general factorization theorems for collinear singularities [8], it allows to justify the so-called QCD-improved parton model whereby experimental cross sections can be computed by convoluting some uncalculable, but process independent, quantities with process-dependent, but calculable, elementary cross sections. The best known case of this type is undoubtfully that of structure functions, which can be measured in deep inelastic lepton–hadron collisions in some kinematical regime and then used to compute either the same process or a completely new hard reaction at a different scale. Besides this utilitaristlc value, structure functions have also provided, for many years, an invaluable source of information [6] about the structure of hadrons in terms of valence and sea quarks and gluons together with interesting information on their polarization state. Another much studied set of uncalculable, universal functions is that of the so-called fragmentation functions, providing the probability that a given hadron is produced (inclusively) in a jet initiated by a given parton. A typical use of factorization resides here in the possibility of computing multihadron final states in jet physics, by convoluting the above fragmentation functions with the calculable perturbative jet evolution [9]. With the advent of the new powerful electron–proton collider HERA at DESY, more phase space is becoming available together with a richer variety of channels. One may thus wonder if the only QCD-inspired use of the machine should be the refined measurements of structure and fragmentation functions together with tests of their predictable evolution and factorization properties. There seems to be some widespread consensus that this should not be the case and that, on the contrary, the study of hadron structure can be extended at HERA in new directions. Actually, already at hadronic colliders, there have been studies [7] of quantities such as the pomeron structure function, diffractive hard scattering and the like, with stimulating outcomes. The aim of this paper is to give a proper framework in which to talk about these extensions of "bread and butter" QCD physics. We shall argue that, within perturbatlve QCD, it is possible to introduce new uncalculable, but measurable and universal functions, that we call "fracture" functions, which tell us about the structure function of a given target hadron once it has fragmented (hence its name) into another given final state hadron. Fracture functions (besides exhibiting a mild, calculable Q^2 dependence) depend upon two hadronic and one partonic label and on two momentum fractions, a Bjorken x and a Feynman z variable $M = M^j_{p,h}(x, z, Q^2)$. One can also say that M measures the parton distribution of the object exchanged between the target and the final hadron, without making a (possibly doubtful) model about what that object actually is, a single particle, a Regge trajectory, a multiparticle continuum, or else. As for ordinary structure functions, the importance of measuring such an object will be twofold: (i) it will teach us about the structure of hadronic systems other than the usual targets and (ii) it can be used as input for computing other hard semi-inclusive processes at other machines, such as some future hadronic colliders. By a judicious choice of the final hadron and of its momentum, one will be able, for instance, to enrich the gluonic component of the partonic flux and thus to enhance signal to background ratios for interesting gluon-induced processes in hadron–hadron collisions".

The intent with the predictable evolution and the factorization properties we had in mind at that time were pursued by the experiments almost literally as stated, and, with a series of comparisons with the HERA deep inelastic data it appeared that was possible to verify features and properties of fracture functions.

From the theoretical side it is also useful to remind here that, as already stated above, the basic formalism, without which, this straightforward definition of the fracture functions could not have been given, is the one of the "jet calculus" of Konishi, Ukawa and Veneziano [9]. The properties of fracture functions, according to the original formulation were possible in terms of the typical jet calculus variables by using, for instance, the evolution variables Y, y, y_0 as in the jet calculus, the properties of real $\hat{P}^j_i(u)$ and regularized $P^j_i(u)$ Altarelli–Parisi vertices as well as the "evolution functions" $E^j_i(x, Y-y)$. Jet calculus formalism, as for entire "jet physics", did constitute the proper rich and fruitful background for defining them.

In this report we review the idea of fracture functions as was originally proposed. We then discuss some of the theoretical developments it has further received and some of the applications made in the course of the years. The paper is organized as follows: In Sect. 2 we recall original definitions and the evolutions equations together with the relevant properties of the fracture functions. The complete proof of the factorization of fracture functions, by using the cut vertex formalism, is then given. In Sect. 2.3 extended fracture functions are defined. Two-loop next-to-leading fracture functions are then introduced in Sect. 2.4. In Sect. 2.5 transverse momentum fracture functions are obtained. The formalism of fracture functions for diffractive processes is shown in Sect. 2.6. In Sect. 3.1 the phenomenological application to diffractive processes is discussed. Further applications to Higgs production, polarized processes and heavy quark production are shown in Sects. 3.2–3.4. The extension of the fracture function concept to the case of multiple hadronic inclusive processes is sketched in Sect. 3.5. A new formulation, via fracture functions, of initial state jets can be found in Sect. 4.

2 Formalism and Definitions

To define fracture functions let us follow again [5] : "In any hard process with at least one hadron in the initial state the question arises on how to describe both target and current fragmentation in perturbative QCD. For the current fragmentation, according to factorization theorems [8] in the QCD-improved parton model, inclusive single particle distributions can be accounted for by factorized convolution of structure and fragmentation functions with the hard point-like cross section, i.e.

$$\sigma_{current} \simeq \int F_p^i \, \hat{\sigma}_i^j \, D_j^h . \tag{1}$$

For target fragmentation, however, distributions have to be defined differently. To this purpose, fracture functions have been introduced. A fracture function $M_{p,h}^i(x, z, Q^2)$ describes the distribution of a given final state hadron h in a process with a target hadron p with the parton-like state i exchanged with an hard process at the scale Q^2. A fracture function depends on three labels: two hadronic and a partonic one and on three variables: two momentum fractions x and z the Bjorken and Feynman variables, respectively, of the i and h states and the hard process scale Q^2. A new kind of factorization is conjectured to take place. The factorized form of the target fragmentation cross section can be written in terms of fracture functions as follows:

$$\sigma_{target} \simeq \int M_{p,h}^i \, \hat{\sigma}_i . \tag{2}$$

Fracture functions are not calculable but measurable and universal functions as structure function are. As for ordinary structure functions, measuring them will give us informations about hadronic systems and dynamics. These informations can be used, as for structure and fragmentation functions, as an input for different hard, semi-inclusive processes, also, eventually, at different energies, by means of evolution equations. By identifying the final h hadron, one can costrain the exchanged, parton-like, state. As an example the i parton-like state in $M_{p,h}^i(x, z, Q^2)$ will be a gluon-rich state for $h = p$ and a quark-rich state, possibly a pion-like object for $h = n$ where n is a neutron".

From the point of view of a consistent formulation in terms of factorized amplitudes of the theoretical inputs and assumptions, fracture functions do represent a further step toward the control of the singularities within the perturbation theory.

It has been shown with a one-loop evaluation in [10] that an entire class of collinear divergencies, due to the configurations corresponding to hadrons emitted along the initial state directions are naturally absorbed within fracture functions. This observation extends the validity of the factorization theorems [8] also to the initial state mass singularities within the target fragmentation region.

Two separate contributions can be isolated in the target cross section [5]:

$$\sigma_{target} \simeq \int M^i_{p,h}\,\hat{\sigma}_i + \int F^i_p\,D^h_k\,\hat{\sigma}_i\,. \tag{3}$$

Correspondingly, one can associate to the cross section two terms, i.e. $\sigma_{target} = M^{NP} + M^P$. The first is a non-perturbative contribution and the second a perturbative one. They can be defined at a given scale Q_0^2 by requiring that $M^P|_{Q^2=Q_0^2} = 0$. It is possible to obtain an evolution equation to determine the fracture function $M^i_{p,h}(x,z,Q^2)$ at any other scale Q^2. The evolution equation has the form

$$\frac{\partial M^j_{p,h}(x,z,Q^2)}{\partial ln\,Q^2} = \frac{\alpha_s(Q^2)}{2\pi}\int_{\frac{x}{1-z}}^{1}\frac{du}{u}P^j_i(u)\,M^i_{p,h}(\frac{x}{u},z,Q^2) + \frac{\alpha_s(Q^2)}{2\pi}$$
$$\int_x^{\frac{x}{x+z}}\frac{u\,du}{x(1-u)}\hat{P}^{j,l}_i(u)\,D^h_l(\frac{zu}{x(1-u)},Q^2)\,F^i_p(\frac{x}{u},Q^2) \tag{4}$$

$P^j_i(u)$ and $\hat{P}^{j,l}_i(u)$ being the regularized and real Altarelli–Parisi vertices, respectively [9]. $D^h_l(z,Q^2)$ represents the fragmentation function of the parton l into hadron h and $F^i_p(x,Q^2)$ is the ordinary deep inelastic proton structure function. The evolution equation can be solved and the solution reads

$$M^j_{p,h}(x,z,Q^2) = \frac{\alpha_s(Q^2)}{2\pi}\int_x^{1-z}\frac{dw}{w}\,E^j_i(\frac{x}{w},Q^2,Q_0^2)\,M^i_{p.h}(w.z.Q_0^2) + \frac{\alpha_s(Q^2)}{2\pi}$$
$$\int_{Q_0^2}^{Q^2}\frac{dk^2}{k^2}\int_{x+z}^{1}\frac{dw}{w^2}\int_{\frac{x}{w}}^{1-\frac{z}{w}}\frac{du}{u(1-u)}E^j_k(\frac{x}{wu},Q^2,k^2)\,\hat{P}^{kl}_i(u) \tag{5}$$
$$\times\,D^k_l(\frac{z}{w(1-u)},k^2)\,F^i_p(w,k^2).$$

The first term describes the hadron distribution at a given arbitrary scale Q_0^2 evolving it to a scale Q^2 by means of the perturbative evolution function $E^j_i(\frac{x}{w},Q^2,Q_0^2)$ which satisfies the equation [9]

$$Q^2\frac{\partial}{\partial\,Q^2}E^j_i(x,Q^2,Q_0^2) = \frac{\alpha_s(Q^2)}{2\pi}\int_x^1\frac{du}{u}\,P^j_k(u)\,E^k_i(\frac{x}{u},Q^2). \tag{6}$$

The second term describes the perturbative evolution from Q_0^2 to Q^2 of the active exchanged parton i. The perturbatively generated partonic shower accompanying the evolution of the parton i contains an inclusive distribution for an additional parton l which finally fragments into the hadron h. Fracture functions do satisfy several properties [5] as follows:

- Do not depend on the arbitrary choosen scale Q_0^2, i.e.

$$\frac{\partial}{\partial Q_0^2} M^j_{p,h}(x,z,Q^2) = 0 \tag{7}$$

- Both $D_l^h(x,Q^2)$ and $F_p^i(x,Q^2)$ satisfy the usual Altarelli–Parisi evolution equations and $\sum_h \int_0^1 dz\ z\ D_l^h(x,Q^2) = 1$ and $\sum_i \int_0^1 dx\ x\ F_p^i(x,Q^2) = 1$ with

$$\sum_i \int_0^1 du\ u\ P_i^j(u) = 0 \tag{8}$$

$M^j_{p,h}(x,z,Q^2)$ satisfies the momentum sum rule:

$$\sum_h \int_0^1 dz\ z\ M^j_{p,h}(x,z,Q^2) = (1-x)\ F_p^j(x,Q^2) \tag{9}$$

accounting for the s-channel unitarity constraint.
- In terms of moments, by defining

$$\int_0^1 dz\ z^m \int_0^{1-z} dx\ x^n\ M^i_{p,h}(x,z,Q^2) = M^{i,ph}_{m,n}(Q^2) \tag{10}$$

with $\int duu^n P_i^j(u) = A_n^{ij}$, one obtains for the moments of the fracture functions the evolution equation

$$\frac{\partial M^{i,ph}_{m,n}(Q^2)}{\partial lnQ^2} = \frac{\alpha_s(Q^2)}{2\pi} A_n^{ij}\ M^{i,ph}_{m,n}(Q^2) + \frac{\alpha_s(Q^2)}{2\pi}\ P^{k,ij}_{mn}\ F^{p,j}_{m+n}(Q^2)\ D^{kh}_m(Q^2). \tag{11}$$

2.1 Factorization of Fracture Functions: a Proof

Here we give the proof of the factorization of fracture functions. We follow closely the line of reasoning of [11] and give the proof by using a generalized operator product expansion (OPE)–cut vertex [12] formalism.

The definition of cut vertices is used in the simple case of $(\phi^3)_6$ theory. This toy model, despite its simpler structure, shares several important properties with QCD. It has a dimensionless, asymptotically free coupling constant and the diagrams with leading mass singularities have the same topology as in (light cone gauge) QCD. For these reasons, $(\phi^3)_6$ is an excellent theoretical framework for the study of factorization properties [15].

To define cut vertices, we consider the inclusive deep inelastic process

$$p + J(q) \to X$$

off the current $J = \frac{1}{2}\phi^2$. We define as usual Q^2 and x as

$$Q^2 = -q^2 \qquad x = \frac{Q^2}{2pq} \,. \tag{12}$$

Let us choose a frame in which $p = (p_+, p_-, \mathbf{0})$ with $p_+ \gg p_-$ and $pq \simeq p_+ q_-$. Given a vector $k = (k_+, k_-, \mathbf{k})$ define $\hat{k} = (k_+, 0, \mathbf{0})$. The structure function, defined as

$$F(p,q) = \frac{Q^2}{2\pi} \int d^6y \; e^{iqy} < p|J(y)J(0)|p >, \tag{13}$$

describes the interaction of the far off-shell current $J(q)$ with an elementary quantum of momentum p through the discontinuity of the forward scattering amplitude (see Fig. 1).

The leading contribution to the structure function comes from the decomposition shown in Fig. 2. Here, with the notations of [12], τ is the hard part of the diagram, i.e. the one in which the large momentum flows, while λ is the soft part. Decompositions with more than two legs connecting the hard to the soft part are suppressed by powers of $1/Q^2$.

Such decomposition can be written in formulae as

$$F(p,q) = \sum_{\tau} \int V_\lambda(p,k) \; H_\tau(\hat{k},q) \; \frac{d^6k}{(2\pi)^6} \,, \tag{14}$$

where $V_\lambda(p,k)$ and $H_\tau(k,q)$ are the discontinuities of the long and short distance parts, respectively.

Moreover, in order to pick up the leading contribution in (14), the momentum k which enters τ is taken to be collinear to the external momentum p. Neglecting renormalization, let us define for a given decomposition into a λ and a τ subdiagram

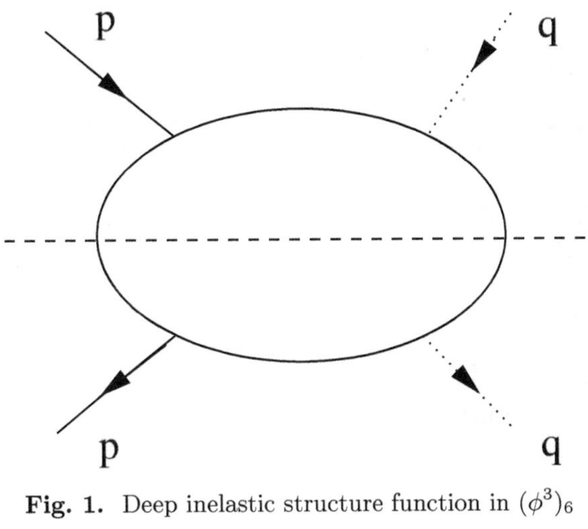

Fig. 1. Deep inelastic structure function in $(\phi^3)_6$

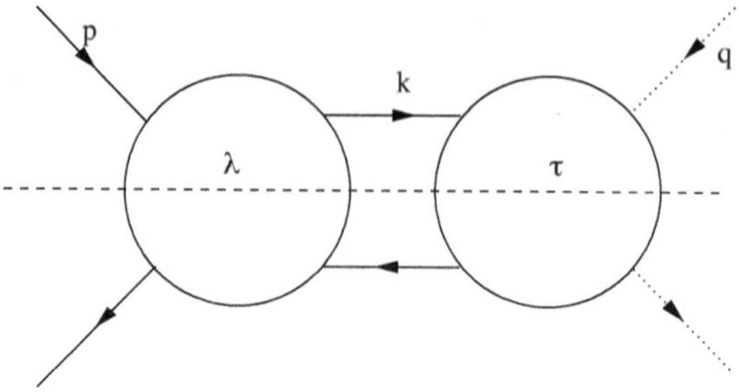

Fig. 2. Relevant decomposition for the deep inelastic structure function in $(\phi^3)_6$

$$v_\lambda(p^2,x) = \int V_\lambda(p,k)\, x\, \delta\left(x - \frac{k_+}{p_+}\right) \frac{d^6k}{(2\pi)^6} \tag{15}$$

and

$$C_\tau(x,Q^2) = H_\tau(k^2=0,x,q^2). \tag{16}$$

Here $v_\lambda(p^2,x)$ represents the contribution of λ when the hard part is contracted to a point, while $C_\tau(x,Q^2)$ is the hard part in which one neglects the virtuality of the incoming momentum with respect to Q^2. Since

$$x \simeq \frac{Q^2}{2p_+q_-} \tag{17}$$

and

$$H_\tau(\hat{k},q) = H_\tau(0, \frac{Q^2}{2k_+q_-}, q^2), \tag{18}$$

using the definition of $\hat{k}$ and (15)–(19), we can write

$$\begin{aligned} F(p,q) &= \sum_\tau \int V_\lambda(p,k)\, H_\tau(\hat{k},q)\, \frac{d^6k}{(2\pi)^6} \\ &= \sum_\tau \int V_\lambda(p,k)\, \delta\left(u - \frac{k_+}{p_+}\right) du\, C_\tau(x/u,Q^2)\, \frac{d^6k}{(2\pi)^6} \\ &= \sum_\tau \int v_\lambda(p^2,u)\, C_\tau(x/u,Q^2)\, \frac{du}{u} \equiv \int v(p^2,u)\, C(x/u,Q^2)\, \frac{du}{u}. \end{aligned} \tag{19}$$

The last integral defines the space-like *cut vertex* $v(p^2,x)$ and the corresponding coefficient function $C(x,Q^2)$. As usual, a simpler factorized expression for the structure function is obtained by taking moments with respect to x. Defining the Mellin transform as

$$f_\sigma = \int_0^1 dx\, x^{\sigma-1} f(x), \tag{20}$$

we find immediately

$$F_\sigma(p^2, Q^2) = v_\sigma(p^2)\, C_\sigma(Q^2). \tag{21}$$

It was shown in [13, 14] that the cut vertex represents the analytic continuation in the spin variable of a matrix element of operators of minimal twist. This correspondence has been confirmed up to two loops by direct calculation of the anomalous dimensions of cut vertices and leading twist operators [13]. Hence, in the case of DIS, the factorized expression (21) can be identified with the one given by OPE

$$F_n(p^2, Q^2) = A_n(p^2)\, C_n(Q^2)\,, \tag{22}$$

where $A_n(p^2)$ are now matrix elements of local operators. Thus, for integer values of σ, the coefficient function which appears in (21) is the same as in (22). This fact will be used in the next section where the evolution of the extended fracture function will be shown to be driven by the anomalous dimension of the same set of local operators.

2.2 A Cut Vertex Approach to Semi-inclusive Processes

Let us consider now, still within $(\phi^3)_6$, a deep inelastic reaction in which a particle with momentum p' is inclusively observed in the final state, i.e. the process

$$p + J(q) \to p' + X.$$

By using the same line of reasoning as for the inclusive case we may define a semi-inclusive structure function as (see Fig. 3)

$$W(p, p', q) = \frac{Q^2}{2\pi} \sum_X \int d^6x\, e^{iqx} < p|J(x)|p'X >< X\ p'|J(0)|p > \tag{23}$$

in terms of matrix elements of the current operator between the incoming hadron with momentum p and the outgoing hadron with momentum p' plus anything.

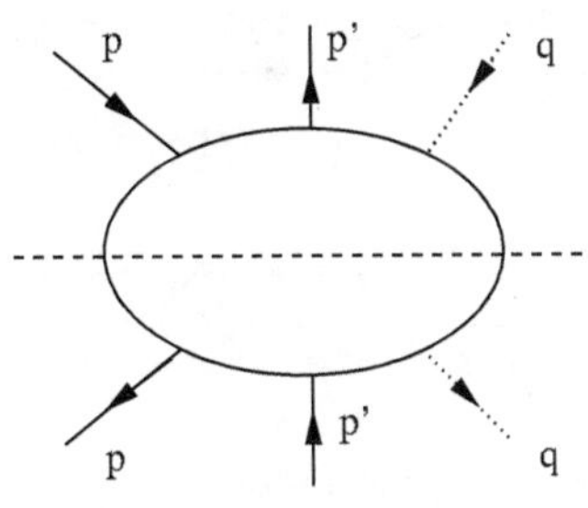

Fig. 3. Deep inelastic semi-inclusive structure function in $(\phi^3)_6$

When the observed particle has transverse momentum $p'^2_\perp$ of order Q^2 the cross section is dominated by the current fragmentation mechanism and can be written in the usual factorized way [16]

$$W(p,p',q) = \int \frac{dx'}{x'}\frac{dz'}{z'}\, f_A(x',Q^2)\,\hat{\sigma}(x/x',z',Q^2)\,D_{A'}(z/z',Q^2), \qquad (24)$$

where

$$z = \frac{pp'}{pq} \simeq \frac{p'_-}{q_-}\,. \qquad (25)$$

In the language of cut vertices, (24) is a convolution of a space-like and a time-like cut vertex through a coefficient function [17]

$$W(p,p',q) = \int \frac{dx'}{x'}\frac{dz'}{z'} v(p^2,x')\,C(x/x',z',Q^2)\,v'(p'^2,z/z'). \qquad (26)$$

By contrast, the limit $t = -(p-p')^2 \ll Q^2$ is dominated by the target fragmentation mechanism and has not been considered in either approach. In particular, it has been shown [10] at one loop that in the limit $t \to 0$ a new collinear singularity appears in the semi-inclusive cross section, which cannot be absorbed into parton densities and fragmentation functions, and so must be lumped into a new phenomenological distribution, i.e. the fracture function.

Following the same steps as before, one can argue that, in the region $t \ll Q^2$, the leading contribution to the semi-inclusive cross section is given by the decomposition shown in Fig. 4.

Such a decomposition implies that an expansion similar to the one in (19) holds, in terms of a new function $v(p,p',\bar{x})$ and a coefficient function $C(\bar{x},Q^2)$

$$W(p,p',q) = \int v(p,p',u)\,C(\bar{x}/u,Q^2)\,\frac{du}{u} \qquad (27)$$

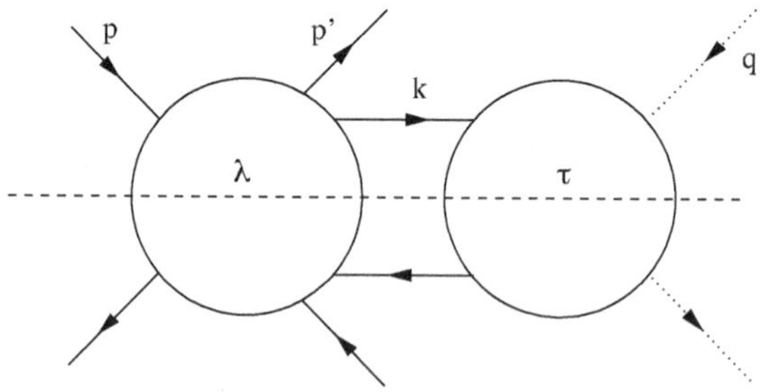

Fig. 4. Relevant decomposition for the semi-inclusive structure function in $(\phi^3)_6$ in the limit $t \ll Q^2$

where we have defined a new variable z as

$$z = \frac{p'q}{pq} \simeq \frac{p'_+}{p_+} \tag{28}$$

and a rescaled variable $\bar{x} = x/(1-z)$. The new function $v(p, p', \bar{x})$ is given by

$$v(p, p', \bar{x}) = \int T(p, p', k)\, \bar{x}\, \delta\left(\bar{x} - \frac{k_+}{p_+ - p'_+}\right) \frac{d^6 k}{(2\pi)^6}, \tag{29}$$

where $T(p, p', k)$ is the discontinuity of a six-point amplitude in the channel $(p - p' - k)^2$. The function $v(p, p', \bar{x})$ is a new object that we will call a *generalized cut vertex*, which depends both on p and p' and embodies all the leading mass singularities of the cross section. By taking moments with respect to $\bar{x}$ as in (20), (27) becomes

$$W_\sigma(p, p', q) = v_\sigma(p, p')\, C_\sigma(Q^2) \tag{30}$$

that is a completely factorized expression analogous to (21).

We are now going to show that this expansion holds up to corrections suppressed by powers of $1/Q^2$. In order to do so, it can be used the method of infrared power counting [18, 19] applied to our process.

In order to get insight into the large Q^2 limit of the semi-inclusive cross section, let us look at the singularities in the limit p^2, p'^2, $t \to 0$. The infrared power counting technique can predict the strength of such singularities. Starting from a given diagram, its *reduced* form in the large Q^2 limit is constructed by simply contracting to a point all the lines whose momenta are not on shell. The general reduced diagrams in the large Q^2 limit for the process under study involve a jet subdiagram J, composed by on-shell lines collinear to the incoming particle, from which the detected particle emerges in the forward direction, since in the large Q limit p and p' can be taken as parallel, and a hard subgraph H in which momenta of order Q circulate, which is connected to the jet by an arbitrary number of collinear lines. Soft connections between J and H can be possibly collected into a soft blob S which is connected to the rest of the diagram by an arbitrary number of lines (see Fig. 5). In $(\phi^3)_6$, by using power counting [19], we find that the leading contributions come from graphs with no soft lines and the minimum number of collinear lines connecting the hard to the jet subdiagram, as in Fig. 6. This fact has been verified by an explicit one-loop calculation in [20].

Any other diagram containing additional collinear lines between J and H is suppressed by powers of $1/Q^2$. It follows that $W(p, p', q)$ is of the following form:

$$W(p, p', q) = \int \frac{d^6 k}{(2\pi)^6}\, T(p, p', k)\, H(\hat{k}, q) + \mathcal{O}(1/Q^2). \tag{31}$$

It is now straightforward to show that (31) is equivalent to (27) with the substitution $H(0, x, q^2) = C(x, Q^2)$. Thus the expansion (27) corresponds to taking the leading part of the semi-inclusive cross section.

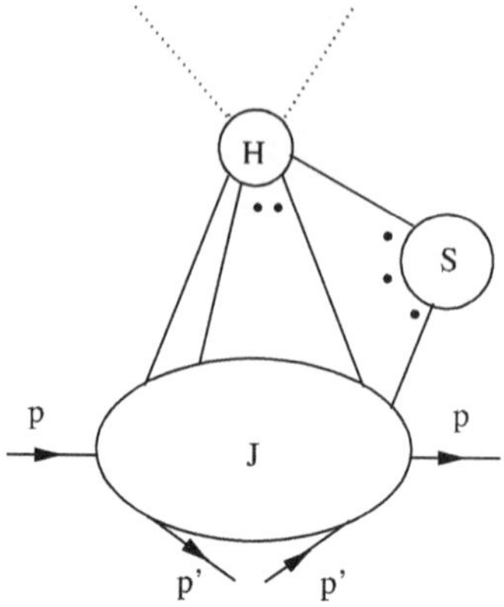

Fig. 5. General reduced graphs which contribute to the semi-inclusive structure function in $(\phi^3)_6$

2.3 Extended Fracture Functions

In the previous section we have given arguments for the validity of a generalized cut vertex expansion for the process $p + J(q) \to p' + X$ in the region $t \ll Q^2$. Let us now investigate the consequences of such a result.

The coefficient function which appears in (27) is the same as that of (19) since it comes from the hard part of the graphs which is exactly the same as in DIS. So we can draw the important conclusion that the evolution of the coefficient function appearing in (27) is directly related to the anomalous dimension of the leading twist local operator which drives the evolution of the DIS coefficient function.

Despite the fact that the theoretical framework in which we have been working is the model field theory $(\phi^3)_6$ we expected the main consequences

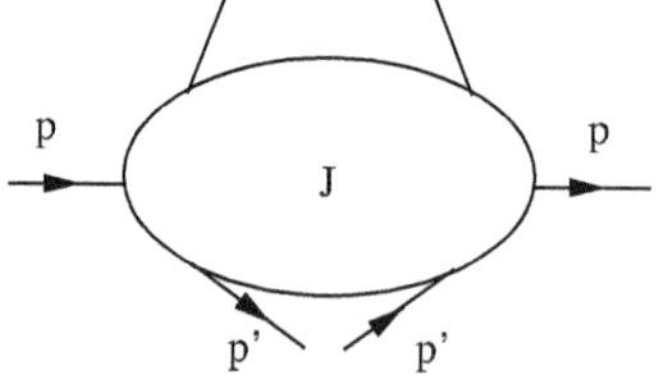

Fig. 6. Leading contributions to the semi-inclusive structure function in $(\phi^3)_6$

expressed in (27) remain valid also in a gauge theory such as QCD [11]. The only further complication which are expected to arise are due to soft gluon lines connecting the hard to the jet subdiagrams. Unlike in $(\phi^3)_6$, in QCD these diagrams are not suppressed by power counting. The only way to get rid of such contributions is to show that they cancel out. As already argued in [21], we did not expect that this complication would destroy factorization. The issue of a complete factorization proof in QCD has been later considered in [22]. In QCD, by using renormalization group, we have

$$C_n^i(Q^2) \equiv C_n^i(Q^2/Q_0^2, \alpha_s) = \left[e^{\int_{\alpha_s}^{\alpha_s(Q^2)} d\alpha \frac{\gamma^{(n)}(\alpha)}{\beta(\alpha)}} \right]_{ij} C_n^j(1, \alpha_s(Q^2)), \quad (32)$$

where Q_0 is the renormalization scale, $\alpha_s \equiv \alpha_s(Q_0^2)$, $\gamma^{(n)}$ is the anomalous dimension matrix of the relevant operators and an ordered exponential is to be understood. Thus we can write the analogue of (30) in QCD as

$$W_n(z, t, Q^2) = \sum_i \mathcal{M}_n^i(z, t, Q^2)\; C_n^i(1, \alpha_s(Q^2)) \quad (33)$$

where, by following [13], we have defined a t-dependent fracture function

$$\mathcal{M}_n^j(z, t, Q^2) \equiv V_n^i(z, t, Q_0^2) \left[e^{\int_{\alpha_s}^{\alpha_s(Q^2)} d\alpha \frac{\gamma^{(n)}(\alpha)}{\beta(\alpha)}} \right]_{ij} \quad (34)$$

just in terms of a cut vertex $V_n^i(z, t, Q_0^2)$ (see Fig. 7).

Inverting the moments and expressing the extended fracture function in terms of the usual Bjorken variable x, one finds that $\mathcal{M}_{A,A'}^i(x, z, t, Q^2)$ obeys the simple homogeneous evolution equation

$$Q^2 \frac{\partial}{\partial Q^2} \mathcal{M}_{A,A'}^i(x, z, t, Q^2) = \sum_j \int_{\frac{x}{1-z}}^1 \frac{du}{u} K_{ij}(u, \alpha_s(Q^2))\; \mathcal{M}_{A,A'}^j(x/u, z, t, Q^2) \quad (35)$$

where $K_{ij}(u, \alpha)$, defined as

$$K_{ij}(u, \alpha) \equiv \frac{1}{2\pi i} \int_{\frac{1}{2}-i\infty}^{\frac{1}{2}+i\infty} dn\; \gamma_{ij}^{(n)}(\alpha)\; u^{-n}, \quad (36)$$

is the same Dokshitzer–Gribov–Lipatov–Altarelli–Parisi (DGLAP) kernel, which controls the evolution of the ordinary parton distribution functions. This result looks particularly appealing since it means that the evolution of the extended fracture function follows the usual perturbative behaviour. One may ask at this point how this result matches with the peculiar equation, which drives the evolution of ordinary fracture functions [5]

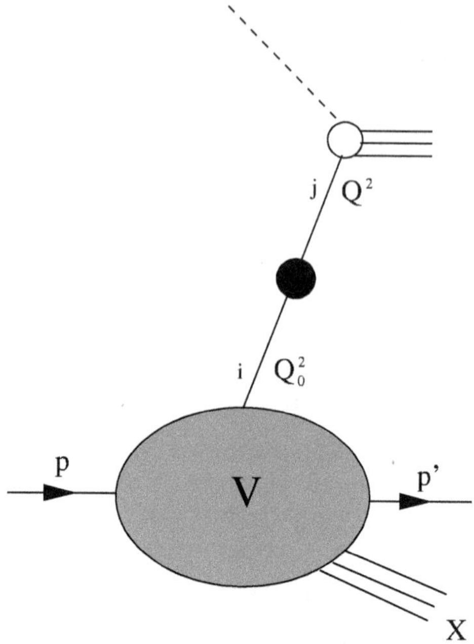

Fig. 7. Extended fracture function

$$Q^2\frac{\partial}{\partial Q^2}M^j_{A,A'}(x,z,Q^2)=\frac{\alpha_s(Q^2)}{2\pi}\int_{\frac{x}{1-z}}^{1}\frac{du}{u}\,P^j_i(u)\,M^i_{A,A'}(x/u,z,Q^2)$$
$$+\frac{\alpha_s(Q^2)}{2\pi}\int_x^{\frac{x}{x+z}}\frac{du}{x(1-u)}\,F^i_A(x/u,Q^2)\,\hat{P}^{jl}_i(u)\,D_{l,A'}\left(\frac{zu}{x(1-u)},Q^2\right). \quad (37)$$

The evolution equation for $M^j_{A,A'}(x,z,Q^2)$ contains two terms: a homogeneous term describing the non-perturbative production of a hadron coming from target fragmentation and an inhomogeneous term whose origin is the perturbative fragmentation due to initial state bremsstrahlung. As discussed in [5], the separation between perturbative and non-perturbative fragmentation introduces an arbitrary scale but the fracture function itself $M^j_{A,A'}(x,z,Q^2)$ does not depend on it.

It can be shown [23] that the evolution equation (37) can be derived by an explicit calculation using (35) together with jet calculus rules [9]. By defining in fact the ordinary fracture function as an integral over t up to a cut-off of order Q^2, e.g. ϵQ^2 with $\epsilon < 1$:

$$M^j_{A,A'}(x,z,Q^2)=\int^{\epsilon Q^2}dt\,\mathcal{M}^j_{A,A'}(x,z,t,Q^2), \quad (38)$$

the inhomogeneous term in the evolution equation is obtained by taking into account the Q^2 dependence of the integration cut-off.

Moreover, the interplay between the scales Q^2 and t has a sizeable effect in terms of a new class of perturbative corrections of the form $\log Q^2/t$. Such corrections are large and potentially dangerous in the region $t \ll Q^2$ since they can ruin a reliable perturbative expansion. Those terms are naturally resummed into (34). For the extended fracture function, these corrections do play an important role for understanding the dynamics of semi-inclusive processes in the kinematic region we have been considering here [23].

Despite the proof of factorization in deep inelastic scattering [11, 22] the one that fracture functions factorize in hadron–hadron scattering has not yet been given.

2.4 Two-loop Next-to-leading Fracture Functions

Daleo, Garcia-Canal and Sassot [24, 25] have considered the extension of the fracture function formalism to include $O(\alpha_s^2)$ QCD corrections and to evaluate amplitudes and evolution equations to next-to-leading (NLO) accuracy. The factorization of fracture functions has been also explicitly checked. The main features related to fracture functions, have been studied in these works up to NLO accuracy, as it is standard in the inclusive case. In particular, there were neither explicit checks of factorization at $O(\alpha_s^2)$ nor indications of how relevant the non-homogeneous evolution might be at NLO.

The evaluation of the NLO corrections in the semi-inclusive channel and the explicit check of the factorization of the collinear singularities need a careful treatment. With respect to the inclusive case, where after a convenient integration over final states singularities may be written as distributions in only one variable times a regular function, in the semi-inclusive one at $O(\alpha_s^2)$, it is necessary to keep additional variables unintegrated. Consequently, entangled singularities in more than one variable have to be dealt with. In order to check factorization, it has to be kept track of the kinematical origin or configuration, which gives rise to the singularity [24]. This requires [24, 25] a detailed analysis of the singularity structure characteristic of the process. In the paper [24] the case where the initial state parton is a gluon is addressed. After obtaining the explicit expressions for the renormalized fracture functions, the explicit evolutions equations can be derived [24]:

$$\begin{aligned}\frac{\partial M^j_{p,h}(x,z,Q^2)}{\partial\, lnQ^2} &= \frac{\alpha_s(Q^2)}{2\pi}\int_{\frac{x}{1-z}}^{1}\frac{du}{u}\,[P^j_i(u)+\frac{\alpha_s(Q^2)}{2\pi}P^{(1)j}_i(u)]\,M^i_{p,h}(\frac{x}{u},z,Q^2)\\ &+\frac{\alpha_s(Q^2)}{2\pi}\frac{1}{x}\int_x^{\frac{x}{x+z}}\frac{du}{u}\int_{\frac{z}{x}}^{\frac{1-u}{u}}\frac{dv}{v}\,[\hat{P}^{j,k}_i(u,v)+\frac{\alpha_s(Q^2)}{2\pi}\hat{P}^{(1)j,k}_i(u,v)]\\ &\cdot D^h_k(\frac{z}{xv},Q^2)\,F^i_p(\frac{x}{u},Q^2).\end{aligned} \quad (39)$$

Here $P^{j,k}_i(u)$, $P^{(1)j,k}_i(u)$, $\hat{P}^{j,k}_i(u)$ and $\hat{P}^{(1)j,k}_i(u)$ are the leading and NLO complete and real kernel, respectively. The corresponding expressions allow the complete determination of the non-homogeneous evolution as for the ordinary DGLAP [26] evolution equations. These therefore allow to verify the factorization of collinear singularities up to $O(\alpha_s^2)$. The relevance of next-to-leading corrections may be also explicitely shown when the effects of the new evolution kernels are compared with the leading order corrections.

The case where the initial state parton is a quark has been addressed in [25]. Here a more complex singularity structure is present which implies a corresponding more involved pattern of factorization. The explicit check of the factorization and a corresponding evolution equation analogous to the one obtained for the gluon case has been explicitly derived [25]. The comparison with the leading order non-homogeneous equation shows for the quark-initiated semi-inclusive hadronic distributions that the impact of next-to-leading corrections depends on the kinematical region of the final hadron emission. Next-to-leading corrections result larger for smaller values of the Bjorken variable x and longitudinal momentum fraction z of the hadrons and of the parent partons and the impact, at fixed values of z, becomes larger as x decreases. In diffractive processes the inhomogeneous term is kinematically suppressed ($z \to 1$) while for forward hadron production, where the inhomogeneous term becomes important, the impact of non-leading contributions increases [27].

2.5 Transverse Momentum-dependent Fracture Functions

In this section we discuss the explicit inclusion of transverse momenta for the semi-inclusive distributions by using fracture functions. We follow the work of [28]. In the current fragmentation, transverse momentum of the detected hadron is taken into account through the following DGLAP time-like equation [29]:

$$Q^2 \frac{\partial \mathcal{D}_i^h(z_h, Q^2, \boldsymbol{p}_\perp)}{\partial Q^2} = \frac{\alpha_s(Q^2)}{2\pi} \int_{z_h}^1 \frac{du}{u} P_{ij}(u, \alpha_s(Q^2)) \cdot \frac{d^2 \boldsymbol{q}_\perp}{\pi}$$
$$\delta(\,u(1-u)Q^2 - q_\perp^2)\, \mathcal{D}_j^h\Big(\frac{z_h}{u}, Q^2, \boldsymbol{p}_\perp - \frac{z_h}{u}\boldsymbol{q}_\perp\Big) . \quad (40)$$

The corresponding space-like equation can be derived as follows:

$$Q^2 \frac{\partial \mathcal{F}_P^i(x_B, Q^2, \boldsymbol{k}_\perp)}{\partial Q^2} = \frac{\alpha_s(Q^2)}{2\pi} \int_{x_B}^1 \frac{du}{u^3} P_j^i(u, \alpha_s(Q^2)) \cdot \frac{d^2 \boldsymbol{q}_\perp}{\pi}$$
$$\delta(\,(1-u)Q^2 - q_\perp^2)\, \mathcal{F}_P^j\Big(\frac{x_B}{u}, Q^2, \frac{\boldsymbol{k}_\perp - \boldsymbol{q}_\perp}{u}\Big) . \quad (41)$$

Perturbative evolution is, however, at work even in target fragmentation region and we expect that a non-negligible amount of transverse momentum is also produced there. We thus generalize fracture function distributions to contain also transverse degrees of freedom. By definition, fracture functions $\mathcal{M}_{p,h}^i(x, \boldsymbol{k}_\perp, z, \boldsymbol{p}_\perp, Q^2)$ give the conditional probability to find in a proton P, at a scale Q^2, a parton with momentum fraction x and transverse momentum

$\boldsymbol{k}_\perp$ while a hadron h, with momentum fraction z and transverse momentum $\boldsymbol{p}_\perp$, is detected. Under these assumptions the following evolution equations can thus be derived [28]:

$$Q^2 \frac{\partial \mathcal{M}^i_{p,h}(x, \boldsymbol{k}_\perp, z, \boldsymbol{p}_\perp, Q^2)}{\partial Q^2} = \frac{\alpha_s(Q^2)}{2\pi} \left\{ \int_{\frac{x}{1-z}}^1 \frac{du}{u^3} P^i_j(u) \right.$$
$$\cdot \int \frac{d^2 \boldsymbol{q}_\perp}{\pi} \delta(\,(1-u)Q^2 - q_\perp^2) \mathcal{M}^j_{p,h}\Big(Q^2, \frac{x}{u}, \frac{\boldsymbol{k}_\perp - \boldsymbol{q}_\perp}{u}, z, \boldsymbol{p}_\perp\Big)$$
$$+ \int_x^{\frac{x}{x+z}} \frac{du}{x(1-u)u^2} \hat{P}^{i,l}_j(u) \int \frac{d^2 \boldsymbol{q}_\perp}{\pi} \delta(\,(1-u)Q^2 - q_\perp^2)$$
$$\left. \cdot \mathcal{F}^j_p\Big(\frac{x}{u}, \frac{\boldsymbol{k}_\perp - \boldsymbol{q}_\perp}{u}, Q^2\Big) \mathcal{D}^h_l\Big(\frac{zu}{x(1-u)}, \boldsymbol{p}_\perp - \frac{zu}{x(1-u)} \boldsymbol{q}_\perp, Q^2\Big) \right\}. \quad (42)$$

As in the longitudinal case, two terms contribute to the evolution of transverse momentum fracture functions as displayed in Fig. 8. The homogeneous one has a pure non-perturbative nature since involves the fragmentation of the proton remnants into the hadron h. The inhomogeneous one takes into account the production of the hadron h from a time-like cascade of parton j and thus is dubbed *perturbative*. The transverse momentum fracture functions fulfil the normalization condition

$$\int d^2 \boldsymbol{k}_\perp \int d^2 \boldsymbol{p}_\perp \mathcal{M}^i_{P,h}(x, \boldsymbol{k}_\perp, z, \boldsymbol{p}_\perp, Q^2) = \mathcal{M}^i_{P,h}(x, z, Q^2)\,, \quad (43)$$

as direct consequence of the kinematics of both terms in the evolution equations, (42). The proof of factorization, i.e. that all singularities occuring in the target remnant direction can be properly renormalized by the less inclusive transverse momentum quantity $\mathcal{M}^i_{P,h}(x, \boldsymbol{k}_\perp, z, \boldsymbol{p}_\perp, Q^2)$, is still lacking at present. In the following we assume such a factorization to hold. Once

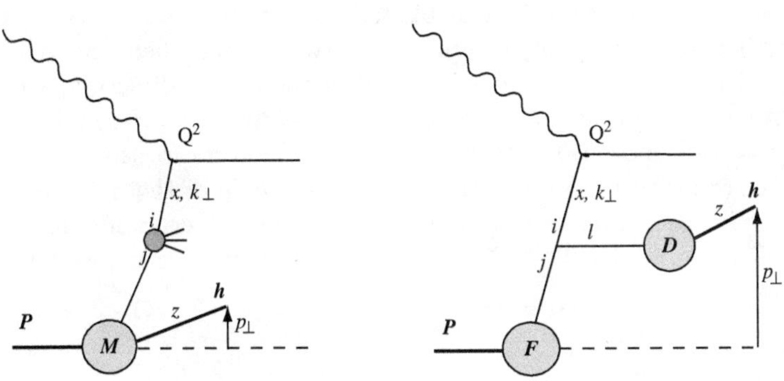

Fig. 8. Evolution of fracture functions M: (**a**) homogeneous term; (**b**) inhomogeneous one

transverse momentum evolution equations are solved, these predictions can be compared with semi-inclusive DIS data, as for the longitudinal case, provided that a factorization theorem holds even for transverse momentum distributions. Such a theorem has been shown to hold in the current fragmentation region for the structure function H_2 in [30]:

$$H_2(x_B, z_h, \boldsymbol{P}_{h\perp}, Q^2) = \sum_{i=q,\bar{q}} e_q^2 \int d^2\boldsymbol{k}_\perp d^2\boldsymbol{p}_\perp \delta^{(2)}(z_h \boldsymbol{k}_\perp + \boldsymbol{p}_\perp - \boldsymbol{P}_{h\perp})$$
$$\times \mathcal{F}_P^i(x_B, \mu_F^2, \boldsymbol{k}_\perp,)\, \mathcal{D}_i^h(z_h, \mu_D^2, \boldsymbol{p}_\perp)\, C(Q^2, \mu_F^2, \mu_D^2)\,, \tag{44}$$

where the standard semi-inclusive variables are defined as follows:

$$z_h = \frac{P \cdot P_h}{P \cdot q}, \quad x_B = \frac{Q^2}{2P \cdot q}, \tag{45}$$

and μ_F^2 and μ_D^2 are the factorization scales. The above results are accurate up to powers in $(P_{h\perp}^2/Q^2)^n$ for soft transverse momenta $P_{h\perp} \simeq \Lambda_{QCD}$. Evolution equations for $\mathcal{F}$ and $\mathcal{D}$ are given in (40) and (41). The factor C is the process-dependent hard coefficient function computable in perturbative QCD and to leading logarithmic accuracy(LLA) we can set C=1. Provided that factorization holds for the transverse momentum fracture functions, we may add, according to (42), their contributions to H_2:

$$H_2(x_B, z_h, \boldsymbol{P}_{h\perp}, Q^2) = \sum_{i=q,\bar{q}} e_q^2 \int d^2\boldsymbol{k}_\perp d^2\boldsymbol{p}_\perp \Big\{ \delta^2(z_h \boldsymbol{k}_\perp + \boldsymbol{p}_\perp - \boldsymbol{P}_{h\perp})$$
$$\cdot \mathcal{F}_P^i(x_B, Q^2, \boldsymbol{k}_\perp)\, \mathcal{D}_i^h(z_h, Q^2, \boldsymbol{p}_\perp) A(0) \tag{46}$$
$$+(1 - x_B)\mathcal{M}_{p,h}^i(x_B, \boldsymbol{k}_\perp, z, \boldsymbol{p}_\perp, Q^2)\, \delta^2(\boldsymbol{p}_\perp - \boldsymbol{P}_{h\perp}) A(1) \Big\}$$

where we have identified all the three factorization scales with the hard scale, $Q^2 = \mu_F^2 = \mu_D^2 = \mu_M^2$. Although, formally, the two contributions are simply added in (46), at LLA and in photon–proton centre of mass frame, the produced hadrons are mainly distributed in two opposite hemispheres. Target fragmented hadrons are produced mainly in the $\theta = \pi$ direction, while current fragmented hadrons mainly along the $\theta = 0$ direction. Here θ is the angle of the produced hadron h with respect to the photon direction, as shown in Fig. 8b. In order to keep track of the emission angle of the detected hadron h, we supplement current and target fragmentation terms in (46) with an angular distribution $A(v)$ [10]. The angular and energy variables v and z are defined as

$$z = \frac{E_h}{E_p(1 - x_B)}, \qquad v = \frac{1 - \cos\theta}{2}, \qquad z_h = z\, v\,. \tag{47}$$

In (47), E_h and E_p denote, respectively, the energies of the detected hadron and of the incoming proton in the photon–proton centre of mass frame. The variables z and v are a useful frame-dependent representation for the hadronic invariant z_h in two respects: z reduces to z_h in the current fragmentation region so that we recover the standard definitions, while for low z_h-values we can distinguish soft hadrons ($z \to 0$) from the ones produced in the target remnant direction ($\theta \to \pi$). Since to LLA all sources of transverse momenta contributing to $\boldsymbol{P}_{h\perp}$ have been taken into account we may pictorially represent (46) as in Fig. 9. This figure shows the sources of transverse momenta in the current and target fragmentation regions extending the description of semi-inclusive processes transverse degrees of freedom.

2.6 Diffraction

Diffractive reactions in photon–hadron interactions can be defined as as the reactions where, within the final state, an isolated hadron can be observed as separated from the rest of the process by a large rapidity gap.

When compared with an inclusive deep inelastic process $\gamma^* p \to X$ a diffractive channel can be represented by the reaction $\gamma^* p \to p^* X$, where the final inclusive collection of hadrons X is well separated from the final eventually excited hadron p^*. Differently from the totally inclusive deep inelastic scattering, a diffractive one might be considered as a combination of hard photon virtualness Q^2 scale and another either hard or eventually soft scale t transferred momentum between p and p^*. Therefore, in the case of a perturbative approach, a combination between perturbative evolutions leading to peculiar final state consisting in a hard hadronic output together with an unbroken proton or a slightly excited final hadronic state.

The peculiar signature of the large rapidity gap events suggests furthermore that between the lower and the upper parts of the reactions the exchange of a colourless object does take place. The mechanism of colour

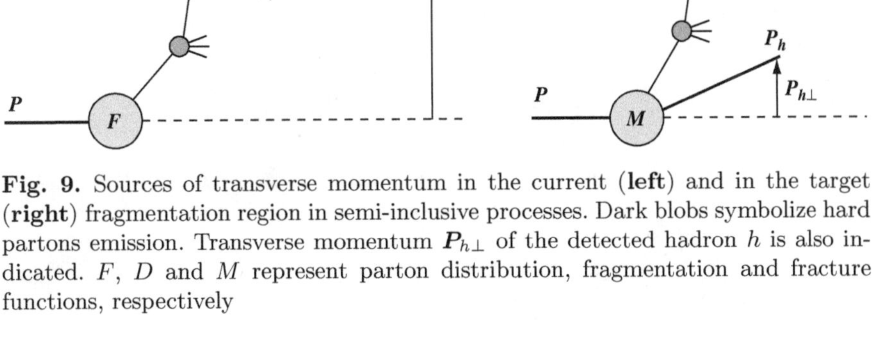

Fig. 9. Sources of transverse momentum in the current (**left**) and in the target (**right**) fragmentation region in semi-inclusive processes. Dark blobs symbolize hard partons emission. Transverse momentum $\boldsymbol{P}_{h\perp}$ of the detected hadron h is also indicated. F, D and M represent parton distribution, fragmentation and fracture functions, respectively

screening shows itself directly in diffraction. The dynamics of the deep inelastic diffractive reactions has been already studied long time ago [31] in terms of space–time evolution of a composite photon. Analogously to the case of the totally inclusive deep inelastic scattering, also in the case of diffractive reactions it is possible to define particular distributions in terms of suitable structure and fragmentation functions.

The typical diffractive reaction $\gamma^* p \to p^* X$ can be written in terms of a new kind of structure function $F_2^{D(3)}(x, Q^2, \xi)$, i.e.

$$\frac{d\sigma}{dx dQ^2 d\xi} = \frac{4\pi\alpha^2}{xQ^4}(1 - y + \frac{y^2}{2})F_2^{D(3)}(x, Q^2, \xi)$$

where the process is fully defined by the variables

$$x = \frac{Q^2}{2p\cdot q}\ ;\ \xi = \frac{q\cdot(p-p')}{q\cdot q}\ ;\ \beta = \frac{Q^2}{2q\cdot(p-p')} = \frac{x}{\xi}\ ;\ y = \frac{Q^2}{xs} \tag{48}$$

with q and p the photon and initial proton momenta, p' the momenta of the final hadron, and x and Q^2 the Bjorken variable and the hard scale. ξ and β characterize the intermediate state of the process. According to the Ingelman–Schlein [32] approach to diffactive processes, a diffractive distribution can be written as

$$\frac{d\sigma}{dt d\xi} = f_P(\xi, t)\ \hat{\sigma}_P(M^2) \tag{49}$$

where $\xi \simeq \frac{Q^2+M_x^2}{Q^2+W^2}$ and M_x^2 and W^2 are the final hadron invariant masses. $\hat{\sigma}_P(M^2)$ is the point-like hard cross section and f_P the pomeron partonic distribution. The fully differential diffractive distribution can be written as

$$\frac{d\sigma^D(Q^2, x, \xi, t)}{dx dQ^2 d\xi dt} = \frac{4\pi\alpha^2}{xQ^4}\frac{dF_2^D}{d\xi dt}(x, Q^2, \xi, t) \tag{50}$$

where $F_2^D(x, Q^2, \xi, t)$ is the diffractive structure function given by the factorized expression

$$\frac{dF_2^D}{d\xi dt}(x, Q^2, \xi, t) = \sum_a \int dx' \frac{df_{a/P}}{d\xi dt}(x', Q^2, \xi, t)\ \hat{F}_{2a} \tag{51}$$

If one uses for the differential parton distribution the Regge parametrization it can be written as

$$\frac{df_{a/P}}{d\xi dt}(x', Q^2, \xi, t) \simeq \xi^{1-2\alpha_P}\frac{g(t)^2}{8\pi^2}\ f_{a/P} \tag{52}$$

with $f_{a/P}$ the parton a pomeron structure function. In perturbative QCD, a more direct model-independent expression can be given

$$\frac{d\sigma}{dt d\xi}(x, Q^2, \xi, t) = \sum_i \int_x^\xi dy\ \hat{\sigma}_i(x, Q^2, y)\cdot\frac{df_i(y, \xi, t)}{d\xi dt} \tag{53}$$

where the parton distributions [21] $\frac{df_i(y,\xi,t)}{d\xi dt}$ are just fracture functions

$$\frac{df_i(y,\xi,t)}{d\xi dt} = M^i_{p,p}(x, 1-\xi, Q^2, t). \qquad (54)$$

This factorized expression does not require any Regge or any alternative parametrization for the parton distributions. Fracture functions do represent a natural continuation of the Ingelman–Schlein [32] approach to describe diffractive processes in the sense that fracture functions allow the perturbative QCD evolution of the distributions in terms of the variable Q^2.

An interesting description of diffractive scattering and factorization has been proposed by Hautmann, Kunszt and Soper [33] in terms of diffractive parton distributions. According to this formulation in hadronic systems with small transverse size, diffraction occurs predominantly at short distances and the diffractive parton distributions can be studied by perturbative methods. For larger systems it is discussed the possibility that diffractive parton distributions are controlled essentially by semi-hard physics at a scale of the order of giga electron volt. The authors find that this possibility accounts for important qualitative aspects of the diffractive data from HERA as the flat behaviour in β and the delay in the fall-off with Q^2.

Arguments have been given in [39] against the diffractive factorization in hadron–hadron scattering.

3 Applications and Phenomenology

3.1 Diffraction

The phenomenological description of the diffraction dynamics in terms of fracture functions has been extensively used to analyse HERA data.

Here we follow the work of De Florian and Sassot [34] where a fracture function-based QCD analysis of the first data produced by the H1 and ZEUS collaborations was given. Both the diffractive and the leading proton deep inelastic lepton–proton scattering structure functions $F_2^{D(3)}$ and $F_2^{LP(3)}$ have been considered. The aim is to verify if the QCD framework for semi-inclusive processes, based on fracture functions, is able to allow a unified treatment of diffractive and leading proton processes, with a detailed perturbative QCD description for them, alternative to those that rely on model-dependent assumptions.

By defining the leading proton structure function $F_2^{LP(3)}$ from the corresponding triple-differential deep inelastic scattering cross section

$$\frac{d^3\sigma^{LP}}{dx\, dQ^2\, d\xi} \equiv \frac{4\pi\alpha^2}{x\, Q^4}\left(1-y+\frac{y^2}{2}\right) F_2^{LP(3)}(x, Q^2, \xi)\,, \qquad (55)$$

with the usual kinematical variables. Even though the processes accounted for are of a semi-inclusive nature, the formulation based on the leading proton structure function is used instead of the usual approach for semi-inclusive deep inelastic scattering

$$\frac{d^3\sigma^p_{current}}{dx\, dQ^2\, dz} \simeq \frac{4\pi\alpha^2}{xQ^4}\left(1-y+\frac{y^2}{2}\right) x\sum_i e_i^2\, F_p^i(x,Q^2)\, D_i^h(z,Q^2) \tag{56}$$

in terms of parton distributions and fragmentation functions, since the last one only takes into account hadrons produced in the current fragmentation region and thus not contributing to the forward leading hadron observables. In terms of fracture functions, for very forward protons with $1-\xi \simeq z$

$$\frac{d^3\sigma^p_{target}}{dx\, dQ^2\, dz} = \frac{4\pi\alpha^2}{x\, Q^4}\left(1-y+\frac{y^2}{2}\right)\sum_i e_i^2 x M_p^{i,p}(x,z,Q^2)\,, \tag{57}$$

where $M_p^{i,p}(x,z,Q^2)$ is the fracture function that accounts for target fragmentation processes and obeys the evolution equation (4). Defining the equivalent to F_2 for fracture functions, i.e.

$$M_{2p}^p(x,z,Q^2) \equiv x\sum_i e_i^2 M_p^{i,p}(x,z,Q^2), \tag{58}$$

and taking into account the shift from z to ξ, the relation between this function and the leading proton structure function is quite apparent.

Similarly the differential cross section for diffractive deep inelastic scattering is usually written in terms of the diffractive structure function $F_2^{D(3)}$

$$\frac{d^3\sigma^D}{d\beta\, dQ^2\, dx_{I\!P}} \equiv \frac{4\pi\alpha^2}{\beta\, Q^4}\left(1-y+\frac{y^2}{2}\right) F_2^{D(3)}(\beta,Q^2,x_{I\!P})\,, \tag{59}$$

where $x_{I\!P} \equiv \xi$, and the variable β is used instead of x. To collect the data, the integration over the small transverse momentum of the final state proton is implied, i.e. on the variable $t=(P-P')^2$. When the integration over the variable t is not performed, then the "extended fracture functions" [11], with an explicit dependence on that variable and obeying homogeneous evolution equations can be used. These do correspond to the diffractive structure functions $F_2^{D(4)}(\beta,Q^2,x_{I\!P},t)$.

The diffractive region is given by small values of $x_{I\!P}$ ($x_{I\!P} < 0.1$), whereas leading proton data are associated with larger values of $x_{I\!P}$ ($x_{I\!P} > 0.1$).

As a suitable parametrization for the proton-to-proton fracture function $M_{2p}^p(\beta,Q_0^2,x_{I\!P})$ at a given initial scale Q_0^2, is choosen [34] by selecting a simple functional dependence in the variables β and $x_{I\!P}$. The quark singlet component ($M_p^{q,p} \equiv 3M_u^{p/p} = 3M_d^{p/p} = 3M_s^{p/p}$) of the fracture function is parametrized as [34]

$$xM_q^{p/p}(\beta, Q_0^2, x_{I\!P}) = N_s\, \beta^{a_s} (1-\beta)^{b_s}$$
$$\left\{ C_{I\!P}\, \beta\, x_{I\!P}^{\alpha_{I\!P}} + C_{LP}\, (1-\beta)^{\gamma_{LP}}\, (1 + a_{LP} (1 - x_{I\!P})^{\beta_{LP}}) \right\}, \quad (60)$$

and similarly for gluons with the corresponding parameters N_g, a_g and b_g. The normalization constants N_s, $C_{I\!P}$ and C_{LP}, are also properly set. Concerning the evolution the choosen values are $Q_0^2 = 2.5\,\mathrm{GeV}^2$ and $\Lambda_{QCD} = 0.232\,\mathrm{GeV}^2$ in a scheme with a variable number of flavours, where charm and bottom distributions are radiatively generated from their corresponding thresholds. The data of the H1 and ZEUS collaborations [35, 36, 37] have been analysed in [34]. The results can be listed here in Figs. 10–12. The parametrization in terms of fracture functions describes the data over the entire range of Q^2 and β when compared with the diffractive H1 (Figs. 10 and 11) and ZEUS data.

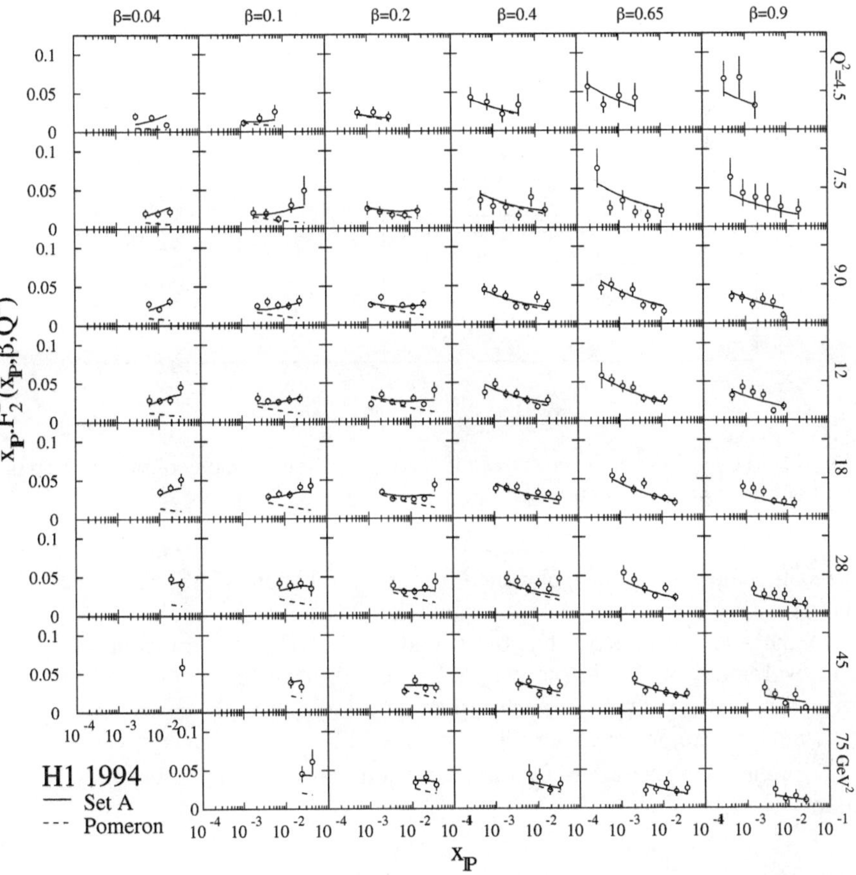

Fig. 10. H1 diffractive data against the outcome of the fracture function parametrization (*solid lines*) and its pomeron-like component (*dashed lines*). From [34]

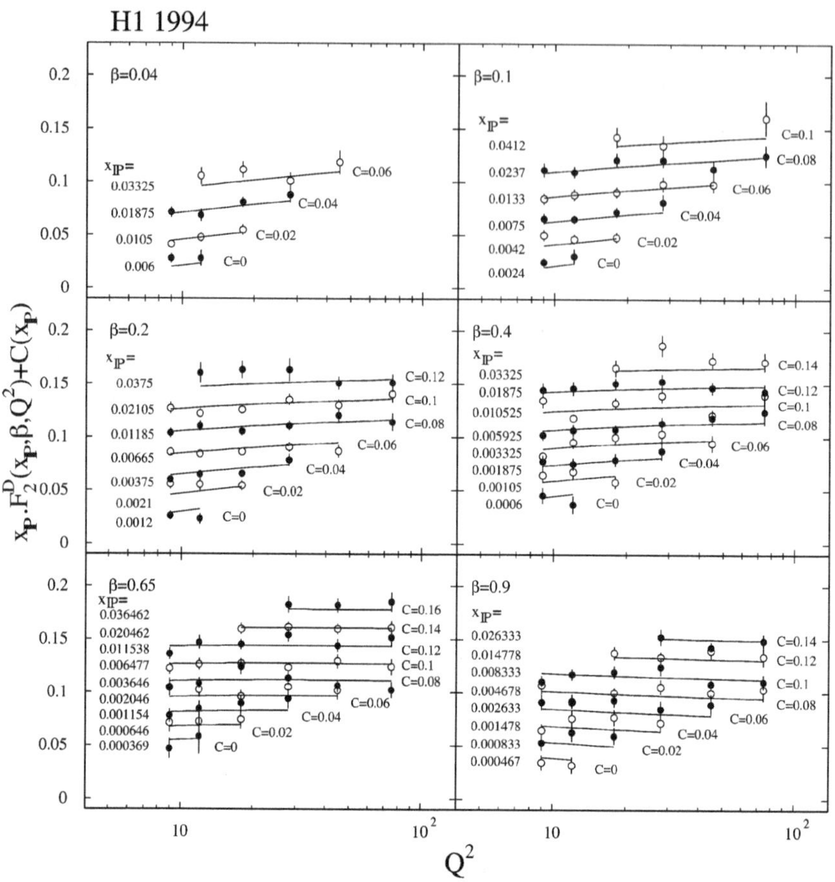

Fig. 11. H1 scale dependence of H1 diffractive data and the one obtained evolving the fracture functions. From [34]

Also the scale dependence, obtained from the evolution, shows agreement with the data over all the range of the values of Q^2 and β.

A more recent analysis has been made by the H1 collaboration [38]. Here a measurement of the diffractive parton distribution functions has been performed by using diffractive parton distribution functions.

The data are presented in the form of a "diffractive reduced cross section" $\sigma_r^{D(3)}$, related to the differential cross section measured experimentally by the equation [38]

$$\frac{d^3\sigma^{ep\to eXY}}{dx_{I\!P}\,dx\,dQ^2} = \frac{2\pi\alpha^2}{xQ^4} \cdot Y_+ \cdot \sigma_r^{D(3)}(x_{I\!P}, x, Q^2) \ , \tag{61}$$

where $Y_+ = 1+(1-y)^2$. Similarly to what done for the inclusive deep inelastic case [46], the reduced e^+p cross section depends on the diffractive structure

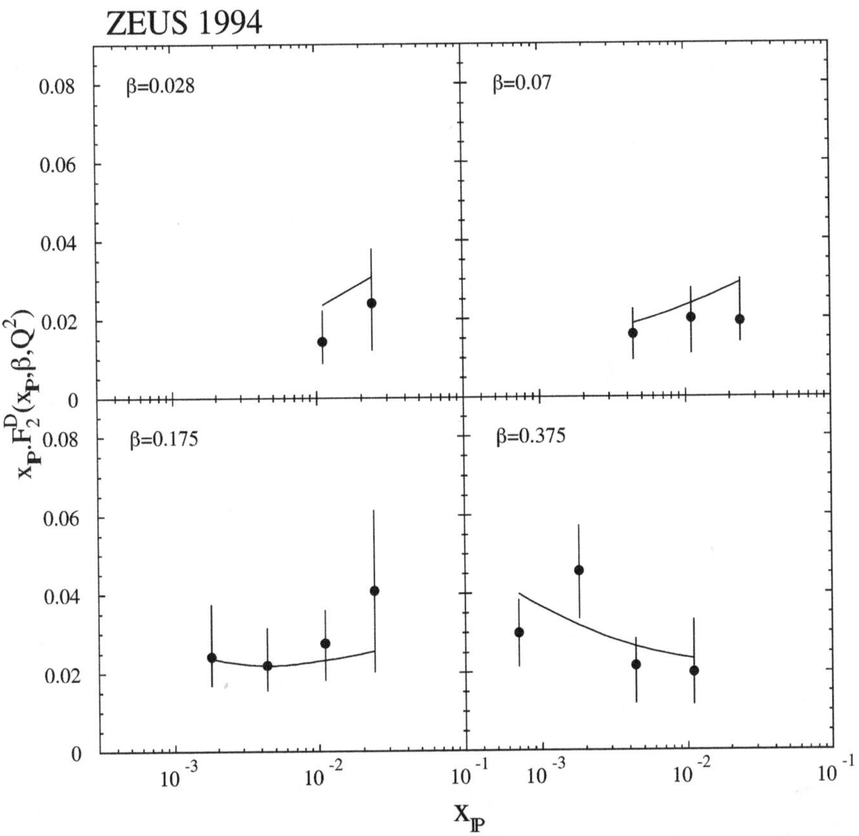

Fig. 12. ZEUS diffractive data against the expectation coming from the fracture function parametrization. From [34]

functions $F_2^{D(3)}$ and $F_L^{D(3)}$ in the one-photon exchange approximation according to the relation

$$\sigma_r^{D(3)} = F_2^{D(3)} - \frac{y^2}{Y_+} F_L^{D(3)} . \tag{62}$$

Since for y not too close to unity, $\sigma_r^{D(3)} = F_2^{D(3)}$ holds to very good approximation. differently from the previous measurements of inclusive diffractive deep inelastic scattering at HERA, where the data were presented in terms of $F_2^{D(3)}$ in [38] are given in terms of $\sigma_r^{D(3)}$.

The charged current measurements of the data are integrated over some or all of the kinematic variables. They are presented as a total cross section and single differentially in either $x_{I\!P}$, β or Q^2.

The Q^2 dependence is quantified by fitting the data at fixed $x_{I\!P}$ and β to the form

$$\sigma_r^{D(3)}(x_{I\!P}, Q^2, \beta) = a_{\mathrm{D}}(\beta, x_{I\!P}) + b_{\mathrm{D}}(\beta, x_{I\!P}) \ln Q^2 , \tag{63}$$

such that $b_{\mathrm{D}}(\beta, x_{I\!P}) = \left[\partial\sigma_r^{D(3)}/\partial \ln Q^2\right]_{\beta, x_{I\!P}}$ is the first logarithmic Q^2 derivative of the reduced cross section.

As discussed before QCD hard scattering collinear factorization, when applied to diffractive deep inelastic scattering implies that the cross section for the process $ep \to eXY$ can be written in terms of convolutions of partonic cross sections $\hat{\sigma}^{ei}(x, Q^2)$ with diffractive parton distribution functions f_i^D as

$$\mathrm{d}\sigma^{ep\to eXY}(x, Q^2, x_{I\!P}, t) = \sum_i f_i^D(x, Q^2, x_{I\!P}, t) \otimes \mathrm{d}\hat{\sigma}^{ei}(x, Q^2) \,. \tag{64}$$

The partonic cross sections are the same as those of inclusive deep inelastic scattering, and the functions f_i^D represent probability distributions for the i-th parton in the proton, under the constraint that the proton is scattered to a particular system Y with specified four-momentum. They are not known from first principles, but can be determined from fits to the data using the evolution equations [26].

The analysis is carried by using input parameters describing the diffractive parton distribution functions at a starting scale Q_0^2 for QCD evolution are adjusted to obtain the best description of the data after NLO DGLAP [47] evolution to $Q^2 > Q_0^2$ and convolution of the diffractive parton distribution functions with coefficient functions. The fit is performed in the $\overline{MS}$ renormalization scheme. The strong coupling is set via $\Lambda_{QCD}^{(3)} = 399 \pm 37$ MeV for three flavours. The evolution of the diffractive reduced cross section with Q^2 is compared with that of the inclusive deep inelastic reduced cross section σ_r by forming the ratio

$$\left.\frac{\sigma_r^{D(3)}(x_{I\!P}, x, Q^2)}{\sigma_r(x, Q^2)}\right|_{x, x_{I\!P}} , \tag{65}$$

at fixed x and $x_{I\!P}$, by using parameterizations of the σ_r previously analysed data. This ratio is shown multiplied by $x_{I\!P}$ in Figs. 13–17 as a function of Q^2 for all measured $x_{I\!P}$ and $x = \beta x_{I\!P}$ values.

From the figures it appears a well-defined pattern of scaling violation showing that diffractive deep inelastic scattering do obey evolution equations as for ordinary inclusive deep inelastic scattering. At large values of $x_{I\!P}$ it is apparent the presence of partonic degrees of freedom with a perturbative QCD evolution. Diffractive parton distribution functions are dominated by the gluon distribution [36].

Factorization in diffractive deep inelastic scattering has been checked by using diffractive parton distribution functions by comparing diffractive events with final state cross sections for jets [53, 54] and heavy quarks [55]. These comparisons show consistency among them.

On the other hand, the apparent failure of the factorization in hadron–hadron collider data [56] when deep inelastic parton diffractive distributions are used, leaves still open the question if the input of deep inelastic data can

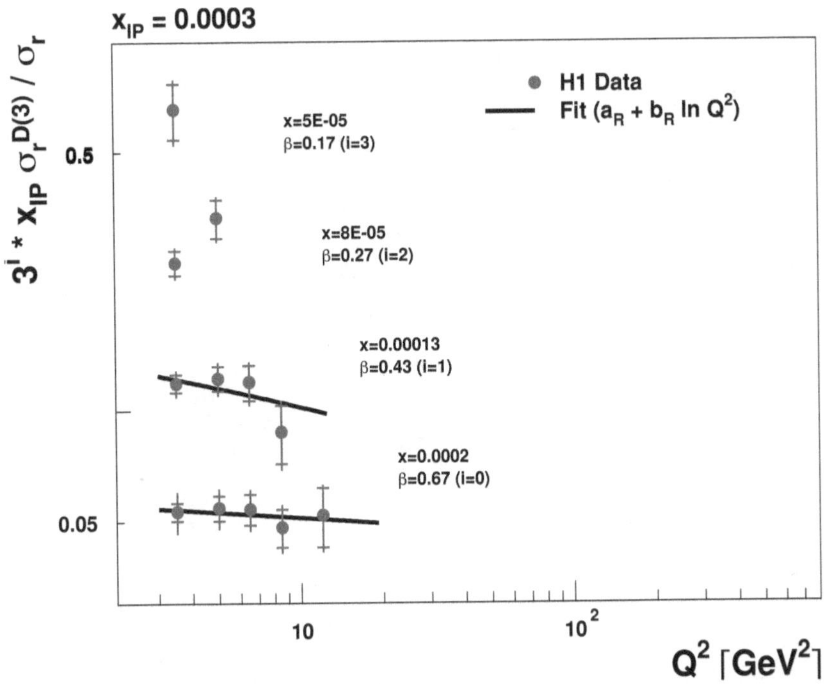

Fig. 13. The ratio of the diffractive to the inclusive reduced cross section, multiplied by $x_{\mathbb{P}}$ and shown as a function of Q^2 for fixed x and fixed $x_{\mathbb{P}} = 0.0003$. The data are multiplied by a further factor of 3^i for visibility, with i as indicated. The inner and outer error bars represent the statistical and total uncertainties, respectively. Normalization uncertainties are not shown. The results of fits of a linear dependence on $\log Q^2$ to the data are also shown. Picture taken from [38]

be used at the hadron colliders. These and other related issues have been recently discussed [57] and have been investigated by the diffraction working group at the HERA–LHC Workshop [58].

From the analyses of the H1 [38] and ZEUS [52] collaborations it clearly emerges the picture of a perturbative QCD description of the diffraction via factorized diffractive parton distributions as indicated by the fracture function approach.

3.2 Higgs Production

The possibility of producing the Higgs boson via a diffractive reaction by using fracture functions has been proposed by Graudenz and Veneziano [48]. This rests on the factorization hypothesis for semi-inclusive hard processes in QCD at the hadronic colliders. In principle, the diffractive production of the Standard Model Higgs boson at LHC can be studied by using only, as input, diffractive hard-processes data of the type recently collected and analysed

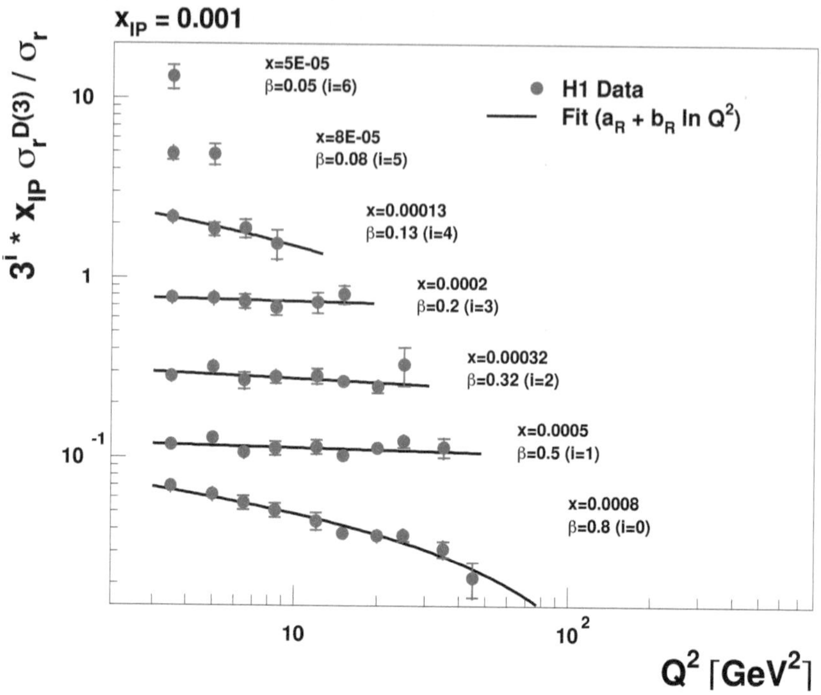

Fig. 14. As in Fig.13 with $x_{I\!P} = 0.001$. From [38]

by the H1 and ZEUS collaborations at HERA. In [48] the existing HERA data have been combined with a simple pomeron exchange picture. A large spread in the Higgs boson production cross section is found, depending on the input parametrization of the pomerons' parton content. In particular, if the pomeron gluon density $f_g x_{I\!P}(x)$ is peaked at large β for small scales, single diffractive events can represent a sizeable fraction of all produced Higgs bosons with an expected better-than-average signal-to-background ratio. Different analyses are also possible, since, as the more precise HERA data have shown, a hard perturbative QCD approach to diffractive processes, in defined kinematical regions has to be preferred to the hadron–pomeron vertex relying in a parametrized diffractive parton distribution. It would be interesting to test the possibility of diffractive Higgs production (modulo the factorization problem) by using the present available H1 and ZEUS data, as well as the future (and probably more precise) data.

3.3 Polarized Processes

The application of fracture functions to describe polarized processes has been studied by De Florian, Garcia Canal, Sampayo and Sassot [50, 51]. The aim is to extend to the target fragmentation region the description of polarized processes. They discuss the factorization of the collinear singularities

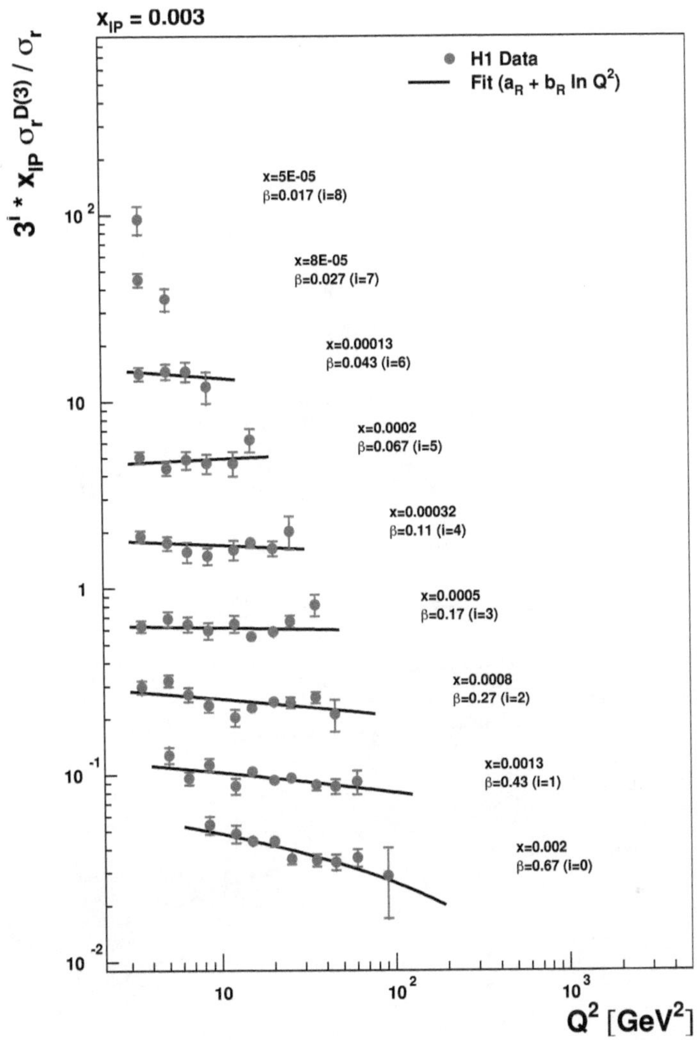

Fig. 15. As in Fig. 13 with $x_{IP} = 0.003$. From [38]

related to the polarized processes, particularly those which are absorbed in the redefinition of the spin-dependent analogue of fracture functions.[1] In [50] they show that, with the inclusion of polarized fracture functions, it is possible to consistently factorize all the collinear singularities that occur and that the

[1] An extensive discussion on the role of the $U(1)_A$ anomaly in QCD phenomenology can be found in the contribution. In this review the issues related to spin physics and to the theoretical implications as well as the use of fracture functions in experiments on semi-inclusive polarized deep inelastic scattering are also discussed.

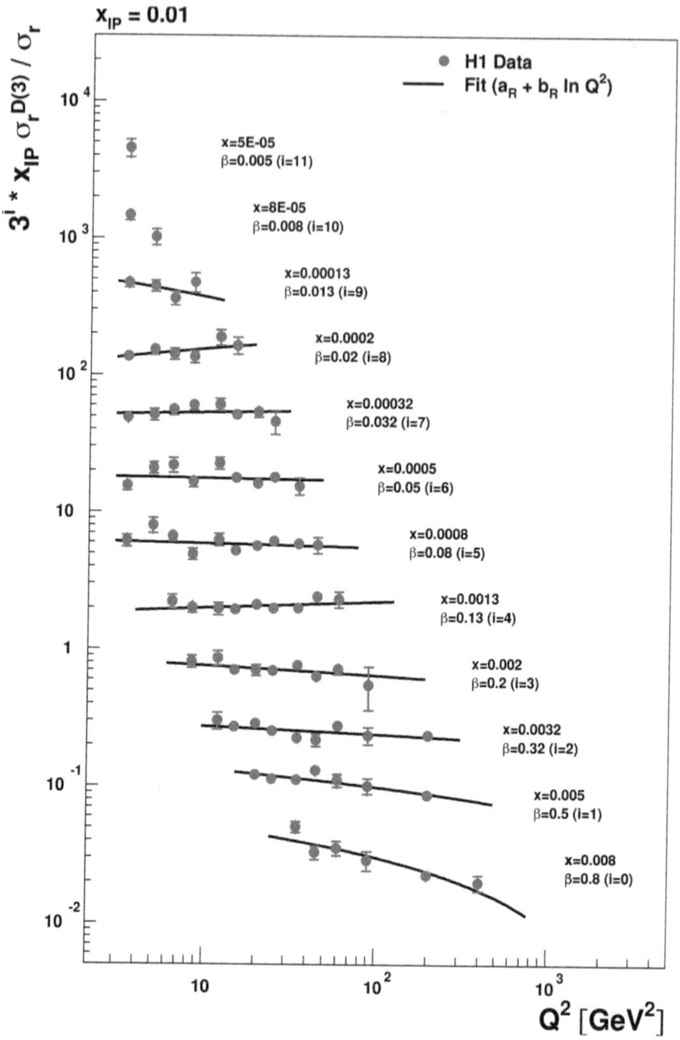

Fig. 16. The ratio of the diffractive to the inclusive reduced cross section, multiplied by $x_{\mathbb{P}}$ and shown as a function of Q^2 for fixed x and fixed $x_{\mathbb{P}} = 0.01$. See the caption of Fig. 13 for further details. From [38]

formalism can be straightforwardly applied in order to factorize unwanted finite soft contributions. In this way the conservation of the non-singlet axial current and the non-conservation of the singlet one, as dictated by the anomaly result is preserved. This requirement allows the definition of polarized parton and fracture distributions intimately related to the fraction of the nucleon spin carried by partons. The definition of an universal and physically meaningful factorization scheme for both current and target fragmentation, consistent with those used in totally inclusive spin-dependent deep inelastic scattering and unpolarized electron–proton annihilation, allows to perform an

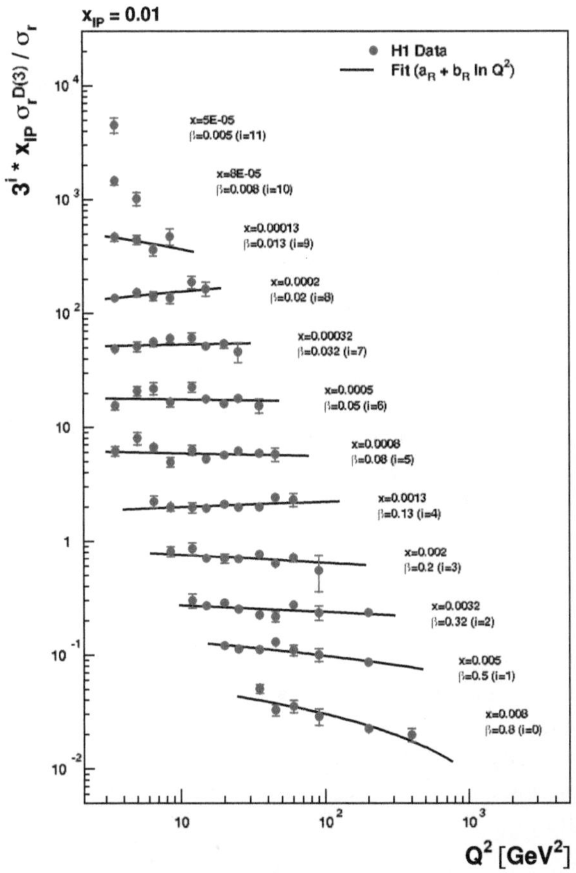

Fig. 17. As in Fig. 16 with $x_{I\!P} = 0.03$. See the caption of Fig. 13 for further details. From [38]

unambiguous $O(\alpha_s)$ analysis of inclusive experiments. A phenomenological analysis has been carried in [51], where results of different experiments have been discussed.

The potential relevance of the fracture functions to describe spin-dependent distributions has been advocated by Teryaev [42]. They may be applied at fixed target energies and may also include interference and final state interaction, providing a source for azimuthal asymmetries at HERMES and polarization at NOMAD. Accordingly, the work of [44] can be rephrased in terms of fracture functions (see also [43]).

Kotzinian [45] has recently discussed the role of hadronization mechanism in polarization phenomena in semi-inclusive deep inelastic scattering and a purity method for extraction of polarized distribution functions. By using a Monte Carlo event generator producing hadrons via current quark as well as target diquark fragmentation or light cluster decays. Since the purity method assumes that only quark fragmentation gives contribution to hadron production in the current fragmentation region, it turns out that the ignorance of contributions from target diquark fragmentation and cluster decays to asymmetry can be the source of incorrect values of polarized quark distributions extracted by the purity method.

3.4 Fracture Functions and Heavy Quark Production

The possibility of using fracture function formalism to describe the production of heavy quarks in the target fragmentation region has been studied by Graudenz [59]. Several interesting features are advocated in favour of this possibility:

i) Fixed-target experiments permit the study of hadron production in the target fragmentation region.
ii) The tagging of specific particles in the target fragments can be employed to introduce a bias in the hard scattering process towards a specific flavour content. The case of hadrons containing a heavy quark is particularly attractive because of the clear experimental signatures and the applicability of perturbative QCD. One of such cases, modulo factorization, can be considered also the production of heavy quarks at the hadron colliders.
iii) The standard approach to one-particle inclusive processes based on fragmentation functions is valid in the current fragmentation region and for large transverse momenta p_t in the target fragmentation region, but it fails for particle production at small p_t in the target fragmentation region. A collinear singularity, which cannot be absorbed in the standard way into the phenomenological distribution functions, prohibits the application of this procedure.

This situation, remedied by the introduction of fracture functions which describe particle production in the target fragmentation region, and can be viewed as correlated distribution functions in the momentum fractions of the observed particle and of the parton initiating the hard scattering process. In [59] it is shown, in a next-to-leading-order calculation for the case of deep inelastic lepton–nucleon scattering, that the additional singularity can be consistently absorbed into the renormalized target fragmentation functions on the one-loop level. The formalism is applied to the production of heavy quarks. The renormalization group equation of the target fragmentation functions for the perturbative contribution is solved numerically, and the results of a case study for deeply inelastic lepton–nucleon scattering at DESY (H1 and ZEUS at HERA), at CERN (NA47) and at Fermilab (E665) are discussed.

Higgs production and the possible association with heavy quark fracture functions have been recently considered in [49].

3.5 Multiple Inclusive Processes

By extending the definition of single hadron fracture function to include a second hadron one has that for a generic double inclusive DIS process, $l+P \to$

$l' + h_1 + h_2 + X$ the corresponding cross sections at leading logarithm level is [40]

$$\frac{d\sigma}{dx dz_1 dz_2 dQ^2} = (1-x) \sum_{i=q,\bar{q}} e_i^2 \Big[\frac{x}{1-x} F_p^i(x,Q^2) D_2^{h_1 h_2}(z_1,z_2,Q^2)$$
$$+ M^i_{2,h_1h_2/p}(x,z_1,z_2,Q^2) + \Big(M^i_{1,h_1/p}(x,z_1,Q^2) D^{h2}_{1,i}(x,z_2)$$
$$+ M^i_{1,h_2/p}(x,z_2,Q^2) D^{h_1}_{1,i}(z_1,Q^2) \Big) \Big] \tag{66}$$

The first term takes into account the current fragmentation of h_1 and h_2, the second the target fragmentation, and for completeness we also added mixed terms in which one hadron is produced in the opposite fragmentation region with respect to the other.

In the following we will discuss evolution equations for $M^i_{2,h_1h_2,p}$ (x, z_1, z_2, Q^2). M_2 gives the conditional probability of finding an active quark i with fraction x of the incoming hadron while two secondary hadrons are produced with momentum fraction z_1 and z_2 with respect to the incoming hadron momentum. The evolution equation for M_2 can be obtained [40]

$$Q^2 \frac{\partial M_2(x,z_1,z_2,Q^2)}{\partial Q^2} = \frac{\alpha_s(Q^2)}{2\pi} \Bigg\{ \int_{\frac{x}{1-z_1-z_2}}^1 \frac{du}{u} P(u) M_2(x/u,z_1,z_2,Q^2)$$
$$+ \int_{\frac{x}{1-z_2}}^{\frac{x}{x+z_1}} \frac{du}{u} \frac{u}{x(1-u)} \hat{P}(u) M_1(x/u,z_2,Q^2) D\Big(\frac{z_1 u}{x(1-u)},Q^2\Big) +$$
$$+ \int_x^{\frac{x}{x+z_1+z_2}} \frac{du}{u} \frac{u^2}{x^2(1-u)^2} \hat{P}(u) F(x/u,Q^2) D_2\Big(\frac{z_1 u}{x(1-u)}, \frac{z_2 u}{x(1-u)}, Q^2\Big) \Bigg\}. \tag{67}$$

It can be extended to M_n

$$Q^2 \frac{\partial M_n(x,z_1,..,z_n,Q^2)}{\partial Q^2} = \frac{\alpha_s(Q^2)}{2\pi} \Bigg\{ \int_{\frac{x}{1-\sum_i^n z_i}}^1 \frac{du}{u} P(u) M_n(x/u,z_1,..,z_n,Q^2)$$
$$+ \sum_{j=1}^{n-1} \int_{\frac{x}{1-\sum z_i}}^{\frac{x}{x+\sum z_i}} \frac{du}{u} \Big[\frac{u}{x(1-u)}\Big]^j \hat{P}(u) M_{n-j}(x/u,z_{n-j},..,z_n,Q^2)$$
$$D_j\Big(\frac{z_1 u}{x(1-u)},..,\frac{z_j u}{x(1-u)},Q^2\Big)$$
$$+ \int_x^{\frac{x}{x+\sum_i^n z_i}} \frac{du}{u} \Big[\frac{u}{x(1-u)}\Big]^n \hat{P}(u) F(x/u,Q^2) D_n\Big(\frac{z_1 u}{x(1-u)},..,\frac{z_n u}{x(1-u)},Q^2\Big) \Bigg\} \tag{68}$$

where the numeric subscript represents the number of hadrons described by a given distribution and the remaining partonic and hadronic indexes have been suppressed for simplicity.

4 Jet Cross sections and Fracture Functions

In the previous section we have shown that theoretical predictions for inclusive multi-particle distributions, although available, these become more involved as we increase the number of identified particles in the final state. The situation is also worsened, at the experimental level, by the high-multiplicity nature of events at present and future colliders. Even in the next-to-simplest case, M_2, an analysis seems to be prohibitive. Outgoing hadrons however emerge as clusters of particles in defined portions of momentum space, a signature of the dominant collinear branching scheme of QCD dynamics. For this reason jets become the natural representation of hadronic activity. Perturbative calculations with an arbitrary number of partons in the final state and experimental jet observables can be quantitatively compared only once a common jet-algorithm is chosen and used on both the theoretical and experimental level. Let us focus now on DIS jet cross sections. Inclusive DIS structure functions can be decomposed in terms of n-particles exclusive structure functions $F_2^{(n)}$ as [61]

$$\frac{1}{\sigma}\frac{d\sigma}{dxdQ^2} \equiv F_2(x,Q^2) = x\sum_{i=q,\bar{q}} e_i^2 q_i(x,Q^2) = \sum_{n=1}^{\infty} F_2^{(n)} = \sum_{n=1}^{\infty}\frac{d\sigma^{(n)}}{dxdQ^2} \tag{69}$$

Analogously for jet cross sections, in terms of suitable jet algorithm, as for example the one defined in [41], the jet exclusive structure functions show a factorized structure of the type

$$F_2^{(n)}(x,Q^2;E_t^2,y_{\rm cut}) = \sum_{i=q,\bar{q}}\int_x^1 \frac{dz}{z} F_p^i(x/z,\mu_F^2) R_{2,i}^{(n)}\left(z,\alpha_s,\frac{Q^2}{E_t^2},y_{\rm cut}\right) \tag{70}$$

where $y_{\rm cut}$ represents the jet structure resolution parameter [41] and is defined in terms of an arbitrary perturbative scale E_t^2, with $\Lambda^2 \ll E_t^2 \le Q^2$. In (70), initial state collinear divergences are factorized into parton distributions functions. The jet coefficients $R_{2,i}^{(n)}$ are calculable in perturbation theory and again depend on the particular jet algorithm chosen, as indicated by the dependence on E_t^2 and y_{cut}. Since we are interested in initial state jets, i.e. jets originating by the space-like struck parton, we briefly recall some features of the approach of [41]. Initial state jets in arbitrary number have been accounted for by using a generating functional method [41]. The n-jet cross sections were then constructed with an iterative block structure. Given the Sudakov form factor $\Delta(Q_i^2,Q_j^2)$, which inhibites emissions off the struck parton lines i in between the two scales Q_i^2 and Q_j^2 as

$$\Delta_i(Q_i^2,Q_j^2) \equiv \exp\Big[-\sum_j\int_{Q_i^2}^{Q_j^2}\frac{dt}{t}\int_{\frac{Q_i^2}{Q_j^2}}^{1-\frac{Q_i^2}{Q_j^2}} dz\frac{\alpha_s(t)}{2\pi}\hat{P}_{ji}(z)\Big] \tag{71}$$

it guarantees that no hadronic activity takes place between each couple of jets. The emission of the real partons is then controlled by real splitting functions $\hat{P}$ [9]. Their subsequent decays are taken into account via the jet function [60] $J(Q^2, k^2)$

$$J(Q^2, k^2) = \sum_h \int_0^1 dz\, d^h(z, Q^2, k^2)\,, \tag{72}$$

where $d^h(z, Q^2, k^2)$ is the probability that an initial parton with mass Q^2 decays into a parton with a longitudinal momentum fraction z with respect to the parent parton and with a mass $k^2 \ll Q^2$. The d's functions satisfy the properties

$$\int_0^{Q^2} dk^2 d(z; Q^2, k^2) \equiv D(z; Q^2); \qquad \int_0^{Q^2} dk^2 J(Q^2, k^2) = 1\,. \tag{73}$$

In the construction of the n-jet cross sections, an iterative block-like structure $\mathcal{G}$ is associated to each jet insertion

$$\mathcal{G}_{ik}^{n_{jet}=1}(u, Q_i^2, Q_j^2) \equiv \Delta_{ij}(Q_i^2, Q_j^2)\, \hat{P}_{lj}^m(u)\, J_m(Q_j^2, Q_0^2)\, \Delta_{lk}(Q_j^2, Q_k^2) \tag{74}$$

Along the struck parton line with ordered virtualities $Q_0^2 < \ldots < Q_i^2 < Q_j^2 < Q_k^2 < \ldots < Q^2$ representing the scales where real parton emissions are allowed. Here Q_0^2 is intended to be the factorization scale, while Q^2 is the virtuality of the parton which directly interacts with the photon. Let us discuss differences between this approach and the jet calculus one. The structure of (74) is obtained by construction at the exclusive level. The evolution function $E(u, Q_i^2, Q_k^2)$ as defined in (6), can be regarded the analogous of as the inclusive level of the function $G_{ik}^{n_{jet}=1}(u, Q_i^2, Q_j^2)$. It takes into account the corresponding of (74) at the inclusive level. It inclusively sums all the radiated partons between Q_i^2 and Q_k^2 and does not describe jet production. We may define a new inclusive distribution [40] that gives the probability of detecting hadrons in a portion of phase space defined by z and t. The sum over the partons i, specified by x and Q^2, and struck by the virtual photon, is understood as in the totally inclusive case (69)

$$\frac{1}{\sigma_{tot}} \frac{d\sigma}{dx dQ^2 dz dt} \equiv \sum_{i=q,\bar{q}} e_i^2 \mathcal{M}^i(x, Q^2, z, t) \tag{75}$$

We interpret such hadronic distributions as characterized only by the partonic indexes, where x and Q^2 are determined by the scattered lepton variables, as for the inclusive case in (69). The variables z and t are the fraction of the longitudinal momentum and the invariant transferred momentum squared of the final state hadrons h_i, respectively. The hadrons that are contained in a portion of phase space $\mathcal{R}$ limited by the constraints

$$\mathcal{R}: \; t_i = -(P - h_i)^2 < t, \quad t_0 \le t \le Q^2\,. \tag{76}$$

Once this procedure is followed, we may obtain z by summing the fractional longitudinal momenta of the hadrons satisfying the phase space constraint of (76):

$$z = \sum_i z_i, \quad h_i \in \mathcal{R} \tag{77}$$

By using n-particle exclusive cross sections,

$$\Sigma^{(n)}_{excl} \equiv \frac{1}{n!} \frac{d^{2n+2}\sigma^{(n)}}{dxdQ^2 \prod_{m=1}^{n} dz_m dt_m} \tag{78}$$

which may be obtained directly from experiments we may derive, in analogy with (69), the hadronic distributions in (75) as

$$\frac{1}{\sigma_{tot}} \frac{d\sigma}{dxdQ^2dzdt} \equiv \frac{1}{\sigma_{tot}} \sum_{k=1}^{\infty} \Big\{ \prod_{m=1}^{k} \int_{t_0}^{t} dt_m \int_0^1 dz_m \Big\} \Sigma^{(k)}_{excl}\, \delta\Big(z - \sum_{k=1}^{n} z_k\Big). \tag{79}$$

with the $\mathcal{R}$ phase space constraints already implemented in the cross sections. The minimum value t_0 corresponds to the beam pipe acceptance where hadrons, being not measured, have not to be counted in $\mathcal{M}^i$. We may recover the structure function F_2 by simply integrating over the hadronic variables

$$\frac{d\sigma}{dxdQ^2} = \int_0^1 dz \int_{t_0}^{Q^2} dt \frac{d\sigma}{dxdQ^2dzdt} \tag{80}$$

The dynamics of the evolution along the struck parton line, as seen from a leading logarithmic accurate evolution equation, can be sketched as follows: partons are emitted strongly ordered in t, with increasing values of t along the line towards the virtual photon, while softest k_t-emissions are closest to the proton remnant. In such configurations, planar diagrams give, as is well known in the inclusive case, the leading logarithmic contributions to the cross sections, which are actually resummed by parton evolution functions $E(x, Q_i^2, Q_j^2)$. Let us define, by using jet calculus rules as in the previous sections, the semi-inclusive extended functions $\mathcal{M}^i$ in terms of parton evolution functions $E(x, Q_i^2, Q_j^2)$

$$\mathcal{M}^j(x, Q^2, z, t) = \int_x^{1-z} \frac{dw}{w} \mathcal{M}^i(w, t, z, t) E_i^j(x/w, t, Q^2) \tag{81}$$

The $\mathcal{M}^i(w, t, z, t)$ are intended as the distributions corresponding to a parton with longitudinal momentum w and scale t and a final state hadrons configuration specified by (76) and (77). The convolution limits are determined by using momentum conservation. Once $\mathcal{M}^i(w, t, z, t)$ is measured, the evolution can be obtained by differentiating (81) with respect to Q^2

$$Q^2 \frac{\partial}{\partial Q^2} \mathcal{M}^j(x, Q^2, z, t) = \frac{\alpha_s(Q^2)}{2\pi} \int_{\frac{x}{1-z}}^{1} \frac{du}{u} P_k^j(u) \mathcal{M}^k(x/u, Q^2, z, t) \tag{82}$$

This evolution equation actually resums large logarithm of the type $\alpha_s \log \frac{Q^2}{t}$. In reality, however, t-ordering is only partially realized. Higher-order corrections together with large angle emissions, i.e. fixed order matrix element, produce partons that, even if originated by parent parton with a hard t-scale, end up along the time-like shower in final state hadrons with soft values of t. Since experiments and also the clustering procedure do not distinguish the origin of such hadrons, the description becomes increasingly reliable as much as the accuracy in describing the partonic shower increases. This can be achieved, for instance, by inserting appropriate higher loop splitting functions

$$P_k^j(u) = P_k^{j\,(0)}(u) + \frac{\alpha_s}{2\pi} P_k^{j\,(1)}(u) + \dots \tag{83}$$

The coefficients corresponding to the two-loop vertex functions, are the ones given in [24, 25] allowing a next-to-leading logarithmic accuracy evolution, do provide a space-like jet calculus formulation via fracture functions.

At variance with inclusive case, $\mathcal{M}^i$ is more sensitive to the details of the struck parton evolution since a *portion* of final state hadrons is observed semi-inclusively and not just summed over. The evolution equations, (82), is formally equivalent to the one for one-particle inclusive extended fracture functions, [23]. The reason for this similarity is that, in (82), is actually the parton with the hardest $t_i = \bar{t}$ in the region $\mathcal{R}$ which pilots the evolution of $\mathcal{M}$, as in the one-particle case. An illustrative example is found in the diffractive data analysis at HERA [38, 52]. Whenever we perform a semi-inclusive measurement with a leading baryon detected in the forward spectrometer, we can of course describe the cross sections in terms of one-particle extended fracture functions M. On the other hand, a diffractive event is also specified by observing a gap in the forward hadronic activity while the proton or its low-mass excitation escape undetected. This ensemble of particles, all of which have a $t_i < \bar{t}$ is what we call, collectively, $\mathcal{M}$.

5 Conclusions

Fracture functions represent a new approach and a useful theoretical tool to describe initial state radiation in QCD semi-inclusive processes. A series of successful applications have been already explored. Theoretical and phenomenological developments are underway. Higher statistics data from HERA and from the higher energy experiments at hadron colliders will constitute further important tests for the fracture function idea.

Acknowledgements

This note has been written to celebrate the 65th birthday of Gabriele Veneziano. It is a great privilege to work with Gabriele, to share with him the bright intuition, the vivid imagination, the sharp reasoning and the profound knowledge of physics. I would like to express to Gabriele also the deep gratitude for the generosity, for the enthusiasm and, sometimes, for the encouragement he has been able to transmit, unchanged in the course of the years, as a mentor and as a friend and for the continuing enjoyable collaboration. I wish to Gabriele to be happy and to continue to do physics, in his extraordinary way, for many more years to come, for his pleasure and ours. I have much benefited from conversations and discussions with several friends and colleagues. In addition to Gabriele I would like also to thank Gianni Camici, Federico Ceccopieri, Dirk Graudenz and Massimiliano Grazzini, for the collaboration we have had on the topics discussed here.

References

1. R.P. Feynman: Phys. Rev. Lett. **23**, (1969) 1415
   J.D. Bjorken, E.A. Paschos: Phys. Rev. **185**, (1969) 1975
2. H. Fritzsch, M. Gell-Mann, H. Leutwyler: Phys. Lett. B **47**, 365 (1973)
3. H. D. Politzer: Phys. Rev. Lett. **30**, 1346 (1973); D. J. Gross, F. Wilczek: Phys. Rev. Lett. **30**, 1343 (1973)
4. R. P. Feynman: *Photon-Hadron Interactions* (W. A. Benjamin Advanced Book Program, New York, 1972)
5. L. Trentadue, G. Veneziano: Phys. Lett. B **323**, 201(1994)
6. R . Taylor: *An Historical Review of Lepton Proton Scattering*, SLAC-PUB-5832 (June 1992); G Altarelli, Phys. Rep. **81**, 1 (1992)
7. P. V. Landshoff: in Proc. 27th Rencontre de Moriond on *Perturbative QCD and Hadronic Interactions* (22–28 March, 1992), ed. by J. Tran Thanh Van (Editions Frontieres), p. 393 and references therein, Gif-sur-Yvette, France
8. D. Amati, R. Petronzio, G. Veneziano: Nucl. Phys. B **140** ,54 (1978), B **146** 29(1978); R. K. Ellis, H. Georgl, M. Machacek, H. D. Politzer, G. G. Ross: Phys. Lett. B **78** 281(1978); Nucl. Phys. B **152**, 285 (1979)
9. K. Konishi, A. Ukawa, G. Veneziano: Phys. Lett. B **78**, 243 (1978) Phys. Lett. B **80**, 259 (1979); Nucl. Phys. B **157**, 45 (1979)
10. D. Graudenz: Nucl. Phys. B **432**, 351 (1994)
11. M. Grazzini, L. Trentadue, G. Veneziano: Nucl. Phys. B **519**, 394 (1998)
12. A. H. Mueller: Phys. Rev. D **18**, 3705 (1978)
13. L. Baulieu, E.G. Floratos, C. Kounnas: Nucl. Phys. B **166**, 321 (1980)
14. T. Munehisa: Prog. Theor. Phys. **67**, 882 (1982)
15. J.C. Taylor: Phys. Lett. B **73**, 85 (1978); Y. Kazama, Y.P. Yao: Phys. Rev. Lett. **41**, 611 (1978); Phys. Rev. D **19**, 3111 (1979); T. Kubota: Nucl. Phys. B**165**, 277 (1980); L. Baulieu, E.G. Floratos, C. Kounnas: Phys. Rev. D **23**, 2464 (1981)
16. G. Altarelli, R.K. Ellis, G. Martinelli, S.Y. Pi: Nucl. Phys. B **160**, 301 (1979)
17. S. Gupta, A.H. Mueller: Phys. Rev. D **20**, 118 (1979)
18. G. Sterman: Phys. Rev. D **17**, 2773 (1978)
19. J.C. Collins, D.E. Soper, G. Sterman: in *Perturbative Quantum Chromodynamics*, ed. by A.H. Mueller (World Scientific, Singapore, 1989)
20. M. Grazzini: Nucl. Phys. B **518**, 303 (1998); see also M. Grazzini: Phys. Rev. D **57**, 4352 (1998)
21. A. Berera, D.E. Soper: Phys. Rev. D **50**, 4328 (1994)

22. J. Collins: Phys. Rev. D **57**, 305 (1998); Erratum ibid. D **61**, 019902 (2000)
23. G. Camici, M. Grazzini, L. Trentadue: Phys. Lett. B **439**, 382 (1998)
24. A. Daleo, C.A. Garcia-Canal, R. Sassot: Nucl. Phys. B **662**, 334 (2003)
25. A. Daleo, R. Sassot: Nucl. Phys. B **673**, 357 (2003)
26. V. Gribov, L. Lipatov: Sov. J. Nucl. Phys. **15**, 438 (1972) [Yad. Fiz. **15**, 781 (1972)]; ibid. **15**, 675 (1972) [Yad. Fiz. **15**, 1218 (1972)]; Yu. L. Dokshitzer: Sov. Phys. JETP **46**, 641 (1977) [Zh. Eksp. Teor. Fiz. **73**, 1216 (1977)]; G. Altarelli, G. Parisi: Nucl. Phys. B **126**, 298 (1977)
27. A. Daleo, De Florian, R. Sassot: Phys. Rev. D **71**, 034013 (2005); A. Daleo, R. Sassot: Phys. Rev. D **73**, 054014 (2006)
28. F. A. Ceccopieri, L. Trentadue: Phys. Lett. B **636**, 310 (2006)
29. A. Bassetto, M. Ciafaloni, G. Marchesini: Nucl. Phys. B **163**, 477 (1980)
30. X. Ji, J. Ma, F. Yuan: Phys. Rev. D **71**, 034005 (2005)
31. J.D. Bjorken, J.B. Kogut, Phys. Rev. D **8**, 1341 (1973)
32. G. Ingelman,P. Schlein: Phys. Lett. B **152**, 256 (1985)
33. F. Hautmann, Z. Kunszt, D. E. Soper: Nucl. Phys. B **563**, (1999) 153; Phys. Rev. Lett. **81**, (1998) 3333
34. D. de Florian, R. Sassot: Phys. Rev. D **58**, 054003 (1998)
35. A. Prinias [H1 and ZEUS Collaborations], Talk given at International Europhysics Conference on High-Energy Physics (HEP 97), Jerusalem, Israel, 19–26 August 1997
36. C. Adloff et al [H1 Collaboration]: Z. Phys. C **76**, 613 (1997)
37. J. Breitweg et al [ZEUS Collaboration]: Eur. Phys. J. C **1**, 81 (1998)
38. A. Aktas et al [H1 Collaboration]: Eur. Phys. J. C **48**, 715 (2006)
39. J. C. Collins, L. Frankfurt, M. Strikman, Phys. Lett. B **307**, 161 (1993)
40. F. Ceccopieri, L. Trentadue, arXiv:0706.4242 [hep-ph], Phys. Lett. B (in press).
41. S. Catani, Yu. L. Dokshitzer, B.R. Webber: Phys. Lett. B **285**, 291 (1992)
42. O. V. Teryaev: Acta Phys. Polon. B **33**, 3749 (2002)
43. O. V. Teryaev: Phys. Part. Nucl. **35**, 524 (2004)
44. S. J. Brodsky, D. S. Hwang, I. Schmidt: Phys. Lett. B **530**, 99 (2002)
45. A. Kotzinian: Phys. Lett. B **552**, 172 (2003)
46. C. Adloff et al [H1 Collaboration]: Eur. Phys. J. C **30**, 1 (2003)
47. W. Furmanski, R. Petronzio: Z. Phys. C **11**, 293 (1982)
48. D. Graudenz, G. Veneziano: Phys. Rev. D **66**, 010001 (2002)
49. F. Maltoni, T. McElmurry, S. Willenbrock: Phys. Rev. D **72**, 074024 (2005)
50. D. de Florian, C. A. Garcia Canal, R. Sassot: Nucl. Phys. B **470**, 195 (1996)
51. D. de Florian, O. A. Sampayo, R. Sassot: Phys. Rev. D **66**, 010001 (2002)
52. S. Chekanov et al. [ZEUS Collaboration]: Nucl. Phys. B **713**, 3 (2005)
53. S. Chekanov et al. [ZEUS Collaboration]: Eur. Phys. J. C **38**, 43 (2004); J. Breitweg et al. [ZEUS Collaboration]: Eur. Phys. J. C **5**, 41 (1998); K. Golec-Biernat, J. Kwiecinski: Phys. Lett. B **353**, 329 (1995); C. Royon et al.: Phys. Rev. D **63**, 074004 (2001)
54. C. Adloff et al. [H1 Collaboration]: Eur. Phys. J. C **6**, 421 (1999); C. Adloff et al. [H1 Collaboration]: Eur. Phys. J. C **20**, 29 (2001)
55. C. Adloff et al. [H1 Collaboration]: Phys. Lett. B **520**, 191 (2001)
56. F. Abe et al. [CDF Collaboration]: Phys. Rev. Lett. **79**, 2636 (1997); F. Abe et al. [CDF Collaboration]: Phys. Rev. Lett. **78**, 2698 (1997); B. Abbott et al. [D0 Collaboration]: Phys. Lett. B **531**, 52 (2002); T. Affolder et al. [CDF Collaboration]: Phys. Rev. Lett. **84**, 5043 (2000); T. Affolder et al. [CDF Collaboration]: Phys. Rev. Lett. **85**, 4215 (2000); V. Abazov et al. [D0 Collaboration]: Phys. Lett. B **574**, 169 (2003)

57. J. Bjorken, Phys. Rev. D **47**, (1993) 101;
E. Gotsman, E. Levin, U. Maor, Phys. Lett. B **309**, 199 (1993);
E. Gotsman, E. Levin, U. Maor, Phys. Lett. B **438**, 229 (1998);
B. Cox, J. Forshaw, L. Lönnblad, JHEP **9910**, 023 (1999);
A. Kaidalov, V. Khoze, A. Martin, M. Ryskin, Phys. Lett. B **567**, 61 (2003)
58. M. Arneodo et al.: in *Proc. of the HERA-LHC Workshop*, ed. by A. De Roeck, H. Jung (CERN-2005-014, 2005), p. 417
59. D. Graudenz: Fortsch. Phys. **45**, 629 (1997)
60. S. Catani, L. Trentadue, Nucl. Phys. B **327**, 323 (1989)
61. C.E. Detar, D.Z. Freedman, G. Veneziano: Phys. Rev. D **4**, 906 (1971)

Part IV

Non-perturbative QCD

Coherence and Incoherence in QCD Jets Dynamics (QCD Jets and Branching Processes)

A. Giovannini[1] and R. Ugoccioni[2]

[1] Theoretical Physics Department, Torino University, Italy, and INFN, Sezione di Torino, Italy
giovannini@to.infn.it
[2] Theoretical Physics Department, Torino University, Italy, and INFN, Sezione di Torino, Italy
ugoccioni@to.infn.it

Abstract. The interpretation of QCD jets as Markov branching processes obtained by solving Konishi–Ukawa–Veneziano equations [1] in the leading logarithmic approximation with a fixed cut-off regularization prescription [2] is reviewed, and its impact in multiparticle dynamics critically examined. Independent intermediate gluon sources (clans) are generated through quark–bremsstrahlung, each source then decays into final partons according to a cascading mechanism dominated by gluon self-interaction. At the hadron level, approximate universal regularities are expected in the different components (or substructures) of the various classes of high-energy collisions. The general behavior of collective variables of final multiplicity distributions is reproduced in terms of the weighted superposition of the above-mentioned regularities controlling the component behaviors of each collision. Predictions of signals of new physics at LHC [3] are reviewed, and perspective of the $1/\mathfrak{N}$ expansion approach [4] indicated.

1 Introduction

The research activity of Gabriele Veneziano is very wide and covers different fields, but usually the emphasis is on his discoveries in dual resonances models and on his contribution to the understanding of string theory as the correct quantum theory of gravity. On the occasion of his 65th anniversary, which motivated the present volume, we would like to point out also the impact of his work on the search in multiparticle production and correlations, in a region that is by definition far from the perturbative sector of QCD. In this paper we will focus our attention on the influence of Gabriele Veneziano in this sector of physics, starting from his results on jet-calculus and their further applications to a probabilistic description of parton showers in the leading logarithmic

A. Giovannini and R. Ugoccioni: *Coherence and Incoherence in QCD Jets Dynamics (QCD Jets and Branching Processes)*, Lect. Notes Phys. **737**, 223–234 (2008)
DOI 10.1007/978-3-540-74233-6_11

approximation (LLA). These developments, together with the idea of local parton-hadron duality [13] and its related generalization [6], is at the basis of quite successful models of event generators and of many lines of research in this field.

The Konishi–Ukawa–Veneziano (KUV) evolution equations opened indeed a new horizon in the basic understanding of the partonic sector of multiparticle production processes. They provided a QCD basis for the approximate description of observed universal regularities in the final charged-particle multiplicity distributions, in all classes of collisions—both in full phase space and in (pseudo) rapidity windows—in terms of independent intermediate gluon sources (or partonic clan ancestors) formation, which then decay into final partons through cascading mechanism dominated by gluon self-interactions. This result is a consequence of the simplified description of quark and gluon QCD jets as Markov branching processes, as obtained from the above-mentioned KUV equations. One is led indeed to a sound QCD framework for the approximate description of the observed final charged particle multiplicity distributions, and of the properties of the related collective variables discovered in high energy collisions. These observations are the subject of the next section.

2 Elementary Models and Unexplained Facts in Multiparticle Dynamics in the Early 1970s

The first regular behavior to be recalled concerns the description of the multiplicity distributions (MD) of all available final particles in high-energy hadronic collisions, in the accelerator region, in terms of a two-parameter MD which we will call from now on the Pascal (NB) regularity.[1]

The two mentioned parameters are the average charged-particle multiplicity $\bar{n}$ of the distribution itself, and the k parameter, which is linked to the dispersion D of the distribution by the simple relation $k = \bar{n}^2/(D^2 - \bar{n})$. The accelerator results confirmed an earlier discovery (in the 1960s) of the multiplicity distribution for the pion component in cosmic ray physics, in a variety of observations performed with different primary nucleon energies [9].

The motivation of this successful phenomenological search in the accelerator region (57 experiments were examined) has to be found in the statistical generalization of the multiperipheral model of multiparticle production (proposed since 1972), which through the Poissonian superposition of properly

[1] It should be noticed that the Pascal (NB) distribution appears with different names in the literature on multiparticle dynamics. When introduced for the first time it was called Polya–Eggenberger [7] multiplicity distribution (a due tribute to biology, where it has been widely applied), while in more recent times the name Negative binomial (NB) was used (in memory of its statistical origin). However, as the first historical appearance of the multiplicity distribution in science goes back to Blaise Pascal [8], the name Pascal distribution was finally proposed, a choice which will be followed in the present paper.

weighted multiperipheral diagrams, led to the Pascal (NB) description of the final particle multiplicity distributions [7, 10].

The parameter k was interpreted, in this context, as the ratio of the reggeon–reggeon particle vertex to the pomeron–reggeon particle vertex; in its high-energy limit it predicted Koba–Nielsen–Olesen (KNO) scaling violations in multiparticle production for hadronic collisions, and provided an explanation of observed deviations [11] from the standard expectations of the multiperipheral model in terms of the onset of the pomeron coupling.[2]

It should be recalled that the experimental occurrence of the Pascal distributions in the MD of the final charged particles suggested also another possible interpretation of the data in terms of a stochastic cell model [12]. This model assumed stimulated emission of identical bosons by identical cells, where each cell was producing a Bose–Einstein distribution. The parameter k is here an integer number ≥ 1, and was interpreted as the number of identical cells involved in the collision, according to an old idea of Max Planck [13]. Such an interpretation was disproved by the data, since k was found to be in general a non-integer number, and even smaller than one in some cases.

3 KUV Differential Evolution Equations and the Advent of QCD in the Late 1970s

It should be pointed out that KUV parton evolution equations [1, 2] are an application of jet calculus in the leading log approximation (LLA) which allows—as already mentioned—a probabilistic description of parton shower processes originated by a quark or a gluon. Since both collinear and infrared singularities are present in the LLA expression of the DGLAP kernels [14], the new problem was how to cure such singularities.

It turns out that collinear singularities can be avoided by imposing a soft cutoff to the evolution of the parton population, whereas infrared divergences are cured by imposing a fixed cutoff on the variable z in the Dokshitzer–Gribov–Lipator–Altarelli–Parisi (DGLAP) elementary kernel $P_{jk}(z)$ describing emission of parton k from parton j, with parton k carrying a fraction z of j's momentum, i.e., $z_{min} = \epsilon' = 1 - z_{max}$. The integrals of the regularized kernels are then interpreted as elementary splitting probabilities,

$$A \equiv \int_{\epsilon'}^{1-\epsilon'} P_{gg}(z)dz = \frac{C_a}{\epsilon} = \frac{\mathcal{N}_c}{\epsilon}; \tag{1}$$

$$\tilde{A} \equiv \int_{\epsilon'}^{1-\epsilon'} P_{gq}(z)dz = \frac{C_F}{\epsilon} = \frac{\mathcal{N}_c^2 - 1}{2\epsilon\mathcal{N}_c}; \tag{2}$$

[2] The paper [7] was in part done during a stay at MIT of one of the present authors, and profited of many discussions with G. Veneziano, as witnessed in the acknowledgments at the end of the paper itself.

$$B \equiv \mathcal{N}_f \int_{\epsilon'}^{1-\epsilon'} P_{qg}(z)dz = \frac{\mathcal{N}_f}{3}, \tag{3}$$

where $\epsilon = (-2\ln\epsilon')^{-1}$, and $\mathcal{N}_f$, $\mathcal{N}_c$ are the number of flavors and colors, respectively.

The jet thickness Y can be used as evolution variable from the virtual scale W down to the scale Q,

$$Y = \frac{1}{2\pi b}\log\left(\frac{\alpha_s(Q^2)}{\alpha_s(W^2)}\right) = \frac{1}{2\pi b}\log\left(\frac{\log(W^2/\Lambda^2)}{\log(Q^2/\Lambda^2)}\right), \tag{4}$$

with $b = (11\mathcal{N}_c - 2\mathcal{N}_f)/12\pi$. The jet thickness Y contains the dependence on the running coupling constant (leading order), and is the mixture of three scales: W (the virtual mass of the primary parton), Q (the splitting scale) and the QCD scale Λ. Y is a small number (< 1) during the early stages of the shower evolution ($Q \lesssim W$). The probability $P_q(Q|W)dQ$ that a quark q of virtuality W splits in the range $[Q, Q + dQ]$, by emitting a gluon, is then given by [2]

$$P_q(Q|W)dQ = e^{-\tilde{A}Y}\tilde{A}dY, \tag{5}$$

and the probability $P_g(Q|W)dQ$ that a gluon g splits (by either emitting another gluon or a quark–antiquark pair) by

$$P_g(Q|W)dQ = e^{-(A+B)Y}(A + B)dY. \tag{6}$$

Neglecting conservation laws, the last two equations imply that the splitting is constant for each dY interval. This simplified assumption allows us to classify the process as Markovian, and therefore to write the corresponding approximate (forward and backward) Kolmogorov equations to create n_q quarks and n_g gluons, starting from an initial quark, $P_q(n_q, n_g; Y)$, or an initial gluon, $P_g(n_q, n_g; Y)$, at thickness Y. The corresponding non-zero transition probabilities in the interval dY are

$$\begin{aligned}(n_q, n_g) \to (n_q, n_g) &= 1 - An_g dY - \tilde{A}n_q dY - Bn_g dY;\\ (n_q, n_g) \to (n_q, n_g + 1) &= An_g dY + \tilde{A}n_q dY;\\ (n_q, n_g) \to (n_q + 2, n_g - 1) &= Bn_g dY.\end{aligned} \tag{7}$$

It is simpler to use the generating functions, calculated from the corresponding transition probabilities

$$G_a(u, v; Y) \equiv \sum_{n_q, n_g} u^{n_q} v^{n_g} P_a(n_q, n_g; Y), \tag{8}$$

where $a = q, g$. Accordingly, the following differential equations are then obtained:

$$\frac{dG_g}{dY} = A(G_g^2 - G_g) + B(G_q^2 - G_g), \tag{9}$$

$$\frac{dG_q}{dY} = \tilde{A}G_q(G_g - 1)\,. \tag{10}$$

When the production of quark–antiquark pairs can be neglected (i.e., when $B = 0$) the above equations decouple; by looking only at the gluon population generating function at Y one has

$$G_g(u, v; Y) = v[v + (1 - v)e^{AY}]^{-1}\,, \tag{11}$$

$$G_q(u, v; Y) = u[v + (1 - v)e^{AY}]^{-\tilde{A}/A}\,. \tag{12}$$

It turns out that the gluon multiplicity distribution, in a gluon-initiated shower, is a shifted geometric distribution with average gluon multiplicity e^{AY} and parameter $k \approx 1$. The gluon multiplicity, in a quark-initiated shower, is instead a Pascal (NB) multiplicity distribution, with average gluon multiplicity $\bar{n} = \tilde{A}(e^{AY} - 1)/A$ and parameter $k = \tilde{A}/A(\approx 4/9)$: k is then the ratio between the gluon self-interaction ($g \to g + g$), with vertex $\tilde{A}$, and the gluon bremsstrahlung initiated by a quark ($q \to q + g$), with vertex A.

KUV evolution equations revealed in this way the approximate QCD skeleton in the early stages of multi-parton production: they single out the essentials of QCD dynamics to be taken into account in its application to the exploration of the partonic sector. The evolution is characterized at this stage by the dominance of the $g \to g + g$ vertex over the $g \to q + \bar{q}$ vertex, and by the weak effects of coherence and conservation laws.

We can summarize the situation after Sects. 2 and 3 as follows.

(a) From Sect. 2 one learns that the Pascal (NB) multiplicity distribution appears experimentally as the natural candidate for describing the final pion multiplicity distributions in cosmic ray physics; the parameter k is decreasing going from low to high energy of the primary-hadron, whereas the average charged particle multiplicity $\bar{n}$ is increasing in the same energy range.
The mentioned regularity appears also in the accelerator region (it has been tested in 57 experiments), although the general trend of its parameters is not so spectacular as in cosmic rays, in view of the relatively low p_{lab} of the incident particle on the fixed target experiments. It also appears in theoretical work on the statistical generalization of the multiperipheral model, where k is interpreted as the ratio of reggeon to pomeron couplings, or, more generally, in terms of a coherent production mechanism over an incoherent one.

(b) From Sect. 3 one notices the occurrence of the Pascal (NB) multiplicity distribution also in the approximate description of QCD parton showers, originated by an initial quark and an initial gluon according to the corresponding KUV evolution equation under the simplified assumption $B = \mathcal{N}_f/3 \to 0$.

To the common wisdom the occurrence of the Pascal multiplicity distribution in so many (apparently different) theoretical and experimental situations was considered to be not interesting enough. Very few people in the field had an opposite point of view; for them, the wide occurrence of the Pascal MD in hadronic reactions was a signature of the approximately unified nature of multiparticle production processes, and of the universality of QCD thanks to the Markov branching nature of the quark and gluon showers in the early stages of their evolution .

4 The Collaboration with Léon Van Hove, and the UA5 Collaboration Results at CERN $p\bar{p}$ Collider on Multiplicity Distributions, in Full Phase Space and in Restricted Pseudo-rapidity Windows

Results summarized in points (a) and (b) of Sect. 3 were not overlooked by Léon Van Hove. His interest on the subject was enhanced by the discovery by the UA5 Collaboration (in the 1980s) that the Pascal (NB) regularity was describing quite well in $p\bar{p}$ collisions at CERN Collider at various c.m. energies (200, 560 and 900 GeV) the MD of final charged particles, not only in full phase space, but also in restricted pseudo-rapidity windows [15].

The result by the UA5 Collaboration was independent from the previous results obtained in the full phase space in cosmic ray physics and in hadronic collisions in the accelerator region, and initially was not related to the theoretical work which had led to the introduction of the Pascal (NB) MD in high-energy phenomenology. The UA5 results were confirmed by the NA 22 Collaboration data on MD's in pp and $\pi^{\pm}$p collisions at 22 GeV c.m. energy [16].

The characteristic experimental trend of the parameters of the Pascal (NB) MD was that the k parameter was decreasing in full phase space as the c.m. energy was increasing, while the opposite occurred to the charged particle multiplicity $\bar{n}$, as expected from previous experiments at lower energies and in cosmic ray physics; in addition, at different fixed c.m. energies, k and $\bar{n}$ were both decreasing from large to restricted (pseudo-) rapidity windows.

All these facts led Léon Van Hove to look for a new interpretation of the occurrence of the regularity in hadronic collisions. A first paper was then produced [17].

One question was still to be answered, concerning the e^+e^- annihilation and deep inelastic scattering. More precisely, is the regularity found in hadron–hadron collisions also present in other classes of collisions? assuming the answer is positive, what is the trend of its parameters with respect to the energy and rapidity variables?

A positive reply would had been remarkable for exploring the partonic sector controlled in its early stages by QCD through KUV equations, where the Pascal (NB) MD had also been discovered in the form discussed in Sect. 2.

This search was motivated by the conviction that the complex structures which we observe in experimental data have often a simple origin at the parton level, and are revealed by universal approximate regularities at the hadron level. The intimate conviction was indeed that an eventual satisfactory explanation of the observed regularity at the final hadron level should be found in a QCD framework at the parton level. In order to proceed in our program, we had to solve the following two problems:

(a) calculate final parton MD from KUV and DGLAP equations or, in more general terms, join the non-perturbative to the perturbative sector of parton showers;
(b) look for final charged particle MD in e^+e^- annihilation and deep inelastic scattering experiments.

In a region where QCD had no predictions the answer to the point (a) was found by following an encouraging result provided by W. Kittel [18]. It led us to rely on the Monte Carlo model of event generator JETSET 7.2 [19], which is based on DGLAP equations in the partonic sector, and has a hadronization prescription based on the string model for the transfer of information from the partonic to the hadronic sector [20]. All event generators have indeed one feature in common, namely the DGLAP or KUV equations, and differ by the type of hadronization model, which is the string model in JETSET, and the cluster model in HERWIG [21]; more recently, a statistical hadronization model has been proposed by F. Becattini [22].

In order to answer to point (b), the HRS and EMC Collaborations were asked to produce the requested data in final hadron multiplicity distributions. Remarkably, all replies were positive [23, 24]. The new facts were the following.

(a) The final parton multiplicity distributions originated, at various virtualities, by a quark–antiquark system and by a gluon–gluon system were all approximate Pascal (NB) MD, either in full phase space or in restricted windows in rapidity [6].
In addition, the parameter k was approximately the same after the hadronization, and the average number of particles was varying by a constant factor $\rho \approx 2$, i.e. $\bar{n}_{hadron} = \rho\bar{n}_{parton}$ and $k_{hadron} = k_{parton}$ (the generalized local hadron–parton duality (GLPHD) prescription).
Notice that LPHD [5] is based on preconfinement (i.e., it says that $\bar{n}_{hadron} = \rho\bar{n}_{parton}$), whereas GLPHD is expressed in terms of n-particle inclusive rapidity distributions of partons, $Q_{n,p}(y_1,\ldots,y_n)$, and hadrons, $Q_{n,h}(y_1,\ldots,y_n)$, by the equations

$$Q_{n,h}(y_1,\ldots,y_n) = \rho^n Q_{n,p}(y_1,\ldots,y_n)\,, \qquad (13)$$

ρ being constant. In addition, by assuming that one of the two (partonic or hadronic) MD is of the Pascal (NB) type in a rapidity domain, then the other one is again a Pascal (NB) multiplicity distribution, and the corresponding NB parameters are linked by the above relations with constant ρ [6].

(b) The Pascal (NB) distribution was describing the MD of the final charged particles in all classes of examined collisions, representing with specific parameter the trends at various c.m. energies in full phase space and in restricted (pseudo-) rapidity windows.

Since then an avalanche of data was produced leading to the same mentioned results in all experiments available at that time. All these facts suggested the interpretation of the Pascal (NB) MD in terms of a two-step process [25]. In the first step, independent objects (the clan ancestors) are produced according to a Poisson MD, in the second step, each ancestor decays, following a logarithmic MD (the clan MD). No correlations exist among particles produced in different clans, and each clan contains at least one particle.

This new perspective led to introducing two new variables in the production process: the average number of clans, $\bar{N}$, and the average number of particles per clan, $\bar{n}_c$, which are linked to the standard parameters $\bar{n}$ and k by the relations

$$\bar{N} = k \ln\left(1 + \frac{\bar{n}}{k}\right) , \tag{14}$$

$$\bar{n}_c = \frac{\bar{n}}{\bar{N}} . \tag{15}$$

The introduction of the clan concept in the interpretation of the approximate Pascal (NB) universal regularity, in the mentioned experiments, led to very suggestive results. The average number of clans is larger in e^+e^- annihilation than in hadron–hadron collisions, whereas just the opposite occurs for the average number of particles per clan. Clan bremsstrahlung is stronger and clan size smaller in the former case than in the latter. An intermediate situation occurs for deep inelastic scattering, where clans are less numerous than in e^+e^-, but the average number of particles per clan is much larger. In addition, pumping energy into a collision does not increase the average number of clans but only the average length of the showers, which are larger in more central than in more peripheral rapidity intervals (see Fig. 1).

Clan formalism can be applied also at parton level, where it disentangles the two above- mentioned QCD vertices; in fact, by recalling the results described in Sect. 3, one obtains

$$\bar{N} = \tilde{A}Y , \tag{16}$$

$$\bar{n}_c = \frac{e^{AY} - 1}{AY} , \tag{17}$$

i.e., gluon production from a quark is controlled by parameter $\tilde{A}$ (average clan production), and gluon emission from a gluon is controlled by parameter A (average gluon shower production inside clans). Clans in this context can be approximately understood (under the assumption $Y < 1$) as bremsstrahlung gluon jets. These can be considered indeed as the building blocks of a unified, although approximate, description of multiparticle production in all classes of collisions.

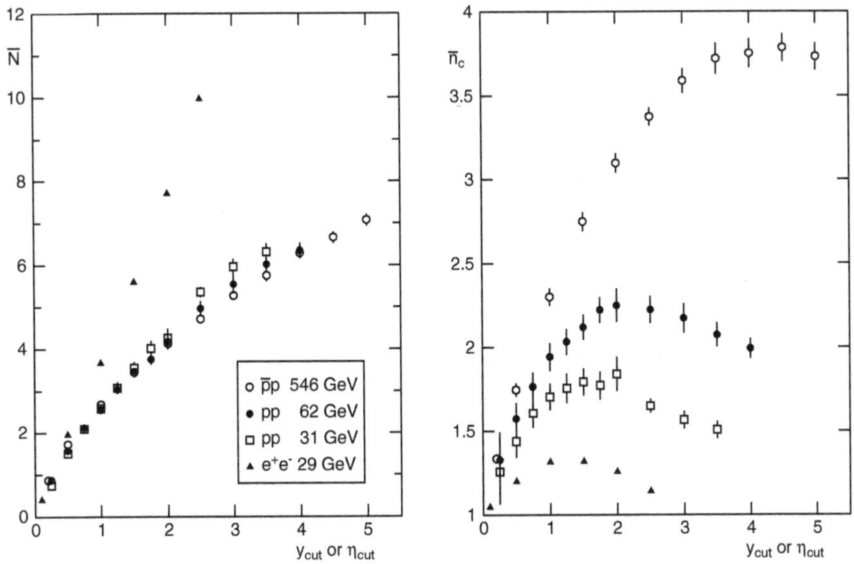

Fig. 1. Average number of clans, $\bar{N}$ (**left panel**), and average number of particles per clan, $\bar{n}_c$ (**right panel**), versus the half-width of the pseudo-rapidity (for 546 GeV data) or rapidity (for the other energies) interval [26]

The picture one has in mind goes as follows: independent, intermediate gluon sources are produced at the parton level via bremsstrahlung, and they later decay into gluon showers dominated by gluon self-interactions. The GLHPD prescription allows us to determine the germane evolution of the showers at the hadron level, with specific different behavior in the various classes of collisions.

The next step to be performed, in order to overcome the approximate description of multiparticle production processes, was to build up a parton shower model based on essentials of QCD in a correct kinematical framework, including conservation laws and coherence effects.

5 New Experimental Findings on Final Charged Particle MD in e^+e^- Annihilation at LEP c.m. Energy, and More Precise Measurements on Final Particle MD at $p\bar{p}$ Collider Top c.m. Energy. The Occurrence of Substructures or Components in the Various Collisions

The situation described in Sect. 4 was simple and quite satisfactory. As already pointed out, however, including in such a framework coherence effects and conservation laws was still an open problem. This search was in part accomplished. But since its natural goal was to build up a Monte Carlo event generator model, to be added to the already existing (and successful) ones, we decided to pay more attention to the new experimental facts in multiparticle production processes which requested a deeper level of investigation than previously thought. We sketch below the relevant steps; a complete review can be found in [27].

The Shoulder Effect

A shoulder structure in the multiplicity distribution ("shoulder effect") was seen experimentally in $p\bar{p}$ collisions at 900 GeV c.m. energy and in e^+e^- annihilation at LEP c.m. energies. This was explained with a weighted superposition mechanism of two classes of events [28], each described by Pascal (NB) MD, and identified, respectively, with soft (without mini-jets) and semihard (with mini-jets) events in $p\bar{p}$, and with two-jet and three-or-more-jet events in e^+e^-.

Cumulant and Factorial Moments

When computing higher-order cumulant and factorial moments of experimental MD, their ratio H_q, when plotted versus the order q, shows sign oscillations (both in $p\bar{p}$ and e^+e^-): the weighted superposition mechanism of two classes of events described by Pascal (NB) MD again is able to explain this feature [29].

Forward–Backward Multiplicity Correlations

Forward–backward multiplicity correlations (FBMC) in e^+e^- annihilation and in $p\bar{p}$ collisions appear different: barely visible in the first case, rather stronger and increasing with c.m. energy in the second one. Both behaviors can be explained combining together the weighted superposition mechanism and particle production via clans [30], the differences in clan behavior being just the key for correctly describing the features of FBMC in the different reactions. Notice that this makes FBMC a very relevant characteristic to investigate in future experiments at CERN, because they can be used to explore the "color" landscape of very high-energy collisions.

The point to be stressed is that all the above-mentioned experimental facts can be explained in terms of the weighted superposition mechanism of two classes of events (or components), each described by a Pascal MD with characteristic parameters. This explanation allows us to maintain the original simple interpretation of the regularity, which was violated when applied to the full sample of events: the regularity is not a property of the full sample of events of the various collisions (except at lower energies, when the full sample essentially coincides with a single class); instead, it is a property of the different components or substructures (classes of events) in which each reaction could be eventually disentangled, and whose properly weighted superposition should reproduce observed experimental data for the general behavior of collective variables.

6 New Physics at CERN. The Weighted Superposition of Three Classes of Events (Soft, Semihard, and Hard) in pp Collisions at LHC

A new class of hard events to be added to the soft class of events (no mini-jets), and to the semihard class of events (with mini-jets), has been envisioned to exist at LHC [3]. It has been proposed to be described by a Pascal (NB) MD with $k \ll 1$ and $\bar{n}$ very large (or, in the clan language, by $\bar{N} \approx 1$, $\bar{n}_c$ very large). The total MD of the final charged particles is given by its weighted superposition with the soft and semihard classes of events. The problem is how to distinguish the three classes of events.

In addition to the seminal work of Gabriele and coworkers on KUV equations (which led to the understanding of parton showers and of the occurrence of the Pascal (NB) regularity in multiparticle production), there is another work (among his many papers) which might have—in our opinion—a new interesting application. We are referring to the article on the $1/\mathfrak{N}$ expansion (with $\mathfrak{N}$ the number of chains) and on Bose–Einstein (BE) interferometry [4].

Although the predictions presented in that paper were disproved by the data shortly after the paper was produced, we recall that the paper foresaw indeed no BE interference, contrary to experimental findings, in case of a coherent reaction like e^+e^- annihilation (the number of chain is here just one). The interference was expected to be large in *pp* (one pomeron exchange, $\mathfrak{N} = 2$), and even larger in $p\bar{p}$ ($\mathfrak{N} = 3$). This trend of a coherent versus an incoherent reaction is today what one should expect for disentangling eventual substructures in the minimum bias sample of events in pp collisions at LHC [31]. The three classes of events, each described by a Pascal (NB) MD with different parameters, would correspond to:

(a) one parton to one parton scattering (soft class of coherent events);
(b) two partons to two partons scattering (semihard class of partly coherent events);
(c) three partons to three partons scattering (hard class of incoherent events).

As the number of parton–parton scatterings (chains) becomes larger the collision becomes harder, the temperature higher and the parton density larger. Future experiments at LHC will provide data to be confronted with this speculative thought.

References

1. K. Konishi, A. Ukawa, G. Veneziano: Nucl. Phys. B **157**, 45 (1979)
2. A. Giovannini, Nucl. Phys. B **161**, 429 (1979).
3. A. Giovannini, R. Ugoccioni: Phys. Rev. D **59**, 094020 (1999); Phys. Rev. D **60**, 074027 (1999)
4. A. Giovannini, G. Veneziano: Nucl. Phys. B **130**, 61 (1977)

5. D. Amati, G. Veneziano: Phys. Lett. B **83**, 87 (1979)
6. L. Van Hove, A. Giovannini: Acta Phys. Pol. B **19**, 917 (1988)
7. A. Giovannini: Il Nuovo Cimento A **10**, 713 (1972); Il Nuovo Cimento A **15**, 543 (1973)
8. B. Pascal: *Varia Opera Mathematica, D. Petri de Fermat* (Tolossae, France, 1679)
9. P.K. MacKeown, A.W. Wolfendale: Proc. Phys. Soc. **89**, 553 (1966)
10. A. Giovannini, P. Antich, E. Calligarich, G. Cecchet, R. Dolfini, F. Impellizzeri, S. Ratti: Il Nuovo Cimento A **24**, 421 (1974); M. Garetto, A. Giovannini, E. Calligarich, G. Cecchet, R. Dolfini, S. Ratti: Il Nuovo Cimento A **38**, 38 (1977)
11. M. Derrick et al.: Phys. Rev. Lett. **29**, 515 (1972)
12. L. Mandel: Proc. Phys. Soc. London, 233 (1959)
13. M. Planck: Sitzungber. Deutsch. Akad. Wiss. Berlin **33**, 355 (1923)
14. Yu. L. Dokshitzer, V. A. Khoze, A. H. Mueller, S. I. Troyan: *Basics of Perturbative QCD* (Editions Frontières, Gif-sur-Yvette, 1991)
15. G. J. Alner et al. (UA5 Collaboration): Phys. Lett. B **160**, 193 (1985)
16. M. Adamus et al. (NA22 Collaboration): Phys. Lett. B **177**, 239 (1986)
17. A. Giovannini, L. Van Hove: Z. Phys. C **30**, 391 (1986)
18. W. Kittel: in *Workshop on Physics with Future Accelerators*, ed. by J. Mulvay (CERN, Yellow Rep. 87-7, 1987), Vol. II, p. 424
19. T. Sjöstrand, M. Bengstsson: Computer Physics Commun. **82**, 74 (1994)
20. B. Andersson: *The Lund Model* (Cambridge University Press, Cambridge, 1996)
21. G. Marchesini, B. R. Webber: Nucl. Phys. B **310**, 461 (1988)
22. F. Becattini: Z. Phys. C **69**, 485 (1996); F. Becattini, U. W. Heinz: Z. Phys. C **76**, 269 (1997)
23. M. Derrick et al. (HRS Collaboration): Phys. Lett. B **168**, 299 (1986)
24. M. Arneodo et al. (EMC Collaboration): Z. Phys. C **35**, 335 (1987)
25. L. Van Hove, A. Giovannini: in *XVII International Symposium on Multiparticle Dynamics*, ed. by M. Markitan et al. (World Scientific, Singapore, 1987), p. 561
26. A. Breakstone et al.: Il Nuovo Cimento A **102**, 1199 (1989)
27. A. Giovannini, R. Ugoccioni: Int. J. Mod. Phys. A **20**, 3897 (2005)
28. A. Giovannini, S. Lupia, R. Ugoccioni: Nucl. Phys. B (Proc. Suppl.) **25**, 115 (1992); Phys. Lett. B **374**, 231 (1996)
29. R. Ugoccioni, A. Giovannini, S. Lupia: Phys. Lett. B **342**, 387 (1995)
30. A. Giovannini, R. Ugoccioni, Phys. Rev. D **66**, 034001 (2002)
31. W.D. Walker: Phys. Rev. D **69**, 034007 (2004)

The $U(1)_A$ Anomaly and QCD Phenomenology

G. M. Shore

Department of Physics, University of Wales, Swansea, Swansea SA2 8PP, UK
g.m.shore@swansea.ac.uk

Abstract. The role of the $U(1)_A$ anomaly in QCD phenomenology is reviewed, focusing on the relation between quark dynamics and gluon topology. Topics covered include a generalisation of the Witten–Veneziano formula for the mass of the η', the determination of pseudoscalar meson decay constants, radiative pseudoscalar decays and the $U(1)_A$ Goldberger–Treiman relation. Sum rules are derived for the proton and photon structure functions g_1^p and g_1^γ measured in polarised deep inelastic scattering (DIS). The first moment sum rule for g_1^p (the 'proton spin' problem) is confronted with new data from COMPASS and HERMES on the deuteron structure function and shown to be quantitatively explained in terms of topological charge screening. Proposals for experiments on semi-inclusive DIS and polarised two-photon physics at future ep and high-luminosity e^+e^- colliders are discussed.

1 Introduction

The $U(1)_A$ anomaly has played an important historical role in establishing QCD as the theory of the strong interactions. The description of radiative decays of the pseudoscalar mesons in the framework of a gauge theory requires the existence of the electromagnetic axial anomaly and determines the number of colours to be $N_c = 3$. The compatibility of the symmetries of QCD with the absence of a ninth light pseudoscalar meson – the so-called $U(1)_A$ problem – in turn depends on the contribution of the colour gauge fields to the anomaly. More recently, it has become clear how the anomaly-mediated link between quark dynamics and gluon topology (the non-perturbative dynamics of topologically non-trivial gluon configurations) is the key to understanding a range of phenomena in polarised QCD phenomenology, most notably the 'proton spin' sum rule for the first moment of the structure function g_1^p.

In this paper, based on original research performed in a long-standing collaboration with Gabriele Veneziano, we review the role of the $U(1)_A$ anomaly in describing a wide variety of phenomena in QCD, ranging from the low-energy dynamics of the pseudoscalar mesons to sum rules in polarised deep-inelastic scattering. The aim is to show how these experiments reveal subtle

G. M. Shore: *The $U(1)_A$ Anomaly and QCD Phenomenology*, Lect. Notes Phys. **737**, 235–288 (2008)
DOI 10.1007/978-3-540-74233-6_12

aspects of quantum field theory, in particular topological gluon dynamics, which go beyond simple current algebra or parton model interpretations.

We begin in Sect. 2 with a brief review of the essential theoretical toolkit: anomalous chiral Ward identities, Zumino transforms, the renormalisation group and the range of expansion schemes associated with large N_c, notably the (Okubo–Zweig–Iizuki) (OZI) approximation. Then, in Sect. 3, we build on Veneziano's seminal 1979 paper [1] to describe how the pseudoscalar mesons saturate the Ward identities in a way compatible with both the renormalisation group and large-N_c constraints and derive a generalisation of the famous Witten–Veneziano mass formula for the η' which incorporates, but goes beyond, the original large-N_c derivation [2, 3].

In Sect. 4, we turn to QCD phenomenology and describe how this intuition on the resolution of the $U(1)_A$ problem allows a quantitative description of low-energy pseudoscalar meson physics, especially radiative decays, the determination of the pseudoscalar decay constants, and meson–nucleon couplings. We review the $U(1)_A$ extension of the Goldberger–Treiman formula first proposed by Veneziano [4] as the key to understanding the 'proton spin' problem and test an important hypothesis on the origin of OZI violations and their relation to the renormalisation group. Low-energy η and η' physics is currently an active experimental field and we explain the importance of an accurate determination of the couplings $g_{\eta NN}$ and $g_{\eta' NN}$ in elucidating the role of gluon topology in QCD.

All of these low-energy phenomena have counterparts in high-energy, polarised deep inelastic scattering. This enables us to formulate a new sum rule for the first moment of the polarised photon structure function g_1^γ (Sect. 6). The dependence of this sum rule on the invariant momentum of the off-shell target photon measures the form factors of the three-current AVV Green function and encodes a wealth of information about the realisation of chiral symmetry in QCD, while its asymptotic limit reflects both the electromagnetic and colour $U(1)_A$ anomalies. We show how this sum rule, which we first proposed in 1992 [5, 6], may soon be tested if the forthcoming generation of high-luminosity e^+e^- colliders, currently conceived as B factories, are run with polarised beams [7].

The most striking application of these ideas is, however, to the famous 'proton spin' problem, which originated with the observation of the violation of the Ellis–Jaffe sum rule for the first moment of the polarised proton structure function g_1^p by the EMC collaboration at CERN in 1988. This experiment, and its successors at SLAC, DESY (HERMES) and CERN (SMC, COMPASS) determined the axial charge a^0 of the proton. In the simple valence-quark parton model, this can be identified with the quark spin and its observed suppression led to an intense experimental and theoretical search over two decades for the origin of the proton spin. In fact, as Veneziano was the first to understand [4], a^0 does not measure spin in QCD itself and its suppression is related to OZI violations induced by the $U(1)_A$ anomaly.

In a series of papers, summarised in Sect. 5, we have shown how a^0 decouples from the real angular momentum sum rule for the proton (the form factors for this sum rule are given by generalised parton distributions (GPDs) which can be extracted from less inclusive measurements such as deeply virtual Compton scattering) and is instead related to the gluon topological susceptibility [8, 9]. The experimentally observed suppression is a manifestation of topological charge screening in the QCD vacuum. In a 1994 paper with Narison [10], using QCD spectral sum rule methods, we were able to compute the slope of the topological susceptibility and give a quantitative prediction for a^0. Our prediction, $a^0 = 0.33$, has within the past few months been spectacularly confirmed by the latest data on the deuteron structure function from the COMPASS and HERMES collaborations.

Hopefully, this impressive new evidence for topological charge screening will provide fresh impetus to experimental 'spin' physics – first, to verify the real angular momentum sum rule by measuring the relevant GPDs, and second, to pursue the programme of target-fragmentation studies in semi-inclusive DIS at polarised *ep* colliders which we have proposed as a further test of our understanding of the g_1^p sum rule [11].

This review has been prepared in celebration of the 65th birthday of Gabriele Veneziano. I first met Gabriele when I came to Geneva as a CERN fellow in 1981. In fact, our first interaction was across a tennis court, in a regular Friday doubles match with Daniele Amati and Toine Van Proeyen. I like to think that in those days I could show Gabriele a thing or two about tennis – physics, of course, was a different matter. It has been my privilege through these ensuing 25 years to collaborate with one of the most brilliant and innovative physicists of our generation. But it has also been fun. As all his collaborators will testify, his good humour, generosity to younger colleagues, and enthusiasm in thinking out solutions to the deepest and most fundamental problems in particle physics and cosmology make working with Gabriele not only intellectually rewarding but hugely enjoyable.

In his contribution to the 'Okubofest' in 1990 [12], Gabriele concluded an account of the relevance of the OZI rule to g_1^p by hoping that he had 'made Professor Okubo happy'. In turn, I hope that this review will make Gabriele happy: happy to recall how his original ideas on the $U(1)_A$ problem have grown into a quantitative description of anomalous QCD phenomenology, and happy at the prospect of new discoveries from a rich programme of experimental physics at future polarised colliders. It is my pleasure to join all the contributors to this volume in wishing him a happy birthday.

2 The $U(1)_A$ Anomaly and the Topological Susceptibility

We begin by reviewing some essential features of the $U(1)_A$ anomaly, chiral Ward identities and the renormalisation group, placing particular emphasis on the role of the gluon topological susceptibility. As we shall see, the anomaly provides the vital link between quark dynamics and gluon topology which is essential in understanding a range of phenomena in polarised QCD phenomenology.

2.1 Anomalous Chiral Ward Identities

An anomaly arises when a symmetry which is present in the classical limit cannot be consistently imposed in a quantum field theory. The original example of an anomaly, and one which continues to have far-reaching implications for the phenomenology of QCD, is the famous Adler–Bell–Jackiw axial anomaly [13, 14, 15], which was first understood in its present form in 1969. In fact, calculations exhibiting what we now recognise as the anomaly had already been performed much earlier by Steinberger in his analysis of meson decays [16] and by Schwinger [17].

Anomalies manifest themselves in a number of ways. The original derivations of the axial anomaly involved the impossibility of simultaneously imposing conservation of both vector and axial currents due to regularisation issues in the AVV triangle diagram in QED. More generally, they arise as anomalous contributions to the commutation relations in current algebra. A modern viewpoint, due to Fujikawa [18], sees anomalies as due to the non-invariance of the fermionic measure in the path integral under transformations corresponding to a symmetry of the classical Lagrangian. In this approach, the result of a chiral transformation $q \to e^{i\alpha^a T^a \gamma_5} q$ on the quark fields in the QCD generating functional $W[V^a_{\mu 5}, V^a_\mu, \theta, S^a_5, S^a]$ defined as[1]

$$e^{iW} = \int \mathcal{D}A\mathcal{D}\bar{q}\mathcal{D}q \, \exp\Big[i\int dx \big(\mathcal{L}_{QCD} + V_5^{\mu a} J^a_{\mu 5} + V^{\mu a} J^a_\mu + \theta Q + S^a_5 \phi^a_5 + S^a \phi^a\big)\Big] \tag{1}$$

[1] Our notation follows that of [3]. The currents and pseudoscalar fields $J^a_{\mu 5}$, Q, ϕ^a_5 together with the scalar ϕ^a are defined by

$$J^a_{\mu 5} = \bar{q}\gamma_\mu\gamma_5 T^a q \qquad J^a_\mu = \bar{q}\gamma_\mu T^a q \qquad Q = \frac{\alpha_s}{8\pi}\mathrm{tr} G_{\mu\nu}\tilde{G}^{\mu\nu}$$
$$\phi^a_5 = \bar{q}\gamma_5 T^a q \qquad \phi^a = \bar{q} T^a q$$

where $G_{\mu\nu}$ is the field strength for the gluon field. Here, $T^i = \frac{1}{2}\lambda^i$ are flavour $SU(n_f)$ generators, and we include the singlet $U(1)_A$ generator $T^0 = \mathbf{1}/\sqrt{2n_f}$ and let the index $a = 0, i$. With this normalisation, $\mathrm{tr} T^a T^b = \frac{1}{2}\delta^{ab}$ for all the generators T^a. This accounts for the rather unconventional factor $\sqrt{2n_f}$ in the anomaly equation but has the advantage of giving a consistent normalisation to the full set of decay constants including the flavour singlets $f^{0\eta'}$ and $f^{0\eta}$.

We will only need to consider fields where i corresponds to a generator in the Cartan sub-algebra, so that $a = 3, 8, 0$ for $n_f = 3$ quark flavours. We define d-symbols by $\{T^a, T^b\} = d_{abc}T^c$. For $n_f = 3$, the explicit values are $d_{000} = d_{033} = d_{088} = d_{330} = d_{880} = \sqrt{2/3}$, $d_{338} = d_{383} = -d_{888} = \sqrt{1/3}$.

is

$$\int \mathcal{D}A\mathcal{D}\bar{q}\mathcal{D}q \left[\partial^\mu J^a_{\mu 5} - \sqrt{2n_f}\delta^{a0}Q - d_{abc}m^b\phi^c_5 - \delta\Big(\int d^4x\mathcal{L}_{\rm QCD}\Big)\right] \exp\Big[\ldots\Big] = 0 \tag{2}$$

The terms in the square bracket are simply those arising from Noether's theorem, including soft breaking by the quark masses, with the addition of the anomaly involving the gluon topological charge density Q. Re-expressing the chiral variation of the elementary fields in terms of a variation with respect to the sources $V^a_{\mu 5}, V^a_\mu, \theta, S^a_5, S^a$ then gives the functional form of the anomalous chiral Ward identities:

$$\begin{aligned}\partial_\mu W_{V^a_{\mu 5}} &- \sqrt{2n_f}\delta_{a0}W_\theta - d_{abc}m^b W_{S^c_5} \\ &+ f_{abc}V^b_\mu W_{V^c_{\mu 5}} + f_{abc}V^b_{\mu 5}W_{V^c_\mu} + d_{abc}S^b W_{S^c_5} - d_{abc}S^b_5 W_{S^c} = 0\end{aligned} \tag{3}$$

where we have abbreviated functional derivatives as suffices. This is the key to all the results derived in this section. It makes precise the familiar statement of the anomaly as

$$\partial^\mu J^a_{\mu 5} - \sqrt{2n_f}Q\delta_{a0} - d_{abc}m^b\phi^c_5 \sim 0 \tag{4}$$

The chiral Ward identities for two- and higher-point Green functions are found by taking functional derivatives of (3) with respect to the sources. The complete set of identities for two-point functions is given in our review [19]. As an example, we find[2]

$$\partial_\mu W_{V^a_{\mu 5}S^b_5} - \sqrt{2n_f}\delta_{a0}W_{\theta S^b_5} - M_{ac}W_{S^c_5 S^b_5} - \Phi_{ab} = 0 \tag{5}$$

which in more familiar notation reads

$$\partial^\mu\langle 0|T\ J^a_{\mu 5}\ \phi^b_5|0\rangle - \sqrt{2n_f}\delta_{a0}\langle 0|T\ Q\ \phi^b_5|0\rangle - d_{adc}m^d\langle 0|T\ \phi^c_5\ \phi^b_5|0\rangle - d_{abc}\langle\phi^c\rangle = 0 \tag{6}$$

[2] We use the following $SU(3)$ notation for the quark masses and condensates:

$$\begin{pmatrix} m_u & 0 & 0 \\ 0 & m_d & 0 \\ 0 & 0 & m_s \end{pmatrix} = \sum_{a=0,3,8} m^a T^a$$

and

$$\begin{pmatrix} \langle\bar{u}u\rangle & 0 & 0 \\ 0 & \langle\bar{d}d\rangle & 0 \\ 0 & 0 & \langle\bar{s}s\rangle \end{pmatrix} = 2\sum_{0,3,8}\langle\phi^a\rangle T^a$$

where $\langle\phi^a\rangle$ is the VEV of $\phi^a = \bar{q}\ T^a q$. It is also convenient to use the compact notation

$$M_{ab} = d_{acb}m^c \qquad\qquad \Phi_{ab} = d_{abc}\langle\phi^c\rangle$$

The anomaly breaks the original $U(n_f)_L \times U(n_f)_R$ chiral symmetry to $SU(n_f)_L \times SU(n_f)_R \times U(1)_V/Z^V_{n_f}$ and the quark condensate spontaneously breaks this further to the coset $SU(n_f)_L \times SU(n_f)/SU(n_f)_V$. Goldstone's theorem follows immediately. In the chiral limit, there are $(n_f^2 - 1)$ massless Nambu–Goldstone bosons, which acquire masses of order $\sqrt{m}$ for non-zero quark mass. There is no flavour singlet Nambu–Goldstone boson since the corresponding current is anomalous.

The zero-momentum Ward identities are especially important here, since they control the low-energy dynamics. With the assumption that there are no exactly massless particles coupling to the currents, we find

$$\begin{aligned} \sqrt{2n_f}\delta_{a0}W_{\theta\theta} + M_{ac}W_{S_5^c\theta} &= 0 \\ \sqrt{2n_f}\delta_{a0}W_{\theta S_5^b} + M_{ac}W_{S_5^c S_5^b} + \Phi_{ab} &= 0 \end{aligned} \tag{7}$$

Another key element of our analysis will be the chiral Ward identities for the effective action $\Gamma[V^a_{\mu5}, V^a_\mu, Q, \phi^a_5, \phi^a]$, defined as the generating functional for vertices which are 1PI with respect to the set of fields Q, ϕ^a_5 and ϕ^a but *not* the currents $J^a_{\mu5}$, J^a_μ. This is achieved using the partial Legendre transform (or Zumino transform):

$$\Gamma[V^a_{\mu5}, V^a_\mu, Q, \phi^a_5, \phi^a] = W[V^a_{\mu5}, V^a_\mu, \theta, S^a_5, S^a] - \int dx \left(\theta Q + S^a_5\phi^a_5 + S^a\phi^a\right) \tag{8}$$

The chiral Ward identities for Γ are

$$\begin{aligned} \partial_\mu\Gamma_{V^a_{\mu5}} & - \sqrt{2n_f}\delta_{a0}Q - d_{abc}m^b\phi^c_5 \\ & + f_{abc}V^b_\mu\Gamma_{V^c_{\mu5}} + f_{abc}V^b_{\mu5}\Gamma_{V^c_\mu} - d_{abc}\phi^c_5\Gamma_{\phi^b} + d_{abc}\phi^c\Gamma_{\phi^b_5} = 0 \end{aligned} \tag{9}$$

Again, the zero-momentum identities for the two-point vertices play an important role:

$$\begin{aligned} \Phi_{ac}\Gamma_{\phi^c_5 Q} - \sqrt{2n_f}\delta_{a0} &= 0 \\ \Phi_{ac}\Gamma_{\phi^c_5\phi^b_5} - M_{ab} &= 0 \end{aligned} \tag{10}$$

These will be used in Sect. 3 to construct an effective action which captures the low-energy dynamics of QCD in the pseudoscalar sector.

2.2 Topological Susceptibility

The connection with topology arises through the identification of the gluon operator Q in the anomaly with a topological charge density. Q is a total divergence:

$$Q = \frac{\alpha_s}{8\pi} \,\mathrm{tr}\, G_{\mu\nu}\tilde{G}^{\mu\nu} = \partial^\mu K_\mu \tag{11}$$

where K_μ is the Chern–Simons current,

$$K_\mu = \frac{\alpha_s}{4\pi}\epsilon_{\mu\nu\rho\sigma}\mathrm{tr}\big(A^\nu G^{\rho\sigma} - \frac{1}{3}gA^\nu[A^\rho, A^\sigma]\big) \tag{12}$$

Nevertheless, the integral over (Euclidean) spacetime of Q need not vanish. In fact, for gauge field configurations such as instantons which become pure gauge at infinity,

$$\int d^4x\ Q \ = \ n \in \mathbf{Z} \tag{13}$$

where the integer n is the topological winding number, an element of the homotopy group $\pi_3(SU(N_c))$.

The form of the anomaly is then understood as follows. Under a chiral transformation, the fermion measure in the path integral transforms as (for one flavour)

$$\mathcal{D}\bar{q}\mathcal{D}q \ \rightarrow \ e^{-2i\alpha\int dx \varphi_i^\dagger\gamma_5\varphi_i}\ \mathcal{D}\bar{q}\mathcal{D}q \ = \ \exp^{-2i\alpha(n_+ - n_-)}\ \mathcal{D}\bar{q}\mathcal{D}q \tag{14}$$

where φ_i is a basis of eigenfunctions of the Dirac operator $\not{D}$ in the background gauge field. The non-zero eigenvalues are chirality paired, so the Jacobian only depends on the difference $(n_+ - n_-)$ of the positive and negative chirality zero modes of $\not{D}$. Finally, the index theorem relates the anomaly to the topological charge density:

$$\mathrm{ind}\not{D} \ = \ n_+ - n_- \ = \ \int d^4x\ Q \tag{15}$$

The topological susceptibility $\chi(p^2)$ is defined as the two-point Green function of Q, viz.

$$\chi(p^2) = i\int dx\ e^{ipx}\langle 0|T\ Q(x)\ Q(0)|0\rangle \tag{16}$$

We are primarily concerned with the zero-momentum limit $\chi(0) = W_{\theta\theta}(0)$. Combining (7) gives the crucial Ward identity satisfied by $\chi(0)$:

$$2n_f\chi(0) = M_{0a}W_{S_5^a S_5^b}M_{b0} + (M\Phi)_{00} \tag{17}$$

that is,

$$n_f^2\int dx\ \langle 0|T\ Q(x)\ Q(0)|0\rangle \ = \ \int dx\ m^a m^b\langle 0|T\ \phi_5^a(x)\ \phi_5^b(0)|0\rangle \ + \ m^a\langle\phi^a\rangle \tag{18}$$

Determining exactly how this is satisfied in QCD is at the heart of the Witten–Veneziano approach to the $U(1)_A$ problem [1, 20].

The zero-momentum Ward identities allow us to write a precise form for the topological susceptibility in QCD in terms of just one unknown dynamical constant [21]. To derive this, recall that the matrix of two-point vertices is simply the inverse of the two-point Green function matrix, so in the pseudoscalar sector we have the following inversion formula:

$$\Gamma_{QQ} = -\Big(W_{\theta\theta} - W_{\theta S_5^a}(W_{S_5S_5})^{-1}_{ab}W_{S_5^b\theta}\Big)^{-1} \tag{19}$$

Using the identities (7) and (17), this implies that at zero momentum

$$\Gamma_{QQ}^{-1} = -\chi\Big(1 - 2n_f\chi(M\Phi)_{00}^{-1}\Big)^{-1} \tag{20}$$

and inverting this relation gives

$$\chi = -\Gamma_{QQ}^{-1}\Big(1 - 2n_f\Gamma_{QQ}^{-1}(M\Phi)_{00}^{-1}\Big)^{-1} \tag{21}$$

Finally, substituting for $(M\Phi)_{00}^{-1}$ using the definitions above, we find the following important identity which determines the quark mass dependence of the topological susceptibility in QCD:

$$\chi(0) = -A\Big(1 - A\sum_q \frac{1}{m_q\langle\bar{q}q\rangle}\Big)^{-1} \tag{22}$$

where we identify the non-perturbative coefficient A as Γ_{QQ}^{-1}.

Notice immediately how this expression exposes the well-known result that $\chi(0)$ vanishes if any quark mass is set to zero. In Sect. 3, we will see how it also clarifies the role of the $1/N_c$ expansion in the $U(1)_A$ problem.

2.3 Renormalisation Group

The conserved current corresponding to a non-anomalous symmetry is not renormalised and has vanishing anomalous dimension. However, an anomalous current such as the flavour singlet axial current $J_{\mu5}^0$ is renormalised. The composite operator renormalisation and mixing in the $J_{\mu5}^0, Q$ sector is as follows [22]:

$$\begin{aligned} J_{\mu5R}^0 &= ZJ_{\mu5B}^0 \\ Q_R &= Q_B - \frac{1}{\sqrt{2n_f}}(1-Z)\partial^\mu J_{\mu5B}^0 \end{aligned} \tag{23}$$

Notice the form of the mixing of the operator Q with $\partial^\mu J_{\mu5}^0$ under renormalisation. This ensures that the combination $\big(\partial^\mu J_{\mu5}^0 - \sqrt{2n_f}Q\big)$ occurring in the $U(1)_A$ anomaly equation is RG invariant. The chiral Ward identities therefore take precisely the same form expressed in terms of the bare or renormalised operators, making precise the notion of 'non-renormalisation of the anomaly'. We may therefore interpret the above Ward identities, which were derived in terms of the bare operators, as identities for the renormalised composite operators (and omit the suffix $_R$ for notational simplicity).

The renormalisation group equation (RGE) for the generating functional $W[V_{\mu5}^a, V_\mu^a, \theta, S_5^a, S^a]$ follows immediately from the definitions (23) of the renormalised composite operators. Including also a standard multiplicative renormalisation $Z_\phi = Z_m^{-1}$ for the pseudoscalar and scalar operators ϕ_5^a and

ϕ^a and denoting the anomalous dimensions corresponding to Z and Z_ϕ by γ and γ_ϕ, respectively, we find[3]

$$\mathcal{D}W = \gamma\Big(V^0_{\mu 5} - \frac{1}{\sqrt{2n_f}}\partial_\mu\theta\Big)W_{V^0_{\mu 5}} + \gamma_\phi\Big(S^a_5 W_{S^a_5} + S^a W_{S^a}\Big) + \ldots \qquad (24)$$

where $\mathcal{D} = \left(\mu\frac{\partial}{\partial\mu} + \beta\frac{\partial}{\partial g} - \gamma_m\sum_q m_q\frac{\partial}{\partial m_q}\right)\Big|_{V,\theta,S_5,S}$.

The RGEs for Green functions are found by functional differentiation of (24) and can be simplified using the Ward identities. For example, for $W_{\theta\theta}$ we find

$$\mathcal{D}W_{\theta\theta} \;=\; 2\gamma W_{\theta\theta} + 2\gamma\frac{1}{\sqrt{2n_f}}M_{0b}W_{\theta S^b_5} + \ldots \qquad (25)$$

At zero momentum, we can then use the first identity in (7) to prove that the topological susceptibility $\chi(0)$ is RG invariant,

$$\mathcal{D}\chi(0) \;=\; 0 \qquad (26)$$

which is consistent with its explicit expression (22).

A similar RGE holds for the effective action $\Gamma[V^a_{\mu 5}, V^a_\mu, Q, \phi^a_5, \phi_5]$, which allows the scaling behaviour of the proper vertices involving Q and ϕ^a_5 to be determined [9, 23, 24]. This reads

$$\mathcal{D}\Gamma = \gamma\Big(V^0_{\mu 5} - \frac{1}{\sqrt{2n_f}}\Gamma_Q\partial_\mu\Big)\Gamma_{V^0_{\mu 5}} - \gamma_\phi\Big(\phi^a_5\Gamma_{\phi^a_5} + \phi^a\Gamma_{\phi^a}\Big) + \ldots \qquad (27)$$

An immediate consequence is that $\mathcal{D}\Gamma_{QQ} = 0$ at zero momentum, which ensures the compatibility of (22) with the RG invariance of $\chi(0)$.

2.4 $1/N_c$, the Topological Expansion and OZI

The final theoretical input into our analysis of the $U(1)_A$ problem and phenomenological implications of the anomaly concerns the range of dynamical approximation schemes associated with the large-N_c limit. At various points we will refer either to the original large-N_c expansion of 't Hooft [25], the topological expansion introduced by Veneziano [26] and the OZI limit [27, 28, 29]. A very clear summary of the distinction between them is given in Veneziano's 'Okubofest' review [12], which we follow here.

In terms of Feynman diagrams, the leading order in the large N_c, fixed n_f ('t Hooft) limit is the most restrictive of these approximations, including only planar diagrams with sources on a single quark line and no further quark loops (Fig. 1).

[3] The notation $+\ldots$ refers to additional terms which are required to produce the contact term contributions to the RGEs for n-point Green functions and vertices of composite operators. These are discussed fully in [9, 23, 24], but will be omitted here for simplicity. They vanish at zero momentum.

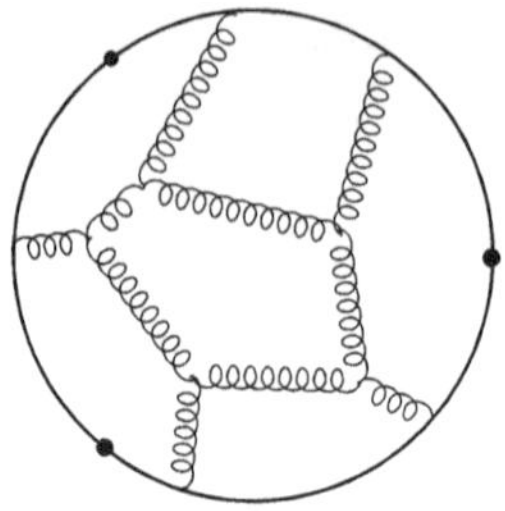

Fig. 1. A typical Feynman diagram allowed in the large-N_c limit. The dots on the quark loop represent external sources

A better approximation to QCD is the quenched approximation familiar in lattice gauge theory. This is a small n_f expansion at fixed N_c, i.e. excluding quark loops but allowing non-planar diagrams (Fig. 2).

An alternative is the topological expansion, which allows any number of internal quark loops, but restricts to planar diagrams at leading order. Provided the sources remain attached to the same quark line, this corresponds to taking large N_c at fixed n_f/N_c. This means that quarks and gluons are treated democratically and the order of approximation is determined solely by the topology of the diagrams (Fig. 2).

Finally, the OZI approximation is a still closer match to full QCD with dynamical quarks than either the leading order quenched or topological expansions. Non-planar diagrams and quark loops are retained, but diagrams in which the external sources are connected to different quark loops are still excluded (Fig. 3). This means that amplitudes which involve purely gluonic intermediate states are suppressed. This is the field-theoretic basis for the original empirical OZI rule.

In each of these large-N_c expansions, except the topological expansion where n_f/N_c is fixed, the $U(1)_A$ anomaly does not contribute at leading order. More precisely, the anomalous contribution $\langle 0|T\ Q\ \phi_5^b|0\rangle$ in the chiral Ward identity (6) is suppressed by $O(1/N_c)$ relative to the current term

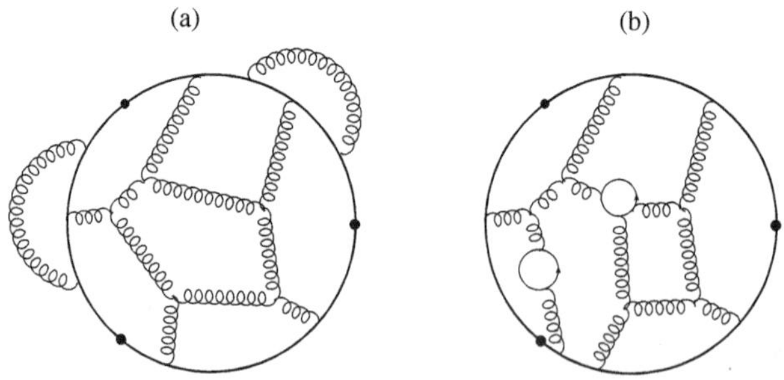

Fig. 2. Feynman diagrams allowed in the quenched approximation (**left**) or leading order in the topological expansion (**right**)

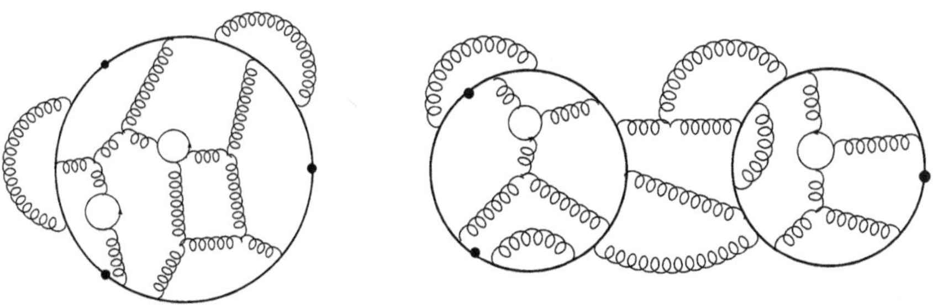

Fig. 3. Feynman diagrams allowed (**left**) and forbidden (**right**) by the OZI rule

$\langle 0|T\ J^0_{\mu 5}\ \phi^b_5|0\rangle$. This means that the flavour singlet current is conserved, Goldstone's theorem applies, and conventional PCAC methods can be used to understand the dynamics of the Green functions with a full set of $(n_f^2 - 1)$ massless bosons in the chiral limit. Taking this as a starting point, we can then learn about the spectral decomposition of the actual QCD Green functions as we relax from the leading-order limits. In particular, this leads us to the famous Witten–Veneziano mass formula for the η' meson [1, 20].

The behaviour of the topological susceptibility at large N_c is central to this analysis. It is clear from looking at planar diagrams that at leading order in $1/N_c$, $\chi(0)$ in QCD coincides with the topological susceptibility $\chi(0)|_{\rm YM}$ in the corresponding pure Yang–Mills theory. Referring now to the explicit expression (22) for $\chi(0)$, large-N_c counting rules give $A = O(1)$ while $\langle \bar{q}q\rangle = O(N_c)$. It follows that *for non-zero quark masses*,

$$\chi(0) \;=\; -\,A \,+\, O(n_f/N_c) \tag{28}$$

where $A = \Gamma^{-1}_{QQ}$ is identified as $-\chi(0)_{\rm YM} + O(1/N_c)$. On the other hand, if we consider the limit of $\chi(0)$ for $m_q \to 0$ at finite N_c, then we have

$$\chi(0)|_{m_q\to 0} \;=\; 0 \tag{29}$$

The 't Hooft large-N_c limit is therefore not smooth in QCD; the $N_c \to \infty$ and $m_q \to 0$ limits do not commute [1, 20, 21]. This is remedied in the topological expansion, where quark loops are retained and the $O(n_f/N_c)$ contribution in (28) allows the smooth chiral limit $\chi(0) \to 0$ even for large N_c.

3 '$U(1)_A$ Without Instantons'

The $U(1)_A$ problem has a long history, pre-dating QCD itself, and has been an important stimulus to new theoretical ideas involving anomalies and gluon topology.

At its simplest, the original '$U(1)_A$ problem' in current algebra is relatively straightforwardly resolved by the existence of the anomalous contributions to the chiral Ward identities (anomalous commutators in current algebra) and the consequent absence of a ninth light Nambu–Goldstone boson in $n_f = 3$ QCD.[4] However, a full resolution requires a much more detailed understanding of the dynamics of the pseudoscalar sector and the role of topological fluctuations in the anomalous Green functions.

In this section, we review the analysis of the $U(1)_A$ problem presented by Veneziano in his seminal 1979 paper, '$U(1)_A$ without instantons' [1].[5] As well as deriving the eponymous mass formula relating the η' mass to the topological susceptibility, the essential problem resolved in [1] is how to describe the dynamics of the Green functions of the pseudoscalar operators in QCD in terms of a spectral decomposition compatible with the n_f, N_c, θ and quark mass dependence imposed by the anomalous Ward identities.

First, recall that in the absence of the anomaly, there will be light pseudoscalar mesons η^α coupling derivatively to the currents with decay constants $f^{a\alpha}$, i.e. $\langle 0|J^a_{\mu 5}|\eta^\alpha\rangle = ip_\mu f^{a\alpha}$. (We use the notation η^α to denote the physical mesons π^0, η and η', while the $SU(3)$ index $a = 3, 8, 0$.) The mass matrix $\mu^2_{\alpha\beta}$ satisfies the Dashen, Gell-Mann–Oakes–Renner (DGMOR) relation [35, 36]

$$f^{a\alpha}\mu^2_{\alpha\beta}f^{T\beta b} \quad = \quad -(M\Phi)_{ab} \qquad \text{(no anomaly)} \tag{30}$$

The consequences of the anomaly are determined by the interaction of the pseudoscalar fields ϕ^a_5 with the topological charge density Q and the subsequent mixing. This gives rise to an additional contribution to the masses. Moreover, we can no longer identify the flavour singlet decay constant by the coupling to $J^0_{\mu 5}$ since this is not RG invariant. Instead, the physical decay constants $f^{a\alpha}$ are defined in terms of the couplings of the η^α to the pseudoscalar fields through the relation $f^{a\alpha}\langle 0|\phi^b_5|\eta^\alpha\rangle = d_{abc}\langle\phi^c\rangle$. This coincides with the usual definition except in the flavour singlet case.

The most transparent way to describe how all this works is to use an effective action $\Gamma[Q, \phi^a_5]$ constructed to satisfy the anomalous chiral Ward identities. It is important to emphasise from the outset that this is an effective action in the sense of Sect. 2.1, i.e. the generating functional for vertices which are 1PI with respect to the set of fields Q, ϕ^a_5 only. The choice of fields is designed to capture the degrees of freedom essential for the dynamics.[6] A different choice (or linear combination) redefines the physical meaning of the

---

[4] The existence of a light flavour-singlet Nambu–Goldstone boson would produce a rapid off-shell variation in the $\eta \to 3\pi$ decay amplitude, in contradiction with the experimental data [30].

[5] For reviews of the instanton approach to the resolution of the $U(1)_A$ problem, see, e.g. [31, 32, 33, 34].

[6] Note especially the frequently misunderstood point that the choice of fields in Γ is not required to be in any sense a complete set, nor does the restriction to a given set of fields constitute an approximation. Before imposing dynamical simplifications, the identities derived from Γ are *exact* – increasing the set of basis fields simply changes the definitions of the 1PI vertices. The effective action considered here is therefore different from the non-linear chiral Lagrangians incorporating

vertices, so it is important that the final choice of fields in Γ results in vertices which are most directly related to physical couplings.

The simplest effective action consistent with the anomalous Ward identities and the renormalisation group is

$$\Gamma[Q,\phi_5^a] = \int dx \left(\frac{1}{2A}Q^2 + Q(\sqrt{2n_f}\delta_{0a} - B_a\partial^2)\Phi^{-1}_{ab}\phi_5^b + \frac{1}{2}\phi_5^a\Phi^{-1}_{ac}((M\Phi)_{cd} - C_{cd}\partial^2)\Phi^{-1}_{db}\phi_5^b\right) \tag{31}$$

The constants C_{ab} and B_a are related to $\Gamma_{V^a_{\mu 5}V^b_{\nu 5}}$ and $\Gamma_{V^a_{\mu 5}Q}$, respectively. The inclusion of the term with B_a is unusual, but is required for consistency with the RGEs derived from (27) beyond zero momentum.

This form of $\Gamma[Q,\phi_5^a]$ encodes three key dynamical assumptions:

- Pole dominance. We assume that the Green functions are dominated by the contribution of single-particle poles associated with the pseudoscalar mesons *including* the flavour singlet. This extends the usual PCAC assumption to the singlet sector.
- Smoothness. We assume that pole-free dynamical quantities such as the decay constants and couplings (1PI vertices) are only weakly momentum-dependent in the range from $p=0$ to their on-shell values. This allows us to impose relations derived from the zero-momentum Ward identities, provided this is compatible with the renormalisation group.
- Topology. There must exist topologically non-trivial fluctuations which can give a non-vanishing value to $\chi(0)|_{\rm YM}$ in pure gluodynamics. This is required to give the non-vanishing coefficient in the all-important $\frac{1}{2A}Q^2$ term in $\Gamma[Q,\phi_5^a]$. Notice that we do not require a kinetic term for Q, which would be associated with a (presumed heavy) pseudoscalar glueball.

The second derivatives of $\Gamma[Q,\phi_5^a]$ are

$$\begin{pmatrix} \Gamma_{QQ} & \Gamma_{Q\phi_5^b} \\ \Gamma_{\phi_5^a Q} & \Gamma_{\phi_5^a\phi_5^b} \end{pmatrix} = \begin{pmatrix} A^{-1} & (\sqrt{2n_f}\delta_{0d} + B_d p^2)\Phi^{-1}_{db} \\ \Phi^{-1}_{ac}(\sqrt{2n_f}\delta_{c0} + B_c p^2) & \Phi^{-1}_{ac}((M\Phi)_{cd} + C_{cd}p^2)\Phi^{-1}_{db} \end{pmatrix} \tag{32}$$

The corresponding Green functions (composite operator propagators) are given by inversion:

$$\begin{pmatrix} W_{\theta\theta} & W_{\theta S_5^b} \\ W_{S_5^a\theta} & W_{S_5^a S_5^b} \end{pmatrix} = -\begin{pmatrix} \Gamma_{QQ} & \Gamma_{Q\phi_5^b} \\ \Gamma_{\phi_5^a Q} & \Gamma_{\phi_5^a\phi_5^b} \end{pmatrix}^{-1} \tag{33}$$

and we find, to leading order in p^2,

---

the large-N_c approach to the pseudoscalar mesons constructed by a number of groups. See, for example, [21, 37, 38, 39, 40, 41, 42].

$$\begin{aligned} W_{\theta\theta} &= -A\,\tilde{\Delta}^{-1} \\ W_{\theta S_5^b} &= W_{S_5^b\theta} \simeq \sqrt{2n_f}A\Delta_{0d}^{-1}\Phi_{db} \\ W_{S_5^a S_5^b} &= -\Phi_{ac}\,\Delta_{cd}^{-1}\,\Phi_{db} \end{aligned} \tag{34}$$

where

$$\tilde{\Delta} = 1 - \Big(2n_f A\delta_{a0}\delta_{0b} + \sqrt{2n_f}A(\delta_{a0}B_b + B_a\delta_{0b})p^2\Big)\big(M\Phi + Cp^2\big)^{-1}_{ab} \tag{35}$$

and

$$\Delta_{ab} = \Big(C_{ab} - \sqrt{2n_f}A(\delta_{a0}B_b + B_a\delta_{0b})\Big)p^2 + (M\Phi)_{ab} - 2n_f A\,\delta_{a0}\delta_{0b} \tag{36}$$

In this form, however, the propagator matrix is not diagonal and the operators are not normalised so as to couple with unit decay constants to the physical states. It is therefore convenient to make a change of variables in Γ so that it is written in terms of operators which are more closely identified with the physical states. We do this is in two stages, since the intermediate stage allows us to make direct contact with the discussion in [1] and will play an important role in some of the phenomenological applications considered later.

First, we define rescaled fields $\hat{\eta}^\alpha$ whose kinetic terms, before mixing with Q, are canonically normalised. That is, we set

$$\hat{\eta}^\alpha = \hat{f}^{T\alpha a}\Phi_{ab}^{-1}\phi_5^b \tag{37}$$

with the 'decay constants' $\hat{f}^{a\alpha}$ defined such that $\frac{d}{dp^2}\Gamma_{\hat{\eta}^\alpha\hat{\eta}^\beta}|_{p=0} = \delta_{\alpha\beta}$. This implies

$$(\hat{f}\hat{f}^T)_{ab} = C_{ab} = \frac{d}{dp^2}W_{S_D^a S_D^b}\Big|_{p=0} \tag{38}$$

where $D^a = \sqrt{2n_f}\delta_{a0}Q + M_{ab}\phi_5^b$ is the divergence of the current $J_{\mu 5}^a$. In the chiral limit, this reduces in the flavour singlet sector to

$$(\hat{f}\hat{f}^T)_{00} = \frac{d}{dp^2}\chi(p^2)\Big|_{p=0} = \chi'(0) \tag{39}$$

a result which plays a vital role in understanding the 'proton spin' problem. Notice however that the $\hat{f}^{a\alpha}$ are *not* RG invariant: in fact, $\mathcal{D}\hat{f}^{a\alpha} = \gamma\delta_{a0}\hat{f}^{a\alpha}$. The effective action $\Gamma[Q,\hat{\eta}^\alpha]$ is

$$\begin{aligned} \Gamma[Q,\hat{\eta}^a] = \int dx\,\bigg(&\frac{1}{2A}Q^2 + Q\big(\sqrt{2n_f}\delta_{0a} - B_a\partial^2\big)(\hat{f}^{-1})^{a\alpha}\hat{\eta}^\alpha \\ &+ \frac{1}{2}\hat{\eta}^\alpha\big(-\partial^2 + \hat{f}^{-1T}M\Phi\hat{f}^{-1}\big)_{\alpha\beta}\hat{\eta}^\beta\bigg) \end{aligned} \tag{40}$$

In this form, the $\hat{\eta}^\alpha$ are the canonically normalised fields corresponding to the 'would-be Nambu–Goldstone bosons' in the absence of the anomaly, before they acquire an additional anomaly-induced mass. In the framework of the large-N_c or OZI approximations, they would correspond to true Nambu-Goldstone bosons. The singlet $\hat{\eta}^0$ is what we have therefore referred to in our previous papers as the 'OZI boson' η'_{OZI}. As we see later, the naive current algebra relations hold when expressed in terms of the $\hat{\eta}^\alpha$ and $\hat{f}^{a\alpha}$, though these do *not* correspond to physical states or decay constants.

The physical particle masses are identified with the poles in the two-point Green functions (34). We immediately see that due to mixing with the topological charge density Q, the physical pseudoscalar meson mass $m^2_{\alpha\beta}$ is shifted from its original value. From the pole in (36), we immediately find

$$f^{a\alpha} m^2_{\alpha\beta} f^{T\beta b} = -(M\Phi)_{ab} + 2n_f A \delta_{a0}\delta_{b0} \tag{41}$$

where we identify the physical, RG-invariant decay constants as

$$(ff^T)_{ab} = (\hat{f}\hat{f}^T)_{ab} - \sqrt{2n_f} A(\delta_{a0}B_b + B_a\delta_{0b}) \tag{42}$$

Equation (41) is the key result. It generalises the original DGMOR relations (30) to the flavour-singlet sector with the anomaly properly incorporated and the renormalisation group constraints satisfied. It represents a generalisation of the Witten–Veneziano mass formula which makes no direct reference to large-N_c arguments but depends only on the three dynamical assumptions stated above [2].

With this clarification of the distinction between the physical decay constants $f^{a\alpha}$ and the RG non-invariant $\hat{f}^{a\alpha}$, we can rewrite (35) for the topological susceptibility $\chi(p^2) = W_{\theta\theta}(p^2)$ as

$$\chi(p^2) = -A\Big[1 - \mathrm{tr}\big((\hat{f}\hat{f}^T - ff^T)p^2 + 2n_f A\mathbf{1}_{00}\big)\big(\hat{f}\hat{f}^T p^2 + M\Phi\big)^{-1}\Big]^{-1} \tag{43}$$

It is clear that in the zero-momentum limit, this expression successfully reproduces (22) for $\chi(0)$. For one flavour, the formula simplifies to

$$\chi(p^2) = -A(\hat{f}\hat{f}^T p^2 + M\Phi)\Big[ff^T p^2 + M\Phi + 2n_f A\Big]^{-1} \qquad (n_f = 1) \tag{44}$$

showing clearly the pole at the shifted mass m^2 of (41). The occurrence of both $\hat{f}^{a\alpha}$ and $f^{a\alpha}$ in these expressions allows them to satisfy the RGE (25) for the topological susceptibility, which requires $\mathcal{D}\chi(p^2) = O(p^2)$.

The second stage is to make a change of variable which diagonalises the propagator matrix, so as to give the most direct possible relation between the operators and the physical states. Choosing

$$\begin{aligned} G &= Q - W_{\theta S_5^a} W^{-1}_{S_5^a S_5^b}\phi_5^b \simeq Q + \sqrt{2n_f} A\Phi^{-1}_{0b}\phi_5^b \\ \eta^\alpha &= f^{T\alpha a}\Phi^{-1}_{ab}\phi_5^b \end{aligned} \tag{45}$$

defines the effective action $\Gamma[G, \eta^\alpha]$ as

$$\Gamma[G, \eta^\alpha] = \int dx \left(\frac{1}{2A} G^2 + \frac{1}{2} \eta^\alpha (-\partial^2 - m^2)_{\alpha\beta} \eta^\beta \right) \tag{46}$$

with $m^2_{\alpha\beta}$ given by (41). The corresponding propagators are

$$\begin{aligned} \langle 0|T\ G\ G|0\rangle &= -A \\ \langle 0|T\ \eta^\alpha\ \eta^\beta|0\rangle &= \frac{-1}{p^2 - m^2_{\eta^\alpha}} \delta^{\alpha\beta} \end{aligned} \tag{47}$$

where with no loss of generality we have taken $m^2_{\alpha\beta}$ diagonal.

Notice also that the states mix in the complementary way to the operators. In particular, the mixing for the states corresponding to (45) for the fields G and η^α is

$$\begin{aligned} |G\rangle &= |Q\rangle \\ |\eta^\alpha\rangle &= (f^{-1})^{\alpha a} (\Phi_{ab}|\phi_5^b\rangle - \sqrt{2n_f} A\delta_{a0}|Q\rangle) \end{aligned} \tag{48}$$

In this sense, we see that we can regard the physical η' (and, with $SU(3)$ breaking, the η) as an admixture of quark and gluon components, while the unphysical state $|G\rangle$ is purely gluonic.

An immediate corollary is the following relation, which we will use repeatedly in deriving alternative forms of the current algebra identities for the pseudoscalar mesons:

$$\Phi_{ab} \frac{\delta}{\delta\phi_5^b} = \hat{f}^{a\alpha} \frac{\delta}{\delta\hat{\eta}^\alpha} = f^{a\alpha} \frac{\delta}{\delta\eta^a} + \sqrt{2n_f} A\delta_{a0} \frac{\delta}{\delta G} \tag{49}$$

The formulation in terms of $\Gamma[G, \eta^\alpha]$ is exactly what we need to construct a simple '$U(1)_A$ PCAC' with which to interpret the low-energy phenomenology of the pseudoscalar mesons. We turn to this in the next section.

Here, we focus on the intermediate formulation $\Gamma[Q, \hat{\eta}^\alpha]$ in order to describe Veneziano's analysis of the $U(1)_A$ problem in the framework of the large-N_c and topological expansions. The starting point is the anomalous Ward identity (18) for the topological susceptibility:

$$n_f^2 \int dx\ \langle 0|T\ Q(x)\ Q(0)|0\rangle = \int dx\ m^a m^b \langle 0|T\ \phi_5^a(x)\ \phi_5^b(0)|0\rangle + m^a\langle\phi^a\rangle \tag{50}$$

The essential problem is how to understand this relation in terms of a spectral decomposition in the context of the $1/N_c$ expansion.

Assuming that $\chi(0)_{YM} = -A + O(1/N_c)$ is non-vanishing at $O(1)$, the l.h.s. is $O(n_f^2)$ in leading order in $1/N_c$. On the other hand, the r.h.s. includes the condensate term of $O(n_f N_c m)$. To resolve this apparent paradox, we have to go beyond leading order in $1/N_c$ and consider the quark loop contributions which are included in the topological expansion. Although these are formally suppressed by powers of (n_f/N_c), they contain light intermediate states that can enhance the order of the Green function. As we have seen above, these light states are just the 'OZI bosons' $|\hat{\eta}^\alpha\rangle$ with masses $\mu^2_{\alpha\beta}$ of $O(n_f m)$. Inserting these intermediate states, we therefore find that:

$$\chi(p^2) \;=\; \chi(p^2)|_{\rm YM} - \langle 0|Q|\hat{\eta}^\alpha\rangle \frac{1}{(p^2-\mu^2)_{\alpha\beta}} \langle \hat{\eta}^\beta|Q|0\rangle \;+\; \ldots \qquad (51)$$

where the coupling $\langle 0|Q|\hat{\eta}^\alpha\rangle$ is $O(\sqrt{n_f/N_c})$.

Approximating $\chi(p^2)_{\rm YM} \sim -A$ (a low-momentum smoothness assumption) and $\langle 0|Q|\hat{\eta}^\alpha\rangle \sim \sqrt{2n_f}A(f^{-1})^{0\alpha}$, then summing the series of intermediate state contributions, we find

$$\chi(p^2) \;\simeq\; -\frac{A}{1 - 2n_f A\Big(f(p^2-\mu^2)f^T\Big)^{-1}_{00}} \qquad (52)$$

This expression reproduces (13) of [1]. Clearly, it is dominated by the physical pseudoscalar pole with anomaly-induced mass given by (41). It does not completely recover our more precise expression (43) because of the approximation for the coupling of Q to the $|\hat{\eta}^\alpha\rangle$, which misses the subtleties related to the introduction of B_a in the effective action $\Gamma[Q,\hat{\eta}^\alpha]$ and the distinction of $\hat{f}^{a\alpha}$ and $f^{a\alpha}$. These are effects of higher order in $1/N_c$ but, as we have seen, they are necessary to establish full RG consistency and will prove to be important for phenomenology.

To see how a term with the $O(n_f N_c m)$ dependence of the condensate can arise in $n_f^2\chi(0)$, notice from (41) that the physical pseudoscalar mass squared $m^2_{\eta^\alpha}$ has two contributions, the first of $O(m)$ from the conventional quark mass term and the new, anomaly-induced contribution of $O(n_f/N_c)$. If we are in a regime where the anomaly contribution dominates ($m < \Lambda_{\rm QCD}/N_c$), then it follows that the above expression for $\chi(0)$ indeed becomes of $O(n_f^{-1}N_c m)$.

The original Witten–Veneziano mass formula for the η' is the large-N_c limit of (41). In the chiral limit there is no flavour mixing and the singlet mass is given by

$$m^2_{\eta'} \;=\; \frac{1}{(f^{0\eta'})^2}2n_f A \;=\; -\frac{2n_f}{f^2_\pi}\chi(0)_{\rm YM} + O((n_f/N_c)^2) \qquad (53)$$

This formula provided the first link between the η' mass and gluon topology. For an alternative recent derivation in the context of an n_f/N_c expansion, see also [43].

What we learn from all this is that the Green functions in the anomalous chiral Ward identities admit a consistent spectral decomposition in terms of a full set of (n_f^2-1) pseudoscalar mesons, provided they satisfy the generalised DGMOR mass formula (41) *including* the all-important anomaly term. The presence of these light poles can enhance the apparent order of the Green functions, as is familiar with Nambu–Goldstone bosons, and the anomaly-induced

$O(n_f/N_c)$ contribution to $m^2_{\eta^\alpha}$ is critical in ensuring complete consistency with the Ward identities.

Similar considerations apply to the resolution of apparent paradoxes in the θ-dependence of some Green functions. For example [1], we can show from the anomalous Ward identities that the condensate satisfies

$$\sum_q m_q \langle \bar{q}q \rangle|_\theta \equiv m^a \langle \phi^a \rangle = \cos(\theta/n_f)\, m^a \langle \phi^a \rangle|_{\theta=0} \tag{54}$$

This implies

$$\begin{aligned} \tfrac{\partial^2}{\partial\theta^2} m^a \langle \phi^a \rangle|_{\theta=0} &= -\, m^a \int dx \int dy\ \langle 0|T\ Q(x)\ Q(y)\ \phi^a(0)|0\rangle \\ &= -\frac{1}{n_f^2}\ m^a \langle \phi^a \rangle|_{\theta=0} \end{aligned} \tag{55}$$

Here, the Green function is superficially of $O(n_f/N_c)$ while the r.h.s. is $O(N_c/n_f)$. The resolution is simply that it contains pseudoscalar intermediate states contributing two light poles with $m^2 \sim O(n_f/N_c)$. So once again we see how the spectral decomposition in terms of the full set of pseudoscalar mesons, including the flavour singlet, ensures consistency with the anomalous Ward identities.

4 Pseudoscalar Mesons

This theoretical analysis provides the basis for an extension of the conventional PCAC or chiral Lagrangian description of the phenomenology of the pseudoscalar mesons to the flavour singlet sector. In this section[7] we describe the role of the $U(1)_A$ anomaly in the radiative decays of π^0, η and η' and derive the $U(1)_A$ Goldberger–Treiman relation, first proposed by Veneziano as a resolution of the 'proton spin' problem.

4.1 $U(1)_A$ Dashen, Gell-Mann–Oakes–Renner relations

The extension of the DGMOR relations to the $U(1)_A$ sector follows from the application of the three key dynamical assumptions used above (viz. pole dominance by the nonet of pseudoscalar mesons, smoothness of decay constants and couplings over the range from zero to on-shell momentum, and the existence of topologically non-trivial gluon dynamics) to the anomalous chiral Ward identities.

The fundamental $U(1)_A$ DGMOR relation

$$f^{a\alpha} m^2_{\alpha\beta} f^{T\beta b} = -M_{ac}\Phi_{cb} + 2n_f A \delta_{a0}\delta_{b0} \tag{56}$$

[7] This section is based on the presentation in [3], where we extend and update our original work [2, 44] to include a detailed comparison with experimental data.

has been derived above in the course of the general discussion of the $U(1)_A$ problem. In order to make this section self-contained, we give a brief and direct derivation here.

Recall that the physical meson fields are given as $\eta^\alpha = f^{T\alpha a}\Phi^{-1}_{ab}\phi^b_5$, with the decay constants defined so that the propagator $W_{S^\alpha_\eta S^\beta_\eta} = -1/(p^2 - m^2_\eta)_{\alpha\beta}$. It follows immediately that at zero momentum,

$$f^{a\alpha}m^2_{\alpha\beta}f^{T\beta b} = \Phi_{ac}(W_{S_5S_5})^{-1}_{cd}\Phi_{db} \tag{57}$$

Using the chiral Ward identities of Sect. 2 together with the identification (21) of the topological susceptibility, we can then show

$$\begin{aligned}\Phi_{ac}(W_{S_5S_5})^{-1}_{cd}\Phi_{db} &= (\Phi M)_{ac}(MW_{S_5S_5}M)^{-1}_{cd}(M\Phi)_{db} \\ &= (M\Phi)_{ac}\Big(-(M\Phi) + 2n_f\chi(0)\mathbf{1}_{00}\Big)^{-1}_{cd}(M\Phi)_{db} \\ &= -(M\Phi)_{ab} + 2n_f\Gamma^{-1}_{QQ}\,\delta_{a0}\delta_{b0}\end{aligned} \tag{58}$$

proving the result (56).

Expanding this out, and assuming the mixed decay constants $f^{0\pi}$, $f^{8\pi}$, $f^{3\eta}$, $f^{3\eta'}$ are all negligible, we have

$$(f^{0\eta'})^2m^2_{\eta'} + (f^{0\eta})^2m^2_\eta \;=\; -\frac{2}{3}\big(m_u\langle\bar{u}u\rangle + m_d\langle\bar{d}d\rangle + m_s\langle\bar{s}s\rangle\big) + 6A \tag{59}$$

$$f^{0\eta'}f^{8\eta'}m'^2_\eta + f^{0\eta}f^{8\eta}m^2_\eta \;=\; -\frac{\sqrt{2}}{3}\big(m_u\langle\bar{u}u\rangle + m_d\langle\bar{d}d\rangle - 2m_s\langle\bar{s}s\rangle\big) \tag{60}$$

$$(f^{8\eta'})^2m'^2_\eta + (f^{8\eta})^2m^2_\eta \;=\; -\frac{1}{3}\big(m_u\langle\bar{u}u\rangle + m_d\langle\bar{d}d\rangle + 4m_s\langle\bar{s}s\rangle\big) \tag{61}$$

$$f^2_\pi m^2_\pi \;=\; -\left(m_u\langle\bar{u}u\rangle + m_d\langle\bar{d}d\rangle\right) \tag{62}$$

and we can add the standard DGMOR relation for the K^+,

$$f^2_K m^2_K \;=\; -\left(m_u\langle\bar{u}u\rangle + m_s\langle\bar{s}s\rangle\right) \tag{63}$$

We emphasise that these formulae, as well as the radiative decay and $U(1)_A$ Goldberger–Treiman relations derived below, do not depend at all on the $1/N_c$ expansion. In particular, the constant A appearing in the flavour singlet formula is defined as the non-perturbative parameter determining the topological susceptibility $\chi(0)$ in QCD according to the exact identity (22). As explained above, large-N_c ideas do indeed provide a rationale for extending the familiar PCAC assumptions of pole dominance and smoothness to the flavour singlet channel, but these assumptions can be tested independently against experimental data.

The most useful form of these relations for phenomenology is to assume exact $SU(2)$ flavour symmetry and eliminate the quark masses and condensates in favour of f_π, f_K, m_π^2 and m_K^2 in the DGMOR relations for the η and η'. This gives

$$(f^{0\eta'})^2 m_{\eta'}^2 + (f^{0\eta})^2 m_\eta^2 = \frac{1}{3}(f_\pi^2 m_\pi^2 + 2 f_K^2 m_K^2) + 6A \tag{64}$$

$$f^{0\eta'} f^{8\eta'} m_\eta'^2 + f^{0\eta} f^{8\eta} m_\eta^2 = \frac{2\sqrt{2}}{3}(f_\pi^2 m_\pi^2 - f_K^2 m_K^2) \tag{65}$$

$$(f^{8\eta'})^2 m_\eta'^2 + (f^{8\eta})^2 m_\eta^2 = -\frac{1}{3}(f_\pi^2 m_\pi^2 - 4 f_K^2 m_K^2) \tag{66}$$

We can also now clarify the precise relation of these results to the Witten–Veneziano formula for the mass of the η' in its non-vanishing quark mass form, viz.

$$m_{\eta'}^2 + m_\eta^2 - 2m_K^2 = -\frac{6}{f_\pi^2}\chi(0)|_{YM} \tag{67}$$

Of course, only the $m_{\eta'}^2$ term on the l.h.s. is present in the chiral limit. Substituting in the explicit values for the masses in this formula gives a prediction [1] for the topological susceptibility, $\chi(0)|_{\mathrm{YM}} \simeq -(180\ \mathrm{MeV})^4$, which as we see below is remarkably close to the subsequently calculated lattice result.

If we now add the DGMOR formulae (64) and (66), we find

$$(f^{0\eta'})^2 m_{\eta'}^2 + (f^{0\eta})^2 m_\eta^2 + (f^{8\eta})^2 m_\eta^2 + (f^{8\eta'})^2 m_{\eta'}^2 - 2 f_K^2 m_K^2 = 6A \tag{68}$$

which we repeat is valid to all orders in $1/N_c$. To reduce this to its Witten–Veneziano approximation, we impose the large-N_c limit to approximate the QCD topological charge parameter A with $-\chi(0)|_{YM}$ as explained in Sect. 2.4. We then set the 'mixed' decay constants $f^{0\eta}$ and $f^{8\eta'}$ to zero and all the remaining decay constants $f^{0\eta'}$, $f^{8\eta}$ and f_K equal to f_π. With these approximations, we recover (67). Eventually, after we have found explicit experimental values for all these quantities, we will be able to demonstrate quantitatively how good an approximation the large-N_c Witten–Veneziano formula is to the generalised $U(1)_A$ DGMOR relation in full QCD.

4.2 Radiative Decay Formulae for $\pi^0, \eta, \eta' \to \gamma\gamma$

Radiative decays of the pseudoscalar mesons are of particular interest as they are controlled by the electromagnetic $U(1)_A$ anomaly,

$$\partial^\mu J_{\mu 5}^a - M_{ab}\phi_5^b - \sqrt{2n_f}Q\delta_{a0} - a_{\mathrm{em}}^a \frac{\alpha}{8\pi} F^{\mu\nu}\tilde{F}_{\mu\nu} = 0 \tag{69}$$

where $F_{\mu\nu} = \partial_\mu A_\nu - \partial_\nu A_\mu$ is the usual electromagnetic field strength and the anomaly coefficients $a^a_{\rm em}$ are determined by the quark charges. The generating functional $\Gamma[V^a_{\mu 5}, V^a_\mu, Q, \phi^a_5, \phi^a, A_\mu]$ of 1PI vertices including the photon satisfies the Ward identity

$$\partial_\mu \Gamma_{V^a_{\mu 5}} - \sqrt{2n_f}\delta_{a0} Q - a^a_{\rm em}\frac{\alpha}{8\pi} F^{\mu\nu}\tilde{F}_{\mu\nu} - d_{abc} m^b \phi^c_5$$
$$+ f_{abc} V^b_\mu \Gamma_{V^c_{\mu 5}} + f_{abc} V^b_{\mu 5}\Gamma_{V^c_\mu} - d_{abc}\phi^c_5 \Gamma_{\phi^b} + d_{abc}\phi^c \Gamma_{\phi^b_5} = 0 \quad (70)$$

To derive the radiative decay formulae, we first differentiate this identity with respect to the photon field A_μ. This gives

$$ip_\mu \Gamma_{V^a_{\mu 5} A^\lambda A^\rho} + \Phi_{ab}\Gamma_{\phi^b_5 A^\lambda A^\rho} = - a^a_{\rm em}\frac{\alpha}{\pi}\,\epsilon_{\mu\nu\lambda\rho} k_1^\mu k_2^\nu \quad (71)$$

where k_1, k_2 are the momenta of the two photons. Notice that the mass term does not contribute directly to this formula. From its definition as 1PI w.r.t. the pseudoscalar fields, the vertex $\Gamma_{V^a_{\mu 5} A^\lambda A^\rho}$ does not have a pole at $p^2 = 0$, even in the massless limit, so we find simply

$$\Phi_{ab}\Gamma_{\phi^b_5 A^\lambda A^\rho}\Big|_{p=0} = - a^a_{\rm em}\frac{\alpha}{\pi}\,\epsilon_{\mu\nu\lambda\rho} k_1^\mu k_2^\nu \quad (72)$$

The radiative couplings $g_{\eta^\alpha\gamma\gamma}$ for the physical mesons $\eta^\alpha = \pi^0, \eta, \eta'$ are defined as usual from the decay amplitude $\langle\gamma\gamma|\eta^\alpha\rangle$. With the PCAC assumptions already discussed, they can be identified with the 1PI vertices as follows:

$$\langle\gamma\gamma|\eta^\alpha\rangle = - i g_{\eta^\alpha\gamma\gamma}\,\epsilon_{\mu\nu\lambda\rho} k_1^\mu k_2^\nu \epsilon^\lambda(k_1)\epsilon^\rho(k_2) = i\Gamma_{\eta^\alpha A_\lambda A_\rho}\epsilon^\lambda(k_1)\epsilon^\rho(k_2) \quad (73)$$

Re-expressing (72) in terms of the canonically normalised 'OZI bosons' $\hat{\eta}^\alpha$, we therefore have the first form of the decay formula,

$$\hat{f}^{a\alpha} g_{\hat{\eta}^\alpha\gamma\gamma} = a^a_{\rm em}\frac{\alpha}{\pi} \quad (74)$$

Then, rewriting this in terms of the physical pseudoscalar couplings $g_{\eta^\alpha\gamma\gamma}$ and decay constants according to the relation (49) gives the final form for the generalised $U(1)_A$ PCAC formula describing radiative pseudoscalar decays, incorporating both the electromagnetic and colour anomalies:

$$f^{a\alpha} g_{\eta^\alpha\gamma\gamma} + \sqrt{2n_f} A g_{G\gamma\gamma}\delta_{a0} = a^a_{\rm em}\frac{\alpha}{\pi} \quad (75)$$

Expanding this formula, we have

$$f^{0\eta'} g_{\eta'\gamma\gamma} + f^{0\eta} g_{\eta\gamma\gamma} + \sqrt{6} A g_{G\gamma\gamma} = a^0_{\rm em}\frac{\alpha}{\pi} \quad (76)$$

$$f^{8\eta'} g_{\eta'\gamma\gamma} + f^{8\eta} g_{\eta\gamma\gamma} = a^8_{\rm em}\frac{\alpha}{\pi} \quad (77)$$

$$f_\pi g_{\pi\gamma\gamma} \;=\; a^3_{\rm em}\frac{\alpha}{\pi} \tag{78}$$

where $a^0_{\rm em} = \frac{2\sqrt{2}}{3\sqrt{3}}N_c$, $a^8_{\rm em} = \frac{1}{3\sqrt{3}}N_c$ and $a^3_{\rm em} = \frac{1}{3}N_c$.

The new element in the flavour singlet decay formula is the gluonic coupling parameter $g_{G\gamma\gamma}$. It takes account of the fact that because of the anomaly-induced mixing with the gluon topological density Q, the physical η' is not a true Nambu–Goldstone boson so the naive PCAC formulae must be modified. $g_{G\gamma\gamma}$ is *not* a physical coupling and must be regarded as an extra parameter to be fitted to data, although in view of the identifications in (48) it may reasonably be thought of as the coupling of the photons to the gluonic component of the η'.

The renormalisation group properties of these relations are readily derived from the RGE (27) for Γ. In the 'OZI boson' form, the unphysical coupling $g_{\hat{\eta}^\alpha\gamma\gamma}$ satisfies the complementary RGE to the decay constant $\hat{f}^{a\alpha}$ so the combination is RG invariant:

$$\mathcal{D}\hat{f}^{a\alpha} = \gamma\delta_{a0}\hat{f}^{a\alpha} \qquad\qquad \mathcal{D}\big(\hat{f}^{a\alpha}g_{\hat{\eta}^\alpha\gamma\gamma}\big) = 0 \tag{79}$$

In contrast, *all* the decay constants and couplings in the relation (75) can be shown to be separately RG invariant, including the gluonic coupling $g_{G\gamma\gamma}$ [24, 44].

4.3 The Renormalisation Group, OZI and $1/N_c$: a Conjecture

Although these $U(1)_A$ PCAC relations have been derived purely on the basis of the pole dominance and smoothness assumptions, we will nevertheless find it useful in practical applications to exploit their OZI or large-N_c behaviour, in conjunction with the renormalisation group.

The basic idea is that violations of the OZI rule, or equivalently anomalous large-N_c behaviour, are generally related to the existence of the $U(1)_A$ anomaly. Moreover, we can identify the quantities which will be particularly sensitive to the anomaly as those which have RGEs involving the anomalous dimension γ. We therefore conjecture that the dependence of Green functions and 1PI vertices on γ will be an important guide in identifying propagators and couplings which are likely to show violations of the OZI rule and those for which the OZI (or large-N_c) limit should be a good approximation [9, 24].

As an example, the large-N_c order of the quantities in the flavour singlet decay relation (76) is as follows: $f^{a\alpha} = O(\sqrt{N_c})$ for all the decay constants, $g_{\eta^\alpha\gamma\gamma} = O(\sqrt{N_c})$, $g_{G\gamma\gamma} = O(1)$, $a^a_{\rm em} = O(N_c)$ and the topological susceptibility parameter $A = O(1)$. The renormalisation group behaviour is especially simple, with both the meson and gluonic couplings $g_{\eta^\alpha\gamma\gamma}$ and $g_{G\gamma\gamma}$ as well as the decay constants being RG invariant. Putting this together, we find that all the terms in the decay formula are of $O(N_c)$ except the anomalous contribution $Ag_{G\gamma\gamma}$ which is $O(1)$. Since it is RG invariant and independent of the anomalous dimension γ, we conjecture that it is a quantity for which the OZI (or large-N_c) approximation should be reliable so we expect it to be numerically small compared with the other contributions. In the next section, we test this against experiment.

As we shall see later, this conjecture has far-reaching implications for a range of predictions related to the anomaly, particularly in the interpretation of the $U(1)_A$ Goldberger–Treiman relation and associated ideas on the first moment sum rules for g_1^p and g_1^γ in deep-inelastic scattering.

4.4 Phenomenology

After all this theoretical development, we finally turn to experiment and use the data on the radiative decays $\eta, \eta' \to \gamma\gamma$ to deduce values for the pseudoscalar meson decay constants $f^{0\eta'}$, $f^{0\eta}$, $f^{8\eta'}$ and $f^{8\eta}$ from the set of decay formulae (76), (77) and $U(1)_A$ DGMOR relations (64)–(66). We will also find the value of the unphysical coupling parameter $g_{G\gamma\gamma}$ and test the realisation of the $1/N_c$ expansion in real QCD.

The two-photon decay widths are given by

$$\Gamma(\eta'(\eta) \to \gamma\gamma) \;=\; \frac{m^3_{\eta'(\eta)}}{64\pi} |g_{\eta'(\eta)\gamma\gamma}|^2 \tag{80}$$

The current experimental data, quoted in the Particle Data Group tables [45], are

$$\Gamma(\eta' \to \gamma\gamma) \;=\; 4.28 \pm 0.19 \text{ KeV} \tag{81}$$

which is dominated by the 1998 L3 data [46] on the two-photon formation of the η' in $e^+e^- \to e^+e^-\pi^+\pi^-\gamma$, and

$$\Gamma(\eta \to \gamma\gamma) \;=\; 0.510 \pm 0.026 \text{ KeV} \tag{82}$$

which arises principally from the 1988 Crystal Ball [47] and 1990 ASP [48] results on $e^+e^- \to e^+e^-\eta$. From this data, we deduce the following results for the couplings $g_{\eta'\gamma\gamma}$ and $g_{\eta\gamma\gamma}$:

$$g_{\eta'\gamma\gamma} \;=\; 0.031 \pm 0.001 \text{ GeV}^{-1} \tag{83}$$

and

$$g_{\eta\gamma\gamma} \;=\; 0.025 \pm 0.001 \text{ GeV}^{-1} \tag{84}$$

which may be compared with $g_{\pi\gamma\gamma} = 0.024 \pm 0.001$ GeV.

We also require the pseudoscalar meson masses:

$$\begin{aligned} m_{\eta'} &= 957.78 \pm 0.14 \text{ MeV} \qquad & m_\eta &= 547.30 \pm 0.12 \text{ MeV} \\ m_K &= 493.68 \pm 0.02 \text{ MeV} \qquad & m_\pi &= 139.57 \pm 0.00 \text{ MeV} \end{aligned} \tag{85}$$

and the decay constants f_π and f_K. These are defined in the standard way, so we take the following values (in our normalisations) from the PDG [45]:

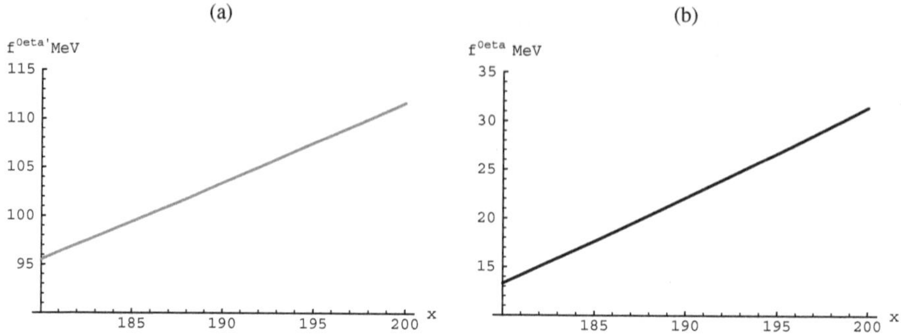

Fig. 4. The decay constants $f^{0\eta'}$ and $f^{0\eta}$ as functions of the non-perturbative parameter $A = (x \text{ MeV})^4$ which determines the topological susceptibility in QCD

$$f_K = 113.00 \pm 1.03 \text{ MeV} \qquad f_\pi = 92.42 \pm 0.26 \text{ MeV} \tag{86}$$

giving $f_K/f_\pi = 1.223 \pm 0.012$.

The octet decay constants $f^{8\eta}$ and $f^{8\eta'}$ are obtained from (66) and (77). This leaves three remaining equations which determine the singlet decay constants $f^{0\eta'}$, $f^{0\eta}$ and the gluonic coupling $g_{G\gamma\gamma}$ in terms of the QCD topological susceptibility parameter A. This dependence is plotted in Figs. 4 and 5.

To make a definite prediction, we need a theoretical input value for the topological susceptibility. In time, lattice calculations in full QCD with dynamical fermions should be able to determine the parameter A. For the moment, however, only the topological susceptibility in pure Yang–Mills theory is known accurately. The most recent value [49] is

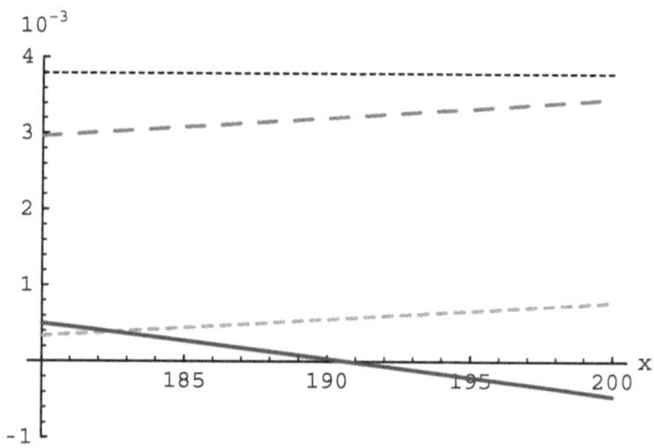

Fig. 5. This shows the relative sizes of the contributions to the flavour singlet radiative decay formula (76) expressed as functions of the topological susceptibility parameter $A = (x \text{ MeV})^4$. The dotted (black) line denotes $\frac{2\sqrt{2}}{\sqrt{3}} \frac{\alpha_{\rm em}}{\pi}$. The dominant contribution comes from the term $f^{0\eta'} g_{\eta'\gamma\gamma}$, denoted by the long-dashed (green) line, while the short-dashed (blue) line denotes $f^{0\eta} g_{\eta\gamma\gamma}$. The contribution from the gluonic coupling, $\sqrt{6} A g_{G\gamma\gamma}$, is shown by the solid (red) line

$$\chi(0)|_{YM} = -(191 \pm 5\ \text{MeV})^4 = -(1.33 \pm 0.14) \times 10^{-3}\ \text{GeV}^4 \tag{87}$$

This supersedes the original value $\chi(0)|_{YM} \simeq -(180\ \text{MeV})^4$ obtained some time ago [50]. Similar estimates are also obtained using QCD spectral sum rule methods [51]. At this point, therefore, we have to make an approximation and so we assume that the $O(1/N_c)$ corrections in the identification

$$A = \chi(0)_{\text{YM}} + O(1/N_c) \tag{88}$$

are numerically small. With this provisional input for A, we can then determine the full set of decay constants:

$$\begin{aligned} f^{0\eta'} &= 104.2 \pm 4.0\ \text{MeV} & f^{0\eta} &= 22.8 \pm 5.7\ \text{MeV} \\ f^{8\eta'} &= -36.1 \pm 1.2\ \text{MeV} & f^{8\eta} &= 98.4 \pm 1.4\ \text{MeV} \end{aligned} \tag{89}$$

and

$$g_{G\gamma\gamma} = -0.001 \pm 0.072\ \text{GeV}^{-4} \tag{90}$$

It is striking how close both the diagonal decay constants $f^{0\eta'}$ and $f^{8\eta}$ are to f_π. Predictably, the off-diagonal ones $f^{0\eta'}$ and $f^{8\eta'}$ are strongly suppressed.

It is also useful to quote these results in the two-angle parametrisation normally used in phenomenology. Defining,

$$\begin{pmatrix} f^{0\eta'} & f^{0\eta} \\ f^{8\eta'} & f^{8\eta} \end{pmatrix} = \begin{pmatrix} f_0 \cos\theta_0 & -f_0 \sin\theta_0 \\ f_8 \sin\theta_8 & f_8 \cos\theta_8 \end{pmatrix} \tag{91}$$

we find

$$\begin{aligned} f_0 &= 106.6 \pm 4.2\ \text{MeV} & f_8 &= 104.8 \pm 1.3\ \text{MeV} \\ \theta_0 &= -12.3 \pm 3.0\ \text{deg} & \theta_8 &= -20.1 \pm 0.7\ \text{deg} \end{aligned} \tag{92}$$

that is

$$\frac{f_0}{f_\pi} = 1.15 \pm 0.05 \qquad \frac{f_8}{f_\pi} = 1.13 \pm 0.02 \tag{93}$$

Given these results, we can now investigate how closely our expectations based on OZI or $1/N_c$ reasoning are actually realised by the experimental data. With the input value (87) for A, the numerical magnitudes and $1/N_c$ orders of the terms in the flavour singlet decay relation are as follows (see Fig. 5):

$$\begin{aligned} &f^{0\eta'} g_{\eta'\gamma\gamma}\ [N_c;\ 3.23] + f^{0\eta} g_{\eta\gamma\gamma}\ [N_c;\ 0.57] + \sqrt{6} A g_{G\gamma\gamma}\ [1;\ -0.005 \pm 0.23] \\ &= a^0_{\text{em}} \frac{\alpha_{\text{em}}}{\pi}\ [N_c;\ 3.79] \end{aligned} \tag{94}$$

The important point is that the gluonic contribution $g_{G\gamma\gamma}$, which is suppressed by a power of $1/N_c$ compared to the others, is also experimentally small. The near-vanishing for the chosen value of A is presumably a coincidence, but we see from Fig. 5 that across a reasonable range of values of the topological susceptibility it is still contributing no more than around 10%, in line with our expectations for a RG-invariant, OZI-suppressed quantity.

It is also interesting to see how the $1/N_c$ approximation is realised in the $U(1)_A$ DGMOR generalisation (68) of the Witten–Veneziano formula (67). Here we find

$$\begin{aligned}(f^{0\eta'})^2 m_{\eta'}^2\ [N_c;\ 9.96]\ +\ (f^{0\eta})^2 m_\eta^2\ [N_c;\ 0.15]\ +\ (f^{8\eta'})^2 m_\eta'^2\ [N_c;\ 1.19]\\ +\ (f^{8\eta})^2 m_\eta^2\ [N_c;\ 2.90] - 2f_K^2 m_K^2 [N_c; -6.22]\ =\ 6A\ [1;\ 7.98]\end{aligned} \tag{95}$$

This confirms the picture that the anomaly-induced contribution of $O(1/N_c)$ to $m_{\eta'}^2$, which gives a sub-leading $O(1)$ effect in $(f^{0\eta'})^2 m_{\eta'}^2$, is in fact numerically dominant and matched by the $O(1)$ topological susceptibility term $6A$. Away from the chiral limit, the conventional non-anomalous terms are all of $O(N_c)$ and balance as expected. The surprising numerical accuracy of the Witten–Veneziano formula (18) is seen to be in part due to a cancellation between the underestimates of $f^{8\eta'}$ (taken to be 0) and f_K (set equal to f_π). This emphasises, however, that great care must be taken in using the formal order in the $1/N_c$ expansion as a guide to the numerical importance of a physical quantity, especially in the $U(1)_A$ channel.

Nevertheless, the fact that the RG-invariant, OZI-suppressed coupling $g_{G\gamma\gamma}$ is experimentally small is a very encouraging result. It increases our confidence that we are able to identify quantities where the OZI, or leading $1/N_c$, approximation is likely to be numerically good. It also shows that $g_{G\gamma\gamma}$ gives a contribution to the decay formula which is entirely consistent with its picturesque interpretation as the coupling of the photons to the anomaly-induced gluonic component of the η'. A posteriori, the fact that its contribution is at most 10% explains the general success of previous theoretically inconsistent phenomenological parametrisations of η' decays in which the naive current algebra formulae omitting the gluonic term are used.

However, while the flavour singlet decay formula is well-defined and theoretically consistent, it is necessarily non-predictive. To be genuinely useful, we would need to find another process in which the same coupling enters. The problem here is that, unlike the decay constants which are universal, the coupling $g_{G\gamma\gamma}$ is process-specific just like $g_{\eta'\gamma\gamma}$ or $g_{\eta\gamma\gamma}$. There are of course many other processes to which our methods may be applied such as $\eta'(\eta) \to V\gamma$, where V is a flavour singlet vector meson ρ, ω, ϕ, or $\eta'(\eta) \to \pi^+\pi^-\gamma$. The required flavour singlet formulae may readily be written down, generalising the naive PCAC formulae. However, each will introduce its own gluonic coupling, such as $g_{GV\gamma}$. Although strict predictivity is lost, our experience with the two-photon decays suggests that these extra couplings will give relatively small, at most $O(10-20\%)$, contributions if like $g_{G\gamma\gamma}$ they can be identified as RG invariant and $1/N_c$ suppressed. This observation restores at least a reasonable degree of predictivity to the use of PCAC methods in the $U(1)_A$ sector.

4.5 $U(1)_A$ Goldberger–Treiman Relation

A further classic application of PCAC is to the pseudoscalar couplings of the nucleon. For the pion, the relation between the axial-vector form factor of the nucleon and the pion–nucleon coupling $g_{\pi NN}$ is the famous Goldberger–Treiman relation. Here, we present its generalisation to the flavour singlet sector, which involves the anomaly and gluon topology. This $U(1)_A$ Goldberger–Treiman relation was first proposed by Veneziano [4] in an investigation of the 'proton-spin' problem and further developed in [3, 8, 9, 52].

The axial-vector form factors are defined from

$$\langle N|J^a_{\mu 5}|N\rangle = 2m_N\Big(G^a_A(p^2)s_\mu + G^a_P(p^2)p.sp_\mu\Big) \tag{96}$$

where $s_\mu = \bar{u}\gamma_\mu\gamma_5 u/2m_N$ is the covariant spin vector. In the absence of a massless pseudoscalar, only the form factors $G^a_A(0)$ contribute at zero momentum.

Expressing the matrix element in terms of the 1PI vertices derived from the generating functional $\Gamma[V^a_{\mu 5}, V^a_\mu, Q, \phi^a_5, \phi^a]$, including spectator fields $N, \bar{N}$ for the nucleon, we have

$$\langle N|J^a_{\mu 5}|N\rangle = \bar{u}\Big(\Gamma_{V_5^{\mu a}\bar{N}N} + W_{V_5^{\mu a}\theta}\Gamma_{Q\bar{N}N} + W_{V_5^{\mu a}S_5^b}\Gamma_{\phi_5^b\bar{N}N}\Big)u \tag{97}$$

Note that this expansion relies on the specific definition(8) of Γ as a partial Legendre transform.

We also need the following relation, valid for all momenta, which is derived directly from the fundamental anomalous chiral Ward identity (9) for Γ:

$$\partial_\mu\Gamma_{V^a_{\mu 5}\bar{N}N} = -\Phi_{ab}\Gamma_{\phi^b_5\bar{N}N} \tag{98}$$

Now, taking the divergence of (97), using this Ward identity and then[8] taking the zero-momentum limit, noting that the propagators vanish at zero momentum since there is no massless pseudoscalar, gives

$$2m_N G^a_A(0)\ \bar{u}\gamma_5 u = i\bar{u}\ \Phi_{ab}\Gamma_{\phi^b_5\bar{N}N}\big|_{p=0}u \tag{99}$$

The meson–nucleon couplings are related to the 1PI vertices by

$$\langle N|\eta^\alpha N\rangle = g_{\eta^\alpha NN}\ \bar{u}\gamma_5 u = i\bar{u}\Gamma_{\eta^\alpha\bar{N}N}u \tag{100}$$

[8] The $p \to 0$ limit is delicate, as is the case for the derivation of the conventional Goldberger–Treiman relation, and should be taken in this order. Literally at $p = 0$, both sides vanish since $\bar{u}\gamma_5 u = 0$.

Re-expressing (99) in terms of the canonically normalised 'OZI boson' field $\hat{\eta}^\alpha$, we therefore derive

$$2m_N G_A^a(0) = \hat{f}^{a\alpha} g_{\hat{\eta}^\alpha NN} \tag{101}$$

This relation will be useful to us when we consider the 'proton spin' problem.

All that now remains to cast this into its final form is to make the familiar change of variables from $Q, \hat{\eta}^\alpha$ to G, η^α, where η^α are interpreted as the physical mesons. We therefore find the generalised $U(1)_A$ Goldberger–Treiman relation:

$$2m_N G_A^a(0) = f^{a\alpha} g_{\eta^\alpha NN} + \sqrt{2n_f} A g_{GNN} \delta_{a0} \tag{102}$$

For the individual components, this is

$$2m_N G_A^3 = f_\pi g_{\pi NN} \tag{103}$$
$$2m_N G_A^8 = f^{8\eta'} g_{\eta' NN} + f^{8\eta} g_{\eta NN} \tag{104}$$
$$2m_N G_A^0 = f^{0\eta'} g_{\eta' NN} + f^{0\eta} g_{\eta NN} + \sqrt{6} A g_{GNN} \tag{105}$$

The renormalisation group properties of these relations are described in great detail in [9]. It is clear that the flavour singlet axial coupling G_A^0 satisfies a homogeneous RGE and scales with the anomalous dimension γ corresponding to the multiplicative renormalisation of $J_{\mu 5}^0$. In the form (101), RG consistency is simply achieved by

$$\mathcal{D}\hat{f}^{a\alpha} = \gamma \delta_{a0} \hat{f}^{a\alpha} \qquad \mathcal{D} g_{\hat{\eta}^\alpha NN} = 0 \tag{106}$$

All the scale dependence is in the decay constant $\hat{f}^{0\alpha}$ while the the coupling $g_{\hat{\eta}^\alpha NN}$ of the 'OZI boson' to the nucleon is RG invariant (in contrast to $g_{\hat{\eta}^\alpha \gamma\gamma}$). In the final form (102) involving the physical decay constants, a careful analysis shows that apart from $G_A^0(0)$ the only other non RG-invariant quantity is the gluonic coupling g_{GNN}, which is required to satisfy the following non-homogeneous RGE to ensure the self-consistency of (105):

$$\mathcal{D} g_{GNN} = \gamma \Big(g_{GNN} + \frac{1}{\sqrt{2n_f}} \frac{1}{A} f^{0\alpha} g_{\eta^\alpha NN} \Big) \tag{107}$$

The large-N_c behaviour in the flavour singlet relation is as follows: $G_A^0 = O(N_c)$, $f^{0\eta}, f^{0\eta'} = O(\sqrt{N_c})$, $A = O(1)$, $g_{\eta NN}, g_{\eta' NN} = O(\sqrt{N_c})$, $g_{GNN} = O(1)$. So the final term $A g_{GNN}$ is $O(1)$, suppressed by a power of $1/N_c$ compared to all the others, which are $O(N_c)$.

We see that, like $g_{G\gamma\gamma}$, the gluonic coupling g_{GNN} is suppressed at large N_c relative to the corresponding meson couplings. However, unlike $g_{G\gamma\gamma}$ that is RG invariant, g_{GNN} has a complicated RG non-invariance and depends on the anomaly-induced anomalous dimension γ. The conjecture in Sect. 4.3 then suggests that while the OZI or large-N_c approximation should be a good guide to the value of $g_{G\gamma\gamma}$, we may expect significant OZI violations for g_{GNN}. We would therefore not be surprised to find that g_{GNN} makes a sizeable numerical contribution to the $U(1)_A$ Goldberger–Treiman relation.

We now try to test these expectations against the experimental data. We first introduce a notation that has become standard in the literature on deep-inelastic scattering. There, the axial couplings are written as

$$G_A^3 = \frac{1}{2}\, a^3 \qquad G_A^8 = \frac{1}{2\sqrt{3}}\, a^8 \qquad G_A^0 = \frac{1}{\sqrt{6}}\, a^0 \tag{108}$$

where the a^a have a simple interpretation in terms of parton distribution functions.

Experimentally,

$$a^3 = 1.267 \pm 0.004 \qquad a^8 = 0.585 \pm 0.025 \tag{109}$$

from low-energy data on nucleon and hyperon beta decay. The latest result[9] for a^0 quoted by the COMPASS collaboration [53] from deep inelastic scattering data is

$$a^0|_{Q^2\to\infty} = 0.33 \pm 0.06 \tag{110}$$

with a similar result from HERMES [54].

The OZI expectation is that $a^0 = a^8$. In the context of DIS, this is a prediction of the simple quark model, where it is known as the Ellis–Jaffe sum rule [57]). We return to this in the context of the 'proton spin' problem in Sect. 5 but for now we concentrate on the low-energy phenomenology of the pseudoscalar meson–nucleon couplings.

The original Goldberger–Treiman relation (103) gives the following value for the pion–nucleon coupling:

$$g_{\pi NN} = 12.86 \pm 0.06 \tag{111}$$

consistent to within about 5% with the experimental value 13.65(13.80)$\pm$0.12 (depending on the data set used [58]). In an ideal world where $g_{\eta NN}$ and $g_{\eta' NN}$ were both known, we would now verify the octet formula (104) then determine the gluonic coupling g_{GNN} from the singlet Goldberger–Treiman relation (105). However, the experimental situation with the η and η'-nucleon couplings is far less clear. One would hope to determine these couplings from the near threshold production of the η and η' in nucleon–nucleon collisions, i.e. $pp \to pp\eta$ and $pp \to pp\eta'$, measured for example at COSY-II [59, 60, 61]. However, the η production is dominated by the $N(1535)S_{11}$ nucleon resonance which decays to $N\eta$, and as a result very little is known about $g_{\eta NN}$ itself. The detailed production mechanism of the η' is not well understood. However,

[9] This supersedes the result $a^0|_{Q^2=4\mathrm{GeV}^2} = 0.237^{+0.024}_{-0.029}$ quoted by COMPASS in 2005 [55, 56], which we used as input into our analysis of the phenomenology of the $U(1)_A$ GT relation in [3]. The fits presented here are updated from those of [3] to take account of this. For a further discussion of the experimental situation, see Sect. 5.

since there is no known baryonic resonance decaying into $N\eta'$, we may simply assume that the reaction $pp \to pp\eta'$ is driven by the direct coupling supplemented by heavy-meson exchange. This allows an upper bound to be placed on $g_{\eta' NN}$ and on this basis [62] quotes $g_{\eta' NN} < 2.5$. This is supported by an analysis [63] of very recent data from CLAS [64] on the photoproduction reaction $\gamma p \to p\eta'$. Describing the cross section data with a model comprising the direct coupling together with t-channel meson exchange and s and u-channel resonances, it is found that equally good fits can be obtained for several values of $g_{\eta' NN}$ covering the whole region $0 < g_{\eta' NN} < 2.5$.

In view of this experimental uncertainty, we shall use the octet and singlet Goldberger–Treiman relations to plot the predictions for $g_{\eta NN}$ and g_{GNN} as a function of the ill-determined η'-nucleon coupling in the experimentally allowed range $0 < g_{\eta' NN} < 2.5$. The results (again taking the value (87) for A) are given in Fig. 6. In Fig. 7 we have shown the relative magnitudes of the various contributions to the flavour-singlet formula.

What we learn from this is that for values of $g_{\eta' NN}$ approaching the upper end of the experimentally allowed range, the contribution of the OZI-suppressed gluonic coupling g_{GNN} is quite large. The variation of $f^{0\eta'} g_{\eta' NN}$ over the allowed range is compensated almost entirely by the variation of $\sqrt{6} g_{GNN}$, with the $f^{0\eta} g_{\eta NN}$ contribution remaining relatively constant.

For example, if experimentally we found $g_{\eta' NN} \simeq 2.5$, which corresponds to the cross sections for $pp \to pp\eta'$ and $\gamma p \to p\eta'$ being almost entirely determined by the direct coupling, then we would have $g_{\eta NN} \simeq 4.14$ and $g_{GNN} \simeq -31.2\ \mathrm{GeV}^{-3}$. In terms of the contributions to the $U(1)_A$ Goldberger–Treiman relation, this would give (in GeV)

$$\begin{aligned} 2m_N G_A^0[N_c;\ 0.25] &= f^{0\eta'} g_{\eta' NN}[N_c;\ 0.26] + f^{0\eta} g_{\eta NN}[N_c;\ 0.09] \\ &\quad + \sqrt{6} A g_{GNN}[O(1);\ -0.10] \end{aligned} \tag{112}$$

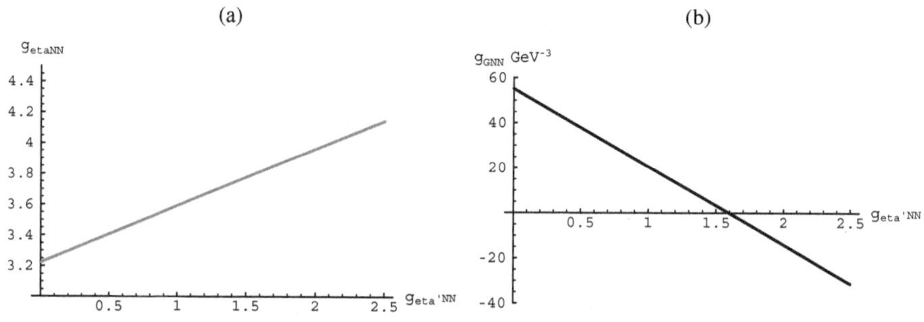

Fig. 6. These figures show the dimensionless η-nucleon coupling $g_{\eta NN}$ and the gluonic coupling g_{GNN} in units of GeV^{-3} expressed as functions of the experimentally uncertain η'-nucleon coupling $g_{\eta' NN}$, as determined from the flavour octet and singlet Goldberger–Treiman relations (104) and (105)

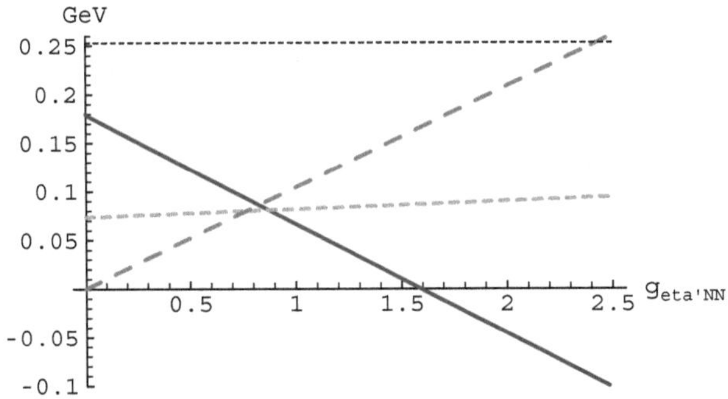

Fig. 7. This shows the relative sizes of the contributions to the $U(1)_A$ Goldberger–Treiman relation from the individual terms in (105), expressed as functions of the coupling $g_{\eta' NN}$. The dotted (*black*) line denotes $2m_N G_A^0$. The long-dashed (*green*) line is $f^{0\eta'} g_{\eta' NN}$ and the short-dashed (*blue*) line is $f^{0\eta} g_{\eta NN}$. The solid (*red*) line shows the contribution of the novel gluonic coupling, $\sqrt{6} A g_{GNN}$, where A determines the QCD topological susceptibility

The anomalously small value of G_A^0 compared to its OZI value (the OZI approximation is $2m_N G_A^0\big|_{\mathrm{OZI}} = \sqrt{2}\ 2m_N G_A^8 = 0.45$) is then due to the partial cancellation of the sum of the meson–nucleon coupling terms by the gluonic coupling g_{GNN}. Although formally $O(1/N_c)$ suppressed, numerically it gives a major contribution to the large OZI violation in G_A^0. This would give some support to our conjecture and provide further evidence that we are able to predict the location of large OZI violations using the renormalisation group as a guide.

Of course, it may be that experimentally we eventually find a value for $g_{\eta' NN} \simeq 1.5$, in the region where g_{GNN} contributes only around 10% or less. Although surprising, this would open the possibility that all gluonic couplings of type g_{GXX} are close to zero, which could be interpreted as implying that the gluonic component of the η' wave function is simply small. Clearly, a reliable determination of $g_{\eta' NN}$, or equivalently $g_{\eta NN}$, would shed considerable light on the $U(1)_A$ dynamics of QCD.

5 Topological Charge Screening and the 'Proton Spin'

So far, we have focused on the implications of the $U(1)_A$ anomaly for low-energy QCD phenomenology. However, the anomaly also plays a vital role in the interpretation of high-energy processes, in particular polarised deep-inelastic scattering.

In this section, we discuss one of the most intensively studied topics in QCD of the past two decades – the famous, but misleadingly named, 'proton spin' problem. We review the interpretation initially proposed by Veneziano [4] and developed by us in a series of papers exploring the relation with the

$U(1)_A$ GT relation and gluon topology [8, 9, 65]. In a subsequent work with Narison, we were able to quantify our prediction by using QCD spectral sum rules to compute the slope $\chi'(0)$ of the topological susceptibility [10, 52]. Remarkably, the most recent experimental data from the COMPASS [53] and HERMES [54] collaborations, released in September 2006, now confirms our original 1994 numerical prediction [10].

5.1 The g_1^p and Angular Momentum Sum Rules

The 'proton spin' problem concerns the sum rule for the first moment of the polarised proton structure function g_1^p. This is measured in polarised DIS experiments through the inclusive processes $\mu p \to \mu X$ (EMC, SMC, COMPASS at CERN) or $ep \to eX$ (SLAC, HERMES at DESY) together with similar experiments on a deuteron target. The polarisation asymmetry of the cross-section is expressed as

$$x\frac{d\Delta\sigma}{dxdy} = \frac{Y_P}{2}\frac{16\pi^2\alpha^2}{s} g_1^p(x,Q^2) \ + \ O\Big(\frac{M^2x^2}{Q^2}\Big) \tag{113}$$

with conventional notation: $Q^2 = -q^2$ and $x = Q^2/2p_2.q$ are the Bjorken variables, where p_2, q are the momenta of the target proton and incident virtual photon, respectively, $y = Q^2/xs$ and $Y_p = (2-y)/y$.

According to standard theory, g_1^p is determined by the proton matrix element of two electromagnetic currents carrying a large spacelike momentum. The sum rule for the first moment of g_1^p is derived from the twist 2, spin 1 terms in the operator product expansion for the currents:

$$J^\lambda(q)J^\rho(-q) \underset{Q^2\to\infty}{\sim} 2\epsilon^{\lambda\rho\nu\mu}\frac{q_\nu}{Q^2}\Big[\Delta C_1^{NS}(\alpha_s)\Big(J^3_{\mu5}+\frac{1}{\sqrt{3}}J^8_{\mu5}\Big)+\frac{2\sqrt{2}}{\sqrt{3}}\Delta C_1^S(\alpha_s)J^0_{\mu5}\Big] \tag{114}$$

where ΔC_1^{NS} and ΔC_1^S are Wilson coefficients and $J^a_{\mu5}$ $(a = 3, 8, 0)$ are the renormalised axial currents, with the normalisations defined in Sect. 2. It is the occurrence of the axial currents in this OPE that provides the link between the $U(1)_A$ anomaly and polarised DIS. The sum rule is therefore:

$$\Gamma_1^p(Q^2) \equiv \int_0^1 dx\ g_1^p(x,Q^2) = \frac{1}{12}\Delta C_1^{NS}\Big(a^3+\frac{1}{3}a^8\Big)+\frac{1}{9}\Delta C_1^S a^0(Q^2) \tag{115}$$

where the axial charges a^3, a^8 and $a^0(Q^2)$ are defined in terms of the forward proton matrix elements as in (108). Here, we have explicitly shown the Q^2 scale dependence associated with the RG non-invariance of $a^0(Q^2)$.

Since the flavour non-singlet axial charges are known from low-energy data, a measurement of the first moment of g_1^p amounts to a determination of the flavour singlet $a^0(Q^2)$. At the time of the original EMC experiment in 1988 [66] the theoretical expectation based on the quark model was that $a^0 = a^8$. The resulting sum rule for g_1^p is known as the Ellis–Jaffe sum rule [57].

The great surprise of the EMC measurement was the discovery that in fact a^0 is significantly suppressed relative to a^8, and indeed the earliest results suggested it could even be zero. However, the reason the result sent shockwaves through both the theoretical and experimental communities (to date, the EMC paper has over 1300 citations) was the interpretation that this implies that the quarks contribute only a fraction of the total spin of the proton.

In fact, this interpretation relies on the simple valence quark model of the proton and is *not* true in QCD, where the axial charge decouples from the real angular momentum sum rule for the proton. Rather, as we shall show, the suppression of $a^0(Q^2)$ reflects the dynamics of gluon topology and appears to be largely independent of the structure of the proton itself. Precisely, it is a manifestation of *topological charge screening* in the QCD vacuum.

The angular momentum sum rule is derived by taking the forward matrix element of the conserved angular momentum current $M^{\mu\nu\lambda}$, defined in terms of the energy-momentum tensor as

$$M^{\mu\nu\lambda} = x^{[\nu}T^{\lambda]\mu} + \partial_\rho X^{\rho\mu\nu\lambda} \tag{116}$$

The inclusion of the arbitrary tensor $X^{\rho\mu\nu\lambda}$ just reflects the usual freedom in QFT of defining conserved currents. This gives us some flexibility in attempting to write $M^{\mu\nu\lambda}$ as a sum of local operators, suggesting interpretations of the total angular momentum as a sum of 'components' of the proton spin. In fact, however, it is not possible to write $M^{\mu\nu\lambda}$ as a sum of operators corresponding to quark and gluon spin and angular momentum in a gauge-invariant way. The best decomposition is [67, 68, 69]

$$M^{\mu\nu\lambda} = O_1^{\mu\nu\lambda} + O_2^{\mu[\lambda}x^{\nu]} + O_3^{\mu[\lambda}x^{\nu]} + \ldots \tag{117}$$

where the dots denote terms whose forward matrix elements vanish. Here,

$$\begin{aligned}
O_1^{\mu\nu\lambda} &= \frac{1}{2}\epsilon^{\mu\nu\lambda\sigma}\bar{q}\gamma_\sigma\gamma_5 q = \frac{1}{2}\epsilon^{\mu\nu\lambda\sigma}\sqrt{2n_f}J^0_{\sigma 5} \\
O_2^{\mu\lambda} &= i\bar{q}\gamma^\mu \overleftrightarrow{D}^\lambda q \\
O_3^{\mu\lambda} &= F^{\mu\rho}F_\rho{}^\lambda
\end{aligned} \tag{118}$$

At first sight, $O_1^{\mu\nu\lambda}$ looks as if it could be associated with 'quark spin', since for *free* Dirac fermions the spin operator coincides with the axial vector current. $O_2^{\mu[\lambda}x^{\mu]}$ would correspond to 'quark orbital angular momentum', leaving $O_3^{\mu[\lambda}x^{\nu]}$ as 'gluon total angular momentum'. Any further decomposition of the gluon angular momentum is necessarily not gauge invariant.

The forward matrix elements of these operators may be expressed in terms of form factors and, as we showed in [68], this exhibits an illuminating cancellation. After some analysis, we find

$$\langle p,s|O_1^{\mu\nu\lambda}|p,s\rangle = a^0 m_N \epsilon^{\mu\nu\lambda\sigma}s_\sigma$$

$$\begin{aligned}\langle p,s|O_2^{\mu[\lambda}x^{\nu]}|p,s\rangle &= J_q\frac{1}{2m_N}p_\rho p^{\{\mu}\epsilon^{[\lambda\}\nu]\rho\sigma}s_\sigma - a^0 m_N\epsilon^{\mu\nu\lambda\sigma}s_\sigma \\ \langle p,s|O_3^{\mu[\lambda}x^{\nu]}|p,s\rangle &= J_g\frac{1}{2m_N}p_\rho p^{\{\mu}\epsilon^{[\lambda\}\nu]\rho\sigma}s_\sigma \end{aligned} \tag{119}$$

The angular momentum sum rule for the proton is then just

$$\frac{1}{2} = J_q + J_g \tag{120}$$

where the Lorentz and gauge-invariant form factors J_q and J_g may reasonably be thought of as representing quark and gluon total angular momentum. However, even this interpretation is not at all rigorous, not least because J_q and J_g mix under renormalisation and scale as

$$\frac{d}{d\ln Q^2}\begin{pmatrix} J_q \\ J_g \end{pmatrix} = \frac{\alpha_s}{4\pi}\begin{pmatrix} -\frac{8}{3}C_F & \frac{2}{3}n_f \\ \frac{8}{3}C_F & -\frac{2}{3}n_f \end{pmatrix}\begin{pmatrix} J_q \\ J_g \end{pmatrix} \tag{121}$$

Only the total angular momentum is Lorentz, gauge and scale invariant.[10]

The crucial observation, however, is that the axial charge a^0 explicitly *cancels* from the angular momentum sum rule. a^0 is an important form factor, which relates the first moment of g_1^p to gluon topology via the $U(1)_A$ anomaly, but it is *not* part of the angular momentum sum rule for the proton.

Just as a^0 can be measured in polarised inclusive DIS, the form factors J_q and J_g can be extracted from measurements of unpolarised generalised parton distributions (GPDs) in processes such as deeply virtual Compton scattering $\gamma^* p \to \gamma p$. These can also in principle be calculated in lattice QCD. The required identifications with GPDs are given in [68].

5.2 QCD Parton Model

Before describing our resolution of the 'proton spin' problem, we briefly review the parton model interpretation of the first moment sum rule for g_1^p.

In the simplest form of the parton model, the proton structure at large Q^2 is described by parton distributions corresponding to free valence quarks only. The polarised structure function is given by

$$g_1^p(x) = \frac{1}{2}\sum_{i=1}^{n_f} e_i^2\ \Delta q_i(x) \tag{122}$$

where $\Delta q_i(x)$ is defined as the difference of the distributions of quarks (and antiquarks) with helicities parallel and antiparallel to the nucleon spin. It is

[10] For a careful discussion of the parton interpretation of longitudinal and transverse angular momentum sum rules, see [70]. This confirms our assertion that the axial charge a^0 is not to be identified with quark helicities in the parton model.

convenient to work with the conventionally defined flavour non-singlet and singlet combinations Δq^{NS} and Δq^S (often also written as $\Delta\Sigma$).

In this model, the first moment of the singlet quark distribution $\Delta q^S = \int_0^1 dx\ \Delta q^S(x)$ can be identified as the sum of the helicities of the quarks. Interpreting the structure function data *in this model* then leads to the conclusion that the quarks carry only a small fraction of the spin of the proton. There is indeed a real contradiction between the experimental data and the *free valence quark* parton model.

However, this simple model leaves out many important features of QCD, the most important being gluons, RG scale dependence and the chiral $U_A(1)$ anomaly. When these effects are included, in the QCD parton model, the naive identification of Δq^S with spin no longer holds and the experimental results for g_1^p can be accommodated, though not predicted.

In the QCD parton model, the polarised structure function is written in terms of both quark and gluon distributions as follows:

$$g_1^p(x,Q^2) = \int_x^1 \frac{du}{u}\,\frac{1}{9}\Big[\Delta C^{NS}\Big(\frac{x}{u}\Big)\Delta q^{NS}(u,t) + \Delta C^S\Big(\frac{x}{u}\Big)\Delta q^S(u,t) + \Delta C^g\Big(\frac{x}{u}\Big)\Delta g(u,t)\Big] \tag{123}$$

where $\Delta C^S, \Delta C^g$ and ΔC^{NS} are perturbatively calculable functions related to the Wilson coefficients and the quark and gluon distributions have a priori a $t = \ln Q^2/\Lambda^2$ dependence determined by the RG evolution, or DGLAP, equations. The first moment sum rule is therefore

$$\Gamma_1^p(Q^2) = \frac{1}{9}\Big[\Delta C_1^{NS}\Delta q^{NS} + \Delta C_1^S \Delta q^S + \Delta C_1^g \Delta g\Big] \tag{124}$$

Comparing with (115), we see that the axial charge $a^0(Q^2)$ is identified with a linear combination of the first moments of the singlet quark and gluon distributions. It is often, though not always, the case that the moments of parton distributions can be identified in one-to-one correspondence with the matrix elements of local operators. The polarised first moments are special in that two parton distributions correspond to the same local operator.

The RG evolution equations for the first moments of the parton distributions are derived from the matrix of anomalous dimensions for the lowest spin, twist 2 operators. This introduces an inevitable renormalisation scheme ambiguity in the definitions of Δq and Δg, and their physical interpretation is correspondingly nuanced. The choice closest to our own analysis is the 'AB' scheme [71] where the parton distributions have the following RG evolution:

$$\tfrac{d}{d\ln Q^2}\Delta q^{NS} = 0 \qquad\qquad \tfrac{d}{d\ln Q^2}\Delta q^S = 0$$

$$\tfrac{d}{d\ln Q^2}\tfrac{\alpha_s}{2\pi}\Delta g(Q^2) = \gamma\Big(\tfrac{\alpha_s}{2\pi}\Delta g(Q^2) - \tfrac{1}{3}\Delta q^S\Big) \tag{125}$$

which requires $\Delta C_1^g = \frac{3\alpha_s}{2\pi}\Delta C_1^S$. It is then possible to make the following identifications with the axial charges:

$$\begin{aligned} a^3 &= \Delta u - \Delta d \\ a^8 &= \Delta u + \Delta d - 2\Delta s \\ a^0(Q^2) &= \Delta u + \Delta d + \Delta s - \frac{3\alpha_s}{2\pi}\Delta g(Q^2) \end{aligned} \tag{126}$$

with $\Delta q^S = \Delta u + \Delta d + \Delta s$. Notice that in the AB scheme, all the scale dependence of the axial charge $a^0(Q^2)$ is assigned to the gluon distribution $\Delta g(Q^2)$.

This was the identification originally introduced for the first moments by Altarelli and Ross [72], and resolves the 'proton spin' problem in the context of the QCD parton model. In this scheme, the Ellis–Jaffe sum rule follows from the assumption that in the proton both Δs and $\Delta g(Q^2)$ are zero, which is the natural assumption in the free valence quark model. This is equivalent to the OZI approximation $a^0(Q^2) = a^8$. However, in the full QCD parton model, there is no reason why $\Delta g(Q^2)$, or even Δs, should be zero in the proton. Indeed, given the different scale dependence of $a^0(Q^2)$ and a^8, it would be unnatural to expect this to hold in QCD itself.

An interesting conjecture [72] is that the observed suppression in $a^0(Q^2)$ is due overwhelmingly to the gluon distribution $\Delta g(Q^2)$ alone. Although by no means a necessary consequence of QCD, this is a reasonable expectation given that it is the anomaly (which is due to the gluons and is responsible for OZI violations) which is responsible for the scale dependence in $a^0(Q^2)$ and $\Delta g(Q^2)$ whereas the Δq are scale invariant. This would be in the same spirit as our conjecture on OZI violations in low-energy phenomenology in Sect. 4.3. To test this, however, we need to find a way to measure $\Delta g(Q^2)$ itself, rather than the combination $a^0(Q^2)$. The most direct option is to extract $\Delta g(x, Q^2)$ from processes such as open charm production, $\gamma^* g \to c\bar{c}$, which is currently being intensively studied by the COMPASS [73], STAR [74] and PHENIX [75] collaborations.

5.3 Topological Charge Screening

We now describe a less conventional approach to deep inelastic scattering based entirely on field-theoretic concepts. In particular, the role of parton distributions is taken over by the 1PI vertices of composite operators introduced above (for a review, see [76]).

Once again, the starting point is the use of the OPE to express the moments of a generic structure function $F(x, Q^2)$ as

$$\int_0^1 dx\, x^{n-1} F(x, Q^2) \;=\; \sum_A C_A^n(Q^2)\, \langle p|\mathcal{O}_A^n(0)|p\rangle \tag{127}$$

where $\mathcal{O}_A^n$ denotes the set of lowest twist, spin n operators and $C_A^n(Q^2)$ are the corresponding Wilson coefficients. The next step is to introduce a new set of composite operators $\tilde{\mathcal{O}}_B$, chosen to encompass the physically relevant degrees of freedom, and write the matrix element as a product of two-point Green functions and 1PI vertices as follows:

$$\int_0^1 dx\, x^{n-1} F(x,Q^2) = \sum_A \sum_B C_A^n(Q^2)\, \langle 0|T\, \mathcal{O}_A^n\, \tilde{\mathcal{O}}_B|0\rangle\, \Gamma_{\tilde{\mathcal{O}}_B pp} \qquad (128)$$

This decomposition splits the structure function into three parts – first, the Wilson coefficients $C_A^n(Q^2)$ which can be calculated in perturbative QCD; second, non-perturbative but *target independent* Green functions that encode the dynamics of the QCD vacuum; third, non-perturbative vertex functions that characterise the target by its couplings to the chosen operators $\tilde{\mathcal{O}}_B$.[11]

Now specialise to the first moment sum rule for g_1^p. For simplicity, we first present the analysis for the chiral limit, where there is no flavour mixing. Using the anomaly (4), we can express the flavour singlet contribution to the sum rule as

$$\Gamma_1^p(Q^2)_{singlet} \equiv \int_0^1 dx\, g_1^p(x,Q^2)_{singlet} = \frac{2}{3}\frac{1}{2m_N}\, \Delta C_1^S(\alpha_s)\, \langle p|Q|p\rangle \qquad (129)$$

The obvious choice for the operators $\tilde{\mathcal{O}}_B$ in this case are the flavour singlet pseudoscalars and it is natural to choose the 'OZI boson' field $\hat{\eta}^0 = \hat{f}^{00}\frac{1}{\langle\bar{q}q\rangle}\phi_5^0$, which is normalised so that $d/dp^2\, \Gamma_{\hat{\eta}^0\hat{\eta}^0}\big|_{p=0} = 1$. As we have seen in (106), the corresponding 1PI vertex is then RG invariant. Writing the 1PI vertices in terms of nucleon couplings as in (100), we find (see Fig. 8)

$$\Gamma_1^p(Q^2)_{singlet} = \frac{2}{3}\frac{1}{2m_N}\Delta C_1^S(\alpha_s)\Big(\langle 0|T\, Q\, Q|0\rangle\, g_{QNN} + \langle 0|T\, Q\, \hat{\eta}^0|0\rangle\, g_{\hat{\eta}^0 NN}\Big) \qquad (130)$$

Recalling that the matrix of two-point Green functions is given by the inversion formula

$$\begin{pmatrix} W_{\theta\theta} & W_{\theta S_{\hat{\eta}^0}} \\ W_{S_{\hat{\eta}^0}\theta} & W_{S_{\hat{\eta}^0}S_{\hat{\eta}^0}} \end{pmatrix} = -\begin{pmatrix} \Gamma_{QQ} & \Gamma_{Q\hat{\eta}^0} \\ \Gamma_{\hat{\eta}^0 Q} & \Gamma_{\hat{\eta}^0\hat{\eta}^0} \end{pmatrix}^{-1} \qquad (131)$$

---

[11] We emphasise again that this decomposition of the matrix elements into products of Green functions and 1PI vertices is *exact*, independent of the choice of the set of operators $\tilde{\mathcal{O}}_B$. In particular, it is not necessary for $\tilde{\mathcal{O}}_B$ to be in any sense a complete set. If a different choice is made, the vertices $\Gamma_{\tilde{\mathcal{O}}_B pp}$ themselves change, becoming 1PI with respect to a different set of composite fields. In practice, the set of operators $\tilde{\mathcal{O}}_B$ should be as small as possible while still capturing the essential degrees of freedom. A good choice can also result in vertices $\Gamma_{\tilde{\mathcal{O}}_B pp}$ which are both RG invariant and closely related to low-energy physical couplings.

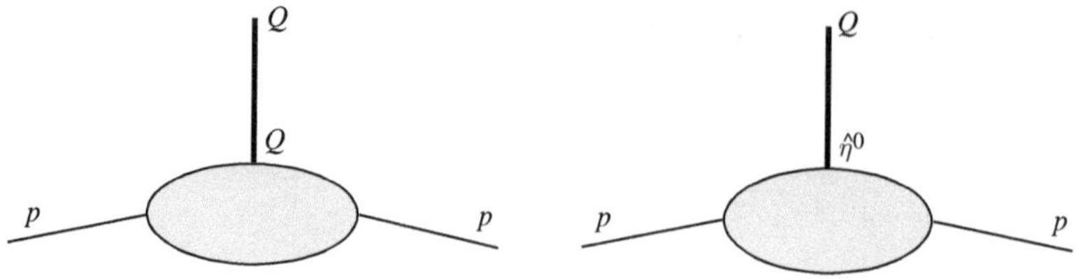

Fig. 8. Illustration of the decomposition of the matrix element $\langle p|Q|p\rangle$ into two-point Green functions and 1PI vertices. The Green function in the first diagram is $\chi(0)$; in the second it is $\sqrt{\chi'(0)}$

and using the normalisation condition for $\hat{\eta}^0$, we can easily show that at zero momentum,

$$W^2_{\theta S_{\hat{\eta}^0}} = \frac{d}{dp^2} W_{\theta\theta}\big|_{p=0} \tag{132}$$

Finally, therefore, we can represent the first moment of g_1^p in the following, physically intuitive form:

$$\Gamma_1^p(Q^2)_{singlet} = \frac{2}{3}\frac{1}{2m_N} \Delta C_1^S(\alpha_s) \Big(\chi(0)\; g_{QNN} + \sqrt{\chi'(0)}\; g_{\hat{\eta}^0 NN}\Big) \tag{133}$$

This shows that the first moment is determined by the gluon topological susceptibility in the QCD vacuum as well as the couplings of the proton to the pseudoscalar operators Q and $\hat{\eta}^0$. In the chiral limit, $\chi(0) = 0$ so the first term vanishes. The entire flavour singlet contribution is therefore simply

$$\Gamma_1^p(Q^2)_{singlet} = \frac{2}{3}\frac{1}{2m_N} \Delta C_1^S(\alpha_s)\; \sqrt{\chi'(0)}\; g_{\hat{\eta}^0 NN} \tag{134}$$

The 1PI vertex $g_{\hat{\eta}^0 NN}$ is RG invariant, and we see from (25) that *in the chiral limit* the slope of the topological susceptibility scales with the anomalous dimension γ, viz.

$$\frac{d}{d\ln Q^2}\sqrt{\chi'(0)} = \gamma\, \sqrt{\chi'(0)} \tag{135}$$

ensuring consistency with the RGE for the flavour singlet axial charge.

The formulae (133) and (134) are our key result. They show how the first moment of g_1^p can be factorised into couplings g_{QNN} and $g_{\hat{\eta}^0 NN}$, which carry information on the proton structure, and Green functions that characterise the QCD vacuum. In the case of g_1^p, the Green functions reduce simply to the topological susceptibility $\chi(0)$ and its slope $\chi'(0)$. We now argue that the experimentally observed suppression in the first moment of g_1^p is due *not* to a suppression in the couplings, but to the vanishing of the topological susceptibility $\chi(0)$ and an anomalously small value for its slope $\chi'(0)$. This is what we refer to as *topological charge screening* in the QCD vacuum.

The justification follows our now familiar conjecture on the relation between OZI violations and RG scale dependence. We expect the source of OZI violations to be in those quantities which are sensitive to the anomaly, as identified by their scaling dependence on the anomalous dimension γ, in this case $\chi'(0)$. In contrast, it should be a good approximation to use the OZI value for the RG-invariant vertex $g_{\hat{\eta}^0 NN}$, that is $g_{\hat{\eta}^0 NN} \simeq \sqrt{2} g_{\hat{\eta}^8 NN}$. The corresponding OZI value for $\sqrt{\chi'(0)}$ would be $f_\pi/\sqrt{6}$. This gives our key formula for the flavour singlet axial charge:

$$\frac{a^0(Q^2)}{a^8} \simeq \frac{\sqrt{6}}{f_\pi} \sqrt{\chi'(0)} \tag{136}$$

The corresponding prediction for the first moment of g_1^p is

$$\Gamma_1^p(Q^2)_{singlet} = \frac{1}{9} \Delta C_1^S(\alpha_s)\, a^8 \frac{\sqrt{6}}{f_\pi} \sqrt{\chi'(0)} \tag{137}$$

The final step is to compute the slope of the topological susceptibility. In time, lattice gauge theory should provide an accurate measurement of $\chi'(0)$. However, this is a particlarly difficult correlator for lattice methods since it requires a simulation of QCD with light dynamical fermions and algorithms that implement topologically non-trivial configurations in a sufficiently fast and stable way. Instead, we have estimated the value of $\chi'(0)$ using the QCD spectral sum rule method. Full details and discussion of this computation can be found in [10, 52]. The result is

$$\sqrt{\chi'(0)} = 26.4 \pm 4.1 \text{ MeV} \tag{138}$$

This gives our final prediction for the flavour singlet axial charge and the complete first moment of g_1^p:

$$a^0\big|_{Q^2=10\text{GeV}^2} = 0.33 \pm 0.05 \tag{139}$$

$$\Gamma_1^p\big|_{Q^2=10\text{GeV}^2} = 0.144 \pm 0.009 \tag{140}$$

Topological charge screening therefore gives a suppression factor of approximately 0.56 in a^0 compared to its OZI value $a^8 = 0.585$.

In the decade since we made this prediction, the experimental measurement has been somewhat lower than this value, in the range $a^0 \simeq 0.20 - 0.25$. This would have suggested there is also a significant OZI violation in the nucleon coupling $g_{\hat{\eta}^0 NN}$ itself, implicating the proton structure in the anomalous suppression of Γ_1^p. Very recently, however, the COMPASS and HERMES collaborations have published new results on the deuteron structure function which spectacularly confirm our picture that topological charge screening in the QCD vacuum is the dominant suppression mechanism.

These new data are shown in Fig. 9. This is based on data collected by COMPASS at CERN in the years 2002–2004 and has only recently been published. The accuracy compared to earlier SMC data at small x is significantly improved and the dip in xg_1^d around $x \sim 10^{-2}$ suggested by the SMC data is no longer present (Fig. 9). This explains the significantly higher value for a^0 found by COMPASS compared to SMC. From this data, COMPASS quote the first moment for the proton–neutron average $g_1^N = (g_1^p + g_1^n)/2$ as [53]

$$\Gamma_1^N\Big|_{Q^2=3\mathrm{GeV}^2} = 0.050 \pm 0.003(stat) \pm 0.002(evol) \pm 0.005(syst) \qquad (141)$$

Extracting the flavour singlet axial charge from the analogue of (115) for Γ_1^N then gives

$$a^0\Big|_{Q^2=3\mathrm{GeV}^2} = 0.35 \pm 0.03(stat) \pm 0.05(syst) \qquad (142)$$

or evolving to the $Q^2 \to \infty$ limit,

$$a^0\Big|_{Q^2\to\infty} = 0.33 \pm 0.03(stat) \pm 0.05(syst) \qquad (143)$$

Similar results are found by HERMES, who quote [54]

$$a^0\Big|_{Q^2=5\mathrm{GeV}^2} = 0.330 \pm 0.011(th) \pm 0.025(exp) \pm 0.028(evol) \qquad (144)$$

The agreement with our prediction (139) is striking.

To close this section, we briefly comment on the extension of our analysis beyond the chiral limit. In this case, the operator $\sqrt{2n_f}Q$ in (129) is replaced by the full divergence of the flavour singlet axial current, viz. $D^0 = \sqrt{2n_f}Q + d_{0bc}m^b\phi_5^c$. Separating the matrix element $\langle p|D^0|p\rangle$ into Green functions and 1PI vertices, we find from the zero-momentum Ward identities that $\langle 0|T\ D^0\ Q|0\rangle = 0$ so the contribution from g_{QNN} still vanishes. The

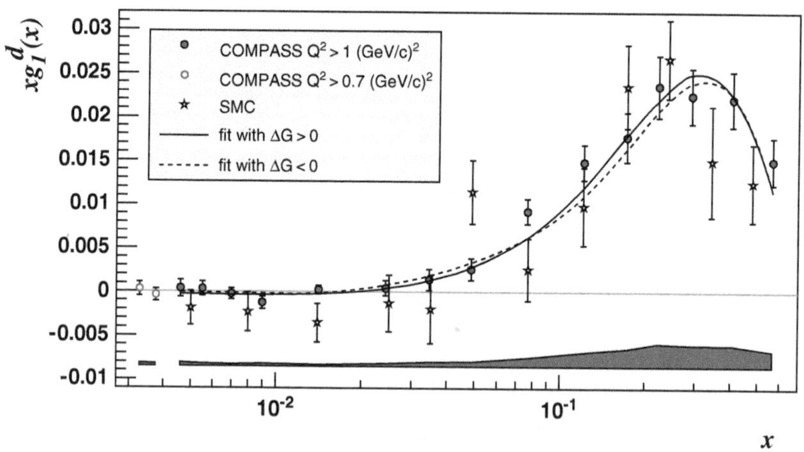

Fig. 9. COMPASS and SMC data for the deuteron structure function $g_1^d(x)$. Statistical error bars are shown with the data points. The shaded band shows the systematic error

other Green function is $\langle 0|T\ D^0\ \hat{\eta}^\alpha|0\rangle = -\hat{f}^{0\alpha}$, so the first moment sum rule becomes

$$\Gamma_1^p(Q^2)_{singlet} = \frac{1}{9}\frac{1}{2m_N}\ \Delta C_1^S(\alpha_s)\ \sqrt{6}\ \hat{f}^{0\alpha} g_{\hat{\eta}^\alpha NN} \tag{145}$$

It is clear that this is simply an alternative derivation of the $U(1)$ GT relation (101) for a^0. We could equally use the alternative form (102) to write

$$\Gamma_1^p(Q^2)_{singlet} = \frac{1}{9}\frac{1}{2m_N}\ \Delta C_1^S(\alpha_s)\ \sqrt{6}\ \left(f^{0\alpha} g_{\eta^\alpha NN} + \sqrt{6} A g_{GNN}\right) \tag{146}$$

Recalling the RGE (107) for g_{GNN}, we see that this bears a remarkable similarity to the expression for a^0 in terms of parton distributions in the AB scheme (126). This was first pointed out in [8, 9].

Manipulating the zero-momentum Ward identities in a similar way to that explained above in the chiral limit now shows that we can express the decay constants $\hat{f}^{a\alpha}$ in terms of vacuum Green functions as follows (see (38)):

$$(\hat{f}\hat{f}^T)_{ab} = \frac{d}{dp^2}\langle 0|T\ D^a\ D^b|0\rangle\big|_{p=0} \tag{147}$$

However, for non-zero quark masses there is flavour mixing amongst the 'OZI bosons' $\hat{\eta}^\alpha$ and we cannot extract the decay constants simply by taking a square root, as was the case in writing $\hat{f}^{00} = \sqrt{\chi'(0)}$ in the chiral limit. Nevertheless, in [52] we estimated the decay constants and form factors in the approximation where we use (147) with the full divergence D^a but neglect flavour mixing. Assuming OZI for the couplings, this gives the estimate

$$\frac{a^0(Q^2)}{a^8} \simeq \sqrt{6}\ \frac{\hat{f}^{00}}{\hat{f}^{88}} \tag{148}$$

where we take

$$\hat{f}^{00} \simeq \sqrt{\frac{d}{dp^2}\langle 0|T\ D^0\ D^0|0\rangle\big|_{p=0}} \qquad \hat{f}^{88} \simeq \sqrt{\frac{d}{dp^2}\langle 0|T\ D^8\ D^8|0\rangle\big|_{p=0}} \tag{149}$$

Evaluating the Green functions using QCD spectral sum rules gives

$$a^0\big|_{Q^2=10\mathrm{GeV}^2} = 0.31 \pm 0.02 \tag{150}$$

$$\Gamma_1^p\big|_{Q^2=10\mathrm{GeV}^2} = 0.141 \pm 0.005 \tag{151}$$

As we have seen in the last section, flavour mixing can be non-negligible in the phenomenology of the pseudoscalar mesons, so we should be a little cautious in overestimating the accuracy of these estimates. (The quoted errors do not include this systematic effect.) Nevertheless, the fact that they are consistent with those obtained in the chiral limit reinforces our confidence that the flavour singlet axial charge is relatively insensitive to the quark masses and that (139) and (140) indeed provide an accurate estimate of the first moment of g_1^p.

The observation that the 'proton spin' sum rule could be explained in terms of an extension of the Goldberger–Treiman relation to the flavour singlet sector was made in Veneziano's original paper [4]. This pointed out for the first time that the suppression in a^0 was an OZI-breaking effect. Since the Goldberger–Treiman relation connects the pseudovector form factors with the pseudoscalar channel, where it is known that there are large OZI violations for the flavour singlet, it becomes natural to expect similar large OZI violations also in a^0. This is the fundamental intuition which we have developed into a quantitative resolution of the 'proton spin' problem.

5.4 Semi-inclusive Polarised DIS

While the agreement between our prediction for the first moment of g_1^p and experiment is now impressive, it would still be interesting to find other experimental tests of topological charge screening. A key consequence of this mechanism is that the OZI violation observed in a^0 is not a property specifically of the proton, but is *target independent*. This leads us to look for ways to make measurements of the polarised structure functions of other hadronic targets besides the proton and neutron. We now show how this can effectively be done by studying semi-inclusive DIS $eN \to ehX$ in the target fragmentation region (see Fig. 10).

The differential cross section in the target fragmentation region can be written analogously to (113) in terms of fracture functions:

$$x\frac{d\Delta\sigma^{target}}{dxdydzdt} = \frac{Y_P}{2}\frac{4\pi\alpha^2}{s}\Delta M_1^{hN}(x,z,t,Q^2) \tag{152}$$

where $x = Q^2/2p_2.q$, $x_{\mathcal{B}} = Q^2/2k.q$, $z = p_2'.q/p_2.q$ so that $1-z = x/x_{\mathcal{B}}$, and the invariant momentum transfer $t = K^2 = -k^2$, where k is the momentum

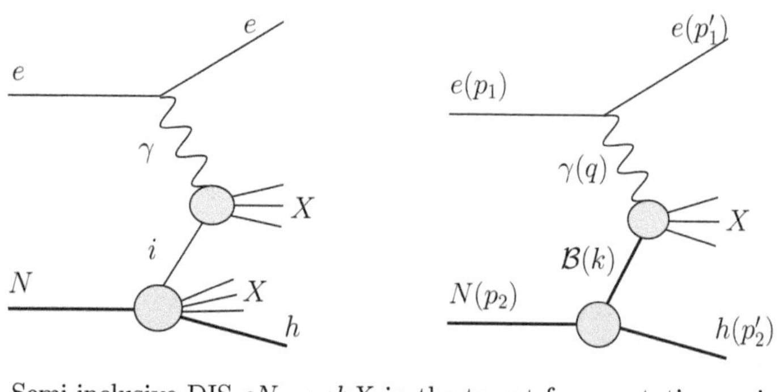

Fig. 10. Semi-inclusive DIS $eN \to ehX$ in the target fragmentation region. In the equivalent current fragmentation process, the detected hadron h is emitted from the hard collision with γ. The right-hand figure shows a simple Reggeon exchange model valid for $z \sim 1$, where h carries a large target energy fraction

of the struck parton. For $K^2 \ll Q^2$, $z \simeq E_h/E_N$ (in the photon–nucleon CM frame) is the energy fraction of the target nucleon carried by the detected hadron h.

ΔM_1^{hN} is the fracture function [77] equivalent of the inclusive structure function g_1^N, so in the same way as in (122) we have

$$\Delta M_1^{hN}(x,z,t,Q^2) = \frac{1}{2}\sum_i e_i^2 \Delta M_i^{hN}(x,z,t,Q^2) \qquad (153)$$

Here, $\Delta M_i^{hN}(x,z,t,Q^2)$ is an extended fracture function, introduced by Grazzini, Trentadue and Veneziano [78], which carries an explicit dependence on t. One of the advantages of these fracture functions is that they satisfy a simple, homogeneous RG evolution equation analogous to the usual inclusive parton distributions.

Our proposal [11, 79] (see also [80]) is to study semi-inclusive DIS in the kinematical region where the detected hadron h (π, K or D) carries a large target energy fraction, i.e. z approaching 1, with a small invariant momentum transfer t. In this region, it is useful to think of the target fragmentation process as being simply modelled by a single Reggeon exchange (see Fig. 10), i.e.

$$\Delta M_1^{hN}(x,z,t,Q^2)\big|_{z\sim 1} \simeq F(t)(1-z)^{-2\alpha_{\mathcal{B}}(t)} g_1^{\mathcal{B}}(x_{\mathcal{B}},t,Q^2) \qquad (154)$$

If we consider ratios of cross sections, the dynamical Reggeon emission factor $F(t)(1-z)^{-2\alpha_{\mathcal{B}}(t)}$ will cancel and we will be able to isolate the ratios of $g_1^{\mathcal{B}}(x_{\mathcal{B}},t,Q^2)$ for different effective targets $\mathcal{B}$. Although single Reggeon exchange is of course only an approximation to the more fundamental QCD description in terms of fracture functions (see [81] for a more technical discussion), it shows particularly clearly how observing semi-inclusive processes at large z with particular choices of h and N amounts in effect to performing inclusive DIS on virtual hadronic targets $\mathcal{B}$. Since our predictions will depend only on the $SU(3)$ properties of $\mathcal{B}$, together with target independence, they will hold equally well when $\mathcal{B}$ is interpreted as a Reggeon rather than a pure hadron state.

The idea is therefore to make predictions for the ratios $\mathcal{R}$ of the first moments of the polarised fracture functions $\int_0^{1-z} dx \Delta M_1^{hN}(x,z,t,Q^2)$ or equivalently $\int_0^1 dx_{\mathcal{B}} g_1^{\mathcal{B}}(x_{\mathcal{B}},t,Q^2)$ for various reactions. The first moments $\Gamma_1^{\mathcal{B}}$ are calculated as in (115) in terms of the axial charges a^3, a^8 and $a^0(Q^2)$ for a state with the $SU(3)$ quantum numbers of $\mathcal{B}$. We then use topological charge screening to say that $a^0(Q^2) \simeq s(Q^2)a^0\big|_{\rm OZI}$, i.e. the flavour singlet axial charge is suppressed relative to its OZI value by a universal, target-independent, suppression factor $s(Q^2)$. From our calculation of $\sqrt{\chi'(0)}$ and the experimental results for g_1^p, we have $s|_{Q^2=10{\rm GeV}^2} \simeq 0.33/0.585 = 0.56$.

Some of the more interesting predictions obtained in [11] are as follows. The ratio

$$\mathcal{R}\left(\frac{en \to e\pi^+ X}{ep \to e\pi^- X}\right)_{z\sim 1} \simeq \frac{2s-1}{2s+2} \qquad (155)$$

is calculated by comparing Γ_1 for the Δ^- and Δ^{++}. It is particularly striking because the physical value of $s(Q^2)$ is close to one half, so the ratio becomes very small. For strange mesons, on the other hand, the ratio depends on whether the exchanged object is in the **8** (where the reduced matrix elements involve the appropriate F/D ratio) or **10** representation, so the prediction is less conclusive, viz.

$$\mathcal{R}\left(\frac{en \to eK^+ X}{ep \to eK^0 X}\right)_{z\sim 1} \simeq \frac{2s-1-3(2s-1)F/D}{2s-1-3(2s+1)F/D} \;(\mathbf{8}) \quad \text{or} \quad \frac{2s-1}{2s+1} \;(\mathbf{10}) \qquad (156)$$

which we find by comparing Γ_1 for either the Σ^- and Σ^+ in the **8** representation or Σ^{*-} and Σ^{*+} in the **10**. For charmed mesons, we again find

$$\mathcal{R}\left(\frac{en \to eD^0 X}{ep \to eD^- X}\right)_{z\sim 1} \simeq \frac{2s-1}{2s+2} \qquad (157)$$

corresponding to the ratio for Σ_c^0 to Σ_c^{++}.

At the other extreme, for z approaching 0, the detected hadron carries only a small fraction of the target nucleon energy. In this limit, the ratio $\mathcal{R}$ of the fracture function moments becomes simply the ratio of the structure function moments for n and p, i.e. using the current experimental values, $\mathcal{R}_{z\sim 0} \simeq \Gamma_1^n/\Gamma_1^p = -0.30$. This is to be compared with the corresponding OZI or Ellis–Jaffe value of -0.12.

The differences between the OZI, or valence quark model, expectations and our predictions based on topological charge screening can therefore be quite dramatic and should give a very clear experimental signal. In [79], together with De Florian, we analysed the potential for realising these experiments in some detail. Since we require particle identification in the target fragmentation region, fixed-target experiments such as COMPASS or HERMES are not appropriate. The preferred option is a polarised ep collider.

The first requirement is to measure particles at extremely small angles ($\theta \leq$ 1 mrad), corresponding to t less than around 1 GeV^2. This has already been achieved at HERA in measurements of diffractive and leading proton/neutron scattering using a forward detection system known as the Leading Proton Spectrometer (LPS). The technique for measuring charged particles involves placing detectors commonly known as 'Roman Pots' inside the beam pipe itself.

The next point is to notice that the considerations above apply equally to ρ as to π production, since the ratios $\mathcal{R}$ are determined by flavour quantum numbers alone. The particle identification requirements will therefore be less stringent, especially as the production of leading strange mesons from protons or neutrons is strongly suppressed. However, we require the forward detectors to have good acceptance for both positive and negatively charged mesons $M = \pi, \rho$ in order to measure the ratio (155).

The reactions with a neutron target can be measured if the polarised proton beam is replaced by polarised ^3He. In this case, if we assume that $^3He = Ap + Bn$, the cross section for the production of positive hadrons h^+ measured in the LPS is given by

$$\sigma(^3He \to h^+) \simeq A\sigma(p \to h^+) + B\sigma(n \to p) + B\sigma(n \to M^+) \qquad (158)$$

The first contribution can be obtained from measurements with the proton beam. However, to subtract the second one, the detectors must have sufficient particle identification at least to distinguish protons from positively charged mesons.

Finally, estimates of the total rates [79] suggest that around 1% of the total DIS events will contain a leading meson in the target fragmentation region where a LPS would have non-vanishing acceptance ($z > 0.6$) and in the dominant domain $x < 0.1$. The relevant cross sections are therefore sufficient to allow the ratios $\mathcal{R}$ to be measured.

The conclusion is that while our proposals undoubtedly pose a challenge to experimentalists, they are nevertheless possible. Given the theoretical importance of the 'proton spin' problem and the topological charge screening mechanism, there is therefore strong motivation to perform target fragmentation experiments at a future polarised *ep* collider [82].

6 Polarised Two-photon Physics and a Sum Rule for g_1^γ

The $U(1)_A$ anomaly plays a vital role in another sum rule arising in polarised deep inelastic scattering, this time for the polarised photon structure function $g_1^\gamma(x_\gamma, Q^2; K^2)$. For real photons, the first moment of g_1^γ vanishes as a consequence of electromagnetic current conservation [83]. For off-shell photons, we proposed a sum rule in 1992 [5, 6] whose dependence on the virtual momentum of the target photon encodes a wealth of information about the anomaly, chiral symmetry breaking and gluon dynamics in QCD. This is of special current interest since, given the ultra-high luminosity of proposed e^+e^- colliders designed as B factories, a detailed measurement of our sum rule is about to become possible for the first time.

6.1 The First Moment Sum Rule for g_1^γ

The polarised structure function g_1^γ is measured in the process $e^+e^- \to e^+e^-X$, which at sufficiently high energy is dominated by the two-photon interaction shown in Fig. 11. The deep inelastic limit is characterised by $Q^2 \to \infty$ with $x = Q^2/2p_2.q$ and $x_\gamma = Q^2/2k.q$ fixed, where $Q^2 = -q^2$, $K^2 = -k^2$ and $s = (p_1 + p_2)^2$. The target photon is assumed to be relatively soft, $K^2 \ll Q^2$.

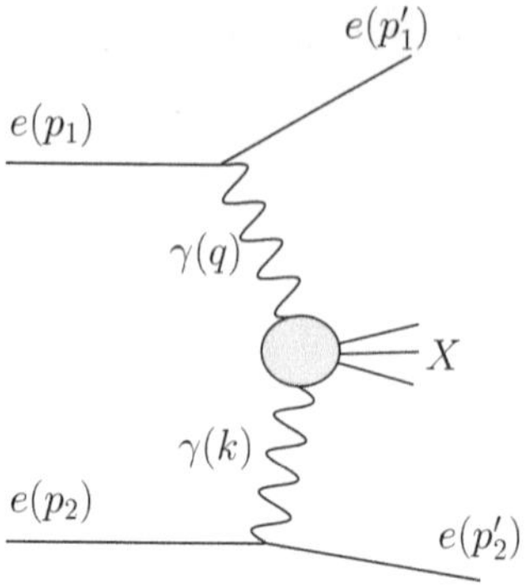

Fig. 11. Kinematics for the two-photon DIS process $e^+e^- \to e^+e^-X$

We are interested in the dependence of the photon structure function $g_1^\gamma(x_\gamma, Q^2; K^2)$ on the invariant momentum K^2 of the target photon. Experimentally, this is given by $K^2 \simeq EE_2'\theta_2^2$ where E_2' and θ_2 are the energy and scattering angle of the target electron. For the values $K^2 \sim m_\rho^2$ of interest in the sum rule, the target electron is nearly forward and θ_2 is very small. If it can be tagged, then the virtuality K^2 is simply determined from θ_2; otherwise K^2 can be inferred indirectly from a measurement of the total hadronic energy.

The total cross section σ and the spin asymmetry $\Delta\sigma$ can be expressed formally in terms of 'electron structure functions' as follows [5]:

$$\sigma = 2\pi\alpha^2 \frac{1}{s} \int_0^\infty \frac{dQ^2}{Q^2} \int_0^1 \frac{dx}{x^2} \left[F_2^e \frac{1}{y}\left(1 - y + \frac{y^2}{2}\right) - F_L^e \frac{y}{2}\right] \tag{159}$$

$$\Delta\sigma = 2\pi\alpha^2 \frac{1}{s} \int_0^\infty \frac{dQ^2}{Q^2} \int_0^1 \frac{dx}{x}\, g_1^e \left(1 - \frac{y}{2}\right) \tag{160}$$

where $\sigma = \frac{1}{2}(\sigma_{++} + \sigma_{+-})$ and $\Delta\sigma = \frac{1}{2}(\sigma_{++} - \sigma_{+-})$ with $+,-$ referring to the electron helicities. The parameter $y = Q^2/xs \ll 1$ and only the leading order terms are retained below.

These electron structure functions can be expressed as convolutions of the photon structure functions with appropriate splitting functions. In particular, we have

$$g_1^e(x, Q^2) = \frac{\alpha}{2\pi} \int_0^\infty \frac{dK^2}{K^2} \int_x^1 \frac{dx_\gamma}{x_\gamma} \Delta P_{\gamma e}\left(\frac{x}{x_\gamma}\right) g_1^\gamma(x_\gamma, Q^2; K^2) \tag{161}$$

where $\Delta P_{\gamma e}(x) = (2 - x)$. This allows us to relate the x_γ-moments of the photon structure functions to the x-moments of the cross sections. For the first moment of g_1^γ, we find

$$\int_0^1 dx\, x \frac{d^3\Delta\sigma}{dQ^2 dx dK^2} = \frac{3}{2}\alpha^3 \frac{1}{sQ^2K^2} \int_0^1 dx_\gamma\, g_1^\gamma(x_\gamma, Q^2; K^2) \tag{162}$$

The first moment sum rule follows, as for the proton, by using the OPE (114) to express the product of electromagnetic currents for the incident photon in terms of the axial currents $J^a_{\mu 5}$. The matrix elements $\langle\gamma^*(k)|J^a_{\mu 5}|\gamma^*(k)\rangle$ with the target photon are then expressed in terms of the three-current AVV Green function involving one axial and two electromagnetic currents. We define form factors for this fundamental correlator as follows:

$$\begin{aligned}-i\langle 0|J^a_{\mu 5}(p)J_\lambda(k_1)J_\rho(k_2)|0\rangle \;&=\; A^a_1\; \epsilon_{\mu\lambda\rho\alpha}k_1^\alpha \;+\; A^a_2\; \epsilon_{\mu\lambda\rho\alpha}k_2^\alpha \\ &\quad + A^a_3\; \epsilon_{\mu\lambda\alpha\beta}k_1^\alpha k_2^\beta k_{2\rho} \;+\; A^a_4\; \epsilon_{\mu\rho\alpha\beta}k_1^\alpha k_2^\beta k_{1\lambda} \\ &\quad + A^a_5\; \epsilon_{\mu\lambda\alpha\beta}k_1^\alpha k_2^\beta k_{1\rho} \;+\; A^a_6\; \epsilon_{\mu\rho\alpha\beta}k_1^\alpha k_2^\beta k_{2\lambda}\end{aligned} \tag{163}$$

where the six form factors are functions of the invariant momenta, i.e. $A^a_i = A^a_i(p^2, k_1^2, k_2^2)$. We also abbreviate $A^a_i(0, k^2, k^2) = A^a_i(K^2)$.

The first moment sum rule for g_1^γ is then [5]:

$$\int_0^1 dx_\gamma\; g_1^\gamma(x_\gamma, Q^2; K^2) \;=\; 4\pi\alpha \sum_{a=3,8,0} \Delta C_1^a(Q^2)\Big(A_1^a(K^2) - A_2^a(K^2)\Big) \tag{164}$$

where the Wilson coefficients are related to those in (115) by $\Delta C_1^3 = \Delta C_1^{NS}$, $\Delta C_1^8 = \frac{1}{\sqrt{3}}\Delta C_1^{NS}$ and $\Delta C_1^0 = \frac{2\sqrt{2}}{\sqrt{3}}\Delta C_1^S$.[12]

Now, just as the sum rule for the proton structure function g_1^p could be related to low-energy meson–nucleon couplings via the $U(1)_A$ Goldberger–Treiman relations, we can relate this sum rule for g_1^γ to the pseudoscalar meson radiative decays using the analysis in Sect. 4.2. Introducing the *off-shell* radiative pseudoscalar couplings for photon virtuality K^2, we define form factors

$$F^a(K^2) \;=\; 1 - \left(a^a_{\rm em}\frac{\alpha}{\pi}\right)^{-1} \hat{f}^{a\alpha} g_{\hat{\eta}^\alpha\gamma\gamma}(K^2) \tag{165}$$

or alternatively,

$$\begin{aligned}F^3(K^2) \;&=\; 1 - \left(a^3_{\rm em}\frac{\alpha}{\pi}\right)^{-1} f_\pi g_{\pi\gamma\gamma}(K^2) \\ F^8(K^2) \;&=\; 1 - \left(a^8_{\rm em}\frac{\alpha}{\pi}\right)^{-1} \left(f^{8\eta} g_{\eta\gamma\gamma}(K^2) + f^{8\eta'} g_{\eta'\gamma\gamma}(K^2)\right) \\ F^0(K^2) \;&= 1 - \left(a^0_{\rm em}\frac{\alpha}{\pi}\right)^{-1} \left(f^{0\eta} g_{\eta\gamma\gamma}(K^2) + f^{0\eta'} g_{\eta'\gamma\gamma}(K^2) + \sqrt{6}A g_{G\gamma\gamma}(K^2)\right)\end{aligned} \tag{166}$$

---

[12] Explicitly,

$$\Delta C_1^{NS} = \frac{1}{3}\Big(1 - \frac{\alpha_s(Q^2)}{\pi}\Big), \qquad \Delta C_1^S = \frac{1}{3}\Big(1 - \frac{\alpha_s(Q^2)}{\pi}\Big)\; \exp\int_0^{t(Q)} dt'\; \gamma(\alpha_s(t'))$$

at leading order, where $t(Q) = \frac{1}{2}\ln\frac{Q^2}{\mu^2}$ and $\gamma = -\frac{3}{4}\frac{\alpha_s^2}{(4\pi)^2}$ is the anomalous dimension corresponding to the $U(1)_A$ current renormalisation.

where the $a^a_{\rm em}$ are the electromagnetic $U(1)_A$ anomaly coefficients defined earlier. We may then rewrite the sum rule as

$$\int_0^1 dx_\gamma\; g_1^\gamma(x_\gamma, Q^2; K^2) \;=\; \frac{1}{2}\frac{\alpha}{\pi} \sum_{a=3,8,0} \Delta C_1^a(Q^2)\; a^a_{\rm em}\; F^a(K^2) \qquad (167)$$

The dependence of the g_1^γ on the invariant momentum K^2 of the target photon reflects many key aspects of both perturbative and non-perturbative QCD dynamics. For on-shell photons, $K^2 = 0$, we have simply [5, 83]

$$\int_0^1 dx_\gamma\; g_1^\gamma(x_\gamma, Q^2; K^2 = 0) \;=\; 0 \qquad (168)$$

This is a consequence of electromagnetic current conservation. This follows simply by taking the divergence of (163) and observing that in the limit $p \to 0$, both A_1 and A_2 are of $O(K^2)$.[13]

In the asymptotic limit where $K^2 \ll m_\rho^2$, a relatively straightforward renormalisation group analysis combined with the anomaly equation shows that, for the flavour non-singlets, the A_i^a tend to the value $\frac{1}{2}\frac{\alpha}{\pi} a^a_{\rm em}$. while in the flavour singlet sector, A_i^0 has an additional factor depending on the anomalous dimension γ. Using the explicit expressions for the Wilson coefficients, we find

$$\begin{aligned}
&\int_0^1 dx_\gamma\; g_1^\gamma(x_\gamma, Q^2; K^2 \ll m_\rho^2) \\
&= \frac{1}{6}\frac{\alpha}{\pi}\Big(1 - \frac{\alpha_s(Q^2)}{\pi}\Big)\Big(a^3_{\rm em} + \frac{1}{\sqrt{3}} a^8_{\rm em} + \frac{2\sqrt{2}}{\sqrt{3}} a^0_{\rm em} \exp\Big[\int_{t(K)}^{t(Q)} dt'\; \gamma(\alpha_s(t'))\Big]\Big) \\
&= \frac{1}{3}\frac{\alpha}{\pi}\Big[1 \;-\; \frac{4}{9}\frac{1}{\ln Q^2/\Lambda^2} \;+\; \frac{16}{81}\Big(\frac{1}{\ln Q^2/\Lambda^2} - \frac{1}{\ln K^2/\Lambda^2}\Big)\Big]
\end{aligned} \qquad (169)$$

The asymptotic limit is therefore determined by the electromagnetic $U(1)_A$ anomaly, with logarithmic corrections reflecting the anomalous dimension of the flavour singlet current due to the colour $U(1)_A$ anomaly. (See also [84] for a NNLO analysis.)

In between these limits, the first moment of g_1^γ provides a measure of the form factors defining the three-current AVV Green function, which encodes a great deal of information about the dynamics of QCD, especially the non-perturbative realisation of chiral symmetry [6]. Equivalently, in the form

---

[13] Electromagnetic current conservation in (163) implies

$$A_1^a = A_3^a k_2^2 + A_5^a \frac{1}{2}(p^2 - k_1^2 - k_2^2), \qquad A_2^a = A_4^a k_1^2 + A_6^a \frac{1}{2}(p^2 - k_1^2 - k_2^2)$$

The chiral limit is special since the form factors can have massless poles and is considered in detail in [6]. The sum rule (168) still holds.

(167), it measures the momentum dependence of the off-shell radiative couplings of the pseudoscalar mesons as the form factors $F^a(K^2)$ vary from 0 to 1.

Just as for g_1^p, we can again isolate a dependence on the topological susceptibility through the identification of the flavour singlet decay constant $\hat{f}^{00}$ in (165) with $\sqrt{\chi'(0)}$ in the chiral limit. This time, however, it is unlikely to be a good approximation to set the corresponding coupling $g_{\hat{\eta}^0\gamma\gamma}$ equal to its OZI value since it is not RG invariant. A more promising approximation is to recall from Sect. 4 that the RG-invariant gluonic coupling $g_{G\gamma\gamma}(0)$ is OZI suppressed and likely to be small. This was confirmed by the phenomenological analysis. If we assume this is also true of the off-shell coupling, then we may approximate the sum rule for g_1^γ entirely in terms of the off-shell couplings of the physical mesons π^0, η and η'.

In general, the momentum dependence of the form factors $(A_1^a - A_2^a)$ or F^a will depend on the fermions contributing to the AVV Green function [6]. In the case of leptons, or heavy quarks, the crossover scale as the form factors $F^a(K^2)$ rise from 0 to 1 with increasing K^2 will be given by the fermion mass. For the light quarks, however, we expect the crossover scale to be a typical hadronic scale $\sim m_\rho$ rather than $m_{u,d,s}$. This can be justified by a rough OPE argument and is consistent with old ideas of vector meson dominance [6, 85]. This behaviour would be an interesting manifestation of the spontaneous breaking of chiral symmetry.

Once again, therefore, we see a close relation between the realisation of sum rules in high-energy deep inelastic scattering and low-energy meson physics. All these issues are discussed at some length in our earlier papers, but here we now turn our attention to the vital question of whether the g_1^γ sum rule can be measured in current or future collider experiments [7].

6.2 Cross Sections and Spin Asymmetries at Polarised B Factories

The spin-dependent cross sections for the two-photon DIS process $e^+e^- \to e^+e^-X$ were analysed in [5, 7] taking account of the experimental cuts on the various kinematical parameters. Keeping the lower cut on Q^2 as a free parameter, we found the following results for the total cross section and spin asymmetry:

$$\sigma \simeq 0.5 \times 10^{-8} \frac{1}{Q^2_{\rm min}} \log \frac{Q^2_{\rm min}}{\Lambda^2} \left(\log \frac{s}{Q^2_{\rm min}}\right)^2 \tag{170}$$

and

$$\frac{\Delta\sigma}{\sigma} = \frac{1}{2} \frac{Q^2_{\rm min}}{s} \log \frac{s}{4Q^2_{\rm min}} \left[1 + \log \frac{s}{4\Lambda^2} \left(\log \frac{Q^2_{\rm min}}{\Lambda^2}\right)^{-1}\right] \tag{171}$$

In order to measure the g_1^γ sum rule, we need to find collider parameters such that the spin asymmetry is significant in a kinematic region where the total cross section is still large. A useful statistical measure of the significance of the asymmetry is that $\sqrt{L\sigma}\Delta\sigma/\sigma \gg 1$, where L is the luminosity.

When we first proposed the first moment sum rule for g_1^γ, the luminosity available from the then current accelerators was inadequate to allow it to be studied. For example, for a polarised version of LEP operating at $s = 10^4$ GeV^2 with an annual integrated luminosity of $L = 100$ pb^{-1}, and optimising the cut at $Q^2_{\rm min} = 10$ GeV^2, we only have $\sigma \simeq 35$ pb and $\Delta\sigma/\sigma \simeq 0.01$. The corresponding annual event rate would be 3.5×10^3 and the statistical significance $\sqrt{L\sigma}\Delta\sigma/\sigma \simeq 0.5$, so even a reliable measurement of the spin asymmetry could not be made.

Clearly, a hugely increased luminosity is required and this has now become available with proposals for machines with projected annual integrated luminosities measured in inverse attobarns. However, as noted in [5], if this increased luminosity is associated with increased CM energy, then the $1/s$ factor in the spin asymmetry (171) sharply reduces the possibility of extracting g_1^γ. There is also a competition as $Q^2_{\rm min}$ is varied between increasing spin asymmetry and decreasing total cross section. This is particularly evident when we analyse the potential of the ILC [86, 87] for measuring the sum rule [7]. We find that even optimising the $Q^2_{\rm min}$ cut, the spin asymmetry is still only of order $\Delta\sigma/\sigma \simeq 0.002$ when σ itself has fallen to around 15 pb. While, given the high luminosity, this would allow a measurement of the first moment of g_1^γ integrated over K^2, a detailed study of the K^2-dependence of the sum rule requires a much greater spin asymmetry.

This leads us to consider instead the new generation of ultra-high luminosity e^+e^- colliders. Although these are envisaged as B factories, these colliders operating with polarised beams would, as we now show, be extremely valuable for studying polarisation phenomena in QCD. As an example of this class, we take the proposed SuperKEKB collider. (The analysis for PEPII is very similar, the main difference being the additional 10-fold increase in luminosity in the current SuperKEKB proposals.)

SuperKEKB is an asymmetric e^+e^- collider with $s = 132$ GeV^2, corresponding to electron and positron beams of 8 and 3.5 GeV, respectively. The design luminosity is 5×10^{35} cm^{-2}s^{-1}, which gives an annual integrated luminosity of 5 ab^{-1} [88]. To see the effects of the experimental cut on $Q^2_{\rm min}$ in this case, we have plotted the total cross section and the spin asymmetry in Fig. 12, in the range of $Q^2_{\rm min}$ from 1 to 10 GeV^2. In this range σ is falling like $1/Q^2_{\rm min}$ while $\Delta\sigma/\sigma$ rises to what is actually a maximum at $Q^2_{\rm min} = 10$ GeV^2.

Taking $Q^2_{\rm min} = 5$ GeV^2, we find $\sigma \simeq 12.5$ pb with spin asymmetry $\Delta\sigma/\sigma \simeq 0.1$. The annual event rate is therefore 6.25×10^7, with $\sqrt{L\sigma}\Delta\sigma/\sigma \simeq 750$. This combination of a very high event rate and the large 10% spin asymmetry means that SuperKEKB has the potential not only to measure $\Delta\sigma$ but to access the full first moment sum rule for g_1^γ itself. Recall from (162) that to measure $\int_0^1 dx\ g_1^\gamma(x, Q^2; K^2)$ we need not just $\Delta\sigma$ but the fully differential cross section w.r.t. K^2 as well as x and Q^2 if the interesting non-perturbative QCD physics is to be accessed. To measure this, we need to divide the data

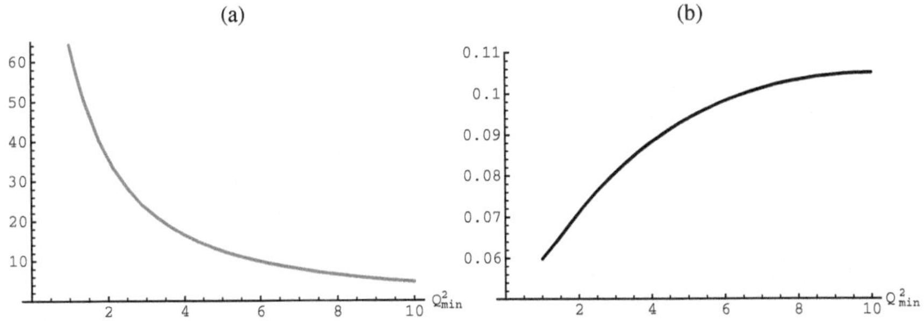

Fig. 12. The left-hand graph shows the total cross section σ (in pb) at SuperKEKB as the experimental cut $Q^2_{\rm min}$ is varied from 1 to 10 GeV^2. The right-hand graph shows the spin asymmetry $\Delta\sigma/\sigma$ over the same range of $Q^2_{\rm min}$

into sufficiently fine K^2 bins in order to plot the explicit K^2 dependence of g_1^γ, while still maintaining the statistical significance of the asymmetry. The ultra-high luminosity of SuperKEKB ensures that the event rate is sufficient, while its moderate CM energy means that the crucial spin asymmetry is not overly suppressed by its $1/s$ dependence.

Our conclusion is that the new generation of ultra-high luminosity, moderate energy e^+e^- colliders, currently conceived as B factories, could also be uniquely sensitive to important QCD physics if run with polarised beams. In particular, they appear to be the only accelerators capable of accessing the full physics content of the sum rule for the first moment of the polarised structure function $g_1^\gamma(x, Q^2; K^2)$. The richness of this physics, in particular the realisation of chiral symmetry breaking, the manifestations of the axial $U(1)_A$ anomaly and the role of gluon topology, provides a strong motivation for giving serious consideration to an attempt to measure the g_1^γ sum rule at these new colliders.

Acknowledgements

In addition to Gabriele, I would like to thank Daniel de Florian, Massimiliano Grazzini, Stephan Narison and Ben White for their collaboration on the original research presented here. This paper has been prepared with the partial support of PPARC grant PP/D507407/1.

References

1. G. Veneziano: Nucl. Phys. B **159**, 213 (1979)
2. G. M. Shore: Nucl. Phys. B **569**, 107 (2000)
3. G. M. Shore: Nucl. Phys. B **744**, 34 (2006)
4. G. Veneziano: Mod. Phys. Lett. A **4**, 1605 (1989)

5. S. Narison, G. M. Shore, G. Veneziano: Nucl. Phys. B **391**, 69 (1993)
6. G. M. Shore, G. Veneziano: Mod. Phys. Lett. A **8**, 373 (1993)
7. G. M. Shore: Nucl. Phys. B **712**, 411 (2005)
8. G. M. Shore, G. Veneziano: Phys. Lett. B **244**, 75 (1990)
9. G. M. Shore, G. Veneziano: Nucl. Phys. B **381**, 23 (1992)
10. S. Narison, G. M. Shore, G. Veneziano: Nucl. Phys. B **433**, 209 (1995)
11. G. M. Shore, G. Veneziano: Nucl. Phys. B **516**, 333 (1998)
12. G. Veneziano: in *From Symmetries to Strings: Forty Years of Rochester Conferences*, ed. by A. Das (World Scientific, Singapore, 1990), pp. 13–26
13. S. L. Adler: Phys. Rev. **177**, 2426 (1969)
14. J. S. Bell, R. Jackiw: Nuovo Cimento A **60**, 47 (1969)
15. S. L. Adler, W.A. Bardeen: Phys. Rev. **182**, 1517 (1969)
16. J. Steinberger: Phys. Rev. **76**, 1180 (1949)
17. J. Schwinger: Phys. Rev. **82**, 664 (1951)
18. K. Fujikawa: Phys. Rev. Lett. **42**, 1195 (1979); Phys. Rev. D **21**, 2848 (1980), erratum-ibid. D **22**, 1499 (1980)
19. G. M. Shore: in *Hidden Symmetries and Higgs Phenomena*, Zuoz Summer School, Switzerland, 1998, pp. 201–223; arXiv:hep-ph/9812354
20. E. Witten: Nucl. Phys. B **156**, 269 (1979)
21. P. Di Vecchia, G. Veneziano: Nucl. Phys. B **171**, 253 (1980)
22. D. Espriu, R. Tarrach: Z. Phys. C **16**, 77 (1982)
23. G. M. Shore: Nucl. Phys. B **362**, 85 (1991)
24. G. M. Shore, G. Veneziano: Nucl. Phys. B **381**, 3 (1992)
25. G. 't Hooft: Nucl. Phys. B **72**, 461 (1972)
26. G. Veneziano: Phys. Lett. B **52**, 220 (1974); Nucl. Phys. B **117**, 519 (1976)
27. S. Okubo: Phys. Lett. **5**, 165 (1963)
28. G. Zweig: CERN report 8419/TH412 (1964)
29. J. Iizuka: Prog. Theor. Phys. Suppl. **37–38**, 21 (1966)
30. S. Weinberg: Phys. Rev. D **11**, 3583 (1975)
31. G. 't Hooft: Phys. Rev. D **14**, 3432 (1976); [Erratum-ibid. D **18**, 2199 (1978)]
32. R. J. Crewther: Riv. Nuovo Cimento **2N8**, 63 (1979)
33. G. A. Christos: Phys. Rep. **116**, 251 (1984)
34. G. 't Hooft: Phys. Rept. **142**, 357 (1986)
35. M. Gell-Mann, R. J. Oakes, B. Renner: Phys. Rev. **175**, 2195 (1968)
36. R. F. Dashen: Phys. Rev. **183**, 1245 (1969)
37. C. Rosenzweig, J. Schechter, C. G. Trahern: Phys. Rev. D **21**, 3388 (1980)
38. P. Di Vecchia, F. Nicodemi, R. Pettorino, G. Veneziano: Nucl. Phys. B **181**, 318 (1981)
39. K. Kawarabayashi, N. Ohta: Nucl. Phys. B **175**, 477 (1980)
40. P. Herrera-Siklody, J. I. Latorre, P. Pascual, J. Taron: Nucl. Phys. B **497**, 345 (1997); Phys. Lett. B **419**, 326 (1998)
41. H. Leutwyler: Nucl. Phys. Proc. Suppl. **64**, 223 (1998)
42. R. Kaiser, H. Leutwyler: Eur. Phys. J. C **17**, 623 (2000)
43. L. Giusti, G. C. Rossi, M. Testa, G. Veneziano: Nucl. Phys. B **628**, 234 (2002)
44. G. M. Shore: Phys. Scr. T **99**, 84 (2002)
45. Particle Data Group: Review of Particle Properties, Phys. Lett. B **592**, 1 (2004)
46. M. Acciarri et al., L3 Collaboration: Phys. Lett. B **418**, 399 (1998)
47. D. A. Williams et al., Crystal Ball Collaboration: Phys. Rev. D **38**, 1365 (1988)
48. N. A. Roe et al., ASP Collaboration: Phys. Rev. D **41**, 17 (1990)

49. L. Del Debbio, L. Giusti, C. Pica: Phys. Rev. Lett. **94**, 032003 (2005)
50. A. Di Giacomo: Nucl. Phys. Proc. Suppl. **23**, 191 (1991)
51. S. Narison: Phys. Lett. B **255**, 101 (1991); Z. Phys. C **26**, 209 (1984)
52. S. Narison, G. M. Shore, G. Veneziano: Nucl. Phys. B **546**, 235 (1999)
53. V. Y. Alexakhin et al. [COMPASS Collaboration]: "The deuteron spin-dependent structure function g1(d) and its first moment," arXiv:hep-ex/0609038
54. A. Airapetian et al. [HERMES Collaboration]: "Precise determination of the spin structure function g(1) of the proton, deuteron and neutron," arXiv:hep-ex/0609039
55. G. Mallot, S. Platchkov, A. Magnon: CERN-SPSC-2005-017; SPSC-M-733
56. E. S. Ageev et al. [COMPASS Collaboration]: Phys. Lett. B **612**, 154 (2005)
57. J. R. Ellis, R. L. Jaffe: Phys. Rev. D **9**, 1444 (1974), [Erratum-ibid. D **10**, 1669 (1974)]
58. D. V. Bugg: Eur. Phys. J. C **33**, 505 (2004)
59. P. Moskal: "Hadronic interaction of eta and eta′ mesons with protons," arXiv:hep-ph/0408162
60. S. D. Bass: Phys. Scr. T **99**, 96 (2002)
61. P. Moskal et al.: Int. J. Mod. Phys. A **20**, 1880 (2005)
62. P. Moskal et al.: Phys. Rev. Lett. **80**, 3202 (1998)
63. K. Nakayama, H. Haberzettl: "Analyzing eta′ photoproduction data on the proton at energies of 1.5GeV–2.3GeV," arXiv:nucl-th/0507044.
64. M. Dugger [CLAS Collaboration]: "S=0 pseudoscalar meson photoproduction from the proton," arXiv:nucl-ex/0512005.
65. G. M. Shore: Nucl. Phys. Proc. Suppl. **39BC**, 101 (1995)
66. J. Ashman et al: Phys. Lett. B **206**, 364 (1988); Nucl. Phys. B **328**, 1 (1990)
67. R. L. Jaffe, A. Manohar: Nucl. Phys. B **337**, 509 (1990)
68. G. M. Shore, B. E. White: Nucl. Phys. B **581**, 409 (2000)
69. G. M. Shore: Nucl. Phys. Proc. Suppl. **96**,171 (2001)
70. B. L. G. Bakker, E. Leader, T. L. Trueman: Phys. Rev. D **70**, 114001 (2004)
71. R. D. Ball, S. Forte, G. Ridolfi: Phys. Lett. B **378**, 255 (1996)
72. G. Altarelli, G. G. Ross: Phys. Lett. B **212**, 391 (1988)
73. S. Procureur [COMPASS Collaboration]: "New measurement of Delta(G)/G at COMPASS," arXiv:hep-ex/0605043
74. R. Fatemi [STAR Collaboration]: "Using jet asymmetries to access Delta(G), the gluon helicity distribution of the proton at STAR," arXiv:nucl-ex/0606007.
75. Y. Fukao [PHENIX Collaboration]: "The overview of the spin physics at RHIC-PHENIX experiment," AIP Conf. Proc. **842**, 321 (2006)
76. G. M. Shore: in *From the Planck Length to the Hubble Radius*, Erice 1998, ed. by A. Zichichi (World Scientific, Singapore, 1998), pp. 79–105
77. L. Trentadue, G. Veneziano: Phys. Lett. B **323**, 201 (1994)
78. M. Grazzini, L. Trentadue, G. Veneziano: Nucl. Phys. B **519**, 394 (1998)
79. D. de Florian, G. M. Shore, G. Veneziano: in *Proceedings of the 1997 Workshop with Polarized Protons at Hera*, ed. by A. de Roeck, T. Gehrmann (Hamburg/Zeuthen, 1997) pp. 696–703; arXiv:hep-ph/9711353
80. G. M. Shore: Nucl. Phys. Proc. Suppl. **64**, 167 (1998)
81. M. Grazzini, G. M. Shore, B. E. White: Nucl. Phys. B **555**, 259 (1999)
82. A. De Roeck: Nucl. Phys. Proc. Suppl. **105**, 40 (2002)
83. S. D. Bass: Int. J. Mod. Phys. A **7**, 6039 (1992)

84. K. Sasaki, T. Ueda, T. Uematsu: Phys. Rev. D **73**, 094024 (2006)
85. T. Ueda, T. Uematsu, K. Sasaki: Phys. Lett. B **640**, 188 (2006)
86. F. Richard et al.: "TESLA: The Superconducting electron positron linear collider with an integrated X-ray laser laboratory. Technical Design Report, Part I"; hep-ph/0106314
87. M. Woods et al.: in *Proc. 5th International Workshop on Electron–Electron Interactions at TeV Energies*, Santa Cruz, 2003; physics/0403037
88. A. G. Akeroyd et al (SuperKEKB Physics Working Group): "Physics at Super B Factory"; hep-ex/0406071

Planar Equivalence 2006*

A. Armoni[1] and M. Shifman[2]

[1] Department of Physics, Swansea University, Singleton Park, Swansea SA2 8PP, UK
A.Armoni@swansea.ac.uk
[2] William I. Fine Theoretical Physics Institute, University of Minnesota, Minneapolis, MN 55455, USA
shifman@umn.edu

Abstract. Planar equivalence between supersymmetric Yang–Mills theory and its orientifold daughters is a promising tool for explorations of nonperturbative aspects of quantum chromodynamics. Taking our 2004 review as a starting point, we summarize some recent developments in this issue.

The most interesting processes in quantum chromodynamics (QCD) are those occurring at large distances, at strong coupling. The large distance dynamics determining such salient features as chiral symmetry breaking and color confinement are the realm of nonperturbative phenomena. Despite the practical importance of the issue and the fact that this is a very deep theoretical problem, very few analytic methods of calculations (of a limited scope) were developed over the years, for a recent review see [1].

The situation is much better in supersymmetric (SUSY) theories: certain quantities (which go under the name of F terms) can be calculated *exactly*, due to holomorphic dependences on various parameters. In particular, it is possible to calculate the exact value of the gluino condensate [2] in pure $\mathcal{N}=1$ super-Yang–Mills (SYM) theory (we will also refer to this theory as supersymmetric gluodynamics).

* A mini-review in honor of Gabriele Veneziano's 65th birthday.

A. Armoni and M. Shifman: *Planar Equivalence 2006*, Lect. Notes Phys. **737**, 289–300 (2008)
DOI 10.1007/978-3-540-74233-6_13

The basic idea behind planar equivalence is to approximate QCD by a supersymmetric theory!

The history of planar equivalence is as follows. In 1998, soon after the seminal AdS/CFT paper of Maldacena [3], Kachru and Silverstein [4] suggested a class of nonsupersymmetric large-N conformal gauge theories. The candidate theories were the duals of $AdS_5 \times S^5/\Gamma$ and, therefore, named "orbifold field theories." Although it turns out that these theories are in fact not conformal (not even in perturbation theory, see [5, 6]) Kachru–Silverstein's conjecture led to a more subtle conjecture by Strassler [7]. A refined version of Strassler's conjecture is *planar equivalence for orientifold field theories.* In contrast to various other conjectures, the latter can be proven [8, 9] under rather mild assumptions, see Sect. 6. The orientifold daughter of SUSY gluodynamics is a nonsupersymmetric Yang–Mills theory with one Dirac fermion in the two-index antisymmetric (or symmetric) representation of SU(N). In this mini-review written on the occasion of Gabriele Veneziano's 65th birthday we focus on recent developments in the issue of planar equivalence—those that took place after our detailed review on this subject [10] was published.

The statement of planar equivalence for (the minimal) orientifold field theory is as follows: *at large N, in a certain well defined bosonic sector, SU(N) $\mathcal{N} = 1$ SYM theory is equivalent to an SU(N) gauge theory with a Dirac fermion in the two-index antisymmetric representation.* The same statement holds for Dirac fermions in the two-index symmetric representation.

Although planar equivalence is an extremely interesting theoretical statement per se, its practical importance goes far beyond since it relates a supersymmetric gauge theory to a nonsupersymmetric one. Thus, potentially, it is a very useful tool for QCD. Let us make a simple observation: for $\mathrm{SU}(3)_{\mathrm{color}}$ a Dirac fermion in the antisymmetric representation is equivalent to a Dirac fermion in the fundamental representation. Therefore, the SU(3) version of the orientifold field theory is in fact one-flavor QCD! Thus, we can approximate one-flavor QCD by supersymmetric Yang–Mills and in this way evaluate some nonperturbative quantities in QCD. In particular, planar equivalence will enable us to calculate the quark condensate in one-flavor QCD by using the value of the gluino condensate in supersymmetric gluodynamics.

1 Planar Equivalence: a Refined Proof

Originally, the idea of planar equivalence between supersymmetric gluodynamics and its orientifold daughter was formulated in 2003. Since then we refined the proof and made it more rigorous [9]. Let us briefly outline the main ingredients of the proof.

It is instructive to start from a perturbative analysis. We want to show that all planar graphs of the two theories coincide. To this end it is useful to use 't Hooft's notation. In this notation the adjoint representation is denoted

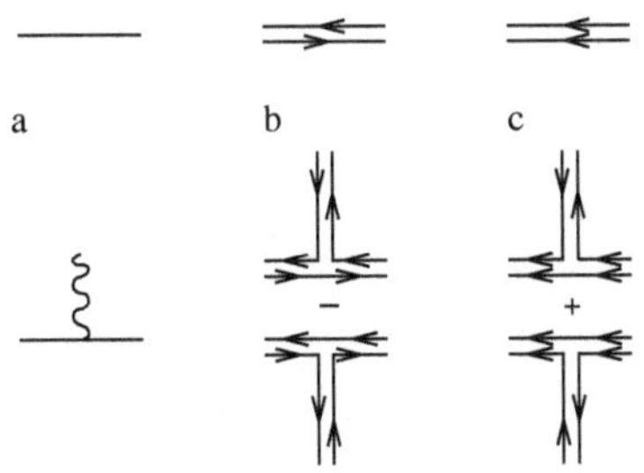

Fig. 1. (**a**) The quark-gluon vertex; (**b**) In $\mathcal{N} = 1$ SYM theory; (**c**) In the orientifold field theory

by two parallel lines with color flow arrows pointing in the opposite directions, whereas the antisymmetric (symmetric) representation is denoted by two parallel lines with the arrows pointing in the same direction. The Feynman rules of the two theories are depicted in Fig. 1.

Next, we observe that the direction of the color flow arrows does not affect the value of the planar graphs under consideration. To see that this is indeed the case, imagine that we paint every pair of the fermionic lines in blue and red colors, respectively. Accordingly, the gluon lines will be either both red or both blue. A planar graph then will be divided into blue regions and red regions separated by fermionic loops. A typical example is given in Fig. 2.

Now imagine that we reverse the arrows attached to the red lines. In this way we map a planar graph of one theory onto a planar graph of the other theory. This action does not change the value of the graph. *Quod Erat Demonstrandum.*

The complete nonperturbative proof [9] is more involved, of course. The main ingredients are as follows. First, define, for a generic Dirac fermion in the representation r, the generating functional

$$e^{-\mathcal{W}_r(J_{\rm YM},\,J_\Psi)} = \int DA_\mu\, D\Psi\, D\bar{\Psi}\, e^{-S_{\rm YM}[A,J_{\rm YM}]}\, \exp\left\{\bar{\Psi}\left(i\,\partial\!\!\!/ + A\!\!\!/^a\, T_r^a + J_\Psi\right)\Psi\right\}. \tag{1}$$

Next, integrate out fermions to arrive at

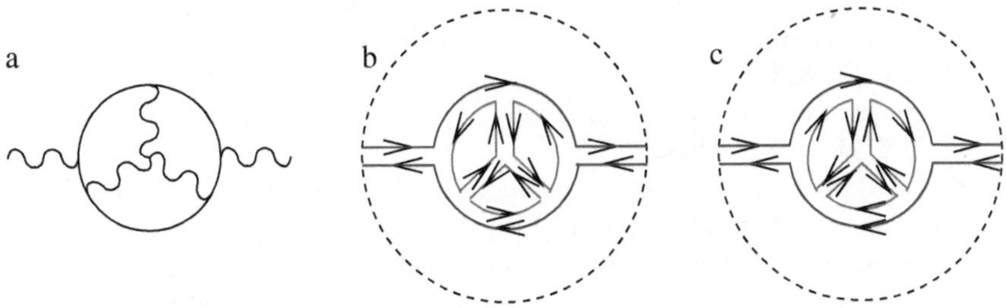

Fig. 2. A typical planar graph in SYM and the orientifold field theory

$$e^{-\mathcal{W}_r(J_{\rm YM},\, J_\Psi)} = \int DA_\mu\, e^{-S_{\rm YM}[A, J_{\rm YM}] + \Gamma_r[A, J_\Psi]}\,, \tag{2}$$

where

$$\Gamma_r[A, J_\Psi] = \log\det\left(i\,\not{\partial} + \not{A}^a\, T^a_r + J_\Psi\right)\,. \tag{3}$$

For what follows it is convenient to write the effective action $\Gamma_r[A, J_\Psi]$ in the world-line formalism [11], as an integral over (super-)Wilson loops

$$\Gamma_r[A, J_\Psi] = -\frac{1}{2}\int_0^\infty \frac{dT}{T}$$

$$\times \int \mathcal{D}x\mathcal{D}\psi\, \exp\left\{-\int_\epsilon^T d\tau\, \left(\frac{1}{2}\dot{x}^\mu\dot{x}^\mu + \frac{1}{2}\psi^\mu\dot{\psi}^\mu - \frac{1}{2}J_\Psi^2\right)\right\}$$

$$\times\, {\rm Tr}\,\mathcal{P}\exp\left\{i\int_0^T d\tau\, \left(A^a_\mu\dot{x}^\mu - \frac{1}{2}\psi^\mu F^a_{\mu\nu}\psi^\nu\right) T^a_r\right\}\,. \tag{4}$$

Thus, the generating functionals of theories with matter in the antisymmetric/adjoint are very similar. The dependence on the representation enters through the Wilson loops. The latter can be written as follows:

$$W_{\rm AS} = \frac{1}{2}\left(({\rm Tr}\,U)^2 - {\rm Tr}\,U^2\right) + (U \to U^\dagger), \tag{5}$$

$$W_{\rm adjoint} = {\rm Tr}\,U\,{\rm Tr}\,U^\dagger - 1 + (U \to U^\dagger) = 2\left({\rm Tr}\,U\,{\rm Tr}\,U^\dagger - 1\right)\,, \tag{6}$$

where U (respectively $U^\dagger$) represents the same group element in the *fundamental* (respectively *antifundamental*) representation of SU(N).

To complete the proof [9], one must show that at large N one can use

$$W_{\rm AS} \sim \frac{1}{2}({\rm Tr}\,U)^2 + \frac{1}{2}({\rm Tr}\,U^\dagger)^2\,, \tag{7}$$

$$W_{\rm adjoint} \sim 2{\rm Tr}\,U\,{\rm Tr}\,U^\dagger\,, \tag{8}$$

and that U can be replaced by $U^\dagger$ everywhere.[2] The factor 2 in (8) is canceled by the factor $\frac{1}{2}$, since the adjoint representation is realized by the Majorana rather than Dirac fermions.

A remarkable consequence of nonperturbative planar equivalence is that (non-SUSY) orientifold field theories exhibit some feature of supersymmetric theories. This is surprising since the spectrum of the large-N theory consists of bosons only— it is impossible to form finite-mass fermionic color singlets. As a "remnant" of SUSY they are predicted to have an even/odd parity degeneracy, as in supersymmetric gluodynamics. More generally, two bosons

[2] See Sect. 6 for a more detailed discussion.

from one and the same would-be supermultiplet, must be degenerate in mass at $N \to \infty$. In addition, the quark condensate $\langle \bar{\Psi}\Psi \rangle$ will form, and its value will be identical to that of the gluino condensate in $\mathcal{N} = 1$ SYM theory. Other important properties are the NSVZ β function, the domain wall spectrum and gluonic Green functions [8, 10].

2 The Orientifold Large-N Expansion

Let us forget for a short while about supersymmetry and look at planar equivalence from a broader perspective. Assume we are interested in the large-N limit of multiflavor QCD. There are various options of generalizing SU(3) QCD to SU(N) gauge theory. In the original 't Hooft large-N expansion [12] both g^2N and the number of flavors N_f (quarks in the fundamental representation) is kept fixed (this is realized in the modern gauge/string duality by keeping the number of flavor branes fixed). In the Veneziano large-N expansion (the topological expansion [13]) the ratio N_f/N is kept fixed, together with g^2N (it can be achieved by placing branes on orbifold singularities, in a certain region of the moduli space). The advantage of the latter expansion is that the quark loops are not suppressed at large N and, hence, flavor physics is better captured in this approximation. In particular, the η' mass does not vanish even when $N \to \infty$, that is to say, a massive η' is a part of the planar theory.

While both expansions are interesting and useful, there is no full quantitative solution to either. It is tempting to say that large-N QCD is dual to a string theory, and there was a significant progress along these lines [3], but it would be certainly wrong to say that an accurate and well-developed description of QCD has been already attained. Therefore, alternative large-N limits may well prove to be very useful.

Let us discuss a new *orientifold* large-N expansion [14]. It will lead to certain *quantitative* predictions for QCD. We start from $\mathrm{SU}(3)_{\mathrm{color}}$ Yang–Mills theory with N_f quark flavors in the fundamental representation (to be referred to as multiflavor QCD). Since for SU(3) the Dirac fermion in the fundamental representation is equivalent to the Dirac fermion in the antisymmetric two-index representation, we have the option of generalizing the theory to $\mathrm{SU}(N)_{\mathrm{color}}$ treating N_f fermions as antisymmetric Dirac fermions, see Fig. 3.

The next step is to consider the large-N limit of this theory while keeping N_f fixed. This large-N approximation, to be referred to as the *orientifold large-N* approximation, is somewhat similar to the topological expansion since the quark loops are not suppressed with respect to the gluon loops.

Through planar equivalence the theory with N_f Dirac quarks in the two-index antisymmetric approximation is related to the theory with N_f adjoint Majorana quarks (in the common sector).

While phenomenological consequences of the orientifold large-N limit so far remain essentially unexplored, in purely theoretical aspect planar

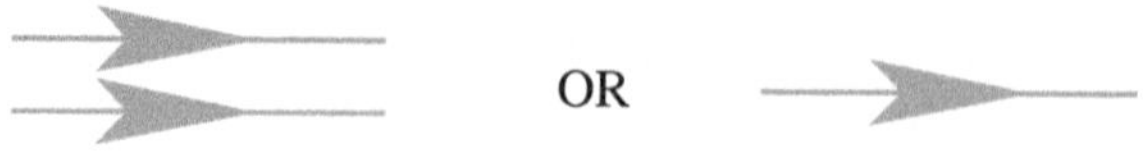

Fig. 3. Antisymmetric/fundamental representation in SU(3)

equivalence of these theories revived interest in gauge theories with quarks in higher representations, other than the fundamental representation. In particular, one can ask the question as to the form of the chiral Lagrangian in the Yang–Mills theory with antisymmetric or adjoint quarks. The chiral Lagrangian of QCD with fundamental quarks supports skyrmions which can be identified with baryons [17]. And what about the chiral Lagrangian in the theory with antisymmetric quarks? The pattern of the spontaneous breaking of the chiral symmetry in this *gedanken* case is well-known. The corresponding chiral Lagrangian is not drastically different from that of QCD. It supports skyrmions too. However, the mass of the skyrmions in this case scales as N^2 rather than N, as is the case in the 't Hooft limit. At first sight, there is no apparent match between skyrmions and baryons. It turns out [18] that N-quark hadrons built of antisymmetric fermions are unstable with regards to fusion of N species into a huge compound object built of N^2 quarks. It is the latter which is an analog of the baryon! For subsequent discussions see [19].

Moreover, chiral Lagrangians were found in theories with the adjoint quarks [20]. The issue of baryon analogs and skyrmions in this case is intriguing and subtle. There is no conservation of fermion number; rather it is $(-1)^F$ which is conserved. It was argued [20] that an analog of the baryon is a compound object built of N^2 quarks with an abnormal assignment of $(-1)^F$. On the skyrmion side, it is seen as a Hopf skyrmion whose topological stability is associated with a nontrivial Hopf invariant.

3 Applications for One-flavor QCD

As we explained in Sects. 1 and 2, we can approximate one-flavor QCD by a planar theory with one Dirac two-index antisymmetric fermion. This theory is planar-equivalent to $\mathcal{N} = 1\,$SYM theory. We can therefore make several *quantitative* predictions about the nonperturbative regime of the one-flavor QCD.

The first prediction concerns the spectrum of the theory. As we discussed at the end of Sect. 1, the color-singlet spectrum of the orientifold field theory exhibits an odd/even parity degeneracy. Thus, we expect a similar degeneracy in the spectrum of one-flavor QCD, within a $1/N$ error,

$$\frac{M_-^S}{M_+^S} = 1 + \mathcal{O}(1/N)\,, \tag{9}$$

where M_-^S is a color-singlet bosonic degree of freedom with spin S and *odd* parity and M_+^S is a color-singlet bosonic degree of freedom with spin S and *even* parity. In particular the η' and the σ mesons should be approximately degenerate. This prediction was supported by lattice QCD analyses, see [15].

Another prediction is the value of the quark condensate in one-flavor QCD. The analysis carried out in [16] was recently tested in a lattice simulation by DeGrand et al. [21]. A comment on this issue is in order here. It is convenient to deal with a renormalization group invariant definition of the gluino condensate and the quark condensate,

$$\langle \bar{\Psi}\Psi \rangle_{\rm RGI} \equiv (g^2)^{\gamma/\beta} \langle \bar{\Psi}\Psi \rangle \, . \tag{10}$$

The renormalization group invariant value of the gluino condensate is

$$\langle \lambda\lambda \rangle = -\frac{N^2}{2\pi^2} \Lambda^3 \, . \tag{11}$$

Nonperturbative planar equivalence implies the equality of the orientifold quark condensate and the gluino condensate at infinite N. Moreover, since we know that for $N = 2$ the antisymmetric representation is equivalent to a color-singlet, we can make an educated guess that the value of the quark condensate at any N is

$$\langle \bar{\Psi}\Psi \rangle = -\left(1 - \frac{2}{N}\right) \frac{N^2}{2\pi^2} \Lambda^3 \, . \tag{12}$$

The evaluation of the quark condensate for $N = 3$ (one-flavor QCD) at 2 GeV (assuming the 't Hooft coupling is 0.115) yields

$$\langle \bar{q}q \rangle_{2\ \rm GeV}^{\rm orientifold} = -(262\ {\rm MeV})^3\ \pm 30\% \, . \tag{13}$$

This value can be compared with a recent lattice evaluation by DeGrand et al. [21]

$$\langle \bar{q}q \rangle_{2\ \rm GeV}^{\rm lattice} = -(269(9)\ {\rm MeV})^3 \, . \tag{14}$$

The agreement is more than satisfactory.

4 Applications for Three-flavor QCD

Is it possible to use planar equivalence to calculate nonperturbative quantities in real three-flavor QCD? In a bid to answer this question positively, a "mixed" approach has been suggested.

Consider an SU(N) gauge theory with one Dirac fermion Ψ in the antisymmetric representation and two extra Dirac fermions χ^i in the fundamental representation. For SU(3) this model reduces to three-flavor QCD. When $N \to \infty$ the fundamental flavors can be neglected and our model is planar equivalent to $\mathcal{N} = 1$ SYM theory. Thus, the model at hand interpolates between QCD for SU(3) and SYM theory at large N.

Several subtleties arise while considering this model. Because of a chiral symmetry breaking Goldstone bosons occur in this model, at any finite N. Therefore, in the attempt to match quantities of this theory and $\mathcal{N}=1$ SYM theory, one has to choose sources which do not couple to these Goldstone particles.

A detailed analysis of the model [22] leads to the estimate

$$\langle \bar{\Psi}\Psi \rangle_{\rm RGI}/\Lambda^3 = -\left(1-\frac{2}{N}\right)\frac{N^2}{2\pi^2}\,, \tag{15}$$

just as in the previous case. Note, however, that in this model $\beta_0 = 3N$ (as in SYM theory), and, as a result, the running coupling is different than in one-flavor QCD. As a result, we find, instead of (13),

$$\langle \bar{q}q \rangle^{\rm orientifold}_{2\,\rm GeV} \quad = \quad -\,(317\ \pm\ 30\ \pm\ 36\ {\rm MeV})^3\,. \tag{16}$$

The errors here are due to the 30% uncertainty of the $1/N$ formula and the experimental uncertainty in the 't Hooft coupling at 2 GeV. The above prediction should be compared with a recent lattice analysis by McNeile [23]

$$\langle \bar{q}q \rangle^{\rm lattice}_{2\,\rm GeV} \quad = \quad -\,(259\ \pm\ 27\ {\rm MeV})^3\,. \tag{17}$$

The orientifold prediction and the lattice simulation result are confronted in Fig. 4.

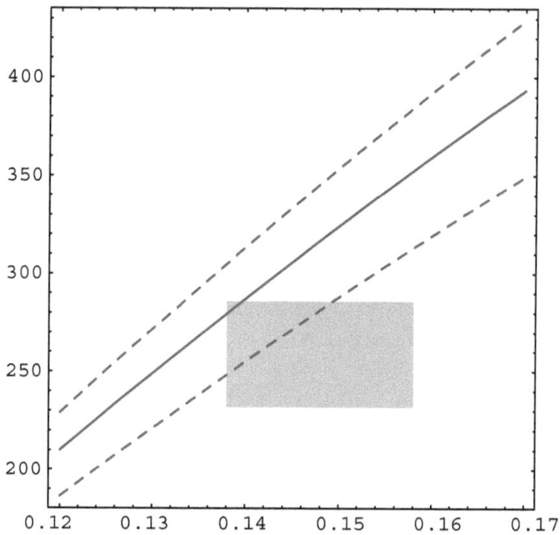

Fig. 4. The quark condensate expressed as $-(y\ {\rm MeV})^3$ as a function of the 't Hooft coupling λ. The solid line represents the prediction of planar equivalence. The two dashed lines represent the $\pm 30\%$ error. The $\pm 1\sigma$ range of the coupling, $0.138 < \lambda < 0.158$ and the lattice estimate $-(259 \pm 27\ {\rm MeV})^3$ define the shaded region

5 Sagnotti's Model and the Gauge/String Correspondence

Orientifold field theories originate in string theory. The starting point is 10D type-0B string theory. By adding the orientifold

$$\Omega' \equiv \Omega(-1)^{f_R}$$

and 32 D9 branes we end up with a nonsupersymmetric nontachyonic string theory [24, 25]. The low-energy spectrum of the closed string modes consists of the dilaton, the graviton, and a set of the Ramond–Ramond (RR) fields. There are no fermions (the Neveu–Schwarz–Ramond sector). The open string sector consists of a 10-dimensional $U(32)$ gauge theory with an antisymmetric fermion. The model is free of RR tadpoles.

In order to obtain a realization of the 4D orientifold field theory one can use a Hanany–Witten brane configuration in type 0A, namely a set of N D4 branes and $O'4$ plane suspended between rotated NS5 branes [8]. An alternative realization [26] is via fractional D3 branes placed on a $C^3/Z_2 \times Z_2$ orbifold singularity in type $0'B$. The latter description is useful for the gauge/gravity correspondence [27]. Since at $g_{\rm st} = 0$ the bosonic gravity modes of type $0'B$ and their interactions are identical to those of type IIB, the gauge/gravity correspondence (provided that it holds) provides an additional evidence in favor of planar equivalence: if the bosonic sectors of two gauge theories are described by the same bosonic sectors of two string theories at $g_{st} = 0$ then the two gauge theories must be equivalent at infinite N.

The gauge/gravity correspondence for the orientifold field theories was used recently [27] to make predictions regarding the theories at finite N. In contrast to the supersymmetric type IIB background which contains N units of the RR flux, the type-0B background contains $N - 2$ units of the RR flux, due to the presence of the $O'5$ plane that shifts the flux by -2. Certain quantities are sensitive to this shift. This is in agreement with results from the effective action approach presented in [28].

6 Charge Conjugation and the Validity of Planar Equivalence

Recently it was pointed out [29] that a necessary and *sufficient* condition for orientifold planar equivalence to hold is the absence of the spontaneous breaking of charge conjugation symmetry (for earlier work related to planar equivalence between SYM theories and *orbifold* daughters, see [30]). This assumption was implicit in our refined proof [9]. It is clear that this issue deserves a separate discussion in Yang–Mills theory per se, not necessarily in association with supersymmetry or planar equivalence.

Motivated by [29] we argued [31] that C parity does not break spontaneously in any vector-like gauge theory on R^4. We first argued that charge conjugation is not broken in pure Yang–Mills theory. Our reasoning is based on the uniqueness of the Yang–Mills vacuum. Being physically compelling our arguments, unfortunately, stop short of a rigorous mathematical proof of the type given in [32] regarding P parity. There is a deep distinction between these two aspects of QCD. While the spatial parity conservation is essentially nondynamical and is based on a general feature of vector-like gauge theories with spinor quarks, the C-parity conservation versus nonconservation is a dynamical question. The uniqueness of the Yang–Mills vacuum provides us with the necessary dynamical information.

Then we prove [31] that if the charge conjugation is unbroken in pure Yang–Mills it is not broken in any vector-like theory.

The above arguments are general and apply to QCD as well as to any other vector-like theory. The absence of the spontaneous breaking of C parity is sufficient for planar equivalence to be valid. It is instructive to return to the proof [9] and heck where exactly we assume charge conjugation to hold.

In fact, as was noted in Sect. 1, we need to assume the expectation values of traces of all Wilson loops to coincide with those of their conjugated, being evaluated in the pure Yang–Mills vacuum. This requires unbroken C parity of pure Yang–Mills theory. Once it is established, it automatically covers the theories with vector-like quarks provided that the expansion in quark loops is convergent.

7 Other Developments

Planar equivalence was used in both formal works and in phenomenology. Papers on the subject appeared on all theoretical high-energy archives: hep-th, hep-ph, and hep-lat. Here we would like to mention a few.

The lattice works are mainly devoted to verification of planar equivalence. A formal strong coupling and large mass proof was given by Patella [33]. The paper by DeGrand et al. [21] confirms our prediction for the quark condensate in one-flavor QCD. The prediction regarding the mass ratio $m_{\eta'}^2/m_\sigma^2$ was confirmed by Keith-Hynes and Thacker [15].

Phenomenological papers, mainly by Sannino and collaborators [34, 35] were devoted to constructing of technicolor models based on the orientifold field theories with symmetric matter. In another recent work [36], predictions about one-flavor QCD were used for "beyond the standard model phenomenology."

Among the more formal aspects, it is worth mentioning the work by Di Vecchia et al. [26] who studied realizations of the orientifold field theories in type-0' string theory as well as tree level string amplitudes in these models.

A partial list of other related works is given in [37, 38, 39, 40, 41].

Summarizing, planar equivalence is a new useful tool in a very limited toolkit available at present for calculations of nonperturbative quantities in

QCD. It has already resulted in a few promising applications, both in QCD, string theory, AdS/CFT, lattice gauge theory and beyond the standard model phenomenology. We believe that further studies are needed in order to exploit the potential of this method. In particular, it seems promising to search for new planar-equivalent pairs with the aim of learning about one of them from the other.

Acknowledgments

We are happy to thank Gabriele Veneziano for a fruitful and enjoyable collaboration. We are grateful to Courtney Davis for the kind permission to use her cartoon of Gabriele Veneziano from *(M)agazine*^{one} 2006.

A.A. is supported by the PPARC advanced fellowship award. The work of M.S. is supported in part by DOE grant DE-FG02-94ER408.

References

1. M. Shifman: Int. J. Mod. Phys. A **21**, 5695 (2006)
2. M. A. Shifman, A. I. Vainshtein: Nucl. Phys. B **296**, 445 (1988)
3. J. M. Maldacena: Adv. Theor. Math. Phys. **2**, 231 (1998); also S. S. Gubser, I. R. Klebanov, A. M. Polyakov: Phys. Lett. B **428**, 105 (1998); E. Witten: Adv. Theor. Math. Phys. **2**, 253 (1998)
4. S. Kachru, E. Silverstein: Phys. Rev. Lett. **80**, 4855 (1998)
5. A. Armoni, E. Lopez, A. M. Uranga: JHEP **0302**, 020 (2003)
6. A. Dymarsky, I. R. Klebanov, R. Roiban: JHEP **0511**, 038 (2005)
7. M. J. Strassler: *On Methods for Extracting Exact Nonperturbative Results in Nonsupersymmetric Gauge Theories,* hep-th/0104032
8. A. Armoni, M. Shifman, G. Veneziano: Nucl. Phys. B **667**, 170 (2003)
9. A. Armoni, M. Shifman, G. Veneziano: Phys. Rev. D **71**, 045015 (2005)
10. A. Armoni, M. Shifman, G. Veneziano: in *From Fields to Strings: Circumnavigating Theoretical Physics*, ed. by M. Shifman, A. Vainshtein, J. Wheater (World Scientific, Singapore, 2005), Vol. 1, p. 353
11. R. Casalbuoni, J. Gomis, G. Longhi: Nuovo Cimento A **24** 249 (1974); R. Casalbuoni: Nuovo Cimento A **33**, 389 (1976); A. Barducci, F. Bordi, R. Casalbuoni: Nuovo Cimento B **64**, 287 (1981); L. Brink, P. Di Vecchia, P. S. Howe: Nucl. Phys. B **118**, 76 (1977); M. J. Strassler: Nucl. Phys. B **385**, 145 (1992); E. D'Hoker, D. G. Gagne: Nucl. Phys. B **467**, 272 (1996)
12. G. 't Hooft: Nucl. Phys. B **72**, 461 (1974)
13. G. Veneziano: Nucl. Phys. B **117**, 519 (1976)
14. A. Armoni, M. Shifman, G. Veneziano: Phys. Rev. Lett. **91**, 191601 (2003)
15. P. Keith-Hynes, H. B. Thacker: *Double Hairpin Diagrams and the Planar Equivalence of $\mathcal{N} = 1$ Supersymmetric Yang–Mills Theory and One-Flavor QCD,* hep-lat/0610045; *Relics of Supersymmetry in Ordinary One-Flavor QCD: Hairpin Diagrams and Scalar-Pseudoscalar Degeneracy,* hep-th/0701136
16. A. Armoni, M. Shifman, G. Veneziano: Phys. Lett. B **579**, 384 (2004)

17. E. Witten: Nucl. Phys. B **223**, 422 (1983); Nucl. Phys. B **223**, 433 (1983) [reprinted in S. Treiman et al., *Current Algebra and Anomalies* (Princeton University Press, Princeton, 1985), p. 515]
18. S. Bolognesi: *Baryons and Skyrmions in QCD with Quarks in Higher Representations,* hep-th/0605065
19. A. Cherman, T. D. Cohen: JHEP **0612**, 035 (2006); Phys. Lett. B **641**, 401 (2006)
20. R. Auzzi, M. Shifman: *Low-Energy Limit of Yang–Mills with Massless Adjoint Quarks: Chiral Lagrangian and Skyrmions,* hep-th/0612211; S. Bolognesi, M. Shifman: *The Hopf Skyrmion in QCD with Adjoint Quarks,* hep-th/0701065
21. T. DeGrand, R. Hoffmann, S. Schaefer, Z. Liu: *Quark Condensate in One-Flavor QCD,* hep-th/0605147
22. A. Armoni, G. Shore, G. Veneziano: Nucl. Phys. B **740**, 23 (2006)
23. C. McNeile: Phys. Lett. B **619**, 124 (2005)
24. A. Sagnotti: *Some Properties of Open String Theories,* hep-th/9509080.
25. A. Sagnotti: Nucl. Phys. Proc. Suppl. **56B**, 332 (1997)
26. P. Di Vecchia, A. Liccardo, R. Marotta, F. Pezzella: JHEP **0409**, 050 (2004)
27. A. Armoni, E. Imeroni: Phys. Lett. B **631**, 192 (2005)
28. F. Sannino, M. Shifman: Phys. Rev. D **69**, 125004 (2004)
29. M. Ünsal, L. G. Yaffe: Phys. Rev. D **74**, 105019 (2006)
30. P. Kovtun, M. Ünsal, L. G. Yaffe: JHEP **0312**, 034 (2003); JHEP **0507**, 008 (2005)
31. A. Armoni, M. Shifman, G. Veneziano: *A note on C-Parity Conservation and the Validity of Orientifold Planar Equivalence,* hep-th/0701229
32. C. Vafa, E. Witten: Phys. Rev. Lett. **53**, 535 (1984).
33. A. Patella: *A Proof of Orientifold Planar Equivalence on the Lattice,* hep-lat/0511037
34. F. Sannino, K. Tuominen: Phys. Rev. D **71**, 051901 (2005)
35. D. K. Hong, S. D. H. Hsu, F. Sannino: Phys. Lett. B **597**, 89 (2004)
36. M. J. Strassler, K. M. Zurek: *Echoes of a Hidden Valley at Hadron Colliders,* hep-ph/0604261
37. A. Feo, P. Merlatti, F. Sannino: Phys. Rev. D **70**, 096004 (2004)
38. J. L. F. Barbon, C. Hoyos: JHEP **0601**, 114 (2006)
39. G. Veneziano, J. Wosiek: JHEP **0601**, 156 (2006)
40. T. DeGrand, R. Hoffmann, *QCD with One Compact Spatial Dimension,* hep-lat/0612012
41. T. J. Hollowood, A. Naqvi: *Phase Transitions of Orientifold Gauge Theories at Large N in Finite Volume,* hep-th/0609203

Part V

Supersymmetric Gauge Theories

Instantons and Supersymmetry

M. Bianchi[1], S. Kovacs[2], and G. Rossi[3]

[1] University of Rome Tor Vergata and INFN, Sez. di Roma Tor Vergata, Via della Ricerca Scientifica-00133 Roma, Italy
Massimo.Bianchi@roma2.infn.it
[2] Trinity College Dublin Dublin 2, Ireland
kovacs@maths.tcd.ie
[3] University of Rome Tor Vergata and INFN, Sez. di Roma Tor Vergata, Via della Ricerca Scientifica-00133 Roma, Italy
giancarlo.rossi@roma2.infn.it

Abstract. The role of instantons in describing non-perturbative aspects of globally supersymmetric gauge theories is reviewed. The cases of theories with $\mathcal{N}=1$, $\mathcal{N}=2$ and $\mathcal{N}=4$ supersymmetry are discussed. Special attention is devoted to the intriguing relation between instanton solutions in field theory and branes in string theory.

1 Introduction

In this review we discuss the role of instantons in describing non-perturbative effects in globally supersymmetric gauge theories[1].

Instantons (anti-instantons) are non-trivial self-dual (anti-self-dual) solutions of the equations of motion of the pure non-abelian Yang–Mills theory when the latter is formulated in a compactified S_4 Euclidean manifold. Instanton solutions are characterised by a topological charge (the Pontrjagin number, K) which takes integer values, $K \in$. The integer K represents the number of times the (sub)group $SU(2)$ of the gauge group is wrapped by the classical solution, while its space–time location spans the S_3-sphere at infinity.

The presentation of the material of this review can be naturally split into three parts according to the number of supersymmetries endowed by the theory. In the first part (Sects. 2–5) we discuss (weak and strong coupling) computations of instanton-dominated correlators in pure $\mathcal{N}=1$ super Yang–Mills (SYM) and in some super QCD-like (SQCD) and chiral extensions of it. A careful analysis of the latter case shows that with a suitable choice of the chiral matter flavour representations the interesting phenomenon of dynamical

[1] See the contribution by G. Shore to this book for applications to the non-supersymmetric case of QCD.

M. Bianchi et al.: *Instantons and Supersymmetry*, Lect. Notes Phys. **737**, 303–470 (2008)
DOI 10.1007/978-3-540-74233-6_14

breaking of supersymmetry occurs in the theory, as a consequence of the constraints imposed by the Konishi anomaly equation.

In the second part (Sects. 6–11) we move to the $\mathcal{N}=2$ super Yang–Mills case. We will show how the highly sophisticated instanton calculus, developed in the years, is able to produce the correct coefficients that determine the exact expression of the $\mathcal{N}=2$ prepotential, derived in the famous Seiberg–Witten (SW) construction. We will also review the construction of the instanton solution in terms of branes with the purpose of illustrating the intriguing relation with string theory.

In the third part (Sects. 12–18) we discuss the role of instantons in $\mathcal{N}=4$ super Yang–Mills. Although there are no anomalous $U(1)$'s in this theory (with the consequence that there exist no chiral $U(1)$ selection rules that would limit the value of Pontrjagin number of the instanton solutions contributing to correlators, as instead happens in the $\mathcal{N}=1$ and $\mathcal{N}=2$ cases), instantons are crucial to check the validity of the Maldacena conjecture beyond the realm of perturbation theory. Furthermore, their correspondence with IIB string D-instantons gives us hope to understand the yet elusive Montonen–Olive duality between the weak and strong coupling regimes of $\mathcal{N}=4$ super Yang–Mills.

A detailed outline of this review is as follows. In Sect. 2 we start with some introductory remarks about instantons and their interpretation as field configurations interpolating between classical Euclidean vacua (quantum tunnelling) and we discuss in detail how semi-classical calculations are performed in the instanton background with special reference to the notion of collective coordinates. We also illustrate the simplifications that occur in supersymmetric theories. In Sect. 3 we derive from supersymmetry and holomorphicity the general structure of the Green functions with only insertions of lowest (highest) components of chiral (anti-chiral) superfields. We show that these Green functions do not depend on the operator insertion points and have a completely fixed dependence upon the parameters of the theory (like masses and coupling constant). We then move in Sect. 4 to the explicit semi-classical instanton computation of constant Green functions in pure SYM and in massive SQCD finding perfect agreement between the theoretical expectations spelled out in the previous section and the results of actual calculations. The main result of this analysis is that the perturbative non-renormalisation theorems of supersymmetry are violated in the semi-classical instanton approximation. The instanton calculus is then extended to encompass the more delicate cases of massless SQCD and to Georgi–Glashow-type theories with matter in suitable non-anomalous chiral representations. In the first case, certain inconsistencies are found between results obtained in the massless limit of the massive theories and what can be directly computed in the strictly massless situation. The problem is discussed in detail and the issue of "strong coupling" *vs* "weak coupling" instanton calculation strategy is addressed. In the second case, conflicting results with the constraints imposed by the Konishi anomaly equation lead to the conclusion that in certain supersymmetric chiral theories supersymmetry is dynamically broken by non-perturbative instanton effects.

In Sect. 5 we give the expression of the effective action for all the cases for which we have obtained results in the semi-classical instanton approximation. A nice agreement between these two approaches is found, which gives support to instanton-based computations.

We then pass to discuss instanton effects in $\mathcal{N} = 2$ SYM theories. After the introduction to the subject contained in Sect. 6, we present some general discussion of their properties in Sect. 7. We start by recalling their supermultiplet content. We then describe the coupling of vector multiplets to hypermultiplets and the structure of the classical and effective actions. In Sect. 8 we review the celebrated analysis of Seiberg and Witten and the derivation of the analytic prepotential in the case of pure $\mathcal{N} = 2$ SYM with $SU(2)$ gauge group. $\mathcal{N} = 2$ instanton calculus is argued to provide a powerful check of the SW prepotential in Sect. 9. In Sect. 9.1 we describe Matone's non-linear recursion relations for the expansion coefficients. The validity of the recursion relations and thus of the expression of the analytic prepotential itself is checked against instanton calculations for winding numbers $K = 1$ and $K = 2$ in Sect. 9.2. In order to go beyond these two cases, we follow the strategy advocated by Nekrasov and collaborators which is based on the possibility of topologically twisting $\mathcal{N} = 2$ SYM theories and turning on a non-commutative deformation that localises the integration over instanton moduli spaces. After reviewing the strategy in Sect. 10, we describe how to couple hypermultiplets in Sect. 10.1. We then sketch the mathematical arguments that lead to the localisation of the measure in Sect. 10.2 and the computation of the residues that allows a non-perturbative check of the correctness of the SW prepotential for arbitrary winding number in Sect. 10.3. In Sect. 11 we change gear and exploit the Veneziano model and D-branes in order to embed (supersymmetric) YM theories in string theory. In particular, we outline the emergence of the Atiyah–Drinfeld–Hitchin–Manin (ADHM) data and the ADHM equations as a result of the introduction of lower dimensional D-branes in a given configuration with maximal $\mathcal{N} = 4$ supersymmetry. Finally, we describe in Sect. 11.1 the truncation to $\mathcal{N} = 2$ supersymmetry and the derivation of the SW prepotential within this framework in Sect. 11.2.

In the final part of the review, starting in Sect. 12, we discuss instanton effects in $\mathcal{N} = 4$ SYM, focussing in particular on the role of instantons in the context of the anti-de Sitter space/conformal field theory (AdS/CFT) correspondence. $\mathcal{N} = 4$ SYM is the maximally extended (rigid) supersymmetric theory in four dimensions and is believed to be exactly conformally invariant at the quantum level. The main properties of the model are reviewed in Sect. 13. We give explicitly the form of the action and the supersymmetry transformations and we discuss the basic implications of conformal invariance on the physics of the theory, highlighting some of the features which make it special compared to the $\mathcal{N} = 1$ and $\mathcal{N} = 2$ theories considered in previous sections. General aspects of instanton calculus in $\mathcal{N} = 4$ SYM are presented in Sect. 14. We describe the general strategy for the calculation of instanton contributions to correlation functions of gauge-invariant composite operators in the semi-classical approximation, emphasising again the essential differences with respect to the $\mathcal{N} = 1$ and $\mathcal{N} = 2$ cases. In Sect. 15 we focus on the case of the $SU(N_c)$ gauge group, which is relevant for the AdS/CFT duality, and we discuss in detail the calculation of correlation functions in the one-instanton sector. We first construct a generating function which facilitates a systematic analysis of instanton contributions to gauge-invariant correlators and we then present some explicit examples. The generalisation of these results to multi-instanton sectors in the large-N_c limit is briefly outlined in Sect. 16. At this point we change somewhat perspective and we move to a discussion of the remarkable gauge/gravity duality conjectured by Maldacena, explaining how instanton calculus allows to test its validity beyond perturbation theory. In Sect. 17 we recall the basic aspects of the duality which relates $\mathcal{N} = 4$ SYM to type IIB superstring theory in an $\mathrm{AdS}_5 \times S^5$ background. Instanton effects in $\mathcal{N} = 4$ SYM are in correspondence with the effects of D-instantons in string theory. More precisely instanton contributions to correlators in $\mathcal{N} = 4$ SYM are related to D-instanton-induced scattering amplitudes in the $\mathrm{AdS}_5 \times S^5$ string theory. In Sects. 18.1 and 18.2 we present the calculation of D-instanton contributions to the string amplitudes dual to the SYM correlation functions studied in Sect. 15. The remarkable agreement between gauge and string theory calculations provides a rather stringent test of the conjectured duality. Finally in Sect. 18.3 we review the role of instantons in a particularly interesting limit of the AdS/CFT correspondence, the so-called BMN limit, in which the gravity side of the correspondence is under a better quantitative control beyond the low-energy supergravity approximation. We show how instanton effects provide again important insights into the non-perturbative features of the duality.

Our notation and various technical details are discussed in a number of appendices.

Given the pedagogic nature of this review we refrain from drawing any conclusion or present any speculation. In Sect. 19 we try, instead, to summarise the crucial contributions given by Gabriele to the subject both as the father of open string theory and as one of the deepest and most original investigators of the non-perturbative aspects of gauge theories. We thus simply list a few lines of research activity where Gabriele's profound insight was precious to put existing problems in the correct perspective and help in solving them.

2 Generalities about Instantons

Instantons (anti-instantons) are self-dual ($F_{\mu\nu} = \tilde{F}_{\mu\nu}$, $\tilde{F}_{\mu\nu} = \frac{1}{2}\epsilon_{\mu\nu\rho\sigma}F_{\rho\sigma}$) (anti-self-dual, $F_{\mu\nu} = -\tilde{F}_{\mu\nu}$) solutions of the classical non-abelian Yang–Mills (YM) Euclidean equations of motion (e.o.m.) [1, 2][2]. They are classified by a topological number, the Pontrjagin (or winding) number, $K \in$, which

[2] There are very many good reviews on the subject of instantons and their role in field theory. Some are listed in [3, 4, 5, 6].

represents the number of times the (sub)group $SU(2)$ of the gauge group is wrapped by the classical solution, $A_\mu^{aI}(x)$, when x spans the S_3-sphere at the infinity of the compactified S_4 Euclidean space–time[3]. Homotopy theory shows, in fact, that the homotopically inequivalent mappings $S_3 \to SU(2)$ are classified by integers since $\Pi_3(SU(2)) \sim \Pi_3(S_3) =$ [7].

2.1 The Geometry of Instantons

In the Feynman gauge ($\partial_\mu A_\mu^a = 0$) the explicit expression of the gauge instanton with winding number $K = 1$ for the $SU(2)$ gauge group (to which case we now restrict) is

$$A_\mu^{aI} = \frac{2}{g}\bar{\eta}_{\mu\nu}^a \frac{(x-x_0)_\nu}{(x-x_0)^2} \frac{\rho^2}{(x-x_0)^2+\rho^2}\,, \tag{1}$$

where $\bar{\eta}_{\mu\nu}^a$ are the 't Hooft symbols [2][4]. In (1) x_0 and ρ are the so-called location and size of the instanton, respectively. They are not fixed by the YM classical e.o.m., neither is the orientation of the instanton gauge field in colour space. Consequently, the derivatives of the instanton solution with respect to each one of these parameters (collective coordinates [8]) will give rise to zero modes of the operator associated with the quadratic fluctuations of the gauge field in the instanton background [2, 9, 10] (see Appendix B for details).

The winding number of a gauge configuration can be expressed in terms of the associated field strength through the (gauge invariant) formula

$$K = \frac{g^2}{32\pi^2}\int \mathrm{d}^4x\, F_{\mu\nu}^a \tilde{F}_{\mu\nu}^a\,. \tag{2}$$

For the action of a self-dual (or anti-self-dual) instanton configuration one then gets

$$S^I = \frac{8\pi^2}{g^2}|K|\,. \tag{3}$$

The topological nature of (1) can be better enlightened by first recasting it in the form ($y \equiv x - x_0$)

$$A_\mu^I = \frac{i}{g}\frac{\rho^2}{y^2+\rho^2}\left[\Omega^{(1)\dagger}\partial_\mu \Omega^{(1)}\right](y)\,, \tag{4}$$

---

[3] In the following adjoint gauge (colour) indices will be indicated with early Latin letters, $a, b, c, \ldots$, and vector indices by middle Latin letters, $i, j, k, \ldots$. Thus for an $SU(N_c)$ gauge group we will have $a, b, c, \ldots = 1, 2, \ldots, N_c^2 - 1$ and $i, j, k, \ldots = 1, 2, \ldots, N_c$. Further notations are summarised in Appendix A.

[4] They relate the generators of one of the two $SU(2)$ groups, in which the Euclidean Lorentz group, $SO(4)$, can be decomposed ($SO(4) \sim SU_L(2) \times SU_R(2)$), to the generators of the latter through the formula $\Sigma_{\mu\nu}^L = \frac{1}{2}\bar{\eta}_{\mu\nu}^a \sigma_a$ with σ_a the Pauli matrices. The similar coefficients for other $SU(2)$ group are the $\eta_{\mu\nu}^a$ symbols with $\Sigma_{\mu\nu}^R = \frac{1}{2}\eta_{\mu\nu}^a \sigma_a$.

where (see (A.14))

$$\left[\Omega^{(1)\dagger}\partial_\mu\Omega^{(1)}\right](x) = -\bar{\sigma}_{\mu\nu}\frac{x_\nu}{\sqrt{x^2}} \,. \tag{5}$$

In the previous equations

$$\Omega^{(1)}(x) = \sigma_\mu \frac{x_\mu}{\sqrt{x^2}} \tag{6}$$

is a topologically non-trivial $SU(2)$ gauge transformation, since it does not tend to the group identity as $\sqrt{x^2}$ tends to infinity. To compute the winding number of the gauge configuration (4), it is convenient to gauge transform it by the transformation $\Omega^{(1)}$ itself. One gets in this way

$$\begin{aligned}(A^I_\mu)^{N.S.} &= \left[A^I_\mu\right]^{\Omega^{(1)}} = \frac{i}{g}\frac{y^2}{y^2+\rho^2}\left[\Omega^{(1)}\partial_\mu\Omega^{(1)\dagger}\right](y) \\ &= -\frac{i}{g}\sigma_{\mu\nu}\frac{y_\nu}{\sqrt{y^2}}\frac{y^2}{y^2+\rho^2} \,. \end{aligned} \tag{7}$$

From the second equality we see that $(A^I_\mu)^{N.S.}$ tends at infinity to a non-trivial pure gauge. Inserting (7) into (2), one gets the expected result, $K=1$.

The form (1) (or (4)) of the one-instanton field is called "singular" because the point where the non-vanishing contribution to the action integral comes from is at $x = x_0$, unlike the "non-singular" form (7) in which this point has been brought to infinity.

We recall in this context that the $F^a_{\mu\nu}\tilde{F}^a_{\mu\nu}$ density can be locally rewritten as the divergence of a gauge non-invariant vector through the formula

$$F^a_{\mu\nu}\tilde{F}^a_{\mu\nu} = 2\partial_\mu K_\mu \,, \qquad K_\mu = \epsilon_{\mu\nu\rho\sigma}\mathrm{Tr}\left[A_\nu F_{\rho\sigma} + \frac{2ig}{3}A_\nu A_\rho A_\sigma\right] . \tag{8}$$

Thus (2) can be written

$$K = \frac{g^2}{16\pi^2}\int_{\mathrm{S}_3^\infty} \mathrm{d}S_\mu K_\mu \,, \tag{9}$$

where S_3^∞ is the three-sphere at infinity in S_4. Since, as we noticed above, $(A^I_\mu)^{N.S.}$ tends to a pure gauge at infinity and hence its field strength vanishes, (9) can be cast in the very expressive form

$$K = \frac{1}{24\pi^2}\int_{\mathrm{S}_3^\infty} \mathrm{d}S_\mu \epsilon_{\mu\nu\rho\sigma}\mathrm{Tr}\left[\Omega^\dagger\partial_\nu\,\Omega\Omega^\dagger\partial_\rho\Omega\,\Omega^\dagger\partial_\sigma\Omega\right] , \tag{10}$$

in which we recognise the Cartan–Maurer formula. In general terms this quantity is an integer, which represents the winding number of the $SU(2)$ gauge transformation, Ω[5].

[5] It can be explicitly proved that, setting $\Omega_h = \exp[iT^a h^a]$, K is invariant under the infinitesimal deformations $h^a \to h^a + \delta h^a$. Thus K only depends on the homotopy class to which Ω_h belongs and can always be normalised so as to be an integer.

2.2 Quantum Tunnelling

The existence of instanton solutions in YM can be interpreted as an indication of quantum tunnelling between different vacua, the latter being pure gauge configurations characterised by their winding numbers [2, 11, 12]. This fact can be illustrated in quite a number of ways. An easy, but heuristic argument is described below. A more sophisticated analysis is presented in Appendix C.

Consider again the asymptotic formula (9), in which, however, the closed surface S_3^∞ has been (smoothly) deformed to an other closed surface, which we take as a hyper-cylinder of length T, bounded at $t = -T/2$ and $t = T/2$ by three-dimensional compact spatial manifolds, S_3. In the limit $T \to \infty$ (9) can then be rewritten as the sum of three contributions

$$K = \frac{ig^3}{24\pi^2} \lim_{T\to\infty} \Big(\int_{S_3} \mathrm{d}^3x\, \epsilon_{4ijk} \mathrm{Tr}\big[A_i A_j A_k\big]\big|_{t=T/2} \\ - \int_{S_3} \mathrm{d}^3x\, \epsilon_{4ijk} \mathrm{Tr}\big[A_i A_j A_k\big]\big|_{t=-T/2} \\ + \int_{-T/2}^{T/2} \int_{S_L} \mathrm{d}S_i\, \epsilon_{i\nu\rho\sigma} \mathrm{Tr}\big[A_\nu A_\rho A_\sigma\big]\Big) , \tag{11}$$

where S_L is the three-dimensional lateral surface of the cylinder. It can be proved [13] that one can always find a gauge where (1) $A_0 = 0$ on the lateral surface S_L and (2) the (time-independent) gauge transformations $\Omega_\pm(\boldsymbol{x})$ at $t = \pm T/2$ are such that at large $|\boldsymbol{x}|$ they are independent on the direction $\boldsymbol{x}/|\boldsymbol{x}|$. Under these conditions the third term in the r.h.s. of (11) vanishes. The other two terms take integer values because they represent the winding number of the mapping $\boldsymbol{x} \to \Omega_\pm(\boldsymbol{x})$ from the (manifold ^{3} compactified to the) S_3 sphere onto $SU(2)$[6]. The net result of these considerations is that (11) can be cast in the form

$$K = n_+ - n_- , \tag{12}$$

$$n_\pm = -\frac{1}{24\pi^2} \epsilon_{ijk} \int_{S_3} \mathrm{d}^3x\, \mathrm{Tr}\big[\Omega_\pm^\dagger \partial_i \Omega_\pm\, \Omega_\pm^\dagger \partial_j \Omega_\pm\, \Omega_\pm^\dagger \partial_k \Omega_\pm(\boldsymbol{x})\big] , \tag{13}$$

which shows that the instanton solution ($K \neq 0$) interpolates between vacuum states (pure gauge configurations) with different winding numbers.

We refer the reader to Appendix C for a more rigorous discussion of instanton-induced tunnelling effects in a YM theory.

---

[6] As an example of such gauge transformations one can take $\Omega(\boldsymbol{x}) = \exp\big[i\pi\boldsymbol{\sigma}\cdot\boldsymbol{x}/\sqrt{\boldsymbol{x}^2+1}\big]$, in which all the point at large $|\boldsymbol{x}|$ are mapped into the group element $-$. In this way the three-dimensional space manifolds at $t = \pm T/2$ become topologically equivalent to S_3.

2.3 Introducing Fermions

In this subsection we want to briefly recall some elementary facts on how to deal with fermions in the functional language in general and in the semi-classical approximation in particular.

Fermionic Functional Integration

When fermions are introduced it is necessary to define integration rules for Grassmann variables. This is a beautiful piece of mathematics of which a simple account can be found in [14]. The well-known results of this analysis can be summarised as follows.

(1) The functional integration over the degrees of freedom of a Dirac fermion belonging to the representation $\mathbf{R}$ of the gauge group has the effect of adding to the gauge action a contribution which is formally given by

$$\log\left\{\int \mathcal{D}\mu[\mathbf{R}](\psi,\bar\psi)\exp\left[\int \mathrm{d}^4x\,(\bar\psi i\gamma_\mu D_\mu[\mathbf{R}]\psi)(x)\right]\right\}$$
$$= \log\left\{\det\left[i\gamma_\mu D_\mu[\mathbf{R}]\right]\right\} = \mathrm{Tr}\log\left[i\gamma_\mu D_\mu[\mathbf{R}]\right], \tag{14}$$

where

$$iD_\mu[\mathbf{R}]\gamma_\mu = iD_\mu[\mathbf{R}]\begin{pmatrix} 0 & \sigma_\mu \\ \bar\sigma_\mu & 0 \end{pmatrix} \tag{15}$$

and the matrices σ_μ and $\bar\sigma_\mu$ are defined in Appendix C.

(2) For a Weyl fermion, which has half the degrees of freedom of a Dirac fermion (cf. (15) and (A.12)), there is the subtlety that the Dirac–Weyl operator maps dotted indices into undotted ones, thus making problematic the definition of a determinant for such an operator. In the literature many prescriptions have been proposed to address this issue in a rigorous way (see [15] and works quoted therein). Actually this difficulty is not relevant in practice, because one can always imagine to factor out the free operator and compute the determinant of the resulting operator which is perfectly well defined [16]. The contribution from the free part is obviously irrelevant in the computation of Green functions as it will cancel with an identical contribution from the normalisation factor (see (A.1)). Loosely speaking, looking at (15) and (A.12), we may say that ceteris paribus the contribution of a Weyl fermion to the functional integral is the "square root" of that of a Dirac fermion.

Fermionic Zero Modes

In computing the fermionic functional integral one is led to consider the decomposition of the associated spinor fields in eigenstates of the fermionic kinetic operator. As is well known, the existence of zero modes in certain background gauge fields, such as instantons, is of particular relevance for non-perturbative calculations both in ordinary and supersymmetric field theories [2, 4, 5].

The number of zero modes of the Dirac operator in an external field is controlled by the famous Atiyah–Singer index theorem [17]. The theorem states that "the index of the Dirac operator (15), i.e. the difference between the number of left-handed (n_L) and right-handed (n_R) zero modes, is equal to twice the Dynkin index of the representation **R** times the Pontrjagin number of the background gauge field". In formulae we write

$$\mathrm{ind}(D_\mu[\mathbf{R}]\gamma_\mu) \equiv n_L - n_R = 2\ell[\mathbf{R}]K\,, \tag{16}$$

where $\ell[\mathbf{R}]$ is the Dynkin index of the representation **R**. Let us now consider some interesting applications of this theorem.

1. Weyl fermion in the adjoint representation, **Adj**, of the gauge group. We must distinguish between the left-handed ($D_\mu[\mathbf{Adj}]\bar{\sigma}_\mu$) and the right-handed ($D_\mu[\mathbf{Adj}]\sigma_\mu$) Weyl operator. In the first case $n_R = 0$ and the formula (16) becomes

$$\mathrm{ind}(D_\mu[\mathbf{Adj}]\bar{\sigma}_\mu) = n_L = 2N_c K\,, \tag{17}$$

because $\ell[\mathbf{Adj}] = N_c$. Since obviously n_L is a non-negative number, there can exist zero modes of the left-handed Weyl operator only if the classical background instanton field has positive winding number, $K > 0$. Similarly for the right-handed Weyl operator one gets

$$\mathrm{ind}(D_\mu[\mathbf{Adj}]\sigma_\mu) = -n_R = 2N_c K\,, \tag{18}$$

implying that there can be zero modes only if $K < 0$.

2. Actually the number of zero modes in the adjoint representation of any compact Lie group, G, is always given by twice the value of the quadratic Casimir operator, $2c_2(\mathbf{Adj}(G))$. This result follows from the formula [18]

$$\mathbf{Adj}(G) = 4\mathbf{Adj}(SU(2)) + n(G)\,\mathbf{2} + s(G)\,\mathbf{1}\,, \tag{19}$$

which expresses how the adjoint representation of G can be decomposed into irreducible representations of $SU(2)$. In (19) we have introduced the definitions

$$n(G) = 2(c_2(\mathbf{Adj}(G)) - 2)\,, \tag{20}$$
$$s(G) = d(\mathbf{Adj}(G)) - 4c_2(\mathbf{Adj}(G)) + 5\,, \tag{21}$$

where $d(\mathbf{Adj}(G))$ is the dimension of the adjoint representation of G. The number of zero-modes is then $4 + 2(c_2(\mathbf{Adj}(G)) - 2) = 2c_2(\mathbf{Adj}(G))$

3. Dirac fermion in the fundamental representation, $\mathbf{N_c}$, of the gauge group. Equation (16) with $\ell[\mathbf{N_c}] = 1/2$ gives

$$\mathrm{ind}(D_\mu[\mathbf{N_c}]\gamma_\mu) = n_L - n_R = K\,. \tag{22}$$

Again in the classical background instanton field there can be either left-handed fermionic zero modes, if $K > 0$, or right-handed ones, if $K < 0$.

4. Fermion in the rank two anti-symmetric representation $\mathbf{N_c(N_c-1)/2}$. The number of zero modes (of definite chirality) is $(N_c-2)K$.

Deriving the explicit expression of all these fermionic zero modes is beyond the scope of this review and we refer the interested reader to the general methods that, starting from the seminal paper of [19], have been developed in the literature [6, 20]. However, for completeness we give their explicit expression for a few cases more relevant for this review in Appendix A.

2.4 Putting Together Fermion and Boson Contributions

As we said, we are interested in computing expectation values of (multi-local) gauge-invariant operators, by dominating the functional integral with the semi-classical contributions coming from the non-trivial minima (instantons) of the Euclidean action. The obvious question is whether this computational strategy leads to a reliable estimate of $\langle O\rangle$.

The General Case

In order to prepare ourselves for this analysis, let us write down the formal result obtained by performing the integration over the quadratic fluctuations (semi-classical approximation, s.c.) around an instanton solution with winding number K. Including also the fermionic contribution in (B.17) and assuming for simplicity that there are no scalar fields in the theory[7], one gets

$$\langle O\rangle\Big|_{s.c.} = \mu^{n_B-\kappa_F n_F}\frac{e^{-\frac{8\pi^2}{g^2}|K|}}{Z|_{s.c.}}\int\prod_{j=1}^{n_F}\mathrm{d}c_j \tag{23}$$

$$\times\int\prod_{i=1}^{n_B}\mathrm{d}\beta_i\frac{||a^{(i)}||}{\sqrt{2\pi}}\frac{(\det{}'[\mathcal{M}^{g.f.}_{\mu\nu}])^{-\frac{1}{2}}\det[-D^2(A^I)]}{(\det[\mathcal{M}^{g.f.}_{0;\mu\nu}])^{-\frac{1}{2}}\det[-\partial^2]}\frac{\det{}'[\not{D}(A^I)]}{\det[\not{\partial}]}O(c;A^I)\,.$$

where $\not{D}$ is the fermionic kinetic operator appropriate to the kind of fermion one is dealing with (Dirac or Weyl) and the prime on the determinants is there to mean that obviously only non-zero eigenvalues are to be included. Further observations about this formula are the following.

• The factor κ_F is 1 for a Dirac fermion and $\frac{1}{2}$ for a Weyl fermion.

• The residual fermionic integration $\prod_{j=1}^{n_F} dc_j$ is over the Grassmannian coefficients associated with the n_F zero modes of the fermionic kinetic operator. We stress that in order not to get a trivially vanishing result, the Berezin [14] integration rules require a perfect matching in the number of fermionic zero modes between those of the fermion operators in the action and those contained in the operator O.

[7] The extension of the formulae of this section to the more general case where also elementary scalar fields are present is possible, but not completely trivial. See below and [4, 6, 21, 22, 23].

• The extra μ dependence in front of the r.h.s. of (23) (with respect to (B.17)) is due (similarly to the case of the bosonic functional integration, see Appendix B), to unmatched μ factors coming from the determinant of the fermion Pauli–Villars (PV) regulators.[8]

• The power $\kappa_F n_F$ is dictated by the way in which zero modes contribute to the fermionic mass term in the action and the nature of the Grassmannian integration rules.

• No further factor comes from dealing with the fermionic zero modes, provided they are normalised to one, which we will always do (this is at variance with what happens for bosonic zero modes, each of which contributes a factor $||\mathrm{norm}||/\sqrt{2\pi}$ to (23)).

• Generally speaking, the ratio of determinants in (23) will be a function of the instanton collective coordinates as well as μ.

The computational strategy outlined above can be safely used if it can be convincingly argued that the classical minima (instantons) really dominate the integral. This is a delicate issue which can only be settled on a case by case basis. For instance, for the instanton contribution to dominate the functional integral one can imagine considering Green functions that are zero in perturbation theory. The argument here is that otherwise the non-perturbative instanton contribution, which is proportional to $\exp(-8\pi^2|K|/g^2)$, would represent a completely negligible correction with respect to any perturbative term. This is the situation one is usually dealing with in $\mathcal{N}=1$ supersymmetric theories.

In the $\mathcal{N}=2$ and $\mathcal{N}=4$ cases it is, instead, interesting to consider more general correlators which do not necessarily vanish in perturbation theory. In these cases instanton contributions, though comparatively exponentially small, can always be "tracked" if the theory ϑ-dependence is followed (see Appendix C).

A second crucial question concerns the finiteness of the r.h.s. of (23). In his beautiful paper 't Hooft [2] has shown that in QCD the integration over the instanton collective coordinates around the classical instanton solution

$$A_\mu = A^I_\mu \,, \qquad \text{all other fields equal to zero} \tag{24}$$

does not lead to a finite result. The reason behind this fact is that the integration over the size of the instanton, ρ, which comes from the ratios of determinant in (23) as well as from the norm of the bosonic zero modes, diverges in the infrared limit, i.e. for large values of ρ (the integration near $\rho = 0$ is, instead, convergent thanks to asymptotic freedom). This problem is not present in the supersymmetric case which we discuss next.

The Supersymmetric Case

Something really surprising indeed happens in the case of a supersymmetric theory. There, irrespective of the details of the theory (number of

[8] In order to have a more readable formula we have not shown in (23) the determinants of the various PV regulators.

supersymmetries, gauge group, matter content, etc.), the whole ratio of (regularised) determinants is always exactly equal to 1 [24]. This is because the eigenvalues of the various kinetic operators in the instanton background are, up to multiplicities, essentially all equal and, due to supersymmetry, there is a perfect matching between bosonic and fermionic degrees of freedom, leading to contributions that are one the inverse of the other. The formula (23) thus becomes

$$\langle O\rangle\Big|_{s.c.}^{\mathrm{SUSY}} = \mu^{n_B-\frac{1}{2}n_F}\,\frac{\mathrm{e}^{-\frac{8\pi^2}{g^2}|K|}}{Z|_{s.c.}} \times \int\prod_{i=1}^{n_B}\mathrm{d}\beta_i\frac{||a^{(i)}||}{\sqrt{2\pi}}\sum_{\{j_k\}}(-1)^{P_{\{j_k\}}}O(\prod_k f_{j_k};A^I)\,, \tag{25}$$

where we have explicitly carried out the final integration over the Grassmannian variables c_j, $j=1,2,\ldots,n_F$. As a result the product of the n_F fermionic fields contained in O is simply replaced by the product of the wave functions, $f_j(x,\beta)$, of the n_F zero modes. The sum over permutations is weighted by alternating signs because of Fermi statistics. Finally, we have set $\kappa_F=\frac{1}{2}$, because in supersymmetric theories fermions are always introduced as Weyl particles.

Actually in the case $K=1$ (25) can be made even more explicit, because for a gauge invariant operator the only dependence on the collective coordinates is that on the size and position of the instanton. Using (B.18) of Appendix B and the coset integration formula (derived in [9]) necessary for the generalisation to the case of the $SU(N_c)$ group, one gets to leading order in g (where $Z|_{s.c.}=1$)

$$\langle O\rangle\Big|_{s.c.}^{\mathrm{SUSY}} = V_{N_c}\mu^{4N_c-\frac{1}{2}n_F}\,\frac{\mathrm{e}^{-\frac{8\pi^2}{g^2}}}{(g^2)^{2N_c}} \times \int\frac{\mathrm{d}\rho}{\rho^5}\,\mathrm{d}^4x_0(\rho^2)^{2N_c}\sum_{\{j_k\}}(-1)^{P_{\{j_k\}}}O(\prod_k f_{j_k};A^I)\,, \tag{26}$$

with

$$V_{N_c}=\frac{4}{\pi^2}\frac{(4\pi^2)^{2N_c}}{(N_c-1)!(N_c-2)!}\,. \tag{27}$$

Supersymmetry has in store another surprise for us. Recalling the multiplicity of the fermionic zero modes as given by the Atiyah–Singer theorem (see Appendix A), one finds that for a supersymmetric theory

$$4N_c-\frac{1}{2}n_F=b_1\,, \tag{28}$$

$$\beta=-\frac{g^3}{16\pi^2}b_1+\mathrm{O}(g^5)\,, \tag{29}$$

where $b_1 > 0$ is the first coefficient of the Callan–Symanzik β-function. To prove (28) we recall the general formula

$$b_1 = \frac{11}{3}\ell[\mathbf{Adj}] - \frac{2}{3}\sum_{\mathbf{R_F}} n_{\mathbf{R_F}}\ell[\mathbf{R_F}] - \frac{1}{3}\sum_{\mathbf{R_B}} n_{\mathbf{R_B}}\ell[\mathbf{R_B}]\,, \tag{30}$$

where $n_{\mathbf{R_F}}$ and $n_{\mathbf{R_B}}$ are the numbers of fermions and bosons in the representations $\mathbf{R_F}$ and $\mathbf{R_B}$, respectively. Since in a supersymmetric theory each fermion is accompanied by a bosonic partner belonging to the same representation $\mathbf{R}$, (30) simplifies to

$$b_1 = 3\ell[\mathbf{Adj}] - \sum_{\mathbf{R}} n_{\mathbf{R}}\ell[\mathbf{R}] = 3N_c - \sum_{\mathbf{R}} n_{\mathbf{R}}\ell[\mathbf{R}]\,, \tag{31}$$

with $n_{\mathbf{R}}$ the number of chiral superfields in the representation $\mathbf{R}$. To be able to compare b_1 in the above equation with the combination that appears in the l.h.s. of (28) we make use of the Atiyah–Singer theorem (see (16)). Separating out the contribution due to gluinos (the fermions in the gauge supermultiplet) which accounts for a $2N_c/2$ contribution, we can write the l.h.s. of (28) in the form

$$4N_c - \frac{1}{2}n_F = 4N_c - N_c - \frac{1}{2}2\sum_{\mathbf{R}} n_{\mathbf{R}}\ell[\mathbf{R}] = 3N_c - \sum_{\mathbf{R}} n_{\mathbf{R}}\ell[\mathbf{R}]\,, \tag{32}$$

in agreement with (31).

The interesting consequence of this equality is that we can combine the exponential of the instanton action with the explicit μ dependence to form the renormalisation group-invariant Λ-parameter of the theory. Introducing the running coupling $g(\mu)$, we can thus write

$$\mu^{4N_c - \frac{1}{2}n_F}\mathrm{e}^{-\frac{8\pi^2}{g(\mu)^2}} = \Lambda^{b_1}\,. \tag{33}$$

We will exploit this key observation in the following, making it more precise (Sect. 4.1).

3 Chiral and Supersymmetric Ward–Takahashi Identities

Before embarking in explicit instantonic calculations of correlators, we want to spell out the constraints imposed on correlators by chiral and supersymmetric Ward–Takahashi identities (WTIs). We will show that in some interesting cases, when these "geometric" constraints are coupled to the requirement of renormalisability, the expression of certain Green functions is (up to multiplicative numerical constants) completely fixed.

The special Green functions which enjoy this amazing property are the n-point correlation functions of lowest (highest) components of chiral (anti-chiral) gauge-invariant composite superfields. Although this is a very limited set of correlators, we will see that their knowledge, when used in conjunction with clustering, is sufficient to draw interesting non-perturbative information about the structure of the vacuum and of its symmetry properties. For this reason, in this section we will limit our consideration to such correlators. We will in particular concentrate on the case of $\mathcal{N}=1$ super QCD (SQCD) (see Appendix A for notations) with the purpose of exploring the properties of a sufficiently general theory in which also mass terms can be present.

WTIs provide relations among different Green functions. They will be worked out under the assumption that supersymmetry is not spontaneously (or explicitly) broken, i.e. under the assumption that the vacuum of the theory is annihilated by all the generators of supersymmetry. Our philosophy will be that, if we find that some dynamical calculation turns out to be in contradiction with constraints imposed by supersymmetry, then this should be interpreted as evidence for spontaneous supersymmetry breaking.

As we explained above, we are now going to consider the n-point Green functions

$$G(x_1,\ldots,x_n) = \langle 0|T\big(\chi_1(x_1)\ldots\chi_n(x_n)\big)|0\rangle\,, \tag{34}$$

where each $\chi_k(x_k)$ is a local gauge-invariant operator made out of a products of lowest components of the fundamental chiral superfields of the theory. Thus the operators χ_k are themselves lowest components of some composite chiral super field, X_k, for which we formally have the expansion

$$X_k(x) = \chi_k(y) + \sqrt{2}\theta^\alpha\psi_{\alpha k}(y) + \theta^2 F_k(y)\,, \tag{35}$$

$$y_\mu = x_\mu + i\theta^\alpha\sigma_\mu^{\alpha\dot\alpha}\bar\theta_{\dot\alpha}\,. \tag{36}$$

On these fields the Q and $\bar Q$ generators of supersymmetry act as "raising" and "lowering" operators according to the (anti-)commutation rules

$$[\bar Q^{\dot\alpha},\chi_k(x)] = 0\,, \qquad \{\bar Q^{\dot\alpha},\psi_k^\alpha(x)\} = \sqrt{2}\,\bar\sigma_\mu^{\dot\alpha\alpha}\partial_\mu\chi_k(x)\,, \tag{37}$$

$$[Q_\alpha, F_k(x)] = 0\,, \qquad \{Q_\alpha,\psi_{k\beta}(x)\} = \sqrt{2}\epsilon_{\alpha\beta}F_k(x)\,. \tag{38}$$

3.1 Space–time Dependence

The independence of the correlators of the form (34) from space–time arguments immediately follows from the (anti-)commutation relations (37). Taking, in fact, the derivative of G with respect to x_ℓ and contracting with $\sqrt{2}\,\bar\sigma_\mu^{\dot\alpha\alpha}$, one gets

$$\begin{aligned}&\sqrt{2}\,\bar\sigma_\mu^{\dot\alpha\alpha}\frac{\partial}{\partial x_{\ell\mu}}G(x_1,\ldots,x_n)\\ &= \langle 0|T\big(\chi_1(x_1)\ldots\{\bar Q^{\dot\alpha},\psi_\ell^\alpha(x_\ell)\}\ldots\chi_n(x_n)\big)|0\rangle = 0\,.\end{aligned} \tag{39}$$

The last equality is a consequence of the fact that $\bar Q$ can be freely (first commutation rule in (37)) brought to act on the vacuum state at the beginning and at the end of the string of χ_k operators and that, under the assumption that supersymmetry is unbroken, $\bar{Q}|0\rangle = 0$. Contributions coming from the derivative acting on the θ-functions that prescribe the time ordering of operators in G are zero because they give rise to the vanishing equal time commutators, $[\chi_\ell(\boldsymbol{x}_\ell, t_\ell), \chi_k(\boldsymbol{x}_k, t_k)]\delta(t_k - t_\ell) = 0$. Equation (39) proves the constancy of G. A similar result clearly holds for n-point correlators, G^*, where only lowest components of anti-chiral superfields are inserted.

We end this section with the important observation that all these correlators vanish identically in perturbation theory. Only non-perturbative instanton-like contributions can make them non-zero.

3.2 Mass and g Dependence

The following further properties hold for correlators of lowest components of chiral, G, (or anti-chiral, G^*) superfields [25, 26, 4]:

(a) G is an analytic function of the complex mass parameters m_f, i.e. it does not depend on m_f^* (the opposite being true for G^*).

(b) The mass dependence of G (and G^*) is completely fixed.

(c) When renormalisation group-invariant operators are inserted, the dependence upon the coupling constant is, in a mass-independent renormalisation scheme, fully accounted for by the renormalisation group-invariant (RGI) quantities Λ and $[m_f]_{\rm inv} \equiv \hat{m}_f$ (see below (51)).

It is important to remark that properties of this kind can be readily exported to the generating functional of Green functions, as they only follow from symmetry principles. They provide strong constraints on the form of the associated effective action. A celebrated example of application of this observation, though in a different context, can be found in the construction of the low-energy effective action that describes the interaction of pions in QCD [27, 28]. In supersymmetric theories the invariances are so tight that often the full expression of the effective superpotential is completely determined [21, 29, 30, 31] (see Sect. 5).

(a) Mass Analyticity

The statement (a) follows from the supersymmetric relation (no sum over f)

$$m_f^* \frac{\partial}{\partial m_f^*} G(x_1, \dots, x_n) = m_f^* \langle 0| T\big(\chi_1(x_1), \dots, \chi_n(x_n) \int \mathrm{d}^4 x\, F_f^{*f}(x)\big)|0\rangle$$

$$= m_f^* \int \mathrm{d}^4 x\, \langle 0| T\big(\chi_1(x_1) \dots \chi_n(x_n) \{\bar{Q}_{\dot\alpha} \bar{\psi}_f^{f\dot\alpha}(x)\}\big)|0\rangle = 0\,, \tag{40}$$

with F_f^{*f} the auxiliary field of the anti-chiral superfield $T_f^{*f} = (\chi_f^{*f}, \bar{\psi}_f^{*f}, F_f^{*f})$. The first equality follows from the fact that, before the auxiliary field is eliminated by the e.o.m., F_f^{*f} is the coefficient of m_f^*. The second is a consequence of the complex conjugate of the anti-commutation relation in (38). Finally, since $\bar{Q}$ commutes with the χ_k's, it can be brought in contact with the vacuum state which is thus annihilated.

(b) Mass Dependence

In order to simplify this analysis we restrict to Green functions where only the gauge-invariant composite operators

$$\frac{g^2}{32\pi^2}\lambda^{\alpha a}(x)\lambda^a_\alpha(x) \equiv \frac{g^2}{32\pi^2}\lambda\lambda(x)\,, \tag{41}$$

$$\tilde{\phi}^{fr}(x)\phi_{hr}(x) \equiv \tilde{\phi}^f\phi_h(x) \tag{42}$$

are inserted. They are the lowest components of chiral superfields which will be called S and T^f_h, respectively. Besides their obvious complex conjugate fields, we will sometimes also consider the composite operators $\tilde{\psi}^{fr}_\alpha(x)\psi^\alpha_{hr}(x) = \tilde{\psi}^f\psi_h(x)$. In general terms we will then consider correlators of the kind

$$\begin{aligned}&G^{(p,q)f_1,\ldots,f_p}_{h_1,\ldots,h_p}(x_1,\ldots,x_p;x_{p+1},\ldots,x_{p+q}) \qquad (43)\\ &= \langle 0|T\big(\tilde{\phi}^{f_1}\phi_{h_1}(x_1)\ldots\tilde{\phi}^{f_p}\phi_{h_p}(x_p)\frac{g^2}{32\pi^2}\lambda\lambda(x_{p+1})\ldots\frac{g^2}{32\pi^2}\lambda\lambda(x_{p+q})\big)|0\rangle\,.\end{aligned}$$

The dependence of (43) upon the mass parameters can be established in the following way. First of all we notice that from (40) we have

$$m_f\frac{\partial}{\partial m_f}G^{(p,q)} = \left(m_f\frac{\partial}{\partial m_f} - m^*_f\frac{\partial}{\partial m^*_f}\right)G^{(p,q)} = \frac{1}{i}\frac{\partial}{\partial\alpha_f}G^{(p,q)}\,, \tag{44}$$

where we have set

$$m_f = |m_f|\mathrm{e}^{i\alpha_f}\,. \tag{45}$$

In order to compute the derivative in the r.h.s. of (44) we perform the non-anomalous $U^f_A(1)$ transformation (see Appendix A)

$$\begin{aligned}&(\tilde{\psi}^h,\psi_h) \to \mathrm{e}^{i\delta_{fh}\alpha_f/2}(\tilde{\psi}^h,\psi_h)\,, \quad (\tilde{\phi}^h,\psi_h) \to \mathrm{e}^{i(\delta_{fh}-1/N_c)\alpha_f/2}(\tilde{\phi}^h,\psi_h)\,,\\ &\lambda \to \mathrm{e}^{-i\alpha_f/2N_c}\lambda\,, \qquad (46)\end{aligned}$$

by means of which the α_f dependence of the action is eliminated, but it is brought in the fields appearing in $G^{(p,q)}$. This allows to carry out in an explicit way the α_f derivative, leading to the result

$$m_f\frac{\partial}{\partial m_f}G^{(p,q)f_1,\ldots,f_p}_{h_1,\ldots,h_p} = q^{(f),f_1,\ldots,f_p}_{h_1,\ldots,h_p}G^{(p,q)f_1,\ldots,f_p}_{h_1,\ldots,h_p}\,, \tag{47}$$

$$q^{(f),f_1,\ldots,f_p}_{h_1,\ldots,h_p} = \frac{p+q}{N_c} - \frac{1}{2}\sum_{\ell=1}^{p}(\delta_{f_\ell,f} + \delta_{h_\ell,f})\,, \tag{48}$$

where $q^{(f)}$ is the sum of all the $U_A^f(1)$ charges of the operators contained in $G^{(p,q)}$. The above differential equation is easily integrated and yields

$$\prod_{\ell=1}^{p}(m_{f_\ell}m_{h_\ell})^{\frac{1}{2}}G^{(p,q)f_1,\dots,f_p}_{h_1,\dots,h_p} = C^{(p,q)f_1,\dots,f_p}_{h_1,\dots,h_p}(\mu,g)\Big(\prod_{\ell=1}^{N_f}m_\ell\Big)^{\frac{(p+q)}{N_c}} . \qquad (49)$$

The μ dependence of $C^{(p,q)f_1,\dots,f_p}_{h_1,\dots,h_p}(\mu,g)$ is trivially fixed by dimensional analysis and one finds

$$C^{(p,q)f_1,\dots,f_p}_{h_1,\dots,h_p}(\mu,g) \propto \mu^{(p+q)(3-N_f/N_c)} . \qquad (50)$$

(c) g Dependence

The g dependence of $C^{(p,q)f_1,\dots,f_p}_{h_1,\dots,h_p}(\mu,g)$ is completely determined by renormalisability. In fact, having factorised in the l.h.s. of (49) the mass factor $\prod_{\ell=1}^{p}(m_{f_\ell}m_{h_\ell})^{\frac{1}{2}}$, which precisely serves the purpose of making the $\tilde{\phi}^f\phi_h$ operators in $G^{(p,q)}$ behave like RGI insertions, the rest of the g dependence must all be expressed through the RGI quantities

$$\Lambda = \mu\exp\Big(-\int^g \frac{1}{\beta(g')}\mathrm{d}g'\Big), \quad \hat{m} = m\exp\Big(-\int^g \frac{\gamma_m(g')}{\beta(g')}\mathrm{d}g'\Big), \qquad (51)$$

where $\beta \neq 0$ and $\gamma_m(g)$ are the Callan–Symanzik function of the theory and the mass anomalous dimension of the matter superfield, respectively. This implies that the g dependence must be of the form

$$C^{(p,q)f_1,\dots,f_p}_{h_1,\dots,h_p}(\mu,g) \propto \exp\Big\{-\int^g \frac{\mathrm{d}g'}{\beta(g')}(p+q)\big[(3-\frac{N_f}{N_c})+\gamma_m(g')\frac{N_f}{N_c}\big]\Big\}, \qquad (52)$$

in order to have

$$\prod_{\ell=1}^{p}(m_{f_\ell}m_{h_\ell})^{\frac{1}{2}}G^{(p,q)f_1,\dots,f_p}_{h_1,\dots,h_p} = \Lambda^{(p+q)(3-N_f/N_c)}\Big(\prod_{\ell=1}^{N_f}\hat{m}_\ell\Big)^{\frac{(p+q)}{N_c}} t^{(p,q)f_1,\dots,f_p}_{h_1,\dots,h_p} , \qquad (53)$$

with $t^{(p,q)f_1,\dots,f_p}_{h_1,\dots,h_p}$ a dimensionless constant tensor in flavour space.

The form of $t^{(p,q)f_1,\dots,f_p}_{h_1,\dots,h_p}$ is strongly constrained (and sometimes completely determined) by the pattern of unbroken flavour symmetries of the theory. Its explicit computation will be one of the main subjects of the next sections.

3.3 The Anomalous $U_\lambda(1)$ R-symmetry

The integrated WTI associated with the anomalous $U_\lambda(1)$ R-symmetry (see Appendix A, (A.28), (A.29) and (A.17)) reads

$$2iKN_c\langle O(x_1,\dots,x_n)\rangle = \sum_{i=1}^{n}\langle\frac{\partial O^{(\alpha)}}{\partial\alpha(x_i)}(x_1,\dots,x_n)\Big|_{\alpha=0}\rangle\,, \tag{54}$$

where $O^{(\alpha)}$ is the operator which is obtained by performing on O a $U_\lambda(1)$ rotation of an angle α. For the special Green function $G^{(p,q)f_1,\dots,f_p}_{h_1,\dots,h_p}$ (see (43)) and (54) simply becomes

$$\begin{aligned}
&2KN_cG^{(p,q)f_1,\dots,f_p}_{h_1,\dots,h_p}(x_1,\dots,x_p;x_{p+1},\dots,x_{p+q})\\
&= 2(p+q)G^{(p,q)f_1,\dots,f_p}_{h_1,\dots,h_p}(x_1,\dots,x_p;x_{p+1},\dots,x_{p+q})\,,
\end{aligned} \tag{55}$$

because the $U_\lambda(1)$ rotation of $G^{(p,q)f_1,\dots,f_p}_{h_1,\dots,h_p}$ is proportional to $G^{(p,q)f_1,\dots,f_p}_{h_1,\dots,h_p}$ itself through the factor $2(p+q)$. As a result only if

$$p+q = KN_c\,, \tag{56}$$

we can get a non-vanishing result. Notice that (56) implies $K>0$ consistently with the fact that we are dealing with lowest components of chiral superfields. Negative values of K will come into play in correlators with insertions of highest components of anti-chiral superfields.

A particularly interesting situation arises if we insist that each flavour should appear exactly K times. Then (56) requires $N_f \le N_c$. At this point to simplify our treatment, we restrict ourselves to the case $K=1$. Since in this situation $p=N_f$, the whole dependence on the bare mass parameters drops out from the Green function we are considering and we get

$$G^{(N_f,N_c-N_f)f_1,\dots,f_{N_f}}_{h_1,\dots,h_{N_f}}(x_1,\dots,x_{N_f};x_{N_f+1},\dots,x_{N_c}) \propto \Lambda^{3N_c-N_f}\,. \tag{57}$$

We expressly note that the exponent to which Λ is raised in (57) is not only the physical dimension of $G^{(N_f,N_c-N_f)}$, but it also coincides with the first coefficient of the β-function of SQCD (see the discussion and the formulae in Sect. 2.4).

Among the Green functions of the type (43) which fulfil the further requirements spelled out in this subsection, we wish to specially mention here the one relevant in pure SYM where one gets the famous correlator [32, 33]

$$G^{(0,N_c)}(x_1,\dots,x_{N_c}) = \langle\frac{g^2}{32\pi^2}\lambda\lambda(x_1)\dots\frac{g^2}{32\pi^2}\lambda\lambda(x_{N_c})\rangle\,. \tag{58}$$

3.4 The Konishi Anomaly

The general need to regularise products of operator fields at the same point is at the origin of the axial anomaly [34] (see Appendix A) and of the anomalous contribution that appears in certain supersymmetric anti-commutators. Starting from the supersymmetry graded algebra summarised in (37) and (38), it has been shown in [35] that, after regularisation, in massive SQCD the following (anomalous) anti-commutation relation holds:

$$\frac{1}{2\sqrt{2}}\{\bar{Q}^{\dot{\alpha}}, \bar{\psi}^f_{\dot{\alpha}}\phi_h(x)\} = -m_f\tilde{\phi}^f\phi_h(x) + \frac{g^2}{32\pi^2}\lambda\lambda(x)\delta^f_h\,, \tag{59}$$

where besides the naive $m_f\tilde{\phi}^f\phi_h(x)$ term an extra contribution appears. This relation is what usually goes under the name of "Konishi anomaly". Clearly, if the vacuum of the theory is supersymmetric, by taking the vacuum expectation value (v.e.v.) of (59) a proportionality relation between gluino and scalar condensates emerges, namely

$$m_f\langle\tilde{\phi}^f\phi_f\rangle = \frac{g^2}{32\pi^2}\langle\lambda\lambda\rangle\,, \quad \text{no sum over } f\,, \tag{60}$$

besides

$$\langle\tilde{\phi}^f\phi_h\rangle = 0\,, \quad f \neq h\,. \tag{61}$$

4 Instanton Calculus

We want to show in this section that the Green functions considered in (57) receive a non-vanishing computable one-instanton contribution. In other words, although zero in perturbation theory, they can be exactly evaluated in the semi-classical approximation by dominating the functional integral with the one-instanton saddle point. A non-trivial result is obtained because the number of fermionic fields that are inserted in $G^{(N_f,N_c-N_f)}$ (either at face value or at the appropriate order in g) is precisely equal to the number of fermionic zero modes present in the $K=1$ instanton background.

4.1 Instanton Calculus in SYM

The computation of the correlator (58) in the semi-classical one-instanton approximation is not too difficult by using the results we have recollected in Appendix B (about bosonic zero modes and collective coordinate integration) and the explicit expression of the $2N_c$ gluino zero modes that can be found in Appendix A [4, 20].

We will consider the case of a pure SYM theory with gauge group $SU(N_c)$.[9] The striking outcome of the calculation (which is based on equations from (25)

[9] For SYM theories with other compact Lie group see [36].

to (27)) is that the apparently extremely complicated dependence of the correlator upon the space–time location of the inserted operators is completely washed out by the bosonic collective coordinate integration and, as expected, a space–time-independent (constant) result is obtained in agreement with the supersymmetric WTI (39). Explicitly one finds [32, 33]

$$G^{(0,N_c)}(x_1,\ldots,x_{N_c}) = C_{N_c}\left(\Lambda_{\rm SYM}^{2-\rm loops}\right)^{3N_c}, \tag{62}$$

where[10]

$$C_{N_c} = \frac{2^{N_c}}{(3N_c-1)(N_c-1)!}, \tag{63}$$

$$\Lambda_{\rm SYM}^{2-\rm loops} = \mu\, e^{-8\pi^2/3N_c g^2}(g^2)^{-1/3}. \tag{64}$$

Equation (64) follow from the known value of the two-loop coefficient of the β-function of the theory and shows that dominating the functional integral by the semi-classical one-instanton saddle point gives a (two-loop) RGI answer.

From this result two important consequences can be derived, one concerning the form of the β-function of the theory and the second the structure of the vacuum.

The SYM β-function

One can argue that the result (64) is valid to all loops in the sense that higher-order power corrections in g are indeed all vanishing. The argument goes as follows. As we remarked in Appendix B just at the end of the first subsection, one can go on with perturbation theory around the instanton background by expanding in powers of g terms cubic and quartic in the fluctuations, as well as terms coming from the Faddeev–Popov procedure. One should be finding in this way logarithmically divergent contributions which would be interpreted as higher order terms in the Callan–Symanzik β-function. In the present case, however, no such term can arise because there is no dimensionful quantity with which we might scale the (would-be) logarithmically divergent μ dependence. In fact, the only other dimensionful quantities are the relative distances $x_i - x_j$ of the operator insertion points. But the supersymmetric WTI (39) prohibits any such dependence.

We must conclude that in the regularisation and renormalisation scheme we work and in the background gauge, the Λ parameter is "two-loop exact". This observation is equivalent to the result of β-exactness first put forward in [39], which amounts to say that one has the exact formula

$$\beta_{\rm SYM}(g) = -\frac{g^3}{16\pi^2}\frac{3N_c}{1-2g^2N_c/16\pi^2}. \tag{65}$$

[10] The constant C_{N_c} differs from the similar constant appearing in (4.9) of [4] by a factor 2^{N_c}. This mistake was pointed out by various authors [23, 37, 38] and was the consequence of an erroneous normalisation of the gluino zero modes.

Introducing (65) in

$$\int_{g(\mu_0)}^{g(\mu)} \frac{\mathrm{d}g'}{\beta_{\mathrm{SYM}}(g')} = \int_{\mu_0}^{\mu} \frac{\mathrm{d}\mu'}{\mu'} \,. \tag{66}$$

one gets, in fact, by a straightforward integration precisely (64). We stress that no approximation (no expansion in powers of g) has been performed in the step from (65) to the formula (64).

Equation (65) can be generalised [39] to encompass the case of extended supersymmetry with $\mathcal{N} = 1, 2, 4$ supercharge multiplets through the simple formula

$$\beta_{\mathcal{N}}(g) = -\frac{g^3}{16\pi^2} \frac{(4-\mathcal{N})N_c}{1-2(2-\mathcal{N})g^2 N_c/16\pi^2} \,, \tag{67}$$

which incorporates the known facts that the $\mathcal{N} = 2$ β-function is one-loop exact and the $\mathcal{N} = 4$ β-function just vanishes.

The Structure of the SYM Vacuum

The space–time constancy of the result (64) allows us to compute the expectation value of the composite operator $g^2\lambda\lambda/32\pi^2$ by simply imagining that the separations $|x_i - x_j|$ are very large. Using clustering, it will be possible to write $G^{(0,N_c)}$ as the product of the v.e.v.'s of such operators (gluino condensate, in the following).

The computation is straightforward if the vacuum of the theory is unique. Here the situation is more complicated because of the very fact that the gluino condensate is not vanishing. This means, in fact, that the residual $_{2N_c}$ symmetry of the theory (see Appendix A) is actually spontaneously broken down to $_2$ with the consequence that there are N_c degenerate vacua in which the theory can live, related by $_{N_c}$ transformations. Incidentally, we note that this result is perfectly consistent with the prediction based on the Witten index calculation [40].

In the presence of many equivalent vacua the functional integral yields non-perturbative results where contributions coming from different vacua are averaged out. Thus in order to extract useful information from the clustering properties of the theory, one has to take into account this phenomenon and go through a procedure called "vacuum disentangling" [4, 33]. All this simply means that we should write for $G^{(0,N_c)}$ the formula

$$G^{(0,N_c)} = \frac{1}{N_c} \sum_{k=1}^{N_c} \left[\langle \Omega_k | \frac{g^2}{32\pi^2} \lambda\lambda | \Omega_k \rangle\right]^{N_c} , \tag{68}$$

with the gluino condensates transforming under $_{2N_c}$ as

$$\langle\Omega_k|\frac{g^2}{32\pi^2}\lambda\lambda|\Omega_k\rangle = \mathrm{e}^{\frac{2\pi i k}{N_c}}\langle\Omega_0|\frac{g^2}{32\pi^2}\lambda\lambda|\Omega_0\rangle\,, \quad k=1,2,\ldots,N_c\,. \tag{69}$$

This equation is telling us that the average in (68) is trivial and we get in the k-th vacuum

$$\langle\Omega_k|\frac{g^2}{32\pi^2}\lambda\lambda|\Omega_k\rangle = \mathrm{e}^{\frac{2\pi i k}{N_c}}\left(C_{N_c}\right)^{1/N_c}\left(\Lambda_{\mathrm{SYM}}^{2-\mathrm{loops}}\right)^3\,, \quad k=1,\ldots,N_c\,. \tag{70}$$

Discussion of the Results

The picture we got from the calculation presented in the previous section looks rather convincing and physically sound. It perfectly matches all our expectations and it has been carried out in a clean and rigorous mathematical way. It is uniquely based on the assumption that Green functions which can receive contributions only from the $K=1$ sector of the theory can be reliably computed by dominating the functional integral by the one-instanton saddle point. We have also argued that the two-loop RGI result obtained in the semi-classical approximation is exact in the sense that it does not get further perturbative corrections.

Despite all these nice features, it has been argued in the literature that the method employed to get the result (62) cannot be right because it seems to encounter a number of problems with other considerations.

(1) The N_c dependence of C_{N_c} leads in the 't Hooft limit ($N_c \to \infty$ with $g^2 N_c$ fixed) to an N_c dependence of the gluino condensate (70) that it is not what one would expect from the fact that the gluinos (together with the gauge field, A_μ) belong to the adjoint representation of the gauge group. Taking into account the g^2 factor that was introduced in front of the gluino bilinear, one would naively expect $\langle g^2\lambda\lambda\rangle \sim \mathrm{O}(g^2 N_c^2) \sim \mathrm{O}(N_c)$ in the 't Hooft limit. From (62) and (63), one finds instead $\langle g^2\lambda\lambda\rangle \sim \mathrm{O}(g^2 N_c) \sim \mathrm{O}(1)$.

(2) In [38] the calculation of $G^{(0,N_c)}$ has been repeated in a fully supersymmetric formalism and the result summarised in (62), (63) and (64) was confirmed (up to the correction for the factor 2^{N_c} that we already mentioned). Interestingly, these authors have also been able to extend, in the large N_c limit, the semi-classical instanton calculation to Green functions which receive contributions from topological sectors with winding numbers $K>1$, i.e. to Green functions with KN_c insertions of the gluino bilinear. The result of this calculation, when clustering is used, is inconsistent with it, because it leads to a value of the condensate which is not independent of K.

(3) The computation of $G^{(0,N_c)}$ can be indirectly done starting from the more complicated case where extra massive matter supermultiplets are added [21, 31, 41] and then exploit the notion of decoupling [42, 43]. We recall that in a nutshell decoupling is the property of a local field theory according to which when some mass becomes large, the corresponding matter field disappears from the low energy physics (see Appendix D), modulo possible consistency conditions resulting from the requirement of anomaly cancellation [44].

From symmetry arguments it is often possible to determine, up to a multiplicative constant, the form of the effective superpotential of the enlarged theory in terms of the relevant composite operators (see Sect. 5). Then consistency arguments, following from sending to infinity each mass successively, supplemented by "constrained instanton" calculations (see the next paragraph), can be used to determine this constant (see Sect. 5.2). Clearly checking its value is important for the self-consistency and the reliability of the various approaches (see point (1) above). One finds that, if computed by looking at the effective superpotential calculations, the value of this constant does not agree with what one can deduce from the formulae (62)–(64). Of course, the comparison was done after having properly matched the RGI parameters associated with the different regularisations employed in the various calculations [23]. For instance, in the $SU(2)$ case, one finds from the equations in Sect. 4.1 $C_2 = 4/5$ instead of the result $C_2 = 1$ one would obtain from decoupling arguments.

Despite a lot of work in the years that followed these findings, there is no clear understanding of why there are such discrepancies and where they come from. One line of arguments [21, 31, 45, 46], first prompted by the results described in (3), relies on the observation that when scalar fields are present other quasi-saddle points exist, in which scalar fields get a non-vanishing v.e.v., which should be taken as background configurations in the semi-classical calculation of Green functions. In fact, the (partial or full) breaking of the gauge symmetry leads in the limit of very large v.e.v.'s to a weakly coupled theory, where semi-classical instanton calculations are expected to be reliable. In this context the non-trivial problem arising from the fact that in the presence of non-vanishing scalar v.e.v.'s the SQCD e.o.m. have no solution (owing precisely to the nature of the scalar boundary conditions) is circumvented by making recourse to the so-called "constrained instanton" method [47][11].

It is not completely clear to us whether the constrained instanton method (sometimes also called the "weak coupling instanton" (WCI) method, from which the nickname "strong coupling instanton" (SCI) method was in opposition attributed to the approach described in Sect. 4.1 and further employed in Sect. 4.2 below) can be considered as a completely satisfactory solution to the problems listed above. We now want to briefly discuss this question by illustrating some pro's and con's of the two approaches.

• Certainly, if one accepts the WCI computational strategy, the problem mentioned in (1) disappears. As for the question of consistency with clustering (point (2) above), to date no check of the kind done in [38] was carried out.

[11] The theoretical foundation of the method is somewhat delicate (it relies on introducing in the functional integral a suitable "constraint" which breaks the integration measure into sectors of well-defined instanton scale size) and its technical implementation requires a number of non-trivial mathematical steps. Its presentation is beyond the scope of this review, but can be found in the original literature. We recommend to the reader the nice work of [23].

Finally we do not see a really rigorous way to decide on the basis of the present knowledge whether $\sqrt{C_2} = 2/\sqrt{5}$ or $\sqrt{C_2} = 1$ is the correct answer for the constant in front of the gluino condensate. One possibility to settle this question could be to make recourse to a lattice formulation of SYM [48] and directly measure in Monte Carlo simulations the gluino condensate. Up to now, unfortunately, severe technical difficulties have prevented such a measurement. For a recent review on the subject of supersymmetry on the lattice see [49].

• The whole idea of working in the Higgs phase of SQCD comes from the key observation that, in the massless limit, the superpotential possesses a complicated vacuum manifold (see Appendix E). It is customary in the literature to speak about "flat directions" [21, 31, 50], i.e. constant values of the scalar fields along which the D-term vanishes. It is in this situation that all the explicit WCI calculations have been carried out.[12]. Despite the fact that explicit instanton calculations have been carried out in the massless limit, their results and implications have been employed in the massive case. In this context it should be noted that the massless limit of the massive SQCD theory is a very delicate one. For instance, as we shall see, the SCI approach gives results in the massive case that, when extrapolated to vanishing mass, are not consistent with results directly obtained in the massless theory. This feature finds a natural explanation in the infrared structure of the theory which is such that the massless limit of the massive theory does not coincide with the strictly massless situation [43].

• The WCI approach has found its most successful application in predicting the non-perturbative expansion coefficients of the SW [51] expression for the effective prepotential of the $\mathcal{N} = 2$ SYM theory (see Sect. 9).

• On the other hand in $\mathcal{N} = 4$ SYM, despite the fact that there are flat directions for the scalar potential, no scalar v.e.v.'s are assumed to be generated (as a non-vanishing v.e.v. would break the (super)conformal invariance of the theory) and all instanton calculations are performed in the SCI way we described in Sect. 4.1. Actually, in $\mathcal{N} = 4$ there is no running of the gauge coupling (67) and one can always think that calculations are done at infinitesimally small values of g. Thus non-perturbative instanton calculations in $\mathcal{N} = 4$ SYM do not seem to fall under the criticisms raised for the $\mathcal{N} = 1$ and $\mathcal{N} = 2$ cases (see Sects. 14 and 15).

4.2 Instanton Calculus in SQCD

In this section we move to SQCD. The action of SQCD is obtained by coupling (in a gauge invariant and supersymmetric way) to the SYM supermultiplet N_f pairs of matter chiral superfields, Φ^r_f and $\tilde{\Phi}^f_r$ ($f = 1, 2, \ldots, N_f$, $r = 1, 2, \ldots, N_c$) belonging, respectively, to the $\mathbf{N_c}$ and $\bar{\mathbf{N}}_\mathbf{c}$ representation of the gauge group (see Appendix A for some detail and [36] for an extension

[12] Besides the original papers in [45], the basic work from which all the old WCI calculations make reference to is the paper quoted in [23].

of these considerations to theories with different gauge groups and matter content).

We want to identify and compute, according to the strategy developed in Sect. 4.1 to deal with SYM, the Green functions that, besides being space–time constant, can be reliably evaluated by dominating the functional integral with the one-instanton saddle point. We shall start by analysing the massive case, where the further information provided to us by the Konishi anomaly relation [35] can be exploited and will allow to determine both the gluino and the scalar matter condensates and check the internal consistency of our calculations. In Sect. 4.2 we will discuss the puzzling features that arise when the limit $m \to 0$ is taken.

Massive SQCD

Already looking at the general results derived in Sect. 3.2 about the mass dependence of Green functions with only lowest components of chiral superfields, we see that their small mass limit is rather delicate, as infrared divergences seem to arise. To avoid hitting this difficulty, we start by limiting the use of instanton calculus to the computation of the correlators that according to (53) are mass independent. Among those we will concentrate here on the following three (see (57)):

$$(A) \quad F^{(0,N_c)}(x_1,\dots,x_{N_c}) \prod_{f=1}^{N_f} \frac{\partial}{\partial m_f} G^{(0,N_c)}(x_1,\dots,x_{N_c})$$

$$= \prod_{f=1}^{N_f} \frac{\partial}{\partial m_f} \langle \frac{g^2}{32\pi^2}\lambda\lambda(x_1)\dots\frac{g^2}{32\pi^2}\lambda\lambda(x_{N_c})\rangle \,, \tag{71}$$

$$(B) \quad G^{(N_f,N_c-N_f)}(x_1,\dots,x_{N_c})$$

$$= \langle \tilde{\phi}^1\phi_1(x_1)\dots\tilde{\phi}^{N_f}\phi_{N_f}(x_{N_f})\frac{g^2}{32\pi^2}\lambda\lambda(x_{N_f+1})\dots\frac{g^2}{32\pi^2}\lambda\lambda(x_{N_c})\rangle \,, \tag{72}$$

and, in the particular case $N_f = N_c$,

$$(C) \quad D(x,x') = \langle \mathrm{det}\phi(x)\mathrm{det}\tilde{\phi}(x')\rangle \,, \tag{73}$$

$$\det\phi = \frac{1}{N_c!}\epsilon^{f_1,\dots,f_{N_f}}\epsilon_{r_1,\dots,r_{N_c}}\phi^{r_1}_{f_1},\dots,\phi^{r_{N_c}}_{f_{N_f}} \,, \tag{74}$$

$$\det\tilde{\phi} = \frac{1}{N_c!}\epsilon_{f_1,\dots,f_{N_f}}\epsilon^{r_1,\dots,r_{N_c}}\tilde{\phi}^{f_1}_{r_1},\dots,\tilde{\phi}^{f_{N_f}}_{r_{N_c}} \,, \tag{75}$$

(A) Let us start the discussion with $F^{(0,N_c)}$. We notice that it contains exactly the number of gluino fields necessary to match the number of zero modes that the theory possesses in the $K = 1$ sector. We recall that, since at the moment we are considering the case in which the matter is massive, no zero modes associated with matter Weyl operators exist. In this situation, we can safely compute the functional integral which defines the above correlator by dominating it with the one-instanton saddle point. The calculation goes through the following steps.

(1) Every factor $\partial/\partial m_f$ can be replaced by the insertion of the action mass term

$$\frac{\partial}{\partial m_f} \to \int \mathrm{d}^4x \left[\tilde{\psi}^f \psi_f(x) + m_f^*(\phi^{*f}\phi_f(x) + \tilde{\phi}^f \tilde{\phi}_f^*(x))\right], \tag{76}$$

(2) which, after integration over the matter supermultiplets, becomes

$$\frac{|m_f|^2}{\mu} \mathrm{Tr}\left[\frac{2}{D^2 - |m_f|^2} - \mathrm{tr}\left(\frac{1}{\not{D}\not{D} - |m_f|^2}\right)\right]. \tag{77}$$

It is understood that the covariant operators $\not{D}$ and D^2 in (77) are computed in the one-instanton background field. The multiplicative mass factors in front of the trace have the following origin: (i) the term m_f^* comes from the expression of the matter propagators and (ii) the ratio m_f/μ comes from what is left out from the ratio between the determinant of the matter Weyl operator and its regulator, after the supersymmetric cancellation of the non-zero mode contribution has been taken care of.

(3) A cancellation of modes also takes place between the two terms in (77). Only the fermionic mode with eigenvalue m_f (i.e. the zero mode in the massless limit) contributes and one simply gets

$$\frac{\partial}{\partial m_f} \to \frac{|m_f|^2}{\mu} \frac{1}{|m_f|^2} = \frac{1}{\mu}. \tag{78}$$

(4) At this point, the functional integration with respect to the gauge supermultiplet fields remains to be done. Since the sole effect of the matter integration is to yield the factor μ^{-N_f}, we are left with exactly the same calculation we did in Sect. 4.1. We thus get

$$F^{(0,N_c)}(x_1,\ldots,x_{N_c}) = C_{N_c} \frac{(\Lambda_{\mathrm{SQCD}}^{1-\mathrm{loop}})^{3N_c-N_f}}{g^{2N_c}}, \tag{79}$$

from which, by integrating with respect to m_f, $f = 1,\ldots,N_f$, we obtain

$$G^{(0,N_c)}(x_1,\ldots,x_{N_c}) = C_{N_c} \frac{(\Lambda_{\mathrm{SQCD}}^{1-\mathrm{loop}})^{3N_c-N_f}}{g^{2N_c}} \prod_{f=1}^{N_f} m_f. \tag{80}$$

Two observations are in order here. First of all, by taking the N_c-th root of the above expression, one can determine the value of the gluino condensate in massive SQCD. One finds

$$\langle\Omega_k|\frac{g^2}{32\pi^2}\lambda\lambda|\Omega_k\rangle = \mathrm{e}^{\frac{2\pi ik}{N_c}}\left(C_{N_c}(\Lambda_{\mathrm{SQCD}}^{1-\mathrm{loop}})^{3N_c-N_f}\frac{1}{g^{2N_c}}\prod_{f=1}^{N_f} m_f\right)^{1/N_c}, \tag{81}$$

which shows that, as in SYM, the discrete ${}_{2N_c}$ symmetry is spontaneously broken down to ${}_2$, living behind an N_c-fold vacuum degeneracy. This is the expected result, since the presence of massive fields cannot modify the value of the Witten index [40].

Secondly, it can be checked that the formulae (80) and (81) define two-loop RGI quantities, as it follows from the known expressions of the β and γ_m functions of the theory. Through $\mathrm{O}(g^5)$ and $\mathrm{O}(g^2)$ they read, respectively,

$$\beta_{\mathrm{SQCD}} = -\frac{g^3}{16\pi^2}(3N_c - N_f) + \frac{g^5}{(16\pi^2)^2}(-6N_c^2 + 4N_cN_f - 2\frac{N_f}{N_c}) + \mathrm{O}(g^7)\,, \quad (82)$$

$$\gamma_m = -\frac{g^2}{8\pi^2}\frac{N_c^2-1}{N_c} + \mathrm{O}(g^4)\,. \quad (83)$$

Actually it has been argued [39] that the following "exact" formula holds.

$$\beta_{\mathrm{SQCD}}(g) = -\frac{g^3}{16\pi^2} - \frac{3N_c - N_f[1-\gamma_m(g)]}{1-2g^2N_c/16\pi^2}\,, \quad (84)$$

which generalises (65) to the SQCD case. Formula (84) perfectly fits with the previous ones to the order they are known and renders the expressions (80) and (81) RGI quantities to all orders.

(B) The computation of the correlator (72) is much more subtle. First of all, one notices that it vanishes to lowest order in g because at the instanton saddle point $\phi_f = \tilde{\phi}^f = 0$. Secondly, the number of inserted gluino fields does not appear to match the number of the existing zero modes. Finally, the matter functional integration requires the knowledge of the massive fermion and scalar propagators, $(\not{D}\not{D} - |m|^2)^{-1}$ and $(D^2 - |m|^2)^{-1}$, in the instanton background which is not available in closed form.

The first and second problems are solved by observing that the integration over the scalar matter fields amounts to substituting ϕ_f and $\tilde{\phi}^f$ with the solutions of their classical e.o.m., which schematically read

$$\phi_f = -i\sqrt{2}g\,(D^2 - |m_f|^2)^{-1}\lambda\psi_f\,, \quad (85)$$

$$\tilde{\phi}^f = i\sqrt{2}g\,(\tilde{D}^2 - |m_f|^2)^{-1}\tilde{\psi}_f\lambda\,. \quad (86)$$

One easily checks that, at the expenses of going to higher order in g, in this fashion one ends up having the right number of inserted gluino fields.

As for the last problem, we start by observing that the integration over the matter fermions has the effect of replacing for each flavour the $\tilde{\psi}^f(x)\psi_f(x')$ product with the corresponding fermionic propagator in the instanton background. After the matter integration one thus arrives at an extremely complicated integral over the collective instanton coordinates, where the unknown fermion and scalar background propagators appear. In order to proceed with the calculation, we notice that the instanton semi-classical approximation respects supersymmetry and that consequently the correlators we are considering will come out to be constant in space–time and mass independent, as shown in Sect. 3. The idea is then to perform the residual computation in the limit of very large masses (more precisely in the limit $m_f \gg |x_i - x_j|^{-1} \gg \Lambda_{\text{SQCD}}$), where the fermion and scalar background propagators tend to their free-field expression. One ends up in this way with feasible integrals which yield the result (Ref. (80))

$$G^{(N_f,N_c-N_f)}(x_1,\ldots,x_{N_c}) = C_{N_c}\frac{(\Lambda^{1-\text{loop}}_{\text{SQCD}})^{3N_c-N_f}}{g^{2N_c}}\,. \tag{87}$$

We remark that this quantity is not RGI as it stands. To make it RGI we must renormalise the scalar fields. One way of doing this is to multiply both sides of (87) by the factor $\prod_f m_f$.

(C) The computational strategy outlined above leads for the correlator (73) to the simple result

$$D(x,x') = 0\,. \tag{88}$$

From the results (81) and (87) one can compute both the gluino and the scalar condensates. Recalling (82) and (83), one gets

$$\begin{aligned}\langle\Omega_k|m_f\tilde{\phi}^f\phi_f|\Omega_k\rangle &= \langle\Omega_k|\frac{g^2}{32\pi^2}\lambda\lambda|\Omega_k\rangle\\ &= \mathrm{e}^{\frac{2\pi i k}{N_c}}\Big(C_{N_c}\big(\Lambda^{2-\text{loop}}_{\text{SQCD}}\big)^{3N_c-N_f}\prod_{f=1}^{N_f}\hat{m}_f^{1-\text{loop}}\Big)^{1/N_c}\,.\end{aligned} \tag{89}$$

Furthermore, one can derive the relations [26, 4]

$$\langle\Omega_k|\tilde{\phi}^f\phi_h|\Omega_k\rangle = 0\,,\quad f\neq h\,, \tag{90}$$

$$\langle\Omega_k|\det\tilde{\phi}|\Omega_k\rangle = \langle\Omega_k|\det\phi|\Omega_k\rangle = 0\,. \tag{91}$$

All these results (see (89)–(91)) are fully consistent with the WTIs of supersymmetry and with (60) and (61) implied by the Konishi anomaly relation [4].

The important conclusion of this thorough analysis is that the non-renormalisation theorems [52] of supersymmetry are violated by instanton effects as it results from the fact that chiral (composite) operators acquire non-vanishing v.e.v.'s, while they are identically zero at the perturbative level. One way of understanding this surprising finding in the language of the effective theory approach of Sect. 5 is to say that instantons generate a contribution to the effective superpotential which is non-perturbative in nature.

Massless SQCD

We now consider the strictly massless ($m_f = 0$, $f = 1,\ldots,N_f$) SQCD theory. From the formulae we derived in the previous sections it should be already clear that the limit $m_f \to 0$ is not smooth. Indeed, we will see that a straightforward application of the instanton calculus rules, that we have developed in the massive case, to massless SQCD leads to results that do not agree with the massless limit of the massive formulae.

The origin of this discrepancy is not completely clear. As we said, one possibility is that the $m_f \to 0$ limit of the massive theory does not coincide with the strictly massless theory, as a consequence of the fact that the small mass limit of massive SQCD is plagued by infrared divergences. Besides the divergences encountered if the massless limit of (89) is taken, a simple analysis shows, in fact, that a (naive) small $|m_f|^2$ Taylor expansion gives raise to $|m_f|^2 \times 1/|m_f|^2$ contributions that would be absent in the strictly massless SQCD theory. Another possibility, strongly advocated in [32] and [21, 31], is related to the observation that in the absence of mass terms the matter superpotential has a huge manifold of flat directions along which the exponential of the action does not provide any damping. In this situation it is not at all clear that the instanton solution (24) can be taken as the configuration that dominates the functional integral. Other types of quasi-saddle points, where scalar fields take a non-zero v.e.v., may be also relevant. The strategy suggested by these authors to deal with this situation will be discussed in Sect. 5. Here we want to first show what sort of results follow when the massive instanton calculus developed in Sect. 4.2 is blindly applied to massless SQCD.

The Green functions that have the correct number of fermionic zero modes in the one-instanton background are restricted to

$$
\begin{aligned}
&G^{(N_f,N_c-N_f)\{f\}}_{\{h\}}(x_1,\ldots,x_{N_c}) \qquad\qquad \text{for } N_c \geq N_f \qquad (92)\\
&= \langle \tilde{\phi}^{f_1}\phi_{h_1}(x_1)\ldots\tilde{\phi}^{f_{N_f}}\phi_{h_{N_f}}(x_{N_f})\frac{g^2}{32\pi^2}\lambda\lambda(x_{N_f+1})\ldots\frac{g^2}{32\pi^2}\lambda\lambda(x_{N_c})\rangle\,,\\
&D(x,x') = \langle \det[\tilde{\phi}(x)]\det[\tilde{\phi}(x')]\rangle\,, \qquad\qquad \text{for } N_c = N_f\,, \qquad (93)
\end{aligned}
$$

because now there exist zero modes also for the matter fermions, $\tilde{\psi}^f$ and ψ_f. A non-vanishing result is obtained if for each scalar field an appropriate Yukawa interaction term is brought down from the action. In this way $2N_c$ gluino zero modes, λ_0, together with the fermionic matter zero modes, $\tilde{\psi}^f_0$ and ψ_{0f}, $f = 1,\ldots,N_f$, will appear simultaneously. At the same time, when scalars are contracted in pairs, the scalar propagator in the instanton background, $(D^2)^{-1}$ or $(\tilde{D}^2)^{-1}$, is generated which will act on the product $\lambda_0\psi_0$ or $\tilde{\psi}_0\lambda_0$, respectively. Unlike the massive case, closed expressions for $(D^2)^{-1}$ and $(\tilde{D}^2)^{-1}$ exist which allows to explicitly compute the form of the "induced scalar modes", by solving the field equations $D^2\phi + ig\sqrt{2}\lambda_0\psi_0 = 0$ and $\tilde{D}^2\tilde{\phi} - ig\sqrt{2}\tilde{\psi}_0\lambda_0 = 0$, respectively.

The problem with the SCI computational strategy we have briefly described can already be seen by taking, for simplicity the case $N_c = 2$ and $N_f = 1$. In massless SQCD (after correcting for the usual factor 2^{N_c} with respect to result quoted in [4]), one gets

$$
\langle \tilde{\phi}\phi(x_1)\frac{g^2}{32\pi^2}\lambda\lambda(x_2)\rangle\Big|_{m=0} = \frac{1}{2}\frac{(\Lambda_{2,1}^{1-\text{loop}})^5}{g^4}\,, \qquad (94)
$$

while for the same Green function in the massive case we got (see (87))

$$\langle\tilde{\phi}\phi(x_1)\frac{g^2}{32\pi^2}\lambda\lambda(x_2)\rangle\big|_{m\neq 0} = \frac{4}{5}\frac{(\Lambda_{2,1}^{1-\text{loop}})^5}{g^4} \,. \tag{95}$$

Apart from the numerical discrepancy visible between (94) and (95), what is more disturbing is that (94) is in conflict with the Konishi anomaly relation (60), which in the massless regime (and using clustering) implies the vanishing of the gluino condensate. An alternative to this conclusion would be to say that the scalar condensate can be infinite in massless SQCD (see the discussion in Sect. 5.2).

Notice that for $N_f > 1$ the massless SQCD action possesses a non-anomalous $SU_L(N_f) \times SU_R(N_f) \times U_V(1) \times U_{\hat{A}}(1)$ symmetry (see (A.36)). This means that in extracting the scalar condensates a vacuum disentangling step analogous to the one performed in Sect. 4.1 is necessary. Proceeding in this way, one again finds results for the condensates that do not agree with what was found in the massive case.

Also the result for the correlator (93) is at variance with (88). We now find $D(x, x') \neq 0$, which implies (no disentangling is necessary here, as $\det\phi$ and $\det\tilde{\phi}$ are invariant under the chiral flavour group)

$$\langle \det\tilde{\phi}\rangle \neq 0\,, \qquad \langle\det\phi\rangle \neq 0\,, \tag{96}$$

signalling the spontaneous breaking of the $U_V(1)$ symmetry.

4.3 The case of Chiral Theories

In this section we wish to discuss the very interesting case of supersymmetric theories of the Georgi–Glashow type [53], where matter fermions are chiral. There is a quite remarkable literature on the subject. A selection of useful papers can be found in [4, 31, 54, 55, 56].

In this review we will limit to consider $SU(N_c)$ gauge theories with matter in the fundamental, $\mathbf{N_c}$, and anti-symmetric, $\mathbf{N_c(N_c-1)/2}$, representation. We recall that gauge anomaly cancellation requires the number of fundamentals, n_{fund}, and anti-symmetric, $n_{\text{anti}} \equiv M$, representations to be related by

$$n_{\text{fund}} = M(N_c - 4)\,. \tag{97}$$

The resulting β-function

$$\beta_{\text{GG}} = -\frac{g^2}{8\pi^2}[3(N_c + M) - MN_c] + \mathrm{O}(g^5) \tag{98}$$

implies asymptotic freedom if $M < 3N_c/(N_c - 3)$.

The composite operators that, besides $g^2\lambda\lambda/32\pi^2$, come into play are generically constructed in terms of the lowest components of the chiral matter superfields for which we introduce the notation

$$\begin{aligned} &\Phi_r^I \quad &, \quad &I = 1, 2, \ldots, n_{\text{fund}} \,, \\ &X_i^{rs} = -X_i^{sr} \quad &, \quad &r, s = 1, 2, \ldots, N_c \,, \; i = 1, 2, \ldots, M \,. \end{aligned} \tag{99}$$

Non-perturbative calculations are of special importance here, as Witten index arguments have so far been unable to make any definite statement about the nature of the vacua of the theory. Actually a variety of scenarios turn out to be realised according to the specific matter content of the action that can be summarised as follows.

(I) *Unbroken supersymmetry with well-defined vacua* [54, 31]. One such example is the $SU(6)$ case with $M = 1$ and correspondingly $n_{\text{fund}} = 2$. The allowed superpotential possesses no flat directions and the unique perturbative vacuum is at vanishing values of the scalar fields. There exist instanton-dominated (constant) Green functions, which upon using clustering give results in perfect agreement with the constraints coming from the Konishi anomaly relations. One finds that the discrete $_{30}$ symmetry is spontaneously broken down to $_{6}$, leaving behind 30/6=5 well-defined supersymmetric vacua. We remark that here, unlike SYM and SQCD, the number of vacua is not equal to N_c. It should be noted that in this example vacuum disentangling can be trivially carried out.

A more delicate situation occurs if we double the number of families, i.e. if we take $M = 2$ [56], because a non-trivial vacuum disentangling over the transformations of the complexification of the global symmetry group $SU(4) \times SU(2)$ is necessary here. When this is done, results from instanton calculations allow to determine all the condensates. In particular, one finds that the discrete $_{12}$ symmetry group is spontaneously broken down to $_{3}$, leaving behind 12/3=4 well-defined supersymmetric vacua. The interesting observation is that in this theory also the relations entailed by the Konishi anomaly equations allow to completely compute all the condensates. Reassuringly, the two sets of results turn out to be in perfect agreement. For a discussion of these results from the complementary effective action point of view, see Sect. 5.3.

In both the above cases when the superpotential is switched off the vacuum becomes ill defined, because in this limit necessarily some of the condensates must "run away" to infinity. This is due to the fact that the relations among condensates involve (inverse) factors of the Yukawa couplings.

(II) *Unbroken supersymmetry with ill-defined vacua.* This situation occurs in theories based on a $SU(N_c)$ gauge group with N_c even and larger than 8. Also in the presence of a non-vanishing superpotential, one finds that, in order to reconcile instanton results with the implication of the Konishi anomaly relations, one has to assume that some of the scalar condensates run away to infinity [56]. Such a result is seen to be related to the existence of flat directions in the superpotential. In this respect the situation is similar to massless SQCD, where we had at the same time flat directions in the superpotential and infinite scalar condensate in order to avoid contradictory results between instanton calculations and the Konishi anomaly equation.

(III) *Spontaneously broken supersymmetry.* This conclusion indirectly arises in Georgi–Glashow-type models in which the gauge group is $SU(N_c)$ with N_c odd, because of conflicting constraints between instanton calculations and relations implied by the Konishi anomaly equation. Several specific cases have been considered in the literature. We list here some interesting examples.

1. $N_c = 5$, $M = 1$ and consequently $n_{\text{fund}} = 1$. Although in this case no superpotential can be constructed, it can be shown that the theory does not admit any perturbative flat direction [31]. Because of the absence of superpotential the Konishi anomaly relations imply that the gluino condensate must vanish. When this result is put together with the non-perturbative calculations of certain instanton-dominated Green functions [54, 4], one is led to the conclusion that the involved scalar condensate cannot take a finite value, if clustering is used and the vacuum is supersymmetric. The wandering to infinity of the scalar condensate in the absence of flat directions looks highly implausible and one should rather conclude that there is a dynamical breaking of supersymmetry owing to non-perturbative instanton effects.
2. $N_c = 5$, $M = 2$ and consequently $n_{\text{fund}} = 2$. This time the theory admits a superpotential but still no flat directions exist. Unlike the previous case, from the Konishi anomaly equations one can prove that all condensates must vanish. This is in contradiction with the non-vanishing result given by the instanton calculation of the Green function where the product of these condensates appear, if clustering is invoked and the vacuum is supersymmetric. The most tempting conclusion is that supersymmetry is dynamically broken. One might object to this conclusion that actually the instanton calculation is performed in the absence of a superpotential, i.e. in a situation where disentangling is necessary and flat directions are present. This should not be a problem, however, because, unlike the case of the mass dependence, one expects the limit in which the superpotential coupling vanishes to be a smooth one. For a discussion of these results from the complementary effective action point of view, see Sect. 5.3.
3. $N_c \geq 7$, $M = 1$. There exist many instanton-dominated Green functions which, after vacuum disentangling, yield an overdetermined set of relations for the condensates [56]. One can solve the resulting equations finding full consistency with clustering. However, the Konishi anomaly equations are such as to imply the vanishing of several condensates and thus through instanton results the run away of others. Because no (perturbative) flat directions exist in these models, one is led again to conclude that supersymmetry is spontaneously broken.

5 The Effective Action Approach

Symmetry properties of the action in the form of anomalous and non-anomalous WTI's together with explicit dynamical (instanton) calculations have taught us a lot about the nature of supersymmetric $\mathcal{N} = 1$ theories. A very useful and elegant way to recollect all these results is to make recourse to the notion of "effective" or "low-energy", action (sometimes also referred to as "effective Lagrangian"). This notion, though with slightly different meanings and realm of application, has a long history. It was first introduced in the papers of [27], as a way of compactly deriving the soft pion theorems of current algebra, then fully developed for QCD with the inclusion of the η' meson and the $U_A(1)$ anomaly in the works quoted in [57] and [28]. A parallel road was opened by Symanzik [58] to deal with the lattice regularisation of QCD, which turned out to be crucial for understanding the approach to the continuum of the lattice theory.

The extension of these ideas to supersymmetric theories was first proposed in refs. [29, 30], where the cases of $\mathcal{N} = 1$ pure SYM and SQCD were considered, and then expanded to a field of investigation of its own in [21, 37, 41]. Some review papers on the subject can be found in [59].

Effective actions for all the theories we have discussed in the previous sections have been constructed and many interesting results have been obtained. In the following we want to briefly review what was done with the main purpose of comparing with instanton results.

5.1 The Effective Action of SYM

The first step along the way of constructing the effective action, Γ_{eff}, describing the low-energy dynamics of a theory is to identify the degrees of freedom relevant in the energy regime $E \ll \Lambda$, where Λ is the theory RGI mass scale. In the pure SYM case, where confinement seems to hold, the obvious degrees of freedom can be collected in the (dimension three) superfield

$$S = \frac{g^2}{16\pi^2}\mathrm{Tr}(W^\alpha W_\alpha)\,, \tag{100}$$

whose lowest component is precisely the gluino composite operator (41).

The second step is the observation that the interesting piece of $\Gamma_{\text{eff}}^{\text{SYM}}$ is not so much its kinetic contribution (a D-term which is non-holomorphic in S and reduces to the standard kinetic terms as $g \to 0$), but rather the F-term which provides the correct anomalous transformation properties of the effective action. In the present case, it is enough and convenient to make reference to the $U_R(1)$ symmetry (see (A.33)) to fix the form of this term, which is often referred to with the name of "effective superpotential" in the literature.

Recalling the $U_R(1)$ transformation properties of the superfield S (see the table in Appendix A)

$$S(x,\theta) \to \mathrm{e}^{3i\alpha} S(x, \theta \mathrm{e}^{-3i\alpha/2})\,, \tag{101}$$

we are led to write for the full effective action the formula

$$\Gamma^{\rm SYM}_{{\rm eff};N_c} = \Gamma^{\rm SYM}_{\rm kin}(S,S^*) + \left[W^{\rm SYM}_{{\rm eff};N_c}(S) + {\rm h.c.}\right] . \qquad (102)$$

where

$$\Gamma^{\rm SYM}_{\rm kin}(S,S^*) = k\big[(S^*S)^{1/3}\big]_D , \qquad (103)$$

$$W^{\rm SYM}_{{\rm eff};N_c}(S) = -\Big[S\big(\log \frac{S^{N_c}}{(c\Lambda_{\rm SYM})^{3N_c}} - N_c\big)\Big]_F . \qquad (104)$$

In the above equations we used the shorthand notation

$$[(\dots)]_F \equiv \int {\rm d}^4x\, {\rm d}^2\theta\, (\dots) , \qquad [(\dots)]_D \equiv \int {\rm d}^4x\, {\rm d}^2\theta\, {\rm d}^2\bar\theta\, (\dots) . \qquad (105)$$

The expression of $\Gamma^{\rm SYM}_{\rm kin}(S,S^*)$ in (103) is in no way unique. It is only an example of a functional having the property that (with a suitable choice of the constant k) it reproduces the standard form of the kinetic term in the limit $g \to 0$. The other constant c cannot be fixed by symmetry considerations only. A way of determining its value will be discussed in Sect. 5.2. Equation (104) is the famous Veneziano–Yankielowicz effective action [29].

It is not difficult to prove that the second term in (102) has the desired transformation properties under $U_R(1)$ (see (A.34)). From (101) we get in fact (the x-dependence and the corresponding space–time integration is understood)

$$\bullet \int {\rm d}^2\theta\, S(\theta) \to {\rm e}^{3i\alpha} \int {\rm d}^2\theta\, S(x,\theta {\rm e}^{-3i\alpha/2})$$
$$= \int {\rm d}^2(\theta {\rm e}^{-3i\alpha/2}) S(\theta {\rm e}^{-3i\alpha/2}) \int {\rm d}^2\theta\, S(\theta) , \qquad (106)$$

$$\bullet \int {\rm d}^2\theta\, S(\theta) \log S(x,\theta) \to {\rm e}^{3i\alpha} \int {\rm d}^2\theta\, S(\theta {\rm e}^{-3i\alpha/2})\big(3i\alpha + \log S(\theta {\rm e}^{-3i\alpha/2})\big)$$
$$= 3i\alpha \int {\rm d}^2(\theta {\rm e}^{-3i\alpha/2}) S(\theta {\rm e}^{-3i\alpha/2}) + \int {\rm d}^2(\theta {\rm e}^{-3i\alpha}) S(\theta {\rm e}^{-3i\alpha}) \log S(\theta {\rm e}^{-3i\alpha/2})$$
$$= 3i\alpha \int {\rm d}^2\theta\, S(\theta) + \int {\rm d}^2\theta\, S(\theta) \log S(\theta) . \qquad (107)$$

Useful information about the non-perturbative properties of the theory can be obtained from the formula (102) by determining the values of S which make $\Gamma^{\rm SYM}_{\rm eff}$ stationary. These are constant field configurations which minimise the effective action. Thus they yield the values of the v.e.v. of the gluino composite operator (gluino condensate). From (102) one gets the result

$$\langle S \rangle = (c\Lambda_{\rm SYM})^3 {\rm e}^{2i\pi k/N_c} , \quad k = 1, \dots, N_c . \qquad (108)$$

If c^3 is identified with $(C_{N_c})^{1/N_c}$, then (108) becomes identical to (70). However, in connection with the comments we made in Sect. 4.1, we must remark here that there is a discrepancy between the number given by the above identification and the choice $c = 1$ made in [23, 31, 59]. The latter can be justified in the framework of the SQCD effective action approach, if in conjunction with WCI calculations, certain consistency relations following from decoupling (see Sect. 5.2) are employed.

A crystal clear way to resolve this puzzling discrepancy would be to arrive at an evaluation of the SYM effective action from first principles, i.e. à la Wilson–Polchinski [60]. Many efforts have been made in this direction and a lot of interesting results have been obtained [61] insisting on the role of anomalies [35, 62] in the construction of the Wilsonian action. A different and perhaps more promising road has been recently undertaken which uses the matrix model formulation of SYM [63]. In this framework the form of the effective action could be derived [64, 65] and the result $c = 1$ was obtained.

5.2 The Effective Action of SQCD

For the purpose of extending the previous considerations to SQCD it is convenient to distinguish among the three cases $N_c > N_f$, $N_c = N_f$ and $N_c < N_f$ and separately discuss the massive and the massless situation.

SQCD with $N_c > N_f$

The Massive Case

The generalisation of the previous formulae to massive SQCD is almost immediate if one includes among the degrees of freedom that describe the low energy dynamics of the theory also the composite operators (42)

$$T^f_h = \tilde{\phi}^f_r(x)\phi^r_h(x)\,. \tag{109}$$

Apart from the unessential (for this discussion) kinetic terms, one finds that the formula which extends (102) is

$$\begin{aligned} &\Gamma^{\rm SQCD}_{{\rm eff};N_c,N_f}(S,S^*;T,T^*) &(110)\\ &= \Gamma^{\rm SQCD}_{\rm kin}(S,S^*;T,T^*) + \left[W^{\rm SQCD}_{{\rm eff};N_c,N_f}(S;T) + {\rm h.c.}\right], \\ &W^{\rm SQCD}_{{\rm eff};N_c,N_f}(S;T) &(111)\\ &= \Big[-S\Big(\log\frac{S^{N_c-N_f}\,{\rm det}T}{(c'\Lambda_{\rm SQCD})^{3N_c-N_f}} - (N_c-N_f)\Big) + \sum_f m_f T^f_f\Big]_F\,. \end{aligned}$$

As before, the constant c' cannot be fixed by symmetry considerations only. We will discuss the important issue of determining its precise value below. The minimisation conditions (also called F-flatness conditions) allow to determine all the condensates and one finds[13]

[13] We recall the elementary formula $d\log[{\rm det}T]/dT^f_h = (T^{-1})^h_f$.

$$\langle S \rangle = \mathrm{e}^{\frac{2i\pi k}{N_c}} (c' \Lambda_{\mathrm{SQCD}})^3 \prod_{f=1}^{N_f} \Big(\frac{m_f}{c' \Lambda_{\mathrm{SQCD}}} \Big)^{1/N_c} , \quad k = 1, \dots, N_c , \tag{112}$$

$$\langle T_h^f \rangle = \delta_h^f \frac{\langle S \rangle}{m_f} , \tag{113}$$

viz. the same results that were obtained in (89) up to the normalisation of the Λ parameter.

Equation (111) has a number of nice properties.

(1) It is mathematically meaningful not only for $N_c > N_f$, where instanton calculations are feasible, but for any value of N_c and N_f, except $N_c = N_f$. Actually in the last case a further composite operator has to come into play, as already hinted at by the results of Sects. 4.2 and 4.2. We will discuss in detail this case below.

(2) It obeys the expected decoupling theorem in the sense that when one of the flavours gets infinitely massive, (111) precisely turns into the effective action for the theory with one less flavour, in which that particular flavour is absent, provided the Λ parameters of the two theories, $\Lambda_{\mathrm{SQCD}}^{N_f}$ and $\Lambda_{\mathrm{SQCD}}^{N_f-1}$, are matched as described in Appendix D.

(3) The massive S field can be "integrated out", leaving a pure matter effective superpotential

$$\begin{aligned} &W_{\mathrm{eff};N_c,N_f}^{\mathrm{SQCD}}(T) \\ &= \Big[(N_c - N_f) \Big(\frac{(c' \Lambda_{\mathrm{SQCD}})^{3N_c - N_f}}{\det T} \Big)^{\frac{1}{(N_c - N_f)}} + \sum_f m_f T_f^f \Big]_F , \end{aligned} \tag{114}$$

which coincides with the Affleck–Dine–Seiberg [21] effective superpotential. One has to remark, however, that the formula is only meaningful for $N_c > N_f$.

The Massless Case

The discussion of the massless case is quite delicate (and controversial). This is to be expected looking at the results of the dynamical instanton calculations presented in Sect. 4.2, which showed conflicting findings for the condensates when the massless limit of the massive results were compared to what one gets in the strictly massless case.

Let us now see what are the implications of the massless SQCD effective action from using the formulae (111) or (114). If we start from (111), we conclude that, when any of the masses is sent to zero, the gluino condensate must be taken to vanish for consistency with (112), while (113) does not contain sufficient information to determine any of the scalar condensates $\langle T_h^f \rangle$. If on the contrary all the masses are set to zero from the beginning, as was done in [31], and (114) is used, one must conclude that the scalar condensates run away to infinity as the potential generated by the massless effective action (114) monotonically decreases to zero in this limit.

SQCD with $N_c = N_f$

When $N_f = N_c$ new composite operators can be constructed which must appear among the fields of the effective action. They are the determinants, X and $\tilde{X}$, over colour and flavour indices of the matter fields, whose lowest components are shown in (74) and (75). The need for new field operators is confirmed by the observation that the action (110) does not contain a sufficiently large number of massless fermions to satisfy the 't Hooft anomaly matching conditions [41, 44, 59] associated with the non-anomalous $U_{\hat{A}}(1)$ symmetry defined by (A.35) and (A.36) (see also the table in Appendix A).

The Massive Case

The most general form of effective action is

$$\Gamma_{\rm eff}^{\rm SQCD}(S,S^*;T,T^*;X,\tilde{X}) = \Gamma_{\rm kin}^{\rm SQCD}(S,S^*;T,T^*;X,\tilde{X}) + \Big[-S\big(\log\frac{{\rm det}T}{(c''\Lambda_{\rm SQCD})^{2N_c}} + f(Z)\big) + \sum_f m_f T_f^f + {\rm h.c.}\Big]_F , \tag{115}$$

where $f(Z)$ is a function of the ratio $Z = X\tilde{X}/{\rm det}T$. For $m_f \neq 0$ (all f) the stationarity conditions lead to the equations

$$\log\frac{{\rm det}T}{(c''\Lambda_{\rm SQCD})^{2N_c}} + f(Z) = 0\,, \tag{116}$$

$$m_f\delta_h^f = S\left[1 - Z\frac{\partial f(Z)}{\partial Z}\right](T^{-1})_f^h\,, \tag{117}$$

$$S\,\frac{\partial f(Z)}{\partial Z}\,\frac{\tilde{X}}{{\rm det}T} = 0\,, \tag{118}$$

$$S\frac{\partial f(Z)}{\partial Z}\,\frac{X}{{\rm det}T} = 0\,, \tag{119}$$

which have the solution

$$\langle X\rangle = \langle\tilde{X}\rangle = 0\,, \tag{120}$$

$$m_f\langle T_h^f\rangle = \delta_h^f\langle S\rangle\,, \tag{121}$$

$$\langle{\rm det}T\rangle = {\rm e}^{-f(0)}c''\big(\Lambda_{\rm SQCD}\big)^{2N_c}\,, \tag{122}$$

a result with exactly the same structure as (89)–(91).

The Massless Case

The massless case is, as usual, more subtle. Besides $\langle S\rangle = 0$ (as implied by the massless limit of the Konishi anomaly equation (121)), by varying the effective action with respect to S one still gets the constraint

$$\log\frac{{\rm det}T}{(c''\Lambda_{\rm SQCD})^{2N_c}} + f(Z) = 0\,, \tag{123}$$

which (if $f(Z) \neq 0$) only fixes one combination of $\langle X \rangle$, $\langle \tilde{X} \rangle$ and $\langle \det T \rangle$, leaving the other two undetermined. This can be interpreted as the equivalent of the statement that for $N_f = N_c$ and $m_f = 0$ the perturbative flat directions are not (all) removed, so vacua with arbitrarily large values of these condensates can occur. The effective action vanishes at the minimum and one is only left with the constraint (123).

The explicit form of this constraint was worked out in [41] with the conclusion that the classical relation $\det T = X\tilde{X}$ is lifted by quantum correction to the formula

$$\det T - X\tilde{X} = (\Lambda_{\rm SQCD})^{2N_c} \,. \tag{124}$$

This was the first example of a by now well-known phenomenon (see Sect. 8) according to which the quantum theory can display a whole manifold of (degenerate) vacuum states where supersymmetry is unbroken. It is a complex Kähler manifold (often called the "quantum moduli space", $\mathcal{M}$) to which the point representing the classical vacuum not always belongs. We end this discussion by noticing that the constraint (124) can also be derived by a massless effective action of the type (115) if one simply takes

$$f(Z) = \log(1 - Z) \,. \tag{125}$$

SQCD with $N_f > N_c$

In this case neither dynamical instanton calculations are possible (see our discussion in Sect. 4) nor the general considerations of [31] apply. In principle, one can imagine to go on with the effective action approach, guided by information on the relevant low-energy degrees of freedom provided by the 't Hooft anomaly matching conditions.

- For instance, in the case $N_f = N_c + 1$ the two baryon-like superfields

$$B^f = \epsilon^{f f_1 f_2 \dots f_{N_c}} \epsilon_{r_1 r_2 \dots r_{N_c}} \Phi^{r_1}_{f_1} \Phi^{r_2}_{f_2} \dots \Phi^{r_{N_c}}_{f_{N_c}} \,, \tag{126}$$

$$\tilde{B}_f = \epsilon_{f f_1 f_2 \dots f_{N_c}} \epsilon^{r_1 r_2 \dots r_{N_c}} \bar{\Phi}^{f_1}_{r_1} \bar{\Phi}^{f_2}_{r_2} \dots \bar{\Phi}^{f_{N_c}}_{r_{N_c}} \,. \tag{127}$$

must come into play in order to fulfil such conditions. They can combine with T^f_h to give the term $B^f T^h_f \bar{B}_h$ in the effective action. The whole expression of the latter can then be argued to have the form

$$\Gamma^{\rm SQCD}_{\rm eff} = \Gamma^{\rm SQCD}_{\rm kin} + \Big[\frac{\det T - B^f T^h_f \tilde{B}_h}{(\Lambda_{\rm SQCD})^{b_1}} + {\rm h.c.} \Big]_F \,, \tag{128}$$

where $b_1 = 3N_c - N_f = 2N_c - 1$. As a consistency check, it can be shown that, if the $(N_f + 1)$-th flavour is given a mass and decoupled, then the situation we described in the previous subsection where we had $N_f = N_c$ is recovered.

It is interesting to remark that by solving the F-flatness equations implied by the effective action (128), one finds that, unlike the case $N_f = N_c$, the point corresponding to the classical vacuum $\langle T^h_f \rangle = \langle B^f \rangle = \langle \tilde{B}_h \rangle = 0$ belongs to the moduli space of the theory. At this point the (non-anomalous) symmetry of the classical action, $SU_L(N_f) \times SU_R(N_f) \times U_V(1) \times U_{\hat{A}}(1)$, is fully unbroken.

• For larger values of N_f ($N_f > N_c + 1$, but smaller than $3N_c$, where UV asymptotic freedom is lost) the above formulae have an obvious generalisation. One introduces the chiral superfields

$$B^{f_1 \dots f_{\tilde{N}_c}} = \epsilon^{f_1 \dots f_{\tilde{N}_c} f_{\tilde{N}_c+1} \dots f_{N_f}} \epsilon_{r_1 \dots r_{N_c}} \Phi^{r_1}_{f_{\tilde{N}_c+1}} \dots \Phi^{r_{N_c}}_{f_{N_f}} , \tag{129}$$

$$\tilde{B}_{f_1 \dots f_{\tilde{N}_c}} = \epsilon_{f_1 \dots f_{\tilde{N}_c} f_{\tilde{N}_c+1} \dots f_{N_f}} \epsilon^{r_1 \dots r_{N_c}} \tilde{\Phi}^{f_{\tilde{N}_c+1}}_{r_1} \dots \tilde{\Phi}^{f_{N_f}}_{r_{N_c}} , \tag{130}$$

where, following [41], we have set

$$\tilde{N}_c = N_f - N_c \, . \tag{131}$$

In terms of the operators (129), (130) and T^h_f one may construct the effective action

$$\begin{aligned} \Gamma^{\rm SQCD}_{\rm eff} &= \Gamma^{\rm SQCD}_{\rm kin} \\ &+ \Big[\frac{\det T - B^{f_1 f_2 \dots f_{\tilde{N}_c}} T^{h_1}_{f_1} T^{h_2}_{f_2} \dots T^{h_{\tilde{N}_c}}_{f_{\tilde{N}_c}} \tilde{B}_{h_1 h_2 \dots h_{\tilde{N}_c}}}{(\Lambda_{\rm SQCD})^{b_1}} + {\rm h.c.} \Big]_F , \end{aligned} \tag{132}$$

where b_1 is the first coefficient of the β-function of the theory.

The trouble with this analysis is that neither the 't Hooft anomaly conditions are fulfilled, if only the above set of composite operators is considered, nor the superpotential has the correct quantum numbers to fit the anomalous symmetries of the theory.

An inspiring and physically compelling interpretation of the situation was given in [41], where it was argued that the theory admits also a "dual" description in terms of a SQCD-like action with the same global "flavour" symmetries, hence with quark fields Q^f and $\tilde{Q}_f$ ($f = 1, 2, \dots, N_f$), but with gauge group $SU(\tilde{N}_c)$ with $\tilde{N}_c = N_f - N_c$. This conclusion follows from the observation that the moduli space of the theory remains unmodified quantum mechanically for all values of $N_f > N_c+1$, at least up to $N_f = 3N_c$. In turn this means that the classical vacuum at the origin (where all the expectation values of the composite fields which represent the degrees of freedom of the low energy theory vanish) preserves the original $SU_L(N_f) \times SU_R(N_f) \times U_V(1) \times U_{\hat{A}}(1)$ symmetry. Consequently, in the dual theory there must necessarily be N_f quark fields, though coupled to a different gauge group. In this theory the operators (129) and (130) are interpreted as composite operators of the form

$$B^{f_1 \dots f_{\tilde{N}_c}} = \epsilon^{r_1 \dots r_{\tilde{N}_c}} \tilde{Q}^{f_1}_{r_1} \tilde{Q}^{f_2}_{r_2} \dots \tilde{Q}^{f_{\tilde{N}_c}}_{r_{\tilde{N}_c}} , \tag{133}$$

$$\tilde{B}_{f_1 \dots f_{\tilde{N}_c}} = \epsilon_{r_1 \dots r_{\tilde{N}_c}} Q^{r_1}_{f_1} Q^{r_2}_{f_2} \dots Q^{r_{\tilde{N}_c}}_{f_{\tilde{N}_c}} . \tag{134}$$

An additional chiral, gauge invariant, supermultiplet, M^h_f, is assumed to exist, which is necessary for matching the 't Hooft anomaly conditions. In terms of the above composite fields an effective superpotential can be written down. It reads $Q^f M_f^h \tilde{Q}_h$. The relation between this theory and the original theory is referred to as "non-abelian electric–magnetic duality" (or more simply as "Seiberg duality") and indeed it can be argued to be a duality relation in the sense that the dual of the dual is the original theory with the quarks and gluons of one description interpreted as solitons (magnetic monopoles) of the other.

Summarising, according to [41, 59], we can briefly describe what happens to SQCD when N_f increases at fixed N_c beyond $N_c + 1$ as follows.

- - For $N_c + 2 \leq N_f < 3N_c/2$ the asymptotic particles of the theory are the the dual quark fields Q^f and $\tilde{Q}_f$ and the mesons M_f^h, which interact through an IR-free ($b_1 = 3\tilde{N}_c - N_f = 2N_f - 3N_c < 0$) supersymmetric theory with gauge group $SU(\tilde{N}_c)$. This how the SQCD theory we started with looks in terms of "magnetic" variables dual to the original "electric" variables (which are instead strongly coupled in this range of N_f values).
- - As soon N_f goes through the value $3N_c/2$, the first coefficient of the dual theory β-function changes sign and the theory is expected to flow to a non-trivial IR fixed point. This continues to be true for the whole range of values $3N_c/2 < N_f < 3N_c$. Both the original and the dual theory can be argued to be conformal theories of interacting quarks and gluons (we remark that $3N_c/2 < N_f < 3N_c \iff 3\tilde{N}_c/2 < N_f < 3\tilde{N}_c$). However, as N_f increases the electric variables tend to become more and more weakly coupled and the opposite happens for the dual magnetic variables.
- - At $N_f = 3N_c$, where the original theory loses asymptotic freedom, the IR fixed fixed point comes to zero coupling.
- - For even larger values of $N_f > 3N_c$ the original electric theory is an IR free theory of quarks and gluons.

Normalising the SQCD and SYM Effective Action

As we have seen, the interesting piece of the SYM and SQCD effective actions can be fixed by symmetry arguments only up to a constant rescaling of their Λ parameter. We want to show in this section how, exploiting the self-consistency requirement implicit in the decoupling theorem, one can fix these constants, if at the same time a dynamical (e.g. instanton based) information is available. We will develop the argument along the line of reasoning advocated in [31, 23, 37] and summarised in [59].

The Case of SQCD

Starting from (114) with just one massive flavour, say the N_f-th one, we require that the effective superpotential of the theory with N_f flavours, namely,

$$W^{\rm SQCD}_{{\rm eff};N_c,N_f}(T)$$
$$= \Big[(N_c - N_f)\eta(N_f)\Big(\frac{(\Lambda^{(N_f)}_{\rm SQCD})^{3N_c-N_f}}{{\rm det}T}\Big)^{\frac{1}{(N_c-N_f)}} + m_{N_f}T^{N_f}_{N_f}\Big]_F\,, \quad (135)$$

goes over to the effective superpotential of the theory with one flavour less when m_{N_f} gets large after using (D.3). To simplify and clarify notations, we have introduced the new constant $\eta(N_f) = (c')^{3N_c-N_f/N_c-N_f}$ with respect to what we had in (114) and we have attached the extra superscript N_f to the Λ parameter of SQCD in order to trace the number of "active" flavours in each theory.

We now proceed to eliminate $T^{N_f}_{N_f}$ by using the F-flatness condition for $T^{N_f}_{N_f}$, which amounts to the stationarity equation $\partial W^{\rm SQCD}_{{\rm eff};N_c,N_f}(T)/\partial T^{N_f}_{N_f} = 0$. We also notice that the analogous conditions for the $T^f_{N_f}$ and $T^{N_f}_h$ components imply the vanishing of their expectation value. After some algebra one finds that the r.h.s. of (135) becomes

$$\Big[\big(\eta(N_f)\big)^{\frac{N_c-N_f}{N_c-N_f+1}}(N_c - N_f + 1)\Big(\frac{m_{N_f}(\Lambda^{(N_f)}_{\rm SQCD})^{3N_c-N_f}}{{\rm det}\tilde{T}}\Big)^{\frac{1}{(N_c-N_f+1)}}\Big]_F\,, \quad (136)$$

where ${\rm det}\tilde{T}$ is the matter determinant with the N_f-th flavour missing. Since the decoupling condition (D.3) implies

$$m_{N_f}(\Lambda^{(N_f)}_{\rm SQCD})^{3N_c-N_f} = (\Lambda^{(N_f-1)}_{\rm SQCD})^{3N_c-N_f+1}\,, \quad (137)$$

we see that the expression (136) becomes the formula for the effective superpotential of SQCD with $N_f - 1$ flavours if $\eta(N_f)$ satisfies the equation

$$\big(\eta(N_f)\big)^{\frac{N_c-N_f}{N_c-N_f+1}} = \eta(N_f - 1)\,. \quad (138)$$

The most general solution of (138) is

$$\eta(N_f) = \eta_0^{\frac{1}{N_c-N_f}}\,, \quad (139)$$

with η_0 a quantity which does not depend on N_f. The last observation is rather important as it can be exploited to simplify the calculation of η_0. In practice, one can proceed in two ways. One is based on an explicit dynamical computation which was done in the WCI approach with the result $\eta_0 = 1$ [21, 31, 45, 46]. The calculation is performed in the especially simple case of SQCD with $N_f = N_c - 1$ flavours, where the $SU(N_c)$ gauge symmetry is completely broken by non-vanishing scalar v.e.v.'s (see Appendix E). In this situation the theory is weakly coupled for sufficiently large v.e.v.'s, thus constrained [47] instanton calculations are expected to be fully reliable.

The second strategy [37] consists in determining η_0 by means of a self-consistency constraint that fixes the value of the gluino condensate in an

$SU(2)_1 \times SU(2)_2$ gauge theory with matter in the $(\mathbf{2}, \mathbf{2})$ representation. The argument, which is quite elegant, exploits the knowledge of the effective superpotential of the theory, derived in [66], and confirms the result $\eta_0 = 1$.

The Case of SYM

Already the result mentioned at the end of the previous section is telling us that the normalising constant, c, in (104) is to be taken equal to one, at variance with the direct SCI calculation which, if a factor $N_c = 2$ is divided out, gave $c = 1/\sqrt{5}$ (see (63)–(70) and the discussion in Sect. 4.1).

There are other similar indirect ways to determine c. An elegant one is to start from a pure $\mathcal{N} = 2$ SYM theory with the addition of a mass term for the chiral (matter) superfield which breaks supersymmetry down to $\mathcal{N} = 1$. Decoupling the massive multiplet by sending the mass parameter to infinity leaves behind a pure $\mathcal{N} = 1$ SYM theory. The reason to reach $\mathcal{N} = 1$ in this somewhat complicated way is that for the effective action of pure $\mathcal{N} = 2$ SYM we have the beautiful SW formula [51] and a simple description of the theory in terms of low energy degrees of freedom (see Sect. 8).

To illustrate the method we wish to present here a simplified adaptation of the original argument given in [38]. The starting point is the formula

$$\langle g^2 \lambda^{\alpha a} \lambda^a_\alpha \rangle = -16\pi i \, \frac{\partial W^{\mathcal{N}=1}_{\text{eff}}}{\partial \tau} = \frac{32\pi^2}{b_1} \Lambda \frac{\partial W^{\mathcal{N}=1}_{\text{eff}}}{\partial \Lambda} , \tag{140}$$

where for the $SU(2)$ gauge group $b_1 = 3 \times 2$ and we have set

$$\tau = \frac{4\pi i}{g^2} + \frac{\vartheta}{2\pi} , \tag{141}$$

$$\Lambda = \mu \mathrm{e}^{2\pi i \tau(\mu)} . \tag{142}$$

We need to compute $W^{\mathcal{N}=1}_{\text{eff}}$ with its correct normalisation. In principle $W^{\mathcal{N}=1}_{\text{eff}}$ could be obtained from (104), after integrating out the S superfield. Precisely because at this stage the normalisation of the $\mathcal{N} = 1$ effective action is unknown, we shall start from the well-defined expression of the $\mathcal{N} = 2$ effective superpotential which in the relevant strong coupling regime takes the form

$$W^{\mathcal{N}=2}_{\text{eff}}(A_D, M, \bar{M}) = \sqrt{2} \bar{M} A_D M + m U(A_D) , \tag{143}$$

where the chiral superfields $M, \bar{M}$ describe the monopole multiplet and A_D is the dual Higgs superfield. In this regime the quantum modulus, U, is naturally expressed in terms of A_D (and not of A). Solving the F-flatness conditions leads to the v.e.v.'s

$$a_D \equiv \langle A_D \rangle = 0 , \qquad \langle M \rangle = \langle \bar{M} \rangle = \Big(- \frac{m}{\sqrt{2}} \frac{\mathrm{d}U(x)}{\mathrm{d}x} \Big|_{x=0} \Big)^{\frac{1}{2}} . \tag{144}$$

In this vacuum configuration, one finds

$$\begin{aligned} W_{\text{eff}}^{\mathcal{N}=2}(0, \langle M \rangle, \langle \bar{M} \rangle) &= m\, u(0) + \dots \\ &= m\big(\Lambda_{\text{SW}}\big)^2 + \dots = 2m\big(\Lambda_{\text{PV}}^{\mathcal{N}=2}\big)^2 + \dots , \end{aligned} \tag{145}$$

where $u(0) = \langle U(0) \rangle = \Lambda_{\text{SW}}^2$ (as it follows from the known relation between a_D and $u = \langle U \rangle$[14]) and Λ_{SW} is the SW dynamical scale. In the last equality for the purpose of comparing with the rest of our formulae, we have introduced the more standard Pauli–Villars scale (which we consistently used throughout this review) related to the former by the relation [37] $\Lambda_{\text{PV}}^{\mathcal{N}=2} = \Lambda_{\text{SW}}/\sqrt{2}$.

The last step of this quite elaborate argument consists in decoupling the matter superfield by sending m to infinity while keeping fixed the combination (see (D.3) and (67))

$$\Lambda^6 = m^2 \left(\Lambda_{\text{PV}}^{\mathcal{N}=2}\right)^4 . \tag{146}$$

Inserting this relation in the last equality of (145) gives

$$W_{\text{eff}}^{\mathcal{N}=1} = 2\Lambda^3 , \tag{147}$$

from which the equation

$$\langle \frac{g^2}{32\pi^2} \lambda^{\alpha a} \lambda_{\alpha}^{a} \rangle = \Lambda^3 \tag{148}$$

follows. This calculation again yields the so-called WCI result $c = 1$.

Although, as we have developed the argument, this computation is enough to fix the normalisation of the effective potential for any number of colours, it would be nice to repeat a similar reasoning in the generic case of an $SU(N_c)$ gauge group in order to explicitly check the N_c behaviour of the gluino condensate. This issue is of relevance for the interesting question of relating non-supersymmetric QCD-like gauge theories with supersymmetric ones in the large N_c limit as proposed in the nice papers of [67].

5.3 The Effective Action of Georgi–Glashow-type Models

A number of interesting results have been obtained in the literature [55, 68] for supersymmetric theories with chiral matter. Here, for brevity, we will only discuss two specific cases (1) $SU(6)$ with two matter superfields in the **6** and one in the $\bar{\mathbf{15}}$ representation and (2) $SU(5)$ with one matter superfields in the **5** and another one in the $\bar{\mathbf{10}}$ representation, as prototypes of two different typical situations, namely unbroken supersymmetry with well-defined vacua and dynamically broken supersymmetry, respectively (see the corresponding discussion in Sect. 4.3).

[14] We recall the SW formula $a_D = \frac{\sqrt{2}}{\pi} \int_{\Lambda_{\text{SW}}^2}^{u} \mathrm{d}x \sqrt{\frac{x-u}{x^2 - \Lambda_{\text{SW}}^4}}$, see Sect. 8.

$SU(6)$ with Matter in $2 \times 6 + \bar{15}$

The construction of the effective action of this theory requires, besides the chiral composite superfields (see (99))

$$S = \frac{g2}{32\pi 2} W^\alpha W_\alpha \,, \quad T = \epsilon_{IJ} \Phi^I_r \Phi^J_s X^{rs} \,, \tag{149}$$

$$U = \epsilon_{r_1 s_1 r_2 s_2 r_3 s_3} X^{r_1 s_1} X^{r_2 s_2} X^{r_3 s_3} \,, \tag{150}$$

the real (vector) ones

$$R^J_I = \Phi^\dagger_I \mathrm{e}^{2gV(\mathbf{6})} \Phi^J \,, \quad Q = X^\dagger \mathrm{e}^{2gV(\bar{\mathbf{15}})} X \,. \tag{151}$$

The expression of the effective action which fulfils all the relevant anomalous and non-anomalous WTIs of the microscopic theory reads [55]

$$\begin{aligned} \Gamma^{\mathrm{GG}-SU(6)}_{\mathrm{eff}} &= \Big[\mathrm{Tr} R + Q + S^\star S + \xi_R \mathrm{Tr} \log R + \xi_Q \log Q \\ &+ \mathrm{Tr}(T^\dagger R^{-1} T)(\det R)^{-1} Q^{-1} + U^\star U Q^{-3} \Big]_D \\ &+ \Big[\mathrm{Tr} W_R^2 + W_Q^2 + S \log \frac{S^3 X T}{\Lambda^{15}_{\mathrm{GG}}} + hT + h'U + \mathrm{h.c.} \Big]_F \,. \end{aligned} \tag{152}$$

where ξ_R, ξ_Q are (in principle) calculable constants, Λ_{GG} is the RGI scale parameter of the theory and

$$W_{R\alpha} = -\frac{1}{4} \bar{D}^2 R^{-1} D_\alpha R \,, \qquad W_{Q\alpha} = -\frac{1}{4} \bar{D}^2 Q^{-1} D_\alpha Q \,. \tag{153}$$

Despite its quite complicated form, the consequences of (152) are rather simple. One gets for the v.e.v.'s of the composite operators (149) and (150)

$$\langle S \rangle_k = (hh')^{1/5} (\Lambda_{\mathrm{GG}})^3 \mathrm{e}^{2\pi i k/5} \,, \quad k = 1, 2, \ldots, 5 \,, \tag{154}$$

$$\langle T \rangle_k = \frac{\langle S \rangle_k}{h} \,, \qquad \langle U \rangle_k = \frac{\langle S \rangle_k}{h'} \,, \tag{155}$$

in perfect agreement with instanton results and the constraints imposed by the Konishi anomaly equations. One finds that the discrete $_{15}$ symmetry group is spontaneously broken down to $_3$, leaving behind 15/3=5 well-defined supersymmetric vacua. As we noticed in Sect. 4.3 point (I), for non-vanishing value of the Yukawa couplings h, h' supersymmetry is unbroken and the vacuum states are well defined. Only when either h or h' go to zero and flat directions appear in the superpotential, some of the condensates run away to infinity.

$SU(5)$ with Matter in $5 + \bar{10}$

This case is more interesting as the phenomenon of dynamical breaking of supersymmetry is seen to occur [68]. The construction of the effective action which fulfils all the anomalous and non-anomalous WTIs of the microscopic theory again requires the introduction of the two real composite (vector) superfields

$$R = \Phi^\dagger e^{2gV(\mathbf{5})}\Phi\,, \quad Q = X^\dagger e^{2gV(\overline{\mathbf{10}})}X\,. \tag{156}$$

Furthermore, besides the chiral composite operators (149) and (150), the chiral superfields

$$Y = W_s^{r_1} W_r^s \Phi_t X^{tr} X^{r_2 r_3} X^{r_4 r_5} \epsilon_{r_1 r_2 r_3 r_4 r_5}\,, \quad A = \frac{g^2}{16\pi^2}(W^2)_s^r \Phi_r \Phi_{s'} X^{s's} \tag{157}$$

must come into play in order to fulfil the 't Hooft anomaly conditions. Finally, the requirement of the absence of flat directions in the microscopic theory implies a judicious choice of the invariant kinetic terms. An expression of the effective action which satisfies all the above constraints is

$$\begin{aligned}\Gamma_{\text{eff}}^{\text{GG}-SU(5)} &= \Big[R + Q + S^\star S + \xi_1 \log R + \xi_2 \log Q \\ &+ (Y^\star R^{-1} T Q^{-3} Y)^{-1} + A^\star R Q A\Big]_D \\ &+ \Big[\kappa_1 W_R^2 + \kappa_2 W_Q^2 + S \log \frac{S^2 Y}{\Lambda_{\text{GG}}^{13}} + \text{h.c.}\Big]_F\,.\end{aligned} \tag{158}$$

The minimisation of $\Gamma_{\text{eff}}^{\text{GG}-SU(5)}$, displays the phenomenon of dynamical breaking of supersymmetry. One finds, in fact, that the minimum occurs at finite non-vanishing values of all the condensates (with the exception of A for which a vanishing result is obtained) and that at this point the effective superpotential is positive.

It is interesting to look at the spectrum of the low-lying states that emerges from the analysis of the effective potential (158). Together with supersymmetry, a non-anomalous $U(1)$ is spontaneously broken by the v.e.v. of Y. Another anomalous $U(1)$ remains instead unbroken and its triangle anomaly is saturated by the composite fermion in A, which remains massless. The only other massless fermion in the spectrum is the Goldstino associated with the spontaneous breaking of supersymmetry. The latter partially lies in the real vector fields R and Q. In the spin zero sector we find the massless Goldstone boson of the spontaneously broken $U(1)$ mentioned above. Two more would-be Goldstone bosons are eaten up *a la Higgs* to give mass to the vector bosons belonging to R and Q. It is the fact that the Goldstino partially lies in the real superfields R and Q that prevents integrating out their massive degrees of freedom, because if one does so the manifest supersymmetry of the effective action is lost.

The overall picture that is coming out is completely consistent with the symmetry breaking pattern that emerges from the dynamical instanton computation of the Green functions with only insertions of lowest components of chiral composite superfields (see Sect. 4.3, point (III) 1).

6 $\mathcal{N} = 2$ SYM: Introduction

As shown in the previous discussion of $\mathcal{N} = 1$ SYM theories, the combination of instanton calculus with holomorphy of the F-terms in the (low-energy) effective action proves to be very powerful in that it allows to determine non-perturbative corrections to the superpotential and argue for dynamical supersymmetry breaking in a class of models. Unfortunately, the spectrum of bound states in supersymmetric vacuum configurations, if present, depends not only on the F-terms, encompassing the superpotential and gauge kinetic terms, but also on D-terms, encoding the kinetic terms for chiral multiplets and their couplings to vector multiplets. D-terms are determined by the Kähler potential $K(\Phi, \Phi^\dagger, V)$, a real non-holomorphic "function" of the (light) chiral multiplets and the vector multiplets, that in principle receives both perturbative and non-perturbative corrections.[15]

The situation significantly improves for $\mathcal{N} = 2$ SYM theories, since the extra supersymmetry relates what in the $\mathcal{N} = 1$ description would be unrelated, i.e. the Kähler potential, the superpotential and the gauge kinetic function [69]. This is true not only when $\mathcal{N} = 2$ vector multiplets are present but also when one couples the resulting $\mathcal{N} = 2$ SYM to "matter" fields belonging to so-called hypermultiplets, or hypers for short.[16] $\mathcal{N} = 2$ supersymmetry allows only ($\mathcal{N} = 2$) minimal couplings of hyper to vector multiplets, coded in "tri-holomorphic moment maps", and the hypers are known to have vanishing anomalous dimensions [71]. The (low-energy) effective theory is thus determined by an analytic prepotential $\mathcal{F}$, which only depends on the $\mathcal{N} = 2$ vector multiplets, and a choice of gauging of tri-holomorphic isometries of the hyperkähler manifold described by the hypers [72]. Vector multiplets are "chiral" in the $\mathcal{N} = 2$ description. In turn the analytic prepotential is known to receive only one-loop and non-perturbative corrections. In their seminal paper [51], Seiberg and Witten were able to determine the exact form of $\mathcal{F}$ for pure $\mathcal{N} = 2$ SYM with gauge group $SU(2)$ by a series of elegant arguments based on electric-magnetic duality [73]. In a subsequent paper [74], they extended their arguments to the case of $\mathcal{N} = 2$ SQCD with gauge group $SU(2)$ that arise after minimal coupling of N_f hypermultiplets belonging to the pseudo-real fundamental representation of $SU(2)$. Later on, these results have been generalised to other gauge groups with hypers in various representations both in the Coulomb branch, corresponding to turning on v.e.v.'s of scalars in vector multiplets thus preserving the rank of the gauge group,

[15] For this reason the "exact" β-function of [32] should be properly seen as an elegant way to hide one's ignorance of the anomalous dimensions γ of chiral multiplets.

[16] In fact it can even improve if the hypermultiplets belong to some special representation of the gauge group, whereby the theory becomes exactly superconformal and thus UV finite so that the two derivative effective action does not receive any correction either perturbatively or non- perturbatively. For instance, this is the case when one extra hypermultiplet is added that belongs to the adjoint representation, leading to the $\mathcal{N} = 4$ SYM theory [70].

and in the Higgs branch, corresponding to turning on v.e.v.'s of scalars in hypers thus generically reducing the rank of the gauge group. The possibility of having new and mixed branches has also been widely explored.

Our aim is to first describe the structure of $\mathcal{N} = 2$ SYM theories both at the microscopic level and at the macroscopic one, when they are described in terms of Wilsonian low-energy effective actions. We then review the arguments of Seiberg and Witten leading to the identification of an auxiliary Riemann surface, i.e. a "complex curve", encoding the complexified gauge coupling τ in (the ratio of) the derivatives of its two "period" integrals, eventually arriving at the determination of $\mathcal{F}$ in the simplest case of $SU(2)$. We then discuss the non-linear recursion relations satisfied by the coefficients of the instanton expansion following the work of Matone's [75] and check the consistency of Matone's relations and, thus, of the SW prepotential, with instanton calculus in sectors with $K = 1, 2$ [76]. In order to tackle the general case, i.e. arbitrary K and generic gauge group (with $SU(N_c)$ in mind), one may exploit the "topological twist" of $\mathcal{N} = 2$ SYM theories [77] that, combined with some "non-commutativity" parameter [78], in the form of the so-called Ω-background, allowed Nekrasov and collaborators [79, 80, 81, 82] to localise the integrals over instanton moduli spaces and compute recursively the expansion coefficients of the non-perturbative series. To this end we sketch these beautiful, but at the same time rather technical, mathematical arguments underlying the ADHM construction. We then turn to the string description of the ADHM construction and its ramifications [83, 84]. The astonishing feature of string theory is that the sophisticated algebro-geometric ADHM construction becomes rather transparent and intuitive once D-branes and their open string excitations are taken into account [85, 86]. In particular, (supersymmetric) gauge theories emerge as the low-energy effective theories governing the dynamics of stacks of D-branes [87]. In this setting instantons can be realised as lower dimensional D-branes within higher dimensional ones [83, 84]. The structure of the ADHM data emerge naturally from the set of open strings connecting the various stacks of branes. Even Nekrasov's Ω-background admits a natural description in terms of the closed string graviphoton that couples to D-branes and their open string excitations [85, 86]. Last but not least, the long sought for duality between gauge fields and strings turns out to emerge quite naturally, at least in the maximally supersymmetric case ($\mathcal{N} = 4$ SYM), in the form of Maldacena's holographic correspondence [88, 89, 90, 91, 92]. We will return to this unprecedented achievement of string theory in Sects. 17 and 18. Here we only give a schematic description of how instanton effects can be computed within string theory in a particular double scaling limit [85, 86].

7 $\mathcal{N} = 2$ SYM: Generalities

$\mathcal{N} = 2$ SYM theories admit two kinds of massless multiplets, both containing four bosonic and as many fermionic degrees of freedom. Vector multiplets are described by chiral $\mathcal{N}=2$ superfields that comprise a vector boson, two Weyl fermions (the gauginos) and a complex scalar all in the adjoint representation of the gauge group. $\mathcal{N}=2$ vector superfields will be denoted by A and their θ expansion schematically reads

$$A(x,\theta) = a(x) + \theta^r_\alpha \lambda^\alpha_r(x) + \frac{1}{2}\theta^r_\alpha \sigma^{\mu\nu\alpha}{}_\beta \theta^\beta_r F_{\mu\nu}(x) + \dots . \tag{159}$$

Higher-order terms in θ^r with $r = 1,2$ can be expressed as derivatives of the lower ones. In terms of $\mathcal{N}=1$ supersymmetry, a $\mathcal{N}=2$ vector multiplet can be decomposed into a vector superfield V and a chiral superfield Φ, both in the adjoint representation of the gauge group (see Appendix A for notation).

$\mathcal{N}=2$ massless "matter" appears in hypermultiplets that consist of four real scalars and two Weyl fermions (the hyperinos), belonging to a real representation of the gauge group. In terms of $\mathcal{N}=1$ supersymmetry, they can be decomposed into a chiral superfield Q in an a priori complex representation $\mathbf{R}$ of the gauge group and a chiral superfield $\tilde{Q}$ in the conjugate representation $\bar{\mathbf{R}}$. Among the massive representations, a special role is played by the 1/2 BPS representations that are shorter than generic massive representations in that they only involve eight bosonic and as many fermionic degrees of freedom (see Appendix G for a brief explanation of this and related notions). Thanks to the relation between mass and "central" charge

$$M = |Z| , \tag{160}$$

1/2 BPS states are indeed annihilated by half of the supersymmetry charges.

The structure of classical $\mathcal{N}=2$ SYM theories is tightly constrained by the large amount of (super)symmetry they are endowed with. The most general two-derivative classical action is completely determined in terms of an "analytic" prepotential $\mathcal{F}$ that is a priori an analytic function of the $\mathcal{N}=2$ vector multiplets A and the complex coupling constant

$$\tau = \frac{\vartheta}{2\pi} + \frac{4\pi i}{g^2} . \tag{161}$$

The $\mathcal{N}=2$ hypermultiplet dynamics is described by a non linear σ model on a hyperkähler space (see Appendix J). The coupling of $\mathcal{N}=2$ hypermultiplets to vector multiplets is minimal in that the vector fields "gauge" (make local) the global hyperkähler isometries of the hypermultiplet metric that preserve the three Kähler structures

$$\omega^I = \frac{1}{2}\,\omega^I_{ij}\, \mathrm{d}q^i \wedge \mathrm{d}q^j , \tag{162}$$

where $\omega^I_{ij} = -\omega^I_{ji}$ with $I = 1,2,3$ and $i,j = 1,\dots,4n_H$ are anti-symmetric tensors such that $\mathrm{d}\omega^I = 0$, where d denotes the exterior differential in field space, i.e. with respect to the scalar components q^i of the n_H hypermultiplets.

In the simple case of constant ω^I_{ij}, writing $i = f + 4r$ with $f = 1, ..., 4$ and $r = 0, ..., n_H - 1$, one can choose

$$\omega^I_{f+4r,f'+4r'} = \eta^I_{ff'}\delta_{rr'}, \tag{163}$$

where $\eta^I_{ff'}$ are the 't Hooft symbols [2].

The effect of "gauging" (hyperkähler) isometries can be elegantly expressed through the minimal substitution

$$\partial_\mu q^i \to D_\mu q^i = \partial_\mu q^i + gA^a_\mu \xi^i_a(q), \tag{164}$$

where $a = 1, ..., n_V$. A tri-holomorphic isometry generated by the vector field $\xi_a = \xi^i_a \partial/\partial q^i$ satisfies

$$\mathcal{L}_\xi \omega^I \equiv \iota_{\xi_a} \mathrm{d}\omega^I + \mathrm{d}(\iota_{\xi_a}\omega^I) = \mathrm{d}(\iota_{\xi_a}\omega^I) = 0, \tag{165}$$

where ι_{ξ_a} denotes contraction with the Killing vector field $\xi_a(q)$. As a consequence, $\iota_{\xi_a}\omega^I = \mathrm{d}\mu^I_a$ a tri-holomorphic Killing vector $\xi_a(q)$ admits hyperkähler moment maps $\mu^I_a(q)$ since locally $\xi^i_a(q)\omega^I_{ij}(q) = \partial_j \mu^I_a(q)$. The $\mu^I_a(q)$ may be thought of as some sort of $\mathcal{N} = 2$ auxiliary fields. In the $\mathcal{N} = 1$ notation, whereby a hypermultiplet with scalar components q^f is described by two chiral multiplets with scalar components $\phi = q^1 + iq^2$ and $\tilde{\phi} = q^3 + iq^4$, one has

$$\begin{aligned}
\mu^3_a(q) &= D_a(\phi, \tilde{\phi}; \phi^\dagger, \tilde{\phi}^\dagger) \\
\mu^+_a(q) &= \mu^1_a(q) + i\mu^2_a(q) = F_a(\phi, \tilde{\phi}) \\
\mu^-_a(q) &= \mu^1_a(q) - i\mu^2_a(q) = \bar{F}_a(\phi^\dagger, \tilde{\phi}^\dagger).
\end{aligned} \tag{166}$$

Indeed the contribution of the hypermultiplets to the potential is exactly given by

$$V_H(q) = \frac{1}{2}\delta_{IJ} \mathrm{Im}\, \tau_a(a) \mu^I_a(q) \mu^J_a(q). \tag{167}$$

Notice that except for the minimal coupling (164) and its $\mathcal{N} = 2$ completion, entailing (167) and various Yukawa-type interactions, there is no neutral coupling between the vector and the hyper multiplets. In particular, as indicated, the complexified gauge couplings $\tau_a(a)$ can only depend on the lowest scalar components a of the n_V vector multiplets A.

Quantum renormalisability drastically restricts the choice of $\mathcal{F}(A)$ and $\xi(q)$, or equivalently of $\mu(q)$. In the microscopic fundamental theory, $\mathcal{F}(A)$ is at most a quadratic function of A, while $\xi^i_a(q)$ are linear in the q's, viz.

$$\xi^i_a(q) = (T_a)^i{}_j q^j, \tag{168}$$

where T_a are the generators of the gauge group in the (a priori reducible) representation spanned by the scalars in the hypermultiplets. Moreover the hyperkähler metric is flat, up to global tri-holomorphic identifications ${}^{4n}/\Gamma$

(examples are the ALE spaces $^4/\Gamma_{ADE}$ where $n = 1$ and Γ_{ADE} is one of the Kleinian discrete subgroups of $SU(2)$, in the ADE classification, see e.g. [93]). As a result the tri-holomorphic moment maps $\mu^I_a(q)$ are completely determined in case of semi-simple gauge groups. When abelian factors are present in the gauge group, one can add constant tri-holomorphic Fayet–Iliopoulos terms ζ^I_a, so that $\mu^I_a(q) = \hat{\mu}^I_a(q) + \zeta^I_a$, where $\hat{\mu}^I_a(q)$ is such that $\hat{\mu}^I_a(q = 0) = 0$.

A different story applies to the Wilsonian effective action[17] for the light (massless) modes that survive, i.e. do not acquire a mass after, partial or complete gauge symmetry breaking below the scale Λ [60]. Here Λ is the explicit cut-off in the Wilsonian effective action, such that all modes with mass or energy above this scale have been integrated out. It is not known how to explicitly perform this task, but the outcome of the "integrating out" procedure is severely constrained by symmetries and one can often "guess" the correct result to lowest order approximation, which is nothing else but a bookkeeping of the relevant degrees of freedom and the symmetries of the theory.

In addition to the Coleman–Weinberg [94]-type logarithmic correction at one-loop

$$\mathcal{F}_{1-\mathrm{loop}}(A) = \frac{i}{8} b_1 A^2 \log \frac{A^2}{\Lambda^2} , \tag{169}$$

where b_1 is the β function coefficient, i.e. $b_1 = 2N_c - N_f$ for $SU(N_c)$ with N_f hypers in the fundamental representation, the prepotential can and in fact must acquire an infinite number of non-perturbative corrections. Indeed $\mathcal{N} = 2$ supersymmetry prevents further perturbative corrections, but the one-loop term violates positivity of the imaginary part of the effective gauge coupling

$$\tau(a) = \frac{\partial^2 \mathcal{F}(a)}{\partial a^2} = \frac{\vartheta(a)}{2\pi} + \frac{4\pi i}{g^2(a)} , \tag{170}$$

where a denotes the lowest (scalar) component of the chiral superfield A that describes the $\mathcal{N} = 2$ vector multiplet and has been defined in (159).

8 Seiberg–Witten Analysis

In their seminal paper [51], Seiberg and Witten have shown how one can exactly compute the analytic prepotential $\mathcal{F}(A)$ in the case of an $SU(2)$ gauge theory without hypermultiplets. In another closely related paper [74] they have shown how to incorporate $2N_f$ half-hypermultiplets in the fundamental representation of $SU(2)$, leading to a theory that deserves to be called $\mathcal{N} = 2$

[17] M.B. would like to thank M. Bochicchio for first pointing out the important difference between the "non-local" 1PI effective action, with an arbitrary number of derivatives, and the Wilsonian low-energy effective action, often considered only up to two derivatives.

SQCD. The case $N_f = 2N_c = 4$ is special since it corresponds to an exact quantum $\mathcal{N} = 2$ superconformal theory.

Clearly, in the Coulomb phase we are focussing on, higher derivative terms are generated by quantum effects with finite coefficients and are suppressed by inverse powers of the v.e.v.'s of the scalar fields. In the superconformal phase with vanishing v.e.v.'s the situation is much subtler. The relevant observables are correlation functions of gauge invariant operators. The modern tool to tackle this interesting issue is the AdS/CFT correspondence proposed by Maldacena that will be discussed in Sects. 17 and 18.

After briefly reviewing the arguments of Seiberg and Witten's, based on symmetries, monodromies and duality, we will describe how to check the result by (constrained) instanton calculus.

As we already mentioned, classical $\mathcal{N} = 2$ SYM admits an infinite tower of BPS-saturated monopoles and dyons thanks to the existence of a complex scalar central charge Z in the $\mathcal{N} = 2$ extended superalgebra. In the simple case of $SU(2)$, they are the supersymmetric analogue of the classical solutions found by 't Hooft [95] and Polyakov [96] and by Julia and Zee [97]. Indeed, we recall that the potential of pure $\mathcal{N} = 2$ SYM, which reads

$$V(a, a^\dagger) = \frac{1}{2}\,\mathrm{Tr}([a, a^\dagger]^2)\,, \tag{171}$$

has flat directions identified by the condition $[a, a^\dagger] = 0$. Up to gauge transformations, this means that both a and $a^\dagger$ belong to the Cartan subalgebra generated by, e.g. $T_3 = \sigma_3/2$. In modern terms, one says the theory admits a one complex dimensional moduli space of classical vacua parametrised by a_3 or rather by the gauge-invariant composite

$$u = \mathrm{Tr}(a^2) = \frac{1}{2}a_3^2\,. \tag{172}$$

Henceforth, we denote a_3 by a for simplicity. Along the flat direction the gauge group is broken to $U(1)$ and one automatically realises the Prasad–Sommerfield condition[18] without the need of sending the scalar self-coupling to zero [98]. Monopole solutions that saturate the Bogomol'nyi bound

$$M_M = |p\tau_0 a|\,, \tag{173}$$

where τ_0 denotes the classical "complexified" coupling, that we have already encountered many times by now, and p is the magnetic charge, can be explicitly constructed solving Nahm's equations [99]. Notice the striking analogy with the Higgs formula for the mass of a $U(1)$ W-like boson with charge q

$$M_w = |q_e a|\,. \tag{174}$$

[18] The Prasad–Sommerfield condition of non vanishing scalar v.e.v. with zero potential is usually achieved by setting the scalar self-coupling λ to zero in the potential $V(\varphi) = \lambda(\varphi^2 - \varphi_0^2)^2$ while keeping $|\varphi| = |\varphi_0|$ at infinity.

In fact one can do better and show that 1/2 BPS-saturated dyons have a mass spectrum given by the formula

$$M_D = |q_e a + q_m a_D| = |Z|, \tag{175}$$

where we have introduced the notation $a_D = \tau_0 a$ and Z is the "central" extension of the $\mathcal{N} = 2$ superalgebra, that being central, i.e. commuting with all the remaining generators, has to be a c-number by Schur's Lemma [100, 101]. In terms of the analytic prepotential $\mathcal{F}(a) = \tau_0 a^2/2 + \cdots$, one is led to the identification

$$a_D = \frac{\partial \mathcal{F}}{\partial a} = \tau_0 a + \cdots, \tag{176}$$

where the dots take into account quantum corrections to $\mathcal{F}(a)$ and thus to $\tau(a) = \tau_0 + \cdots$. The exact (quantum) identification $a_D = \frac{\partial \mathcal{F}}{\partial a}$ is tantamount to assuming that the classical formula (175) for the central charge retains the same form in the quantum theory, as strongly suggested by consideration of $\mathcal{N} = 2$ supersymmetry. Actually, the formula $|Z| = |q_e a + q_m a_D|$ displays a remarkable symmetry under $SL(2, \;)$ transformations acting on the electric and magnetic charges q and p. In fact, under

$$q_e \to k q_e + l q_m \qquad q_m \to m q_e + n q_m \tag{177}$$

Z is invariant if one simultaneously performs a "monodromy" transformation

$$a \to n a - l a_D \qquad a_D \to -m a + k a_D\,. \tag{178}$$

In this way a, a_D are seen as components of a section of an $SL(2, \;)$ bundle over the moduli space of vacua parametrised by the gauge-invariant composite u, for which we write $a = a(u), a_D = a_D(u)$. This geometrical description implies that the components $a = a(u)$ and $a_D = a_D(u)$ undergo non-trivial transformations, i.e. acquire non-trivial monodromy, when one parallel transports them as functions of u around some special points. As a result of their dependence of $a = a(u), a_D = a_D(u)$ on u, the complexified coupling that has been so far considered a function of a can be considered a function of u given by the ratio of the derivatives of $a_D(u)$ and $a(u)$ through the chain rule

$$\tau(a) = \tau(a(u)) = \tau(u) = \frac{\partial^2 \mathcal{F}(a)}{\partial a^2} = \frac{\partial}{\partial a}\frac{\partial \mathcal{F}(a)}{\partial a} = \frac{\partial a_D}{\partial a} = \frac{\frac{\mathrm{d}a_D}{\mathrm{d}u}}{\frac{\mathrm{d}a}{\mathrm{d}u}}, \tag{179}$$

with $\mathrm{Im}\,\tau(u) > 0$ (for vacuum stability). Remarkably, at this point, the complexified effective coupling $\tau(u)$ can be considered as the modular parameter (the "period") of an auxiliary torus, a Riemann surface of genus one. The latter is also known as an "elliptic curve", i.e. a complex dimension one manifold whose periods are determined in terms of elliptic integrals. In fact determining this auxiliary elliptic curve, the so-called "Seiberg–Witten curve", allows one to compute its periods from the equations

$$a'_{D}(u) = \frac{da_{D}}{du}, \quad a'(u) = \frac{da_{D}}{du} \tag{180}$$

and, after integration w.r.t. u, $\mathcal{F}(a)$ itself, since $a_{D} = \frac{\partial \mathcal{F}(a)}{\partial a}$.

In order to determine the SW curve one starts by computing the monodromy of the section $(a = a(u), a_{D} = a_{D}(u))$ at infinity where the theory, being asymptotically free ($b_1 = 2N_c = 4$ for $SU(2)$), is weakly coupled and (see (169))

$$\mathcal{F}(a) \approx \frac{i}{2} a^2 \log \frac{a^2}{\Lambda^2} \tag{181}$$

with Λ the RGI scale

$$\Lambda = M \exp(-8\pi^2 / b_1 g^2(M)) \,. \tag{182}$$

In this way one gets

$$a_{D} = \mathcal{F}'(a) \approx \frac{i}{2} [2a \log \frac{a^2}{\Lambda^2} + 2a] \,. \tag{183}$$

Under $u \to e^{2\pi i} u$, one has $a \to -a$ and $a_{D} \to -a_{D} + 2a$. These considerations fix the monodromy of the section $(a = a(u), a_{D} = a_{D}(u))$ around the branch point at infinity.

Perturbatively, the other branch point of $\mathcal{F}(a)$ is at $a = 0$. If this were the full story, the theory would be inconsistent since $\mathrm{Im}\,\tau$ could not possibly be positive throughout the moduli space, being τ holomorphic and thus $\mathrm{Im}\,\tau$ harmonic. Seiberg and Witten argued that the non-abelian symmetry ($a = 0$) is never restored at the quantum level and that this is consistent with assuming the existence of only two more singular points. They interpreted the singularities as due to the fact that some massive states become massless at each of the two additional singular points in the moduli space. In fact, the two relevant states are a monopole ($q_e = 0, q_m = 1$) and a dyon ($q_e = 1, q_m = -1$).[19] In order to identify the location of these extra singularities, it is crucial to exploit a discrete $_4$ symmetry of the quantum theory for $N_c = 2$, which is a remnant of the anomalous $U(1)_R$ subgroup of the $U(2)$ R-symmetry of the classical theory. Indeed, classically $\mathcal{N} = 2$ SYM is invariant under global $SU(2) \times U(1)_R$ transformations under which the gauge field is invariant, the gaugini rotate as a $\mathbf{2}_{1/2}$ and the complex boson is a charge $+1$ $SU(2)$ singlet. The $U(1)_R$ symmetry is broken by the quantum anomaly that preserves a $_{4N_c} \approx {}_{2N_c} \times {}_2$ where the latter factor is fermion parity and the former is the above-mentioned $_4$ under which $u \to -u$. This is enough to completely determine the SW curve

$$\mathcal{E}_{\mathrm{SW}}: \quad y^2 = (x^2 - \Lambda^2)(x^2 - u^2) \,, \tag{184}$$

[19] This particular choices of electric and magnetic charges simplify the notation in the following. Other choices are possible but require performing $SL(2, \ )$ transformations.

which is indeed singular when $u = \pm\Lambda$. A generic elliptic curve can be written as a double cover of the sphere $y^2 = (x - x_1)(x - x_2)(x - x_3)(x - x_4)$ with branch points at $x = x_i$. By the $SL(2, \ )$ symmetry of the sphere, one can always put three of the branch points at, say, $0, 1$ and ∞ so that the remaining complex parameter determines the shape of the torus (actually the ratio of the two periods). In order not to spoil the $_2$ symmetry of quantum $\mathcal{N} = 2$ SYM theory, it is however more convenient to fix only two branch points at, say, $\pm\Lambda$ (or ± 1 after rescaling the variables). The remaining two branch points are set at $\pm u$. When u reaches $\pm\Lambda$ the curve (184) representing the torus degenerates, i.e. one of the two cycles and the corresponding period become zero signalling the presence of a singularity in the theory.

The periods of $\mathcal{E}_{\rm SW}$ can be expressed in terms of elliptic integrals and after identifying the cycles that correspond to $a'_{_D}(u)$ and $a'(u)$, one can eventually compute $\mathcal{F}$.

Making more precise the above geometrical considerations, we can say that the vector $(a_{_D}, a)$ is a section of a flat bundle over the moduli space parametrised by u with monodromy group $\Gamma(2) \subset SL(2, \ )$ generated by

$$M_{-1} = \begin{pmatrix} -1 & 2 \\ -2 & 3 \end{pmatrix} \qquad M_1 = \begin{pmatrix} 1 & 0 \\ -2 & 1 \end{pmatrix} \qquad M_\infty = \begin{pmatrix} -1 & 2 \\ 0 & -1 \end{pmatrix} \tag{185}$$

such that $M_1 M_{-1} = M_\infty$. It can be checked that the monodromy transformations of $(a_{_D}, a)$ around $\pm\Lambda$ are indeed represented by the matrices $M_{\pm 1}$.

If the modular parameter τ of $\mathcal{E}_{\rm SW}$ were the ratio $a_{_D}/a$ we would have completed our task, since $a_{_D}$ and a would have coincided with the two "canonical" periods of the unique holomorphic[20] differential $\mathrm{d}\omega = \mathrm{d}x/y$. However $\tau = a'_{_D}(u)/a'(u)$ and this means that $a_{_D}$ and a are periods of a meromorphic differential whose derivative w.r.t. u is the unique holomorphic differential $\mathrm{d}\omega$. Seiberg and Witten identified the "natural" meromorphic differential $\mathrm{d}\lambda(u)$ that prior to monodromy considerations is only determined up to a meromorphic differential independent of u. Setting $\Lambda = 1$ for simplicity, one finds that

$$\mathrm{d}\lambda(u) = \frac{\sqrt{x-u}}{\sqrt{x^2-1}}\,\mathrm{d}x \tag{186}$$

is the unique differential that satisfies all the requirements, i.e. $\mathrm{d}\lambda'(u) = \mathrm{d}\omega$ and it is such that

$$a_{_D}(u) = \frac{\sqrt{2}}{\pi}\int_1^u \frac{\sqrt{x-u}}{\sqrt{x^2-1}}\,\mathrm{d}x\,, \quad a(u) = \frac{\sqrt{2}}{\pi}\int_{-1}^1 \frac{\sqrt{x-u}}{\sqrt{x^2-1}}\,\mathrm{d}x \tag{187}$$

have the correct monodromy when parallel transported around ± 1 and ∞. Consistently with the surviving $_4$ symmetry, one can write

$$\mathcal{F}(a) = \mathcal{F}_{\rm pert}(a) + \mathcal{F}_{\rm non-pert}(a)\,, \tag{188}$$

[20] Although it does not look holomorphic, ω is in fact holomorphic!.

where

$$\mathcal{F}_{\text{pert}}(a) = \mathcal{F}_{\text{tree}}(a) + \mathcal{F}_{1-\text{loop}}(a) = \frac{i}{2} a^2 \log \frac{a^2}{\Lambda^2} \tag{189}$$

and

$$\mathcal{F}_{\text{non-pert}}\ (a) = a^2 \sum_{K=1}^{\infty} \mathcal{F}_K \frac{\Lambda^{4K}}{a^{4K}}, \tag{190}$$

with the latter incorporating the contribution of instantons of increasing winding number K.

A few comments are in order here. First $a = 0$ is excised from the moduli space, i.e. there is no value of u such that $a(u) = 0$. Second the singular points $u = \pm\Lambda$ correspond to $a_D(u) = 0$ and $a_D(u) = a$, respectively, and lie on the so-called surface of marginal stability where $\text{Im}\,\tau = 0$. This is the locus where the lattice of BPS states collapses and transitions of the form $|Z = Z_1 + Z_2\rangle \to |Z_1\rangle + |Z_2\rangle$ are allowed by both charge and mass conservation.

The effective coupling $g^2(a) = 4\pi^2/\text{Im}\,\tau(u)$ is always semi-positive definite and never grows too much. At large a this is due to asymptotic freedom. In the interior of the moduli space all charged vector bosons become extremely massive and the theory is essentially abelian. Near the singular points, one better switches to a dual magnetic or dyonic description, whereby the abelian magnetic or dynonic photons are coupled to light monopoles and dyons. The effective coupling decreases with the renormalisation scale μ in the IR until it reaches the value

$$\tilde{g}^2(\mu)|_{\mu=m} \approx \frac{1}{|\log(m/\Lambda)|}, \tag{191}$$

with m the mass of the lightest charged state, be it a monopole or a dyon. As stressed above, these states are arranged in hypermultiplets. Due to the presence of light charged particles in hypermultiplets coupled to abelian vector multiplets, the dual magnetic or dyonic theory is different from the original electric theory, that only involved non-abelian vector multiples. Yet electric-magnetic duality led us for quite a long way. Only for $\mathcal{N} = 4$ SYM and for other exactly (super)conformal-invariant theories the dual magnetic or dyonic theory is expected to coincide with the electric theory.[21]

Finally, at the singular points $u = \pm\Lambda^2$ new branches of the moduli space open up where monopoles or dyons can condense, i.e. acquire a v.e.v., thus inducing (oblique) confinement of the chromo-electric charges and flux tubes due to the dual Meissner effect [103]. Adding $\mathcal{N} = 1$ supersymmetric mass terms to the adjoint chiral multiplet induces dynamical chiral symmetry breaking and confinement in a controllable way.

[21] More precisely in these cases the duality maps the electric theory into a magnetic theory with the same action but dual gauge group [102], G^*. The latter is obtained from the original gauge group of the electric theory, G, exchanging the role of the weight and root lattices. Therefore, in the case of groups with simply laced Lie algebras, G and G^* are isomorphic. For groups with non-simply laced algebras this is not the case and one has the following pairs: $G \leftrightarrow G^*$: $SO(2n+1) \leftrightarrow Sp(n)$, $F_4 \leftrightarrow F_4'$ and $G_2 \leftrightarrow G_2'$.

9 Checking the SW Formula by Instanton Calculations

Our next task is to check the SW prepotential against explicit instanton computations. A one-instanton check is not enough because of ambiguities in the definition of Λ that can rescale it by a finite constant. The perspective for a two-instanton check seems a priori daunting but the calculation turns out to be feasible [37, 76]. In fact one can do much better. Matone has shown that the coefficients of the SW prepotential satisfy non linear recurrence relations that can be checked to hold in instanton calculus [75]. Fucito and Travaglini have shown that multi-instanton calculus precisely reproduces the desired relations [76]. More recently the problem has been attacked once again in a beautiful series of papers by Nekrasov and collaborators [79, 80, 81, 82]. Introducing suitable deformations parameters (Ω-background), one can localise the measure over the multi-instanton (super)moduli space reducing the calculation of the $\mathcal{F}_K$ coefficients to a mere, though certainly not trivial, combinatorial problem. We will have limited space to discuss this fascinating issue and we refer the reader to the original literature as well as to the accessible reviews [79, 80, 81, 82]. We cannot resist saying immediately that the somewhat obscure deformation parameters introduced by Nekrasov admit a very natural explanation in a string setting for the problem whereby open string excitations of D-branes account for the gauge and matter light degrees of freedom [83, 84, 85]. The Ω-deformation is equivalent to turning on a background for a closed string state in the so-called Ramond–Ramond (R–R) sector, the graviphoton, effectively producing a non (anti-)commutative superspace [86], i.e. a superspace where the θ variables do not anti-commute very much like the x variables do not commute. From this vantage point, higher-order terms in the deformation, receiving instanton corrections, are associated with higher derivative gravitational F-terms that appear in the type II low-energy effective actions after compactification on Calabi–Yau threefolds [104]. A similar approach for the calculation of the SW prepotential, based on localisation on the instanton moduli space, was proposed in [105].

9.1 Matone Relations

Exploiting powerful results in the theory of uniformisation of Riemann surfaces, it was shown in [75] that the non-perturbative coefficients in the expansion of $\mathcal{F}(a)$ satisfy certain recursion relations known as Matone's relations. In order to achieve this result, it is convenient to consider the auxiliary function

$$\mathcal{G}(a) = \mathcal{F}(a) - \frac{a}{2}\frac{\partial \mathcal{F}(a)}{\partial a}, \tag{192}$$

where $\mathcal{F}(a)$ is defined in (188) and consider the expansion

$$\mathcal{G}(a) = a^2 \sum_{K=0}^{\infty} \mathcal{G}_K \frac{\Lambda^{4K}}{a^{4K}}. \tag{193}$$

We will momentarily see that $\mathcal{G}(a) = u$. The expansion coefficients $\mathcal{G}_K$ and $\mathcal{F}_K$ are related by

$$\mathcal{G}_K = 2K\mathcal{F}_K\,, \tag{194}$$

for $K \neq 0$ while $\mathcal{G}_0 = 1/2$.

As previously discussed, for $SU(2)$ the moduli space is parameterised by $u = \mathrm{Tr}(\phi^2)$ and turns out to be a Riemann sphere with three punctures at $u_1 = -\Lambda$, $u_2 = \Lambda$ and $u_3 = \infty$ with a symmetry $u \leftrightarrow -u$. We recall that $(a_D(u), a(u))$ is a section of a flat bundle over the moduli space with monodromy group $\Gamma(2) \subset SL(2, \quad)$ generated by the three matrices M_{-1}, M_{+1} and M_∞ in (185) with $M_{-1}M_{+1} = M_\infty$.

Using the obvious integrability of the differential

$$W(u)\,\mathrm{d}u = a\,\mathrm{d}a_D - a_D\mathrm{d}a\,, \tag{195}$$

that being a complex function of the single variable u is necessarily an exact differential, one can define the auxiliary function

$$g(u) = \int_1^u \mathrm{d}z\, W(z)\,. \tag{196}$$

This helps determining the behaviour of $\mathcal{F}$ under monodromy (modular transformations). In fact by integrating

$$\partial_u\mathcal{F} = a_D\partial_u a = \frac{1}{2}[\partial_u(a_D a) - W(u)] \tag{197}$$

one finds

$$\mathcal{F}(u) = \frac{1}{2}[a_D a - g(u)] + \mathcal{F}_0\,. \tag{198}$$

One can check that, under

$$\begin{aligned} a_D &\to \tilde{a}_D = k a_D - m a \\ a &\to \tilde{a} = -l a_D + n a\,, \end{aligned}$$

with $kn - lm = 1$, one has

$$\tilde{\mathcal{F}}(\tilde{a}) = \mathcal{F}(a) + \frac{1}{2}[lm a_D a - kl a_D^2 - mn a^2]\,, \tag{199}$$

while $\mathcal{G}(a)$, conveniently defined as above, turns out to be modular invariant, i.e.

$$\tilde{\mathcal{G}}(\tilde{a}) = \mathcal{G}(a)\,, \tag{200}$$

since u and hence $g(u)$ are invariant. By taking the ratio of $a'_D(u)$ and $a'(u)$ and keeping in mind that u is invariant, one also finds that

$$\tau(a) = \frac{\partial^2\mathcal{F}(a)}{\partial a^2} = \frac{a'_D(u)}{a'(u)} \to \tilde{\tau}(\tilde{a}) = \frac{\partial^2\tilde{\mathcal{F}}(\tilde{a})}{\partial\tilde{a}^2} = \frac{k\tau(a) - m}{-l\tau(a) + n}\,, \tag{201}$$

which is the expected projective transformation of the complexified coupling.

By uniformisation arguments, i.e. monodromy invariance and asymptotic behaviour at large a, Matone eventually showed that [75]

$$\mathcal{G}(a) = -i\pi g(u)/2 = u\,. \tag{202}$$

The linear u dependence of $g(u)$ is tantamount to saying that $W(u)$ is a constant, independent of u. In fact $W(u) = a(u)a'_{D}(u) - a_{D}(u)a'(u)$ is nothing but the Wronskian of the solutions of the second-order differential equation satisfied by $a(u)$ and $a_{D}(u)$, which in canonical (Schrödinger-like) form reads

$$(1-u^2)\,\frac{\mathrm{d}^2\psi(u)}{\mathrm{d}u^2} - \frac{1}{4}\psi(u) = 0\,. \tag{203}$$

As a result of the uniformisation theorem of the moduli space of Riemann surfaces, $\mathcal{G}(a)$ obeys a non-linear differential equation of the form

$$(1-\mathcal{G}^2)\,\frac{\mathrm{d}^2\mathcal{G}}{\mathrm{d}a^2} + \frac{1}{4}a\left(\frac{\mathrm{d}\mathcal{G}}{\mathrm{d}a}\right)^3 = 0\,, \tag{204}$$

so that the coefficients of the expansion (193) satisfy the sought for recursion relation

$$\begin{aligned}\mathcal{G}_{K+1} = {} & \frac{1}{8\mathcal{G}_0^2(K+1)^2} \times \Bigg\{(2K-1)(4K-1)\mathcal{G}_K \\ & + 2\mathcal{G}_0 \sum_{N=0}^{K-1} c(N,K)\,\mathcal{G}_{K-N}\,\mathcal{G}_{N+1} \\ & - 2\sum_{L=0}^{K-1}\sum_{N=0}^{L+1} d(L,N,K)\,\mathcal{G}_{K-L}\,\mathcal{G}_{L+1-N}\,\mathcal{G}_N\Bigg\}\,,\end{aligned} \tag{205}$$

where

$$\begin{aligned} c(N,K) &= 2N(K-N-1) + K - 1 \\ d(L,N,K) &= [2(K-L)-1][2K-3L-1+2N(L-N+1)] \end{aligned} \tag{206}$$

and $\mathcal{G}_0 = 1/2$. The first few coefficients read

$$\mathcal{G}_1 = \frac{1}{2^2}\,, \quad \mathcal{G}_2 = \frac{5}{2^6}\,, \quad \mathcal{G}_3 = \frac{9}{2^7}\,, \tag{207}$$

in perfect agreement with the results of Seiberg and Witten. Moreover since $u = \mathcal{G}(a)$ using the asymptotic behaviour of $\mathcal{G}$, one can determine the constant value of W that reads

$$W = a'_{D}a - a_{D}a' = \frac{2i}{\pi}\,. \tag{208}$$

This relation is very useful in order to determine the "critical" curve where $\mathrm{Im}(a_{D}/a) = 0$. On this curve the lattice of BPS states collapses to a line, as already observed.

9.2 (Constrained) Instanton Checks for $K = 1, 2$

Following Matone [75], Fucito and Travaglini [76] have been able to check the non-perturbative relation

$$\langle \mathrm{Tr}(\phi^2) \rangle (a) = u(a) = \mathcal{G}(a) = \left(\mathcal{F}(a) - \frac{a}{2} \frac{\partial \mathcal{F}(a)}{\partial a} \right) \tag{209}$$

for $K = 1, 2$ and show agreement with the SW prepotential.

Using the relation between $\mathcal{F}$ in (188) and $\mathcal{G}$ in (193), one finds

$$\langle \mathrm{Tr}(\phi^2) \rangle (a) = -\frac{a^2}{2} - \sum_K \mathcal{G}_K \frac{\Lambda^{4K}}{a^{4K-2}} \, . \tag{210}$$

The calculation was carried out by making use of the ADHM construction, which we now briefly review in the $SU(2)$ case [19]. In the ADHM approach [19], the gauge connection is written in the form

$$A_\mu(x) = U^\dagger(x) \partial_\mu U(x) \, . \tag{211}$$

The key observation is that $U(x)$ is not a unitary $SU(2)$ matrix but rather a $(1 + K) \times 1$ "array" of quaternions, satisfying

$$\Delta^\dagger(x) U(x) = 0 \, , \tag{212}$$

where

$$\Delta(x) = a + bx \, , \tag{213}$$

with $x = x^\mu \sigma_\mu$ the position quaternion. Self-duality requires

$$\Delta^\dagger(x) \Delta(x) = f^{-1} \otimes \quad , \tag{214}$$

with f an invertible $K \times K$ matrix and the 2×2 identity matrix. The projector on the kernel of $\Delta^\dagger(x)$, spanned by $U(x)$, reads

$$P(x) = U(x) U^\dagger(x) = \quad - \Delta f \Delta^\dagger(x) \, . \tag{215}$$

Gauge field zero modes, that we here denote by a_μ, are orthogonal to the gauge orbit and can be parametrised as [20, 76]

$$a_\mu(x) = U^\dagger(x) [C \bar{\sigma}_\mu f b^\dagger - b f \sigma_\mu C^\dagger] U(x) \, , \tag{216}$$

with C a $(1 + K) \times K$ "matrix" of quaternions satisfying

$$\Delta^\dagger(x) C = (\Delta^\dagger(x) C)^T \, . \tag{217}$$

These conditions reduce the number of independent (quaternionic) components of C from $(1 + K) \times K$ to $(1 + K) \times K - (K - 1) \times K = 2K$, i.e. $8K$ zero modes as expected for $SU(2)$ instantons. Modulo symmetries, which are local $SU(2)$ and global $SO(K)$, the components of C can be identified with the fluctuations of Δ, $\delta\Delta$, i.e. variations of the ADHM data, satisfying the self-duality condition

$$(\Delta + \delta\Delta)^{\dagger}(x)(\Delta + \delta\Delta)(x) = f^{-1} \otimes \tag{218}$$

if non-linear terms are neglected. Since $\delta\Delta = C$ is linear in the gauge field zero modes parametrised by C, one can identify zero modes of the gauge fields with solutions of the linearised ADHM equations around a given self-dual solution. This is equivalent to identifying the bosonic zero modes as solutions of the equation $S[A_\mu + a_\mu] = S[A_\mu]$ up to cubic terms. One can similarly determine the fermionic zero modes that in the case of $\mathcal{N} = 2$ are as many as the bosonic zero modes and are given by [20, 76]

$$\lambda^{(i)}_{\beta\dot{\alpha}} = \sigma^{\mu}_{\beta\dot{\alpha}} a^{(i)}_{\mu} , \tag{219}$$

with $i = 1, ..., 8$. In the presence of flat directions of the classical scalar potential, the constrained instanton method entails an expansion around a solution of the (approximate) coupled equations[22]

$$D_\mu F^{\mu\nu} = 0 , \quad D^2 \phi = 0 \tag{220}$$

with boundary condition at infinity, $\phi \to \phi_{\text{flat}}$. For $SU(2)$ $\phi_{\text{flat}} = a\sigma_3/2i$ modulo gauge transformations.

For $K = 1$, everything simplifies drastically. As discussed in detail in Sects. 2.3 and 2.4 and Appendix B, the bosonic measure ("integrated" over $SU(2)/\ {}_2$) reads

$$\mathrm{d}\mu_B = \frac{4}{\pi^2}\left(\frac{2\pi\rho\mu}{g^2}\right)^8 \frac{\mathrm{d}^4 x_0\, \mathrm{d}\rho}{\rho^5} . \tag{221}$$

Using the fermionic zero modes, that are not normalised, the fermionic measure is given by

$$\mathrm{d}\mu_F = \mathrm{d}^4\eta\, \mathrm{d}^4\bar{\xi}\left(\frac{g^2}{32\pi^2\mu}\right)^4 . \tag{222}$$

Due to the presence of the scalar v.e.v., a, the classical action consists of various terms

$$S_{\text{cl}} = S_{\text{YM}} + S_{\text{scal}} + S_{\text{ferm}} + S_{\text{Yuk}} + S_{\text{pot}} . \tag{223}$$

---

[22] The attentive reader may notice that these are not the classical equations since the scalar induced source $J^\nu = \phi^\dagger(D^\nu\phi) - (D^\nu\phi^\dagger)\phi$ is being neglected. Exact topologically non-trivial solutions in the presence of non-zero v.e.v.'s for the scalars are not known [106]. The standard approach, which allows to control the fluctuations around the approximate solution, consists in adding to the action a "constraint" on the instanton size. The resulting "solution" is thus known as a "constrained instanton" [47].

After integration over the fluctuations of ϕ and $\phi^\dagger$ around their v.e.v., the Yukawa couplings produce an additional (to ϕ_{harmonic}) inhomogeneous term in ϕ of the form

$$\phi^a_{\text{inhom}} = \sqrt{2}[D^{-2}]^a{}_b \epsilon^{bcd} \lambda^\alpha_c \lambda_{d\alpha} = \sqrt{2}\zeta^\alpha \lambda^a_\alpha \,, \tag{224}$$

where $\zeta^\alpha = \eta^\alpha + x^\mu \sigma^{\alpha\dot\alpha}_\mu \bar\xi_{\dot\alpha}$. The absence of zero modes with "wrong" chirality leads to

$$S_{\text{Yuk}} = \bar a^b \bar\xi_{\dot\alpha} \sigma^{\dot\alpha\dot\beta}_b \bar\xi_{\dot\beta} \frac{g}{\sqrt{2}} \left(\frac{g^2}{32\pi^2\mu}\right)^{-1} . \tag{225}$$

Moreover

$$S_{\text{scal}} = 4\pi^2 |a|^2 \rho^2 \tag{226}$$

and we set

$$\Lambda^4 = \mu^4 \mathrm{e}^{-8\pi^2/g^2} . \tag{227}$$

The explicit computation of u (the v.e.v. of $\mathrm{Tr}(\phi^2)$) then yields

$$\begin{aligned} u = \langle \phi^a \phi_a \rangle_{K=1} &= \Lambda^4 \int \frac{4}{\pi^2} \left(\frac{2\pi}{g}\right)^8 \mathrm{d}^4 x_0 \mathrm{d}\rho \, \rho^3 \, \mathrm{e}^{-4\pi^2|a|^2\rho^2} F^a_{\mu\nu} F^{\mu\nu}_a \\ &\times \int \mathrm{d}^4\eta \mathrm{d}^4\bar\xi \left(\frac{g^2}{32\pi^2}\right)^4 (\eta\eta)^2 \exp\left[-\bar a^b \bar\xi \sigma_b \bar\xi \frac{g}{\sqrt{2}} \left(\frac{g^2}{32\pi^2\mu}\right)^{-1}\right] . \end{aligned} \tag{228}$$

Performing the integrations over the collective coordinates yields

$$\langle \phi^a \phi_a \rangle_{K=1} = \frac{2}{g^4} \frac{\Lambda^4}{a^2} \tag{229}$$

in agreement with $\mathcal{G}_1$.

For $K = 2$, the (constrained) instanton calculus is more laborious. The off-diagonal component d of the lower sub-block of Δ is of the form

$$d = \frac{1}{2} \frac{y}{y^2} (\bar v_2 v_1 - \bar v_1 v_2) \,, \tag{230}$$

where $y = x_1 - x_2 \equiv 2e$ and $x_0 = (x_1 + x_2)/2$, with x_1 and x_2 denoting the two instanton "centres" and v_1 and v_2 the two extra quaternionic collective coordinates. Similar restrictions as before apply to $C = \delta\Delta$ so that the off-diagonal component γ of the lower sub-block of C is of the form

$$\gamma = \frac{y}{y^2} (2\bar d \eta + \bar v_2 \nu_1 - \bar v_1 \nu_2) \,, \tag{231}$$

where η, ν_1 and ν_2 are quaternions that parametrise the independent fluctuations of the fermions. Separating the four collective coordinates associated with translations, x_0, and the four broken Poincaré supersymmetries, η_0, the relevant correlator reads

$$\langle \phi^a \phi_a \rangle_{K=2} = \frac{\Lambda^8}{16} \int \mathrm{d}^4 v\, \mathrm{d}^4 e\, \mathrm{d}^4 \xi\, \mathrm{d}^4 \nu_1\, \mathrm{d}^4 \nu_2 \left(\frac{J_B}{J_F}\right)^{1/2} \mathrm{e}^{-S_{\mathrm{Yuk}} - S_{\mathrm{scal}}}$$
$$\times \int \mathrm{d}^4 x_0 \mathrm{d}^4 \eta_0\, (\eta_0 \eta_0)^2 F^a_{\mu\nu} F^{\mu\nu}_a \tag{232}$$

where

$$\left(\frac{J_B}{J_F}\right)^{1/2} = \frac{2^{10}}{\pi^8} \frac{|\,|e|^2 - |d|^2\,|}{|v_1|^2 + |v_2|^2 + 4(|d|^2 + |c|^2)} \,. \tag{233}$$

Performing all the many necessary integrations yields

$$\langle \phi^a \phi_a \rangle_{K=2} = -\frac{5}{4g^8} \frac{\Lambda^8}{a^6} \tag{234}$$

in agreement with $\mathcal{G}_2$.

Actually, one can formally prove that Matone relations are satisfied by instanton calculus for any K [76].

Another elegant approach to derive the SW prepotential from first principles is based on the so-called $\mathcal{N} = 2^*$ theory. This is nothing else but $\mathcal{N} = 4$ SYM theory deformed by the addition of a mass M for the hypermultiplet in the adjoint representation, or equivalently the same mass $M_1 = M_2 = M$ for two of the three adjoint chiral multiplets in the $\mathcal{N} = 1$ description of the $\mathcal{N} = 4$ theory. Quite remarkably the hypermultiplet, $H = \{\Phi_1, \Phi_2\}$, appears quadratically in the microscopic action,

$$S[\Phi_{I=1,2}; \Phi_3, V] = \int \mathrm{d}^2\theta \mathrm{d}^2\bar\theta\, \mathrm{Tr}(\Phi_I^\dagger \mathrm{e}^{gV} \Phi^I) \tag{235}$$
$$+ \int \mathrm{d}^2\theta\, g \mathrm{Tr}([\Phi_1, \Phi_2]\Phi_3) + \frac{1}{2} M\, \mathrm{Tr}(\Phi_I)^2 + \mathrm{h.c.}\,,$$

and can be integrated out in a Gaussian fashion. One ends up with an effective action à la Wilson–Polchinski where M plays the role of an UV cut-off. The advantage of the approach is the UV finiteness of $\mathcal{N} = 4$ SYM theory which persists after the inclusion of the $\mathcal{N} = 2$ supersymmetric mass terms. The resulting low-energy effective action is expected to coincide with the one resulting from the SW prepotential. As we said in Sect. 5, this has been partially checked by means of the exact renormalisation group in [61].

10 Topological Twist and Non-commutative Deformation

$\mathcal{N} = 2$ SYM theories admit an interesting reformulation which goes under the name of "topological twist" [77]. Although the topologically twisted version is not fully equivalent to the original (dynamical) theory, some of the observables coincide. In particular, one can suspect that the analytic prepotential $\mathcal{F}$ could be one of these observables thanks to holomorphy. As we will see later on, this is not completely true. The topological theory cannot reproduce the logarithmic term generated by one-loop corrections. Yet, a properly defined partition function of the topological theory captures all the non-perturbative corrections to $\mathcal{F}$ and more. Indeed, higher derivative "gravitational" F-terms can be reliably computed by means of its topologically twisted version if a suitable background inducing "non-commutativity" is turned on. After briefly reviewing the topological twist formalism, we will sketch the arguments leading to the derivation of $\mathcal{F}_{\text{non-pert}}$ from the topological partition function.

The topological twist consists in bringing bosons and fermions to transform in the same way under the subgroup $SU(2)_L \times SU(2)_D \subset SU(2)_L \times SU(2)_R \times SU(2)_I$, where D stands for the diagonal subgroup of $SU(2)_R \times SU(2)_I$, which is not to be confused with the (Euclidean) Lorentz group $SU(2)_L \times SU(2)_R$, since $SU(2)_I$ is part of the R-symmetry group $U(1) \times SU(2)_I$. Under $SU(2)_L \times SU(2)_D$ the two Weyl gaugini transform as a four-vector, $\psi_\mu \in (1/2, 1/2)$, a singlet, $\bar{\eta} \in (0,0)$, and a self-dual tensor, $\bar{\chi}^+_{\mu\nu} \in (0,1)$, where, adhering to standard notation, (j_L, j_D) refers to the $SU(2)_L \times SU(2)_D$ spins rather than the dimension of the representation. Similarly the superspace variables dual to the eight supercharges are $\theta \to \theta_\mu, \bar{\theta}^+_{\mu\nu}, \bar{\theta}$, so that the chiral superfield Φ admits the newly looking decomposition

$$\Phi = \phi + \theta^\mu \psi_\mu + \frac{1}{2}\theta^\mu \theta^\nu F_{\mu\nu} + \cdots \,, \tag{236}$$

where

$$\theta^\mu = \frac{1}{2}\sigma^\mu_{\alpha,r}\theta^{\alpha,r} \,. \tag{237}$$

The supercharge $\bar{Q} = \varepsilon^{\dot{\alpha}r}\bar{Q}_{\dot{\alpha}r}$ is a scalar and plays the role of topological Becchi, Rouet, Stora and Tyutin (BRST) charge. In the topologically twisted version, which we would like to stress is only a reformulation of $\mathcal{N}=2$ SYM theories, the action reads

$$S_{\text{top}} = \int F \wedge F + \{\bar{Q}, \Psi\} \,, \tag{238}$$

where $\Psi = \phi \partial_\mu \psi^\mu + F_{\mu\nu}\chi^{\mu\nu} + \eta[\bar{\phi}, \phi]$ is the "topological gauge fermion".

For hyperkähler manifolds, i.e. manifolds with three closed Kähler forms, the supercharges $\bar{Q}^+_{\mu\nu}$ can also be exploited in order to perform the topological twist [77]. Nekrasov [81, 82] proposed to also use Q_μ or better deform $\bar{Q}$ to

$$\bar{Q}_E = \bar{Q} + E_a V^a_{\mu\nu} x^\mu Q^\nu \,, \tag{239}$$

where $V^a_{\mu\nu} = -V^a_{\nu\mu}$ are the six generators of the Euclidean rotation group $SO(4)$ and E_a are constant parameters. This allows to define equivariant forms $\Omega(E) = \sum_p \Omega_p(E) = \sum_p \sum_{i_1,\dots,i_p} \frac{1}{p!}\Omega_{i_1,\dots,i_p} \mathrm{d}x^{i_1} \wedge \dots \mathrm{d}x^{i_p}$ such that

$$R\Omega(E) = \Omega(R^{-1}ER) \tag{240}$$

for any $R \in SO(4)$. $\Omega(E)$ are naturally acted on by the equivariant exterior derivative

$$\mathrm{d}_E = \mathrm{d} + \iota_{V(E)}\,, \tag{241}$$

where $\iota_{V(E)}$ denotes contraction with the vector field $V(E) = E_a V^a_{\mu\nu} x^\mu \partial^\nu$, i.e.

$$\mathrm{d}_E \Omega_p = \mathrm{d}\Omega_p + \iota_{V(E)} \Omega_p\,. \tag{242}$$

As a result, acting with d_E on a p-form generically yields both a $(p+1)$-form $\mathrm{d}\Omega_p$ and a $(p-1)$-form $\iota_{V(E)}\Omega_p$.

One can check that the topological observable

$$\mathcal{O}_P^{\Omega(E)} = \int_4 \Omega(E) \wedge P(\Phi) \tag{243}$$

is $\bar{Q}_E$-closed iff $\Omega(E)$ is "equivariantly" closed, i.e. iff $\mathrm{d}_E\Omega(E) = 0$. For generic choices of E^a the set of equivariantly closed forms is empty. However one can consider $E^a \in U(2)_\omega \subset SO(4)$, where $U(2)_\omega$ is the stability group of a "reference" symplectic (Kähler and thus closed) form,

$$\omega = \mathrm{d}x^1 \wedge \mathrm{d}x^2 + \mathrm{d}x^3 \wedge \mathrm{d}x^4\,, \tag{244}$$

that by definition satisfies $\mathrm{d}\omega = 0$. In this way from the condition of equivariance, that can be phrased in terms of the vanishing of the following Lie derivative

$$\mathcal{L}_{V(E)}\omega = 0 = \mathrm{d}(\iota_{V(E)}\omega) + \iota_{V(E)}\mathrm{d}\omega\,, \tag{245}$$

it follows that, at least locally,

$$\iota_{V(E)}\omega = \mathrm{d}\mu(E)\,, \tag{246}$$

or in other terms

$$\mathrm{d}_E(\omega - \mu(E)) = 0\,. \tag{247}$$

Decomposing $\mu(E)$ along the four generators of the stability group $U(2)_\omega$, one finds

$$h(x) \equiv \mu^0 = \delta_{\mu\nu} x^\mu x^\nu \quad \mu^a = \sum_{\mu<\nu} \eta^a_{\mu\nu} x^\mu x^\nu\,, \tag{248}$$

where $\eta^a_{\mu\nu}$ are 't Hooft symbols. Since ω defines a complex structure one can introduce complex coordinates z_1, z_2 such that $\omega = \mathrm{d}z_1 \wedge \mathrm{d}\bar{z}_1 + \mathrm{d}z_2 \wedge \mathrm{d}\bar{z}_2$. We also define

$$H = \mu_R(E) = \epsilon_1 |z_1|^2 + \epsilon_2 |z_2|^2\,, \tag{249}$$

where $\mu_R(E) = \frac{1}{2}(\epsilon_1 + \epsilon_2)\mu^0(z,\bar{z}) + \frac{1}{2}(\epsilon_1 - \epsilon_2)\mu^3(z,\bar{z})$ is an arbitrary linear combination of the "real" moment maps, the complex part being $\mu_C = \mu^1 + i\mu^2$.

Relying on the equivariance properties of ω and H, one can define the generating function of $\bar{Q}_E$-closed observables by the formula

$$
Z(a,\epsilon) = \left\langle \exp\left\{ \frac{1}{(2\pi i)^2} \int_{{}^4\equiv\ {}^2} \left[\omega \wedge \mathrm{Tr}(\phi F + \frac{1}{2}\psi \wedge \psi) \right.\right.\right.
$$

$$
\left.\left.\left. -\frac{1}{2} H(x) \mathrm{Tr}(F \wedge F) \right] \right\} \right\rangle_a , \qquad (250)
$$

where the suffix a denotes the dependence on the scalar v.e.v., a. Supersymmetry, which in this context is tantamount to topological invariance since $\bar{Q}_E$ is a linear combination of the supercharges, guarantees a perfect cancellation of all perturbative contributions between bosons and fermions. As a result, $Z(a,\epsilon)$ is saturated by instantons, viz.

$$
Z(a,\epsilon) = \sum_K q^K Z_K(a,\epsilon) , \qquad (251)
$$

where $q = \exp(2\pi i \tau)$. Moreover, the presence of H suppresses the contribution of widely separated instantons and can be combined with ω into

$$
\mathcal{H}(x,\theta) = H(x) + \frac{1}{2}\omega_{\mu\nu}\theta^\mu\theta^\nu . \qquad (252)
$$

$\mathcal{H}(x,\theta)$ represents a manifestly supersymmetric regulator for the holomorphic function $\mathcal{F}(a,\Lambda)$, where the explicit presence of Λ as an argument is to denote the dependence of $\mathcal{F}$ on the renormalisation group-invariant scale. In turn, the latter gets effectively replaced by $\mathcal{F}(a,\Lambda e^{-\mathcal{H}})$. Indeed rescaling the metric of ${}^4 \equiv {}^2$ by a factor λ and taking the limit $\lambda \to \infty$, only the last term survives in the partition function, since all the other terms are suppressed by inverse powers of λ that appear in the propagators needed for the contractions. Taking into account that derivatives of H with respect to x^μ or, equivalently, z_1 and z_2, are proportional to $\epsilon_{1,2}$ one finds

$$
\begin{aligned}
Z(a,\epsilon) &= \exp\left\{ \frac{1}{2(2\pi i)^2} \int_{4} \omega \wedge \omega \frac{\partial^2 \mathcal{F}(a,\Lambda e^{-H})}{\partial \log \Lambda^2} \right\} + \mathcal{O}(\epsilon) \\
&\approx \exp\left\{ \frac{\mathcal{F}_{\text{inst}}(a,\Lambda) + \mathcal{O}(\epsilon)}{\epsilon_1\epsilon_2} \right\}
\end{aligned} \qquad (253)
$$

where

$$
\mathcal{F}_{\text{inst}}(a,\Lambda) = \int_0^\infty \frac{\partial^2 \mathcal{F}(a,\Lambda e^{-H})}{\partial H^2} H \,\mathrm{d}H . \qquad (254)
$$

Equation (254) makes the analytic properties of Z and $\mathcal{F}$ manifest.

10.1 Including Hypermultiplets

In the presence of N_f hypermultiplets in the fundamental representation with masses m_f, a possible parametrisation[23] of the SW curve is [107]

[23] In order to make contact with the parametrisation used previously one has to perform the transformation $y = w - \frac{1}{2}P(z)$ and set $Q(z)$ to zero in the absence of hypers.

$$w + \frac{\Lambda^{2N_c - N_f} Q(z)}{w} = P(z)\,, \tag{255}$$

where

$$Q(z) = \prod_{f=1}^{N_f} (z + m_f) \tag{256}$$

and

$$P(z) = \prod_{l=1}^{N_c} (z - \alpha_l)\,. \tag{257}$$

The α_l's are related to the v.e.v.'s of the adjoint scalars belonging to the Cartan subalgebra and are such that $\sum_l \alpha_l = 0$. The space of monic polynomials $P(z)$, i.e. polynomials where the coefficient of the monomial of highest degree is 1, so that the coefficient of the monomial of next to highest degree is 0, is $\mathcal{U} = {}^{N_c-1}$ and can thus be parametrised by the $N_c - 1$ variables $u_n = \mathrm{Tr}(a^n)$, with $n = 1, ..., N_c - 1$. The latter are symmetric polynomials in the α_l that can be identified with the $N_c - 1$ Casimirs of $SU(N_c)$. The first two symmetric polynomials are 1 and $u_2 = \sum_l \alpha_l^2$ or, equivalently, $\sum_{l<l'} \alpha_l \alpha_{l'}$ since $\sum_l \alpha_l = 0$. The relation between u_n and α_l can be similarly determined. We now discuss how to determine the relation between u_n and the periods a_l and a_l^D of the SW curve.

In the perturbative region, where $|\alpha_l|, |\alpha_l - \alpha_n| \gg |\Lambda|, |m_f|$, one can choose local coordinates

$$a_l = \frac{1}{2\pi i} \oint_{A_l} \frac{z\, \mathrm{d}w}{w} \tag{258}$$

and

$$a_l^D = \frac{1}{2\pi i} \oint_{B_l} \frac{z\, \mathrm{d}w}{w}\,, \tag{259}$$

where the A_l cycles encircle the cuts in the z-plane from the point α_l^+ to the point α_l^-, where the points $\alpha_l^\pm$ are such that

$$P(z = \alpha_l^\pm) = \pm 2 \Lambda^{N_c - \frac{N_f}{2}} \sqrt{Q(z = \alpha_l^\pm)}\,. \tag{260}$$

Not all A_l cycles are homologically independent since $\sum_l A_l \approx 0$ can be shrunk to zero. The B_l cycles go through the cuts from α_l^+ to $\alpha_{l+1(\mathrm{mod}\,N)}^-$. Once again $\sum_l B_l \approx 0$ in homology. As a result, $\sum_l \mathrm{d}a_l \wedge \mathrm{d}a_l^D = 0$ and on local patches one can introduce the prepotential $\mathcal{F}(a; m, \Lambda)$ such that

$$\mathrm{d}\mathcal{F}(a; m, \Lambda) = \sum_l a_l^D \mathrm{d}a_l\,. \tag{261}$$

$\mathcal{F}(a; m, \Lambda)$ admits an expansion of the form

$$\mathcal{F}(a; m, \Lambda) = \mathcal{F}_{\mathrm{pert}}(a; m, \Lambda) + \mathcal{F}_{\mathrm{inst}}(a; m, \Lambda)\,, \tag{262}$$

where $\mathcal{F}_{\text{inst}}(a;m,\Lambda)$ encompasses the instanton contribution, that can be computed by the localisation techniques outlined below, and

$$\mathcal{F}_{\text{pert}}(a;m,\Lambda) = \frac{1}{2}\sum_{l\neq l'}(a_l - a_{l'})^2 \log\left(\frac{a_l - a_{l'}}{\Lambda}\right) - \sum_{l,f}(a_l + m_f)^2 \log\left(\frac{a_l + m_f}{\Lambda}\right) \tag{263}$$

encodes the logarithmic running of the gauge coupling with the mass scales at play. Indeed $a_l - a_{l'}$ are the masses of the W-bosons and $a_l + m_f$ are the masses of the charged hypermultiplets.

10.2 Instanton Measure and Localisation for Arbitrary *K*

Following the ADHM construction [19], the moduli space $\mathcal{M}_{K,N_c}$ of K instantons in $SU(N_c)$ with fixed framing (i.e. orientation in colour space) at infinity is a $4KN_c$ dimensional variety and can be viewed as the hyperkähler quotient of the ADHM data (B_1, B_2, I, J), where $B_{1,2} \in \text{End}(V_K)$, $I \in \text{Hom}(W_{N_c}, V_K)$ and $J \in \text{Hom}(V_K, W_{N_c})$, with respect to the action of $U(K)$. The corresponding formulae

$$\mu_{\mathbf{C}} = [B_1, B_2] + IJ = 0 \tag{264}$$

and

$$\mu_{\mathbf{R}} = [B_1, B_1^\dagger] + [B_2, B_2^\dagger] + II^\dagger - J^\dagger J = 0 \tag{265}$$

are the celebrated ADHM equations [19] that indeed enjoy invariance under $U(K)$ transformations.

As a result $\mathcal{M}_{K,N_c}$ is neither compact in the UV (due to small size instantons) nor in the IR (due to the non-compactness of ^{4}).

Various compactifications of $\mathcal{M}_{K,N_c}$ have been proposed [108]. The Uhlenbeck compactification $\mathcal{M}^U_{K,N_c}$ corresponds to the construction of a hyperkähler orbifold where the UV problem is cured by including point-like instantons, e.g. gluing subspaces of the form $\mathcal{M}_{K-1,N_c} \times {}^4$, $\mathcal{M}_{K-2,N_c} \times {}^8$, $\mathcal{M}_{K-3,N_c} \times {}^{12}$ and so on. Alternatively, according to Nekrasov and Schwarz the singularities of $\mathcal{M}^U_{K,N_c}$ can be blown up to a smooth space $\mathcal{M}^{NS}_{K,N_c}$ which includes "exceptional divisors" in place of the original singularities [79, 80]. This blowing up relies on a non-commutative extension of the gauge theory that translates in the possibility of deforming the ADHM equations (264) and (264) to[24]

$$\mu_{\mathbf{R}} = \zeta_{\mathbf{R}} \ {}_K \,, \quad \mu_{\mathbf{C}} = 0\,. \tag{266}$$

Deformed instanton calculus then boils down to computing equivariant volumes of $\mathcal{M}^{NS}_{K,N_c}$, provided one uses in the definition of the integration measure

[24] In principle one can deform $\mu_{\mathbf{C}}$ as well. But this deformation is irrelevant as it can always be eliminated by a non analytic change of coordinates.

the closed symplectic two-form[25] lifted from $\mathcal{M}^{U}_{K,N_c}$, where the relevant symplectic form is the reference Kähler form. Since this symplectic form vanishes when restricted to the exceptional divisors, it does not add contributions "extraneous" to the original "commutative" gauge theory. In order to localise the measure, i.e. reduce the integrals to contour integrals that are calculable by the residue theorem, it is convenient to consider the combined action of $U(K)$, $G = SU(N_c)/\;{}_{N_c}$ and $\mathbf{T}^2$, the latter representing the maximal torus, i.e. the exponential of the Cartan subalgebra, of $SO(4)$. The use of this combined action is instrumental in deforming the symplectic Kähler form ω of ^{4} by the moment maps $\mu_G = \delta_G \mathcal{A}^{\Lambda} \omega^{\mathrm{ADHM}}_{\Lambda\Lambda'} \mathcal{A}^{\Lambda'}$, where $\mathcal{A}^{\Lambda'}$ collectively denote the ADHM data, and $\mu_{T^2} = \epsilon_a x^i (V^a_{T^2})_i{}^j \omega_{jk} x^k$ and in constructing an equivariant form that localises the integrals on point-like abelian instantons.

The partition function over the compactified instanton moduli space reads

$$Z(a, \epsilon_1, \epsilon_2; q) = \sum_K q^K \oint_{\mathcal{M}_K} 1 \tag{267}$$

where $q = \mathrm{e}^{2\pi i \tau}$ and $\oint 1$ denotes the localisation of the integral to point-like instantons while $a = (a_1, ..., a_{N_c})$ parametrise the Cartan subalgebra of $G = SU(N_c)$, i.e. $\sum_{i=1}^{N_c} a_i = 0$ and ϵ_1, ϵ_2 are deformation parameters corresponding to the Ω background, defined below. For the purpose of computing, the integral it is convenient to rewrite the contour integral in the form

$$Z_K = \oint_{\mathcal{M}_K} 1 = \int_{\mathcal{M}_K} \exp(\omega + \mu_G(a) + \mu_{T^2}(\epsilon)), \tag{268}$$

due to topological BRST invariance. The non-perturbative contributions to the prepotential, but not the perturbative ones, are proportional to the logarithm of the topological partition function

$$Z(a, \epsilon_1, \epsilon_2; q) = \exp\left(\frac{1}{\epsilon_1 \epsilon_2} \mathcal{F}_{\mathrm{non-pert}}(a, \epsilon_1, \epsilon_2; q)\right), \tag{269}$$

as previously shown. Here we are only describing an efficient way to explicitly compute the contour integrals that yield Z_K, the coefficients of the expansion of the topological partition function. Using localisation, one can indeed derive an explicit expression for $Z(a, \epsilon_1, \epsilon_2; q)$. Taking for simplicity $\epsilon_1 = -\epsilon_2 = \hbar$ (the notation $\hbar$ suggests that some quantum non-commutativity is switched on as we will see!), one finds

$$Z(a, \hbar, -\hbar; q) = \sum_{\boldsymbol{K}} q^{|\boldsymbol{K}|} \prod_{(m,n) \neq (i,j)} \frac{a_{mi} + \hbar(K_{m,n} - K_{i,j} + j - n)}{a_{mi} + \hbar(j - n)}, \tag{270}$$

[25] A symplectic 2-form is the generalisation of the familiar 2-form $\omega = \sum_i \mathrm{d}p_i \wedge \mathrm{d}q^i$ in phase space.

where $a_{mi} = a_m - a_i$ and the sum is over the "coloured" partitions of the instanton numbers among the N_c abelian factors $U(1)^{N_c}$ of the Cartan subalgebra of $U(N_c)$

$$\boldsymbol{K} = (\boldsymbol{K}_1, \ldots, \boldsymbol{K}_{N_c}) \tag{271}$$

with

$$\boldsymbol{K}_n = \{K_{n,1} \geq K_{n,2} \geq \cdots \geq K_{n,l_n} \geq K_{n,l_n+1} = K_{n,l_n+2} = \cdots = 0\}, \tag{272}$$

while the product in (270) is over $1 \leq m, i \leq N_c$ and $n, j \geq 1$.

The theory can be enlarged by the addition of N_f hypermultiplets in the fundamental $\mathbf{N_c} + \mathbf{N_c}^*$ with masses $m_1, ..., m_{N_f}$. The explicit expression of $Z(a_i, m_f, \hbar, -\hbar; q)$ in this case becomes

$$Z(a_i, m_f, \hbar, -\hbar; q) = \sum_{\boldsymbol{K}} (q\hbar^{N_f})^{|\boldsymbol{K}|} \prod_{(m,n)} \prod_{f=1}^{N_f} \frac{\Gamma(\frac{1}{\hbar}(a_m + m_f) + 1 + K_{m,n} - n)}{a_{mi} + \hbar(j-n)}$$
$$\times \prod_{(m,n)\neq(i,j)} \frac{a_{mi} + \hbar(K_{m,n} - K_{i,j} + j - n)}{a_{mi} + \hbar(j-n)} . \tag{273}$$

10.3 Computing the Residues and Checking the Instanton Contributions

In a remarkable paper, Moore, Nekrasov and Shatashvili [80] have indeed been able to reduce the computation of Z_K in the K-instanton sector to contour integrals of the form

$$Z_K(a; \epsilon_i) = \frac{1}{K!} \frac{(\epsilon_1 + \epsilon_2)^K}{(2\pi i \epsilon_1 \epsilon_2)^K} \oint \prod_{I=1}^{K} \frac{d\phi_I Q(\phi_I)}{P(\phi_I) P(\phi_I + \epsilon_1 + \epsilon_2)}$$
$$\times \prod_{1 \leq I < J \leq K} \frac{\phi_{IJ}^2 (\phi_{IJ}^2 - (\epsilon_1 + \epsilon_2)^2)}{(\phi_{IJ}^2 - \epsilon_1^2)(\phi_{IJ}^2 - \epsilon_2^2)} , \tag{274}$$

where the complex variables $\phi_{IJ} = \phi_I - \phi_J$ can be thought of as entries of a $K \times K$ matrix, $P(z)$ and $Q(z)$ were defined before and the integration contours run along the real axis.

The variables ϕ_I, a_l and $\epsilon_{1,2}$ represent an infinitesimal deformation of the ADHM equations such that

$$[B_1, \phi] = \epsilon_1 B_1 \quad [B_2, \phi] = \epsilon_2 B_2$$
$$-\phi I + Ia = 0 \quad -aJ + J\phi = -(\epsilon_1 + \epsilon_2) J . \tag{275}$$

In the bases of the K-dimensional vector space V_K and the N_c-dimensional vector space W_{N_c} of the ADHM construction, where the $K \times K$ matrix ϕ and the $N_c \times N_c$ matrix a, representing the scalar v.e.v.'s, are diagonal, one has

$$(\phi_{IJ}+\epsilon_1)B_{1,IJ}=0 \qquad (\phi_{IJ}+\epsilon_2)B_{2,IJ}=0$$
$$(\phi_I - a_l)I_{I,l}=0 \qquad (\phi_I+\epsilon_1+\epsilon_2-a_l)J_{l,I}=0\,. \tag{276}$$

The poles at $\phi_{IJ} = \pm\epsilon_{1,2}$ should be avoided by deforming the contour or setting $\epsilon_{1,2} \to \epsilon_{1,2} + i\delta$. Similarly $a_l \to a_l + i\delta'$ in order to avoid the zeroes of P in the denominator. The origin of the poles at $\phi_{IJ} = \pm\epsilon_{1,2}$ can be understood by means of the Duistermaat–Heckman (DH) formula

$$\frac{1}{n!}\int_{X^{2n}} \omega^n \, \mathrm{e}^{-\mu[\xi]} = \sum_{P_f:V_\xi(P_f)=0} \frac{\mathrm{e}^{-\mu[\xi](P_f)}}{\prod_{i=1}^n W_i[\xi](P_f)}\,, \tag{277}$$

where X^{2n} is a symplectic manifold with symplectic form ω and μ is the moment map of a "torus" action generated by ξ and represented by V_ξ, that has fixed points P_f with "exponents" $W_i[\xi](P_f)$. In the case of $X = \mathcal{M}^{NS}_{K,N_c}$, the relevant torus action, that consists of the geometric transformations that form an abelian group, is $U(1)^{N_c-1} \subset SU(N_c)$ and $U(1)^2 \subset U(2)_\omega$. Indeed $U(1)^{N_c-1}$ is the maximal torus of the gauge group, generated by the Cartan subalgebra and one cannot hope to get any larger torus action from the gauge group generators. Similarly $U(1)^2$ is the maximal abelian subgroup of the stability group of the symplectic Kähler form and one cannot get anything more from the Euclidean rotation group. For generic ADHM data, the deformed ADHM equations have solutions only in correspondence with the poles of the integrand, this means that ϕ_I and $\phi_{IJ} = \phi_I - \phi_J$ are uniquely specified in terms of a_l, $\epsilon_{1,2}$ and P_f. The last ingredient, $\prod_{i=1}^n W_i[\xi](f)$, in the DH formula can be related to the Chern character of the tangent bundle of $\mathcal{M}^{NS}_{K,N_c}$ at the point P_f.

Another important step in the computation of the contour integral is the classification of the residues in terms of Young tableaux[26] $\boldsymbol{Y} = (Y_1, ..., Y_{N_c})$, such that $\sum_l |Y_l| = K$. Indeed to each Y_l with $0 < K_l \le K$ boxes corresponds a partition

$$K_{l,1} \ge \cdots \ge K_{l,n_l} \ge K_{l,n_l+1} = K_{l,n_l+2} = \cdots = 0\,. \tag{278}$$

Then the pole corresponding to a given $\boldsymbol{Y}$ is located at

$$\phi_I^{(r,s)} = a_l + \epsilon_1(r-1) + \epsilon_2(s-1)\,, \tag{279}$$

with the integers r and s such that $0 \le r \le n_l$ and $0 \le s \le K_{l,r}$.

In more physical terms the fixed points of the action of $G \times T^2$ on the "resolved" $\mathcal{M}^{NS}_{K,N_c}$ correspond to $U(N_c)$ non-commutative instantons that split into $U(1)^{N_c}$ non-commutative instantons such that the instanton charge K is

[26] Young tableaux are sets of boxes. The number of columns is N_c for $U(N_c)$. Starting from the first column the number of boxes should not increase. Boxes in the same column correspond to anti-symmetrised indices. Boxes in the same row correspond to symmetrised indices.

split into $K = \sum_l K_l$ with K_l in the lth subgroup. The non-commutativity induced by the ϵ-deformation prevents the instantons from coalescing one on top of the other.

For a given $\boldsymbol{Y}$ the residue of the contour integral reads

$$R(\boldsymbol{Y}) = \frac{1}{(\epsilon_1\epsilon_2)^K} \times \tag{280}$$

$$\prod_{l=1}^{N_c}\prod_{r=1}^{n_l}\prod_{s=1}^{K_{l,r}} \frac{T_l(\epsilon_1(r-1)+\epsilon_2(s-1))}{(\epsilon(\ell_l(r,s)+1)-\epsilon_2 h_l(r,s))(\epsilon_2 h_l(r,s)-\epsilon\ell_l(r,s))} \prod_{l<m}^{1,N_c}\prod_{r=1}^{n_l}\prod_{p=1}^{n_m}$$

$$\prod_{s=1}^{K_{l,r}}\prod_{t=1}^{K_{m,p}} \left[\frac{(a_{lm}+\epsilon_1(t-K_{m,p})-\epsilon_2(s-1))(a_{lm}+\epsilon_1 t-\epsilon_2(s-1-K_{l,r}))}{(a_{lm}+\epsilon_1(t-K_{m,p})-\epsilon_2(s-1-K_{l,r}))(a_{lm}+\epsilon_1 t-\epsilon_2(s-1)}\right]^2$$

where $\epsilon = \epsilon_1+\epsilon_2$, $a_{lm} = a_l - a_m$, $\ell_l(r,s) = K_{l,r} - s$, $h_l(r,s) = K_{l,r} + K_{l,s} - r - s + 1$ and

$$T_l(z) = \frac{Q(z+a_l)}{\prod_{m\neq l}(z+a_{lm})(z+\epsilon+a_{lm})} \,. \tag{281}$$

For future use it is convenient to define

$$S_l(z) = \frac{Q(z+a_l)}{\prod_{m\neq l}(z+a_{lm})^2} \,, \tag{282}$$

in terms of which the first two coefficients of the instanton expansion of the topological partition function are given by

$$Z_1 = \frac{1}{\epsilon_1\epsilon_2}\sum_l S_l(0) \,, \tag{283}$$

$$Z_2 = \frac{1}{(\epsilon_1\epsilon_2)^2}\left[\frac{1}{4}\sum_l S_l(0)[S_l(\hbar)+S_l(-\hbar)] + \frac{1}{2}\sum_{l\neq m}\frac{S_l(0)S_m(0)a_{lm}^4}{(a_{lm}^2-\hbar^2)^2}\right]$$

and so on. Using the known relation between the topological partition function and the non-perturbative contribution to the holomorphic prepotential (190) one gets

$$\mathcal{F}_1 = \sum_l S_l(0) \,,$$

$$\mathcal{F}_2 = \frac{1}{4}\sum_l S_l(0)S_l''(0) + \sum_{l\neq m}\frac{S_l(0)S_m(0)}{a_{lm}^2} + \mathcal{O}(\hbar^2) \tag{284}$$

and so on. Formulae tend soon to become unwieldy but Nekrasov has been able to check agreement with previous results for the holomorphic prepotential up to five instantons [81, 82]. The consistency among various independent approaches confirms the correctness of the result for the SW prepotential.

11 (Constrained) Instantons from Open Strings

One of the most astonishing features of critical strings is the presence of a massless vector boson in the open string spectrum and of a massless symmetric tensor in the closed string spectrum. The latter can be interpreted as the graviton. The former can be interpreted as the photon in the abelian case or as a gauge boson in the non-abelian one. Originally, a Yang–Mills group was introduced ad hoc through Chan–Paton (CP) factors. They respect the cyclicity of the Veneziano amplitude [109], that requires insertions of open string vertex operators on the boundary of a disk. In modern terms the group theory structure emerges from certain configuration of Dp-branes (D standing for Dirichelet, p for the number of spatial dimensions of the brane), i.e. hypersurfaces where open strings can end [110].

In the supersymmetric case, i.e. after GSO projection, the low-energy world-volume dynamics of N_c coincident Dp-branes is governed by the dimensional reduction from $d = 10$ to $d = p + 1$ of the $\mathcal{N} = 1$ SYM theory with gauge group $U(N_c)$ [111]. In particular, $p = 3$ corresponds to the celebrated $\mathcal{N} = 4$ SYM in $d = 4$, some (non-)perturbative properties of which will be discussed later on. From a macroscopic viewpoint, Dp-branes are 1/2 BPS solitons of type II or type I supergravities, in that they preserve one-half of the supersymmetries of the parent theory. Configurations with different kinds of Dp-branes are generically non-supersymmetric except for very special choices of embeddings, i.e. dimensions and orientation of the various branes w.r.t. one another. For our purposes of relating strings to instanton calculus, it is crucial that a configuration with K D$(p-4)$-branes lying within a stack of N_c Dp-branes, i.e. such that the branes have $p-4$ dimensions in common, preserves 1/4 of the original supersymmetries. In fact this configuration is a "bound state" at threshold [111], i.e. the mass of the bound state is the sum of the masses of the constituent branes. Moreover, the D$(p-4)$-branes have all the right to be considered as a "gas" of instantons within the Dp-branes [83].

We will exploit the fact that D$(p-4)$-branes behave as a gas of instantons within the Dp-branes for the case $p = 3$ that corresponds to $\mathcal{N} = 4$ SYM and will indicate how to get instantons in gauge theories with less or no supersymmetries. We will also discuss how to tune the parameters, i.e. the string tension $T = 1/2\pi\alpha'$ and the string coupling g_s (related to the v.e.v. of the massless scalar dilaton) in order to decouple heavy string modes. We will not consider the cases $p \neq 3$.

In the presence of N_c D3 and K D(-1)-branes there are three sectors of the open string spectrum. Strings that start and end on D3-branes provide the $U(N_c)$ gauge fields and their superpartners. Strings that start and end on D(-1)-branes yield $U(K)$ non-dynamical (background) gauge fields and their superpartners. Together they provide a subset of the (super) ADHM data, e.g. the centre of mass $x_{CM} = \sum_i M_i x_i / \sum_i M_i$, where M_i are the masses of the brane constituents, and global SUSY parameters. Strings that start on

D3-branes and end on D(−1)-branes or that start on D(−1)-branes and end on D3-branes provide the remaining (super) ADHM data.

Suppressing CP factors for the moment, the vertex operators for gauge bosons, that belong to the Neveu–Schwarz (NS) sector, read

$$V_A = A_M(p)\Psi^M \mathrm{e}^{-\varphi}\mathrm{e}^{ip\cdot X} , \tag{285}$$

where X^M and Ψ^M, with $M = 0,...,9$, denote the bosonic and fermionic string coordinates, respectively, and φ the superghost boson. BRST invariance requires $p^2 = 0$ and $p \cdot A(p) = 0$, which is the form that the linearised Yang–Mills equations for $A^M(p)$ take in the transverse gauge. Vertex operators for gauginos, belonging to the Ramond (R) sector, read

$$V_\Lambda = \Lambda^a(p) S_a \mathrm{e}^{-\varphi/2}\mathrm{e}^{ip\cdot X} , \tag{286}$$

where S_a, with $a = 1,...16$, is a chiral spin field that creates a cut for Ψ^M, i.e. a line connecting two branch points of the polydromous fields Ψ^M. This means that the operator product expansion (OPE) of $\Psi^M(z)$ with $S_a(w)$ contains half integer powers of $z-w$. BRST invariance requires $p^2 = 0$ and $p\cdot\Gamma_{ab}\Lambda^b(p) = 0$, which is the massless Dirac equation for $\Lambda^b(p)$.

After reduction to $d = 4$, relevant for D3-branes, the gauge bosons in $d = 10$ yield gauge bosons A_μ as well as six real scalars $A_i = \phi_i$. The $d = 10$ gauginos yield four Weyl gauginos Λ^A_α and their anti-particles $\bar\Lambda^{\dot\alpha}_A$. The structure of the on-shell effective action can be extracted from the knowledge of the scattering amplitudes on the disk with D3-brane boundary conditions. In the low-energy limit, $\alpha' \to 0$ with the Yang–Mills coupling $g^2 = 4\pi g_s$ fixed, the effective action coincides with $\mathcal{N} = 4$ SYM theory.

After reduction to $d = 0$, relevant for D(−1)-branes, also known as D-instantons, the gauge field vertex operator V_A defined above yields 10 non-dynamical "fields", i.e. matrices whose dynamics is governed by an action in 0 dimensions. Due to the breaking of the (Euclidean) Lorentz symmetry $SO(10)$ to $SO(4) \times SO(6)$ in the presence of D3-branes, it turns out to be convenient to split the ten "gauge bosons", a_M, into four gauge bosons, a_μ, and six real "scalars", χ_i. Similarly, the $d = 10$ gauginos, V_Λ, produce four non-dynamical Weyl "gauginos", Θ^A_α, and their anti-particles, $\bar\Theta^{\dot\alpha}_A$. The structure of the on-shell effective action can be extracted from the scattering amplitudes on the disk with D(−1)-brane boundary conditions. In the low-energy limit, $\alpha' \to 0$ with the zero-dimensional Yang–Mills coupling $g_0^2 = g_s/4\pi^3(\alpha')^2$ fixed, the effective action for the low lying excitations of the D(−1)-brane reads

$$\mathcal{S}_{\mathrm{D}(-1)} = \mathcal{S}_{\mathrm{cub}} + \mathcal{S}_{\mathrm{quart}} , \tag{287}$$

where

$$\mathcal{S}_{\mathrm{cub}} = \frac{i}{g_0^2}\,\mathrm{Tr}_K\left(\bar\Theta_A\sigma^\mu[a_\mu,\Theta^A] - \frac{1}{2}\tau_i^{AB}\bar\Theta_A[\chi^i,\bar\Theta_B] \right.$$
$$\left. -\frac{1}{2}\bar\tau^i_{AB}\Theta_A[\chi_i,\Theta_B]\right) , \tag{288}$$

with Tr_K denoting the trace in the K-dimensional representation of $U(K)$ and

$$\mathcal{S}_{\mathrm{quart}} = \frac{1}{4g_0^2}\,\mathrm{Tr}_K\left([a_\mu, a_\nu][a^\mu, a^\nu] + 2[a_\mu, \chi_i][a^\mu, \chi^i] + [\chi_i, \chi_j][\chi^i, \chi^j]\right) . \quad (289)$$

In what follows, it is crucial to replace $\mathcal{S}_{\mathrm{quart}}$ with a cubic action $\mathcal{S}'_{\mathrm{cub}}$, through the Hubbard–Stratonovich procedure, that entails the introduction of auxiliary fields $X_{\mu\nu}$, $Y_{\mu i}$ and Z_{ij}. Their vertex operators, bilinear in the fermions Ψ's, are not BRST invariant and a priori one should not insert them as vertices in scattering amplitudes. Nevertheless, three-point amplitudes with one auxiliary field insertion are consistent and yield the correct interactions, because the BRST non-invariant part decouples. In the end one replaces $\mathcal{S}_{\mathrm{quart}}$ with

$$\mathcal{S}'_{\mathrm{cub}} = \frac{1}{2g_0^2}\,\mathrm{Tr}_K\Big(\frac{1}{2}X_{\mu\nu}X^{\mu\nu} + Y_{\mu i}Y^{\mu i} + \frac{1}{2}Z_{ij}Z^{ij} + X_{\mu\nu}[a^\mu, a^\nu] + 2Y_{\mu i}[a^\mu, \chi^i] + Z_{ij}[\chi^i, \chi^j]\Big) . \quad (290)$$

We now pass to consider open strings connecting D3-branes to D(-1)-branes. Vertex operators in this sector involve ${}_2$ bosonic twist fields, $\sigma_{(\mu)}$, because one is changing the boundary conditions of the four (Euclidean) "spacetime" coordinates from Neumann (D3) to Dirichelet (D(-1)). Twist fields are local conformal primary operators that generate a cut in the bosonic coordinate field X very much like spin fields, already encountered above, generate cuts in the fermionic coordinates Ψ. In the canonical superghost picture $q = -1$ for bosons, the vertex operators read

$$V_w^{(-1)} = w_{\dot\alpha}\Sigma C^{\dot\alpha}\mathrm{e}^{-\varphi}T_{K,N_c} , \qquad V_{\bar w}^{(-1)} = \bar w_{\dot\alpha}\Sigma C^{\dot\alpha}\mathrm{e}^{-\varphi}T_{N_c,K} , \quad (291)$$

where $\Sigma = \prod_\mu \sigma_{(\mu)}$ is a bosonic twist field of dimension $1/4 = 4 \times 1/16$ and $C^{\dot\alpha}$ is an $SO(4)$ spin field of dimension $1/4$. $T_{N_c,K}$ denote the $K \times N_c$ Chan–Paton "matrices". The supersymmetry partners, in the canonical $q = -1/2$ picture for fermions, have vertex operators of the form

$$V_\nu^{(-1/2)} = \nu^A\Sigma C_A\mathrm{e}^{-\varphi/2}T_{K,N_c} , \qquad V_{\bar\nu}^{(-1/2)} = \bar\nu^A\Sigma C_A\mathrm{e}^{-\varphi/2}T_{N_c,K} , \quad (292)$$

where C_A is an $SO(6)$ spin field.

Computing amplitudes on disks with mixed boundary conditions allows one to extract the effective action for the "twisted" sector. Defining the $K \times K$ matrices

$$W^a = (w\sigma^a\bar w)_{K\times K} , \quad (293)$$

the action that governs the dynamics of the light modes (or moduli) of the system of D(-1)-branes in the presence of D3-branes, takes the form

$$\mathcal{S}_{\rm twist} = \frac{2i}{g_0^2}\,{\rm Tr}_K\Big((w_{\dot\alpha}\bar\nu^A + \nu^A \bar w_{\dot\alpha})\bar\Theta^{\dot\alpha}_A - X_a W^a + \frac{1}{2}\chi_i \tau^i_{AB}\nu^A\bar\nu^B - i\chi_i w^{\dot\alpha}\bar w_{\dot\alpha}\chi^i \Big)\,, \tag{294}$$

where we have set $X_{\mu\nu} = X_a\bar\eta^a_{\mu\nu} + \bar X_a \eta^a_{\mu\nu}$, so that the three components $\bar X_a$ actually decouple because $\eta^a_{\mu\nu}\bar\eta^{\mu\nu}_b = 0$.

Combining with the previous terms and rescaling appropriately the fields, so as to get a non-trivial field theory limit, one finds for the *complete* action that governs the dynamics of the light modes (or moduli) of the system of D(−1)-branes in the presence of D3-branes

$$\mathcal{S}_{\rm moduli} = \mathcal{S}_{\rm cub} + \mathcal{S}'_{\rm cub} + \mathcal{S}_{\rm twist}\;. \tag{295}$$

One can check that

$$x_0^\mu = {\rm Tr}_K(a^\mu) \quad \text{and} \quad \theta_0^{\alpha A} = {\rm Tr}_K(\Theta^{\alpha A})\,, \tag{296}$$

drop from the action, while varying w.r.t. X_a and $\bar\Theta^{\dot\alpha}_A$ yields the super ADHM equations. The latter consist in $3K\times K$ real bosonic equations

$$W^a + i\bar\eta^a_{\mu\nu}[a^\mu, a^\nu] = 0 \tag{297}$$

that, taking into account $U(K)$ invariance, impose $4K\times K$ constraints on the ADHM data which implement the hyperkähler quotient, and $8K\times K$ fermionic constraints (for $\mathcal{N}=4$ supersymmetry)

$$w_{\dot\alpha}\bar\nu^A + \nu^A\bar w_{\dot\alpha} + [\Theta^{\alpha A}, a_\mu]\sigma^\mu_{\alpha\dot\alpha} = 0\,, \tag{298}$$

that reduce the number of independent fermionic zero modes. These ingredients, i.e. the constrained ADHM superdata encoded in the various open string vertex operators and their interactions encoded in the scattering amplitudes, are sufficient to reconstruct the classical super instanton profile as well as to compute instanton contributions to correlation functions. In particular

$$A^{\rm inst}_\mu(p; w, \bar w) = \langle\langle V^{(-1)}_{\bar w} U^{(0)}_\mu(-p) V^{(-1)}_w \rangle\rangle = (\bar w \sigma_a w)_{N_c\times N_c}\bar\eta^a_{\mu\nu}p^\nu {\rm e}^{-ip\cdot x_0}\,, \tag{299}$$

where $U^{(0)}_\mu$ is the "amputated" vertex operator

$$U^{(0)}_\mu(-p) = 2i(\partial X_\mu - ip\cdot\Psi\Psi_\mu){\rm e}^{-ip\cdot X} \tag{300}$$

in the $q=0$ superghost picture. After Fourier transforming to x space one obtains

$$A^{\rm inst}_\mu(x; w, \bar w) = \int \frac{{\rm d}^4p}{4\pi^2 p^2} A^{\rm inst}_\mu(p; w, \bar w){\rm e}^{ip\cdot x} = (\bar w\sigma_a w)_{N_c\times N_c}\bar\eta^a_{\mu\nu}\frac{(x-x_0)^\nu}{(x-x_0)^4}\,, \tag{301}$$

which should coincide with the asymptotic behaviour of the *unconstrained* instanton at large distance in the *singular* gauge. Indeed, focussing on $K = 1$ and $N_c = 2$, if one sets $2\rho^2 = \bar{w}w$ by a global $SU(2)$ rotation, one finds

$$A_\mu^{\mathrm{inst},a}(x;\rho) \approx 2\rho^2 \bar{\eta}^a_{\mu\nu} \frac{(x-x_0)^\nu}{(x-x_0)^4} , \tag{302}$$

which is the large distance term in the expansion of the celebrated BPST solution. To make contact with (1) one clearly has to extract a factor g from (302). Higher-order terms in $\rho^2 = \bar{w}w/2$ are sub-dominant at large distances and are anyway determined by solving the YM equations with the given asymptotic behaviour. By similar methods one can compute the classical asymptotic profiles of the other elementary fields (gauginos and scalars) that involve the 16 supersymmetry (8 Poincaré and 8 superconformal) parameters broken by the D-instanton but preserved by the D3-branes (in the near horizon limit). These profiles enter the computation of instanton contributions to amplitudes.

One can then embark in the computation of instanton-dominated correlators. Denoting by $U_{\mathcal{O}}(p)$ the unintegrated open string vertex operators corresponding to the SYM fields $\mathcal{O}(-p)$, one schematically has to compute

$$\begin{aligned} &\langle \mathcal{O}_1(p_1)...\mathcal{O}_n(p_n)\rangle|^{\mathrm{D-inst}}_{\mathrm{amp}} \\ &= \int \mathrm{d}\mathcal{M} \langle\langle U_{\mathcal{O}_1}(-p_1)\rangle\rangle_{\mathcal{D}(\mathcal{M})} ... \langle\langle U_{\mathcal{O}_n}(-p_n)\rangle\rangle_{\mathcal{D}(\mathcal{M})} \mathrm{e}^{-\mathcal{S}(\mathcal{M})} . \end{aligned} \tag{303}$$

The simple "product" form of the integrand is due to the fact that the amplitude is dominated by disconnected disks with mixed boundary conditions $\mathcal{D}(\mathcal{M})$ obtained by inserting the non-dynamical (super)moduli fields, which must include at least the 16 exact fermionic zero modes. This is the most interesting part of the string construction of instantons. We have only devoted few lines to it because, once the "super-instanton" profile has been generated and the "supermoduli" have been correctly identified, one can repeat word by word what has been pedagogically said and carefully done in the discussion of $\mathcal{N} = 1$ SYM.

11.1 $\mathcal{N} = 2$ SYM from Open Strings

There are various ways to realise $d = 4$ $\mathcal{N} = 2$ SYM in string theory. The easiest way is to put a stack of D3-branes at an orbifold point,[27] let us say the origin of $^6/\Gamma$, such that the holonomy group[28] Γ is a discrete subgroup

[27] Another possibility is to consider intersecting branes or brane with internal magnetic fluxes preserving $\mathcal{N} = 2$ supersymmetry. Other configurations are possible in M-theory, e.g. by wrapping M5-branes around Riemann surfaces producing SW curves, etc.

[28] If the holonomy group $\Gamma \subset SU(3)$ one has $\mathcal{N} = 1$ SYM, when $\Gamma = 1$ (trivial holonomy group) one has $\mathcal{N} = 4$ SYM.

of $SU(2)$ of dimension r. As discussed in [93, 112], in the context of ALE instantons in string theory, there are essentially two kinds of branes one can consider. Regular branes are those that transform in the "regular" representation of Γ, i.e. the (usually reducible) representation of dimension $r = \sum_i n_i^2$ equal to the dimension of Γ. For instance for the cyclic group $_n$, the regular n-dimensional reducible representation is simply the direct sum of the n one-dimensional irreducible representations. The D3-branes can be moved away from the orbifold point, where the curvature is concentrated, to the flat bulk in such a way that the r images in the covering space actually correspond to one physical brane in $^6/\Gamma$. There can be other branes that transform under smaller (irreducible) representation of Γ, e.g. any of the $n-1$ non-trivial one-dimensional irreps of $_n$, and are called "fractional" branes in that they carry fractional R–R charge in $^6/\Gamma$, corresponding to integer charge in the covering space. Branes of this kind cannot be moved away from the orbifold point and give rise to gauge theories with lower supersymmetry than branes that can be moved into the flat bulk (here "moving" has exactly the same meaning as above). In orbifolds the curvature is concentrated at the singularity. If a (stack of) branes is displaced from the singular (orbifold) point and placed in the bulk, the effective field theory governing the dynamics of the light modes enjoys $\mathcal{N} = 4$ SUSY.

For definiteness, let us consider the case of $\Gamma = {}_n \subset SU(2)$, corresponding to the A-series in the ADE classification of discrete subgroups of $SU(2)$ and thus the case of ALE instantons, see, e.g. [93]. The regular representation is n-dimensional and reducible. One starts with n stacks of N_c-branes each. The reduction of SUSY from $\mathcal{N} = 4$ to $\mathcal{N} = 2$ is achieved by truncating the parent theory with gauge group $U(nN_c)$ to the sector which is invariant under the action of $_n$. The natural action of $_n \subset SU(2)$ on the gauge fields and complex scalars is given by

$$A_\mu \to A_\mu \quad , \quad \phi_3 \to \phi_3 \quad , \quad \phi_1 \to \omega\phi_1 \quad , \quad \phi_2 \to \bar{\omega}\phi_2 \quad , \tag{304}$$

where $\omega = \exp(2\pi i/n)$. Furthermore, $_n$ is taken to act on the gauge group $U(nN_c)$ via a discrete Wilson line

$$W_{\text{reg}} = (\ {}_{n\times n}, \omega \ {}_{n\times n}, \ldots, \omega^{n-1} \ {}_{n\times n}) \,, \tag{305}$$

in such a way that

$$T^a \to W T^a W^{-1} \,. \tag{306}$$

Taking into account the combined action of $_n$ in (304)–(306), one concludes that the condition $W\Phi W^{-1} = \omega_\Phi \Phi$, where Φ collectively denotes the (bosonic) fields, truncates the theory to one with a vector boson and a complex scalar ϕ_3 in the adjoint of $U(N_c)^n$ and two complex bosons $\phi_{1,2}$ in the bi-fundamental of adjacent $U(N_c)$'s. Since we have chosen precisely $_n \subset SU(2)$, out of the 16 supersymmetry parameters associated with the $\mathcal{N} = 4$ Poincaré supersymmetry, 8 are invariant and generate the $\mathcal{N} = 2$ Poincaré supersymmetry.

Indeed the $(\mathbf{2}_L, \mathbf{4})$ and $(\mathbf{2}_R, \mathbf{4}^*)$ spinors (that arise from dimensional reduction of the $\mathbf{16}$ of $\mathcal{N}=1$ SYM in $d=10$) give rise to $(\mathbf{2}_L, \mathbf{2}, \mathbf{1})$ and $(\mathbf{2}_R, \mathbf{2}, \mathbf{1})$ spinors that are invariant under $SU(2)_H$ as well as to $(\mathbf{2}_L, \mathbf{1}, \mathbf{2})$ and $(\mathbf{2}_R, \mathbf{1}, \mathbf{2})$ spinors that are not invariant under $SU(2)_H$. The resulting $\mathcal{N}=2$ Poincaré supersymmetry implies that each of the above bosons is accompanied by its fermion superpartner that promote the theory to $\mathcal{N}=2$ SYM coupled to hypermultiplets in the $(\mathbf{N}_i, \mathbf{N}^*_{i+1}) \oplus (\mathbf{N}^*_i, \mathbf{N}_{i+1})$ representation. The one-loop β-function of $SU(N_c)^n$ turns out to be zero, because $2N_c - 2N_c = 0$, while the $U(1) \subset U(N_c)^n$ are IR free (as for any abelian gauge theory coupled to charged matter) and thus the $U(1)$ vector multiplets decouple at low energies. One is dealing with an exact $\mathcal{N}=2$ superconformal theory in the IR. In fact, one can turn on v.e.v.'s of the adjoint scalar (Coulomb branch) or of the bi-fundamentals (Higgs branch). The former generically breaks the group to $U(1)^{nN_c}$, the latter to $U(N_c)_{\rm diag}$ realising the expected simultaneous motion of the n stacks of N_c branes away from the fixed point into the bulk, where supersymmetry is enhanced to $\mathcal{N}=4$, since the hypermultiplets in the bi-fundamentals produce the extra adjoint of $U(N_c)_{\rm diag}$ needed to promote a $\mathcal{N}=2$ vector multiplet to a $\mathcal{N}=4$ vector multiplet. The diagonal $U(1) \subset U(N_c)_{\rm diag}$ is free and corresponds to the centre of mass motion of the bound state of the various stacks of D-branes.

If instead of choosing the "regular" embedding of $\;_n$ in $U(nN_c)$ one takes another representation for W, one gets non-superconformal theories that live on fractional branes. In the extreme case where $W = W_k$ with

$$W_k = (\omega^k \;\; _{M\times M}) , \tag{307}$$

and $\omega = e^{2\pi i/n}$ for any $k = 1, ..., n-1$ one gets pure $\mathcal{N}=2$ SYM with gauge group $U(M)$ where M is not necessarily a multiple of n, i.e. $M \neq nN_c$ generically. Fractional branes are stuck at the fixed point, conventionally put at the origin of $\;^6/\;_n$ and cannot move away from it. Referring to our previous notation, out of the six real ϕ_i's only two (one complex), ϕ and $\phi^\dagger$, survive the orbifold projection. The precise linear combination of the six original real scalar fields is determined by the choice of the embedding of $SU(2)$ into the rotation group of $\;^6$, $SO(6) \approx SU(4)$. Similarly, out of the four gaugini only two survive the projection, i.e. the ones that are singlets of $SU(2) \supset \Gamma$ and transform as a charged doublet under the $SU(2) \times U(1)$ subgroup of $SO(6) \approx SU(4)$ commuting with Γ. The complexified gauge coupling of the surviving $\mathcal{N}=2$ SYM theory with gauge group $U(M)$ is determined by the closed string background, i.e. the v.e.v.'s of the so-called blowing up modes of the orbifold fixed point. The blowing up modes are nothing but twist fields for the closed string coordinates, this means that the OPE of the bosonic coordinates $X(z,\bar{z})$ with the bosonic twist fields $\sigma(w,\bar{w})$ contains fractional powers. We have already encountered twist fields for the open string coordinates. Since closed string vertex operators are given by combinations of open string vertex operators for the left- and right-moving excitations of the closed string, blowing up modes are described by products of twist fields for the left and right movers, schematically $\sigma(z,\bar{z}) = \sigma_L(z)\sigma_R(\bar{z})$.

Indeed one may regard fractional D3-branes as D5-branes wrapped around homologically non-trivial cycles, sometimes called "exceptional divisors", that are complex varieties of codimension one[29] in ${}^4/\Gamma \equiv {}^2/\Gamma$, that shrink to zero size, i.e. to zero area in one's preferred units, at the fixed point in the orbifold limit, i.e. prior to resolution of the singularity. For a ${}_n$ singularity there are $n-1$ two-spheres that intersect according to the Cartan matrix of A_{n-1}. The complexified coupling is given by the "period integrals" of the 2-form $B_2 + iC_2$, with B_2 belonging to the Neveu–Schwarz–Neveu–Schwarz (NS–NS) sector and C_2 belonging to the R–R sector. For regular branes, the gauge coupling of the diagonal subgroup, the one surviving when the branes move to the bulk, is given by $\Phi + iC_0$, where Φ is the NS–NS dilaton and C_0 is the R–R scalar "axion". Indeed one can show that the corresponding tadpoles precisely match the one-loop running of the couplings [113, 114]!

Essentially the same analysis applies to open strings with both ends on D(−1)-branes (D-instantons). Taking K fractional D-instantons with

$$W_l^{\mathrm{D-inst}} = (\omega^l \quad {}_{K\times K}) \,, \tag{308}$$

produces the truncation of the world-volume low-energy theory to pure (zero-dimensional!) $\mathcal{N}=2$ SYM with gauge group $U(K)$. The surviving adjoint scalars will be denoted by χ and $\chi^\dagger$. The two associated non-dynamical fermions will be denoted by Θ^r_α with $r=1,2$ and their conjugates by $\bar{\Theta}^{\dot\alpha}_r$. Setting

$$a^\mu_{K\times K} = x^\mu_0 \quad {}_{K\times K} + y^\mu_g \mathbf{T}^g_{K\times K} \,, \tag{309}$$

where $\mathbf{T}^g_{K\times K}$ are the generators of $SU(K)$, and

$$\Theta^{r\alpha}_{K\times K} = \theta^{r\alpha}_0 \quad {}_{K\times K} + \zeta^{r\alpha}_g \mathbf{T}^g_{K\times K} \tag{310}$$

one can regard x^μ_0 and $\theta^{r\alpha}_0$ as coordinates in $\mathcal{N}=2$ superspace.

Open strings connecting N_c fractional D3-branes to K fractional D-instantons belong to the bi-fundamental $(\mathbf{N_c}, \bar{\mathbf{K}})$ representation of $U(N_c) \times U(K)$. The bosonic modes $w_{\dot\alpha}$ and $\bar{w}_{\dot\alpha}$ are as in the $\mathcal{N}=4$ case, while the fermionic modes are halved and will be consistently denoted by ν^r and $\bar{\nu}^r$.

In the double scaling limit $\alpha' \to 0$, $g_0 \to \infty$ (g_0 has mass dimension $+2$) with $(4\pi^2\alpha' g_0)^2 = 4\pi g_s = g^2$ fixed, the non-dynamical moduli fields are governed by the action

$$\mathcal{S}^{\mathcal{N}=2}_{\mathrm{moduli}} = \mathcal{S}_{\mathrm{bose}} + \mathcal{S}_{\mathrm{fermi}} + \mathcal{S}_{\mathrm{ADHM}} \,, \tag{311}$$

where

$$\begin{aligned}
\mathcal{S}^{\mathcal{N}=2}_{\mathrm{bose}} &= \mathrm{Tr}_K\big(-2[\chi^\dagger, a^\mu][\chi, a_\mu] + \chi w^{\dot\alpha}\bar{w}_{\dot\alpha}\chi^\dagger + \chi^\dagger w^{\dot\alpha}\bar{w}_{\dot\alpha}\chi\big) \\
\mathcal{S}^{\mathcal{N}=2}_{\mathrm{fermi}} &= i\frac{\sqrt{2}}{2}\varepsilon_{rs}\mathrm{Tr}_K\big(\nu^r\bar{\nu}^s\chi^\dagger - \Theta^r[\chi, \Theta^s]\big) \\
\mathcal{S}^{\mathcal{N}=2}_{\mathrm{ADHM}} &= -i\mathrm{Tr}_K[\bar{\Theta}^{\dot\alpha}_r(w_{\dot\alpha}\bar{\nu}^r + \nu^r\bar{w}_{\dot\alpha} + [\Theta^{\alpha r}, a_\mu]\sigma^\mu_{\alpha\dot\alpha}) - X_a(W^a + i\bar{\eta}^a_{\mu\nu}[a^\mu, a^\nu])] .
\end{aligned} \tag{312}$$

[29] Recall that $\Gamma \subset SU(2)$ only acts on ${}^2 \equiv {}^4 \subset {}^3 \equiv {}^6$.

Varying the action w.r.t. X_a and $\bar{\Theta}^{\dot{\alpha}}_A$ yields the $\mathcal{N}=2$ super ADHM constraints

$$W^a + i\bar{\eta}^a_{\mu\nu}[a^\mu, a^\nu] = 0 \tag{313}$$

and

$$w_{\dot{\alpha}}\bar{\nu}^r + \nu^r \bar{w}_{\dot{\alpha}} + [\Theta^{\alpha r}, a_\mu]\sigma^\mu_{\alpha\dot{\alpha}} = 0\,. \tag{314}$$

As before, one can perform a Hubbard–Stratonovich transformation and replace the quartic couplings in $\mathcal{S}^{\mathcal{N}=2}_{\text{bose}}$ with trilinear couplings to auxiliary fields $Y^\mu_{K\times K}$ and $U^{\dot{\alpha}}_{N_c\times K}$ and $\bar{U}^{\dot{\alpha}}_{K\times N_c}$ and their conjugates. As a result one gets

$$\begin{aligned}\mathcal{S}^{\mathcal{N}=2}_{\text{bose}} = \mathrm{Tr}_K\big(&2Y^\mu Y^\dagger_\mu - 2Y_\mu[\chi^\dagger, a^\mu] - 2Y^\dagger_\mu[\chi, a_\mu] \\ &+ U^\dagger_{\dot{\alpha}}U^{\dot{\alpha}} + \bar{U}^\dagger_{\dot{\alpha}}\bar{U}^{\dot{\alpha}} + \bar{U}^\dagger_{\dot{\alpha}}\bar{w}^{\dot{\alpha}}\chi + \bar{U}_{\dot{\alpha}}\bar{w}^{\dot{\alpha}}\chi^\dagger + \chi w^{\dot{\alpha}}U^\dagger_{\dot{\alpha}} + \chi^\dagger w^{\dot{\alpha}}U_{\dot{\alpha}}\Big)\,.\end{aligned} \tag{315}$$

Computing amplitudes with insertions of the scalar field vertex operator

$$V^{(-1)}_\phi(p) = \phi(p)\mathrm{e}^{-\varphi}\mathrm{e}^{ip\cdot X} \tag{316}$$

at $p=0$, that correspond to turning on a v.e.v. for ϕ in the Cartan subalgebra of $U(N_c)$, one can construct the relevant action for the moduli fields. By the invariance of the scattering amplitudes under the exchange of the dynamical field ϕ with the non-dynamical field χ, the effect of the presence of a constant ϕ in the computation of instanton effects simply amounts to the replacements

$$\chi_{K\times K} \otimes \;{}_{N_c\times N_c} \to \chi_{K\times K} \otimes \;{}_{N_c\times N_c} - \;{}_{K\times K} \otimes \phi_{N_c\times N_c} \tag{317}$$

and

$$\chi^\dagger_{K\times K} \otimes \;{}_{N_c\times N_c} \to \chi^\dagger_{K\times K} \otimes \;{}_{N_c\times N_c} - \;{}_{K\times K} \otimes \phi^\dagger_{N_c\times N_c}\,. \tag{318}$$

It is crucial to observe at this point that ϕ and $\phi^\dagger$ do not enter the fermionic action in the same way, indeed the additional terms in the fermionic action of the $\mathcal{N}=2$ supermoduli read

$$\Delta\mathcal{S}^{\mathcal{N}=2}_{\text{fermi}} = i\frac{\sqrt{2}}{2}\,\varepsilon_{rs}\,\mathrm{Tr}_K\left(\nu^r\phi^\dagger\bar{\nu}^s\right)\,. \tag{319}$$

As a consequence all $2K(N_c-2)$ zero modes associated with $\bar{\nu}^r$ and ν^s are lifted.

11.2 SW Prepotential from String Instantons

Let us now specialise to the case of a $SU(2)$ gauge group. We are ready to accomplish the task of checking the SW prepotential, $\mathcal{F}_{\text{SW}}$, by means of Veneziano's open string theory! The Wilsonian effective action for the light neutral modes is

$$S_{\text{eff}}[\Phi] = \int \mathrm{d}^4x\,\mathrm{d}^4\theta\,\mathcal{F}(\Phi) + \text{h.c.}\,, \tag{320}$$

where $\Phi = \Phi_3 \sigma^3/2$ is the $\mathcal{N}=2$ vector superfield,

$$\Phi(x,\theta) = \phi(x) + \theta^\alpha_r \lambda^r_\alpha(x) + \frac{1}{2}\theta^\alpha_r \theta^\beta_s (\varepsilon^{rs}\sigma^{\mu\nu}_{\alpha\beta}F_{\mu\nu}(x) + \sigma^{rs}_a \varepsilon_{\alpha\beta}X^a(x)) + \cdots \,. \quad (321)$$

In (321) $\cdots$ stands for higher-order terms in θ's that can be expressed in terms of the lowest components. We hope the reader does not get confused by the notation. In this section, Φ denotes an $\mathcal{N}=2$ chiral superfield (previously denoted by A), ϕ is its lowest component and v denote the v.e.v. of ϕ, while a or more precisely a_μ are the non-dynamical moduli fields.

The contribution of the K-instanton sector to $S_{\text{eff}}[\Phi]$ is given by

$$S^{(K)}_{\text{eff}}[\Phi] = \int_{\mathcal{M}_K} \mathrm{d}\mu_K \, \mathrm{e}^{-\mathcal{S}_K(\Phi,\mu)} \,, \quad (322)$$

where μ collectively denotes the supermoduli parametrising $\mathcal{M}_K$. Separating the collective coordinates x^μ_0 and $\theta^{r\alpha}_0$ and, dropping the subscript 0 for simplicity, one gets

$$S^{(K)}_{\text{eff}}[\Phi] = \int \mathrm{d}^4x \, \mathrm{d}^4\theta \int_{\hat{\mathcal{M}}_K} \mathrm{d}\hat{\mu}_K \, \mathrm{e}^{-\mathcal{S}_K(\Phi,\hat{\mu})} \,, \quad (323)$$

so that comparison with the formula (323) yields

$$\mathcal{F}_K(\Phi) = \int_{\hat{\mathcal{M}}_K} \mathrm{d}\hat{\mu}_K \, \mathrm{e}^{-\mathcal{S}_K(\Phi,\hat{\mu})} \,, \quad (324)$$

where $\hat{\mathcal{M}}_K$ denotes the supermoduli space of "centred" instantons. $\hat{\mathcal{M}}_K$ describes configurations with fixed position of the centre of mass of the various instantons, which in turn are parameterised by $\hat{\mu}$'s, i.e. by the collective coordinates that do not move the position of the centre of mass. Since $\Phi(x,\theta)$ may be taken to be a constant (slowly varying) superfield $\Phi(x,\theta) = \phi$ independent of the $\hat{\mu}$'s, one can compute $\mathcal{F}_K(\phi)$ and then promote the argument ϕ to a chiral superfield by holomorphy in the low-energy approximation. Indeed higher (super)derivatives would contribute to the 1PI effective action. Resumming the infinite number of such contributions should reveal the spectrum of stable particles (BPS monopoles and dyons), expected on the basis of the SW analysis. The study of this feature is beyond the scope of the present analysis.

Following this strategy till the end, one finds

$$\mathcal{F}_K(\phi) = \mathcal{C}_K \phi^2 \frac{\Lambda^{4K}}{\phi^{4K}} \,, \quad (325)$$

where $\Lambda = v \exp(-8\pi^2/g^2(v)b_1)$ is the RG 1 scale, dynamically generated by dimensional transmutation, and v is an arbitrary scale that can be taken to coincide with the v.e.v. of ϕ. The coefficients of the ϕ expansion of $\mathcal{F}$ can be computed by setting ϕ to any convenient value, including $\phi = 0$, and are given by

$$\mathcal{C}_K = \int_{\hat{\mathcal{M}}_K} d\hat{\mu}_K \, e^{-\mathcal{S}_K(\phi=0,\hat{\mu})} \,. \tag{326}$$

The coefficients $\mathcal{C}_K$ are also known as Gromov–Witten invariants. They have been explicitly computed for $K = 1$ and $K = 2$ by performing the integral over $\hat{\mathcal{M}}_K$ and shown to match with the SW proposal and to reproduce Matone's relations, as previously reviewed.

In fact, as previously shown in Sect. 10.2, they can all be computed by exploiting powerful localisation properties of the integral over the (super) moduli space. Nekrasov and collaborators [79, 80, 81, 82] have been able to localise the integrals over instanton moduli spaces by turning on the so-called Ω-background, characterised by a constant self-dual anti-symmetric tensor $\Omega_{\mu\nu} = \varepsilon_a \eta^a_{\mu\nu}$. From the string vantage point, the Ω-background amounts to a constant R–R graviphoton field strength in the (Euclidean) spacetime directions $f_{\mu\nu} = f_a \eta^a_{\mu\nu}$. The precise numerical factor is $\frac{1}{2}$ so that $f_{\mu\nu} = \frac{1}{2}\Omega_{\mu\nu}$ or $f_a = \frac{1}{2}\varepsilon_a$. In the presence of such a background, the D3-brane action gets modified to[30]

$$S_{\rm D3} = S_{\rm SYM} - \int d^4x \, [2ig f_{\mu\nu} {\rm Tr}_{N_c}(\bar{\phi} F^{\mu\nu}) + g^2 f_{\mu\nu} f^{\mu\nu} \, {\rm Tr}_{N_c}(\bar{\phi}^2)] \,, \tag{327}$$

where ${\rm Tr}_{N_c}$ denotes the trace over the N_c-dimensional representation of the $U(N_c)$ Chan–Paton group associated with the D3-branes. The modification of the effective action of the D3-branes after switching on the Ω-background can be derived by the procedure of computing open string scattering amplitudes on the disk with an insertion of a closed string vertex operator for the R–R graviphoton. In the canonical $(-1/2, -1/2)$ superghost picture the vertex operator for the R–R graviphoton reads

$$V_f = f_{\mu\nu} S^\alpha \sigma^{\mu\nu}_{\alpha\beta} \tilde{S}^\beta \Sigma \tilde{\Sigma}^\dagger e^{-(\varphi+\tilde{\varphi})/2} \,. \tag{328}$$

We observe that the only relevant amplitude is

$$\langle\langle V_A^{(-1)} V_{\bar{\phi}}^{(0)} V_f^{(-1/2,-1/2)} \rangle\rangle \,. \tag{329}$$

In fact all other amplitudes, including the one with $V_{\bar{\phi}}^{(0)}$ replaced by $V_\phi^{(0)}$, either vanish or are irrelevant in the low energy limit, i.e. produce higher derivative terms. The combined effect of the Ω-background and the non-vanishing v.e.v. for ϕ is to replace the standard ADHM matrix $\Delta_{(N_c+2K)\times 2K}$ with

$$\Delta_{(N_c+2K)\times 2K} \to \Delta_{(N_c+2K)\times 2K} + i\mathcal{A}_{(N_c+2K)\times 2K}(v, \varepsilon) + \cdots \,. \tag{330}$$

[30] In order to expose the relative strength of the various terms in the action, we henceforth switch to the perturbative normalisation, whereby we drop the overall $1/g^2$.

It is important to note hat the upper block of $\mathcal{A}_{(N_c+2K)\times 2K}(v,\varepsilon)$ is given by

$$\mathcal{A}^{\rm up}_{N_c\times 2K}(\phi,\varepsilon) = \phi^u{}_v w^v_{i\dot\alpha} - w^u_{j\dot\alpha}\chi^j{}_i\,, \qquad (331)$$

where $u,v = 1,...N_c$, $i,j = 1,...,K$, and the lower block is

$$\mathcal{A}^{\rm low}_{2K\times 2K}(v,\varepsilon) = [\chi, a_\mu]\sigma^\mu_{\alpha\dot\alpha} + \varepsilon_a\sigma^{a\,\beta}_\alpha a_\mu\sigma^\mu_{\beta\dot\alpha}\,. \qquad (332)$$

As in the standard (commutative, in the absence of graviphoton background, i.e. for $\Omega = 0$) case, the gauge field can be written in the convenient form $gA_\mu = U^\dagger\partial_\mu U$, which, as before, is not a pure gauge because U is not an $N_c \times N_c$ matrix.

The fermionic zero modes can be parametrised as

$$g^{1/2}\lambda^{\alpha r} = U^\dagger(\mathcal{L}^r f(w,x)\bar b^\alpha + b^\alpha f(w,x)\bar{\mathcal{L}}^r)U\,, \qquad (333)$$

where $\mathcal{L}^r, \bar{\mathcal{L}}^r$ are $K\times K$ spinor matrices satisfying the super ADHM constraints and $b,\bar b$ are $(N_c+2K)\times 2K$ constant spinor matrices with vanishing upper block and diagonal lower block. Moreover, one has

$$f_{K\times K}(w,x) = (\bar w_{\dot\alpha}w^{\dot\alpha} + (a-x\ \ )^2)^{-1} \qquad (334)$$

as a consequence of the ADHM constraints. Finally, the scalar field profile in the presence of the non-commutative Ω background is given by

$$\phi = i\sqrt{2}\epsilon_{rs}U^\dagger\mathcal{L}^r f(w,x)\bar{\mathcal{L}}^s U + U^\dagger\mathcal{J}U \qquad (335)$$

and correctly satisfies

$$D^2\phi = -i\sqrt{2}\epsilon_{rs}\lambda^{\alpha r}\lambda^s_\alpha - ig\Omega_{\mu\nu}F^{\mu\nu} \qquad (336)$$

to lowest order in g with $F_{\mu\nu} = \tilde F_{\mu\nu}$. We note that $\mathcal{J}$ is an $(N_c+2K)\times(N_c+2K)$ block diagonal matrix with the upper block $\mathcal{J}^{\rm up}_{N_c\times N_c} = v_{N_c\times N_c}$, where $v_{N_c\times N_c}$ represents the v.e.v. of the dynamical scalar fields $\phi_{N_c\times N_c}$, and the lower block $\mathcal{J}^{\rm low}_{2K\times 2K} = \chi_{K\times K}\otimes \quad + \quad \otimes\,\varepsilon_a\sigma^a$, where $\chi_{K\times K}$ represents the non-dynamical scalar fields (moduli).

In principle one can analyse by similar means gauge theories with lower ($\mathcal{N}=1$) or no supersymmetry. This analysis is only in its infancy and goes beyond the scope of this review. It is the subject of very intense research activity at present, see e.g. [115]. We hope we have provided the interested and proficient reader with the necessary tools to enter the *arena* of this fascinating endeavour.

12 Instanton Effects in $\mathcal{N}=4$ SYM

In the following sections we shall review the calculation of instanton effects in $\mathcal{N}=4$ SYM [70]. This is the maximally extended (rigid) supersymmetric theory in four dimensions and possesses a number of remarkable properties. It is ultraviolet finite [116] and provides an example of four-dimensional quantum field theory with exact conformal invariance at the quantum level. The theory is also believed to be invariant under a strong↔weak coupling duality, known as S-duality, which generalises the Montonen–Olive electric-magnetic duality [73, 102]. Originally, the interest in the theory was driven by the discovery of its finiteness properties. In recent years it has been extensively studied in the context of the AdS/CFT duality [88, 89, 90], which relates it to type IIB superstring theory in an $AdS_5 \times S^5$ background.

As a conformal field theory $\mathcal{N} = 4$ SYM has rather different physical properties from those of the $\mathcal{N} = 1$ and $\mathcal{N} = 2$ theories previously discussed. However, instanton effects play a decisive role also here and the methods of supersymmetric instanton calculus described in the previous sections have recently been extensively applied to the study of non-perturbative aspects of the model. In particular, instantons are expected to be instrumental to the realisation of S-duality in $\mathcal{N} = 4$ SYM and the study of their contributions has led to some of the most striking tests of the validity of the AdS/CFT correspondence.

After a brief overview of the structure of $\mathcal{N} = 4$ SYM and its notable properties, in Sects. 14–16 we shall describe the calculation of instanton contributions to correlation functions, essentially following the method that in Sect. 2 was called the SCI method. In Sects. 17 and 18 we shall then discuss the rôle of instantons in the AdS/CFT duality.

13 $\mathcal{N} = 4$ Supersymmetric Yang–Mills Theory

The $\mathcal{N} = 4$ supersymmetric Yang–Mills theory was originally constructed in [70] as the dimensional reduction of the 10-dimensional $\mathcal{N} = 1$ supersymmetric Yang–Mills theory on a six torus. The field content of the theory consists of a gauge field, A_μ, four Weyl fermions, λ^A_α $(A = 1, \ldots, 4)$ and six real scalars, φ^i $(i = 1, \ldots, 6)$. In terms of $\mathcal{N} = 1$ multiplets these fields combine into one vector and three chiral multiplets. All the fields are in the adjoint representation of the gauge group, which in most of the following will be taken to be $SU(N_c)$. The global supergroup of symmetries of the theory is $PSU(2,2|4)$, whose maximal bosonic subgroup is $SO(2,4) \times SU(4)$. The $SO(2,4)$ factor is the four-dimensional conformal group and $SU(4)$ is the R-symmetry group under which the fermions transform in the $\mathbf{4}$ (and their conjugates in the $\bar{\mathbf{4}}$), the scalars in the $\mathbf{6}$ and the gauge field is a singlet. It is often convenient to label the scalars by an anti-symmetric pair of indices in the $\mathbf{4}$, as φ^{AB}, subject to the reality constraint

$$\bar{\varphi}_{AB} = (\varphi^{AB})^\dagger = \frac{1}{2}\varepsilon_{ABCD}\varphi^{CD} . \tag{337}$$

The two parametrisations are related by

$$\varphi^i = \frac{1}{\sqrt{2}}\, \Sigma^i_{AB}\varphi^{AB}\,, \qquad \varphi^{AB} = \frac{1}{\sqrt{8}}\, \bar{\Sigma}^{AB}_i \varphi^i\,, \tag{338}$$

where Σ^i_{AB} ($\bar{\Sigma}^{AB}_i$) are Clebsch–Gordan coefficients projecting the product of two **4**'s (two $\bar{\mathbf{4}}$'s) onto the **6**. They can be expressed in terms of the 't Hooft symbols [2] as

$$\begin{aligned} \Sigma^i_{AB} &= (\Sigma^a_{AB}, \Sigma^{a+3}_{AB}) = (\eta^a_{AB}, i\bar{\eta}^a_{AB})\,, \\ \bar{\Sigma}^{AB}_i &= (\bar{\Sigma}^a_{AB}, \bar{\Sigma}^{a+3}_{AB}) = (-\eta^{AB}_a, i\bar{\eta}^{AB}_a)\,, \qquad i = 1, \ldots, 6\,,\ a = 1, 2, 3\,. \end{aligned} \tag{339}$$

The elementary fields are conveniently represented as colour matrices and the classical action of the theory, which is uniquely determined (up to the choice of gauge group) by $\mathcal{N} = 4$ supersymmetry, can be written as

$$\begin{aligned} S = \int \mathrm{d}^4x\, \mathrm{Tr} \Big\{ &\frac{1}{2} F_{\mu\nu}F^{\mu\nu} + 2D_\mu\varphi^{AB} D^\mu \bar{\varphi}_{AB} - 2i\lambda^{\alpha A} \not{D}_{\alpha\dot{\alpha}} \bar{\lambda}^{\dot{\alpha}}_A \\ &- 2g\lambda^{\alpha A}[\lambda^B_\alpha, \bar{\varphi}_{AB}] - 2g\bar{\lambda}_{\dot{\alpha}A}[\bar{\lambda}^{\dot{\alpha}}_B, \varphi^{AB}] - 2g^2[\varphi^{AB}, \varphi^{CD}][\bar{\varphi}_{AB}, \bar{\varphi}_{CD}] \Big\}\,. \end{aligned} \tag{340}$$

The action (340) is invariant under the supersymmetry transformations

$$\begin{aligned} \delta_\epsilon \varphi^{AB} &= \frac{1}{2}\left(\lambda^{\alpha A}\epsilon^B_\alpha - \lambda^{\alpha B}\epsilon^A_\alpha\right) + \frac{1}{2}\varepsilon^{ABCD}\bar{\epsilon}_{\dot{\alpha}C}\bar{\lambda}^{\dot{\alpha}}_D \\ \delta_\epsilon \lambda^A_\alpha &= -\frac{1}{2}F_{\mu\nu}\sigma^{\mu\nu}_\alpha{}^\beta \epsilon^A_\beta + 4i\left(\not{D}_{\alpha\dot{\alpha}}\varphi^{AB}\right)\bar{\epsilon}^{\dot{\alpha}}_B - 8g[\bar{\varphi}_{BC}, \varphi^{CA}]\epsilon^B_\alpha \\ \delta_\epsilon A^\mu &= -i\lambda^{\alpha A}\sigma^\mu_{\alpha\dot{\alpha}}\bar{\epsilon}^{\dot{\alpha}}_A - i\epsilon^{\alpha A}\sigma^\mu_{\alpha\dot{\alpha}}\bar{\lambda}^{\dot{\alpha}}_A\,. \end{aligned} \tag{341}$$

Given the gauge group, the action (340) contains a single parameter, the coupling constant g.[31] The absence of divergences in the theory implies that the corresponding β-function vanishes. As discussed in Appendix C, it is possible to add to the action a ϑ-term.

The $\mathcal{N} = 4$ SYM theory has a vacuum manifold parametrised by the v.e.v.'s of the six scalars which make the potential vanish. The resulting moduli space turns out to be

$$\mathcal{M} = \quad {}^{6r}/\mathcal{S}_r\,, \tag{342}$$

where r is the rank of the gauge group and $\mathcal{S}_r$ is the group of permutations of r elements. At a generic point of the moduli space, the theory is in a Coulomb phase and the gauge group is broken down to $U(1)^r$. In this phase and in the presence of a ϑ-term the theory contains BPS-saturated monopole and dyon states characterised by integer quantum numbers, q_e and q_m, associated

[31] In the following we will maintain the notation used in the previous sections, denoting the Yang–Mills coupling by g. The string coupling constant will be denoted by g_s.

with their electric and magnetic charges [100, 101]. The conjectured S-duality of $\mathcal{N}=4$ SYM requires that the spectrum of such states be invariant under the action of $SL(2,\ )$ transformations acting projectively on the complexified coupling, τ (defined in (141)),

$$\tau \to \frac{a\tau + b}{c\tau + d}\,, \qquad a,b,c,d \in \ \,, \qquad ad - bc = 1\,, \tag{343}$$

while simultaneously rotating the electric and magnetic quantum numbers according to

$$\begin{pmatrix} q_e \\ q_m \end{pmatrix} \to \begin{pmatrix} -a & b \\ c & -d \end{pmatrix} \begin{pmatrix} q_e \\ q_m \end{pmatrix}. \tag{344}$$

Significant evidence in support of this conjecture has been obtained using semi-classical methods [117].

The conformal phase of the theory corresponds to the origin of the moduli space where all the scalar v.e.v.'s vanish. As already observed, at this point the classical (super)conformal symmetry is preserved at the quantum level, resulting in a non-trivial interacting conformal field theory. In this phase the fundamental observables are correlation functions of gauge-invariant composite operators constructed from the elementary fields in (340). Such operators are classified according to their transformation under the global symmetries and are organised in multiplets of the superconformal group, $PSU(2,2|4)$. Some properties of the $\mathcal{N}=4$ superconformal group and its multiplets are reviewed in Appendix C. Each operator is characterised by its quantum numbers with respect to the bosonic subgroup $SO(2,4)\times SU(4)$. These can be chosen to be two spins, (j_1, j_2), and the scaling dimension, Δ, identifying the transformation under the conformal group together with three Dynkin labels, $[k,l,m]$, identifying the $SU(4)$ representation under which the operator transforms. $\mathcal{N}=4$ composite operators can be broadly divided into two classes, protected operators belonging to short or semi-short "BPS" multiplets of the superconformal group and unprotected ones belonging to long multiplets [118, 92]. Correlation functions of protected operators satisfy special non-renormalisation properties. A notable example of BPS multiplet is the one comprising the $PSU(2,2|4)$ conserved currents, i.e. the energy–momentum tensor, $\mathcal{T}_{\mu\nu}$, and the supersymmetry and R-symmetry currents, $\Sigma^{\mu}_{\alpha A}$ and $\mathcal{J}^{\mu\,B}_{A}$, respectively. We give here the explicit form of the first few components of the $\mathcal{T}_{\mu\nu}$ multiplet

$$\begin{aligned}
\mathcal{Q}^{[A_1B_1][A_2B_2]} &= \mathrm{Tr}\left(2\varphi^{A_1B_1}\varphi^{A_2B_2} + \varphi^{A_1A_2}\varphi^{B_1B_2} + \varphi^{A_1B_2}\varphi^{A_2B_1}\right) \\
\mathcal{X}^{A_1[A_2B_2]}_{\alpha} &= \mathrm{Tr}\left(2\lambda^{A_1}_{\alpha}\varphi^{A_2B_2} + \lambda^{A_2}_{\alpha}\varphi^{A_1B_2} - \lambda^{B_2}_{\alpha}\varphi^{A_1A_2}\right) \\
\mathcal{E}^{(A_1A_2)} &= \mathrm{Tr}\left(-\lambda^{\alpha A_1}\lambda^{A_2}_{\alpha} + g\, t^{(A_1A_2)}_{CDEFGH}\,\varphi^{CD}\varphi^{EF}\varphi^{GH}\right) \\
\mathcal{B}^{[A_1A_2]}_{\mu\nu} &= \mathrm{Tr}\left(\lambda^{\alpha A_1}\sigma_{\mu\nu\,\alpha}{}^{\beta}\lambda^{A_2}_{\beta} + 2iF_{\mu\nu}\varphi^{A_1A_2}\right) \\
\Lambda^{A}_{\alpha} &= \mathrm{Tr}\left\{\sigma^{\mu\nu\,\beta}_{\alpha}F_{\mu\nu}\lambda^{A}_{\beta} + g[\bar{\varphi}_{BC},\varphi^{CA}]\lambda^{B}_{\alpha} + \left(\not{D}_{\alpha\dot{\alpha}}\bar{\lambda}^{\dot{\alpha}}_{B} + g[\lambda^{C}_{\alpha},\bar{\varphi}_{BC}]\right)\varphi^{AB}\right\}.
\end{aligned} \tag{345}$$

The operator $\mathcal{Q}$ is the lowest component of the multiplet and transforms in the $\mathbf{20}'$ of the $SU(4)$ R-symmetry, $\mathcal{E}$ and $\mathcal{B}_{\mu\nu}$ are respectively in the $\mathbf{10}$ and $\mathbf{6}$ and the fermionic operators $\mathcal{X}_\alpha$ and Λ_α transform in the $\mathbf{20}$ and $\mathbf{4}$, respectively. The tensor $t^{(A_1 A_2)}_{CDEFGH}$ in $\mathcal{E}$ projects the product of three $\mathbf{6}$'s onto the $\mathbf{10}$. In the next sections we shall study various examples of correlation functions involving the operators in (345). We shall also consider other BPS multiplets in the same class whose lowest component is a dimension ℓ scalar operator which, in terms of the φ^i scalars, takes the form[32]

$$\mathcal{Q}_\ell^{\{i_1 i_2 \cdots i_\ell\}} = \mathrm{Tr}\left[\varphi^{\{i_1}\varphi^{i_2}\cdots\varphi^{i_\ell\}}\right] . \tag{346}$$

The first example of long (non-BPS) multiplet is the $\mathcal{N}=4$ Konishi multiplet, whose lowest component is the dimension 2, $SU(4)$ singlet scalar

$$\mathcal{K}_1 = \varepsilon_{ABCD}\,\mathrm{Tr}\left(\varphi^{AB}\varphi^{CD}\right) . \tag{347}$$

Conformal invariance implies that, given a complete basis of operators in the theory, any correlation function is fully determined, via the operator product expansion (OPE), by two sets of numbers, the scaling dimensions of the operators and the Wilson coefficients that couple triplets of operators. Both scaling dimensions and Wilson coefficients receive perturbative and non-perturbative quantum corrections, so they are non-trivial functions of the coupling, g, and the ϑ-angle.

The spectrum of scaling dimensions in $\mathcal{N}=4$ SYM has been the subject of extensive study in the context of the AdS/CFT correspondence. The scaling dimensions of composite operators are determined by their two-point correlation functions. For a primary operator, $\mathcal{O}(x)$, conformal invariance fixes the form of the two-point function $\langle \mathcal{O}(x)\mathcal{O}^\dagger(y)\rangle$ to be

$$\langle \mathcal{O}(x)\mathcal{O}^\dagger(y)\rangle = \frac{c}{(x-y)^{2\Delta}} , \tag{348}$$

where Δ is the scaling dimension and c is a constant which may, in general, depend on g and the ϑ-angle.[33] In general Δ receives quantum corrections, $\Delta = \Delta_0 + \gamma$, where Δ_0 is the bare or engineering dimension and γ the anomalous part. The latter has an expansion of the form

$$\gamma(g,\vartheta) = \sum_{n=1}^{\infty} \gamma_n^{\mathrm{pert}} g^{2n} + \sum_{K>0}\sum_{m=0}^{\infty}\left[\gamma_m^{(K)} g^{2m}\,\mathrm{e}^{(-8\pi^2/g^2 + i\vartheta)K} + \mathrm{c.c.}\right] , \tag{349}$$

---

[32] We use curly brackets to denote symmetrisation with subtraction of traces. For simple symmetrisation and anti-symmetrisation we use parentheses and square brackets, respectively.

[33] This is actually an oversimplification. In general, in a given sector characterised by certain quantum numbers, one needs to consider a complete set of operators and resolve their mixing. The resolution of the operator mixing diagonalises the matrix of two-point functions. Only after this step equations of the form (348) determine the physical scaling dimensions.

where the first series contains the perturbative contributions and the second double series the instanton and anti-instanton contributions. The two-point functions of protected operators are not renormalised implying that their bare dimensions are not corrected ($\Delta = \Delta_0$).

The behaviour (349) of the anomalous dimensions illustrates an important feature of $\mathcal{N} = 4$ observables. In general, physical quantities receive contributions at all orders in perturbation theory and from all instanton sectors. Moreover, in each instanton sector there exists an infinite series of perturbative corrections arising from fluctuations around the leading instanton semi-classical contribution. This is the consequence of the absence of chiral selection rules and marks an important difference with respect to the cases of $\mathcal{N} = 1$ and $\mathcal{N} = 2$ theories considered in the previous sections. Indeed, in $\mathcal{N} = 4$ SYM there are no anomalous $U(1)$'s. As a consequence, as will be discussed in the next sections, there exist no correlation functions which are dominated by the contribution of specific instanton sectors.

In the conformal phase the field equations of $\mathcal{N} = 4$ SYM admit no (non-singular) monopole or dyon solutions and the conjectured S-duality has a different realisation. Specifically, it requires that the spectrum of scaling dimensions of gauge-invariant operators be invariant under the $SL(2, \ )$ transformations (343). This suggests that the scaling dimensions should naturally be written as functions of τ and $\bar{\tau}$ in the form

$$\Delta = \Delta(\tau, \bar{\tau}) = \Delta_0 + \gamma(\tau, \bar{\tau}) \,. \tag{350}$$

We then conclude that instanton effects, which are the source of the ϑ dependence in (349), must play a crucial role here. Similarly, it can be argued that instantons are important in determining the behaviour of correlation functions under S-duality. As will be discussed in Sects. 17 and 18, the arguments outlined here also resonate with what is understood about the role of D-instantons in the dual type IIB string theory compactified on $\mathrm{AdS}_5 \times S^5$. Unfortunately, little is known beyond these qualitative considerations and the details of how the S-duality of $\mathcal{N} = 4$ SYM is implemented in the superconformal phase remain largely elusive, see, however, [119] and [120] for recent progress.

14 Instanton Calculus in $\mathcal{N} = 4$ SYM

In this section we discuss some general features of instanton calculus in the $\mathcal{N} = 4$ SYM theory, highlighting the differences with respect to the $\mathcal{N} = 1$ and $\mathcal{N} = 2$ cases. In the next section we analyse in more detail one-instanton contributions to some specific correlation functions, focussing on the case of $SU(N_c)$ gauge group which is particularly important for the applications to the AdS/CFT correspondence. Multi-instanton configurations are described using the ADHM formalism [19] (see also Sect. 9.2). A full account of the technical aspects of the ADHM construction is beyond the scope of the present work. A comprehensive review of multi-instanton calculus in supersymmetric gauge theories can be found in [6]. The calculation of multi-instanton contributions is extremely involved and their direct evaluation is only possible for small instanton numbers (although remarkable progress was made in [81, 82]. See the discussion in Sect. 10). However, dramatic simplifications occur in the large N_c limit of relevance for the AdS/CFT duality. A brief review of the computation of multi-instanton corrections to $\mathcal{N}=4$ correlators in this limit will be given in Sect. 16 following the work of [121].

The calculation of correlation functions in $\mathcal{N}=4$ SYM in the semi-classical approximation proceeds as described in a general setting in Sect. 2. The path integral is evaluated using a saddle point approximation around the instanton configuration and thus reduced to a finite dimensional integral over the collective coordinate manifold associated with bosonic and fermionic zero modes. However, the form of the interaction terms in the $\mathcal{N}=4$ action (340), and in particular the coupling to the scalar fields, requires a modification of the previous analysis.

In principle, as done in Sect. 2, it is possible to use as saddle point configuration the solution in which the gauge field is given by the standard bosonic instanton with all the other fields vanishing, i.e.

$$A_\mu = A^I_\mu\,, \qquad \lambda^A_\alpha = \bar\lambda^{\dot\alpha}_A = \varphi^{AB} = 0\,. \tag{351}$$

The fluctuations of the various fields in the background of this configuration, including fermions, should then be treated perturbatively. This results in an expansion which, in the case under consideration of $\mathcal{N}=4$ SYM, is somewhat hard to handle beyond leading order. A more efficient approach consists in utilising as saddle point configuration a finite action solution of the complete set of coupled field equations of the theory including higher order corrections in g. This also provides a natural framework for implementing the supersymmetric generalisation of the ADHM construction.

A generic n-point correlation function computed in the semi-classical approximation around such a saddle point has an expression analogous to (23)–(25) which we schematically rewrite in the form

$$\langle \mathcal{O}_1(x_1)\cdots\mathcal{O}_n(x_n)\rangle = \int \mathrm{d}\mu(\beta,c)\,\mathrm{e}^{-S_{\rm inst}}\,\hat{\mathcal{O}}_1(x_1;\beta,c)\cdots\hat{\mathcal{O}}_n(x_n;\beta,c)\,, \tag{352}$$

where $\mathrm{d}\mu(\beta,c)$ is the integration measure over the bosonic (β) and fermionic (c) collective coordinates arising from the zero-mode fluctuations around the classical solution and $S_{\rm inst}$ is the action evaluated on the solution. With $\hat{\mathcal{O}}_i$ we denote the classical expressions of the operators at the saddle point. The latter depend on the insertion points of the operators and the collective coordinates.

For pure $\mathcal{N}=1$ SYM there exists a whole manifold of saddle points (including the one in (351)) which correspond to field configurations with $A_\mu = A^I_\mu$, a gaugino solution of the Weyl–Dirac equation

$$\bar{\not{D}}^{\dot\alpha\alpha}\lambda_\alpha = 0\,, \tag{353}$$

and the anti-chiral fermion, $\bar{\lambda}_{\dot{\alpha}}$, identically zero. The resulting semi-classical expectation values involve integrals over the n_B bosonic collective coordinates as well as the n_F fermion zero modes resulting from the index theorem and discussed in previous sections.

The generalisation of this analysis to the $\mathcal{N} = 4$ case incurs in a serious obstacle: no exact topologically non-trivial solution of the coupled field equations is known except (351). The $\mathcal{N} = 4$ SYM field equations read

$$\begin{aligned} & D_\mu F^{\mu\nu} + i\,g\{\lambda^{\alpha A}\sigma^{\nu}_{\alpha\dot{\alpha}}, \bar{\lambda}^{\dot{\alpha}}_{A}\} + \frac{1}{2}g[\bar{\varphi}_{AB}, D^\nu\varphi^{AB}] = 0 \\ & D^2\varphi^{AB} + \sqrt{2}\,g[\{\lambda^{\alpha A}, \lambda^{B}_{\alpha}\} + \frac{1}{2}\,\varepsilon^{ABCD}\{\bar{\lambda}_{\dot{\alpha}C}, \bar{\lambda}^{\dot{\alpha}}_{D}\}] - \frac{g^2}{2}[\bar{\varphi}_{CD}, [\varphi^{AB}, \varphi^{CD}]] = 0 \\ & \bar{\not{D}}^{\dot{\alpha}\alpha}\lambda^{A}_{\alpha} + i\sqrt{2}\,g[\varphi^{AB}, \bar{\lambda}^{\dot{\alpha}}_{B}] = 0 \\ & \not{D}_{\alpha\dot{\alpha}}\bar{\lambda}^{\dot{\alpha}}_{A} - i\sqrt{2}\,g[\bar{\varphi}_{AB}, \lambda^{B}_{\alpha}] = 0\,. \end{aligned} \tag{354}$$

In the following discussion we denote by $\Phi^{(n)}$ a solution of the classical equations of motion for the generic field Φ, which depends on n zero modes of the Dirac operator in the instanton background. As already observed, the equations (354) are solved by the purely bosonic configuration (351), where $A^{I}_{\mu} = A^{(0)}_{\mu}$ is a charge-K instanton, with n_B associated collective coordinates. One can try to do better and solve iteratively the full set of coupled equations (354). Upon substituting the instanton solution (351), the equation for λ^A is of the form (353) and admits n_F independent solutions for each "flavour" $A = 1, \ldots, 4$. After this first step of the iteration, which generates a non-trivial solution, $\lambda^{A(1)}_{\alpha}$, for the gauginos, one notices that the configuration

$$A_\mu = A^{(0)}_{\mu}\,, \quad \lambda^{A}_{\alpha} = \lambda^{A(1)}_{\alpha}\,, \quad \varphi^{AB} = \bar{\lambda}^{\dot{\alpha}}_{A} = 0\,, \tag{355}$$

unlike what happens in the cases of $\mathcal{N} = 1$ and $\mathcal{N} = 2$ SYM, is not a solution of (354) and the process has to continue. The equation for the scalar fields, obtained inserting back $\lambda^{A(1)}_{\alpha}$, admits a solution which is bilinear in the fermion modes, $\varphi^{AB(2)}$. Again the resulting configuration

$$A_\mu = A^{(0)}_{\mu}\,, \quad \lambda^{A}_{\alpha} = \lambda^{A(1)}_{\alpha}\,, \quad \varphi^{AB} = \varphi^{AB(2)}\,, \quad \bar{\lambda}^{\dot{\alpha}}_{A} = 0\,, \tag{356}$$

is not an exact solution of (354). A further iteration generates a non-trivial field configuration, $\bar{\lambda}^{\dot{\alpha}(3)}_{A}$, for the anti-chiral fermions involving three zero-modes. At this point also the first equation in (354) gets an extra term, so that at the next step a modification, $A^{(4)}_{\mu}$, of the standard bosonic instanton is necessary. One may notice that the field strength associated with $A^{(4)}_{\mu}$ is no longer (anti-)self-dual.

In principle this recursive procedure can only stop when, after a number of successive iterations, a field configuration involving a number of fermion modes exceeding n_F is generated. The first few steps of the construction described above were explicitly carried out in [122]. However, a complete super-instanton multiplet, which would exactly solve (354) in closed form is not known. Indeed it has been argued in [6] that for generic gauge group such an exact solution may not exist. In spite of this obstacle, it is possible to consistently compute the semi-classical contribution to correlation functions expanding the path integral around an appropriately approximate solution. The crucial observation is that successive steps in the iterative procedure outlined above produce corrections to the solution which are suppressed by increasing powers of g. Therefore, in the weak coupling limit, it is consistent to employ as saddle point configuration an approximate truncated solution of the equations of motion in which only terms up to a certain power of g are retained. Thus the idea is to solve the field equations to leading order and include in the integration in (352) all the zero modes of the truncated equations. For the purpose of computing correlators in the semi-classical approximation, the relevant saddle point is determined by solving the system

$$\begin{aligned} &F_{\mu\nu} = \widetilde{F}_{\mu\nu}, \\ &\bar{\not{D}}^{\dot{\alpha}\alpha} \lambda^A_\alpha = 0, \\ &D^2 \varphi^{AB} = g\sqrt{2}\left(\lambda^{\alpha A}\lambda^B_\alpha - \lambda^{\alpha B}\lambda^A_\alpha\right), \end{aligned} \tag{357}$$

with the integration measure in (352) including all the associated fermion zero modes. The action evaluated on the solution of (357) is not simply given by $8\pi^2/g^2$, but manifestly depends on a subset of the collective coordinates

$$S_{\text{inst}} = \frac{8\pi^2}{g^2} - i\vartheta + \tilde{S}_{\text{inst}}(\tilde{\beta}, \tilde{c}), \tag{358}$$

where we have denoted with $\tilde{\beta}$ and $\tilde{c}$ the collective coordinates associated with the "non-exact" zero modes, i.e. those which are zero modes of the truncated equations (357), but not of the full coupled equations (354). These non-exact zero modes are said to be "lifted" by the interactions. As will be discussed more explicitly in the next section, in the case of gauge group $SU(N_c)$, the only fermion modes which remain unlifted are those associated with the Poincaré and special supersymmetries which are broken in the instanton background. All the remaining modes are lifted by the coupling to the scalars.

The lifting of fermion zero modes has important consequences for the properties of correlation functions which receive instanton contributions. Since some of the fermionic collective coordinates ($\tilde{c}$ in the formula above) appear explicitly in the action, it is not necessary to saturate the corresponding fermionic integrations with the operator insertions in order to obtain a non-vanishing result. This implies that in the $\mathcal{N} = 4$ theory there are no (strict) selection rules determining which correlation functions receive contributions from which winding number sector, unlike what happens in the $\mathcal{N} = 1$ and $\mathcal{N} = 2$ cases. In particular, non-vanishing correlators receive contributions from configurations with arbitrary instanton number. This result emerging from explicit calculations has its origin in the absence of an anomalous $U(1)$ R-symmetry in $\mathcal{N}=4$ SYM.

It must be stressed that the approach described here is consistent only if restricted to the calculation of the leading order contributions in g (see (357)). The reason is that in order to go to higher orders in a consistent way one should also take systematically into account all the quantum fluctuations that beyond the semi-classical approximation have been neglected.

15 One-instanton in $\mathcal{N}=4$ SYM with $SU(N_c)$ Gauge Group

In the one-instanton sector, the approach described in the previous section can be implemented in a straightforward way. We focus here on the case of $SU(N_c)$ gauge group, but the generalisation to orthogonal and symplectic groups is not too difficult. As explained above, we shall use as saddle point for the calculation of correlation functions in the semi-classical approximation the solution to the truncated equations (357).[34] The resulting saddle point field configuration is characterised by $4N_c$ bosonic collective coordinates. As we know they are the position, x_0, and size, ρ, of the instanton and its global gauge orientation parameters. The latter can be conveniently identified with the set of variables, $w_{u\dot{\alpha}}$ and $\bar{w}^{\dot{\alpha}u}$ (where $u=1,\ldots,N_c$ is a colour index and $\dot{\alpha}=1,2$ is a spinor index), parametrising the coset $SU(N_c)/SU(N_c-2)\times U(1)$ describing the $SU(2)$ colour orientation of the instanton and its embedding into $SU(N_c)$ [9]. Moreover, as we will be working with the approximate solution of (357), all the $8N_c$ fermionic collective coordinates associated with the zero modes of the Dirac operator will be included in the integration measure. These comprise the 16 moduli associated with the Poincaré and special supersymmetries broken by the bosonic instanton and denoted, respectively, by η^A_α and $\bar{\xi}^A_{\dot{\alpha}}$ $(A=1,2,3,4)$. For brevity we shall refer collectively to these modes as superconformal modes. The remaining fermion moduli, which can be thought of superpartners of the gauge orientation parameters, are described by $8N_c$ parameters, ν^A_u and $\bar{\nu}^{Au}$, subject to the 2×8 constraints

$$\bar{w}^{\dot{\alpha}u}\nu^A_u=0\,,\quad \bar{\nu}^{Au}w_{u\dot{\alpha}}=0\,, \tag{359}$$

which effectively reduce their number to $8(N_c-2)$.

As discussed in the previous section, only the 16 superconformal modes remain exact zero modes to all orders in g. The ν^A_u and $\bar{\nu}^{Au}$ modes are lifted

[34] In preparation for the forthcoming discussion on the AdS/CFT duality we shall from now on work with fields rescaled by a factor of g. Consequently, the action of the $\mathcal{N}=4$ theory will have an overall factor $1/g^2$ in front. This is the normalisation which arises naturally in the dual string theory and therefore this rescaling will simplify the comparison of string and gauge theory results.

and appear explicitly in the instanton action. The solution of the coupled equations (357) generates a non-trivial configuration for the scalars, φ^{AB}, which is bilinear in the $8N_c$ fermion zero modes. Substituting this solution, together with $\bar\lambda^{\dot\alpha}_A = 0$, in the action (truncated to the cubic couplings for consistency with our iterative procedure) gives

$$S_{\rm inst} = -2\pi i\tau + S_{4F} = -2\pi i\tau + \frac{\pi^2}{2g^2\rho^2}\,\varepsilon_{ABCD}\,\mathcal{F}^{AB}\mathcal{F}^{CD}\,, \qquad (360)$$

where τ was defined in (141) and

$$\mathcal{F}^{AB} = \frac{1}{2\sqrt{2}}\left(\bar\nu^{Au}\nu^B_u - \bar\nu^{Bu}\nu^A_u\right)\,. \qquad (361)$$

It is the four-fermion term, S_{4F}, arising from the Yukawa couplings $\lambda^A[\lambda^B, \bar\varphi_{AB}]$ which is responsible for the lifting the of the $8(N_c-2)$ ν^A_u and $\bar\nu^{Au}$ modes.

15.1 A Generating Function

In the following we only consider correlators of gauge-invariant operators for which the integration over the moduli describing the global orientation of the instanton simply produces a volume factor which can be absorbed in the measure [9]. We denote by $\mathrm{d}\mu_{\rm phys}$ the gauge-invariant, or physical, integration measure on the instanton moduli space obtained after integration over the gauge orientation parameters. The physical measure to be used in the calculation of expectation values in semi-classical approximation in the one-instanton sector is[35]

$$\int \mathrm{d}\mu_{\rm phys}\,\mathrm{e}^{-S_{\rm inst}} = \frac{\pi^{-4N_c}g^{4N_c}\mathrm{e}^{2\pi i\tau}}{(N_c-1)!(N_c-2)!} \qquad (362)$$
$$\times \int \mathrm{d}\rho\,\mathrm{d}^4x_0 \prod_{A=1}^{4} \mathrm{d}^2\eta^A \mathrm{d}^2\bar\xi^A\,\mathrm{d}^{N_c-2}\nu^A \mathrm{d}^{N_c-2}\bar\nu^A\,\rho^{4N_c-13}\mathrm{e}^{-S_{4F}}\,,$$

where the instanton action is given in (360) and (361) and the ρ, g and N_c dependence is the result of the normalisation of the collective coordinates as explained in Sect. 2.

Following [121] the integration over the $8(N_c-2)$ non-exact modes, ν^A_u and $\bar\nu^{Au}$, can be reduced to a Gaussian form introducing auxiliary bosonic coordinates, χ^i, $i=1,\ldots,6$, and rewriting the r.h.s. of (362) in the form

$$\frac{\pi^{-4N_c}g^{4N_c}\mathrm{e}^{2\pi i\tau}}{(N_c-1)!(N_c-2)!}\int \mathrm{d}\rho\,\mathrm{d}^4x_0\,\mathrm{d}^6\chi \prod_{A=1}^{4} \mathrm{d}^2\eta^A \mathrm{d}^2\bar\xi^A\,\mathrm{d}^{N_c-2}\nu^A \mathrm{d}^{N_c-2}\bar\nu^A$$
$$\times\rho^{4N_c-7}\exp\left[-2\rho^2\chi^i\chi^i + \frac{4\pi i}{g}\chi_{AB}\mathcal{F}^{AB}\right]\,, \qquad (363)$$

[35] Here and in the following formulae we omit a (N_c-independent) numerical constant that will be reinstated in the final expressions.

where $\chi_{AB} = \frac{1}{\sqrt{8}}\Sigma^i_{AB}\chi^i$ and $\mathcal{F}^{AB}$ is defined in (361).

The semi-classical contribution to a correlation function of local gauge-invariant operators is obtained by integrating the product of their profiles in the instanton background with the above measure. The integration over the Grassmann variables in (363) requires that the classical expressions of the operators soak up the 16 fermion modes η^A_α and $\bar{\xi}^A_{\dot{\alpha}}$ for the result to be non-zero. The Grassmann integrals over the ν^A and $\bar{\nu}^A$ modes are non-vanishing even if the operators do not contain any dependence on these variables. In the following we shall use the terminology introduced in [123] and refer to correlation functions in which the operator insertions soak up only the 16 exact modes as "minimal". Correlators in which the operators contain a dependence on more than sixteen modes will be referred to as "non-minimal". In order to systematically study the instanton contributions to generic correlation functions, it is convenient to construct a generating function. This allows to drastically simplify the combinatorics associated with the ν^A and $\bar{\nu}^A$ integrations in the non-minimal cases [123]. For this purpose, we introduce sources, $\bar{\vartheta}^u_A$ and ϑ_{Au}, in (363) coupled to those fermionic variables and define

$$Z[\vartheta,\bar{\vartheta}] = \frac{\pi^{-4N_c} g^{4N_c} \mathrm{e}^{2\pi i\tau}}{(N_c-1)!(N_c-2)!}\int \mathrm{d}\rho\, \mathrm{d}^4x_0\, \mathrm{d}^6\chi \prod_{A=1}^{4} \mathrm{d}^2\eta^A\, \mathrm{d}^2\bar{\xi}^A\, \mathrm{d}^{N_c-2}\bar{\nu}^A\, \mathrm{d}^{N_c-2}\nu^A$$
$$\times \rho^{4N_c-7} \exp\left[-2\rho^2\chi^i\chi^i + \frac{\sqrt{8}\pi i}{g}\bar{\nu}^{Au}\chi_{AB}\nu^B_u + \bar{\vartheta}^u_A\nu^A_u + \vartheta_{Au}\bar{\nu}^{Au}\right]. \tag{364}$$

Since the integrals over $\bar{\nu}$ and ν are Gaussian, they can be immediately computed. Introducing polar coordinates

$$\chi^i \to (r,\Omega)\,, \quad \sum_{i=1}^{6}(\chi^i)^2 = r^2\,, \tag{365}$$

we find

$$Z[\vartheta,\bar{\vartheta}] = \frac{2^{-29}\pi^{-13} g^8 \mathrm{e}^{2\pi i\tau}}{(N_c-1)!(N_c-2)!}\int \mathrm{d}\rho\, \mathrm{d}^4x_0\, \mathrm{d}^5\Omega \prod_{A=1}^{4}\mathrm{d}^2\eta^A\, \mathrm{d}^2\bar{\xi}^A\, \rho^{4N_c-7}$$
$$\times \int_0^\infty \mathrm{d}r\, r^{4N_c-3}\mathrm{e}^{-2\rho^2 r^2}\, \mathcal{Z}(\vartheta,\bar{\vartheta};\Omega,r)\,, \tag{366}$$

where all the numerical coefficients omitted in previous expressions have been reinstated and we have introduced the density

$$\mathcal{Z}(\vartheta,\bar{\vartheta};\Omega,r) = \exp\left[-\frac{ig}{\pi r}\,\bar{\vartheta}^u_A\Omega^{AB}\vartheta_{Bu}\right], \tag{367}$$

where the symplectic form Ω^{AB} is given by (see (365))

$$\Omega^{AB} = \bar{\Sigma}^{AB}_i\Omega^i\,, \qquad \sum_{i=1}^{6}\left(\Omega^i\right)^2 = 1\,. \tag{368}$$

Notice that the angular variables, Ω^i, introduced in the polar representation of the auxiliary coordinates, χ^i, parametrise a five-sphere. This will play a very important role in comparing with string theory results in the context of the AdS/CFT correspondence.

A n-point correlation function in the semi-classical approximation takes now the form (see (352))

$$\langle \mathcal{O}_1(x_1) \cdots \mathcal{O}_n(x_n) \rangle = \int \mathrm{d}\mu_{\mathrm{phys}}\, \mathrm{e}^{-S_{\mathrm{inst}}}\, \hat{\mathcal{O}}_1 \cdots \hat{\mathcal{O}}_n \,, \tag{369}$$

where

$$\hat{\mathcal{O}}_i = \hat{\mathcal{O}}_i(x_i; x_0, \rho, \eta^A, \bar{\xi}^A, \nu^A, \bar{\nu}^A) \tag{370}$$

denotes the classical instantonic profile of the operator $\mathcal{O}_i$ and generically depends on all the bosonic and fermionic moduli. In particular, the ν^A and $\bar{\nu}^A$ modes appear in gauge-invariant operators only in colour singlet bilinears, in either symmetric or anti-symmetric combinations belonging to the representation **10** or **6** of $SU(4)$, respectively, i.e. in the combinations

$$(\bar{\nu}^A \nu^B)_{\mathbf{10}} \equiv \bar{\nu}^{u(A} \nu_u^{B)} = (\bar{\nu}^{Au} \nu_u^B + \bar{\nu}^{Bu} \nu_u^A) \,, \tag{371}$$

$$(\bar{\nu}^A \nu^B)_{\mathbf{6}} \equiv \bar{\nu}^{u[A} \nu_u^{B]} = (\bar{\nu}^{Au} \nu_u^B - \bar{\nu}^{Bu} \nu_u^A) \,. \tag{372}$$

The strategy is then to rewrite the dependence on these collective coordinates in each insertion in (369) in terms of derivatives with respect to the sources $(\vartheta_A, \bar{\vartheta}_A)$. In this way the dependence on the ν^A's and $\bar{\nu}^A$'s is traded for a dependence on the angular variables Ω^{AB}. After this step the integration over the radial parameter, r, can be computed and one is left with an integration over the bosonic coordinates, x_0, ρ, Ω^{AB}, and the 16 coordinates associated with the exact modes, η^A_α and $\bar{\xi}^A_{\dot{\alpha}}$. In the next subsections we present some examples of such calculations.

15.2 Minimal Correlation Functions

A class of correlation functions which have been extensively studied in the context of the AdS/CFT correspondence are those involving the operators in the $\mathcal{N} = 4$ supercurrent multiplet (345). This is a 1/2-BPS supermultiplet and thus all its components are protected operators. Their two- and three-point functions are not renormalised and in particular do not receive instanton contributions. However, their four- and higher-point functions can be non-zero in an instanton background and turn out to contain interesting dynamical information. The first examples of correlators in this class were considered in [124] in the case of $SU(2)$ gauge group. The calculations have then been generalised to $SU(N_c)$ in [125] and to multi-instantons in the large N_c limit in [121].

• The simplest minimal correlation function involves 16 insertions of the fermionic operator[36]

$$\Lambda_\alpha^A = \frac{1}{g^2}\,\mathrm{Tr}\,\{\sigma^{\mu\nu\,\beta}_{\alpha}F_{\mu\nu}\lambda^A_\beta + [\bar\varphi_{BC},\varphi^{CA}]\lambda^B_\alpha + (\not{D}_{\alpha\dot\alpha}\bar\lambda^{\dot\alpha}_B + [\lambda^C_\alpha,\bar\varphi_{BC}])\,\varphi^{AB}\}\,, \qquad (373)$$

transforming in the **4** of the $SU(4)$ R-symmetry group

$$G_{16}(x_1,x_2,\ldots,x_{16}) = \langle \Lambda^{A_1}_{\alpha_1}(x_1)\,\Lambda^{A_2}_{\alpha_2}(x_2)\cdots\Lambda^{A_{16}}_{\alpha_{16}}(x_{16})\rangle\,. \qquad (374)$$

For the calculation of (374) in the semi-classical approximation only the contribution of the first term in (373) to the classical profile of Λ^A_α is relevant. In fact by reinserting for a moment the powers of g that were absorbed in the redefinition of the fields, it is immediately seen that all the other terms are of higher order in g.

Substituting the solution for $A^{(0)}_\mu$ and $\lambda^{A(1)}_\alpha$ one obtains for the classical profile of Λ^A_α

$$\hat\Lambda^A_\alpha(x) = \frac{96}{g^2}\,\rho^4\,[f(x)]^4\,\zeta^A_\alpha(x)\,, \qquad (375)$$

where the function $f(x)$ is defined in (A.40) and $\zeta^A_\alpha(x)$ is the combination

$$\zeta^A_\alpha(x) = \frac{1}{\sqrt\rho}\left[\rho\,\eta^A_\alpha - (x-x_0)_\mu\sigma^\mu_{\alpha\dot\alpha}\bar\xi^{\dot\alpha A}\right]. \qquad (376)$$

We explicitly note that $\hat\Lambda^A_\alpha$ is linear in the superconformal collective coordinates and does not depend on the ν^A and $\bar\nu^A$ modes. This means that in evaluating the correlator (374) the sources $\bar\vartheta^u_A$ and ϑ_{Au} can be set to zero. We thus get

$$G_{16}(x_1,\ldots,x_{16}) = \frac{2^{51}\,3^{16}\,\pi^{-13}\,\mathrm{e}^{2\pi i\tau}}{g^{24}\,(N_c-1)!(N_c-2)!}\int \mathrm{d}\rho\,\mathrm{d}^4x_0\,\mathrm{d}^5\Omega\,\prod_{A=1}^4 \mathrm{d}^2\eta^A\,\mathrm{d}^2\bar\xi^A$$
$$\times \int_0^\infty \mathrm{d}r\,r^{4N_c-3}\mathrm{e}^{-2\rho^2r^2}\,\rho^{4N_c-7}\prod_{i=1}^{16}\rho^4\,[f(x_i)]^4\,\zeta^{A_i}_{\alpha_i}(x_i)\,. \qquad (377)$$

The r integral is elementary and yields

$$\int_0^\infty \mathrm{d}r\,r^{4N_c-3}\mathrm{e}^{-2\rho^2r^2} = 2^{-2N_c}\,\rho^{2-4N_c}\,\Gamma(2N_c-1)\,, \qquad (378)$$

so that

[36] Here and in the following we use for the composite operators the normalisation appropriate in the context of the AdS/CFT correspondence, which requires that their tree-level correlation functions be proportional to N_c^2 (see (422) and (423)).

$$G_{16}(x_1,\dots,x_{16}) = \frac{c_1(N_c)\,2^{51}\,3^{16}\,\pi^{-13}\,\mathrm{e}^{2\pi i\tau}}{g^{24}} \int \frac{\mathrm{d}\rho\,\mathrm{d}^4x_0}{\rho^5}\,\mathrm{d}^5\Omega \prod_{A=1}^{4} \mathrm{d}^2\eta^A\,\mathrm{d}^2\bar{\xi}^A$$
$$\times \prod_{i=1}^{16} \frac{\rho^4}{[(x_i-x_0)^2+\rho^2]^4}\,\zeta_{\alpha_i}^{A_i}(x_i)\,, \tag{379}$$

where

$$c_1(N_c) = \frac{2^{-2N_c}\Gamma(2N_c-1)}{(N_c-1)!(N_c-2)!} \underset{N_c\to\infty}{\sim} N_c^{1/2}\,. \tag{380}$$

The integration over the fermion modes selects terms with eight η^A's and eight $\bar{\xi}^A$'s in (379) and results in a fully anti-symmetric tensor in the $SU(4)$ and spinor indices. As will be discussed in Sect. 18, the unintegrated expression (379) suffices for the comparison with the associated dual process in string theory.

• As a second example of minimal correlation function, we consider the four-point function

$$G_4(x_1,\dots,x_4) = \langle \mathcal{Q}^{A_1B_1C_1D_1}(x_1)\cdots\mathcal{Q}^{A_4B_4C_4D_4}(x_4)\rangle\,, \tag{381}$$

where the scalar operators $\mathcal{Q}^{ABCD}$ belong to the $\mathbf{20'}$ of $SU(4)$ and are given by

$$\mathcal{Q}^{ABCD} = \frac{1}{g^2}\,\mathrm{Tr}\left(2\varphi^{AB}\varphi^{CD} + \varphi^{AC}\varphi^{BD} - \varphi^{AD}\varphi^{BC}\right)\,. \tag{382}$$

When evaluated on the solution of the saddle point equations (357), the $\mathcal{Q}^{ABCD}$'s contain four fermion modes. Unlike the fermions Λ_α^A they also involve the ν^A and $\bar{\nu}^A$ modes. However, in the minimal correlator (381) this dependence can be neglected as all the fermion modes need to be of type η^A and $\bar{\xi}^A$ to saturate the corresponding Grassmann integrals. The relevant terms giving the profile of $\mathcal{Q}^{ABCD}$ are

$$\hat{\mathcal{Q}}^{ABCD} = \frac{96}{g^2}\,\rho^4\,[f(x)]^4\,\left[(\zeta^{\alpha A}\zeta_\alpha^C)(\zeta^{\beta B}\zeta_\beta^D) - (\zeta^{\alpha A}\zeta_\alpha^D)(\zeta^{\beta B}\zeta_\beta^C)\right]\,, \tag{383}$$

with ζ_α^A defined in (376).

Proceeding as for the 16-point function (374), one finds

$$G_4(x_1,\dots,x_4) = c_1(N)\,2^{-9}\,3^4\,\pi^{-13}\,\mathrm{e}^{2\pi i\tau} \int \frac{\mathrm{d}\rho\,\mathrm{d}^4x_0}{\rho^5}\,\mathrm{d}^5\Omega \prod_{A=1}^{4} \mathrm{d}^2\eta^A\,\mathrm{d}^2\bar{\xi}^A$$
$$\times \prod_{i=1}^{4} \frac{\rho^4}{[(x-x_0)^2+\rho^2]^4}\,\left[(\zeta^{A_i}\zeta^{C_i})(\zeta^{B_i}\zeta^{D_i}) - (\zeta^{A_i}\zeta^{D_i})(\zeta^{B_i}\zeta^{C_i})\right](x_i)\,, \tag{384}$$

with $c_1(N)$ given in (380).

In the case of the four-point function (381) the final integrals in (384) have been explicitly computed in [126]. The result is a very complicated function of the distances $x_{ij}^2 = (x_i - x_j)^2$, which, however, can be used to extract information about instanton contributions to the anomalous dimensions of certain operators via the OPE analysis. As discussed in [126], the result shows, in particular, that the Konishi operator (347) does not acquire an instanton induced anomalous dimension. The study of the OPE also shows that there are $SU(4)$ singlet operators with $\Delta_0 = 4$, which do receive an instanton contribution (as well as possibly a perturbative one) to their anomalous dimension. We shall briefly return to the calculation of instanton corrections to the scaling dimensions of composite operators at the end of the next subsection and to the interpretation of (379) and (384) in Sect. 18.2.

15.3 Non-minimal Correlation Functions

The minimal correlators considered above are not dominated by the contribution of the one-instanton sector. Apart from ordinary perturbative corrections, they receive contributions from $K > 1$ instanton configurations as well as from perturbative fluctuations in each instanton sector. The non-minimal correlation functions, in which the operator insertions soak up more than the minimal number of fermion zero modes, have similar properties in this respect. However, their calculation presents new complications that will now be illustrated with explicit examples. In general, one can distinguish two classes of such non-minimal correlation functions, based on the features that differentiate them from related minimal cases, i.e. those involving additional insertions and those involving higher-dimensional operators.

- An example of the first type is the 20-point function

$$G_{20}(x_1, \dots, x_{20}) = \langle \Lambda_{\alpha_1}^{A_1}(x_1) \cdots \Lambda_{\alpha_{16}}^{A_{16}}(x_{16})\, \mathcal{E}^{B_1C_1}(y_1) \cdots \mathcal{E}^{B_4C_4}(y_4) \rangle\,, \quad (385)$$

where Λ_α^A is defined in (373) and

$$\mathcal{E}^{BC} = \frac{1}{g^2}\,\mathrm{Tr}\left(-\lambda^{\alpha B}\lambda_\alpha^C + t_{DEFGHL}^{(BC)}\,\varphi^{DE}\varphi^{FG}\varphi^{HL}\right). \quad (386)$$

For the calculation of (385) in the semi-classical approximation only the first term in (386) is relevant. Its contribution to the classical profile of $\mathcal{E}^{BC}$ is

$$\hat{\mathcal{E}}^{BC} = -\frac{96}{g^2}\,\rho^4\,[f(x)]^4\,\zeta^{\alpha B}\zeta_\alpha^C - \frac{2}{g^2}\,\rho^2\,[f(x)]^3\,(\bar{\nu}^{Bu}\nu_u^C + \bar{\nu}^{Cu}\nu_u^B)\,. \quad (387)$$

As explained in the previous subsection, the operator Λ_α^A does not depend on the fermion modes of type ν^A and $\bar{\nu}^A$ and thus in evaluating (385) we need to use for each $\mathcal{E}^{BC}$ insertion the second term in (387), as the Λ_α^A insertions already soak up all the superconformal modes. In this way we get

$$G_{20}(x_1, \dots, x_{20}) = \frac{1}{g^{40}} \int \mathrm{d}\mu_{\mathrm{phys}}\, \mathrm{e}^{-S_{\mathrm{inst}}} \prod_{i=1}^{16} 96\rho^4\,[f(x_i)]^4\,\zeta^{\alpha_i A_i}(x_i)$$
$$\times \prod_{j=1}^{4} 2\rho^2\,[f(y_j)]^3\,(\bar{\nu}^{(B_j}\nu^{C_j)})\,. \quad (388)$$

Using the generating function (366), this formula can be rewritten in the form

$$G_{20}(x_1,\ldots,x_{20}) = \frac{2^{55}3^{16}\pi^{-13}\,\mathrm{e}^{2\pi i\tau}}{g^{32}(N_c-1)!(N_c-2)!}\int \mathrm{d}\rho\,\mathrm{d}^4x_0\,\mathrm{d}^5\Omega\prod_{A=1}^{4}\mathrm{d}^2\eta^A\,\mathrm{d}^2\bar{\xi}^A$$
$$\times\int_0^\infty \mathrm{d}r\, r^{4N_c-3}\mathrm{e}^{-2\rho^2r^2}\,\rho^{4N_c-7}\prod_{i=1}^{16}\rho^4\left[f(x_i)\right]^4\zeta^{\alpha_iA_i}(x_i)\prod_{j=1}^{4}\rho^2\left[f(y_j)\right]^3$$
$$\times\left[\frac{\delta^8\mathcal{Z}(\vartheta,\bar\vartheta,\Omega,r)}{\delta\vartheta_{u_1(B_1}\delta\bar\vartheta^{u_1}_{C_1)}\cdots\delta\vartheta_{u_4(B_4}\delta\bar\vartheta^{u_4}_{C_4)}}\right]\Bigg|_{\vartheta=\bar\vartheta=0}. \qquad (389)$$

After evaluating the derivatives and eliminating the sources the integral over r can be performed and one gets

$$G_{20}(x_1,\ldots,x_{20}) = \frac{2^{57}3^{16}\pi^{-17}\,c_2(N_c)\,\mathrm{e}^{2\pi i\tau}}{g^{28}}\int\frac{\mathrm{d}\rho\,\mathrm{d}^4x_0}{\rho^5}\prod_{A=1}^{4}\mathrm{d}^2\eta^A\,\mathrm{d}^2\bar\xi^A$$
$$\times\prod_{i=1}^{16}\frac{\rho^4}{[(x_i-x_0)^2+\rho^2]^4}\zeta^{\alpha_iA_i}(x_i)\prod_{j=1}^{4}\frac{\rho^3}{[(y_j-x_0)^2+\rho^2]^3}$$
$$\times\int\mathrm{d}^5\Omega\,\left[\Omega^{B_1C_2}\Omega^{B_2C_1}\Omega^{B_3C_4}\Omega^{B_4C_3}+\cdots\right], \qquad (390)$$

where the ellipsis in the last line stands for permutations of the B_i, C_i indices and

$$c_2(N_c) = \frac{2^{-2N_c}(N_c-2)^2\Gamma(2N_c-3)}{(N_c-1)!(N_c-2)!}$$
$$\underset{N_c\to\infty}{\sim} N_c^{1/2}\left[1-\frac{25}{8N_c}+\mathrm{O}(1/N_c^2)\right]. \qquad (391)$$

The factor of $(N_c-2)^2$ in the numerator of $c_2(N_c)$ comes from the contraction of colour indices in the ϑ_{Au}'s and $\bar\vartheta^u_A$'s sources. The integration over the five sphere in (390) gives the $SU(4)$ tensor

$$t^{B_1C_1\cdots B_4C_4} = \varepsilon^{B_1C_2B_2C_1}\varepsilon^{B_3C_4B_4C_3} + \text{permutations}. \qquad (392)$$

The main difference to be noted with respect to the minimal cases is the non-trivial dependence on the angular variables parametrising the five-sphere. In general, as in the above expression, the five-sphere integral factorises and gives rise to $SU(4)$ selection rules. Specifically, a correlation function can receive a non-zero instanton contribution only if the $SU(4)$ flavour indices carried by the non-exact modes, ν^A and $\bar\nu^A$, appear in a combination containing the $SU(4)$ singlet representation. We shall re-examine the results (390) and (391) in Sect. 18.2 in connection with the corresponding processes in the dual string theory. In particular, we will see that the calculation of non-minimal correlators such as (385) leads to a puzzle: the N-dependence in the SYM result does not agree with that of the amplitudes which are naturally identified as their dual. The resolution of the puzzle will require taking into account further types of contributions which do not arise in the minimal case (see Sect. 18.2).

From the previous example and the form of the generating function (366) we can deduce some general features of non-minimal correlation functions. The insertion of each $(\bar{\nu}^A \nu^B)$ bilinear in a correlator corresponds to two derivatives of (367) with respect to the sources. This, besides producing a factor of g, also modifies the r dependence of the integrand, thus affecting the overall N_c-dependence, see (378). Moreover additional factors of N_c are associated with the contraction of the colour indices carried by the ν_u^A's and $\bar{\nu}^{Au}$'s variables. From (366) and (367) one checks that in general the insertion of any $(\bar{\nu}^A \nu^B)_{\mathbf{10}}$ pair yields a factor g and the insertion of each $(\bar{\nu}^A \nu^B)_{\mathbf{6}}$ pair a factor $g\sqrt{N_c}$.

Schematically, for a generic non-minimal n-point correlation function containing q $(\bar{\nu}^A \nu^B)_{\mathbf{10}}$ factors and p $(\bar{\nu}^A \nu^B)_{\mathbf{6}}$ bilinears one finds

$$\begin{aligned}\langle \mathcal{O}_1(x_1)\cdots\mathcal{O}_n(x_n)\rangle \sim g^{8+p+q}\, \mathrm{e}^{2\pi i\tau} \alpha(N_c) \int \frac{\mathrm{d}\rho\, \mathrm{d}^4 x_0}{\rho^5}\, \mathrm{d}^5\Omega \prod_{A=1}^{4} \mathrm{d}^2\eta^A\, \mathrm{d}^2\bar{\xi}^A \\ \times \rho^{p+q} \prod_{i=1}^{n} \tilde{\mathcal{O}}_i(x_i; x_0, \rho, \eta, \bar{\xi}, \Omega)\,, \end{aligned} \tag{393}$$

where $\tilde{\mathcal{O}}_i$ denote the profiles of the operators after the dependence on the non-exact modes has been re-expressed in terms of the Ω^{AB}'s of (368). For future use we give the expression of the coefficient $\alpha(N_c)$ at large N_c

$$\begin{aligned}\alpha(N_c) &= \frac{2^{-2N_c}\Gamma\left(2N_c - 1 - \frac{p+q}{2}\right)}{(N_c-1)!(N_c-2)!} N_c^{p+\frac{q}{2}} \left[1 + \mathrm{O}(\frac{1}{N_c})\right] \\ &\sim N_c^{\frac{1}{2}+\frac{p}{2}} \left[1 + \mathrm{O}(\frac{1}{N_c})\right]. \end{aligned} \tag{394}$$

• All the operators in the supercurrent multiplet considered so far only involve the $(\bar{\nu}^A \nu^B)_{\mathbf{10}}$ bilinears. Anti-symmetric bilinears occur in higher dimension operators such as those in multiplets having as lowest component the scalars (346) with $\ell \geq 3$. An example of correlation function containing such insertions is

$$G_{16}(x_1,\ldots,x_{16}) = \langle \Lambda^{A_1}_{\alpha_1}(x_1)\cdots \Lambda^{A_{14}}_{\alpha_{14}}(x_{14}) \tilde{\Lambda}^{B_1B_2B_3}_{\beta_1}(y_1) \tilde{\Lambda}^{C_1C_2C_3}_{\beta_2}(y_2)\rangle\,, \tag{395}$$

where the operator $\tilde{\Lambda}^{B_1B_2B_3}_{\alpha}$, which belongs to the same multiplet as $\mathcal{Q}_{\ell=3}$ and transforms in the $\overline{\mathbf{20}}$ of $SU(4)$, is

$$\begin{aligned}\tilde{\Lambda}^{B_1B_2B_3}_{\alpha} = &\frac{1}{g^3 N_c^{1/2}} \mathrm{Tr}\left[2\lambda^{B_1}_{\alpha}\left(\lambda^{\beta\, B_2}\lambda^{B_3}_{\beta} + \lambda^{\beta\, B_3}\lambda^{B_2}_{\beta}\right)\right. \\ &+\lambda^{B_2}_{\alpha}\left(\lambda^{\beta\, B_1}\lambda^{B_3}_{\beta} + \lambda^{\beta\, B_3}\lambda^{B_1}_{\beta}\right) + \lambda^{B_3}_{\alpha}\left(\lambda^{\beta\, B_1}\lambda^{B_2}_{\beta} + \lambda^{\beta\, B_2}\lambda^{B_1}_{\beta}\right) \\ &\left.+F_{mn}\sigma^{mn\,\beta}_{\alpha}\left(\{\lambda^{B_2}_{\beta}, \varphi^{B_1B_3}\} + \{\lambda^{B_3}_{\beta}, \varphi^{B_1B_2}\}\right) + \cdots\right], \end{aligned} \tag{396}$$

with the ellipsis referring to terms which are negligible in the semi-classical approximation.

The profile of the operator (396) in the one-instanton background is

$$\hat{\tilde{\Lambda}}_{\alpha}^{B_1B_2B_3} = \frac{24}{g^3 N_c^{1/2}}\, \rho^4\, [f(x)]^5 \left[\zeta_{\alpha}^{B_2}(\bar{\nu}^{[B_1}\nu^{B_3]}) + \zeta_{\alpha}^{B_3}(\bar{\nu}^{u[B_1}\nu_u^{B_2]})\right] . \tag{397}$$

The correlation function (395) is computed by replacing each inserted operator with its classical profile and integrating over the moduli space. In particular, the normalisation of the operators $\tilde{\Lambda}_{\alpha}^{ABC}$ is such as to compensate the additional factors N_c associated with the $(\bar{\nu}^A\nu^B)_{\mathbf{6}}$ bilinears and the final result behaves again like $N_c^{1/2}$ in the large-N_c limit. Proceeding as in the previous cases, one finds

$$\begin{aligned} G_{16}(x_1,\ldots,x_{16}) &= \frac{c_3(N_c)\, 2^{48}\, 3^{16}\, \mathrm{e}^{2\pi i\tau}}{\pi^{17}\, g^{24}} \int \frac{\mathrm{d}\rho\, \mathrm{d}^4x_0}{\rho^5}\, \mathrm{d}^5\Omega \prod_{A=1}^{4} \mathrm{d}^2\eta^A\, \mathrm{d}^2\bar{\xi}^A \\ &\times \prod_{i=1}^{14} \frac{\rho^4}{[(x_i-x_0)^2+\rho^2]^4}\, \zeta_{\alpha_i}^{A_i}(x_i) \\ &\times \left[\frac{\rho^5}{[y_1-x_0)^2+\rho^2]^5}\, \zeta_{\beta_1}^{B_2}(y_1)\, \Omega^{B_1B_3}\, \frac{\rho^5}{[y_2-x_0)^2+\rho^2]^5}\, \zeta_{\beta_2}^{C_2}(y_2)\, \Omega^{C_1C_3} + \cdots\right] , \end{aligned} \tag{398}$$

where the $\cdots$ refers to symmetrisation in (B_2, B_3) and (C_2, C_3). The N_c dependence is contained in the coefficient $c_3(N_c)$, where

$$c_3(N_c) = \frac{2^{-2N_c}(N_c-2)^2\Gamma(2N_c-2)}{N_c(N_c-1)!(N_c-2)!} \underset{N_c\to\infty}{\sim} N_c^{1/2} . \tag{399}$$

The integration over the five-sphere in this case gives a single ε-tensor

$$\int \mathrm{d}^5\Omega\, \Omega^{AB}\Omega^{CD} = \frac{\pi^3}{6}\, \varepsilon^{ABCD} . \tag{400}$$

The example of (395) allows to illustrate another feature of non-minimal correlators. Since not all the fields are employed to soak up the 16 exact superconformal modes, there are contributions to the expectation value in which pairs of fields are contracted with an instantonic propagator. In the case of (395), for instance, it is possible to contract pairs of scalars in the two $\tilde{\Lambda}_a^{ABC}$ operators. Contributions of this type are of the same order in g as those in which the extra insertions soak up ν^A and $\bar{\nu}^A$ modes, since with the normalisations we are using ($S \propto 1/g^2$) the scalar propagator is proportional to g^2. They are, however, sub-leading with respect to terms containing $(\bar{\nu}^A\nu^B)_{\mathbf{6}}$ pairs at large N_c. The evaluation of the contributions with contractions is rather involved because they require the use of the propagator in the instanton background, which has a complicated expression [20, 127]. We shall not discuss further these effects, but we stress that they are essential for the comparison with certain string theory amplitudes [123].

There are many other interesting examples of non-minimal correlation functions in $\mathcal{N}=4$ SYM which could be discussed. For lack of space we conclude this section with a brief list of some other notable cases, referring the reader to the original literature for further details. A comprehensive study of non-minimal correlators can be found in [123].

- A special class of correlation functions in $\mathcal{N}=4$ SYM are the so-called extremal correlators. These are n-point functions of operators of the type (346) in which the dimension, ℓ_1, of one of the operators equals the sum of the dimensions, ℓ_i, $i=2,\ldots,n$, of the others ($\ell_1=\sum_{i=2}^n \ell_i$). The analysis of the associated dual amplitudes in supergravity led to the prediction that such correlation functions should not be renormalised [128]. This was then confirmed by field theory calculations in [129, 130]. In particular, an argument for the absence of instanton corrections to extremal correlators, based on the analysis of fermion zero modes, was given in [129]. Similar results have been shown to hold for next-to-extremal correlation functions for which $\ell_1=\sum_{i=2}^n \ell_i-2$ [131]. A more complicated class are the near extremal correlators, characterised by the condition $\ell_1=\sum_{i=2}^n \ell_i-m$ with $m \leq n-3$. These satisfy certain partial non-renormalisation properties [132], which have been argued in [123] to survive the inclusion of instanton corrections.
- The Wilson loop is a particularly important operator in non-abelian gauge theories since it plays the rôle of order parameter characterising confinement. In pure Yang–Mills theory the Wilson loop is the expectation value

$$W[\mathcal{C}]=\frac{1}{N_c}\langle \mathrm{Tr}_{N_c}\mathcal{P}\exp\left[i\oint_{\mathcal{C}}\mathrm{d}x^\mu\, A_\mu\right]\rangle \tag{401}$$

of the holonomy associated with the closed contour $\mathcal{C}$. A generalisation of this quantity in $\mathcal{N}=4$ SYM has been constructed in [133] together with a proposal for the dual quantity in string theory to be associated with it. A special class of Wilson loops in $\mathcal{N}=4$ SYM are circular BPS loops, which are annihilated by 16 linear combinations of Poincaré and special supersymmetries. These Wilson loops are defined as

$$W[\mathcal{C}_R]=\frac{1}{N_c}\langle \mathrm{Tr}_{N_c}\mathcal{P}\exp\left[i\oint_{\mathcal{C}_R}\mathrm{d}s\,\left(A_\mu\dot{x}^\mu+i\varphi_i n^i|\dot{x}|\right)\right]\rangle\,, \tag{402}$$

where n^i is a constant unit vector on the five-sphere and $\mathcal{C}_R$ is a circle of radius R. An elegant method for computing instanton corrections to (402) in the case of $SU(2)$ gauge group was devised in [134]. The BPS Wilson loop is a non-minimal correlator since it is non-polynomial in the fields. In the $SU(N_c)$ case its calculation in the instanton background is a formidable task and the $SU(2)$ analysis of [134] has not been extended to this more general case.

- In Sect. 15.2 we mentioned that certain results concerning instanton corrections to the anomalous dimensions of gauge-invariant composite operators can be obtained from the OPE analysis of four-point functions such as (381). On the other hand, as discussed in Sect. 13, the anomalous dimensions can be computed directly from two-point functions after resolving the operator mixing. Depending on the bare dimension of the operators one is considering two-point functions can be minimal or non-minimal. A systematic study of instanton contributions to two-point functions of scalar operator was initiated in [135]. As discussed in Sect. 13, general considerations, and in particular arguments based on S-duality, suggest that generically anomalous dimensions in $\mathcal{N}=4$ SYM should receive both perturbative and non-perturbative contributions. A rather surprising result found in [135] is the absence of instanton corrections to the majority of scalar operators of bare dimensions $\Delta_0 \leq 5$.
- Finally an important class of non-minimal correlation functions are those relevant for the so-called BMN limit, which is the subject of Sect. 18.3.

16 Generalisation to Multi-instanton Sectors

The generalisation of the analysis presented in the previous section to multi-instanton sectors is technically very involved and requires the full machinery of the ADHM construction [19]. A detailed description of this formalism and its generalisation to supersymmetric theories, as well as references to the original literature can be found in [6]. Due to space limits we shall only report an important result of [121], where multi-instanton contributions to $\mathcal{N}=4$ SYM correlation functions were explicitly evaluated in the large N_c limit.

In the generic K-instanton sector and with gauge group $SU(N_c)$ an instanton configuration in pure Yang–Mills theory is characterised by $4KN_c$ collective coordinates parametrising a hyper-Kähler manifold, $\mathcal{M}_K$. A description of the moduli space associated with general (anti-)self-dual gauge configurations can be given using the ADHM construction [19]. This is based on the introduction of an overcomplete set of matrix-valued parameters on $\mathcal{M}_K$, satisfying non-linear constraints, which can be shown to be equivalent to the self-duality condition for the Yang–Mills field strength. The constraints can be implemented describing the moduli space, $\mathcal{M}_K$, and the associated metric by means of what is referred to as a hyper-Kähler quotient construction [136]. In the case of the $SU(N_c)$ $\mathcal{N}=4$ SYM theory there are also $8KN_c$ fermionic collective coordinates in the generic K instanton sector. These can be included in the ADHM formalism as matrix-valued generalisations of the collective coordinates introduced in the one-instanton sector, subject to suitable constraints.

As usual, the calculation of instanton contributions to correlation functions involves the integration over the instanton moduli space. This can be formally achieved integrating over the redundant set of bosonic and fermionic ADHM matrices and imposing the constraints via δ-functions. However, as already observed, the calculations are not feasible for generic K, since an explicit solution to the constraints is not known.

In the case of $\mathcal{N}=4$ SYM, these calculations are also not particularly enlightening since, in general, correlators receive non-vanishing contributions from all instanton sectors. However, a dramatic simplification occurs in the large N_c limit, making the calculation of K-instanton corrections to correlation functions feasible for arbitrary K [121]. The reason for this simplification is that in the large N_c limit the integration over the (super) moduli space is dominated by a very special configuration and can be evaluated using a saddle point approximation. After the introduction of a matrix generalisation of the auxiliary variables χ^i, the saddle point that dominates the K-instanton moduli space integration corresponds to a configuration in which the K instantons have the same size and share the same location both in space–time and in the five-sphere directions parametrised by the χ^i's.[37] As in the one-instanton sector only the 16 exact fermion zero modes associated with the broken superconformal symmetries remain exact. The physical moduli space integration measure obtained using the saddle point approximation is

$$\int \mathrm{d}\mu^{(K)}_{\mathrm{phys}}\, \mathrm{e}^{-S_{\mathrm{inst}}} \tag{403}$$
$$\underset{N_c\to\infty}{\longrightarrow} \frac{N_c^{1/2}\, g^8\, \mathrm{e}^{2\pi i K\tau}}{K^3\, 2^{17K^2/2-K/2+25}\, \pi^{9K^2/2+9}} \int \frac{\mathrm{d}^4x_0\, \mathrm{d}\rho}{\rho^5}\, \mathrm{d}^5\Omega \prod_{A=1}^{4} \mathrm{d}^2\eta^A\, \mathrm{d}^2\bar{\xi}\, Z_K\,,$$

where Z_K contains the integration over the fluctuations around the saddle point. These can be expressed in terms of $[K]\times[K]$ bosonic and fermionic matrices, A_M, $M=0,\ldots,9$ and Ψ_r, $r=1,\ldots,16$, by means of which the Z_K factor in (403) takes the form of the partition function of a $SU(K)$ supersymmetric matrix model[38], i.e.

$$Z_K = \frac{1}{\mathrm{Vol}\, SU(K)} \int \mathrm{d}^{10}A\, \mathrm{d}^{16}\Psi\, \mathrm{e}^{-S(A,\Psi)}\,, \tag{404}$$

where

$$S(A,\Psi) = -\frac{1}{2}\,\mathrm{Tr}_K \left([A_M, A_N]^2 + \bar{\Psi}[\not{A}, \Psi] \right) . \tag{405}$$

The partition function Z_K was computed in [80, 137] with the result

$$Z_K = 2^{17K^2/2-K/2-8}\, \pi^{9K^2-9/2}\, K^{-1/2} \sum_{m|K} \frac{1}{m^2}\,, \tag{406}$$

---

[37] In the analysis of the fluctuations around the saddle point it is also important that, as far as the global gauge orientation is concerned, the K instantons lie in mutually orthogonal $SU(2)$ subgroups of $SU(N_c)$.

[38] This is the dimensional reduction to zero dimensions of 10-dimensional $\mathcal{N}=1$ SYM.

where the sum is over the positive integer divisors of K.

Correlation functions of composite operators in the large N_c limit are computed integrating their profiles in the K-instanton background with the measure (403). In particular, if the operator profiles do not depend on the collective coordinates parametrising the matrix model, the partition function Z_K factors out. This is the case for minimal correlation functions of gauge invariant operators such as those considered in Sect. 15.2. As an example of this type we consider the K-instanton contribution to (374). In the large N_c limit the profile of the operator Λ^A_α in the K-instanton background does not depend on the matrix model coordinates. It is proportional to its one-instanton expression, namely

$$\hat{\Lambda}^A_\alpha\Big|_{K-\text{inst}} = \frac{96\,K}{g^2}\,\frac{\rho^4}{[(x-x_0)^2+\rho^2]^4}\,\zeta^A_a \equiv K\,\hat{\Lambda}^A_\alpha\Big|_{1-\text{inst}}\,. \tag{407}$$

Therefore one finds [121]

$$\langle \Lambda^{A_1}_{\alpha_1}(x_1)\cdots\Lambda^{A_{16}}_{\alpha_{16}}(x_{16})\rangle_{K-\text{inst}} = \frac{N_c^{1/2}\,K^{25/2}\,2^{47}\,3^{16}\,\pi^{-27/2}\,\mathrm{e}^{2\pi i K\tau}}{g^{24}}$$
$$\times \sum_{m|K}\frac{1}{m^2}\int\frac{\mathrm{d}\rho\,\mathrm{d}^4x_0}{\rho^5}\,\mathrm{d}^5\Omega\prod_{A=1}^{4}\mathrm{d}^2\eta^A\,\mathrm{d}^2\bar{\xi}^A\prod_{i=1}^{16}\frac{\rho^4}{[(x_i-x_0)^2+\rho^2]^4}\,\zeta^{A_i}_{\alpha_i}(x_i)\,. \tag{408}$$

The calculation of multi-instanton contributions to non-minimal correlation functions, even in the large N_c limit, is much more complicated. In the non-minimal case the operator insertions depend on the one-instanton moduli and also on the matrix model variables and thus one cannot simply factor out Z_K. It is natural to expect that in these cases, instead of the partition function, the integration over the A_M and Ψ_r variables should be related to certain correlation functions in the matrix model giving rise to generalisations of (406).

17 AdS/CFT Correspondence: a Brief Overview

As already mentioned, the recent renewed interest in $\mathcal{N}=4$ SYM stems from the conjecture about the AdS/CFT correspondence [88, 89, 90]. In this section we provide a brief overview of the main concepts at the basis of this conjecture and in the following sections we review the rôle of instantons in this context.

The idea of the AdS/CFT correspondence was presented in [88] and a more concrete formulation was given in [89, 90]. Reviews can be found in [91, 92]. In [88] Maldacena proposed a remarkable duality relation connecting two completely different theories, $\mathcal{N}$=4 SYM with $SU(N_c)$ gauge group and type IIB superstring theory in an $\mathrm{AdS}_5\times S^5$ background.

The type IIB superstring theory has $\mathcal{N} = (2,0)$ supersymmetry in 10 dimensions, i.e. it is invariant under 32 supersymmetries. Its spectrum contains a finite number of massless states and an infinite tower of massive states. The massless spectrum is chiral. The bosonic degrees of freedom are divided into the so-called Neveu–Schwarz–Neveu–Schwarz (NS–NS) and Ramond–Ramond (R–R) sectors. The massless NS–NS sector contains a traceless rank-two symmetric tensor (the graviton, g_{MN}, $M, N = 0, \ldots, 9$), an anti-symmetric two-form (B_{MN}) and a scalar (the dilaton, ϕ). The massless R–R sector contains a scalar ($C_{(0)}$), an anti-symmetric two-form (C_{MN}) and an anti-symmetric four-form (C_{MNPQ}), with self-dual field strength. The massless fermions are the spin 1/2 dilatino (λ) and the spin 3/2 gravitino (ψ_M). These are complex Weyl spinors of opposite chiralities. The theory has two parameters, the coupling constant, related to the v.e.v. of the dilaton, $g_s = e^{\langle\phi\rangle}$, and the inverse string tension, α'. The latter sets the scale for the massive states in the spectrum which have masses proportional to $1/\sqrt{\alpha'}$.

The background relevant for the correspondence with $\mathcal{N} = 4$ SYM, $\mathrm{AdS}_5 \times S^5$, is the product of a five-dimensional anti-de Sitter space and a five-sphere. The non-compact factor, Lorentzian AdS_5, can be described as a hyperboloid embedded in six dimensions, i.e. in terms of six Cartesian coordinates, X^i, $i = 0, \ldots, 5$, satisfying the constraint

$$X_0^2 - X_1^2 - \cdots - X_4^2 + X_5^2 = L^2 \tag{409}$$

where L is the (constant) radius of curvature. This definition immediately shows that the AdS_5 space has isometry group $SO(2,4)$. The so-called global coordinates for AdS_5 are introduced setting

$$\begin{aligned} &X_0 = L\cosh\rho\,\cos t\,, \qquad X_5 = L\cosh\rho\,\sin t\,, \\ &X_r = L\sinh\rho\,\Omega_r\,, \qquad r = 1,2,3,4\,, \qquad \sum_r \Omega_r^2 = 1\,. \end{aligned} \tag{410}$$

In terms of these coordinates the metric reads

$$\mathrm{d}s^2 = L^2(-\cosh^2\rho\,\mathrm{d}t^2 + \mathrm{d}\rho^2 + \sinh^2\rho\,\mathrm{d}\Omega^2)\,. \tag{411}$$

Another convenient set of coordinates for AdS_5 are the so-called Poincaré coordinates, (z_μ, z_0). The z_0 coordinate parametrises the radial direction of AdS_5 and the four z_μ coordinates parametrise the directions parallel to the boundary located at $z_0 = 0$. In terms of these coordinates the metric is

$$\mathrm{d}s^2 = \frac{L^2}{z_0^2}\left(\mathrm{d}z_\mu^2 + \mathrm{d}z_0^2\right)\,. \tag{412}$$

The $\mathrm{AdS}_5 \times S^5$ space is maximally supersymmetric if the two factors have the same radius of curvature, L. The non-vanishing components of the Ricci tensor are

$$R_{mn} = -\frac{4}{L^2}\,g_{mn}\,, \qquad R_{ab} = \frac{4}{L^2}\,g_{ab}\,, \tag{413}$$

where the indices m, n span the AdS_5 directions and the indices a, b the S^5 directions. Moreover the self-dual R–R five-form field strength has a non-vanishing background value

$$F_{mnpqr} = \frac{1}{L}\,\varepsilon_{mnpqr}\,, \qquad F_{abcde} = \frac{1}{L}\,\varepsilon_{abcde}\,. \tag{414}$$

The conjectured duality has a holographic nature in that it relates the physics described by the string theory in the bulk of $\mathrm{AdS}_5 \times S^5$ to that of a gauge theory, $\mathcal{N} = 4$ SYM, living on the four-dimensional boundary of AdS_5.

The first ingredient of the correspondence is a dictionary relating the parameters of the two theories. In $\mathcal{N} = 4$ SYM the parameters are the coupling, g, and the rank of the gauge group. In the string theory, besides the coupling constant, g_s, and the inverse string tension, α', the radius of curvature, L, of the AdS_5 and S^5 spaces enters as an additional dimensionful parameter. The relations among the gauge and string theory parameters are

$$g^2 = 4\pi g_\mathrm{s}\,, \qquad L^4 = 4\pi g_\mathrm{s}{\alpha'}^2 N_c\,. \tag{415}$$

The second equation can be used to relate the dimensionless ratio $L^4/{\alpha'}^2$ to the 't Hooft coupling, $\lambda = g^2 N_c$,

$$\frac{L^4}{{\alpha'}^2} = \lambda\,. \tag{416}$$

The ϑ-angle, that can be turned on in the gauge theory, is related to the expectation value of the R–R scalar

$$\frac{\vartheta}{2\pi} = \langle C_{(0)} \rangle\,. \tag{417}$$

Given this dictionary for the parameters of the two theories, the correspondence is formulated in terms of two additional basic ingredients:

- A map between the fundamental degrees of freedom of the two theories.
- A prescription for the computation the observables of one theory in terms of those of the other.

The map between degrees of freedom is dictated by the symmetries. The (super)isometries of the string background, under which the states in the string spectrum are classified, coincide with the (super)group of global symmetries of the gauge theory, which, as already discussed, is $PSU(2,2|4)$. The duality associates states in the string spectrum with gauge-invariant composite operators in $\mathcal{N} = 4$ SYM, which have the same quantum numbers under the $SO(2,4) \times SO(6)$ maximal bosonic subgroup of $PSU(2,2|4)$. Specifically, (1) the Lorentz quantum numbers are identified, (2) the masses of the string states are related to the scaling dimensions of the dual operators and (3) the $SO(6)$ quantum numbers arising in the Kaluza–Klein (KK) reduction on S^5

of the string theory are related to the Dynkin labels characterising the transformation of the dual gauge theory operators under the $SU(4)$ R-symmetry. Supersymmetry then implies that entire multiplets are related. The simplest case of this relation is represented by the correspondence between the supergravity multiplet, which contains the graviton and its superpartners, and the $\mathcal{N}=4$ supercurrent multiplet discussed in Sect. 13.

The prescription relating observables on the two sides of the correspondence is based on the identification of properly defined partition functions. The string partition function in $\mathrm{AdS}_5 \times S^5$ is a functional of the boundary values of the fields. The latter play the rôle of sources for the dual operators in the boundary gauge theory [89, 90] and one is led to propose the holographic formula

$$Z_{\mathrm{IIB}}[\Phi|_{\partial\mathrm{AdS}} = J] = \int [\mathrm{d}A][\mathrm{d}\lambda][\mathrm{d}\bar{\lambda}][\mathrm{d}\varphi]\, \exp\Big(-S_{\mathcal{N}=4} + \int \mathcal{O}_{\Phi} J\Big)\,. \tag{418}$$

Here Φ denotes a generic field in the string theory and $\mathcal{O}_{\Phi}$ is the dual composite operator in $\mathcal{N}=4$ SYM according to the map previously described.

The quantisation of string theory in an $\mathrm{AdS}_5 \times S^5$ background is not understood well enough to make really operative use of (418). However, interesting results can be obtained in certain limits. In particular in the weak coupling and small curvature limit on the gravity side, where

$$g_{\mathrm{s}} \ll 1\,, \qquad \frac{L^2}{\alpha'} \gg 1\,, \tag{419}$$

classical supergravity becomes a good approximation. Based on the dictionary (415), this limit corresponds to the limit of large N_{c} and large 't Hooft coupling, λ, in the gauge theory. Since in the $N_c \to \infty$ limit λ plays effectively the rôle of coupling constant, one obtains a duality between classical type IIB supergravity in $\mathrm{AdS}_5 \times S^5$ and the strong coupling limit of $\mathcal{N}=4$ SYM in the planar approximation. This observation illustrates the strong/weak nature of the duality, which on the one hand makes it difficult to test, but on the other makes it a powerful tool for the study of strongly coupled gauge theories.

In the limit (419) the IIB partition function in (418) is well approximated by

$$Z_{\mathrm{IIB}}[\Phi|_{\partial\mathrm{AdS}} = J] \sim \mathrm{e}^{-S_{\mathrm{IIB}}[\Phi|_{\partial\mathrm{AdS}}=J]}\,, \tag{420}$$

where S_{IIB} is the classical type IIB supergravity action in the $\mathrm{AdS}_5 \times S^5$ background.

In this limit the relation (418) has a simple and intriguing interpretation. Correlation functions in the gauge theory are obtained taking functional derivatives with respect to the sources on the r.h.s. of (418). Using the approximation (420), one finds that differentiating with respect to the sources is equivalent to solving the supergravity equations of motion with boundary conditions $\Phi|_{\partial\mathrm{AdS}} = J$. Therefore the correspondence states that an n-point

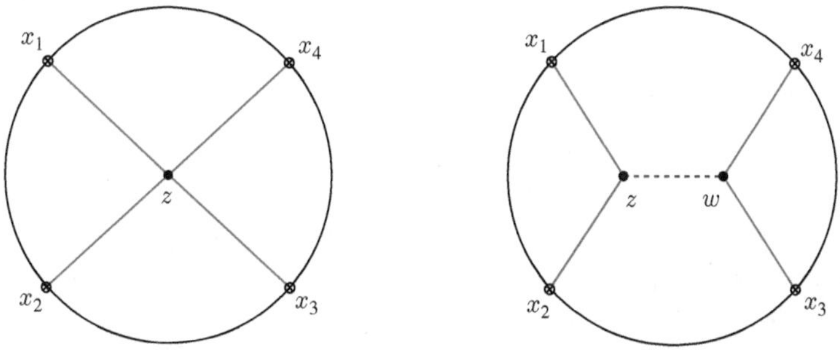

Fig. 1. Contact and exchange contributions to a four-point amplitude in $AdS_5 \times S^5$

correlation function, $\langle \mathcal{O}_1(x_1) \cdots \mathcal{O}_n(x_n) \rangle$, in $\mathcal{N} = 4$ SYM is equal to an amplitude in which, for each $\mathcal{O}_i$ insertion, the dual supergravity state, Φ_i, is propagated from the bulk to the boundary point x_i. An intuitive graphical representation of this prescription was proposed in [90].[39] Figure 1 represents the supergravity amplitudes contributing to the process dual to a SYM four-point function, $\langle \mathcal{O}_1(x_1) \cdots \mathcal{O}_4(x_4) \rangle$. The interior of the circle in Fig. 1 represents the bulk of $AdS_5 \times S^5$ and the circle itself is the four-dimensional boundary where the gauge theory lives.

A normalisable solution of the free supergravity equations of motions satisfying the boundary condition $\Phi|_{\partial \mathrm{AdS}} = J$ can be written as

$$\Phi^{(i)}(z_\mu, z_0) = \int \mathrm{d}^4 x \, K_i(z_\mu, z_0; x_\mu^{(i)}) \, J_\Phi(x_\mu^{(i)}) \,, \tag{421}$$

where the function $K_i(z_\mu, z_0; x_\mu^{(i)})$ is a so-called bulk-to-boundary propagator, i.e. the kernel that allows to express a supergravity field, Φ, at the bulk point (z_μ, z_0) in terms of its boundary value, J_Φ, at $(z_\mu = x_\mu^{(i)}, z_0 = 0)$. Substituting the solution (421) into the generating functional (420) one obtains the following expressions for the two contributions in Fig. 1:

$$\mathcal{A}_{\mathrm{cont}}(x_1, \ldots, x_4) = N_c^2 \int \frac{\mathrm{d}^4 z \, \mathrm{d}z_0}{z_0^5} \, \mathrm{d}^5\omega \prod_{i=1}^{4} K_i(z_\mu, z_0; x_\mu^{(i)}) \tag{422}$$

$$\mathcal{A}_{\mathrm{exc}}(x_1, \ldots, x_4) = N_c^2 \int \frac{\mathrm{d}^4 z \, \mathrm{d}z_0}{z_0^5} \, \mathrm{d}^5\omega \sum_m \prod_{i=1}^{2} \prod_{j=3}^{4} K_i(z_\mu, z_0; x_\mu^{(i)}) \times G_m(z, w) \, K_j(z_\mu, z_0; x_\mu^{(j)}) \,, \tag{423}$$

where $G_m(z, w)$ in the second amplitude, corresponding to the exchange diagram, represents a bulk-to-bulk propagator in AdS_5 and the index m runs over the set of all allowed intermediate states. In (422) and (423) we have

[39] In the following we shall refer to processes of this type as scattering amplitudes in AdS.

used Poincaré coordinates, (z_μ, z_0), to parametrise AdS_5 and angles, ω_i, for the five-sphere. In terms of these parameters the $\mathrm{AdS}_5 \times S^5$ metric becomes

$$\mathrm{d}s^2 = \frac{L^2}{z_0^2}\left(\mathrm{d}z_\mu^2 + \mathrm{d}z_0^2 + z_0^2\,\mathrm{d}\omega_5^2\right) . \tag{424}$$

In (422) and (423) the overall factor of N_c^2 is obtained rewriting the coefficient in front of the IIB supergravity action in the string frame, namely $L^8/\alpha'^4 g_s^2$, in terms of Yang–Mills parameters using (415).

In the next sections we shall discuss the inclusion of instanton effects in this picture. In Sect. 18.3 we shall consider another notable limit, i.e. the BMN limit, in which the string theory ↔ field theory correspondence is under control beyond the supergravity approximation.

18 Instanton Effects in the AdS/CFT Duality

In the AdS/CFT correspondence, the effects of Yang–Mills instantons in $\mathcal{N}=4$ SYM are related to non-perturbative effects induced by D-instantons in the dual IIB string theory [138]. In the low-energy supergravity limit of string theory, D-instantons arise as non-trivial solutions of the Euclidean field equations. In the 10-dimensional Euclidean space they correspond to configurations in which the metric (in the Einstein frame) is flat and the dilaton and the R–R scalar have non-constant profiles while all the other fields vanish. As the ordinary Yang–Mills instantons, the supergravity D-instantons are characterised by their integer-valued charge. The supergravity action evaluated on a charge-K D-instanton configuration is proportional to K, as in the Yang–Mills case, and inversely proportional to the string coupling, g_s. The D-instanton solution of the type IIB field equations in $\mathrm{AdS}_5 \times S^5$ has similar properties and can be obtained from the flat space solution [124]. In string theory D-instantons are identified with D(−1)-branes, i.e. point-like objects in Euclidean 10-dimensional space. Their world-volume is zero-dimensional and therefore open strings ending on D(−1)-branes carry no propagating degrees of freedom. They describe instead zero-modes associated with the D-instantons as discussed in Sect. 11. D-branes, and D-instantons in particular, can also be described in terms of closed string modes as collective excitations using the so-called boundary state formalism. This will be utilised in a special case in Sect. 18.3.

The discussion of the general principles of the AdS/CFT correspondence in Sect. 17 and specifically the fundamental relation (418) indicate that, in order to make contact with the calculation of instanton contributions to $\mathcal{N}=4$ correlation functions, one should study D-instanton induced contributions to string scattering amplitudes in $\mathrm{AdS}_5 \times S^5$. In principle, this involves including in the genus expansion of the closed string amplitudes the contribution of world-sheets with boundaries associated with the presence of D(−1)-branes.

However, as already explained, in the $\mathrm{AdS}_5 \times S^5$ background such calculations are not under control and one is restricted to a low-energy supergravity analysis. In the supergravity approximation, the inclusion of the effect of D-instantons requires a refinement of (420) in which the classical supergravity action is replaced by the low-energy effective action which incorporates the effect of the infinite tower of massive string excitations on the dynamics of the massless modes.

18.1 The Type IIB Effective Action

The type IIB string theory effective action is expressed as a powers series in the inverse string tension, α'. It takes the form

$$S_{\mathrm{IIB}}^{\mathrm{eff}} = \frac{1}{\alpha'^4}\left(S^{(0)} + \alpha'^3 S^{(3)} + \alpha'^4 S^{(4)} + \cdots + \alpha'^r S^{(r)} + \cdots\right), \qquad (425)$$

where $S^{(0)}$ denotes the classical action and the subsequent terms contain higher derivative couplings, which receive D-instanton contributions. The inclusion of such vertices in supergravity amplitudes in $\mathrm{AdS}_5 \times S^5$ gives rise to contributions which are in correspondence with the correlation functions discussed in Sects. 15 and 16.

The form of (425) is in principle determined by supersymmetry. The terms appearing in the leading correction, $S^{(3)}$, have been extensively studied. The couplings arising at this level include the well known $\mathcal{R}^4$ term and a large number of other terms related to it by supersymmetry. Schematically, in the string frame, the form of $S^{(3)}$ is

$$\begin{aligned}\alpha'^3\, S^{(3)} = \frac{1}{\alpha'}\int \mathrm{d}^{10}X\,\sqrt{-g}\,\mathrm{e}^{-\phi/2}\Big[f_1^{(0,0)}(\tau,\bar\tau)\left(\mathcal{R}^4 + (G\bar G)^4 + \cdots\right) + \cdots \\ + f_1^{(8,-8)}(\tau,\bar\tau)\left(G^8 + \cdots\right) + \cdots + f_1^{(12,-12)}(\tau,\bar\tau)\lambda^{16}\Big]\,. \qquad (426)\end{aligned}$$

The precise form of many of these couplings has been determined, see for instance [139] where the $\mathcal{R}^4$ coupling was studied. In the following we shall further discuss certain vertices which are relevant for the comparison with the Yang–Mills calculations of the previous sections. The coefficients in (426) are functions of the complex scalar, $\tau = \tau_1 + i\tau_2 = C_{(0)} + i\mathrm{e}^{-\phi}$, where ϕ is the dilaton and $C_{(0)}$ the R–R scalar. The effective action is invariant under $SL(2, \ )$ transformations acting on τ as

$$\tau \to \frac{a\tau + b}{c\tau + d}, \qquad (427)$$

where the integers a, b, c, d satisfy $ad - bc = 1$. Under such transformations any supergravity field, Φ, acquires a phase

$$\Phi \to \left(\frac{c\tau + d}{c\bar\tau + d}\right)^{q_\Phi} \Phi, \qquad (428)$$

where q_Φ is the charge of Φ under the local $U(1)$ symmetry of (425) which also rotates the two chiral supersymmetries [140]. In particular, the metric and the IIB self-dual five-form are not charged, the complex combination $G_{(3)} = (\tau \mathrm{d}B_{(2)} + \mathrm{d}C_{(2)})/\sqrt{\tau_2}$ (where $B_{(2)}$ and $C_{(2)}$ are the NS–NS and R–R two forms) has charge 1, the fluctuation of the complex scalar, $\delta\tau \equiv \hat{\tau}$, has charge 2, the dilatino, λ, and the gravitino, ψ_M, have charge 3/2 and 1/2, respectively. The coefficient functions in (426) transform as modular forms with holomorphic and anti-holomorphic weights $(w, -w)$, so that

$$f_1^{(w,-w)}(\tau,\bar{\tau}) \to \left(\frac{c\tau+d}{c\bar{\tau}+d}\right)^w f_1^{(w,-w)}(\tau,\bar{\tau}) \,. \tag{429}$$

Invariance under $SL(2, \ )$ requires that the weight w of the modular form in each term in the effective action be equal to half the sum of the $U(1)$ charges of the fields in the vertex.

The modular forms $f_1^{(w,-w)}(\tau,\bar{\tau})$ are obtained acting on $f_1^{(0,0)}(\tau,\bar{\tau})$ with modular covariant derivatives

$$f_1^{(w,-w)}(\tau,\bar{\tau}) = D_{w-1}D_{w-2}\cdots D_0 f_1^{(0,0)}(\tau,\bar{\tau})\,, \tag{430}$$

where $D_w = \tau_2 \frac{\partial}{\partial\tau} - i\frac{w}{2}$.

The modular form in front of the $\mathcal{R}^4$ term, $f_1^{(0,0)}(\tau,\bar{\tau})$, is given by a non-holomorphic Eisenstein series

$$f_1^{(0,0)}(\tau,\bar{\tau}) = \sum_{(m,n)\neq(0,0)} \frac{\tau_2^{3/2}}{|m+n\tau|^3}\,. \tag{431}$$

It can be expanded in Fourier modes as

$$\begin{aligned} f_1^{(0,0)}(\tau,\bar{\tau}) &= \sum_{K=-\infty}^{\infty} \mathcal{F}_K^1(\tau_2)\,\mathrm{e}^{2\pi i K\tau_1} = 2\zeta(3)\tau_2^{\frac{3}{2}} + \frac{2\pi^2}{3}\tau_2^{-\frac{1}{2}} \\ &+4\pi\sum_{K\neq 0}|K|^{\frac{1}{2}}\mu(K,1)\,\mathrm{e}^{-2\pi(|K|\tau_2 - iK\tau_1)}\sum_{j=0}^{\infty}(4\pi K\tau_2)^{-j}\frac{\Gamma(j-1/2)}{\Gamma(-j-1/2)j!}\,, \end{aligned} \tag{432}$$

where the r.h.s. is the result of a further weak coupling (large τ_2) expansion.

The non-zero Fourier modes are interpreted as D-instanton contributions with instanton number K ($K > 0$ terms are D-instanton contributions while $K < 0$ terms are anti-D-instanton contributions). The measure factor, $\mu(K,1)$, is

$$\mu(K,1) = \sum_{m|K}\frac{1}{m^2}\,, \tag{433}$$

where the sum is over the positive integer divisors of K. The coefficients of the D-instanton terms in (432), include an infinite series of perturbative fluctuations around any charge-K D-instanton. The leading term in this series is the one of relevance for the comparison with the semi-classical Yang–Mills instanton calculations. In the case of $f_1^{(0,0)}(\tau,\bar{\tau})$ this term is independent of τ_2. From (430) it follows that the leading D-instanton term in the modular form $f_1^{(w,-w)}(\tau,\bar{\tau})$ behaves as $\tau_2^w = g_s^{-w}$. The zero D-instanton term, $\mathcal{F}_0^1$, contains only two power-behaved contributions that arise in string perturbation theory as tree-level and one-loop contributions, with no higher-loop terms.

Much less is known about higher-order terms beyond $S^{(3)}$ in the string effective action, but various terms in $S^{(5)}$ are known and certain classes of terms at higher orders have been studied. Among the interactions at order α'^5 we have the following:

$$\alpha'^5 S^{(5)} = \alpha' \int \mathrm{d}^{10}X\,\sqrt{-g}\,\mathrm{e}^{\phi/2}\Big[f_2^{(0,0)}(\tau,\bar{\tau})\,D^4\mathcal{R}^4 + f_2^{(2,-2)}(\tau,\bar{\tau})\,G^4\mathcal{R}^4 + f_2^{(12,-12)}(\tau,\bar{\tau})\,\mathcal{R}^2\lambda^{16} + f_2^{(12,-12)}(\tau,\bar{\tau})\,\mathcal{R}^2\lambda^{16} + \cdots\Big]\,. \quad (434)$$

The modular forms, $f_2^{(w,-w)}(\tau,\bar{\tau})$, appearing here are generalisations of those previously defined. More generally at higher orders in the α' expansion one expects modular forms of the type

$$f_l^{(0,0)}(\tau,\bar{\tau}) = \sum_{(m,n)\neq(0,0)} \frac{\tau_2^{l+\frac{1}{2}}}{|m+n\tau|^{2l+1}}\,. \quad (435)$$

All these functions satisfy relations similar to (430). The weak coupling expansion of $f_2^{(0,0}(\tau,\bar{\tau})$ is

$$f_2^{(0,0)}(\tau,\bar{\tau}) = 2\zeta(5)\tau_2^{\frac{5}{2}} + \frac{4\pi^4}{135}\tau_2^{-\frac{3}{2}} + \frac{8\pi^2}{3}\sum_{K\neq 0}|K|^{\frac{3}{2}}\mu(K,2)\,\mathrm{e}^{-2\pi(|K|\tau_2 - iK\tau_1)}\left(1+\frac{3}{16\pi K}\tau_2^{-1}+\cdots\right), \quad (436)$$

where $\mu(K,2) = \sum_{m|K} 1/m^4$.

In the next subsection we shall discuss how the D-instanton induced terms appearing in the IIB effective action are related to instanton contributions to $\mathcal{N}=4$ correlation functions. In the analysis of processes dual to non-minimal correlators it will also be important to include the effect of the fluctuations, $\hat{\tau}$, of the complex scalar in the modular forms $f_l^{(w,-w)}(\tau,\bar{\tau})$. For instance, rewriting the complex scalar as $\tau = \tau_0 + \hat{\tau}$ (where τ_0 is the constant background value of τ), the expansion of the D-instanton exponential factor in $f_1^{(0,0)}(\tau,\bar{\tau})$ gives rise to a series of the form

$$\mathrm{e}^{2\pi iK\tau} = \mathrm{e}^{2\pi iK\tau_0}\sum_r \frac{(2\pi iK)^r}{r!}\,\hat{\tau}^r\,. \quad (437)$$

Equations (437) and (426) show that at order α' in the string low-energy action there are effective vertices of the form $\hat{\tau}^r\,\mathcal{R}^4$, which can contribute to scattering amplitudes in the $\mathrm{AdS}_5 \times S^5$ background.

18.2 D-instantons in $AdS_5 \times S^5$ and Comparison with Yang–Mills Instantons

The discussion in the previous subsection provides the background necessary to analyse the processes dual to the correlation functions computed in Sects. 15 and 16. These are dual to supergravity amplitudes involving the D-instanton induced vertices in the IIB effective action. In order to make contact with $\mathcal{N} = 4$ SYM, one needs to specialise the general expressions of the vertices in (426) to the case of the $AdS_5 \times S^5$ background. For this purpose, we shall expand the 10-dimensional supergravity fields in harmonics on the five-sphere [141]

$$\Phi(X) = \sum_{\ell} \Phi^{I_\ell}(z)\, \mathcal{Y}^{(\ell)}_{I_\ell}(\omega)\,, \tag{438}$$

where the $\mathcal{Y}^{(\ell)}_{I_\ell}(\omega)$'s are spherical harmonics, with ℓ denoting the level and I_ℓ a set of $SO(6)$ indices. After expanding the supergravity fields in $S^{\rm eff}_{\rm IIB}$ in this way the amplitudes dual to SYM correlators are computed using the prescription described in Sect. 17.

In studying AdS amplitudes we distinguish again between minimal and non-minimal cases, characterising an amplitude as (non-)minimal if it is dual to a (non-)minimal Yang–Mills correlator.

Minimal AdS Amplitudes

The simplest minimal amplitude is the one dual to the 16-point correlation function (374). The operator Λ^A_α in (373) is dual to the type IIB dilatino, λ, and thus according to the prescription explained in Sect. 17 we need to consider an amplitude with 16 dilatini propagating to the boundary. The vertex in the effective action which contributes to such process is

$$\frac{1}{\alpha'} \int {\rm d}^{10}X\, \sqrt{-g}\, {\rm e}^{-\phi/2} f_1^{(12,-12)}(\tau,\bar\tau)\, t_{16}\, \lambda^{16}\,, \tag{439}$$

where t_{16} is a 16-index anti-symmetric tensor contracting the spinor indices of the 16 dilatini. The amplitude dual to (374) involves the leading D-instanton term in (439) (see (430)–(432)), i.e.

$$\frac{1}{\alpha'} \int {\rm d}^{10}X\, \sqrt{-g}\, 2^{14}\pi^{13} \sum_{K>0} K^{25/2} \sum_{m|K} \frac{1}{m^2}\, {\rm e}^{2\pi i K\tau}\, {\rm e}^{-25\phi/2}\, t_{16}\, \lambda^{16}\,. \tag{440}$$

The amplitude induced by this interaction is depicted on the l.h.s. of Fig. 2: it is a contact amplitude in which the 16 dilatini interact via the vertex (440) and propagate to the boundary points $x_1, \ldots, x_{16}$.

After introducing an explicit parametrisation for $AdS_5 \times S^5$ and rewriting the string theory parameters, g_s and α', in terms of Yang–Mills parameters using the dictionary (415), the amplitude in Fig. 2 becomes

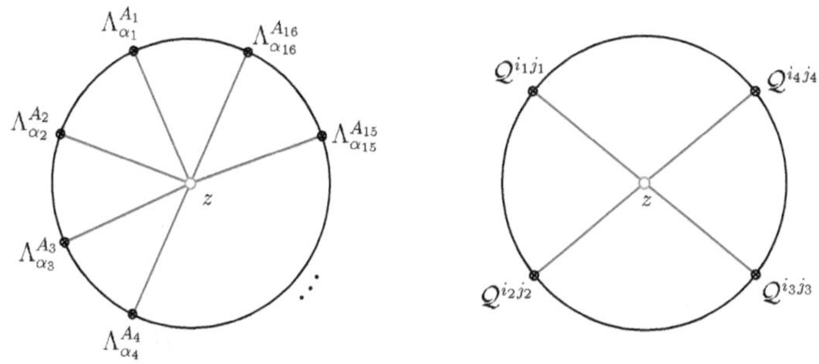

Fig. 2. D-instanton induced minimal amplitudes in $\mathrm{AdS}_5 \times S^5$

$$\frac{N_c^{1/2}}{g^{24}} \sum_{K>0} K^{\frac{25}{2}} \sum_{m|K} \frac{1}{m^2}\, \mathrm{e}^{2\pi i K\tau} \int \frac{\mathrm{d}^4 z\, \mathrm{d}z_0}{z_0^5}\, \mathrm{d}^5\omega\, t_{16}$$
$$\times \prod_{i=1}^{16} \mathcal{Y}_F^{(0)}(\omega)\, K^F_{7/2}(z, z_0; x_i) \ , \tag{441}$$

where overall numerical constants have been dropped and no indices have been indicated explicitly. In (441) we have used Poincaré coordinates, (z_μ, z_0), for AdS_5, with z_μ parametrising the directions parallel to the boundary and z_0 the radial direction. In terms of these coordinates and five angular variables for the S^5 factor, the $\mathrm{AdS}_5 \times S^5$ metric has the form (424). The 10-dimensional dilatino has been expanded in spherical harmonics. In the expansion we have retained the ground state component, dual to the SYM operator Λ^A_α. $\mathcal{Y}_F^{(0)}(\omega)$ denotes the corresponding harmonic function. In (441) $K^F_{7/2}$ denotes the bulk-to-boundary propagator for the dilatino, i.e. a spin 1/2 fermion with AdS mass $-\frac{3}{2L}$

$$K^F_{7/2}(z, z_0; x) = K_4(z, z_0; x) \left[\sqrt{z_0}\, \gamma_5 - \frac{1}{\sqrt{z_0}}\, (x-z)_\mu \gamma^\mu \right] , \tag{442}$$

where

$$K_\Delta(z, z_0; x) = \frac{z_0^\Delta}{[(z-x)^2 + z_0^2]^\Delta} \,. \tag{443}$$

Remarkably, the result (441), in its unintegrated form, is in exact agreement with the multi-instanton contribution to the correlation function (374) (cf. (379) and its multi-instanton generalisation (408)) after the integration over the 16 exact fermion zero-modes in the latter. To compare the two results, one identifies the AdS_5 coordinates, z_μ, z_0, with the position and size of the instanton and the S^5 angles with the auxiliary angular variables, Ω^{AB}, introduced in the gauge theory calculation. The integration over the position of the interaction point in the supergravity amplitude reproduces the integration over the $\mathcal{N} = 4$ moduli space, which, in the large N_c limit and with the inclusion of the auxiliary variables, is precisely one copy of $\mathrm{AdS}_5 \times S^5$. The bulk-to-boundary propagators reconstruct the dependence on the moduli contained in the profiles of the Yang–Mills operators. Finally, although we have not kept track of all the numerical factors, the dependence on the parameters, g and N_c, as well as on the instanton number, K, are in perfect agreement. The factors of τ_2 in the weak coupling expansion of the modular form $f_1^{(12,-12)}(\tau, \bar{\tau})$ give rise to the same g dependence as in the Yang–Mills result. Similarly the matrix model partition function is reproduced by the measure factor, $\mu(K, 1)$, in the modular form, see (433). The power of N_c in (441) follows from the application of the AdS/CFT dictionary (415), which gives

$$\frac{\mathrm{e}^{-\phi/2} L^2}{\alpha'} = 2\pi^{1/2} \sqrt{N_c} \, . \tag{444}$$

The calculation of the amplitude dual to the four-point function (381) is completely analogous. The $\mathcal{N} = 4$ scalar operator $\mathcal{Q}^{ABCD}$ in the $\mathbf{20}'$ of SU(4) is dual to a linear combination of the trace part of the metric in the S^5 directions and the S^5 components of the R–R four-form potential. The scalar in the supergravity multiplet, corresponding to $\mathcal{Q}^{ABCD}$, arises at level $\ell = 2$ in the expansion in spherical harmonics. An amplitude contributing to the process dual to (381) involves the $\mathcal{R}^4$ interaction in the bulk. This is depicted on the r.h.s in Fig. 2. Proceeding as in the case of the 16-point amplitude one finds that this four-point amplitude is

$$N_c^{1/2} \sum_{K>0} K^{1/2} \sum_{m|K} \frac{1}{m^2} \, \mathrm{e}^{2\pi i K \tau} \int \frac{\mathrm{d}^4 z \, \mathrm{d} z_0}{z_0^5} \, \mathrm{d}^5 \omega \times \prod_{i=1}^{4} \mathcal{Y}_B^{(2)}(\omega) \, K_4(z, z_0; x_i) \ , \tag{445}$$

where the bulk-to-boundary propagator, K_4, is now the one appropriate for a scalar of mass squared $-4/L^2$ in AdS and the $\mathcal{Y}_B^{(2)}$'s are $\ell = 2$ scalar spherical harmonics. Again the result agrees perfectly with the Yang–Mills calculation.

The examples described here illustrate the striking agreement between instanton contributions to Yang–Mills correlation functions and D-instanton induced supergravity amplitudes. The agreement found represents one of the most convincing tests of the validity of the AdS/CFT correspondence. The majority of the explicit tests of the Maldacena conjecture compare protected quantities which do not depend on the coupling constant and thus coincide with their free theory expressions. In these cases, the comparison is not affected by the strong/weak coupling nature of the correspondence, but the calculations simply test that the same non-renormalisation properties are valid on both sides. The calculations reviewed above represent instead one of the few instances in which a precise comparison is possible for quantities which do receive non-trivial quantum corrections.[40] Such a precise agreement is remarkable and somewhat unexpected, since the calculations in this section and those in Sect. 15 appear to have different regimes of validity. It is natural to interpret the result of the comparison as due to an underlying partial non-renormalisation property [142], whose origin, however, remains unexplained.

Non-minimal AdS Amplitudes

The discussion in the previous subsection has a natural generalisation to the case of amplitudes dual to the non-minimal correlation functions of Sect. 15.3. In the non-minimal case, the study of SYM correlators has not been generalised to multi-instanton sectors. In this section we show how the supergravity analysis gives results which are in qualitative agreement with those of the $\mathcal{N}=4$ calculations in the one-instanton sector. We consider supergravity amplitudes related to the two main types of non-minimal correlators discussed in Sect. 15.3.

Correlators such as (395) correspond to amplitudes involving KK excited states in the spectrum of type IIB supergravity in $\mathrm{AdS}_5 \times S^5$. The $\mathcal{N}=4$ operators in 1/2 BPS multiplets with lowest component a scalar of the form (346) with $\ell > 2$ are dual to KK excited states. In particular, the fermionic operator $\tilde{\Lambda}_{\alpha}^{ABC}$ in (396) is dual to the first KK excitation of the dilatino. Therefore, the amplitude dual to the correlation function (395) is similar to the 16-point amplitude considered in the previous subsection, but with two of the dilatini in the first KK excited level.

The other class of non-minimal correlators presented in Sect. 15.3 comprises higher-point functions which have a natural interpretation as dual to amplitudes induced by vertices in the string effective action at order α'^5 and higher. However, in these cases the comparison is less straightforward and as will be shown shortly there are subtleties that need to be taken into account.

The amplitude dual to the 16-point correlator (395) involves the same interaction as in (439) and (440) with the only difference that upon reduction on the five-sphere one selects for two of the dilatini the first excited state instead of the KK ground state. The diagrammatic representation of the amplitude is given in Fig. 3, where the double lines indicate bulk-to-boundary propagators for the KK excited states. The dilatini in the first KK level are spin 1/2 fermions of mass $-\frac{5}{2L}$ for which the bulk-to-boundary propagator is

$$K^F_{9/2}(z,z_0;x) = K_5(z,z_0;x)\left[\sqrt{z_0}\,\gamma_5 - \frac{1}{\sqrt{z_0}}\,(x-z)_\mu\gamma^\mu\right]. \tag{446}$$

[40] The BPS Wilson loops mentioned in Sect. 15.3 provide another notable example in perturbation theory.

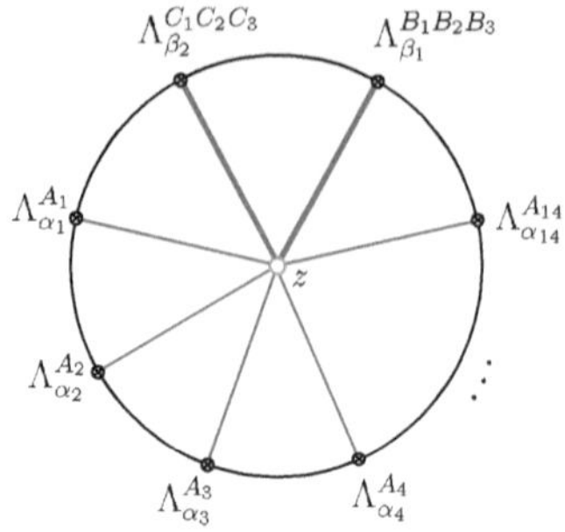

Fig. 3. Supergravity amplitude dual to the non-minimal 16-point function (395)

The resulting amplitude is

$$\frac{N_c^{1/2}}{g^{24}} \sum_{K>0} K^{\frac{25}{2}} \sum_{m|K} \frac{1}{m^2} \mathrm{e}^{2\pi i K\tau} \int \frac{\mathrm{d}^4 z\, \mathrm{d}z_0}{z_0^5} \mathrm{d}^5\omega\, t_{16}$$
$$\times \prod_{i=1}^{14} \mathcal{Y}_F^{(0)}(\omega)\, K^F_{7/2}(z, z_0; x_i) \prod_{j=1}^{2} \mathcal{Y}_F^{(1)}(\omega)\, K^F_{9/2}(z, z_0; y_i)\,. \qquad (447)$$

where $\mathcal{Y}_F^{(1)}(\omega)$ denotes the first excited fermionic spherical harmonic. The result (447) is again in agreement with the corresponding Yang–Mills calculation (398).

Non-minimal amplitudes of this type, involving KK excited states, are generalisations of the analogous minimal ones. Apart from the appearance of bulk-to-boundary propagators for fields of the appropriate mass, the only difference is in the five-sphere integrals, because of the presence of higher harmonics, which are necessary to reproduce the Ω-dependence of the corresponding Yang–Mills expressions.

The study of the other class of non-minimal amplitudes is more complicated and yields some surprises. In order to describe the main features of these amplitudes, we focus on the example of the process dual to the correlation function (385) involving 16 fermionic operators, Λ^A_α, in the **4** of SU(4) and four scalar operators, $\mathcal{E}^{AB}$, in the **10**. As already observed, it is natural to expect that in these cases the amplitudes should involve couplings of order α'^5 and beyond. The operator $\mathcal{E}^{AB}$ in (386) is dual to a linear combination of the NS–NS and R–R two forms with indices in the internal directions. The corresponding field strength, $G_{(3)}$, was defined in Sect. 18.1. Therefore a contact amplitude involving the vertex

$$\alpha' \int \mathrm{d}^{10}X\, \sqrt{-g}\, \mathrm{e}^{\phi/2} f_2^{(14,-14)}(\tau,\bar\tau)\, G^4\, \lambda^{16} \qquad (448)$$

represents an obvious candidate for the dual of the correlator (385). This process is represented in the second diagram of Fig. 4.

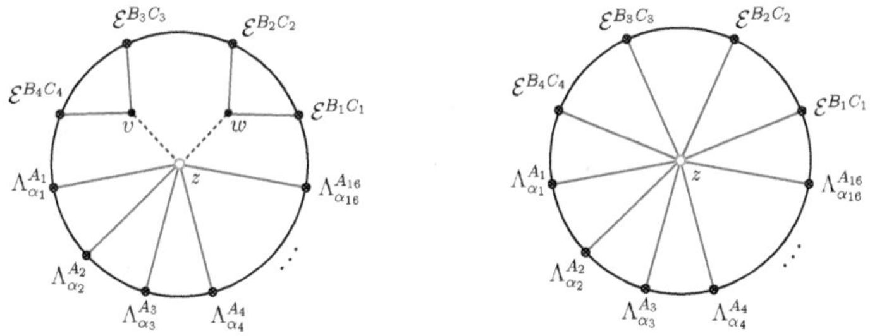

Fig. 4. Contributions to the 20-point amplitude dual to the correlation function (385)

This interpretation, however, leads to a puzzle. Using the dictionary (415) the 12-derivative couplings at order α'^5 give rise to contributions of order $N_c^{-1/2}$, in fact

$$\frac{\alpha' \, \mathrm{e}^{\phi/2}}{L} \sim N_c^{-1/2} \,, \tag{449}$$

which is not the behaviour expected from the Yang–Mills analysis. The leading contribution to the correlation function (385) is in fact of order $N_c^{1/2}$, as follows from (390) and (391).

The resolution of this mismatch requires the inclusion in the supergravity analysis of contributions of a type not encountered in the calculation of minimal amplitudes. These are exchange diagrams involving a D-instanton-induced vertex as well as additional perturbative couplings. The relevant D-instanton vertices are those of order $1/\alpha'$ in the expansion of the effective action, so that the resulting amplitudes give rise to contributions of order $N_c^{1/2}$, see (444). In order to generate contributions to non-minimal amplitudes one needs to include in the D-instanton vertices the fluctuations of the complex scalar, $\tau = \tau_0 + \hat{\tau}$, as described at the end of Sect. 18.1.

In the case of the amplitude under consideration one needs to consider the vertex coupling 16 dilatini with two additional insertions of $\hat{\tau}$ coming from the expansion of the exponential factor in the modular form $f_1^{(12,-12)}(\tau,\bar{\tau})$. The non-perturbative part of the effective vertex is

$$\frac{1}{\alpha'} \int \mathrm{d}^{10}X \, \sqrt{-g} \, \mathrm{e}^{-25\phi/2} \, \mathrm{e}^{2\pi i \tau} \, \hat{\tau}^2 \, t_{16} \, \lambda^{16} \,, \tag{450}$$

where only the $K = 1$ contribution relevant for the comparison with the one-instanton sector in $\mathcal{N} = 4$ SYM has been included. The amplitude contributing to the dual of the 20-point correlator (385) is depicted on the l.h.s. of Fig. 4. The two bulk-to-bulk lines joining the D-instanton vertex at point z to the points v and w are $\langle \hat{\tau} \hat{\bar{\tau}} \rangle$ propagators and the two cubic vertices are $\hat{\bar{\tau}} GG$ couplings from the classical type IIB action.

In evaluating this diagram, upon using dimensional reduction-like formulae as (438), one has to sum over all the contributions associated with the exchange of the KK excitations of the complex scalar. The coupling to the external three forms restricts this sum to the states allowed by the $SO(6)$ selection rules enforced by the integration over the five-sphere. In the present case there is only one allowed contribution, corresponding to the exchange of a complex scalar in the second KK excited level, i.e. a state in the representation $\mathbf{20}'$ of $SO(6) \sim SU(4)$.

At first sight the resulting amplitude does not resemble the $\mathcal{N} = 4$ SYM result: it is an exchange amplitude requiring integrations over three bulk points. However, because of the specific coupling involved the integrations over the positions of the two cubic couplings can be performed. This is because, after using expressions such as (438), one can integrate by parts the derivatives in each of the three-form field strengths onto the $\hat{\tau}$ scalar, and the cubic couplings schematically reduce to the form

$$(\partial^2 + m_{\tau_{\mathbf{20}'}})\hat{\tau}\, B_{ij}B^{ij}\,, \tag{451}$$

where B is the complex combination of the NS–NS and R–R two forms and the mass term comes from the derivatives in the S^5 directions. After the integration by parts one thus reconstructs the AdS$_5$ wave operator acting on the internal bulk-to-bulk propagators which then yield five-dimensional δ-functions. The integrations at the points v and w in Fig. 4 can thus be computed and the exchange diagram reduces to a contact contribution. Therefore, the net effect of the exchange diagram is to give rise to a new coupling in the AdS$_5$ effective action of the form

$$\frac{1}{g_s^2\alpha'}\int\frac{\mathrm{d}^4z\,\mathrm{d}z_0}{z_0^5}\,\mathrm{e}^{-25\phi/2}\,\mathrm{e}^{2\pi i\tau}\,t_{16}\,\lambda^{16}\,B^4\,, \tag{452}$$

where the factor of g_s^{-2} arises from the rescaling of the complex scalar, $\hat{\tau}$, needed to make its kinetic term canonically normalised.

The amplitude induced by this vertex (expressed in terms of SYM parameters) takes schematically the form

$$\frac{\sqrt{N_c}\,\mathrm{e}^{2\pi i\tau}}{g^{28}}\int\frac{\mathrm{d}^4z\,\mathrm{d}z_0}{z_0^5}\prod_{i=1}^{16}K^F_{7/2}(z,z_0;x_i)\prod_{j=1}^{4}K_3(z,z_0;y_j)\,, \tag{453}$$

which reproduces the leading large N_c term in the $\mathcal{N} = 4$ result (390) with the correct space–time dependence.

The amplitude involving the order α'^5 vertex $\lambda^{16}G^4$ in (448) gives rise to a contribution with the same space–time dependence, but of order $N_c^{-1/2}$. This sub-leading contribution is interpreted as corresponding to the $1/N_c$ correction in the SYM result (390).

The example of the above 20-point function illustrates some features common to many non-minimal amplitudes. In general, unlike in the minimal cases, the amplitudes dual to non-minimal $\mathcal{N} = 4$ correlation functions receive several contributions. Various effects such as those described in the previous example need to be taken into account to show agreement between the Yang–Mills and supergravity calculations. More details and other non-minimal examples are discussed in [123].

18.3 Beyond Supergravity: the BMN Limit

The AdS/CFT correspondence discussed in the previous sections is a very remarkable duality and the study of instanton effects has led to some of the most successful tests of its validity. In the formulation presented so far the duality has, however, some limitations. Because of our present limited understanding of the quantisation of string theory in non-trivial backgrounds such as $\mathrm{AdS}_5 \times S^5$, the study of the gravity side of the correspondence is restricted to the supergravity approximation. Moreover, even in this regime, the strong–weak coupling nature of the duality makes the direct comparison of the two sides problematic. In this section we briefly review a very interesting limit of the correspondence, the so-called BMN limit [143], which allows to overcome both the above limitations.[41] The idea is to consider string theory in a background obtained from $\mathrm{AdS}_5 \times S^5$ via a special procedure known as Penrose limit [146]. The result of the limit is a background with the geometry of a maximally supersymmetric gravitational plane wave [147]. Remarkably, despite the non-flatness of the metric and the presence of a R–R background, it is possible to quantise string theory in this geometry [148, 149]. In [143] it has been proposed that strings propagating in this particular plane-wave background are dual to a certain sector of $\mathcal{N}=4$ SYM. The latter, usually referred to as the BMN sector, comprises operators of large scaling dimension, Δ, and large charge, J, with respect to a $U(1)$ subgroup of the $SU(4)$ R-symmetry group. The possibility of quantising string theory in the plane-wave background has made the comparison between string and gauge theory possible beyond the supergravity approximation, albeit only in a specific sector of $\mathcal{N}=4$ SYM. Moreover, in this limit there exists a regime in which both sides of the correspondence are weakly coupled, so that the strong–weak coupling problem is also avoided.

At the heart of the correspondence proposed in [143] is a relation between the energy, E, of plane-wave string states and a combination of the scaling dimension and R-charge of the dual operators, which reads

$$\frac{1}{\mu} E = \Delta - J \,, \tag{454}$$

where the parameter μ is related to the value of the R–R self-dual five-form present in the background, see (459). The validity of (454) has been successfully tested in perturbation theory in a number of cases. Reviews of these results can be found in [150]. In this subsection we present a brief overview of the non-perturbative tests carried out in [151, 152, 153].

---

[41] Another limit that has attracted some attention is the highly "stringy" regime $\lambda \to 0$ where the theory exposes higher spin symmetry enhancement [144]. The bulk counterpart of the recombination of semi-short multiplets into long ones and the emergence of anomalous dimensions in the boundary theory is a pantagruelic Higgs mechanism termed *La Grande Bouffe* [145].

In order to take the Penrose limit that gives rise to the plane-wave background we start with the $\mathrm{AdS}_5 \times S^5$ metric written in global coordinates

$$\mathrm{d}s^2 = L^2 \left[-\cosh^2\rho\, \mathrm{d}t^2 + \mathrm{d}\rho^2 + \sinh^2\rho\, \mathrm{d}\Omega_3^2 \right.$$
$$\left. + \cos^2\theta\, \mathrm{d}\psi^2 + \mathrm{d}\theta^2 + \sin^2\theta\, \mathrm{d}\tilde{\Omega}_3^2 \right], \tag{455}$$

where Ω_3 and $\tilde{\Omega}_3$ refer to angles parametrising the two three spheres inside AdS_5 and S^5, respectively. In this coordinate system one can choose

$$\tilde{x}^{\pm} = \pm \frac{1}{\sqrt{2}}(t \pm \psi) \tag{456}$$

as light-cone variables and define the new coordinates

$$x^+ = \frac{1}{\mu}\tilde{x}^+, \qquad x^- = \mu L^2 \tilde{x}^-, \qquad \rho = \frac{r}{L}, \qquad \theta = \frac{y}{L}, \tag{457}$$

where μ is an arbitrary scale. The Penrose limit is obtained sending L to infinity while keeping $x^{\pm}, \rho$ and y "fixed". The resulting metric is that of a plane wave

$$\mathrm{d}s^2 = 2\mathrm{d}x^+\mathrm{d}x^- - \mu^2 x^I x^I (\mathrm{d}x^+)^2 + \mathrm{d}x^I \mathrm{d}x^I, \tag{458}$$

where x^I, $I = 1, \ldots, 8$, are Cartesian coordinates such that $x^I x^I = r^2 + y^2$. The original $\mathrm{AdS}_5 \times S^5$ background has also a non-zero self-dual R–R five-form (414), which after the limit has non-vanishing components

$$F_{+1234} = F_{+5678} = 2\mu, \tag{459}$$

with indices $1, 2, \ldots, 8$ corresponding to the x^I directions.[42]

The plane-wave background preserves the same (maximal) amount of supersymmetry as the original $\mathrm{AdS}_5 \times S^5$. In fact at the level of the super-isometries the Penrose limit corresponds to an Inönü–Wigner contraction.[43] The supergroup of isometries resulting from the contraction of $PSU(2,2|4)$ is $PSU(2|2) \times PSU(2|2) \times U(1) \times U(1)$ with maximal bosonic subgroup $H(4)^2 \rtimes SO(4) \times SO(4) \times U(1) \times U(1)$, where $H(4)$ denotes the four-dimensional Heisenberg group [147]. In the following we shall denote the two $SO(4)$ factors with $SO(4)_C$ and $SO(4)_R$, where the subscript refers to the fact that they are subgroups, respectively, of the conformal group and the R-symmetry group of the dual $\mathcal{N} = 4$ SYM theory. The states in the string spectrum can therefore be labelled by quantum numbers characterising their transformation under $SO(4)_C \times SO(4)_R \times U(1) \times U(1)$. These are identified with two pairs of spins, $(s_1, s_2; s_1', s_2')$, and the light-cone energy and momentum, (p_+, p_-), associated with translations in the x^+ and x^- directions.

---

[42] Notice that the metric (458) is $SO(8)$ invariant, but the background value of the five-form breaks this symmetry down to $SO(4) \times SO(4)$.

[43] More precisely this contraction should be called a Saletan contraction [154].

A very remarkable feature of the plane-wave background is that it allows the quantisation of string theory in the Green–Schwarz (GS) formalism.[44] As shown in [148], in the light-cone gauge the plane-wave GS string is described by a (massive) free world-sheet theory. This allows to carry out the quantisation essentially in same fashion as in flat space. The string action constructed in [148] is

$$S=\frac{1}{2\pi\alpha'}\int_{-\infty}^{+\infty}\mathrm{d}\tau\int_0^{2\pi p_-}\mathrm{d}\sigma\left[\frac{1}{2}\partial_+X^I\partial_-X^I-\frac{1}{2}m^2X^IX^I\right.$$
$$\left.+i\left(S^a\partial_+S^a+\tilde{S}^a\partial_-\tilde{S}^a-2mS^a\Pi_{ab}\tilde{S}^b\right)\right], \qquad (460)$$

where the mass parameter $m=\mu p_-\alpha'$ has been introduced. In (460) the X^I's denote the transverse coordinates of the string and the index $I=1,\dots,8$ is in the $\mathbf{8}_\mathrm{v}$ of $SO(8)$. The S^a's and $\tilde{S}^a$'s, $a=1,\dots,8$, are GS fermions. These are $SO(8)$ spinors of the same chirality in the $\mathbf{8}_\mathrm{s}$. The matrix Π is a product of $SO(8)$ γ-matrices, $\Pi=\gamma^1\gamma^2\gamma^3\gamma^4$. As usual in the light-cone gauge the non-physical components have been eliminated setting $X^+(\sigma,\tau)=2\pi\alpha' p_-\tau$, whereas $X^-(\sigma,\tau)$ is expressed in terms of the X^I's using the so-called Virasoro constraints which follow from consistency with the equation of motion for the world-sheet metric.

Since (460) describes a free theory the equations of motions lead to a standard mode expansion. For instance, for the transverse bosons one finds

$$X^I(\sigma,\tau)=\cos(m\tau)\,x_0^I+\frac{1}{m}\sin(m\tau)\,p_0^I$$
$$+i\sum_{n\neq 0}\frac{1}{\omega_n}\left(\mathrm{e}^{-i\omega_n\tau+2i\pi n\sigma}\alpha_n^I+\mathrm{e}^{-i\omega_n\tau-2i\pi n\sigma}\tilde{\alpha}_n^I\right), \qquad (461)$$

where

$$\omega_n=\mathrm{sign}(n)\sqrt{m^2+n^2}\,. \qquad (462)$$

The GS fermions have a similar mode expansion with coefficients which will be denoted by $S^a_{\pm n}$ and $\tilde{S}^a_{\pm n}$.

Upon quantisation, the coefficients in the expansion of the world-sheet fields give rise to creation and annihilation operators for the states in the string spectrum. In order to construct the physical creation and annihilation operators, the $SO(8)$ oscillators need to be decomposed under $SO(4)_C\times SO(4)_R$. Massive excitations of the string are associated with non-zero oscillators. The bosonic ones, α_n^I and $\tilde{\alpha}_n^I$, are in the $\mathbf{8}_\mathrm{v}$ which decomposes as $\mathbf{8}_\mathrm{v}\to(\mathbf{4};\mathbf{1})\oplus(\mathbf{1};\mathbf{4})$, and one obtains

[44] By "quantisation" we here mean the determination of the spectrum of states with $p_-\neq 0$, with $p_->0$ for incoming and $p_-<0$ for outgoing states. Interactions and the spectrum of states with $p_-=0$ are much subtler and not fully known even at tree level.

$$\alpha_n^I \to \alpha_n^i\,,\ \alpha_n^{\mu+5}\,, \qquad \tilde{\alpha}_n^I \to \tilde{\alpha}_n^i\,,\ \tilde{\alpha}_n^{\mu+5}\,, \qquad n \in \ ,\quad n \neq 0\,, \tag{463}$$

where $i = 1,2,3,4$ and $\mu = 0,1,2,3$ are vector indices in $SO(4)_R$ and $SO(4)_C$, respectively. The fermions are in the $\mathbf{8}_s$ which decomposes into $(\mathbf{2}_L;\mathbf{2}_L) \oplus (\mathbf{2}_R;\mathbf{2}_R)$. The fermionic oscillators are decomposed using the projectors $P_\pm = \frac{1}{2}(1 \pm \Pi)$

$$S_n^a \to S_n^\pm = P_\pm S_n^a\,, \qquad \tilde{S}_n^\pm = P_\pm \tilde{S}_n^a\,, \tag{464}$$

which yield spinors with $SO(4)_C \times SO(4)_R$ chiralities $(+,+)$ and $(-,-)$.

The zero modes in the expansion of the world-sheet fields are treated similarly. The bosonic ones, x_0^I and p_0^I, associated with the transverse position and momentum of the string, are combined into

$$a^I = \frac{1}{\sqrt{2|m|}}(p_0^I - i|m|x_0^I) \quad \text{and} \quad a^{I\dagger} = \frac{1}{\sqrt{2|m|}}(p_0^I + i|m|x_0^I)\,. \tag{465}$$

These are then decomposed as in (463). The fermion zero modes, S_0 and $\tilde{S}_0$, are combined into

$$\theta = \frac{1}{\sqrt{2}}(S_0 + i\tilde{S}_0) \quad \text{and} \quad \bar{\theta} = \frac{1}{\sqrt{2}}(S_0 - i\tilde{S}_0) \tag{466}$$

and then further decomposed as

$$\theta_{L,R} = P_{+,-}\theta\,, \qquad \bar{\theta}_{L,R} = P_{+,-}\bar{\theta}\,. \tag{467}$$

The Fock space of states is built on a vacuum, $|0\rangle_h = |0,p_-\rangle_h$, defined as the state annihilated by θ_R, $\bar{\theta}_L$, a^I and all the non-zero oscillators of positive frequency. This is a non-degenerate bosonic state of zero mass usually referred to as the BMN vacuum. The fermionic zero modes θ_L and $\bar{\theta}_R$ are creation operators and generate the supergravity multiplet acting on $|0\rangle_h$ [149]. The bosonic zero modes $a^{I\dagger}$ create the Kaluza–Klein-like excitations. The massive string modes are created by combinations of non-zero oscillators with negative frequencies, α_{-n}^i, $\tilde{\alpha}_{-n}^i$, $\alpha_{-n}^{\mu+5}$, $\tilde{\alpha}_{-n}^{\mu+5}$, $S_{-n}^\pm$ and $\tilde{S}_{-n}^\pm$, acting on the BMN vacuum. Physical states, $|s\rangle_{\text{phys}}$, are subject to the level-matching condition

$$\left(N - \tilde{N}\right)|s\rangle_{\text{phys}} = 0\,, \tag{468}$$

where the left and right moving number operators are defined by

$$\begin{aligned} N &= \sum_{n=1}^{\infty}\left(\frac{n}{\omega_n}\,\alpha_{-n}^I\alpha_n^I + n\,S_{-n}^a S_n^a\right) \\ \tilde{N} &= \sum_{n=1}^{\infty}\left(\frac{n}{\omega_n}\,\tilde{\alpha}_{-n}^I\tilde{\alpha}_n^I + n\,\tilde{S}_{-n}^a\tilde{S}_n^a\right)\,. \end{aligned} \tag{469}$$

The string theory Hamiltonian can be expressed in terms of the above oscillators as

$$
\begin{aligned}
2p_- H = & m\left(a^{I^\dagger} a^I + \theta^a_L \bar{\theta}^a_L + \bar{\theta}^a_R \theta^a_R\right) \\
& + \sum_{k=1}^{\infty} \left[\alpha^I_{-k}\alpha^I_k + \tilde{\alpha}^I_{-k}\tilde{\alpha}^I_k + \omega_k \left(S^a_{-k} S^a_k + \tilde{S}^a_{-k}\tilde{S}^a_k\right)\right] . \qquad (470)
\end{aligned}
$$

From the form of the Hamiltonian (470) it is straightforward to compute the free string spectrum. The mass of a generic massive string excitation is

$$
\frac{1}{\mu} M = \frac{1}{m} \sum_{n=1}^{\infty} \left(N + \tilde{N}\right) |\omega_n| , \qquad (471)
$$

with ω_n defined in (462) and $m = \mu p_- \alpha'$.

In the following our discussion will focus on massive string states, which are more interesting in the context of the duality with $\mathcal{N} = 4$ SYM since their masses receive quantum corrections. For simplicity, we shall restrict our attention to states created by the α^i_{-n} and $\tilde{\alpha}^i_{-n}$ oscillators

$$
|s\rangle = \alpha^{i_1}_{-n_1} \cdots \alpha^{i_r}_{-n_r} \cdots \tilde{\alpha}^{j_1}_{-m_1} \cdots \tilde{\alpha}^{j_s}_{-m_s} \cdots |0\rangle_h , \qquad (472)
$$

with $\sum_r n_r = \sum_s m_s$ to satisfy level matching.

As has been mentioned before, the states in the spectrum are characterised by the their $SO(4)_C \times SO(4)_R$ quantum numbers, besides their mass and light-cone momentum. As in the original formulation of the AdS/CFT correspondence, the quantum numbers associated with the symmetries on the two sides also dictate the map relating string states to composite operators in $\mathcal{N} = 4$ SYM. Equation (457) leads to the following identifications:

$$
\begin{aligned}
\frac{1}{\mu} H \equiv \frac{1}{\mu} p_+ &= -i\partial_+ = -i(\partial_t + \partial_\psi) \to \mathcal{D} - \mathcal{J} \\
p_- = -i\partial_- &= \frac{i}{2\mu L^2}(\partial_t - \partial_\psi) \to \frac{1}{2\mu L^2}(\mathcal{D} + \mathcal{J}) ,
\end{aligned} \qquad (473)
$$

where, as usual, $L^4 = 4\pi g_s N_c \alpha'^2$. Equations (473) relate the string light-cone energy and momentum to linear combinations of the dilation operator, $\mathcal{D}$, and the generator, $\mathcal{J}$, of the $U(1)$ subgroup of the $SU(4)$ R-symmetry singled out in taking the Penrose limit. From the above relations it follows that in the $L \to \infty$ limit string states with finite energy and light-cone momentum correspond to SYM operators with values of $\mathcal{D}$ and $\mathcal{J}$ satisfying

$$
\Delta \to \infty , \qquad J \to \infty , \qquad \Delta - J \quad \text{finite} . \qquad (474)
$$

Operators with these properties form the BMN sector of $\mathcal{N} = 4$ SYM. The explicit form of the operators dual to states in the plane-wave string spectrum was proposed in [143]. The starting point for the construction of such operators is the definition of the dual to the BMN vacuum, which is identified with the operator

$$\mathcal{O} = \frac{1}{\sqrt{J N_c^J}} \operatorname{Tr}\left(Z^J\right) , \tag{475}$$

where Z is the complex combination of the $\mathcal{N} = 4$ scalar fields with $J = 1$ for which we choose $Z = 2\varphi^{14}$ (see (337) and (338)). The operator (475) has $\Delta - J = 0$ as expected for the dual of a zero energy state. Operators corresponding to the other states in the string spectrum are obtained inserting in the trace in (475) "impurities", i.e. other elementary fields in the $\mathcal{N} = 4$ fundamental multiplet. The action of each creation operator on the string side corresponds to the insertion of an impurity of a certain type.[45] In particular, the operators dual to states of the form (472) are

$$\begin{aligned} \mathcal{O}^{i_1 \dots i_k}_{J; n_1 \dots n_k} = & \frac{1}{\sqrt{J^{k-1} \left(\frac{g^2 N_c}{8\pi^2}\right)^{J+k}}} \\ & \times \sum_{\substack{p_1, \dots, p_{k-1}=0 \\ p_1 + \dots + p_{k-1} \le J}}^{J} \mathrm{e}^{2\pi i [(n_1 + \dots + n_{k-1}) p_1 + (n_2 + \dots + n_{k-1}) p_2 + \dots + n_{k-1} p_{k-1}]/J} \\ & \times \operatorname{Tr}\left(Z^{J-(p_1+\dots+p_{k-1})} \varphi^{i_1} Z^{p_1} \varphi^{i_2} \dots Z^{p_{k-1}} \varphi^{i_k}\right) , \end{aligned} \tag{476}$$

where the integers n_i correspond to the mode numbers of the string state[46] and the action of the creation operators in (472) is in correspondence with the insertion of impurities, φ^i, for which we have the definitions

$$\begin{aligned} \varphi^1 &= \frac{1}{\sqrt{2}} \left(-\varphi^{13} + \varphi^{24}\right) , \quad \varphi^2 = \frac{1}{\sqrt{2}} \left(\varphi^{12} + \varphi^{34}\right) , \\ \varphi^3 &= \frac{i}{\sqrt{2}} \left(-\varphi^{13} - \varphi^{24}\right) , \quad \varphi^4 = \frac{i}{\sqrt{2}} \left(\varphi^{12} - \varphi^{34}\right) , \end{aligned} \tag{477}$$

The four real scalars, φ^i, transform in the $(\mathbf{1};\mathbf{4})$ of $SO(4)_C \times SO(4)_R$ and the map between the operators (476) and the string states (472) is determined by the $SO(4)_R$ quantum numbers.

To make the comparison between the string theory in the plane-wave background and the BMN sector of $\mathcal{N} = 4$ SYM possible at a quantitative level, one has to consider the large N_c limit. The combination of the large N_c limit with the limit of large Δ and J, implies that new effective parameters, λ' and g_2, arise [143, 155, 156], which are related to the ordinary 't Hooft parameters, λ and $1/N_c$, by a rescaling

$$\lambda' = \frac{g^2 N_c}{J^2} , \qquad g_2 = \frac{J^2}{N_c} . \tag{478}$$

[45] For this reason the string excitations are also often referred to as impurities.

[46] Conventionally, left-moving modes correspond to $n_i > 0$ and right-moving ones to $n_i < 0$.

These in turn are related to the parameters of the plane-wave string theory by

$$m^2 = (\mu p_- \alpha')^2 = \frac{1}{\lambda'}, \qquad 4\pi g_s m^2 = g_2 . \tag{479}$$

The double scaling limit defined by (474) and $N_c \to \infty$ with J^2/N_c fixed connects the weak coupling regime of the gauge theory to string theory at small g_s and large m. The property that in this limit physical quantities can be expanded in powers of the effective parameters, λ' and g_2, is referred to as BMN scaling.

Instanton Effects in the BMN Limit

Tests of the BMN limit of the AdS/CFT correspondence consist in verifying the validity of the relation

$$\frac{1}{\mu} H = \mathcal{D} - \mathcal{J} . \tag{480}$$

This is an operator relation and it requires that the eigenvalues of the two sides be equal, i.e. masses of states in the plane-wave string spectrum, rescaled by a factor of μ, should equal the combination $\Delta - J$ for the dual operators. In general the comparison requires the resolution of a mixing problem, i.e. the diagonalisation of the operators in (480).

The quantum corrections to the string mass spectrum are extracted from two-point amplitudes. At the perturbative level, calculations of such amplitudes have been performed using string field theory methods. The non-perturbative corrections that we are interested in are induced by two-point amplitudes in which the external states are coupled to D-instantons. These quantum corrections to the string masses should be compared to instanton corrections to the eigenvalues of the operator $\mathcal{D} - \mathcal{J}$, i.e. to the anomalous dimensions of BMN operators since the charge J is not renormalised.

The leading D-instanton contribution to a two-point amplitude is obtained coupling the external states to two disconnected disks, with Dirichlet boundary conditions, localised at the same space–time point. This is schematically depicted in Fig. 5.

The D-instanton in the plane-wave string theory can be described as a collective excitation of elementary closed string oscillators using the boundary state formalism [157]. The construction of the D-instanton boundary state in

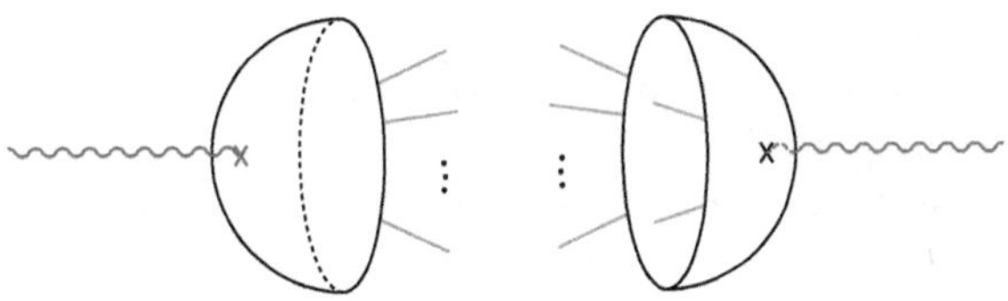

Fig. 5. Leading D-instanton contribution to a two-point scattering amplitude

the plane-wave background follows closely the approach used in [158] for the light-cone GS string theory in flat space. The boundary state describing a D-instanton with transverse position z^I will be denoted by $||z^I\rangle\rangle$. It is defined by the following gluing conditions:

$$\begin{aligned} &\left(\alpha_n^I - \tilde{\alpha}_{-n}^I\right) ||z^I\rangle\rangle = 0\,, \\ &\left(S_n^a + iM_n^{ab}\tilde{S}_{-n}^b\right) ||z^I\rangle\rangle = 0\,, \qquad n \in \quad , \end{aligned} \tag{481}$$

where the matrix M_n is

$$M_n = \frac{1}{n}\left(\omega_n \quad - m\Pi\right)\,, \tag{482}$$

with ω_n given in (462). The explicit expression for $||z^I\rangle\rangle$ is [157]

$$||z^I\rangle\rangle = (4\pi m)^2 \exp\left(\sum_{k=1}^{\infty} \frac{1}{\omega_k}\alpha_{-k}^I\tilde{\alpha}_{-k}^I - iS_{-k}^a M_k \tilde{S}_{-k}^a\right) ||z^I\rangle\rangle_0\,, \tag{483}$$

where $||z^I\rangle\rangle_0$ denotes the zero-mode part

$$||z^I\rangle\rangle_0 = \mathrm{e}^{-|m|(z^I)^2/2}\,\mathrm{e}^{i\sqrt{2|m|}z^I a^{I\dagger}}\,\mathrm{e}^{\frac{1}{2}a^{I\dagger}a^{I\dagger}}|0\rangle_D \tag{484}$$

and $|0\rangle_D = \theta_L^1\theta_L^2\theta_L^3\theta_L^4|0\rangle_h$.

The plane-wave background is maximally supersymmetric, i.e. it is invariant under 32 supersymmetries. These are divided into 16 kinematical supersymmetries, which do not commute with the string Hamiltonian, and 16 dynamical ones, which commute with the Hamiltonian. The boundary state (483) and (484) is annihilated by eight kinematical and eight dynamical supersymmetries as a consequence of the conditions (481). The other half of the supersymmetries acting on $||z^I\rangle\rangle$ generate the fermion zero modes of the D-instanton. These are represented by the open strings attached to the boundary of the disks in Fig. 5. We shall denote the combinations of kinematical and dynamical supersymmetries which act non-trivially on the boundary state by $(\bar{q}_L, \bar{q}_R)$ and Q^-, respectively. The bosonic collective coordinates of the D-instanton correspond to its position in the 10-dimensional plane-wave geometry.

In order to compute a two-point amplitude of the type represented in Fig. 5, one needs to construct a state that includes the full dependence on the collective coordinates of the D-instanton and couples to two external states. We shall refer to such a state as a "dressed two-boundary state". The latter describes the two disks in Fig. 5 and is obtained considering the product of two boundary states associated with distinct Fock spaces but located at the same position, z^I. The "dressing" corresponding to the inclusion of the dependence on the bosonic and fermionic collective coordinates is achieved acting with the bosonic and fermionic generators of the broken symmetries. Denoting the two Fock spaces with indices 1 and 2, the dressed two-boundary state can be written as

$$
\begin{aligned}
||V_2; \boldsymbol{z}, \eta, \epsilon\rangle\rangle &= \mathrm{e}^{iz^+(p_{1+}+p_{2+})}\mathrm{e}^{iz^-(p_{1-}+p_{2-})} \\
&\times \left[\eta\left(Q_1^- + Q_2^-\right)\right]^8 \left[\epsilon_L\left(q_{1L}+q_{2L}\right)\right]^4 \left[\epsilon_L\left(q_{1L}+q_{2L}\right)\right]^4 ||z^I\rangle\rangle_1 \otimes ||z^I\rangle\rangle_2 \,,
\end{aligned} \tag{485}
$$

where $\boldsymbol{z} = (z^I, z^+, z^-)$ is the 10-dimensional location of the D-instanton and the $SO(8)$ spinors η and $\epsilon = (\epsilon_L, \epsilon_R)$ denote the fermionic collective coordinates associated with the dynamical and kinematical supersymmetries broken by the D-instanton.

The two-point amplitude that we are interested in is obtained coupling the dressed two-boundary state to a pair of external physical states and integrating over the bosonic and fermionic collective coordinates of the D-instanton. Denoting the incoming and outgoing states by $|s_1\rangle$ and $|s_2\rangle$, the one-particle irreducible part of the amplitude is

$$
\mathcal{A} = c\, g_{\mathrm{s}}^{7/2}\, \mathrm{e}^{2\pi i \tau} \int \mathrm{d}^8 z\, \mathrm{d}z^+\, \mathrm{d}z^-\, \mathrm{d}^8\eta\, \mathrm{d}^8\epsilon \left(\langle s_1| \otimes \langle s_2|\right) ||V_2; \boldsymbol{z}, \eta, \epsilon\rangle\rangle \,, \tag{486}
$$

where c is a numerical constant which has not been explicitly computed and the measure factor, $g_{\mathrm{s}}^{7/2}\, \mathrm{e}^{2\pi i \tau}$, follows from the comparison with the D-instanton induced contributions to the low-energy effective action.

As an example of amplitude of the type (486) we consider the leading D-instanton contribution to the case in which $|s_1\rangle$ and $|s_2\rangle$ are particular states in the class (472). Specifically, we consider $SO(4)_R$ singlet states with four impurities

$$
\begin{aligned}
\langle s_1| \otimes \langle s_2| &= \frac{1}{\omega_{n_1}\omega_{n_2}\omega_{m_1}\omega_{m_2}}\, \varepsilon_{ijkl}\, \varepsilon_{i'j'k'l'} \\
&\times {}_h\langle 0|\alpha_{n_1}^{(1)i}\alpha_{n_2}^{(1)j}\tilde{\alpha}_{n_1}^{(1)k}\tilde{\alpha}_{n_2}^{(1)l} \otimes {}_h\langle 0|\alpha_{m_1}^{(2)i'}\alpha_{m_2}^{(2)j'}\tilde{\alpha}_{m_1}^{(2)k'}\tilde{\alpha}_{m_2}^{(2)l'} \,,
\end{aligned} \tag{487}
$$

where the prefactors ensure that the states are normalised to one. A generic four impurity state has three independent mode numbers, after imposing the level matching condition. In (487) we have made a special choice: each of the states contains two left- and two right-moving excitations with pairwise equal mode numbers. This is because only states of this type couple to a D-instanton at leading order in g_{s}. This is a general property of D-instanton-induced amplitudes: because of the way in which the creation operators enter in (483) and (484), the boundary state couples only to states with the same number of left- and two right-moving oscillators with pairwise matched mode numbers.

To proceed with the calculation of the leading D-instanton contribution we insert (487) into (486). The general strategy for the calculation of such amplitudes consists in expanding the $||z^I\rangle\rangle$ factors in the dressed boundary state in a power series retaining only the terms which do not annihilate the product $\langle s_1| \otimes \langle s_2|$ on the left. The integration over the collective coordinates

z^+ and z^- imposes conservation of light-cone energy and momentum. Because of the non-linear dispersion relation (462) energy conservation requires that the mode numbers in the incoming and outgoing states be equal. Of the remaining integrations, those over the eight transverse z^I's and over the eight ϵ fermionic moduli are trivial in the case of external states of the type we are considering.[47] The non-trivial part of the calculation is the integration over the eight fermion moduli η

$$\langle s_1| \otimes \langle s_2| \int \mathrm{d}^8\eta \, \left[\eta(Q_1^- + Q_2^-)\right]^8 \exp\left[\sum_n \frac{1}{\omega_n}(\alpha_{-n}^{(1)}\tilde{\alpha}_{-n}^{(1)} + \alpha_{-n}^{(2)}\tilde{\alpha}_{-n}^{(2)}) - i(S_{-n}^{(1)}M_n\tilde{S}_{-n}^{(1)} + S_{-n}^{(2)}M_n\tilde{S}_{-n}^{(2)})\right] |0\rangle_1 \otimes |0\rangle_2 \,. \tag{488}$$

These integrations induce a coupling between the two disks, since the dynamical supercharges, Q^-, which couple to the η's depend non-trivially on the non-zero string oscillators. The calculation is greatly simplified when one considers the large m limit relevant for the comparison with $\mathcal{N}=4$ SYM at weak coupling. Since in this limit $M_n \sim m$, the dominant contribution to the amplitude with external states (487) is obtained retaining in the expansion of the boundary state two $SM\tilde{S}$ factors on each disk and distributing the eight Q^-'s evenly on the two disks. After some lengthy but straightforward algebra one obtains

$$\mathcal{A}(n_1,n_2) = \varepsilon_{ijkl}\,\varepsilon_{i'j'k'l'}\,\mathrm{e}^{2\pi i\tau}\,g_s^{7/2}\,m^8\,\frac{1}{n_1^2 n_2^2}\,I_\eta^{ijkl,i'j'k'l'} \tag{489}$$

where

$$\begin{aligned} I_\eta^{ijkl,i'j'k'l'} &= \int \mathrm{d}^8\eta\,\eta^+\gamma^{ij}\eta^+\eta^+\gamma^{kl}\eta^+\eta^-\gamma^{i'j'}\eta^-\eta^-\gamma^{k'l'}\eta^- \\ &= (\varepsilon^{ijkl} + \delta^{ik}\delta^{jl} - \delta^{il}\delta^{jk})(\varepsilon^{i'j'k'l'} - \delta^{i'k'}\delta^{j'l'} + \delta^{i'l'}\delta^{j'k'})\,. \end{aligned} \tag{490}$$

In the case of the amplitude (489) the only contribution comes from the term containing the product of two ε-tensors in (490) and one gets

$$\mathcal{A}(n_1,n_2) = 576\,\mathrm{e}^{2\pi i\tau}\,g_s^{7/2}\,m^8\,\frac{1}{n_1^2 n_2^2}\,. \tag{491}$$

The same integral (490) arises in the calculation of two-point amplitudes between other $SO(4)_C \times SO(4)_R$ singlet four-impurity operators and in these cases the terms involving Kronecker δ's in $I_\eta^{ijkl,i'j'k'l'}$ can contribute.

The result (489) is the leading non-perturbative correction to the one particle irreducible part of the two-point amplitude and thus it yields the D-instanton correction to the mass matrix for states of the form (487)

[47] They give rise to δ-functions that in the present case simply integrate to one.

$$\frac{1}{\mu}\,\delta M \sim \frac{\mathrm{e}^{2\pi i\tau}\, g_s^{7/2}\, m^7}{(n_1 n_2)^2} = \frac{\mathrm{e}^{-\frac{8\pi^2}{g_2\lambda'}+i\vartheta}\, g_2^{7/2}}{(n_1 n_2)^2}\,. \tag{492}$$

The SYM operators dual to the states in (487) are a special case of (476), i.e. four impurity $SO(4)_C \times SO(4)_R$ singlets. They are given by

$$\mathcal{O}_{n_1,n_2,n_3} = \frac{\varepsilon_{ijkl}}{\sqrt{J^3 (g^2 N_c)^{J+4}}} \sum_{\substack{p,q,r=0 \\ p+q+r\leq J}}^{J} \mathrm{e}^{2\pi i[(n_1+n_2+n_3)p+(n_2+n_3)q+n_3 r]/J} \times \mathrm{Tr}\left[Z^{J-(p+q+r)}\varphi^i Z^p \varphi^j Z^q \varphi^k Z^r \varphi^l\right]. \tag{493}$$

In order to compute the one instanton contribution to the matrix of anomalous dimensions for such operators, one considers the two-point function

$$G(x_1,x_2) = \langle \mathcal{O}_{n_1 n_2 n_3}(x_1)\, \bar{\mathcal{O}}_{m_1 m_2 m_3}(x_2)\rangle\,. \tag{494}$$

The calculation proceeds as in the case of the correlators discussed in Sect. 15. In the semi-classical approximation one needs to compute the classical profiles of the operators $\mathcal{O}_{n_1 n_2 n_3}$ and $\bar{\mathcal{O}}_{m_1 m_2 m_3}$ and integrate them over the instanton moduli space. The profiles of the operator (493) and of its conjugate contain $2J+8$ fermion zero modes each and thus (494) is non-minimal according to the terminology introduced in Sect. 15.

Although the calculation of the two-point function (494) presents no new conceptual difficulties, it involves rather complicated combinatorics associated with the distribution of the exact and non-exact fermion zero modes in the two operators. Each of the two operators should soak up eight of the 16 superconformal modes in the combination $(\zeta^1)^2(\zeta^2)^2(\zeta^3)^2(\zeta^4)^2$, while the remaining modes are of type ν^A and $\bar{\nu}^A$. Expanding the trace in (493) and in the conjugate operator one obtains a large number of terms satisfying this requirement. The double limit $N_c \to \infty$, $J \to \infty$, with J^2/N_c fixed, simplifies somewhat the analysis. The dominant contributions in this limit come from certain specific distributions of the fermion modes. The large N_c limit requires that all the $\bar{\nu}^A \nu^B$ bilinears be in the **6**, see (394). Moreover at large J the leading contributions to the operator profiles come from terms in which as many of the superconformal modes as possible are provided by the Z's and $\bar{Z}$'s rather than by the impurities. This is because one gets roughly a multiplicity factor of J associated with every Z or $\bar{Z}$ providing one such mode. Taking into account these simplifications the calculation of the profiles of $\mathcal{O}_{n_1 n_2 n_3}$ and $\bar{\mathcal{O}}_{m_1 m_2 m_3}$, albeit rather tedious, is feasible. Eventually, the dependence on the collective coordinates in all the relevant terms in the profile of the operator $\mathcal{O}_{n_1 n_2 n_3}$ reduces to

$$\frac{\rho^8}{[(x_1-x_0)^2+\rho^2]^{J+8}} \left(\bar{\nu}^{[1}\nu^{4]}\right)^J \left[\left(\zeta^1\right)^2\left(\zeta^2\right)^2\left(\zeta^3\right)^2\left(\zeta^4\right)^2\right](x_1)\,. \tag{495}$$

Similarly, all the terms in the classical profile of $\bar{\mathcal{O}}_{m_1m_2m_3}$, which contribute in the BMN limit contain the following factor:

$$\frac{\rho^8}{[(x_2-x_0)^2+\rho^2]^{J+8}}\left(\bar{\nu}^{[2}\nu^{3]}\right)^J\left[\left(\zeta^1\right)^2\left(\zeta^2\right)^2\left(\zeta^3\right)^2\left(\zeta^4\right)^2\right](x_2)\,. \tag{496}$$

After factoring out the dependence on the collective coordinates the dependence on the mode numbers, n_i and m_i, is determined by sums of the form

$$\begin{aligned}K(n_1,n_2,n_3;J) = &\sum_{\substack{p,q,r=0\\ p+q+r\leq J}}^{J} \mathrm{e}^{2\pi i[(n_1+n_2+n_3)p+(n_2+n_3)q+n_3r]/J} \\ &\times\left[\frac{c_1}{4!}p(p-1)(p-2)(p-3)+\frac{c_2}{3!}qp(p-1)(p-2)+\cdots\right],\end{aligned} \tag{497}$$

where each term contains combinatorial factors and $c_1, c_2, \ldots$ are numerical coefficients.

The two-point function (494) is thus

$$\begin{aligned}&\langle\mathcal{O}(x_1)\,\bar{\mathcal{O}}(x_2)\rangle_{\text{inst}} \\ &= c(g,N_c,J)\int\frac{\mathrm{d}^4x_0\,\mathrm{d}\rho}{\rho^5}\frac{\rho^{2J+16}}{[(x_1-x_0)^2+\rho^2]^{J+8}[(x_2-x_0)^2+\rho^2]^{J+8}} \\ &\times\int\mathrm{d}^8\eta\,\mathrm{d}^8\bar{\xi}\prod_{A=1}^{4}\left[\zeta^A(x_1)\right]^2\left[\zeta^A(x_2)\right]^2 \\ &\times\int\mathrm{d}^5\Omega\,\left(\Omega^{14}\right)^J\left(\Omega^{23}\right)^J\left[K(n_1,n_2,n_3;J)\,K(m_1,m_2,m_3,J)\right]\end{aligned} \tag{498}$$

where $c(g,N_c,J)$ contains the dependence on the parameters arising from the normalisation of the operators and the moduli space integration measure, as well as the factors of $g\sqrt{N_c}$ obtained rewriting the $(\bar{\nu}^A\nu^B)_{\mathbf{6}}$ bilinears in terms of the angular variables Ω^{AB}. In the large J limit the sums in (497) can be approximated with integrals. For instance, the first term becomes

$$\begin{aligned}&\sum_{p,q,r=0}^{J}\mathrm{e}^{2\pi i[(n_1+n_2+n_3)p+(n_2+n_3)q+n_3r]/J}p(p-1)(p-2)(p-3) \\ &\to J^7\int_0^1\mathrm{d}x\int_0^{1-x}\mathrm{d}y\int_0^{1-x-y}\mathrm{d}z\,\mathrm{e}^{2\pi i[(n_1+n_2+n_3)x+(n_2+n_3)y+n_3z]}\,x^4\end{aligned} \tag{499}$$

From (498) and (499), recalling the analysis in Sect. 15, one can deduce the dependence of the two-point function on the parameters. There are numerous sources of powers of g, N_c and J in the calculation, but remarkably the final result can be expressed only in terms of the parameters g_2 and λ', as required by BMN scaling. In detail one gets

$$\underbrace{\left(\frac{1}{\sqrt{J^3(g^2N_c)^{J+4}}}\right)^2}_{\text{norm. operators}} \times \underbrace{e^{2\pi i\tau} g^8\sqrt{N_c}}_{\text{measure}} \times \underbrace{\frac{(g\sqrt{N_c})^{2J}}{J^2}}_{\substack{\nu,\bar\nu \\ \text{integrals}}} \times \underbrace{\frac{1}{J^2}}_{\substack{x_0,\rho \\ \text{integrals}}} \times \underbrace{(J^7)^2}_{\text{sums}}$$

$$\sim \frac{J^7}{N_c^{7/2}}\, e^{2\pi i\tau} = g_2^{7/2}\, e^{-\frac{8\pi^2}{g_2\lambda'}+i\vartheta}, \tag{500}$$

which is in agreement with the λ' and g_2 dependence of the string theory result (489).

The simple mode number dependence of the string two-point amplitude is more complicated to reproduce. In the SYM two-point function the dependence on the integers n_i and m_i is contained in the functions $K(n_1,n_2,n_3;J)$ and $K(m_1,m_2,m_3,J)$ defined in (497). Each term in these sums receives a large number of contributions resulting in very complicated expressions. However, combining all the contributions leads to impressive cancellations and a very simple result. In conclusion, the one-instanton contribution to the two-point function (494) can be written in the form

$$G(x_1,x_2) = \frac{3^2\,(g_2)^{7/2}\, e^{-\frac{8\pi^2}{g_2\lambda'}+i\vartheta}}{2^{41}\,\pi^{13/2}} \frac{1}{(n_1n_2m_1m_2)}\frac{1}{(x_{12}^2)^{J+4}} \log\left[\Lambda^2 x_{12}^2\right], \tag{501}$$

where Λ is a scale that appears as a consequence of the logarithmic divergence in the x_0 and ρ integrals, which signals a contribution to the matrix of anomalous dimensions. Notably the result is only non-zero if the mode numbers in the two operators are equal in pairs, again in agreement with string theory.

From the coefficient of (501), one can read off the contribution to the matrix of anomalous dimensions. The above calculation is not sufficient to determine the actual anomalous dimension of the operator (493) since this requires the diagonalisation of the matrix of two-point functions of all the operators with the same quantum numbers. However, all such two-point functions are expected to have the same dependence on λ' and g_2 found in (501). Therefore, one can conclude that the behaviour of the leading instanton contribution to the anomalous dimension of four impurity $SO(4)_C \times SO(4)_R$ singlet operators is

$$\gamma_{\text{inst}} \sim \frac{g_2^{7/2}\, e^{-\frac{8\pi^2}{g_2\lambda'}+i\vartheta}}{(n_1n_2)^2}, \tag{502}$$

in agreement with (492). In view of the complexity of the calculation, this result provides a striking test of the BMN proposal.

A number of other two-point string amplitudes and their dual correlation functions have been studied in [151, 152, 153]. The many interesting results obtained in these papers can be summarised in the following statements.

- Four impurity operators in other representations of $SO(4)_R$ and the corresponding string states have two-point functions which behave as

$(\lambda')^2 (g_2)^{7/2} \exp(-8\pi^2/g_2\lambda' + i\vartheta)$, i.e. they are suppressed by two powers of λ' with respect to those in the singlet sector.

- Two impurity operators have the same suppression. The calculation of instanton contributions to two-point functions of two impurity operators in $\mathcal{N} = 4$ SYM is rather subtle because in order to saturate the integrations over the superconformal modes one needs to use the classical solution for the scalar fields involving six fermion modes, $\varphi^{(6)\,AB}$.
- Supergravity states and their KK excitations do not couple to the D-instanton boundary state and thus, as expected, their masses do not receive non-perturbative corrections. This result is far from obvious in the gauge theory and requires non-trivial cancellations which have not been explicitly verified.
- (D-)Instantons contribute to the mixing of states in the NS–NS and R–R sectors of the plane string theory.[48]
- Instanton contributions to two-point functions of certain operators dual to R–R string states, i.e. operators with an even number of fermionic impurities, involve inverse powers of λ'. Although this behaviour is rather surprising, it is not pathological in the $\lambda' \to 0$ limit because the inverse powers of λ' are accompanied by the instanton weight $\exp(-8\pi^2/\lambda' g_2)$. These two-point functions vanish in perturbation theory.

It is notable that many of these results can be straightforwardly obtained in string theory where they are easily deduced from properties of the D-instanton boundary state, whereas they are much more complicated to obtain from a field theoretical calculation in $\mathcal{N} = 4$ SYM.

19 Conclusions

We would like to conclude this long review by highlighting the many topics where Gabriele's contributions along the years have been at the heart of the theoretical developments that have made our understanding of non-perturbative effects of field theory so deep and powerful.

Conceptually, perhaps the most important contributions in this direction have been his works on the foundation of the notion of effective action in a supersymmetric framework. The effective action for the $\mathcal{N} = 1$ SYM theory [29] and its extension to SQCD [30] are milestones along the way of dealing with the non-perturbative structure of field theory. These works appear as an immediate extension and generalisation of the approach established for the description of the low-energy degrees of freedom of QCD [27, 28], as soon as the fundamental rôle of anomalies was recognized [57, 159, 160]. The validation of the famous Witten–Veneziano formula [161] for the η' mass, yielded by lattice simulations [162], and the explicit instanton calculations, carried out in

[48] Unlike in flat space, in the plane-wave background this mixing occurs also in perturbation theory beyond tree-level [153].

various instances in supersymmetric theories [4], have beautifully confirmed the predictive power of the effective action approach both in a supersymmetric and in a non-supersymmetric context.

Together with many other important, independently derived, results [21], these ideas have proved to be of enormous impact on the way we think today of possible extensions of the Standard Model.

We cannot end this review without mentioning what we consider the most important step of modern physics beyond field theory, namely the construction of the dual Veneziano amplitude [109], which is expressed by the remarkably simple formula

$$A(s,t) = \int_0^1 \mathrm{d}x\, x^{-\alpha' s-1}(1-x)^{-\alpha' t-1} \,. \tag{503}$$

It is unanimously recognized that (503) represents the founding paper of String Theory. It took some time to realise that the infinite tower of "resonances" exchanged in the s and t channel are the excitations of an open bosonic string living in 26 dimensions. Planar duality, $A(s,t) = A(t,s)$, and the UV softness of the amplitude are exposed quite neatly by its geometric interpretation in terms of vertex operators inserted on the boundary of a disk. The presence of a massless vector excitation has brought String Theory to be the most credited candidate for the unification of all interactions, including gravity. In this respect, the graviton comes in as the massless excitation of the closed string spectrum and its vertex operator is a sort of "square" of the vertex operator for the massless vector of the Veneziano amplitude. That open strings might be considered more fundamental than closed strings is something which seems to emerge in all modern approaches, where D-branes and their open string excitations are used to describe interactions mediated by gauge bosons. We want also to recall that in a somewhat more distant context string excitations have been shown to be able to account for the microscopic degrees of freedom of black holes, thus yielding what is considered today the only satisfactory solution to the holographic puzzle of black hole thermodynamics [163].

In the present review we have briefly sketched the enormous simplification that open strings bring into the ADHM construction of instantons. However, for lack of space we had no chance to stress the far-reaching consequences of ideas underlying the Veneziano amplitude in the quest for unification and in the process of clarification of the many puzzles of quantum gravity. We dare to conclude by saying that we expect the Veneziano amplitude to be among the basic blocks of any consistent formulation of the fundamental laws of Nature.

Acknowledgements

Discussions with Massimo Testa, Yassen Stanev and especially Michael Green are gratefully acknowledged. The work of S.K. was supported in part by a

Marie Curie Intra-European Fellowship and by the EU-RTN network *Constituents, Fundamental Forces and Symmetries of the Universe* (MRTN-CT-2004-005104).

Appendix A – Notations

A.1 Generalities

We work in Euclidean metric with $g^E_{\mu\nu} = \delta_{\mu\nu}$. Factors of the gauge coupling constant, g, will be explicit everywhere. We are interested in computing expectation values of gauge-invariant (possibly multi-local) renormalisable, composite operators, i.e. functional integrals of the type

$$\langle O\rangle = \frac{1}{Z}\int \mathcal{D}\mu(\psi,\bar{\psi})\mathcal{D}A_\mu \exp\left[-S_{\mathrm{YM}} + \int \mathrm{d}^4x\bar{\psi}(\not{D}+m)\psi\right]O[\psi,\bar{\psi},A_\mu], \quad \text{(A.1)}$$

where $\not{D}$ can be either a Dirac or a Weyl–Dirac operator (see below) and Z is a similar functional integral with O replaced by the identity operator.

A.2 Yang–Mills Action

The pure Yang–Mills action has the form

$$S_{\mathrm{YM}} = \frac{1}{2}\int \mathrm{d}^4x\,\mathrm{Tr}[F_{\mu\nu}F_{\mu\nu}] = \frac{1}{4}\int \mathrm{d}^4x \sum_a F^a_{\mu\nu}F^a_{\mu\nu}, \quad \text{(A.2)}$$

$$F_{\mu\nu} = T^a F^a_{\mu\nu}, \qquad F_{\mu\nu} = \partial_\mu A_\nu - \partial_\nu A_\mu + ig[A_\mu, A_\nu]. \quad \text{(A.3)}$$

A.3 Some Group Theory Formulae

In (A.2) the matrices $T^a \equiv T^a_{\mathbf{N_c}}$, $a = 1, 2, \ldots, N_c^2 - 1$ are the $SU(N_c)$ generators in the fundamental representation, $\mathbf{N_c}$. In general, the generators, $T_{\mathbf{R}}$, in the (irreducible) representation $\mathbf{R}$ are normalised according to the formula

$$\mathrm{Tr}\left[T^a_{\mathbf{R}}T^b_{\mathbf{R}}\right] = \ell[\mathbf{R}]\delta^{ab}, \quad \text{(A.4)}$$

with $\ell[\mathbf{R}]$ the Dynkin index of the representation. It is customary to normalise generators in the $\mathbf{N_c}$, $\bar{\mathbf{N}}_{\mathbf{c}}$ and **Adj** representations so that

$$\ell[\mathbf{N_c}] = \ell[\bar{\mathbf{N}}_{\mathbf{c}}] = \frac{1}{2}, \qquad \ell[\mathbf{Adj}] = N_c. \quad \text{(A.5)}$$

Taking the trace of the equation which defines the quadratic Casimir operator of the representation $\mathbf{R}$

$$\sum_a T^a_{\mathbf{R}}T^a_{\mathbf{R}} = c_2[\mathbf{R}]\ \mathbb{1}_{\dim(\mathbf{R})\times\dim(\mathbf{R})}, \quad \text{(A.6)}$$

and using (A.4), one gets the useful relation

$$c_2[\mathbf{R}]\dim(\mathbf{R}) = \ell[\mathbf{R}]\dim(G). \quad \text{(A.7)}$$

A.4 Dirac Fermions

The Euclidean action of a Dirac fermion, ψ, $\bar{\psi}$, in the representation $\mathbf{R}$ of the gauge group $SU(N_c)$ takes the form

$$S_{\rm DF} = \int {\rm d}^4x\, \bar{\psi}_r D^r_{\mu s}[\mathbf{R}]\gamma_\mu \psi^s\,, \quad r,s = 1,\dots,\dim(\mathbf{R})\,, \tag{A.8}$$

where Dirac indices are understood and

$$D^r_{\mu s}[\mathbf{R}] = \partial_\mu \delta^r_s - g(T^a_{\mathbf{R}})^r_s A^a_\mu\,. \tag{A.9}$$

In eq. (A.8) hermitean γ-matrices are used, satisfying the anti-commutation relations $\{\gamma_\mu, \gamma_\nu\} = 2\delta_{\mu\nu}$.

A.5 Weyl Fermions

The Euclidean action of a Weyl fermion, $\lambda^a_\alpha, \bar{\lambda}^a_{\dot{\alpha}}$ $(\alpha, \dot{\alpha} = 1,2)$ belonging to the adjoint representation of the gauge group $SU(N_c)$ takes the form

$$S_{\rm WF} = \int {\rm d}^4x\, \bar{\lambda}^a_{\dot{\alpha}} D^{ab}_\mu[\mathbf{Adj}]\bar{\sigma}^{\dot{\alpha}\alpha}_\mu \lambda^b_\alpha\,, \tag{A.10}$$

where

$$D^{ab}_\mu[\mathbf{Adj}] = \partial_\mu \delta^{ab} - g f^{abc} A^c_\mu\,, \tag{A.11}$$

$$\bar{\sigma}_\mu = (\ , -i\sigma_k)\,, \tag{A.12}$$

with σ_k the Pauli matrices. It is also useful to introduce the matrices $(\sigma_\mu)_{\alpha\dot{\alpha}}$

$$\sigma_\mu = (\ , i\sigma_k)\,. \tag{A.13}$$

and the definitions

$$(\sigma_{\mu\nu})^\beta_\alpha = \frac{1}{2}(\sigma_{\mu\alpha\dot{\alpha}}\bar{\sigma}^{\dot{\alpha}\beta}_\nu - \sigma_{\nu\alpha\dot{\alpha}}\bar{\sigma}^{\dot{\alpha}\beta}_\mu)\,, \tag{A.14}$$

$$(\bar{\sigma}_{\mu\nu})^{\dot{\beta}}_{\dot{\alpha}} = \frac{1}{2}(\bar{\sigma}^{\dot{\beta}\alpha}_\mu \sigma_{\nu\alpha\dot{\alpha}} - \bar{\sigma}^{\dot{\beta}\alpha}_\nu \sigma_{\mu\alpha\dot{\alpha}})\,. \tag{A.15}$$

A.6 The SYM Action

The Euclidean action of the minimal $\mathcal{N} = 1$ supersymmetric gauge theory (super Yang–Mills, SYM), $S_{\rm SYM}$, when written in components, is simply given by the sum of (A.2) and (A.10). The classical action is invariant under the $U_\lambda(1)$ R-symmetry [164]

$$\lambda \to {\rm e}^{i\alpha}\lambda\,, \qquad A_\mu \text{ untouched}\,. \tag{A.16}$$

Quantum–mechanically this symmetry is anomalous with

$$\partial_\mu J^{(\lambda)}_\mu = 2i\ell[\mathbf{Adj}]\frac{g^2}{32\pi^2}F^a_{\mu\nu}\tilde{F}^a_{\mu\nu} = 2iN_c\frac{g^2}{32\pi^2}F^a_{\mu\nu}\tilde{F}^a_{\mu\nu}\,, \tag{A.17}$$

$$J^{(\lambda)}_\mu = \bar{\lambda}^a_{\dot{\alpha}}\bar{\sigma}^{\dot{\alpha}\alpha}_\mu \lambda^a_\alpha\,, \tag{A.18}$$

but has a non-anomalous discrete subgroup

$$ _{2N_c} = \{z_k = \mathrm{e}^{i\alpha_k}\,, \alpha_k = 2\pi k/2N_c\,, k = 1, 2, \ldots, 2N_c\}\,. \tag{A.19}$$

This important statement can be proved in different ways. An elegant proof makes use of the (natural) extension of SYM in which a ϑ term is added to the action (see the last section of Appendix C). In this situation under a $U_\lambda(1)$ rotation of the gluino fields in the functional integral, we get the (anomalous) WTI

$$\langle O_1(x_1)\ldots O_n(x_n)\rangle^{(\vartheta)} = \langle O_1(x_1)\ldots O_n(x_n)\rangle^{(\vartheta+2N_c\alpha)}\,. \tag{A.20}$$

Since the theory is classically invariant under such a rotation (the transformation $u(\alpha)O_k u^\dagger(\alpha) = \exp(i\eta_k\alpha)O_k$, with η_k the $U_\lambda(1)$ charge of O_k leaves invariant the correlator if the vacuum is annihilated by the unitary operator $u(\alpha)$), the only effect of the transformation is to change the value of the ϑ angle. The change does not affect physics if

$$2N_c\alpha = 2n\pi\,, \quad n \in \quad . \tag{A.21}$$

Clearly this result holds for any value of ϑ, thus also at $\vartheta = 0$.

When written in superfields, the SYM action takes the form [164]

$$S_{\mathrm{SYM}} = \int \mathrm{d}^4x\,\mathrm{d}^2\theta\,\mathrm{Tr}[W_\alpha W^\alpha]\,, \tag{A.22}$$

$$W_\alpha = -\frac{1}{4}\bar{D}^2\left(\mathrm{e}^{-gV}D_\alpha\mathrm{e}^{gV}\right)\,, \tag{A.23}$$

$$\begin{aligned} V(x,\theta) &= C(x) + \theta\mu(x) + \bar\theta\bar\mu(x) + \frac{1}{2}\theta^2 S(x) + \frac{1}{2}\bar\theta^2\bar S(x) \\ &+ \theta\sigma^\mu\bar\theta A_\mu(x) + \frac{1}{2}\bar\theta^2\theta\lambda(x) + \frac{1}{2}\theta^2\bar\theta\bar\lambda(x) + \frac{1}{4}\theta^2\bar\theta^2 D(x) + \ldots\,, \end{aligned} \tag{A.24}$$

where dots stand for terms that can be expressed as derivatives of the fields already present in (A.24).

A.7 The SQCD Action

The Euclidean action of the $\mathcal{N} = 1$ supersymmetric theory which more closely resembles QCD is obtained by coupling in a supersymmetric and gauge invariant way to the SYM supermultiplet N_f pairs of matter chiral superfields fields ($f = 1, \ldots, N_f$, $r = 1, \ldots, N_c$)

$$\Phi^r_f(x) = \phi^r_f(y) + \sqrt{2}\theta^\alpha\psi^r_{\alpha f}(y) + \theta 2F^r_f(y)\,, \quad y_\mu = x_\mu + i\theta\sigma_\mu\bar\theta\,, \tag{A.25}$$

$$\tilde\Phi^f_r(x) = \tilde\phi^f_r(y) + \sqrt{2}\theta^\alpha\tilde\psi^f_{\alpha r}(y)\,, + \theta 2\tilde F^f_r(y)\,, \tag{A.26}$$

belonging to the representations $\mathbf{N_c}$ and $\bar{\mathbf{N}}_\mathbf{c}$, respectively, of the gauge group. In this way the gauge-invariant mass term

$$S_{\mathrm{SQCD}}^{\mathrm{mass}} = \sum_f \Big[m_f \int \mathrm{d}^4x \sum_{\alpha r} \tilde{\psi}_r^{\alpha f} \psi_{\alpha f}^r + m_f^* \int \mathrm{d}^4x \sum_{\dot{\alpha} r} \bar{\psi}_{\dot{\alpha} r}^f \bar{\tilde{\psi}}_f^{\dot{\alpha} r} + |m_f|^2 \int \mathrm{d}^4x \sum_r (\phi_r^{*f} \phi_f^r + \tilde{\phi}_r^f \tilde{\phi}_f^{*r}) \Big] \tag{A.27}$$

can be constructed. The rest of the action is completely standard and can be found in any textbook or, for instance, in [4].

A.8 The "Flavour" Symmetries of the SQCD Action

(I) The classical SQCD action is invariant under the $U_\lambda(1)$ R symmetry [164] [49] which now transforms in a non-trivial way gluinos and scalars according to

$$\begin{aligned} &\lambda \to \mathrm{e}^{i\alpha}\lambda\,, \quad \phi \to \mathrm{e}^{i\alpha}\phi\,, \quad \tilde{\phi} \to \mathrm{e}^{i\alpha}\tilde{\phi}\,, \quad + \quad \text{complex conjugate}\,, \\ &A_\mu\,, \quad \psi, \bar{\psi}\,, \quad \tilde{\psi}, \bar{\tilde{\psi}}\,, \quad \text{untouched}\,. \end{aligned} \tag{A.28}$$

The $U_\lambda(1)$ R-symmetry of SQCD is anomalous with the same anomaly as in SYM (see (A.17)). Again only the $_{2N_c}$ subgroup is unbroken. With respect to the SYM case $J_\mu^{(\lambda)}$ must now be augmented with the inclusion of the matter contribution and reads

$$J_\mu^{(\lambda)} = \bar{\lambda}_{\dot{\alpha}}^a \bar{\sigma}_\mu^{\dot{\alpha}\alpha} \lambda_\alpha^a + \sum_f \Big\{ \big[i\phi^{f*} \overleftrightarrow{\partial}_\mu \phi_f \big] + \big[\phi_f \to \tilde{\phi}^f \big] \Big\}\,. \tag{A.29}$$

(II) The massless theory with N_f flavours possesses a global $SU(N_f) \times SU(N_f) \times U_V(1) \times U_A(1)$ symmetry. The chiral $SU(N_f) \times SU(N_f)$ symmetry is broken by matter mass terms. For instance, if all masses are equal ($m_f = m$), the unbroken subgroup is the diagonal vector group $U_V(N_f)$, while, if all masses are different ($m_1 \neq m_2 \neq \ldots \neq m_f$), the leftover unbroken subgroup is $U(1)^{N_f}$.

(III) The $U_A(1)$ transformation

$$\begin{aligned} &(\phi, \psi) \to \mathrm{e}^{i\alpha}(\phi, \psi)\,, \quad (\tilde{\phi}, \tilde{\psi}) \to \mathrm{e}^{i\alpha}(\tilde{\phi}, \tilde{\psi})\,, \quad + \quad \text{complex conjugate}\,, \\ &\lambda, \quad A_\mu\,, \quad \text{untouched}\,. \end{aligned} \tag{A.30}$$

is classically a symmetry at vanishing masses, but it is quantum–mechanically anomalous with

$$\partial_\mu J_\mu^{(A)} = 2iN_f \frac{g^2}{32\pi^2} F_{\mu\nu}^a \tilde{F}_{\mu\nu}^a\,, \tag{A.31}$$

$$J_\mu^{(A)} = \sum_f \Big\{ \big[\bar{\psi}_{\dot{\alpha}}^f \bar{\sigma}_\mu^{\dot{\alpha}\alpha} \psi_{\alpha f} + i\phi^{*f} \overleftrightarrow{\partial}_\mu \phi_f \big] + \big[(\phi_f, \psi_f) \to (\tilde{\phi}^f, \tilde{\psi}^f) \big] \Big\}. \tag{A.32}$$

[49] $U_\lambda(1)$ is sometimes also called $U_A^{PQ}(1)$ [25], where PQ stands for Peccei–Quinn [165], because it is anomalous and classically unbroken even at non-vanishing masses.

(IV) Often, instead of $U_\lambda(1)$, the linear combination

$$U_R(1) = \frac{3}{2}U_\lambda(1) - \frac{1}{2}U_A(1) \tag{A.33}$$

is introduced because the associated current belongs to the current supermultiplet [166]. This classical symmetry is anomalous and from (A.17) and (A.31) we see that the associated current obeys the anomaly equation

$$\partial_\mu J_\mu^{(R)} = i(3N_c - N_f)\frac{g^2}{32\pi^2} F^a_{\mu\nu}\tilde{F}^a_{\mu\nu}\,. \tag{A.34}$$

(V) It is possible to construct a non-anomalous, exactly conserved current, $J_\mu^{(\hat{A})}$, out of the two anomalous currents $J_\mu^{(A)}$ and $J_\mu^{(\lambda)}$ (see (A.32) and (A.29)). One finds

$$J_\mu^{(\hat{A})} = -N_f J_\mu^{(\lambda)} + N_c J_\mu^{(A)}\,. \tag{A.35}$$

The transformations induced on the fields by the associated $U_{\hat{A}}(1)$ symmetry (A.35) are

$$(\phi,\tilde{\phi}) \to \mathrm{e}^{i(N_c-N_f)\alpha}(\phi,\tilde{\phi})\,,\quad (\psi,\tilde{\psi}) \to \mathrm{e}^{iN_c\alpha}(\psi,\tilde{\psi})\,,\quad + \text{ complex conjugate}\,,$$
$$\lambda \to \mathrm{e}^{-iN_c\alpha}\lambda\,,\quad + \quad \text{complex conjugate}\,. \tag{A.36}$$

(VI) In Table A.1 we recollect for convenience the charges of elementary and composite gauge and matter fields under the various $U(1)$'s we have introduced. In the last two rows we report the coefficient of the anomaly of the associated current in units of $Q = \frac{g^2}{32\pi^2}F^a_{\mu\nu}\tilde{F}^a_{\mu\nu}$ and we indicate whether the conservation of the current is broken by mass terms or not.

Table A. 1. The $SU(N_f)\times SU(N_f)$ quantum numbers and $U(1)$-charges of elementary and composite fields of SQCD. The anomaly associated with each $U(1)$ current is given in units of $g^2/32\pi^2 F^a_{\mu\nu}\tilde{F}^a_{\mu\nu}$. In the last line by "Yes" ("No") we mean that the corresponding symmetry is (is not) broken by the presence of mass terms

Field	$SU(N_f)$	$SU(N_f)$	$U_A(1)$	$U_R(1)$	$U_{\hat{A}}(1)$	$U_\lambda(1)=U_A^{PQ}(1)$	$U_V(1)$
λ	**1**	**1**	0	3/2	$-N_f$	1	0
ψ	$\mathbf{N_f}$	**1**	1	1	N_c	0	1
ϕ	$\mathbf{N_f}$	**1**	1	-1/2	$N_c - N_f$	1	0
$\tilde{\psi}$	**1**	$\mathbf{N_f}$	1	1	N_c	0	-1
$\tilde{\phi}$	**1**	$\mathbf{N_f}$	1	-1/2	$N_c - N_f$	1	0
$S\|_{\theta=0}$	**1**	**1**	0	3	$-2N_f$	2	0
$T_h^f\|_{\theta=0}$	$\mathbf{N_f}$	$\mathbf{N_f}$	2	2	$2(N_c - N_f)$	2	0
Anomaly	/	/	$2N_f$	$3N_c - N_f$	0	$2N_c$	0
Mass term	Yes	Yes	Yes	Yes	Yes	No	No

A.9 Gluino Zero Modes

The explicit expression of the $2N_c$ gluino zero modes endowed with the correct normalisation $\int d^4x \sum_a \lambda^{a\alpha}(x)\lambda^{a*}_{\alpha}(x) = 1$ is (the index counting the $2N_c$ zero modes is indicated in parentheses)

$$(\lambda^{\alpha}_{(k)})^s_r(x) = \frac{1}{\pi}(\delta^{\alpha}_r\delta^s_k - \epsilon^{\alpha s}\epsilon_{rk})\rho^2(f(x))^2\,, \quad k = 1, 2\,, \tag{A.37}$$

$$(\lambda^{\alpha(\dot{\alpha})})^s_r(x) = -\frac{i}{\sqrt{2}\pi}\bar{\sigma}^{\dot{\alpha}\beta}_{\mu}(x - x_0)_{\mu}(\delta^{\alpha}_r\delta^s_{\beta} - \epsilon^{\alpha s}\epsilon_{r\beta})\rho(f(x))^2\,, \quad \dot{\alpha} = 1, 2\,, \tag{A.38}$$

$$(\lambda^{\alpha}_{(\pm i)})^s_r(x) = \frac{1}{\sqrt{2}\pi}(\delta^{\alpha}_r\delta^s_i \pm \epsilon^{\alpha s}\delta_{ri})\rho(f(x))^{3/2}\,, \quad i = 1, \ldots, N_c - 2\,, \tag{A.39}$$

with

$$f(x) = \frac{1}{(x - x_0)^2 + \rho^2}\,. \tag{A.40}$$

The first four modes are $SU(2)$ triplets, while the last $2(N_c - 2)$ are doublets. The four triplets can be directly generated starting from the expression (24) and acting on it with anyone of two supersymmetric and two superconformal transformations that are unbroken in the background instanton field. They are often called "exact zero modes" in the literature, see, for instance, [121] and references therein. This name originates from the following observation. Effectively the overall field configuration which is relevant for the kind of computations we have presented in Sect. 4 (see Sect. 14 for further applications) is given by the gauge instanton solution, the associated set of fermionic zero modes and the expression of the scalar fields that are obtained by solving their linearised classical e.o.m., i.e. the e.o.m. that result upon neglecting the quartic scalar self-interaction terms. The reason for neglecting such terms is that the latter would give rise to contributions of higher order in g compared to the leading ones we have been keeping. When the action of the theory is computed in this approximation and on the above field configuration, it just happens that the result does not depend on the fermionic collective coordinates associated with the four $SU(2)$ triplet zero modes of (A.37) and (A.38). The other fermionic zero modes will give origin to quartic terms in the remaining $2(N_c - 2)$ fermionic collective coordinates (see Sect. 14).

Appendix B – Bosonic Collective Coordinates and Functional Integration

In this appendix we want to explain how one can compute the pure gauge part of the functional integration in the semi-classical approximation around a non-trivial instantonic background. We will follow the method of [10], which neatly explains how to deal with the problem of bosonic zero modes and the consequent need of introducing collective coordinates.

In the semi-classical approximation one starts by expanding the (gauge) action around the (instanton) classical solution, keeping only terms up to quadratic fluctuations. Setting

$$A_\mu = A^I_\mu + Q_\mu \,, \tag{B.1}$$

one gets in this way

$$S_{\rm YM} = S^I - \frac{1}{2}\int \mathrm{d}^4x \,\mathrm{Tr}\big[Q_\mu \mathcal{M}_{\mu\nu}(A^I)Q_\nu\big] + \mathrm{O}(Q^3)\,, \tag{B.2}$$

where

$$S^I = \frac{8\pi^2}{g^2}|K|\,, \tag{B.3}$$

$$\mathcal{M}_{\mu\nu}(A^I) = -D^2(A^I)\delta_{\mu\nu} + D_\mu(A^I)D_\nu(A^I) \\ -2[F^I_{\mu\nu},\cdot]\,, \tag{B.4}$$

$$D_\mu(A^I) = \partial_\mu + g[A^I_\mu,\cdot]\,. \tag{B.5}$$

The operator $\mathcal{M}_{\mu\nu}(A^I)$ has quite a large manifold of (normalisable and non-normalisable) zero modes. Not only it is annihilated by all the functions of the form $D_\nu(A^I)F(x)$, as a consequence of gauge invariance, but also by the $4|K|N_c$ normalisable vectors that are obtained by differentiating the instanton field configuration with respect to the $4|K|N_c$ parameters, β_i, $i=1,\ldots,4|K|N_c$ (bosonic collective coordinates in the following), upon which the most general classical solution depends.[50] The existence of such zero modes is immediately proved by noticing that by differentiating the classical instanton e.o.m., $(\delta S_{\rm YM}/\delta A_\mu)_{A^I_\mu}=0$, with respect to β_i, one gets

$$\int \mathrm{d}^4x' \Big(\frac{\delta^2 S_{\rm YM}}{\delta A_\mu(x)\delta A_\nu(x')}\Big)_{A^I_\mu} \frac{\partial A^I_\nu(x',\beta)}{\partial\beta_i} \\ = \mathcal{M}_{\mu\nu}(A^I)\frac{\partial A^I_\nu(x,\beta)}{\partial\beta_i} = 0\,, \\ i=1,\ldots,4|K|N_c\,. \tag{B.6}$$

The most elegant way to deal with an operator with such a kernel was worked out some time ago in [10]. The idea is to functionally integrate over all fluctuations, $Q_\mu(x) = A_\mu(x) - A^I_\mu(x,\beta)^U$, that are orthogonal to the manifold described by $A^I_\mu(x,\beta)^U$ when U spans the space of topologically trivial gauge transformations, $\mathcal{G}_0$, and the parameters β_i are let to move in their allowed

[50] We are referring here to the $SU(N_c)$ gauge group case. In the one-instanton sector, $|K|=1$, the collective coordinates are the size and the location of the instanton and its $4N_c-5$ "orientation angles" in colour space [9, 20, 23].

range of variation. In more mathematical terms the latter manifold is called the "instanton moduli space".

The orthogonality conditions (B.6) are imposed by a straightforward generalisation of the usual Faddeev–Popov (FP) procedure [167] which consists in introducing in the functional integral the identity

$$1 = \Delta_{\rm FP} \int_{\mathcal{G}_0} \prod_{a,x} \delta h^a(x) \int_{\mathcal{M}} \prod_i \mathrm{d}\beta_i\, \delta\Big(< (A_\mu - A^I_\mu(\beta)^{U_h}), \frac{\delta A^I_\mu(\beta)^{U_h}}{\delta h^a(x)} > \Big)$$
$$\times \delta\Big(< (A_\nu - A^I_\nu(\beta)^{U_h}), \frac{\partial A^I_\nu(\beta)^{U_h}}{\delta \beta_i} > \Big), \qquad \text{(B.7)}$$

where $\Delta_{\rm FP}$ is the FP determinant. In (B.7) we have used the shorthand notation

$$< f_\mu, g_\mu > = \frac{1}{2} \int \mathrm{d}^4x\, {\rm Tr}_{\mathbf{Adj}}[f_\mu(x) g_\mu(x)] \qquad \text{(B.8)}$$

for the scalar product $< \cdot, \cdot >$ induced in the space of functions by the form of the gauge action. After some algebra (see [168] for details) (B.7) can be cast in the more expressive form

$$1 = \Delta_{\rm FP} \int_{\mathcal{G}_0} \prod_{a,x} \mathcal{D}\mu[h^a(x)] \int_{\mathcal{M}} \prod_i \mathrm{d}\beta_i\, \delta\Big({\rm Tr}[T^a D_\mu(A^I)(A_\mu(x)^{U_h^\dagger} - A^I_\mu(x,\beta))] \Big)$$
$$\times \delta\Big(< (A_\nu^{U_h^\dagger} - A^I_\nu(\beta)), \frac{\partial A^I_\nu(\beta)}{\delta \beta_i} > \Big), \qquad \text{(B.9)}$$

which shows that we are naturally brought to work in the instanton background gauge. $\Delta_{\rm FP}$ can be shown to have in the semi-classical approximation the expression

$$\Delta_{\rm FP} = {\rm det}^{x,y}_{a,b}\big[- D^2(A^I)^{ab} \delta(x-y) \big] {\rm det}_{i,j}\big[< a^{(i)}(\beta), a^{(j)}(\beta) > \big], \qquad \text{(B.10)}$$

where the $a^{(i)}$'s ($i = 1, 2, \ldots, 2|K|N_c$) are the mutually orthogonal (see the next subsection) vectors

$$a^{(i)}_\mu(x,\beta) = \Big[\delta_{\mu\nu} - D_\mu(A)[D^2(A)]^{-1} D_\nu(A)\big|_{A_\mu = A^I_\mu} \Big] \frac{\partial A^I_\nu(x,\beta)}{\partial \beta_i}. \qquad \text{(B.11)}$$

We will indicate by $||a^{(i)}||$ their norm in the metric induced by the scalar product (B.8). The vectors $a^{(i)}_\mu(x,\beta)$ are not exactly the functions $\partial A^I_\mu(x,\beta)/\partial\beta_i$. They differ from the latter by a term which makes them to fulfil the equation

$$D_\mu(A^I) a^{(i)}_\mu(x,\beta) = 0, \qquad \text{(B.12)}$$

i.e. which makes them transverse with respect to the covariant derivative in the instanton background.

Putting everything together and noticing that the orthogonality condition among the vectors (B.11) makes immediate the computation of the factor

$\det_{i,j}[< a^{(i)}(\beta), a^{(j)}(\beta) >]$ in $\Delta_{\rm FP}$, one finally gets for the v.e.v. of a gauge invariant operator, $O(A)$, in the semi-classical approximation around an instanton configuration with winding number $|K|$ the expression

$$\langle O\rangle\Big|_{s.c.} = \frac{\mathrm{e}^{-\frac{8\pi^2}{g^2}|K|}}{Z|_{s.c.}} \int \mathcal{D}Q_\mu \prod_i \mathrm{d}\beta_i \frac{||a^{(i)}||}{\sqrt{2\pi}} \times \mathrm{e}^{-\frac{1}{2}\int \mathrm{d}^4x\mathrm{d}^4y\, Q_\mu \mathcal{M}^{g.f.}_{\mu\nu} Q_\nu} \det[-D^2(A^I)]\delta(D^{ab}_\mu(A^I)Q^b_\mu) O(A^I)\,, \tag{B.13}$$

where

$$Z|_{s.c.} = \int \mathcal{D}Q_\mu \mathrm{e}^{-\frac{1}{2}\int \mathrm{d}^4x\mathrm{d}^4y\, Q_\mu \mathcal{M}^{g.f.}_{0;\mu\nu} Q_\nu} \det[-\partial^2]\delta(\partial_\mu Q^a_\mu)\,, \tag{B.14}$$

$$\mathcal{M}^{g.f.}_{\mu\nu} = -D^2(A^I)\delta_{\mu\nu} - 2\,[F^I_{\mu\nu}, \cdot\,]\,, \tag{B.15}$$

$$\mathcal{M}^{g.f.}_{0;\mu\nu} = -\partial^2\delta_{\mu\nu}\,. \tag{B.16}$$

$Z|_{s.c.}$ is the necessary normalisation factor which, in order to be consistent with the approximation we are working in, must be evaluated by expanding the action around the trivial solution of the field e.o.m. keeping only terms quadratic in the fluctuations. Note that to make more transparent analogies and differences between (B.13) and (B.14) we have named Q_μ the integration variable also in (B.14). $\mathcal{M}^{g.f.}_{\mu\nu}$ ($\mathcal{M}^{g.f.}_{0\mu\nu}$) is the gauge fixed operator that governs the quadratic fluctuations of the gauge field in the instanton (trivial) background and $\det[-D^2(A^I)]$ ($\det[-\partial^2]$) is the associated FP determinant.

One can formally perform the gauge functional integrations in the r.h.s. of (B.13), getting

$$\langle O\rangle\Big|_{s.c.} = \mu^{n_B} \frac{\mathrm{e}^{-\frac{8\pi^2}{g^2}|K|}}{Z|_{s.c.}} \int \prod_{i=1}^{n_B} \mathrm{d}\beta_i \frac{||a^{(i)}||}{\sqrt{2\pi}} \times \frac{(\det{}'[\mathcal{M}^{g.f.}_{\mu\nu}])^{-\frac{1}{2}}\det[-D^2(A^I)]}{(\det[\mathcal{M}^{g.f.}_{0;\mu\nu}])^{-\frac{1}{2}}\det[-\partial^2]} O(A^I)\,, \tag{B.17}$$

where $n_B = 4|K|N_c$ is the number of bosonic zero modes and μ is the subtraction point (see below). The prime on $\det{}'[\mathcal{M}^{g.f.}_{\mu\nu}]$ is to mean that the determinant should be taken in the space orthogonal to the manifold spanned by the zero modes (B.11).

A number of observations are in order here.

(1) As is seen from the above equations, by the method of [10] one is naturally led to the background gauge fixing condition $D^{ab}_\mu(A^I)Q^b_\mu = 0$.

(2) One must imagine that the above functional integral has been computed in some regularisation. In these instantonic computations it is customary to work in the Pauli–Villars (PV) regularisation [2], where a ghost-like field with mass μ (but opposite statistics) is introduced for each fundamental field in the action (gluons, FP ghosts and, if present, fermions). The net effect

of the presence of PV regulators is that the result of the functional integration over quadratic fluctuations will have the form of a product of factors, with each term being the ratio of the determinant of each particle quadratic operator divided by the associated PV ghost determinant (raised to the appropriate power according to multiplicity and statistics).

(3) When the limit $\mu \to \infty$ is taken, the only left-over μ dependence is the multiplicative factor μ^{n_B}, $n_B = 4|K|N_c$. This factor comes about because of the following reason. There is a one-to-one correspondence between the eigenvectors (and the eigenvalues) of analogous operators in each ratio of determinants, except for the zero modes. There is, in fact, a mismatch between the numerator and the denominator in the sense that there are some (actually $n_B = 4|K|N_c$) eigenvalues in the PV denominator that do not have their counterpart in the "primed" determinant in the numerator. This leaves out precisely a factor $(\mu^2)^{\frac{1}{2}}$ for each bosonic collective coordinate (and actually a factor $\mu^{-\frac{1}{2}}$ for each Weyl fermion zero mode, see (23) in Sect. 2.3).

(4) The factor $1/\sqrt{2\pi}$ for each bosonic zero mode appears for a similar reason. In fact, the integrations that give rise to the product of eigenvalues finally leading to the various bosonic determinants are all Gaussian in the (quadratic, i.e. semi-classical) approximation in which we are working. Since, as we noticed above, there is a one-to-one correspondence between physical modes and PV modes, all the factors $\sqrt{2\pi}$ will compensate between the numerator and the denominator, except for the factors coming from the integration over the PV modes that are in correspondence with the bosonic zero modes. The reason is that no Gaussian integration is associated with the bosonic zero modes, as the latter were replaced by integrations over the related collective coordinates. In this way a factor $1/\sqrt{2\pi}$ for each bosonic zero mode will be left in the denominator.

(5) In principle, one can go beyond the semi-classical formulae (B.13) and (B.17), including perturbatively $\mathrm{O}(Q_\mu^3) \sim \mathrm{O}(g)$ and $\mathrm{O}(Q_\mu^4) \sim \mathrm{O}(g^2)$ corrections that were neglected before. As is well known, perturbation theory in an external field is perfectly well defined and fully renormalisable.

B.1 Bosonic Zero Modes

We close this appendix by reporting in the case $N_c = 2$ and $K = 1$ the explicit expression of the $4|K|N_c = 8$ "transverse" bosonic zero modes and of their norms. One finds ($y = x - x_0$)

$$
\begin{aligned}
a_\mu^{(\nu)} &= F_{\mu\nu}^I(y)\,, & ||a_\mu^{(\nu)}|| &= \tfrac{2\sqrt{2}\pi}{g}\,, \\
a_\mu^{(\mathrm{dil.})} &= A_\mu^I(y)\tfrac{2y^2}{\rho(y^2+\rho^2)}\,, & ||a_\mu^{(\mathrm{dil.})}|| &= \tfrac{4\pi}{g}\,, \\
a_\mu^{(a)} &= D_\mu(A^I)\left[\tfrac{T^a}{g}\tfrac{y^2}{y^2+\rho^2}\right], & ||a_\mu^{(a)}|| &= \tfrac{2\pi\rho}{g}\,.
\end{aligned}
\qquad \text{(B.18)}
$$

One can check that these vectors are mutually orthogonal.

Appendix C – Quantum Tunnelling

The emergence of the quantum tunnelling phenomenon in the presence of instantons is most easily and rigorously explained in the Schrödinger functional formalism [169], where the theory is formulated in terms of a "propagation kernel" which expresses the probability amplitude to find the gauge field configuration $\boldsymbol{A}^{(2)}(\boldsymbol{x}) = (A_i^{(2)}(\boldsymbol{x}), i = 1,2,3)$ at the final time $t = T/2$, if the gauge field configuration at the initial time $t = -T/2$ was $\boldsymbol{A}^{(1)}(\boldsymbol{x}) = (A_i^{(1)}(\boldsymbol{x}), i = 1,2,3)$.

The Schrödinger Functional in the Temporal Gauge

The Schrödinger kernel is most expressively written in the temporal gauge [170]. As a result of making use of the Faddeev–Popov procedure, one can show that in the formal continuum language it takes the form[51]

$$\mathcal{K}[\boldsymbol{A}^{(2)}, \boldsymbol{A}^{(1)}; T] = \int_{\hat{\mathcal{G}}_0} \prod_{\boldsymbol{x}} \mathcal{D}\mu[h(\boldsymbol{x})] \int_{\boldsymbol{A}^{(1)}(\boldsymbol{x})}^{[\boldsymbol{A}^{(2)}(\boldsymbol{x})]^{U_0[h(\boldsymbol{x})]}} \prod_{a,\boldsymbol{x}} \times \prod_{-\frac{T}{2}<t<\frac{T}{2}} \mathcal{D}A^a(\boldsymbol{x},t) \exp\left[-S_{YM}[\boldsymbol{A}, A_0 = 0]\right], \tag{C.1}$$

where $U_0[h] = \exp(iT^a h^a) \in \hat{\mathcal{G}}_0$ with $\hat{\mathcal{G}}_0$ the group of the time-independent, topologically trivial gauge transformations (i.e. those that tend to the group identity at spatial infinity) and $\mathcal{D}\mu[h(\boldsymbol{x})]$ is the invariant Haar measure over $SU(N_c)$ at each spatial point $\boldsymbol{x}$. The integration over the spatial components of the gauge field is extended to all configurations that satisfy the boundary conditions $\boldsymbol{A}(\boldsymbol{x},T/2) = [\boldsymbol{A}^{(2)}(\boldsymbol{x})]^{U_0[h(\boldsymbol{x})]}$ and $\boldsymbol{A}(\boldsymbol{x},-T/2) = \boldsymbol{A}^{(1)}(\boldsymbol{x})$.

The gauge integration over $\hat{\mathcal{G}}_0$ plays a crucial role in the formalism as it has the effect of projecting out from the kernel all the states that do not satisfy the Gauss' law constraint. In fact, since the Gauss' law operator is the generator of the time-independent topologically trivial gauge transformations, only the states annihilated by it will appear in the spectral decomposition of $\mathcal{K}[\boldsymbol{A}^{(2)}, \boldsymbol{A}^{(1)}; T]$ [170], for which we can then formally write

$$\mathcal{K}[\boldsymbol{A}^{(2)}, \boldsymbol{A}^{(1)}; T] = \sum_n \mathrm{e}^{-E_n T} \Psi_n[\boldsymbol{A}^{(2)}](\Psi_n[\boldsymbol{A}^{(1)}])^*, \tag{C.2}$$

where

$$\mathcal{H}\Psi_n[\boldsymbol{A}] = E_n \Psi_n[\boldsymbol{A}], \tag{C.3}$$

$$D_i(\boldsymbol{A})^{ab} \frac{\delta}{\delta A_i^b(\boldsymbol{x})} \Psi_n[\boldsymbol{A}] = 0. \tag{C.4}$$

[51] For the lattice regularised formulation of the Schrödinger kernel – more commonly called Schrödinger functional in that context – see [171].

The last equation is indeed the statement that the eigenstates of the Hamiltonian appearing in the spectral decomposition (C.2) are left untouched by time-independent gauge transformations that tend to the identity at infinity. In fact, from the invariance property

$$\mathcal{U}_0[h]\Psi_n[\boldsymbol{A}] = \Psi_n[\boldsymbol{A}^{U_0[h]}] = \Psi_n[\boldsymbol{A}]\,, \tag{C.5}$$

$$\mathcal{U}_0[h] = \exp\Big(- \int \mathrm{d}^3x\, (D_i^{ab} h^b(\boldsymbol{x})) \frac{\delta}{\delta A_i^a(\boldsymbol{x})} \Big)\,, \tag{C.6}$$

the Gauss' law (C.4) follows by expanding (C.5) in powers of $h(\boldsymbol{x})$, if the latter function vanishes as $|\boldsymbol{x}| \to \infty$, i.e. precisely if $U_0[h] \in \hat{\mathcal{G}}_0$. An equivalent way to prove this statement is to observe that the Schrödinger kernel enjoys the invariance properties

$$\mathcal{K}[(\boldsymbol{A}^{(2)})^{U_0}, (\boldsymbol{A}^{(1)}); T] = \mathcal{K}[\boldsymbol{A}^{(2)}, \boldsymbol{A}^{(1)}; T] = \mathcal{K}[\boldsymbol{A}^{(2)}, (\boldsymbol{A}^{(1)})^{U_0}; T]\,. \tag{C.7}$$

The first equality follows from the invariance of the Haar measure, as the U_0 gauge transformation can be reabsorbed in the integration measure over $\hat{\mathcal{G}}_0$. The second equality is an immediate consequence of the previous equation and the invariance property

$$\mathcal{K}[(\boldsymbol{A}^{(2)})^{U}, (\boldsymbol{A}^{(1)})^{U}; T] = \mathcal{K}[\boldsymbol{A}^{(2)}, \boldsymbol{A}^{(1)}; T]\,, \quad U \in \hat{\mathcal{G}}_0 \tag{C.8}$$

which in turn follows from the observation that any time-independent gauge transformation acting on the boundary fields can be reabsorbed by the change of variables $\boldsymbol{A} \to \boldsymbol{A}' = \boldsymbol{A}^U$ in (C.1). The invariance property (C.8) can be used to show that the $\hat{\mathcal{G}}_0$ gauge integration in (C.1) can be equally well performed over the time-independent gauge transformations acting on the boundary field $\boldsymbol{A}^{(1)}$ at $t = -T/2$.

C.2 Emergence of the ϑ Angle

We finally notice that the states Ψ_n also support a unitary representation, $\mathcal{U}_K$, of the abelian homotopy group $\Pi_3(SU(2)) \sim \Pi_3(\mathrm{S}_3) =$. Since the Hamiltonian, $\mathcal{H}$, and $\mathcal{U}_K$ commute, they can be simultaneously diagonalised. Thus on their common eigenvectors (for a while we will keep calling them Ψ_n) we have

$$\mathcal{U}_K \Psi_n[\boldsymbol{A}] = \Psi_n[\boldsymbol{A}^{U_K}] = \mathrm{e}^{-i\vartheta_K} \Psi_n[\boldsymbol{A}]\,. \tag{C.9}$$

Consistency with the group property

$$\mathcal{U}_K \mathcal{U}_{K'} = \mathcal{U}_{K+K'} \tag{C.10}$$

implies

$$\vartheta_K = K\vartheta\,, \tag{C.11}$$

naturally leading to the emergence of a ϑ-angle. States should (and will) then be indicated by $\Psi_n^{(\vartheta)}[\boldsymbol{A}]$ in the following.

C.3 Classical Vacua and Quantum Tunnelling

The classical vacua of the theory are immediately identified as the gauge configurations for which the classical Hamiltonian

$$H = \int d^3x \left(\frac{1}{2}\dot{A}_i^a \dot{A}_i^a + \frac{1}{4}F_{ij}^a F_{ij}^a\right) \tag{C.12}$$

vanishes, thus as time-independent ($\dot{A}_i^a = 0$) pure gauges ($F_{ij}^a = 0$). This simple argument shows that there are infinitely many "vacua" labelled by an integer, $K \in$, which is telling us which homotopy class the K-th vacuum belongs to.

In Euclidean time the one-instanton ($K = 1$) solution interpolates between adjacent minima, i.e. between pure gauge configurations with winding number differing by one unit.[52] The formulae (12) and (13) can then be immediately proved. Since $A_0 = 0$, one successively gets, in fact

$$\begin{aligned} K &= \frac{g^2}{32\pi^2}\int d^4x\, F_{\mu\nu}^a \tilde{F}_{\mu\nu}^a(x) = \frac{g^2}{16\pi^2}\int d^4x\, \partial_\mu K_\mu(x) = \frac{g^2}{16\pi^2}\int d^4x\, \partial_0 K_0(\boldsymbol{x}, t) \\ &= \frac{g^2}{16\pi^2}\left[\int d^3x\, K_0(\boldsymbol{x}, +\infty) - \int d^3x\, K_0(\boldsymbol{x}, -\infty)\right] = n_+ - n_- \,. \end{aligned} \tag{C.13}$$

The last equality follows remembering that at very large (positive and negative) times $K_0 \propto \epsilon_{ijk}\mathrm{Tr}[A_i A_j A_k]$ with $\boldsymbol{A}$ a pure gauge.

The classical vacuum degeneracy is removed by the quantum mechanical tunnelling between adjacent minima occurring with an amplitude $\Gamma^I \propto \exp(-S^I) = \exp(-8\pi^2/g^2)$. A band spectrum is generated with the lowest energy eigenstates and eigenvalues given by

$$\Psi_0^{(\vartheta)}[\boldsymbol{A}] = \sum_{K\in} e^{-iK\vartheta}\Psi_0^{(K)}[\boldsymbol{A}]\,, \tag{C.14}$$

$$E_0(\vartheta) = \alpha_0 + \beta_0 \cos\vartheta\,, \tag{C.15}$$

where, at the leading order in Γ^I, $\Psi_0^{(K)}[\boldsymbol{A}]$ is the perturbative vacuum state functional "centred" around the K-th minimum of the energy, i.e. around a pure gauge field with winding number K and α_0, β_0 are computable constants proportional to the spatial volume of the system.

It is not too difficult to prove the result (C.15). We start by observing that, once quantum tunnelling has been recognised to take place, the spectral

[52] Multi-instanton solutions with $|K| > 1$ connect vacua with winding numbers differing by exactly $|K|$ units. They have an action (see (B.3)) which is exponentially small with respect to the one-instanton action. In the approximation we are working, their contribution is automatically taken care of by the exponentiation of the one-instanton contribution implicit in the spectral formula (C.2). Incidentally this is the way in which within the Schrödinger functional formalism the "dilute gas" approximation [3, 12] is recovered.

decomposition of the Schrödinger kernel can be written in more informative form ($\sum_n \to \int \mathrm{d}\vartheta \sum_\ell$)

$$\mathcal{K}[\boldsymbol{A}^{(2)}, \boldsymbol{A}^{(1)}; T] = \int_0^{2\pi} \mathrm{d}\vartheta \sum_\ell \mathrm{e}^{-E_\ell(\theta)T} \Psi_\ell^{(\theta)}[\boldsymbol{A}^{(2)}](\Psi_\ell^{(\theta)}[\boldsymbol{A}^{(1)}])^* . \tag{C.16}$$

For the purpose of our calculation, it is enough to take $\boldsymbol{A}^{(2)}$ and $\boldsymbol{A}^{(1)}$ as pure gauges. At this point only their winding number matters and we can simplify our notation by writing the Schrödinger kernel and the associated state functionals in the form $\mathcal{K}[K^{(2)}, K^{(1)}; T]$ and $\Psi_\ell^{(\vartheta)}[K]$, respectively. In this notation (C.9) becomes

$$\mathcal{U}_K \Psi_\ell^{(\vartheta)}[0] = \Psi_\ell^{(\vartheta)}[K] = \mathrm{e}^{-i\vartheta K} \Psi_\ell^{(\vartheta)}[0] , \tag{C.17}$$

where, we stress, "0" means a pure gauge configuration with $K = 0$.

To leading order in the instanton tunnelling amplitude, we only need to evaluate the kernels $\mathcal{K}[K, K; T]$, $\mathcal{K}[K, K+1; T]$ and $\mathcal{K}[K+1, K; T]$, as all the others should be considered exponentially small to this order

$$\mathcal{K}[K, K'; T] = 0 , \qquad |K - K'| > 1 . \tag{C.18}$$

In order to proceed further we first note the relation

$$\begin{aligned}
&\mathcal{K}[K, K+1; T] = (\mathcal{K}[K+1, K; T])^* \\
&= \int_0^{2\pi} \mathrm{d}\vartheta \sum_\ell \Psi_\ell^{(\vartheta)}[K](\Psi_\ell^{(\vartheta)}[K+1])^* \mathrm{e}^{-E_\ell(\vartheta)T} \\
&= \int_0^{2\pi} \mathrm{d}\vartheta \, \mathrm{e}^{i\vartheta} \sum_\ell \Psi_\ell^{(\vartheta)}[0](\Psi_\ell^{(\vartheta)}[0])^* \mathrm{e}^{-E_\ell(\vartheta)T} ,
\end{aligned} \tag{C.19}$$

that follows from (C.17). Since we are interested in computing the energy of the lowest lying state, we shall take T very large, keeping however $T \exp(-8\pi^2/g^2) < 1$. Expanding the exponential of the energy up to terms linear in T, one finds

$$\begin{aligned}
&\mathcal{K}[K, K+1; T] = \int_0^{2\pi} \mathrm{d}\vartheta \mathrm{e}^{i\vartheta} |\Psi_0^{(\vartheta)}[0]|^2 (1 - E_0(\vartheta)T + \mathrm{O}(T^2)) \\
&= |\Psi_0^{\mathrm{P.T.}}[0]|^2 \int_0^{2\pi} \frac{\mathrm{d}\vartheta}{2\pi} \mathrm{e}^{i\vartheta} (1 - E_0(\vartheta)T + \mathrm{O}(T^2)) \\
&= -T |\Psi_0^{\mathrm{P.T.}}[0]|^2 \int_0^{2\pi} \frac{\mathrm{d}\vartheta}{2\pi} \mathrm{e}^{i\vartheta} E_0(\vartheta) + \mathrm{O}(T^2) ,
\end{aligned} \tag{C.20}$$

$$\begin{aligned}
&\mathcal{K}[K+1, K; T] = -T |\Psi_0^{\mathrm{P.T.}}[0]|^2 \int_0^{2\pi} \frac{\mathrm{d}\vartheta}{2\pi} \mathrm{e}^{-i\vartheta} E_0(\vartheta) \\
&+ \mathrm{O}(T^2) ,
\end{aligned} \tag{C.21}$$

$$\mathcal{K}[K,K;T] = |\Psi_0^{\mathrm{P.T.}}[0]|^2 - T|\Psi_0^{\mathrm{P.T.}}[0]|^2 \cdot \int_0^{2\pi} \frac{\mathrm{d}\vartheta}{2\pi} E_0(\vartheta) + \mathrm{O}(T^2)\,, \tag{C.22}$$

where the first equality in (C.20) follows from the fact that in the semi-classical approximation one has

$$|\Psi_0^{(\vartheta)}[0]|^2 = \frac{1}{2\pi} |\Psi_0^{\mathrm{P.T.}}[0]|^2\,. \tag{C.23}$$

We conclude from (C.18), (C.20), (C.21) and (C.22) that the coefficient of the terms linear in T has only the three non-vanishing Fourier components of order $\pm 1, 0$. Thus $E_0(\vartheta)$ is precisely of the form (C.15).

C.4 Adding a ϑ-term

It is instructive to see what happens if a ϑ-term is added to the gauge action. In this case the contribution

$$i\vartheta \frac{g^2}{32\pi^2} \int \mathrm{d}^4x\, F^a_{\mu\nu} \tilde{F}^a_{\mu\nu}(x) \tag{C.24}$$

should be included in (A.2).[53] It is easy to prove that such an action describes a world with a well-defined ϑ-angle (obviously equal to the value appearing in (C.24)). From the formula (see (C.1))

$$\mathcal{K}^{(\vartheta)}[\boldsymbol{A}^{(2)}, \boldsymbol{A}^{(1)}; T] = \int_{\hat{\mathcal{G}}_0} \prod_{\boldsymbol{x}} \mathcal{D}\mu[h(\boldsymbol{x})] \tilde{\mathcal{K}}^{(\vartheta)}[(\boldsymbol{A}^{(2)})^{U_0[h]}, \boldsymbol{A}^{(1)}; T]\,, \tag{C.25}$$

$$\tilde{\mathcal{K}}^{(\vartheta)}[\boldsymbol{A}^{(2)}, \boldsymbol{A}^{(1)}; T] = \int_{\boldsymbol{A}^{(1)}(\boldsymbol{x})}^{\boldsymbol{A}^{(2)}(\boldsymbol{x})} \prod_{a,\boldsymbol{x}} \prod_{-\frac{T}{2}<t<\frac{T}{2}} \mathcal{D}A^a(\boldsymbol{x},t)$$

$$\exp\Big[- S_{YM}[\boldsymbol{A}, A_0 = 0] - i\vartheta \frac{g^2}{32\pi^2} \int \mathrm{d}^4x\, F^a_{\mu\nu} \tilde{F}^a_{\mu\nu}(x)\Big]\,, \tag{C.26}$$

one checks, in fact, that under a homotopically non-trivial (time-independent) gauge transformation with winding number K, acting, say, on the boundary gauge field at $T/2$, the Schrödinger kernel is not invariant (recall the situation in the absence of a ϑ-term, (C.8)), rather one has

$$\mathcal{K}^{(\vartheta)}[(\boldsymbol{A}^{(2)})^{U_K}, \boldsymbol{A}^{(1)}; T] = \mathrm{e}^{-iK\vartheta} \mathcal{K}^{(\vartheta)}[\boldsymbol{A}^{(2)}, \boldsymbol{A}^{(1)}; T]\,. \tag{C.27}$$

This result (which incidentally implies that physics is invariant if we replace ϑ with $\vartheta + 2\pi$) follows from the fact that the exponent in (C.26) is not invariant

[53] Notice the presence of the imaginary unit in front of this term even in Euclidean metric.

under such gauge transformation. Obviously, the YM action is invariant, but the second term is not. The reason can be traced back to the fact that the vector K_μ in (8) is not gauge invariant. Under the time-independent gauge transformation $U_{\scriptscriptstyle K}$ in (C.27) one finds, in fact (recall that we are in the temporal gauge)

$$\begin{aligned} \frac{g^2}{32\pi^2}\int \mathrm{d}^4x\,\big[F^a_{\mu\nu}\tilde F^a_{\mu\nu}\big]^{U_K} &= \frac{g^2}{16\pi^2}\int \mathrm{d}^3x\,\big[K_0[(\boldsymbol{A}^{(2)})^{U_K}] - K_0[\boldsymbol{A}^{(1)}]\big] \\ &= \frac{g^2}{16\pi^2}\int \mathrm{d}^3x\,\big[K_0[\boldsymbol{A}^{(2)}] - K_0[\boldsymbol{A}^{(1)}]\big] \\ &+\frac{\epsilon_{ijk}}{24\pi^2}\int_{\mathrm{S}_3}\mathrm{d}^3x\,\mathrm{Tr}\big[U^\dagger_K\partial_i U_K\, U^\dagger_K\partial_j U_K\, U^\dagger_K\partial_k U_K\big] = \frac{g^2}{32\pi^2}\int \mathrm{d}^4x\,\big[F^a_{\mu\nu}\tilde F^a_{\mu\nu}\big] + K\,. \end{aligned} \tag{C.28}$$

From the spectral decomposition of $\mathcal{K}^{(\vartheta)}$, one concludes that (C.17) holds for each state appearing in it, thus proving the announced statement.

Appendix D – Decoupling

The physical content of the Applequist–Carazzone theorem [42] is that in an asymptotically free theory a heavy particle (i.e. a particle with $m_f \gg \Lambda$) should "decouple", that is to say, it should not influence physics at energies $E \ll m_f$.

The most important (for us) consequence of this statement is that one can relate the Λ parameter of an $SU(N_c)$ gauge theory with N_f flavours to that of the theory with $N_f - 1$ dynamically active flavours, which is obtained after the mass of one of the flavours (say the N_f-th one) has been sent to infinity.

The running of the coupling constant of the two theories is guided at one loop by the evolution equations (the dependence of b_1 on N_c is understood)

$$\frac{g^2_{N_f}(\mu)}{8\pi^2} = \frac{1}{b_{1,N_f}\log\mu/\Lambda^{(N_f)}}\,, \tag{D.1}$$

$$\frac{g^2_{N_f-1}(\mu)}{8\pi^2} = \frac{1}{b_{1,N_f-1}\log\mu/\Lambda^{(N_f-1)}}\,. \tag{D.2}$$

A necessary implication of decoupling is that for $m_f \gg \Lambda^{(N_f-1)}, \Lambda^{(N_f)}$ the running of g^2 in the theory with N_f flavour must change from the behaviour in (D.1) – when μ is sufficiently larger than m_f – to that in (D.2) – when μ is well below it. The equality of the coupling constants at $\mu \sim m_f$ (required by smoothness) leads to the sought relation

$$\left(\frac{m_f}{\Lambda^{(N_f)}}\right)^{b_{1,N_f}} = \left(\frac{m_f}{\Lambda^{(N_f-1)}}\right)^{b_{1,N_f-1}}\,. \tag{D.3}$$

Notice that, since we are assuming that m_f is larger than both $\Lambda^{(N_f-1)}$ and $\Lambda^{(N_f)}$, from (D.3) it follows $\Lambda^{(N_f-1)} > \Lambda^{(N_f)}$. This relation is phenomenologically quite important. It is telling us that, when the energy scale, E, of a process goes through the production threshold of a particle of mass m_f, since the running of the coupling constant switches from that of (D.1) to that of (D.2), it just happens that the value taken by the effective coupling constant, $g^2_{\rm eff}(E)$, that controls the process is always the largest between $g^2_{N_f-1}(E)$ and $g^2_{N_f}(E)$ for all values of E.

Appendix E – Flat Directions of Massless SQCD

In this appendix we want to elucidate the structure of the vacuum manifold of massless SQCD. The theory possesses the (non-anomalous) symmetry group (see Table A.1)

$$G = SU_L(N_f) \times SU_R(N_f) \times U_V(1) \times U_{\hat{A}}(1)\,. \tag{E.1}$$

Any field configuration of the type

$$A_\mu = \lambda = \psi = \bar{\psi} = \tilde{\psi} = \bar{\tilde{\psi}} = 0\,, \tag{E.2}$$

$$D^a = \phi^{r\,\dagger}_f (T^a)^{r'}_r \phi^f_{r'} - \tilde{\phi}^r_f (T^a)^{r'}_r \tilde{\phi}^{f\,\dagger}_{r'} = 0 \tag{E.3}$$

has vanishing energy, thus it is to be interpreted as a classical vacuum state. Non-renormalisation theorems ensure that this configuration is stable against perturbative corrections (but, as we have seen, not against non-perturbative instantonic corrections).

For the applications it is important to determine the solutions of (E.3). In order to simplify the discussion, it is convenient to separately examine the case $N_f < N_c$ and $N_f \geq N_c$.

• For $N_f < N_c$ it can be easily seen that (up to symmetry operations) the most general solution of (E.3) is given by

$$\langle \phi^f_r \rangle = \langle \tilde{\phi}^{f\,\dagger}_r \rangle = \begin{cases} v_r \delta^{rf} & 1 \leq r \leq N_f \\ 0 & \text{otherwise}\,. \end{cases} \tag{E.4}$$

If the v's are all non-vanishing the gauge symmetry is broken from the original $SU(N_c)$ group down to $SU(N_c - N_f)$. In the special case $N_f = N_c - 1$ the gauge symmetry is completely broken. Among the quark superfields, $(2N_c - N_f)N_f$ of them become heavy owing to the super-Higgs mechanism, while the remaining N_f^2 will contain the Goldstone bosons of the various broken global symmetries, as well as their superpartners. The pattern of surviving symmetries will depend upon the detailed values assumed by the v_r's in (E.4).

• For $N_f \geq N_c$ the analysis of (E.3) is a bit more involved. The result is that (up to symmetry operations) the most general pattern of scalar v.e.v.'s that makes D^a vanish is

$$\langle \phi_r^f \rangle = \begin{cases} v_r \delta^{rf} & 1 \le f \le N_c \\ 0 & \text{otherwise} \end{cases} \tag{E.5}$$

$$\langle \tilde{\phi}_r^{f\,\dagger} \rangle = \begin{cases} (|v_r|^2 - b^2)^{\frac{1}{2}} \delta^{rf} & 1 \le f \le N_c \\ 0 & \text{otherwise} \end{cases} \tag{E.6}$$

where b is an arbitrary constant. For non-zero v_r the gauge symmetry is completely broken and $N_c^2 - 1$ quarks become massive by the super-Higgs mechanism. Again the detailed pattern of surviving symmetries depend on the particular values taken by the scalar v.e.v.'s (E.5) and (E.6).

We wish to conclude with a comment. As we have seen, the vacuum manifold is not compact. This is due to the fact that the symmetry of the set of supersymmetric vacua is a certain complexification of the symmetry group of the Lagrangian [172]. In fact, any rescaling of the massless scalar fields, although not a symmetry of the theory, when applied to a vacuum configuration leads to another acceptable, physically inequivalent, vacuum.

Appendix F – $\mathcal{N} = 2$ Lagrangian and Supersymmetry Transformations

Rigid $\mathcal{N} = 2$ supersymmetric theories consist of two kinds of massless multiplets. Vector multiplets and hypermultiplets.

Vector multiplets contain a vector A_μ, two spinor gaugini λ_α^r and a complex scalar ϕ all transforming in the adjoint representation of the gauge group. Vector multiplets are described by chiral superfields usually denoted by $\mathcal{A}$ whose θ expansion schematically reads

$$\mathcal{A}(x,\theta) = a(x) + \theta_\alpha^r \lambda_r^\alpha(x) + \frac{1}{2} \theta_\alpha^r \sigma^{\mu\nu\alpha}{}_\beta \theta_r^\beta F_{\mu\nu}(x) + \cdots . \tag{F.1}$$

Higher order terms in θ^r with $r = 1, 2$ can be expressed as derivatives of the lower ones.

The Lagrangian of pure $\mathcal{N} = 2$ SYM theory is given by

$$L = \int \mathrm{d}^4\theta \, \mathcal{F}(\mathcal{A}) , \tag{F.2}$$

where $\mathcal{F}(\mathcal{A})$ is a group invariant function of the chiral superfields. Renormalisability restricts $\mathcal{F}(\mathcal{A})$ to be quadratic

$$\mathcal{F}(\mathcal{A}) = \frac{1}{2} \tau_0 \mathcal{A}^2 , \tag{F.3}$$

so that

$$\tau_0 \delta_{ab} = \frac{\partial^2 \mathcal{F}(\mathcal{A})}{\partial \mathcal{A}_a \partial \mathcal{A}_b} \tag{F.4}$$

and

$$L_{\mathcal{N}=2} = \mathrm{Im}\,\tau_0\,\mathrm{Tr}\left(\frac{1}{4}F_{\mu\nu}F^{\mu\nu} + i\lambda^r\sigma^\mu D_\mu\bar{\lambda}_r + D_\mu\phi D^\mu\phi^\dagger + [\phi,\phi^\dagger]^2 + \lambda^r[\phi^\dagger,\lambda_r] + \bar{\lambda}^r[\phi,\bar{\lambda}_r]\right) . \tag{F.5}$$

The Lagrangian $L_{\mathcal{N}=2}$ is invariant under Poincaré transformations (up to total derivatives), under $U(2)_R$ R-symmetry transformations and under the global $\mathcal{N}=2$ supersymmetry transformations

$$\begin{aligned} \delta A_\mu &= \eta^r\sigma_\mu\bar{\lambda}_r + \bar{\eta}^r\bar{\sigma}_\mu\lambda_r \\ \delta\lambda^r &= (\frac{1}{2}F_{\mu\nu}\sigma^{\mu\nu} + [\phi,\phi^\dagger])\eta^r + i\sigma^\mu D_\mu\phi\bar{\eta}^r \\ \delta\phi &= \eta^r\lambda_r\,, \end{aligned} \tag{F.6}$$

where A_μ is the gauge potential associated with the field strength $F_{\mu\nu}$. R-symmetry indices are raised and lowered with the symplectic matrix ε^{rs}.

Appendix G – BPS Configurations

The acronym BPS, for Bogomol'nyi–Prasad–Sommefield, was initially introduced to designate certain solitonic solutions in non-supersymmetric quantum field theories. The simplest configuration of this type is a symmetric monopole arising in the Georgi–Glashow model [173] describing a $SU(2)$ gauge field coupled to a scalar field in the adjoint representation. The Lagrangian of the model is

$$\mathcal{L} = -\frac{1}{4}F^a_{\mu\nu}F^{a\mu\nu} + \frac{1}{2}D_\mu\Phi^a D^\mu\Phi^a - \frac{\lambda}{4}(\Phi^a\Phi^a - v^2)^2\,, \qquad a = 1,2,3\,, \tag{G.1}$$

with gauge coupling constant e. As shown by 't Hooft [95] and Polyakov [96], in the Coulomb phase, i.e. in the presence of a v.e.v. for the adjoint scalar, the theory admits monopole solutions characterised by an integer-valued topological charge. Static finite energy configurations have vanishing scalar potential at spatial infinity. The condition for the vanishing of the potential defines a two-sphere, $\sum_a \Phi^a\Phi^a = v^2$. Therefore, for such configurations, the scalar field provides a map from the two-sphere at spatial infinity into the two-sphere of the Higgs vacuum. This map defines the second homotopy group of S^2, $\Pi_2(S^2) \equiv$. As a result, the magnetic charge, g, associated with a solution of the field equations satisfies a Dirac quantisation condition. Denoting by $\boldsymbol{B}$ the non-abelian magnetic field with components $B^{ai} = -\frac{1}{2}\varepsilon^{ijk}F^a_{jk}$, one finds

$$g = \int_{S^2_\infty} \boldsymbol{B}\cdot \mathrm{d}\boldsymbol{\Sigma} = \frac{1}{2ev^3}\int \mathrm{d}\Sigma^i\,\varepsilon^{ijk}\varepsilon^{abc}\Phi^a\partial^j\Phi^b\partial^k\Phi^c = \frac{4\pi n}{e}\,, \tag{G.2}$$

where the integer n is the winding number determined by the behaviour of the Higgs field at spatial infinity.

No exact solution to the complete non-linear field equations is know explicitly, even in the simplest case of gauge group $SU(2)$. However, the analysis can be drastically simplified taking advantage of the implications of a general bound on the mass of field configurations with non-vanishing winding number known as the *Bogomol'nyi bound*. For a static field configuration with vanishing electric field, $E^{ai} = -F^{a0i} = 0$, the energy (mass) satisfies

$$\begin{aligned} M &= \int \mathrm{d}^3 r \, \frac{1}{2} \left[\boldsymbol{B}^a \cdot \boldsymbol{B}^a + \boldsymbol{D}\Phi^a \cdot \boldsymbol{D}\Phi^a + V(\Phi) \right] \\ &\geq \frac{1}{2} \int \mathrm{d}^3 r \, (\boldsymbol{B}^a - \boldsymbol{D}\Phi^a)^2 + vg \, , \end{aligned} \tag{G.3}$$

implying the bound

$$M \geq vg \, . \tag{G.4}$$

Minimal energy configurations saturate the bound and thus should have vanishing potential and should satisfy the first-order Bogomol'nyi equation

$$\boldsymbol{B}^a = \boldsymbol{D}\Phi^a \, . \tag{G.5}$$

The first explicit example of solution to (G.5) with $M = vg$ is the spherically symmetric one constructed by Prasad and Sommerfield [98]. Following their analysis, the expression *BPS saturated* has been used to designate solutions to the field equations saturating the Bogomol'nyi bound. For dyons with electric and magnetic charges e and g, respectively, the bound generalises to

$$M \geq v(e^2 + g^2)^{1/2} \, . \tag{G.6}$$

A comprehensive review of the physics of solitons and monopoles in gauge theory can be found in [174].

Appendix H – Extended Superalgebras, Central Charges and Multiplet Shortening

In the context of supersymmetric theories, certain short multiplets which correspond to special representations of the supersymmetry algebra (see also Appendix A) are referred to as *BPS multiplets*. States in such multiplets saturate a generalisation of the Bogomol'nyi bound (G.4), which relates their mass, M, to their "central charge".

The $\mathcal{N} = 1$ supersymmetry algebra in $D = 4$,

$$\{Q_\alpha, \bar{Q}_{\dot\alpha}\} = i\sigma^\mu_{\alpha\dot\alpha} P_\mu \, , \qquad \{Q_\alpha, Q_\beta\} = 0 \, , \tag{H.1}$$

admits no central extension. Generalised (non-scalar) central charges associated with the existence of domain wall configurations may appear, but they carry Lorentz indices.

Extended supersymmetry algebras, on the other hand, admit non-trivial *bona fide* central charges, usually denoted by Z. The $\mathcal{N}=1$ superalgebra (H.1) can be generalised to

$$\{Q^A_\alpha, \bar{Q}_{\dot\alpha B}\} = i\delta^A{}_B \sigma^\mu_{\alpha\dot\alpha} P_\mu \,, \qquad \{Q^A_\alpha, Q^B_\beta\} = Z^{AB}\epsilon_{\alpha\beta} \,, \tag{H.2}$$

where $A, B = 1, ..., \mathcal{N}$ are supersymmetry indices and the central charges, Z^{AB}, satisfy $Z^{AB} = -Z^{BA}$. In particular, for $\mathcal{N}=2$, there is only one complex central charge, $Z \equiv Z^{12}$, while for $\mathcal{N}=4$ there are six complex central charges satisfying a (self) duality condition $Z^{AB} = \frac{1}{2}\epsilon^{ABCD}\bar{Z}_{CD}$, very much as the elementary scalar fields in the theory. By means of a R-symmetry transformation, the central charges can be skew diagonalised and shown to satisfy

$$M \geq |Z_1| \geq |Z_2| \geq \cdots \geq |Z_i| \geq \cdots \geq 0 \,, \tag{H.3}$$

where $M^2 = P_\mu P^\mu$ is one of the Casimirs of the representation one is considering, for a proper ordering of the skew eigenvalues Z_i, $i = 1, ..., [\mathcal{N}/2]$. For $\mathcal{N}$ odd, one eigenvalue is necessarily zero by Binet's theorem. The relation (H.3) represents a generalisation of the Bogomol'nyi bound.

Irreducible "massive" representations of the supersymmetry algebra are indeed constructed by going to the rest frame. This reduces the form of the algebra to that of a Clifford algebra and one can split the $4\mathcal{N}$ supercharges into $2\mathcal{N}$ creation operators and $2\mathcal{N}$ annihilation operators by considering suitable linear combinations [164]. This means that, in general, a massive multiplet consists of $2^{2\mathcal{N}}$ states, $2^{2\mathcal{N}-1}$ bosonic and as many fermionic. However, if some of the central charges coincide with M the multiplet shortens, since some of the creation operators annihilate the ground state.

In the case of $\mathcal{N}=2$, this happens when $M = |Z|$. The corresponding multiplet is said to be 1/2 BPS since half the creation operators (four out of eight) act trivially. As a result the supermultiplet contains only half as many states, i.e. eight (four bosons and four fermions) instead of $16 = 2^4$.

In the case of the $\mathcal{N}=4$ superalgebra, one has two sub-cases $M = |Z_1| = |Z_2|$ (1/2 BPS) and $M = |Z_1| \neq |Z_2|$ (1/4 BPS). The corresponding multiplets are, respectively, 1/2 and 3/4 the length of ordinary $\mathcal{N}=4$ multiplets.

As discussed in Sect. 13, in the conformal phase the $\mathcal{N}=4$ SYM theory has a larger group of symmetries, the $\mathcal{N}=4$ superconformal group, $PSU(2,2|4)$, see Appendix T.

In this situation the fundamental degrees of freedom of the theory are gauge-invariant composite operators which are organised into multiplets forming unitary irreducible representations (UIRs) of $PSU(2,2|4)$. Each composite operator in a multiplet can be labelled by the quantum numbers associated with the maximal bosonic sub-group of $PSU(2,2|4)$, $SO(2,4) \times SO(6)$, i.e. two spins, j_1 and j_2, the scaling dimension, Δ, and three $SO(6)$ Dynkin labels, $[k, l, m]$.

A further generalisation of the concept of BPS multiplet arises in this context. The UIRs of $PSU(2,2|4)$ have been classified in [175]. For a review and applications to the AdS/CFT correspondence, see [118, 92]. Ordinary long representations contain a number of states proportional to 2^{16}, with the proportionality constant related to the dimension of the representation of the bosonic sub-group under which the lowest component transforms. Shorter representations arise when specific relations occur among the $SO(2,4) \times SO(6)$ quantum numbers of the lowest component of the multiplet. The correlation functions considered in Sect. 15 involve operators belonging to multiplets classified as 1/2 BPS. These are characterised by the fact that their lowest component is a Lorentz scalar operator of dimension $\Delta = \ell$, with $\ell \geq 2$, transforming in the $[0, \ell, 0]$ representation of the $SO(6)$ R-symmetry group. Generic multiplets of this type have $\frac{1}{12}\ell^2(\ell^2 - 1)\, 2^8$ components. The cases $\ell = 2, 3$ are special in that they are characterised by a further accidental shortening and are sometimes referred to as ultra-short. Many other shortening conditions can be identified. For instance, 1/4 BPS multiplets arise when the lowest component is a Lorentz scalar, double trace operator with $\Delta = 2k + l$ transforming in the representation $[k, l, k]$ of $SO(6)$. In the case of 1/2 and 1/4 BPS multiplets the range of spin is, respectively, 4 and 6 units, whereas long multiplets have a spin range of 8 units.

Appendix I – The $\mathcal{N} = 4$ Superconformal Group

In this appendix we summarise some basic properties of the four-dimensional $\mathcal{N} = 4$ superconformal group, $PSU(2, 2|4)$. More details and references can be found in the reviews [91]. The maximal bosonic subgroup of $PSU(2, 2|4)$ is the direct product of the four-dimensional conformal group, $SO(2, 4) \sim SU(2, 2)$, and of the R-symmetry group of the $\mathcal{N} = 4$ superalgebra, $SO(6) \sim SU(4)$. The conformal group is the group of transformations which preserve the form of the metric up to a (position dependent) scale factor. In four-dimensional Minkowski space, with metric $\eta_{\mu\nu} = \mathrm{diag}(-, +, +, +)$, it is generated by translations, Lorentz transformations, dilations and special conformal transformations. We denote the corresponding generators respectively by P_μ, $L_{\mu\nu}$, D and K_μ, $\mu, \nu = 0, 1, 2, 3$. For the generators of the $SU(4)$ R-symmetry we use T^a, $a = 1, 2, \ldots, 15$.

The action of infinitesimal conformal transformations on the coordinates, $x_\mu \to x'_\mu(x) = x_\mu + \delta x_\mu(x)$, is the following:

$$\begin{aligned}
&\delta x_\mu(x) = a_\mu \quad \text{(translations)} \\
&\delta x_\mu(x) = \Lambda_\mu{}^\nu x_\nu \quad \text{(Lorentz transformations)} \\
&\delta x_\mu(x) = \lambda x_\mu \quad \text{(dilations)} \\
&\delta x_\mu(x) = 2 b_\nu x^\nu x_\mu - x_\nu x^\nu b_\mu \quad \text{(special conformal transformations)}\,,
\end{aligned} \tag{I.1}$$

where a_μ and b_μ are constant vectors, $\lambda \in \;^{+}$ and the constant matrix $\Lambda_\mu{}^\nu$ satisfies $\eta^{\rho\sigma} \Lambda_\rho{}^\mu \Lambda_\sigma{}^\nu = \eta^{\mu\nu}$.

The fermionic symmetries in $PSU(2,2|4)$ comprise 16 Poincaré supersymmetries, generated by the supercharges Q^A_α and $\bar{Q}^{\dot\alpha}_A$, with $A = 1,2,3,4$ and $\alpha, \dot\alpha = 1,2$, and 16 special (or conformal) supersymmetries, generated by the supercharges S^α_A and $\bar{S}^A_{\dot\alpha}$.

As explained in Appendix G the superconformal algebra also admits six complex scalar central charges as well as additional generalised central charges which carry Lorentz indices.

Neglecting the central extensions the superconformal algebra reads

$$\begin{aligned}
&[L_{\mu\nu}, P_\rho] = -i(\eta_{\mu\rho}P_\nu - \eta_{\nu\rho}P_\mu)\,, \quad [L_{\mu\nu}, K_\rho] = -i(\eta_{\mu\rho}K_\nu - \eta_{\nu\rho}K_\mu)\,,\\
&[L_{\mu\nu}, L_{\rho\sigma}] = -i\eta_{\mu\rho}L_{\nu\sigma} + \text{permutations}\,, \quad [P_\mu, K_\nu] = 2iL_{\mu\nu} - 2i\eta_{\mu\nu}D\,,\\
&[D, L_{\mu\nu}] = 0\,, \quad [D, P_\mu] = -iP_\mu\,, \quad [D, K_\mu] = iK_\mu\,,\\
&\{Q^A_\alpha, Q^B_\beta\} = \{S^\alpha_A, S^\beta_B\} = \{Q^A_\alpha, \bar{S}^B_{\dot\alpha}\} = \{\bar{Q}^{\dot\alpha}_A, S^\beta_B\} = 0\,,\\
&\{Q^A_\alpha, \bar{Q}_{\dot\alpha B}\} = 2\sigma^\mu_{\alpha\dot\alpha}P_\mu\delta^A{}_B\,, \quad \{S_{\alpha A}, \bar{S}^B_{\dot\alpha}\} = 2\sigma^\mu_{\alpha\dot\alpha}K_\mu\delta_A{}^B\,,\\
&\{Q^A_\alpha, S_{\beta B}\} = \varepsilon_{\alpha\beta}(\delta^A{}_B D + T^A{}_B) + \frac{1}{2}\delta^A{}_B L_{\mu\nu}\varepsilon_{\beta\gamma}\sigma^{\mu\nu\,\gamma}_{\alpha}\,.
\end{aligned} \tag{I.2}$$

The R-symmetry group of automorphisms of the generic $\mathcal{N}$-extended supersymmetry algebra is $U(\mathcal{N})$. The $\mathcal{N} = 4$ case under consideration is special in that the $U(1)$ factor in the decomposition $U(\mathcal{N}) = SU(\mathcal{N}) \times U(1)$ of the R-symmetry becomes an outer automorphism: none of the other generators in the algebra is charged under this $U(1)$ symmetry [175, 176] and all the fields and composite operators in $\mathcal{N} = 4$ SYM are neutral under this central $U(1)$. The absence of the abelian factor in the R-symmetry is reflected in the notation $PSU(2,2|4)$ as opposed to $SU(2,2|4)$.

As has been discussed in Sect. 13, the observables in the $\mathcal{N} = 4$ SYM theory are correlation functions of local gauge-invariant composite operators. Such operators are labelled by the quantum numbers characterising their transformation under the bosonic subgroup $SO(2,4) \times SO(6)$. A class of operators playing a special role in a conformal field theory such as $\mathcal{N} = 4$ SYM are the *conformal primary operators*. These are defined by the condition of being annihilated by special conformal transformations acting at the origin,

$$[K_\mu, \mathcal{O}(x)]\big|_{x=0} = 0\,. \tag{I.3}$$

The existence of such operators is associated with the presence of a lower bound on the dimension of fields and operators in a unitary conformal field theory. Since the action of K_μ lowers the dimension of an operator, the existence of the unitarity bound implies that in every representation of the conformal group there must be an operator satisfying (I.3). The action of the generators of the conformal group on primary operators is as follows:

$$\begin{aligned}
&[P_\mu, \mathcal{O}(x)] = i\partial_\mu\mathcal{O}(x)\\
&[L_{\mu\nu}, \mathcal{O}(x)] = [i(x_\mu\partial_\nu - x_\nu\partial_\mu) + M_{\mu\nu}]\mathcal{O}(x)\\
&[D, \mathcal{O}(x)] = -i(\Delta - x^\mu\partial_\mu)\mathcal{O}(x)\\
&[K_\mu, \mathcal{O}(x)] = [i(x^2\partial_\mu - 2x_\mu x^\nu\partial_\nu + 2x_\mu\Delta) - 2x^\nu M_{\mu\nu}]\mathcal{O}(x)\,.
\end{aligned} \tag{I.4}$$

The functional form of two- and three-point functions of primary operators is fixed by conformal invariance. In the case of Lorentz scalars, for instance, the two-point function vanishes unless the two operators have the same scaling dimension, in which case it takes the form

$$\langle \mathcal{O}_i(x_1)\, \mathcal{O}_j(x_2)\rangle = \frac{c_{ij}}{|x_{12}|^{2\Delta}}\,, \tag{I.5}$$

where $x_{12} = x_1 - x_2$, c_{ij} are constants and Δ is the common dimension of $\mathcal{O}_i$ and $\mathcal{O}_j$. For three-point functions conformal invariance implies

$$\langle \mathcal{O}_i(x_1)\mathcal{O}_j(x_2)\mathcal{O}_k(x_3)\rangle = \frac{c_{ijk}}{|x_{12}|^{\Delta_i+\Delta_j-\Delta_k}|x_{13}|^{\Delta_i+\Delta_k-\Delta_j}|x_{23}|^{\Delta_j+\Delta_k-\Delta_i}}\,, \tag{I.6}$$

where c_{ijk} are numerical coefficients. The form of four- and higher-point functions is not completely determined by the conformal symmetry. Four-point functions, for instance, are determined up a function of two conformally invariant cross-ratios constructed from the four insertion points, e.g. $r = x_{12}^2 x_{34}^2/x_{13}^2 x_{24}^2$ and $s = x_{14}^2 x_{23}^2/x_{13}^2 x_{24}^2$. The scaling dimensions and the coefficients, c_{ij} and c_{ijk}, in (I.5) and (I.6) are in general functions of the Yang–Mills coupling constant, g, and the ϑ-angle.

Local composite operators in $\mathcal{N} = 4$ SYM are organised in multiplets of the superconformal group. The bottom component of any such multiplet, i.e. the operator of lowest dimension, is referred to as a *superconformal primary operator*. Superconformal primary operators are annihilated by the special supersymmetry generators acting at the origin,

$$\{S^A_\alpha, \mathcal{O}(x)]\big|_{x=0} = 0\,, \tag{I.7}$$

where the symbol $\{S, \mathcal{O}]$ indicates a commutator if $\mathcal{O}$ is bosonic and an anti-commutator if $\mathcal{O}$ is fermionic. Notice that superconformal primary operators are always also conformal primaries, but the opposite is not true.

As discussed in Appendix G there are special UIRs of the superconformal group corresponding to short BPS multiplets. Operators in such multiplets are protected and their two- and three-point functions do not receive quantum corrections. This implies that their scaling dimensions and three-point couplings are not renormalised.

Appendix J – Compendium of Differential Geometry

An n-dimensional topological manifold is a set of points such that the neighbourhood of a point P (any open set containing the point P) looks like ^{n}. In order to describe a manifold, one needs an atlas made of many patches that are related to one another by transition functions. If the transition functions are continuous, then the manifold is continuous. If the transition functions are differentiable, then the manifold is differentiable. If the transition functions are complex analytic, then the manifold is complex.

One can add further structures. On a differentiable manifold, one can define a metric which is a symmetric bilinear form on the vector fields such that $g(U,V) = g(V,U) = g_{ij}U^iV^j$ in a local coordinate patch. Parallel transport is achieved by means of a connection Γ^i_{jk} that can be fixed to be the Christoffel connection imposing that the metric be covariantly constant, i.e. $0 = D_i g_{jk} \equiv \partial_i g_{jk} - \Gamma^l_{ik} g_{jl} - \Gamma^l_{ji} g_{lk}$. One can then construct the Riemann curvature tensor R^k_{ijl} and its contractions, the Ricci curvature tensor R_{il} and the scalar curvature R. After parallel transport a vector gets transformed by means of a $SO(n)$ rotation. Transformations along closed paths form the holonomy group of the manifold, which is necessarily a subgroup of $SO(n)$.

On differentiable manifolds, one can define p-forms with $p < n$. A 0-form is a function, a 1-form is combination of the differentials of the local coordinates $A = A_i(x)\mathrm{d}x^i$. In a local coordinate patch

$$A_p = \frac{1}{p!} \sum_{i_1,\ldots,i_p} A_{i_1,\ldots,i_p} \mathrm{d}x^{i_1} \wedge \ldots \wedge \mathrm{d}x^{i_p} \,. \tag{J.1}$$

where the (anti-)symmetric wedge product satisfies $A_p \wedge B_q = (-)^{pq} B_q \wedge A_p$. On forms one can define an exterior differential $\mathrm{d}A_p = B_{p+1}$ that satisfies the (graded) Leibniz rule. On a Riemannian manifold, one can also define a Hodge star operator $*A_p = \tilde{A}_{n-p}$. Combining d and $*$ one can define a differential operator δ such that $\delta A_p \equiv *\,\mathrm{d}*\,A_p = C_{p-1}$, that generalises the divergence. The Lie derivative of a p-form along a vector field V is defined by

$$\mathcal{L}_V A_p = \iota_V \mathrm{d}A_p + \mathrm{d}(\iota_V A) \,, \tag{J.2}$$

where ι_V denotes contraction with V. In a local coordinate patch one has

$$\mathcal{L}_V A_{i_1,\ldots,i_p} = V^i \partial_i A_{i_1,\ldots,i_p} - \sum_{k=1}^{p} A_{i_1,\ldots,i,\ldots,i_p} \partial_{i_k} V^i \,. \tag{J.3}$$

Both d and δ are nilpotent, i.e. $\mathrm{d}^2 = 0$ and $\delta^2 = 0$. The generalised Laplacian is given by $\Delta A_p = (\mathrm{d}\delta + \delta\mathrm{d})A_p$. It coincides with the standard Laplacian $\Delta = ||g||^{-1/2}\partial_i(||g||^{1/2}g^{ij}\partial_j)$ on 0-forms (scalars). A form is closed if $\mathrm{d}A = 0$ and exact if $A = \mathrm{d}C$. A form is co-closed if $\delta A = 0$ and co-exact if $A = \delta C$. A form which is closed and co-closed is harmonic, i.e. $\Delta A = 0$. The equivalence classes of closed forms $\mathcal{C}^p$ that differ by exact forms $\mathcal{E}^p$ define the cohomology groups $\mathcal{H}^p = \mathcal{C}^p/\mathcal{E}^p$. De Rham has shown that one can always find a harmonic representative in each cohomology class.[54]

[54] Given a closed p-form such that $\mathrm{d}A = 0$ but $\delta A \neq 0$, one can always find a cohomologous form $A' = A + \mathrm{d}\Lambda$ such that $\mathrm{d}A' = 0$, by construction, and $\delta A' = 0$ by requiring that $\delta\mathrm{d}\Lambda = -\delta A$, i.e. inverting the elliptic operator $\Lambda = -(\delta\mathrm{d})^{-1}\delta A$.

A symplectic manifold is an even dimensional manifold that admits a closed 2-form, known as symplectic form, e.g. for the phase space of a point in ^{n} one has $\Omega = \sum_i \mathrm{d}p_i \wedge \mathrm{d}x^i$.

On complex manifolds, one can decompose d as $d = \partial + \bar{\partial}$, where both ∂ and $\bar{\partial}$ are nilpotent. By a complex coordinate change, one can always put the metric in Hermitean form $\mathrm{d}s^2 = g_{i\bar{j}}\mathrm{d}z^i\mathrm{d}\bar{z}^{\bar{j}}$ at least locally.

The Kähler form on a complex Riemannian manifold is $\omega = g_{i\bar{j}}\mathrm{d}z^i \wedge \mathrm{d}\bar{z}^{\bar{j}}$. If ω is closed, $\mathrm{d}\omega = 0$, which implies $\partial\omega = 0 = \bar{\partial}\omega$, the manifold is Kähler. Locally $\omega = \partial\bar{\partial}K$, where $K(z,\bar{z})$ is the Kähler potential. If the manifold has real dimension $4n$ and admits three closed Kähler forms, $\mathrm{d}\omega^I = 0$, $I = 1, 2, 3$, whose components satisfy the algebra of quaternions the manifold is said to be hyper-Kähler. If the three Kähler forms are not closed but rather $\mathrm{d}\omega^I = c_n\epsilon^{IJK}\omega_J \wedge \omega_K$, the manifold is said to be quaternionic.

An isometry of the metric is a coordinate transformation that leaves the metric invariant and is thus generated by a vector field V that satisfies

$$0 = \mathcal{L}_V g_{ij} = V^k\partial_k g_{ij} - g_{ik}\partial_j V^k - g_{ik}\partial_j V^k \equiv -\nabla_i V_j - \nabla_j V_i \,, \qquad \text{(J.4)}$$

where indices are raised and lowered with the metric.

A holomorphic isometry is such that $\mathcal{L}_V\omega = 0$. Thanks to the closure of ω, V admits a prepotential because $\mathrm{d}(\iota_V\omega) = 0$ implies $\iota_V\omega = \mathrm{d}\mu_V$ locally. The prepotential is known also as the holomorphic Kähler map. A tri-holomorphic isometry is such that $\mathcal{L}_V\omega^I = 0$. Thanks to the closure of ω^I, V admits three prepotentials because $\mathrm{d}(\iota_V\omega^I) = 0$ implies $\iota_V\omega^I = \mathrm{d}\mu_V^I$ locally. The prepotentials are known also as the tri-holomorphic Kähler maps.

References

1. A.A. Belavin, A.M. Polyakov, A.M. Schwartz, Yu.S. Tyupkin: Phys. Lett. B **59**, 85 (1975)
2. G. 't Hooft: Phys. Rev. Lett. **37**, 8 (1976); Phys. Rev. D **14**, 3432 (1976); ibid. **18**, 2199 (1978); Phys. Rep. **142**, 357 (1986)
3. S.R. Coleman: *The Uses of Instantons*, Lecture delivered at 1977 International School of Subnuclear Physics, Erice, Italy, 23 July–10 August, 1977 (Plenum Press, New York, 1978)
4. D. Amati, K. Konishi, Y. Meurice, G.C. Rossi, G. Veneziano: Phys. Rep. **162**, 169 (1988)
5. A.I. Vainshtein, V.I. Zakharov, V.A. Novikov, M.A. Shifman: Sov. Phys. Usp. **25** 195 (1982) [Usp. Fiz. Nauk **136** 553 (1982)], revised and updated version published in *ITEP Lectures on Particle Physics and Field Theory* (World Scientific, Singapore, 1999), Vol. 1, pp. 201–299;
   M.A. Shifman: Lectures given at the International School of Physics "Enrico Fermi", Varenna, Italy, 3–6 July 1995, `hep-th/9704114`;
   M.A. Shifman, A.I. Vainhstein: `hep-th/9902016`
6. N. Dorey, T.J. Hollowood, V.V. Khoze, M.P. Mattis: Phys. Rep. **371**, 231 (2002), and references therein

7. M. Nakahara: *Geometry, Topology and Physics*, Graduate Student Series in Physics, Gen. Ed. D.F. Brewer (Institute of Physics Publishing, Bristol and Philadelphia, 1990)
8. J.L. Gervais, B. Sakita: Phys. Rev. B **11**, 2943 (1975);
E. Tomboulis: Phys. Rev. B **12**, 1678 (1975)
9. C. Bernard: Phys. Rev. D **19**, 3013 (1979)
10. L.G. Yaffe: Nucl. Phys. B **151**, 247 (1979).
11. R. Jackiw, C. Rebbi: Phys. Rev. Lett. **37**, 172 (1976); Phys. Rev. D **14**, 517 (1976)
12. C. Callan, R. Dashen, D. Gross: Phys. Lett. B **63**, 334 (1976); Phys. Lett. B**66**, 375 (1977)
13. D.I. Olive, R.J. Crewther, S. Sciuto: Riv. Nuovo Cimento **2N8**, 1 (1979)
14. F.A. Berezin: *The Method of Second Quantization* (Academic Press, New York, 1966);
L.D. Faddeev: *Introduction to Functional Methods*, in *Methods in Field Theory*, Les Houches 1975, Eds. R. Balian, J. Zinn-Justin (North-Holland, Amsterdam, 1976);
P. Ramond: *Field Theory: a Modern Primer* (Benjamin-Cumming, Reading, Mass., 1981)
15. Y. Meurice: Phys. Lett. B **164**, 141 (1985)
16. L. Maiani, G.C. Rossi and M. Testa: Phys. Lett. B **292**, 397 (1992)
17. M.F. Atiyah, I.M. Singer: Bull. Amer. Math. Soc. **69**, 422 (1963); Ann. Math. **87**, 485; 546 (1968);
M.F. Atiyah, G.B. Segal: Ann. Math. **87**, 531 (1968);
L. Alvarez-Gaumé: Comm. Math. Phys. **90**, 161 (1983);
D. Friedan, P. Windey: Nucl. Phys. B **235**, 395 (1984)
18. C.W. Bernard, N.H. Christ, A.H. Guth, E.J. Weinberg: Phys. Rev. D **16**, 2967 (1977)
19. M.F. Atiyah, V. Drinfeld, N. Hitchin, Y. Manin: Phys. Lett. A **65**, 185 (1978)
20. E. Corrigan, D. Fairlie, S. Templeton, P. Goddard: Nucl. Phys. B **140**, 31 (1978);
N.H. Christ, E.G. Weinberg, N.K. Stanton: Phys. Rev. D **18**, 2013 (1978);
E. Corrigan, P. Goddard, S. Templeton: Nucl. Phys. B **151**, 63 (1979)
21. I. Affleck, M. Dine, N. Seiberg: Phys. Rev. Lett. **51**, 1026 (1983); Nucl. Phys. B**241**, 493 (1984); Nucl. Phys. B **256**, 557 (1985)
22. V.A. Novikov, M.A. Shifman, A.I. Vainshtein, V.I. Zakharov: JETP Lett. **39** 601 (1984)
23. S.F. Cordes: Nucl. Phys. B **273**, 629 (1986)
24. A. D'Adda, P. Di Vecchia: Phys. Lett. B **73**, 162 (1978)
25. G. Veneziano: Phys. Lett. B **124**, 357 (1983)
26. D. Amati, G.C. Rossi, G. Veneziano: Nucl. Phys. B **249**, 1 (1985)
27. S. Coleman, J. Wess, B. Zumino: Phys. Rev. **177**, 2239 (1969);
C.G. Callan, S. Coleman, J. Wess, B. Zumino: Phys. Rev. **177**, 2247 (1969);
S. Weinberg: Physica A **96**, 327 (1979)
28. J. Gasser, H. Leutwyler: Phys. Rep. **87**, 77 (1982); Ann. Phys. **158**, 142 (1984); Nucl. Phys. B **250**, 465 (1985)
29. G. Veneziano, S. Yankielowicz: Phys. Lett. B **113**, 321 (1982)
30. T. Taylor, G. Veneziano, S. Yankielowicz: Nucl. Phys.B **218**, 493 (1983)
31. I. Affleck, M. Dine, N. Seiberg: Phys. Lett. B **137**, 187 (1983); Phys. Rev. Lett. **52**, 1677 (1984); Phys. Lett. B **140**, 59 (1984)

32. V.A. Novikov, M.A. Shifman, A.I. Vainshtein, V.I. Zakharov: Nucl. Phys. B **223**, 445 (1983); **229**, 407 (1983)
33. G.C. Rossi, G. Veneziano: Phys. Lett. B **138**, 195 (1984).
34. J. Schwinger: Phys. Rev. **82**, 664 (1951);
S. Adler: Phys. Rev. **177**, 2426 (1969);
J.S. Bell, R. Jackiw: Nuovo Cimento A **60**, 47 (1969)
35. K. Konishi: Phys. Lett. B **135**, 439 (1984);
K. Konishi, K. Shizuya: Nuovo Cimento A **90**, 111 (1985);
T.E. Clark, O. Piguet, K. Sibold: Nucl. Phys. B **143**, 445 (1978); ibid. **159**, 1 (1979); *ibidem* **169**, 77 (1980);
S. Gates, Jr., M.T. Grisaru, M. Roček, W. Siegel: *Superspace* (Benjamin/Cummings, New York, 1983)
36. K. Konishi, H. Panagopoulos: Phys. Lett. B **191**, 290 (1987)
37. D. Finnell, P. Pouliot: Nucl. Phys. B **453**, 227 (1995)
38. T.J. Hollowood, V.V. Khoze, W.J. Lee, M.P. Mattis: Nucl. Phys. B **570**, 241 (2000)
39. V.A. Novikov, M.A. Shifman, A.I. Vainshtein, V.I. Zakharov: Nucl. Phys. B **229**, 381 (1983); Phys. Lett. B **166** 329 (1986] [Sov. J. Nucl. Phys. **43**, 294 (1986); Yad. Fiz. **43**, 459 (1986)];
T. Morris, D. Ross, C. Sachrajda: Phys. Lett. B **172**, 40 (1986)
40. E. Witten: Nucl. Phys. B **202**, 253 (1982)
41. N. Seiberg: Phys. Rev. D **49**, 6857 (1994); Nucl. Phys. B **431** 484 (1995)
42. T. Appelquist, J. Carazzone: Phys. Rev. D **11**, 2856 (1975)
43. D. Amati, Y. Meurice, G.C. Rossi, G. Veneziano: Nucl. Phys. B **263**, 591 (1986)
44. G. 't Hooft: in *Proceedings of the 1979 Cargése Summer School*, Eds. G. 't Hooft et al. (Plenum Press, New York, 1980)
45. V.A. Novikov, M.A. Shifman, A.I. Vainshtein, V.I. Zakharov: JETP Lett. **39**, 601 (1984); Nucl. Phys. B **260**, 157 (1985);
M.A. Shifman, A.I. Vainshtein, V.I. Zakharov: Usp. Fiz. Nauk **146**, 683 (1985) [Sov. Phys. Usp. **28**, 709 (1985)]
46. J. Fuchs, M.G. Schmidt: Z. Phys. C **30**, 161 (1986);
J. Fuchs: Nucl. Phys. B **272**, 677 (1986); ibid. **282**, 437 (1987)
47. I. Affleck: Nucl. Phys. B **191**, 429 (1981)
48. G. Curci, G. Veneziano: Nucl. Phys. B **292**, 555 (1987)
49. I. Montvay: Int. J. Mod. Phys. A **17**, 2377 (2002)
50. F. Buccella, J.P. Derendiger, S. Ferrara, C. Savoy: Phys. Lett. B **115**, 375 (1982)
51. N. Seiberg, E. Witten: Nucl.Phys. B **426**, 19 (1994); Erratum: ibid. **430**, 485 (1994)
52. J. Wess, B. Zumino: Phys. Lett. B **49**, 52 (1974);
J. Iliopoulos, B. Zumino: Nucl. Phys. B **76**, 310 (1974);
S. Ferrara, J. Iliopoulos, B. Zumino: Nucl. Phys. B **77**, 413 (1974);
S. Weinberg: Phys. Lett. B **62**, 111 (1976);
M.T. Grisaru, M. Roček, W. Siegel: Nucl. Phys. B **159**, 429 (1979)
53. H. Georgi, S. Glashow: Phys. Rev. Lett. **32**, 438 (1974)
54. Y. Meurice, G. Veneziano: Phys. Lett. B **141**, 69 (1984)
55. E. Guadagnini, K. Konishi: Nuovo Cimento A **90**, 400 (1985);
A. Bicci, K. Konishi: Europhys. Lett. **1**, 275 (1986)
56. K. Konishi: Nucl. Phys. B **289**, 253 (1987)

57. C. Rosenzweig, J. Schechter, G. Trahern: Phys. Rev. D **21**, 3388 (1980);
P. Di Vecchia, G. Veneziano: Nucl. Phys. B **171**, 253 (1980);
E. Witten: Ann. Phys. **128**, 363 (1980);
K. Kawarabayashi, N. Ohta: Nucl. Phys. B **175** 477 (1980);
K. Kawarabayashi, N. Ohta: Prog. Theor. Phys. **66** 1789 (1981);
P. Nath, A. Arnowitt: Phys. Rev. D **23**, 473 (1981)
58. K. Symanzik: in *New Developments in Gauge Theories*, Eds. G. 't Hooft et al. (Plenum, New York, 1980), p. 313;
K. Symanzik: "Some topics in quantum field theory" in *Mathematical Problems in Theoretical Physics*, Eds. R. Schrader et al., Lectures Notes in Physics, Vol. 153 (Springer, New York, 1982);
K. Symanzik: Nucl. Phys. B **226**, 187; Nucl. Phys. B **226**, 205 (1983)
59. G. Shore, G. Veneziano: Int. J. Mod. Phis. A **1**, 499 (1986);
M. Peskin: Proc. of the 1996 Theoretical Advanced Study Institute on *Fields, String and Duality* (Boulder, CO, 2–28 June 1996), `hep-th/9702094`
60. K.G. Wilson: Phys. Rev. B **4**, 3174 (1971);
J. Polchinski: Nucl. Phys. B **231**, 269 (1984);
G. Gallavotti: Rev. Mod. Phys. **57**, 471 (1985)
61. S. Arnone, C. Fusi, K. Yoshida: JHEP **9902**, 022 (1999);
S. Arnone, S. Chiantese, K. Yoshida: Int. J. Mod. Phys. A **16**, 1811 (2001);
S. Arnone, D. Francia, K. Yoshida: Mod. Phys. Lett. A **17**, 1191 (2002);
J. Ambjörn, R.A. Janik: Phys. Lett. B **569**, 81 (2003);
S. Arnone, K. Yoshida: Int. J. Mod. Phys. B **18**, 469 (2004);
S. Arnone, F. Guerrieri, K. Yoshida: JHEP **0405**, 031 (2004)
62. N. Arkani-Hamed, H. Murayama: Phys. Rev. D **57**, 6638 (1998); JHEP **0006**, 030 (2000)
63. R. Dijkgraaf, C. Vafa: `hep-th/0208048`;
R. Dijkgraaf, M.T. Grisaru, C.S. Lam, C. Vafa, D. Zanon: Phys. Lett. B **573**, 138 (2003)
64. G. Hailu, H. Georgi: JHEP **0402**, 038 (2004)
65. H. Kawai, T. Kuroki, T. Morita, K. Yoshida: Phys. Lett. B **611**, 269 (2005)
66. K.A. Intriligator, R.G. Leigh, N. Seiberg: Phys. Rev. D **50**, 1092 (1994)
67. A. Armoni, M.A. Shifman, G. Veneziano: Nucl. Phys. B **667**, 170 (2003); Phys. Rev. Lett. **91**, 191601 (2003); Phys. Lett. B **579**, 384 (2004);
A. Armoni, G. Shore, G. Veneziano: Nucl. Phys. B **740**, 23 (2006)
68. K. Konishi, G. Veneziano: Phys. Lett. B **187**, 106 (1987)
69. S. Ferrara, B. Zumino: Nucl. Phys. B **79**, 413 (1974);
M. Sohnius, K.S. Stelle, P.C. West: Phys. Lett. B **92**, 123 (1980);
W. Lerche: "Lecture on $N = 2$ supersymmetric gauge theory", given at the *NATO Advanced Study Institute: Les Houches Summer School on Theoretical Physics, Session 64: Quantum Symmetries*, Les Houches, France, 1 August–8 September 1995;
A. Bilal: `hep-th/9601007`
70. L. Brink, J.H. Schwarz, J. Scherk: Nucl. Phys. B **121**, 77 (1977);
F. Gliozzi, J. Scherk, D. Olive: Nucl. Phys. B **122**, 256 (1977)
71. P.S. Howe, K.S. Stelle, P.C. West: Phys. Lett. B **124**, 55 (1983).
72. L. Andrianopoli, M. Bertolini, A. Ceresole, R. D'Auria, S. Ferrara, P. Fre, T. Magri: J. Geom. Phys. **23**, 111 (1997)
73. C. Montonen, D.I. Olive: Phys. Lett. B **72**, 117 (1977)

74. N. Seiberg, E. Witten: Nucl. Phys. B **431**, 484 (1994)
75. M. Matone: Phys. Lett. B **357**, 342 (1995);
M. Matone: JHEP **0104**, 041 (2001)
76. F. Fucito, G. Travaglini: Phys. Rev. D **55**, 1099 (1997)
77. E. Witten: Commun. Math. Phys. **117**, 353 (1988)
78. N. Seiberg, E. Witten: JHEP **9909**, 032 (1999)
79. N. Nekrasov, A.S. Schwarz: Commun. Math. Phys. **198**, 689 (1998)
80. G. W. Moore, N. Nekrasov, S. Shatashvili: Commun. Math. Phys. **209**, 77 (2000)
81. N.A. Nekrasov: Commun. Math. Phys. **241**, 143 (2003)
82. N.A. Nekrasov: Adv. Theor. Math. Phys. **7**, 831 (2004)
83. M.R. Douglas: `hep-th/9512077`
84. E. Witten: Nucl. Phys. B **460**, 335 (1996);
E. Witten: JHEP **0204**, 012 (2002)
85. M. Billo, M. Frau, I. Pesando, F. Fucito, A. Lerda, A. Liccardo: JHEP **0302**, 045 (2003)
86. M. Billo, M. Frau, F. Fucito, A. Lerda: `hep-th/0606013`
87. A. Giveon, D. Kutasov: Rev. Mod. Phys. **71**, 983 (1999)
88. J.M. Maldacena: Adv. Theor. Math. Phys. **2**, 231 (1998) [Int. J. Theor. Phys. **38**, 1113 (1999)]
89. S.S. Gubser, I.R. Klebanov, A.M. Polyakov: Phys. Lett. B **428**, 105 (1998)
90. E. Witten: Adv. Theor. Math. Phys. **2**, 253 (1998)
91. O. Aharony, S.S. Gubser, J.M. Maldacena, H. Ooguri, Y. Oz: Phys. Rep. **323**, 183 (2000);
E. D'Hoker, D.Z. Freedman: Lectures given at *Theoretical Advanced Study Institute in Elementary Particle Physics on Strings, Branes and Extra Dimensions* (Boulder, CO, 3–29 June 2001), `hep-th/0201253`
92. M. Bianchi: Nucl. Phys. Proc. Suppl. **102**, 56 (2001);
M. Bianchi: Fortsch. Phys. **53**, 665 (2005)
93. M. Bianchi, F. Fucito, G.C. Rossi, M. Martellini: Nucl. Phys. B **440**, 129 (1995);
M. Bianchi, F. Fucito, G.C. Rossi, M. Martellini: Nucl. Phys. B **473**, 367 (1996)
94. S.R. Coleman, E. Weinberg: Phys. Rev. D **7**, 1888 (1973)
95. G. 't Hooft: Nucl. Phys. B **79**, 276 (1974).
96. A.M. Polyakov: JETP Lett. **20**, 194 (1974) [Pisma Zh. Eksp. Teor. Fiz. **20**, 430 (1974)]
97. B. Julia, A. Zee: Phys. Rev. D **11**, 2227 (1975)
98. M.K. Prasad, C.M. Sommerfield: Phys. Rev. Lett. **35**, 760 (1975)
99. W. Nahm: Phys. Lett. B **90**, 413 (1980)
100. E. Witten, D.I. Olive: Phys. Lett. B **78**, 97 (1978)
101. H. Osborn: Phys. Lett. B **83**, 321 (1979)
102. P. Goddard, J. Nuyts, D.I. Olive: Nucl. Phys. B **125**, 1 (1977)
103. G. 't Hooft: Nucl. Phys. B **153**, 141 (1979)
104. R. Dijkgraaf, M.T. Grisaru, H. Ooguri, C. Vafa, D. Zanon: JHEP **0404**, 028 (2004)
105. T.J. Hollowood: JHEP **0203**, 038 (2002) and Nucl. Phys. B **639**, 66 (2002).
106. D. Anselmi, P. Fré: Nucl. Phys. B **404**, 288 (1993) ; Phys. Lett. B **347**, 247 (1995);
E. Witten: Math. Res. Lett. **1**, 769 (1994)
107. A. Klemm, W. Lerche, S. Theisen: Int. J. Mod. Phys. A **11**, 1929 (1996)

108. S.K. Donaldson, P.B. Kronheimer: *The Geometry of Four-Manifolds*, Oxford Mathematical Monographs (Oxford University Press, Oxford, New York, 1997).
109. G. Veneziano: Nuovo Cimento A **57**, 190 (1968)
110. J. Dai, R.G. Leigh, J. Polchinski: Mod. Phys. Lett. A **4**, 2073 (1989);
R.G. Leigh: Mod. Phys. Lett. A **4**, 2767 (1989);
J. Polchinski: Phys. Rev. Lett. **75**, 4724 (1995);
C. Angelantonj, A. Sagnotti: Phys. Rept. **371**, 1 (2002) [Erratum: ibid. **376**, 339 (2003)]
111. E. Witten: JHEP **9807**, 006 (1998)
112. M.R. Douglas, G.W. Moore: hep-th/9603167
113. C. Angelantonj, A. Armoni: Phys. Lett. B **482**, 329 (2000)
114. M. Bianchi, J.F. Morales: JHEP **0008**, 035 (2000)
115. F. Fucito, J.F. Morales, R. Poghossian, A. Tanzini: JHEP **0601**, 031 (2006);
R. Blumenhagen, M. Cvetic, T. Weigand: hep-th/0609191;
M. Haack, D. Krefl, D. Lust, A. Van Proeyen, M. Zagermann: hep-th/0609211;
L.E. Ibanez, A.M. Uranga: hep-th/0609213.
B. Florea, S. Kachru, J. McGreevy, N. Saulina: hep-th/0610003;
N. Akerblom, R. Blumenhagen, D. Lust, E. Plauschinn, M. Schmidt-Sommerfeld: hep-th/0612132;
M. Bianchi, E. Kiritsis: hep-th/0702015
116. L.V. Avdeev, O.V. Tarasov, A.A. Vladimirov: Phys. Lett. B **96**, 94 (1980);
M.T. Grisaru, M. Roček, W. Siegel: Phys. Rev. Lett. **45**, 1063 (1980);
M.F. Sohnius, P.C. West: Phys. Lett. B **100**, 245 (1981);
W.E. Caswell, D. Zanon: Nucl. Phys. B **182**, 125 (1981);
S. Mandelstam: Nucl. Phys. B **213**, 149 (1983);
L. Brink, O. Lindgren, B.E.W. Nilsson: Phys. Lett. B **123**, 323(1983);
P.S. Howe, K.S. Stelle, P.K. Townsend: Nucl. Phys. B **236**, 125 (1984)
117. A. Sen: Phys. Lett. B **329**, 217 (1994)
118. M. Bianchi, F.A. Dolan, P.J. Heslop, H. Osborn: hep-th/0609179
119. C. Vafa, E. Witten: Nucl. Phys. B **431**, 3 (1994)
120. A. Kapustin, E. Witten: hep-th/0604151;
S. Gukov, E. Witten: hep-th/0612073;
A. Kapustin: hep-th/0612119
121. N. Dorey, T.J. Hollowood, V.V. Khoze, M.P. Mattis, S. Vandoren: Nucl. Phys. **B552**, 88 (1999)
122. A.V. Belitsky, S. Vandoren, P. van Nieuwenhuizen: Class. Quant. Grav. **17**, 3521 (2000)
123. M.B. Green, S. Kovacs: JHEP **0304**, 058 (2003)
124. M. Bianchi, M.B. Green, S. Kovacs, G.C. Rossi: JHEP **9808**, 013 (1998)
125. N. Dorey, V.V. Khoze, M.P. Mattis, S. Vandoren: Phys. Lett. B **442**, 145 (1998)
126. M. Bianchi, S. Kovacs, G.C. Rossi, Ya.S. Stanev: JHEP **9908**, 020 (1999)
127. L.S. Brown, R.D. Carlitz, D.B. Creamer, C. Lee: Phys. Rev. D **17**, 1583 (1978)
128. E. D'Hoker, D.Z. Freedman, S.D. Mathur, A. Matusis, L. Rastelli: hep-th/9908160
129. M. Bianchi, S. Kovacs: Phys. Lett. B **468**, 102 (1999)
130. B. Eden, P.S. Howe, C. Schubert, E. Sokatchev, P.C. West: Phys. Lett. B **472**, 323 (2000)
131. J. Erdmenger, M. Perez-Victoria: Phys. Rev. D **62**, 045008 (2000);
B.U. Eden, P.S. Howe, E. Sokatchev, P.C. West: Phys. Lett. B **494**, 141 (2000)

132. E. D'Hoker, J. Erdmenger, D.Z. Freedman, M. Perez-Victoria: Nucl. Phys. B **589**, 3 (2000)
133. S.J. Rey, J.T. Yee: Eur. Phys. J. C **22**, 379 (2001);
J.M. Maldacena: Phys. Rev. Lett. **80**, 4859 (1998)
134. M. Bianchi, M.B. Green, S. Kovacs: JHEP **0204**, 040 (2002); hep-th/0107028
135. S. Kovacs: Nucl. Phys. B **684**, 3 (2004)
136. N.J. Hitchin, A. Karlhede, U. Lindstrom, M. Roček: Commun. Math. Phys. **108**, 535 (1987)
137. W. Krauth, H. Nicolai, M. Staudacher: Phys. Lett. B **431**, 31 (1998);
W. Krauth, M. Staudacher: Phys. Lett. B **435**, 350 (1998)
138. T. Banks, M.B. Green: JHEP **9805**, 002 (1998)
139. M.B. Green, M. Gutperle: Nucl. Phys. B **498**, 195 (1997)
140. J.H. Schwarz: Nucl. Phys. B **226**, 269 (1983)
141. H.J. Kim, L.J. Romans, P. van Nieuwenhuizen: Phys. Rev. D **32**, 389 (1985)
142. R. Gopakumar, M.B. Green: JHEP **9912**, 015 (1999)
143. D. Berenstein, J.M. Maldacena, H. Nastase: JHEP **0204**, 013 (2002)
144. B. Sundborg: Nucl. Phys. B **573**, 349 (2000);
S.E. Konstein, M.A. Vasiliev, V.N. Zaikin: JHEP **0012**, 018 (2000);
E. Witten: *Spacetime Reconstruction*, Talk at *JHS 60 Conference*, California Institute of Technology, 3–4 November 2001 http://quark.caltech.edu/jhs60/witten/1.html;
E. Sezgin, P. Sundell: JHEP **0109**, 036 (2001);
E. Sezgin, P. Sundell: JHEP **0109**, 025 (2001);
A.M. Polyakov: Int. J. Mod. Phys. A **17S1**, 119 (2002)
145. M. Bianchi, J.F. Morales, H. Samtleben: JHEP **0307**, 062 (2003);
N. Beisert, M. Bianchi, J.F. Morales, H. Samtleben: JHEP **0402**, 001 (2004);
N. Beisert, M. Bianchi, J.F. Morales, H. Samtleben: JHEP **0407**, 058 (2004)
146. R. Penrose: in *Differential Geometry and Relativity*, eds. M. Cahen, M. Flato (Reidel, Dordrecht, Netherlands, 1976);
R. Gueven: Phys. Lett. B **482**, 255 (2000)
147. M. Blau, J. Figueroa-O'Farrill, C. Hull, G. Papadopoulos: Class. Quant. Grav. **19**, L87 (2002); JHEP **0201**, 047 (2002)
148. R.R. Metsaev: Nucl. Phys. B **625**, 70 (2002)
149. R.R. Metsaev, A.A. Tseytlin: Phys. Rev. D **65**, 126004 (2002)
150. A. Pankiewicz: Fortsch. Phys. **51**, 1139 (2003);
J.C. Plefka: Fortsch. Phys. **52**, 264 (2004);
J.M. Maldacena: Lectures given at the *Theoretical Advanced Study Institute in Elementary Particle Physics (TASI 2003)on Recent Trends in String Theory* (Boulder, CO, 1–27 June 2003), hep-th/0309246;
M. Spradlin, A. Volovich: Lectures given at *ICTP Spring School on Superstring Theory and Related Topics* (Trieste, Italy, 31 March–8 April 2003), hep-th/0310033;
D. Sadri, M.M. Sheikh-Jabbari: Rev. Mod. Phys. **76**, 853 (2004);
R. Russo, A. Tanzini: Class. Quant. Grav. **21**, S1265 (2004)
151. M.B. Green, S. Kovacs, A. Sinha: JHEP **0505**, 055 (2005)
152. M.B. Green, S. Kovacs, A. Sinha: JHEP **0512**, 038 (2005)
153. M.B. Green, S. Kovacs, A. Sinha: Phys. Rev. D **73**, 066004 (2006)
154. E.J. Saletan: J. Math. Phys. **7**, 53 (1961)
155. C. Kristjansen, J. Plefka, G.W. Semenoff, M. Staudacher: Nucl. Phys. B **643**, 3 (2002)

156. N.R. Constable, D.Z. Freedman, M. Headrick, S. Minwalla, L. Motl, A. Postnikov, W. Skiba: JHEP **0207**, 017 (2002)
157. M.R. Gaberdiel, M.B. Green: Ann. Phys. **307**, 147 (2003)
158. M.B. Green, M. Gutperle: Nucl. Phys. B **476**, 484 (1996)
159. E. Witten: Nucl. Phys. B **223**, 422 (1983)
160. P. Di Vecchia, F. Nicodemi, R. Pettorino, G. Veneziano: Nucl. Phys. B **181**, 318 (1981)
161. E. Witten: Nucl. Phys. B **156**, 269 (1979);
G. Veneziano: Nucl. Phys. B **159**, 213 (1979)
162. For a recent review see, L. Del Debbio, L. Giusti, C. Pica: Nucl. Phys. (Proc. Suppl) B **140**, 603 (2005) [hep-lat/0409100] and references therein
163. C. Vafa, A. Strominger: Phys. Lett. **B379**, 99 (1996)
164. J. Wess, B. Zumino: Nucl. Phys. B **70**, 39 (1974);
P. Fayet, S. Ferrara: Phys. Rept. **32**, 250 (1977);
J. Wess, J. Bagger: *Supersymmetry and Supergravity*, 2nd Edition (Princeton University Press, Princeton, 1992)
165. R. Peccei, H. Quinn: Phys. Rev. Lett. **38**, 1440 (1977); Phys. Rev. D **16**, 1791 (1977)
166. S. Ferrara, B. Zumino: Nucl. Phys. B **87**, 207 (1975);
O. Piguet, K. Sibold: Nucl. Phys. B **196**, 428 (1982); ibid. 447 (1982); Helv. Phys. Acta **63**, 71 (1990)
167. L. Faddeev, V. Popov: Phys. Lett. B **25**, 29 (1967)
168. G. Travaglini: Ph.D. lectures, University of Rome “Tor Vergata”, unpublished
169. R.P. Feynman, A.R. Hibbs: *Quantum Mechanics and Path Integrals* (McGraw-Hill, New York, 1977)
170. G.C. Rossi, M. Testa: Nucl. Phys. B **163**, 109 (1980); ibid. **B176**, 477 (1980); ibid. **B237**, 442 (1984);
K. Symanzik: Nucl. Phys. B **190**, 1 (1981);
J.P. Leroy, J. Micheli, G.C. Rossi, K. Yoshida: Z. Phys. C **48**, 653 (1990)
171. M. Lüscher: Comm. Math. Phys. **54**, 283 (1977);
G. Marchesini, E. Onofri: Nuovo Cimento A **65**, 298 (1981);
M. Lüscher, R. Narayanan, P. Weisz, U. Wolff: Nucl. Phys. B **384**, 168 (1992);
M. Lüscher, R. Sommer, P. Weisz, U. Wolff: Nucl. Phys. B **389**, 247 (1993); ibid. **413**, 481 (1994);
S. Sint: Nucl. Phys. B **421**, 135 (1994); Nucl. Phys. B **451**, 416 (1995)
172. B.A. Ovrut, J. Wess: Phys. Rev. D **25**, 409 (1982);
W.E. Lerche: Nucl. Phys. B **238**, 582 (1984);
T. Kugo, I. Ojima, T. Yanagida: Phys. Lett. B **135**, 402 (1984)
173. H. Georgi, S. L. Glashow: Phys. Rev. D **6**, 2977 (1972)
174. D. Tong: Lectures given at *Theoretical Advanced Study Institute in Elementary Particle Physics: Many Dimensions of String Theory* (Boulder, CO, 5 June–1 July 2005), hep-th/0509216
175. V.K. Dobrev, V.B. Petkova: Phys. Lett. B **162**, 127 (1985)
176. K. Intriligator: Nucl. Phys. B **551**, 575 (1999)

The Magnetic Monopoles Seventy-five Years Later

K. Konishi

Dipartimento di Fisica, "E. Fermi", Università degli Studi di Pisa, Largo Pontecorvo, 3, Ed. C, 56127 Pisa, Italy, and INFN, Sezione di Pisa, Pisa, Italy
konishi@df.unipi.it

Abstract. Non-Abelian monopoles are present in the fully quantum–mechanical low-energy effective action of many solvable supersymmetric theories. They behave perfectly as point-like particles carrying non-Abelian dual magnetic charges. They play a crucial role in confinement and in dynamical symmetry breaking in these theories. There is a natural identification of these excitations within the semiclassical approach, which involves the flavor symmetry in an essential manner. We review in an introductory fashion the recent development which has led to a better understanding of the nature and definition of non-Abelian monopoles, as well as of their role in confinement and dynamical symmetry breaking in strongly interacting theories.

Three quarters of a century have passed since the introduction of magnetic monopoles in quantum field theory by Dirac [1]. Our understanding of the soliton sector of spontaneously broken gauge theories [2] is still largely unsatisfactory. In particular, the development in our understanding of *non-Abelian* versions of monopoles [3, 4, 5, 6, 7, 8, 9, 10, 11, 12, 13, 14] and vortices [18] have been very slow, in spite of many articles written on these subjects, and in spite of the important role these topological excitations are likely to play in various areas of physics. For instance, they might hold the key to the mystery of the quark confinement in QCD. Their quantum–mechanical properties are gradually emerging, however, thanks to an ever improving grasp on the non-perturbative dynamics in the context of supersymmetric gauge theories. Some of the ingredients of this development include the Seiberg–Witten solution of $\mathcal{N} = 2$ supersymmetric gauge theories and exact instanton summations, better understanding of the properties of (super-) conformal field theories, exact results on the chiral condensates and symmetry breaking pattern in a wide class of $\mathcal{N} = 1$ supersymmetric gauge theories, and so on. Also, many new results on non-Abelian *vortices* and *domain walls* are now available, which are closely related to the problems concerning the monopoles.

It is the author's opinion that a serious discussion about confinement and non-Abelian monopoles today cannot ignore these basic results from

K. Konishi: *The Magnetic Monopoles Seventy-five Years Later*, Lect. Notes Phys. **737**, 471–521 (2008)
DOI 10.1007/978-3-540-74233-6_15

supersymmetric gauge theories. This lecture presents a review of what the author believes to be some of the most relevant aspects of this development, which should serve as an introduction to this very exciting area of research.

1 Color Confinement

One of the profound unsolved problems in the elementary particle physics today is quark confinement. A popular idea, due to 't Hooft and Mandelstam [19] holds that the ground state of QCD (quantum chromodynamics) is a dual superconductor: the quarks are confined by the chromo-electric vortices, analogous to the magnetic Abrikosov–Nielsen–Olesen vortex in the usual type II superconductors in solid. The Lagrangian of QCD

$$L = -\frac{1}{4} F^a_{\mu\nu} F^{\mu\nu\, a} + \bar{\psi}\,(i\gamma_\mu\, \mathcal{D}^\mu + m)\,\psi, \tag{1}$$

however, describes the dynamics of quarks and gluons, and it is not obvious from (1) how magnetic (dual) degrees of freedom appear and how they interact. One way to detect such degrees of freedom is 't Hooft's Abelian gauge fixing. One chooses the gauge so that a given field (perhaps some composite of $F^a_{\mu\nu}$) in the adjoint representation to take an Abelian form

$$X = \begin{pmatrix} \lambda_1 & 0 & 0 \\ 0 & \lambda_2 & 0 \\ 0 & 0 & \lambda_3 \end{pmatrix}, \qquad \lambda_1 > \lambda_2 > \lambda_3. \tag{2}$$

For a generic gauge-field configuration $A_\mu(x)$, however, it is not possible to keep the above diagonal form everywhere in $\mathbf{R}^4$. Near a singularity $\lambda_1 = \lambda_2$, diagonalization of the matrix

$$X = X|_{\lambda_1=\lambda_2} + \begin{pmatrix} C(x) & 0 \\ 0 & 0 \end{pmatrix} \tag{3}$$

where C is a 2×2 matrix, for instance, of the form,

$$C = \tau^i (x - x_0)^i, \tag{4}$$

by a gauge transformation $U(x)$, introduces a magnetic monopole, $A_i \simeq U(x)\,\partial\, U(x)^{-1}$.

Another possibility is to use the Cho–Faddeev–Niemi decomposition [20] of the gauge fields (for $SU(2)$)

$$A^a_\mu = C_\mu \mathbf{n}^a + \tilde{\sigma}(x)(\partial_\mu \mathbf{n} \times \mathbf{n})^a + \rho\, \partial_\mu \mathbf{n}^a; \qquad \tilde{\sigma}(x) = 1 + \sigma(x), \tag{5}$$

in terms of the unit vector field $\mathbf{n}$ and the Abelian gauge field C_μ which live on S^2 and S^1 factors, respectively, of $SU(2)$, and a charged "scalar" field

$$\phi = \rho(x) + i\sigma(x). \tag{6}$$

The Wu–Yang singular monopole solution [21], for instance, corresponds to

$$n^a = \frac{x^a}{r}, \quad C_\mu = \phi = 0. \tag{7}$$

It is possible that these singularities, regularized, e.g., by the zero of $1 - |\phi|^2$, somehow manage to behave as dominant degrees of freedom in the ground state of QCD.

Whichever way, a central question is whether the magnetic monopoles of QCD is of Abelian or non-Abelian type. The 't Hooft–Mandelstam scenario is essentially Abelian. By assuming that the relevant infrared degrees of freedom are those which signal the singularities of Abelian gauge fixing, one tacitly makes a highly nontrivial dynamical assumption.

In this respect, the $SU(2)$ gauge theory is an exception, though. It is quite possible that in this particular case 't Hooft's (or related) Abelian gauge-fixing procedure allows us to "detect" the correct magnetic degrees of freedom, even if the system does not dynamically Abelianize.[1] The singularities of the Abelian gauge-fixing would signal the presence of the magnetic degrees of freedom, which correspond [22] just to the Wu–Yang monopoles, (7). As the Cho–Faddeev–Niemi $n_a(x)$ field parametrizes $\pi_2(S^2) \sim \pi_2(SU(2)/U(1)) = \mathbf{Z}$, the magnetic charge of the Wu–Yang monopoles are the same, and quantized in the same way, as the 't Hooft–Polyakov monopoles of the Georgi–Glashow model. In more general $SU(N)$ theories with $N \geq 3$, however, one does not expect such a lucky situation. If the system does not dynamically Abelianize to $U(1)^{N-1}$ effective system at some low-energy scales, it would not be appropriately described by an effective Lagrangian describing the Abelian monopoles which signal the singularities of the Abelian gauge fixing.[2]

Actually, there is no hint that such a dynamical scenario (dynamical Abelianization) is realized in Nature. We must seriously consider the much more subtle possibility that somehow non-Abelian, magnetic degrees of freedom play a role in the physics of confinement and chiral symmetry breaking. Are there models in which the low-energy dynamics is known and in which non-Abelian magnetic degrees of freedom play a central role?

It does not seem to be widely known that not only do such systems exist, but that in a sense this (occurrence of light non-Abelian monopoles) is a most typical dynamical phenomenon in a wide class of supersymmetric gauge systems. The class of models in question is $N = 2$ supersymmetric theories

---

[1] This could explain the mysterious success of the Abelian dominance idea in lattice simulations of the pure $SU(2)$ gauge theory, even if there are no other indications for dynamical Abelianization. The author thanks T. Suzuki for useful discussions.

[2] Vice versa, in a system where Abelianization does take place, as in a class of supersymmetric models mentioned in Sect. 5.4 below, 't Hooft's Abelian gauge fixing should be a perfectly valid tool for extracting and studying the relevant infrared degrees of freedom.

with SU, SO or USp gauge groups with quark hypermultiplets in various representations [23, 24, 25, 26, 27, 28, 29, 30, 31]. Moreover, the class of models in which one can make reliable analysis about their low-energy behavior, have increased enormously thanks to a more recent work on certain $N = 1$ models [32] with scalar multiplets in the adjoint representation. Again, the appearance of massless, non-Abelian monopoles in their low-energy effective action is a rule, rather than an exception, in these models.

Of course, in the context of superconformal theories there are famous examples of non-Abelian dualities such as the Montonen–Olive duality in $N = 4$ supersymmetric theories [33] or the Seiberg duality in the $N = 1$ supersymmetric models [34] with nontrivial infrared fixed points.

These problems (conformal invariance and confinement) are closely related, as the confinement and dynamical symmetry breaking can often be seen as the result of breaking of (nontrivial) conformal invariance near an infrared fixed-point theory.

Evidently, supersymmetric theories are trying to tell us something important about the non-Abelian monopoles and confinement. In what follows we review briefly the old difficulties associated with the semiclassical concepts of non-Abelian monopoles. It will be argued that the dual group properties of non-Abelian monopoles occurring in a system with gauge symmetry breaking $G \longrightarrow H$ are best defined by setting the low-energy H system in Higgs phase, so that the dual system is in confinement phase. The transformation law of the monopoles follows from that of monopole–vortex mixed configurations in the system with a large hierarchy of energy scales, $v_1 \gg v_2$,

$$G \xrightarrow{v_1} H \xrightarrow{v_2} \emptyset, \tag{8}$$

under an unbroken, exact color–flavor diagonal symmetry H_{C+F} This last symmetry is broken by individual soliton vortex, so the latter develops continuous moduli. The transformation law among the regular monopoles, which appear at the end point of the vortex, follows from that of the vortices. This defines, once rewritten in the dual, magnetic variables, the dual group $\tilde{H}$ under which the monopoles transform as a multiplet.

2 Difficulties with the Semiclassical "Non-Abelian Monopoles"

2.1 Abelian Monopoles

A system in which the gauge symmetry is spontaneously broken

$$G \stackrel{\langle\phi_1\rangle\neq 0}{\longrightarrow} H \tag{9}$$

where H is some non-Abelian subgroup of G, possesses a set of regular magnetic monopole solutions in the semiclassical approximation. They are natural generalizations of the Abelian 't Hooft–Polyakov monopoles [2], found originally in the $G = SO(3)$ theory broken to $H = U(1)$ by a Higgs mechanism. In that theory, the field content is just the $SU(2)$ gauge fields and a scalar field in the adjoint representation of the gauge group; the energy of a static field configuration has an expression

$$E = \int d^3x [\frac{1}{4} F_{ij}^{a\,2} + \frac{1}{2}(D_i\phi^a)^2 + \frac{\lambda}{8}(\phi^{a\,2} - F^2)^2] \tag{10}$$

where

$$F_{ij}^a = \partial_i\, A_j - \partial_j\, A_i - g\, \epsilon^{abc}\, A_i^b\, A_j^c;$$

while $D_i\phi^a$ is a covariant derivative,

$$D_i\phi^a = \partial_i\phi^a - g\, \epsilon^{abc} A_i^b\, \phi^c.$$

Now the static finite energy solution of the equation of motion must behave asymptotically as

$$\phi^a \to n^a(x)\, F, \qquad n^a(x)^2 = 1, \tag{11}$$

where the vector field $n^a(x)$ clearly label the winding of the map $S^2 \to S^2$, the first sphere being the space sphere surrounding the monopole, the second sphere representing the vacuum orientation in the group space. One possibility is n^a has a fixed orientation, such as $n^a(x) = (0, 0, 1)$ everywhere: this represents a vacuum. Another possibility is that n^a makes a nontrivial winding in the group space as x_i goes around the sphere, e.g.,

$$n^a(x) = (\sin\theta\, \cos m\phi, \sin\theta\, \sin m\phi, \cos\theta), \qquad m = \pm 1, \pm 2, \ldots.$$

This integer labels the homotopy classes

$$\pi_2(SU(2)/U(1)) \sim \pi_2(S^2) \sim \mathbf{Z}$$

of the scalar field configurations. The gauge fields must reduce to the pure gauge,

$$A_i^a \to \frac{1}{g} \epsilon^{abc}\, n^b(x)\, \partial_i\, n^c(x)$$

in order for the energy to be finite.

The solution of the equation of motion in the nontrivial sectors can be found by rewriting (10) as

$$E = \int d^3x [\frac{1}{4}(F_{ij}^a - \epsilon_{ijk}\, D_k\phi^a)^2 + \frac{1}{2}\epsilon_{ijk}\, F_{ij}^a\, D_k\, \phi^a + \frac{\lambda}{8}(\phi^{a\,2} - F^2)^2. \tag{12}$$

The crucial observation is that while the first and third terms are semipositive definite, the second term is a total derivative,

$$\frac{1}{2}\epsilon_{ijk}\, F_{ij}^a\, D_k\, \phi^a = \partial_k\, B_k, \qquad B_k = \frac{1}{2}\epsilon_{ijk}\, F_{ij}^a\, \phi^a.$$

We used above a useful identity for the derivatives for gauge invariant products

$$\partial_k \operatorname{Tr}(A\, B\, \ldots) = \operatorname{Tr}(D_k A\, B\, \ldots) + \operatorname{Tr}(A\, D_k B\, \ldots) + \ldots .$$

Thus the second term of (12) represents F times the "magnetic" charge

$$\int dv\, \nabla \cdot \mathbf{B} = \int dS \cdot \mathbf{B} = 4\,\pi g_m, \qquad \mathbf{B} \sim \frac{g_m}{r^3}\,\mathbf{r}.$$

If $\lambda = 0$, $|\phi^{a\,2}| \to F^2$ (Bogomol'nyi–Prasad–Sommerfield (BPS) limit) the mass is proportional to the magnetic charge, $4\,\pi\, g_m\, F = \frac{\langle\phi\rangle}{g}$, while the field configuration satisfies the linear BPS equation

$$F_{ij}^a - \epsilon_{ijk}\, D_k \phi^a = 0,$$

with an explicit (BPS) solution [2]

$$A_i^a = \epsilon_{aij} r_j \frac{1 - K(r)}{g\, r^2}, \qquad K(r) = \frac{gFr}{\sinh gFr}, \tag{13}$$

$$\phi^a = r^a\, \frac{H(r)}{g\, r^2}, \qquad H(r) = gFr \coth gFr - 1. \tag{14}$$

2.2 Non-Abelian Unbroken Group

When the "unbroken" gauge group is non-Abelian, the asymptotic gauge field can be written as

$$F_{ij} = \epsilon_{ijk} B_k = \epsilon_{ijk} \frac{r_k}{r^3} (\beta \cdot \mathbf{H}), \tag{15}$$

in an appropriate gauge, where $\mathbf{H}$ are the diagonal generators of H in the Cartan subalgebra. A straightforward generalization of the Dirac's quantization condition leads to

$$2\,\beta \cdot \alpha \in \mathbf{Z} \tag{16}$$

where α are the root vectors of H.[3] It is not difficult to write down explicit classical solutions [5, 6] by generalizing (13) and (14).

The constant vectors β (with the number of components equal to the rank of the group H) label possible monopoles. It is easy to see that the solution of (16) is that β is any of the *weight vectors* of a group whose nonzero roots are given by

$$\alpha^* = \frac{\alpha}{\alpha \cdot \alpha}. \tag{17}$$

---

[3] This is most easily seen by considering $\operatorname{Tr} e^{ig \oint A_i\, dx^i}$ along an infinitesimal closed curve on the surface of a sphere surrounding the monopole. By enlarging the loop and reclosing it at the other side of the sphere, one ends up with

$$e^{ig \int d\mathbf{S} \cdot \mathbf{B}} = e^{4\pi i \beta \cdot \mathbf{H}}.$$

This should be an identity operator: commuting the above with nondiagonal generators E_α yields (16).

This is just a standard group theory theorem: (16) can in fact be rewritten as the well-known relation between a weight vector and a root vector of any group, $2\,\beta\cdot\alpha^*/(\alpha^*\cdot\alpha^*)\in\mathbf{Z}$.

The group generated by (17) is known as the *dual* (we shall call it Goddard–Nuyts–Olive–Weinberg (GNOW) dual below) of H, let us call $\tilde{H}$. One is thus led to a set of semiclassical *degenerate* monopoles, with multiplicity equal to that of a representation of $\tilde{H}$; this has led to the so-called GNOW conjecture, i.e., that they form a multiplet of the group $\tilde{H}$, dual of H [4, 5, 6]. For simply laced groups (with the same length of all nonzero roots) such as $SU(N)$, $SO(2N)$, the dual of H is basically the same group, except that the allowed representations tell us that

$$U(N)\leftrightarrow U(N);\qquad SO(2N)\leftrightarrow SO(2N), \tag{18}$$

while

$$SU(N)\leftrightarrow \frac{SU(N)}{\mathbf{Z}_N};\qquad SO(2N+1)\leftrightarrow USp(2N). \tag{19}$$

There is no difficulty in explicitly constructing these degenerate set of monopoles [6]. The basic idea is to embed the 't Hooft–Polyakov monopoles in various broken $SU(2)$ subgroups. The main results are summarized in Appendixes 9 and 9. These set of monopoles constitute the prime candidates for the members of a multiplet of the dual group $\tilde{H}$.

There are however well-known difficulties in such an interpretation. The first concerns the topological obstruction discussed in [11]: in the presence of the classical monopole background, it is not possible to define a globally well-defined set of generators isomorphic to H. As a consequence, no "colored dyons" exist. In a simplest case with the breaking

$$SU(3)\stackrel{\langle\phi_1\rangle\neq 0}{\longrightarrow} SU(2)\times U(1), \tag{20}$$

this means that

$$\textit{no monopoles with charges}\quad (\underline{2},1^*)\quad \textit{exist}, \tag{21}$$

where the asterisk indicates the dual, magnetic $U(1)$ charge.

The second can be regarded as an infinitesimal version of the same difficulty: certain bosonic zero modes around the monopole solution, corresponding to H gauge transformations, are non-normalizable (behaving as $r^{-1/2}$ asymptotically). Thus the standard procedure of quantization leading to H multiplets of monopoles, does not work. Some progress on the check of GNOW duality along this orthodox line of thought, has been reported nevertheless [14], in the context of $\mathcal{N}=4$ supersymmetric gauge theories. Their approach, however, requires the consideration of particular class of multi monopole systems, neutral with respect to the non-Abelian group (more precisely, non-Abelian part of) H only.

Both of these difficulties concern the transformation properties of the monopoles under the subgroup H, while the relevant question should be how they transform under the dual group, $\tilde{H}$. As field transformation groups, H and $\tilde{H}$ are relatively nonlocal, the latter should look like a nonlocal transformation group in the original, electric description.

Another related question concerns the *multiplicity* of the monopoles: Take again the case of the system with a breaking pattern, (20). One might argue that there is only one monopole, as all the degenerate solutions are related by the unbroken *gauge* group $H = SU(2)$.[4] Or one might say that there are two monopoles, in the sense that according to the semiclassical GNO classification they are supposed to belong to a doublet of the dual $SU(2)$ group. Or, perhaps, one should conclude that there are infinitely many, continuously related solutions, as the two solutions obtained by embedding the 't Hooft solutions in $(1, 3)$ and $(2, 3)$ subspaces, are clearly part of the continuous set of (moduli of) solutions. In short, what is the multiplicity ($\mathcal{N}$) of the monopoles:

$$\mathcal{N} = 1, \quad 2, \quad \text{or} \quad \infty \, ? \tag{22}$$

Formulated perhaps more adequately:

- What is the dual group? How do the degenerate magnetic monopoles transform among themselves under the dual group? Which of the semiclassical aspects of monopoles survive quantum effects?

In the attempt to answer these questions, some general considerations seem to be unavoidable. The first is the fact since H and $\tilde{H}$ groups are non-Abelian the dynamics of the system should enter the problem in essential way. For instance, the non-Abelian H interactions can become strongly-coupled at low energies and can break itself dynamically. This indeed occurs in pure $\mathcal{N} = 2$ super Yang–Mills theories (i.e., theories without quark hypermultiplets), where the exact quantum–mechanical result is known in terms of the Seiberg–Witten curves [23, 24, 25]: see below. Consider for instance, a pure $\mathcal{N} = 2$, $SU(N+1)$ gauge theory. Even though partial breaking, e.g., $SU(N+1) \to SU(N) \times U(1)$ looks perfectly possible semiclassically, in an appropriate region of classical degenerate vacua, no such vacua exist quantum mechanically. In *all* vacua the light monopoles are abelian, the effective, magnetic gauge group being $U(1)^N$.

Generally speaking, the concept of a dual group multiplet is well-defined when $\tilde{H}$ interactions are weak (or at worst, conformal). This however means that one must study the original, electric theory in the regime of strong coupling, which would usually make the task exceedingly difficult. Fortunately, in $\mathcal{N} = 2$ supersymmetric gauge theories, exact Seiberg–Witten curves describe the fully quantum–mechanical consequences of the strong-interaction

[4] This interpretation however encounters the difficulties mentioned above. Also there are cases in which degenerate monopoles occur, which are not simply related by the group H, see below.

dynamics in terms of weakly coupled dual magnetic variables. This is how we know that the non-Abelian monopoles exist in fully quantum theories [27]: in the so-called r-vacua of softly broken $\mathcal{N} = 2$, $SU(N)$ gauge theory, the light monopoles appear as the dominant infrared degrees of freedom and interact as point-like particles having the charges of a fundamental multiplet $\underline{r}$ of an effective, dual $SU(r)$ gauge group. In an $SU(3)$ gauge theory broken to $SU(2) \times U(1)$ as in (20), with an appropriate number of quark multiplets ($N_f \geq 4$), for instance, light magnetic monopoles carrying the charges

$$(\underline{2}^*, 1^*) \tag{23}$$

under the dual $SU(2) \times U(1)$ appear in the low-energy effective action. (Dual) colored dyons do exist! The distinction between H and $\tilde{H}$ is crucial (cf. (21)).

In $\mathcal{N} = 2$, $SU(N)$ SQCD with N_f flavors, light non-Abelian monopoles with $SU(r)$ dual gauge group appear for $r \leq \frac{N_f}{2}$ only. Such a limit clearly reflects the dynamics of the soliton monopoles under renormalization group: the effective low-energy gauge group must be either infrared free or conformally invariant, in order for the monopoles to emerge as recognizable low-energy degrees of freedom [28, 29, 30].

A closely related point concerns the phase of the system. If the dual group were in Higgs phase, the multiplet structure among the monopoles would get lost, generally. Therefore one must study the dual ($\tilde{H}$) system in confinement phase.[5] But then, according to the standard electromagnetic duality argument, *one must analyze the electric system in Higgs phase.* The monopoles will appear confined by the vortices of the H system, which can be naturally interpreted as confining string of the dual system $\tilde{H}$.

We are thus led to study the system with a hierarchical symmetry breaking,

$$G \xrightarrow{v_1} H \xrightarrow{v_2} \emptyset, \tag{24}$$

where

$$v_1 \gg v_2, \tag{25}$$

instead of the original system (9). The smaller VEV breaks H completely. However, in order for the degeneracy among the monopoles not to be broken by the breaking at the scale v_2, we require that some global color–flavor diagonal group

$$H_{C+F} \subset H_{color} \otimes G_F \tag{26}$$

remains unbroken (see below).

As we shall see, such a scenario is very naturally realized in $\mathcal{N} = 2$ supersymmetric theories. An important lesson one learns from these considerations (and from the explicit models), is that the role of the massless flavor is fundamental. This manifests itself in more than one ways.

[5] Non-Abelian monopoles in the *Coulomb phase* suffer from the difficulties already discussed.

(i) H must be nonasymptotically free, this requires that there be sufficient number of massless flavors: otherwise, H interactions would become strong at low energies and H group can break itself dynamically;
(ii) The physics of the r vacua [28, 30] indeed shows that the non-Abelian dual group $SU(r)$ appear only for $r \leq \frac{N_f}{2}$. This limit can be understood from the renormalization group: in order for a nontrivial r vacuum to exist, there must be at least $2r$ massless matter flavor in the original, electric theory;
(iii) Non-Abelian vortices [35, 37], which as we shall see are closely related to the concept of non-Abelian monopoles, require also an exact flavor group. The non-Abelian flux moduli arise as a result of an exact color–flavor diagonal symmetry of the system, broken by individual soliton vortices.

3 Non-Abelian Monopoles from Vortex Moduli

It turns out that the properties of the monopoles induced by the breaking

$$G \to H \tag{27}$$

are closely related to the properties of the vortices, which develop when the low-energy H gauge theory is put in Higgs phase by a set of scalar VEVs, $H \to \emptyset$. The crucial instrument is the exact homotopy sequence,

$$\cdots \to \pi_2(G) \to \pi_2(G/H) \to \pi_1(H) \to \pi_1(G) \to \cdots \tag{28}$$

But first a few words on homotopy groups and on the use of these relations to characterize the semiclassical monopoles. We shall come back to consider monopole–vortex mixed configurations later.

$\pi_1(M)$ and $\pi_2(M)$ are the first and second homotopy groups, respectively, representing the distinct classes of maps from S^1 or S^2 to the (group) manifold M. Now "products" among such equivalent classes can be defined and they turn out to form a group structure [39, 8]. The definition of "the relative homotopy groups" such as $\pi_2(G/H)$ and the proof of the exactness of the sequence (28) can be found in the first reference. An exact sequence is a useful tool for studying the structure of different groups through their correspondences (group homomorphisms). "Exact" means that the kernel of the map at any point of the chain is equal to the image of the preceding map. Such relations are shown pictorially in Fig. 1. These sequences can be used, for instance, as follows. Assume for simplicity that $\pi_2(G)$ and $\pi_1(G)$ are both trivial. In this case it is clear that each element of $\pi_1(H)$ is an image of a corresponding element of $\pi_2(G/H)$: all monopoles are regular, 't Hooft–Polyakov monopoles.

Consider now the case $\pi_1(G)$ is nontrivial. Take for concreteness $G = SO(3)$, with $\pi_1(SO(3)) = \mathbf{Z}_2$, and $H = U(1)$, with $\pi_1(U(1)) = \mathbf{Z}$. For any compact Lie groups $\pi_2(G) = \emptyset$. The exact sequence illustrated in Fig. 1 in this case implies that the monopoles, classified by $\pi_1(U(1)) = \mathbf{Z}$ can further by divided into two classes, one belonging to the image of $\pi_2(SO(3)/U(1))$—'t Hooft–Polyakov monopoles!—and those which are not related to the breaking—the singular, Dirac monopoles. The correspondence is two-to-one: the monopoles of magnetic charges $2n$ times ($n = 1, 2, \ldots$) the Dirac unit are regular monopoles while those with charges $2n+1$ are Dirac monopoles. In other words, the regular monopoles correspond to the kernel of the map $\pi_1(H) \to \pi_1(G)$ [8].

The exact sequence (28) assumes an important significance when we consider the system with a hierarchical symmetry breaking (24),

$$G \xrightarrow{v_1} H \xrightarrow{v_2} \emptyset.$$

As H is now completely broken the low-energy theory has vortices, classified by $\pi_1(H)$. If $\pi_1(G) = \emptyset$, however, the full theory cannot have vortices. This apparent paradox is solved when one realizes that there is another related paradox: monopoles representing $\pi_2(G/H)$ cannot be stable, because in the full theory the gauge group is completely broken, $G \to \emptyset$, and because for any Lie group, $\pi_2(G) = \emptyset$. These paradoxes solve themselves: the vortices of the low-energy theory end at the monopoles, which have large but finite masses. Or they are broken in the middle by (though suppressed) monopole–antimonopole pair production. Vice versa, the monopoles are not stable as its flux is carried away by the vortex (see Fig. 2).

Applied to the case of $SO(3) \to U(1) \to \emptyset$, this was precisely the logic used by 't Hooft in his pioneering paper on the monopoles. As is seen from Fig. 1,

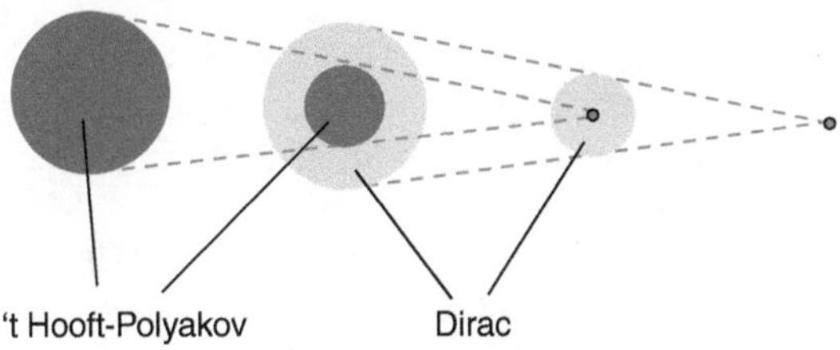

Fig. 1. A pictorial representation of the exact homotopy sequence, (28), with the leftmost figure corresponding to $\pi_2(G/H)$

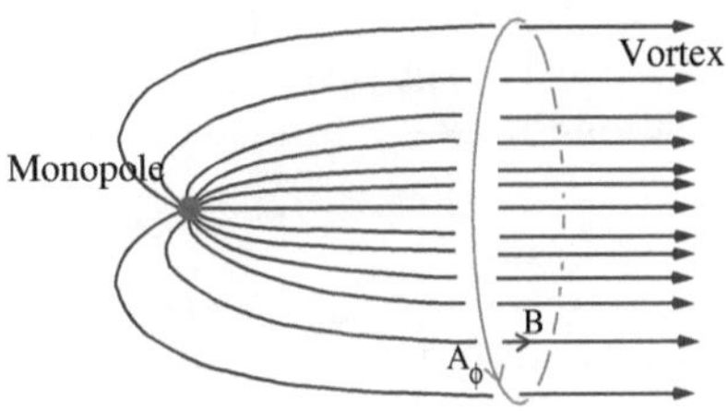

Fig. 2. The monopole as a sink of the magnetic flux lines

the vortices ($\pi_1(U(1)) = \mathbf{Z}$) of the winding number two, corresponding to the trivial element of $\pi_1(SO(3)) = \mathbf{Z}_2$, should not be stable in the full theory: there must be a regular monopole-like configuration, having the magnetic charge twice the Dirac unit, $g_m = 4\pi/g$, where g is the the gauge coupling constant of the $SO(3)$ theory, acting as a source or a sink of the magnetic flux (Fig. 2). [6]

An important new aspect we have here, as compared to the case discussed by 't Hooft [2] is that now the unbroken group H is non-Abelian and that the low-energy vortices carry continuous, non-Abelian flux moduli. As the color–flavor diagonal symmetry is an exact unbroken symmetry of the full theory, and the non-Abelian moduli among the low-energy vortices is a consequence of it, it follows that the the monopoles appearing as the end points of such vortices carry the same continuous moduli.

The monopole transformation properties follow from those of the vortices, which can be studied in the low-energy approximation.

4 $\mathcal{N} = 2$ Supersymmetric Gauge Theories and Light Non-Abelian Monopoles

It is always a healthy attitude to try to test one's general idea against a concrete model. For various reasons it turns out that $\mathcal{N} = 2$ models provides a good testing ground, as the results of strong infrared dynamics are known in the form of exact Seiberg–Witten curves. Another advantage is that by varying certain parameters upon which the system depends holomorphically, as is usual in supersymmetric theories, one can study the system (8) in different regimes.

In the regions of parameters where $v_1 \gg v_2 \gg \Lambda$, semiclassical analysis in the original electric theory is justified, and one can study monopoles (in the effective theory at mass scales much higher than v_2) and separately, the vortices (in the effective theory valid at mass scales much lower than v_1). The symmetry and homotopy-map argument allows to obtain the missing information about the non-Abelian transformation properties of the monopoles, from the known properties of the vortices. We come back to this discussion in Sect. 6.2. In the concrete models studied there the breaking mass scales are given by $m_i = m \sim v_1$; $\sqrt{\mu m} \sim v_2$, so the parameter regions explored correspond to $|m_i| \gg |\mu| \gg \Lambda$.

These results are then checked against the fully quantum–mechanical results on the monopoles appearing as the massless degrees of freedom in the

[6] The relation appears to violate the Dirac quantization condition: actually, the minimum electric charge which could be introduced in the theory is that of a quark, $e = g/2$, and which satisfies $g_m\, e = 2\pi$, in accordance with Dirac's condition.

magnetic dual theory, in the region $v_1 \sim v_2 \sim \Lambda$. This regime will be discussed first. In the following Sect. 4.2, in fact, the parameters are chosen to be $m_i, \mu \sim \Lambda$, and in particular, $m_i \to m$.

We shall return later (Sect. 6.2) to see that how our ideas on non-Abelian duality based on the hierarchical symmetry breaking and on color–flavor diagonal symmetry can be studied in the same model reliably and see that the results found match the full quantum results.

4.1 Seiberg–Witten Solution of Pure $\mathcal{N} = 2$ Yang–Mills

$\mathcal{N} = 2$ supersymmetric $SU(2)$ Yang–Mills theory is described by the Lagrangian,

$$L = \frac{1}{8\pi} \mathrm{Im}\, \tau_{cl} \left[\int d^4\theta\, \Phi^\dagger e^V \Phi + \int d^2\theta\, \frac{1}{2} W W \right] \tag{29}$$

where

$$\tau_{cl} \equiv \frac{\theta_0}{2\pi} + \frac{4\pi i}{g_0^2} \tag{30}$$

is the bare θ parameter and coupling constant. $\Phi = \phi + \sqrt{2}\,\theta\,\psi + \dots$, and $W_\alpha = -i\lambda + \frac{i}{2}\,(\sigma^\mu\,\bar{\sigma}^\nu)_\alpha^\beta\, F_{\mu\nu}\,\theta_\beta + \dots$ are $\mathcal{N} = 1$ chiral and gauge superfields, both in the adjoint representation of the gauge group. The theory possesses $\mathcal{N} = 2$ supersymmetry as there are two gauginos, λ and ψ.

The scalar potential in this case is just the so-called D term

$$V_D = \frac{g^2}{8} ||[\Phi^\dagger, \Phi]||^2, \tag{31}$$

only, and the system has a continuous vacuum degeneracy (CMS—classical moduli space), parametrized by a complex number a,

$$\langle \Phi \rangle = \begin{pmatrix} a & 0 \\ 0 & -a \end{pmatrix}. \tag{32}$$

At any given a the gauge symmetry is broken by Higgs mechanism to $U(1)$. The low-energy theory is a $U(1)$ theory, describing the photon and photino λ, and the $\mathcal{N} = 2$ partners, $A = (A, \psi)$.

The general requirement of $\mathcal{N} = 2$ supersymmetry implies that the Lagrangian has the form

$$L_{eff} = \frac{1}{4\pi} \mathrm{Im}\, [\int d^4\theta \frac{dF(A)}{dA}\, \bar{A} + \int \frac{1}{2} \frac{d^2F(A)}{dA^2}\, W^\alpha\, W_\alpha\,], \tag{33}$$

with $F(A)$ is holomorphic in A. $F(A)$ is known as prepotential. Going to component fields, the fermionic and gauge parts take the form

$$L_{ferm} = \frac{1}{8\pi^2} [\mathrm{Im} \frac{d^2F(A)}{dA^2}] (i\bar{\psi}\bar{\sigma}^\mu \bar{D}_\mu \psi + i\bar{\lambda}\bar{\sigma}^\mu \bar{D}_\mu \lambda + \dots),$$

$$L_{gauge} = \frac{1}{16\pi^2}[\mathrm{Im}\frac{d^2F(A)}{dA^2}](F_{\mu\nu}^2 + i\,F_{\mu\nu}\,\tilde{F}^{\mu\nu} + \ldots),$$

which shows clearly ψ and λ have the same properties as the adjoint fermions ($SU_R(2)$ global symmetry of $\mathcal{N} = 2$ supersymmetry); the second formula shows that

$$\tau_{eff} = \frac{dA_D}{dA} = \frac{d^2F(A)}{dA^2}, \qquad A_D \equiv \frac{dF(A)}{dA},$$

acts as the low-energy effective (complex) coupling constant

$$\tau_{eff} = \frac{\theta_{eff}}{2\pi} + \frac{4\pi i}{g_{eff}^2}\,. \tag{34}$$

Let us recall that in general $4D$ supersymmetric sigma model, with a set of scalar multilpets Φ, the kinetic term is given by a (real) Kähler potential

$$L = \int d^4\theta\, K(\Phi, \bar{\Phi}) = \frac{\partial^2 K}{\partial\phi_i\,\partial\bar{\phi}_j}\,\partial_\mu\phi_i\,\partial^\mu\bar{\phi}_j + \ldots.$$

Here the Kähler potential has a special form, determined by the prepotential,

$$K = \frac{1}{2i}[\frac{dF(A)}{dA_i}\,\bar{A}_i - \frac{dF(\bar{A})}{d\bar{A}_i}\,A_i\,]$$

(termed special geometry).

Coming back to the $SU(2)$ $\mathcal{N} = 2$ Yang–Mills theory where there is only one scalar multiplet A, the bosonic part of the Lagrangian has the form

$$L_{bos} = \frac{1}{2\,i}\,(\partial_\mu\,a_D\,\partial^\mu\bar{a} - \partial_\mu\,a\,\partial^\mu\bar{a}_D) + \mathrm{Im}\tau(a)(F_{\mu\nu}^+)^2, \qquad F_{\mu\nu}^+ = F_{\mu\nu} + i\,\tilde{F}_{\mu\nu}.$$

Now this model has a nice property of (form) invariance under the generalized electromagnetic duality transformation [40]

$$\begin{pmatrix} a_D \\ a \end{pmatrix} \to M\begin{pmatrix} a_D \\ a \end{pmatrix}, \qquad \begin{pmatrix} F_{\mu\nu}^+ \\ G_{\mu\nu}^+ \end{pmatrix} \to M\begin{pmatrix} F_{\mu\nu}^+ \\ G_{\mu\nu}^+ \end{pmatrix}; \tag{35}$$

where

$$G_{\mu\nu}^+ \equiv \frac{1}{2}\frac{\partial}{\partial F_{\mu\nu}^+}\,[\tau(a)\,F_{\mu\nu}^{+\,2}]$$

and M is an $SL(2, Z)$ matrix,

$$M = \begin{pmatrix} A\;B \\ C\;D \end{pmatrix}, \qquad A\,D - B\,C = 1.$$

Such an invariance group includes the electromagnetic duality transformation $F_{\mu\nu} \leftrightarrow \tilde{F}_{\mu\nu}$, together with $a \leftrightarrow a_D$.

Since $F(A)$ is holomorphic, so is $\tau(A)$: it is harmonic, $\nabla\tau = \nabla\mathrm{Im}\tau = 0$. Thus $\mathrm{Im}\tau$ cannot be everywhere positive. This means that A cannot be a good global variable everywhere in the field space: there must be some singularities where the description in terms of a, $F_{\mu\nu}$ fails.

The beautiful argument by Seiberg and Witten [23, 24] that the singularity be related to the point where the magnetic monopole of the theory—as the bosonic part of the model is just the Giorgi–Glashow model the soliton monopoles found by 't Hooft and Polyakov are part of the spectrum—becomes *massless* due to quantum effects, and the consequent determination of the the prepotential $F(A)$ are by now well known. For completeness we summarize the main points of the solution in Appendix .4. Let us recall the main result here: by introducing an auxiliary torus (whose genus 1 corresponds to the rank of the gauge group $SU(2)$), described by the algebraic *curve*

$$y^2 = (x^2 - \Lambda^4)(x-u) = (x+\Lambda^2)(x-\Lambda^2)(x-u), \qquad u \equiv \langle \mathrm{Tr}\Phi^2 \rangle, \tag{36}$$

the solution is expressed as

$$\frac{da_D}{du} = \oint_\beta \frac{dx}{y}, \qquad \frac{da}{du} = \oint_\alpha \frac{dx}{y}, \tag{37}$$

where α and β are the two canonical cycles on the torus, Fig. 3. Explicitly,

$$a_D(u) = \frac{\sqrt{2}}{\pi} \int_{\Lambda^2}^{u} \left(\frac{x-u}{x^2-\Lambda^4} \right)^{1/2} = i\, \frac{u-\Lambda^2}{2}\, F(\frac{1}{2}, \frac{1}{2}; 2; \frac{\Lambda^2-u}{2}),$$

$$a(u) = \frac{\sqrt{2}}{\pi} \int_{\Lambda^2}^{\Lambda^2} \left(\frac{x-u}{x^2-\Lambda^4} \right)^{1/2} = \sqrt{2}\,(u+\Lambda^2)^{1/2}\, F(-\frac{1}{2}, \frac{1}{2}; 1; \frac{2}{u+\Lambda^2}). \tag{38}$$

The key step of the solution (37) was the theorem in algebraic geometry that the integrals of the holomorphic differential ($\frac{dx}{y}$ in our case of the genus one torus (36)) along the canonical cycles α and β (they are called period integrals) satisfy

$$\mathrm{Im} \frac{\oint_\alpha \frac{dx}{y}}{\oint_\beta \frac{dx}{y}} > 0,$$

independently of the way canonical cycles are redefined. According to the identification of the period integrals with the physical quantities as (37) this guarantees that

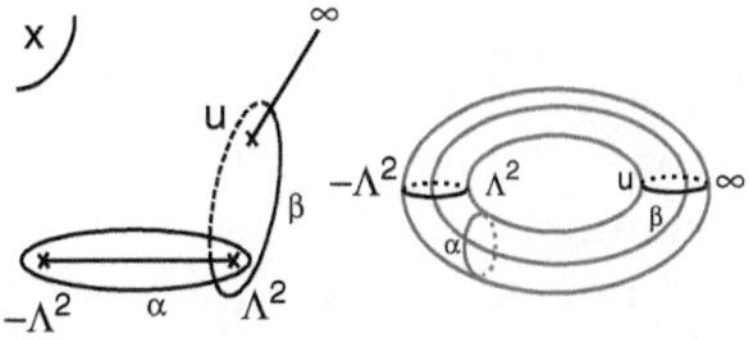

Fig. 3. The torus (36) represented as a two-sheeted Riemann surfaces, with two branch cuts (**left**). Note that two Riemann spheres attached at two cuts are equivalent to a torus (**figure on the right**)

$$\text{Im}\,\tau_{eff} = \text{Im}\frac{da_D}{da} = \frac{4\pi}{g_{eff}^2} > 0.$$

Let us add several remarks.

(i) Another key observation by Seiberg–Witten is that the $\mathcal{N} = 2$ supersymmetry implies an exact mass formula for BPS saturated states with magnetic and electric charges n_m, n_e:

$$M_{n_m,n_e} = \sqrt{2}\,|n_m\, a_D + n_e\, a|. \tag{39}$$

This is a consequence of the fact that the system has an underlying $\mathcal{N} = 2$ supersymmetry with a central extension (see Appendix .4). This formula generalizes the standard Higgs formula, $M_{0,n_e} = g\, n_e\langle\phi\rangle$, as $a \sim g\langle\phi\rangle$ semiclassically, and at the same time, the 't Hooft–Polyakov monopole mass formula, $M_{n_m,0} = 4\,\pi\, n_m\, \langle\phi\rangle/g$ (semiclassically $a_D \sim 4\,\pi\,\langle\phi\rangle/g$). Note that in the fully quantum formula (39) the magnetic and electric charges appear symmetrically. Indeed the mass formula is invariant under the generalized duality transformations (35), modulo appropriate relabeling of magnetic and electric charges.

(iii) Quite remarkably the low-energy effective action thus determined contains quantum effects in its entirety, the one-loop perturbative effects plus the sum of infinite instanton contributions. Indeed, the Seiberg–Witten curves have been checked against direct instanton calculations [41], and more recently, have been rederived by an explicit instanton resummation [42].

(iii) The Seiberg–Witten solution nicely solves an old (apparent) paradox related to the Dirac quantization versus renormalization group [8]: how can the relation $g_m\, g_e = 2\,\pi\, n$, $n = 0, 1, \ldots$ be compatible with the fact that both the electric and magnetic charges are Abelian $U(1)$ coupling constants, expected to get renormalized in the same direction? In the Seiberg–Witten solution, $g_m(\mu)$ gets renormalized as in (magnetic version of) QED, though monopole loops, with monopoles replacing the role of the electron. The same infrared behavior is explained, in the original electric picture, as due to instanton-induced nonperturbative renormalization of the electric coupling constant $g_e(\mu)$. As a consequence $g_m(\mu)\, g_e(\mu) = 4\,\pi$ holds [23] at any infrared cutoff $\mu = \langle a_D\rangle$. For other subtle issues related to renormalization group properties of Seiberg–Witten solution, see [43].

(iv) How do we know that these massless monopoles are related to the 't Hooft–Polyakov monopoles? That they *are* indeed them, can be verified by studying the electric and quark (in the cases with $N_f = 1, 2, 3$) number charges. As is well known the 't Hooft–Polyakov monopoles acquire these $U(1)$ charges quantum mechanically, via a beautiful phenomenon of charge fractionalization [44], which in this specific situation are the Witten's [45] and Jackiw–Rebbi's effects [46]. By moving within the space of vacua (QMS) and going into the regions where semiclassical approximation is valid (where $u = \langle \text{Tr}\,\Phi^2\rangle \gg \Lambda^2$), one can compare these fractional

$U(1)$ charges read off from the leading terms of the exact Seiberg–Witten solution with the ones obtained many years earlier by standard quantization of fermion fields around the semiclassical monopole backgrounds [47]. The results exactly match [48, 49].

(v) The low-energy effective Lagrangian near one of the singularities, e.g., $u = \Lambda^2$, looks like a (dual) QED with a massless monopole, whose Lagrangian has the standard $\mathcal{N} = 2$ QED form,

$$L = \frac{1}{4\pi}\mathrm{Im}\,[\int d^4\theta \frac{dF(A_D)}{dA_D}\,\bar{A}_D + \int \frac{1}{2}\frac{d^2F(A_D)}{dA_D^2}\,W_D^{\alpha}\,W_{D\,\alpha}\,] + \int d^4\theta(\bar{M}e^{V_D}M + \tilde{M}e^{-V_D}\bar{\tilde{M}}) + \int d^2\theta\sqrt{2}\tilde{M}A_D M, \tag{40}$$

where the gauge terms are just the dual of (33); the third and fourth terms describe the monopole.

(vi) Addition of a $\mathcal{N} = 1$ perturbation, the adjoint scalar mass term, $\mu\,\mathrm{Tr}\Phi^2$ in the original electric theory induces $\Delta L = \mu\,U(A_D)$, where the function $U(A_D)$ is the inverse of the solution $a_D(u)$. By minimizing the potential, the degeneracy (quantum moduli space—QMS) is eliminated leaving just two vacua, where

$$a_D = 0, \quad u = \langle \mathrm{Tr}\Phi^2 \rangle = \pm\Lambda^2, \qquad \langle M \rangle = \langle \tilde{M} \rangle = \mu\,\frac{\partial U}{\partial A_D} \sim \mu\,\Lambda.$$

The first result says that the magnetic monopole is massless in this vacuum (see (40)), the third states that the magnetic monopole condenses, leading to confinement à la 't Hooft–Mandelstam. This is perhaps the first example of nontrivial 4D system where this phenomenon has been demonstrated explicitly and analytically.

4.2 Seiberg–Witten Solutions for $\mathcal{N} = 2$ Models with Quarks

A general enthusiasm (alarm?) caused by the news that the $SU(2)$ Seiberg–Witten model with a small $\mathcal{N} = 1$ perturbation exhibited the 't Hooft–Mandelstam mechanism of confinement, was followed by a widespread delusion (relief?) among theoretical physicists when it was realized that the light monopoles appearing in the low-energy theory were Abelian and at the same time confinement was accompanied by dynamical Abelianization. This surely was not a good model of QCD! The fact that in the $SU(2)$ models with $N_f = 1, 2, 3$ hypermultiplets of quarks, studied in the (quite remarkable) second paper by Seiberg and Witten [24], as well as in pure $\mathcal{N} = 2$ Yang–Mills theories with more general gauge groups [25], the low-energy monopoles were always Abelian, did not help.

What was not realized at the time, however, was the fact that there was a clear reason for the Abelianization in these simplest models (see Sect. 4.3 below), and that, in the context of a more general class of $\mathcal{N} = 2$ theories with quark multiplets, Abelian confinement belonged to the exceptional cases. In fact, confinement is more typically caused by condensation of non-Abelian monopoles, as the subsequent analyses have revealed. We shall below briefly summarize the main features of these models, with technical aspects kept at its minimum.

The systems we consider are simple generalization of the $\mathcal{N}=2$ models with "quark" multiplets. The $N=1$ chiral and gauge superfields $\Phi = \phi + \sqrt{2}\,\theta\,\psi + \ldots$, and $W_\alpha = -i\lambda + \frac{i}{2}\,(\sigma^\mu\,\bar{\sigma}^\nu)_\alpha^\beta\,F_{\mu\nu}\,\theta_\beta + \ldots$ are both in the adjoint representation of the gauge group, while the hypermultiplets are taken in the fundamental representation of the gauge group. The Lagrangian takes the form,

$$L = \frac{1}{8\pi}\mathrm{Im}\,\tau_{cl}\left[\int d^4\theta\,\Phi^\dagger e^V \Phi + \int d^2\theta\,\frac{1}{2}WW\right] + L^{(quarks)} + \Delta L + \Delta' L, \quad (41)$$

where

$$L^{(quarks)} = \sum_i [\int d^4\theta\,\{Q_i^\dagger e^V Q_i + \tilde{Q}_i^\dagger e^{\tilde{V}}\tilde{Q}_i\} + \int d^2\theta\,\{\sqrt{2}\tilde{Q}_i\Phi Q^i + m_i\tilde{Q}_i Q^i\} \quad (42)$$

describes the n_f flavors of hypermultiplets ("quarks"), and

$$\tau_{cl} \equiv \frac{\theta_0}{\pi} + \frac{8\pi i}{g_0^2} \quad (43)$$

is the bare θ parameter and coupling constant. The $N=1$ chiral and gauge superfields $\Phi = \phi + \sqrt{2}\,\theta\,\psi + \ldots$, and $W_\alpha = -i\lambda + \frac{i}{2}\,(\sigma^\mu\,\bar{\sigma}^\nu)_\alpha^\beta\,F_{\mu\nu}\,\theta_\beta + \ldots$ are both in the adjoint representation of the gauge group, while the hypermultiplets are taken in the fundamental representation of the gauge group.

We consider small *generic* nonvanishing bare masses m_i for the hypermultiplets ("quarks"), which is consistent with $\mathcal{N}=2$ supersymmetry. Furthermore, it is convenient to introduce the mass for the adjoint scalar multiplet

$$\Delta L = \int d^2\theta\,\mu\,\mathrm{Tr}\,\Phi^2 \quad (44)$$

which breaks supersymmetry to $\mathcal{N}=1$. An advantage of doing so is that all flat directions are eliminated and one is left with a finite number of isolated vacua; keeping track of this number (and the symmetry breaking pattern in each of them) allows us to make highly nontrivial check of our analyses at various stages.

Below we summarize the physical results on these systems. To solve the system (41), the first step is the generalization of the curve (36) to the case of general group G. When the breaking is maximum, $G \to U(1)^{r_G}$ where r_G is the rank of the group G, we set $\mu = 0$ and consider vacua

$$\langle\Phi\rangle = \mathrm{diag}(\phi_1, \phi_2, \ldots), \qquad \phi_1 \neq \phi_2, \quad etc. \quad (45)$$

The auxiliary genus $g = N_c - 1$ (or N_c) curves for $SU(N_c)$ ($USp(2N_c)$) theories corresponding to these classical vacua (called Coulomb branch of the moduli space) are given by

$$y^2 = \prod_{k=1}^{n_c} (x - \phi_k)^2 + 4\Lambda^{2n_c - n_f} \prod_{j=1}^{n_f} (x + m_j), \qquad SU(N_c), \ N_f \leq 2N_c - 2, \tag{46}$$

and

$$y^2 = \prod_{k=1}^{n_c} (x - \phi_k)^2 + 4\Lambda \prod_{j=1}^{n_f} \left(x + m_j + \frac{\Lambda}{N_c} \right), \qquad SU(N_c), \ N_f = 2N_c - 1, \tag{47}$$

with ϕ_k subject to the constraint $\sum_{k=1}^{n_c} \phi_k = 0$, and

$$xy^2 = \left[x \prod_{a=1}^{n_c} (x - \phi_a^2)^2 + 2\Lambda^{2n_c + 2 - n_f} m_1 \cdots m_{n_f} \right]^2 - 4\Lambda^{2(2n_c + 2 - n_f)} \prod_{i=1}^{n_f} (x + m_i^2) \tag{48}$$

for $USp(2N_c)$. Analogous results for $SO(N_c)$ theories are also known.

The connection between these genus g hypertori and physics is made [23, 24, 25, 26, 27, 28, 29] through the identification of various period integrals of the holomorphic differentials on the curves with $(da_{Di}/du_j, da_i/du_j)$, where the gauge invariant parameters u_j's are defined by the standard relation

$$\prod_{a=1}^{n_c} (x - \phi_a) = \sum_{k=0}^{N_c} u_k \, x^{N_c - k}, \qquad u_0 = 1, \quad u_1 = 0, \qquad SU(N_c); \tag{49}$$

$$\prod_{a=1}^{n_c} (x - \phi_a^2) = \sum_{k=0}^{N_c} u_k \, x^{N_c - k}, \qquad u_0 = 1, \qquad USp(2N_c), \tag{50}$$

and $u_2 \equiv \langle \mathrm{Tr}\, \Phi^2 \rangle$, $u_3 \equiv \langle \mathrm{Tr}\, \Phi^3 \rangle$, etc. The VEVs of a_{Di}, a_i, which are directly related to the physical masses of the BPS particles through the exact Seiberg–Witten mass formula [23, 24]

$$M^{n_{mi}, n_{ei}, S_k} = \sqrt{2} \left| \sum_{i=1}^{g} (n_{mi} \, a_{Di} + n_{ei} \, a_i) + \sum_k S_k m_k \right|, \tag{51}$$

are constructed as integrals over the nontrivial cycles of the meromorphic differentials on the curves. S_k are the i-th quark number charge of the monopole under consideration, which enters the formula for the central charges (hence the mass).

(i) These formulae naturally generalize those of the pure $SU(2)$ theory, (37) and (39). The singularities of the curves (46)–(48) are the points in the space of vacua (QMS) where various particles become massless.

(ii) When $m_i \gg \Lambda$ these singularities are at the points where $\phi \sim m_i$ (where the quarks become massless—see (42)) and at the points where monopoles of pure Yang–Mills theory become massless. The latter are the points the curve of the Yang–Mills theory,

$$y^2 = \prod_{k=1}^{n_c} (x - \phi_k)^2 + 4\Lambda_{YM}^{2n_c}$$

become maximally singular, $\sim \prod_{i=1}^{n_c-1} (x - x_i)^2 (x - \alpha)(x - \beta)$.

(iii) It is the property of these curves that when $m_i \sim \Lambda$ all singularities are found to correspond to magnetic degrees of freedom (massless monopoles and dyons). To trace how, as m_i are varied, the original "electric" singularities (massless quarks) make a metamorphosis into magnetic monopoles, due to the movement of certain branch points (or branch cuts) sliding under other branch cuts (branch surfaces), is a rather complicated business, and has been analyzed satisfactorily only in the $SU(2)$ theories with matter [24, 51].

(iv) The particular form of the curve specific to different groups reflect different global symmetries. A nice discussion is given in [26].

4.3 Exact Quantum Behavior of Light Non-Abelian Monopoles

Physics of confining vacua and properties of light monopoles in these theories are studied by identifying all of the $\mathcal{N} = 1$ vacua (the points in the QMS—quantum moduli space, that is, the space of vacua—which survive the $\mathcal{N} = 1$ perturbation) and studying the low-energy action for each of them. The underlying $\mathcal{N} = 2$ theory, especially with $m_i = 0$ or with equal masses $m_i = m$, has a large continuous degeneracy of vacua (flat directions), which has been studied by using the Seiberg–Witten curves, nonrenormalization of Higgs branch metrics, superconformal points and their universality, their moduli structure and symmetries, etc. [28, 29]. For the purpose of this section, however, we are most interested in the set of vacua which are picked up when the small generic bare quark masses m_i and a small nonzero adjoint mass μ are present. At the roots of these different branches of $\mathcal{N} = 2$ vacua where the Higgs branches meet the Coulomb branch, lie all these vacua (see Fig. 4), which survives the $\mathcal{N} = 1$ perturbation, (44). In $SU(N_c)$ theories with N_f flavors with generic masses, all $\mathcal{N} = 1$ vacua arising this way have been completely classified [30, 31].

For nearly equal quark masses they fall into classes $r = 0, 1, \ldots, \frac{N_f}{2}$ groups of vacua near the "roots of nonbaryonic Higgs branches," and for $N_f \geq N_c$, there are special vacua at the "roots of baryonic Higgs branches." These names reflect the fact that in the respective Higgs branch nonbaryonic or baryonic squark VEV,

$$\langle Q_i^a \, \tilde{Q}_a^j \rangle, \qquad \langle \epsilon_{a_1 \, a_2 \, \ldots, a_{N_c}} Q_{i_1}^{a_1} Q_{i_2}^{a_2} \ldots Q_{i_{N_c}}^{a_{N_c}} \rangle, \tag{52}$$

are formed (see Fig. 4). Each group of vacua coalesce in single vacua where the gauge symmetry is enhanced into non-Abelian gauge groups, as in Table 1.

The vacua at the root of the baryonic branch are in "free-magnetic" phase; the light non-Abelian magnetic monopoles appear as asymptotic states; they do not condense, no confinement and no symmetry breaking occur. Although the appearance of the Seiberg dual gauge group, $SU(\tilde{N}_c)$, $\tilde{N}_c \equiv N_f - N_c$ is certainly intriguing [28], these are not type of vacua we are interested in.

Our main interest is the first classes of the so-called r-vacua, where the magnetic gauge group is

$$U(r) \times U(1)^{N_c - r},$$

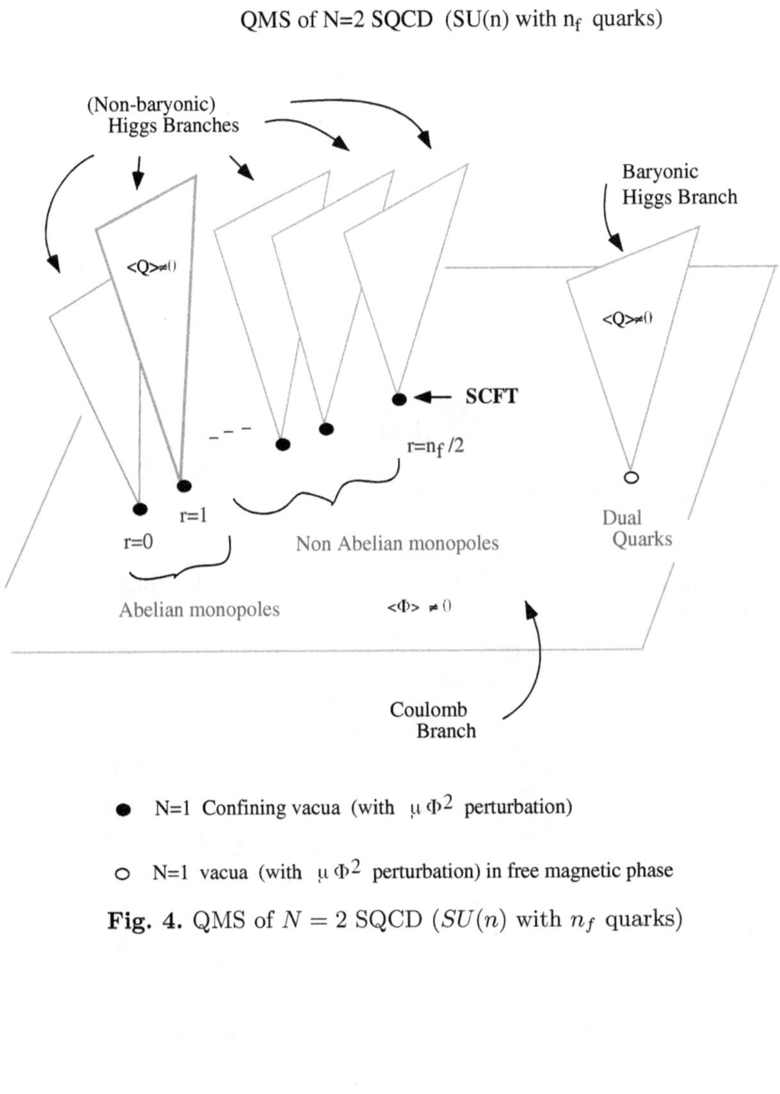

Fig. 4. QMS of $N = 2$ SQCD ($SU(n)$ with n_f quarks)

and the massless matter multiplets consist of N_f monopoles in the fundamental representation of $U(r)$, and flavor-singlet Abelian monopoles carrying a single charge, each with respect to one of the $U(1)$ factors (Table 1). [7]

Once the gauge group and the quantum numbers of the matter fields are all known, the $\mathcal{N}=2$ supersymmetry uniquely fixes the structure of the effective action. We find that

(i) We see the non-Abelian monopoles in action, in the generic r ($2 \le r \le \frac{N_f}{2}$) vacua (see Table 2 taken from [30]). They behave perfectly as point-like particles, albeit in a dual, magnetic gauge system. Upon $\mathcal{N}=1$ perturbation they condense (confinement phase) $\langle q_a^i \rangle \sim \delta_a^i \sqrt{\mu\,\Lambda}$ and induces flavor symmetry breaking

$$SU(N_f) \times U(1) \to U(r) \times U(N_f - r).$$

(ii) The upper limit $r \le \frac{N_f}{2}$ is a manifestation of monopole dynamics: only in this range of r the non-Abelian monopoles can appear as *recognizable infrared degrees of freedom.* We now see why in the $SU(2)$ Seiberg–Witten models, as well as in pure $\mathcal{N}=2$ Yang–Mills (i.e., $N_f=0$) models with different gauge groups, the low-energy monopoles were found to be always Abelian: in all these cases, non-Abelian monopoles would interact too strongly, not enough of them being there. We remind the reader that the beta function in $\mathcal{N}=2$ SU theories has the pure one-loop form with $\beta_0 \propto 2\,r - N_f$.

(iii) Indeed, there are homotopy and symmetry arguments [30, 52] which suggest that non-Abelian monopoles appearing in the r-vacua are "baryonic constituents" of an Abelian ('t Hooft–Polyakov) monopople,

$$\textit{Abelian monopole} \sim \epsilon^{a_1 \dots a_r} q_{a_1}^{i_1} q_{a_2}^{i_2} \dots q_{a_r}^{i_r}, \tag{53}$$

a_i being the dual color indices and i_m the flavor indices. The $SU(r)$ gauge ineractions, being infrared-free, are unable to keep the Abelian monopole bound: they disintegrate into non-Abelian monopoles.

(iv) That the effective degrees of freedom in the r vacua are non-Abelian rather than Abelian monopoles, is actually required also by symmetry of

Table 1. The effective degrees of freedom and their quantum numbers at the "non-baryonic root"

	$SU(r)$	$U(1)_0$	$U(1)_1$	$\dots$	$U(1)_{n_c-r-1}$	$U(1)_B$
$n_f \times q$	$\underline{\mathbf{r}}$	1	0	$\dots$	0	0
e_1	$\underline{\mathbf{1}}$	0	1	$\dots$	0	0
$\vdots$	$\vdots$	$\vdots$	$\vdots$	$\ddots$	$\vdots$	$\vdots$
e_{n_c-r-1}	$\underline{\mathbf{1}}$	0	0	$\dots$	1	0

[7] We shall use the notation $N_c = n_c$ indistinguishably, and analogously $N_f = n_f$.

Table 2. Phases of $SU(n_c)$ gauge theory with n_f flavors. $\tilde{n}_c \equiv n_f - n_c$

Label (r)	Deg. freed.	Eff. gauge group	Phase	Global symmetry
0	Monopoles	$U(1)^{n_c-1}$	Confinement	$U(n_f)$
1	Monopoles	$U(1)^{n_c-1}$	Confinement	$U(n_f-1)\times U(1)$
$\leq [\frac{n_f-1}{2}]$	NA Monopoles	$SU(r)\times U(1)^{n_c-r}$	Confinement	$U(n_f-r)\times U(r)$
$n_f/2$	Rel. nonloc.	-	Confinement	$U(n_f/2)\times U(n_f/2)$
BR	NA monopoles	$SU(\tilde{n}_c)\times U(1)^{n_c-\tilde{n}_c}$	Free magnetic	$U(n_f)$

the system [30, 53], not only from the dynamics. If the Abelian monopoles of the r-th tensor flavor representation were the correct degrees of freedom, the low-energy effective theory would have too large an accidental symmetry – $SU(\binom{N_f}{r})$. The condensation of such monopoles would produce far-too-many Nambu–Goldstone bosons than expected from the symmetry of the underlying theory. The system prevents such an awkward situation from being realized in an elegant manner, introducing smaller solitons, non-Abelian monopoles, in the fundamental representation of the $SU(N_f)$ so that the low-energy theory has the right symmetry.

(v) An analogous argument might be used in the standard QCD, to exclude Abelian picture of confinement, though admittedly this is not a very rigorous one. We know from lattice simulations of $SU(3)$ theory that confinement and chiral symmetry breaking are closely related. If Abelian 't Hooft–Monopole–Mandelstam monopoles were the right degrees of freedom describing confinement, their condensation would somehow have to describe chiral symmetry breaking as well. We would then be led to assume that they carry flavor quantum numbers of $SU(N_f)_L \times SU(N_f)_R$, e.g.,

$$Monopoles \sim M_i^j, \qquad \langle M_i^j \rangle \propto \delta_i^j \, \Lambda_{QCD},$$

where i, j are $SU(N_f)_L \times SU(N_f)_R$ indices. But such a system would have a far too large accidental symmetry. Confinement would be accompanied by a large number of unexpected (and indeed unobserved) light Nambu–Goldstone bosons.

(vi) The limiting case of r vacua, with $r = \frac{N_f}{2}$, as well as the massless ($m_i \to 0$) limit of $USp(2N_c)$ and $SO(N_c)$ theories, are of great interest (see Fig. 5 and Table 3). The low-energy effective theory in these cases turn out to be conformally invariant (nontrivial infrared fixed-point) theories. This is an analogue of an Abelian superconformal vacuum found first in the pure

Table 3. Phases of $USp(2n_c)$ gauge theory with n_f flavors with $m_i \to 0$. $\tilde{n}_c \equiv n_f - n_c - 2$

	Deg. freed.	Eff. gauge group	Phase	Global symmetry
First group	Rel. nonloc.	–	Confinement	$U(n_f)$
Second group	Dual quarks	$USp(2\tilde{n}_c)\times U(1)^{n_c-\tilde{n}_c}$	Free magnetic	$SO(2n_f)$

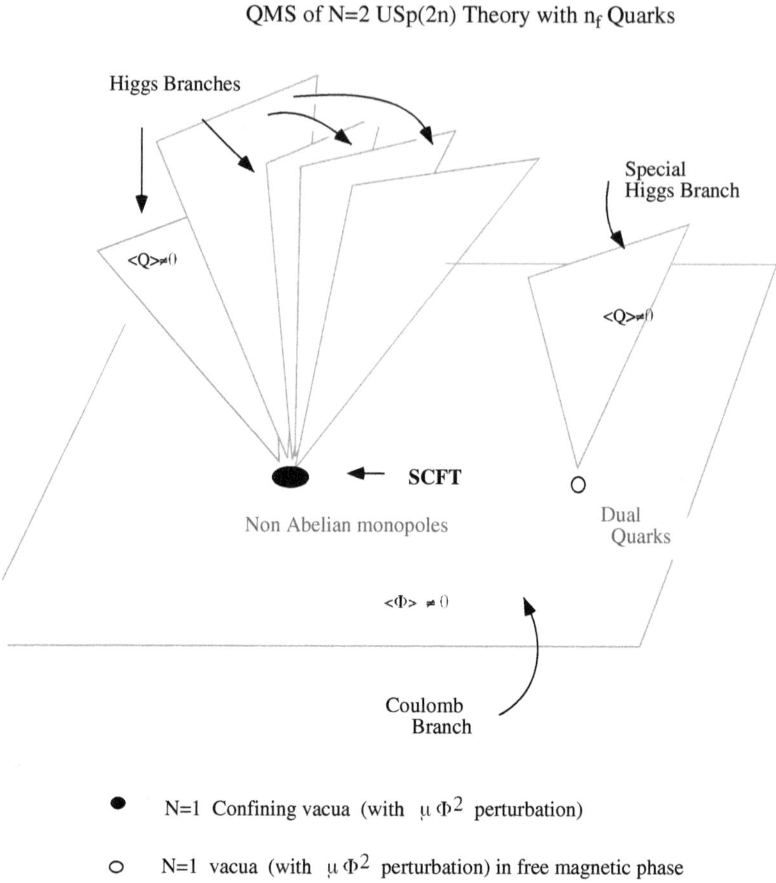

Fig. 5. QMS of $N = 2$ SQCD $USp(2n)$ theory with n_f quarks

$SU(3)$ Yang–Mills theory by Argyres and Douglas [54]. It can be explicitly checked that the low-energy degrees of freedom include relatively nonlocal monopoles and dyons [30, 53, 55]. There are no local effective Lagrangians describing the infrared dynamics. These are the most difficult cases to analyze, but are potentially the most interesting ones, from the point of view of understanding QCD. We shall come back to these (perhaps, crucial) cases at the end of the lecture, Sect. 7.

5 Vortices

The moral of the story is that the non-Abelian monopoles do exist in fully quantum–mechanical systems. In typical confining vacua in supersymmetric gauge theories they are the relevant infrared degrees of freedom. Their condensation induces confinement and dynamical symmetry breaking. This brings us back to the problem of *understanding* these light, magnetic degrees of freedom as quantum solitons:

- What are their semiclassical counterparts?
- Are they Goddard–Nuyts–Olive–Weinberg monopoles?
- In which sense condensation of non-Abelian monopoles imply confinement?
- How has the difficulty related to the dual group mentioned earlier been avoided?

These are the questions we wish to answer. The idea is to take advantage of the fact that in supersymmetric theories there are parameters which can be varied, upon which the physical properties of the system depend in a holomorphic fashion. As m_i and μ are varied, there cannot be phase transition at some $|\mu|$ or at $|m_i|$: the number of Nambu–Goldstone bosons and hence the pattern of the symmetry breaking, must be invariant.

5.1 Abrikosov–Nielsen–Olesen Vortex

Topologically stable vortices arise when the ground states of a system have a nontrivial moduli space which is not simply connected. The best-known case [56] is the Abelian gauge theory with a charged complex matter field in Higgs phase (superconductor), where the static configurations have energy density

$$H = \frac{1}{4}F_{ij}^2 + |D_i\phi|^2 + V(|\phi|), \qquad D_i = \partial_i - i\,e\,A_i.$$

The potential V is assumed to attain its minimum at $|\phi| = v \neq 0$. The asymptotic gauge and scalar fields must be such that the field energy be finite,

$$|\phi(x)| \to v, \qquad D_i\phi \to 0, \qquad F_{ij}^2 \to 0.$$

These allow for nontrivial configurations classified by an integer,

$$\pi_1(U(1)) = \mathbf{Z},$$

i.e., by an integer winding number n,

$$\phi \to e^{in\varphi}\,v, \qquad A_\varphi \to \frac{n}{e\rho},$$

where ρ, ϕ, and z are the position variables of the cylindrical coordinate system. At the center of the vortex $\phi(\rho = 0, \varphi) = 0$ in order for $\phi(\rho, \phi)$ to be a smooth configuration: the gauge symmetry is restored along the vortex core.

Depending on the potential, the vacuum can be superconductor of type II where single isolated (Abrikosov–Nielsen–Olesen) vortices are stable, type I systems where vortices stick together to form the regions of normal ground state, and finally there is the critical case between them (BPS) where vortices has no net interaction and the tension of winding number k vortex is equal to k times that of the minimum-winding vortex.

5.2 $\mathbf{Z}_N$ Vortices

In pure $SU(N)$ theory with all matter fields in adjoint representation, the true gauge group is $\frac{SU(N)}{\mathbf{Z}_N}$. When the gauge group is completely broken the vacuum manifold has nontrivial structure,

$$\pi_1(\frac{SU(N)}{\mathbf{Z}_N}) = \mathbf{Z}_N. \tag{54}$$

The asymptotic behavior of the fields, required by finiteness of the tension is

$$A_i \sim \frac{i}{g} U(\phi)\partial_i U^\dagger(\phi); \quad \phi_A \sim U\phi_A^{(0)}U^\dagger, \qquad U(\phi) = \exp i \sum_j^r \beta_j T_j \phi$$

where T_j are the generators of the Cartan subalgebra of H, $\phi_A^{(0)}$ are the (set of) VEVs of the adjoint scalar fields which break the $SU(N)$ group completely. The smoothness of the configurations requires the quantization condition: (α = root vectors of H)

$$U(2\pi) \in \mathbf{Z}_N, \qquad \alpha \cdot \beta \in \mathbf{Z}. \tag{55}$$

The second condition of (55) appears to imply that these vortices be characterized by the *weight vectors* of the group $\tilde{H} = SU(N)$, dual of $H = SU(N)/\mathbf{Z}_N$ [4]: one vortex for each irreducible representation of $\tilde{H}$. Actually, (54) shows that there is just one stable vortex with a given $\mathbf{Z}_N$ charge (N-ality)[8].

An interesting model of this sort is the so-called $N = 1^*$ theory [57, 58, 59] defined as the $\mathcal{N} = 4$ supersymmetric theory with addition of mass terms for the three adjoint scalar multiplets,

$$\Delta L = \sum_{i=1}^3 m_i \, \Phi_i^2|_{\theta\theta},$$

which break supersymmetry to $\mathcal{N} = 1$. The general properties of chiral condensates,

$$\langle W W \rangle, \quad \langle \Phi_1^2 \rangle, \quad \langle \Phi_2^2 \rangle, \quad \langle \Phi_3^2 \rangle,$$

in all possible types of vacua (confinement vacua, Coulomb vacua, Higgs vacua) have been analyzed exactly in a series of papers [60].

This model is based on the underlying $\mathcal{N} = 4$ model, which is believed to display exact Olive–Montonen duality. In spite of the relative simplicity of the model, the properties of $\mathbf{Z}_N$ monopoles in the Higgs (or partially Higgs) vacua in the $\mathcal{N} = 1^*$ are not very well known, except for the $SU(2)$ [61] or $SU(3)$ cases.

[8] That an excitation in a theory in which all fields are neutral with respect to $\mathbf{Z}_N$ is characterized by a fractional $\mathbf{Z}_N$ charge, may be thought of as an analogue of a very general behavior of solitons: charge fractionalization.

5.3 Non-Abelian Vortices in a $U(N)$ Model

The $\mathbf{Z}_N$ vortice discussed in the preceding section at first sight appears to carry a non-Abelian charge, being labeled by the weight vector of a non-Abelian dual group $\tilde{H}$: actually, they do not [62]. It is just a single solution, which can be transformed by Weyl transformations of H. There are no continuous moduli associated to it.

Truly non-Abelian vortices have been constructed [35, 37] in the context of a $\mathcal{N}=2$ supersymmetric $U(N)$ gauge theory, with N_f flavors, where the gauge group is broken by the VEVs of a set of scalar fields in the fundamental representations. The model Lagrangian has the form

$$\begin{aligned}\mathcal{L} = \mathrm{Tr}&\left[-\frac{1}{2g^2}F_{\mu\nu}F^{\mu\nu} - \frac{2}{g^2}\mathcal{D}_\mu \phi^\dagger \mathcal{D}^\mu \phi - \mathcal{D}_\mu H \mathcal{D}^\mu H^\dagger - \lambda\left(c\,\mathbf{1}_N - H H^\dagger\right)^2\right] \\ &+\mathrm{Tr}\left[(H^\dagger\phi - M H^\dagger)(\phi H - H M)\right] \end{aligned} \tag{56}$$

where $F_{\mu\nu} = \partial_\mu W_\nu - \partial_\nu W_\nu + i\,[W_\mu, W_\nu]$ and $\mathcal{D}_\mu H = (\partial_\mu + i\,W_\mu)\,H$, and H represents the fields in the fundamental representation of $SU(N)$, written in a color–flavor $N \times N_f$ matrix form, $(H)^i_\alpha \equiv q^i_\alpha$, and M is a $N_f \times N_f$ mass matrix. Here, g is the $U(N)_{\mathrm{G}}$ gauge coupling, λ is a scalar coupling. For

$$\lambda = \frac{g^2}{4} \tag{57}$$

the system is BPS saturated. For such a choice, (56) can be regarded as a truncation of the bosonic sector of an $\mathcal{N}=2$ supersymmetric $U(N)$ gauge theory, and with $(H)^i_\alpha$ representing the half of the squark fields,

$$(H)^i_\alpha \equiv q^i_\alpha, \qquad \tilde{q}^\alpha_i \equiv 0 \tag{58}$$

In the supersymmetric context the parameter c is the Fayet–Iliopoulos parameter. In the following we set $c > 0$ so that the system be in Higgs phase, and so as to allow stable vortex configurations. For generic, unequal quark masses,

$$M = diag\,(m_1, m_2, \ldots, m_{N_f}), \tag{59}$$

the adjoint scalar VEV takes the form,

$$\langle\phi\rangle = M = \begin{pmatrix} m_1 & 0 & 0 & 0 \\ 0 & m_2 & 0 & 0 \\ & & & \\ 0 & 0 & \ddots & 0 \\ 0 & 0 & 0 & m_N \end{pmatrix}, \tag{60}$$

which breaks the gauge group to $U(1)^N$.

In order to have a non-Abelian vortex, it is necessary to choose masses equal,

$$M = diag\,(m, m, \ldots, m), \tag{61}$$

the adjoint and squark fields have the vacuum expectation value (VEV)

$$\langle \phi \rangle = m\,\mathbf{1}_N, \qquad \langle H \rangle = \sqrt{c}\begin{pmatrix} 1 & 0 & 0 \\ 0 & \ddots & 0 \\ 0 & 0 & 1 \end{pmatrix} \tag{62}$$

where only the first N flavors are left explicit. The squark VEV breaks the gauge symmetry completely, while leaving an unbroken $SU(N)_{C+F}$ color–flavor diagonal symmetry (the flavor group acts on H from the right while the $U(N)_{\rm G}$ gauge symmetry acts on H from the left). The global symmetry group associate with the other $N_f - N$ flavors also remains unbroken. The BPS vortex equations are

$$(\mathcal{D}_1 + i\mathcal{D}_2)\, H = 0, \quad F_{12} + \frac{g^2}{2}\left(c\,\mathbf{1}_N - H\,H^\dagger\right) = 0. \tag{63}$$

The matter equation can be solved [65, 66, 67] by use of the $N \times N$ moduli matrix $H_0(z)$ whose components are holomorphic functions of the complex coordinate $z = x^1 + ix^2$,

$$H = S^{-1}(z,\bar{z})\, H_0(z), \quad W_1 + i\,W_2 = -2\,i\,S^{-1}(z,\bar{z})\,\bar{\partial}_z S(z,\bar{z}). \tag{64}$$

The gauge field equations then take the simple form ("master equation")

$$\partial_z\,(\Omega^{-1}\partial_{\bar{z}}\,\Omega) = \frac{g^2}{4}\,(c\,\mathbf{1}_N - \Omega^{-1}\,H_0\,H_0^\dagger). \tag{65}$$

The moduli matrix and S are defined up to a redefinition,

$$H_0(z) \to V(z)\,H_0(z), \qquad S(z,\bar{z}) \to V(z)\,S(z,\bar{z}), \tag{66}$$

where $V(z)$ is any nonsingular $N \times N$ matrix which is holomorphic in z. This class of model has been extensively studied recently [65, 66, 67, 68, 69, 70, 71]. In particular, in the context of these models, a considerable attention was given to the system in which $U(N)$ gauge symmetry is either explicitly or dynamically broken to $U(1)^N$, producing *Abelian* monopoles. As the terminology used and concepts involved, though physically distinct, are often similar to the concept of non-Abelian monopoles discussed in this note, and could be misleading.

5.4 Dynamical Abelianization

As should be clear from what we said so far, it is crucial that the color–flavor diagonal symmetry $SU(N)$ remains exactly conserved, for the emergence of non-Abelian dual gauge group (see the next section). Consider, instead, the cases in which the gauge $U(N)$ (or $SU(N) \times U(1)$) symmetry is broken to Abelian subgroup $U(1)^N$, either by small quark mass differences ((60)) or dynamically, as in the $\mathcal{N} = 2$ models with $N_f < 2N$ [36, 69]. From the breaking of various $SU(2)$ subgroups to $U(1)$ there appear light 't Hooft–Polyakov monopoles of mass $O(\frac{\Delta m}{g})$ (in the case of an explicit breaking) or $O(\Lambda)$ (in the case of dynamical breaking). As the $U(1)^N$ gauge group is further broken by the squark VEVs, the system develops ANO vortices. The light magnetic monopoles, carrying magnetic charges of two different $U(1)$ factors, look confined by the two vortices (Fig. 6). These cases have been discussed extensively [67, 68, 69, 70], within the context of $U(N)$ model of Sect. 5.3.

The dynamics of the fluctuation of the orientational modes along the vortex turns out to be described by a two-dimensional $\mathbf{C}P^{N-1}$ model [35, 37]. It has been shown [35, 36, 69, 70], that the kinks of the two-dimensional sigma model precisely correspond to these light monopoles, to be expected in the underlying $4D$ gauge theory. In particular, it was noted that there is an elegant matching between the dynamics of two-dimensional sigma model (describing the dynamics of the vortex orientational modes in the Higgs phase of the $4D$ theory) and the dynamics of the $4D$ gauge theory in the Coulomb phase, including the precise matching of the coupling constant renormalization [36, 68, 69].

Note that these cases are analogue of *what would occur* in QCD if the color $SU(3)$ symmetry were to dynamically break itself to $U(1)^2$. Confinement would be described in this case by the condensation of magnetic monopoles carrying the Abelian charges Q^1, or Q^2, and the resulting ANO vortices will be of two types, 1 and 2 carrying the related fluxes.

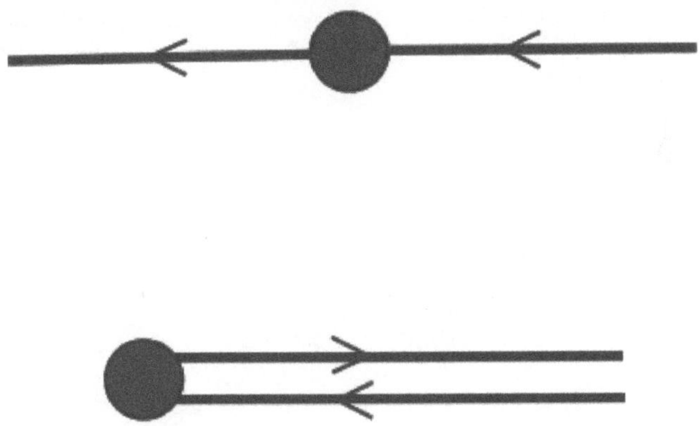

Fig. 6. Monopoles in $U(N)$ systems with abelianization are confined by two Abelian vortices

6 The Model

Actually the model we need here is not exactly the model of Sect. 5.3, but is a model which contains it as a low-energy approximation. It is the same model already discussed in Sect. 4.2, but now we analyze it in the region, $m_i \gg \mu \gg \Lambda$, so that the semiclassical reasoning of Sect. 3 makes sense. For concreteness, we take as our model the standard $\mathcal{N}=2$ SQCD with N_f quark hypermultiplets, with a larger gauge symmetry, e.g., $SU(N+1)$, which is broken at a much larger mass scale ($v_1 \sim |m_i|$) as

$$SU(N+1) \stackrel{v_1 \neq 0}{\longrightarrow} \frac{SU(N)\times U(1)}{\mathbf{Z}_N}. \tag{67}$$

The unbroken gauge symmetry is completely broken at a lower mass scale, $v_2 \sim |\sqrt{\mu m}|$, as in (78) below.

Clearly, one can attempt a similar embedding of the model (56) in a larger gauge group broken at some higher mass scale, in the context of a nonsupersymmetric model, even though in such a case the potential must be judiciously chosen and the dynamical stability of the scenario would have to be carefully monitored. Here we choose to study the softly broken $\mathcal{N}=2$ SQCD for concreteness, and above all because the dynamical properties of this model are well understood: this will provide us with a nontrivial check of our results. Another motivation is purely of convenience: it gives a definite potential with desired properties.[9]

We are hereby back to our argument on the duality and non-Abelian monopoles, defined through a better-understood non-Abelian *vortices* presented in general terms in Sect. 2.2, but now in the context of a concrete model, where the fully quantum–mechanical answer is known.

The underlying theory is thus

$$\mathcal{L} = \frac{1}{8\pi}\mathrm{Im}\, S_{cl}\left[\int d^4\theta\, \Phi^\dagger e^V \Phi + \int d^2\theta\, \frac{1}{2} W W\right] + \mathcal{L}^{(\mathrm{quarks})} + \int d^2\theta\, \mu\, \mathrm{Tr}\Phi^2 + h.c.; \tag{68}$$

$$\mathcal{L}^{(\mathrm{quarks})} = \tag{69}$$
$$\sum_i \left[\int d^4\theta\, \{Q_i^\dagger e^V Q_i + \tilde{Q}_i e^{-V} \tilde{Q}_i^\dagger\} + \int d^2\theta\, \{\sqrt{2}\tilde{Q}_i \Phi Q^i + m_i\, \tilde{Q}_i\, Q^i\} + h.c.\right]$$

where m_i are the bare masses of the quarks and we have defined the complex coupling constant

$$S_{cl} \equiv \frac{\theta_0}{\pi} + \frac{8\pi i}{g_0^2}. \tag{70}$$

[9] Recent developments [32, 77] allow us actually to consider systems of this sort within a much wider class of $\mathcal{N}=1$ supersymmetric models, whose infrared properties are very much under control.

We also added the parameter μ, the mass of the adjoint chiral multiplet, which breaks the supersymmetry softly to $\mathcal{N}=1$. The bosonic sector of this model is described, after elimination of the auxiliary fields, by

$$\mathcal{L} = \frac{1}{4g^2} F_{\mu\nu}^2 + \frac{1}{g^2} |\mathcal{D}_\mu \Phi|^2 + |\mathcal{D}_\mu Q|^2 + \left|\mathcal{D}_\mu \bar{\tilde{Q}}\right|^2 - V_1 - V_2, \tag{71}$$

where

$$V_1 = \frac{1}{8} \sum_A \left(t^A_{ij} \left[\frac{1}{g^2} (-2) \, [\Phi^\dagger, \Phi]_{ji} + Q^\dagger_j Q_i - \tilde{Q}_j \tilde{Q}^\dagger_i \right] \right)^2 ; \tag{72}$$

$$\begin{aligned} V_2 = & \, g^2 |\mu \, \Phi^A + \sqrt{2} \, \tilde{Q} \, t^A Q|^2 + \tilde{Q} \, [m + \sqrt{2} \Phi] \, [m + \sqrt{2} \Phi]^\dagger \, \tilde{Q}^\dagger \\ & + Q^\dagger \, [m + \sqrt{2} \Phi]^\dagger \, [m + \sqrt{2} \Phi] \, Q. \end{aligned} \tag{73}$$

In the construction of the approximate monopole and vortex solutions, we shall consider only the VEVs and fluctuations around them which satisfy

$$[\Phi^\dagger, \Phi] = 0, \qquad Q_i = \tilde{Q}^\dagger_i, \tag{74}$$

and hence the D-term potential V_1 can be set identically to zero throughout.

In order to keep the hierarchy of the gauge symmetry breaking scales, (24), we choose the masses such that

$$m_1 = \ldots = m_{N_f} = m, \tag{75}$$

$$m \gg \mu \gg \Lambda. \tag{76}$$

Although the theory described by the above Lagrangian has many degenerate vacua, we are interested in the vacuum where (see [30] for the detail)

$$\langle \Phi \rangle = - \frac{1}{\sqrt{2}} \begin{pmatrix} m & 0 & 0 & 0 \\ 0 & \ddots & \vdots & \vdots \\ 0 & \ldots & m & 0 \\ 0 & \ldots & 0 & -N m \end{pmatrix} ; \tag{77}$$

$$Q = \tilde{Q}^\dagger = \begin{pmatrix} d & 0 & 0 & 0 & \ldots \\ 0 & \ddots & 0 & \vdots & \ldots \\ 0 & 0 & d & 0 & \ldots \\ 0 & \ldots & 0 & 0 & \ldots \end{pmatrix}, \qquad d = \sqrt{(N+1) \, \mu \, m}. \tag{78}$$

This is a particular case of the so-called r vacuum, with $r = N$. Although such a vacuum certainly exists classically, the existence of the quantum $r = N$ vacuum in this theory requires $N_f \geq 2\,N$, which we shall assume.[10]

[10] This might appear to be a rather tight condition as the original theory loses asymptotic freedom for $N_f \geq 2\,N + 2$. This is not so. An analogous discussion can be made by considering the breaking $SU(N) \to SU(r) \times U(1)^{N-r}$. In this case the condition for the quantum non-Abelian vacuum is $2\,N > N_f \geq 2\,r$, which is a much looser condition.

To start with, ignore the smaller squark VEV, (78). As $\pi_2(G/H) \sim \pi_1(H) = \pi_1(U(1)) = \mathbf{Z}$, the symmetry breaking (77) gives rise to regular magnetic monopoles with mass of order of $O(\frac{v_1}{g})$, whose continuous transformation property is our main concern here.

The semiclassical formulas for their mass and fluxes [6, 52] are summarized in Appendix 9.

6.1 Low-energy Approximation and Vortices

At scales much lower than $v_1 = m$ but still neglecting the smaller squark VEV $v_2 = d = \sqrt{(N+1)\,\mu\, m} \ll v_1$, the theory reduces to an $SU(N) \times U(1)$ gauge theory with N_f light quarks $q_i, \tilde{q}^i$ (the first N components of the original quark multiplets $Q_i, \tilde{Q}^i$). By integrating out the massive fields, the effective Lagrangian valid between the two mass scales has the form,

$$\mathcal{L} = \frac{1}{4g_N^2}(F_{\mu\nu}^a)^2 + \frac{1}{4g_1^2}(F_{\mu\nu}^0)^2 + \frac{1}{g_N^2}|\mathcal{D}_\mu \phi^a|^2 + \frac{1}{g_1^2}|\mathcal{D}_\mu \phi^0|^2 + |\mathcal{D}_\mu q|^2 + |\mathcal{D}_\mu \bar{\tilde{q}}|^2$$
$$-g_1^2 \left| -\mu\, m \sqrt{N(N+1)} + \frac{\tilde{q}\, q}{\sqrt{N(N+1)}} \right|^2 - g_N^2 |\sqrt{2}\, \tilde{q}\, t^a q\,|^2 + \dots \tag{79}$$

where $a = 1, 2, \dots N^2-1$ labels the $SU(N)$ generators, t^a; the index 0 refers to the $U(1)$ generator $t^0 = \frac{1}{\sqrt{2N(N+1)}}\, diag(1, \dots, 1, -N)$. We have taken into account the fact that the $SU(N)$ and $U(1)$ coupling constants (g_N and g_1) get renormalized differently towards the infrared.

The adjoint scalars are fixed to its VEV, (77), with small fluctuations around it,

$$\Phi = \langle \Phi \rangle (1 + \langle \Phi \rangle^{-1} \tilde{\Phi}), \qquad |\tilde{\Phi}| \ll m. \tag{80}$$

In the consideration of the vortices of the low-energy theory, they will be in fact replaced by the constant VEV. The presence of the small terms (80), however, makes the low-energy vortices not strictly BPS (and this will be important in the consideration of their stability below).[11]

The quark fields are replaced, consistently with (74), as

$$\tilde{q} \equiv q^\dagger, \qquad q \to \frac{1}{\sqrt{2}}\, q, \tag{81}$$

where the second replacement brings back the kinetic term to the standard form.

---

[11] In the terminology used in Davis et al. [63] in the discussion of the Abelian vortices in supersymmetric models, our model corresponds to an F model while the models of [68, 69, 66] correspond to a D model. In the approximation of replacing Φ with a constant, the two models are equivalent: they are related by an $SU_R(2)$ transformation [64, 78].

We further replace the singlet coupling constant and the $U(1)$ gauge field as

$$e \equiv \frac{g_1}{\sqrt{2N(N+1)}}; \qquad \tilde{A}_\mu \equiv \frac{A_\mu}{\sqrt{2N(N+1)}}, \qquad \tilde{\phi}^0 \equiv \frac{\phi^0}{\sqrt{2N(N+1)}}. \quad (82)$$

The net effect is

$$\mathcal{L} = \frac{1}{4g_N^2}(F_{\mu\nu}^a)^2 + \frac{1}{4e^2}(\tilde{F}_{\mu\nu})^2 + |\mathcal{D}_\mu q|^2 - \frac{e^2}{2}\,|\,q^\dagger\, q - c\,\mathbf{1}\,|^2 - \frac{1}{2}\,g_N^2\,|\,q^\dagger\, t^a q\,|^2, \quad (83)$$

$$c = N(N+1)\sqrt{2\,\mu\, m}. \quad (84)$$

Neglecting the small terms left implicit, this is identical to the $U(N)$ model (56), except for the fact that $e \neq g_N$ here. The transformation property of the vortices can be determined from the moduli matrix, as was done in [76]. Indeed, the system possesses BPS saturated vortices described by the linearized equations

$$(\mathcal{D}_1 + i\mathcal{D}_2)\; q = 0, \quad (85)$$

$$F_{12}^{(0)} + \frac{e^2}{2}\left(c\,\mathbf{1}_N - q\,q^\dagger\right) = 0; \qquad F_{12}^{(a)} + \frac{g_N^2}{2}\, q_i^\dagger\, t^a\, q_i = 0. \quad (86)$$

The matter equation can be solved exactly as in [65, 66, 67] ($z = x^1 + ix^2$) by setting

$$q = S^{-1}(z,\bar{z})\, H_0(z), \quad A_1 + i\, A_2 = -2\, i\, S^{-1}(z,\bar{z})\, \bar{\partial}_z S(z,\bar{z}), \quad (87)$$

where S is an $N \times N$ invertible matrix over whole of the z plane, and H_0 is the moduli matrix, holomorphic in z.

The gauge field equations take a slightly more complicated form than in the $U(N)$ model (56):

$$\partial_z\,(\Omega^{-1}\partial_{\bar{z}}\,\Omega) = -\frac{g_N^2}{2}\,\mathrm{Tr}\,(\,t^a\,\Omega^{-1}\,q\,q^\dagger\,)\,t^a - \frac{e^2}{4N}\,\mathrm{Tr}\,(\,\Omega^{-1} q\,q^\dagger - \mathbf{1}), \quad \Omega = S\,S^\dagger. \quad (88)$$

The last equation reduces to the master equation (65) in the $U(N)$ limit, $g_N = e$.

The advantage of the moduli matrix formalism is that all the moduli parameters appear in the holomorphic, moduli matrix $H_0(z)$. Especially, the transformation property of the vortices under the color–flavor diagonal group can be studied by studying the behavior of the moduli matrix.

6.2 Dual Gauge Transformation from the Vortex Moduli

The concepts such as the low-energy BPS vortices or the high-energy BPS monopole solutions are thus only approximate: their explicit forms are valid only in the lowest-order approximation, in the respective kinematical regions.

Nevertheless, there is a property of the system which is exact and does not depend on any approximation: the full system has an exact, global $SU(N)_{C+F}$ symmetry, which is neither broken by the interactions nor by both sets of VEVs, v_1 and v_2. This symmetry is broken by individual soliton vortex, endowing the latter with non-Abelian orientational moduli, analogous to the translational zero modes of a kink. Note that the vortex breaks the color–flavor symmetry as

$$SU(N)_{C+F} \to SU(N-1) \times U(1), \tag{89}$$

leading to the moduli space of the minimum vortices, which is

$$\mathcal{M} \simeq \mathbf{C}P^{N-1} = \frac{SU(N)}{SU(N-1) \times U(1)}. \tag{90}$$

The fact that this moduli coincides with the moduli of the quantum states of an N-state quantum–mechanical system, is a first hint that the monopoles appearing at the end point of a vortex, transform as a fundamental multiplet $\underline{N}$ of a group $SU(N)$.

The moduli space of the vortices is described by the moduli matrix (we consider here the vortices of minimal winding, $k = 1$)

$$H_0(z) \simeq \begin{pmatrix} 1 & 0 & 0 & -a_1 \\ 0 & \ddots & 0 & \vdots \\ 0 & 0 & 1 & -a_{N-1} \\ 0 & \dots & 0 & z \end{pmatrix}, \tag{91}$$

where the constants a_i, $i = 1, 2, \dots, N-1$ are the coordinates of $\mathbf{C}P^{N-1}$. Under $SU(N)_{C+F}$ transformation, the squark fields transform as

$$q \to U^{-1} q\, U, \tag{92}$$

but as the moduli matrix is defined *modulo* holomorphic redefinition (66), it is sufficient to consider

$$H_0(z) \to H_0(z)\, U. \tag{93}$$

Now, for an infinitesimal $SU(N)$ transformation acting on a matrix of the form (91), U can be taken in the form

$$U = \mathbf{1} + X, \qquad X = \begin{pmatrix} \mathbf{0} & \boldsymbol{\xi} \\ -(\boldsymbol{\xi})^\dagger & 0 \end{pmatrix}, \tag{94}$$

where $\boldsymbol{\xi}$ is a small $N-1$ component constant vector. Computing $H_0\, X$ and making a V transformation from the left to bring back H_0 to the original form, we find

$$\delta a_i = -\xi_i - a_i\, (\boldsymbol{\xi})^\dagger \cdot \boldsymbol{a}, \tag{95}$$

which shows that a_i's indeed transform as the inhomogeneous coordinates of $\mathbf{C}P^{N-1}$. In other words, the vortex represented by the moduli matrix (91) transforms as a fundamental multiplet of $SU(N)$.[12]

As an illustration consider the simplest case of $SU(2)$ theory. In this case, the moduli matrix is simply [72]

$$H_0^{(1,0)} \simeq \begin{pmatrix} z-z_0 & 0 \\ -b_0 & 1 \end{pmatrix}; \qquad H_0^{(0,1)} \simeq \begin{pmatrix} 1 & -a_0 \\ 0 & z-z_0 \end{pmatrix}. \tag{96}$$

with the transition function between the two patches:

$$b_0 = \frac{1}{a_0}. \tag{97}$$

The points on this $\mathbf{C}P^1$ represent all possible $k=1$ vortices. Note that points on the space of a quantum–mechanical two-state system,

$$|\Psi\rangle = a_1|\psi_1\rangle + a_2\,|\psi_2\rangle, \qquad (a_1,a_2) \sim \lambda\,(a_1,a_2), \quad \lambda \in \mathbf{C}, \tag{98}$$

can be put in one-to-one correspondence with the inhomogeneous coordinate of a $\mathbf{C}P^1$,

$$a_0 = \frac{a_1}{a_2}, \qquad b_0 = \frac{a_2}{a_1}. \tag{99}$$

In order to make this correspondence manifest, note that the minimal vortex (96) transforms under the $SU(2)_{C+F}$ transformation, as

$$H_0 \to V\,H_0\,U^\dagger, \qquad U = \begin{pmatrix} \alpha & \beta \\ -\beta^* & \alpha^* \end{pmatrix}, \quad |\alpha|^2+|\beta|^2 = 1, \tag{100}$$

where the factor $U^\dagger$ from the right represents a flavor transformation, V is a holomorphic matrix which brings H_0 to the original triangular form [76]. The action of this transformation on the moduli parameter, for instance, a_0, can be found to be

$$a_0 \longrightarrow \frac{\alpha\,a_0+\beta}{\alpha^*-\beta^*\,a_0}. \tag{101}$$

But this is precisely the way a doublet state (98) transforms under $SU(2)$,

$$\begin{pmatrix} a_1 \\ a_2 \end{pmatrix} \to \begin{pmatrix} \alpha & \beta \\ -\beta^* & \alpha^* \end{pmatrix} \begin{pmatrix} a_1 \\ a_2 \end{pmatrix}. \tag{102}$$

The fact that the vortices (seen as solitons of the low-energy approximation) transform as in the $\underline{N}$ representation of $SU(N)_{C+F}$, implies that there exist a set of monopoles which transform accordingly, as $\underline{N}$. The existence of such a set follows from the exact $SU(N)_{C+F}$ symmetry of the theory, broken by the individual monopole–vortex configuration.

---

[12] Note that, if a $\underline{N}$ vector $\mathbf{c}$ transforms as $\mathbf{c} \to (\mathbf{1}+X)\,\mathbf{c}$, the inhomogeneous coordinates $a_i = c_i/c_N$ transform as in (95).

This answers some of the questions formulated earlier (below (22)) unambiguously [76]. Note that in our derivation of continuous transformations of the monopoles, the explicit, semiclassical form of the latter is not used.

A subtle point is that in the high-energy approximation, and to lowest order of such an approximation, the semiclassical monopoles are just certain nontrivial field configurations involving $\phi(x)$ and $A_i(x)$ fields only, and therefore apparently transform under the color part of $SU(N)_{C+F}$ only. When the full monopole–vortex configuration $\phi(x), A_i(x), q(x)$ (Fig. 2) is considered, however, only the combined color–flavor diagonal transformations keep the energy of the configuration invariant. In other words, the monopole transformations must be regarded as part of more complicated transformations involving flavor, when higher-order effects in $O(\frac{v_1}{v_2})$ are taken into account. And this means that the transformations are among *physically distinct states*, as the vortex moduli describe obviously physically distinct vortices [37].

This discussion highlights the crucial role played by the (massless) flavors in the underlying theory as has been already summarized at the end of Sect. 2. There is, however, another important independent effect due to the massless flavors. Due to the zero modes of the fermions, the semiclassical monopoles are converted to some irreducible multiplets in the *flavor* group $SU(N_f)$ [46]. The "clouds" of the fermion zero-mode fluctuation fields surrounding the monopole have an extension of $O(\frac{1}{v_1})$, which is much smaller than the distance scales associated with the infrared effects discussed here. We conclude that there was one more crucial role of the flavor on non-Abelian monopoles: it allows to generate the dual magnetic gauge group on the one hand, and to "dress" the monopoles and endow them with global, flavor quantum numbers à la Jackiw–Rebbi, on the other. They should be regarded as two, distinct effects.

Our construction has been generalized to the symmetry breaking $SO(2N+1) \to U(N) \to \emptyset$, $SO(2N+1) \to U(r) \times U(1)^{N-r} \to \emptyset$, in the concrete context of softly broken $\mathcal{N} = 2$ models. There is an interesting difference in the quantum fate of the semiclassical monopoles in the case the unbroken SU factor has the maximum rank and in the cases where $r \le N-1$. The semiclassical (vortex–monopole complex) argument of Sect. 3 and in this section and the fully quantum-mechanical results (of Sects. 4.2 and 4.3) agree qualitatively, quite nontrivially [76].

The fact that the vortices of the low-energy theory are BPS saturated, which allows us to analyze their moduli and transformation properties elegantly as discussed above, while in the full theory there are corrections which make them non-BPS (and unstable), might cause some concern. Actually, the rigor of our argument is not affected by those terms which can be treated as perturbation. The attributes characterized by integers such as the transformation property of certain configurations as a multiplet of a non-Abelian group which is an exact symmetry group of the full theory, cannot receive renormalization. This is similar to the current algebra relations of Gell–Mann, which are not renormalized. CVC of Feynman and Gell–Mann also hinges upon an analogous situation.[13] The results obtained in the BPS limit (in the limit $v_2/v_1 \to 0$) are thus valid at any finite values of v_2/v_1 [79]. Thus

The dual group $\tilde{H}$ is the transformation group H_{C+F}, seen in the dual magnetic description.

6.3 Other Symmetry Breaking Patterns

The cases such as $SO(2N+3) \to SO(2N+1) \times U(1)$ or $USp(2N+2) \to USp(2N) \times U(1)$, are particularly interesting, as the groups $SO(2N+1)$ and $USp(2N)$ are interchanged by the GNOW duality. In the first case, for instance, the GNOW conjecture states that the monopoles belong to multiplets of the dual group $USp(2N)$. Although there are some hints how such GNOW dual monopoles might emerge naturally in the semiclassical approximations [80], there is a strong argument (based on $\mathcal{N}=2$ supersymmetry and global symmetry [30, 53]) as well as clear evidence [30], against the appearance of these GNOW monopoles as the light degrees of freedom. In other words, even if they might emerge in a semiclassical approximation, they do not survive quantum effects.

It is perhaps not a coincidence that the Seiberg duals of $\mathcal{N}=1$ supersymmetric theories do not coincide always with GNOW duals.

The systems $USp(2N) \to U(r) \times U(1)^{N-r} \to \emptyset$ also is known to possess light non-Abelian monopoles in the fundamental representation of the dual group $SU(r)$ [30], which can be nicely understood by our definition of the dual group.

7 Confinement Near Conformal Vacua

A particular class of confining vacua, in which confinement and dynamical symmetry breaking are described by non-Abelian magnetic monopoles *interacting strongly*, are of great interest. The vacua we are talking about are known as non-Abelian Argyres–Douglas vacua. These are found as a particular case of r vacua, with $r = N_f/2$ of $SU(N)$ SQCD, as well as in the massless limit ($m_i \to 0$) of all of confining vacua of $SO(N)$ and $USp(2N)$ theories. Many other examples of vacua with analogous properties can be found in the context of wider class of $\mathcal{N}=1$ supersymmetric gauge theories [32].

Although the details (the global symmetry, the light–degrees of freedom) depend on the model, there is a common feature in this class of systems which makes these particularly interesting. Because of dynamics and for symmetry requirement the system chooses to produce non-Abelian (rather than Abelian) magnetic monopoles as the low-energy degrees of freedom, but cannot produce quite as many of them as to make the effective theory infrared-free.

[13] The absence of "colored dyons" [11] mentioned earlier can also be interpreted in this manner.

As a consequence, confinement is caused by the condensation of certain monopole composites rather than by the condensation of single monopoles [53]. As non-Abelian monopoles carry flavor quantum numbers of the original quarks (this is necessary for the low-energy theory to have the correct symmetry of the underlying theory), the pattern of the symmetry breaking reflects such a mechanism. These considerations have been distilled from studies on this class of systems and on the problem of understanding non-Abelian monopoles discussed in various parts of this lecture.

8 Quantum Chromodynamics

What does all this teach about QCD? That the Abelian superconductor picture is probably not the correct picture of real-world QCD ($SU(3)$) has been already pointed out. In particular, the fact that the deconfinement and chiral restoration transitions occur at exactly the same temperatures in $SU(3)$ lattice measurement, appears to make the assumption that Abelian $U(1)^2$ monopoles are responsible for confinement and chiral symmetry breaking, rather awkward (the remark (v) of Sect. 4.3). On the other hand, in ordinary (non-supersymmetric) gauge theories, the "sign flip" of the beta function needed to make the non-Abelian monopoles recognizable infrared (or intermediate-scale) degrees of freedom, is much more difficult to achieve. If the dual "magnetic" group were again $SU(3)$, the magnetic monopoles of such a theory (regularized Z_3 monopoles?) would probably interact too strongly and would form composite monopoles (cf. the point (iii) of Sect. 4.3). A small number of light flavors, dressing these monopoles with flavor quantum numbers, would not be sufficient.

We might speculate that the dynamics of QCD lies somewhere between. The dual theory could be an

$$SU(2) \times U(1) \qquad \text{or} \qquad U(2) \tag{103}$$

theory, with magnetic monopoles in $\underline{2}$ of the $SU(2)$ group and moreover we expect them to carry flavor $SU_L(2) \times SU_R(2)$ quantum numbers. We expect them to interact strongly, but not too much, and it is possible that the system is close to a nontrivial infrared fixed point, with relatively nonlocal dyons present at the same time, as in the SCFT effective low-energy theories of the supersymmetric models discussed in the previous subsection.

Let us assume that they are M_a^i, $\tilde{M}_j^b$, with the (dual) color a, b and flavor indices i, j, and carrying opposite $U(1)$ charges. A condensate of the form

$$\langle M_a^i \, \tilde{M}_j^a \rangle \sim \Lambda^2 \, \delta_j^i \tag{104}$$

might form, inducing confinement and chiral symmetry breaking $SU_L(2) \times SU_R(2) \to SU_V(2)$ simultaneously. It could be that the standard quark condensate

$$\langle \psi_L^i \, \bar{\psi}_{Rj} \rangle \sim \Lambda^3 \, \delta_j^i \tag{105}$$

is closely related dynamically to or induced by the monopole condensation, (104), for instance, via the Rubakov effect [81].

It is interesting that in such a picture, there should be a considerable difference between a theory with quarks in the fundamental representation and a (unrealistic) theory with quarks in the adjoint representation. The Jackiw–Rebbi effect works diffrently in the two cases. In the former case the fermion zero modes give rise to *bosonic* multiplet of degenerate monopoles, while in the latter case some of the monopoles become fermions. In the theory with adjoint quarks, then, there can be considerable difference between the phenomenon of confinement and that of chiral symmetry breaking. There is an ample evidence for such a difference (e.g., different transition temperatures) in lattice gauge theory, as is well known.

9 Conclusive Remarks

Non-Abelian monopoles are present in the fully quantum–mechanical low-energy effective action of many solvable supersymmetric theories. They behave perfectly as point-like particles carrying non-Abelian dual magnetic charges. They play a crucial role in confinement and in dynamical symmetry breaking in these theories. There is a natural identification of these excitations within the semiclassical approach, which involves the flavor symmetry in an essential manner. It is hoped that such an improved grasp on the nature of non-Abelian monopoles would one day lead to a better understanding of confinement in QCD.

Acknowledgments

It is a great pleasure for me to present these notes in honor of the 65th birthday of my friend Gabriele Veneziano. With his deep understanding of physics, brilliant intuition, elegance of his logics, and inexhaustible fantasy, as well as with his exemplary human quality, he has been a guide to many of us contemporary and younger generations of theoretical physicists for so many years. It is not easy to emulate such a high standard, but I present these lecture notes, with the best of my efforts and with a deep sense of gratitude to Gabriele. Finally, I wish to thank many friends and collaborators who contributed at various stages of this investigation.

Appendix A—Semiclassical "Non-Abelian" Monopoles

In this appendix we review some general formulae [6, 4]. These degenerate monopoles appear in a system with the gauge symmetry breaking

$$G \stackrel{\langle\phi\rangle\neq 0}{\longrightarrow} H \tag{A.1}$$

with a nontrivial $\pi_2(G/H)$ and non-Abelian H.

The normalization of the generators can be chosen [4] so that the metric of the root vector space is[14]

$$g_{ij} = \sum_{roots} \alpha_i \alpha_j = \delta_{ij}. \tag{A.4}$$

The Higgs field vacuum expectation value (VEV) is taken to be of the form

$$\phi_0 = \mathbf{h} \cdot \mathbf{H}, \tag{A.5}$$

where $\mathbf{h} = (h_1, \ldots, h_{\mathrm{rank}(G)})$ is a constant vector representing the VEV. The root vectors orthogonal to $\mathbf{h}$ belong to the unbroken subgroup H.

The monopole solutions are constructed from various $SU(2)$ subgroups of G that do not commute with H,

$$S_1 = \frac{1}{\sqrt{2\alpha^2}}(E_\alpha + E_{-\alpha}); \qquad S_2 = -\frac{i}{\sqrt{2\alpha^2}}(E_\alpha - E_{-\alpha}); \qquad S_3 = \alpha^* \cdot \mathbf{H}, \tag{A.6}$$

where α is a root vector associated with a pair of *broken* generators $E_{\pm\alpha}$. α^* is a dual root vector defined by

$$\alpha^* \equiv \frac{\alpha}{\alpha \cdot \alpha}. \tag{A.7}$$

The symmetry breaking (A.1) induces the Higgs mechanism in such an $SU(2)$ subgroup, $SU(2) \to U(1)$. By embedding the known 't Hooft–Polyakov monopole [2, 38] lying in this subgroup and adding a constant term to ϕ so that it behaves correctly asymptotically, one easily constructs a solution of the equation of motion [6, 27]:

$$A_i(\mathbf{r}) = A_i^a(\mathbf{r}, \mathbf{h}\cdot\alpha)\, S_a; \qquad \phi(\mathbf{r}) = \chi^a(\mathbf{r}, \mathbf{h}\cdot\alpha)\, S_a + [\,\mathbf{h} - (\mathbf{h}\cdot\alpha)\,\alpha^*]\cdot \mathbf{H}, \tag{A.8}$$

where

$$A_i^a(\mathbf{r}) = \epsilon_{aij}\frac{r^j}{r^2}A(r); \qquad \chi^a(\mathbf{r}) = \frac{r^a}{r}\chi(r), \qquad \chi(\infty) = \mathbf{h}\cdot\alpha \tag{A.9}$$

is the standard 't Hooft–Polyakov–BPS solution. Note that $\phi(\mathbf{r} = (0,0,\infty)) = \phi_0$.

---

[14] In the Cartan basis, the Lie algebra of the group G takes the form

$$[H_i, H_k] = 0, \qquad (i,k = 1,2,\ldots,r); \qquad [H_i, E_\alpha] = \alpha_i\, E_\alpha;$$
$$[E_\alpha, E_{-\alpha}] = \alpha^i\, H_i; \tag{A.2}$$

$$[E_\alpha, E_\beta] = N_{\alpha\beta}\, E_{\alpha+\beta} \qquad (\alpha + \beta \neq 0). \tag{A.3}$$

$\alpha_i = (\alpha_1, \alpha_2, \ldots)$ are the root vectors.

The mass of a BPS monopole is then given by

$$M = \int d\mathbf{S} \cdot \mathrm{Tr}\,\phi\,\mathbf{B}, \qquad \mathbf{B} = \frac{r_i(\mathbf{S}\cdot\mathbf{r})}{r^4}. \tag{A.10}$$

This can be computed by going to the gauge in which

$$\mathbf{B} = \frac{\mathbf{r}S_3}{r^3} = \frac{\mathbf{r}}{r^3}\,\alpha^* \cdot \mathbf{H}, \tag{A.11}$$

to be

$$M = \frac{4\pi h_i \alpha_j^*}{g}\,\mathrm{Tr}\,H_i\,H_j. \tag{A.12}$$

For instance, the mass of the minimal monopole of $SU(N{+}1) \to SU(N){\times}U(1)$ can be found easily by using (B.4)–(B.10)

$$M = \frac{2\pi\,v\,(N+1)}{g}. \tag{A.13}$$

For the cases $SO(N+2) \to SO(N) \times U(1)$ and $USp(2N+2) \to USp(2N) \times U(1)$, where $\mathrm{Tr}H_i\,H_j = C\,\delta_{ij}$, one finds

$$M = \frac{4\pi\,C\,\mathbf{h}\cdot\alpha^*}{g} = \frac{4\,\pi\,v}{g}, \tag{A.14}$$

while for $SO(2N) \to SU(N) \times U(1)$, $SO(2N+1) \to SU(N) \times U(1)$, and $USp(2N) \to SU(N) \times U(1)$, the mass is

$$M = \frac{8\pi\,C\,\mathbf{h}\cdot\alpha^*}{g} = \frac{8\,\pi\,v}{g}. \tag{A.15}$$

In order to get the $U(1)$ magnetic charge,[15] we first divide by an appropriate normalization factor in the mass formula (A.10)

$$F_m = \int d\mathbf{S} \cdot \frac{\mathrm{Tr}\,\phi\,\mathbf{B}}{N_\phi} = \int d\mathbf{S} \cdot \mathbf{B}^{(0)}, \qquad \mathbf{B} = \frac{r_i(\mathbf{S}\cdot\mathbf{r})}{r^4}. \tag{A.16}$$

The result, which is equal to $4\pi g_m$ by definition, gives the magnetic charge. The latter must then be expressed as a function of the minimum $U(1)$ electric charge present in the given theory, which can be easily found from the normalized (such that $\mathrm{Tr}\,T^{(a)}\,T^{(a)} = \frac{1}{2}$) form of the relevant $U(1)$ generator.

For example, in the case of the symmetry breaking, $SO(2N) \to U(N)$, the adjoint VEV is of the form, $\phi = \sqrt{4N}\,v\,T^{(0)}$, where $T^{(0)}$ is a $2N \times 2N$ block-diagonal matrix with N nonzero submatrices $\frac{i}{\sqrt{4N}}\begin{pmatrix} 0 & 1 \\ -1 & 0 \end{pmatrix}$. Dividing the mass (A.15) by $\sqrt{N}\,v$ and identifying the flux with $4\pi g_m$ one gets $g_m = \frac{2}{\sqrt{N}\,g}$.

[15] In this calculation it is necessary to use the generators normalized as $\mathrm{Tr}\,T^{(a)}\,T^{(b)} = \frac{1}{2}\delta_{ab}$, such that $\mathbf{B} = \mathbf{B}^{(0)}\,T^{(0)} + \ldots$.

Finally, in terms of the minimum electric charge of the theory $e_0 = \frac{g}{\sqrt{4N}}$ (which follows from the normalized form of $T^{(0)}$ above) one finds

$$g_m = \frac{2}{\sqrt{N}\,g} = \frac{2}{N} \cdot \frac{1}{2\,e_0}. \tag{A.17}$$

The calculation is similar in other cases.

The asymptotic gauge field can be written as

$$F_{ij} = \epsilon_{ijk} B_k = \epsilon_{ijk} \frac{r_k}{r^3} (\beta \cdot \mathbf{H}), \qquad \beta = \alpha^* \tag{A.18}$$

in an appropriate gauge ((A.10)). The Goddard–Nuyts–Olive quantization condition [4]

$$2\,\beta \cdot \alpha \in \mathbf{Z} \tag{A.19}$$

then reduces to the well-known theorem that for two root vectors α_1, α_2 of any group,

$$\frac{2\,(\alpha_1 \cdot \alpha_2)}{(\alpha_1 \cdot \alpha_1)} \tag{A.20}$$

is an integer.

Appendix B—Root Vectors and Weight Vectors

.1 $A_N = SU(N+1)$

It is sometimes convenient to have the root vectors and weight vectors of the Lie algebra $SU(N+1)$ as vectors in an $(N+1)$-dimensional space rather than an N-dimensional one. The root vectors are then simply

$$(\cdots, \pm 1, \cdots, \mp 1, \cdots). \tag{B.1}$$

($\cdots$ stand for zero elements) which all lie on the plane

$$x_1 + x_2 + \ldots + x_{N+1} = 0, \tag{B.2}$$

while the weight vectors are projections in this plane of the orthogonal vectors

$$\boldsymbol{\mu} = (\cdots, \pm 1, \cdots) \tag{B.3}$$

where the dots represent zero elements.

In order to use the general formulas of Weinberg and Goddard–Olive–Nuyts we normalize these vectors so that the diagonal (Cartan) generators may be written

$$\mathbf{H}_{\mathrm{i}} = \mathrm{diag}\,(w_1^i,\, w_2^i \ldots,\, w_N^i,\, w_{N+1}^i\,), \qquad i = 1, 2 \ldots N \tag{B.4}$$

where w_k represents the k-th weight vector of the fundamental representation of $SU(N+1)$, satisfying

$$\mathbf{w}_k \cdot \mathbf{w}_l = -\frac{1}{2(N+1)^2}; \quad (k \neq l);$$
$$\mathbf{w}_k \cdot \mathbf{w}_k = \frac{N}{2(N+1)^2}, \qquad k,l = 1,2,\ldots,N+1; \tag{B.5}$$

and $\sum_{k=1}^{N+1} \mathbf{w}_k = 0$. They are vectors lying in an N-dimensional space (B.2): in the coordinates of the $(N+1)$-dimensional space,

$$\mathbf{w}_i = \frac{1}{\sqrt{2(N+1)^3}}(-1,\ldots,-1,N,-1,-1,\ldots). \tag{B.6}$$

The root vectors are simply

$$\alpha = \mathbf{w}_i - \mathbf{w}_j = \frac{1}{\sqrt{2(N+1)}}(\cdots,\pm 1,\cdots,\mp 1,\cdots) \tag{B.7}$$

with the norm

$$\alpha \cdot \alpha = \frac{1}{N+1}. \tag{B.8}$$

Note that for $i \neq j$

$$\mathrm{Tr}\,(H_i\,H_j) = w_1^i w_1^j + \ldots + w_{N+1}^i w_{N+1}^j = \frac{-2N+N-1}{2(N+1)^3} = -\frac{1}{2(N+1)^2}, \tag{B.9}$$

while

$$\mathrm{Tr}\,(H_i\,H_i) = \frac{N^2+N}{2(N+1)^3} = \frac{N}{2(N+1)^2}. \tag{B.10}$$

The adjoint VEV causing the symmetry breaking $SU(N+1) \to SU(N) \times U(1)$ is of the form,

$$\phi = \mathbf{h} \cdot \mathbf{H}, \qquad \mathbf{h} = v\sqrt{2(N+1)^3}\,(0,0,\ldots,1). \tag{B.11}$$

.2 $B_N = SO(2N+1)$

The N generators in the Cartan subalgebra of the Lie algebra $SO(2N+1)$ can be taken to be

$$H_i = \begin{pmatrix} -iw_1^i\mathbf{J} & & & & \\ & -iw_2^i\mathbf{J} & & & \\ & & \ddots & & \\ & & & -iw_N^i\mathbf{J} & \\ & & & & 0 \end{pmatrix}, \qquad \mathbf{J} = \begin{pmatrix} & 1 \\ -1 & \end{pmatrix} \tag{B.12}$$

where $\mathbf{w}_k$ $(k=1,2,\ldots,N)$ are the weight vectors of the fundamental representation, which are vectors in an N-dimensional Euclidean space

$$\mathbf{w}_k \cdot \mathbf{w}_l = 0; \quad k \neq l; \qquad \mathbf{w}_k \cdot \mathbf{w}_k = \frac{1}{2(2N-1)} : \tag{B.13}$$

they form a complete set of orthogonal vectors. The root vectors of $SO(2N+1)$ group are $\alpha = \{\pm \mathbf{w}_i, \ \pm \mathbf{w}_i \pm \mathbf{w}_j\}$; their duals are:

$$\alpha^* = \pm 2(2N-1)\, \mathbf{w}_i, \qquad (2N-1)[\pm \mathbf{w}_i \pm \mathbf{w}_j]. \tag{B.14}$$

The diagonal generators satisfy

$$\mathrm{Tr}\, H_i\, H_j = \frac{1}{2N-1}\, \delta_{ij}. \tag{B.15}$$

In the system with symmetry breaking $SO(2N+1) \to SO(2N-1) \times U(1)$ the adjoint scalar VEV is

$$\phi = \mathbf{h} \cdot \mathbf{H}, \qquad \mathbf{h} = iv\sqrt{2(2N-1)}\,(0,0,\ldots,1). \tag{B.16}$$

.3 $C_N = USp(2N)$

The N generators in the Cartan subalgebra of $USp(2N)$ are the following $2N \times 2N$ matrices:

$$\mathbf{H}_i = \begin{pmatrix} \mathbf{B_i} & \mathbf{0} \\ \mathbf{0} & -\mathbf{B_i}^{\mathrm{t}} \end{pmatrix}, \quad i = 1, 2, \ldots, N, \tag{B.17}$$

where

$$\mathbf{B_i} = \begin{pmatrix} w^i_1 & & & & \\ & w^i_2 & & & \\ & 0 & \ddots & 0 & \\ & & & w^i_{N-1} & \\ & & & & w^i_N \end{pmatrix}, \qquad i = 1, 2 \ldots N. \tag{B.18}$$

The weight vectors $\mathbf{w}_k$ $(k = 1, 2, \ldots, N)$ form a complete set of orthogonal vectors in an N-dimensional Euclidean space and satisfy

$$\mathbf{w}_k \cdot \mathbf{w}_l = 0; \quad k \neq l; \qquad \mathbf{w}_k \cdot \mathbf{w}_k = \frac{1}{4(N+1)}. \tag{B.19}$$

The root vectors of $USp(2N)$ group are $\alpha = \{\pm 2\, \mathbf{w}_i, \ \pm \mathbf{w}_i \pm \mathbf{w}_j\}$. The diagonal generators satisfy

$$\mathrm{Tr}\, H_i\, H_j = \frac{1}{2(N+1)}\, \delta_{ij}. \tag{B.20}$$

For the breaking $USp(2N) \to USp(2(N-1)) \times U(1)$ the adjoint scalar VEV is

$$\phi = \mathbf{h} \cdot \mathbf{H}, \qquad \mathbf{h} = v\sqrt{4(N+1)}\,(0,0,\ldots,1). \tag{B.21}$$

.4 $D_N = SO(2N)$

The N generators in the Cartan subalgebra of the $SO(2N)$ group can be chosen to be

$$H_i = \begin{pmatrix} -iw_1^i \begin{pmatrix} & 1 \\ -1 & \end{pmatrix} & & & \\ & -iw_2^i \begin{pmatrix} & 1 \\ -1 & \end{pmatrix} & & \\ & & \ddots & \\ & & & -iw_N^i \begin{pmatrix} & 1 \\ -1 & \end{pmatrix} \end{pmatrix}, \tag{B.22}$$

where $\mathbf{w}_k$ $(k = 1, 2, \ldots, N)$ are the weight vectors of the fundamental representation, living in an N-dimensional Euclidean space and satisfying

$$\mathbf{w}_k \cdot \mathbf{w}_l = 0; \quad k \neq l; \qquad \mathbf{w}_k \cdot \mathbf{w}_k = \frac{1}{4(N-1)} : \tag{B.23}$$

they form a complete set of orthogonal vectors. The root vectors of $SO(2N)$ are $\alpha = \{\pm\mathbf{w}_i \pm \mathbf{w}_j\}$. The diagonal generators satisfy

$$\mathrm{Tr}\, H_i\, H_j = \frac{1}{2(N-1)}\, \delta_{ij}. \tag{B.24}$$

In the system with symmetry breaking $SO(2N) \to SO(2N-2) \times U(1)$ the adjoint scalar VEV takes the form

$$\phi = \mathbf{h} \cdot \mathbf{H}, \qquad \mathbf{h} = iv\sqrt{4(N-1)}\,(0, 0, \ldots, 1). \tag{B.25}$$

Appendix C—Seiberg–Witten Curves for $SU(2)$ $\mathcal{N} = 2$ Super Yang-Mills Theory

The variable a and a_D are to be considered as local variables, describing the low-energy effective action in a particular patch of the space of vacua (QMS). On the other hand, the variable $u = \mathrm{Tr}\,\langle \Phi^2 \rangle$ is a gauge invariant and apparently unique and global variable describing the QMS. The space (a_D, a) is the covering space $\tilde{\mathcal{M}}$ of the space $\mathcal{M}$ whose coordinate is the complex VEV u. If the base space were simply connected, the map $\tilde{\mathcal{M}} \to \mathcal{M}$ would be trivial. In general, a closed loop of the point u in the base space induces a discrete transformation, called monodromy group, among the inverse images of the point u in the covering group.

The fact that the space $\mathcal{M}$ is nontrivial follows from the one-loop beta function,

$$\tau_{eff} = da_D/da = \frac{\theta_{eff}}{2\pi} + \frac{4\pi i}{g^2_{eff}} \sim \frac{i}{2\pi} \log a + \ldots$$

so

$$F(A) \simeq \frac{i}{2\pi} A^2 \log \frac{A^2}{\Lambda^2}, \qquad a_D = \frac{dF(a)}{da} \simeq \frac{i}{2\pi}\,(a \,\log a + \frac{a}{2}).$$

The effect of a loop at large $u \sim a^2/2$, $u \to e^{2\pi i} u$ is $a \to e^{\pi i}$, so

$$a_D \to -a_D + 2a, \quad a \to -a,$$

or

$$\begin{pmatrix} a_D \\ a \end{pmatrix} \to M_\infty \begin{pmatrix} a_D \\ a \end{pmatrix}, \qquad M_\infty = \begin{pmatrix} -1 & 2 \\ 0 & -1 \end{pmatrix}.$$

A singularity at ∞ in the u space implies the presence of at least one more singularity at finite u. As the theory possesses an invariance under spontaneously broken discrete $\mathbf{Z}_2$, under which $u \to -u$, it is natural to assume a pair of singularirties at $u = \pm\Lambda^2$. The key idea of Seiberg and Witten is that these singularities correspond to the points of u where the 't Hooft–Polyakov monopole becomes massless due to quantum effects. Near $u \sim \Lambda^2$ then

$$a_D(u = \Lambda^2) = 0, \qquad \tau_D = -\frac{da}{da_D} \simeq -\frac{i}{\pi} \log a_D, \tag{C.1}$$

and

$$a_D \sim c_0(u - \Lambda^2),$$

where (C.1) is the standard beta function of $N = 2$ supersymmetric QED. Thus a closed loop in u around the point Λ^2 induces the monodromy transformation

$$a \to a - a - 2a_D; \quad a_D \to a_D, \qquad M_{\Lambda^2} = \begin{pmatrix} 1 & 0 \\ -2 & 1 \end{pmatrix}.$$

The monodromy transformation around $-\Lambda^2$ follows from the consistency condition,

$$M_{\Lambda^2} \cdot M_{\Lambda^2} = M_\infty.$$

The map $a_D(u)$, $a(u)$, with the desired properties is precisely the one given in (36)–(38).

Appendix D—One-particle Representations of $\mathcal{N} = 1$ and $\mathcal{N} = 2$ Supersymmetry Algebra

(i) For a massive $\mathcal{N} = 1$ supersymmetric particle states, one has ($P^\mu = (M, 0, 0, 0)$)

$$\{Q_\alpha, \bar{Q}_{\dot\alpha}\} = \delta_{\alpha\dot\alpha}\, 2M, \qquad \alpha, \dot\alpha = 1, 2, \tag{D.1}$$

or, by defining

$$b_\alpha^\dagger = \frac{1}{\sqrt{2M}} Q_\alpha, \qquad b_{\dot\alpha} = \frac{1}{\sqrt{2M}} \bar{Q}_{\dot\alpha}. \tag{D.2}$$

These can be regarded as two pairs of annihilation and creation operators, $\{b_{\dot{\alpha}}, b^\dagger_\alpha\} = \delta_{\alpha\dot{\alpha}}$. The complete set of one particle states can then be constructed by defining the vacuum state by ($i = 1, 2$)

$$b_i\,|0\rangle = 0; \tag{D.3}$$

the full set of states are

$$|0\rangle, \quad b_1^\dagger|0\rangle, \quad b_2^\dagger|0\rangle, \quad b_1^\dagger b_2^\dagger|0\rangle, \tag{D.4}$$

they form a degenerate supersymmetry multiplet (two bosons and two fermions). For $\mathcal{N}$ supersymmetry, the same argument shows that the multiplicity of a massive multiplet is

$$\sum_{n=0}^{2\mathcal{N}} \binom{2\mathcal{N}}{n} = 2^{2\mathcal{N}}. \tag{D.5}$$

(ii) Massless $\mathcal{N} = 1$ supersymmetric particle states: In this case it is not possible to go to the rest frame but the momentum can be chosen as $P^\mu = (p, 0, 0, p)$. Then

$$\{Q_\alpha, \bar{Q}_{\dot{\alpha}}\} = \begin{pmatrix} 2p & 0 \\ 0 & 0 \end{pmatrix}_{\alpha\dot{\alpha}} \tag{D.6}$$

The state $b_2^\dagger\,|0\rangle$ have a zero norm. The particle states are given by the positive norm states, half of (D.4),

$$|0\rangle, \quad b_1^\dagger|0\rangle. \tag{D.7}$$

The multiplicity of a massless $\mathcal{N} = 1$ supersymmetry multiplet is

$$\sum_{n=0}^{\mathcal{N}} \binom{2\mathcal{N}}{n} = 2^{\mathcal{N}}. \tag{D.8}$$

(iii) Massive $\mathcal{N} = 2$ supersymmetric particle states with central charges. In the rest frame ($P^\mu = (M, 0, 0, 0)$) the supersymmetry algebra reduces to

$$\{Q^i_\alpha, \bar{Q}^j_{\dot{\alpha}}\} = \delta^{ij}\,\delta_{\alpha\dot{\alpha}}\,2M, \qquad \alpha, \dot{\alpha} = 1, 2, \quad i, j = 1, 2, \tag{D.9}$$

$$\{Q^i_\alpha, Q^j_\beta\} = \epsilon_{\alpha\beta}\,\epsilon^{ij}\,(U + iV) \tag{D.10}$$

Within an irreducible representation U and V are just numbers (electric and magnetic charges of these particles). There are three cases:

1. $2M < \sqrt{U^2 + V^2}$: It is not possible to find a positive-norm representation of the algebra;
2. $2M = \sqrt{U^2 + V^2}$: A representation exists with multiplicity $2^{\mathcal{N}} = 4$ (short multiplet) (these are the so-called BPS saturated case);

3. $2M > \sqrt{U^2+V^2}$: A representation exists with multiplicity $2^{2\mathcal{N}} = 16$ (long multiplet).

Proof. Define

$$\frac{Q_1^1}{\sqrt{2M}} = b_1 \qquad \frac{Q_2^1}{\sqrt{2M}} = b_2 \qquad \frac{Q_1^2}{\sqrt{2M}} = b_3 \qquad \frac{Q_2^2}{2M} = b_4 \tag{D.11}$$

$$-\frac{U}{\sqrt{2M}} = u \qquad -\frac{V}{\sqrt{2M}} = v \tag{D.12}$$

then

$$\{b_i, b_j^\dagger\} = \delta_{ij} \qquad \{b_1, b_4\} = u + iv \qquad \{b_2, b_3\} = -u - iv \tag{D.13}$$

$$\{b_1^\dagger, b_4^\dagger\} = u - iv \qquad \{b_2^\dagger, b_3^\dagger\} = -u + iv \tag{D.14}$$

Now make the change of variables

$$Q_\alpha^1 \longrightarrow e^{i\gamma}\, Q_\alpha^1 \qquad Q_\alpha^2 \longrightarrow Q_\alpha^2 \tag{D.15}$$

$$b_1 \longrightarrow e^{i\gamma}\, b_1 \qquad b_2 \longrightarrow e^{i\gamma}\, b_2 \tag{D.16}$$

to have $\{b_1, b_4\}$ real and positive:

$$\{b_1, b_4\} = \{b_1^\dagger, b_4^\dagger\} = \alpha = \frac{\sqrt{U^2+V^2}}{2M} \tag{D.17}$$

$$\{b_2, b_3\} = \{b_2^\dagger, b_3^\dagger\} = -\alpha \tag{D.18}$$

In order to see the spectrum, it is convenient to set

$$A = b_1 \cos\vartheta + b_4^\dagger \sin\vartheta, \qquad B = -b_1 \sin\vartheta + b_4^\dagger \cos\vartheta. \tag{D.19}$$

The condition $\{A, B\} = \{A, B^\dagger\} = 0$ yields $\vartheta = \frac{\pi}{4}$: A and B satisfy disjoint anticommutators

$$\{A, B\} = 0, \qquad \{A, A^\dagger\} = 1 + \alpha, \qquad \{B, B^\dagger\} = 1 - \alpha. \tag{D.20}$$

Thus if $|\alpha| < 1$ there are two creation operators $A^\dagger$, $B^\dagger$; while if $\alpha = \pm 1$ $B^\dagger$ (or $A^\dagger$) creates zero-norm states. The same passages for b_2 and $b_3^\dagger$ lead to a similar result. The net result is that particles with mass $M > \frac{\sqrt{U^2+V^2}}{2}$ come in "long multiplets", with multiplicity $2^{2\mathcal{N}} = 8$, while the BPS particles with mass $M = \frac{\sqrt{U^2+V^2}}{2}$ come in "short multiplets" of multiplicity $2^{\mathcal{N}} = 4$.

References

1. P.A.M. Dirac: Proc. Roy. Soc. (1931) A **133**, 60; Phys. Rev. 74, 817 (1948)
2. G. 't Hooft: Nucl. Phys. B **79**, 817 (1974), A.M. Polyakov: JETP Lett. **20**, 194 (1974); M.K. Prasad, C.M. Sommerfield: Phys. Rev. Lett. **35**, 760 (1975); W. Nahm: Phys. Lett. B **90**, 413 (1980)
3. E. Lubkin: Ann. Phys. **23**, 233 (1963); E. Corrigan, D.I. Olive, D.B. Fairlie: J. Nuyts, Nucl. Phys. B **106**, 475 (1976)
4. P. Goddard, J. Nuyts, D. Olive: Nucl. Phys. B **125**, 1 (1977)
5. F.A. Bais: Phys. Rev. D **18**, 1206 (1978)
6. E.J. Weinberg: Nucl. Phys. B **167**, 500 (1980); Nucl. Phys. B **203**, 445 (1982); K. Lee, E. J. Weinberg, P. Yi: Phys. Rev. D **54** , 6351 (1996)
7. C.H. Taubes: Commun. Math. Phys. **80**, 343 (1980)
8. S. Coleman: "The Magnetic Monopole Fifty Years Later", Lectures given at International School of Subnuclear Physics, Erice, Italy (1981)
9. R.S. Ward: Commun. Math. Phys. **86**, 437 (1982)
10. N. Manton: Phys. Lett. B **154**, 397 (1985), Erratum ibid. B **157**, 475 (1985)
11. A. Abouelsaood: Nucl. Phys. B **226**, 309 (1983); P. Nelson, A. Manohar: Phys. Rev. Lett. **50**, 943 (1983); A. Balachandran, G. Marmo, M. Mukunda, J. Nilsson, E. Sudarshan, F. Zaccaria: Phys. Rev. Lett. **50**, 1553 (1983); P. Nelson, S. Coleman: Nucl. Phys. B **227**, 1 (1984)
12. C. Rebbi, G. Soliani: *Soliton and Particles* (World Scientific, Singapore, 1984). Many earlier references on the solitons are collected in this book
13. P.A. Horvathy, J.H. Rawnsley: Phys. Rev. D **32**, 968 (1985); J. Math. Phys. **27**, 982 (1986)
14. N. Dorey, C. Fraser, T.J. Hollowood, M.A.C. Kneipp: "NonAbelian duality in $N = 4$ supersymmetric gauge theories" [arXiv: hep-th/9512116]; Phys.Lett. B **383**, 422 (1996)
15. C.J. Houghton, P.M. Sutcliffe: J. Math. Phys. **38**, 5576 (1997)
16. B.J. Schroers, F.A. Bais: Nucl. Phys. B **512**, 250 (1998); Nucl. Phys. B **535**, 197 (1998)
17. M. Strassler: Prog. Theor. Phys. Suppl. **131**, 439 (1998)
18. H.J. de Vega: Phys. Rev. D **18**, 2932 (1978); H.J. de Vega, F.A. Shaposnik: Phys. Rev. Lett. **56**, 2564 (1986); Phys. Rev. D34, 3206 (1986); J. Heo, T. Vachaspati: Phys. Rev. D **58**, 065011 (1998), P. Suranyi: hep-lat/9912023; F.A. Shaposnik, P. Suranyi: Phys. Rev. D **62**, 125002 (2000); J. Edelstein, W. Fuertes, J. Mas, J. Guilarte: Phys. Rev. D **62**, 065008 (2000); M. Kneipp, P. Brockill: Phys. Rev. D **64**, 125012 (2001)
19. G. 't Hooft: Nucl. Phys. B **190**, 455 (1981); S. Mandelstam: Phys. Lett. **53B**, 476 (1975); Phys. Rep. C **23**, 245 (1976)
20. Y.M. Cho: Phys. Rev. D **21**, 1080 (1980); L.D. Faddeev and A.J. Niemi: Phys. Rev. Lett. **82**, 1624 (1999); Phys. Lett. B **449**, 214 (1999)
21. T.T. Wu, C.N. Yang: in *Properties of Matter Under Unusual Conditions*, ed. by H. Mark, S. Fernbach (Interscience, New York, 1969)
22. K. Konishi, K. Takenaga: Phys. Lett. B **508**, 392 (2001)
23. N. Seiberg, E. Witten: Nucl. Phys. B **426**, 19 (1994); Erratum ibid. B **430**, 485 (1994)
24. N. Seiberg, E. Witten: Nucl. Phys. B **431**, 484 (1994)

25. P. C. Argyres, A. F. Faraggi: Phys. Rev. Lett **74**, 3931 (1995); A. Klemm, W. Lerche, S. Theisen, S. Yankielowicz: Phys. Lett. B **344**, 169 (1995); Int. J. Mod. Phys. A **11**, 1929 (1996), A. Hanany, Y. Oz: Nucl. Phys. B **452**, 283 (1995)
26. P. C. Argyres, M. R. Plesser, A. D. Shapere: Phys. Rev. Lett. **75**, 1699 (1995); P. C. Argyres, A. D. Shapere: Nucl. Phys. B **461**, 437 (1996); A. Hanany: Nucl. Phys. B **466**, 85 (1996)
27. S. Bolognesi, K. Konishi: Nucl. Phys. B **645**, 337 (2002)
28. P. C. Argyres, M. R. Plesser, N. Seiberg: Nucl. Phys. B **471**, 159 (1996); P.C. Argyres, M.R. Plesser, A.D. Shapere: Nucl. Phys. B **483**, 172 (1997); K. Hori, H. Ooguri, Y. Oz: Adv. Theor. Math. Phys. **1**, 1 (1998)
29. A. Hanany, Y. Oz: Nucl. Phys. B **466**, 85 (1996)
30. G. Carlino, K. Konishi, H. Murayama: JHEP **0002**, 004 (2000); Nucl. Phys. B **590**, 37 (2000)
31. G. Carlino, K. Konishi, S. P. Kumar, H. Murayama: Nucl. Phys. B **608**, 51 (2001)
32. F. Cachazo, M. R. Douglas, N. Seiberg, E. Witten: JHEP **0212**, 071 (2002); F. Cachazo, N. Seiberg, E. Witten: JHEP **0302**, 042 (2003); F. Cachazo, N. Seiberg, E. Witten: JHEP **0304**, 018 (2003); for a review and further references, see: R. Argurio, G. Ferretti, R. Heise: Int. J. Mod. Phys. A **19**, 2015 (2004)
33. C. Montonen, D. Olive: Phys. Lett. B **72**, 117 (1977)
34. N. Seiberg: Nucl. Phys. B **435**, 129 (1995)
35. A. Hanany, D. Tong: JHEP **0307**, 037 (2003)
36. A. Hanany, D. Tong: JHEP **0404**, 066 (2004)
37. R. Auzzi, S. Bolognesi, J. Evslin, K. Konishi, A. Yung: Nucl. Phys. B **673**, 187 (2003)
38. E.B. Bogomolnyi: Sov. J. Nucl. Phys. **24**, 449 (1976)
39. B.A. Dubrovin, A.T. Fomenko, S.P. Novikov: *Modern Geometry—Methods and Applications, Part II. The Geometry and Topology of Manifolds*, translated by R.G. Burns (Graduate Text in Mathematics, Springer, Berlin, 1985)
40. M.K. Gaillard, B. Zumino: Nucl. Phys. B **193**, 221 (1981)
41. D. Finnell, P. Pouliot: Nucl. Phys. B **453**, 225 (1995); N. Dorey, V.V. Khoze, M.P. Mattis: Nucl. Phys. B **492**, 607 (1997)
42. N. A. Nekrasov: Adv. Theor. Math. Phys. **7** , 831 (2004)
43. K. Konishi: Int. J. Mod. Phys. A **16**, 1861 (2001)
44. J. Goldstone, F. Wilczek: Phys. Rev. Lett. **47**, 986 (1981)
45. E. Witten: Phys. Lett. B **86**, 283 (1979)
46. R. Jackiw, C. Rebbi: Phys. Rev. D **13**, 3398 (1976)
47. A. J. Niemi, Manu B. Paranjape, G. W. Semenoff: Phys. Rev. Lett. **53**, 515 (1984)
48. F. Ferrari: Phys. Rev. Lett. **78**, 795 (1997)
49. K. Konishi, H. Terao: Nucl. Phys. B **511**, 264 (1998); G. Carlino, K. Konishi, H. Terao: JHEP **9804**, 003 (1998)
50. A. Rebhan, P. van Nieuwenhuizen, R. Wimmer: Phys. Lett. B **594**, 234 (2004); Phys. Lett. B **632**, 145 (2006); JHEP **0606**, 056 (2006)
51. A. Bilal, F. Ferrari: Nucl. Phys. B **516**, 175 (1998); A. Cappelli, P. Valtancoli, L. Vergnano: Nucl. Phys. B **524**, 469 (1998)
52. R. Auzzi, S. Bolognesi, J. Evslin, K. Konishi, H. Murayama: Nucl. Phys. B **701**, 207 (2004)

53. G. Marmorini, K. Konishi, N. Yokoi: Nucl. Phys. B **741**, 180 (2006)
54. P. C. Argyres, M. R. Douglas: Nucl. Phys. B **448**, 93 (1995); P. C. Argyres, M. R. Plesser, N. Seiberg, E. Witten: Nucl. Phys. B **461**, 71 (1996); T. Eguchi, K. Hori, K. Ito, S.-K. Yang: Nucl. Phys. B **471**, 431 (1996)
55. R. Auzzi, R. Grena, K. Konishi: Nucl. Phys. B **653**, 204 (2003)
56. A. Abrikosov: Sov. Phys. JETP **32**, 1442 (1957); H. Nielsen, P. Olesen: Nucl. Phys. B **61**, 45 (1973)
57. R. Donagi, E. Witten: Nucl. Phys. B **460**, 299 (1996)
58. M. J. Strassler: "Messages for QCD from the Superworld" [arXiv: hep-th/9803009]
59. A. Hanany, M. Strassler, A. Zaffaroni: Nucl. Phys. B **513**, 87 (1998)
60. N. Dorey: JHEP **9811**, 005 (1998); N. Dorey, T.J. Hollowood, S.P. Kumar: Nucl. Phys. B **624**, 95 (2002)
61. V. Markov, A. Marshakov, A. Yung: Nucl. Phys. B **709**, 267 (2005)
62. K. Konishi, L. Spanu: Int. J. Mod. Phys. A **18**, 249 (2003)
63. S. C. Davis, A-C. Davis, M. Trodden: Phys. Lett. B **405**, 257 (1997)
64. A.I. Vainshtein, A. Yung: Nucl. Phys. B **614**, 3 (2001)
65. Y. Isozumi, M. Nitta, K. Ohashi, N. Sakai: Phys. Rev. D **71**, 065018 (2005)
66. M. Eto, Y. Isozumi, M. Nitta, K. Ohashi, N. Sakai: Phys. Rev. Lett. **96**, 161601 (2006)
67. M. Eto, Y. Isozumi, M. Nitta, K. Ohashi, N. Sakai: J. Phys. A **39**, 315 (2006)
68. D. Tong: "TASI lectures on solitons: Instantons, monopoles, vortices and kinks" [arXiv: hep-th/0509216]
69. M. Shifman and A. Yung: Phys. Rev. D **66**, 045012 (2002); M. Shifman and A. Yung: Phys. Rev. D **70**, 045004 (2004)
70. A. Gorsky, M. Shifman, A. Yung: Phys. Rev. D **71**, 045010 (2005)
71. M. Shifman, A. Yung: Phys. Rev. D **73**, 125012 (2006)
72. M. Eto, Y. Isozumi, M. Nitta, K. Ohashi, N. Sakai: Phys. Rev. D **72**, 025011 (2005)
73. S. Bolognesi, K. Konishi, G. Marmorini: Nucl. Phys. B **718**, 134 (2005)
74. N. Seiberg: Nucl. Phys. B **435**, 129 (1994)
75. R. Auzzi, S. Bolognesi, J. Evslin, K. Konishi: Nucl. Phys. B **686**, 119 (2004)
76. M. Eto, K. Konishi, G. Marmorini, M. Nitta, K. Ohashi, W. Vinci, N. Yokoi: Phys. Rev. D **74**, 065021 (2006)
77. R. Dijkgraaf, C. Vafa: Nucl. Phys. B **644**, 3 (2002); Nucl. Phys. B **644**, 21 (2002)
78. R. Auzzi, S. Bolognesi, J. Evslin: JHEP **0502**, 046 (2005)
79. M. Eto, L. Ferretti, K. Konishi, G. Marmorini, M. Nitta, K. Ohashi, W. Vinci, N. Yokoi: "Non-abelian duality from vortex moduli: a dual model of color-confinement", [arXiv: hep-th/0611313]
80. L. Ferretti, K. Konishi: in *Sense of Beauty in Physics,* Festschrift in honor of the 70th birthday of A. Di Giacomo, (Edizioni PLUS, University of Pisa Press, Pisa, 2006)
81. V. A. Rubakov: Nucl. Phys. B **203**, 311 (1982)

Part VI

String dualities and symmetries

Novel Symmetries of String Theory

J. Maharana

Institute of Physics, Bhubaneswar-751005, India
maharana@iopb.res.in

Abstract. The evolution of a closed bosonic string in its massless background is considered. The target space local symmetries associated with the graviton and the two-form antisymmetric tensor are exposed through Ward identities using a path-integral formalism in the Hamiltonian phase space. Similar Ward identities are derived for nonabelian gauge bosons that appear as massless excitation of compactified strings. It is proposed that excited massive stringy states might be endowed with hidden local symmetries. We find evidence in support of this conjecture when we examine string evolution in the background of some of the low-lying massive states.

1 Introduction

It is recognized that string theory holds the promise to unify the fundamental forces of Nature. There have been important developments to achieve this goal [1]. One of the wonders of string theories is the rich strove of their symmetries. The symmetry contents have been unraveled over the decades. It is not obvious whether we have exhausted and comprehended all the stringy symmetries so far. There are symmetries associated with the worldsheet such as the invariance under Weyl rescaling and the reparametrization invariance. It is well known that the quantum constraints determine the critical dimensions of the target space. On the other hand, if one envisages evolution of a string on the background of its massless excitations, the quantum constraints severely restrict the configurations of these backgrounds [2, 3]. The so-called β-function equations govern the evolutions of the backgrounds. The effective action constructed from the β-function equations reveal the local symmetries associated with the target space. For example, the massless spectrum of closed strings contains a spin-2 state identified with the graviton. Therefore, it is expected that string theory will encode general coordinate transformation invariance in its dynamics when we consider interaction of the graviton with other states. The open string spectrum incorporates a gauge boson, so

J. Maharana: *Novel Symmetries of String Theory*, Lect. Notes Phys. **737**, 525–552 (2008)
DOI 10.1007/978-3-540-74233-6_16

that gauge invariance is naturally expected to be manifested in string theory where such states are involved. Furthermore, it is well-known that nonabelian gauge bosons appear when a suitable compactification scheme is adopted. The resulting effective action is found to be invariant under those nonabelian gauge symmetries. Furthermore, dualities have played a cardinal role in our understanding of the string dynamics. The extended nature of string, and of other associated objects like branes, is responsible for the novel features which are not encountered in field theory describing point particles. The richness of the spectrum, the appearance of a limiting temperature due to an enormous degeneracy of the excited levels, and the presence of symmetries, are at the root of some of the mysteries of the theory which are yet to be fully understood.

The article presents attempts of Gabriele Veneziano and of the author to understand symmetries of string theory, and our endeavours to explore presence of new symmetries which might be associated with higher excited states of strings [4, 5]. It is worthwhile to ask whether one could unravel any local symmetries for the excited massive levels, just as massless states are endowed with 'gauge' symmetries [5]. Moreover, one might contemplate if the higher levels acquire their masses due to a Higgs-like mechanism so that there is a phase where all excited levels are massless. Whereas string field theory might be the appropriate arena to address these issues, the first quantized approach might provide a more intuitive picture where one studies the attributes of S-matrix elements. We mention in passing that most of our results are based on computations of S-matrix elements, and there is not enough headway to evaluate off-shell amplitudes. Of course, ideally, a complete theoretical framework should provide us tools to derive Green's functions. The relevance of the preceding comments will be alluded to later, in the context of symmetries associated with massive excited states of the strings.

The path-integral Hamiltonian formulation is adopted in order to exhibit symmetries associated with stringy states. Let us first consider the evolution of a closed bosonic string in the background of its massless excitations. The action corresponds to that of a σ-model where the backgrounds play the role of coupling constants. The quantum constraints lead to β-function equations when we carry out the computation by the standard prescriptions in the weak field approximation. In order to expose the underlying local symmetries associated with the massless states, we resort to the phase-space BRS Hamiltonian framework in the path-integral formalism. As will be demonstrated in sequel, the local symmetries are exhibited through the Ward identities satisfied by the S-matrix elements. We shall arrive at the Ward identities through the following steps: (i) obtain the canonical Hamiltonian, (ii) derive the algebra of constrains associated with the worldsheet symmetries, (iii) present the gauge fixed Hamiltonian with the relevant ghost terms and (iv) formally construct the generating functional for the S-matrix. We shall adopt the same prescription when we explore the presence of symmetries associated with excited massive stringy states [6, 7].

The article is organized as follows. We present a pedagogical introduction to BRS Hamiltonian formalism in the context of the problem at hand. In the next section, a closed bosonic string is envisaged in the presence of graviton and of the antisymmetric tensor field, and the Hamiltonian constraint analysis is presented. Next we discuss the modifications necessary in the σ-model action when nonabelian gauge bosons are present due to adoption of some compactification schemes. The fourth section is devoted to the derivation of Ward identities for the S-matrix elements with massless external states. We discuss how to incorporate the coupling of a dilaton background to the string in our approach. We present an illustrative example to show how anomalies creep in to viciate the Ward identities. Next, we present our approach to study presence of symmetries for massive stringy states. This section is based on unpublished works [6, 7]. We elaborate on the difficulties encountered by us while attempting to generalize our formalism for the massless case to the higher levels, and point out why we believe that the excited states are endowed with symmetries. The summary and conclusions are presented in the final section.

2 Hamiltonian Formalism and BRS Quantization

The quantization of theories with local symmetries such as gauge theories, Einstein's general theory of relativity and string theory pose problems due to the presence of first class constraints. Therefore, we are required to choose a gauge fixing prescription. The starting point is to construct a covariant theory and in the process we introduce more degrees of freedom. For example, in the relativistically covariant description of free electrodynamics, one deals with the vector potential A_μ with four degrees of freedom, whereas the photon is endowed with two physical degrees of freedom. The procedure of gauge fixing allows us to eliminate undesirable components. However, in this process, we might loose the covariance character of the theory. On the other hand, introducing a gauge fixing term like $\frac{1}{2}(\partial_\mu A^\mu)^2$ requires enlargement of the Hilbert space, as is well-known. The covariant gauge fixed descriptions of nonabelian gauge theory and string theory is best described in the framework of the BRS formalism.

The constrained Hamiltonian dynamics due to Dirac provides a very powerful and elegant technique to study theories with local symmetries and to quantize them. It is well-known that such theories possess first class constraints (they might have additional second class constraints). These constraints satisfy Poisson bracket algebra describing the underlying local symmetries. The gauge fixing conditions are introduced to eliminate the redundant degrees of freedom of the theories. The set of gauge fixing conditions, together with the originally derived first class constraints, constitute a set of second class constraints. Next, one may follow Dirac's prescriptions to quantize the gauge fixed theories. One of the most attractive features of

Dirac's formulation is that we can keep accounts of the degrees of freedoms of theory in the Hamiltonian phase space. Subsequently, the procedures of canonical quantization may be followed. The BRS formalism is best suited for quantization of such theories with covariant gauge fixing. It is customary to start from the gauge invariant Lagrangian, supplement it with a covariant gauge fixing term, and finally add the ghost part. The resulting new Lagrangian is no longer invariant under the local symmetry; however, it is invariant under the global BRS transformation. The BRS charge is nilpotent and it annihilates all physical states. Furthermore, it commutes with all operators corresponding to observables of the theory. The Balatin–Fradkin–Vilkovisky (BFV) [8, 9, 10] Hamiltonian phase-space approach lays down procedures to construct the Hamiltonian action, incorporating the constrained Hamiltonian formalism of Dirac. Let $\{\mathcal{F}_a, a = 1,..N\}$ be a set of first class constraints identified for the given theory. They satisfy the algebra

$$\{\mathcal{F}_a, \mathcal{F}_b\}_{PB} = f^c_{ab}\mathcal{F}_c, \tag{1}$$

where $\approx$ stands for weak equality, i.e. first the the PB are evaluated and then the constraints are to be set to zero in all such computations. Here f^c_{ab} are the structure 'constants' which usually do not depend on the phase-space variables; however, in general, they could carry dependence on such variables. Indeed, they coincide with structure constants of the underlying Lie algebra for nonabelian gauge theories. Whereas, in the case of Einstein–Hilbert action, these are derivatives of appropriate δ-functions as they appear in the algebra of Hamiltonian and momentum constraints of the theory. Furthermore, the set $\{\mathcal{F}_a\}$ also satisfy

$$\{\mathcal{H}_C, \mathcal{F}_a\}_{PB} \approx V^b_a \mathcal{F}_b. \tag{2}$$

Here $\mathcal{H}_C$ is the canonical Hamiltonian density derived from the gauge invariant action and V^b_a are constants determined for the theory under investigation. Equation (2) conveys that the time evolution of a first class constraint can be expressed as a linear combination of the first class constrains.

The

BFV provide a prescription to construct the BRS charge

$$\mathcal{Q} = \mathcal{F}_a \eta^a + \frac{1}{2}\mathcal{P}^a f^c_{ab}\eta^b\eta_c, \tag{3}$$

where $\mathcal{P}^a, \eta^a$ are sets of Grassmann odd ghosts (in quantum theory these are anticommuting objects) satisfying the PB relation

$$\{\mathcal{P}^a, \eta_b\}_{PB} = \delta^a_b, \tag{4}$$

with appropriate definition of PB brackets for such objects (note that there will be a δ-function on the *RHS* for fields). Moreover, $\mathcal{Q}$ is nilpotent by construction

$$\{\mathcal{Q}, \mathcal{Q}\}_{PB} = 0. \tag{5}$$

It is to be emphasized that (5) is a nontrivial statement for a fermionic charge like $\mathcal{Q}$. Furthermore, this condition imposes severe constraints on the underlying Hilbert space of the corresponding quantum theory. The next step is to construct the gauge fixed action. Recall that the PB of $\mathcal{Q}$ vanishes with $\mathcal{H}_C$ by construction, since $\mathcal{F}_a$ are first class. Then a Grassmann odd fermionic object χ is introduced which is a function of the fields, of their conjugate momenta, and of the set of ghosts $\{\mathcal{P}_a, \eta_a\}$. The gauge fixed Hamiltonian density is constructed to be

$$\mathcal{H}_\chi = \mathcal{H}_C + \mathcal{P}^a V_a^b \eta_b - \{\chi, \mathcal{Q}\}_{PB}. \tag{6}$$

Therefore, choice of the gauge fixing function, χ, determines the effective gauge fixed Hamiltonian density. Now the Hamiltonian action is

$$S_H = \int d^d x \left(\phi_i \pi^i_\phi - \mathcal{H}_\chi \right), \tag{7}$$

where ϕ_i are generic fields (gauge fields, scalars, fermions), and π^i_ϕ their conjugate momenta. The expression (7) is for d-dimensional spacetime.

Let us consider the evolution of a closed string in d-dimensional target space. The string traces out a cylinder on the worldsheet surface during its evolution. The underlying action is required to be invariant under the reparametrization of the worldsheet coordinates. The Polyakov action is the most convenient form to describe and quantize open and closed strings [11],

$$S = -\frac{1}{2} \int d^2\sigma \sqrt{-\gamma}\gamma^{ab} \partial_a X^\mu \partial_b X^\nu \eta_{\mu\nu}, \tag{8}$$

where γ_{ab} is the worldsheet metric, γ^{ab} is its inverse, γ is determinant of worldsheet
metric and $\eta_{\mu\nu}$ is the flat space metric of the target space. The variation of the action with respect to γ^{ab} results in the worldsheet energy–momentum tensor,

$$T_{ab} = \partial_a X^\mu \partial_b X^\nu \eta_{\mu\nu} - \frac{1}{2}\gamma_{ab}\gamma^{cd}\partial_c X^\mu \partial_d X^\nu \eta_{\mu\nu}. \tag{9}$$

Note that $T_{ab} = 0$, since there is no kinetic term for the worldsheet metric, as the analogue of Einstein–Hilbert piece, $\int d^2\sigma\sqrt{-\gamma}\ R^{(2)}$, is a topological term. We can solve for γ_{ab} from the above equation. If we insert the above expression for the worldsheet metric into the Polyakov action, then we recover the string action as proposed by Nambu and Goto. An important point to note is that the equivalence between the Polyakov and the Nambu–Goto action will hold when the equation of motion for the worldsheet metric is utilized in (8).

The action (8) has the following symmetry properties.
(a) Two-dimensional reparametrization invariance,

$$\delta\gamma_{ab} = \xi^c\partial_c\gamma_{ab} + \partial_a\xi^c\gamma_{bc} + \partial_b\xi^c\gamma_{ac}, \tag{10}$$

and hence $\delta\sqrt{-\gamma} = \partial_a(\xi^a\sqrt{-\gamma})$. The string coordinate transforms as

$$\delta X^\mu = \xi^a\partial_a X^\mu. \tag{11}$$

(b) Weyl invariance

$$\delta\gamma_{ab} = 2\Omega\gamma_{ab}, \qquad \delta X^\mu = 0, \tag{12}$$

where the parameter Ω depends on the worldsheet variables.
(c) Poincare invariance (in target space)

$$\delta X^\mu = \omega^\mu_\nu X^\nu + a^\mu, \qquad \delta\gamma_{ab} = 0, \tag{13}$$

where $\omega_{\mu\nu}$ are antisymmetric parameters associated with the Lorentz transformation and a^μ are the parameters of translation.

Note that the Weyl invariance implies tracelessness of the two-dimensional energy momentum tensor for the classical theory. The quantum invariance of this symmetry has far reaching consequences in string theory.

If we make the orthonormal gauge choice for the worldsheet metric, $\gamma_{ab} = e^{2\Omega(\sigma,\tau)}\eta_{ab}$ with $\eta_{ab} = \mathrm{diag}(-1,+1)$, the form of Polyakov action simplifies since $\sqrt{-\gamma}\gamma^{ab} = \eta^{ab}$ in this gauge. The condition of the vanishing of T_{ab} reduces to two constraints

$$(\dot{X} \pm X')^2 = 0. \tag{14}$$

These are the Virasoro constraints. They take the following form in the Hamiltonian formalism:

$$P_\mu X'^\mu = 0, \qquad H = \frac{1}{2}(P^2 + X'^2) = 0, \tag{15}$$

where P_μ is momentum conjugate to X^μ derived from the Polyakov action. It is easy to check that the first constraint generates σ translations on the worldsheet, whereas the latter, being the canonical Hamiltonian, generates τ translations. It is more convenient to define the constraints $L_\pm = \frac{1}{4}(P^\mu \pm X'^\mu)\eta_{\mu\nu}(P^\nu \pm X'^\nu)$ whose 'equal time' algebra takes an elegant form

$$\{L_\pm(\sigma), L_\pm(\sigma')\}_{PB} = \pm\Big(L_\pm(\sigma) + L_\pm(\sigma')\Big)\partial_\sigma\delta(\sigma-\sigma') \tag{16}$$

and

$$\{L_\pm(\sigma), L_\mp(\sigma')\}_{PB} = 0. \tag{17}$$

It is obvious from the constraint algebra (16) and (17) that the $L_\pm$ represent a pair of first class constraints. The classical BRS charge is

$$\mathcal{Q} = \int d\sigma (L_+ \eta_+ + L_- \eta_- + \mathcal{P}_+ \eta_+ \eta'_+ - \mathcal{P}_- \eta_- \eta'_-). \tag{18}$$

The gauge fixed Hamiltonian density is

$$H_\chi = \{\chi, \mathcal{Q}\}. \tag{19}$$

Note that χ is the gauge fixing function; for ON gauge choice $\chi = \mathcal{P}_+ + \mathcal{P}_-$. As discussed above, the ON gauge choice corresponds to $H = \frac{1}{2}(P^2 + X'^2) = L_+ + L_-$ involving string coordinates. Therefore, this choice of χ gives us the full ON gauge Hamiltonian,

$$H_{ON} = L_+ + L_- + 2\mathcal{P}_+ \eta'_+ + \mathcal{P}'_+ \eta_+ - 2\mathcal{P}_- \eta'_- - \mathcal{P}'_- \eta_-. \tag{20}$$

The classical BRS charge $\mathcal{Q}$ is nilpotent by construction, i.e. $\{\mathcal{Q}, \mathcal{Q}\}_{PB} = 0$. The quantum BRS charge is defined with a normal ordering prescription. The string coordinate X^μ, its canonical momenta P_μ and the ghost fields $\mathcal{P}_\pm, \eta_\pm$ are expanded in Fourier series with creation and annihilation operators. The quantum constraint $\hat{\mathcal{Q}}^2 = 0$ implies the number of spacetime dimensions, $D = 26$, and the 'intercept' $\alpha_0 = 2$ (we are discussing closed bosonic string) [12]. We may remind the reader that the massless excitations of a closed bosonic string contain a scalar $\varPhi$, called dilaton, a symmetric tensor, $G_{\mu\nu}$, identified with the graviton, and an antisymmetric tensor, $B_{\mu\nu}$. One of our principal goals is to unveil the target space symmetries of the theory. The first step in this direction is to adopt the first quantized framework and envisage the evolution of the string in the background of these massless excitations, whose worldsheet action takes the form

$$S = -\frac{1}{2} \int d^2\sigma \Big(\sqrt{\gamma} \gamma^{ab} \partial_a X^\mu \partial_b X^\nu G_{\mu\nu}(X) + \epsilon^{ab} \partial_a X^\mu \partial_b X^\nu B_{\mu\nu}(X) \Big). \tag{21}$$

The generators

$$L_\pm = \frac{1}{4}(P_\mu \pm X'^\rho G_{\mu\rho} + X'^\rho B_{\mu\rho}) G^{\mu\nu} (P_\nu \pm X'^\lambda G_{\nu\lambda} + X'^\lambda B_{\nu\lambda}) \tag{22}$$

satisfy the same Poisson bracket algebra as in (16) and (17). Next, the generating functional can be formally defined using the phase-space path-integral formalism in order to implement BRS quantization,

$$\Sigma[G, B] = \int d[X^\mu] d[P_\mu] d[\eta_\pm] d[\mathcal{P}_\pm] \exp \left[i \int d\sigma \Big(\dot{X}^\mu P_\mu + \mathcal{P}_\pm \dot{\eta}_\pm - H_\chi \Big) \right]. \tag{23}$$

Here $H_\chi = L_+ + L_- + 2\mathcal{P}_+ \eta'_+ + \mathcal{P}'_+ \eta_+ - 2\mathcal{P}_- \eta'_- - \mathcal{P}'_- \eta_-$ and H_χ is the gauge fixed Hamiltonian density for the case at hand. In the context of the BRS quantization, the following remarks are to be borne in mind. The nilpotency of the quantum BRS charge $\hat{\mathcal{Q}}$ imposes stringent constraints on the admissible backgrounds in the form of differential equations. These are precisely the β-function equations computed while adopting the conformal field theory techniques.

The generating functional (23) was introduced by Fradkin and Tseytlin [2] in order to study string dynamics in the presence of nontrivial background fields in their Hamiltonian path-integral approach. Notice that Σ plays the role of generating functional for S-matrix elements in the following sense. Let us collectively denote the backgrounds as

$$\mathcal{B}(X(\sigma)) = (\phi(x), G_{\mu\nu}(x), B_{\mu\nu}(x), M), \tag{24}$$

where M collectively stands for massive stringy states such as the tachyon, T and higher excited levels. If we consider the worldsheet action in the presence of $\mathcal{B}$, the resulting σ-model action is required to be conformally invariant. In the simplest case we consider massless backgrounds $\mathcal{B} = (\phi(x), G_{\mu\nu}(x), B_{\mu\nu}(x),\ M = 0)$, and such that $\mathcal{B}$ fluctuates around a trivial vacuum configuration.

$$\mathcal{B} = \mathcal{B}_0 + \tilde{\mathcal{B}}, \tag{25}$$

where $\mathcal{B}_0 =$ (const, $\eta_{\mu\nu}$, const) and $\tilde{\mathcal{B}} = (e^{ik_\phi.x}, \alpha_{\mu\nu}e^{ik.x}, ...)$, where $\alpha_{\mu\nu}$ is identified with the polarization tensor of graviton. The fluctuating fields are required to satisfy the vanishing 'β-function' conditions $k_\phi^2 = 0$, $k^2 = 0$, $k_\mu \alpha_{\mu\nu} = 0, ...$). If we have to introduce the tachyon background, the corresponding constraint is $\tilde{\mathcal{B}}_T = e^{ik_T.x}$, $k_T^2 = 4/\alpha'$. Note that for such trivial backgrounds Σ generates the S-matrix elements for the scattering of those states. It is hoped that we shall be able to obtain the S-matrix elements for the massless excitations even when nontrivial vacuum configurations are envisaged. Furthermore, it is expected that the S-matrix elements can be derived also when higher level massive modes are included in the corresponding σ-model action, and consequently Σ is treated as a functional of $(\phi, G, B, M...)$.

Our starting point is to construct Σ in a formal sense, include the ghost fields, and define it through the phase-space path integral. We introduce a set of generating functionals for canonical transformation which will play an important role in exhibiting the local symmetries associated with the massless states of the string in the target space. Indeed, when one constructs the string effective action in target space starting from the β-function equations, the effective action is invariant under local target space symmetries. However, it is not obvious to explain how the two-dimensional σ-model encodes the local symmetries of the target space. Note that we have not introduced the string coupling to dilaton background in (21). The additional term is

$$\frac{1}{4\pi}\int d^2\sigma\sqrt{-\gamma}R^{(2)}\Phi(X.) \tag{26}$$

Here $R^{(2)}$ is the scalar curvature of the two-dimensional worldsheet. The dilaton coupling (26) is not conformally invariant. However, we demand that the sum of (21) and (26) be conformally invariant, which is equivalent to the vanishing of the β-function conditions [3]. Consequently, one obtains three sets of coupled differential equations involving backgrounds $G_{\mu\nu}, B_{\mu\nu}$ and Φ (the total number of equations being d^2). In the context of BRS Hamiltonian formalism, the coupling of a string to the dilaton background needs careful discussions, and we shall return to this issue later.

It is well-known that massless gauge bosons appear in the spectrum of closed strings when some of the spatial dimensions are compactified [13]; of special interests are the toroidally compactified coordinates. Indeed, the hope of unifying all fundamental forces in string theory gained strong support with the construction of heterotic string theory [14]. This ten-dimensional theory not only contained the spectrum of $N = 1$ supergravity, but also admitted nonabelian gauge theories with $SO(32)$ or $E_8 \times E_8$ groups. We may recall that the seminal work of Green and Schwarz [15], which generated the so-called 1984 superstring revolution, had shown that these were the only two admissible gauge groups in order to satisfy anomaly-free conditions for theories in ten dimensions. Therefore, when we examine the presence of local symmetries for a string in its massless backgrounds, we are expected to address the relevant issues for compactified string coordinates.

Let us consider the case where d spatial coordinates are compactified (we focus on a bosonic string in critical dimensions $D = 26$). These coordinates are denoted by $X^I, I = 1, 2, ..d$, and for a closed string (in absence of nontrivial backgrounds) are separated into left and right movers. Let us recall that there will be d Kaluza–Klein gauge bosons if we were to consider a higher-dimensional theory of gravity with compact coordinates. Our goal is to couple the string to the nonabelian gauge boson background. It is more convenient to fermionize the compact bosonic coordinates [16]. Essentially, each compact bosonic coordinate corresponds to two fermionic degrees of freedom. Thus for a set of $\{X^I\}, I = 1, 2..d$ bosonic coordinates, one introduces $\psi_i, i = 1, 2, ..2d$ two-dimensional Majorana fermions. Therefore, for a free closed string (in critical $D = 26$ dimensions) with 16 compact coordinates, there are 32 left moving and 32 right moving chiral fermions. Moreover, the nonabelian gauge group is $SO(32)_L \times SO(32)_R$, or two $E_8 \times E_8$ gauge groups when appropriate boundary conditions are imposed on the worldsheet fermions (corresponding to fermionic representations of the compactified bosonic coordinates). For the sake of simplicity, let us couple the right-handed sector to the corresponding background gauge fields. The worldsheet action is

$$S = -\frac{1}{2}\int d^2\sigma \Big[\sqrt{-\gamma}\gamma^{ab}\partial_a X^\mu \partial_b X^\nu G_{\mu\nu} + \epsilon^{ab}\partial_a X^\mu \partial_b X^\nu B_{\mu\nu} + ie\psi^i_L e^a_- \partial_a \psi^i_L$$

$$+ ie\psi^i_R (e^a_\alpha \partial_a \delta_{ij} + \Gamma^M_{ij} A^M_\mu \partial_a X^\mu)\psi^j_R\Big], \qquad (27)$$

where e^a_α is the zweibein associated with the worldsheet and e is its determinant; Γ^M_{ij} are the generators of the gauge group $SO(32)$ for $d = 16$ in the fundamental representation to which $\psi^i_{L,R}$ belong. The above action (27) is conformally invariant at the classical level. The two constraints, $L_{\pm}$, can be easily computed after some lengthy calculations.

$$0 = L_+ - L_- = P_\mu X'^\mu + \Pi^i_R \partial_1 \psi^i_R + \Psi_L \partial_1 \psi^i_L, \tag{28}$$

and

$$0 = L_+ + L_- = \frac{1}{2}\tilde{P}_\mu \tilde{P}_\nu G^{\mu\nu}(X) + \frac{1}{2} X'^\mu X'^\nu G_{\mu\nu}(X) - \Pi^i_L \partial_1 \psi^i_L \\ + \Pi^i_R \partial_1 \psi^i_R - \Pi^i_R T^M_{ij} A^M_\mu(X) \psi^j_R X'^\mu, \tag{29}$$

where the conjugate momenta of the chiral worldsheet fermions are $\Pi^i_{L,R} = \frac{1}{2} ie\psi^i_{L,R}$ and

$$\tilde{P}_\mu = P_\mu + X'^\lambda B_{\mu\lambda} + \Pi^i_R T^M_{ij} A^M_\mu \psi^j_R. \tag{30}$$

It is straightforward to show that the computation of the classical brackets $\{L_+, L_+\}_{PB}$, $\{L_-, L_-\}_{PB}$ and $\{L_+, L_-\}_{PB}$ take the same form as in (16) and (17). The BRS charge and the Hamiltonian can be obtained for the case under study (i.e. with backgrounds G, B and A) following the prescriptions described above for the case where the string evolves in the presence of the graviton and of the antisymmetric tensor fields. Notice that the massless sector, for a string with compact coordinates, also contains scalars belonging to the adjoint representations of the (left $\times$ right) gauge groups. The corresponding action in the presence of these states can be written down easily.

3 Canonical Transformations and Invariance Properties of Σ

We show that a chosen set of canonical transformations on the phase-space path integral (23) brings out the local symmetries of the generating functional [5], Σ, under transformations of the background fields, $G_{\mu\nu}$, $B_{\mu\nu}$ and A^M_μ. Note, however, that this will be demonstrated at a formal level. We are aware that careful computations might lead to anomalies. We shall present an example where such an anomaly can be explicitly computed. In what follows, we shall focus on exhibiting the underlying local symmetries. First, we shall deal with the case on noncompact string in backgrounds of G and B, and then take up the case of a string with compact coordinates.

(i) String with noncompact coordinates:
Let Φ_F be a generator of canonical transformation in the phase space such that $\delta z = \{z, \Phi_F\}_{PB}$, where $\{z\}$ collectively stand for the phase-space variables X^μ and P_μ. We introduce a generator

$$\Phi_G = \int d\sigma P_\mu \xi^\mu(X). \tag{31}$$

The transformations induced on z are

$$\delta_G X^\mu = \xi^\mu(X), \quad \delta_G P_\mu = -P_\nu \xi^\nu{}_{,\mu}(X), \quad \text{and} \quad \delta_G(\eta, \mathcal{P}) = 0. \tag{32}$$

Here the comma stands for ordinary derivative. The backgrounds, $G_{\mu\nu}(X)$ and $B_{\mu\nu}(X)$, are functions of X^μ and therefore, under Φ_G, the coordinates shift according to the rules (32) leading to

$$\delta_G G_{\mu\nu} = G_{\mu\nu},_\lambda \xi^\lambda, \quad \delta_G B_{\mu\nu} = B_{\mu\nu},_\lambda \xi^\lambda. \tag{33}$$

The Hamiltonian action

$$S_H = \int d^2\sigma \Big(\dot{X}^\mu P_\mu + \mathcal{P}_\pm \dot{\eta}_\pm - H_{ON} \Big) \tag{34}$$

exhibits the property

$$\delta_G S_H = -\delta^{GTC} S_H. \tag{35}$$

The *LHS* of the above equation is well defined from the rules described above, while δ^{GCT} is to interpreted as follows: noting that $G_{\mu\nu}$ and $B_{\mu\nu}$ are second rank tensors in the target space, their transformation rules under general coordinate transformations (GCT) are dictated from their tensorial properties, namely

$$\delta^{GCT} G_{\mu\nu} = -G_{\mu\lambda}\xi^\lambda{}_{,\nu} - G_{\nu\lambda}, \xi^\lambda_\mu - G_{\mu\nu},_\lambda \xi^\lambda. \tag{36}$$

A similar expression follows for $\delta^{GCT} B_{\mu\nu}$. If we introduce another generator of canonical transformation,

$$\Phi_B = \int d\sigma X'^\mu \Lambda_\mu(X), \tag{37}$$

then the variation of the phase-space variables, and consequently the variations of the background, can be computed according to the prescriptions already given. The analogous transformation property of S_H is

$$\delta_B S_H = -\delta^{B-gauge} S_H. \tag{38}$$

We define the B-gauge transformation to be

$$\delta^{B-gauge} = \partial_\mu \Lambda_\nu(X) - \partial_\nu \Lambda_\mu(X). \tag{39}$$

Note that the target space metric, $G_{\mu\nu}$ is not affected by $\delta^{B-gauge}$ transformation. We assume that the phase-space measure remains invariant under the canonical transformations induced by the generators Φ_G and Φ_B. We arrive at the following conclusion, after taking into account the relations (35) and (38):

$$\Sigma(G, B) = \Sigma\Big(G + \delta^{GCT} G, B + \delta^{GCT} B + \delta^{B-gauge} B \Big). \tag{40}$$

(ii) Compact case:
The massless sector of a closed bosonic string, with compact coordinates,

consists of nonabelian gauge fields in addition to G and B (we ignore the coupling of the string to the massless scalars arising due to compactification). The generator of canonical transformations, which exposes the nonabelian gauge symmetry, turns out to be

$$\Phi_A = \int d\sigma \psi_R^i \psi_R^j T_{ij}^M \xi^M(X). \tag{41}$$

The corresponding Hamiltonian action satisfies the relation

$$\delta_A S_H = -\frac{i}{2} \delta^{gauge} S_H. \tag{42}$$

Here δ^{gauge} is the nonabelian gauge transformation acting on $A_\mu^M(X)$ such that

$$\delta^{gauge} A_\mu^M = \xi^M{}_{,\mu}(X) + f^{MNP} A_\mu^N \xi^P(X), \tag{43}$$

where f^{MNP} are the structure constants defined by the relation

$$\left[T^M, T^N\right] = i f^{MNP} T^P. \tag{44}$$

Taking into account (42), and repeating the arguments for the non-compact string case, we conclude that

$$\Sigma(G, B, A) = \Sigma(G, B, A + \delta^{gauge} A). \tag{45}$$

We may remind the reader that the relations (40) and (45) hold modulo the anomalies alluded to earlier. We shall briefly address this issue at the end of this section.

The invariance properties of the generating functional allows us to derive Ward identities for the S-matrix elements, as they are generated by Σ in an elegant manner. Let us first focus on GCT. We arrive at

$$0 = \delta^{GCT} \Sigma = \left\langle \int d^D x \left[\frac{\delta S_H}{\delta G_{\mu\nu}} \delta^{GCT} G_{\mu\nu} + \frac{\delta S_H}{\delta B_{\mu\nu}} \delta^{GCT} B_{\mu\nu} \right] \right\rangle_{G,B}. \tag{46}$$

The symbol $<>_{G,B}$ stands for an average in the sense of functional integration in the Hamiltonian phase space with the weight factor $\exp(iS_H(G,B))$. The backgrounds $G_{\mu\nu}$ and $B_{\mu\nu}$ are finally set to those configurations enforced by the β-function equations. The generic form of S_H is

$$S_H = \int d^2\sigma \mathcal{L}(X, P, \eta, \mathcal{P}, G(X(\sigma)), B(X(\sigma))). \tag{47}$$

Therefore, the functional derivative of S_H with respective to a generic background gives us the corresponding vertex operator. As an illustrative example consider the variation of S_H with respect to $G_{\mu\nu}$:

$$\frac{\delta S_H}{\delta G_{\mu\nu}(x)} = \int d^2\sigma \delta(x - X(\sigma)) \frac{\delta \mathcal{L}_H}{\delta G_{\mu\nu}(X)} = \int d^2\sigma \delta(x - X(\sigma)) V_G^{\mu\nu}(X). \quad (48)$$

We can similarly obtain the corresponding vertex operator associated with $B_{\mu\nu}$. We define $V_G^{\mu\nu}$ and $V_B^{\mu\nu}$ as the graviton and antisymmetric tensor vertex operators in a background. These operators can be obtained explicitly as

$$V_G^{\mu\nu} = \frac{1}{2} X'^{\mu} X'^{\nu} - \frac{1}{2} G^{\mu\rho} G^{\nu\lambda} P_\rho P_\lambda - P_\rho G^{\rho\mu} G^{\nu\sigma} B_{\sigma\tau} X'^{\tau} - \frac{1}{2} X'^{\rho} B_{\alpha\rho} B_{\beta\sigma} G^{\mu\alpha} G^{\nu\beta} \quad (49)$$

and

$$2V_B^{\mu\nu} = P_\rho G^{\rho\mu} X'^{\nu} + X'^{\nu} B\rho\sigma X'^{\sigma} G^{\rho\mu} - (\mu \leftrightarrow \nu). \quad (50)$$

We can use in (46) the expression (36) for $\delta^{GCT} G_{\mu\nu}$, and the definition of vertex operator (49), to arrive at

$$0 = \Bigg\langle \int d^2\sigma \Bigg[V_G^{\mu\nu} \Big(G_{\mu\lambda}(X) \xi^{\lambda},_{\nu}(X) + G_{\nu\lambda}(X) \xi^{\lambda}(X),_{\mu} + G_{\mu\nu},_{\lambda}(X) \xi^{\lambda}(X) \Big)$$
$$+ V_B^{\mu\nu}(X) \Big(B_{\mu\lambda}(X) \xi^{\lambda},_{\nu}(X) - B_{\nu\lambda}(X) \xi^{\lambda},_{\mu}(X) + B_{\mu\nu},_{\lambda}(X) \xi^{\lambda}(X) \Big) \Bigg] \Bigg\rangle_{G,B}. \quad (51)$$

Now we are in a position to derive the desired Ward identities (WI) for processes with multigravitons and antisymmetric tensor field. Notice that (51) holds for arbitrary infinitesimal parameters $\xi^{\alpha}(X)$. Therefore, we are permitted to take the functional derivative of the right-hand side of (51) with respect to $\xi^{\alpha}(X)$ at $\xi^{\alpha} = 0$. Furthermore, we can take an arbitrary number of derivatives of the above equation with respect to $G_{\mu\nu}(Y)$ and $B_{\mu\nu}(Z)$ at the ground state values of $G_{\mu\nu}$ and $B_{\mu\nu}$ (i.e. backgrounds corresponding to a string vacuum configuration). As an illustrative example, let us envisage the case of a closed bosonic string in critical dimensions, $D = 26$, in which the background metric describes the Minkowski space. If we consider the n-graviton amplitude (ignore the B-field, for the moment), then (51) can be expressed as

$$\frac{\delta^n}{\delta G_{\mu_1\nu_1}(y_1) \ldots \delta G_{\mu_n\nu_n}(y_n)} \Bigg\langle \int d^2\sigma \Big(G_{\mu\lambda} \partial_\nu \delta(x - X) + (\mu \leftrightarrow \nu) + G_{\mu\nu},_{\lambda} \delta(x - X) \Big) \Bigg\rangle = 0. \quad (52)$$

This form of WI is familiar in multigraviton scattering amplitudes. Let us examine (52) a little carefully. The chain of functional derivatives, applied to the expression for the path-integral average, acts in following three ways: the derivative with respect to metric $G_{\mu_i\nu_i}(y_i)$ can act (i) on the vertex operator $V_G^{\mu\nu}$, or (ii) on $G_{\mu\lambda}$, $G_{\nu\lambda}$, $G_{\mu\nu},_{\lambda}$ or (iii) on the Hamiltonian action S_H, which is hidden in the definition of the path integral average $< \ldots >_G$ itself. The action of the G-functional derivatives in case (i) and (ii) kills the presence of any metric and produces a $\delta(y_i - X)$ type of terms which are the contact terms. On the other hand, each G-functional derivative acting on S_H produces an additional V_G. Note that WI presented above is in the x-space representation. If we Fourier transform (52), the familiar form WI can be recovered:

$$\left\langle \prod_{i=1}^{n} \int d^2\sigma_i V_G^{\mu_i\nu_i}(X(\sigma_i)) e^{ik_i X(\sigma_i)} \int d^2\sigma 2k_\nu V_G^{\lambda\nu}(X(\sigma)) e^{ikX(\sigma)} \right\rangle_{G=\eta}$$
$$= \sum_{i=1}^{n} k_i^\lambda \left\langle \int d^2\sigma_i V_G^{\mu_i\nu_i}(X(\sigma_i)) e^{(k+k_i)X(\sigma_i)} \right.$$
$$\left. \prod_{j\neq i} \int d^2\sigma_j V_G^{\mu_j\nu_j}(X(\sigma_j)) e^{k_j X(\sigma_j)} \right\rangle + (\text{other contact terms}). \quad (53)$$

The above equation tells us that the divergence of an $(n+1)$-graviton amplitude is related to the sum of lower-point S-matrix elements. It is to be kept in mind that there could be potential anomaly terms in the WI. We arrived at these results through formal manipulations. We expect WI to hold good for backgrounds which are consistent with BRS invariance. Furthermore, the fluctuations should correspond to on-shell external particles. Naturally, if want to derive WI for off-shell Green functions, there could be additional terms in the RHS of the above equation.

We can extend the aforementioned arguments to derive Ward identities for amplitudes with gravitons, antisymmetric tensor fields, and nonabelian gauge bosons, in the presence of generic backgrounds compatible with BRS invariance. The vertex operator for the gauge bosons is

$$V_{AM}^{\mu} = \frac{\delta S_H}{\delta A_\mu^M} = \psi_R^i T_{ij}^M \psi_R^j (G^{\mu\nu}\tilde{P}_\nu - X'^\mu), \quad (54)$$

and $\tilde{P}_\mu$ is given by (30). Note that $G_{\mu\nu}$ and $B_{\mu\nu}$ are all buried in the expression for $\tilde{P}_\mu$ derived from an action which describes the evolution of the compactified closed bosonic string in a background with the graviton, the antisymmetric tensor field and a nonabelian gauge boson. The corresponding WI can then be derived from the basic equation

$$\left\langle \int d^2\sigma V_{AM}^{\mu}(X(\sigma)) \Big[\partial_\mu \delta(X-x)\delta_{MP} \right.$$
$$\left. + f^{NMP} A_\mu^N(X(\sigma))\delta(X-x) \Big] \right\rangle_{G,B,A} = 0 \quad (55)$$

We briefly discuss the coupling of a string to the dilaton background in the context of the BRS Hamiltonian formalism. The well-known coupling $\sim \int d^2\sigma \sqrt{-\gamma} R^{(2)} \Phi(X)$ is to be converted into a term involving ghost fields. This issue has been addressed in the Lagrangian BRS approach in [17]. We noticed that an additional piece contributes to the gauge-fixed Hamiltonian density involving ghosts and the dilaton background,

$$\begin{aligned} H_D = &\frac{8}{3}\mathcal{P}_+\eta_+(P_\mu + X'^\rho B_{\mu\rho} - X'^\rho G_{\mu\rho})G^{\mu\nu}\partial_\nu\Phi(X) \\ &+ \frac{8}{3}\mathcal{P}_-\eta_-(P_\mu + X'^\rho B_{\mu\rho} + X'^{rho})G^{\mu\nu}\partial_\nu\Phi(X) \\ &+ \frac{64}{9}\mathcal{P}_+\eta_+\mathcal{P}_-\eta_-\partial_\mu\Phi G^{\mu\nu}\partial_\nu\Phi. \end{aligned} \tag{56}$$

The total Hamiltonian density is $H = H_{ON} + H_D$. The phase-space variables transform as follows under the canonical transformation induced by the generator by $\mathbf{\Phi}_{ghost} = \int d^2\sigma(\mathcal{P}_+\eta_+ + \mathcal{P}_-\eta_-)\epsilon(X)$:

$$\begin{aligned} &\delta\eta_\pm = \epsilon(X)\eta_\pm, \quad \delta\mathcal{P}_\pm = -\epsilon(X)\mathcal{P}_\pm, \quad \delta X^\mu = 0, \\ &\delta P_\mu = -(\partial_\mu\epsilon)(\mathcal{P}_+\eta_+ + \mathcal{P}_-\eta_-). \end{aligned} \tag{57}$$

These variations of the phase-space variables induce a change in the Hamiltonian action, which is equivalent to shifting the dilaton by a parameter ϵ, i.e.

$$\delta_{ghost} = -\frac{1}{8}\epsilon(X). \tag{58}$$

Therefore, one might naively argue that $\Phi(X)$ can be rotated away by a field redefinition. However, it is well-known that such classical arguments are invalid when quantum–mechanical considerations are invoked.

Let us focus our attention on a simple scenario when the left moving sector of a closed bosonic string is compactified on a d-dimensional torus. We may choose d such that it corresponds to even self-dual lattice, if we desire. If we fermionize the compact coordinates, we must introduce $2d$ chiral fermions, and the massless sector will have nonabelian gauge bosons A_μ^M. The generator of canonical transformation given by (41) is instrumental for deriving the WI associated with gauge invariance. We consider backgrounds where $G_{\mu\nu} = \eta_{\mu\nu}$ and $B_{\mu\nu} = 0$, and we are interested in investigating whether the canonical transformation introduces anomalies, i.e. whether the phase measure remains invariant or not [18]. We must deal carefully with the transformed two-dimensional chiral fermion measure, which is essential in the definition of $\Sigma(A)$. The anomaly, if any, will arise from the noninvariance of this measure [19] under the transformations induced by (41). We start from the action

$$S = -\frac{1}{2}\int d^2\sigma(\partial_\alpha X^\mu\partial^\alpha X^\nu\eta_{\mu\nu} + i\psi^i\rho_\alpha\partial^\alpha\psi^i), \tag{59}$$

where ψ^i, $i = 1, ..2d$ are the two-dimensional Majorana fermions on the world-sheet, and ρ_α are the two-dimensional 'γ'-matrices. Here we have already adopted the orthonormal gauge. Let us couple only the left-moving fermions to the corresponding gauge field background, so that the action is schematically written as

$$S = -\frac{1}{2}\int d^2\sigma\Big(\partial_\alpha X^\mu \partial^\alpha X_\mu \psi^i \Big[i\partial_\alpha \delta_{ij} + \frac{(1-\rho_5)}{2}\partial_\alpha X^\mu T^M_{ij} A^M_\mu \Big]\psi^j\Big), \quad (60)$$

where ρ_5 is the product of the three ρ-matrices. The action is invariant under the following transformations: (i) $\delta\psi = \frac{i}{2}(1-\rho_5)\theta(X)T\psi$, (ii) $\delta A^M_\mu \partial_\mu\theta(X) + A\wedge\theta$ and (iii) $\delta X^\mu = 0$. Here θ is the gauge parameter. In order to check the presence of any anomaly, in the form of noninvariance of the fermionic path-integral measure, we define the Euclidean Dirac operator

$$\rho^\alpha D_\alpha = \rho_\alpha \left[\partial_\alpha - \frac{i}{2}(1-\rho_5)a_\alpha\right], \quad (61)$$

where $a_\alpha = \partial_\alpha X^\mu T^M A^M_\mu$, suppressing the indices with the understanding that a_α is a matrix. Note that caution has to be exercised to compute anomalies in presence of 'γ_5' couplings. First write

$$\rho_\alpha D_\alpha = \rho_\alpha \left[\partial_\alpha - \frac{i}{2}v_\alpha + \frac{i}{2}\rho_5 a_\alpha\right], \quad (62)$$

with the prescription that we set $v_\alpha = a_\alpha$ at the end of the calculations. It is necessary to analytically continue $a_\alpha \to i a_\alpha$ in order to make the operator hermitian; however, we rotate it back at the end of the computation. The standard trick is to expand both ψ and $\bar\psi$ in terms of a complete set of the eigenfunctions of the hermitian Dirac operator defined by

$$\psi = \sum_n c_n \phi_n, \quad \bar\psi = \sum_n d_n \phi_n^\dagger, \quad (63)$$

where ϕ_n correspond to complete set of functions satisfying the Dirac equation $\rho_\alpha D_\alpha \phi_n = \lambda_n \phi_n$, and the Dirac operator is the one which is obtained after going through the aforementioned steps. Now, the path-integral measure becomes

$$D\bar\psi D\psi = \prod_n dc_n dd_n. \quad (64)$$

The gauge transformation introduces a change in the fermionic measure,

$$D\bar\psi' D\psi = [(\det\tilde{C}_{nm})(\det C_{nm})]^{-1} D\bar\lambda D\lambda, \quad (65)$$

where

$$C_{nm} = \delta_{nm} + \int d^2\sigma \phi_n^\dagger(\sigma)\frac{i(1-\rho_5)}{2}(\theta T)\phi_m, \quad (66)$$

$$\tilde{C}_{nm} = \delta_{nm} - \int d^2\sigma \phi_n^\dagger(\sigma)\frac{i(1+\rho_5)}{2}(\theta T)\phi_m. \quad (67)$$

In order to evaluate $\det C_{nm}$ we use (66), and write it in the following form for infinitesimal gauge transformation:

$$\det C_{nm} = \exp\left[\mathrm{Tr}\sum_n \int d^2\sigma \phi_n^\dagger(\sigma)\frac{i}{2}(1-\rho_5)\theta T\phi_n(\sigma)\right]. \quad (68)$$

This equation can be rewritten as

$$\begin{aligned}\det C_{nm} &= \lim_{M^2\to\infty} \exp\left[\mathrm{Tr}\sum_n \int d^2\sigma {\phi_n}^\dagger(\sigma)\frac{i}{2}(1-\rho_5)\theta T\phi_n(\sigma)e^{-\frac{{\lambda_n}^2}{M^2}}\right] \\ &= \lim_{M^2\to\infty} \exp\left[\mathrm{Tr}\sum_n \int d^2\sigma {\phi_n}^\dagger(\sigma)\frac{i}{2}(1-\rho_5)\theta T\phi_n(\sigma)e^{-\frac{\mathcal{D}^2}{M^2}}\right],\end{aligned} \tag{69}$$

where $\mathcal{D} = \rho_\alpha D^\alpha$. The expression for det C_{nm} can be evaluated by using the completeness of states, with the aid of some identities specific to two dimensions. At the end, we analytically continue back $a_\alpha \to -ia_\alpha$ and eventually set $v_\alpha = a_\alpha$. The determinant (69) becomes

$$\det C_{nm} = \exp\left[\frac{i}{8\pi}\int d^2\sigma\theta^M(X)(-i\partial^\alpha {a^M}_\alpha - \epsilon_{\alpha\beta}\partial^\alpha a^{M\beta})\right]. \tag{70}$$

A rather simple and analogous calculation shows that

$$\det \tilde{C}_{nm} = 1. \tag{71}$$

The change in the path-integral measure assumes the form

$$S_{\text{anomalous}} = -\frac{1}{16\pi}\int d^2\sigma\theta^M(X)(\partial_\alpha a^{M\alpha} - \epsilon^{\alpha\beta}\partial_\alpha a^M_\beta)\Big]. \tag{72}$$

We use the freedom to add a local counter term

$$S_{CT} = -\frac{1}{32\pi}\int d^2\sigma {a^M}_\alpha a^{M\alpha}, \tag{73}$$

whose gauge variation will cancel the first term of $S_{\text{anomalous}}$. We use the definition of a_α and then after some algebra we arrive at

$$S_{\text{anomalous}} = -\frac{1}{16\pi}\int d^2\sigma\epsilon^{\alpha\beta}\partial_\alpha X^\mu\partial_\beta X^\nu\partial_\mu\theta^M A^M_\nu. \tag{74}$$

We conclude that the gauge coupling for such a theory will be inconsistent since the path-integral measure is affected by an anomaly. We recall that the string also couples to the antisymmetric two-form, $B_{\mu\nu}(X)$:

$$S_B = \frac{1}{32\pi}\int d^2\sigma B_{\mu\nu}(X)\epsilon^{\alpha\beta}\partial_\alpha X^\mu\partial_\beta X^\nu. \tag{75}$$

Notice the slightly different normalization factor on the *RHS*. The gauge invariance is restored if we demand that the two-form B-field transforms as

$$\delta B_{\mu\nu} = \partial_{[\mu}\theta^M(X)A^M_{\nu]} \tag{76}$$

under nonabelian gauge transformation, as was implemented on A_μ^M. We recall that the field strength of $B_{\mu\nu}$, $H_{\mu\nu\lambda}$ will require the addition of a gauge Chern–Simons (C–S) term in its transformation (76). Thus, we notice that the coupling of a gauge background alone to a chiral sector (in the worldsheet action) introduces an anomalous term in the fermionic path-integral measure. This piece can be removed by introducing a local counter term, and demanding that the two-form B-field transforms in a specific manner under the nonabelian gauge transformation associated with the gauge background. In turn, the field strength $\mathbf{H}$ of the B-field is required to be modified by a C–S term. As emphasizes earlier, our Ward identities are valid modulo anomalies, and the above illustrative example shows how such anomalies could be computed in the path-integral context.

4 Symmetries of Massive String Excitations

We have elucidated how the local symmetries in target space could be unraveled by introducing canonical transformations in phase space. This was achieved within the first-quantized approach to string theory. There are reasons to believe that we are yet to unveil hidden symmetries (higher symmetries) of string theory. There are hints about existence of such symmetries from the exponential degeneracy of excited string states, and from the description of very high-energy collision processes in their string theoretic descriptions [20, 21]. Therefore, it is argued that discovering and understanding such higher symmetries will provide us with deeper insight of string theory. It is natural to ask how does one go about exploring such symmetries. We adopt the conventional approach, in the sense that we look for classical symmetries. These are easy to understand. Subsequently, we examine these symmetries in a quantum context. As we illustrated earlier, the symmetry might be affected by anomalies, leading to the breakdown of the symmetry. There are reasons to explore in these directions. It is recognized that string-theory vacua are embarrassingly rich. It is not unreasonable to speculate that the discovery of new stringy symmetries might provide us a way to identify the vacuum which describes the low-energy standard model, as well as the Universe we live in. It is quite natural to presume that string field theory will encode the symmetries of string theory in their totality. Therefore, this seems to be the right setting to seek answers to the questions raised earlier. There are some hints that string field theory could be the right forum to address these issues. However, string field theory has not fully developed efficient techniques to carry out practical calculations. Therefore, our approach is based on a more pragmatic first-quantized formulation. Indeed, we envisage the evolution of a string in the background of its massive states, generalizing the two-dimensional worldsheet σ-model action for massless backgrounds.

We proceed to explore the higher symmetries following our experience with massless excitations. We work in a classical framework where, quite interestingly, we can get glimpses of the underlying symmetries. Moreover, even in this simple approach, we can safely understand some of the features of these symmetries. Let us first briefly recall some salient and relevant features of the local symmetries we have studied so far. We constructed the Hamiltonian action, S_H, for the string in the background of its massless excitations $G_{\mu\nu}, B_{\mu\nu}$ and A_μ^M. We introduced generators of canonical transformations associated with general coordinate transformation, gauge symmetry of $B_{\mu\nu}$ and nonabelian gauge symmetries. Note that, when implementing transformations associated with GCT, the vertex involved both $B_{\mu\nu}$ and A_μ^M, since S_H has pieces where $G_{\mu\nu}$ couples to these backgrounds, and the vertex is a functional derivative of S_H with respect to $G_{\mu\nu}$. The same arguments can be repeated when we derive WI for the other two massless excitations. The point to be emphasized is that, if we perform any one of the three canonical transformations introduced in Sect. 3, we always obtain only the three backgrounds and their vertex functions. In other words, the canonical transformations are such that we do not have to introduce additional terms in the Hamiltonian action, corresponding to new vertex functions, when we look for its invariance properties. Let us consider infinitesimal general coordinate transformations of the form

$$\delta_\xi X^\mu = \xi^\mu(X), \quad \text{and } \delta_\eta X^\nu = \eta^\nu(X), \tag{77}$$

such that a transformation of this kind acting on a function $f(X)$ produce the following variation:

$$\delta_\xi f(X) = f(X+\xi) - f(X) = f_{,\lambda}\, \xi^\lambda. \tag{78}$$

Thus, two transformations involving the shift of X^μ will result in

$$\delta_\eta\delta_\xi f(X) = f_{,\lambda\rho}\, \xi^\lambda \eta^\rho + f_{,\lambda}\, \xi^\lambda{}_{,\rho}\, \eta^\rho. \tag{79}$$

Therefore, it is straightforward to verify that

$$(\delta_\eta\delta_\xi - \delta_\xi\delta_\eta) = \delta_{\xi\eta}, \tag{80}$$

where $\delta_{\xi\rho} = \xi^\lambda{}_{,\rho}\, \eta^\rho - \eta^\rho{}_{,\lambda}\, \xi^\lambda$. Therefore, two such coordinate transformations result in another coordinate transformation operation as easily demonstrated.

Let consider a generalized form of shift of the string coordinates,

$$\tilde\delta X^\mu = \xi^\mu + \partial X^\nu \bar\partial X^\rho \zeta^\mu_{\{\nu\rho\}}, \tag{81}$$

where ∂ and $\bar\partial$ are derivatives defined in terms of worldsheet complexified coordinates. The notation $\{\mu\nu\}$ is to be interpreted as symmetrization with respect to the two indices μ and ν, here and everywhere. We consider a worldsheet action in the presence of a tachyon and graviton background only (and ignore all others):

$$S = \int d^2z \Big[T(X) + (\partial X^\mu \bar\partial X^\rho + \partial X^\rho \bar\partial X^\mu) G_{\mu\rho}(X)\Big]. \tag{82}$$

Under this new shift transformation (81), the tachyon part simply gets transformed to

$$\tilde{\delta}T(X) = T(X)_{,\mu}\,\xi^{\mu} + T(X)_{,\mu}\,\partial X^{\nu}\bar{\partial}X^{\rho}\zeta^{\mu}_{\{\nu\rho\}}. \tag{83}$$

Notice the form of the last term in the above equation: when the tachyon background is varied, the produced extra piece looks like a graviton vertex, since $\partial X^{\nu}\bar{\partial}X^{\rho}$ couples to the graviton $G_{\nu\rho}(X)$. At this stage, we can already notice an interesting feature. Suppose we had consider string in a tachyonic background alone, and implemented the new shift (81). Then the variation of the action will generate a graviton-like vertex. If we want to apply the arguments of the previous section, then we shall have to introduce a graviton vertex to see if we can compensate this shift by a generalized form of GCT, and obtain a relation $\delta^{new}S_H = -\delta^{new}_{GCT}S_H$, just as we had $\delta_G S_H = -\delta^{GCT}S_H$. We shall show that the variation of the graviton vertex under (81) yields some interesting features. Note that δX^{μ}, in this context, has two parts: one that corresponds to the usual GCT, and another piece which we have introduced. It will be argued that the second piece could be associated with a (local) higher symmetry transformation.

Let us consider the variation of the graviton vertex,

$$\tilde{\delta}\left[(\partial X^{\mu}\bar{\partial}X^{\rho} + \partial X^{\rho}\bar{\partial}X^{\mu})G_{\mu\rho}\right], \tag{84}$$

with $\tilde{\delta}X^{\lambda} = \xi^{\lambda}(X) + \zeta^{\lambda}_{\{\nu\rho\}}(X)\partial X^{\nu}\bar{\partial}X^{\rho}$. Under the new coordinate transformations, partial derivatives of X^{μ} will transform in two ways: they will get shifted in the usual form due the ξ-shift, and a new shift will be added due to ζ-part in those derivatives. The graviton background will also vary as its argument X undergoes the shifts. It is quite obvious that the ζ-part is going to add some extra pieces to (84), since it contains derivatives of the string coordinates (crudely speaking these are (1,1) operators). Thus, we can see that there will be terms like $\partial X^{\mu}\partial X^{\nu}\bar{\partial}X^{\rho}$ contracted with ζ, and so on. We shall discuss about them later. Let us look at the coefficients of $\partial X^{\mu}\bar{\partial}X^{\nu}$ which appear after the variation operation has been performed in (84). Note that the usual shift of the form $X^{\mu} \to X^{\mu} + \xi^{\mu}(X)$ gives rise to a variation $\delta_{\xi}(\partial X^{\mu}) = \xi^{\mu}{}_{,\lambda}(X)\partial X^{\lambda}$. Therefore, under this variation,

$$\begin{aligned}\delta_{\xi}[(\partial X^{\mu}\bar{\partial}X^{\nu} + \partial X^{\nu}\bar{\partial}X^{\mu})G_{\mu\nu}] &= \xi^{\mu}{}_{,\lambda}\,(\partial X^{\lambda}\bar{\partial}X^{\rho} + \partial X^{\rho}\bar{\partial}X^{\lambda})G_{\mu\rho} \\ &\quad + (\partial X^{\mu}\bar{\partial}X^{\rho} + \partial X^{\rho}\bar{\partial}X^{\mu})G_{\mu\rho,\lambda}\,\xi^{\lambda}. \end{aligned} \tag{85}$$

This is the usual variation which was the starting point for the derivation of gravitational WI, and which we obtained through the generator of canonical transformation, Φ_G. However, there is another piece in (84), $\zeta^{\mu}_{\{\nu\rho\}}\partial X^{\nu}\bar{\partial}X^{\rho}$. We have argued earlier that the variation induced by δ_{ζ} on $\partial X^{\nu}\bar{\partial}X^{\rho}G_{\nu\rho}$ takes it away from the form of the graviton vertex. However, recall that the variation of the tachyon background under this shift is of the form of a graviton vertex.

To briefly summarize, the effect of the $\tilde{\delta} = \delta_\xi + \delta_\zeta$ shift transformation is that there is a piece in the transformed graviton vertex which is in the form of a graviton vertex multiplied by $\xi,^\mu_\lambda$ (coming from δ_ξ variation), and another piece coming from the δ_ζ variation of tachyon. This variation in the action can be compensated by the following variation of the metric:

$$\tilde{\delta}^{GCT} G_{\mu\rho} = G_{\mu\lambda}\xi^\lambda,_\rho + \xi^\lambda,_\mu G_{\rho\lambda} + \xi^\lambda G_{\mu\rho},_\lambda + \zeta^\alpha_{\{\mu\rho\}} T,_\alpha\,, \tag{86}$$

in the sense that if we ignore the presence of other terms arising due to the $\tilde{\delta}$ variation, we could derive a new WI. We can use the relation $\tilde{\delta} S_H = -\tilde{\delta}^{GCT} S_H + ...$ where ellipses stand for the terms we have ignored.

Let us now look at the coefficients of the parameter $\zeta^\mu_{\{\nu\rho\}}$, which will be obtained from the variation of the graviton vertex under δ_ζ:

$$\begin{aligned}
&\left[\partial X^\nu \bar{\partial} X^\eta \zeta^\mu_{\{\nu\eta\}},_\lambda \partial X^\lambda + \partial\partial X^\nu \bar{\partial} X^\eta \zeta^\mu_{\{\nu\eta\}} + \partial X^\nu \partial\bar{\partial} X^\eta \zeta^\mu_{\{\nu\eta\}}\right] \bar{\partial} X^\rho G_{\mu\rho} \\
&+ \partial X^\mu \left[\partial X^\nu \bar{\partial} X^\eta \zeta^\rho_{\{\nu\eta\}},_\lambda \bar{\partial} X^\lambda + \partial\bar{\partial} X^\nu \bar{\partial} X^\eta \zeta^\rho_{\{\nu\eta\}} + \partial X^\nu \bar{\partial}\bar{\partial} X^\eta \zeta^\rho_{\{\nu\eta\}}\right] G_{\mu\rho} \\
&+ \mu \leftrightarrow \rho + (\partial X^\mu \bar{\partial} X^\rho + \bar{\partial} X^\mu \partial X^\rho) G_{\mu\rho},_\lambda \zeta^\lambda_{\{\eta\alpha\}} \partial X^\eta \bar{\partial} X^\alpha .
\end{aligned} \tag{87}$$

Let us examine the structure of the terms appearing in the above equation. Suppressing target space indices we may write them as (i) $\partial X \bar{\partial} X \partial X$, (ii) $\partial\bar{\partial} X$ and (iii) terms where we interchange $\partial \leftrightarrow \bar{\partial}$ in (i) and (ii); and there is finally the term (iv) $\partial X \bar{\partial} X \partial X \bar{\partial} X$. We recall that we use equations of motion when we derive Ward identities: therefore, pieces appearing in the category (ii) will vanish due to the on-shell condition. The appearance of these types of terms (after equations of motion are implemented) forces us to think that we must add additional vertex operators to the worldsheet action if we are to use our technique to derive WI associated with the δ_ζ shift of the coordinates. Therefore, we include the vertex operators corresponding to the first excited massive states, which assume the form

$$\begin{aligned}
&F^{(1)}_{\mu\nu\rho'} \partial X^\mu \partial X^\nu \bar{\partial}\bar{\partial} X^{\rho'} + F^{(2)}_{\mu\nu'\rho'} \partial\partial X^\mu \bar{\partial} X^{\nu'} \bar{\partial} X^{\rho'} \\
&+ S_{\{\mu\nu\}\{\rho'\eta'\}} \partial X^\mu \partial X^\nu \bar{\partial} X^{\rho'} \bar{\partial} X^{\eta'} .
\end{aligned} \tag{88}$$

Let us now examine how the terms appearing in the above equation should transform under the combined shifts δ_ξ and δ_ζ. We know the rules for transformations of ∂X and $\bar{\partial} X$ already. The three index background undergoes a variation $\delta F^{(1)}_{\mu\nu\rho'} = F^{(1)}_{\mu\nu\rho'},_\lambda \delta X^\lambda$,

$$\delta F^{(1)}_{\mu\nu\rho'} = F^{(1)}_{\mu\nu\rho'},_\lambda \xi^\lambda + F^{(1)}_{\mu\nu\rho'},_\lambda \partial X^\kappa \bar{\partial} X^\eta \zeta^\lambda_{\{\kappa\eta\}} . \tag{89}$$

Similarly, we can obtain the variation of $F^{(2)}_{\mu\nu'\rho'}$. Let us look at the vertex operator associated with $F^{(1)}$, and note what will be its transformed form after the variations of ∂X, $\bar\partial X$ and $F^{(1)}$. It is easy to see that there will be some pieces coming from the δ_ξ variation that will look like the ones coming from the δ_ζ variation of the graviton vertex. Thus a symmetry associated with local ζ-type shift needs the introduction of $F^{(1),(2)}$ backgrounds. Note, however, that the ζ-variation of F's already tells us that we need introducing vertex operators associated with still higher-level states.

We argue more qualitatively below, rather than presenting our explicit lengthy algebra. Notice the first term of (88), which gives the coupling of a string to the $F^{(1)}$ background. When we consider the variation of this background due to the ξ-shift, there will be one term which will be associated with the variation of a the massive background with four indices (the S-field appearing in (88)). Under usual GCT in the target space, $F^{(1)}$ transforms as a tensor. Now we focus attention on the vertex involving the four-index background: $S_{\{\mu\nu\}\{\rho'\eta'\}}\partial X^\mu \partial X^\nu \bar\partial X^{\rho'} \bar\partial X^{\eta'}$. As before, the δ_ξ-variation of this vertex will consists of several terms: there will be terms which will have pieces like $\partial X \partial X \bar\partial X \bar\partial X$, and also terms which are the product of five pieces involving $\partial X \bar\partial X$..... Again we see that the usual δ_ξ-shift already is seeking the presence of higher massive level vertex. If we consider the consequences of δ_ζ-shift on the aforementioned vertex, we immediately realize that we have to add more vertices corresponding to even higher massive levels. Therefore, a simple δ_ξ-shift already requires presence of higher states, and hints at a hidden symmetry. It is quite interesting to explore the consequences of inducing the shifts we have considered so far. However, we cannot make any headway beyond a certain limit, since testing our proposition through explicit computations becomes unmanageable. Moreover, there is no reason to consider only δ_ξ and δ_ζ shifts. One could consider a more general form such as $\delta_\Sigma X^\mu = \partial X^\rho \partial X^\alpha \bar\partial X^{\eta'} \Sigma^\mu_{\{\rho\alpha\}\eta'}$ which is generic. Obviously, one has to add other suitable terms to this expression. We can conclude that the presence of δ_Σ will transform tachyon in such a way that we shall need higher massive vertices by looking at the tachyon background variation alone. Therefore, there are two different avenues opening up if we generalize our original prescription (successfully utilized for massless backgrounds), to investigate the symmetries of string theory. (i) We add vertex functions corresponding to massive string states and generalize the δ_ξ-shift by adding one extra piece, i.e. δ_ζ shift. The consequences have been already discussed. (ii) We can generalize the δ_ξ shift by adding all possible allowed terms, and immediately note that such generalization also requires addition of additional vertex functions to the worldsheet action. Both paths lead to the conclusion that string theory is endowed with higher symmetries.

It is worthwhile exploring additional properties of the transformations δ_ξ and δ_ζ. The former is associated with the GCT, and we have seen that two such successive GCT transfomations correspond to another GCT transformation. Let us closely look at δ_ξ and δ_ζ transformations. A simple explicit calculation will illustrate the point:

$$\left(\delta_\zeta\delta_\xi - \delta_\xi\delta_\zeta\right)X^\mu = \xi^\mu{}_{,\alpha}\,\partial X^\kappa\bar{\partial}X^{\sigma'}\zeta^\alpha_{\{\kappa\sigma'\}} - \xi^\alpha{}_{,\beta}\,\zeta^\mu_{\{\alpha\eta'\}}\partial X^\beta\bar{\partial}X^{\eta'}$$
$$- \xi^{\eta'}{}_{,\beta}\,\zeta^\mu_{\{\alpha\eta'\}}\bar{\partial}X^\beta\partial X^\alpha - \zeta^\mu_{\{\alpha\eta'\}}{}_{,\beta}\,\xi^\beta\partial X^\alpha\bar{\partial}X^{\eta'}. \quad (90)$$

It is obvious that the above rule of operation is applicable to any function of X, $f(X)$, or to any tensor which depends on X. Therefore, we conclude that

$$\left[\delta_\zeta\delta_\xi, \delta_\xi\delta_\zeta\right] = \delta_{\hat{\zeta}}, \quad (91)$$

where $\hat{\zeta}^\mu_{\{\rho\sigma\}} = \zeta^\lambda_{\{\rho\sigma\}}\xi^\mu{}_{,\lambda} - \zeta^\mu_{\{\lambda\rho\}}\xi^\lambda{}_{,\sigma} - \zeta^\mu_{\{\lambda\sigma\}}\xi^\lambda{}_{,\rho} - \zeta^\mu_{\{\rho\sigma\}}{}_{,\lambda}\,\xi^\lambda$, and it is to be understood in the light of (90). Therefore, a usual GCT followed by a ζ-shift is still a combined operation of these two shift. However, two combined ζ-shift operations will take us out of a ζ-shift, and will signal that we have to introduce higher transformations. If we continue to repeat this process, we will have to introduce a hierarchy of higher and higher shifts.

Our discussion has been confined to the classical level only. The generators of canonical transformations that we have introduced for deriving WI associated with massless backgrounds have not been restricted. Therefore, they take us from the phase-space manifold of a string to another domain which is huge indeed. There is a way to constraint the choice of the generators, to some extent. We argued than the canonical transformations are associated with undelying symmetries. In the context of string theory, we interpret symmetry transformation as existence of physically indistinguishable solutions to the string equations of motion of the backgrounds. Thus, the transformed backgrounds and phase-space variables correspond to isomorphic conformal field theories. Rephrased in another way, with each solution of the string equations of motion there is a two-dimensional conformally invariant theory. Moreover, this theory is defined by specifying the phase-space variables and the generators $L_\pm$. As mentioned earlier, the spacetime fields are the coupling constants of the σ-model. The couplings of the backgrounds to the string are identified as vertex operators. We note that the vertex should be BRS invariant in order to fulfill the requirements of conformal invariance of the theory. When we implement canonical transformation, not only phase-space variables but also vertex operators are transformed in a specified way. However, these vertex operators must be BRS invariant too. In turn, this condition imposes constraints on the choice of the genators.

Let us look at the graviton vertex operator $V_G = \partial X^\mu\bar{\partial}X^\nu G_{\mu\nu}(X)$, which will get transformed according to the rules we have given. For infinitesimal transformations we have $\tilde{V}_G = V_G + \delta V_G$, where $\delta V_G = \{V_G, \Phi_G\}_{PB}$ and $\delta Q = \{Q, \Phi\}_{PB}$. However, $\{Q, V_G\}_{PB} = 0$. If a generator corresponds to a symmetry of the theory, then it should commute with the BRS charge, i.e. $\{Q, \Phi_G\}_{PB} = 0$. Thus, $\{\delta V_G, Q\}_{PB} = 0$. The generator is already constrainted by such requirements. Let us consider the case of weak graviton background, i.e. $G_{\mu\nu} = \eta_{\mu\nu} + h_{\mu\nu}$. It is well-known that the BRS invariance yields the equation of motion, $\nabla^\mu\nabla_\mu h_{\alpha\beta}=0$, and the transversality condition, $h_{\mu\nu},_\mu = h_{\mu\nu},_\nu = 0$. We mention that a lot of care must be taken in actual computations since the quantum operators must be defined keeping in mind issues like operator ordering and other delicate points. It is relatively simple nowadays to derive equations of motion for massless backgrounds using standard techniques.

It is a more difficult task to fully analyse the consequences of conformal invariance when excited massive levels couple to the string. There have been attempts to study such cases [22] from the perspective of conformal field theory (CFT). As in case of the graviton coupling we obtain equation of motion and transversality condition by demanding that the vertex operator be of dimension $(1,1)$. To simplify matter further, let us look at a subset of coupling of the first excited massive levels. The $F^{(1)}$ coupling in (88) is slightly modified [22] if we look for a vertex of the type $(1,0)$. The desired piece is

$$\mathbf{F}^{(1)} = F^{(1)}_{\mu\nu\lambda'}\partial X^\mu \partial X^\nu \bar\partial X^{\lambda'} + \partial_\mu F^{(1)\mu}_{\nu\lambda'}\partial\partial X^\nu \bar\partial X^{\lambda'}, \tag{92}$$

and here ∂_μ is a partial derivative with respect to the spacetime coordinates. This vertex is $(1,0)$ if

$$F^{(1)\mu}_{\mu\lambda'} + 2\partial_\mu\partial_\nu F^{(1)\mu\nu}_{\lambda'} = 0, \tag{93}$$

$$\partial_{\lambda'} F^{(1)\lambda'}_{\mu\nu} = 0, \tag{94}$$

and

$$\nabla^\mu\nabla_\mu F^{(1)}_{\mu\nu\lambda'} = 0. \tag{95}$$

A similar argument is to be adopted to include additional terms in (88), when we demand that the corresponding operators should be of the $(0,1)$ type. If we want to derive 'equations of motion' for other backgrounds, we could proceed along these lines. It is assumed that a string is evolving in the flat Minkowski background and that the vertex operators are to be added to the worldsheet action in the weak field approximation. In a more general setting, the 'beta' function equations for massive excited backgrounds can be computed by adopting the technique of Riemann normal coordinate expansion (as is customary for massless tensor background fields). It is quite obvious that such computations are not easy, although can be carried out, in principle, using the conformal field theory techniques.

We would like to draw attention to one more important feature of our approach in deriving the Ward identities. It will be neccessary to introduce a prescription to define off-shell amplitudes within the first quantized framework. Let us condider a process such as the scattering graviton + graviton → tachyon + tachyon. Note that we shall have to go off-shell for this reaction. However, if we consider the $(n+1)$-point amplitude with a single graviton and n-tachyons, then we can write down WI in a field-theoretic framework. The amplitude will require an off-shell description. If we are interested in computing such an amplitude in string theory, going off-shell will amount to introducing a conformal factor, with a specific prescription, in the amplitude, and this prescription is not necessarily unique. We already know (from our experiences in quantum field theory with gauge symmetries) that a gauge fixing parameter appears in the general n-point functions. It is the S-matrix element which is required to be gauge invariant. In the context of string theory, a conformal factor (signaling the presence of a gauge fixing parameter) is to be introduced. Kubota and Veneziano [23] have suggested a prescription to construct off-shell amplitudes, and one of their motivations is to get a handle on the n-point amplitudes involving multi tachyons and massless stringy excitations. This programme is far from being complete, although its importance is recognized. We may stress that the importance of string field theory is realized at every stage when we venture to address the issues alluded to earlier.

5 Summary and Conclusions

The goal of this article is to unveil and investigate symmetries of string theory in its first quantized formulation. We adopted the point of view that the worldsheet action for a string in the background of its massless excitations encodes the local symmetries of the theory. The background fields may be envisaged as coupling constants of the σ-model. The conformal invariance of the theory, leading to the equations of motion, already provides us with some clue about the local symmetries of the target space. It becomes more transparent when we construct the effective action from the equations of motion which is manifestly invariant under target space local symmetries. It is worth while to note that higher-order corrections force us to add higher derivative terms in the background fields; however, the local symmetries are maintained at each order of the perturbation theory. One may intuitively claim that the worldsheet description inherently contains those target space symmetries.

The Hamitonian path integral formulation in phase space, adopted here, provides an elegant technique to expose the target space symmetries. Our derivation of the Ward identities is based on formal arguments, and could be interpreted as a classical result. We have followed the traditional avenue, in the sense that our first goal is to study the classical symmetries, and then quantize the theory. We are aware that the WI we derived might be violated due to the presence of anomalies. Indeed, we presented an example where the conservation law is anomalous, and the anomaly was computed within the path integral framework. In the process, we discovered that even the compatified bosonic string requires, under certain circumstances, Green–Schwarz-type Chern–Simons term in the definition of the three-form **H**-field. Our point is that the invariance properties of the measure, a potential source of anomaly, could be analysed using standard prescriptions, at least in principle. We found that a simple generalization of the usual coordinate shift (associated with GCT) leads to a very interesting structure. In the first place, when we include tachyonic background, the generalized shift conveys that GCT gets modified when we adopt our technique to derive WI for graviton–tachyon backgrounds. At the same time, inclusion of δ_ζ in the shift of coordinate already signals the presence of a hierachy of new symmetries, as described in the text. We saw that contributions of excited massive states, in the form of vertex operators, are required in the σ-model action. In fact, it is not possible to truncate the contributions of these terms. Moreover, we analysed the action of the δ_ξ and δ_ζ operating on string coordinates and on functions of string coordinates. We notices that the algebra really does not close. We have already conjectured that the massive stringy states might acquire their masses due to some spontaneously broken gauge symmetry.

It is interesting to note that the symmetries associated with higher excited states of $c = 1$ string theory have been studied a lot in the past. Let us recall that a one-dimensional string coupled to a gravitational background has a two-dimensional target space interpretation. We require that the central charge value be 26. This is achieved by introducing a background charge and, as a consequence, the Virasoro generators and the BRS charge get modified accordingly. In this picture, the general couplings are functions of two variables, one of them being the conformal mode of the two-dimensional target space metric. The ground state is the tachyon. However, this theory also contains an infinite set of discrete states [24]. Therefore, a σ-model action can be constructed involving these states. In this model, we may envisage the possibility of inducing canonical transformations for these states and seek for associated symmetries. Indeed, the $c = 1$ theory is endowed with a W_∞ symmetry [25], and the generators of this transformation are the generators of the W_∞ algebra. In fact, these symmetries have a very nice interpretation as gauge transformations when they are analysed from the perspective of string field theory, suitably formulated for $c = 1$ string theory [26]. This is a very interesting and encouraging result for us. Now the question arises whether, for the bosonic string in critical dimensions, we can unravel higher symmetries from a string field theory perspectives.

We have a partial answer to this question [27]. One sets out with a non-polynomial formulation of string field theory. This string field action is known to be invariant under infintesimal gauge transformations. We recall that the string field theory action can be expanded in terms of component fields; similarly, the corresponding gauge parameter will also have an expansion. It was argued that a specific choice of the gauge function, Λ, can be made, such that this gauge transformation corresponds to a canonical transformation when viewed from the first quantized σ-model description in the worldsheet [27]. Since string field theory can account for off-shell descriptions as well, this gauge transformation is expected to be more general. Of course, a most general gauge transformation could not be implemented due to technical reasons; however, a linearized version was adopted to check the gauge transformation properties of the first few massive levels. Moreover, the gauge functions could be identified for a few levels explicitly. It was possible to identify the gauge transformations, associated with some of the low-lying massive states, with canonical transformations of massive levels in the first quantized descriptions. At this stage one was not able to compute the algebra of the generators and identify the underlying algebra, as was the case with the $c = 1$ theory where W_∞ symmetry was identified as the underlying symmetry. The generators satisfied the corresponding algebra. Nevertheless, it is hearting to unravel the presence of a hierarchy of symmetries in the critical bosonic string; moreover, the presence of W_∞ algebra in $c = 1$ strings also gives us a clue that the critical bosonic string might be endowed with a rich underlying symmetry.

Acknowledgments

I would like to thank Ashok Das and Nick Mavromatos for useful discussions over the years. I especially thank T. Kubota for sharing his unpublished work and his valuable notes with me. It has been a rewarding experience to know Gabriele as a collaborator, as a collegue and as a friend. I have immensely benefited from long discussions with him, from his very deep insights in physics and from his human values. I wish him many more happy, prosperous and productive years ahead.

References

1. M. B. Green, J. H. Schwarz, E. Witten: *Superstring Theory* (Cambidge University Press, Cambridge, 1987); J. Polchinski: *String Theory* (Cambridge University Press, Cambridge, 1998)
2. E. S. Fradkin, A. A. Tseytlin: Phys. Lett. B **158**, 316 (1985); Nucl. Phys. B **261**, 1 (1985)
3. C. Lovelace: Phys. Lett. B **135** , 75 (1984); Nucl. Phys. B **273**, 413 (1986); A. Sen: Phys. Rev. D **32**, 2102 (1985) C. G. Callan, D. Friedan, E. J. Martinec, M. J. Perry: Nucl. Phys. B **262**, 593 (1985)
4. G. Veneziano: Phys. Lett. B **167**, 387 (1986)
5. J. Maharana, G. Veneziano: Phys. Lett. B **169**, 177 (1985); Nucl. Phys. B **283**, 126 (1987)
6. J. Maharana, G. Veneziano (unpublished work 1986)
7. J. Maharana, G. Veneziano (unpublished work 1991 and 1993)
8. I. A. Batalin, G. A. Vilkovisky: Phys. Lett. B **69**, 309 (1977)
9. I. A. Batalin, E. S. Fradkin: Phys. Lett. B **122**, 157 (1983)
10. For a review see M. Henneaux: Phys. Rep. **126**, 1 (1985)
11. A. M. Polyakov: Phys. Lett. B **103**, 207 (1981)
12. M. Kato, K. Ogawa: Nucl. Phys. B **212** , 443 (1983); S. Hwang: Phys. Rev. D **28**, 2614 (1983)
13. I. B. Frenkel, V. G. Kac: Inv. Math. **62**, 23 (1981); I. B. Frenkel: J. Funct. Analysis **44**, 259 (1981); P. Goddard and D. Olive: in *Vertex Operators in Mathematics and Physics*, eds J. Lepowski, S. Mandelstam, I. M. Singer (Springer-Verlag, New York, 1985), p. 51

14. D. J. Gross, J. A. Harvey, E. Martinec, R. Rohm: Nucl. Phys. B **216**, 253 (1985)
15. M. B. Green, J. H. Schwarz: Phys. Lett. B **149**, 117 (1984)
16. E. Witten: Commun. Math. Phys. **92**, 451 (1984)
17. T. Banks, D. Nemshansky, A. Sen: Nucl. Phys. B **277**, 67 (1986)
18. A. Das, J. Maharana, P. Panigrahi: Mod. Phys. Lett. A **8**, 759 (1988)
19. K. Fujikawa: Phys. Rev. Lett. **42**, 1195 (1979); Phys. Rev. D **21**, 2848 (1980); Phys. Rev. D **D29**, 285 (1984)
20. D. J. Gross, P. Mende: Phys. Lett. B **197**, 129 (1987); Nucl. Phys.B **303**, 407 (1988); D. J. Gross: Phys. Rev. Lett **60**, 1229 (1988)
21. D. Amati, M. Ciafaloni, G. Veneziano: Phys. Lett. B **197**, 81 (1987); Int. J. Mod. Phys. A **3**, 1615 (1988); Phys. Lett. B **216**, 41 (1989); Phys. Lett. B **289**, 87 (19989); Nucl. Phys. B **403**, 707 (1993)
22. M. Evans, B. Ovrut: Phys. Rev. D **39**, 3016 (1989); Phys. Rev. D **41**, 3149 (1990); R. Akhoury, Y. Okada: Nucl. Phys. B **318**, 176 (1989)
23. T. Kubota, G. Veneziano: Phys. Lett. B **207**, 419 (1988); and unpublished results.
24. A. M. Polyakov: Mod. Phys. Lett. A **6**, 635 (1991); S. Mukherji, S. Mukhi, A. Sen: Phys. Lett. B **266**, 337 (1991); B. Lian, G. Zuckerman: Phys. Lett. B **266**, 21 (1991; S. Mukherji, S. Mukhi, A. Sen: Phys. Lett. B **266**, 337 (1991)
25. J. Avan, A. Jevicki: Phys. Lett. B **266**, 35 (1991); G. Moore, N. Seiberg: Int. J. Mod. Phys. A **7**, 2634 (1992); S. Das. A. Dhar, G. Mandal, S. R. Wadia: Int. J. Mod. Phys. A **7**, 5165 (1992); I. Klebanov, A. M. Polyakov: Mod. Phys. Lett. A **6**, 3373 (1991); E. Witten: Nucl. Phys. B **373**, 187 (1992); D. Minic, J. Polchinski, Z. Yang: Nucl. Phys. B **369**, 324 (1992)
26. S. Mukherji, S. Mukhi, A. Sen: Phys. Lett. B **275**, 39 (1991)
27. J. Maharana, S. Mukherji: Phys. Lett. B **284**, 36 (1992)

Threshold Effects Beyond the Standard Model

T. R. Taylor

Department of Physics, Northeastern University, Boston, MA 02115, USA
taylor@neu.edu

Abstract. In this contribution to the Festschrift celebrating Gabriele Veneziano on his 65th birthday, I discuss the threshold effects of extra dimensions and their applications to physics beyond the standard model, focusing on superstring theory.

1 Introduction

I am very happy to contribute to the Festschrift celebrating Gabriele Veneziano on his 65th birthday. I have known Gabriele for more than 25 years and worked with him on many projects, learning not only physics, but how to *enjoy* physics. "Amusing" is the word that he often uses to describe interesting ideas, and that single word characterizes best a unique style of *joyful* research that led to his pioneering work on string theory, particle physics and cosmology described in this book. When researching Gabriele's original work on running coupling constants, preceding our 1988 collaboration [1] described below, I ran into a write-up of his lectures on "Topics in String Theory" delivered in 1987 in China and in India [2]. His paper concludes with: "But my moral, I hope, is a clear one for the young string theorist: If string math. is lots of fun, string phys. is no less." Indeed, I had much fun working on string physics over the following 20 years. In this contribution, I discuss the threshold effects of extra dimensions and their applications to physics beyond the standard model, focusing on superstring theory.

2 Threshold Effects of Extra Dimensions

At a given time in the history of elementary particle physics, there is always the mystery of higher energies and the hope of building even more powerful accelerators that would take us one step farther in the understanding of short-distance physics. Thirty years ago, discovering yet another quark was a major

T. R. Taylor: *Threshold Effects Beyond the Standard Model*, Lect. Notes Phys. **737**, 553–560 (2008)
DOI 10.1007/978-3-540-74233-6_17

breakthrough; but now, the next round of experiments can hardly satisfy theorists without uncovering extra dimensions or producing black hole fireballs. Threshold effects appear each time a new particle is discovered. They appear in many physical quantities, signaling transition to new energy domains.

As an example, consider the top quark threshold. We want to see how the QCD coupling constant evolves from the region below the top mass scale m_t, across the threshold, to higher energies. In order to determine the corresponding one-loop correction to the effective action, we can consider the vacuum polarization diagram with two external gauge bosons at momentum scale Q, as shown in Fig. 1. Since we are mostly interested in the effects of quark loops, there is no need to use a full-fledged background field method. This two-point function is

$$\Pi^{\mu\nu}(Q) = i(Q^\mu Q^\nu - Q^2 g^{\mu\nu})\Pi(Q) \, , \tag{1}$$

with

$$\Pi(Q) \approx i \sum_{m_n<\Lambda} \beta_n \int \frac{d^4P}{(2\pi)^4} \frac{1}{P^2+m_n^2} \frac{1}{(P+Q)^2+m_n^2} \, . \tag{2}$$

Here, the sum extends over all particles with masses below the ultraviolet cutoff Λ, and β_n denote the respective beta function coefficients:

$$\beta_n = 2(-1)^{F_n}(\lambda_n^2 - \frac{1}{12})C_n \, , \tag{3}$$

where F_n is the fermion number, λ_n the helicity and C_n the quadratic Casimir in the particle's $SU(3)$ color representation. The momentum dependence of the integral (2) changes at the threshold. In a rough approximation,

$$\begin{aligned} Q \lesssim m_t : \ \Pi(Q) &\approx \Big[\sum_{n:\, m_n<m_t} \frac{\beta_n}{(4\pi)^2} \Big] \ln\Big(\frac{Q}{\Lambda}\Big)^2 + \sum_{n:\, \Lambda>m_n\geq m_t} \frac{\beta_n}{(4\pi)^2} \ln\Big(\frac{m_n}{\Lambda}\Big)^2 , \\ Q \gtrsim m_t : \ \Pi(Q) &\approx \Big[\sum_{n:\, m_n\leq m_t} \frac{\beta_n}{(4\pi)^2} \Big] \ln\Big(\frac{Q}{\Lambda}\Big)^2 + \sum_{n:\, \Lambda>m_n>m_t} \frac{\beta_n}{(4\pi)^2} \ln\Big(\frac{m_n}{\Lambda}\Big)^2 . \end{aligned} \tag{4}$$

Below the threshold, the top quark loop does not participate in the logarithmic running of the coupling constant, which is completely determined by the particle spectrum below m_t. However, its contribution ensures a smooth transition to higher energies, where the coupling runs with the beta function

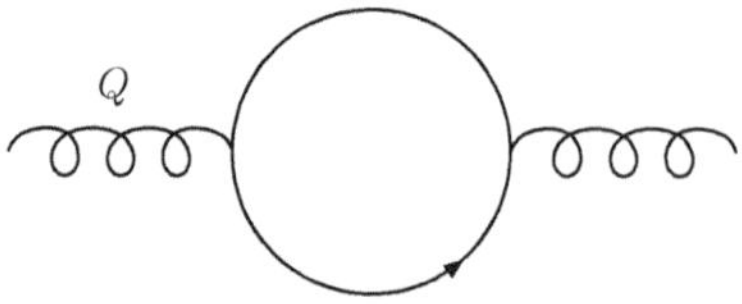

Fig. 1. One-loop contributions to the effective gauge coupling

coefficient including top. While the full renormalization group beta function determines the cutoff dependence of couplings, the finite threshold effects play an important role in the evolution of effective physical couplings. Thus they are very important for all applications involving extrapolations to high energies, in particular in the framework of unification scenarios.

The fact that even a single particle can produce significant threshold effects is very important for grand unification, but it does not excite imagination in a way like the threshold to higher dimensions, envisaged in some Kaluza–Klein (KK) scenarios beyond the standard model [3]. When crossing to higher dimensions, including, say, a circle of radius R, one encounters not just one particle, but an infinite tower of KK excitations with masses $m_n = n/R$ labeled by $n \geq 0$. For each tower with $\beta_n = \beta_0$, the sums in (4), with the threshold mass m_t replaced by $1/R$, split into $0 \leq n < QR$ and $QR < n < \Lambda R$. The latter can be approximated by an integral, giving [1]:

$$\begin{aligned} Q \ll 1/R \quad &: \qquad \Pi(Q) \approx \frac{\beta_0}{(4\pi)^2}\Big[\ln(QR)^2 - 2(\mathcal{N}-1)\Big], \\ 1/R \ll Q \ll \Lambda &: \qquad \Pi(Q) \approx 2\frac{\beta_0}{(4\pi)^2}(RQ - \mathcal{N}), \end{aligned} \tag{5}$$

where $\mathcal{N} = \Lambda R$ is the (large) number of KK excitations below the cutoff Λ. Note that the logarithmic running occurs only below the decompactification scale, and is completely determined by the properties of the massless state at the bottom of the tower. Above the threshold, there is a power (linear) running appropriate to non-compact five dimensions. The logarithmic running is something very special to four dimensions—it is a remnant of infrared divergences that do not appear in higher dimensions. Incidentally, in order to explain why we live in four dimensions, one needs a mechanism that relies on some special properties of $D = 4$; thus infrared divergences are very likely to play a role in such dynamical compactifications [1]. Note that the dominant momentum-independent one-loop threshold correction is of order $\mathcal{O}(\mathcal{N})$.

The computations of one-loop threshold effects can be repeated for more general, possibly anisotropic compactifications, always with the same result that the radius R which determines the scale of logarithmic running should be understood as the largest length scale characterizing the compact space. Thus the onset of power running occurs as soon as the energies approach the first Kaluza—Klein mass.

More recently, large-radius compactifications became quite a popular element of model building beyond the standard model. Although it is a very attractive possibility, it seems to be incompatible with the existence of supersymmetric grand unification suggested by the observed values of gauge coupling constants, which is based on the logarithmic running. However, as shown in [4], it is possible that large threshold corrections can also lead to unification, at lower energy scales, determined by the size of compact dimensions. Here, supersymmetry seems to lose its special appeal; however, it is desirable for another reason. A mechanism based on large one-loop threshold corrections can be reliable only if the higher loop effects are small. Without supersymmetry, there is no reason to expect that this is the case. The common feature of $N = 1$ supersymmetric compactifications is that at heavy Kaluza–Klein levels the spectrum as well as interactions are $N = 2$ supersymmetric. It is well known that $N = 2$ gauge couplings are not renormalized beyond one loop. It can be also shown [5] that the one-loop threshold corrections are dominant, while the higher loop corrections are suppressed, at least by some powers of the tree-level coupling constants.

3 Superstring Threshold Corrections

Threshold corrections appear also in the framework of string theory, which brings two new elements. First, the ultraviolet cutoff is a physical parameter, related to the Regge slope α' that determines the masses of heavy string modes: $\Lambda \approx (\alpha')^{-1/2}$. Thus the cutoff itself becomes a threshold for the production of heavy string modes. Second, the tree-level coupling constants and the radius, as well other geometric quantities characterizing the shape and size of extra dimensions, correspond to vacuum expectation values (VEVs) of certain moduli fields. The moduli parameterize flat directions of the tree-level scalar potential; therefore, the determination of their VEVs is a dynamical problem of "moduli stabilization."

The fact that string theory is ultraviolet finite does not prevent gauge couplings from running which, as explained before, is an infrared effect, and can be studied by using the low-energy effective field theory. A more rigorous, formal treatment of threshold corrections is complicated by the fact that only on-shell amplitudes can be computed by using standard string-theoretical techniques. At the same time when Gabriele was using effective field theory with a string cutoff [2], Kaplunovsky [6] developed a full-fledged formalism for studying threshold corrections in string theory. Then Dixon, Kaplunovsky and Louis [7] studied moduli dependence of string loop corrections in certain heterotic orbifold compactifications.[1] For any untwisted modulus T upon which the threshold corrections Δ do depend non-trivially, the functional form of this dependence is given by

$$\Delta = A \cdot \ln\left(|\eta(T)|^4 \cdot \mathrm{Im}T\right) + T\text{-independent terms} , \tag{6}$$

where A are computable constants determined by the massless spectrum. The Dedekind function is defined by

[1] These computations were later extended to more general orbifolds by Mayr and Stieberger [8]. More recently, Lüst and Stieberger [9] studied gauge threshold corrections in intersecting brane-world models. The formalism for computing threshold corrections to Yukawa couplings has been developed in [10].

$$\eta(T) = e^{\pi i T/12} \prod_{n=1}^{\infty} (1 - e^{2\pi i n T}) \,. \tag{7}$$

It is very interesting to compare (5) and (6). To make it simple, consider a six-dimensional orbifold which is a product of a two-dimensional torus T^2 and "something" four-dimensional, and that T^2 is a product of two circles with radii R_1 and R_2, respectively. For such compactifications, there exists a modulus parameterizing the volume of T^2: $\mathrm{Im}T = R_1 R_2/\alpha' = (\Lambda R_1) \cdot (\Lambda R_2)$. Note that $\mathrm{Im}\, T \sim \mathcal{N}$, measuring also the (approximate) number $\mathcal{N}$ of KK excitations of T^2 with masses below the string cutoff. In the limit of large radii, $\mathcal{N} \to \infty$, and

$$\Delta \sim -A \cdot \frac{\pi}{3} \mathcal{N} \,; \tag{8}$$

thus the string threshold corrections have the same large-radius behavior as a generic sum of KK modes (5), up to a multiplicative constant which is rather ambiguous because it is related to the precise implementation of the mass cut-off on the KK spectrum. However, the coefficients A are non-zero only for the orbifold sectors with $N = 2$ supersymmetry. Furthermore, according to the non-renormalization theorem proved in [11], all higher loop (genus) corrections are zero exactly; thus the full-fledged string computations are perfectly compatible with the effective field theory analysis. A more precise match between the two formalisms has been discussed in [12].

In any closed string theory like the heterotic one, KK modes of each circle are accompanied by strings winding n times around the circle, with masses $m_n = nR/\alpha'$. The spectrum as well as the interactions have a small–large radius symmetry $R \leftrightarrow \alpha'/R$, which is extended in the orbifold compactifications to a full T-duality: $PSL(2, \mathbf{Z})$ modular invariance generated by $T \to -1/T$ and $T \to T + 1$. The threshold correction (6) is $PSL(2, \mathbf{Z})$-invariant. This invariance is realized, however, in quite a non-trivial way. Δ is the coefficient of the kinetic energy terms of gauge bosons so its form is restricted by supersymmetry to be a real part of a holomorphic function of chiral fields. Indeed, at the tree level $g^{-2} = 4\,\mathrm{Re}S$, where S is the dilaton superfield. The presence of $\mathrm{Im}T$ under the logarithm in (6), which is necessary for the modular invariance, is in conflict with that property. Thus the string threshold corrections suffer from a "holomorphic anomaly" [13, 14], which is related to the infrared divergences associated to massless states that *cannot* be described in terms of a *local* effective action [15]. Since then, holomorphic anomalies play an important role in more formal areas of superstring theory, see e.g. [16].

The moduli-dependent threshold corrections have some interesting phenomenological consequences. For example, in some specific orbifold models with the light spectrum below the compactification scale consisting only of the particles belonging to the minimal standard model, a phenomenologically viable gauge coupling unification imposes certain constraints on the modular transformation properties of quark, lepton and Higgs superfields [17].

The fact that gauge (and other) couplings are moduli-dependent may also help is stabilizing the moduli VEVs. In particular, in the context of hidden gaugino condensation mechanism of supersymmetry breaking [18], the scale Λ_{SYM} of gaugino condensation is given, in the two-loop approximation, by

$$\Lambda_{SYM} = \mu\, g^{\beta_1/2\beta_0^2} \exp\left(\frac{-8\pi^2}{\beta_0 g^2}\right), \tag{9}$$

where μ is the scale at which the gauge coupling constant g is defined, and β_0, β_1 are the beta function coefficients of the hidden super Yang–Mills (SYM) sector: $\beta(g) = -\frac{\beta_0}{(4\pi)^2} g^3 - \frac{\beta_1}{(4\pi)^4} g^5 + \ldots$. From Gabriele and Shimon's work [19] we know that gaugino condensation can be described in terms of a simple lagrangian for the composite superfield $\mathcal{W}_\alpha \mathcal{W}^\alpha = \lambda_\alpha \lambda^\alpha + \ldots$, with the coupling constant promoted to a holomorphic function of the dilaton and moduli superfields [20]. In heterotic orbifold compactifications, the moduli dependence is completely determined by modular invariance [21, 22]. As an example, consider a pure SYM hidden sector and focus on the dependence of the Veneziano–Yankielowicz lagrangian on three superfields: $\mathcal{W}_\alpha \mathcal{W}^\alpha$, the dilaton S and one modulus $T = a + iR^2/\alpha'$, where R is the (common) radius of six compact dimensions and a is the associated axion. One finds [21] that the condensation occurs at the expected scale

$$|\langle \lambda_\alpha \lambda^\alpha \rangle| = \Lambda_{SYM}^3\,, \tag{10}$$

with the identification:[2]

$$\mu^2 = \frac{1}{2\,\mathrm{Im}T}\,, \qquad \frac{1}{g^2} = 4\,\mathrm{Re}\big[S + \frac{\beta_0}{(4\pi)^2} \ln \eta(T)\big]\,. \tag{11}$$

The above result can be interpreted by saying that the scale μ is the *infrared* cutoff while g is the Wilsonian coupling constant [23] including the non-anomalous part of threshold corrections (6), due to *massive* KK states with $m_n > \mu$ only. This is the coupling that should be used in the effective four-dimensional field theory at energies below the compactification threshold μ, which from the low-energy point of view becomes an *ultraviolet* cutoff. Indeed, g^{-2} is a real part of a holomorphic function, as required by supersymmetry. However, it is not modular-invariant because the zero mass modes are excluded from loop integrals.

In order to obtain the moduli superpotential generated by gaugino condensation, one integrates out the composite field $\mathcal{W}_\alpha \mathcal{W}^\alpha$. This leads to the following superpotential:

$$W = \exp\Big[-\frac{96\pi^2}{\beta_0}\big(S + \frac{\beta_0}{(4\pi)^2} \ln \eta(T)\big)\Big] = e^{-\frac{96\pi^2}{\beta_0} S} \eta^{-6}(T)\,. \tag{12}$$

The above superpotential transforms under the $PSL(2, \mathbf{Z})$ modular transformations as a form of weight -3, which ensures modular invariance of the

[2] This comparison makes use of the fact that $\beta_1/2\beta_0^2 = -2/3$ in SYM theory.

lagrangian. In fact, its form is determined uniquely by the modular properties and asymptotic behavior, so it can be also derived without necessarily going into details of SYM dynamics. The corresponding scalar potential is modular invariant; therefore, it is symmetric under $R \leftrightarrow \alpha'/R$ and has stationary points at $R^2 = \alpha'$ $(T = i)$. Since its form depends on the details of Kähler potential, which also receives some non-perturbative corrections, it is difficult to prove that T and other moduli are stabilized; however, there are some indications that this is indeed the case in some models. As far as the dilaton is concerned, the scalar potential exhibits a "runaway" behavior at $S \to \infty$, driving the model to its trivial zero-coupling limit. This problem can be circumvented if the hidden SYM sector contains a gauge group consisting of several simple subgroup factors. Then the dilaton VEV can be "locked" by a "racetrack" of the potential [24].

The threshold effects of extra dimensions and the related gaugino condensation mechanism remain as important ingredients of superstring model building, now including not only heterotic strings, but also D-branes and flux compactifications. They may play a major role in connecting superstring theory to the real world.

Acknowledgments

I would like to thank Gabriele for 25 years of enjoyable collaborations, his friendship, guidance and support. I am looking forward to many future projects, as exciting and enjoyable as usual. I am also grateful to my collaborators Ignatios Antoniadis, Pierre Binétruy, Sergio Ferrara, Mary K. Gaillard, Edi Gava, Zurab Kakushadze, Dieter Lüst, Nico Magnoli, Narain and Pran Nath, who worked together with me on the related topics. This work is supported in part by the US National Science Foundation Grant PHY-0600304. Any opinions, findings and conclusions or recommendations expressed in this material are those of the author and do not necessarily reflect the views of the National Science Foundation.

References

1. T. R. Taylor, G. Veneziano: Phys. Lett. B **212**, 147 (1988)
2. G. Veneziano: *Topics in String Theory*, CERN-TH-5019/88 (1988)
3. I. Antoniadis: Phys. Lett. B **246**, 377 (1990)
4. K. R. Dienes, E. Dudas, T. Gherghetta: Phys. Lett. B **436**, 55 (1998)
5. Z. Kakushadze, T. R. Taylor: Nucl. Phys. B **562**, 78 (1999)
6. V. S. Kaplunovsky: Nucl. Phys. B **307**, 145 (1988) [Erratum-ibid. B **382**, 436 (1992)]
7. L. J. Dixon, V. Kaplunovsky, J. Louis: Nucl. Phys. B **355**, 649 (1991)
8. P. Mayr, S. Stieberger: Nucl. Phys. B **407**, 725 (1993)
9. D. Lüst, S. Stieberger: arXiv:hep-th/0302221

10. I. Antoniadis, E. Gava, K. S. Narain, T. R. Taylor: Nucl. Phys. B **407**, 706 (1993)
11. I. Antoniadis, K. S. Narain, T. R. Taylor: Phys. Lett. B **267**, 37 (1991)
12. M. K. Gaillard, T. R. Taylor: Nucl. Phys. B **381**, 577 (1992)
13. J. P. Derendinger, S. Ferrara, C. Kounnas, F. Zwirner: Nucl. Phys. B **372**, 145 (1992)
14. G. Lopes Cardoso, B. A. Ovrut: Nucl. Phys. B **392**, 315 (1993)
15. J. Louis: SLAC-PUB-5527 (1991), published in *Boston PASCOS 1991*, pp. 751–765
16. M. Bershadsky, S. Cecotti, H. Ooguri, C. Vafa: Nucl. Phys. B **405**, 279 (1993)
17. L. E. Ibanez, D. Lüst, G. G. Ross: Phys. Lett. B **272**, 251 (1991); L. E. Ibanez, D. Lüst: Nucl. Phys. B **382**, 305 (1992); P. Mayr, H. P. Nilles, S. Stieberger: Phys. Lett. B **317**, 53 (1993); H. P. Nilles, S. Stieberger: Phys. Lett. B **367**, 126 (1996); Nucl. Phys. B **499**, 3 (1997)
18. H. P. Nilles: Phys. Lett. B **115**, 193 (1982); S. Ferrara, L. Girardello, H. P. Nilles: Phys. Lett. B **125**, 457 (1983); M. Dine, R. Rohm, N. Seiberg, E. Witten: Phys. Lett. B **156**, 55 (1985); C. Kounnas, M. Porrati: Phys. Lett. B **191**, 91 (1987)
19. G. Veneziano, S. Yankielowicz: Phys. Lett. B **113**, 231 (1982)
20. T. R. Taylor: Phys. Lett. B **164**, 43 (1985)
21. S. Ferrara, N. Magnoli, T. R. Taylor, G. Veneziano: Phys. Lett. B **245**, 409 (1990)
22. A. Font, L. E. Ibanez, D. Lüst, F. Quevedo: Phys. Lett. B **245**, 401 (1990); H. P. Nilles, M. Olechowski: Phys. Lett. B **248**, 268 (1990); P. Binétruy, M. K. Gaillard: Phys. Lett. B **253**, 119 (1991); M. Cvetic, A. Font, L. E. Ibanez, D. Lüst, F. Quevedo: Nucl. Phys. B **361**, 194 (1991); D. Lüst, T. R. Taylor: Phys. Lett. B **253**, 335 (1991); P. Binétruy, M. K. Gaillard, T. R. Taylor: Nucl. Phys. B **455**, 97 (1995); P. Nath, T. R. Taylor: Phys. Lett. B **548**, 77 (2002)
23. V. Kaplunovsky, J. Louis: Nucl. Phys. B **444**, 191 (1995)
24. N. V. Krasnikov: Phys. Lett. B **193**, 37 (1987); L. J. Dixon: SLAC-PUB-5229 (1990), published in *DPF Conf. 1990*, pp. 811–822; T. R. Taylor: Phys. Lett. B **252**, 59 (1990)

Dualities in String Cosmology[1]

K. A. Meissner

Institute of Theoretical Physics, Warsaw University, Hoża 69, 00-681 Warsaw, Poland
Krzysztof.Meissner@fuw.edu.pl

Abstract. We describe in this chapter a set of duality symmetries present in the string-inspired theory of gravity coupled to the dilaton. These dualities are the cornerstones of String Cosmology, which provides alternatives to the usual inflation scenario. The crucial role of Prof. Gabriele Veneziano in the discovery and the development of string dualities is described and emphasized.

1 Introduction

Before going over to the description of dualities in String Cosmology and the fundamental role of Prof. Gabriele Veneziano in its discovery I would like to devote a few lines to some personal recollections. I met Gabriele for the first time in 1984 when, as a young experimental physicist, I worked for a few months in the UA2 group at CERN. This was a remarkable year not only for CERN (because of the Z and W discoveries) but also for string theory initiated 17 years earlier by Gabriele (because of the discovery of the Green–Schwarz mechanism of cancellation of anomalies). Obviously at that time a distance between a young experimental physics student and the world famous theoretical physicist was so huge that neither I dared to approach Gabriele nor I imagined that I ever would. Fortunately, a few years later, I gave a seminar at the theory division at CERN and Gabriele was generous enough to encourage me to apply for a 1-year position there. The stay at CERN 1990–1991 was the beginning of our collaboration (marked with publishing two papers that I consider the most important in my life) and the friendship that I am deeply grateful for.

Although general relativity is a theory with an extremely large group of local symmetries (i.e. the group of diffeomorphisms), it is very difficult to find any nontrivial global symmetry not directly linked with diffeomorphisms. The first such symmetry was discovered by Ehlers [1] in 1957 where it was shown

[1] In honour of Prof. Gabriele Veneziano

Krzysztof A. Meissner: *Dualities in String Cosmology*, Lect. Notes Phys. **737**, 561–571 (2008)
DOI 10.1007/978-3-540-74233-6_18

that four-dimensional pure gravity with one Killing vector exhibits $SL(2,\mathbb{R})$ symmetry. The argument is very simple and can be presented in a few lines. One starts with a metric parameterized as (the Killing vector is assumed here to be along the spatial coordinate z)

$$\mathrm{d}s^2 = \left(e^{-\rho} g_{ij} + e^{\rho} A_i A_j\right) \mathrm{d}x^i \mathrm{d}x^j + 2 e^{\rho} A_i \mathrm{d}x^i \mathrm{d}z + e^{\rho} (\mathrm{d}z)^2, \qquad (1)$$

where $i, j = 0, 1, 2$, all fields are real and depend only on x^i. Calculating the scalar curvature we get

$$\sqrt{-g^{(4)}} R^{(4)} = \sqrt{-g^{(3)}} \left(R^{(3)} - \frac{1}{4} e^{2\rho} F_{ij}^2 - \frac{1}{2} (\nabla \rho)^2 + \Box \rho \right). \qquad (2)$$

In three dimensions a vector is dual to a scalar. We perform the dualization by a substitution in the action

$$F_{ij} = e^{-2\rho} \epsilon_{ijk} \partial^k \sigma, \qquad (3)$$

and changing the sign of the resulting expression (which follows from the path integral formulation of dualization). Then introducing a complex field ϕ defined on the upper half-plane

$$\phi = \sigma + \mathrm{i}\, e^{\rho}, \qquad (4)$$

we get the reduced three-dimensional action

$$S = \int \sqrt{-g^{(3)}} \left(R^{(3)} - \frac{\nabla \bar{\phi} \nabla \phi}{2 (\mathrm{Im}\, \phi)^2} \right). \qquad (5)$$

This action is explicitly invariant under the Ehlers $SL(2,\mathbb{R})$ group acting as

$$\phi \to \frac{a\phi + b}{c\phi + d}, \qquad (6)$$

with real numbers a, b, c, d satisfying $ad - bc = 1$ (the Ehlers group was recently found also at higher-derivative orders in the gravitational action [2]).

With two Killing vectors the resulting symmetry is much bigger – in fact it is an infinite symmetry group called the Geroch group [3]. It was found in 1971 as a "solution generating" technique in a set of stationary, axisymmetric solutions of the Einstein's equations (the actual infinite Lie algebra structure was found later, and the group structure of finite transformations still later at the beginning of 1980s in the connection with coset constructions, nonlinear σ-models and Kac–Moody $SL(2,\mathbb{R})$ algebra).

In 1984 (the year of the "first string revolution") there appeared a notion of duality in string theory which in the simplest form says that a propagation of strings on a manifold of radius R is equivalent to a propagation on a manifold of radius $1/R$ [4]. It was later generalized to more complicated fixed background situations and there was an argument that it should be an exact symmetry to all orders in α' [5].

In 1986 Narain discovered a large set of consistent string theories corresponding to toroidal compactifications of the heterotic string [6]. These compactifications are parameterized by points in the coset space $G/H = SO(26-d, 10-d)/SO(26-d) \times SO(10-d)$. Although all these theories were unrealistic phenomenologically (they correspond to $N = 4$ supersymmetry in $D = 4$), the discovery ended the dream of a unique string theory. The points on the Narain's lattice did however, correspond, to different theories and not to different solutions inside the same theory so the construction is more a "theory generating" than "solution generating" technique. The second paper of [6] found a correspondence between the Lorentzian self-dual lattices in the heterotic picture of $(26-d, 10-d)$ compactification and the $(10-d)$ compactification in the presence of gauge and antisymmetric fields (that are of direct concern to subsequent developments).

2 Scale Factor Duality

The crucial observation that later led to extremely fruitful lines of research (especially in cosmology but in many other areas as well) was done by Gabriele in the paper released in April 1991 and published half a year later [7]. As a starting point he took the effective action for fields that are always in the massless spectrum of any closed string theory: gravity and the dilaton ϕ. The action reads

$$\Gamma^{(0)} = \frac{1}{2\kappa^2} \int d^{d+1}x \sqrt{-g}\, e^{-\phi} \left\{R + (\nabla\phi)^2\right\}. \tag{7}$$

Then the assumption is made that all fields depend only on time. With the diagonal ansatz for the metric:

$$g_{\mu\nu} = (-1, a_1^2(t), \ldots, a_d^2(t)), \tag{8}$$

the main observation of the paper states that the equations of motion are invariant under

$$\Phi \to \Phi, \qquad a_i(t) \to a_i^{-1}(t), \tag{9}$$

with

$$\Phi = \phi - \sum \ln(a_i(t)). \tag{10}$$

Such a symmetry is called in the paper scale factor duality (SFD) and it heavily relies on the fact that string theory predicts the relative coefficient of two terms in (7) to be 1 – otherwise the symmetry would not be there! Gabriele showed that starting with some simple solutions one can generate new solutions that still solve the equations of motion. The paper attracted a broad attention and was a basis of an introduction almost 2 years later of String (Pre-Big-Bang) Cosmology by Maurizio Gasperini and Gabriele [8] (each paper has now more than 500 citations).

3 $O(d,d)$ Symmetry to the Lowest Order

Even before the seminal paper on scale factor duality was released, during our discussions with Gabriele the question arose whether it can be generalized to other fields, most notably the antisymmetric tensor $B_{\mu\nu}$ which also is present in the massless spectrum. Soon after it turned out that the symmetry is much bigger than just the discrete Z_2 of SFD – it is actually a large noncompact $O(d,d)$ symmetry with SFD as a small discrete subgroup. The paper released in Autumn 1991 [9] took as a starting point the string massless action

$$\Gamma^{(0)} = \frac{1}{2\kappa^2} \int d^{d+1}x \sqrt{-g}\, e^{-\phi} \left\{ R + (\nabla\phi)^2 - \frac{1}{12} H^2 \right\}, \tag{11}$$

where $H_{\mu\nu\rho} = \partial_\mu B_{\nu\rho} + cyclic$. The actions (7) and (11) are taken as basic ingredients in String Cosmology and the existence of duality plays a fundamental role there (see the contribution of M. Gasperini in this volume [10]).

When fields depend only on time we can write

$$g_{\mu\nu} = \begin{pmatrix} -1 & 0 \\ 0 & G(t) \end{pmatrix}, \qquad B_{\mu\nu} = \begin{pmatrix} 0 & 0 \\ 0 & B(t) \end{pmatrix}. \tag{12}$$

Then the paper introduced a symmetric $2d \times 2d$ matrix (introduced earlier in a different context of fixed backgrounds by Shapere and Wilczek [11])

$$M_0 = \begin{pmatrix} G^{-1} & -G^{-1}B \\ BG^{-1} & G - BG^{-1}B \end{pmatrix} \tag{13}$$

belonging to the $O(d,d)$ group:

$$M_0^T = M_0, \qquad M_0 \eta M_0 = \eta, \qquad \eta = \begin{pmatrix} \mathbf{0} & \mathbf{1} \\ \mathbf{1} & \mathbf{0} \end{pmatrix}. \tag{14}$$

The action (11) can then be rewritten as

$$\Gamma^{(0)} = -\frac{1}{2\kappa^2} \int dt e^{-\Phi} \left(\dot{\Phi}^2 + \frac{1}{8} \mathrm{Tr}[\dot{M}\eta\dot{M}\eta] \right). \tag{15}$$

The action (15) is explicitly symmetric under the action of the $O(d,d)$ group:

$$M_0 \to \Omega^T M_0 \Omega, \qquad \Phi \to \Phi, \tag{16}$$

where Ω belongs to the $O(d,d)$: $\Omega^T \eta \Omega = \eta$.

Soon after our second common paper appeared [12] that discussed the general solutions to the equations of motion using the conserved current connected with the presence of a global continuous $O(d,d)$ symmetry of the action. It is appropriate to describe them here in some detail, as it seems that their potential has not yet been fully exploited.

We start with the description of the group $O(d,d)$. We write the general element of the group as

$$\Omega = \Omega_t \Omega_n = \begin{pmatrix} A_1^{-1} & A_1^{-1}A_2 \\ \mathbf{0} & A_1^T \end{pmatrix} \begin{pmatrix} \mathbf{1} & \mathbf{0} \\ A_3 & \mathbf{1} \end{pmatrix}, \tag{17}$$

where A_2, A_3 are antisymmetric $d \times d$ matrices. The action of Ω_t on M gives trivial constant rescaling and shift of G and B:

$$G \to G' = A_1 G A_1^T, \quad B \to B' = A_1 B A_1^T - A_2. \tag{18}$$

The action of Ω_n, however, gives genuinely new (and complicated) solutions. For example, even if we start from the simplest case $B = 0$, we get

$$G' = (1 - GA_3GA_3)^{-1}G, \quad B' = GA_3(1 - GA_3GA_3)^{-1}G. \tag{19}$$

We now turn to the equations of motion from the action (15). The first equation follows from reintroducing G_{00} into the action and from setting to zero the corresponding variation. This gives directly the "Zero Energy" condition

$$(\dot{\Phi})^2 + \frac{1}{8}\,\mathrm{Tr}\left(\dot{M}\eta\dot{M}\eta\right) - V(\Phi) = 0, \tag{20}$$

where we allow now for a potential $V(\Phi)$. If we assume that the potential does not break the symmetry explicitly, then it can depend on M only through a function of the invariants $\mathrm{Tr}(M\eta)^p$, $p = 1, 2, \ldots$. However, for p odd these traces vanish and for p even they are equal to $2d$; hence, the potential can depend only on Φ.

Such a potential is however rather unusual since Φ is not a scalar under general coordinate transformations (but see [10] for its inclusion into a fully general-covariant formulation). On the other hand, if the presence of $V(\Phi)$ could be justified, then one would have a relatively simple solution to the so-called graceful exit problem in String Cosmology [13, 14]. This point deserves further study.

The variation of the action with respect to Φ yields:

$$(\dot{\Phi})^2 - 2\ddot{\Phi} - \frac{1}{8}\,\mathrm{Tr}\left(\dot{M}\eta\dot{M}\eta\right) - \frac{\partial V(\Phi)}{\partial \Phi} = 0. \tag{21}$$

The variation of the action with respect to M has to be done carefully, since M is subject to several constraints (it is symmetric and belongs to $O(d,d)$). The resulting equation reads

$$\partial_t(M\eta\dot{M}) = \dot{\Phi}(M\eta\dot{M}), \tag{22}$$

which can be integrated to give

$$\mathrm{e}^{-\Phi}(M\eta\dot{M}) = \mathrm{const} = A. \tag{23}$$

The constant matrix A satisfies

$$A^T = -A, \qquad M\eta A = -A\eta M. \tag{24}$$

It is obvious that (20), (21) and (23) are invariant under the full $O(d,d)$ group.

Substituting (23) into (20) we obtain the first-order equation for Φ:

$$(\dot{\Phi})^2 = \frac{\exp(2\Phi)}{8}\ \mathrm{Tr}(A\eta)^2 + V(\Phi), \tag{25}$$

which can be solved by quadratures:

$$t = \int_{\Phi_0}^{\Phi} dy \left[\frac{\exp(2y)}{8}\ \mathrm{Tr}(A\eta)^2 + V(y)\right]^{-1/2} \tag{26}$$

This solution can then be used to define a "dilaton time" τ:

$$\tau = \int_{t_0}^{t} e^{\Phi}\, dt'. \tag{27}$$

In terms of τ the general solution of (23) simply reads

$$M(t) = \exp(-A\eta\tau)M(t_0). \tag{28}$$

The case of vanishing potential $V = 0$ is easy to analyse. In this case we get

$$e^{\Phi} = \frac{C}{T-t}; \qquad C = \sqrt{\frac{8}{\mathrm{Tr}(A\eta)^2}}. \tag{29}$$

We also get

$$\tau = C \ln \frac{T-t}{T-t_0}, \tag{30}$$

so that the solution for M reads

$$M(t) = \exp\left(CA\eta \ln \frac{T-t}{T-t_0}\right), \tag{31}$$

where we assumed that $M(t_0) = \mathbf{1}$. Consider, for instance, the special form of A:

$$A = \begin{pmatrix} 0 & -A_d \\ A_d & 0 \end{pmatrix}, \tag{32}$$

where $A_d = diag(a_1, .., a_d)$. In this case we get

$$M(t) = \begin{pmatrix} diag\left[\left(\frac{T-t}{T-t_0}\right)^{-2\alpha_1}, ..\right] & 0 \\ 0 & diag\left[\left(\frac{T-t}{T-t_0}\right)^{2\alpha_1}, ..\right] \end{pmatrix}, \tag{33}$$

where $\alpha_i = a_i/\sqrt{\Sigma a_i^2}$. These solutions are exactly the ones discussed in [7] (see also [15]).

It is equally easy to analyse the case $V = \Lambda$ =const [12]. The solution then reads

$$e^{\Phi} = C\sqrt{\Lambda}/\sinh(\sqrt{\Lambda}(T-t)) \tag{34}$$

and

$$M(t) = \exp\left(CA\eta \ln \frac{\tanh(\sqrt{\Lambda}(T-t)/2)}{\tanh(\sqrt{\Lambda}(T-t_0)/2)}\right) \tag{35}$$

where we again assumed that $M(t_0) = \mathbf{1}$.

In [12] one amusing solution was found for $d = 9$ (corresponding to the usual uncompactified superstring). As A we take

$$A = \begin{pmatrix} 0 & diag(-a_1,..,-a_9) \\ diag(a_1,..,a_9) & 0 \end{pmatrix}. \tag{36}$$

Then M is equal to

$$M = \begin{pmatrix} diag\left(\tanh^{-2\alpha_1}(\sqrt{\Lambda}(T-t)/2),..\right) & 0 \\ 0 & diag\left(\tanh^{2\alpha_1}(\sqrt{\Lambda}(T-t)/2),..\right) \end{pmatrix} \tag{37}$$

where $\alpha_i = a_i/\sqrt{\Sigma a_i^2}$. The scalar curvature and dilaton are finite for $t \to T$ when

$$\sum \alpha_i = 1. \tag{38}$$

Assuming that all $|\alpha_i|$ are equal, the only solution for α_i is $(-1/3,-1/3,-1/3,+1/3,..,+1/3)$, so that three dimensions expand and six dimensions contract.

4 $O(d,d)$ Symmetry to the Next Order

In the paper [12] an argument was given that $O(d,d)$ symmetry should be present to all orders in the α' expansion. The argument can be summarized as follows: Consider a conformal background (a string vacuum) of massless fields (metric, torsion and dilaton) which do not depend upon a particular set of (possibly noncompact) string coordinates $X^a (a = 1, 2, ..., d)$. The associated nilpotent string BRST operator Q will depend trivially (i.e. quadratically at most) on the $2d$ phase-space variables $Z = (P_a, X'^a)$. We perform now a global, canonical $O(d,d)$ transformation on the Z variables. Since this transformation preserves commutation relations and Wick contractions, the new BRST operator will also be nilpotent. However, the change in Z can be traded for a change in the backgrounds, implying that also the transformed backgrounds define a (generally inequivalent) conformal theory. Thus, $O(d,d)$ should be a symmetry of this particular class of string vacua. The action of

$O(d,d)$ can be very complicated, however, when we go over to higher orders in the α' expansion.

Indeed it was shown later in [16] that to the next order a seemingly very complicated string action

$$\begin{aligned}\Gamma = \frac{1}{2\kappa^2}\int \sqrt{-g}e^{-\phi}\Big\{ R + (\nabla\phi)^2 - \frac{1}{12}H^2 \\ +\alpha'\Big[-R^2_{GB} + 4\left(R^{\mu\nu} - \frac{1}{2}g^{\mu\nu}R\right)\partial_\mu\phi\partial_\nu\phi - 2\Box\phi(\nabla\phi)^2 + (\nabla\phi)^4 \\ +\frac{1}{2}\left(R^{\mu\nu\sigma\rho}H_{\mu\nu\alpha}H_{\sigma\rho}{}^{\alpha} - 2R^{\mu\nu}H^2_{\mu\nu} + \frac{1}{3}RH^2\right) - D^\mu\partial^\nu\phi H^2_{\mu\nu} + \frac{1}{3}\Box\phi H^2 \\ -\frac{1}{6}H^2(\nabla\phi)^2 - \frac{1}{24}H_{\mu\nu\lambda}H^\nu{}_{\rho\alpha}H^{\rho\sigma\lambda}H_\sigma{}^{\mu\alpha} + \frac{1}{8}H^2_{\mu\nu}H^{2\mu\nu} - \frac{1}{144}(H^2)^2\Big]\Big\},\end{aligned} \tag{39}$$

also exhibits the $O(d,d)$ symmetry. To display this symmetry one has to redefine M by $O(d,d)$ rotations of order α'. The redefinition can be written as

$$M \to \omega^T M_0\, \omega, \tag{40}$$

where ω is in the form

$$\Omega = \exp\begin{pmatrix} a_1 & a_2 \\ a_3 & -a_1^T \end{pmatrix}, \quad a_2^T = -a_2, \quad a_3^T = -a_3, \tag{41}$$

with

$$\begin{aligned} a_1 &= -\alpha'\left[-\frac{1}{2}\dot{G}G^{-1}\dot{G}G^{-1} + \frac{1}{2}\dot{B}G^{-1}\dot{B}G^{-1}\right], \\ a_2 &= -\alpha'\Big[-\dot{G}G^{-1}\dot{B} - \dot{B}G^{-1}\dot{G} + \frac{1}{2}(\dot{G}G^{-1}\dot{G} - \dot{B}G^{-1}\dot{B})G^{-1}B \\ &\quad +\frac{1}{2}BG^{-1}(\dot{G}G^{-1}\dot{G} - \dot{B}G^{-1}\dot{B})\Big], \\ a_3 &= 0. \end{aligned} \tag{42}$$

Comparing with (17) we see that this redefinition is a trivial rotation (since $a_3 = 0$).

With this new M, the action (39) reads

$$\begin{aligned}\Gamma = \int dt e^{-\Phi}\Big\{-\dot{\Phi}^2 - \frac{1}{8}\mathrm{Tr}(\dot{M}\eta)^2 \\ -\alpha'\left[\frac{1}{16}\mathrm{Tr}(\dot{M}\eta)^4 - \frac{1}{64}(\mathrm{Tr}(\dot{M}\eta)^2)^2 - \frac{1}{4}(\mathrm{Tr}(\dot{M}\eta)^2)\dot{\Phi}^2 - \frac{1}{3}\dot{\Phi}^4\right]\Big\}.\end{aligned} \tag{43}$$

Since the $O(d,d)$ symmetry is continuous and global, it has an associated conserved current, which means, for a theory depending only on time, that the current should be constant (it is an "integrated once" equation of motion for M). In analogy to (23) we call this constant A:

$$A = e^{-\Phi} \left\{ M\eta\dot{M} + 2\alpha' \left[\frac{1}{2} M(\eta\dot{M})^3 - \frac{1}{8} M\eta\dot{M}\mathrm{Tr}(\dot{M}\eta)^2 - M\eta\dot{M}\dot{\Phi}^2 \right] \right\} \quad (44)$$

where $A^T = -A$ and $A\eta M = -M\eta A$.

The g_{00} equation reads

$$0 = -\dot{\Phi}^2 - \frac{1}{8}\mathrm{Tr}(\dot{M}\eta)^2$$
$$-3\alpha' \left[\frac{1}{16}\mathrm{Tr}(\dot{M}\eta)^4 - \frac{1}{64}(\mathrm{Tr}(\dot{M}\eta)^2)^2 - \frac{1}{4}(\mathrm{Tr}(\dot{M}\eta)^2)\dot{\Phi}^2 - \frac{1}{3}\dot{\Phi}^4 \right]. \quad (45)$$

Equations (44) and (45) cannot be explicitly solved because of their nonlinear structure. Since they are first order in the derivatives, however, they are in principle solvable by quadratures.

5 Discussion

The main result of the above papers consisted in showing that in string theory there exist large symmetries of the dynamical backgrounds (and not only symmetries of propagation of strings in different static backgrounds). These developments led to the idea that the dilaton may play a crucial role in the evolution of the Early Universe. The idea was taken as a cornerstone of the so-called Pre-Big-Bang Cosmology, developed in 1993 by Gabriele and Maurizio Gasperini [8], that has grown into a separate part of early cosmology in itself. In this idea the scale factor duality (or, more generally, $O(d,d)$ duality) plays a crucial role – the history of the Universe for negative times $t < 0$ and positive times $t > 0$ is connected by a duality transformation (9), combined with the time reversal transformation $t \to -t$.

Such a scenario gives a natural possibility for solving all the usual problems of standard cosmology by a phase of superinflation at negative times, driven by the presence of the dilaton field without the need of any extra inflaton field. The scenario gives a spectrum of perturbations which is significantly different from that of the usual inflationary scenario – the power spectra of Pre-Big Bang Cosmology calculated in two papers, written together with Alessandra Buonanno and Carlo Ungarelli [17], show a significant difference in the origin of structure: In this scenario it comes from the axion field fluctuations, and not from the scalar perturbations of the inflaton field. It is however not clear, at present, how to actually realize a connection from negative to positive times ("passing through the singularity"), nor how to stop the dilaton from evolving since, perturbatively, the dilaton does not develop any potential. This is the so-called graceful exit problem [13, 14], for which there is no definite solution, at present (one of the possibilities is to invoke a duality-invariant potential depending on Φ and not on ϕ; this seems to be very much against the spirit of general relativity; see however [10] for a general-covariant, non-local interpretation of such potential).

There was (and still is) quite an intensive research that was initiated by the discovery of dualities in String Cosmology. It is difficult to even list all the different lines of research, so we name just a few of them, here: large-scale magnetic fields [18], cosmological perturbations [19], dilaton production [20], relic gravitational waves [21], noncompact symmetries [22], ekpyrotic [23] and cyclic [24] models of the Universe, phantom duality [25] and triality [26], entropy of the Universe [27], black-hole solutions [28] and many others (for more detailed references the reader may consult recent [29, 30] and earlier [31] reviews).

Although it is not clear yet which of the ideas described in this paper will be part of the future Early Universe Cosmology, one can safely predict that the discovery of duality in the gravi-dilaton system (as well as all other discoveries in Gabriele's monumental set of achievements) will have a profound impact on any future (not necessarily string related) theoretical research on gravity and its symmetries.

References

1. J. Ehlers: *Konstruktionen und Charakterisierung von Lösungen der Einsteinschen Gravitationsfeldgleichungen*, Dissertation (Hamburg, 1957)
2. C. Colonnello, A. Kleinschmidt: *Ehlers symmetry at the next derivative order*, arXiv:0706.2816 [hep-th]
3. R. Geroch: J. Math. Phys. **12**, 918 (1971); J. Math. Phys. **13**, 394 (1972)
4. K. Kikkawa, M. Yamasaki: Phys. Lett. B **149**, 357 (1984);
   N. Sakai, I. Senda: Prog. Theor. Phys. **75**, 692 (1986)
5. A. Giveon, E. Rabinovici, G. Veneziano: Nucl. Phys. B **322** (1989) 167
6. K. S. Narain: Phys.Lett. B **169**, 41 (1986);
   K. S. Narain, M. H. Sarmadi, E. Witten: Nucl. Phys. B **279**, 369 (1987)
7. G. Veneziano: Phys. Lett. B **265**, 287 (1991)
8. M. Gasperini, G. Veneziano: Astropart. Phys. **1**, 317 (1993)
9. K. A. Meissner, G. Veneziano: Phys. Lett. B **267**, 33 (1991)
10. M. Gasperini: *Dilaton cosmology and phenomenology*, this volume
11. A. D. Shapere, F. Wilczek: Nucl. Phys. B **320**, 669 (1989)
12. K. A. Meissner, G. Veneziano: Mod. Phys. Lett. A **6**, 3397 (1991)
13. R. Brustein, G. Veneziano: Phys. Lett. B **329**, 429 (1994)
14. M. Gasperini, J. Maharana, G. Veneziano: Nucl. Phys. B **472**, 349 (1996)
15. M. Mueller: Nucl. Phys. B **337**, 37 (1990)
16. K. A. Meissner: Phys. Lett. B **392**, 298 (1997)
17. A. Buonanno, K. A. Meissner, C. Ungarelli, G. Veneziano: Phys. Rev. D **57**, 2543 (1998); JHEP **9801**, 004 (1998)
18. M. Gasperini, M. Giovannini, G. Veneziano: Phys. Rev. Lett. **75**, 3796 (1995)
19. R. Brustein, M. Gasperini, M. Giovannini, V. F. Mukhanov, G. Veneziano: Phys. Rev. D **51**, 6744 (1995)

20. M. Gasperini, G. Veneziano: Phys. Rev. D **50**, 2519 (1994)
21. R. Brustein, M. Gasperini, M. Giovannini, G. Veneziano: Phys. Lett. B **361**, 45 (1995)
22. J. Maharana, J. H. Schwarz: Nucl. Phys. B **390**, 3 (1993)
23. J. Khoury, B. A. Ovrut, P. J. Steinhardt, N. Turok: Phys. Rev. D **64**, 123522 (2001)
24. P. J. Steinhardt, N. Turok: Phys. Rev. D **65**, 126003 (2002)
25. M. P. Dabrowski, T. Stachowiak, M. Szydlowski: Phys. Rev. D **68**, 103519 (2003)
26. J. E. Lidsey: Phys. Rev. D **70**, 041302 (2004)
27. G. Veneziano: Phys. Lett. B **454**, 22 (1999)
28. A. Sen: Nucl. Phys. B **440**, 421 (1995)
29. M. Gasperini, G. Veneziano: Phys. Rep. **373**, 1 (2003)
30. J. E. Lidsey, D. Wands, E. J. Copeland: Phys. Rep. **337**, 343 (2000)
31. A. Giveon, M. Porrati, E. Rabinovici: Phys. Rep. **244**, 77 (1994)

Spontaneous Breaking of Space–Time Symmetries

E. Rabinovici

Racah Institute of Physics, The Hebrew University of Jerusalem, 91904 Jerusalem, Israel
eliezer@vms.huji.ac.il

Abstract. Kinematical and dynamical mechanisms leading to the spontaneous breaking of space–time symmetries are described. The symmetries affected are space and time translations, space rotations, scale and conformal transformations. Applications are made to solidification, string theory compactifications, the analysis of stable theories with no ground states, supersymmetry breaking and the determination of the value of the vacuum energy.

1 Introduction

This being a contribution to honor Gabriele Veneziano I allow myself to open with some personal words. I have first heard Gabriele's name on the radio when the late Yuval Ne'eman described the great importance of young Gabriele's work. That was in the late 1960s, several years later as a student I had the privilege to learn from a still very young Gabriele about the dual model in full detail. These were outstanding lectures. Over the years I have learned many more things from Gabriele, some of them through direct collaborations, and in parallel we had developed a personal friendship for which I am grateful.

It is not uncommon to young scientists to complain that their teachers didn't educate them appropriately and did not really pass them/point them to the relevant information. I may have some such complaints of my own but not to Gabriele. I would have liked for example to know earlier about the ideas of Kaluza and Klein. So, in order to somewhat reduce the complaints that will be directed at me, I would like to use this opportunity to describe something that it is not taught extensively in particle physics courses, namely the mechanisms to spontaneously break space–time symmetries. The world around us is actually not explicitly invariant under translations or under rotations. It is also not explicitly invariant under scale and conformal symmetries. In this work we will review various mechanisms to break all these space–time symmetries. I think they may yet play an important role in particle physics as

E. Rabinovici: *Spontaneous Breaking of Space–Time Symmetries*, Lect. Notes Phys. **737**, 573–605 (2008)
DOI 10.1007/978-3-540-74233-6_19

well. I will first describe attempts to break translational invariance kinematically by imposing specific boundary conditions. Then I will review the Landau theory of solidification and an attempt to apply it to generate a dynamical mechanism for compactifications. I will discuss both the success and challenges of that approach. Next, in the context of breaking time-translational invariance I will discuss various systems which are well defined but have no ground state. Following a review of the breaking of scale invariance and conformal invariance I will also not miss this opportunity to describe in a Katoish manner that the vacuum energy in conformal/scale invariant theories is very constrained, and its zero value does not depend on the presence or absence of any spontaneously generated scales. This may eventually be recognized as an important ingredient in understanding and explaining the cosmological constant problem.

2 Spontaneous Breaking of Space Symmetries

Space symmetries include space translations and space rotations, and we address here the spontaneous breaking of these space symmetries. This occurs for example when a liquid solidifies and a lattice is formed. The standard manner to identify the ground state of a system is to construct what is called the effective potential. The symmetry properties of the ground state determine whether a spontaneous breaking of symmetries which are manifest in the Lagrangian occurs.

Let's review the manner in which the effective potential is constructed. One first considers all wave functionals which have the same expectation value of the field operator $\tilde{\phi}$,

$$< \Psi(\phi)|\hat{\phi}|\Psi(\phi) >= \tilde{\phi} \, . \tag{1}$$

Out of this subset of wave functionals, one chooses the particular wave functional which minimizes the expectation value of the Hamiltonian. One calls it $V_{eff}(<\phi>)$,

$$V_{eff}(\tilde{\phi}) = min_{\tilde{\phi}} < \Psi(\phi)|\hat{H}|\Psi(\phi) > . \tag{2}$$

Eventually one draws a picture portraying V_{eff} as a function of $\tilde{\phi}$ and one searches for its minimum. The wave functional for which this energy minimum was obtained is the wave functional of the ground-state of the system. However, one usually ignores the possibility that the ground-state wave function would correspond to a non-constant (in x) expectation value $< \phi(x) >$. Of course it makes much easier the drawing of pictures in books; here, however, we will discuss cases where $< \phi(x) >$ actually does depend on x when evaluated in the ground state.

Why does one usually only consider wave functionals with constant values of $< \phi(x) >$?

The reason is expediency—when one wants to pick up the ground state of the system among various candidates, one is interested only in the winner, that is the true ground state. One does not care if one misses out candidate states whose energies are just above that of the ground state of the system. As one generally does not expect spontaneous breakdown of space–time symmetries in the ground state, one considers it enough to search for the ground state only among those candidates for which $< \phi(x) >$ is constant. However, that need not always be the case.

2.1 Kinematics: Attempts to Break Spatial Translational Invariance Through Boundary Conditions

I will first describe an easy way to attempt to break space symmetries—that is to break the symmetries not by the dynamics of the system but kinematically, by imposing certain boundary conditions, which may induce such a breaking. This easy solution is a mirror to what is done in String Theory in several cases, including when one is considering brane sectors. To try and break translational invariance by boundary conditions, one considers for example a system which depends on a scalar field ϕ. Assume the system lives in a box extending from $-L$ to L, and impose the condition of anti-periodicity, namely,

$$\phi(L) = -\phi(-L), \tag{3}$$

where L is the spacial cutoff we put on the system.

If the system at hand is described by an effective potential that has only one minimum, as in Fig. 1, where the expectation value $< \phi >$ vanishes, then there is no effect resulting from imposing the boundary conditions. The

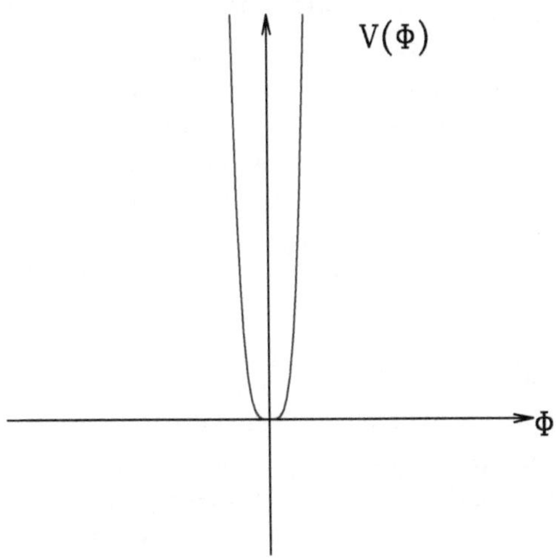

Fig. 1. Unbroken symmetry

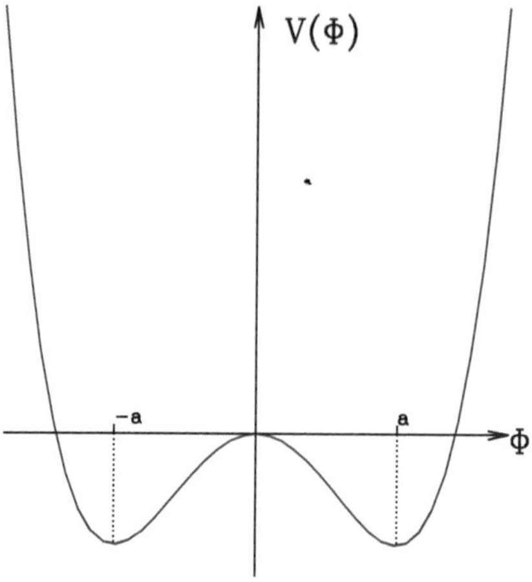

Fig. 2. Broken symmetry

ground state does fulfill the boundary condition, and it remains the one which does not break translational invariance. From the point of view of the wave functional it is concentrated around $\phi = 0$.

However, consider the double-well potential of Fig. 2 (in circumstances where there is no tunneling). In this case the effective potential has two minima, one at $\phi = a$ and the other at $\phi = -a$. Imposing the boundary condition removes both the possible true vacua of the system, because neither the ground state for which $< \phi >= -a$ nor the ground state for which $< \phi >= a$ obeys the boundary condition. One is driven to look for another type of ground state. We know, for example, that in a two-dimensional system composed only of scalar fields there is a finite energy solution, which is a soliton, that at L has a value a and at $-L$ has a value $-a$, see Fig. 3. An anti-soliton will have the opposite values. This is a stable topological configuration, and one may imagine that indeed in such a system there is no translational invariance, because the ground state will have to be such that its spacial expectation value follows the values of the soliton field, and thus is not translational invariant.

It is true that by imposing the boundary conditions one has forced the system into the soliton sector, but one has to remember that this system has a zero mode. Technically, if one solves the small fluctuations of the scalar field in the presence of a background, which is a soliton, one finds that there is a zero mode. This zero mode is a reflection of the underlying translational invariance and it actually tells us that one is not able to determine, by energetic considerations, where the inflection point x_0 (namely, the point from which one turns from one vacuum to the other) is placed (see Fig. 3). Actually there is a valid soliton solution for each value of x_0.

Why is this important? At the case at hand, the zero mode is normalizable. This amounts to saying that the soliton mass is finite. In such a case, there is actually no bulk violation of translational invariance. What one needs to do is

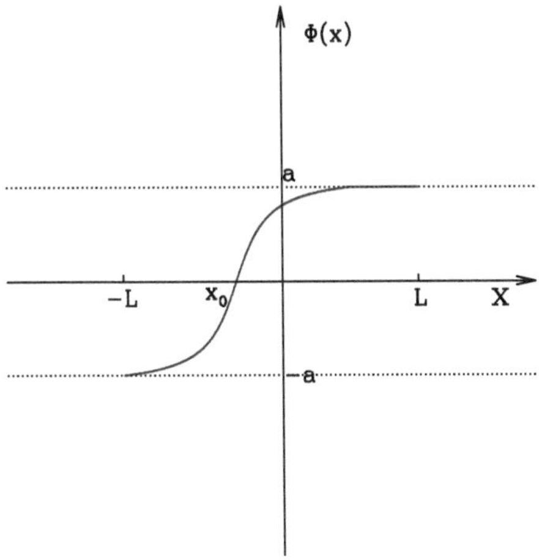

Fig. 3. A soliton attempts to break translational invariance

to construct an eigenstate configuration, which is an eigenstate of the linear momentum operation, a plane wave in terms of the center of mass coordinate of the soliton. The lowest energy state which corresponds to a momentum state has $p = 0$, one has restored in this way translational invariance. The only problem will be to fix the system very near the edges, but in the bulk the symmetry has been restored, and there is no breaking of bulk translational invariance. Could one still have a case where the boundary conditions do cause a spontaneous breaking of translational invariance?

This may occur when one drives the mass of the soliton to infinity by an appropriate choice of parameters. When the mass of the soliton is infinite, physically one cannot form a linear momentum state out of it, and technically the zero mode ceases to be normalizable. In such a case, one does indeed break translational invariance spontaneously by fixing the point where the soliton makes the transition from one vacuum to the other. This occurs for example in String Theory in a sector containing infinite branes: branes which have finite energy do not break translational invariance, and one can build out of them linear momentum states. However, branes which extend up to infinity carry infinite energy, and therefore do lead to the breakdown of translational invariance. I will mention at this point that once upon the time, when people were considering the breakdown of extended global supersymmetries, there was a predominant common wisdom which claimed that one cannot break down extended global supersymmetry to anything but $\mathcal{N} = 0$. That is either all the supersymmetries are manifest together, or they are all broken together. The argument went in the following way: one writes the formula for the Hamiltonian

$$H = \sum_{\alpha} \bar{Q}^I_\alpha Q^I_\alpha, \tag{4}$$

where $I = 1, \ldots, \mathcal{N}$ is *not* summed, and it is a non-trivial constraint to get the same Hamiltonian by summing over different supersymmetry generators. When one can do that, one has an extended supersymmetry. However, it is clear from this that if the Hamiltonian does not vanish on the ground state, then some of the Q^I (for each I independently) do not annihilate the ground state. Therefore, the supersymmetries are either all preserved or all broken.

This type of argument assumed implicitly that Poincaré invariance is present in the system. If one now considers a system of branes (see for example [1, 2]), then part of the Poincaré invariance is preserved and part is broken. This exposes a loop hole in the former argument, and in the absence of full translational invariance (due to the presence of infinite mass branes) one may obtain fractional BPS states, and one may break down $\mathcal{N} = 4$ to $\mathcal{N} = 2$, $\mathcal{N} = 2$ to $\mathcal{N} = 1$, and various other combinations.

This is an example where spontaneous breaking of translation invariance occurs—it has an impact also on the partial breaking of global supersymmetry and, if one wishes, this is a way to break translation invariance by forcing the system, using boundary conditions, to a certain super-selection sector.

This is not what I mainly want to discuss here. I would like to discuss a situation where the dynamics of the system drive the spontaneous breaking of translational and rotational invariance.

2.2 Dynamics: The Landau Theory of Liquid–Solid Phase Transitions

Let us now turn to discuss the transition between a liquid and a solid. This follows the seminal work of Landau [3]. In a monumental paper he simultaneously described spontaneous symmetry breaking of both internal and space–time symmetries. Consider a liquid, a system whose Lagrangian is either relativistic or non-relativistic, and it possesses full rotational and translational invariance. A solid, on the other hand, is a system which maintains only a very small part of the translational invariance and rotational invariance (Fig. 4).

Let us simplify the study by ignoring the point structure at each lattice point which a solid may have. That is, let's not consider the atomic structure at each point. One focuses first on the question of how does the simplest lattice form.

I will describe this following Landau and then, following [4], I am going to describe applications to String Theory. Landau starts by defining the Landau

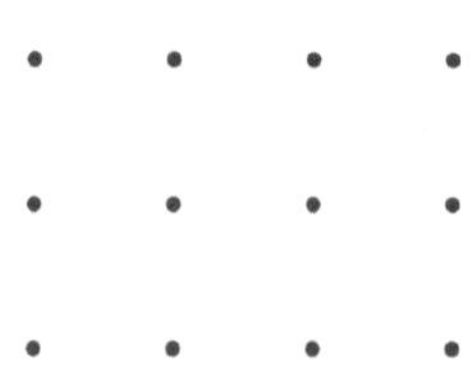

Fig. 4. The solid lattice breaks most of the translational and rotational invariance

order parameter to monitor the transition between a solid and a liquid. It is a scalar order parameter $\varrho(\boldsymbol{x})$,

$$\varrho(\boldsymbol{x}) = \varrho_s(\boldsymbol{x}) - \varrho_0, \tag{5}$$

the difference between the non-translational non-rotational invariant density of the solid $\varrho_s(\boldsymbol{x})$, and the constant density ϱ_0 of the liquid. Next, consider the Fourier decomposition of $\varrho(\boldsymbol{x})$

$$\varrho(\boldsymbol{x}) = \sum \varrho(\boldsymbol{q}) e^{i\boldsymbol{q}\cdot\boldsymbol{x}} + h.c. \tag{6}$$

It is useful to use as order parameters the Fourier components $\varrho(\boldsymbol{q})$.

The question is thus: Does the wave functional of the ground state have support on $\boldsymbol{q} \neq \boldsymbol{0}$? If the answer is positive, then at the very least continuous translational space symmetry would be spontaneously broken. This will be determined by studying the Landau–Ginsburg effective action as expressed in terms of the order parameter $\varrho(\boldsymbol{q})$. The first relevant term of the Landau–Ginsburg action is quadratic in the order parameter and is given by

$$\mathcal{L}_0 = \int d\boldsymbol{q}_1 d\boldsymbol{q}_2 \varrho(\boldsymbol{q}_1)\varrho(\boldsymbol{q}_2) A(|\boldsymbol{q}_1|^2)\delta(\boldsymbol{q}_1 + \boldsymbol{q}_2). \tag{7}$$

The delta function $\delta(\boldsymbol{q}_1 + \boldsymbol{q}_2)$ enforces translational invariance, while rotational invariance is preserved by the dependence on $|\boldsymbol{q}|^2$ of the function $A(|\boldsymbol{q}|^2)$. The function $A(|\boldsymbol{q}|^2)$, like in any Landau–Ginsburg potential, is determined by the microscopic theory. In the particular case at hand, it will depend on the hardcore potential component in the atoms involved and on other possible potentials, as well as on the temperature of the system. In the case of neutron stars, studied in [5], the Pauli exclusion principle plays a role in determining the function $A(|\boldsymbol{q}|^2)$.

Let us treat first an example that we are familiar with, that of a free massive spin-zero particle in a relativistic field theory. In that case the function $A(|\boldsymbol{q}|^2)$ is

$$A(|\boldsymbol{q}|^2) = |\boldsymbol{q}|^2 + m^2. \tag{8}$$

This has a minimum at $|\boldsymbol{q}|^2 = 0$, as shown in Fig. 5, and thus the function $\varrho(\boldsymbol{q})$ should get the support only at $\boldsymbol{q} = \boldsymbol{0}$: There is no spontaneous breakdown of translational invariance, in this case.

In the presence of interactions things may become more complicated; for example, I am not familiar even with a proof that the standard model ground state does not violate space–time symmetry (though most likely it does not). In any case, the microscopic theory may allow a different function for $A(|\boldsymbol{q}|^2)$. In particular, assume that the form of $A(|\boldsymbol{q}|^2)$ is as given in Fig. 6 . In this case, the function $A(|\boldsymbol{q}|^2)$ has a minimum at a value $|\boldsymbol{q}_0|^2 \neq 0$. In such a system the ground state wave functional gives rise to a density concentrated around $|\boldsymbol{q}_0|^2 \neq 0$. In particular, one would expect the support to be concentrated

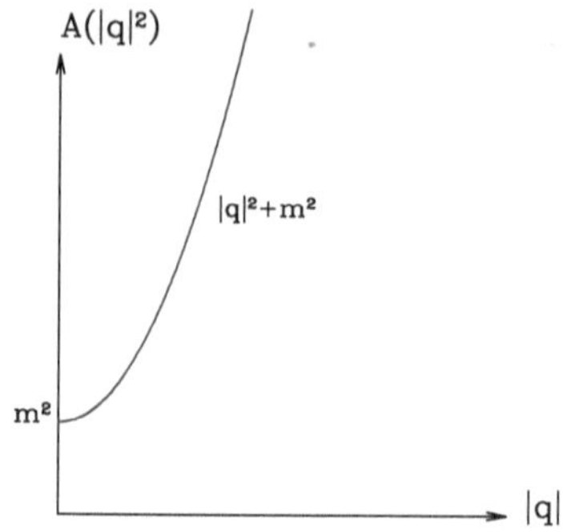

Fig. 5. The form of $A(|\boldsymbol{q}|^2)$ in a free massive relativistic field theory does not lead to spontaneous breaking of translational invariance

around a sphere in $\boldsymbol{q}$-space, whose radius is $|\boldsymbol{q}_0|$. So, given $A(|\boldsymbol{q}|^2)$ of that form, one is in a situation where there is a spontaneous breaking of translational invariance, but not yet also a breaking of rotational invariance, which is what is needed to form a solid. It is good enough to break just translational invariance.

The ground-state density does depend on $\boldsymbol{x}$

$$\varrho(\boldsymbol{x}) = \int_{\mathcal{S}_{|\boldsymbol{q}_0|}} \mathcal{D}\Omega \; \varrho(\boldsymbol{q}) e^{\mathrm{i}\boldsymbol{q}\cdot\boldsymbol{x}} + h.c. \tag{9}$$

In this approximation the wave functional of the ground state is supported on a sphere $\mathcal{S}_{|\boldsymbol{q}_0|}$ whose radius is $\boldsymbol{q}_0$. In particle physics we have become rather sophisticated, and when one writes down the Landau–Ginsburg action, one usually requires that the expansion which one does in the order parameter

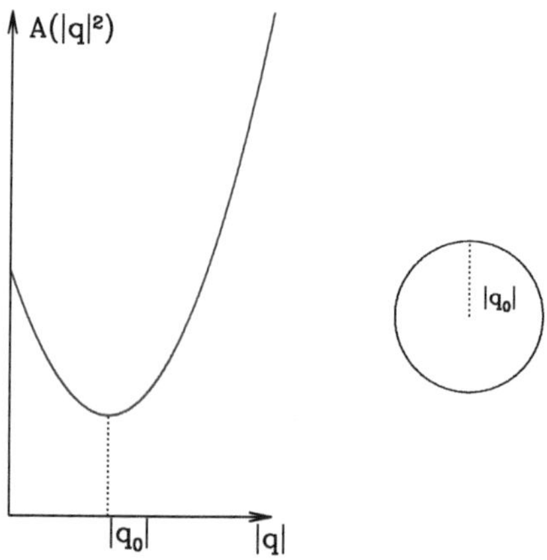

Fig. 6. Example of a function $A(|\boldsymbol{q}^2|)$ which leads to the breaking of translational invariance. An explicit microscopical realization of a such a form appears in neutron stars [5]. The wave functional is concentrated at most on the shell of a sphere of radius $|\boldsymbol{q}_0|$

be under control: That means, for example, that there is a limit in which this expansion becomes exact. In the case at hand this is not the situation, which is actually very complicated; nevertheless, one follows the usual Landau–Ginsburg expansion.

The term which follows the quadratic interaction is a cubic term:

$$\mathcal{L} = \mathcal{L}_2 + \mathcal{L}_3, \tag{10}$$

$$\mathcal{L}_3 = \int d^3\boldsymbol{q}_1 d^3\boldsymbol{q}_2 d^3\boldsymbol{q}_3 \; \varrho(\boldsymbol{q}_1)\varrho(\boldsymbol{q}_2)\varrho(\boldsymbol{q}_3)\delta(\boldsymbol{q}_1 + \boldsymbol{q}_2 + \boldsymbol{q}_3) \times \\ B(|\boldsymbol{q}_1|^2, |\boldsymbol{q}_2|^2, |\boldsymbol{q}_3|^2, \boldsymbol{q}_1 \cdot \boldsymbol{q}_2, \boldsymbol{q}_1 \cdot \boldsymbol{q}_3, \boldsymbol{q}_2 \cdot \boldsymbol{q}_3). \tag{11}$$

For the purpose of illustration I am going to assume, as Landau did, that this is a good perturbation, namely that when one considers $\mathcal{L}_3$ one is going already to assume that the support of ϱ comes from only those values of $\boldsymbol{q}$ such that $|\boldsymbol{q}_1|^2 \cong |\boldsymbol{q}_2|^2 \cong |\boldsymbol{q}_3|^2 \cong |\boldsymbol{q}_0|^2$. This was determined by $\mathcal{L}_2$.

In (11), once again, the delta function $\delta(\boldsymbol{q}_1+\boldsymbol{q}_2+\boldsymbol{q}_3)$ enforces the explicit translational invariance, and the dependence of B on the momentum respects both translational and rotational invariance. The integral in the $\boldsymbol{q}$'s is not over all possible values, but only over those whose lengths is determined by $|\boldsymbol{q}_0|^2$, which in turn was fixed by $\mathcal{L}_2$.

An additional structure emerges due to the effect of the delta function $\delta(\boldsymbol{q}_1+\boldsymbol{q}_2+\boldsymbol{q}_3)$. It restricts the candidates for the ground state to have support on at least three different values for the $\boldsymbol{q}_i$. The three vectors appearing need to sum up to give a triangle, see Fig. 7.

Actually they are six if the field is real since one needs

$$\varrho(\boldsymbol{q}) = \varrho(-\boldsymbol{q}). \tag{12}$$

Thus one has at least six components of $\varrho(\boldsymbol{q})$ which do not vanish. In general, instead of $\varrho(\boldsymbol{q})$ having support on all values of a sphere, they are now broken into triplets where the $\boldsymbol{q}_i$ have to sum together to form triangles (Fig. 7). In this manner also rotational invariance is spontaneously broken.

Let's be even more explicit, because we have used the approximation that all the $\boldsymbol{q}_i$ have the same length, the $\boldsymbol{q}_i$ that tessellate the sphere have to form equilateral triangles, as in Fig. 7. Equilateral triangles single out a specific angle 60°, that is a spontaneous breaking of rotational invariance. One has obtained a non-zero value for $\boldsymbol{q}$, and one has derived that the ground state is built out of objects which have to sum up to form triangles which are equilateral and thus have 60° angles. From energy and combinatorial considerations one then finds that, to be on an extremum, one needs all the values of $\varrho(\boldsymbol{q}_i)$ to be equal,

$$|\varrho(\boldsymbol{q}_i)|^2 = |\varrho(\boldsymbol{q}_0)|^2, \tag{13}$$

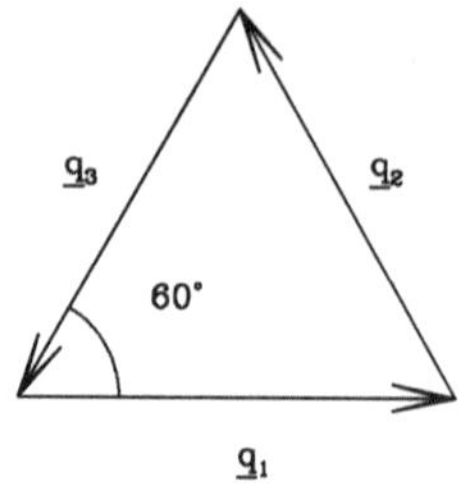

Fig. 7. The sphere $\mathcal{S}_{|q_0|}$ is triangulated due to the presence of a cubic term in the Lagrangian. Since in this approximation all the sides of the triangles have the same length, their angles are determined to be 60°. Rotational invariance is thus spontaneously broken

which leads to

$$|\varrho(\boldsymbol{x})|^2 = n|\varrho(\boldsymbol{q}_0)|^2, \tag{14}$$

where n is the number of non-vanishing components of $\varrho(\boldsymbol{q})$.

There are a couple of general ways to distribute the triplets, one in which each $\boldsymbol{q}_i$ appears in only one of the triplets, and another in which each value of $\boldsymbol{q}_i$ does participate in two triplets. The number of elements (i.e. the number of triplet configurations) is proportional to n in both cases, being either $2n/3$ or $4n/3$. When one does the analysis, and one estimates the value of $\mathcal{L}_3$, one finds that it decreases as the inverse of $\sqrt{n}$:

$$\mathcal{L}_3 \sim \frac{|\varrho(\boldsymbol{q}_0)|^2}{\sqrt{n}}. \tag{15}$$

Thus the ground state will be obtained for some finite value of n. One needs to consider only a finite number of triplet configurations when one searches for the extrema of the free energy. Just three, i.e. six participants (if the field is real), lead to the following density distribution

$$\varrho(x,y) = \pm \left(\frac{2}{3}\right)^{1/2} \varrho_{q_0} \left[\cos(q_0 x) + 2\cos\left(\frac{1}{2}q_0 x\right)\cos\left(\frac{\sqrt{3}}{2}q_0 y\right)\right]. \tag{16}$$

The corresponding free energy is

$$\mathcal{L}_3^{n=3} = \frac{2B\varrho_{q_0}^3}{3\sqrt{3}}. \tag{17}$$

For the case of two spatial dimensions it turns out that if $\varrho(q_0) > 0$ it is advantageous to form a triangular lattice, while if $\varrho(q_0) < 0$, the dual lattice, which is a honeycomb lattice, is formed.

This required only studying the minimal possible configuration. In three spatial dimensions this would be a candidate for a two-dimensional lattice in three dimensions, if one wishes some type of compactification.

In three dimensions one needs to consider also larger configurations to obtain the extrema. The next candidate configuration has six ($n = 6$), i.e. twelve values of $\boldsymbol{q}$. This is a more complicated configuration, whose density distribution is

$$\varrho(x, y, z) = \frac{2}{\sqrt{3}} \varrho_{q_0} \left[\cos\left(\frac{\sqrt{2}}{2} q_0 x\right) \cos\left(\frac{\sqrt{2}}{2} q_0 y\right) + \right.$$
$$\left. + \cos\left(\frac{\sqrt{2}}{2} q_0 x\right) \cos\left(\frac{\sqrt{2}}{2} q_0 z\right) + \cos\left(\frac{\sqrt{2}}{2} q_0 y\right) \cos\left(\frac{\sqrt{2}}{2} q_0 z\right) \right], \quad (18)$$

which is actually that of a BCC lattice (in real space). The value of $\mathcal{L}_3$ is larger than for the former configuration:

$$\mathcal{L}_3^{n=6} = \frac{4B\varrho_{q_0}^3}{3\sqrt{6}} > \mathcal{L}_3^{n=3} , \quad (19)$$

and leads to the extrema of the free energy, corresponding to the most stable configuration.

From amazingly simple considerations, one has a prediction that solids in three dimensions are all BCC lattices—a very universal description of the system. Before confronting this claim with the data one needs to recall that the transitions between solids and liquids are not second-order transitions, they are actually first-order transitions. So, one may question the validity of universality claims in this context. However, it turns out that in many cases one can arrange that the solidifications occur as a weak first-order transitions, in which case approximate universality properties can be present.

Returning to the data and following [6], one discovers that about 40 metals, which are on the left of the periodic table (excluding magnesium (Mg)), form near the solidification point a BCC configuration. I will repeat the difficulties of the analysis and the argumentation to proceed with it nevertheless. The transition is first order—the fact that in many cases it is a weak first-order transition softens this problem. There is no true expansion parameter in the problem. The microscopic theory constructing A and B is very phenomenological, and therefore, the real relative stability of the metal is a very delicate

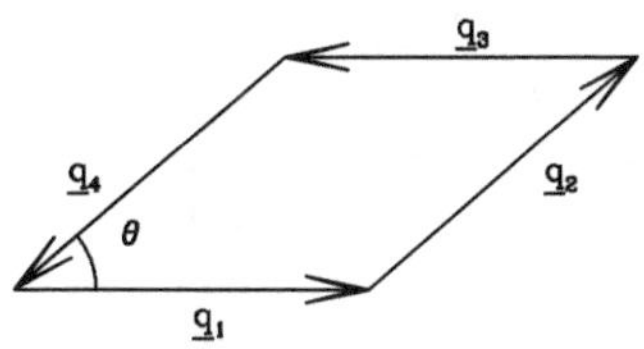

Fig. 8. In the absence of a cubic term, a quartic term would not suffice classically to induce a spontaneous breaking of rotational invariance. A rhombus does not single out a preferred angle θ

matter. Even taking all these into account, the result and its agreement with a large body of the experimental data is striking.

Consider what would have happened without a cubic term. In that case, the term following the quadratic term would be $\mathcal{L}_4$, which schematically would assume the form

$$\begin{aligned}\mathcal{L}_4 = \int d\boldsymbol{q}_1 d\boldsymbol{q}_2 d\boldsymbol{q}_3 d\boldsymbol{q}_4\ \delta(\boldsymbol{q}_1 + \boldsymbol{q}_2 + \boldsymbol{q}_3 + \boldsymbol{q}_4) \times \\ C(|\boldsymbol{q}_1|^2, |\boldsymbol{q}_2|^2 |\boldsymbol{q}_3|^2, |\boldsymbol{q}_4|^2, \boldsymbol{q}_1 \cdot \boldsymbol{q}_2, \boldsymbol{q}_1 \cdot \boldsymbol{q}_3, \dots)\end{aligned} \quad (20)$$

where the delta function enforces translational invariance, and C should be built by such invariants that maintain both rotational and translational invariance.

This does not break rotational invariance because, unlike the case of triangles, the configurations which are enforced now, assuming perturbation theory, are those of quadrilaterals with equal sides. But for a rhombus (Fig. 8) no angle is singled out. The rotational invariance is not broken. Fortunately there is no microscopic symmetry consideration that rules out the cubic term.

Another interesting type of lattices are the Abrikosov lattices formed of vortexes, which we do not discuss here.

String Theory Compactifications

What has been described above has a very solid basis in nature. What we will describe next is of a much more speculative nature, and it is based on work by Elitzur, Forge and myself [4] , in which we try to address the issues of compactification in String Theory. There are several attitudes one might adopt regarding compactification. One, which makes a lot of sense, is to say that the Universe starts up very small, and the issue of compactification is an issue of explaining why four dimensions became very large, while the rest of the dimensions remain small. This is not what I am going to discuss here.

Here, I discuss possible dynamical aspects of compactification taking in account some of the hints learnt from the case of solid-state physics. I don't have much confidence in human imagination when it is totally detached from reality, I would hope that many of the hints available in nature to be useful to understand other phenomena. In particle physics one has learned quite a lot from the dynamics of solid-state physics, and statistical mechanics systems.

Returning to the case at hand we have just reviewed a system which has lost most of its rotational and translational invariance, and we want to see how such a thing could happen in String Theory. One of the key ingredients driving this behavior is the presence of a bulk tachyon.

There are actually at least three types of tachyons/instabilities with which one is familiar right now in String Theory. One is that of the Bosonic String Theory tachyon. This instability could well be an incurable one; nevertheless, let's try and follow it.

The other types of instabilities, which we will discuss later, are an instability in Open String Theory, an open string tachyon, and also localized bulk tachyons.

For the moment we focus on bulk tachyons, which will be one key ingredient. Due to them, it is preferable for a system in String Theory containing a tachyon to have a support on a non-zero value of q^2. One can see this from the form of the tachyon whose vertex operator is the following

$$T(x) = e^{\mathrm{i}q_0 x} + h.c. \tag{21}$$

To obtain a dimension $(1,1)$ operator one needs $q_0 \neq 0$. Tachyons do give us the starting point that appears in Landau theory of solidification (note that here it is not a minimum consideration). The second key ingredient that we need for Landau's theory of solidification, in order to obtain not only the breakdown of translational invariance, but also of rotational invariance, is the presence of a cubic term. We know from the OPE (operator-product-expansion) that three tachyons do couple together (see Fig. 9). In particular, the OPE between two tachyons does contain a third tachyon. So we have in a such a theory a T^3 term. One indeed has the necessary ingredients to try to follow if tachyons could lead to the spontaneous breaking of rotational and translational invariance in String Theory, and maybe also to compactification.

In order to be more concrete, we followed the ideas of [7, 8] and tried to handle in a reliable fashion almost marginal operators. Consider a tachyon which is not an exact $(1,1)$ operator, but one which has $q_0^2 = 2 - \varepsilon$. We will also look at the subset of the full string background, a subset which contains a $c = 2$ sector . We will not deal here with the question of how the total central charge remains at the appropriate value, which is zero, and how to dress operators.

As an illustration, consider the subset of the backgrounds which are string moving in flat space, where the piece of the Lagrangian on which we focus is

$$\mathcal{L} = \partial X^1 \bar{\partial} X^1 + \partial X^2 \bar{\partial} X^2 + T(X^1, X^2). \tag{22}$$

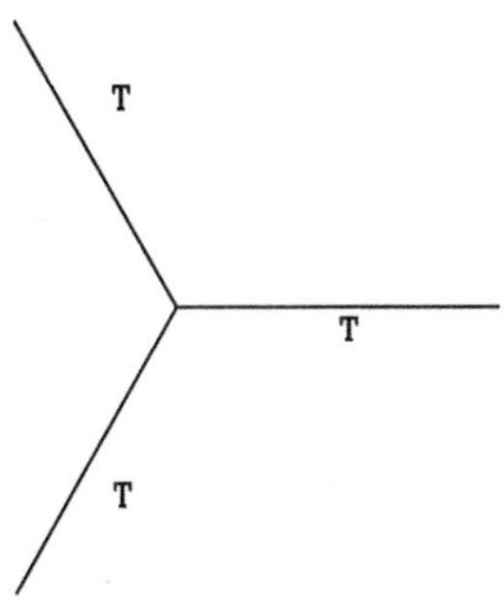

Fig. 9. Tachyonic cubic vertex

From Landau's theory of solidification we know that, because the system has support on a $q_0 \neq 0$, and because the free energy of the system contains a cubic coupling, we can try and build the triplets, which again actually correspond to six vectors, so that they get a support in an appropriate way, i.e. such that they break translational and rotational invariance.

The 60° angle, discussed in the solidification case, manifests itself in a suggested tachyon configuration:

$$T(X^1, X^2) = \sum_{a=1}^{3} T_a \cos(\sum_{i}^{2} k_i^a X^i), \tag{23}$$

where the three momenta $\boldsymbol{k}^1$, $\boldsymbol{k}^2$ and $\boldsymbol{k}^3$ are the following.

$$\boldsymbol{k}^1 = k(1,0) \quad \boldsymbol{k}^2 = k(-1/2, \sqrt{3}/2) \quad \boldsymbol{k}^3 = k(-1/2, -\sqrt{3}/2). \tag{24}$$

All of them have $\boldsymbol{k}^2 = 2 - \varepsilon$, and the structure is very similar to that of the $SU(3)$ root lattice (see Fig. 10), as before: For every k_i there is also the corresponding $-k_i$ contribution.

One can simplify the tachyon potential by taking the ansatz for the amplitudes $T_a = T$. The Lagrangian one needs to solve is the one given in (22), and actually one can show that the beta function of the tachyon alone vanishes to order ε. So (23) is a solution of the approximate tachyon equations of motion. This means that had it been up to the tachyon alone one would have obtained the lattice, perhaps some honeycomb or triangular lattice, which would break both translational and rotational invariance. However, this system contains also gravity so one needs to see what is the influence of the formation of such a lattice on gravity. As shown in [4], the beta function for the graviton $\beta_{G_{\mu\nu}}$ vanishes (at leading order in α') if

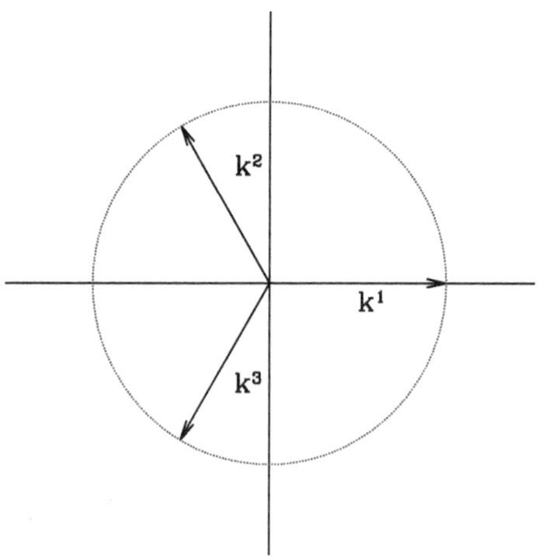

Fig. 10. SU(2) roots

$$\beta_{G_{\mu\nu}} = -R_{\mu\nu} + \nabla_\mu T \nabla_\nu T = -R_{\mu\nu} + \frac{3}{2}\varepsilon^2 \delta_{\mu\nu} = 0. \tag{25}$$

For $D = 2$, due to the Liouville theorem $R_{\mu\nu}$ can be written as $R_{\mu\nu} = a\delta_{\mu\nu}$, so actually one can solve the equation by forming a two-dimensional sphere.

This is actually a highlight of a model for compactification. We started by having just a tachyon. The tachyon would have produced the lattice on its own, but because of the presence of the gravity, the lattice of tachyons actually causes the compactification of space to a sphere.

However, it turns out, and details are presented in [4], that unfortunately this result is not obtained in a desired reliable approximation. The main problem is that, in order to do reliable perturbation theory, we need to perform a plane-wave expansion, with the wave lengths representing a nearly marginal operator. However, when the sphere is formed, the topology changes, and the change of topology means that one should now expand the fields in terms of spherical harmonics $Y_{l,m}$. This topological obstruction takes away the reliability of our calculation. Some defects may form in order to resolve this topological problem, and one conjecture we had at that time was that actually parafermions, which are defects, form to resolve the tension. A more complex form of compactification emerges.

Once again, recall that actually the system, when fully considered, has to be coupled to the dilaton in order to maintain the total central charge. According to the Zamolodchikov theorem [7], once the system starts to flow, the central charge decreases from 2 and this on its own breaks the balance. In a sense, in the case of bulk tachyons we were tantalizingly close to obtain an explicit dynamical mechanism for compactification. However, due to topological obstructions, what was a solution for the beta function locally in space cannot be a global solution without taking into account other effects. We will return to the breaking of translational invariance in the different context of the open string tachyon.

Liquid Crystals

The tachyon is a scalar order parameter, String Theory has additional fields which carry indexes. In particular, one might think that if one looks for a similarity to our universe, maybe one should consider the phase of liquid crystals. Such systems are translational invariant in some directions but not in other (see Fig. 11). We will give now examples of that.

There are various types of liquid crystals and one can ask what is the Landau–Ginsburg theory of them. Actually, one can also ask about vector potential systems which are described, as gauge fields are, by vector-like order parameters. Such systems include detergents which possess a hydrophobic and a hydrophilic pole, and play a crucial role in cleaning our garments. One can try to extract from $\boldsymbol{p}(\boldsymbol{r})$ the various invariants one wants to use in order to describe this system, such as $div\boldsymbol{p}$, $curl\boldsymbol{p}$, $s_{\alpha\beta} = \partial_\alpha p_\beta + \partial_\beta p_\alpha$. It turns out that

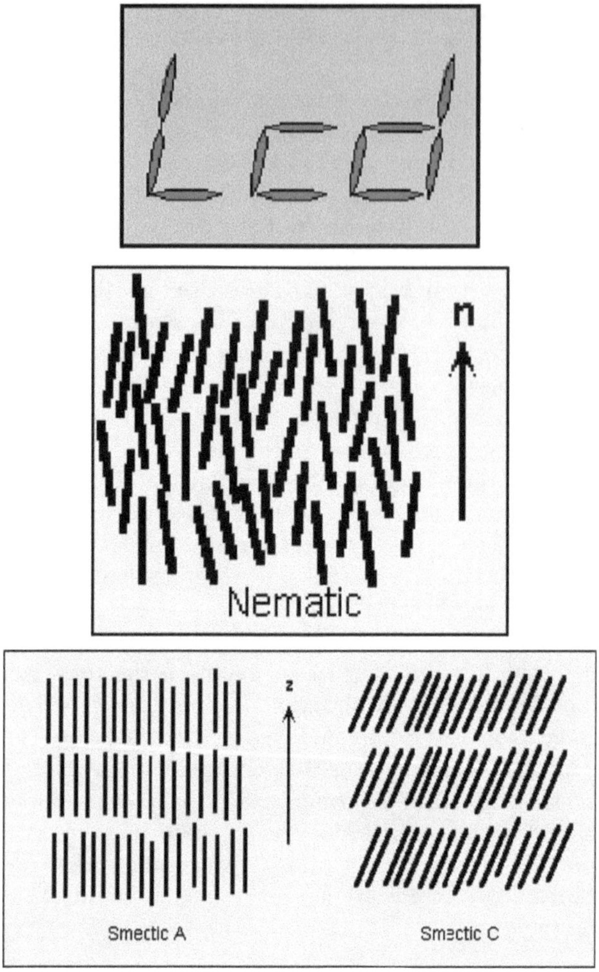

Fig. 11. Various phases of liquid crystals breaking. These systems exhibit asymmetrical breaking of translational and rotational invariance

one can write down a Landau–Ginsburg theory for detergents, which explains many of their very fascinating properties.

Considering the case of liquid crystals, these can also be described by choosing for example particular spherical harmonic functions, and using them as an order parameters.

For illustrative purposes, we give the dependence of the density ϕ on the angles and on the coordinates[1]

[1] The index structure of ϕ has been omitted

$$\phi = \sum_i \mu_i Y_2^2(\theta_i, \phi_i) e^{\mathrm{i}\boldsymbol{k}_i \cdot \boldsymbol{r}_i} + h.c. \tag{26}$$

By assuming the ansatz $\mu_i = \mu$, the effective Landau–Ginsburg free energy is given below

$$F \sim (\alpha_0 + dk + ck^2)\mu^2 - \beta\mu^3 + r\mu^4, \tag{27}$$

from which one can extract the properties of nematic, smectic A and smectic C properties, and many other exciting things for which we refer to the literature [6].

Boundary Perturbations

Next I discuss an example where a breakdown of spatial translational symmetry actually clearly occurs. As mentioned above one can formulate an intuitive theorem in the bulk; the theorem states that under the renormalization group flow, the value of the Virasoro central charge c decreases from its UV value to a smaller IR value. This is due to the integrating out of the degrees of freedom and applies to the unitary sectors of String Theory. In String Theory with its ghosts the total central charge vanishes. One can imagine mechanisms by which the central charge of the ghosts increases [9], but basically one needs to couple the two systems maintaining a total vanishing central charge. This can be done for example with the help of a linear dilaton, and leads to very interesting questions and results . Generically, the matter central charge will decrease to zero leaving one with just a $c = 0$ topological theory, but there are also other possibilities. The central charge is related to the anomaly which exists in the bulk. On the other hand, when one considers the boundary theory, there are no gravitational anomalies in it. Thus in that case one can consider tachyonic open string theory perturbations. In the example given by the action below

$$S = \int_{\Sigma} \mathcal{L}_{CFT} + \int_{\partial\Sigma} g\mathcal{O}_{Rel.}, \tag{28}$$

the bulk theory is defined on the surface Σ, and on its boundary $\partial\Sigma$ one adds a relevant operator $\mathcal{O}_{Rel.}$. There is a boundary renormalization group flow which does not change the bulk central charge, and therefore does not lead to all the problems associated with tachyons in the bulk.

One can associate a term in the boundary which measures the effective number of degrees of freedom, and this has been done by various authors [10, 11].

It can be proved, moreover, that one can define such a function whose value also decreases when the theory flows on the boundary, all this without requiring an adjustment of the total central charge. What happens for example is that the theory flows from Dirichlet(D) to Neuman(N) boundary conditions, so that, in other words, branes may dissolve or may be created under such

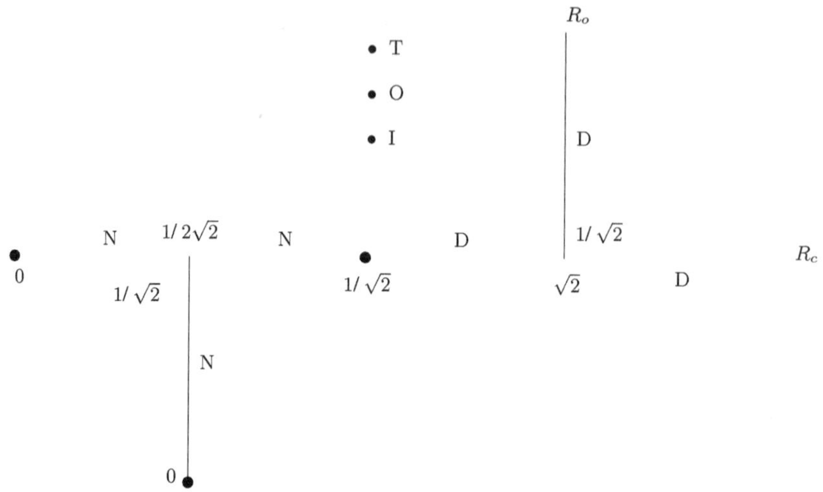

Fig. 12. Map of the preferred boundary conditions in the $c = 1$ moduli space, N stands for Neuman and D for Dirichlet boundary conditions [9]

a flow. In Fig. 12 we give an example of a very simple compactification in which one can identify what are the stable configurations, describing when the system chooses to obey Dirichlet and when the system chooses to obey Neuman boundary conditions [9].

This can be used even further if one changes the relevant operator added on the boundary into a Sine–Gordon one. In that case one can actually has situations where one breaks translational invariance in space–time by a $D-25$ brane, for example dissolving into a lattice of $D-24$ branes [12] (Fig. 13). Again, such a situation will also lead to a reduction of the original amount of supersymmetry. Thus, the idea of spontaneous breaking of spatial translational invariance is demonstratively realized in String Theory by the presence of open string tachyons.

3 Spontaneous Breaking of Time-Translational Invariance and of Supersymmetry

Next I will discuss a somewhat different mechanism which may allow the possibility of a spontaneous breaking of time-translational invariance. For that it is useful to consider conformal and superconformal quantum mechanics. One way to motivate the interest in such systems is to recall some basic facts concerning the validity of a perturbative expansion.

Consider the Hamiltonian,

$$H = \frac{p_q^2}{2m} + \frac{1}{2} g q^n \,. \tag{29}$$

Fig. 13. A lattice of D24 branes is formed from a D25 brane in the presence of a boundary tachyon

One may wonder if it is possible to make a meaningful perturbative expansion in terms of small or large g or small or large m. To answer this one needs to find out if one can remove the g,m dependence from the operators, and relegate it to the total energy scale. This type of rescaling is used for discussing the harmonic oscillator. One attempts to define a new set of dimensionless canonical variables p_x, x that preserve the commutation relations,

$$[p_q, q] = [p_x, x]\hbar \ , \tag{30}$$

and

$$H = h(m,g)\frac{1}{2}(p_x^2 + x^n) \ . \tag{31}$$

The following decomposition

$$q = f(m,g)x \ , \qquad p_q = \frac{1}{f(m,g)}p_x \tag{32}$$

gives

$$2H = \frac{p_x^2}{mf^2(m,g)} + gf(m,g)^n x^n \quad , \tag{33}$$

and so one may choose

$$gf(m,g) = \left(\frac{1}{mf(m,g)^2}\right)^{\frac{1}{n+2}} \ . \tag{34}$$

The Hamiltonian becomes

$$H = g^{1-\frac{n}{n+2}} m^{-\frac{n}{n+2}} \frac{1}{2}(p_q^2 + q^n) \ . \tag{35}$$

The role of g and m is indeed just to determine the overall energy scale. They may not serve as meaningful perturbation parameters. This does not apply to the special case of $n = -2$, the case of conformal quantum mechanics, where g can be a real perturbative parameter.

3.1 Conformal Quantum Mechanics: A Stable System with No Ground State Breaks Time-Translational Invariance

Consider the Hamiltonian

$$H = \frac{1}{2}(p^2 + gx^{-2}) \tag{36}$$

for a positive value of g [13]. H is part of the following algebra:

$$[H, D] = iH \ , \quad [K, D] = iK \ , \quad [H, K] = 2iD \ . \tag{37}$$

It is an SO(2,1) algebra, one representation of which is

$$D = -\frac{1}{4}(xp + px) \ , \quad K = \frac{1}{2}x^2, \tag{38}$$

with H given above. The Casimir is given by

$$\frac{1}{2}(HK + KH) - D^2 = \frac{g}{4} - \frac{3}{16} \ . \tag{39}$$

In the Lagrangian formalism the system is described by

$$\mathcal{L} = \frac{1}{2}(\dot{x}^2 - \frac{g}{x^2}) \ , \quad S = \int dt\mathcal{L} \ . \tag{40}$$

Symmetries of the action S, and not of the Lagrangian $\mathcal{L}$ alone, are given by

$$t' = \frac{at + b}{ct + d} \ , \quad x'(t') = \frac{1}{ct + d}x(t) \ , \tag{41}$$

$$A = \begin{pmatrix} a & b \\ c & d \end{pmatrix} \ , \qquad \mathrm{detA = ad - bc = 1}. \tag{42}$$

H acts as translation

$$A_T = \begin{pmatrix} 1 & 0 \\ \delta & 1 \end{pmatrix} \ , \qquad t' = t + \delta \ . \tag{43}$$

D acts as dilation

$$A_D = \begin{pmatrix} \alpha & 0 \\ 0 & \frac{1}{\alpha} \end{pmatrix} \ , \qquad t' = \alpha^2 t \ . \tag{44}$$

K acts as a special conformal transformation

$$A_K = \begin{pmatrix} 1 & \delta \\ 0 & 1 \end{pmatrix}, \qquad t' = \frac{t}{\delta t + 1} . \tag{45}$$

The spectrum of the Hamiltonian (36) is the open set $(0, \infty)$, the spectrum is therefore continuous and bounded from below. The wave functions are given by

$$\psi_E(x) = \sqrt{x} J_{\sqrt{g+\frac{1}{4}}}(\sqrt{2E}x), \qquad E \neq 0 . \tag{46}$$

The zero-energy state is given by $\phi(x) = x^{\alpha}$:

$$H\phi(x) = \left(-\frac{d^2}{dx^2} + \frac{g}{x^2} \right) x^{\alpha} = 0 . \tag{47}$$

This implies

$$g = -\alpha(\alpha - 1) , \tag{48}$$

and solving this equation gives

$$\alpha = -\frac{1}{2} \pm \frac{\sqrt{1+4g}}{2} . \tag{49}$$

This gives rise to two independent solutions, and by completeness these are all the solutions. The case $\alpha_+ > 0$ does not lead to a normalizable solution since the function diverges at infinity. The case $\alpha_- < 0$ is not normalizable either, since the function diverges at the origin (a result of the scale symmetry).

Thus, there is no normalizable (not even plane-wave normalizable) $E = 0$ solution (Fig. 14)!

Most of the analysis in field theory proceeds by identifying a ground state and the fluctuations around it. How do we deal with a system in the absence of a ground state?

One possibility is to accept this as a fact of life. Perhaps it is possible to view this as similar to cosmological models that also lack a ground state, such

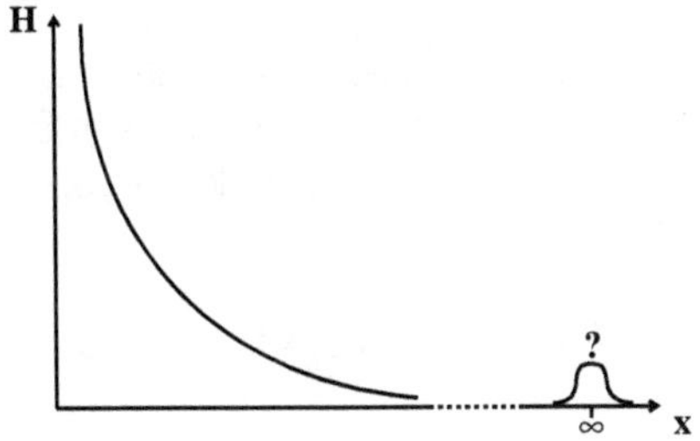

Fig. 14. There is no normalizable ground state for this potential

those with Quintessence. In field theory such systems have no finite-energy states in the spectrum at all. Only time-dependent states are allowed. In the presence of an appropriate cutoff, and in quantum mechanics, it is only the potential lowest energy state which is disallowed.

Another possibility is to define a new evolution operator that does have a ground state

$$G = uH + vD + wK \ . \tag{50}$$

This operator has a ground state if $v^2 - 4uw < 0$. Any choice explicitly breaks scale invariance. Take for example

$$G = \frac{1}{2}\left(\frac{1}{a}K + aH\right) \equiv R \ , \tag{51}$$

where a has the dimension of a length. The eigenvalues of R are

$$r_n = r_0 + n \ , \qquad r_0 = \frac{1}{2}\left(1 + \sqrt{g + \frac{1}{4}}\right) \ . \tag{52}$$

This is a breaking of scale invariance by a dictum and not by the dynamics of the system. Nevertheless, it is very interesting to search for a physical interpretation of this. Surprisingly, this question arises in the context of black-hole physics. Consider a particle of mass m and charge q falling into a charged black hole. The black hole is BPS, meaning that its mass M and charge Q are related, in the appropriate unites, by $M = Q$.

The black hole metric and vector potential are given by

$$ds^2 = -\left(1 + \frac{M}{r}\right)^{-2} dt^2 + \left(1 + \frac{M}{r}\right)^2 (dr^2 + r^2 d\Omega^2) \ , \qquad A_t = \frac{r}{M} . \tag{53}$$

Now consider the near Horizon limit, i.e. $r << M$, which we will reach by taking $M \to \infty$ and keeping r fixed. This produces an $AdS_2 \times S^2$ geometry

$$ds^2 = -\left(\frac{r}{M}\right)^2 dt^2 + \left(\frac{M}{r}\right)^2 dr^2 + M^2 d\Omega^2 \ . \tag{54}$$

We also wish to keep $M^2(m - q)$ fixed as we scale M. This means we must scale $(m - q) \to 0$, that is, the particle itself becomes BPS in the limit.

The Hamiltonian for this falling in particle, in this limit, is given by our old friend:

$$H = \frac{p_r^2}{2m} + \frac{g}{2r^2} \ , \qquad g = 8M^2(m - q) + \frac{4l(l+1)}{M} \ . \tag{55}$$

For $l = 0$ we have $g > 0$, and there is no ground state. This is associated with the coordinate singularity at the Horizon. The change in evolution operator is now associated with a change of time coordinate. One for which the world line of a static particle passes through the black-hole horizon, instead of remaining in the exterior of the space–time. In any case, the consequence of removing the potential lowest energy state of the system from the spectrum can be described as a breaking of time-translational invariance.

3.2 Superconformal Quantum Mechanics: A Stable System with No Ground State Also Breaks Supersymmetry

The bosonic conformal mechanical system had no ground state. The absence of a $E = 0$ ground state in the supersymmetric context leads to the breaking of supersymmetry. This breaking has a different flavor from that which was discussed for the spatial translations. We next examine the supersymmetric version of conformal quantum mechanics [1, 14], to see if indeed supersymmetry is broken. The superpotential is chosen to be

$$W(x) = \frac{1}{2} g \log x^2 , \tag{56}$$

yielding the Hamiltonian:

$$H = \frac{1}{2} \left[\left(p^2 + \left(\frac{dW}{dx} \right)^2 \right) 1 - \frac{d^2 W}{dx^2} \sigma_3 \right] . \tag{57}$$

Representing ψ by $\frac{1}{2}\sigma_-$ and $\psi^*$by $\frac{1}{2}\sigma_+$ gives the supercharges:

$$Q = \psi^+ \left(-ip + \frac{dW}{dx} \right) , \quad Q^+ = \psi \left(ip + \frac{dW}{Dx} \right) . \tag{58}$$

One now has a larger algebra, the superconformal algebra,

$$\begin{aligned} \{Q, Q^+\} &= 2H , \quad \{Q, S^+\} = g - B + 2iD , \\ \{S, S^+\} &= 2K , \quad \{Q^+, S\} = g - B - 2iD . \end{aligned} \tag{59}$$

A realization is

$$B = \sigma_3 , \quad S = \psi^+ x , \quad S^+ = \psi x . \tag{60}$$

The zero-energy solutions are

$$exp(\pm W(x)) = x^{\pm g} , \tag{61}$$

and neither solution is normalizable.

The Hamiltonian H factorizes

$$2H = \begin{pmatrix} p^2 + \frac{g(g+1)}{x^2} & 0 \\ 0 & p^2 + \frac{g(g-1)}{x^2} \end{pmatrix} , \tag{62}$$

and we may solve for the full spectrum:

$$\psi_E(x) = x^{1/2} J_{\sqrt{\nu}}(x\sqrt{2E}) \,, \qquad E \neq 0 \,, \tag{63}$$

where $\nu = g(g-1) + 1/4$ for $N_F = 0$ and $\nu = g(g+1) + 1/4$ for $N_F = 1$.

The spectrum is continuous and there is no normalizable zero-energy state. One must interpret the absence of a normalizable ground state. It is also possible to define a new operator which has a normalizable ground state. By inspection the operator (51) can be used, provided one makes the following identifications:

$$\begin{aligned} N_F &= 1 \qquad g_B = g_{susy}(g_{susy}+1) \,, \\ N_F &= 0 \qquad g_B = g_{susy}(g_{susy}-1) \,. \end{aligned} \tag{64}$$

Thus the spectrum differs between the $N_F = 1$ and $N_F = 0$ sectors, and supersymmetry would be broken. One needs to define a whole new set of operators:

$$\begin{aligned} M &= Q - S \qquad M^+ = Q^+ - S^+ \\ N &= Q^+ + S^+ \qquad N^+ = Q + S^+ \end{aligned} \tag{65}$$

which produces the algebra:

$$\begin{aligned} \frac{1}{4}\{M, M^+\} &= R + \frac{1}{2}B - \frac{1}{2}g \equiv T_1 \,, \\ \frac{1}{4}\{N, N^+\} &= R + \frac{1}{2}B + \frac{1}{2}g \equiv T_2 \,, \\ \frac{1}{4}\{M, N\} &= L_- \qquad \frac{1}{4}\{M^+, N^+\} = L_+ \,, \\ L_\pm &= -\frac{1}{2}(H - K \mp 2iD). \end{aligned} \tag{66}$$

The operators T_1, T_2, H have a doublet spectrum. “Ground states” are given by

$$T_1|0>=0 \,; \quad T_2|0>=0 \,; \quad H|0>=0 \,. \tag{67}$$

In this setup one can also exhibit [1] how in the presence of a breaking of a space–time symmetry, global $N = 2$ can be broken only to $N = 1$. A physical context arises when one considers a supersymmetric particle falling into a black hole [15, 16]. This is the supersymmetric analogue of the situation already discussed.

One should mention again that there is a dictum in the way one has broken scale/conformal invariance in the problem. It is amusing to mention that if one takes the dictated ground state, and decomposes it in terms of the energy eigenstates, then one usually gets that the new ground state looks like a thermal distribution of the old ground states. This looks very attractive and it is related to black holes, which as mentioned above do come up.

Another example where such breakdown of time-translational invariance may occur is the Liouville model. Also, there is no normalizable ground state. For works on the possible breakdown of translational invariance in the two-dimensional Liouville model see [17, 18].

Beyond $d = 2$, we can mention that in four dimensions in $\mathcal{N} = 1$ supersymmetric theories, where the number of flavors N_F is smaller than the number of colors, $0 < N_F < N_C$, one also gets [19, 20] a situation where the spectrum is bounded from below, but there is no ground state. The spectrum is open, and actually in the presence of a cutoff such systems have no finite-energy states at all, which is very interesting as far as Cosmology is concerned.

4 Spontaneous Breaking of Conformal Invariance

Fubini [21] also suggested to discuss such situations in a general number of dimensions. He researched it in a scientific environment which did not yet fully realize that interacting finite theories might exist in various number of dimensions. Therefore, much of his analysis was of a classical nature. He emphasized the conformal features of the system, and we are going to discuss the breakdown of conformal invariance. The discussion of the breakdown of time-translational invariance brought us to conformal theories and now we are discussing also the breakdown of the conformal invariance.

If one considers a theory with only one scalar field, a general classic conformal invariant is given by the following Lagrangian

$$\mathcal{L} = \frac{1}{2}\partial_\mu\phi\partial^\mu\phi - g\phi^{\frac{2d}{d-2}} . \tag{68}$$

The symmetry of the system is the bosonic, $O(d, 2)$ symmetry, and the generator are $M_{\mu\nu}$, P_μ, of the Poincaré group, the special conformal transformation generator K_μ, and the dilatation D. The dictum of Fubini in this case is that the ground state is not translational invariant, and this is not accompanied by any dynamical calculation. The vacuum expectation value $< \phi(x) >$ is x dependent, and actually it looks very much like an instanton

$$< \phi(x) >= b\left(\frac{a^2 + x^2}{2a}\right)^{-\frac{d-2}{2}} , \tag{69}$$

which is a solution of the equation of motion

$$\partial^2\phi(x) - 2g\frac{d}{d-2}\phi^{\frac{d+2}{d-2}}(x) = 0 . \tag{70}$$

By choosing this to be the vacuum, (again I emphasize, this is by dictum), one breaks down the $O(d, 2)$ symmetry (as in Fig. 15) in the following fashion:

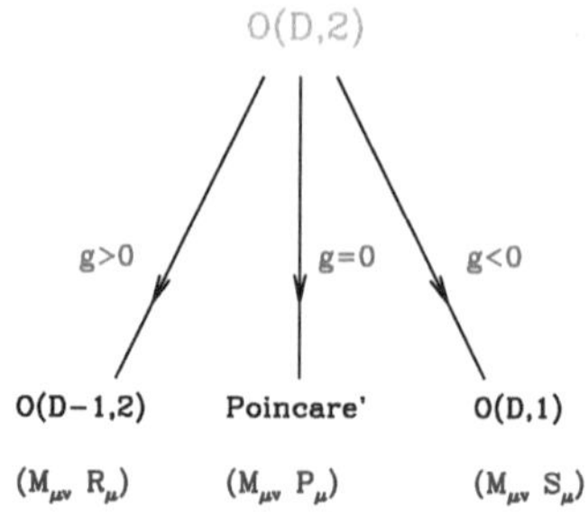

Fig. 15. The sign of the quartic coupling g determines the symmetry breaking patterns of the symmetry group $O(d, 2)$

if the coupling g of the scalar self-interaction is positive, then the theory breaks down to $O(d-1, 2)$ and the resulting symmetries are $M_{\mu\nu}$, R_μ. If $g < 0$, then the symmetry breaks to $O(d-1)$, generated by $M_{\mu\nu}$, S_μ, where

$$S_\mu = \frac{1}{2}\left(aP_\mu - \frac{1}{a}K_\mu\right) . \tag{71}$$

If $g = 0$, one remains with Poincaré invariance (Fig. 15). In the de Sitter example, which occurs for $g > 0$, one can show again that there are signatures of temperature. A question which at the time seemed interesting was: Does a spontaneous breaking of conformal invariance require also the breakdown of translational invariance? Examples were since found where this is not the case. Counter-examples to the idea that the breaking of conformal invariance must drive a breaking of supersymmetry were discovered, and we will discuss in more detail some such examples. One can break scale invariance without breaking rotational or translational invariance. We also mention briefly that conformal invariance and scale invariance are not always equivalent, and in a set of works (see, e.g., [22]) it has been shown that scale invariance leads, under certain conditions, to conformal invariance.

For instance, this occurs in the case where the spectrum of the theory is discrete, such as in a two-dimensional sigma model description in which the target space is compact. But for non-compact target spaces one can find counter-examples [23] in which scale invariance does not lead to conformal invariance. In recent years it has been fully realized that theories which are quantum mechanically scalar invariant and finite may exist in $d = 2, 3, 4, 5, 6$ dimensions. Such theories can exhibit spontaneous breaking, e.g., the $d = 4$, $\mathcal{N} = 4$ super Yang–Mills with $SU(N)$ gauge group which is characterized by the following spectrum

$$(A^a_\mu, \lambda^a, \phi^a + i\varrho^a).$$

The theory is parameterized by the complex parameter $ig + \theta$, where g is the coupling constant and θ is the angle. Such a theory has flat directions which allow phases where either $<\phi>$ vanishes and the theory is realized in a conformal manner, or a phase in which $<\phi>\neq 0$ along flat directions. This is the Coulomb phase, in which the gauge group $SU(N)$ may be reduced all the way to $U(1)^N$, where N is the rank of the gauge group. This is the maximum possible breaking of the gauge group when the fields are in the adjoint representation. In such a case, scale invariance is broken spontaneously and the vacuum energy remains zero, and there is no breakdown of either translational invariance or supersymmetry. Such a theory will have a Goldstone boson, associated with the spontaneous breaking of scale invariance, which is called the dilaton. This is a *true* dilaton worthy of his name. It is interesting to note that in such a system the vacuum energy is not influenced by the value of $<\phi>$, and it vanishes in all the phases.

5 $O(N)$ Vector Models in $d = 3$: Spontaneous Breaking of Scale Invariance and the Vacuum Energy

The next example that we have is related to the spontaneous breaking of scale invariance in a three-dimensional bosonic theory. Such a theory describes the mixing of He_3 and He_4, (see [24] and references therein).

The most general Lagrangian describing such a system is

$$\mathcal{L} = \frac{1}{2}\partial_\mu\phi\,\partial^\mu\phi - \frac{1}{2}\lambda_2(\phi)^2 + \frac{\lambda_4}{4N}(\phi)^4 + \frac{\lambda_6}{N^2}(\phi)^6, \tag{72}$$

and it can be treated at $d = 3 - \varepsilon$. The system has two order parameters, $<(\phi)^2>$ and $<\phi>$.

In a classical analysis performed for $d = 3-\varepsilon$, when the sign of λ_2 changes, $<\phi>$ is produced. However, $<(\phi)^2>\neq 0$ even for $\lambda_2 > 0$, which is exemplified by the diagram shown in Fig. 16.

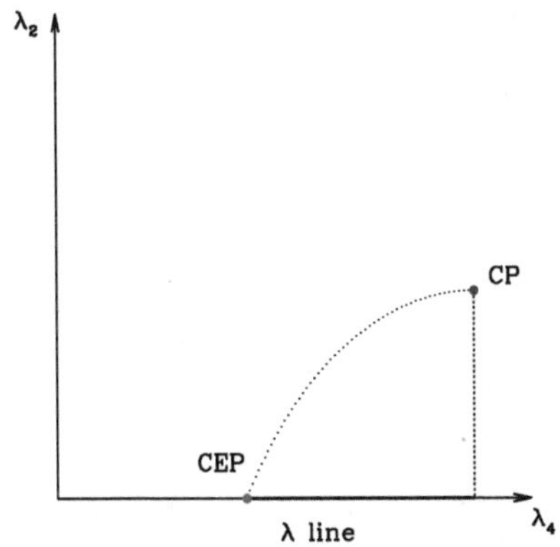

Fig. 16. The phase diagram of a $d = 3 - \varepsilon$ Conformal Theory, in three dimensions the CP and CEP points coincide to produce a flat direction

When one goes to three dimensions, the point which is denoted by CP, which is a critical point, and the point CEP which is the critical end point do actually meet together and lead to a very interesting structure. Going directly to $d = 3$, one can write down the $O(N)$ vector model written below

$$\mathcal{L} = \frac{1}{2}\partial_\mu\phi\,\partial^\mu\phi - \frac{1}{2}\lambda_2(\phi)^2 + \frac{\lambda_4}{4!N}(\phi)^4 + \frac{\lambda_6}{6!N^2}(\phi)^6\,. \tag{73}$$

It should be emphasized that everything said depends on the very specific manner of taking the limit. One first keeps the cutoff Λ fixed and takes $N \to \infty$, by performing a functional integral or selecting a subset of diagrams, and only then does one remove the cutoff, sending it to infinity, setting the renormalized quadratic and quartic couplings to zero. Such a system turns out to be not only classically conformally invariant, but also quantum mechanically, having a vanishing beta function [25]. We next elaborate on such systems.

Let us now review some more known facts about the three-dimensional theory once a classically marginal operator,$(\phi^2)^3$, is added [25]. For any finite value of N, the coupling g_6 of this operator is infrared-free quantum mechanically, as the marginal operator gets a positive anomalous dimension already at one loop. This implies that the theory is only well defined for zero value of the coupling of this operator. In the presence of a cutoff, interacting particles have mass of the order of the cutoff. At its tri-critical point the $O(N)$ model in three dimensions is described by the Lagrangian

$$\mathcal{L} = \frac{1}{2}\partial_\mu\phi\,\partial^\mu\phi + \frac{1}{6N^2}g_6(\phi^2)^3\,, \tag{74}$$

where the fields ϕ are in the vector representation of $O(N)$.

In the limit $N \to \infty$ [25]

$$\beta_{g_6} = 0\,; \tag{75}$$

$1/N$ corrections break conformality. In the large N limit, g_6 is a modulus. It turns out there is no spontaneous breaking of the $O(N)$ symmetry, and it is instructive to write the effective potential in terms of an $O(N)$ invariant field,

$$\sigma = \phi^2\,. \tag{76}$$

The effective potential is [24]

$$V(\sigma) = f(g_6)|\sigma|^3\,, \tag{77}$$

where

$$f(g_6) = g_c - g_6 \tag{78}$$

with

$$g_c = (4\pi)^2\,. \tag{79}$$

The system has various phases. For values of g_6 smaller than g_c, i.e. when $f(g_6)$ is positive, the system consists of N massless non-interacting ϕ particles. These particles do not interact in the infinite N limit; thus, correlation functions do not depend on g_6. For the special value $g_6 = g_c$, $f(g_6)$ vanishes and a flat direction in σ opens up: The expectation value of σ becomes a modulus. For a zero value of this expectation value the theory continues to consist of N massless ϕ fields. For any non-zero value of the expectation value the system has N massive ϕ particles. All have the same mass due to the unbroken $O(N)$ symmetry. Scale invariance is broken spontaneously though the vacuum energy still vanishes. The Goldstone boson associated with the spontaneous breaking of scale invariance, the dilaton, is massless and identified as the $O(N)$ singlet field $\delta\sigma \equiv \sigma - \langle\sigma\rangle$. All the particles are non-interacting in the infinite N limit. This theory is not conformal: In the infrared limit, it flows to another theory containing a single, massless, $O(N)$-singlet particle. For larger values of g_6 the exact potential is unbounded from below. The system is unstable (in the supersymmetric case the potential is bounded from below and the larger g_6 structure is similar to the smaller g_6 structure [26]). Actually this instability is an artifact of the dimensional regularization used above, which does not respect the positivity of the renormalized field σ. In any case a more careful analysis [25] shows that the apparent instability reflects the inability to define a renormalizable interacting theory. All the masses are of the order of the cutoff, and there is no mechanism to scale them down to low mass values. In other words, the theory depends strongly on its UV completion.

This is summarized in Table 1. There, S.B. denotes spontaneous symmetry breaking of scale invariance and V is the vacuum energy. For $f(g_6) < 0$ the theory is unstable. Note that the vacuum energy always vanishes whenever the theory is well defined.

When $\langle\sigma\rangle \neq 0$, and the scale invariance is spontaneously broken, one can write down the effective theory for energy scales below $\langle\sigma\rangle$, and integrate out the degrees of freedom above that scale. The vacuum energy remains zero however, and is not proportional to $\langle\sigma\rangle^3$ as might be expected naively [24, 27, 28, 29, 30].

For completeness we note that by adding more vector fields one has also phases in which the internal global $O(N)$ symmetry is spontaneously broken.

Table 1. Marginal perturbations of the $O(N)$ model

$f(g_6)$	$\lvert\langle\sigma\rangle\rvert$	S.B.	Masses	V
$f(g_6) > 0$	0	No	0	0
$f(g_6) = 0$	0	No	0	0
$f(g_6) = 0$	$\neq 0$	Yes	Massless dilaton, N particles of equal mass	0
$f(g_6) < 0$	∞	Yes, but ill defined	Tachyons or masses of order the cutoff	$-\infty$

An example is the $O(N) \times O(N)$ model [29] with two fields in the vector representation of $O(N)$, with Lagrangian:

$$\mathcal{L} = \partial_\mu \phi_1 \cdot \partial^\mu \phi_1 + \partial_\mu \phi_2 \cdot \partial^\mu \phi_2 + \lambda_{6,0}(\phi_1^2)^3 + \lambda_{4,2}(\phi_1^2)^2(\phi_2^2) + \lambda_{2,4}(\phi_1^2)(\phi_2^2)^2 + \lambda_{0,6}(\phi_2^2)^3 \; . \tag{80}$$

Again, the β functions vanish in the strict $N \to \infty$ limit. There are now two possible scales, one associated with the breakdown of a global symmetry and another with the breakdown of scale invariance. The possibilities are summarized by the table below:

$O(N)$	$O(N)$	Scale	Massless	Massive	V
$+$	$+$	$+$	all	none	0
$-$	$+$	$-$	$(N-1)\pi's, D$	N, σ	0
$+$	$-$	$-$	$(N-1)\pi's, D$	N, σ	0
$-$	$-$	$-$	$2(N-1)\pi's, D$	σ	0

(81)

Again, in all cases, the vacuum energy vanishes. Assume a hierarchy of scales where the scale invariance is broken at a scale much above the scale at which the $O(N)$ symmetries are broken. One would have argued that one would have had a low-energy effective Lagrangian for the massless pions and dilaton, with a vacuum energy given by the scale at which the global symmetry is broken. This is not true, the vacuum energy remains zero. This system has a critical surface, on one patch the deep infrared theory contains only one massless particle: an $O(N) \times O(N)$ singlet. For the other patches the deep infrared theory is described by $O(N)$ massless particles, most of which are not $O(N)$ singlets.

In general, effective field theories should have all possible symmetries of the underlying theory, whether they are realized linearly or non-linearly. In finite scale invariant theories the vacuum energy E_{vac} should be determined by all scales and symmetries involved. It should have the same value (zero in this case), in all phases of the system whether or not expectation values are formed. This punches a hole in Zeldovich-like arguments [31] and offers a different view on the gravity of the cosmological constant problem [32]. If the theory has a global scale invariance, which is spontaneously broken, it will produce a dilaton. The question is: Where is the dilaton? The dilaton should be a massless field. Several authors [33, 34] tried to check the possibilities that the dilatons might exist, noting that the dilaton must be a massless Goldstone boson. Under certain assumptions, one finds out that actually in certain models having a massless dilaton would not violate experimental data. Perhaps it even predicts deviations of the equivalence principle from Galileo famous experiment of $\delta a/a \sim 10^{-12}$. Just below the present experimental sensitivity.

This is done under the assumption that the dilaton couples in the following universal fashion

$$\mathcal{L} = F(\Phi)\left(R - F^2 + 2[\nabla^2\Phi - (\nabla\Phi)^2]\right). \qquad (82)$$

It could also happen that the dilaton gets swallowed in some Higgs-like mechanism. One should also mention that if kinematically a finite-scale invariant is forced by some super-selection rule (such as having a non-trivial monopole number [35]) into a certain solitonic sector, then the rest energy of the system should be accounted for, and the vacuum energy will be slightly lifted from zero.

Let us finish this section by noticing an amusing thing—there are various solutions that go under the name of Randal and Sundrum. One of the constructions contains two types of branes, near the boundary of the space there is a Planck brane with tension T_1, which is fine-tuned so to have zero cosmological Constant. Then at a certain distance, very deep inside the bulk theory, one places the TeV brane, it has negative tension and the tension is again fine tuned, so that the cosmological constant vanishes also on that brane.

The two branes are separated by some distance which in [36] is associated to massless particle, which is the dilaton or the radion (see Fig. 17).

In principle, there are circumstances where this distance is not fixed, and there are several possible situations whose outcome is very similar to that one discussed in the $d = 3$ conformal theory. If the sum of the tensions $T_1 + T_2$ is arranged to vanish, then the system behaves as a spontaneously broken system, the magnitude of the vev of the field is the distance between the two branes.

If $T_1 + T_2 > 0$, the two branes actually are attracted to sit one on top of the other, and when $T_1 + T_2 < 0$, the branes repel, the system is unstable and as a result one of the branes is exiled to infinity.

These three examples are in full correspondence with the conditions on the coefficients of the $(\phi)^6$ theory that we discussed above. The difference between the two theories, and an important difference, is that in case of the

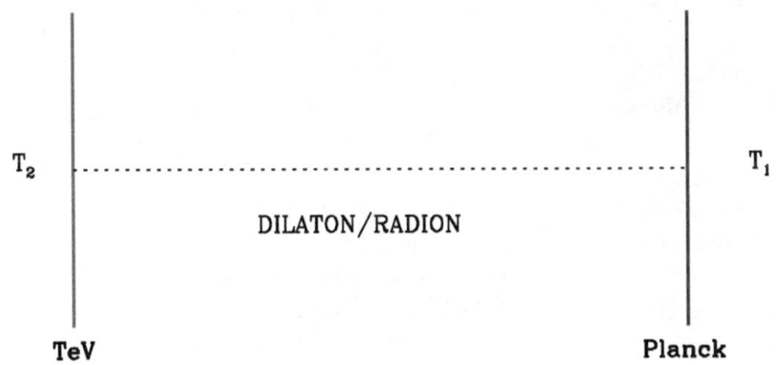

Fig. 17. Planck brane and TeV brane

$(\phi)^6$ theory we are certain that in the large N limit the theory is indeed finite quantum mechanically.

For the case of $(\phi)^4$ we don't have such an assurance, and it would be nice to find a system for which we are guaranteed to be finite also quantum mechanically, which exhibits the same type of behavior.

5.1 Conclusions

• Spontaneous breaking of translational and rotational symmetry are possible. It fits data for many phases of matter, and it may have a manifestation in the dynamics of compactification.

• Conformal/scale invariant theories which are stable but have no ground states indicate a new mechanism of breaking time-translational invariance as well as supersymmetry.

• A finite-scale invariant theory has the same vanishing vacuum energy in all its phases.

• It is a great privilege to recognize Gabriele's outstanding contributions.

Acknowledgements

The author thanks Matteo Cardella for various discussions on this manuscript. The author wishes to thank his various collaborators on these subjects, especially W. Bardeen, S. Elitzur, M. Einhorn, A. Forge, A. Giveon, M. Porrati , A. Schwimmer and G. Veneziano.

References

1. S. Fubini, E. Rabinovici: Nucl. Phys. B **245**, 17 (1984)
2. J. Hughes, J. Polchinski: Nucl. Phys. B **278**, 147 (1986)
3. L. D. Landau: Phys. Z. Soviet II **26**, 545 (1937)
4. S. Elitzur, A. Forge, E. Rabinovici: Nucl. Phys. B **359**, 581 (1991)
5. C. Baym, H. A. Bethe, C. J. Pethick: Nucl. Phys. A **175**, 25 (1971)
6. S. Alexander: *Symmetries and Broken Symmetries in Condensed Matter Physics*, ed by N. Boccara (IDSET, Paris1981) p.141
7. A. B. Zamolodchikov: Sov. Phys. JETP Lett. **43**, 730 (1986)
8. A. W. W. Ludwing, J. L. Cardy: Nucl. Phys. B **285** [FS19], 687 (1987)
9. S. Elitzur, E. Rabinovici, G. Sarkissian: Nucl. Phys. B **541**, 246 (1999)
10. I. Affleck, A. W. W. Ludwig: Phys. Rev. Lett. **67**, 161 (1991)
11. D. Friedan, A. Konechny: Phys. Rev. Lett. **93**, 030402 (2004)
12. J. A. Harvey, D. Kutasov, E. J. Martinec: "On the relevance of tachyons," arXiv:hep-th/0003101

13. V. de Alfaro, S. Fubini, G. Furlan: Nuovo Cim. A **34**, 569 (1976)
14. V. P. Akulov, A. I. Pashnev: Theor. Math. Phys. **56**, 862 (1983) [Teor. Mat. Fiz. **56**, 344 (1983)]
15. P. Claus, M. Derix, R. Kallosh, J. Kumar, P. K. Townsend, A. Van Proeyen: Phys. Rev. Lett. **81**, 4553 (1998)
16. R. Kallosh: "Black holes and quantum mechanics," arXiv:hep-th/9902007
17. E. D'Hoker, R. Jackiw: Phys. Rev. Lett. **50**, 1719 (1983)
18. C. W. Bernard, B. Lautrup, E. Rabinovici: Phys. Lett. B **134**, 335 (1984)
19. I. Affleck, M. Dine, N. Seiberg: Nucl. Phys. B **241**, 493 (1984)
20. T. R. Taylor, G. Veneziano, S. Yankielowicz: Nucl. Phys. B **218**, 493 (1983)
21. Sergio Fubini: Nuovo Cim. A **34**, 521 (1976)
22. J. Polchinski: Nucl. Phys. **303**, 226 (1988)
23. S. Elitzur, A. Giveon, E. Rabinovici, A. Schwimmer, G. Veneziano: Nucl. Phys. B **435**, 147 (1995)
24. D. J. Amit, E. Rabinovici: Nucl. Phys. B **257**, 371 (1985)
25. W. A. Bardeen, M. Moshe, M. Bander: Phys. Rev. Lett. **52**, 1188 (1984)
26. W. A. Bardeen, K. Higashijima, M. Moshe: Nucl. Phys. B **250**, 437 (1985)
27. D. S. Berman, E. Rabinovici: "Supersymmetric gauge theories," arXiv:hep-th/0210044
28. M. B. Einhorn, G. Goldberg, E. Rabinovici: Nucl. Phys. B **256**, 499 (1985)
29. E. Rabinovici, B. Saering, W. A. Bardeen: Phys. Rev. D **36**, 562 (1987)
30. W. M. Alberico, S. Sciuto: *Symmetry and Simplicity in Physics* (Proceedings of the Symposium in occasion of Sergio Fubini's 65th Birthday, Turin, Italy, February 24–26) (World Scientific, Singapore 1994), p 220
31. Ya. B. Zeldovich: Sov. Phys. Uspekhi **11** (1968) 381
32. S. Weinberg: Rev. Mod. Phys. **61** (1989)
33. T. Damour, A. M. Polyakov: Nucl. Phys. B **423**, 532 (1994)
34. T. Damour, F. Piazza, G. Veneziano: Phys. Rev. D **66**, 046007 (2002)
35. E. Rabinovici: unpublished
36. R. Rattazzi, A. Zaffaroni: JHEP **0104**, 021 (2001)

Part VII

String/Quantum Gravity, Black Holes and Entropy

The Information Paradox

D. Amati

International School for Advanced Studies (SISSA), Trieste, Italy
amati@sissa.it

Los muertos que vos matasteis gozan de buena salud[1]

Abstract. The incompatibility between gravity and quantum coherence represented by black holes should be solved by a consistent quantum theory that contains gravity as superstring theory. Despite many encouraging results in that sense, I question here the general feeling of a naïve resolution of the paradox. And indicate non-trivial physical possibilities towards its solution that are suggested by string theory and may be further investigated in its context.

1 Introduction

The fact that black holes represent an apparent contradiction between gravity and quantum mechanics is a too well-known problem to need exhaustive recall. The best way to visualize it is to consider together the formation and evaporation processes. We may envisage a black hole (b.h.) to be formed by a pure quantum state prepared in a distant flat space (an impinging spherical wave, two or 25 particles colliding at high energies and small impact parameter, and so on). If the characteristics of a b.h. – including its evaporation implied by quantum mechanics – depend only on few basic parameters (M, Q, J) as required by general relativity (no hairs), it is clear that quantum coherence of the initial state is totally lost in the process. The contradiction has resisted efforts to doctored modifications (as corrections to the thermal character of Hawking evaporation) and brought distinguished scientists to give up either quantum mechanics [1] or the relevance of classical (general relativity) solutions in a path integral formulation of the quantum theory of gravitation [2].

On the other hand, the advent of string (actually superstring) theory as a consistent quantum theory that contains gravity gave confidence that somehow the paradox should be solved in its framework. Much progress has been

[1] Josè Zorrilla, "Don Juan Tenorio" (1844).

D. Amati: *The Information Paradox*, Lect. Notes Phys. **737**, 609–617 (2008)
DOI 10.1007/978-3-540-74233-6_20

done in studying b.h. regimes in string theories and a remarkable set of coincidences have been revealed. After briefly recalling those results, I will argue that the paradox not only is not trivially solved as often claimed, but manifests its full vitality in compelling some quite novel possibilities in the generalization to the quantum realm of some classical concepts as space–time and its geometry, or in the influence that quantum effects may have on the actual realization of classical geometrical configurations as trapped space–time regions.

2 String Theories and Black Holes

String theories contain arbitrarily massive states within regions characterized by the string length l_s – the basic dimensional parameter of the theory – and thus states that, classically, would represent black holes. The mass beyond which those states should be black holes depends [3] on the string coupling g, the other – dimensionless – basic parameter of the theory. Or, in different words, for every mass (or excitation energy) there is a small coupling string regime, and a large coupling b.h. regime.

In the string regime, D-branes in four [4] and five [5] dimensions with a convenient number of charges have been studied. BPS states have been counted as well as nearly BPS (Bogolmon'y-Prasad-Sommerfeld) states for certain regions of moduli space where perturbative computations are feasible [6]. Decay rates have been computed [7] – by averaging over the many degenerate initial states – and shown to have a typical thermal distribution. The moduli independence of these results allows to conjecture [8] their validity beyond the moduli region where they were computed. And their g independence, also suggested by non-renormalization arguments [9], may imply their possible continuation beyond the weak coupling regime.

An independent treatment – on totally different grounds – of the strong coupling regime substantiates that impression. The large g description of the four- and five-dimensional systems just described is found by solving the 10-d supergravity equations after reduction on the same compact manifold used for the D-brane description. The solution generates a metric [10] that depends on parameters that are related to the charges through the moduli of the compact manifold. The metric shows an event horizon even in the extreme limit in which its area gives the Bekenstein–Hawking entropy of extremal b.h. (see e.g. [11]). This entropy and the ADM (Arnowitt-Deser-Mlsner) mass coincide with the (exponentiated) multiplicity and mass of the BPS states with the same charges, as computed from the D-branes in the small coupling regime. For nearly extremal b.h. the entropy and the evaporation spectrum – obtained by solving wave equations in the corresponding metric background – coincide again [7] with those computed for small g. And, remarkably, even deviations from black body spectrum seem to agree [12].

The microscopic formulation of the 5-d near extremal b.h. has been further studied [13] in terms of the D1–D5 brane system. The AdS/CFT (anti de-Sitter/conformal field theory) correspondence was shown to play a role in the matching between supergravity results and the microscopic (SCFT) formulation of the b.h. thermodynamics and Hawking radiation, the b.h. being defined through a density matrix.

All these agreements among such different computations gave confidence to the g continuation of the theory to a strong coupling regime where b.h. physics is met. This direct connection between the semiclassical black hole picture and a unitary quantum approach, has been considered the sign that the information loss due to b.h. could be somehow recuperated [14]. But how this may be achived is yet far from clear. In the computations just referred it appeared clearly that the thermal Hawking radiation was obtained by the averaging over the degenerate microstates that are counted by the b.h. entropy, while each microstate would have given rise to a complex but non-thermal radiation with well-defined spectra and correlations that carry the precise identity of the microstate from which they would have been originated. This is of course a basic characteristics of a microstate (a pure quantum state) irrespectively of g. In other words, the black hole microstates are not themselves black holes [15]. And this not only because of the absolute specificity of its radiation, but also by not having any signal of an event horizon associated with each of them. This last fact is of course expected by sheer consistency: if a b.h. microstate would be characterized by an event horizon, it would have – itself – a Bekenstein entropy and thus would not be a pure quantum state. The b.h. appears indeed as the macrostate correctly defined by a decoherence procedure – density matrix – over the many non-blackholish microstates of the theory [16].

The obvious consequence of the preceding discussion is that a well prepared quantum state (a spherical shell impinging from large distances, or a two particle scattering at high energy and small impact parameter, etc.) is not expected to give rise to a b.h. even if the classical conditions for a gravitational collapse are apparently satisfied.

The possibility that microstates do not have a horizon has been more recently proposed in a different context [14]: for every wrapping of a D1 brane (whose number defines one of the charges briefly mentioned before) a profile function in transverse space is introduced so to enter into a momentum charge that contributes to the BPS charge. These profile functions then enter into the supergravity solutions that are supposed to hold in the strong coupling regime and change their behaviour at short radius, differently for every different profile function. They are not singular at $r = 0$ and the value of r where they all start resembling the usual b.h. solution outside the horizon is identified as a fuzzy "horizon" of a fuzzball proposal for b.h.. It is unclear, however, if and how a trapped region could emerge for the incoherent superposition characterizing the b.h. macrostate.

3 The Role of Decoherence

If string microstates counted by the entropy of b.h. macrostates are not themselves black holes, it should happen that decoherence, intrinsic to any classical limit, should be critical in building b.h. characteristics as metric singularities and event horizons. It is not surprising that decoherence may have an important role in high excitation string physics due to the very large degeneracy of states in that regime. Indeed, even for $g \to 0$, i.e. for tree diagrams, the non-trivial spectrum of emitted particles in the decay of any high-mass excitation gives rise to a thermal distribution if an average over the very many states with the same mass is performed [17].

Even if effects of this kind may well be at work also for large couplings, decoherence should have much more subtle effects in order to generate b.h. physics from non black-holish microstates. Let me provide some speculative ideas on how a geometric picture could arise from a decoherence procedure in the pregeometric string approach. In this theory, indeed, even space and time are defined through the string; they are operators and not parameters that could be interpreted as coordinates of a space–time that may subtend a dynamical geometry. These are all concepts that may arise in a classical limit of the theory when quantum fluctuations may be neglected. But even in this limit, the theory contains in principle not only the metric and possibly matter fields, but also an infinite number of higher-rank tensor fields whose effect may possibly be ignored only in some conditions. The (infinite number of) equations that these (infinite number of) fields should satisfy, are given by the condition of no conformal anomaly ($\beta = 0$), and it is in the limit of small frequencies (in string length units) that only massless fields appear satisfying Einstein's equations [18]. But in presence of a horizon of a metric solution, the statement of low frequency is not relativistic invariant. Indeed, an arbitrary low-frequency wave for a fixed external observer will be perceived by a free falling one with a blue shift which gets arbitrarily increased when approaching the horizon. This means that to have disregarded contributions with higher derivatives, or fields with higher tensorial character, would have been an unwarranted approximation. And even a small effect of those tensors could have avoided the metric condition that implied the singularity and the trapped region in the usual Einstein equation. There could be many solutions involving different field configurations in which the metric and other tensor fields are classically entangled with relevant phases. And it could well happen that an incoherent superposition of these different background configurations could wash out the higher tensorial fields leaving a geometric description with, eventually, a b.h. metric with its singularity and its event horizon. This could be a hypothetical way in which non-b.h. microstates could give rise to a b.h. macrostate.

In this case, the apparent contradiction between b.h. in classical general relativity and quantum coherence is solved in a conceptually simple way: it is the decoherence procedure, implied in any classical limit, that gives rise – from a consistent quantum theory of gravitation as superstrings – to a classical geometrical space–time description (general relativity) with eventual trapped regions, event horizon and b.h. and, of course, the loss of quantum coherence.

4 High-energy Collisions in String Theory and Metric Back Reaction

Let us now discuss high-energy scattering. Superstrings provide a computational perturbative algorithm for S-matrix amplitudes that, if properly resumed, allows an explicit analysis of the continuation to the strong coupling (b.h.) regime. Therefore, as we shall discuss later, the consistent quantum theory may investigate situations in which, semiclassically, the process should be described by a b.h. formation and subsequent evaporation. Thus, hopefully, the analysis may throw light on how and why may happen that a coherent quantum state would not produce a b.h. even if the classical conditions to form it are met.

Much work has been done to study trans-Planckian collisions in a string approach [18,19,20]. I will recall methods and results that are consistently computable in the string regime and organized in an effective action form [21] to tackle their extention to a strong coupling regime where, semiclassically, b.h. formation and subsequent evaporation should be expected. As already said, string (or actually superstring) theories contain a dimensional scale – the string length l_s – and a dimensionless one g, the string coupling that generates the genus expansion. Gravitational scales, as the Newton constant G or the Schwarzschild radius R_S corresponding to an energy E are given by

$$G = g^2 l_s^2/\hbar, \qquad R_S = GE . \tag{1}$$

For simplicity, (1) and other explicit expressions we shall give will refer to the $d = 4$ case, even if the analysis we recall has been done for an arbitrary number d of non-compactified dimensions. The method used in [18] is to consider a trans-Planckian regime defined by a small coupling large energy

$$g^2 << 1, \qquad El_s/\hbar >> 1 , \tag{2}$$

so that

$$GE^2/\hbar = g^2(El_s/\hbar)^2 > 1 . \tag{3}$$

In the genus expansion of string amplitudes all terms in which g^2 is enhanced by the large factor as in (3) have to be considered, and resumed. Let us notice that in the large energy regime of (2) and (3) $R_S/l_s = g^2 El_s/\hbar$ can be smaller or larger than 1 and, as we shall see, physics will be different on the two sides of the inequality. The computation of the collision amplitude in superstring theory in terms of the energy E and impact parameter b has been organized in powers of R_S^2/b^2. For b larger than both R_S and l_s, the two-particle collision amplitude in the high-energy regime as defined by (2) – obtained by the just discussed all order resummation – has an eikonal form, the eikonal being a Hermitian operator (thus unitary S-matrix) in the Fock space of the two colliding strings. Only for very large values of b – where the amplitude is perturbative and dominated by the graviton pole – the scattering is elastic, while for $b < gEl_s^2/\hbar$ the two colliding gravitons are also excited to other superstring states in the scattering process. The eikonal is large and allows a classical trajectory interpretation through a saddle point in the Bessel transform to transfer momentum. It reproduces the relation between the deflection angle and the impact parameter classically experienced by each particle in the (Aichelburg–Sexl) gravitational field created by the other one. With the extra fact that while deflecting, colliding particles may be excited (in a calculable way) to one of its string recurrences, implying an attenuation of the elastic amplitude (imaginary phase) that increases, together with the deflection angle, for decreasing b. In the $R_S < l_s$ case, b may decrease where string effects become relevant, giving rise to copious inelastic production [22] and thus to a softening that implies an attenuation of the elastic amplitude and a reduced deflection angle. In the $R_S > l_s$ case, when b approaches R_S, new terms appear, as said before, in the form of powers of R_S^2/b^2, that look as classical corrections despite their quantum origin. The first term has been computed in the string framework [16, 23], and an effective action algorithm has been proposed for computing and resuming them all [21].

This may be interpreted as a metric and dilaton background generated by the process or, equivalently, a consistent quantum computation of back reaction on the metric, giving effects that become relevant when approaching situations in which a b.h. formation is classically expected. It could thus represent a way of understanding how and why a b.h. is avoided in a well-defined quantum state as that under discussion. It is perhaps unfortunate that no further effort has been devoted in that direction. I have even a vague recall of a sense of frustration of the scientist to whom we dedicate these contributions, Ciafaloni and myself when – many years ago – some preliminary results could not be forced into the recognition of a horizon. The fact that brought us to give up, while today I would consider it as the expected sign to reveal novel quantum gravitational effects! Furthermore, if this sort of back reaction is efficient in avoiding trapped regions in the well-defined quantum state represented by the two colliding particles, it could perhaps continue to do so in arbitrary collapse situations. Let me also adventure that this possible effect of quantum back reactions on the metric may allow an interpretation of the recent Hawking suggestion [2] that the original classical solution, as the Schwarzschild metric in a gravitational collapse, may give an irrelevant contribution to the path integral for the actual gravitational process.

5 Metric Back Reaction and Possible Avoidance of Black Holes

The idea that standard b.h. may not be the objects realized in nature even at the macroscopic level, has been recently explored within different contexts [24]. In particular, interesting suggestions have been borrowed from geometric acoustical models that can be studied experimentally and show a physics that is associated with classical and quantum fields in curved space–times [25]. Propagation of small disturbances in the flow of even simple fluids are known to behave equivalently to a linear (classical or quantum) field over an acoustic space–time endowed with an acoustic metric [26]. Depending on that endowed metric, acoustic b.h. – trapped region corresponding to a supersonic regime in fluid flow – may be created. It has been however noted [27] that Hawking-like radiation does not necessarily imply the formation of a trapped region; it is sufficient that a sonic point conveniently develops in the asymptotic future. The radiation is then controlled by a temperature that contains both the Hawking one and the rate by which the sonic point is reached for $t \to \infty$. This critical collapse result suggests an alternative scenario for a semiclassical collapse and evaporation of "b.h." objects that – very speculatively – could be exported to semiclassical gravity. Its interpretation would imply that some quantum back reaction on the geometry could prevent the surface of the collapsing star (or impinging matter) from actually crossing the Schwarzschild radius. At later stages, the evaporation process would become more efficient so to induce a chasing of the would be horizon by the surface of the star that could end with the complete evaporation and a flat space–time [27].

6 Conclusions and Outlook

I hope to have substantiated my (probably personal) point of view of why some coincidences between string state multiplicity and average decay spectra, on one hand, and b.h. entropy and evaporation spectra, on the other, are far from having solved in a naive way the apparent paradox of loss of quantum coherence in b.h. formation and evaporation (the information paradox). String microstates, in particular, are not b.h. and well defined quantum states would not generate b.h. even if they would have been expected on classical grounds. I have discussed two ways to resolve this apparent discrepancy, both of them accessible to further investigation in the string framework. The first one starts from the fact that superstring theory is pregeometrical and even the concept of space–time is induced by the string through a classical limit. Thus space–time, geometry, event horizons, black holes and the loss of quantum coherence would all come with the same token, i.e. the decoherence procedure implied in the classical limit that leads to general relativity. Thus no paradox: either bona fide quantum (as superstrings) or classical space–time with dynamical geometry and black holes but no a priori quantum coherence. The other possibility is that the lack of b.h. formation in a quantum state, as two-particle collision, may be due to well-identified quantum contributions that give rise to apparently classical effects that act as quantum back reactions on the metric. Effects that could remain influential even in classical gravitational collapse processes thus avoiding metric singularities, trapped regions and event horizons. Without forming, therefore, even classical b.h. despite the fact that many external observational properties would not look very dissimilar. Thus no paradox because no real black holes: no trapped region or event horizon to spoil quantum coherence or information retrieval.

Recognition

I had the chance to enjoy a lively and fruitful collaboration with Gabriele for many years and on a variety of subjects. Sharing – as also reflected in this paper – the joy of elaborating original physics, the frustration of unexpected obstacles and the persisting challenge of different viewpoints on possible developments. I wish him to keep harvesting success, surrounded by friends and collaborators attracted by his scientific and human qualities. People of all origins and ages ... with me at the oldest end.

References

1. S.W. Hawking: Phys. Rev. D **50**, 3982 (1994)
2. S.W. Hawking: Lecture at the 17th Int. Conf. on General Relativity and Gravitation, 2004 (`http://math.ucr.edu/home/baez/week207.html`)
3. M.J. Bowick, L. Smolin, L.C.R. Wijewardhana: Phys. Rev. Lett. **56**, 424 (1986); G. Veneziano: Europhys. Lett. **2**, 133 (1986); L. Susskind: `hep-th`/9309145
4. J. Maldacena, A. Strominger: Phys. Rev. Lett. **77**, 428 (1966); C. Johnson, R. Khuri, R. Myers: Phys. Lett. B **378**, 78 (1996)
5. A. Strominger, C. Vafa: Phys. Lett. B **379**, 99 (1966)
6. C. Callan, J. Maldacena: Nucl. Phys. B **472**, 591 (1996)
7. S. Das, S. Mathur: Nucl. Phys. B **478**, 561 (1996); S. Gubser, I. Klebanov: Nucl. Phys. B **482**, 173 (1996)
8. G. Horowitz, J. Maldacena, A. Strominger: Phys. Lett. B **383**, 151 (1996)
9. J. Maldacena: `hep-th`/961125
10. M. Cvetic, D. Youm: `hep-th`/9508058, `hep-th`/9512127; G. Horowitz, D. Lowe, J. Maldacena: Phys. Rev. Lett. **77**, 430 (1996)
11. L. Andrianopoli, R. D'Auria, S. Ferrara, M. Trigiante: *Extremal black holes in supergravity*, this volume
12. S. Gubser, I. Klebanov: Phys. Rev. Lett. **77**, 4491 (1996); J. Maldacena, A. Strominger: Phys. Rev. D **55**, 861 (1997)
13. J.R. David, G. Mandal, S.R. Wadia: Phys. Rep. **369**, 549 (2002)
14. G. Horowitz: `gr-qc`/9704072
15. D. Amati: Phys. Lett. B **454**, 203 (1999)

16. S.D. Mathur: hep-th/0502050
17. D. Amati, J. Russo: Phys. Lett. B **454**, 207 (1999)
18. C.G. Callan, E. Martinec, M.J. Perry, D. Friedan: Nucl. Phys. B **262**, 593 (1985)
19. D. Amati, M. Ciafaloni, G. Veneziano: Phys. Lett. B **197**, 81 (1987); Int. J. Mod. Phys. A **3**, 1615 (1988)
20. D.J. Gross, P.F. Mende: Phys. Lett. B **197**, 129 (1987); Nucl. Phys. B **303**, 407 (1988);
P.F. Mende, H. Ooguri: Nuc. Phys. B **339**, 641 (1990)
21. D. Amati, M. Ciafaloni, G. Veneziano: Nucl. Phys. B **403**, 707 (1993)
22. G. Veneziano: JHEP **11**, 001 (2004)
23. E. Gava, R. Iengo, C.-J.Zhu: Nuc. Phys. B **323**, 585 (1989);
D. Amati, M. Ciafaloni, G. Veneziano: Nuc. Phys. B **347**, 550 (1990)
24. T.A. Roman, P.G. Bergmann: Phys. Rev. D **28**, 1265 (1983);
C.R. Stephens, G.'t Hooft, B.F. Whiting: Class. Q. Grav. **11**, 621 (1994);
A. Ashketar, M. Bojowald: gr-qc/050429;
S.Hayward: gr-qc/0506120;
T. Vachaspati, D. Stojkovic, L.M.Krauss: gr-qc/0609024
25. C. Barceló, S. Liberati, M. Visser: Living Rev. Relativity **8**, 12 (2005)
26. W.G. Unruh: Phys. Rev. Lett. **46**, 1351 (1981)
27. C. Barceló, S. Liberati, S. Sonego, M. Visser: Class. Q. Grav. **23**, 5341 (2006); Phys. Rev. Lett. **97**, 171301 (2006)

Cosmological Entropy Bounds

R. Brustein

Department of Physics, Ben-Gurion University, Beer-Sheva 84105, Israel
ramyb@bgu.ac.il

Abstract. I review some basic facts about entropy bounds in general and about cosmological entropy bounds. Then I review the causal entropy bound, the conditions for its validity and its application to the study of cosmological singularities. This article is based on joint work with Gabriele Veneziano and subsequent related research.

1 To Gabriele

On the occasion of your 65th birthday may you continue to find joy in science and life as you have always had, and continue to help us understand our universe with your creative passion and vast knowledge. It is a pleasure and an honor to contribute to this volume and present one of the subjects among your many interests. Thank you for explaining to me why entropy bounds are interesting and for your collaboration on this and other subjects.

2 Introduction

2.1 What Are Entropy Bounds?

The second law of thermodynamics states that the entropy of a closed system tends to grow toward its largest possible value. But what is this maximal value? Entropy bounds aim to answer this question.

Bekenstein [1] has suggested that for a system of energy E whose size R is larger than its gravitational radius $R > R_g \equiv 2G_N E$, entropy is bounded by

$$S \leq ER/\hbar = R_g \, R \, l_P^{-2}. \tag{1}$$

Here l_P is the Planck length. This is known as the Bekenstein entropy bound (BEB).

R. Brustein: *Cosmological Entropy Bounds*, Lect. Notes Phys. **737**, 619–659 (2008)
DOI 10.1007/978-3-540-74233-6_21

Entropy bounds are closely related to black hole (BH) thermodynamics and their interplay with their "normal" environment. They are also probably associated with instabilities to forming BHs; however, this has not been proved in an explicit calculation. The original argument of Bekenstein was based on the Geroch process: a thought experiment in which a small thermodynamic system is moved from infinity into a BH. The small system is lowered slowly until it is just outside the BH horizon, and then falls in. By requiring that the generalized second law (GSL) will not be violated, one gets inequality (1).

A long debate about the relationship between entropy bounds and the GSL has been going on. On one side, Unruh, Wald and others [2, 3] have argued that the GSL holds automatically, so that entropy bounds cannot be inferred from situations where the law seems to be violated. They argue that the microphysics will eventually take care of any apparent violation. Consequently, they argued that the BEB does not have to be postulated as a separate requirement in addition to the GSL. Responding to their arguments Bekenstein [4] has argued that it is not always obvious in a particular example how the system avoids violating the bound and analyzed in detail several of the purported counter-examples of this type and demonstrated in each case the specific mechanism enforcing the bound.

Holography [5] (see below) suggests that the maximal entropy of any system is bounded by $S_{HOL} \leq A l_P^{-2}$, where A is the area of the space-like surface enclosing a certain region of space. For systems of limited gravity $R > R_g$, and since $A = R^2$, the BEB implies the holography bound. Physics up to scales of about 1 TeV is very well described in terms of quantum field theory, which uses, roughly, one quantum mechanical degree of freedom (DOF) for each point in space (the number of DOF is the logarithm of the number of independent quantum states). This seems to imply that $S(V) \sim V$, but the BEB states that $S(V) \leq A$. The BEB does not seem to depend on the detailed properties of the system and can thus be applied to any volume V of space in which gravity is not dominant. The bound is saturated by the Bekenstein–Hawking entropy associated with a BH horizon, stating that no stable spherical system can have a higher entropy than a BH of equal size.

A bold interpretation of the BEB was proposed by 't Hooft and Susskind [5],—that the number of independent quantum DOF contained in a given spatial volume V is bounded by the surface area of the region. In a later formulation by Bousso [6] their conjecture reads "a physical system can be completely specified by data stored on its boundary without exceeding a density of one bit per Planck area." In this sense the world is two-dimensional and not three-dimensional, for this reason their conjecture is called the holographic principle. The holographic principle postulates an extreme reduction in the complexity of physical systems, and is not manifest in a description of nature in terms of quantum field theories on curved space. It is widely believed that quantum gravity has to be formulated as a holographic theory. This point of view has received strong support from the ADS/CFT duality

[7, 8], which defines quantum gravity non-perturbatively in a certain class of space–times and involves only the physical degrees admitted by holography.

One way of viewing entropy bounds is that they are new laws of nature that have to supplement the equations that govern any fundamental theory of quantum gravity. From this perspective the entropy bounds and the holographic principle are presumed to be valid for any physical system and their "true" form has to be unraveled. An alternative perspective is that entropy bounds will be automatically obeyed by any physical system and will be a consequence of the fundamental dynamical equations. As such, entropy bounds will not provide additional independent constraints on the system's evolution. In the final fundamental theory entropy bounds will be tautologically correct. My personal view on this issue at the present time is closer to the second point of view.

My current perspective is that without detailed knowledge of the dynamical equations that govern physics at the shortest distance scales and at the highest energies it is hard to make detailed quantitative use of entropy bounds. They are very useful as qualitative tools in the absence of the final fundamental theory of quantum gravity when one is trying to determine whether a candidate theory is correct by studying its consequences. As I will explain they are particularly useful in discriminating among cosmologies that are suspect of being unphysical for various reasons.

2.2 What Are Cosmological Entropy Bounds?

Is it possible to extend entropy bounds to more general situations, for example, to cosmology? In 1989 Bekenstein [9] proposed that it might be possible to apply the BEB to a region as large as the particle horizon d_p: $d_p(t) = a(t)\int_{t_{\rm initial}}^{t} dt'/a(t')$, $a(t)$ being the scale factor of an Friedman–Robertson–Walker (FRW) universe. If the entropy of a visible part of the universe obeys the usual entropy bound from nearly flat space situations, then Bekenstein suggested that the temperature of the universe is bounded and therefore certain cosmological singularities are avoided. The proposal to apply the holographic bound from nearly flat space to cosmology was first made by Fischler and Susskind [10] and later extended and modified by Bousso [6]. Verlinde [11] proposed an entirely holographic bound on entropy stating that the subextensive component of the entropy (the "Casimir entropy") of a closed universe has to be less than the entropy of a BH of the same size.

To appreciate the necessity to modify the BEB in some situations, let us think [12] about a box of relativistic gas in thermal equilibrium at a temperature T. We assume that the gas consists of $\mathcal{N}$ independent DOF and is confined to a box of macroscopic linear size R. We further assume that R is larger than any fundamental length scale in the system, and in particular that R is much larger than the Planck length $R \gg l_P$. The volume of the box is $V = R^3$. Since the gas is in thermal equilibrium, its energy density is $\rho = \mathcal{N}T^4$ and its entropy density is $s = \mathcal{N}T^3$ (here and in the following we systematically neglect numerical factors). Here we are interested in the case $RT > 1$ which means that the size of the box is larger than the thermal wavelength $1/T$. The case $RT < 1$ has been considered previously in [13]. In this case the temperature is not relevant, rather the field theory cutoff Λ was shown to be the relevant scale.

Under what conditions is this relativistic gas unstable to the creation of BHs? The simplest criterion which may be used to determine whether an instability is present is a comparison of the total energy in the box $E_{\rm Th} = \mathcal{N} T^4 R^3$ to the energy of a BH of the same size $E_{\rm BH} = M_P^2 R$ (M_P is the Planck mass). The two energies are equal when $T^4 = 1/\mathcal{N} M_P^2/R^2$. So thermal radiation in a box has a lower energy than a BH of the same size if

$$(TR)^4 < \frac{1}{\mathcal{N}} M_P^2 R^2. \tag{2}$$

Another way to determine the presence of an instability to creation of BHs is to compare the thermal entropy $S_{\rm Th} = \mathcal{N} T^3 R^3$ to the entropy of the BH $S_{\rm BH} = M_P^2 R^2$. They are equal when $T^3 = 1/\mathcal{N} M_P^2/R$. So thermal radiation in a box has a lower entropy than a BH of the same size if

$$(TR)^3 < \frac{1}{\mathcal{N}} M_P^2 R^2. \tag{3}$$

From (2) and (3) it is possible to conclude the well-known fact that for fixed R and $\mathcal{N}$, if the temperature is low enough, the average thermal free energy is not sufficient to form BHs. For low temperatures the thermal fluctuations are weak and they do not alter the conclusion qualitatively.

Now imagine raising the temperature of the radiation from some low value for which conditions (2) and (3) are comfortably satisfied to higher and higher values such that eventually condition (2) is saturated. Since $TR > 1$, (2) is saturated before (3). We assume that the size of the box R is fixed during this process (recall that the number of species $\mathcal{N}$ is also fixed), and estimate the backreaction of the radiation energy density on the geometry of the box to determine whether the assumption that the geometry of box is fixed is consistent. To obtain a simple estimate, we assume that the box is spherical, homogeneous and isotropic. Then its expansion or contraction rate is given by the Hubble parameter $H = \dot{R}/R$, which is determined by the 00 Einstein equation $H^2 M_P^2 = \mathcal{N} T^4$. However, if (2) is satisfied, then $\frac{1}{R^2} M_P^2 = \mathcal{N} T^4$, and therefore $HR \sim 1$. The conclusion is that if (2) is saturated, then the gravitational time scale is comparable to the light crossing time of the box, and therefore, it is inconsistent to assume that the box has a fixed size which is independent of the energy density inside it.

Thus we have shown that it is not possible to ignore the backreaction of the gas on the geometry under all circumstances. Sometimes the backreaction has to be taken into account. When the BEB is near saturation, we have found that the basic assumptions have to be changed so it has to be modified to adapt to an intrinsically time-dependent situation.

2.3 Why Is It Reasonable to Expect Cosmological Entropy Bounds?

Some have argued incorrectly that it is impossible to discuss entropy bounds in cosmology. They argue that the universe is the whole system and thus one cannot apply thermodynamical arguments that sometimes rely on separating a sub-system from a heat reservoir. This argument is false as the following braneworld thought experiment explicitly demonstrated [12]. Let us consider a brane moving in a higher-dimensional BH background. From the brane point of view it experiences a cosmological evolution and one can imagine that the brane falls into the BH and disappears from an external observer's view into the BH horizon. We are thus in a situation similar to the one envisaged in the Geroch process: the thought experiment in which a thermodynamic system is absorbed by a BH. The aim is to design the process such that the energy absorbed by the BH is minimal. In such a way the entropy that the BH gains will also be minimal, as both the energy and the entropy of the BH depend only on its mass after the absorption. We can make the entropy balance during the process and see under which conditions the GSL is respected.

We can gain some insight by modeling a 4D radiation-dominated (RD) universe as a brane moving in an AdS_5-Schwarzschild space–time. For the BH in AdS to be the dominant configuration over an AdS space filled with thermal radiation as required for our analysis to be relevant, the BH must be large and hot compared to the surrounding AdS_5 [14]. In this limit the closed 4D universe can be treated as flat. The motion of the brane through the bulk spacetime is viewed by a brane observer as a cosmological evolution. According to the prescription of the RS II model [15], the 4D brane is placed at the Z_2 symmetric point of the orbifold. On the other hand, in the so-called mirage cosmology [16, 17], the brane is treated as a test object following a geodesic motion. In both cases the evolution of the brane in the AdS_5-Schwarzschild bulk mimics an FRW RD cosmology. From the 5D perspective one may expect some limits on the entropy of the brane by considering what happens when the BH swallows the brane.

2.4 What Are Cosmological Entropy Bounds Good for?

Our interest in entropy bounds in general and cosmological entropy bounds in particular originated from the interest in determining the fate of cosmological singularities. Specifically, we were interested in finding whether the bounce that is an essential part of the pre-big-bang (PBB) scenario of string cosmology [18] can be physically realized or perhaps there is some principle that requires the solution to be singular. We needed a general principle because string theory could not provide an explicit enough model of the hypothetical bounce transition. The traditional tools for finding such criteria were the energy conditions that are used in the singularity theorems. However, the use of energy conditions is limited because there are examples of cosmologies that do not seem to be problematic in any of their physical properties and for which the singularity theorems are not applicable because some of the energy conditions are violated. On the other hand, there are examples of cosmologies for which we expect some problems while the singularity theorems seem perfectly valid.

Let us consider, for example, the scale factor for a closed deSitter universe. This is a closed universe containing a positive cosmological constant Λ. In $D = 4$ it is given by $a(t) = (\frac{\Lambda}{3})^{-1/2} \cosh \sqrt{\frac{\Lambda}{3}} t$, showing a bounce at $t = 0$. The bounce is not allowed by the classic singularity theorems. This is not surprising since the sources of this model violate the strong energy conditions (SEC). The reliability of the SEC as a criterion of discriminating physical and unphysical solutions is therefore questionable (as is well known in the context of inflationary cosmology). Conversely, in a 4D contracting universe filled with radiation consisting of $\mathcal{N}$ species in thermal equilibrium, the singularity theorems imply that the solution will reach a future singularity. But entropy bounds indicate expected problems already when $T \sim M_{\mathrm{P}}/\mathcal{N}^{1/2}$ as we will show later.

3 The Causal Entropy Bound

3.1 The Hubble Entropy Bound

Motivated by the necessity to resolve the apparent singularity in the lowest order classical PBB scenario, Veneziano has studied the possible role of entropy bounds and proposed the Hubble entropy bound (HEB) [19]. The physical motivations leading to the proposal of the HEB are (i) that in a given region of space the entropy is maximized by the largest BH that can fit in it and (ii) that the largest BH that can hold together without falling apart in a cosmological background has typically the size of the Hubble radius. In the following we review the basic ideas that led to Veneziano's proposal of the HEB.

Veneziano considered the possibility that the BEB or holographic bounds can be applied to an arbitrary sphere of radius R, cut out of a homogeneous cosmological space. Entropy in cosmology is extensive so it grows like R^3, but the boundary's area grows like R^2. Hence, at sufficiently large R, the (naive) holography bound must be violated. On the other hand, $S_{BEB} \sim ER \sim R^4$ appears to be safer at large R.

In order to show how inadequate the naive bounds are in cosmology, Veneziano applied them at the Planck time $t \sim t_P \sim 10^{-43}$ s, within standard FRW cosmology, to the region of space that has become our visible universe today. The size of that region at $t \sim t_P$ was about 10^{30} in units of the Planck length l_P, and the entropy density was of about Planckian. Thus, the actual entropy of the patch is

$$S \sim (10^{30})^3 = 10^{90}, \tag{4}$$

while

$$S_{BEB} \sim \rho R^4/\hbar \sim R^4/l_P^4 \sim 10^{120} \quad , S_{HOL} \sim R^2 l_P^{-2} \sim 10^{60} \ . \tag{5}$$

The actual entropy lies at the geometric mean between the two naive bounds, making one false and the other quite useless. The two bounds differ by a factor $(Hd_p)^2$. While such a factor is of order unity in FRW-type cosmologies, it can be huge after a long period of inflation. For this reason the (naive) holographic entropy bound appears to be stronger than the cosmological version of the BEB, just the opposite of what we argued to be the case for systems of limited gravity.

A sufficiently homogeneous universe has a local time-dependent Hubble expansion rate defined, in the synchronous gauge, by $H = \frac{1}{6}\partial_t \left(\log \det g_{ij}\right)$. If H does not vary much over distances $\sim 1/H$, then the Hubble radius $1/H$ corresponds to the scale of causal connection. If on top of this homogeneous background some isolated lumps of size much smaller than $1/H$ exist, then the expansion of the Universe is irrelevant and the situation should be similar to that of nearly flat space. Veneziano argued that it is possible in this case that a single Hubble patch contains several BHs. The BH can coalesce and in the process their entropy will increase. He argued further that this way of increasing entropy has some limit since it is hard to imagine that a BH of size larger than $1/H$ can form. The different parts of its horizon would be unable to hold together. Strong arguments in this direction were given long ago in the literature [20]. Thus, the largest entropy in a region of space larger than $1/H$ is the one corresponding to one BH per Hubble volume $1/H^3$. Using the Bekenstein–Hawking formula for the entropy of a BH of size $1/H$ leads to the proposal of a "Hubble entropy bound" that the entropy is bounded by $S_{HEB} \equiv n_H S^H$, where n_H is the number of Hubble-size regions within the volume V, each one carrying maximal entropy $S^H = l_P^{-2} H^{-2}$,

$$S(V) < S_{HEB} \equiv n_H S^H = V H^3 l_P^{-2} H^{-2} = V H l_P^{-2} \ . \tag{6}$$

The HEB is partly holographic since S^H scales as an area, and partly extensive since n_H scales as the volume. If the HEB is applied to a region of size d_p, then the bound is the geometric mean of the BEB and the naive holography bound,

$$S_{HEB} = d_p^3 H l_P^{-2} = S_{BEB}^{1/2} \, S_{HOL}^{1/2} \ . \tag{7}$$

3.2 The Causal Entropy Bound

The causal entropy bound (CEB) [21] aims to improve the HEB. It is a covariant bound applicable to entropy on space-like hypersurfaces. We do not insist, a priori, on a holographic bound, but aim at generality of the hypersurface and then investigate how holography may or may not work. For systems of limited gravity Bekenstein's bound is the tightest bound, while, in other situations, the CEB is the strongest one which does not lead to contradictions for space-like regions.

We shall refer to entropy in a region as to a quantity proportional to the number of DOF in that region. To be more precise, we shall exclude from consideration entropy associated with the background gravitational field itself. We will, however, take into account the entropy of the perturbations of the gravitational field. Let us first state our proposal, and then motivate and test it. Consider a generic space-like hypersurface, defined by the equation $\tau = 0$, and a compact region lying within it defined by $\sigma \leq 0$. We have proposed that the entropy contained in this region, $S(\tau = 0,\ \sigma \leq 0)$, is bounded by S_{CEB},

$$S_{CEB} = l_P^{-2} \int\limits_{\sigma<0} d^4x\sqrt{-g}\delta(\tau)\sqrt{\text{Max}_\pm\left[(G_{\mu\nu} \pm R_{\mu\nu})\partial^\mu\tau\partial^\nu\tau\right]} =$$

$$l_P^{-1}\hbar^{-1/2} \int\limits_{\sigma<0} d^4x\sqrt{-g}\delta(\tau)\sqrt{\text{Max}_\pm\left[(T_{\mu\nu} \pm T_{\mu\nu} \mp \frac{1}{2}g_{\mu\nu}\ T)\partial^\mu\tau\partial^\nu\tau\right]}. \quad (8)$$

Here $G_{\mu\nu}$ and $R_{\mu\nu}$ are the Einstein and the Ricci tensor, respectively, $T_{\mu\nu}$ is the energy–momentum tensor and T its trace. To derive the second equality, we have used Einstein equations, $G_{\mu\nu} = 8\pi G_N T_{\mu\nu}$. Note the appearance of the square root of the energy contained in the region and that (8) is manifestly covariant, and invariant under reparametrization of the hypersurface equation: such an invariance requires a square-root of $\partial^\mu\tau\partial^\nu\tau$. Reality of S_{CEB} is assured if sources obey the weak energy condition, $T_{\mu\nu}\partial^\mu\tau\partial^\nu\tau \geq 0$, since then the sum of the two combinations in (8), and thus their maximum, are positive. The weak energy condition is sufficient but not necessary for reality. We expect that for physical systems reality will always be guaranteed.

Since (8) applies to any space-like region, it can be written in a local form rather than in an integrated form by introducing an entropy current s_μ such that $S = \int d^4x\sqrt{-g}\delta(\tau)s_\mu\partial^\mu\tau$. Then (8) becomes equivalent to (λ^μ being an arbitrary time-like vector):

$$s_\mu\lambda^\mu \leq l_P^{-1}\hbar^{-1/2}\sqrt{\text{Max}_\pm\left[(T_{\mu\nu} \pm T_{\mu\nu} \mp \frac{1}{2}g_{\mu\nu}\ T)\lambda^\mu\lambda^\nu\right]}. \quad (9)$$

In the limit in which the hypersurface is light-like, $\partial^\mu\tau\partial_\mu\tau = 0$, (8) and (9) read

$$S_{CEB} = \int_{\sigma<0} d^4x\sqrt{-g}\delta(\tau)\sqrt{T_{\mu\nu}\partial^\mu\tau\partial^\nu\tau}\,,$$
$$s_\mu\lambda^\mu \leq l_P^{-1}\hbar^{-1/2}\sqrt{T_{\mu\nu}\lambda^\mu\lambda^\nu}\,,\quad \lambda_\mu\lambda^\mu = 0\,, \quad (10)$$

and become closely related to the assumptions made in [22] (1.10). We already see signs here that the physics at short scales and high energies is important in determining the value of the maximal entropy because $T_{\mu\nu}$ is generically at least quadratic in the fields.

The physical motivations leading us to the above proposal are similar to those used to motivate the HEB: (*i*) that entropy is maximized, in a given region of space, by the largest BH that can fit in it, and (*ii*) that the largest BH that can hold together without falling apart in a cosmological background has typically the size of the Hubble radius. The second assumption clearly needs to be refined and, possibly, to be defined covariantly. With such a goal in mind, we will proceed as follows: We will start by identifying a critical ("Jeans") length scale above which perturbations are causally disconnected so that BH of larger size, very likely, cannot form. We will first find this causal connection (CC) scale R_{CC} for the simplest cosmological backgrounds, then extend it to more general cases and, finally, guess the completely general expression using general covariance.

In order to identify the CC scale for a homogeneous, isotropic and spatially flat background, let us consider a generic perturbation around such a background in the hamiltonian approach developed in [23]. The Fourier components of the (normalized) perturbation and of its (normalized) conjugate momentum satisfy Schroedinger-like equations $\widehat{\Psi}_k{}''+\left[k^2-(S^{1/2})''S^{-1/2}\right]\widehat{\Psi}_k=0$, $\widehat{\Pi}_k{}''+\left[k^2-(S^{-1/2})''S^{1/2}\right]\widehat{\Pi}_k=0$, where k is the comoving momentum, a prime denotes differentiation w.r.t. conformal time η, and $S^{1/2}$ is the so-called pump field, a combination of the various backgrounds which depends on the specific perturbation under study. The perturbation equations clearly identify a "Jeans-like" CC comoving momentum

$$\begin{aligned} k_{CC}^2 &= \mathrm{Max}\left[(S^{1/2})''S^{-1/2}\ ,\ (S^{-1/2})''S^{1/2}\right] \\ &= \mathrm{Max}\left[\mathcal{K}'+\mathcal{K}^2\ ,\ -\mathcal{K}'+\mathcal{K}^2\right], \end{aligned} \tag{11}$$

where $\mathcal{K}=(S^{1/2})'S^{-1/2}$. Equation (11) always defines a real k_{CC} since the sum of the two quantities appearing on the r.h.s. is positive semi-definite. Since tensor perturbations are always present, let us restrict our attention to them. The "pump field" $S^{1/2}$ is simply given, in this case, by the scale factor $a(\eta)$ so that $\mathcal{K}\to\mathcal{H}=a'/a$. Equation (11) is immediately converted into the definition of a proper "Jeans" CC length $R_{CC}=ak_{CC}^{-1}$. Substituting into (11), and expressing the result in terms of proper-time quantities, we obtain (for tensor perturbations) $R_{CC}^{-2}=\mathrm{Max}\left[\dot{H}+2H^2\ ,\ -\dot{H}\right]$. Before trying to recast this equation in a more covariant form let us remove the assumption of spatial flatness by introducing the usual spatial-curvature parameter κ ($\kappa=0,\pm1$). The study of perturbations in non-flat space is considerably more complicated than in a spatially flat background. The final result, however, appears to be extremely simple [24, 25], and can be obtained from the flat case by the following replacements in (11): $\mathcal{H}^2\to\mathcal{H}^2+\kappa$, $\mathcal{H}'\to\mathcal{H}'$. Using this simple rule we arrive at the following generalization

$$R_{CC}^{-2} = \text{Max}\left[\dot{H} + 2H^2 + \kappa/a^2, \; - \dot{H} + \kappa/a^2\right]. \tag{12}$$

At this point we could have introduced anisotropy in our homogeneous background and study perturbations with or without spatial curvature. Instead, we adopt a shortcut route. We observe that the 00 components of the Ricci and Einstein tensors for our background are given by $R_{00} = -3(\dot{H} + H^2)$ *and* $G_{00} = 3(H^2 + \kappa/a^2)$. Obviously,

$$\begin{aligned} R_{CC}^{-2} &= \frac{1}{3}\text{Max}_{\mp}\left(G_{00} \mp R_{00}\right) \\ &= 4\pi G_N \,\text{Max}\left[\frac{\rho}{3} - p \,,\; \rho + p\right], \end{aligned} \tag{13}$$

where we have inserted Einstein equations using as an example a perfect-fluid energy momentum tensor $T^{\mu}_{\ \nu} = diag(\rho, -p, -p, -p)$. Equation (13) is guaranteed to define a real R_{CC} if the weak energy condition (reading here $\rho > 0$) holds, since the sum of the two combinations is positive in this case. In general, other perturbations may compete with tensor perturbations and define a smaller R_{CC}. In this case, the symbol Max in the above equations also applies to the various types of perturbations. This may help to ensure reality of R_{CC} in all physical situations.

As a final step, let us convert (13) into an explicitly covariant bound on entropy. Using R_{CC} as the maximal scale for BHs, we get a bound on entropy which scales like $S \sim V R_{CC}^{-3}\, R_{CC}^2 l_P^{-2} = V R_{CC}^{-1}\, l_P^{-2}$. We now express R_{CC}^{-1} as in (13) in terms of the components of the Ricci and Einstein tensors in the direction orthogonal to the hypersurface on which the entropy is being computed. This can be done covariantly by defining the hypersurface through the equation $\tau = 0$ and by identifying the normal with the vector $\nabla^{\mu}\tau$. This procedure leads immediately to the proposal (8). The local form (9) clearly follows by shrinking the space-like region to a point. Alternatively, using standard 3 + 1 ADM formalism [26], we can express the relevant components of the Ricci and Einstein tensors in terms of the intrinsic and extrinsic curvature of the hypersurface under study and arrive at the following final formula:

$$S_{CEB} = l_P^{-2} \int d^3x \sqrt{h}\; \left[\text{Max}\left(P \,,\; Q\right)\right]^{1/2}, \tag{14}$$

where $P = \frac{1}{2}\mathcal{R} + \dot{\theta} + \frac{2}{3}\theta^2 + \sigma^2 - \mathcal{A}$ and $Q = \frac{1}{2}\mathcal{R} - \dot{\theta} - 3\sigma^2 + \mathcal{A}$. Using standard notations, we have denoted by $\mathcal{R}$ the intrinsic 3-curvature scalar, by θ the expansion rate, by σ the shear, and by $\mathcal{A}$ the "acceleration" given (for vanishing shifts N_i) in terms of the lapse function N by $\mathcal{A} = N^{-1} N^{,i}_{\ ;i}$.

3.3 The CEB in D Dimensions

In order to generalize the CEB to arbitrary dimension D [27], we generalize the causal-connection scale R_{CC} by looking at perturbation equations in D dimensions. For gravitons, in the case of flat universe, one finds [28]

$$R_{CC}^{-2} = \frac{D-2}{2}\text{Max}\left[\dot{H} + \frac{D}{2}H^2, -\dot{H} + \frac{D-4}{2}H^2\right] . \tag{15}$$

If $H \gg \dot{H}$, $R_{CC} \propto H^{-1}$ and one recovers HEB with a D-dependent prefactor scaling as $\sqrt{D(D-2)}$. The above result generalizes to the case of a spatially curved universe as we have explained previously,

$$R_{CC}^{-2} = \frac{D-2}{2}\text{Max}\left[\dot{H} + \frac{D}{2}H^2 + \frac{D-2}{2}\frac{\kappa}{a^2}, -\dot{H} + \frac{D-4}{2}H^2 + \frac{D-2}{2}\frac{\kappa}{a^2}\right]. \tag{16}$$

A covariant definition of R_{CC} is obtained by expressing (16) in terms of the 00 components of curvature tensors. We find

$$R_{CC}^{-2} = \frac{D-2}{2(D-1)}\text{Max}\left[G_{00} \mp R_{00}\right] = 4\pi G_N\left[\frac{1}{D-1}\rho - p, \frac{2D-5}{D-1}\rho + p\right], \tag{17}$$

where, to derive the second equality, we have used Einstein's equations, $G_{\mu\nu} = 8\pi G_N T_{\mu\nu}$ and a perfect-fluid form for the energy–momentum tensor.

The Bekenstein–Hawking entropy of a Schwarzchild BH of radius R_{BH} in D dimensions is given by $S = \mathcal{A}/4l_P^{D-2}$. The generalization of S_{CEB} for a region of proper volume V is therefore

$$S_{\text{CEB}} = \beta n_H S^{BH} = \beta\frac{V}{V(R_{CC})}\frac{\mathcal{A}}{4l_P^{D-2}}, \tag{18}$$

where $n_H \equiv \frac{V}{V(R_{CC})}$ is the number of causally connected regions in the volume considered, $V(x)$ denotes the volume of a region of size x and β is a fudge factor reflecting current uncertainty on the actual limiting size for BH stability. For a spherical volume in flat space we have $V(x) = \Omega_{D-2}x^{D-1}/(D-1)$, with $\Omega_{D-2} = 2\pi^{(D-1)/2}/\Gamma\left(\frac{D-1}{2}\right)$. But in general the result is different and depends on the spatial-curvature radius.

Following [21], the expression for S_{CEB} in D dimensions can be rewritten in the explicitly covariant form

$$S_{\text{CEB}} = \frac{B}{l_P^{D-2}}\int_{\sigma<0} d^Dx\,\sqrt{-g}\delta(\tau)\sqrt{\text{Max}_{\pm}[(G_{\mu\nu} \pm R_{\mu\nu})\partial^\mu\tau\partial^\nu\tau]} =$$

$$\frac{B(8\pi)^{1/2}}{l_P^{D/2-1}}\int_{\sigma<0} d^Dx\,\sqrt{-g}\delta(\tau)\sqrt{\text{Max}_{\pm}\left[(T_{\mu\nu} \pm T_{\mu\nu} \mp \frac{1}{2}g_{\mu\nu}\,T)\partial^\mu\tau\partial^\nu\tau\right]}, \tag{19}$$

where $\sigma < 0$ defines the spatial region inside the $\tau = 0$ hypersurface whose entropy we are discussing, and T is the trace of the energy–momentum tensor.

The prefactor B can be fixed by comparing (18) and (19). Let us consider the expression (18) in the limit $R_{CC} \ll a$, where a is the radius of the universe. In this case, over a region of size R_{CC} we may neglect spatial curvature and write $V(R_{CC}) = \Omega_{D-2} R_{CC}^{D-1}/(D-1)$, and the area of the BH horizon as $\mathcal{A} = \Omega_{D-2} R_{BH}^{D-2}$, thus giving (apart for negligible terms of order $(R_{CC}/a)^2$)

$$S_{\rm CEB} = \beta \frac{D-1}{4} V R_{CC}^{-1} l_P^{-(D-2)} = B \sqrt{\frac{2(D-1)}{D-2}}\, V R_{CC}^{-1} l_P^{-(D-2)} . \qquad (20)$$

This fixes $B = \sqrt{\frac{(D-1)(D-2)}{32}}\,\beta$.

Since (19) applies to any space-like region, it can be rewritten in a local form as in a 4D case by introducing an entropy current s_μ such that $S = \int d^D x \sqrt{-g}\,\delta(\tau) s_\mu \partial^\mu \tau$. Then (19) becomes equivalent to (with λ^μ an arbitrary time-like vector)

$$s_\mu \lambda^\mu \leq l_P^{-D/2+1} (8\pi)^{1/2} B \sqrt{\mathrm{Max}_\pm \left[(T_{\mu\nu} \pm T_{\mu\nu} \mp \frac{1}{2} g_{\mu\nu}\, T) \lambda^\mu \lambda^\nu \right]} . \qquad (21)$$

In the limit of a light-like vector λ we get one of the conditions proposed by Flanagan et al. [22] in order to recover Bousso's proposal. Their bound corresponds (in $D = 4$) to $B = \frac{1}{4\pi}$ and could be used to fix β (assuming that it is D-independent).

For systems of limited gravity the BEB is tighter than the CEB, $S_{BEB} < S_{CEB}$. Therefore, in all systems for which the BEB is obeyed, the CEB will be obeyed as well. Hence, our bound is most interesting for systems of strong gravity, and in particular in cosmology.

For general collapsing regions we have limited computational power. While the local form (9) looks most appropriate for the study of collapsing regions, most likely the analysis of the general case will need the use of numerical methods. We can qualitatively check cases that are similar to the cosmological ones [29], such as homogeneous, isotropic contracting pressureless regions, or a contracting homogeneous, isotropic region filled with a perfect fluid. The pressureless case can be described by a Friedman interior and a Schwarzschild exterior. Since CEB is valid for the analogue cosmological solution, it is also valid for this case.

A particularly interesting case is that of the (generically non-homogeneous) collapse of a stiff fluid ($p = \rho$) which can be mapped by a simple field redefinition onto the dilaton-driven inflation of string cosmology [18]. In this case one finds a constant S_{CEB} in agreement with the HEB result [19]. Hence, no problem arises in this case, even if one starts from a saturated S_{CEB} at the onset of collapse. For non-stiff equations of state, the situation appears less safe if one starts near saturation. However, care must be taken in this case of perturbations which tend to grow non-linear by and form singularities on rather short time scales. Such cases cannot be described analytically but have been looked at numerically.

3.4 The CEB in Cosmology

The universe is a system of strong self-gravity. The geometry of the universe is determined by self-gravity, and the size of the universe is at least its gravitational radius. The strongest challenges to entropy bounds in general, and to the CEB in particular, come from considering (re)collapsing universes.

In homogeneous and isotropic D-dimensional cosmological backgrounds we have found the dependence of $R_{\rm CC}$ on the Hubble parameter $H(t)$, its time derivative $\dot{H}(t)$ and the scale factor $a(t)$ in (16) and (17),

$$\begin{aligned} R_{\rm CC}^{-2} &= \frac{D-2}{2}{\rm Max}\left[\dot{H}+\frac{D}{2}H^2+\frac{D-2}{2}\frac{\kappa}{a^2}, -\dot{H}+\frac{D-4}{2}H^2+\frac{D-2}{2}\frac{\kappa}{a^2}\right] \\ &= \frac{4\pi G_{\rm N}}{D-1}{\rm Max}\Big[\rho-(D-1)p\,, (2D-5)\rho+(D-1)p\Big], \end{aligned} \tag{22}$$

where $\kappa = 0, \pm 1$ determines the spatial curvature. Notice that $R_{\rm CC}$ is well defined if ρ is positive because the maximum in (22) is larger than the average of the two entries in the brackets, and the average is equal to $2(D-2)\rho$.

The following four cases exhaust all possible types of cosmologies [21, 30]:

1. $|\dot{H}| \sim H^2 \sim |k|/a^2$, or $|\dot{H}| \sim H^2 \gg |k|/a^2$. In this case effective energy density and pressure are of the same order, $\rho \sim p$. All length scales that may be considered in entropy bounds, such as particle horizon, apparent horizon, $R_{\rm CC}$ and the Hubble radius, are parametrically equal. This case includes non-inflationary FRW universes with matter and radiation.
2. $H^2 \gg |k|/a^2, |\dot{H}|$. In this case $|\rho+p| \ll \rho$, and the universe is inflationary. In this case R_{CC} is parametrically equal to $|H|^{-1}$.
3. $|\dot{H}| \gg H^2$, $|k|/a^2$. In this case $|\rho| \ll p$. Since ρ and p are the effective energy density and pressure, there are no problems with causality. This case occurs, for instance, near the turning point of an expanding universe which recollapses, or near a bounce of a contracting universe which re-expands.
4. $k/a^2 \gg |\dot{H}|$, H^2. In this case the spatial curvature determines the causal connection scale. This occurs, for example, when both H and $\dot{H}$ vanish as in a closed Einstein universe.

We will first describe several cosmological models and explain how they satisfy the CEB. Then we will present in a general form the conditions on sources that guarantee the validity of the CEB.

A radiation-dominated Universe

Our first example is a radiation-dominated universe in D dimensions. In this case $\rho = (D-1)p$ and the 00 equation for the scale factor is

$$H^2 + \frac{\kappa}{a^2} = \frac{16\pi G_N}{(D-1)(D-2)}\rho = \frac{16\pi G_N}{(D-1)(D-2)}\rho_0 R_0^D a^{-D}, \quad \kappa = 0, \pm 1 \tag{23}$$

In terms of the conveniently rescaled conformal time η, defined by $a(\eta)d\eta = (D-2)dt$, the solutions can be put in the simple form

$$a(\eta) = A^{\frac{1}{D-2}} \begin{cases} [\sin(\eta/2)]^\alpha & \kappa = 1 \\ (\eta/2)^\alpha & \kappa = 0 \\ [\sinh(\eta/2)]^\alpha & \kappa = -1 \end{cases}, \quad A = \frac{16\pi G_N \rho_0 R_0^D}{(D-1)(D-2)}, \quad \alpha = \frac{2}{D-2}. \tag{24}$$

As can be seen from (24) the qualitative behavior of the solutions does not depend strongly on D. In a (closed, open or flat) RD universe one always has $R_{00} = G_{00}$; therefore, $R_{CC}^{-2} = \frac{D-2}{2}\left(-\dot{H} + \frac{D-4}{2}H^2 + \frac{D-2}{2}\frac{\kappa}{a^2}\right)$. The behavior of S_{CEB} is easily derived from the explicit solution for the scale factor and R_{CC}. In the case D=4 it is shown in Fig. 1.

A related case is when matter can be modeled by a conformal field theory (CFT). Kutasov and Larsen [31] pointed out that for weakly coupled CFTs in a sphere of radius R, the free energy F, the entropy S and the total energy E can be expanded at weak coupling and large $x \equiv 2\pi RT$,

$$-FR = f(x) = \sum_{n\geq 0} a_{D-2n} x^{D-2n} + \dots \tag{25}$$

$$S = 2\pi f'(x)\,, \tag{26}$$

$$ER = (x\partial_x - 1) f(x), \tag{27}$$

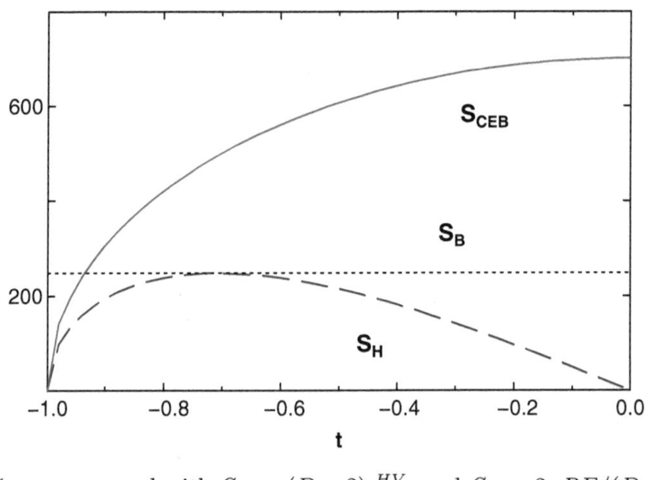

Fig. 1. S_{CEB} compared with $S_H = (D-2)\frac{HV}{4G_N}$ and $S_B \equiv 2\pi RE/(D-1)$ in the expanding phase of a closed $D = 4$, RD Universe. Here we set $\beta = \frac{D-2}{D-1}$

where the dots represent non-perturbative contributions.

We can explicitly check under which conditions the entropy of weakly coupled CFTs obeys the CEB, $S < S_{\mathrm{CEB}} = 4B\sqrt{\pi}\sqrt{EV}l_P^{-(D-2)/2}$. In the limit $TR \gg 1$ we find

$$\frac{S^2}{S^2_{\mathrm{CEB}}} = \frac{\pi a_D D^2}{4B^2(D-1)\Omega_{D-1}} \left(2\pi l_P T\right)^{D-2} . \tag{28}$$

Thus, CEB is obeyed provided that

$$\left(\frac{T}{M_P}\right)^{D-2} < \frac{K(D)}{a_D}, \tag{29}$$

where $K(D)$ is a D-dependent (but CFT-independent) constant. We conclude that CEB is obeyed as long as temperatures are below M_P by a factor $a_D^{-\frac{1}{D-2}}$ Since a_D is proportional to the number $\mathcal{N}$ of CFT-matter species, we obtain a bound on temperature which scales as $\mathcal{N}^{-\frac{1}{D-2}}$ in Planck units.

We can also explicitly check under which conditions strongly coupled CFTs possessing AdS duals as considered by Verlinde [11] obey the CEB. For such CFTs,

$$S = \frac{c}{12}\frac{V}{L^{D-1}}, \tag{30}$$

$$E = \frac{c}{12}\frac{D-1}{4\pi L}\left(1+\frac{L^2}{R^2}\right)\frac{V}{L^{D-1}}, \tag{31}$$

$$T = \frac{1}{4\pi L}\left(D+(D-2)\frac{L^2}{R^2}\right), \tag{32}$$

where c is the central charge of the CFT and $L \sim 1/T$ is the AdS radius.

In this case, in the limit $R/L \sim TR \gg 1$ we find

$$\frac{S^2}{S^2_{\mathrm{CEB}}} = \frac{1}{4(D-1)B^2}\frac{c}{12}\left(\frac{l_P}{L}\right)^{D-2} \tag{33}$$

and thus CEB is obeyed for

$$\frac{1}{4(D-1)B^2}\frac{c}{12}\left(\frac{4\pi T}{DM_P}\right)^{(D-2)} < 1 . \tag{34}$$

Since the central charge c is proportional to the number of CFT fields $\mathcal{N}$, we obtain a bound on temperature which, in Planck units, scales as $\mathcal{N}^{-\frac{1}{D-2}}$, exactly as previously obtained for the weakly coupled case.

For the case $ER \sim a_D$ (which corresponds to $RT \sim 1$) the validity of the CEB guaranteed by a condition similar to (29).

Finally, we would like to show that CEB holds also when $ER \sim 1$. In this case $S_{CEB} \simeq 4B\sqrt{\pi}\sqrt{V/R}l_P^{-(D-2)/2}$ scales as $\left(\frac{R}{l_P}\right)^{\frac{D-2}{2}}$. The appropriate setup for calculating the entropy in this case is the microcanonical ensemble with the result $S \sim \log a_D \sim \log \mathcal{N}$; thus, $S < S_{CEB}$ is guaranteed for a macroscopic universe as long as

$$\left(\frac{R}{l_P}\right)^{\frac{D-2}{2}} > \log \mathcal{N} \,. \tag{35}$$

In a quantum theory of gravity we expect the UV cutoff Λ to be finite and to represent an upper bound on T (as in the example of superstring theory and its Hagedorn temperature) and a lower bound on R (as in the minimal compactification radius). Thus conditions (29) and (34) for the validity of CEB are satisfied as long as $\left(\frac{\Lambda}{M_P}\right)^{D-2} < 1/\mathcal{N}$. A bound of the same form was previously proposed in [9] and [32], and independent arguments in support of bounds of this sort have also been put forward in [13].

The Inflationary Universe

The inflationary universe is completely compatible with the CEB. To a certain extent this is a not such an interesting case, because the CEB is comfortably satisfied.

The entropy balance begins for the inflationary universe after the end of inflation when the energy of the background is converted to matter. This process is historically called reheating and is associated with a large entropy production. In the following we will assume that the reheating process is instantaneous and complete. We will denote by the subscript RH quantities at the instant of reheating.

Since $\dot{H}$ is subleading in this case, it follows from (22) that $R_{CC} \sim 1/H$. In this case the CEB and the HEB are similar,

$$S_{CEB}(t_{RH}) = l_P^{-2} H(t_{RH}) a^3(t_{RH}). \tag{36}$$

Assuming that the energy has been completely converted into radiation, the energy density of the radiation is $\rho(t_{RH}) = T_{RH}^4$. From the 00 Einstein equation $l_P^{-2}H^2 = \rho$, thus

$$\begin{aligned} S_{CEB}(t_{RH}) &= T_{RH}^4 \frac{1}{H(t_{RH})} a^3(t_{RH}) \\ &= \frac{T_{RH}}{H(t_{RH})} T_{RH}^3 a^3(t_{RH}) \\ &= \frac{T_{RH}}{H(t_{RH})} S(t_{RH}). \end{aligned} \tag{37}$$

Here we have used the expression for the radiation entropy $S(t_{RH}) = T_{RH}^3 a^3(t_{RH})$. Since from the 00 Einstein equation $\frac{T_{RH}}{H(t_{RH})} \sim \sqrt{\frac{M_P}{H(t_{RH})}}$, and since we expect that the Hubble parameter at reheat be substantially below the Planck temperature, we conclude that the CEB is comfortable satisfied.

A Universe Near a Turning Point

Let us consider either a flat or closed universe with some perfect fluid in thermal equilibrium and a constant equation of state $p = \gamma\rho$, $1 > \gamma > -1$, and with an additional small negative cosmological constant $\Lambda = -\lambda$. The universe starts out expanding, reaches a maximal size and then contracts toward a singularity. In this case the matter entropy within a comoving volume is constant in time. But near the point of maximal expansion the apparent horizon and the Hubble length diverge causing violation of the HEB. However, for a fixed comoving volume, $S_{CEB} \sim V R_{CC}^{-1}$, and, since R_{CC} is never larger than some maximal value, the CEB has a chance of doing better.

To see this explicitly, let us consider a 4D example. In this case we obtain from (22)

$$R_{CC}^{-2} = \frac{1}{3}\text{Max}\left[\frac{1}{2}\rho_0(1-3\gamma)a^{-3(1+\gamma)} - 2\lambda\ ,\ \ \frac{3}{2}(1+\gamma)\rho_0 a^{-3(1+\gamma)}\right], \quad (38)$$

independently of κ. The initial energy density is ρ_0 and a is the ratio of the scale factor to its initial value. Since the maximum is larger than each of the expressions in the brackets

$$R_{CC}^{-2} \geq \frac{1}{2}(1+\gamma)\rho_0 a^{-3(1+\gamma)}. \quad (39)$$

It follows that in a fixed comoving volume S_{CEB} scales as $\sim a^3 R_{CC}^{-1} \sim a^{3/2(1-\gamma)}$. Since $\gamma < 1$, this means that S_{CEB} grows during the expansion, reaches a maximum at the turning point and then starts decreasing. If the initial conditions are fixed at sufficiently early times when curvature and cosmological constant are negligible, the CEB will be obeyed initially provided energy density and curvature are less than Planckian. But then the evolution of S_{CEB} that we have found will guarantee that the bound is satisfied at all times until Planckian density and curvature is reached in the recollapsing phase. Thus the CEB will be satisfied throughout the classical evolution of our Universe.

A Static Universe

The simplest example of a non-singular cosmology is a static Einstein model in D dimensions which was discussed in [30]. This model requires positive curvature, and two types of sources: cosmological constant and dust; we denote by ρ_Λ and ρ_{m} the energy densities associated with each of the two components. To provide entropy, we need an additional source, which we choose to be radiation consisting of $\mathcal{N}$ species in thermal equilibrium at temperature T. The energy density of the radiation is given by $\rho_{\mathrm{r}} = \mathcal{N} T^D$, and the entropy density of the radiation is given by $s_\mathrm{r} = \mathcal{N} T^{D-1}$ (we ignore here numerical factors since we will be interested in scaling of quantities). The total entropy of the system is given entirely by the entropy of the radiation $S_\mathrm{r} = s_\mathrm{r} V$.

In term of these sources, Einstein equations can be written in the following way:

$$H^2 + \frac{1}{a^2} = \frac{16\pi G_\mathrm{N}}{(D-2)(D-1)}\rho_\mathrm{tot} = \frac{16\pi G_\mathrm{N}}{(D-2)(D-1)}\left(\rho_\Lambda + \rho_\mathrm{m} + \rho_\mathrm{r}\right) \quad (40)$$

$$\begin{aligned} \dot{H} - \frac{1}{a^2} &= -\frac{8\pi G_\mathrm{N}}{(D-2)}\left(\rho_\mathrm{tot} + p_\mathrm{tot}\right) \\ &= -\frac{8\pi G_\mathrm{N}}{(D-2)(D-1)}\left[D\rho_\mathrm{r} + (D-1)\rho_\mathrm{m}\right], \end{aligned} \quad (41)$$

where we have used in (41) the equations of state relating pressure to energy density: $p_\Lambda = -\rho_\Lambda$, $p_\mathrm{m} = 0$ and $(D-1)p_\mathrm{r} = \rho_\mathrm{r}$.

For given ρ_m and ρ_r, one can choose ρ_Λ and the scale factor a such that H and $\dot{H}$ vanish in (40) and (41), and thus obtains a static solution. In particular, the condition given by (41) determines the scale factor in terms of ρ_m and ρ_r,

$$a^2 = \frac{(D-2)(D-1)}{8\pi G_\mathrm{N}}\frac{1}{D\rho_\mathrm{r} + (D-1)\rho_\mathrm{m}}. \quad (42)$$

Since both H and $\dot{H}$ vanish identically, R_CC is determined solely by the scale factor a given in (42), as discussed previously.

We now wish to determine under which conditions (if any) some violations of CEB may occur in this model. Recall that according to (20) the CEB bounds the total entropy of a region contained in a comoving volume V by $S_\mathrm{CEB} = \alpha(D-1)\frac{V}{G_\mathrm{N} R_\mathrm{CC}}$, and that in the static case under consideration $R_\mathrm{CC} = 2a/(D-2)$. The square of the ratio of S_CEB and the entropy of the system S_r is given by

$$\begin{aligned} \left(\frac{S_\mathrm{CEB}}{S_\mathrm{r}}\right)^2 &= \left(\frac{\alpha(D-1)}{s_\mathrm{r} R_\mathrm{CC} G_\mathrm{N}}\right)^2 = \\ &= \left[2\pi\alpha^2(D-1)(D-2)\right]\left[D + (D-1)\frac{\rho_\mathrm{m}}{\rho_\mathrm{r}}\right]\left[\frac{1}{\mathcal{N}}\left(\frac{M_P}{T}\right)^{D-2}\right]. \end{aligned} \quad (43)$$

Since the second factor in expression (43) is larger than unity if ρ_m and ρ_r are positive, and neglecting the overall prefactor which is independent of the sources in the model, we conclude that the CEB is valid provided that

$$\mathcal{N}\left(\frac{T}{M_\mathrm{P}}\right)^{D-2} \leq 1. \quad (44)$$

This is the same condition discussed above which should be interpreted as a requirement that temperatures are sub-Planckian, in the case of many number of species $\mathcal{N}$.

Our conclusion is that as long as the temperature of radiation stays well below Planckian, CEB is upheld. The fact that the model is gravitationally unstable to matter perturbations does not seem to be particularly relevant to the issue of validity of the CEB.

Bekenstein's Non-singular Universe

A time-dependent non-singular cosmological model was found years ago by Bekenstein [33] (see also [34]). This is a 4D Friedman–Robertson–Walker universe which is conformal to the closed Einstein Universe. It contains dust, consisting of N particles of mass μ (N is constant and μ is positive), coupled to a classical conformal massless scalar field ψ, and $\mathcal{N}$ species of radiation in thermal equilibrium. The action for the dust-ψ system is given by

$$\mathcal{S} = -\frac{1}{2}\int \sqrt{-g}\left[(\nabla\psi)^2 + \frac{1}{6}\psi^2 R\right] d^4x - \int (\mu + f\psi)\, d\tau. \tag{45}$$

It includes in addition to the usual action for free point particles of rest mass μ, a dust-scalar field interaction whose strength is determined by the coupling f. Accordingly, we may define the effective mass of the dust particles: $\mu_{\text{eff}} = \mu + f\psi$.

The total energy density and pressure in Bekenstein's Universe are given by

$$\rho_{\text{tot}} = \rho_{\text{r}} + \rho_\psi + \rho_{\text{m}}, \quad p_{\text{tot}} = p_{\text{r}} + p_\psi + p_{\text{m}}, \tag{46}$$

where $\{\rho_{\text{r}}, p_{\text{r}}\}$, $\{\rho_\psi, p_\psi\}$ and $\{\rho_{\text{m}}, p_{\text{m}}\}$ are the energy densities and pressures associated with the radiation, scalar field and dust, respectively. They depend on the scale factor in the following way

$$\begin{aligned} \rho_{\text{r}} &= \mathcal{C}\mathcal{N}a^{-4} = \mathcal{N}T^4, \\ \rho_\psi &= \frac{1}{2}f^2N^2a^{-4}, \\ \rho_{\text{m}} &= N\mu_{\text{eff}}a^{-3} = N\mu a^{-3} - 2\rho_\psi, \end{aligned} \tag{47}$$

and their equations of state $\gamma_{\text{r}} = p_{\text{r}}/\rho_{\text{r}}$, $\gamma_\psi = p_\psi/\rho_\psi$ and $\gamma_{\text{m}} = p_{\text{m}}/\rho_{\text{m}}$ are the following:

$$\begin{aligned} \gamma_{\text{r}} &= 1/3, \\ \gamma_\psi &= -1/3, \\ \gamma_{\text{m}} &= 0. \end{aligned} \tag{48}$$

The dependence of ψ on a $\psi = -fNa^{-1}$ yields $\mu_{\text{eff}} = \mu - f^2Na^{-1}$. $\mathcal{C}$ is an integration constant and the only source of entropy is the radiation whose entropy density is given by $s_{\text{r}} = \mathcal{N}T^3$.

The solution for the scale factor a is given in terms of the conformal time η by

$$a(\eta) = a_0(1 + B \sin \eta). \tag{49}$$

We assume that a_0, the mean value of the scale factor, is macroscopic, so it is large in our Planck units. If $B = 0$, the solution describes a static universe very similar to the closed Einstein Universe discussed previously. For $0 < B < 1$ the solution describes a "bouncing universe": The universe bounces off at $\eta = 3\pi/2$ when the scale factor is minimal $a = a_{\rm min} = a_0(1 - B)$, expands until it turns over at $\eta = 5\pi/2$ when its scale factor is maximal, $a = a_{\rm max} = a_0(1+B)$, and continues to oscillate without ever reaching a singularity. The equations of motion require that the energy densities of the sources obey the following equalities at all times [33]:

$$2\frac{a}{a_0}\left(\frac{\rho_\psi - \rho_{\rm r}}{2\rho_\psi + \rho_{\rm m}}\right) = 1 - B^2 = \frac{a_{\rm min} a_{\rm max}}{{a_0}^2}. \tag{50}$$

Since $2\rho_\psi + \rho_{\rm m} = N\mu a^{-3} > 0$, $\rho_{\rm r} > 0$ and $B^2 < 1$, it follows that a necessary condition for a bounce is that $\rho_{\rm r} < \rho_\psi$. This implies that the total pressure $\frac{1}{3}(\rho_{\rm r} - \rho_\psi)$ is always negative. Moreover, (50) for $a = a_{\rm min}$ implies that $\rho_{\rm m} \leq -2\rho_{\rm r} < 0$ there. But then, the conclusion must be that in order to avoid a singularity, $\mu_{\rm eff} < 0$ at least at the bounce. It is possible, however, to find a range of initial conditions and parameters such that $\mu_{\rm eff}$ is positive near the turnover.

The result that $\rho_{\rm r}$ and ρ_ψ are manifestly positive definite, but $\rho_{\rm m}$ can (and in fact must) be negative some of the time, suggest that it might be possible to parametrically decrease $\rho_{\rm tot}$ by lowering $\mu_{\rm eff}$ (making it large and negative) by increasing the coupling strength f, so that the amounts of radiation and entropy are kept constant. As it turns out this is exactly the case in which the CEB can be potentially violated. Using Einstein equations to express $R_{\rm CC}$ in terms of the total energy density and pressure, we find the ratio $(S_{\rm CEB}/S_{\rm r})^2$:

$$\left(\frac{S_{\rm CEB}}{S_{\rm r}}\right)^2 \sim {G_{\rm N}}^{-2}\left(\frac{\rho_{\rm r}}{\mathcal{N}}\right)^{-3/2}\frac{1}{\mathcal{N}^2} G_{\rm N}\,{\rm Max}\left[\frac{\rho_{\rm tot}}{3} - p_{\rm tot}, \rho_{\rm tot} + p_{\rm tot}\right], \tag{51}$$

a system for which the ratio above is smaller than one would violate the CEB. Recalling that the maximum on the r.h.s. of (51) is always larger than the mean of the two entries and rearranging, we find

$$\left(\frac{S_{\rm CEB}}{S_{\rm r}}\right)^2 \geq \left[\frac{1}{\mathcal{N}}\frac{M_P^2}{T^2}\right]\frac{\rho_{\rm tot}}{\rho_{\rm r}}. \tag{52}$$

Since we assume that the model is sub-Planckian, namely that the first factor is larger than one as in (44), the only way in which CEB could be violated is if somehow the second factor was parametrically small. As discussed above, it does seem that the second term $\rho_{\rm tot}/\rho_{\rm r}$ can be made arbitrarily small by decreasing $\rho_{\rm tot}$ while keeping $\rho_{\rm r}$ constant. Consequently, it is apparently possible to make the ratio $S_{\rm CEB}/S_{\rm r}$ smaller than one and obtain a CEB violating cosmology. But this can be achieved only if the effective mass of the dust particles is negative (and large) as can be seen from (46).

Violations of the CEB (and as a matter of fact, of any other entropy bound) go hand in hand with large negative energy densities in the dust sector. In the model under discussion, this manifests itself in the form of dust particles with highly negative effective masses. Occurrence of such negative energy density would most probably render the model unstable. We argue that any analysis of entropy bounds should be performed for stable models. This is particularly relevant for the CEB, whose definition involves explicitly the largest scale at which stable BHs could be formed. However, the instability does not necessarily lead to violations of the CEB as in the previous case. To support this argument, we have outlined possible instabilities in the dust scalar field system when the dust particles' mass is negative [30].

The Pre-big-bang Scenario

Veneziano [19] was the first to study entropy bounds in the context of the PBB scenario. It has been argued [35, 36] that a form of stochastic PBB is a generic consequence of natural initial conditions corresponding to generic gravitational and dilatonic waves superimposed on the perturbative vacuum of critical superstring theory. In the Einstein-frame metric this can be seen as a chaotic gravitational collapse leading to the formation of BHs of different sizes. For a string frame observer inside each BH this is viewed as a PBB inflationary cosmology. The duration of the inflationary phase is controlled by the size of the BH [35, 36], so from this point of view the observable Universe should be identified with the region of space that was originally inside a sufficiently large BH.

In Veneziano [19] studied a 4D PBB model and followed the evolution of several contributions to the entropy. At time $t = t_i$, corresponding to the first appearance of a horizon, he used the Bekenstein–Hawking formula to evaluate that the entropy in the collapsed region S_{coll}. Then he used the fact [36] that the initial size of the BH horizon determines the initial value of the Hubble parameter and found that

$$S_{coll} \sim (R_{in}/l_{P,in})^2 \sim (H_{in} l_{P,in})^{-2} = S_{HEB}. \tag{53}$$

Thus, initially the entropy is as large as allowed by the HEB (without fine-tuning). Here it was implicitly assumed the initial string coupling is small.

After a short transient phase, dilaton-driven inflation (DDI) should follow [35, 36] and last until t_s, the time at which a string-scale curvature is reached. We expect this classical process not to generate further entropy. During DDI S_{HEB} remains constant and the bound continues to be saturated. This follows from the "conservation law" of string cosmology [18]

$$\partial_t \left(e^{-\phi}\sqrt{g}H\right) = 0; \tag{54}$$

hence,

$$\partial_t \left(\left(\sqrt{g}H^3\right)\ \left(e^{-\phi}H^{-2}\right)\right) = \partial_t \left(n_H S^H\right) = 0\ . \tag{55}$$

Veneziano suggested the following interpretation: At the beginning of the DDI phase the whole entropy is in a single Hubble volume. As DDI proceeds, the same total amount of entropy becomes equally shared between very many Hubble volumes until, eventually, each one of them contributes a small number.

While the coupling is still small, S_{HEB} cannot decrease,

$$\partial_t(e^{-\phi}\sqrt{g}H) \geq 0. \tag{56}$$

It follows that

$$(\dot{\phi} - 3H) \leq \dot{H}/H\ . \tag{57}$$

Veneziano noticed that this constraint may be important. As α' corrections intervene to stop the growth of H, the entropy bound forces $\dot{\phi}-3H$ to decrease and eventually to change sign if H stops growing. But this is just what is needed to convert the DDI solution into the FRW solution [18].

If the initial conditions are such that the string coupling becomes strong while the curvature is still small, then Veneziano argued [19] that the HEB forces a non-singular PBB cosmology as well. This time the entropy production by the squeezing of quantum fluctuations is the important factor. This will be discussed further when we discuss the generalized second law.

3.5 Conditions for the Validity of the CEB in Cosmology

We may summarize the lessons of the previous examples by imposing conditions on sources in a generic cosmological setting such that the CEB is obeyed.

We consider a cosmic fluid consisting of radiation, an optional cosmological constant, and additional unspecified classical dynamical sources which do not include any contributions from the cosmological constant or radiation. For simplicity we assume that the additional sources have negligible entropy. This is the most conservative assumption: If some of the additional sources have substantial entropy our conclusions can be strengthened. We use the previous notations for the total, cosmological and radiation energy densities, $\rho_{\rm tot}$, ρ_Λ and $\rho_{\rm r}$ respectively, and denote by ρ^* the combined energy density of the additional sources. Thus

$$\rho_{\rm tot} = \rho_{\rm r} + \rho_\Lambda + \rho^*. \tag{58}$$

We use the same notation for the relative pressures, and for the equation of state $\gamma^* \equiv \rho^*/p^*$, which may be time dependent.

In terms of these sources, the causal connection scale can be written as

$$R_{\rm CC}^{-2} = \frac{4\pi G_{\rm N}}{D-1}{\rm Max}\left\{ D\rho_\Lambda + \left[1-(D-1)\gamma^*\right]\rho^*\, ,\right.$$

$$(D-4)\rho_\Lambda + \left[(2D-5)+(D-1)\gamma^*\right]\rho^* + 2(D-2)\rho_r \bigg\}. \tag{59}$$

We may now express the ratio of $(S_{\rm CEB}/S_r)^2$, neglecting as usual prefactors of order 1

$$\left(\frac{S_{\rm CEB}}{S_r}\right)^2 \sim \frac{1}{\mathcal{N}}\left(\frac{M_P}{T}\right)^{D-2} \mathrm{Max}\left\{ D\frac{\rho_\Lambda}{\rho_r} + \left[1-(D-1)\gamma^*\right]\frac{\rho^*}{\rho_r},\right.$$

$$\left.(D-4)\frac{\rho_\Lambda}{\rho_r} + \left[(2D-5)+(D-1)\gamma^*\right]\frac{\rho^*}{\rho_r} + 2(D-2)\right\}. \tag{60}$$

Any CEB violation requires that this ratio be parametrically smaller than one. Notice that the first factor is larger than one by our requirement that the radiation energy density be sub-Planckian. Thus the only remaining possibility for violating CEB is that the second factor be parametrically smaller than unity. As we show below, this can occur only if at least one of the additional sources has negative energy density.

The r.h.s. of (60) is larger than the average of the two entries, so that

$$\left(\frac{S_{\rm CEB}}{S_r}\right)^2 \geq \frac{1}{\mathcal{N}}\left(\frac{M_P}{T}\right)^{D-2} (D-2)\frac{\rho_{\rm tot}}{\rho_r}. \tag{61}$$

Therefore, since $\rho_{\rm tot} > 0$, a necessary condition for this expression to be smaller than unity is that $\rho_{\rm tot} \ll \rho_r$, which we may re-express as

$$\frac{\rho_\Lambda}{\rho_r} \sim -\left(1+\frac{\rho^*}{\rho_r}\right). \tag{62}$$

This is not a sufficient condition since the equations of motion could dictate, for example, that the first factor on the r.h.s. of (61) could be parametrically larger than unity at the same time. By substituting condition (62) into (60), we obtain

$$\left(\frac{S_{\rm CEB}}{S_r}\right)^2 \sim \frac{1}{\mathcal{N}}\left(\frac{M_P}{T}\right)^{D-2} \times$$

$$\mathrm{Max}\left\{-\left[(D-1)(1+\gamma^*)\frac{\rho^*}{\rho_r}+D\right], (D-1)(1+\gamma^*)\frac{\rho^*}{\rho_r}+D\right\}. \tag{63}$$

Therefore, an additional necessary condition for S_{CEB}/S_r to be smaller than one is that

$$(1+\gamma^*)\rho^* \simeq -\frac{D}{(D-1)}\rho_r\,. \tag{64}$$

Condition (64) can be satisfied in two ways:

(i) $1+\gamma^* > 0$ and $\rho^* < 0$. This obviously requires that at least one of the sources has negative energy density. In this case (barring pathologies) the magnitude of ρ^* is comparable to that of $\rho_{\rm r}$.

(ii) $1+\gamma^* < 0$ and $\rho^* > 0$. However, for classical dynamical sources, this typically clashes with causality which requires that the pressure and energy density of each of the additional dynamical sources obey $|p_i| < |\rho_i|$; hence, if all $\rho_i > 0$, then necessarily $\gamma^* = (\sum p_i)\,/\,(\sum \rho_i) > -1$.

Consequently, condition (64) cannot be satisfied if all of the dynamical sources have positive energy densities and equations of state $|\gamma_i| \leq 1$. Bekenstein's Universe discussed previously fits well within our framework: The total energy density is positive, but the overall contribution to $\rho_{\rm tot}$ of all the sources, excluding radiation (since the cosmological constant vanishes in this case), is negative and almost cancels the contribution of radiation, leaving a small positive $\rho_{\rm tot}$.

To summarize, if all dynamical sources (different from the cosmological constant) have positive energy densities $\rho_i > 0$ and have causal equations of state ($|\gamma_i| \leq 1$), and if radiation temperatures are sub-Planckian, CEB is upheld.

3.6 The CEB and the Singularity Theorems

The CEB (and entropy bounds in general) refines the classic singularity theorems. It is satisfied by cosmologies for which the singularity theorems are not applicable because some of the energy conditions are violated, but do not seem to be problematic in any of their properties. Conversely, it indicates possible problems when the singularity theorems seem perfectly valid.

In general, the total energy–momentum tensor of a closed "bouncing" universe violates the SEC, but it can obey the CEB. In order to see this explicitly, let us consider the "bounce" condition, i.e. $H=0$, $\dot{H}>0$ for a closed Universe; by using the Einstein equations (40 and 41), we can express this condition in terms of the sources as follows:

$$\rho_{\rm tot} > 0, \quad (D-3)\rho_{\rm tot} + (D-1)p_{\rm tot} < 0. \tag{65}$$

The second of these conditions is (in $D=4$) precisely the condition for violation of the SEC. In terms of $\rho_{\rm r}$, ρ_Λ and ρ^* this reads

$$2\rho_\Lambda - (D-2)\rho_{\rm r} - \left[(D-3)+(D-1)\gamma^*\right]\rho^* > 0\,. \tag{66}$$

In comparison, a necessary condition that the CEB is violated can be obtained from (62) and (64),

$$2\rho_\Lambda - (D-2)\rho_{\rm r} - \left[(D-3)+(D-1)\gamma^*\right]\rho^* \sim 0\,, \tag{67}$$

where the l.h.s. of (67) can be either positive or negative. So we find that there is a range of parameters for which the CEB can be obeyed in some bouncing cosmologies but not in others.

In a spatially flat universe ($\kappa = 0$), the conditions for a bounce are slightly different: $\rho_{\rm tot} = 0$ and $\rho_{\rm tot} + p_{\rm tot} < 0$. At the bounce these conditions imply violation of the null energy condition (NEC). As discussed previously, classical sources are not expected to violate the NEC, but effective quantum sources (such as Hawking radiation) are known to violate the NEC. In terms of $\rho_{\rm r}$, ρ_Λ and ρ^* the condition for a bounce reads

$$\left(1 + \frac{1}{D-1}\right)\rho_{\rm r} + (1+\gamma^*)\rho^* > 0. \tag{68}$$

In comparison, a necessary condition that the CEB is violated can be obtained from (64),

$$\left(1 + \frac{1}{D-1}\right)\rho_{\rm r} + (1+\gamma^*)\rho^* \sim 0\,, \tag{69}$$

where the l.h.s. of (69) can be either positive or negative. So, again, we find that there is a range of parameters for which the CEB can be obeyed in some spatially flat bouncing cosmologies but not in others.

The CEB appears to be a more reliable criterion than energy conditions when trying to decide whether a certain cosmology is reasonable: Taking again the closed deSitter Universe as an example, we can add a small amount of radiation to it, and still have a bouncing model if ρ_Λ is the dominant source, and SEC will not be obeyed (66). Nevertheless, the general discussion in this section shows that in this case the CEB is not violated as long as radiation temperatures remain sub-Planckian, despite the presence of a bounce. This happens, in part, because the CEB is able to discriminate better between dynamical and non-dynamical sources (such as the cosmological constant), and imposes constraints that involve the former ones only, such as (64).

We have reached the following conclusions by studying the validity of the CEB for non-singular cosmologies:

1. Violation of the CEB necessarily requires either high temperatures $\mathcal{N}\left(\frac{T}{M_{\rm P}}\right)^{D-2} \geq 1$, or dynamical sources that have negative energy densities with a large magnitude, or sources with acausal equation of state. Of course, neither of the above is sufficient to guarantee violations of the CEB.
2. Classical sources of this type are suspect of being unphysical or unstable, but each source has to be checked on a case-by-case basis. In the examples that we have discussed the sources were indeed found to be unstable or are strongly suspected to be so.
3. Sources with large negative energy density could allow, in principle, to increase the entropy within a given volume while keeping its boundary

area and the total energy constant. This would lead to violation of all known entropy bounds, and of any entropy bound which depends in a continuous way on the total energy or on the linear size of the system.

4. The CEB is more discriminating than singularity theorems. In the examples we have considered it allows non-singular cosmologies for which singularity theorems cannot be applied, but does not allow them if they are associated with specific dynamical problems.

3.7 Comparison of the CEB to Other Entropy Bounds

Finally, we compare our CEB to other bounds, in particular to Bekenstein's and Bousso's. For systems of limited gravity whose size exceeds their Schwarzschild radius: $R > R_g$, Bekenstein's bound is given by $S < S_{BEB} = l_P^{-2} R\, R_g$, and Bousso's procedure results in the holography bound, $S < S_{HOL} = l_P^{-2} R^2$, but since $R > R_g$, $S_{BEB} < S_{HOL}$, and therefore Bousso's bound is less stringent than Bekenstein's. Consider now the CEB applied to the region of size R containing an isolated system. Expressing CEB in the form (8) one immediately obtains $S_{CEB} = l_P^{-1} R^{3/2} E^{1/2} \hbar^{-1/2} = (S_{HOL}\ S_{BEB})^{1/2}$, implying $S_{BEB} \leq S_{CEB} \leq S_{HOL}$. We conclude that for isolated systems of limited self-gravity the Bekenstein bound is the tightest, followed by our CEB and, finally, by Bousso's holographic bound. Similar scaling properties for the HEB were discussed in [19].

For regions of space that contain so much energy that the corresponding gravitational radius R_g exceeds R, Bekenstein's bound is the weakest, while the naive holographic bound is the strongest (but very often wrong). Bousso's proposal uses the apparent horizon R_{AH}, while CEB uses R_{CC}. For homogeneous cosmologies, $R_{CC} < R_{AH}$, since R_{CC}^{-2}, according to (12), is always larger than the average of the two terms appearing on its r.h.s., which is precisely $R_{AH}^{-2} = H^2 + \kappa/a^2$. Since, for a fixed volume, the bounds scale like R_{AH}^{-1} or R_{CC}^{-1}, we immediately find that CEB is generally more generous. An important difference between our proposal and Bousso's covariant holographic bound [6] that scales as S/A is that there the entropy S is a flux through light-like hypersurfaces. A detailed comparison with Bousso's proposal is therefore more subtle because of his use of the apparent horizon area to bound entropy on light sheets. This can be converted into a bound on the entropy of the space-like region only in special cases.

Verlinde [11] argued that the radiation in a closed, radiation-dominated Universe can be modeled by a CFT, and that its entropy can be evaluated using a generalized Cardy formula. After an appropriate modification of Verlinde's bound which evades the criticism about its validity for weakly coupled CFTs, the new bound is exactly equivalent to CEB within the CFT framework.

4 The Generalized Second Law and the Causal Entropy Bound

4.1 The Generalized Second Law in Cosmology

There seems to be a close relationship between entropy bounds and the GSL. We have proposed a concrete classical and quantum mechanical form of the GSL in cosmology [32], which is valid also in situations far from thermal equilibrium. We discuss various entropy sources, such as thermal, "geometric" and "quantum" entropy, apply GSL to study cosmological solutions and show that it is compatible with entropy bounds. GSL allows a more detailed description of how, and if, cosmological singularities are evaded. The proposed GSL is different from GSL for BHs [37], but the idea that in addition to normal entropy other sources of entropy have to be included has some similarities. We will discuss here only 4D models. Obviously, it should be possible to generalize our analysis to higher dimensions in a straightforward manner along the lines of the generalizations of the CEB to higher dimensions.

The starting point of our classical discussion is the definition of the total entropy of a domain containing more than one cosmological horizon [19]. We have already introduced the number of cosmological horizons within a given comoving volume $V = a(t)^3$. It is simply the total volume divided by the volume of a single horizon, $n_H = a(t)^3/|H(t)|^{-3}$. As usual, we will ignore numerical factors of order unity. Here we use units in which $c = 1, G_N = 1/16\pi, \hbar = 1$ and discuss only flat, homogeneous and isotropic cosmologies. If the entropy within a given horizon is S^H, then the total entropy is given by $S = n_H S^H$. Classical GSL requires that the cosmological evolution, even when far from thermal equilibrium, must obey $dS \geq 0$, in addition to Einstein equations. In particular,

$$n_H \partial_t S^H + \partial_t n_H S^H \geq 0. \tag{70}$$

In general, there could be many sources and types of entropy, and the total entropy is the sum of their contributions. If, in some epoch, a single type of entropy makes a dominant contribution to S^H, for example, of the form $S^H = |H|^\alpha$, α being a constant characterizing the type of entropy source, and therefore $S = (a|H|)^3|H|^\alpha$, (70) becomes an explicit inequality,

$$3H + (3+\alpha)\frac{\dot{H}}{H} \geq 0, \tag{71}$$

which can be translated into energy conditions constraining the energy density ρ, and the pressure p of (effective) sources. Using the FRW equations,

$$\begin{aligned} H^2 &= \frac{1}{6}\rho, \\ \dot{H} &= -\frac{1}{4}(\rho + p), \end{aligned} \tag{72}$$

$$\dot{\rho} + 3H(\rho + p) = 0,$$

and assuming $\alpha > -3$ (which we will see later is a reasonable assumption) and of course $\rho > 0$, we obtain

$$\frac{p}{\rho} \leq \frac{2}{3+\alpha} - 1 \qquad \text{for} \qquad H > 0, \tag{73}$$

$$\frac{p}{\rho} \geq \frac{2}{3+\alpha} - 1 \qquad \text{for} \qquad H < 0. \tag{74}$$

Adiabatic evolution occurs when the inequalities in (73) and (74) are saturated.

A few remarks about the allowed range of values of α are in order. First, the usual adiabatic expansion of a radiation-dominated universe with $p/\rho = 1/3$ corresponds to $\alpha = -3/2$. Adiabatic evolution with $p/\rho < -1$ for which the null energy condition is violated would require a source for which $\alpha < -3$. This is problematic since it does not allow a flat space limit of vanishing H with finite entropy. The existence of an entropy source with $\alpha < -2$ does not allow a finite $\partial_t S$ in the flat space limit and is therefore suspected of being unphysical. Finally, the equation of state $p = -\rho$ (deSitter inflation) cannot be described as adiabatic evolution for any finite α.

Let us discuss in more detail three specific examples. First, as already noted, we have verified that thermal entropy during radiation-dominated evolution can be described without difficulties, as expected. In this case, $\alpha = -\frac{3}{2}$ reproduces the well-known adiabatic expansion, but also allows entropy production. The present era of matter domination requires a more complicated description since in this case one source provides the entropy, and another source the energy.

The second case is that of geometric entropy S_g, whose source is the existence of a cosmological horizon [38, 39]. The concept of geometric entropy is closely related to the holographic principle and to entanglement entropy (see below). For a system with a cosmological horizon S_g^H is given by (ignoring numerical factors of order unity)

$$S_g^H = |H|^{-2} G_N^{-1}. \tag{75}$$

The equation of state corresponding to adiabatic evolution with dominant S_g is obtained by substituting $\alpha = -2$ into (73) and (74), leading to $p/\rho = 1$ for positive and negative H. This equation of state is simply that of a free massless scalar field, also recognized as the two vacuum branches of PBB string cosmology [18] in the Einstein frame. In [19] this was found for the (+) branch in the string frame as an "empirical" observation. In general, for the case of dominant geometric entropy, GSL requires, for positive H, $p \leq \rho$; hence, deSitter inflation is definitely allowed. For negative H, GSL requires $\rho \leq p$, and therefore forbids, for example, a time-reversed history of our universe or a contracting deSitter universe with a negative constant H (unless some additional entropy sources appear).

The third case is that of quantum entropy S_q, associated with quantum fluctuations. This form of entropy was discussed in [40, 41]. Specific quantum entropy for a single physical degree of freedom is approximately given by (again, ignoring numerical factors of order unity)

$$s_q = \int d^3k \ln n_k, \tag{76}$$

where $n_k \gg 1$ are occupation numbers of quantum modes. Quantum entropy is large for highly excited quantum states, such as the squeezed states obtained by amplification of quantum fluctuations during inflation. Quantum entropy does not seem to be expressible in general as $S_q^H = |H|^\alpha$, since occupation numbers depend on the whole history of the evolution. We will discuss this form of entropy in more detail later, when the quantum version of GSL is proposed.

Geometric entropy is related to the existence of a horizon or more generally to the existence of a causal boundary. From my current perspective the geometric entropy corresponds to entanglement entropy of fluctuations whose wavelength is shorter than the horizon, while "quantum" entropy is probably related to entanglement entropy of fluctuations whose wavelength is larger than the horizon (see below).

We would like to show that it is possible to formally define a temperature, and that the definition is compatible with the a generalized form of the first law of thermodynamics (see also [43]). Recall that the first law for a closed system states that $TdS = dE + pdV = (\rho+p)dV + Vd\rho$. Let us now consider the case of single entropy source and formally define a temperature T, $T^{-1} = \left(\frac{\partial S}{\partial E}\right)_V = \frac{\partial s}{\partial \rho}$, since $E = \rho V$ and $S = sV$. Using (72), and $s = |H|^{\alpha+3}$, we obtain $\frac{\partial s}{\partial \rho} = \frac{\alpha+3}{12}|H|^{\alpha+1}$, and therefore,

$$T = \frac{12}{\alpha+3}|H|^{-\alpha-1}. \tag{77}$$

To ensure positive temperatures, $\alpha > -3$, a condition which we have already encountered. Additionally, for $\alpha > -1$, T diverges in the flat space limit, and therefore such a source is suspect of being unphysical, leading to the conclusion that the physical range of α is $-2 \leq \alpha \leq -1$. A compatibility check requires $T^{-1} = \frac{\partial s}{\partial t}/\frac{\partial \rho}{\partial t}$, which indeed yields a result in agreement with (77). Yet another thermodynamic relation $p/T = \left(\frac{\partial S}{\partial V}\right)_E$ leads to $p = sT - \rho$ and therefore to $p/\rho = \frac{2}{\alpha+3} - 1$ for adiabatic evolution, in complete agreement with (73) and (74). For $\alpha = -2$, (77) implies $T_g = |H|$, in agreement with [38], and for ordinary thermal entropy $\alpha = -3/2$ reproduces the known result, $T = |H|^{1/2}$.

Is GSL compatible with entropy bounds? Let us start answering this question by considering a universe undergoing decelerated expansion, i.e. $H > 0$, $\dot{H} < 0$. For entropy sources with $\alpha > -2$, going backward in time, H is prevented by the restriction $S^H \leq S_g^H$ from becoming too large. This requires that at a certain moment in time $\dot{H}$ has reversed sign, or at least vanished. GSL allows such a transition. Evolving from the past toward the future, and looking at (71) we see that a transition from an epoch of accelerated expansion $H > 0$, $\dot{H} > 0$, to an epoch of decelerated expansion $H > 0$, $\dot{H} < 0$, can occur without violation of GSL. But later we discuss a new bound appearing in this situation when quantum effects are included.

For a contracting universe with $H < 0$, and if sources with $\alpha > -2$ exist, the situation is more interesting. Let us check whether in an epoch of accelerated contraction $H < 0$, $\dot{H} < 0$, GSL is compatible with entropy bounds. If an epoch of accelerated contraction lasts, it will inevitably run into a future singularity, in conflict with bound $S^H \leq S^H_g$. This conflict could perhaps have been prevented if at some moment in time the evolution had turned into decelerated contraction with $H < 0$, $\dot{H} > 0$. But a brief look at (71), $\dot{H} \leq -\frac{3}{3+\alpha}H^2$, shows that decelerated contraction is not allowed by GSL. The conclusion is that for the case of accelerated contraction GSL and the entropy bound are not compatible.

To resolve the conflict between GSL and the entropy bound, we propose adding a missing quantum entropy term $dS_{Quantum} = -\mu dn_H$, where $\mu(a, H, \dot{H}, ...)$ is a "chemical potential" motivated by the following heuristic argument. Specific quantum entropy is given by (76), and we consider for the moment one type of quantum fluctuations that preserves its identity throughout the evolution. Changes in S_q result from the well-known phenomenon of freezing and defreezing of quantum fluctuations. For example, quantum modes whose wavelength is stretched by an accelerated cosmic expansion to the point that it is larger than the horizon become frozen ("exit the horizon"), and are lost as dynamical modes, and conversely, quantum modes whose wavelength shrinks during a period of decelerated expansion ("re-enter the horizon") thaw and become dynamical again. Taking into account this "quantum leakage" of entropy requires that the first law should be modified as in open systems $TdS = dE + PdV - \mu dN$.

Consider a universe going through a period of decelerated expansion, containing some quantum fluctuations which have re-entered the horizon (for concreteness, it is possible to think about an isotropic background of gravitational waves). In this case, physical momenta simply redshift, but since no new modes have re-entered, and since occupation numbers do not change by simple redshift, then within a fixed comoving volume, entropy does not change. However, if there are some frozen fluctuations outside the horizon "waiting to re-enter," then there will be a change in quantum entropy, because the minimal comoving wave number of dynamical modes k_{min} will decrease due to the expansion, $k_{min}(t + \delta t) < k_{min}(t)$. The resulting change in quantum entropy, for a single physical degree of freedom, is $\Delta s_q = \int_{k_{min}(t+\delta t)}^{k_{min}(t)} k^2 dk \ln n_k$, and since $k_{min}(t) = a(t)H(t)$, $\Delta S_q = \int_{a(t+\delta t)H(t+\delta t)}^{a(t)H(t)} k^2 dk \ln n_k = -\Delta(aH)^3 \ln n_{k=aH}$, providedln n_k is a smooth enough function. Therefore, for $\mathcal{N}$ physical DOF, and since $n_H = (aH)^3$,

$$dS_q = -\mu \mathcal{N} dn_H, \tag{78}$$

where parameter μ is taken to be positive. Obviously, the result depends on the spectrum n_k, but typical spectra are of the form $n_k \sim k^\beta$, and therefore we may take as a reasonable approximation $\ln n_k \sim constant$ for all $\mathcal{N}$ physical DOF.

We adopt proposal (78) in general

$$\begin{aligned} dS &= dS_{Classical} + dS_{Quantum} \\ &= dn_H S^H + n_H dS^H - \mu \mathcal{N} dn_H, \end{aligned} \tag{79}$$

where S^H is the classical entropy within a cosmological horizon. In particular, for the case that S^H is dominated by a single source $S^H = |H|^\alpha$,

$$\left(3H + 3\frac{\dot{H}}{H}\right) n_H (S^H - \mu \mathcal{N}) + \alpha \frac{\dot{H}}{H} n_H S^H \geq 0. \tag{80}$$

Quantum-modified GSL (80) allows a transition from accelerated to decelerated contraction. As a check, look at $H < 0$, $\dot{H} = 0$, in this case modified GSL requires $3H(S^H - \mu \mathcal{N}) \geq 0$, which, if $\mu \mathcal{N} \geq S^H$, is allowed. If the dominant form of entropy is indeed geometric entropy, the transition from accelerated to decelerated contraction is allowed already at $|H| \sim M_P/\sqrt{\mathcal{N}}$. In models where $\mathcal{N}$ is a large number, such as grand unified theories and string theory where it is expected to be of the order of 1000, the transition can occur at a scale much below the Planck scale, at which classical general relativity is conventionally expected to adequately describe background evolution.

If we reconsider the transition from accelerated to decelerated expansion and require that (80) holds, we discover a new bound derived directly from GSL. It is compatible with, but not relying on, the bound $S^H \leq S_g^H$. Consider the case in which $\dot{H}$ and H are positive, or H positive and $\dot{H}$ negative but $|\dot{H}| \ll H^2$, relevant to whether the transition is allowed by GSL. In this case, (80) reduces to $S^H - \mu \mathcal{N} \geq 0$, that is, GSL puts a lower bound on the classical entropy within the horizon. If geometric entropy is the dominant source of entropy as expected, GSL puts a lower bound on geometric entropy $S_g^H \geq \mu \mathcal{N}$, which yields an upper bound on H,

$$H \leq \frac{M_P}{\sqrt{\mathcal{N}}}. \tag{81}$$

The scale that appeared previously in the resolution of the conflict between entropy bounds and GSL for a contracting universe has reappeared in (81), and remarkably, (81) is the same bound obtained in [9] using different arguments. Bound (81) forbids a large class of singular homogeneous, isotropic, spatially flat cosmologies by bounding the scale of curvature for such a universe.

4.2 The Generalized Second Law in Pre-big-bang String Cosmology

String theory is a consistent theory of quantum gravity, with the power to describe high curvature regions of space–time [44], and as such, we could expect it to teach us about the fate of cosmological singularities, with the expectation that singularities are smoothed and turned into brief epochs of high curvature. However, many attempts to seduce an answer out of string theory regarding cosmological singularities have failed so far in producing a conclusive answer (see for example [45]). The reason is probably that most technical advancements in string theory rely heavily on supersymmetry, but generic time-dependent solutions break all supersymmetries and therefore known methods are less powerful when applied to cosmology.

We have focused [46] on the two sources of entropy defined previously. The first source is the geometric entropy S_g, and the second source is quantum entropy S_q. The entropy within a given horizon is S^H and the total entropy is given by $S = n_H S^H$. We will ignore numerical factors, use units in which $c = 1$, $\hbar = 1$, $G_N = e^{\phi}/16\pi$, ϕ being the dilaton, and discuss only flat, homogeneous and isotropic 4D string cosmologies in the so-called string frame, in which the lowest order effective action is

$$\mathcal{S}_{\rm LO} = \int d^4x \sqrt{-g} e^{-\phi} \left[R + (\partial\phi)^2 \right].$$

Obviously, the discussion can be generalized in a straightforward manner to higher D.

In ordinary cosmology, geometric entropy within a Hubble volume is given by its area $S_g^H = H^{-2} G_N^{-1}$, and therefore, specific geometric entropy is given by $s_g = |H| G_N^{-1}$ [32]. A possible expression for specific geometric entropy in string cosmology is obtained by substituting $G_N = e^{\phi}$, leading to

$$s_g = |H| e^{-\phi}. \tag{82}$$

Reassurance that s_g is indeed given by (82) is provided by the following observation. The action $\mathcal{S}_{\rm LO}$ can be expressed in a $(3+1)$ covariant form, using the 3-metric g_{ij}, the extrinsic curvature K_{ij}, considering only vanishing 3-Ricci scalar and homogeneous dilaton,

$$\mathcal{S}_{\rm LO} = \int d^3x dt \sqrt{g_{ij}} e^{-\phi} \left[-3K_{ij}K^{ij} - 2g^{ij}\partial_t K_{ij} + K^2 - (\partial_t\phi)^2 \right].$$

Now, $\mathcal{S}_{\rm LO}$ is invariant under the symmetry transformation $g_{ij} \to e^{2\lambda} g_{ij}$, $\phi \to \phi + 3\lambda$, for an arbitrary time-dependent λ. From the variation of the action $\delta\mathcal{S} = \int d^3x dt \sqrt{g_{ij}} e^{-\phi} 4K\dot{\lambda}$, we may read off the current and conserved charge $Q = 4a^3 e^{-\phi} K$. The symmetry is exact in the flat homogeneous case, and it seems plausible that it is a good symmetry even when α' corrections are present [42]. With definition (82), the total geometric entropy $S_g = a^3|H|e^{-\phi}$

is proportional to the corresponding conserved charge. Adiabatic evolution, determined by $\partial_t S_g = 0$, leads to a familiar equation, $\frac{\dot{H}}{H} - \dot{\phi} + 3H = 0$, satisfied by the $(\pm)$ vacuum branches of PBB string cosmology.

Quantum entropy for a single field in string cosmology is, as in [40, 41, 32], given by

$$s_q = \int_{k_{min}}^{k_{max}} d^3k f(k), \tag{83}$$

where for large occupation numbers $f(k) \simeq \ln n_k$. The ultraviolet cutoff k_{max} is assumed to remain constant at the string scale. The infrared cutoff k_{min} is determined by the perturbation equation $\psi''_{k_c} + \left(k_c^2 - \frac{\sqrt{s(\eta)}''}{\sqrt{s(\eta)}}\right)\psi_{k_c} = 0$, where η is conformal time $' = \partial_\eta$, and k_c is the comoving momentum related to physical momentum $k(\eta)$ as $k_c = a(\eta)k(\eta)$. Modes for which $k_c^2 \leq \frac{\sqrt{s}''}{\sqrt{s}}$ are "frozen," and are lost as dynamical modes. The "pump field" $s(\eta) = a^{2m}e^{\ell\phi}$ depends on the background evolution and on the spin and dilaton coupling of various fields. We are interested in solutions for which $a'/a \sim \phi' \sim 1/\eta$, and therefore, for all particles $\frac{\sqrt{s}''}{\sqrt{s}} \sim 1/\eta^2$. It follows that $k_{min} \sim H$. In other phases of cosmological evolution our assumption does not necessarily hold, but in standard radiation domination (RD) with frozen dilaton all modes re-enter the horizon. Using the reasonable approximation $f(k) \sim$ constant, we obtain, as in [32],

$$\Delta S_q \simeq -\mu \Delta n_H. \tag{84}$$

Parameter μ is positive, and in many cases proportional to the number of species of particles, taking into account all DOF of the system, perturbative and non-perturbative. The main contribution to μ comes from light DOF, and therefore if some non-perturbative objects such as D branes become light, they will make a substantial contribution to μ.

We now turn to the generalized second law of thermodynamics, taking into account geometric and quantum entropy. Enforcing $dS \geq 0$, and in particular, $\partial_t S = \partial_t S_g + \partial_t S_q \geq 0$, leads to an important inequality,

$$\left(H^{-2}e^{-\phi} - \mu\right)\partial_t n_H + n_H \partial_t \left(H^{-2}e^{-\phi}\right) \geq 0. \tag{85}$$

When quantum entropy is negligible compared to geometric entropy, GSL (85) leads to

$$\dot{\phi} \leq \frac{\dot{H}}{H} + 3H, \tag{86}$$

yielding a bound on $\dot{\phi}$, and therefore on dilaton kinetic energy, for a given H, $\dot{H}$. Bound (86) was first obtained in [19], and interpreted as following from a saturated HEB.

When quantum entropy becomes relevant, we obtain another bound. We are interested in a situation in which the universe expands, $H > 0$, and ϕ and H are non-decreasing, and therefore $\partial_t \left(H^{-2}e^{-\phi}\right) \leq 0$ and $\partial_t n_H > 0$. A necessary condition for GSL to hold is that

$$H^2 \leq \frac{e^{-\phi}}{\mu}, \tag{87}$$

bounding total geometric entropy $He^{-\phi} \leq \frac{e^{-\frac{3}{2}\phi}}{\sqrt{\mu}}$. A bound similar to (87) was obtained in [19] by considering entropy of re-entering quantum fluctuations. We stress that to be useful in analysis of cosmological singularities (87) has to be considered for perturbations that exit the horizon. If the condition (87) is satisfied, then the cosmological evolution always allows a self-consistent description using the low-energy effective action approach.

It is not a priori clear that the form of GSL and entropy sources remains unchanged when curvature becomes large; in fact, we may expect higher-order corrections to appear. For example, the conserved charge of the scaling symmetry of the action will depend in general on higher-order curvature corrections. Nevertheless, in the following we will assume that specific geometric entropy is given by (82), without higher-order corrections, and try to verify that, for some reason yet to be understood, there are no higher-order corrections to (82). Our results are consistent with this assumption.

We now turn to apply our general analysis to the PBB string cosmology scenario, in which the universe starts from a state of very small curvature and string coupling and then undergoes a long phase of dilaton-driven inflation, joining smoothly at later times standard RD cosmology, giving rise to a singularity-free inflationary cosmology. The high-curvature phase joining DDI and RD phases is identified with the "big bang" of standard cosmology. A key issue confronting this scenario is whether and under what conditions can the graceful exit transition from DDI to RD be completed [47]. In particular, it was argued that curvature is bounded by an algebraic fixed point behavior when both H and $\dot{\phi}$ are constants and the universe is in a linear-dilaton deSitter space [42], and coupling is bounded by quantum corrections [48, 49]. But it became clear that another general theoretical ingredient is missing, and we propose that GSL is that missing ingredient.

We have studied numerically examples of PBB string cosmologies to verify that the overall picture we suggest is valid in cases that can be analyzed explicitly. We first consider, as in [42, 50], α' corrections to the lowest order string effective action,

$$\mathcal{S} = \frac{1}{16\pi\alpha'} \int d^4x\sqrt{-g}e^{-\phi}\left[R + (\partial\phi)^2 + \frac{1}{2}\mathcal{L}_{\alpha'}\right], \tag{88}$$

where

$$\mathcal{L}_{\alpha'} = k\alpha'\left[\frac{1}{2}R^2_{GB} + A\,(\partial\phi)^4 + D\partial^2\phi\,(\partial\phi)^2 + C\left(R^{\mu\nu} - \frac{1}{2}g^{\mu\nu}R\right)\partial_\mu\phi\partial_\nu\phi\right], \tag{89}$$

with $C = -(2A + 2D + 1)$, is the most general form of four derivative corrections that lead to equations of motion with at most second (time) derivatives. The rationale for this choice was explained in [50]. k is a numerical factor depending on the type of string theory. Action (88) leads to equations of motion, $-3H^2 + \dot{\bar{\phi}}^2 - \bar{\rho} = 0$, $\bar{\sigma} - 2\dot{H} + 2H\dot{\bar{\phi}} = 0$, $\bar{\lambda} - 3H^2 - \dot{\bar{\phi}}^2 + 2\ddot{\bar{\phi}} = 0$, where $\bar{\rho}$, $\bar{\lambda}$ and $\bar{\sigma}$ are effective sources parameterizing the contribution of α' corrections [50]. Parameters A and D should have been determined by string theory; however, at the moment, it is not possible to calculate them in general. If A and D were determined, we could just use the results and check whether their generic cosmological solutions are non-singular, but since A and D are unavailable at the moment, we turn to GSL to restrict them.

First, we look at the initial stages of the evolution when the string coupling and H are very small. We find that not all the values of the parameters A and D are allowed by GSL. The condition $\bar{\sigma} \geq 0$, which is equivalent to GSL on generic solutions at the very early stage of the evolution, if the only relevant form of entropy is geometric entropy, leads to the following condition on A and D (first obtained by Madden [51]), $40.05A + 28.86D \leq 7.253$. The values of A and D which satisfy this inequality are labeled "allowed," and the rest are "forbidden." In [50] a condition that α' corrections are such that solutions start to turn toward a fixed point at the very early stages of their evolution was found $61.1768A + 40.8475D \leq 16.083$, and such solutions were labeled "turning the right way." Both conditions are displayed in Fig. 2. They select almost the same region of (A, D) space, a gratifying result, GSL "forbids" actions whose generic solutions are singular and do not reach a fixed point.

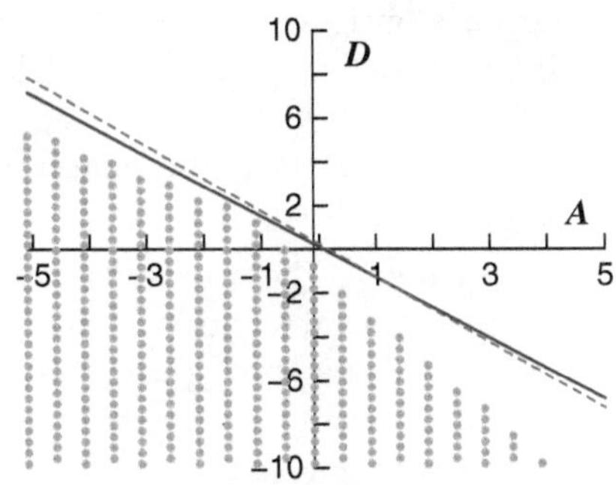

Fig. 2. Two lines, separating actions whose generic solutions "turn the right way" at the early stages of evolution (*red-dashed*), and actions whose generic solutions satisfy classical GSL while close to the (+) branch vacuum (*blue-solid*). The dots represent (A, D) values whose generic solutions reach a fixed point, and are all in the "allowed" region

We further observe that generic solutions which "turn the wrong way" at the early stages of their evolution continue their course in a way similar to the

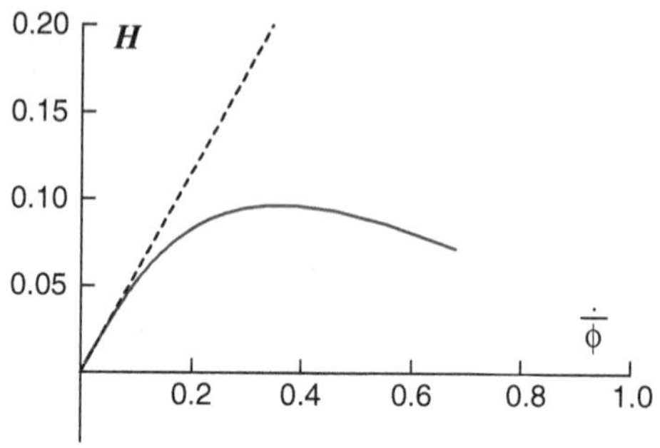

Fig. 3. Typical solution that "turns the wrong way." The dashed line is the (+) branch vacuum

solution presented in Fig. 3. We find numerically that at a certain moment in time H starts to decrease, at that point $\dot{H} = 0$ and particle production effects are still extremely weak, and therefore (86) is the relevant bound, but (86) is certainly violated.

We have scanned the (A, D) plane to check whether a generic solution that reaches a fixed point respects GSL throughout the whole evolution, and conversely, whether a generic solution obeying GSL evolves toward a fixed point. The results are shown in Fig. 2; clearly, the "forbidden" region does not contain actions whose generic solutions go to fixed points. Nevertheless, there are some (A, D) values located in the small wedges near the bounding lines, for which the corresponding solutions always satisfy (86), but do not reach a fixed point, and are singular. This happens because they meet a cusp singularity. Consistency requires adding higher-order α' corrections when cusp singularities are approached, which we will not attempt here.

If particle production effects are strong, the quantum part of GSL adds bound (87), which adds another "forbidden" region in the $(H, \dot{\bar{\phi}})$ plane, the region above a straight line parallel to the $\dot{\bar{\phi}}$ axis. The quantum part of GSL has therefore a significant impact on corrections to the effective action. On a fixed point ϕ is still increasing, and therefore, the bounding line described by (87) is moving downward, and when the critical line moves below the fixed point, GSL is violated. This means that when a certain critical value of the coupling e^{ϕ} is reached, the solution can no longer stay on the fixed point, and it must move away toward an exit. One way this can happen is if quantum corrections, perhaps of the type discussed in [48, 49], exist.

The full GSL therefore forces actions to have generic solutions that are non-singular, classical GSL bounds dilaton kinetic energy and quantum GSL bounds H and therefore at a certain moment of the evolution $\dot{H}$ must vanish (at least asymptotically), and then curvature is bounded. If cusp singularities are removed by adding higher-order corrections, as might be expected, we can apply GSL with similar conclusions also in this case. A schematic graceful exit

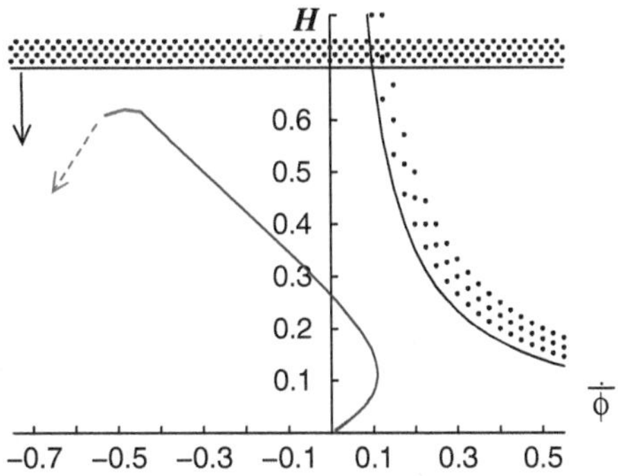

Fig. 4. Graceful exit enforced by GSL on generic solutions. The horizontal line is bound (87) and the curve on the right is bound (86), shaded regions indicate GSL violation

enforced by GSL is shown in Fig. 4. Our result indicates that if we impose GSL in addition to equations of motion, then non-singular PBB string cosmology is quite generic.

5 Area Entropy, Entanglement Entropy and Entropy Bounds

Classical general relativity predicts space–times with event horizons and other causal boundaries, such as apparent horizons, cosmological horizons and acceleration horizons. Observers in space–times with causal boundaries can see very different physics, as demonstrated by comparing the static observer at infinity and a freely falling observer in the Schwarzchild geometry. For the first, the horizon is a very special place: Energies of particles diverge and space–time seems to end there, while for a freely falling observer the horizon and its vicinity do not look special at all. In cosmological space–times with causal boundaries the situation is similar. The existence of causal boundaries is determined by the large-scale properties of space–time, and hence is intrinsically a non-local concept. In cosmology, for example, it is hard for a local observer to determine whether the space–time is de Sitter space that has a cosmological event horizon, or a Robertson–Walker space which looks approximately de Sitter.

The interpretation of the thermodynamic properties of BHs and whether they originate from some underlying, more fundamental, statistical mechanics remains unclear, in spite of the intense efforts and the progress that has been achieved over the last 30 years since the discovery by Bekenstein [37]. Quantum field theory (QFT) in the fixed background of space–times with horizons is a key element in the quantitative understanding of the statistical mechanics of BHs. QFT in such background has several interesting and well-known features. The quantum vacuum states associated with different observers can be very different from each other, leading to strong particle production effects: the Hawking effect and the Unruh effect. In addition, the appearance of large blue shifts of quantum modes near the horizon lead to the trans-Planckian problem [52]. The proposed resolutions include the brick-wall model [53, 54] and the stretched-horizon [55] idea. The entropy and thermodynamics are also observer dependent, as demonstrated by the classic comparison between the Rindler and Minkowski space observers in the Minkowski vacuum. The accelerated observer sees a truly thermal state, while for the Minkowski observer the temperature vanishes. The tension between the possibility of evaluating the entropy and other thermodynamic quantities in the semi-classical approximation and their observer dependence and hence their sensitivity to physics at the highest energy scales is intriguing and is not yet resolved.

My current point of view about the physics of space–times with causal boundaries is the entanglement point of view. I believe that the statistical properties of such space–times arise because classical observers in them have access only to a part of the whole quantum state. When a system is in a pure state, but one cannot access the complete quantum system, and a measurement is performed, one is instructed by the rules of quantum mechanics to trace over the classically inaccessible DOF. This leads to a natural framework for interpreting the physics of spaces with causal boundaries: that it is described by the density matrix which results from tracing over the inaccessible DOF. In the context of BHs the idea was first proposed by 't Hooft [54], and by Sorkin and collaborators [56], and then extended and elaborated by Srednicki [39] and others.

The entanglement approach considers the fundamental physical objects describing the physics of space–times with causal boundaries to be their global quantum state and the unitary evolution operator. The entanglement approach has several obvious advantages: It naturally leads to area-law entropy, it can incorporate the observer dependence of BH thermodynamics and of the thermodynamics of cosmological space–times with causal boundaries. It can naturally accommodate the geometric and quantum entropies—the first resulting from the entanglement entropy of short-wavelength fluctuations and the second resulting from the entanglement entropy of fluctuations whose wavelength is larger than the causal connection scale. This interpretation is also automatically compatible with entropy bounds and the GSL as long as the evolution equations are "physical" because from a global point of view it is clear that nothing special occurs when a horizon develops. Obviously, there are also some unresolved issues that need to be better understood in this context.

The space–times that are traditionally used to explore the entanglement point of view are spaces with bifurcating Killing horizons such as the eternal Schwarzschild BH or Rindler space. Israel [57] has shown that the quantum Hilbert space of fields in space–times with bifurcating Killing horizons has a product structure that is isomorphic to the product structure that arises in thermofield dynamics [58]. In thermofield dynamics one formally doubles the Hilbert space and evaluates quantum expectation values in the thermofield double pure state in order to evaluate expectation values in a thermal state of the original system. In this context the entropy is the entanglement entropy that is obtained from tracing over one of the two spaces.

One of the main unresolved issues confronting the entanglement interpretation is the ultraviolet (UV) divergence of entanglement entropy and other entanglement correlation functions near the horizon, and its dependence on the number of fields [59, 60, 61]. Another issue concerns space–times that do not have non-degenerate bifurcating Killing horizons. For such spaces, it is unclear what is entangled with what, since some of the regions of the extended space–time are missing.

The entanglement point of view has been discussed in the AdS-CFT context by Maldacena [62] who studied eternal BHs in AdS. In 4D, the space has two boundaries that are topologically $S^2 \times S^1$, the dual FT consists of two CFTs "living on the boundary." The product theory in the TFD state defines the string theory in the bulk, whose low energy limit is the AdS-BH. The FT side is completely well defined, and its thermodynamics can obviously be interpreted as entanglement thermodynamics. The low energy state in the bulk is the Hartle–Hawking vacuum. The entanglement point of view suggests the following perspective. Suppose that the universe is in a pure state and that it evolves unitarily. Then the entropy of any sub-system of it is entirely in the eyes of the beholder: a particular classical observer.

We have shown [63] that the entropy resulting from the counting of microstates of non-extremal BHs using field theory duals of string theories can be interpreted as arising from entanglement. The conditions for making such an interpretation consistent were determined. First, we have interpreted the entropy and thermodynamics of space–times with non-degenerate, bifurcating Killing horizons as arising from entanglement. We have used a path integral method to define the Hartle–Hawking vacuum state in such space–times, and reveal explicitly its entangled nature and its relation to the geometry. If string theory on such spacetimes has a field theory dual, then, in the low-energy, weak coupling limit, the field theory state that is dual to the Hartle–Hawking state is a thermofield double state. This allowed us to compare the entanglement entropy to the entropy of the field theory dual, and thus to the Bekenstein–Hawking entropy of the BH.

To further understand the nature of the time evolution of sub-systems in this context, we have considered [64] a collapsing relativistic spherical shell in a free quantum field theory. Once the center of the wavefunction of the shell passes a certain radius r_s, the degrees of freedom inside r_s are traced over. We have found that an observer outside this region will determine that the evolution of the system is non-unitary. The non-unitary evolution occurs only when the wavefunction is in the process of crossing the boundary and the amount of non-unitarity is proportional to the area of the boundary.

Acknowledgments

I would like to thank all the collaborators who participated in the research that is summarized and reviewed in this article. First, I would like to thank Gabriele Veneziano for interesting me in this subject and for collaboration in several related projects. I would like to thank David Eichler, Marty Einhorn, Stefano Foffa, Dick Madden, David Oaknin, Avi Mayo, Riccardo Sturani and Amos Yarom for fruitful collaborations whose results are presented in this article.

References

1. J. D. Bekenstein: Phys. Rev. D **23**, 287 (1981); J. D. Bekenstein: Phys. Rev. D **49**, 1912 (1994)
2. W. G. Unruh, R. M. Wald: Phys. Rev. D **25**, 942 (1982)
3. M. A. Pelath, R. M. Wald: Phys. Rev. D **60**, 104009 (1999); D. Marolf, R. D. Sorkin: Phys. Rev. D **69**, 024014 (2004)
4. J. D. Bekenstein: Phys. Rev. D **70**, 121502 (2004); J. D. Bekenstein: Found. Phys. **35**, 1805 (2005)
5. G. 't Hooft: "Dimensional reduction in quantum gravity", arXiv:gr-qc/9310026; L. Susskind: J. Math. Phys. **36**, 6377 (1995)
6. R. Bousso: Rev. Mod. Phys. **74**, 825 (2002)
7. J. Maldacena: Adv. Theor. Math. Phys. **2**, 231 (1998)
8. O. Aharony, S. S. Gubser, J. M. Maldacena, H. Ooguri, Y. Oz: Phys. Rep. **323**, 183 (2000)
9. J. D. Bekenstein: Int. J. Theor. Phys. **28**, 967 (1989)
10. W. Fischler, L. Susskind: "Holography and cosmology", arXiv:hep-th/9806039
11. E. P. Verlinde: "On the holographic principle in a radiation dominated universe", arXiv:hep-th/0008140; I. Savonije, E. P. Verlinde: Phys. Lett. B **507**, 305 (2001)
12. R. Brustein, D. Eichler, S. Foffa: Phys. Rev. D **71**, 124015 (2005)
13. R. Brustein, D. Eichler, S. Foffa, D. H. Oaknin: Phys. Rev. D **65**, 105013 (2002)
14. E. Witten: Adv. Theor. Math. Phys. **2** (1998) 505
15. L. Randall, R. Sundrum: Phys. Rev. Lett. **83**, 4690 (1999)
16. P. Kraus: JHEP **9912**, 011 (1999)
17. A. Kehagias, E. Kiritsis: JHEP **9911**, 022 (1999)
18. M. Gasperini, G. Veneziano: Phys. Rep. **373**, 1 (2003)
19. G. Veneziano: Phys. Lett. **B454**, 22 (1999)
20. B. J. Carr, S. W. Hawking: Mon. Not. Roy. Astron. Soc. **168**, 399 (1974); B. J. Carr: Astrophys. J. **201**, 1 (1975); I. D. Novikov, A. G. Polnarev, Astron. Zh. **57** (1980) 250 [Sov. Astron. **24** (1980) 147]
21. R. Brustein, G. Veneziano: Phys. Rev. Lett. **84** (2000) 5965
22. E. E. Flanagan, D. Marolf, R. M. Wald: Phys. Rev. D **62**, 084035 (2000)
23. R. Brustein, M. Gasperini, G. Veneziano: Phys. Lett. B **431**, 277 (1998)
24. J. Garriga, X. Montes, M. Sasaki, T. Tanaka: Nucl. Phys. B **513**, 343 (1998)
25. A. Ghosh, G. Pollifrone, G. Veneziano: Phys. Lett. B **440**, 20 (1998)

26. C. W. Misner, K. S. Thorne, J. A. Wheeler: *Gravitation* (Freeman, San Francisco, 1970)
27. R. Brustein, S. Foffa, G. Veneziano: Phys. Lett. B **507** (2001) 270
28. M. Gasperini, M. Giovannini: Phys. Rev. D **47** (1993) 1519
29. C. W. Misner, K. S. Thorne, J. A. Wheeler: Gravitation (Freeman, San Francisco, 1970), pp. 851–859
30. R. Brustein, S. Foffa, A. E. Mayo: Phys. Rev. D **65**, 024004 (2002)
31. D. Kutasov, F. Larsen: JHEP **0101**, 001 (2001)
32. R. Brustein: Phys. Rev. Lett. **84**, 2072 (2000)
33. J. D. Bekenstein: Phys. Rev. D **11**, 2072 (1975)
34. Avraham E. Mayo: "Remarks on Bousso's covariant entropy bound", unpublished
35. G. Veneziano: Phys. Lett. B **406**, 297 (1997); A. Buonanno, K. A. Meissner, C. Ungarelli, G. Veneziano: Phys. Rev. D **57**, 2543 (1998)
36. A. Buonanno, T. Damour, G. Veneziano: Nucl. Phys. B **543**, 275 (1999)
37. J. D. Bekenstein: Phys. Rev. D **7**, 2333 (1973)
38. G. W. Gibbons, S. W. Hawking: Phys. Rev. D **15**, 2738 (1977)
39. M. Srednicki: Phys. Rev. Lett. **71**, 666 (1993)
40. R. H. Brandenberger, V. F. Mukhanov, T. Prokopec: Phys. Rev. Lett. **69**, 3606 (1992); R. H. Brandenberger, T. Prokopec, V. F. Mukhanov: Phys. Rev. D **48**, 2443 (1993)
41. M. Gasperini, M. Giovannini: Phys. Lett. B **301**, 334 (1993)
42. M. Gasperini, M. Maggiore, G. Veneziano: Nucl. Phys. B **494**, 315 (1997)
43. T. Jacobson: Phys. Rev. Lett. **75**, 1260 (1995)
44. J. Polchinski: *String Theory* (Cambridge University Press, Cambridge, 1998)
45. D. Kutasov: Phys. Scripta **T117**, 99 (2005)
46. R. Brustein, S. Foffa, R. Sturani: Phys. Lett. B **471** (2000) 352
47. R. Brustein, G. Veneziano: Phys. Lett. B **329**, 429 (1994); N. Kaloper, R. Madden, K. A. Olive: Nucl. Phys. B **452**, 677 (1995)
48. R. Brustein, R. Madden: Phys. Lett. B **410**, 110 (1997); R. Brustein, R. Madden: Phys. Rev. D **57**, 712 (1998)
49. S. Foffa, M. Maggiore, R. Sturani: Nucl. Phys. B **552**, 395 (1999)
50. R. Brustein, R. Madden: JHEP **9907**, 006 (1999)
51. R. Madden, private communication
52. T. Jacobson: "Introduction to quantum fields in curved spacetime and the Hawking effect", arXiv:gr-qc/0308048
53. G. 't Hooft: Nucl. Phys. B **256**, 727 (1985)
54. G. 't Hooft: Int. J. Mod. Phys. A **11**, 4623 (1996)
55. L. Susskind, L. Thorlacius, J. Uglum: Phys. Rev. D **48**, 3743 (1993)
56. L. Bombelli, R. K. Koul, J. H. Lee, R. D. Sorkin: Phys. Rev. D **34**, 373 (1986)
57. W. Israel: Phys. Lett. A **57**, 107 (1976)
58. Y. Takahasi, H. Umezawa: Collect. Phenom. **2**, 55 (1975)
59. R. M. Wald: Living Rev. Rel. **4**, 6 (2001) [arXiv:gr-qc/9912119]
60. D. Marolf: "On the quantum width of a black hole horizon", arXiv:hep-th/0312059
61. D. Marolf: "A few words on entropy, thermodynamics, and horizons", arXiv:hep-th/0410168
62. J. M. Maldacena: JHEP **0304**, 021 (2003)
63. R. Brustein, M. B. Einhorn, A. Yarom: JHEP **0601**, 098 (2006)
64. R. Brustein, M. B. Einhorn, A. Yarom: "Entanglement and nonunitary evolution", arXiv:hep-th/0609075

Extremal Black Holes in Supergravity*

L. Andrianopoli[1], R. D'Auria[2], S. Ferrara[3] and M. Trigiante[4]

[1] "Centro Enrico Fermi", Compendio Viminale, Via Panisperna 89/A, I-00184 Rome, Italy,
Dipartimento di Fisica, Politecnico di Torino, Corso Duca degli Abruzzi 24, I-10129 Turin, Italy,
Istituto Nazionale di Fisica Nucleare (INFN) Sezione di Torino, Turin, Italy, and CERN PH-TH Division, CH 1211 Geneva 23, Switzerland
laura.andrianopoli@cern.ch

[2] Dipartimento di Fisica, Politecnico di Torino, Corso Duca degli Abruzzi 24, I-10129 Turin, Italy, and Istituto Nazionale di Fisica Nucleare Sezione di Torino, Turin, Italy
riccardo.dauria@polito.it

[3] CERN PH-TH Division, CH 1211 Geneva 23, Switzerland, and Istituto Nazionale di Fisica Nucleare, Laboratori Nazionali di Frascati, Frascati, Italy
sergio.ferrara@cern.ch

[4] Dipartimento di Fisica, Politecnico di Torino, Corso Duca degli Abruzzi 24, I-10129 Turin, Italy, and Istituto Nazionale di Fisica Nucleare Sezione di Torino, Turin, Italy
mario.trigiante@polito.it

Abstract. We present the main features of the physics of extremal black holes embedded in supersymmetric theories of gravitation, with a detailed analysis of the attractor mechanism for BPS and non-BPS black-hole solutions in four dimensions.

1 Introduction: Extremal Black Holes from Classical General Relativity to String Theory

The physics of black holes [1], with its theoretical and phenomenological implications, has a fertile impact on many branches of natural science, such as astrophysics, cosmology, particle physics and, more recently, mathematical physics [2] and quantum information theory [3]. This is not so astonishing in view of the fact that, owing to the singularity theorems of Penrose and Hawking [4], the existence of black holes seems to be an unavoidable consequence of

* One of the authors (Sergio Ferrara) has explored with Gabriele Veneziano the role of "duality" in superstring inspired effective Lagrangians. The same duality plays a central role in the physics of black holes presented in this article.

L. Andrianopoli et al.: *Extremal Black Holes in Supergravity*, Lect. Notes Phys. **737**, 661–727 (2008)
DOI 10.1007/978-3-540-74233-6_22

Einstein's theory of general relativity and of its modern generalizations such as supergravity [5], superstrings, and M-theory [6].

A fascinating aspect of black-hole physics is in their thermodynamic properties that seem to encode fundamental insights of a so far not established final theory of quantum gravity. In this context a central role is played by the Bekenstein–Hawking (in the following, B–H) entropy formula [7]:

$$S_{\text{B–H}} = \frac{k_B}{\ell_P^2}\frac{1}{4}\,\text{Area}_H\,, \tag{1}$$

where k_B is the Boltzman constant, $\ell_P^2 = G\hbar/c^3$ is the squared Planck length while Area_H denotes the area of the horizon surface (from now on we shall use the natural units $\hbar = c = G = k_B = 1$).

This relation between a thermodynamic quantity ($S_{\text{B–H}}$) and a geometric quantity (Area_H) is a puzzling aspect that motivated much theoretical work in the last decades. In fact, a microscopic statistical explanation of the area/entropy formula, related to microstate counting, has been regarded as possible only within a consistent and satisfactory formulation of quantum gravity. Superstring theory is the most serious candidate for a theory of quantum gravity and, as such, should eventually provide such a microscopic explanation of the area law [8]. Since black holes are a typical non-perturbative phenomenon, perturbative string theory could say very little about their entropy: only non-perturbative string theory could have a handle on it. Progress in this direction came after 1995 [9], through the recognition of the role of string dualities. These dualities allow one to relate the strong coupling regime of one superstring model to the weak coupling regime of another. Interestingly enough, there is evidence that the (perturbative and non-perturbative) string dualities are all encoded in the global symmetry group (the U-duality group) of the low-energy supergravity effective action [10].

Let us introduce a particular class of black-hole solutions, which will be particularly relevant to our discussion: the *extremal black holes.* The simplest instance of these solutions may be found within the class of the so-called Reissner–Nordström (R–N) space-time [11], whose metric describes a static, isotropic black hole of mass M and electric (or magnetic) charge Q:

$$ds^2 = dt^2\left(1 - \frac{2M}{\rho} + \frac{Q^2}{\rho^2}\right) - d\rho^2\left(1 - \frac{2M}{\rho} + \frac{Q^2}{\rho^2}\right)^{-1} - \rho^2\, d\Omega^2\,, \tag{2}$$

where $d\Omega^2 = (d\theta^2 + \sin^2\theta\, d\phi^2)$ is the metric on a 2-sphere. The metric (2) admits two Killing horizons, where the norm of the Killing vector $\frac{\partial}{\partial t}$ changes sign. The horizons are located at the two roots of the quadratic polynomial $\Delta \equiv -2M\rho + Q^2 + \rho^2$:

$$\rho_\pm = M \pm \sqrt{M^2 - Q^2}\,. \tag{3}$$

If $M < |Q|$ the two horizons disappear and we have a naked singularity. In classical general relativity people have postulated the so-called *cosmic cen-*

sorship conjecture [5, 12]: space–time singularities should always be hidden inside a horizon. This conjecture implies, in the R–N case, the bound:

$$M \geq |Q| \, . \tag{4}$$

Of particular interest are the states that saturate the bound (4). If

$$M = |Q| \, , \tag{5}$$

the two horizons coincide and, setting: $\rho = r + M$ (where $r^2 = \boldsymbol{x} \cdot \boldsymbol{x}$), the metric (2) can be rewritten as

$$\begin{aligned} ds^2 &= dt^2 \left(1 + \frac{Q}{r}\right)^{-2} - \left(1 + \frac{Q}{r}\right)^2 \left(dr^2 + r^2 \, d\Omega^2\right) \\ &= H^{-2}(r) \, dt^2 - H^2(r) \, d\boldsymbol{x} \cdot d\boldsymbol{x} \end{aligned} \tag{6}$$

in terms of the harmonic function

$$H(r) = \left(1 + \frac{Q}{r}\right) . \tag{7}$$

As (6) shows, the extremal R–N configuration may be regarded as a soliton of classical general relativity, interpolating between two vacua of the theory: the flat Minkowski space–time, asymptotically reached at spatial infinity $r \to \infty$, and the Bertotti–Robinson (B–R) metric [13], describing the conformally flat geometry $AdS_2 \times S^2$ near the horizon $r \to 0$ [5]:

$$ds^2_{\text{B–R}} = \frac{r^2}{M^2_{\text{B–R}}} dt^2 - \frac{M^2_{\text{B–R}}}{r^2} \left(dr^2 + r^2 d\Omega\right) . \tag{8}$$

Last, let us note that the condition $M = |Q|$ can be regarded as a no-force condition between the gravitational attraction $F_g = \frac{M}{r^2}$ and the electric repulsion $F_q = -\frac{Q}{r^2}$ on a unit mass carrying a unit charge.

Until now we have reviewed the concept of extremal black holes as it arises in classical general relativity. However, extremal black-hole configurations are embedded in a natural way in supergravity theories. Indeed supergravity, being invariant under local super-Poincaré transformations, includes general relativity, i.e. it describes gravitation coupled to other fields in a supersymmetric framework. Therefore, it admits black holes among its classical solutions. Moreover, as black holes describe a physical regime where the gravitational field is very strong, a complete understanding of their physics seems to require a theory of quantum gravity, like superstring theory is. In this respect, as anticipated above, extremal black holes have become objects of the utmost relevance in the context of superstrings after 1995 [8, 6, 5, 14]. This interest, which is just part of a more general interest in the p-brane classical solutions of supergravity theories in all dimensions $4 \leq D \leq 11$ [15, 16], stems from the interpretation of the classical solutions of supergravity that preserve a fraction of the original supersymmetries as non-perturbative states, necessary to complete the perturbative string spectrum and make it invariant under the many conjectured duality symmetries [10, 17, 18, 19, 20]. Extremal black holes and their parent p-branes in higher dimensions are then viewed as additional *particle-like* states that compose the spectrum of a fundamental quantum theory. As the monopoles in gauge theories, these non-perturbative quantum states originate from regular solutions of the classical field equations, the same Einstein equations one deals with in classical general relativity and astrophysics. The essential new ingredient, in this respect, is supersymmetry, which requires the presence of *vector fields* and *scalar fields* in appropriate proportions. Hence the black holes we are going to discuss are solutions of generalized Einstein–Maxwell–dilaton equations.

Within the superstring framework, supergravity provides an effective description that holds at lowest order in the string loop expansion and in the limit in which the space–time curvature is much smaller than the typical string scale (string tension). The supergravity description of extremal black holes is therefore reliable when the radius of the horizon is much larger than the string scale, and this corresponds to the limit of large charges. Superstring corrections induce higher derivative terms in the low-energy action and therefore the B–H entropy formula is expected to be corrected as well by terms which are subleading in the small curvature limit. In this paper we will not consider these higher derivative effects.

Thinking of a black-hole configuration as a particular bosonic background of an N-extended locally supersymmetric theory gives a simple and natural understanding at the cosmic censorship conjecture. Indeed, in theories with extended supersymmetry ($N \geq 2$) the bound (4) is just a consequence of the supersymmetry algebra, and this ensures that in these theories the cosmic censorship conjecture is always verified, that is there are no naked singularities. When the black hole is embedded in extended supergravity, the model depends in general also on scalar fields. In this case, as we will see, the electric charge Q has to be replaced by the maximum eigenvalue of the central charge appearing in the supersymmetry algebra (depending on the expectation value of scalar fields and on the electric and magnetic charges). The R–N metric takes in general a more complicated form.

However, extremal black holes have a peculiar feature: even when the dynamics depends on scalar fields, the event horizon loses all information about the scalars; this is true independently of the fact that the solution preserves any supersymmetries or not. Then, as will be discussed extensively in Sect. 4, also if the extremal black hole is coupled to scalar fields, the near-horizon geometry is still described by a conformally flat, B–R-type geometry, with a mass parameter $M_{\mathrm{B-R}}$ depending on the given configuration of electric and magnetic charges, but not on the scalars. The horizon is in fact an attractor point [21, 22, 23]: scalar fields, independently of their boundary conditions at spatial infinity, when approaching the horizon flow to a fixed point given by a certain ratio of electric and magnetic charges. This may be understood in the context of Hawking theory. Indeed quantum black holes are not stable: they radiate a thermic radiation as a black body, and correspondingly lose their energy (mass). The only stable black-hole configurations are the extremal ones, because they have the minimal possible energy compatible with relation (4) and so they cannot radiate. Indeed, physically they represent the limit case in which the black-hole temperature, measured by the surface gravity at the horizon, is sent to zero.

Remembering now that the black-hole entropy is given by the area/entropy B–H relation (1), we see that the entropy of extremal black holes is a topological quantity, in the sense that it is fixed in terms of the quantized electric and magnetic charges, while it does not depend on continuous parameters such as scalars. The horizon mass parameter $M_{\text{B–R}}$ turns out to be given in this case (extremal configurations) by the maximum eigenvalue Z_{max} of the central charge appearing in the supersymmetry algebra, evaluated at the fixed point:

$$M_{\text{B–R}} = M_{\text{B–R}}(p,q) = |Z_{\text{max}}(\phi_{\text{fix}},p,q)| \tag{9}$$

this gives, for the B–H entropy:

$$S_{\text{B-H}} = \frac{A_{\text{B–R}}(p,q)}{4} = \pi |Z_{\text{max}}(\phi_{\text{fix}},p,q)|^2 . \tag{10}$$

A lot of efffort was made in the course of the years to give an explanation for the topological entropy of extremal black holes in the context of a quantum theory of gravity, such as string theory. A particularly interesting problem is finding a microscopic, statistical mechanics interpretation of this thermodynamic quantity. Although we will not deal with the microscopic point of view at all in this paper, it is important to mention that such an interpretation became possible after the introduction of D-branes in the context of string theory [8, 24]. Following this approach, extremal black holes are interpreted as bound states of D-branes in a space–time compactified to four or five dimensions, and the different microstates contributing to the B–H entropy are, for instance, related to the different ways of wrapping branes in the internal directions. Let us mention that all calculations made in particular cases using this approach provided values for the B–H entropy compatible with those obtained with the supergravity, macroscopic techniques. The entropy formula turns out to be in all cases a U-duality-invariant expression (homogeneous of degree 2) built out of electric and magnetic charges and as such it can be in fact also computed through certain (moduli-independent) topological quantities [25], which only depend on the nature of the U-duality groups and the appropriate representations of electric and magnetic charges [26]. We mention for completeness that, as previously pointed out, superstring corrections that take into account higher derivative effects determine a deviation from the area law for the entropy [27, 28]. Recently, a deeper insight into the microscopic description of black-hole entropy was gained, in this case, from the fruitful proposal in [29], describing the microscopic degrees of freedom of black holes in terms of topological strings.

Originally, the attention was mainly devoted to the so-called *BPS-extremal black holes*, i.e. to solutions which saturate the bound in (5). From an abstract viewpoint BPS-saturated states are characterized by the fact that they preserve a fraction, 1/2 or 1/4 or 1/8, of the original supersymmetries. What this actually means is that there is a suitable projection operator $S^2 = S$ acting on the supersymmetry charge Q_{SUSY}, such that:

$$(S \cdot Q_{\mathrm{SUSY}}) \,|\, \mathrm{BPS\ state} \,\rangle = 0\,. \qquad (11)$$

Since the supersymmetry transformation rules of any supersymmetric field theory are linear in the first derivatives of the fields, (11) is actually a system of first-order differential equations, to be combined with the second-order field equations of the theory. Translating (11) into an explicit first-order differential system requires knowledge of the supersymmetry transformation rules of supergravity. The latter have a rich geometric structure whose analysis will be the subject of Sect. 3. The BPS saturation condition transfers the geometric structure of supergravity, associated with its scalar sector, into the physics of extremal black holes. We note that first-order differential equations $\frac{d\Phi}{dr} = f(\Phi)$ have in general fixed points, corresponding to the values of r for which $f(\Phi) = 0$. For the BPS black holes, the fixed point is reached precisely at the black-hole horizon, and this is how the attractor behavior is realized for this class of extremal black holes.

For BPS configurations, non-renormalization theorems based on supersymmetry guarantee the validity of the (BPS) bound $M = |Q|$ beyond the perturbative regime: if the bound is saturated in the classical theory, the same must be true also when quantum corrections are taken into account and the theory is in a regime where the supergravity approximation breaks down. That it is actually an exact state of non-perturbative string theory follows from supersymmetry representation theory. The classical BPS state is by definition an element of a short supermultiplet and, if supersymmetry is unbroken, it cannot be renormalized to a long supermultiplet. For this class of extremal black holes, an accurate agreement between the macroscopic and microscopic calculations was found. For example, in the $N = 8$ theory the entropy was shown to correspond to the unique quartic $E_{7(7)}$-invariant built in terms of the 56-dimensional representation. Actually, topological U-invariants constructed in terms of the (moduli dependent) central charges and matter charges can be derived for all $N \geq 2$ theories; they can be shown, as expected, to coincide with the squared ADM mass at fixed scalars.

Quite recently it has been recognized that the attractor mechanism, which is responsible for the area/entropy relation, has a larger application [30, 31, 32, 33, 34, 35, 36, 37] beyond the BPS cases, being a peculiarity of all *extremal black-hole configurations*, BPS or not. The common feature is that extremal black-hole configurations always belong to some representation of supersymmetry, as will be surveyed in Sect. 2 (this is not the case for non-extremal configurations, since the action of supersymmetry generators cannot be defined for non-zero temperature [38]). Extremal configurations that completely break supersymmetry will belong to long representations of supersymmetry.

Even for these more general cases, because of the topological nature of the extremality condition, the entropy formula turns out to be still given by a U-duality-invariant expression built out of electric and magnetic charges. We will report in Sect. 6 on the classification of all extremal solutions (BPS and non-BPS) of N-extended supergravity in four dimensions.

For all the N-extended theories in four dimensions, the general feature that allows us to find the B–H entropy as a topological invariant is the presence of vectors and scalars in the same representation of supersymmetry. This causes the electric/magnetic duality transformations on the vector field strengths (which for these theories are embedded into symplectic transformations) to also act as isometries on the scalar sectors [39].[1] The symplectic structure of the various σ-models of N-extended supergravity in four dimensions and the relevant relations involving the charges obeyed by the scalars will be worked out in Sect. 3.

As a final remark, let us observe that, since the aim of the present review is to calculate the B-H entropy of extremal black holes, we will only discuss solutions which have $S_{\text{B–H}} \neq 0$. For this class of solutions, known as *large black holes*, the classical area/entropy formula is valid, as it gives the dominant contribution to the black-hole entropy. For these configurations the area of the horizon is in fact proportional to a duality-invariant expression constructed with the electric and magnetic charges, which for these states is not vanishing [41]. This will prove to be a powerful computational tool and will be the subject of Sect. 5.2. As we will see in detail in the following sections, configurations with non-vanishing horizon area in supersymmetric theories preserve at most four supercharges ($N = 1$ supersymmetry) in the bulk of space–time. Black-hole solutions preserving more supercharges do exist, but they do not correspond to classical attractors since in that case the classical area/entropy formula vanishes. These configurations are named *small black holes* and require, for finding the entropy, a quantum attractor mechanism taking into account the presence of higher curvature terms [29, 42, 43].

The paper is organized as follows. Sect. 2 treats the supersymmetry structure of extremal black-hole solutions of supergravity theories, and the black-hole configurations are described as massive representations of supersymmetry. In Sect. 3 we briefly review the properties of four-dimensional extended supergravity related to its global symmetries. A particular emphasis is given to the general symplectic structure characterizing the moduli spaces of these theories. The presence of this structure allows the global symmetries of extended supergravities to be realized as generalized electric–magnetic symplectic duality transformations acting on the electric and magnetic charges

[1] We note that symplectic transformations outside the U-duality group have a non-trivial action on the solutions, allowing one to bring a BPS configuration to a non-BPS one [40]

of dyonic solutions (as black holes). In Sect. 4 we start reviewing extremal regular black-hole solutions embedded in supergravity and, for the BPS case, an explicit solution will be found by solving the Killing spinor equations. In Sect. 5 we give a general overview of extremal and non-extremal solutions showing how the attractor mechanism comes about in the extremal case only. Then a general tool for calculating the Bekenstein–Hawking entropy for both BPS and non-BPS extremal black holes will be given, based on the observation that the black-hole potential takes a particularly simple form in the supergravity case, which is fixed in terms of the geometric properties of the moduli space of the given theory. Moreover, for theories based on moduli spaces given by symmetric manifolds G/H, which is the case of all supergravity theories with $N \geq 3$ extended supersymmetry, but also of several $N = 2$ models, the BPS and non-BPS black holes are classified by some U-duality-invariant expressions, depending on the representation of the isometry group G under which the electric and magnetic charges are classified. Finally in Sect. 6, by exploiting the supergravity machinery introduced in Sect. 3 and 4, we shall give a detailed analysis of the attractor solutions for the various theories of extended supergravity. Section 7 contains some concluding remarks.

Our discussion will be confined to four-dimensional black holes.

2 Extremal Black Holes as Massive Representations of Supersymmetry

We are going to review in the present section the algebraic structure of the massive representations of supersymmetry, both for short and long multiplets, in order to pinpoint, for each supergravity theory, the extremal black-hole configurations corresponding to a given number of preserved supercharges. The condition of extremality is in fact independent on the supersymmetry preserved by the solution, the only difference between the supersymmetric and the non-supersymmetic case being that the configurations preserving some supercharges correspond to short multiplets, while the configurations which completely break supersymmetry will instead belong to long representations of supersymmetry. The highest spin of the configuration [2] depends on the number of supercharges of the theory under consideration [44].

As a result of our analysis we find for example, as far as large BPS black-hole configurations are considered, that for $N = 2$ theories the highest spin of the configuration (which in this case is 1/2-BPS) is $J_{MAX} = 1/2$, for $N = 4$ theories (1/4-BPS) is $J_{MAX} = 3/2$, while for the $N = 8$ case (1/8-BPS) is $J_{MAX} = 7/2$. On the other hand, 1/2-BPS multiplets have maximum spin $J_{MAX} = N/4$ ($N = 2, 4, 8$) as for massless representations. They are given in Tables 2–4. The corresponding black holes (for $N > 2$) have vanishing classical entropy (small black holes) [25].

[2] We confine our analysis here to the *minimal* highest spin allowed for a given theory

The long multiplets corresponding to non-BPS extremal black-hole configurations have $J_{MAX} = 1$ in the $N = 2$ theory, $J_{MAX} = 2$ in the $N = 4$ theory and $J_{MAX} = 4$ in the $N = 8$ theory. However, as we will see in detail in the following, for the non-BPS cases we may have solutions with vanishing or non-vanishing central charge. Since the central charge Z_{AB} is a complex matrix, it is not invariant under CPT symmetry, but transforms as $Z_{AB} \to \bar{Z}_{AB}$.[3] The representation then depends on the charge of the configuration: if the solution has vanishing central charge the long-multiplet will be neutral (real), while if the solution has non-zero central charge the long multiplet will be charged (complex), with a doubled dimension as required for CPT invariance [44].

We have listed in Tables 1–3 all possible massive representations with highest spin $J_{MAX} \leq 3/2$ for $N \leq 8$. The occurence of long spin 3/2 multiplets is only possible for $N = 3, 2$ and of long spin 1 multiplets for $N = 2$. In $N = 1$ there is only one type of massive multiplet (long) since there are no central charges. Its structure is

$$\left[(J_0 + \frac{1}{2}), 2(J_0), (J_0 - \frac{1}{2})\right],$$

except for $J_0 = 0$ where we have $\left[(\frac{1}{2}), 2(0)\right]$.

In the tables we will denote the spin states by (J) and the number in front of them is their multiplicity. In the fundamental multiplet, with spin $J_0 = 0$ vacuum, the multiplicity of the spin $(N - q - k)/2$ is the dimension of the k-fold antisymmetric Ω-traceless representation of $USp(2(N - q))$. For multiplets with $J_0 \neq 0$ one has to make the tensor product of the fundamental multiplet with the representation of spin J_0. We also indicate if the multiplet is long or short.

2.1 Massive Representations of the Supersymmetry Algebra

The $D = 4$ supersymmetry algebra is given by

$$\begin{aligned} \{\bar{Q}_{A\alpha}, \bar{Q}_{B\beta}\} &= -(C\gamma^\mu)_{\alpha\beta} P_\mu \delta_{AB} + \mathrm{i}(C\mathbb{Z}_{AB})_{\alpha\beta} \\ &(A, B = 1, \dots, 2p) \end{aligned} \quad , \tag{12}$$

where the SUSY charges $\bar{Q}_A \equiv Q_A^\dagger \gamma_0 = Q_A^T C$ are Majorana spinors, C is the charge conjugation matrix, P_μ is the four-momentum operator and the antisymmetric tensor $\mathbb{Z}_{AB}$ is defined as

$$\mathbb{Z}_{AB} = \Re(Z_{AB}) + \mathrm{i}\gamma^5 \Im(Z_{AB}), \tag{13}$$

the complex matrix $Z_{AB} = -Z_{BA}$ being the central charge operator. For the sake of simplicity, we shall suppress the spinorial indices in the formulae.

[3] We use here a different definition of central charge with respect to [44]: $Z_{AB} \to \mathrm{i}Z_{AB}$

Table 1. Massive spin 3/2 multiplets

N	Massive spin 3/2 multiplet	Long	Short
8	None		
6	$2\times\left[(\frac{3}{2}),6(1),14(\frac{1}{2}),14'(0)\right]$	No	$q=3$, ($\frac{1}{2}$BPS)
5	$2\times\left[(\frac{3}{2}),6(1),14(\frac{1}{2}),14'(0)\right]$	No	$q=2$, ($\frac{2}{5}$BPS)
4	$2\times\left[(\frac{3}{2}),6(1),14(\frac{1}{2}),14'(0)\right]$	No	$q=1$, ($\frac{1}{4}$BPS)
	$2\times\left[(\frac{3}{2}),4(1),6(\frac{1}{2}),4(0)\right]$	No	$q=2$, ($\frac{1}{2}$BPS)
3	$\left[(\frac{3}{2}),6(1),14(\frac{1}{2}),14'(0)\right]$	Yes	no
	$2\times\left[(\frac{3}{2}),4(1),6(\frac{1}{2}),4(0)\right]$	No	$q=1$, ($\frac{1}{3}$BPS)
2	$\left[(\frac{3}{2}),4(1),6(\frac{1}{2}),4(0)\right]$	Yes	no
	$2\times\left[(\frac{3}{2}),2(1),(\frac{1}{2})\right]$	No	$q=1$, ($\frac{1}{2}$BPS)
1	$\left[(\frac{3}{2}),2(1),(\frac{1}{2})\right]$	Yes	no

Using the symmetries of the theory, it can always be reduced to normal form [45]. For N even it reads

Table 2. Massive spin 1 multiplets

N	Massive spin 1 multiplet	Long	Short
8,6,5	None		
4	$2\times\left[(1),4(\frac{1}{2}),5(0)\right]$	No	$q=2$, ($\frac{1}{2}$BPS)
3	$2\times\left[(1),4(\frac{1}{2}),5(0)\right]$	No	$q=1$, ($\frac{1}{3}$BPS)
2	$\left[(1),4(\frac{1}{2}),5(0)\right]$	Yes	No
	$2\times\left[(1),2(\frac{1}{2}),(0)\right]$	No	$q=1$, ($\frac{1}{2}$BPS)
1	$\left[(1),2(\frac{1}{2}),(0)\right]$	Yes	No

Table 3. Massive spin 1/2 multiplets

N	Massive spin 1/2 multiplet	long	Short
8,6,5,4,3	None		
2	$2 \times [(\frac{1}{2}), 2(0)]$	No	$q = 1$, ($\frac{1}{2}$BPS)
1	$[(\frac{1}{2}), 2(0)]$	Yes	No

$$Z_{AB} = \begin{pmatrix} \epsilon Z_1 & 0 & \dots & 0 \\ 0 & \epsilon Z_2 & \dots & 0 \\ \dots & \dots & \dots & \dots \\ 0 & 0 & \dots & \epsilon Z_p \end{pmatrix}, \tag{14}$$

where ϵ is the 2×2 antisymmetric matrix, (every zero is a 2×2 zero matrix) and the p skew eigenvalues Z_m of Z_{AB} are the central charges. For N odd the central charge matrix has the same form as in (14) with $p = (N-1)/2$, except for one extra zero row and one extra zero column. Note that it is not always possible to reduce Z_{AB} to its normal form with real Z_m by means of symmetries of the theory [45]. This is the case in particular of $N = 8$ supergravity where the $SU(8)$ R-symmetry does not affect the global phase of the skew-eigenvalues Z_m. Therefore, we shall consider the general situation in which Z_m are complex and define for each of them the spinorial matrices which will enter the supersymmetry algebra:

$$\begin{aligned} \mathbb{Z}_m &= \Re(Z_m) + \mathrm{i}\,\gamma^5\, \Im(Z_m)\,, \\ \bar{\mathbb{Z}}_m &= \Re(Z_m) - \mathrm{i}\,\gamma^5\, \Im(Z_m)\,, \quad m = 1, \dots, p\,. \end{aligned} \tag{15}$$

If we identify each index $A, B, \dots$ with the pair of indices

$$A = (a, m) \quad ; \quad a, b, \dots = 1, 2 \quad ; \quad m, n, \dots = 1, \dots, p\,, \tag{16}$$

the matrix $\mathbb{Z}_{AB}$ in the normal frame will have the form

$$\mathbb{Z}_{AB} = \mathbb{Z}_{am,\, bn} = \mathbb{Z}_m\, \delta_{mn}\, \epsilon_{ab}\,, \tag{17}$$

and the superalgebra (12) can be rewritten as

$$\{\bar{Q}_{am}\,,\, \bar{Q}_{bn}\} = -(C\,\gamma^\mu)\; P_\mu\, \delta_{ab}\, \delta_{mn} + \mathrm{i}\, C\, \epsilon_{ab}\, \mathbb{Z}_m\, \delta_{mn}, \tag{18}$$

where ϵ_{ab} is the two-dimensional Levi Civita symbol. Let us consider a generic unit time-like Killing vector ζ^μ ($\zeta^\mu \zeta_\mu = 1$), in terms of which we define the following projectors acting on both the internal (a, m) and Lorentz indices (α, β) of the spinors:

$$S^{(\pm)}_{am,\, bn} = \frac{1}{2}\left(\delta_{ab}\, \delta_{mn} \pm \mathrm{i}\, \zeta_\mu\, \gamma^\mu\, \frac{\bar{\mathbb{Z}}_m}{|Z_m|}\, \delta_{mn}\, \epsilon_{ab}\right),$$

$$\tilde{S}^{(\pm)}_{am,\,bn} = \frac{1}{2}\left(\delta_{ab}\,\delta_{mn} \pm \mathrm{i}\,\zeta_\mu\,\gamma^\mu\,\frac{\mathbb{Z}_m}{|Z_m|}\,\delta_{mn}\,\epsilon_{ab}\right), \tag{19}$$

and define the projected supersymmetry generators:

$$\bar{Q}^{(\pm)} = \bar{Q}\,S^{(\pm)}. \tag{20}$$

The anticommutation relation (18) can be rewritten in the following form:

$$\left\{Q^{(\pm)}_{am}\,,\,\bar{Q}^{(\pm)}_{bn}\right\} = \tilde{S}^{(\pm)}_{am,\,bn}\,\zeta_\mu\,\gamma^\mu\,(\zeta_\nu\,P^\nu \mp |Z_m|)\,. \tag{21}$$

In the case in which $\zeta^\mu = (1,0,0,0)$ and we are in the rest frame ($P^0 = M$) the above relation reads

$$\left\{Q^{(\pm)}_{am}\,,\,Q^{(\pm)\,\dagger}_{bn}\right\} = \tilde{S}^{(\pm)}_{am,\,bn}\,(M \mp |Z_m|)\,. \tag{22}$$

Since the left-hand side of (22) is non-negative definite, we deduce the BPS bound required by unitarity of the representations

$$M \geq |\,Z_m| \qquad \forall Z_m\,,\, m = 1,\ldots,p\,. \tag{23}$$

It is an elementary consequence of the supersymmetry algebra and of the identification between central charges and topological charges [46].

Massive BPS Multiplets

Suppose that on a given state $|BPS\rangle$ the BPS bound (23) is saturated by q of the p eigenvalues Z_m:

$$M = |Z_1| = |Z_2| = \cdots = |Z_q| \quad q \leq p\,, \tag{24}$$

then, from (22) we deduce that

$$Q^{(+)}_{am}\,|BPS\rangle = 0\ \ ,\ \ m = 1,\ldots,q\,, \tag{25}$$

namely q of the pairs of creation–annihilation operators, which have abelian anticommutation relations, annihilate the state. The multiplet obtained by acting on $|BPS\rangle$ with the remaining supersymmetry generators is said to be q/N BPS. Note that $q_{MAX} = N/2$ for N even and $q_{MAX} = (N-1)/2$ for N odd. The $USp(2N)$ symmetry is now reduced to $USp(2(N-q))$. The short multiplet has the same number of states as a long multiplet of the $N-q$ supersymmetry algebra. The fundamental multiplet, with $J = 0$ vacuum, contains $2 \cdot 2^{2(N-q)}$ states with $J_{MAX} = (N-q)/2$. Note the doubling due to CPT invariance. Generic massive short multiplets can be obtained by making the tensor product with a spin J_0 representation of $SU(2)$.

If we write the infinitesimal generator of a supersymmetry in the form

$$\bar{Q}_A\,\epsilon_A = \bar{Q}^{(+)}_A\,\epsilon^{(+)}_A + \bar{Q}^{(-)}_A\,\epsilon^{(-)}_A\,, \tag{26}$$

the supersymmetries preserved by $|BPS\rangle$ are parametrized by $\epsilon^{(+)}_{am}$ with $m \leq q$ and thus defined by the condition

$$\epsilon^{(-)}_{am} = S^{(-)}_{am,\,bn}\,\epsilon_{bn} = 0\ ;\quad m,\,n \leq q\,, \tag{27}$$

$$\epsilon_{am} = 0\ ;\quad m > q\,, \tag{28}$$

which can be written in terms of *Weyl* spinors ϵ_A, ϵ^A in the following form:

$$\epsilon_{am} = \mathrm{i}\,\frac{Z_m}{|Z_m|}\,\zeta_\mu\,\gamma^\mu\,\epsilon_{ab}\,\epsilon^{bm} = \mathrm{i}\,\frac{Z_m}{|Z_m|}\,\epsilon_{ab}\,\gamma^0\,\epsilon^{bm}\quad ;\quad m \leq q\,, \tag{29}$$

$$\epsilon_{am} = 0\ ;\quad m > q\,. \tag{30}$$

If, in a given supergravity theory, the state $|BPS\rangle$ corresponds to a background described by a certain configuration of fields, (25) is translated into the request that the supersymmetry variations of all the fields are zero in the background. We consider extremal black-hole solutions for which the supersymmetry variations of the bosonic fields are identically zero. Then the condition (25) yields a set of first-order differential equations for the bosonic fields, called "Killing spinor" equations, to be satisfied on the given configuration

$$0 = \delta\text{fermions} = \text{SUSY rule}\,(\text{bosons}, \epsilon_{am})\ , \tag{31}$$

where the supersymmetry transformations are made with respect to the residual supersymmetry parameter $\epsilon^{(+)}_{am}$ defined by the conditions (30). These conditions are important in order to be able to recast (31) into differential equations involving only the bosonic fields of the solution.

Massive Non-BPS Multiplets

Massive multiplets with $Z_m = 0$ or $Z_m \neq 0$ but $M > |Z_m|$ are called long multiplets or non-BPS states. They are qualitatively the same, the only difference being that in the first case the supermultiplets are real, while in the second one the representations must be doubled in order to have CPT invariance, since $Z_m \to \bar{Z}_m$ under CPT.

In both cases the supersymmetry algebra can be put in a form with $2N$ creation and $2N$ annihilation operators. It shows explicit invariance under $SU(2)\times USp(2N)$. The vacuum state is now labeled by the spin representation of $SU(2)$, $|\Omega\rangle_J$. If $J = 0$ we have the fundamental massive multiplet with 2^{2N} states. These are organized in representations of $SU(2)$ with $J_{MAX} = N/2$. With respect to $USp(2N)$ the states with fixed $0 < J < N/2$ are arranged in the $(N-2J)$-fold Ω-traceless antisymmetric representation, $[N-2J]$.

The general multiplet with a spin J vacuum can be obtained by tensoring the fundamental multiplet with spin J representation of $SU(2)$. The total number of states is then $(2J+1)\cdot 2^{2N}$.

3 The General Form of the Supergravity Action in Four Dimensions and its BPS Configurations

In this section we begin the study of extremal black-hole solutions of extended supergravity in four space–time dimensions. To this aim we first have to introduce the main features of four-dimensional N-extended supergravities. These theories contain in the bosonic sector, besides the metric, a number n_V of vectors and m of (real) scalar fields. The relevant bosonic action is known to have the following general form:

$$\mathcal{S} = \int \sqrt{-g}\, d^4x \left(-\frac{1}{2} R + \Im \mathcal{N}_{\Lambda\Gamma} F^{\Lambda}_{\mu\nu} F^{\Gamma|\mu\nu} + \frac{1}{2\sqrt{-g}} Re\mathcal{N}_{\Lambda\Gamma} \epsilon^{\mu\nu\rho\sigma} F^{\Lambda}_{\mu\nu} F^{\Gamma}_{\rho\sigma} + \frac{1}{2} g_{rs}(\Phi) \partial_\mu \Phi^r \partial^\mu \Phi^s \right) , \tag{32}$$

where $g_{rs}(\Phi)$ $(r, s, \cdots = 1, \cdots, m)$ is the scalar metric on the σ-model described by the scalar manifold $\mathcal{M}_{scalar}$ of real dimension m and the vectors kinetic matrix $\mathcal{N}_{\Lambda\Sigma}(\Phi)$ is a complex, symmetric, $n_V \times n_V$ matrix depending on the scalar fields. The number of vectors and scalars, namely n_V and m, and the geometric properties of the scalar manifold $\mathcal{M}_{scalar}$ depend on the number N of supersymmetries and are resumed in Table 4. The imaginary part $Im\mathcal{N}$ of the vector kinetic matrix is negative definite and generalizes the inverse of the squared coupling constant appearing in ordinary gauge theories while its real part $Re\mathcal{N}$ is instead a generalization of the *theta*-angle of quantum chromodynamics. In supergravity theories, the kinetic matrix $\mathcal{N}$ is in general not a constant, its components being functions of the scalar fields. However, in extended supergravity $(N \geq 2)$ the relation between the scalar geometry and the kinetic matrix $\mathcal{N}$ has a very general and universal form. Indeed it is related to the solution of a general problem, namely how to lift the action of the scalar manifold isometries from the scalar to the vector fields. Such a lift is necessary because of supersymmetry since scalars and vectors generically belong to the same supermultiplet and must rotate coherently under symmetry operations. This problem has been solved in a general (non-supersymmetric) framework in reference [39] by considering the possible extension of the Dirac electric–magnetic duality to more general theories involving scalars. In the next subsection we review this approach and in particular we show how enforcing covariance with respect to such duality rotations leads to a determination of the kinetic matrix $\mathcal{N}$. The structure of $\mathcal{N}$ enters the black-hole equations in a crucial way so that the topological invariant associated with the hole, that is its entropy, is an invariant of the group of electro-magnetic duality rotations, the U-duality group.

3.1 Duality Rotations and Symplectic Covariance

Let us review the general structure of an abelian theory of vectors and scalars displaying covariance under a group of duality rotations. The basic reference

Table 4. Scalar manifolds of $N > 2$ extended supergravities. In the table, n_V stands for the number of vectors and m for the number of real scalar fields. In all the cases the duality group G is embedded in $Sp(2n_V, \mathbb{R})$

N	Duality group G	Isotropy H	$\mathcal{M}_{scalar}$	n_V	m
3	$SU(3,n)$	$SU(3) \times U(n)$	$\frac{SU(3,n)}{S(U(3)\times U(n))}$	$3+n$	$6n$
4	$SU(1,1) \otimes SO(6,n)$	$U(4) \times SO(n)$	$\frac{SU(1,1)}{U(1)} \otimes \frac{SO(6,n)}{SO(6)\times SO(n)}$	$6+n$	$6n+2$
5	$SU(1,5)$	$U(5)$	$\frac{SU(1,5)}{S(U(1)\times U(5))}$	10	10
6	$SO^\star(12)$	$U(6)$	$\frac{SO^\star(12)}{U(1)\times SU(6)}$	16	30
7, 8	$E_{7(7)}$	$SU(8)$	$\frac{E_{7(7)}}{SU(8)}$	28	70

is the 1981 paper by Gaillard and Zumino [39]. A general presentation in $D = 2p$ dimensions can be found in [47]. Here we fix $D = 4$.

We consider a theory of n_V abelian gauge fields A^Λ_μ, in a $D = 4$ space–time with Lorentz signature (which we take to be mostly minus). They correspond to a set of n_V differential 1-forms

$$A^\Lambda \equiv A^\Lambda_\mu \, dx^\mu \qquad (\Lambda = 1, \dots, n_V) \, . \tag{33}$$

The corresponding field strengths and their Hodge duals are defined by[4]

$$\begin{aligned} F^\Lambda &\equiv d\, A^\Lambda \equiv F^\Lambda_{\mu\nu}\, dx^\mu \wedge dx^\nu , \\ F^\Lambda_{\mu\nu} &\equiv \frac{1}{2}\left(\partial_\mu A^\Lambda_\nu - \partial_\nu A^\Lambda_\mu\right), \\ ({}^\star F^\Lambda)_{\mu\nu} &\equiv \frac{\sqrt{-g}}{2} \varepsilon_{\mu\nu\rho\sigma}\, F^{\Lambda|\rho\sigma} . \end{aligned} \tag{34}$$

The dynamics of a system of abelian gauge fields coupled to scalars in a gravity theory is encoded in the bosonic action (32).

Introducing self-dual and antiself-dual combinations

$$\begin{aligned} F^\pm &= \frac{1}{2}\left(F \pm \mathrm{i}\,{}^\star F\right) , \\ {}^\star F^\pm &= \mp \mathrm{i} F^\pm , \end{aligned} \tag{35}$$

the vector part of the Lagrangian defined by (32) can be rewritten in the form

$$\mathcal{L}_{vec} = \mathrm{i}\left[F^{-T}\bar{\mathcal{N}}F^- - F^{+T}\mathcal{N}F^+\right] . \tag{36}$$

Introducing further the new tensors

$${}^\star G_{\Lambda|\mu\nu} \equiv \frac{1}{2}\frac{\partial \mathcal{L}}{\partial F^\Lambda_{\mu\nu}} = Im\mathcal{N}_{\Lambda\Sigma}\, F^\Sigma_{\mu\nu} + Re\mathcal{N}_{\Lambda\Sigma}\, {}^\star F^\Sigma_{\mu\nu} \leftrightarrow G^\mp_{\Lambda|\mu\nu} \equiv \mp\frac{\mathrm{i}}{2}\frac{\partial \mathcal{L}}{\partial F^{\mp\Lambda}_{\mu\nu}} , \tag{37}$$

[4] We use, for the ϵ tensor, the convention: $\epsilon_{0123} = -1$

the Bianchi identities and field equations associated with the Lagrangian (32) can be written as

$$\begin{aligned} \nabla^{\mu\star} F^{\Lambda}_{\mu\nu} &= 0, \\ \nabla^{\mu\star} G_{\Lambda|\mu\nu} &= 0, \end{aligned} \tag{38}$$

or equivalently

$$\nabla^{\mu} \mathrm{Im} F^{\pm\Lambda}_{\mu\nu} = 0, \tag{39}$$

$$\nabla^{\mu} \mathrm{Im} G^{\pm}_{\Lambda|\mu\nu} = 0\ . \tag{40}$$

This suggests that we introduce the $2n_V$ column vector

$$\mathbf{V} \equiv \begin{pmatrix} {}^\star F \\ {}^\star G \end{pmatrix} \tag{41}$$

and that we consider general linear transformations on such a vector

$$\begin{pmatrix} {}^\star F \\ {}^\star G \end{pmatrix}' = \begin{pmatrix} A & B \\ C & D \end{pmatrix} \begin{pmatrix} {}^\star F \\ {}^\star G \end{pmatrix} . \tag{42}$$

For any constant matrix $\mathcal{S} = \begin{pmatrix} A & B \\ C & D \end{pmatrix} \in GL(2n_V, \mathbb{R})$ the new vector of magnetic and electric field strengths $\mathbf{V}' = \mathcal{S} \cdot \mathbf{V}$ satisfies the same equations (38) as the old one. In a condensed notation we can write

$$\partial\, \mathbf{V} = 0 \quad \Longleftrightarrow \quad \partial\, \mathbf{V}' = 0. \tag{43}$$

Separating the self-dual and antiself-dual parts

$$F = (F^+ + F^-) \quad ; \quad G = (G^+ + G^-)\,, \tag{44}$$

and taking into account that we have

$$G^+ = \mathcal{N} F^+ \quad ; \quad G^- = \bar{\mathcal{N}} F^-, \tag{45}$$

the duality rotation of (42) can be rewritten as

$$\begin{pmatrix} F^+ \\ G^+ \end{pmatrix}' = \begin{pmatrix} A & B \\ C & D \end{pmatrix} \begin{pmatrix} F^+ \\ \mathcal{N} F^+ \end{pmatrix}; \quad \begin{pmatrix} F^- \\ G^- \end{pmatrix}' = \begin{pmatrix} A & B \\ C & D \end{pmatrix} \begin{pmatrix} F^- \\ \bar{\mathcal{N}} F^- \end{pmatrix} . \tag{46}$$

Now, let us note that, since in the system we are considering ((32)) the gauge fields are coupled to the scalar sector via the scalar-dependent kinetic matrix $\mathcal{N}$, when a duality rotation is performed on the vector field strengths and their duals, we have to assume that the scalars get transformed correspondingly, through the action of some diffeomorphism on the scalar manifold $\mathcal{M}_{scalar}$. In particular, the kinetic matrix $\mathcal{N}(\Phi)$ transforms under a duality rotation. Then, a duality transformation ξ acts in the following way on the supersymmetric system:

$$\xi : \begin{cases} V & \to V'^{\mp} = S_\xi V^{\mp} \\ \Phi & \to \quad \Phi' = \xi(\Phi) \\ \mathcal{N}(\Phi) & \to \quad \mathcal{N}'\left(\xi(\Phi)\right) \end{cases} \tag{47}$$

Thus, the transformation laws of the equations of motion and of $\mathcal{N}$, and so also the matrix $\mathcal{S}_\xi$, will be induced by a diffeomorphism of the scalar fields.

Focusing in particular on the first relation in (47), that explicitly reads

$$\begin{pmatrix} F^{\pm\prime} \\ G^{\pm\prime} \end{pmatrix} = \begin{pmatrix} A_\xi F^{\pm} + B_\xi G^{\pm} \\ C_\xi F^{\pm} + D_\xi G^{\pm} \end{pmatrix}, \tag{48}$$

we note that it contains the magnetic field strength $G^{\mp}_\Lambda$ introduced in (37), which is defined as a variation of the kinetic lagrangian. Under the transformations (47) the Lagrangian transforms in the following way:

$$\begin{aligned} \mathcal{L}' = \mathrm{i}\Big[&(A_\xi + B_\xi\mathcal{N})_\Gamma{}^\Lambda \,(A_\xi + B_\xi\mathcal{N})^\Sigma{}_\Delta\, \mathcal{N}'_{\Lambda\Sigma}(\Phi) F^{+\Gamma} F^{+\Delta} \\ &- (A_\xi + B_\xi\bar{\mathcal{N}})_\Gamma{}^\Lambda \,(A_\xi + B_\xi\bar{\mathcal{N}})^\Sigma{}_\Delta\, \bar{\mathcal{N}}'_{\Lambda\Sigma}(\Phi) F^{-\Gamma} F^{-\Delta}\Big]; \end{aligned} \tag{49}$$

Equations (47) must be consistent with the definition of $G^{\mp}$ as a variation of the Lagrangian (49)

$$G'^{+}_\Lambda = (C_\xi + D_\xi\mathcal{N})_{\Lambda\Sigma} F^{+\Sigma} \equiv -\frac{\mathrm{i}}{2}\frac{\partial\mathcal{L}'}{\partial F'^{+\Lambda}} = (A_\xi + B_\xi\mathcal{N})^\Delta{}_\Sigma \mathcal{N}'_{\Lambda\Delta} F^{+\Sigma} \tag{50}$$

that implies

$$\mathcal{N}'_{\Lambda\Sigma}(\Phi') = \left[(C_\xi + D_\xi\mathcal{N})\cdot(A_\xi + B_\xi\mathcal{N})^{-1}\right]_{\Lambda\Sigma}; \tag{51}$$

The condition that the matrix $\mathcal{N}$ is symmetric, and that this property must be true also in the duality transformed system, gives the constraint

$$\mathcal{S} \in Sp(2n_V, \mathbb{R}) \subset GL(2n_V, \mathbb{R}), \tag{52}$$

that is:

$$\mathcal{S}^T\,\mathbb{C}\,\mathcal{S} = \mathbb{C}, \tag{53}$$

where $\mathbb{C}$ is the symplectic invariant $2n_V \times 2n_V$ matrix:

$$\mathbb{C} = \begin{pmatrix} 0 & -1\!\!1 \\ 1\!\!1 & 0 \end{pmatrix}. \tag{54}$$

It is useful to rewrite the symplectic condition (53) in terms of the $n_V \times n_V$ blocks defining $\mathcal{S}$:

$$A^T C - C^T A = B^T D - D^T B = 0 \;;\;\; A^T D - C^T B = 1\!\!1. \tag{55}$$

The above observation has important implications on the scalar manifold $\mathcal{M}_{scalar}$. Indeed, it implies that on the scalar manifold the following homomorphism is defined:

$$Diff(\mathcal{M}_{scalar}) \rightarrow Sp(2n,\mathbb{R})\,. \tag{56}$$

In particular, the presence on the manifold of a function of scalars transforming with a fractional linear transformation under a duality rotation on the scalars, induces the existence on $\mathcal{M}_{scalar}$ of a linear structure (inherited from the vectors). As we are going to discuss in detail in Sect. 3.2, this may be rephrased by saying that the scalar manifold is endowed with a symplectic bundle. As the transition functions of this bundle are given in terms of the *constant* matrix $\mathcal{S}$, the symplectic bundle is flat. In particular, as we will see in Sect. 3.2, for the $N=2$ four-dimensional theory this implies that the scalar manifold be a *special manifold*, that is a Kähler–Hodge manifold endowed with a flat symplectic bundle.

If we are interested in the global symmetries of the theory (i.e. global symmetries of the field equations and Bianchi identities) we will need to restrict the duality transformations, namely the homomorphism in (56), to the isometries of the scalar manifold, which leave the scalar sector of the action invariant. The transformations (47), which are duality symmetries of the system field equations/Bianchi identities, cannot be extended in general to be symmetries of the Lagrangian. The scalar part of the Lagrangian (32) is invariant under the action of the isometry group of the metric g_{rs}, but the vector part is in general not invariant. The transformed Lagrangian under the action of $\mathcal{S}\in Sp(2n_V,\mathbb{R})$ can be rewritten:

$$\begin{aligned} Im\left(F^{-\Lambda}G^-_\Lambda\right) &\rightarrow Im\left(F'^{-\Lambda}G'^-_\Lambda\right)\\ &= Im\big[F^{-\Lambda}G^-_\Lambda + 2(C^TB)_\Lambda{}^\Sigma F^{-\Lambda}G^-_\Sigma\\ &+ (C^TA)_{\Lambda\Sigma}F^{-\Lambda}F^{-\Sigma} + (D^TB)^{\Lambda\Sigma}G^-_\Lambda G^-_\Sigma\big]\,. \end{aligned} \tag{57}$$

It is evident from (57) that only the transformations with $B=C=0$ are symmetries.
If $C\neq 0$, $B=0$ the Lagrangian varies for a topological term

$$(C^TA)_{\Lambda\Sigma}F^\Lambda_{\mu\nu}{}^\star F^{\Sigma|\mu\nu} \tag{58}$$

corresponding to a redefinition of the function $\Re\mathcal{N}_{\Lambda\Sigma}$; such a transformation being a total derivative it leaves classical physics invariant, but it is relevant in the quantum theory. It is a symmetry of the partition function only if $\Delta\Re\mathcal{N}_{\Lambda\Sigma}=\frac{1}{2}(C^TA)$ is an integer multiple of 2π, and this implies that $\mathcal{S}\in Sp(2n_V,\mathbb{Z})\subset Sp(2n_V,\mathbb{R})$.
For $B\neq 0$ neither the action nor the perturbative partition function are invariant. Let us observe that in this case the transformation law (51) of the kinetic matrix $\mathcal{N}$ contains the transformation $\mathcal{N}\rightarrow -\frac{1}{\mathcal{N}}$ that is it exchanges the weak and strong coupling regimes of the theory. One may then think of such

a quantum field theory as being described by a collection of local Lagrangians, each defined in a local patch. They are all equivalent once one defines for each of them what is *electric* and what is *magnetic*. Duality transformations map this set of Lagrangians one into the other.

At this point we observe that the supergravity bosonic Lagrangian (32) is exactly of the form considered in this section as far as the matter content is concerned, so that we may apply the above considerations about duality rotations to the supergravity case. In particular, the U-duality acts in all theories with $N \geq 2$ supersymmetries, where the vector supermultiplets contain both vectors and scalars. For $N = 1$ supergravity, instead, vectors and scalars are still present but they are not related by supersymmetry, and as a consequence they are not related by U-duality rotations, so that the previous formalism does not necessarily apply. [5] In the next subsection we will discuss in a geometric framework the structure of the supergravity theories for $N \geq 2$. In particular, for theories whose σ-model is a coset space (which includes all theories with $N > 2$) we will give the expression for the kinetic vector matrix $\mathcal{N}_{\Lambda\Sigma}$ in terms of the $Sp(2n_V)$ coset representatives embedding the U-duality group. Furthermore we will show that in the $N = 2$ case, although the σ-model of the scalars is not in general a coset space, yet it may be treated in a completely analogous way.

3.2 Duality Symmetries and Central Charges

Let us restrict our attention to N-extended supersymmetric theories coupled to the gravitational field, that is to supergravity theories, whose bosonic action has been given in (32). For each theory we are going to analyze the group theoretical structure and to find the expression of the central charges, together with the properties they obey. As already mentioned, with the exception of the $N = 1$ and $N = 2$ cases, all supergravity theories in four dimensions contain scalar fields whose kinetic Lagrangians are described by σ-models of the form G/H (we have summarized these cases in Table 4). We will first examine the theories with $N > 2$, extending then the results to the $N = 2$ case. Here and in the following, G denotes a non compact group acting as isometry group on the scalar manifold while H, the isotropy subgroup, is of the form

$$H = H_{Aut} \otimes H_{matter}, \tag{59}$$

H_{Aut} being the automorphism group of the supersymmetry algebra while H_{matter} is related to the matter multiplets. (Of course $H_{matter} = \mathbb{1}$ in all cases where supersymmetric matter does not exist, namely $N > 4$).

We will see that in all the theories the fields are in some representation of the isometry group G of the scalar fields or of its maximal compact subgroup

[5] There are however $N = 1$ models where the scalar moduli space is given by a special Kähler manifold. This is the case for example for the compactification of the heterotic theory on Calabi–Yau manifolds

H. This is just a consequence of the Gaillard–Zumino duality acting on the two-form field strengths and their duals, discussed in the preceding section.

The scalar manifolds and the automorphism groups of supergravity theories for any D and N can be found in the literature (see for instance [47, 48, 49, 50]). As it was discussed in the previous section, the group G acts linearly in a symplectic representation on the electric and magnetic field strengths appearing in the gravitational and matter multiplets. Here and in the following the index Λ runs over the dimensions of some representation of the duality group G. Since consistency of the quantum theory requires the electric and magnetic charges to satisfy a quantization condition, the true duality symmetry at the quantum level (U-duality), acting on quantized charges, is a suitable discrete version of the continuous group G [10]. The moduli space of these theories is $G(\mathbb{Z})\backslash G/H$.

All the properties of the given supergravity theories for $N \geq 3$ are completely fixed in terms of the geometry of G/H, namely in terms of the coset representatives L satisfying the relation

$$L(\Phi') = gL(\Phi)h(g,\Phi), \tag{60}$$

where $g \in G$, $h \in H$ and $\Phi' = \Phi'(\Phi)$, Φ being the coordinates of G/H. Note that the scalar fields in G/H can be assigned, in the linearized theory, to linear representations R_H of the local isotropy group H so that dim $R_H =$ dim $G -$ dim H (in the full theory, R_H is the representation which the vielbein of G/H belongs to).

With any field-strength F^Λ we may associate a magnetic charge p^Λ and an electric charge q_Λ given, respectively, by

$$p^\Lambda = \frac{1}{4\pi}\int_{S^2} F^\Lambda, \qquad q_\Lambda = \frac{1}{4\pi}\int_{S^2} G_\Lambda\,, \tag{61}$$

where S^2 is a spatial two-sphere in the space–time geometry of the dyonic solution (for instance, in Minkowski space–time the two-sphere at radial infinity S^2_∞). Clearly the presence of dyonic solutions requires the Maxwell equations (38) to be completed by adding corresponding electric and magnetic currents on the right-hand side. These charges however are not the physical charges of the *interacting theory*; these latter can be computed by looking at the transformation laws of the fermion fields, where the physical field strengths appear dressed with the scalar fields [51, 50]. It is in terms of these interacting dressed field strengths that the field theory realization of the central charges occurring in the supersymmetry algebra (12) is given. Indeed, let us first introduce the central charges: they are associated with the dressed two-form T_{AB} appearing in the supersymmetry transformation law of the gravitino one-form. The physical graviphoton may be identified from the supersymmetry transformation law of the gravitino field in the interacting theory, namely:

$$\delta\psi_A = \nabla\epsilon_A + \alpha T_{AB|\mu\nu}\gamma^a\gamma^{\mu\nu}\epsilon^B V_a + \cdots\,. \tag{62}$$

Here ∇ is the covariant derivative in terms of the space–time spin connection and the composite connection of the automorphism group H_{Aut}, α is a coefficient fixed by supersymmetry, V^a is the space–time vielbein, $A = 1, \cdots, N$ is the index acted on by the automorphism group. Here and in the following the dots denote trilinear fermion terms which are characteristic of any supersymmetric theory but do not play any role in the following discussion. The two-form field strength T_{AB} will be constructed by dressing the bare field strengths F^{Λ} with the coset representative $L(\Phi)$ of G/H, Φ denoting a set of coordinates of G/H.

Note that the same field strength T_{AB} which appears in the gravitino transformation law is also present in the dilatino transformation law in the following way:

$$\delta\chi_{ABC} = P_{ABCD,\ell}\partial_\mu\phi^\ell\gamma^\mu\epsilon^D + \beta T_{[AB|\mu\nu}\gamma^{\mu\nu}\epsilon_{C]} + \cdots . \tag{63}$$

Analogously, when vector multiplets are present, the matter vector field strengths T_I appearing in the transformation laws of the gaugino fields, which are named matter vector field strengths, are linear combinations of the field strengths dressed with a different combination of the scalars:

$$\delta\lambda_{IA} = \mathrm{i}P_{IAB,i}\partial_\mu\Phi^i\gamma^\mu\epsilon^B + \gamma T_{I|\mu\nu}\gamma^{\mu\nu}\epsilon_A + \cdots . \tag{64}$$

Here $P_{ABCD} = P_{ABCD,\ell}d\phi^\ell$ and $P^I_{AB} = P^I_{AB,i}d\Phi^i$ are the vielbein of the scalar manifolds spanned by the scalar fields of the gravitational and vector multiplets, respectively (more precise definitions are given below), and β and γ are constants fixed by supersymmetry.

In order to give the explicit dependence on scalars of T_{AB}, T^I, it is necessary to recall from the previous subsection that, according to the Gaillard–Zumino construction, the isometry group G of the scalar manifold acts on the vector $(F^{-\Lambda}, G^-_\Lambda)$ (or its complex conjugate) as a subgroup of $Sp(2n_V, \mathbb{R})$ (n_V is the number of vector fields) with duality transformations interchanging electric and magnetic field strengths:

$$\mathcal{S}\begin{pmatrix} F^{-\Lambda} \\ G^-_\Lambda \end{pmatrix} = \begin{pmatrix} F^{-\Lambda} \\ G^-_\Lambda \end{pmatrix}' . \tag{65}$$

Let now $L(\Phi)$ be the coset representative of G in the symplectic representation, namely as a $2n_V \times 2n_V$ matrix belonging to $Sp(2n_V, \mathbb{R})$ and therefore, in each theory, it can be described in terms of $n_V \times n_V$ blocks A_L, B_L, C_L, D_L satisfying the same relations (55) as the corresponding blocks of the generic symplectic transformation $\mathcal{S}$.

Since the fermions of supergravity theories transform in a complex representation of the R-symmetry group $H_{Aut} \subset G$, it is useful to introduce a complex basis in the vector space of $Sp(2n_V, \mathbb{R})$, defined by the action of following unitary matrix:[6]

[6] We adopt here and in the following a condensed notation where $\mathbb{1}$ denotes the n_V dimensional identity matrix $\mathbb{1}^M_N = \delta^M_N$. In supergravity calculations, the index

$$\mathcal{A} = \frac{1}{\sqrt{2}} \begin{pmatrix} 1\!\!1 & i\,1\!\!1 \\ 1\!\!1 & -i\,1\!\!1 \end{pmatrix},$$

and to introduce a new matrix $\mathbf{V}(\Phi)$ obtained by complexifying the right index of the coset representative $L(\Phi)$, so as to make its transformation properties under right action of H manifest:

$$\mathbf{V}(\Phi) = \begin{pmatrix} \mathbf{f} & \bar{\mathbf{f}} \\ \mathbf{h} & \bar{\mathbf{h}} \end{pmatrix} = L(\Phi)\mathcal{A}^\dagger, \tag{66}$$

where

$$\mathbf{f} = \frac{1}{\sqrt{2}}(A_L - \mathrm{i}B_L) \; ; \; \mathbf{h} = \frac{1}{\sqrt{2}}(C_L - \mathrm{i}D_L),$$

From the properties of $L(\Phi)$ as a symplectic matrix, it is easy to derive the following properties for $\mathbf{V}$:

$$\mathbf{V}\,\eta\,\mathbf{V}^\dagger = -\mathrm{i}\mathbb{C} \;\; ; \;\; \mathbf{V}^\dagger\,\mathbb{C}\,\mathbf{V} = \mathrm{i}\eta, \tag{67}$$

where the symplectic invariant matrix $\mathbb{C}$ and η are defined as follows:

$$\mathbb{C} = \begin{pmatrix} 0 & -1\!\!1 \\ 1\!\!1 & 0 \end{pmatrix} \; ; \; \eta = \begin{pmatrix} 1\!\!1 & 0 \\ 0 & -1\!\!1 \end{pmatrix}, \tag{68}$$

and, as usual, each block is an $n_V \times n_V$ matrix. The above relations imply on the matrices $\mathbf{f}$ and $\mathbf{h}$ the following properties:

$$\begin{cases} \mathrm{i}(\mathbf{f}^\dagger\mathbf{h} - \mathbf{h}^\dagger\mathbf{f}) = 1\!\!1 \\ (\mathbf{f}^t\mathbf{h} - \mathbf{h}^t\mathbf{f}) \;\; = 0 \end{cases} \tag{69}$$

The $n_V \times n_V$ blocks $\mathbf{f}$, $\mathbf{h}$ of $\mathbf{V}$ can be decomposed with respect to the isotropy group $H_{Aut} \times H_{matter}$ as

$$\begin{aligned} \mathbf{f} &= (f^\Lambda_{AB}, \bar{f}^\Lambda_{\bar{I}}) \equiv (\mathbf{f}^\Lambda{}_M), \\ \mathbf{h} &= (h_{\Lambda AB}, \bar{h}_{\Lambda\bar{I}}) \equiv (\mathbf{h}_{\Lambda M}), \end{aligned} \tag{70}$$

where AB are indices in the antisymmetric representation of $H_{Aut} = SU(N) \times U(1)$, I is an index of the fundamental representation of H_{matter} and $M = (AB, \bar{I})$. Upper $SU(N)$ indices label objects in the complex conjugate representation of $SU(N)$: $(f^\Lambda_{AB})^* = \bar{f}^{\Lambda AB}$, $(f^\Lambda_I)^* = \bar{f}^\Lambda_{\bar{I}} = \bar{f}^{\Lambda I}$, etc.

Let us remark that, in order to make contact with the notation used for the $N = 2$ case, in the definition (70) some of the entries ($\bar{f}^\Lambda_{\bar{I}}$ and $\bar{h}_{\Lambda\bar{I}}$)

M is often decomposed as $M = (AB, I)$, $AB = -BA$ labeling the two-time antisymmetric representation of the R-symmetry group H_{Aut} and I running over the H_{matter} representation of the matter fields. We use the convention that the sum over the antisymmetric couple AB be free and therefore supplemented by a factor 1/2 in order to avoid repetitions. In particular with these conventions, when restricted to the AB indices, the identity reads: $1\!\!1^{AB}_{CD} \equiv 2\,\delta^{AB}_{CD} = \delta^A_C\delta^B_D - \delta^A_D\delta^B_C$.

have been written as complex conjugates of other quantities (f_I^Λ and $h_{\Lambda I}$ respectively). In this way, f_{AB}^Λ and $\bar{f}_{\bar{I}}^\Lambda$ are characterized by having Kähler weight of the same sign. Indeed, for all the matter coupled theories ($N = 2, 3, 4$) we have, as a general feature, that the entries of the blocks $\mathbf{f}$ and $\mathbf{h}$ carrying H_{matter} indices have a Kähler weight with an opposite sign with respect to the corresponding entries with H_{Aut} indices. This may be seen from the supersymmetry transformation rules of the supergravity fields, in virtue the fact that gravitinos and gauginos with the same chirality have opposite Kähler weight. We note however that this notation differs from the one in previous papers, where the upper and lower parts of the symplectic section were defined instead as $(f_{AB}^\Lambda, f_I^\Lambda)$, $(h_{\Lambda\,AB}, h_{\Lambda\,I})$.

It is useful to introduce the following quantities:

$$\begin{aligned}
&\mathbf{V}_M = (V_{AB}, \overline{V}_{\bar{I}}) , \quad \text{where:} \\
&V_{AB} \equiv (f_{AB}^\Lambda, h_{\Lambda AB}) \ ; \ \ V_I \equiv (f_I^\Lambda, h_{\Lambda I}) .
\end{aligned} \tag{71}$$

The vectors $\mathbf{V}_M$ are (complex) symplectic sections of a $Sp(2n_V, \mathbb{R})$ bundle over G/H. As anticipated in the previous subsection, this bundle is actually flat. The real embedding given by $L(\Phi)$ is appropriate for duality transformations of $F^\pm$ and their duals $G^\pm$, according to (46), while the complex embedding in the matrix $\mathbf{V}$ is appropriate in writing down the fermion transformation laws and supercovariant field strengths. The kinetic matrix $\mathcal{N}$, according to Gaillard–Zumino [39], can be written in terms of the sub-blocks $\mathbf{f}$, $\mathbf{h}$, and turns out to be

$$\mathcal{N} = \mathbf{h}\,\mathbf{f}^{-1}, \qquad \mathcal{N} = \mathcal{N}^t , \tag{72}$$

transforming projectively under $Sp(2n_V, \mathbb{R})$ duality rotations as already shown in the previous section. By using (69)and (72) we find that

$$(\mathbf{f}^t)^{-1} = \mathrm{i}(\mathcal{N} - \bar{\mathcal{N}})\bar{\mathbf{f}} , \tag{73}$$

that is

$$(\mathbf{f}^{-1})^{AB}{}_\Lambda = \mathrm{i}(\mathcal{N} - \bar{\mathcal{N}})_{\Lambda\Sigma} \bar{f}^{\Sigma\,AB} , \tag{74}$$

$$(\mathbf{f}^{-1})^{\bar{I}}{}_\Lambda = \mathrm{i}(\mathcal{N} - \bar{\mathcal{N}})_{\Lambda\Sigma} f^{\Sigma\,\bar{I}} . \tag{75}$$

It can be shown [50] that the dressed graviphotons and matter self-dual field strengths appearing in the transformation law of gravitino (62), dilatino (63) and gaugino (64) can be constructed as a symplectic invariant using the $\mathbf{f}$ and $\mathbf{h}$ matrices as follows:

$$\begin{aligned}
T_{AB}^- &= -\mathrm{i}(\bar{\mathbf{f}}^{-1})_{AB\Lambda} F^{-\Lambda} = f_{AB}^\Lambda\, (\mathcal{N} - \bar{\mathcal{N}})_{\Lambda\Sigma}\, F^{-\Sigma} = h_{\Lambda AB}\, F^{-\Lambda} - f_{AB}^\Lambda\, G_\Lambda^- , \\
\bar{T}_{\bar{I}}^- &= -\mathrm{i}(\bar{\mathbf{f}}^{-1})_{\bar{I}\Lambda}\, F^{-\Lambda} = \bar{f}_{\bar{I}}^\Lambda\, (\mathcal{N} - \bar{\mathcal{N}})_{\Lambda\Sigma}\, F^{-\Sigma} = \bar{h}_{\Lambda\bar{I}}\, F^{-\Lambda} - \bar{f}_{\bar{I}}^\Lambda\, G_\Lambda^- , \\
\bar{T}^{+AB} &= (T_{AB}^-)^* , \\
T_I^+ &= (\bar{T}_{\bar{I}}^-)^* ,
\end{aligned} \tag{76}$$

(for $N > 4$, supersymmetry does not allow matter multiplets and $f_I^\Lambda = 0 = T_I$). To construct the dressed charges one integrates $T_{AB} = T_{AB}^+ + T_{AB}^-$ and (for $N = 3,4$) $\bar{T}_{\bar{I}} = \bar{T}^+_{\bar{I}} + \bar{T}^-_{\bar{I}}$ on a large two-sphere. For this purpose we note that

$$T^+_{AB} = h_{\Lambda AB}\, F^{+\Lambda} - f^\Lambda_{AB}\, G^+_\Lambda = 0\,, \tag{77}$$
$$\bar{T}^+_{\bar{I}} = \bar{h}_{\Lambda \bar{I}}\, F^{+\Lambda} - \bar{f}^\Lambda_{\bar{I}}\, G^+_\Lambda = 0\,, \tag{78}$$

as a consequence of (72) and (45). Therefore, we can introduce the central and matter charges as the dressed charges obtained by integrating the two forms T_{AB} and $\bar{T}_{\bar{I}}$:

$$\begin{aligned} Z_{AB} &= -\frac{1}{4\pi}\int_{S^2} T_{AB} = -\frac{1}{4\pi}\int_{S^2}(T^+_{AB} + T^-_{AB}) = -\frac{1}{4\pi}\int_{S^2} T^-_{AB} \\ &= f^\Lambda_{AB}\, q_\Lambda - h_{\Lambda AB}\, p^\Lambda\,, \end{aligned} \tag{79}$$
$$\begin{aligned} \overline{Z}_{\bar{I}} &= -\frac{1}{4\pi}\int_{S^2} \bar{T}_{\bar{I}} = -\frac{1}{4\pi}\int_{S^2}(\bar{T}^+_{\bar{I}} + \bar{T}^-_{\bar{I}}) = -\frac{1}{4\pi}\int_{S^2} \bar{T}^-_{\bar{I}} \\ &= \bar{f}^\Lambda_{\bar{I}}\, q_\Lambda - \bar{h}_{\Lambda \bar{I}}\, p^\Lambda \quad (N \le 4)\,, \end{aligned} \tag{80}$$

where p^Λ and q_Λ were defined in (61) and the sections (f^Λ, h_Λ) on the right-hand side now depend on the v.e.v.'s $\Phi_\infty \equiv \Phi(r=\infty)$ of the scalar fields Φ^r. We see that because of the electric–magnetic duality, the central and matter charges are given in this case by symplectic-invariant expressions.

The scalar field-dependent combinations of fields strengths appearing in the fermion supersymmetry transformation rules have a profound meaning and, as we are going to see in the following, they play a key role in the physics of extremal black holes. The integral of the graviphoton $T_{AB\mu\nu}$ gives the value of the central charge Z_{AB} of the supersymmetry algebra, while by integration of the matter field strengths $T_{I|\mu\nu}$ one obtains the so-called matter charges Z_I.

We are now able to derive some differential relations among the central and matter charges using the Maurer–Cartan equations obeyed by the scalars through the embedded coset representative $\mathbf{V}$. Indeed, let $\Gamma = \mathbf{V}^{-1} d\mathbf{V}$ be the $Sp(2n_V, \mathbb{R})$ Lie algebra left invariant one form satisfying

$$d\Gamma + \Gamma \wedge \Gamma = 0\,. \tag{81}$$

In terms of $(\mathbf{f}, \mathbf{h})$, Γ has the following form:

$$\Gamma \equiv \mathbf{V}^{-1} d\mathbf{V} = \begin{pmatrix} \mathrm{i}(\mathbf{f}^\dagger d\mathbf{h} - \mathbf{h}^\dagger d\mathbf{f}) & \mathrm{i}(\mathbf{f}^\dagger d\bar{\mathbf{h}} - \mathbf{h}^\dagger d\bar{\mathbf{f}}) \\ -\mathrm{i}(\mathbf{f}^t d\mathbf{h} - \mathbf{h}^t d\mathbf{f}) & -\mathrm{i}(\mathbf{f}^t d\bar{\mathbf{h}} - \mathbf{h}^t d\bar{\mathbf{f}}) \end{pmatrix} \equiv \begin{pmatrix} \Omega^{(H)} & \bar{\mathcal{P}} \\ \mathcal{P} & \bar{\Omega}^{(H)} \end{pmatrix}, \tag{82}$$

where the $n_V \times n_V$ sub-blocks $\Omega^{(H)}$ and $\mathcal{P}$ embed the H connection and the vielbein of G/H, respectively. This identification follows from the Cartan decomposition of the $Sp(2n_V, \mathbb{R})$ Lie algebra.

From (66) and (82), we obtain the $(n_V \times n_V)$ matrix equation:

$$D(\Omega)\mathbf{f} = \bar{\mathbf{f}}\,\mathcal{P}\,,$$
$$D(\Omega)\mathbf{h} = \bar{\mathbf{h}}\,\mathcal{P}\,, \tag{83}$$

together with their complex conjugates. Explicitly, if we define the $H_{Aut} \times H_{matter}$-covariant derivative of the $\mathbf{V}_M$ vectors, introduced in (71), as

$$D\mathbf{V}_M = d\mathbf{V}_M - \mathbf{V}_N\,\omega^N{}_M\,, \quad \omega = \begin{pmatrix} \omega^{AB}{}_{CD} & 0 \\ 0 & \omega^I{}_J \end{pmatrix}\,, \tag{84}$$

we have

$$\Omega^{(H)} = \mathrm{i}[\mathbf{f}^\dagger(D\mathbf{h} + \mathbf{h}\omega) - \mathbf{h}^\dagger(D\mathbf{f} + \mathbf{f}\omega)] = \omega\mathbb{1}\,, \tag{85}$$

where we have used

$$D\mathbf{h} = \bar{\mathcal{N}} D\mathbf{f}; \quad \mathbf{h} = \mathcal{N}\mathbf{f}\,, \tag{86}$$

which follow from (83) and the fundamental identity (69). Furthermore, using the same relations, the embedded vielbein $\mathcal{P}$ can be written as follows:

$$\mathcal{P} = -\mathrm{i}(\mathbf{f}^t D\mathbf{h} - \mathbf{h}^t D\mathbf{f}) = \mathrm{i}\mathbf{f}^t(\mathcal{N} - \bar{\mathcal{N}})D\mathbf{f}\,. \tag{87}$$

Using further the definition (70) we have

$$D(\omega)f^\Lambda_{AB} = f^\Lambda_I P^I_{AB} + \frac{1}{2}\bar{f}^{\Lambda CD} P_{ABCD}\,,$$
$$D(\omega)\bar{f}^\Lambda_{\bar{I}} = \frac{1}{2}\bar{f}^{\Lambda AB} P_{AB\bar{I}} + f^{\Lambda\bar{J}} P_{\bar{J}\bar{I}}\,, \tag{88}$$

where we have decomposed the embedded vielbein $\mathcal{P}$ as follows:

$$\mathcal{P} = \begin{pmatrix} P_{ABCD} & P_{AB\bar{J}} \\ P_{\bar{I}CD} & P_{\bar{I}\bar{J}} \end{pmatrix}\,, \tag{89}$$

the sub-blocks being related to the vielbein of G/H, written in terms of the indices of $H_{Aut} \times H_{matter}$. In particular, the component P_{ABCD} is completely antisymmetric in its indices. Note that, since $\mathbf{f}$ belongs to the unitary matrix $\mathbf{V}$, we have: $\overline{\mathbf{V}}^M = (f^\Lambda_{AB}, \bar{f}^\Lambda_{\bar{I}})^\star = (\bar{f}^{\Lambda AB}, f^{\Lambda\bar{I}})$. Obviously, the same differential relations that we wrote for $\mathbf{f}$ hold true for the dual matrix $\mathbf{h}$ as well.

Using the definition of the charges (79) and (80), we then get the following differential relations among charges:[7]

$$D(\omega)Z_{AB} = Z_I P^I_{AB} + \frac{1}{2}\bar{Z}^{CD} P_{ABCD}\,,$$
$$D(\omega)\bar{Z}_{\bar{I}} = \frac{1}{2}\bar{Z}^{AB} P_{AB\bar{I}} + Z^{\bar{J}} P_{\bar{I}\bar{J}}\,. \tag{90}$$

[7] Here we are using for the matter charges a different notation with respect to [50], for instance, in that the quantities Z_I correspond in [50] to $\bar{Z}^I$.

Depending on the coset manifold, some of the sub-blocks of (89) can be actually zero. For example, in $N=3$, the vielbein of $G/H = \frac{SU(3,n)}{SU(3)\times SU(n)\times U(1)}$ [52] is $P_{\bar{I}AB}$ (AB antisymmetric), $I=1,\cdots,n; A,B=1,2,3$ and it turns out that $P_{ABCD} = P_{\bar{I}\bar{J}} = 0$.

In $N=4$, $G/H = \frac{SU(1,1)}{U(1)} \times \frac{O(6,n)}{O(6)\times O(n)}$ [53], and we have $P_{ABCD} = \epsilon_{ABCD}P$, $P_{\bar{I}\bar{J}} = P\delta_{IJ}$, where P is the Kählerian vielbein of $\frac{SU(1,1)}{U(1)}$ ($A,\cdots,D$ $SU(4)$ indices and I,J $O(n)$ indices) and P_{IAB} is the vielbein of $\frac{O(6,n)}{O(6)\times O(n)}$.

For $N>4$ (no matter indices) we have that $\mathcal{P}$ coincides with the vielbein P_{ABCD} of the relevant G/H.

For the purpose of comparison of the previous formalism with the $N=2$ supergravity case, where the σ-model is in general not a coset, it is interesting to note that, if the connection $\Omega^{(H)}$ and the vielbein $\mathcal{P}$ are regarded as data of G/H, then the Maurer–Cartan equations (88) can be interpreted as an integrable system of differential equations for the section (V_{AB}, $\bar{V}_{\bar{I}}$, $\bar{V}^{AB}$, $V^{\bar{I}}$) of the symplectic fiber bundle constructed over G/H. Namely the integrable system (84) that we explicitly write in the following equivalent matrix form:

$$D\begin{pmatrix} V_{AB} \\ \bar{V}_{\bar{I}} \\ \bar{V}^{AB} \\ V^{\bar{I}} \end{pmatrix} = \begin{pmatrix} 0 & 0 & \frac{1}{2}P_{ABCD} & P_{AB\bar{J}} \\ 0 & 0 & \frac{1}{2}P_{\bar{I}CD} & P_{\bar{I}\bar{J}} \\ \frac{1}{2}\bar{P}^{ABCD} & \bar{P}^{AB\bar{J}} & 0 & 0 \\ \frac{1}{2}\bar{P}^{\bar{I}CD} & \bar{P}^{\bar{I}\bar{J}} & 0 & 0 \end{pmatrix} \begin{pmatrix} V_{CD} \\ \bar{V}_{\bar{J}} \\ \bar{V}^{CD} \\ V^{\bar{J}} \end{pmatrix}, \tag{91}$$

has $2n_V$ solutions given by $\mathbf{V}_M$. The integrability condition (81) means that Γ is a flat connection of the symplectic bundle. In terms of the geometry of G/H this in turn implies that the $\mathbb{H}$-curvature associated to the connection $\Omega^{(H)}$ (and hence, since the manifold is a symmetric space, also the Riemannian curvature) is constant, being proportional to the wedge product of two vielbein.

Furthermore, besides the differential relations (90) the charges also satisfy sum rules.

The sum rule has the following form:

$$\frac{1}{2}Z_{AB}\bar{Z}^{AB} + Z_I\bar{Z}^I = -\frac{1}{2}Q^t\mathcal{M}(\mathcal{N})Q\,, \tag{92}$$

where $\mathbb{C}$ is the symplectic metric while $\mathcal{M}(\mathcal{N})$ and Q are

$$\begin{aligned}\mathcal{M}(\mathcal{N}) &= \begin{pmatrix} \mathbb{1} & -\Re\mathcal{N} \\ 0 & \mathbb{1} \end{pmatrix} \cdot \begin{pmatrix} \Im\mathcal{N} & 0 \\ 0 & \Im\mathcal{N}^{-1} \end{pmatrix} \cdot \begin{pmatrix} \mathbb{1} & 0 \\ -\Re\mathcal{N} & \mathbb{1} \end{pmatrix} \\ &= \begin{pmatrix} \Im\mathcal{N} + \Re\mathcal{N}\Im\mathcal{N}^{-1}\Re\mathcal{N} & -\Re\mathcal{N}\Im\mathcal{N}^{-1} \\ -\Im\mathcal{N}^{-1}\Re\mathcal{N} & \Im\mathcal{N}^{-1} \end{pmatrix} = \mathbb{C}\,\mathbf{V}\,\mathbf{V}^\dagger\,\mathbb{C}\,,\end{aligned} \tag{93}$$

and

$$Q = \begin{pmatrix} p^\Lambda \\ q_\Lambda \end{pmatrix} . \tag{94}$$

This result is obtained from the fundamental identities (69) and from the definition of $\mathbf{V}$ and of the kinetic matrix given in (66) and (72). Indeed one can verify that [50, 54]:

$$\begin{aligned} \mathbf{f}\,\mathbf{f}^\dagger &= -\mathrm{i}\left(\mathcal{N} - \bar{\mathcal{N}}\right)^{-1} , \\ \mathbf{h}\,\mathbf{h}^\dagger &= -\mathrm{i}\left(\bar{\mathcal{N}}^{-1} - \mathcal{N}^{-1}\right)^{-1} \equiv -\mathrm{i}\mathcal{N}\left(\mathcal{N} - \bar{\mathcal{N}}\right)^{-1}\bar{\mathcal{N}} , \\ \mathbf{h}\,\mathbf{f}^\dagger &= \mathcal{N}\mathbf{f}\,\mathbf{f}^\dagger , \\ \mathbf{f}\,\mathbf{h}^\dagger &= \mathbf{f}\,\mathbf{f}^\dagger\bar{\mathcal{N}} , \end{aligned} \tag{95}$$

so that, using the explicit expression for the charges in (79) and (80), (92) is easily retrieved.

In the following, studying the applications of these formulas to extremal black holes, other relations coming from the same identities listed above will also be useful, in particular:

$$\begin{aligned} \frac{1}{2}\left(\mathcal{M} + \mathrm{i}\mathbb{C}\right) &= \begin{pmatrix} -\mathbf{h}\,\mathbf{h}^\dagger & \mathbf{h}\,\mathbf{f}^\dagger \\ \mathbf{f}\,\mathbf{h}^\dagger & -\mathbf{f}\,\mathbf{f}^\dagger \end{pmatrix} = \frac{1}{2}\mathbb{C}\,\mathbf{V}\left(\mathbb{1} + \eta\right)\mathbf{V}^\dagger\,\mathbb{C} \\ &= -(\mathbb{C}\,\mathbf{V})_M\,(\mathbb{C}\,\bar{\mathbf{V}})^M , \end{aligned} \tag{96}$$

$$\frac{1}{2}\left(\mathcal{M} + \mathrm{i}\mathbb{C}\right)\mathbf{V}_M = \mathrm{i}\,\mathbb{C}\,\mathbf{V}_M , \tag{97}$$

$$\frac{1}{2}\left(\mathcal{M} - \mathrm{i}\mathbb{C}\right)\mathbf{V}_M = 0 , \tag{98}$$

$$\mathcal{M}\,Q = \mathbb{C}\,\mathbf{V}\,\mathbf{V}^\dagger\,\mathbb{C}\,Q = -2\,Re\left(\mathbb{C}\,\mathbf{V}_M < Q, \bar{\mathbf{V}}^M >\right) , \tag{99}$$

$$\mathbb{C}\,Q = -\mathrm{i}\,\mathbb{C}\,\mathbf{V}\,\eta\,\mathbf{V}^\dagger\mathbb{C}\,Q = -2\,Im\left(\mathbb{C}\,\mathbf{V}_M < Q, \bar{\mathbf{V}}^M >\right) . \tag{100}$$

The symplectic scalar product appearing in (99) and (100) is defined as

$$< V, W > \equiv V^t\,\mathbb{C}\,W , \tag{101}$$

moreover $\bar{\mathbf{V}}^M = (\mathbf{V}_M)^*$. Using (71), (79), and (80) we can use the following short-hand notation for the central charge vector:

$$Z_M = (Z_{AB}, \bar{Z}_{\bar{I}}) =< Q, \mathbf{V}_M > . \tag{102}$$

From the above expression and from (96), (92) follows.

3.3 The $N = 2$ Theory

The formalism we have developed so far for the $D = 4$, $N > 2$ theories is completely determined by the embedding of the coset representative of G/H in $Sp(2n, \mathbb{R})$ and by the embedded Maurer–Cartan equations (88). We want now to show that this formalism, and in particular the identities (69), the differential relations among charges (90) and the sum rules (92) of $N = 2$ matter-coupled supergravity [55, 56] can be obtained in a way completely analogous to the $N > 2$ cases discussed in the previous subsection, where the σ-model was a coset space. This follows essentially from the fact that, though the scalar manifold $\mathcal{M}_{scalar}$ of the $N = 2$ theory is not in general a coset manifold, nevertheless it has a symplectic structure identical to the $N > 2$ theories, as a consequence of the Gaillard–Zumino duality.

In the case of $N = 2$ supergravity, the requirements imposed by supersymmetry on the scalar manifold $\mathcal{M}_{scalar}$ of the theory dictate that it should be the following direct product: $\mathcal{M}_{scalar} = \mathcal{M}^{SK} \otimes \mathcal{M}^{Q}$ where $\mathcal{M}^{SK}$ is a special Kähler manifold of complex dimension n and $\mathcal{M}^{Q}$ a quaternionic manifold of real dimension $4n_H$. Note that n and n_H are, respectively, the number of vector multiplets and hypermultiplets contained in the theory. The direct product structure imposed by supersymmetry precisely reflects the fact that the quaternionic and special Kähler scalars belong to different supermultiplets. In the construction of extremal black holes, it turns out that the hyperscalars are spectators playing no dynamical role. Hence we do not discuss here the hypermultiplets any further and we confine our attention to an $N = 2$ supergravity where the graviton multiplet, containing besides the graviton $g_{\mu\nu}$ also a graviphoton A^0_μ, is coupled to n vector multiplets. Such a theory has an action of type (32) where the number of gauge fields is $n_V = 1 + n$ and the number of (real) scalar fields is $m = 2n$. We shall use capital Greek indices to label the vector fields: $\Lambda, \Sigma \dots = 0, \dots, n$. To make the action (32) fully explicit, we need to discuss the geometry of the manifold $\mathcal{M}^{SK}$ spanned by the vector-multiplet scalars, namely special Kähler geometry. Since $\mathcal{M}^{SK}$ is in particular a complex manifold, we shall describe the corresponding scalars as complex fields: z^i, $\bar{z}^{\bar{\imath}}$, $i, \bar{\imath} = 1, \dots, n$. We refer to [57] for a detailed analysis. A special Kähler manifold $\mathcal{M}^{SK}$ is a Kähler–Hodge manifold endowed with an extra symplectic structure. A Kähler manifold $\mathcal{M}$ is a Hodge manifold if and only if there exists a $U(1)$ bundle $\mathcal{L} \longrightarrow \mathcal{M}$ such that its first Chern class equals the cohomology class of the Kähler two-form K:

$$c_1(\mathcal{L}) = [K]. \tag{103}$$

In local terms we can write

$$K = \mathrm{i}\, g_{i\bar{\jmath}}\, dz^i \wedge d\bar{z}^{\bar{\jmath}}, \tag{104}$$

where z^i are n holomorphic coordinates on $\mathcal{M}^{SK}$ and $g_{i\bar{\jmath}}$ its metric. In this case the $U(1)$ Kähler connection is given by

$$\mathcal{Q} = -\frac{\mathrm{i}}{2}\left(\partial_i \mathcal{K} dz^i - \partial_{\bar{\imath}} \mathcal{K} d\bar{z}^{\bar{\imath}}\right), \tag{105}$$

where $\mathcal{K}$ is the Kähler potential, so that $K = d\mathcal{Q}$.

Let now $\Phi(z, \bar{z})$ be a section of the $U(1)$ bundle of weight p. By definition its covariant derivative is

$$D\Phi = (d + ip\mathcal{Q})\Phi, \tag{106}$$

or, in components,

$$D_i\Phi = (\partial_i + \tfrac{1}{2}p\partial_i\mathcal{K})\Phi \; ; \; D_{\bar{\imath}}\Phi = (\partial_{\bar{\imath}} - \tfrac{1}{2}p\partial_{\bar{\imath}}\mathcal{K})\Phi \, . \tag{107}$$

A covariantly holomorphic section is defined by the equation: $D_{\bar{\imath}}\Phi = 0$. Setting:

$$\tilde{\Phi} = e^{-p\mathcal{K}/2}\Phi \, , \tag{108}$$

we get

$$D_i\tilde{\Phi} = (\partial_i + p\partial_i\mathcal{K})\tilde{\Phi} \; ; \; D_{\bar{\imath}}\tilde{\Phi} = \partial_{\bar{\imath}}\tilde{\Phi} \; , \tag{109}$$

so that under this map covariantly holomorphic sections Φ become truly holomorphic sections.

There are several equivalent ways of defining what a special Kähler manifold is. An intrinsic definition is the following. A special Kähler manifold can be given by constructing a $(2n+2)$-dimensional flat symplectic bundle over the Kähler–Hodge manifold whose generic sections (with weight $p=1$)

$$V = (f^{\Lambda}, h_{\Lambda}) \, , \tag{110}$$

are covariantly holomorphic

$$D_{\bar{\imath}}V = (\partial_{\bar{\imath}} - \frac{1}{2}\partial_{\bar{\imath}}\mathcal{K})V = 0 \, , \tag{111}$$

and satisfy the further condition

$$\mathrm{i} < V, \bar{V} >= \mathrm{i}(\bar{f}^{\Lambda}h_{\Lambda} - \bar{h}_{\Lambda}f^{\Lambda}) = 1 \; , \tag{112}$$

where the $<,>$ product was defined in (101). Defining

$$V_i = D_iV = (f^{\Lambda}_i, h_{\Lambda i}), \tag{113}$$

and introducing a symmetric three-tensor C_{ijk} by

$$D_iV_j = iC_{ijk}g^{k\bar{k}}\bar{V}_{\bar{k}} \; , \tag{114}$$

the set of differential equations

$$\begin{aligned} D_iV &= V_i \, , \\ D_iV_j &= iC_{ijk}g^{k\bar{k}}\bar{V}_{\bar{k}} \, , \\ D_iV_{\bar{\jmath}} &= g_{i\bar{\jmath}}\bar{V} \, , \\ D_i\bar{V} &= 0 \, , \end{aligned} \tag{115}$$

defines a symplectic connection. Requiring that the differential system (115) is integrable is equivalent to requiring that the symplectic connection is flat. Since the integrability condition of (115) gives constraints on the base Kähler–Hodge manifold, we define special-Kähler a manifold whose associated symplectic connection is flat. At the end of this section, we will give the restrictions on the manifold imposed by the flatness of the connection.

It must be noted that, for special Kähler manifolds, the Kähler potential can be computed as a symplectic invariant from (112). Indeed, introducing also the holomorphic sections

$$\begin{aligned} \Omega &= e^{-\mathcal{K}/2} V = e^{-\mathcal{K}/2}(f^\Lambda, h_\Lambda) = (X^\Lambda, F_\Lambda)\,, \\ \partial_{\bar{\imath}}\Omega &= 0\,, \end{aligned} \tag{116}$$

(112) gives

$$\mathcal{K} = -\ln \mathrm{i} < \Omega, \bar{\Omega} >= -\ln \mathrm{i}(\bar{X}^\Lambda F_\Lambda - X^\Lambda \bar{F}_\Lambda)\,. \tag{117}$$

If we introduce the complex symmetric $(n+1)\times(n+1)$ matrix $\mathcal{N}_{\Lambda\Sigma}$ defined through the relations

$$h_\Lambda = \mathcal{N}_{\Lambda\Sigma} f^\Sigma \quad , \quad h_{\Lambda\bar{\imath}} = \mathcal{N}_{\Lambda\Sigma} \bar{f}^\Sigma_{\bar{\imath}}\,, \tag{118}$$

then we have

$$< V, \bar{V} >= (\mathcal{N} - \bar{\mathcal{N}})_{\Lambda\Sigma} f^\Lambda \bar{f}^\Sigma = -\mathrm{i}\,, \tag{119}$$

so that

$$\mathcal{K} = -\ln[\mathrm{i}(\bar{X}^\Lambda\,(\mathcal{N} - \bar{\mathcal{N}})_{\Lambda\Sigma} X^\Sigma)]\,, \tag{120}$$

and

$$g_{i\bar{\jmath}} = -\mathrm{i} < V_i, V_{\bar{\jmath}} >= -2 f_i^\Lambda \mathrm{Im}\mathcal{N}_{\Lambda\Sigma} \bar{f}^\Sigma_{\bar{\jmath}}\,, \tag{121}$$

$$C_{ijk} = < D_i V_j, V_k >= 2\mathrm{i}\mathrm{Im}\mathcal{N}_{\Lambda\Sigma} f_i^\Lambda D_j f_k^\Sigma\,. \tag{122}$$

We shall also use the following identity which follows from the previous ones:

$$f_i^\Lambda\, g^{i\bar{\jmath}}\, \bar{f}^\Sigma_{\bar{\jmath}} = -\frac{1}{2}\,(\mathrm{Im}\mathcal{N})^{-1\,\Lambda\Sigma} - \bar{L}^\Lambda\, L^\Sigma\,. \tag{123}$$

The matrix $\mathcal{N}_{\Lambda\Sigma}$ turns out to be the matrix appearing in the kinetic Lagrangian of the vectors in $N=2$ supergravity. Under coordinate transformations, the sections Ω transform as

$$\tilde{\Omega} = e^{-f_S(z)} \mathcal{S}\Omega\,, \tag{124}$$

where $\mathcal{S} = \begin{pmatrix} A & B \\ C & D \end{pmatrix}$ is an element of $Sp(2n_V, \mathbb{R})$ and the factor $e^{-f_S(z)}$ is a $U(1)$ Kähler transformation. We also note that, from the definition of $\mathcal{N}$, (118):

$$\tilde{\mathcal{N}}(\tilde{X}, \tilde{F}) = [C + D\mathcal{N}(X,F)][A + B\mathcal{N}(X,F)]^{-1}\,. \tag{125}$$

We can now define a matrix $\mathbf{V}$ as in (66) satisfying the relations (67), in terms of the quantities (f^Λ, $\bar{f}^\Lambda{}_{\bar{\imath}}$, h_Λ, $\bar{h}_{\Lambda\bar{\imath}}$) introduced in (110) and (113). In order to identify the blocks $\mathbf{f}$ and $\mathbf{h}$ of $\mathbf{V}$ in (66), we note that in $N=2$ theories $H_{Aut} = SU(2)\times U(1)$, so that the f^Λ_{AB} and $h_{\Lambda\,AB}$ entries in (70) are actually $SU(2)$-singlets. We can therefore consistently write $\mathbf{f}$ and $\mathbf{h}$ as the following $n_V \times n_V$ matrices:

$$\mathbf{f} \equiv \left(f^{\Lambda}_{AB}, \bar{f}^{\Lambda}_{\bar{I}}\right); \quad \mathbf{h} \equiv \left(h_{\Lambda\, AB}, \bar{h}_{\Lambda\, \bar{I}}\right) , \tag{126}$$

where f^{Λ}_{AB}, $h_{\Lambda\, AB}$, and f^{Λ}_{I}, $h_{\Lambda\, I}$ are defined as follows:

$$\begin{aligned} f^{\Lambda}_{AB} &= f^{\Lambda}\, \epsilon_{AB} \;\; ; \;\; h_{\Lambda\, AB} = h_{\Lambda}\, \epsilon_{AB} \, , \\ f^{\Lambda}_{I} &= f^{\Lambda}_{i}\, P^{i}_{I} \;\; ; \;\; h_{\Lambda\, I} = h_{\Lambda\, i}\, P^{i}_{I} \, , \end{aligned} \tag{127}$$

P^{i}_{I}, $P^{\bar{\imath}}_{\bar{I}}$ being the inverse of the Kählerian vielbein P^{I}_{i}, $\bar{P}^{\bar{I}}_{\bar{\imath}}$ defined by the relation:

$$g_{i\bar{\jmath}} = P^{I}_{i}\, \bar{P}^{\bar{J}}_{\bar{\jmath}}\, \eta_{I\bar{J}} \, , \tag{128}$$

and $\eta_{I\bar{J}}$ is the flat metric. From the definition (126) and the properties (119), (121) it is straightforward to verify that the **f** and **h** blocks satisfy the relations (69), or equivalently that the matrix **V** satisfies the conditions (67). The relations (69) therefore encode the set of algebraic relations of special geometry.

Let us now consider the analogous of the embedded Maurer–Cartan equations of G/H. We introduce, as before, the matrix one-form $\Gamma = \mathbf{V}^{-1} d\mathbf{V}$ satisfying the relation $d\Gamma + \Gamma \wedge \Gamma = 0$. We further introduce the covariant derivative of the symplectic section $(f^{\Lambda}, \bar{f}^{\Lambda}_{\bar{I}}, \bar{f}^{\Lambda}, f^{\Lambda}_{I})$ with respect to the $U(1)$-Kähler connection $\mathcal{Q}$ and the spin connection ω^{IJ} of $\mathcal{M}^{SK}$:

$$\begin{aligned} D(f^{\Lambda}, \bar{f}^{\Lambda}_{\bar{I}}, \bar{f}^{\Lambda}, f^{\Lambda}_{I}) &= d(f^{\Lambda}, \bar{f}^{\Lambda}_{\bar{I}}, \bar{f}^{\Lambda}, f^{\Lambda}_{I}) \\ &\quad -(f^{\Lambda}, \bar{f}^{\Lambda}_{\bar{J}}, \bar{f}^{\Lambda}, f^{\Lambda}_{J}) \begin{pmatrix} -\mathrm{i}\mathcal{Q} & 0 & 0 & 0 \\ 0 & \mathrm{i}\mathcal{Q}\delta^{\bar{J}}_{\bar{I}} + \omega^{\bar{J}}{}_{\bar{I}} & 0 & 0 \\ 0 & 0 & \mathrm{i}\mathcal{Q} & 0 \\ 0 & 0 & 0 & -\mathrm{i}\mathcal{Q}\delta^{J}_{I} + \omega^{J}{}_{I} \end{pmatrix} \end{aligned} \tag{129}$$

the Kähler weight of $(f^{\Lambda}, f^{\Lambda}_{I})$ and $(\bar{f}^{\Lambda}, \bar{f}^{\Lambda}_{\bar{I}})$ being $p = 1$ and $p = -1$, respectively. Using the same decomposition as in (82) and (84), (85) we have in the $N = 2$ case:

$$\begin{aligned} \Gamma &= \begin{pmatrix} \Omega & \bar{\mathcal{P}} \\ \mathcal{P} & \bar{\Omega} \end{pmatrix} , \\ \Omega = \omega &= \begin{pmatrix} -\mathrm{i}\mathcal{Q} & 0 \\ 0 & \mathrm{i}\mathcal{Q}\delta^{I}_{J} + \bar{\omega}^{I}{}_{J} \end{pmatrix} . \end{aligned} \tag{130}$$

For the sub-block $\mathcal{P}$ we obtain

$$\mathcal{P} = -\mathrm{i}(f^{t} D h - h^{t} D f) = \mathrm{i} f^{t} (\mathcal{N} - \bar{\mathcal{N}}) D f = \begin{pmatrix} 0 & P_{\bar{I}} \\ P^{J} & P^{J}_{\bar{I}} \end{pmatrix} , \tag{131}$$

where $P^{J} \equiv \eta^{J\bar{I}} P_{\bar{I}}$ is the $(1,0)$-form Kählerian vielbein while

$$P^{J}_{\bar{I}} \equiv \mathrm{i} \left(f^{t} (\mathcal{N} - \bar{\mathcal{N}}) D f\right)^{J}{}_{\bar{I}} \tag{132}$$

is a one-form which in general, in the cases where the manifold is not a coset, represents a new geometric quantity on $\mathcal{M}^{SK}$. Note that we get zero in the first entry of (131) by virtue of the fact that the identity (69) implies $f^{\Lambda}(\mathcal{N}-\bar{\mathcal{N}})_{\Lambda\Sigma}f^{\Sigma}_{\bar{I}}=0$ and that f^{Λ} is covariantly holomorphic. If Ω and $\mathcal{P}$ are considered as data on $\mathcal{M}^{SK}$ then we may interpret $\Gamma=V^{-1}dV$ as an integrable system of differential equations, namely,

$$D(V,\bar{V}_{\bar{I}},\bar{V},V_I)=(V,\bar{V}_{\bar{J}},\bar{V},V_J)\begin{pmatrix}0 & 0 & 0 & \bar{P}_I\\ 0 & 0 & \bar{P}^{\bar{J}} & \bar{P}^{\bar{J}}_{\ I}\\ 0 & P_{\bar{I}} & 0 & 0\\ P^J & P^J_{\ \bar{I}} & 0 & 0\end{pmatrix}, \tag{133}$$

where the flat Kähler indices $I,\bar{I},\cdots$ are raised and lowered with the flat Kähler metric $\eta_{I\bar{J}}$. Note that (133) coincides with the set of relations (115) if we trade world indices $i,\bar{\imath}$ with flat indices $I,\bar{I}$, provided we also identify

$$\bar{P}^{\bar{J}}_{\ I}=\bar{P}^{\bar{J}}_{\ Ik}dz^k=P^{\bar{J},i}P_I^{\ j}C_{ijk}dz^k. \tag{134}$$

Then, the integrability condition $d\Gamma+\Gamma\wedge\Gamma=0$ is equivalent to the flatness of the special Kähler symplectic connection and it gives the following three constraints on the Kähler base manifold

$$d(\mathrm{i}\mathcal{Q})+\bar{P}_I\wedge P^I=0\rightarrow\partial_{\bar{\jmath}}\partial_i\mathcal{K}=P^I_{\ ,i}\bar{P}_{I,\bar{\jmath}}=g_{i\bar{\jmath}}, \tag{135}$$

$$(d\omega+\omega\wedge\omega)^J_{\ \bar{I}}=P_{\bar{I}}\wedge\bar{P}^{\bar{J}}-\mathrm{i}d\mathcal{Q}\delta^{\bar{J}}_{\bar{I}}-\bar{P}^{\bar{J}}_{\ L}\wedge P^L_{\ \bar{I}}, \tag{136}$$

$$DP^J_{\ \bar{I}}=0, \tag{137}$$

$$\bar{P}_J\wedge P^J_{\ \bar{I}}=0. \tag{138}$$

Equation (135) implies that $\mathcal{M}^{SK}$ is a Kähler–Hodge manifold. Equation (136), written with holomorphic and antiholomorphic curved indices, gives

$$R_{\bar{\imath}j\bar{k}l}=g_{\bar{\imath}l}g_{j\bar{k}}+g_{\bar{k}l}g_{\bar{\imath}j}-\bar{C}_{\bar{\imath}\bar{k}\bar{m}}C_{jln}g^{\bar{m}n}, \tag{139}$$

which is the usual constraint on the Riemann tensor of the special geometry. The further special geometry constraints on the three tensor C_{ijk} are then consequences of (137) and (138), which imply

$$\begin{aligned}D_{[l}C_{i]jk}&=0,\\ D_{\bar{l}}C_{ijk}&=0.\end{aligned} \tag{140}$$

In particular, the first of (140) also implies that C_{ijk} is a completely symmetric tensor.

In summary, we have seen that the $N=2$ theory and the higher N theories have essentially the same symplectic structure, the only difference being that since the scalar manifold of $N=2$ is not in general a coset manifold the symplectic structure allows the presence of a new geometric quantity which physically corresponds to the anomalous magnetic moments of the $N=2$ theory. It goes without saying that, when $\mathcal{M}^{SK}$ is itself a coset manifold [58], then the anomalous magnetic moments C_{ijk} must be expressible in terms of the vielbein of G/H.

To complete the analogy between the $N = 2$ theory and the higher N theories in $D = 4$, we also give for completeness the supersymmetry transformation laws, the central and matter charges, the differential relations among them, and the sum rules.

The transformation laws for the chiral gravitino ψ_A and gaugino λ^{iA} fields are

$$\delta\psi_{A\mu} = \nabla_\mu \epsilon_A + \epsilon_{AB} T_{\mu\nu}\gamma^\nu \epsilon^B + \cdots , \tag{141}$$

$$\delta\lambda^{iA} = \mathrm{i}\partial_\mu z^i \gamma^\mu \epsilon^A + \frac{\mathrm{i}}{2} \bar{T}_{\bar{\jmath}\mu\nu}\gamma^{\mu\nu} g^{i\bar{\jmath}} \epsilon^{AB}\epsilon_B + \cdots , \tag{142}$$

where

$$T \equiv h_\Lambda F^\Lambda - f^\Lambda G_\Lambda , \tag{143}$$

$$\bar{T}_{\bar{\imath}} \equiv \bar{T}_{\bar{I}} \bar{P}^{\bar{I}}_{\bar{\imath}} , \text{ with: } \bar{T}_{\bar{I}} \equiv \bar{h}_{\Lambda\bar{I}} F^\Lambda - \bar{f}^\Lambda_{\bar{I}} G_\Lambda , \tag{144}$$

are, respectively, the graviphoton and the matter vectors, and the position of the $SU(2)$ automorphism index A (A,B=1,2) is related to chirality (namely (ψ_A, λ^{iA}) are chiral, $(\psi^A, \lambda^{\bar{\imath}}_A)$ antichiral). In principle only the (anti) self-dual part of F and G should appear in the transformation laws of the (anti)chiral fermi fields; however, exactly as in (77) and (78) for $N > 2$ theories, from (115) it follows that:

$$\begin{aligned} T^+ &= h_\Lambda F^{+\Lambda} - f^\Lambda G^+_\Lambda = 0 , \\ T^-_I &= h_{\Lambda I} F^{-\Lambda} - f^\Lambda_I G^-_\Lambda = 0 , \end{aligned} \tag{145}$$

so that $T = T^-$ and $T_I = T^+_I$ (i.e. $\bar{T} = \bar{T}^+$, $\bar{T}_{\bar{I}} = \bar{T}^-_{\bar{I}}$). Note that both the graviphoton and the matter vectors are symplectic invariant according to the fact that the fermions do not transform under the duality group (except for a possible R-symmetry phase). To define the physical charges let us recall the definition of the moduli-independent charges in (61). The central charges and the matter charges are now defined as the integrals over S^2 of the physical graviphoton and matter vectors

$$\begin{aligned} Z &= -\frac{1}{4\pi}\int_{S^2} T = -\frac{1}{4\pi}\int_{S^2} (h_\Lambda F^\Lambda - f^\Lambda G_\Lambda) = f^\Lambda(z,\bar{z}) q_\Lambda - h_\Lambda(z,\bar{z}) p^\Lambda , \\ Z_I &= -\frac{1}{4\pi}\int_{S^2} T_I = -\frac{1}{4\pi}\int_{S^2} (h_{\Lambda I} F^\Lambda - f^\Lambda_I G_\Lambda) = f^\Lambda_I(z,\bar{z}) q_\Lambda - h_{\Lambda I}(z,\bar{z}) p^\Lambda . \end{aligned} \tag{146}$$

where $z^i, \bar{z}^{\bar{\imath}}$ denote the v.e.v. of the moduli fields in a given background. In virtue of (115) we get immediately:

$$Z_I = P^i_I Z_i \ ; \quad Z_i \equiv D_i Z . \tag{147}$$

As a consequence of the symplectic structure, one can derive two sum rules for Z and Z_I:

$$|Z|^2 + |Z_I|^2 \equiv |Z|^2 + Z_i g^{i\bar{\jmath}} \bar{Z}_{\bar{\jmath}} = -\frac{1}{2} Q^t \mathcal{M} Q \tag{148}$$

where the symmetric matrix $\mathcal{M}$ was defined in (93) and Q is the symplectic vector of electric and magnetic charges defined in (94).

Equation (148) is obtained by using exactly the same procedure as in (92).

4 Supersymmetric Black Holes: General Discussion

We are going to study in this section the peculiarities of extremal black holes that are solutions of extended supergravity theories.

As anticipated in the introduction, for black-hole configurations that are particular bosonic backgrounds of N-extended locally supersymmetric theories, the cosmic censorship conjecture (expressing the request that the space-time singularities are always hidden by event horizons) finds a simple and natural understanding. For the Reissner–Nordstrom black holes this is codified in the bound (4) on the mass M and charge Q of the solution, that we recall here

$$M \geq |Q|. \tag{149}$$

In extended supersymmetric theories this bound is just a consequence of the supersymmetry algebra (21), as a consequence of the fact that

$$\left\{ Q^{(\pm)}_{am} , Q^{\dagger(\pm)}_{am} \right\} \geq 0, \tag{150}$$

so that the cosmic censorship conjecture is always verified.

Another general property of extremal black holes, that will be surveyed in Sect. 5, is encoded in the so-called no-hair theorem. It states that the end point of the gravitational collapse of a black hole is independent of the initial conditions. Then, if one tries to perturb an extremal black hole with whatever additional hair (some slight mass anisotropy, or a long-range field, like a scalar) all these features disappear near the horizon, except for those associated with the conserved quantities of general relativity, namely, for a non-rotating black hole, its mass and charge. When the black hole is embedded in an N-extended supergravity theory, the solution depends in general also on scalar fields. In this case, the electric charge Q has to be replaced by the central charge appearing in the supersymmetry algebra (which is dressed with the expectation value of scalar fields). The black-hole metric takes a generalized form with respect to the Reissner–Nordström one. However, for the extremal case the event horizon looses all information about the scalar 'hair'. As for the Reissner–Nordström case, the near-horizon geometry is still described by a conformally flat, Bertotti–Robinson-type geometry, with a mass parameter $M_{\text{B-R}}$, which only depends on the distribution of charges and not on the scalar fields. As will be discussed extensively in Sect. 5, this follows from the fact that the differential equations on the metric and scalars fields of the extremal black hole (200) and (201) are solved under the condition that the horizon be an attractor point [2] (see (207)). Scalar fields, independently of their boundary conditions at spatial infinity, approaching the horizon flow to a fixed point given by a certain ratio of electric and magnetic charges. Since the dominant contribution to the black-hole entropy is given (at least for large black holes) by the area/entropy Bekenstein–Hawking relation (1), it follows that the entropy of extremal black holes is a topological quantity fixed in terms of the quantized electric and magnetic charges while it does not depend on continuous parameters like scalars.

It will be shown that the request that the scalars Φ^r be regular at the fixed point (reached at the horizon $\tau \to \infty$) implies two important conditions which have both to be satisfied:

$$\left(\frac{d\Phi^r}{d\tau}\right)_{hor} = 0, \tag{151}$$

$$\left(\frac{\partial V_{\text{B–H}}(\Phi)}{\partial \Phi_i}\right)_{hor} = 0. \tag{152}$$

where the function $V_{\text{B–H}}(\Phi, p, q)$, called the black-hole potential, will be introduced in (203).

Exploiting (152), a decade ago a general rule was given [22] for finding the values of fixed scalars, and then the Bekenstein–Hawking entropy, in $N = 2$ theories, through an *extremum principle* in moduli space. This follows from the observation that, when the scalar fields are evaluated at spatial infinity ($\tau = 0$), $V_{\text{B–H}}$ coincides with the squared ADM mass of the black hole. Then, since (152) does not depend explicitly on the radial variable τ (as the extremization is done with respect to the scalar fields at any given point) the expectation values Φ_∞ may be chosen as independent variables. Equation (152) is then reformulated as the statement that the fixed scalars Φ_{fix} are the ones, among all the possible expectation values taken by scalar fields, that extremize the ADM mass of the black hole in moduli space:

$$\Phi_{\text{fix}} : \quad \frac{\partial M_{ADM}(\Phi_\infty)}{\partial \Phi^r_\infty}\Big|_{\Phi_{\text{fix}}} = 0. \tag{153}$$

Correspondingly, the Bekenstein–Hawking entropy is given in terms of that extremum among the possible ADM masses (given by all possible boundary conditions that one can impose on scalars at spatial infinity), this last being identified with the Bertotti–Robinson mass $M_{\text{B–R}}$:

$$M_{\text{B-R}} \equiv M_{ADM}(\Phi_{\text{fix}}). \tag{154}$$

The solutions with the scalar fields constant and everywhere equal to the fixed value Φ_{fix} are called *double extremal black holes.*

The approach outlined above will prove to be a very useful computational tool to calculate the B–H entropy since, as will be explained in Sect. 5, in extended supergravity the explicit dependence of $V_{\text{B–H}}$ on the moduli is given.

4.1 BPS Extremal Black Holes

For the case of BPS extremal black holes, the extremum principle (153) may be explained by means of the Killing spinor equations near the horizon and these are encoded in some relations on the scalars moduli spaces, discussed in detail in Sects. 3.2 and 3.3, which express the embedding of the scalar geometry in a symplectic representation of the U-duality group [59]. For definiteness, to present the argument we will refer, for the sequel of this subsection, to the case $N = 2$, which is the model originally considered in [21, 22].

The Killing spinor equations expressing the existence of unbroken supersymmetries are obtained, for the gauginos in the $N = 2$ case [57], by setting to zero the r.h.s. of (142) that is, using flat indices:

$$\delta\lambda^I_A = P^I_{,i}\partial_\mu z^i\gamma^\mu\varepsilon_{AB}\epsilon^B + \bar{T}^I_{\mu\nu}\gamma^{\mu\nu}\epsilon_A + \cdots = 0. \tag{155}$$

As we will see in detail in the next subsection, approaching the black-hole horizon the scalars z^i reach their fixed values z_{fix}[8] so that

$$\partial_\mu z^i = 0 \tag{156}$$

and (155) is satisfied for

$$T_I = 0, \tag{157}$$

which implies, using integrated quantities:

$$Z_I = Z_i P^i_I = -\frac{1}{4\pi}\int_{S^2} T_I = \left(f^\Lambda_I q_\Lambda - h_{\Lambda I}p^\Lambda\right)|_{\text{fix}} = 0. \tag{158}$$

What we have found is that the Killing spinor equation imposes the vanishing of the matter charges near the horizon. Then, remembering (147), near the horizon we have

$$Z_I = D_I Z = 0 \tag{159}$$

where Z is the central charge appearing in the $N = 2$ supersymmetry algebra, so that:

$$\partial_i|Z| = 0. \tag{160}$$

For an extremal BPS black hole ($|Z| = M_{ADM}$), (160) coincides with (153) giving the fixed scalars $\Phi_{\text{fix}} \equiv z_{\text{fix}}$ at the horizon. We then see that the entropy of the black hole is related to the central charge, namely to the integral of the graviphoton field strength evaluated for very special values of the scalar fields z^i. These special values, the *fixed scalars* z^i_{fix}, are functions solely of the electric and magnetic charges $\{q_\Sigma, p^\Lambda\}$ of the black hole and are attained by the scalars $z^i(r)$ at the black hole horizon $r = 0$.

[8] A point x_{fix} where the phase velocity is vanishing is named *fixed point* and represents the system in equilibrium $v(x_{\text{fix}}) = 0$ [22, 23]. The fixed point is said to be an attractor if $\lim_{t\to\infty} x(t) = x_{\text{fix}}$.

Let us discuss in detail the explicit solution of the Killing spinor equation and the general properties of $N = 2$ BPS-saturated black holes [21, 60, 61, 62]. As our analysis will reveal, these properties are completely encoded in the special Kähler geometric structure of the mother theory.

Let us consider a black-hole ansatz for the metric,[9] restricting the attention to static, spherically symmetric solutions:

$$ds^2 = e^{2U(r)}\, dt^2 - e^{-2U(r)} G_{ij}(r)\, dx^i dx^j; \quad \left(r^2 = G_{ij} x^i x^j\right), \quad i,j = 1,2,3 \tag{161}$$

and for the vector field strengths:

$$F^\Lambda = \frac{p^\Lambda}{2r^3} \epsilon_{abc} x^a dx^b \wedge dx^c - \frac{\ell^\Lambda(r)}{r^3} e^{2U} dt \wedge \boldsymbol{x} \cdot d\boldsymbol{x}. \tag{162}$$

Note that here r parametrizes the distance from the horizon.

It is convenient to rephrase the same ansatz in the complex formalism well-adapted to the $N = 2$ theory. To this effect we begin by constructing a two-form which is anti-self-dual in the background of the metric (161) and whose integral on the two-sphere at infinity S^2_∞ is normalized to 4π. A short calculation yields

$$\begin{aligned} E^- &= \mathrm{i} \frac{e^{2U(r)}}{r^3}\, dt \wedge \boldsymbol{x} \cdot d\boldsymbol{x} + \frac{1}{2} \frac{x^a}{r^3}\, dx^b \wedge dx^c \epsilon_{abc}\,, \\ \int_{S^2_\infty} E^- &= 4\pi, \end{aligned} \tag{163}$$

from which one obtains

$$E^-_{\mu\nu} \gamma^{\mu\nu} = 2\mathrm{i} \frac{e^{2U(r)}}{r^3} \gamma_a x^a \gamma_0 \frac{1}{2} \left[\mathbf{1} + \gamma_5\right], \tag{164}$$

which will simplify the unfolding of the supersymmetry transformation rules. Next, introducing the following complex combination:

$$t^\Lambda(r) = \frac{1}{2} \left(p^\Lambda + \mathrm{i} \ell^\Lambda(r)\right) \tag{165}$$

of the magnetic charges $p^\Lambda = \frac{1}{4\pi} \int_{S^2} F^\Lambda$ and of the functions $\ell^\Lambda(r) = -\frac{1}{4\pi} \int_{S^2} {}^\star F^\Lambda$ introduced in (162), we can rewrite the ansatz (162) as

$$F^{-|\Lambda} = t^\Lambda E^-, \tag{166}$$

and we retrieve the original formulae from

$$\begin{aligned} F^\Lambda &= 2\mathrm{Re} F^{-|\Lambda} = \frac{p^\Lambda}{2r^3} \epsilon_{abc} x^a dx^b \wedge dx^c - \frac{\ell^\Lambda(r)}{r^3} e^{2U} dt \wedge \boldsymbol{x} \cdot d\boldsymbol{x}, \\ {}^\star F^\Lambda &= -2\mathrm{Im} F^{-|\Lambda} = -\frac{\ell^\Lambda(r)}{2r^3} \epsilon_{abc} x^a dx^b \wedge dx^c - \frac{p^\Lambda}{r^3} e^{2U} dt \wedge \boldsymbol{x} \cdot d\boldsymbol{x}. \end{aligned} \tag{167}$$

[9] This ansatz is dictated by the general p-brane solution of supergravity bosonic equations in any dimensions [15].

Before proceeding further, it is convenient to define the electric and magnetic charges of the black hole as it is appropriate in any abelian gauge theory. Recalling the general form of the field equations and of the Bianchi identities as given in (38), we see that on-shell the field strengths $F_{\mu\nu}$ and $G_{\mu\nu}$ are both closed two-forms, since their duals are divergenceless. Hence, for the Gauss theorem, their integral on a closed space-like two-sphere does not depend on the radius of the sphere. These integrals are the (constant) electric and magnetic charges of the black hole defined in (61) that, in a quantum theory, we expect to be quantized. Using the ansatze (167) and the definition (37), we find

$$q_\Lambda = \frac{1}{4\pi}\int_{S^2} G_\Lambda = \Im\mathcal{N}_{\Lambda\Sigma}\ell^\Sigma + \Re\mathcal{N}_{\Lambda\Sigma}p^\Sigma = 2\Re\left(\mathcal{N}_{\Lambda\Sigma}\bar{t}^\Sigma\right). \tag{168}$$

From the above equation we can obtain the field dependence of the functions $\ell^\Lambda(r)$

$$\ell^\Lambda(r) = (\mathrm{Im}\mathcal{N})^{-1\,\Lambda\Sigma}\left(q_\Sigma - \mathrm{Re}\mathcal{N}_{\Sigma\Gamma}\,p^\Gamma\right). \tag{169}$$

Consider now the Killing spinor equations obtained from the supersymmetry transformations rules (141) and (142):

$$0 = \nabla_\mu\,\xi_A + \epsilon_{AB}\,T^-_{\mu\nu}\,\gamma^\nu\xi^B\,, \tag{170}$$

$$0 = \mathrm{i}\,\nabla_\mu\,z^i\,\gamma^\mu\xi^A + \frac{\mathrm{i}}{2}\,g^{i\bar{\jmath}}\,\bar{T}^-_{\bar{\jmath}|\mu\nu}\,\gamma^{\mu\nu}\epsilon^{AB}\,\xi_B\,, \tag{171}$$

where the Killing spinor $\xi_A(r)$ is of the form of a single radial function times a constant spinor satisfying

$$\begin{aligned} \xi_A(r) &= e^{f(r)}\chi_A, \qquad \chi_A = \text{constant},\\ \gamma_0\chi_A &= \mathrm{i}\,\frac{Z}{|Z|}\,\epsilon_{AB}\chi^B \end{aligned} \tag{172}$$

We observe that the condition (172) halves the number of supercharges preserved by the solution. Inserting (143),(144) and (172) into (170) and (171) and using the result (164), with a little work we obtain the first-order differential equations:

$$\begin{aligned}\frac{dz^i}{dr} &= -\left(\frac{e^{U(r)}}{r^2}\right)\frac{Z}{|Z|}\,g^{i\bar{\jmath}}\bar{f}^\Lambda_{\bar{\jmath}}(\mathcal{N}-\bar{\mathcal{N}})_{\Lambda\Sigma}t^\Sigma\\ &= \left(\frac{e^{U(r)}}{r^2}\right)\frac{Z}{|Z|}\,g^{i\bar{\jmath}}D_{\bar{\jmath}}\bar{Z}(z,\bar{z},p,q) = 2\left(\frac{e^{U(r)}}{r^2}\right)g^{i\bar{\jmath}}\,\partial_{\bar{\jmath}}|Z(z,\bar{z},p,q)|\,,\end{aligned} \tag{173}$$

$$\frac{dU}{dr} = \left(\frac{e^{U(r)}}{r^2}\right)|h_\Sigma p^\Sigma - f^\Lambda q_\Lambda| = \left(\frac{e^{U(r)}}{r^2}\right)|Z(z,\bar{z},p,q)|\,, \tag{174}$$

where $\mathcal{N}_{\Lambda\Sigma}(z,\bar{z})$ is the kinetic matrix of special geometry defined by (118), the vector $V = \left(f^{\Lambda}(z,\bar{z}), h_{\Sigma}(z,\bar{z})\right)$, according to (110), is the covariantly holomorphic section of the symplectic bundle entering the definition of a special Kähler manifold. Moreover, according to (146),

$$Z(z,\bar{z},p,q) \equiv f^{\Lambda} q_{\Lambda} - h_{\Sigma} p^{\Sigma}\,, \tag{175}$$

is the local realization on the scalar manifold $\mathcal{SM}$ of the central charge of the $N=2$ superalgebra,

$$\bar{Z}^{i}(z,\bar{z},p,q) \equiv g^{i\bar{\jmath}} D_{\bar{\jmath}} \bar{Z}(z,\bar{z},p,q)\,, \tag{176}$$

are the charges associated with the matter vectors, the so-called matter central charges, written with world indices of the special Kähler manifold. In terms of the complex charge vector t^{Λ} introduced in (165), the central and matter charges have the following useful expressions:

$$Z = -2\mathrm{i}\, f^{\Lambda}\, \mathrm{Im}\mathcal{N}_{\Lambda\Sigma}\, t^{\Sigma}\,, \tag{177}$$

$$\overline{Z}_{\bar{\imath}} = -2\mathrm{i}\, \bar{f}^{\Lambda}_{\bar{\imath}}\, \mathrm{Im}\mathcal{N}_{\Lambda\Sigma}\, t^{\Sigma}\,, \tag{178}$$

In summary, we have reduced the condition that the black hole should be a BPS-saturated state to the pair of first-order differential equations (173), (174) for the metric scale factor $U(r)$ and for the scalar fields $z^{i}(r)$. To obtain explicit solutions, one should specify the special Kähler manifold one is working with, namely the specific Lagrangian model. There are, however, some very general and interesting conclusions that can be drawn in a model-independent way. They are just consequences of the fact that these BPS conditions are first-order differential equations. Because of that there are fixed points (see footnote 171), namely values either of the metric or of the scalar fields which, once attained in the evolution parameter r (= the radial distance), will persist indefinitely. The fixed point values are just the zeros of the right-hand side in either of the coupled equations (174) and (173). The fixed point for the metric equation (174) is $r=\infty$, which corresponds to its asymptotic flatness. The fixed point for the moduli equation (173) is $r=0$. So, independently from the initial data at $r=\infty$ that determine the details of the evolution, the scalar fields flow into their fixed point values at $r=0$, which, as we will show, turns out to be a horizon. Indeed in the vicinity of $r=0$ also the metric takes the universal form of the Bertotti–Robinson $AdS_2 \times S^2$ metric.

Let us see this more closely. To begin with we consider the equations determining the fixed point values for the moduli and the universal form attained by the metric at the moduli fixed point. Using (178), we find

$$0 = g^{i\bar{\jmath}}\, \bar{Z}_{\bar{\jmath}}\big|_{\mathrm{fix}} = -2\mathrm{i}\; g^{i\bar{\jmath}}\, \bar{f}^{\Gamma}_{\bar{\jmath}}\, (\mathrm{Im}\mathcal{N})_{\Gamma\Lambda}\, t^{\Lambda}\big|_{\mathrm{fix}}\,, \tag{179}$$

$$\left(\frac{dU}{dr}\right)\bigg|_{\mathrm{fix}} = \left(\frac{e^{U(r)}}{r^{2}}\right) \left|Z\left(z,\bar{z},p,q\right)\right|\bigg|_{\mathrm{fix}}\,. \tag{180}$$

Multiplying (179) by f_i^{Σ}, using the identity (123) and the definition (177) of the central charge we conclude that at the fixed point the following condition is true:

$$0 = \left(t^{\Lambda} + \mathrm{i}\,\bar{f}^{\Lambda}\,Z\right)\Big|_{\text{fix}}\,. \tag{181}$$

In terms of the previously defined electric and magnetic charges (see (61) and (168)), (181) can be rewritten as

$$p^{\Lambda} = -\mathrm{i}\,\left(Z\,\bar{f}^{\Lambda} - \bar{Z}\,f^{\Lambda}\right)\Big|_{\text{fix}}\,, \tag{182}$$

$$q_{\Sigma} = -\mathrm{i}\,\left(Z\,\bar{h}_{\Lambda} - \bar{Z}\,h_{\Lambda}\right)\Big|_{\text{fix}}\,. \tag{183}$$

Equations (179), or equivalently (182) and (183), can be regarded as algebraic equations determining the value of the scalar fields at the fixed point as functions of the electric and magnetic charges p^{Λ}, q_{Σ}. Note therefore that, at the horizon, also the central charge depends only on the quantized charges: $Z(z,\bar{z},p,q)|_{\text{fix}} \equiv Z(p,q)$.

In the vicinity of the fixed point the differential equation for the metric becomes

$$\frac{dU}{dr} = \frac{|Z(p,q)|}{r^2}\,e^{U(r)} \tag{184}$$

which has the approximate solution:

$$\exp[-U(r)] \stackrel{r\to 0}{\longrightarrow} \frac{|Z(p,q)|}{r}. \tag{185}$$

Hence, near $r=0$ the metric (161) becomes of the Bertotti–Robinson type (see (8)) with Bertotti–Robinson mass given by

$$M^2_{\text{B-R}} = |Z(p,q)|^2. \tag{186}$$

In the metric (8) the surface $r=0$ is light-like and corresponds to a horizon since it is the locus where the Killing vector generating time translations $\frac{\partial}{\partial t}$, which is time-like at spatial infinity $r=\infty$, becomes light-like. The horizon $r=0$ has a finite area given by

$$\text{Area}_H = \int_{r=0} \sqrt{g_{\theta\theta}\,g_{\phi\phi}}\,d\theta\,d\phi \;=\; 4\pi\,M^2_{\text{B-R}}. \tag{187}$$

Hence, independently from the details of the considered model, the BPS-saturated black holes in an N=2 theory have a Bekenstein–Hawking entropy given by the following horizon area:

$$\frac{\text{Area}_H}{4\pi} = |Z(p,q)|^2, \tag{188}$$

where (186) was used, the value of the central charge being determined by (182) and (183). Such equations, as we shall see in the next secton, can also be seen as the variational equations for the minimization of the horizon area as given by (188), if the central charge is regarded as a function of both the scalar fields and the charges:

$$\begin{aligned} \mathrm{Area}_H(z,\bar{z}) &= 4\pi\,|Z(z,\bar{z},p,q)|^2, \\ \frac{\delta\mathrm{Area}_H}{\delta z} &= 0 \longrightarrow z = z_{\mathrm{fix}}\,. \end{aligned} \tag{189}$$

5 BPS and Non-BPS Attractor Mechanism: The Geodesic Potential

Quite recently it was noticed that the attractor behavior of extremal black holes in supersymmetric theories is not peculiar of BPS solutions preserving some supersymmetries [31], and examples of non-supersymmetric extremal black holes exhibiting the attractor phenomenon were found [34, 36, 63, 64, 65, 66].

It is then appropriate to introduce an alternative approach to extremality which does not rely on the existence of supersymmetry [31, 36, 67]. Let us start by writing the space–time metric of a black hole in terms of a new radial parameter τ:

$$ds^2 = e^{2U}\,dt^2 - e^{-2U}\left(\frac{c^4}{\sinh^4(c\tau)}\,d\tau^2 + \frac{c^2}{\sinh^2(c\tau)}\,d\Omega^2\right). \tag{190}$$

The coordinate τ is related to the radial coordinate r by the following relation:

$$\frac{c^2}{\sinh^2(c\tau)} = (r-r_0)^2 - c^2 = (r-r^-)\,(r-r^+)\,. \tag{191}$$

Here $c \equiv 2ST$ is the extremality parameter of the solution, with S the entropy and T the temperature of the black hole. When c is non-vanishing, the black hole has two horizons located at $r^\pm = r_0 \pm c$. The outer horizon is located at $r_H = r^+$ corresponding to $\tau \to -\infty$. The extremality limit at which the two horizons coincide, $r_H = r^+ = r^- = r_0$, is $c \to 0$. In this case the metric (190) takes the simple form in the r coordinate

$$ds^2 = e^{2U}\,dt^2 - e^{-2U}\left(dr^2 + (r-r_H)^2\,d\Omega^2\right). \tag{192}$$

In the general case, if we require the horizon to have a finite area A, the scale function U in the near-horizon limit should behave as follows:

$$e^{-2U} \overset{\tau\to-\infty}{\longrightarrow} \frac{A}{4\pi}\,\frac{\sinh^2(c\tau)}{c^2} = \frac{A}{4\pi}\,\frac{1}{(r-r^-)(r-r^+)}\,, \tag{193}$$

so that the near-horizon metric reads

$$ds^2 = \frac{4\pi}{A}\,(r-r^-)(r-r^+)\,dt^2 - \frac{A}{4\pi}\,\frac{dr^2}{(r-r^-)(r-r^+)} - \frac{A}{4\pi}\,d\Omega^2\,. \tag{194}$$

The above metric coincides with the near-horizon metric of a Reissner–Nordström solution with horizons located at $r^{\pm}$. It is useful to introduce the radial coordinate ρ defined as $\rho = 2\, e^{c\tau}$, in terms of which, in the near-horizon limit, we can write $e^{-2U} \sim \left(\frac{r_H}{\rho c}\right)^2$, where $r_H = \sqrt{A/4\pi}$ is the radius of the (outer) horizon, and the metric becomes

$$ds^2 = \left(\frac{\rho c}{r_H}\right)^2 dt^2 - (r_H)^2\,(d\rho^2 + d\Omega^2)\,. \tag{195}$$

The coordinate ρ measures the *physical distance* from the horizon, which is located at $\rho = 0$, in units of r_H. It is important to note that the distance of a point at some finite ρ_0 from the horizon is finite:

$$d = \int_0^{\rho_0} r_H\, d\rho = r_H\, \rho_0 < \infty\,. \tag{196}$$

Using this feature, in [36] an intuitive argument was given in order to justify the absence of a universal behavior for the scalar fields near the horizon of a non-extremal black hole: the distance from the horizon is not "long enough" in order for the scalar fields to "loose memory" of their initial values at infinity.

Let us now consider the extremal case $c = 0$. The relation between τ and r becomes $\tau = -1/(r - r_H)$. In order to have a finite horizon area, U should behave near the horizon as

$$e^{-2U} \sim \left(\frac{r_H}{r - r_H}\right)^2, \tag{197}$$

The physical distance from the horizon is now measured in units r_H by the coordinate $\omega = \ln(r - r_H)$ in terms of which the near-horizon metric reads

$$ds^2 = \frac{1}{(r_H)^2}\, e^{2\omega}\, dt^2 - (r_H)^2\,(d\omega^2 + d\Omega^2)\,. \tag{198}$$

Since now the horizon is located at $\omega \to -\infty$, the distance of a point at some finite ω_0 from the horizon is always infinite, as opposite to the non-extremal case:

$$d = \int_{-\infty}^{\omega_0} r_H\, d\omega = \infty\,. \tag{199}$$

Therefore, as observed in [36], the infinite distance from the horizon in the extremal case justifies the fact that the scalar fields at the horizon "loose memory" of their initial values at infinity and therefore exhibit a universality behavior. In order to simplify the notation, in the following we shall use the coordinate r to denote the distance from the horizon, consistently with our previous treatment of the BPS black-hole solutions.

Let us consider the field equations for the metric components (see (190)) and for the scalar fields Φ^r coming from the Lagrangian (32). By eliminating the vector fields through their equations of motion, the resulting equations for the metric and the scalar fields, written in terms of the evolution parameter τ, take the following simple form [67]:

$$\frac{d^2U}{d\tau^2} = V_{\text{B–H}}(\Phi, p, q)e^{2U}\,, \tag{200}$$

$$\frac{D^2\Phi^r}{D\tau^2} = g^{rs}(\Phi)\,\frac{\partial V_{\text{B–H}}(\Phi, p, q)}{\partial \Phi^s}e^{2U}\,, \tag{201}$$

with the constraint

$$\left(\frac{dU}{d\tau}\right)^2 + \frac{1}{2}\,g_{rs}(\Phi)\,\frac{d\Phi^r}{d\tau}\frac{d\Phi^s}{d\tau} - V_{\text{B–H}}(\Phi, p, q)e^{2U} = c^2\,, \tag{202}$$

where $V_{\text{B–H}}(\Phi, p, q)$ is a function of the scalars and of the electric and magnetic charges of the theory defined by

$$V_{\text{B–H}} = -\frac{1}{2}\,Q^t\mathcal{M}(\mathcal{N})Q\,, \tag{203}$$

where as usual Q is the symplectic vector of quantized electric and magnetic charges and $\mathcal{M}(\mathcal{N})$ is the symplectic matrix defined in (93) in terms of the matrix $\mathcal{N}_{\Lambda\Sigma}(\Phi)$. Let us note that the field equations (201) can be extracted from the effective one-dimensional Lagrangian:

$$\mathcal{L}_{eff} = \left(\frac{dU}{d\tau}\right)^2 + \frac{1}{2}\,g_{rs}\frac{d\Phi^r}{d\tau}\frac{d\Phi^s}{d\tau} + V_{\text{B–H}}(\Phi, p, q)e^{2U}, \tag{204}$$

constrained with (202). The extremality condition is $c^2 \to 0$.

From (204) we see that the properties of extremal black holes are completely encoded in the metric of the scalar manifold g_{rs} and on the scalar effective potential $V_{\text{B–H}}$, known as black-hole potential or geodesic potential [31, 67]. In particular, as it was shown in [31, 36, 67] and as we shall review below, the area of the event horizon is proportional to the value of $V_{\text{B–H}}$ at the horizon

$$\frac{A}{4\pi} = V_{\text{B-H}}(\Phi_h, p, q) \tag{205}$$

where Φ_h denotes the value taken by the scalar fields at the horizon. [10] This follows from the property that there is an attractor mechanism at work in the extremal case. To see this, let us consider the set of equations (201) at $c = 0$. Regularity of the scalar fields at the horizon, which is located, with respect to the physical distance parameter ω, at $\omega \to -\infty$, implies that at the horizon the first derivative of Φ^r with respect to ω vanishes: $\partial_\omega \Phi^r_{|h} = 0$. Near the horizon a solution to (201), under the hypothesis that $(\partial V_{\text{B–H}}/\partial\Phi^r)_h$ be finite, behaves as follows:

[10] For the sake of clarity in the comparison with equivalent formulas in [36], let us note that in [36] the definition $\Sigma^r = \frac{d\Phi^r}{d\tau}$ has been used.

$$\Phi^r \sim \frac{1}{2\,(r_H)^2}\, g^{rs}(\Phi_h)\, \frac{\partial V_{\text{B–H}}}{\partial \Phi^s}\bigg|_{\Phi_h} \omega^2 + \Phi^r_h. \qquad (206)$$

Regularity of Φ^r at $\omega \to -\infty$ then further requires that $(\partial V_{\text{B–H}}/\partial \Phi^r)_{|h} = 0$, implying that the horizon be an attractor point for the scalar fields. We conclude that in the extremal case the scalar fields tend in the near-horizon limit to some fixed values Φ^r_h, which extremize the potential $V_{\text{B–H}}$:

$$\omega \to -\infty: \quad \Phi^r(\omega) \text{ regular} \Rightarrow \quad \frac{\partial V_{\text{B–H}}}{\partial \Phi^r}\bigg|_{\Phi_h} \to 0 \quad ; \quad \frac{d\Phi^r}{d\omega} \to 0. \qquad (207)$$

These values are functions of the quantized electric and magnetic charges only: $\Phi^r_h = \Phi^r_h(p,q)$. Furthermore, let us consider (202). In the extremal limit $c = 0$, near the horizon it becomes

$$\left(\frac{dU}{d\tau}\right)^2 \sim V_{\text{B–H}}(\Phi_h(p,q),p,q)e^{2U} \qquad (208)$$

from which it follows, for the metric components near the horizon

$$e^{2U} \sim \frac{r^2}{V_{\text{B–H}}(\Phi_h)} = \left(\frac{r}{r_H}\right)^2, \qquad (209)$$

that is:

$$ds^2_{hor} = \frac{r^2}{V_{\text{B–H}}(\Phi_h)} dt^2 - \frac{V_{\text{B–H}}(\Phi_h)}{r^2}\left(dr^2 + r^2 d\Omega\right). \qquad (210)$$

From (208) and (210) we immediately see that the value of the potential at the horizon measures its area, as anticipated in (205). The metric (210) describes a Bertotti–Robinson geometry $AdS_2 \times S^2$, with mass parameter $M^2_{\text{B–R}} = V_{\text{B–H}}(\Phi_h)$.

To summarize, we have just shown that the area of the event horizon of an extremal black hole (and hence its B–H entropy) is given by the black-hole potential evaluated at the horizon, where it gets an extremum. This justifies our assertion at the end of the previous section.

Let us briefly comment on the non-extremal case $c \neq 0$. For these solutions, the physical distance is measured by the coordinate ρ introduced in (193) and the horizon is located at $\rho = 0$. The requirement of regularity of the scalar fields at the horizon is less stringent. It just means that the scalars should admit a Taylor expansion in ρ around $\rho = 0$ and thus it poses no constraints, aside from finiteness, on their derivatives at the horizon:

$$\Phi^r \sim \Phi^r_h + \frac{\partial \Phi^r}{\partial \rho}\bigg|_0 \rho + \frac{1}{2\,(r_H)^2}\, g^{rs}(\Phi_h)\, \frac{\partial V_{\text{B-H}}}{\partial \Phi^s}\bigg|_{\Phi_h} \rho^2 + O(\rho^3). \qquad (211)$$

The horizon is therefore not necessarily an attractor point, since at $\rho = 0$ $(\partial V_{\text{B–H}}/\partial \Phi^r)_{\Phi_h}$ can now be a non-vanishing constant.

5.1 Extremal Black Holes in Supergravity

For supergravity theories, supersymmetry fixes the black-hole potential $V_{\text{B–H}}$ defined in (203) to take a particular form that allows to find its extremum in an easy way. Indeed, an expression exactly coinciding with (203) has been found in Sect. 3 in an apparently different context, as the result of a sum rule among central and matter charges in supergravity theories (93). So, in every supergravity theory, the black-hole potential has the general form

$$V_{\text{B–H}} \equiv -\frac{1}{2}Q^t\mathcal{M}(\mathcal{N})Q = \frac{1}{2}Z_{AB}\bar{Z}^{AB} + Z_I\bar{Z}^I. \tag{212}$$

By making use of the geometric relations of Sect. 3, the value of the charge vector $Q = \begin{pmatrix} p^\Lambda \\ q_\Lambda \end{pmatrix}$ in terms of the moduli Φ is given by (99) and (100). Then, to find the extremum of $V_{\text{B–H}}$ we can apply the differential relations (90) among central and matter charges found in Sect. 3.

Let us now analyze more in detail, for the case of supergravity theories, the extremality condition $c = 0$ as it comes from the constraint (202) which has to be imposed on the solution all over space–time. According to the discussion given in the previous section, the existence of solutions to (202) does not not rely on supersymmetry, therefore also non supersymmetric extremal black holes still exhibit an attractor behavior (207) (found at $c = 0$).

At spatial infinity $\tau \to 0$, where the macroscopic features of the black hole are well defined, we have $U \to M_{ADM}\tau$, as it follows from the general definition of ADM mass in General Relativity (see for example [1]). The metric (161) reduces to the Minkowski one and the constraint (202) becomes

$$M_{ADM}^2 = |Z(\Phi_\infty, p, q)|^2 + |Z_I(\Phi_\infty, p, q)|^2 - \frac{1}{2}\, g_{rs}\frac{d\Phi_\infty^r}{d\tau}\frac{d\Phi_\infty^s}{d\tau}. \tag{213}$$

These solutions do not necessarily saturate the BPS bound, since in general, from (213), $M_{ADM}^2 \neq |Z(\Phi_\infty)|^2$. They then completely break supersymmetry. The behavior at the horizon may nevertheless be easily found thanks to the expression (212) that the black-hole potential takes in supergravity theories, by exploiting the condition (207) and in particular $\frac{\partial V_{\text{B–H}}}{\partial \Phi^r}|_{\Phi_h} \to 0$.

For the cases where the black-hole solution preserves some supersymmetries, we are going to find that the constraint (202) yields the BPS bound on the mass of the solution. Indeed in that case one may apply the results of Sect. 4.1. Let us restrict to the case of $N = 2$ supergravity, where $V_{\text{B–H}} = |Z|^2 + |Z_I|^2$. The Killing spinor equation $\delta_\epsilon \lambda = 0$ gives equation (173) that implies

$$\left|\frac{dz^i}{d\tau}\right|^2 = e^{2U}\,|g^{i\bar{\jmath}}\,D_{\bar{\jmath}}Z|^2. \tag{214}$$

By making use of (214), the constraint (202) reduces in the extremal limit $c = 0$ to the following equation, valid all over space–time:

$$\left(\frac{dU}{d\tau}\right)^2 = e^{2U}\,|Z|^2. \tag{215}$$

At spatial infinity $\tau \to 0$, (214) and (215) become

$$M^2_{ADM} = |Z(\Phi_\infty, p, q)|^2\,; \qquad |Z_I(\Phi_\infty, p, q)|^2 = g_{rs}\frac{d\Phi^r_\infty}{d\tau}\frac{d\Phi^s_\infty}{d\tau}\,. \tag{216}$$

The first equation in (216) may be recognized as the saturation of the BPS bound on the mass of the solution. On the other hand, near the horizon the attractor condition holds

$$\left.\frac{d\Phi^r}{d\tau}\right|_h = 0\,, \tag{217}$$

and from (214) it gives $Z_I|_h = 0$, which may be solved to find $\Phi_{\text{fix}}(p, q)$ leaving, for the mass parameter at the horizon

$$\left(\frac{dU}{d\tau}\right)^2_h = M^2_{\text{B-R}}(p, q) = |Z(\Phi_{\text{fix}}, p, q)|^2. \tag{218}$$

Actually, the extrema of the black-hole potential may be systematically studied, both for the BPS and non-BPS case, by use of the geometric relations (90). One finds that the extrema are given by

$$\begin{aligned} dV_{\text{B-H}} &= \frac{1}{2}DZ_{AB}\bar{Z}^{AB} + DZ_I\bar{Z}^I + c.c. \\ &= \frac{1}{2}\left(\frac{1}{2}\bar{Z}^{AB}\bar{Z}^{CD}P_{ABCD} + \bar{Z}^{AB}\bar{Z}^I P_{ABI} + c.c.\right) \\ &+ \left(\frac{1}{2}\bar{Z}^{AB}\bar{Z}^I P_{ABI} + \bar{Z}^I\bar{Z}^J P_{IJ} + c.c.\right) = 0. \end{aligned} \tag{219}$$

Let us remark that the one introduced in (219) is a covariant procedure, not referring explicitly to the horizon properties for finding the entropy, so it is not necessary to specify explicitly horizon parameters (like the metric and the fixed values of scalars at that point), $V_{\text{B-H}}$ being a well-defined quantity over all the space–time.

The conditions (219), defining the extremum of the black-hole potential and thus the fixed scalars, when restricted to the BPS case have the same content as, and are therefore completely equivalent to, the relations (173) and (174) found in the previous subsection from the Killing-spinor conditions. In particular, extremal black holes preserving one supersymmetry correspond to N-extended multiplets with

$$M_{ADM} = |Z_1| > |Z_2| \cdots > |Z_{[N/2]}| \tag{220}$$

where Z_m, $m = 1, \cdots, [N/2]$, are the skew-eigenvalues of the central charge antisymmetric matrix introduced in (14) [68, 69, 50, 51]: $Z_1 = Z_{12}$,

$Z_2 = Z_{34}, \ldots$. At the attractor point, where M_{ADM} is extremized, supersymmetry requires the vanishing of each term on the right-hand side of (219). In particular, we find $Z_I = 0$ (recall that Z_I does not exist for $N > 4$) and

$$\bar{Z}^{AB}\bar{Z}^{CD}P_{ABCD} = \ \Rightarrow \ \bar{Z}^{[AB}\bar{Z}^{CD]} = 0\,. \tag{221}$$

The above condition is satisfied taking $Z_1 = Z_{12} \neq 0$ and $Z_m = 0$, $m > 1$. A general property of regular BPS black-hole solutions is that supersymmetry doubles at the horizon. This is consistent with the fact that the near horizon geometry is a Bertotti–Robinson space–time of the form $AdS_2 \times S^2$, which is known to have an unbroken $N = 2$ supersymmetry [5]. Let us now give an argument for the vanishing of the supersymmetry variation along ϵ_1, ϵ_2 of the fermion fields at the horizon. As far as the dilatino fields are concerned, it is sufficient to remember that, since $(d\Phi^r/d\tau)_h = 0$, at the horizon the supersymmetry variation is proportional to $Z_{[AB}\epsilon_{C]}$. However, this expression is also zero since the only non-vanishing central charge is $Z_1 \equiv Z_{12}$ and furthermore $Z_{[12}\epsilon_{1]} = Z_{[12}\epsilon_{2]} = 0$. As for the gaugini their supersymmetry variation at the horizon is automatically zero being $Z_I = 0$. Finally, let us remark that the gravitino variation is not actually zero; however, the variation of its field strength along ϵ_1, ϵ_2 vanishes because of the property of the Bertotti–Robinson solution of being conformally flat and the fact that the graviphoton field strength T_{AB} is Lorentz-covariantly constant at the horizon [22].

A case by case analysis of the BPS and non-BPS black holes in the various supergravity models, by inspection of the extrema of $V_{\text{B-H}}$, will be given in Sect.6. As an exemplification of the method, let us anticipate here the detailed study of the BPS solution of $D = 4$, $N = 4$ pure supergravity. The field content is given by the gravitational multiplet, that is by the graviton $g_{\mu\nu}$, four gravitini $\psi_{\mu A}$, $A = 1, \cdots, 4$, six vectors $A_\mu^{[AB]}$, four dilatini $\chi^{[ABC]}$ and a complex scalar $\phi = a + \mathrm{i}e^{\varphi}$ parametrizing the coset manifold $G/H = SU(1,1)/U(1)$. The symplectic $Sp(12)$-sections $(f^\Lambda_{AB}, h_{\Lambda AB})$ ($\Lambda \equiv [AB] = 1, \cdots, 6$) over the scalar manifold are given by

$$\begin{aligned} f^\Lambda_{AB} &= e^{-\varphi/2}\delta^\Lambda_{AB}, \\ h_{\Lambda AB} &= \phi e^{-\varphi/2}\delta_{\Lambda AB}, \end{aligned} \tag{222}$$

so that

$$\mathcal{N}_{\Lambda\Sigma} = (\mathbf{h}\cdot\mathbf{f}^{-1})_{\Lambda\Sigma} = \phi\delta_{\Lambda\Sigma}. \tag{223}$$

The central charge matrix is then given by

$$Z_{AB} = f^\Lambda_{AB}q_\Lambda - h_{\Lambda AB}p^\Lambda = -e^{-\varphi/2}(\phi p_{AB} - q_{AB})\,. \tag{224}$$

The black-hole potential is therefore

$$\begin{aligned} V(\phi, p, q) &= \frac{1}{2}e^{-\varphi}(\phi p_{AB} - q_{AB})(\bar{\phi}p^{AB} - q^{AB}) \\ &= \frac{1}{2}(a^2 e^{-\varphi} + e^{\varphi})p_{AB}p^{AB} + e^{-\varphi}q_{AB}p^{AB} - 2ae^{-\varphi}q_{AB}p^{AB} \end{aligned}$$

$$\equiv \frac{1}{2}(p,q)\begin{pmatrix}1 & 0\\ -a & 1\end{pmatrix}\begin{pmatrix}e^{\varphi} & 0\\ 0 & e^{-\varphi}\end{pmatrix}\begin{pmatrix}1 & -a\\ 0 & 1\end{pmatrix}\begin{pmatrix}p\\ q\end{pmatrix}. \tag{225}$$

By extremizing the potential in the moduli space we get

$$\begin{aligned}\frac{\partial V}{\partial a} &= 0 \to a_h = \frac{q_{AB}p^{AB}}{p_{AB}p^{AB}},\\ \frac{\partial V}{\partial \varphi} &= 0 \to e^{\varphi_h} = \frac{\sqrt{|q_{AB}q^{AB}p_{CD}p^{CD} - (q_{AB}p^{AB})^2|}}{p_{AB}p^{AB}},\end{aligned} \tag{226}$$

from which it follows that the entropy is

$$S_{\text{B-H}} = 4\pi V(\phi_h,p,q) = 4\pi\sqrt{|q_{AB}q^{AB}p_{CD}p^{CD} - (q_{AB}p^{AB})^2|}. \tag{227}$$

As a final observation, let us note, following [31], that the extremum reached by the black-hole potential at the horizon is in particular a minimum, unless the metric of the scalar fields change sign, corresponding to some sort of phase transition, where the effective Lagrangian description (204) of the theory breaks down. This can be seen from the properties of the Hessian of the black-hole potential. It was shown in [31] for the $N = 2$, $D = 4$ case that at the critical point $\Phi = \Phi_{\text{fix}} \equiv \Phi_h$, from the special geometry properties it follows:

$$(\partial_{\bar{\imath}}\partial_j|Z|)_{\text{fix}} = \frac{1}{2}g_{\bar{\imath}j}|Z|_{\text{fix}} \tag{228}$$

and then, remembering, from the above discussion, that $V_{\text{fix}} = |Z_{\text{fix}}|^2$:

$$(\partial_{\bar{\imath}}\partial_j V)_{\text{fix}} = 2g_{\bar{\imath}j}|Z_{\text{fix}}|^2. \tag{229}$$

From (229) it follows, for the $N = 2$ theory, that the minimum is unique. In the next section we will show one more technique for finding the entropy, exploiting the fact that it is a "topological quantity" not depending on scalars. This last procedure is particularly interesting because it refers only to group theoretical properties of the coset manifolds spanned by scalars, and do not need the knowledge of any details of the black-hole horizon.

5.2 B-H Entropy as a U-invariant for Symmetric Spaces

For theories based on moduli spaces given by symmetric manifolds G/H, which is the case of all supergravity theories with $N \geq 3$ extended supersymmetry, but also of several $N = 2$ models, the BPS and non-BPS black holes are classified by some U-duality-invariant expressions depending on the representation of the group G of G/H under which the electric and magnetic charges are classified. In this respect, the classification of the $N = 2$ invariants is entirely similar to the $N > 2$ cases, where all scalar manifolds are symmetric spaces.

For theories that have a quartic invariant I_4 [70] (this includes all $N = 2$ symmetric spaces based on cubic prepotentials [71, 72] and $N = 4, 6, 8$ theories), the B–H entropy turns out to be proportional to its square root

$$S_{\text{B–H}} \propto \sqrt{|I_4|}. \tag{230}$$

The BPS solutions have $I_4 > 0$ while the non-BPS ones (with non-vanishing central charge) have instead $I_4 < 0$. For all the above theories with the exception of the $N = 8$ case, there is also a second non-BPS solution with vanishing central charge and $I_4 > 0$.

For theories based on symmetric spaces with only a quadratic invariant I_2 (this includes $N = 2$ theories with quadratic prepotentials as well as $N = 3$ and $N = 5$ theories), the B–H entropy is

$$S_{\text{B-H}} \propto |I_2|. \tag{231}$$

In these cases, beyond the BPS solution which has $I_2 > 0$ there is only one non-BPS solution, with vanishing central charge and $I_2 < 0$.

All the solutions discussed here give $S_{\text{B–H}} \neq 0$ and then fall in the class of the so-called large black holes, for which the classical area/entropy formula is valid as it gives the dominant contribution to the black-hole entropy. Solutions with $I_4, I_2 = 0$ do exist but they do not correspond to classical attractors since in that case the classical area/entropy formula vanishes. In this case one deals with small black holes, and a quantum attractor mechanism, including higher curvature terms, has to be considered for finding the entropy.

The main purpose of this subsection is to provide particular expressions which give the entropy formula as a moduli-independent quantity in the entire moduli space and not just at the critical points. Namely, we are looking for quantities $S\left(Z_{AB}(\phi), \bar{Z}^{AB}(\phi), Z_I(\phi), \bar{Z}^I(\phi)\right)$ such that $\frac{\partial}{\partial \phi^i} S = 0$, ϕ^i being the moduli coordinates. [11] To this aim, let us first consider invariants I_α of the isotropy group H of the scalar manifold G/H, built with the central and matter charges. We will take all possible H-invariants up to quartic ones for four dimensional theories (except for the $N = 3$ case, where the invariants of order higher than quadratic are not irreducible). Then, let us consider a linear combination $S^2 = \sum_\alpha C_\alpha I_\alpha$ of the H-invariants, with arbitrary coefficients C_α. Now, let us extremize S in the moduli space $\frac{\partial S}{\partial \Phi^i} = 0$, for some set of $\{C_\alpha\}$. Since $\Phi^i \in G/H$, the quantity found in this way (which in all cases turns out to be unique) is a U-invariant, and is indeed proportional to the Bekenstein–Hawking entropy.

These formulae generalize the quartic $E_{7(-7)}$ invariant of $N = 8$ supergravity [70] to all other cases. [12]

[11] The Bekenstein–Hawking entropy $S_{\text{B–H}} = \frac{A}{4}$ is actually πS in our notation

[12] Our analysis is based on general properties of scalar coset manifolds. As a consequence, it can be applied straightforwardly also to the $N = 2$ cases, whenever one considers special coset manifolds.

Let us first consider the theories $N = 3, 4$, where matter can be present [52, 53].

The U-duality groups[13] are, in these cases, $SU(3,n)$ and $SU(1,1) \times SO(6,n)$ respectively. The central and matter charges Z_{AB}, Z_I transform in an obvious way under the isotropy groups

$$H = SU(3) \times SU(n) \times U(1) \qquad (N = 3), \tag{232}$$
$$H = SU(4) \times SO(n) \times U(1) \qquad (N = 4). \tag{233}$$

Under the action of the elements of G/H the charges may get mixed with their complex conjugate. The infinitesimal transformation can be read from the differential relations satisfied by the charges (90) [50] .

For $N = 3$:

$$P^{ABCD} = P_{IJ} = 0, \quad P_{ABI} \equiv \epsilon_{ABC} P^C_I, \quad Z_{AB} \equiv \epsilon_{ABC} Z^C. \tag{234}$$

Then the variations are

$$\delta Z^A = \xi^{AI} Z_I, \tag{235}$$
$$\delta Z_I = \bar{\xi}_{AI} Z^A, \tag{236}$$

where ξ^{AI} are infinitesimal parameters of $K = G/H$.

The possible quadratic H-invariants are

$$I_1 = Z^A \bar{Z}_A, \quad I_2 = Z_I \bar{Z}^I. \tag{237}$$

So, the U-invariant expression is

$$S = |Z^A \bar{Z}_A - Z_I \bar{Z}^I|. \tag{238}$$

In other words, $D_i S = \partial_i S = 0$, where the covariant derivative is defined in [50].

Note that at the attractor point ($Z_I = 0$) it coincides with the moduli-dependent potential (212) computed at its extremum.

For $N = 4$

$$P_{ABCD} = \epsilon_{ABCD} P, \quad P_{IJ} = \eta_{IJ} P, \quad P_{ABI} = \frac{1}{2} \eta_{IJ} \epsilon_{ABCD} \bar{P}^{CDJ}, \tag{239}$$

and the transformations of $K = \frac{SU(1,1)}{U(1)} \times \frac{O(6,n)}{O(6) \times O(n)}$ are

$$\delta Z_{AB} = \frac{1}{2} \epsilon_{ABCD} \, \xi \, \bar{Z}^{CD} + \xi^I_{AB} Z_I, \tag{240}$$

[13] Here we denote by U-duality group the isometry group U acting on the scalars in a symplectic representation, although only a restriction of it to integers is the proper U-duality group [10].

$$\delta Z_I = \bar{\xi}\,\eta_{IJ}\,\bar{Z}^J + \frac{1}{2}\,\bar{\xi}_I^{AB}\,Z_{AB}, \tag{241}$$

with $\bar{\xi}_I^{AB} = \frac{1}{2}\eta_{IJ}\epsilon^{ABCD}\xi_{CD}^J$.

The possible H-invariants are

$$\begin{aligned} I_1 &= Z_{AB}\bar{Z}^{AB} \\ I_2 &= Z_{AB}\bar{Z}^{BC}Z_{CD}\bar{Z}^{DA} \\ I_3 &= \epsilon^{ABCD}Z_{AB}Z_{CD} \\ I_4 &= Z_I Z^I. \end{aligned} \tag{242}$$

There are three $O(6,n)$ invariants given by S_1, S_2, and $\bar{S}_2$ where

$$S_1 = \frac{1}{2}Z_{AB}\bar{Z}^{AB} - Z_I\bar{Z}_I, \tag{243}$$

$$S_2 = \frac{1}{4}\epsilon^{ABCD}Z_{AB}Z_{CD} - Z_I Z_I, \tag{244}$$

and the unique $SU(1,1) \times O(6,n)$ invariant S, $DS = 0$, is given by

$$S = \sqrt{|(S_1)^2 - |S_2|^2|}. \tag{245}$$

At the attractor point $Z_I = 0$ and $\epsilon^{ABCD}Z_{AB}Z_{CD} = 0$ so that S reduces to the square of the BPS mass.
Note that, in absence of matter multiplets, one recovers the expression found in the previous subsecion by extremizing the black hole potential.

For $N = 5, 6, 8$ the U-duality-invariant expression S is the square root of a unique invariant under the corresponding U-duality groups $SU(5,1)$, $O^*(12)$ and $E_{7(-7)}$. The strategy is to find a quartic expression S^2 in terms of Z_{AB} such that $DS = 0$, i.e. S is moduli-independent.

As before, this quantity is a particular combination of the H quartic invariants.

For $SU(5,1)$ there are only two $U(5)$ quartic invariants. In terms of the matrix $A_A^{\ B} = Z_{AC}\bar{Z}^{CB}$ they are $(TrA)^2$, $Tr(A^2)$, where

$$TrA = Z_{AB}\bar{Z}^{BA}, \tag{246}$$

$$Tr(A^2) = Z_{AB}\bar{Z}^{BC}Z_{CD}\bar{Z}^{DA}. \tag{247}$$

As before, the relative coefficient is fixed by the transformation properties of Z_{AB} under $\frac{SU(5,1)}{U(5)}$ elements of infinitesimal parameter ξ^C:

$$\delta Z_{AB} = \frac{1}{2}\xi^C\epsilon_{CABPQ}\bar{Z}^{PQ}. \tag{248}$$

It then follows that the required invariant is

$$S = \frac{1}{2}\sqrt{|4Tr(A^2) - (TrA)^2|}. \tag{249}$$

The $N = 6$ case is the more complicated because under $U(6)$ the left-handed spinor of $O^*(12)$ splits into:

$$32_L \to 15_1 + \bar{15}_{-1} + 1_{-3} + 1_3. \tag{250}$$

The transformations of $\frac{O^*(12)}{U(6)}$ are

$$\delta Z_{AB} = \frac{1}{4}\epsilon_{ABCDEF}\xi^{CD}\bar{Z}^{EF} + \xi_{AB}\bar{X}, \tag{251}$$

$$\delta X = \frac{1}{2}\xi_{AB}\bar{Z}^{AB}, \tag{252}$$

where we denote by X the $SU(6)$ singlet.

The quartic $U(6)$ invariants are

$$I_1 = (TrA)^2 \tag{253}$$

$$I_2 = Tr(A^2) \tag{254}$$

$$I_3 = Re(Pf\, ZX) = \frac{1}{2^3 3!}Re(\epsilon^{ABCDEF} Z_{AB} Z_{CD} Z_{EF} X) \tag{255}$$

$$I_4 = (TrA)X\bar{X} \tag{256}$$

$$I_5 = X^2\bar{X}^2 \tag{257}$$

where the matrix A is, as for the $N = 5$ case, $A_A^{\ B} = Z_{AC}\bar{Z}^{CB}$.

The unique $O^*(12)$ invariant is

$$S = \frac{1}{2}\sqrt{|4I_2 - I_1 + 32I_3 + 4I_4 + 4I_5|} \tag{258}$$

$$DS = 0. \tag{259}$$

Note that at the BPS attractor point $Pf\, Z = 0$, $X = 0$ and S reduces to the square of the BPS mass.

For $N = 8$ the $SU(8)$ invariants are [14]

$$I_1 = (TrA)^2 \tag{260}$$

$$I_2 = Tr(A^2) \tag{261}$$

$$I_3 = Pf\, Z = \frac{1}{2^4 4!}\epsilon^{ABCDEFGH} Z_{AB} Z_{CD} Z_{EF} Z_{GH}. \tag{262}$$

The $\frac{E_{7(-7)}}{SU(8)}$ transformations are

$$\delta Z_{AB} = \frac{1}{2}\xi_{ABCD}\bar{Z}^{CD}, \tag{263}$$

where ξ_{ABCD} satisfies the reality constraint:

[14] The Pfaffian of an $(n \times n)$ (n even) antisymmetric matrix is defined as $PfZ = \frac{1}{2^n n!}\epsilon^{A_1 \cdots A_n} Z_{A_1 A_2} \cdots Z_{A_{N-1} A_N}$, with the property: $|PfZ| = |detZ|^{1/2}$.

$$\xi_{ABCD} = \frac{1}{24}\epsilon_{ABCDEFGH}\bar{\xi}^{EFGH}. \tag{264}$$

One finds the following $E_{7(-7)}$ invariant [70]:

$$S = \frac{1}{2}\sqrt{|4Tr(A^2) - (TrA)^2 + 32Re(Pf\, Z)|}. \tag{265}$$

6 Detailed Analysis of Attractors in Extended Supergravities: BPS and Non-BPS Critical Points

The extremum principle was found originally in the context of $N = 2$ four-dimensional black holes. However, as we have described in Sect. 4, it has a more general validity, being true for all N-extended supergravities in four dimensions (in the cases where the Bekenstein–Hawking entropy is different from zero) [50]. Indeed, the general discussion of Sect. 3.2 shows that the coset structure of extended supergravities in four dimensions (for $N > 2$) induces the existence, in every theory, of differential relations among central and matter charges that generalize the ones existing for the $N = 2$ case. Furthermore, as far as BPS solutions are considered, Killing-spinor equations for gauginos and dilatinos analogous to (90) are obtained by setting to zero the supersymmetry transformation laws of the fermions. Correspondingly, at the fixed point $\partial_\mu \Phi^i = 0$, for any extended supergravity theories one gets some conditions that allow to find the value of fixed scalars and hence of the B–H entropy both for BPS and non-BPS black- hole solutions.

We will first discuss in Sect. 6.1 the case of $N = 2$ supergravity, then in Sect. 6.2 the case of the other extended theories allowing matter couplings to the supergravity multiplet, that is $N = 3, 4$ extended supergravities, and finally we will pass to analyze in Sect. 6.3 $N = 5, 6, 8$ theories, which are pure supergravity models.

For every theory, the strategy adopted to find the extrema will be to solve the equation $dV_{\text{B–H}} = 0$, as given in general in (219), by setting to zero all the independent components in the decomposition on a basis of vielbein of the moduli space [50].

We confine our analysis to large black holes, with finite horizon area.

6.1 $N = 2$ Attractor Equations

In the original paper [31], the $N = 2$ attractor conditions were introduced via an extremum condition on the black-hole potential (203)

$$V_{\text{B–H}} = -\frac{1}{2}Q^T\mathcal{M}Q = |Z|^2 + |D_iZ|^2 \tag{266}$$

discussed in Sect. 5. Indeed, by making use of properties of $N = 2$ special geometry, the extremum condition was written in the form

$$\partial_i V_{\text{B–H}} = 2\bar{Z} D_i Z + \mathrm{i} C_{ijk} g^{j\bar{j}} g^{k\bar{k}} D_{\bar{j}} \bar{Z} D_{\bar{k}} \bar{Z} = 0\,, \tag{267}$$

where use of the special geometry relations (115) was made.

Given (267), it is useful to write the attractor equations in a different form. Indeed, recalling (99) and (100) [73, 74, 33] (which are true all over the moduli space) we may write

$$Q - \mathrm{i}\,\mathbb{C}\,\mathcal{M}(\mathcal{N}) \cdot Q = -2\mathrm{i}\,\bar{\mathbf{V}}^M Z_M = -2\mathrm{i}\left(Z\bar{V} + g^{i\bar{j}} D_{\bar{j}} \bar{Z} D_i V\right)\,, \tag{268}$$

where V is the symplectic section introduced in (110); substituting the extremum condition from (267), (268) gives the value of the charges in terms of the fixed scalars

$$[Q - \mathrm{i}\,\mathbb{C}\,\mathcal{M}(\mathcal{N}) \cdot Q]|_{\text{fix}} = -2\mathrm{i}\left(Z\bar{V} + \frac{\mathrm{i}}{2Z}\bar{C}^{ijk} D_i V\, D_j Z\, D_k Z\right)\Bigg|_{\text{fix}} \quad \text{for } Z_{\text{fix}} \neq 0\,,$$
$$[Q - \mathrm{i}\,\mathbb{C}\,\mathcal{M}(\mathcal{N}) \cdot Q]|_{\text{fix}} = -2\mathrm{i}\,\left(g^{i\bar{j}} D_{\bar{j}} \bar{Z} D_i V\right)\big|_{\text{fix}} \quad \text{for } Z_{\text{fix}} = 0\,. \tag{269}$$

The BPS solution corresponds to set $D_i Z = 0$, in which case, for large black holes ($Z_{\text{fix}} \neq 0$), (269) reduces to (182) and (183).

The attractive nature of the extremum was further seen to come from the fact that the mass matrix at that point is strictly positive since

$$\partial_i\partial_j V_{\text{B-H}}|_{(\partial_i V_{\text{B-H}}=0)} = 0\,; \quad \partial_i\partial_{\bar{j}} V_{\text{B-H}}|_{(\partial_i V_{\text{B-H}}=0)} = 2|Z|^2 g_{i\bar{j}}\,. \tag{270}$$

Non supersymmetric extremal black holes with finite horizon area correspond to solutions of (267) with

$$D_i Z \neq 0\,. \tag{271}$$

These solutions may be divided in two classes

- $D_i Z \neq 0, Z \neq 0\,,$
- $D_i Z \neq 0, Z = 0\,.$

For these more general cases, the horizon mass parameter $M_{\text{B–R}}$ which extremizes the ADM mass in moduli space is then given by

$$M^2_{\text{B–R}} = V_{\text{B–H}}|_{(\partial_i V_{\text{B–H}}=0)} = \left[|Z|^2 + |D_i Z|^2\right]_{(\partial_i V_{\text{B–H}}=0)} > |Z|^2_{(\partial_i V_{\text{B–H}}=0)}\,. \tag{272}$$

Equation (272) is a special case of the BPS bound on the mass.

If the central charge Z vanishes on the extremum, then $D_i Z$ have to satisfy

$$C_{ijk} g^{j\bar{j}} g^{k\bar{k}} D_{\bar{j}} \bar{Z} D_{\bar{k}} \bar{Z} = 0 \qquad \forall i \tag{273}$$

in order to fulfill (267). Solutions to the above equation, for the case of special geometries based on symmetric spaces, have been given in [75].

When $Z \neq 0, D_i Z \neq 0$, one may obtain some further consequences of (267). Let us define

$$Z^{\bar{\imath}} \equiv g^{i\bar{\imath}} D_i Z\,, \quad \bar{Z}^{i} \equiv g^{i\bar{\imath}} D_{\bar{\imath}} \bar{Z}. \tag{274}$$

From (267) we get, by multiplication with $g^{i\bar{\imath}}$

$$Z^{\bar{\imath}} = -\frac{\mathrm{i}}{2\bar{Z}} C^{\bar{\imath}}{}_{jk} \bar{Z}^{j} \bar{Z}^{k} \tag{275}$$

and, by multiplication with $\bar{Z}^{i}$

$$|D_i Z|^2 = -\frac{\mathrm{i}}{2\bar{Z}} N_3(\bar{Z}^k) = \frac{\mathrm{i}}{2Z} N_3(Z^{\bar{k}}) \tag{276}$$

where we have introduced the definition $N_3(\bar{Z}^k) \equiv C_{ijk} \bar{Z}^i \bar{Z}^j \bar{Z}^k$. Note that, if at the attractor point $N_3 = 0$, then $Z = 0$ (or $Z \neq 0$ but then $Z^{\bar{\imath}} = 0$).

The complex conjugate of (267) may be rewritten, using (275) as

$$2Z D_{\bar{\imath}} \bar{Z} = -\frac{\mathrm{i}}{4\bar{Z}^2} C_{\bar{\imath}\bar{\jmath}\bar{k}} C^{\bar{\jmath}}{}_{\ell m} \bar{Z}^{\ell} \bar{Z}^{m} C^{\bar{k}}{}_{pq} \bar{Z}^{p} \bar{Z}^{q}. \tag{277}$$

By making use of the special geometry relation [76, 77, 75]

$$C_{\bar{\imath}\bar{\jmath}\bar{k}} C^{\bar{\jmath}}{}_{(\ell m} C^{\bar{k}}{}_{pq)} = \frac{4}{3} C_{(\ell m p} g_{q)\bar{\imath}} + \bar{E}_{\bar{\imath}\ell m p q}, \tag{278}$$

where the tensor $\bar{E}_{\bar{\imath}\ell mpq}$ defined by this relation is related to the covariant derivative of the Riemann tensor and it exactly vanishes for all symmetric spaces, [15], we may finally rewrite (267) as

$$2\bar{Z} D_i Z = \frac{\mathrm{i}}{3Z^2} D_i Z C_{\bar{\jmath}\bar{k}\bar{\ell}} Z^{\bar{\jmath}} Z^{\bar{k}} Z^{\bar{\ell}} + \frac{\mathrm{i}}{4Z^2} E_{i\bar{\jmath}\bar{k}\bar{\ell}\bar{m}} Z^{\bar{\jmath}} Z^{\bar{k}} Z^{\bar{\ell}} Z^{\bar{m}}. \tag{279}$$

Moreover, using also (276) we obtain

$$\left(|Z|^2 - \frac{1}{3} |D_i Z|^2 \right) D_i Z = \frac{\mathrm{i}}{8Z} E_{i\bar{\jmath}\bar{k}\bar{\ell}\bar{m}} Z^{\bar{\jmath}} Z^{\bar{k}} Z^{\bar{\ell}} Z^{\bar{m}}. \tag{280}$$

For symmetric spaces (280) gives

$$|D_i Z|^2 = 3|Z|^2, \tag{281}$$

implying that for these black holes: $M^2_{\mathrm{B\text{-}R}} = 4|Z|^2_{(\partial_i V_{\mathrm{B\text{-}H}}=0)}$.

This relation, for symmetric spaces, was obtained in [54] and then all the solutions of this type have been classified in [75]. In particular, solutions

[15] In this case equation (278) is a consequence of the special geometry relation $D_{\bar{\imath}} C_{jk\ell} = 0$.

with $C_{ijk} \equiv 0$ correspond to the special series of symmetric special manifolds $\frac{SU(1,1+n)}{U(1)\times SU(1+n)}$ for which only non-BPS solutions with $Z = 0$ may exist.

Solutions of the type in (281) have also been found for non-symmetric spaces based on cubic prepotentials in [34].

However, because of (280), these cannot be the most general solutions. For the generic case of non-symmetric special manifolds, we have instead

$$|D_i Z|^2 = 3|Z|^2 + \Delta, \tag{282}$$

where

$$\Delta = -\frac{3}{4}\frac{E_{i\bar{j}\bar{k}\bar{\ell}\bar{m}} Z^{\bar{j}} Z^{\bar{k}} Z^{\bar{\ell}} Z^{\bar{m}}}{N_3(Z^{\bar{k}})} \tag{283}$$

and the Bekenstein–Hawking entropy is

$$S_{\text{B-H}} = A/4 = \pi\left(4|Z|^2 + \Delta\right). \tag{284}$$

Note that, for these non-BPS black holes, at the attractor point Δ is real and, because of (282), it satisfies $-\Delta < 3|Z|^2$.

In all the cases, the attractive nature of the solution depends on the Hessian matrix, which however may have null directions.

6.2 $N > 2$ Matter Coupled Attractors

The $N = 3$ Case

The scalar manifold for this theory, as discussed in Sect. 3.2, is the coset space

$$G/H = \frac{SU(3,n)}{SU(3)\times SU(n)\times U(1)} \tag{285}$$

and the relations among central and matter charges are (see (90))

$$\begin{aligned} D(\omega) Z_{AB} &= Z_I P^I_{AB}\,, \\ D(\omega) Z_I &= \frac{1}{2} Z_{AB} \bar{P}^{AB}_I\,. \end{aligned} \tag{286}$$

The extremum condition on the black-hole potential is then

$$\begin{aligned} dV_{\text{B-H}} &= \frac{1}{2} DZ_{AB}\bar{Z}^{AB} + \frac{1}{2} Z_{AB} D\bar{Z}^{AB} + DZ_I \bar{Z}^I + Z_I D\bar{Z}^I \\ &= P^I_{AB}\bar{Z}^{AB} Z_I + c.c. = 0, \end{aligned} \tag{287}$$

and allows two different solutions with non-zero area. This is expected from Sect. 5.2 because the isometry group of the symmetric space (285) only has a quadratic invariant

$$I_2 = \frac{1}{2}|Z_{AB}|^2 - |Z_I|^2 . \qquad (288)$$

Then,

- either $Z_{AB} \neq 0$, $Z_I = 0$, in this case we have a BPS attractor and the black-hole potential becomes

$$V_{\text{B–H}}|_{\text{attr}} = I_2|_{\text{attr}} > 0 , \qquad (289)$$

- or $Z_I \neq 0$, $Z_{AB} = 0$, which gives a non-BPS attractor solution with black-hole potential

$$V_{\text{B–H}}|_{\text{attr}} = -I_2|_{\text{attr}} > 0 . \qquad (290)$$

The $N = 4$ Case

In this case the scalar manifold is the coset space

$$G/H = \frac{SU(1,1)}{U(1)} \times \frac{SO(6,n)}{SO(6) \times SO(n)} \qquad (291)$$

and the relations among central and matter charges are (see (90) and the discussion below)

$$\begin{aligned} D(\omega)Z_{AB} &= Z_I P^I_{AB} + \frac{1}{2}\bar{Z}^{CD}\epsilon_{ABCD}\, P , \\ D(\omega)\bar{Z}_I &= \frac{1}{2}\bar{Z}^{AB} P_{ABI} + Z_I\, P . \end{aligned} \qquad (292)$$

We recall that for this theory the vielbein P_{ABI} satisfies the reality condition $\bar{P}^{ABI} \equiv (P_{ABI})^\star = \frac{1}{2}\epsilon^{ABCD} P^I_{CD}$.

The extremum condition on the black-hole potential is then

$$\begin{aligned} dV_{\text{B-H}} &= \frac{1}{2}DZ_{AB}\bar{Z}^{AB} + \frac{1}{2}Z_{AB}D\bar{Z}^{AB} + DZ_I\bar{Z}^I + Z_I D\bar{Z}^I = 0 \\ &= P_{ABI}\left(\bar{Z}^{AB}Z_I + \frac{1}{2}\epsilon^{ABCD}Z_{CD}\bar{Z}_I\right) + P\left(Z_I Z_I + \frac{1}{4}\epsilon_{ABCD}\bar{Z}^{AB}\bar{Z}^{CD}\right) \\ &\quad + \bar{P}\left(\bar{Z}^I\bar{Z}^I + \frac{1}{4}\epsilon^{ABCD}\bar{Z}_{AB}\bar{Z}_{CD}\right) = 0 . \end{aligned} \qquad (293)$$

Equation (293) is satisfied for

$$\begin{cases} \bar{Z}^{AB}Z^I + \frac{1}{2}\epsilon^{ABCD}Z_{CD}\bar{Z}^I &= 0 \\ Z^I Z^J \delta_{IJ} + \frac{1}{4}\epsilon_{ABCD}\bar{Z}^{AB}\bar{Z}^{CD} &= 0 \end{cases} . \qquad (294)$$

Therefore we have, in terms of the proper values Z_1, Z_2 of the central charge antisymmetric matrix Z_{AB} (by means of a $U(1) \subset H$ transformation [45], they may always be chosen real and positive) and of the complex matter charges Z^I

$$\begin{cases} \bar{Z}_1 Z^I + Z_2\bar{Z}^I &= 0 \\ Z^I Z^I + 2\bar{Z}_1\bar{Z}_2 &= 0 \end{cases} . \qquad (295)$$

- The BPS solution with finite area is found, as discussed in general in Sect. 5, for

$$Z_I = 0; \qquad Z_2 = 0 \quad (\text{for } Z_1 > Z_2) \tag{296}$$

and corresponds to the black-hole potential

$$V_{\text{B-H}}|_{\text{attr}} = (Z_1)^2 . \tag{297}$$

This solution partially breaks the symmetry of the moduli space, as

$$\begin{cases} SU(4) \to SU(2) \times SU(2) \times U(1) \\ SO(n) \to \qquad SO(n) \end{cases} .$$

There are also two non-BPS solutions:

- One is found by choosing $Z_I = (z, \mathbf{0})$

$$\begin{cases} Z_1 = Z_2 = \rho \\ z = \sqrt{2}\mathrm{i}\rho \end{cases} \tag{298}$$

which gives, for the black-hole potential

$$V_{\text{B–H}}|_{\text{attr}} = (Z_1)^2 + (Z_2)^2 + |z|^2 = 4\rho^2 . \tag{299}$$

In this case the isotropy symmetry then becomes

$$\begin{cases} SU(4) \to USp(4) \\ SO(n) \to SO(n-1) \end{cases} .$$

- The other is obtained by choosing instead $Z_I = (k_1, k_2, \mathbf{0})$ and $Z_{AB} = 0$. This solves (295) for $k_1^2 + k_2^2 = 0$, that is for $k_2 = \pm ik_1 = ik$, giving

$$V_{\text{B-H}}|_{\text{attr}} = |k_1|^2 + |k_2|^2 = 2|k|^2 . \tag{300}$$

For this case, then, the isotropy symmetry preserved is

$$\begin{cases} SU(4) \to SU(4) \\ SO(n) \to SO(n-2) \end{cases} .$$

The analysis of this section is in accord with the discussion on U-invariants of Sect. 5.2. Indeed, the isometry group of the scalar manifold (291) admits the quartic invariant (245)

$$I_4 = S_1^2 - |S_2|^2, \tag{301}$$

where S_1 and S_2 are the $O(6,n)$ invariants introduced in (243) and (244) and we have $S_{\text{B–H}} = \sqrt{|I_4|}$.

For the BPS case, $I_4 > 0$. For the non-BPS ones we have, in the first case $I_4 = -|S_2|^2 < 0$, in the second case $I_4 = S_1^2 > 0$.

The case of the pure $N = 4$ supergravity model anticipated as an example in Sect. 5 falls in this classification and corresponds to the BPS solution (since in that case $Z_I \equiv 0$). It is however interesting to look at the $N = 2$ reduction of that model, where only two of the six vector fields survive, one as the graviphoton and one inside a vector multiplet whose scalars span the coset $\frac{SU(1,1)}{U(1)}$ (axion–dilaton system). Correspondingly, the two proper values of the $N = 4$ central charge play now two different roles: one, say Z_1, is the $N = 2$ central charge, while the other, Z_2, is the matter charge. Equation (295) has now two distinct solutions (corresponding to the twice degenerate BPS solution in $N = 4$): the BPS one, for $Z_2 = 0$, $M_{ADM} = Z_1$, and a non-BPS one, for $Z_1 = 0$, $Z_2 \neq 0$. This is understood, in terms of invariants, from the fact that $SU(1,1)$ does not have an independent quartic invariant, and in fact, in this case, one finds that I_4 reduces to $I_4 = \left[(Z_1)^2 - (Z_2)^2\right]^2$.

6.3 $N > 4$ Pure Supergravity Attractors

We are going to discuss here the attractor solutions for the extended theories with $N > 4$, where no matter multiplets may be coupled. We will include a discussion of their relation to $N = 2$ BPS and non-BPS black holes, already presented in [71].

The $N = 5$ Case

The moduli space of this model is

$$G/H = \frac{SU(1,5)}{U(5)}, \tag{302}$$

the theory contains 10 graviphotons and the relations among the central charges are

$$D(\omega)Z_{AB} = +\frac{1}{2}\bar{Z}^{CD}P_{ABCD}. \tag{303}$$

Correspondingly, the extremum condition on the black-hole potential is

$$\begin{aligned} dV_{\text{B–H}} &= \frac{1}{2}DZ_{AB}\bar{Z}^{AB} + \frac{1}{2}Z_{AB}D\bar{Z}^{AB} \\ &= \frac{1}{4}P_{ABCD}\bar{Z}^{AB}\bar{Z}^{CD} + c.c. = 0. \end{aligned} \tag{304}$$

This extremum condition allows only one solution with non-zero area, the BPS one. Indeed, in terms of the proper values Z_1, Z_2 of Z_{AB}, (304) becomes

$$Z_1Z_2 + \bar{Z}_1\bar{Z}_2 = 0. \tag{305}$$

However, by means of a $U(5)$ rotation Z_1, Z_2 may always be chosen real and non-negative [45], leaving as the only solution with non-zero area $Z_1 > 0$, $Z_2 = 0$ (or vice versa). The black-hole potential on this solution is

$$V_{\text{B-H}}|_{\text{attr}} = |Z_1|^2 \qquad (\text{or } 1 \leftrightarrow 2). \tag{306}$$

This solution is $\frac{1}{5}$-BPS and breaks the symmetry of the moduli space:

$$U(5) \to SU(2) \times SU(3) \times U(1).$$

However, if we truncate this model $N = 5 \to N = 2$, we have the following decomposition of the 10 vectors:

$$\mathbf{10} \to \mathbf{1} + \overline{\mathbf{3}} + \mathbf{6}\,.$$

The singlet corresponds to the $N = 2$ graviphoton, while $\overline{\mathbf{3}}$ is the representation of the three vectors in the vector multiplets. The 6 extra vectors are projected out in the truncation. Correspondingly, the $N = 5$ central charge Z_{AB} reduces to

$$Z_{AB} \to \begin{pmatrix} Z_{ab} = Z\delta_{ab} & 0 \\ 0 & Z_{IJ} = \epsilon_{IJK}\bar{Z}^K \end{pmatrix}, \quad a,b = 1,2;\; I,J,K = 1,2,3. \tag{307}$$

The two solutions $Z_1 > 0, Z_2 = 0$ and $Z_1 = 0, Z_2 > 0$, which were BPS and degenerate in the $N = 5$ theory, in the $N = 2$ interpretation correspond the first to a BPS solution (if we set $Z_1 \equiv Z$) and the second to a non-BPS solution with $Z = 0$, as for the quadratic series discussed in Sect. 6.1.

Let us inspect these results in terms of the discussion of Sect. 5.2. The $SU(5,1)$ invariant is (in terms of the $U(5)$ invariants introduced in Sect. 5.2):

$$I_4 = 4Tr(A^2) - (TrA)^2 \tag{308}$$

that is, in terms of the proper values of the central charge

$$I_4 = \left[(Z_1)^2 - (Z_2)^2\right]^2 . \tag{309}$$

The solutions $Z_1 \neq Z_2$ are separated by the solution $Z_1 = Z_2$, which corresponds to a small black hole, with $I_4 = 0$. This is the solution which preserves the maximal amount of supersymmetry ($\frac{2}{5}$ unbroken), but it does not come from the attractor equations.

The $N = 6$ Case

The moduli space is

$$G/H = \frac{SO^*(12)}{U(6)}, \tag{310}$$

and the theory contains 16 graviphotons, 15 in the twice-antisymmetric representation of $U(6)$ plus a singlet. The attractor solutions for this theory have already been presented in [75].

The relations among the central charges are

$$D(\omega)Z_{AB} = \frac{1}{2}\bar{Z}^{CD}P_{ABCD} + \frac{1}{4!}\bar{Z}\epsilon_{ABCDEF}\bar{P}^{CDEF}\,,$$
$$D(\omega)Z = \frac{1}{2!4!}\bar{Z}^{AB}\epsilon_{ABCDEF}\bar{P}^{CDEF}\,. \tag{311}$$

The black-hole potential for this theory is

$$V_{\text{B-H}} = \frac{1}{2}Z_{AB}\bar{Z}^{AB} + Z\bar{Z} \tag{312}$$

and the extremum condition is then

$$dV_{\text{B-H}} = \frac{1}{2}DZ_{AB}\bar{Z}^{AB} + \frac{1}{2}Z_{AB}D\bar{Z}^{AB} + DZ\bar{Z} + ZD\bar{Z} = 0$$
$$= \frac{1}{4}P_{ABCD}\left(\bar{Z}^{AB}\bar{Z}^{CD} + \frac{1}{3!}\epsilon^{ABCDEF}Z_{EF}Z\right) + c.c. = 0\,. \tag{313}$$

In terms of the proper-values Z_1, Z_2, Z_3 of Z_{AB}, which may always be chosen real and non negative by a $U(6)$ rotation, the condition to be satisfied on the extremum is

$$Z_1Z_2 + ZZ_3 = 0 \qquad (1 \to 2 \to 3 \to 1 \text{ cyclically})\,. \tag{314}$$

This equation admits one solution $\frac{1}{6}$-BPS with $Z = 0$, and two independent non-BPS solutions, both with $Z \neq 0$.

- The BPS solution is found for

$$Z = 0 \qquad Z_2 = Z_3 = 0, \qquad Z_1 \neq 0\,, \tag{315}$$

if we choose $Z_1 \geq Z_2 \geq Z_3$. In this case the black-hole potential becomes

$$V_{\text{B-H}}|_{\text{attr}} = |Z_1|^2 \qquad (\text{or } 1 \leftrightarrow 2 \leftrightarrow 3) \tag{316}$$

and corresponds to $I_4 > 0$.
This solution breaks the symmetry

$$U(6) \to SU(2) \times U(4)$$

and corresponds to an $\frac{SO^*(12)}{SU(4,2)}$ orbit of the charge vector.

- One non-BPS solution is obtained for

$$Z \neq 0 \qquad Z_1 = Z_2 = Z_3 = 0\,. \tag{317}$$

It gives for the black-hole potential

$$V_{\text{B-H}}|_{\text{attr}} = |Z|^2, \tag{318}$$

and preserves all the $U(6)$ symmetry of the moduli space. This solution corresponds to the orbit $\frac{SO^*(12)}{SU(6)}$. Also for this solution the quartic invariant is positive $I_4 > 0$.

- The third solution is found by setting

$$Z_1 = Z_2 = Z_3 = \rho\,, \qquad Z = -\rho\,. \tag{319}$$

In this case the black-hole potential becomes

$$V_{\text{B-H}}|_{\text{attr}} = 4\rho^2\,. \tag{320}$$

This solution breaks the symmetry $U(6) \to USp(6)$, and corresponds to the charge orbit $\frac{SO^*(12)}{SU^*(6)}$. The quartic invariant for this solution is negative $I_4 < 0$.

It is interesting to note, as already observed in [50, 71, 75], that the bosonic sector of the $N = 6$ is exactly the same as the one of the $N = 2$ model coupled with 15 vector multiplets with scalar sector based on the same coset (310). In the $N = 2$ interpretation of this model, the singlet charge Z plays the role of central charge, while the 15 charges Z_{AB} are interpreted as matter charges.

The interpretation of the three attractor solutions is now different: the first one, which was $\frac{1}{6}$-BPS in the $N = 6$ model, is now non-BPS and breaks supersymmetry, while the second one in this model is $\frac{1}{2}$-BPS. The third solution, where all the proper forms of the dressed charges are different from zero, is non-BPS in both interpretations.

The $N = 8$ Case

This model has been studied in detail in [54]. Its scalar manifold is the coset

$$G/H = \frac{E_{7(7)}}{SU(8)}\,. \tag{321}$$

The relations among the 28 central charges are

$$D(\omega)Z_{AB} = \frac{1}{2}\bar{Z}^{CD}P_{ABCD}\,, \tag{322}$$

where the vielbein P_{ABCD} satisfies the reality condition

$$\bar{P}^{ABCD} = \epsilon^{ABCDEFGH}P_{EFGH}\,. \tag{323}$$

The extremum condition is then

$$\begin{aligned} dV_{\text{B–H}} &= \frac{1}{2}DZ_{AB}\bar{Z}^{AB} + \frac{1}{2}Z_{AB}D\bar{Z}^{AB} = 0 \\ &= \frac{1}{4}P_{ABCD}\left(\bar{Z}^{AB}\bar{Z}^{CD} + \frac{1}{4!}\epsilon^{ABCDEFGH}Z_{EF}Z_{GH}\right) = 0\,. \end{aligned} \tag{324}$$

In terms of the central charge proper values $Z_1, \cdots Z_4$ the condition for the extremum may be written

$$\begin{cases} Z_1 Z_2 + \bar{Z}_3 \bar{Z}_4 = 0 \\ Z_1 Z_3 + \bar{Z}_2 \bar{Z}_4 = 0 \\ Z_2 Z_3 + \bar{Z}_1 \bar{Z}_4 = 0 \end{cases} \tag{325}$$

and admits two independent attractor solutions:

- The BPS solution is found for

$$Z_2 = Z_3 = Z_4 = 0, \qquad Z_1 \neq 0, \tag{326}$$

if we choose $Z_1 \geq Z_2 \geq Z_3 \geq Z_4$. In this case the black hole potential becomes

$$V_{\text{B-H}}|_{\text{attr}} = |Z_1|^2 \qquad (\text{or } 1 \leftrightarrow 2 \leftrightarrow 3 \leftrightarrow 4) \tag{327}$$

and corresponds to $I_4 > 0$. This solution breaks the symmetry

$$SU(8) \to SU(2) \times U(6)$$

and corresponds to an $\frac{E_7}{E_{6(2)}}$ orbit of the charge vector.

- The non-BPS solution is obtained for

$$Z_1 = Z_2 = Z_3 = Z_4 = e^{i\frac{\pi}{4}}\rho, \qquad \rho \in \mathbb{R}^+ . \tag{328}$$

It gives for the black-hole potential

$$V_{\text{B–H}}|_{\text{attr}} = 4\rho^2 . \tag{329}$$

This solution breaks the symmetry $SU(8) \to USp(8)$, and corresponds to the charge orbit $\frac{E_7}{E_{6(6)}}$. The quartic invariant for this solution is negative $I_4 < 0$.

7 Conclusions

This survey has presented the main features of the physics of black holes embedded in supersymmetric theories of gravitation. They have an extremely rich structure and give an interplay between space–time singularities in solutions of Einstein matter coupled equations and the solitonic, particle-like structure of these configurations such as mass, spin and charge.

The present analysis may be extended to rotating black holes and to geometries not necessarily asymptotically flat (such as, for example, asymptotically anti-de Sitter solutions). Furthermore, the concept of entropy may be extended to theories which include higher curvature and higher derivative matter terms [27, 28, 42, 43]. This is important in order to make contact with superstring and M-theory where these terms unavoidably appear. In this context, a remarkable connection has been found between the entropy functional and the topological string partition function, an approach pioneered in [29].

Black-hole attractors fall in the class of possible superstring vacua, which in a wide context have led to the study of the so-called landscape [78].

It is a challenging problem to see which new directions towards a fundamental theory of nature these investigations may suggest in the future.

Acknowledgements

The present review is partly based on the work and discussions with the following people: S. Bellucci, A. Ceresole, M. Duff, P. Fré, E. Gimon, M. Gunaydin, R. Kallosh, M.A. Lledó, J. Maldacena, A. Marrani, and A. Strominger.

Work supported in part by the European Community's Human Potential Program under contract MRTN-CT-2004-005104 "Constituents, fundamental forces and symmetries of the universe", in which L.A., R.D'A., and M.T. are associated to Torino University. The work of S.F. has been supported in part by European Community's Human Potential Program under contract MRTN-CT-2004-005104 'Constituents, fundamental forces and symmetries of the universe" and the contract MRTN- CT-2004-503369 "The quest for unification: Theory Confronts Experiments", in association with INFN Frascati National Laboratories and by D.O.E. grant DE-FG03-91ER40662, Task C.

References

1. B. De Witt, C. De Witt eds.: *Black Holes* (Gordon and Breach, New York, 1973); S. W. Hawking, W. Israel: *General Relativity* (Cambridge University Press, Cambridge, 1979); R. M. Wald: *General Relativity* (University of Chicago Press, Chicago, 1984)
2. G. W. Moore: *Les Houches lectures on strings and arithmetic*, arXiv:hep-th/0401049; M. R. Douglas, R. Reinbacher, S. T. Yau: *Branes, bundles and attractors: Bogomolov and beyond*, arXiv:math.ag/0604597
3. M. J. Duff: *String triality, black-hole entropy and Cayley's hyperdeterminant*, arXiv:hep-th/0601134; R. Kallosh, A. Linde: Phys. Rev. D **73**, 104033 (2006) P. Levay: Phys. Rev. D **74**, 024030 (2006) M. J. Duff, S. Ferrara: *E_7 and the tripartite entanglement of seven qubits*, arXiv:quant-ph/0609227; P. Levay: *Strings, black holes, the tripartite entanglement of seven qubits and the Fano plane*, arXiv:hep-th/0610314
4. S. W. Hawking, R. Penrose: Proc. Roy. Soc. Lond. A **314**, 529 (1970)
5. G. Gibbons: in *Unified theories of Elementary Particles. Critical Assessment and Prospects*, Proceedings of the Heisemberg Symposium, München, Germany, 1981, ed. by P. Breitenlohner, H. P. Dürr, Lecture Notes in Physics, Vol. 160 (Springer-Verlag, Berlin, 1982); G. W. Gibbons: in *Supersymmetry, Supergravity and Related Topics*, Proceedings of the XVth GIFT International Physics (Girona, Spain 1984), ed. by F. del Aguila, J. de Azcárraga, L. Ibáñez, (World Scientific, Singapore, 1985), p. 147; P. Breitenlohner, D. Maison, G. W. Gibbons: Commun. Math. Phys. **120**, 295 (1988); R. Kallosh, A. D. Linde, T. Ortin, A. W. Peet, A. Van Proeyen: Phys. Rev. D **46**, 5278 (1992); R. Kallosh, T. Ortin, A. W. Peet: Phys. Rev. D **47**, 5400 (1993); R. Kallosh: Phys. Lett. B **282**, 80 (1992); R. Kallosh, A. W. Peet: Phys. Rev. D **46**, 5223 (1992); R. R. Khuri, T. Ortin: Nucl. Phys. B **467**, 355 (1996); A. Sen: Nucl. Phys. B **440**, 421 (1995); A. Sen: Phys. Lett. B **303**, 22 (1993); A. Sen: Mod. Phys. Lett. A **10**, 2081 (1995); M. Cvetic, C. M. Hull: Nucl. Phys. B **480**, 296 (1996); M. Cvetic, I. Gaida: Nucl. Phys. B **505**, 291 (1997); M. Cvetic, D. Youm: arXiv:hep-th/9512127; M. Cvetic, A. A. Tseytlin: Phys. Rev. D **53**, 5619 (1996)

6. For reviews on black holes in superstring theory see: J. M. Maldacena: *Black-Holes in String Theory*, hep-th/9607235; A. W. Peet: *TASI lectures on black holes in string theory*, arXiv:hep-th/0008241; B. Pioline: Class. Quant. Grav. **23**, S981 (2006); A. Dabholkar: Class. Quant. Grav. **23**, S957 (2006)
7. S. W. Hawking: Phys. Rev. Lett. **26**, 1344 (1971); J. D. Bekenstein: Phys. Rev. D **7**, 2333 (1973)
8. A. Strominger, C. Vafa: Phys. Lett. B **379**, 99 (1996); C. G. . Callan, J. M. Maldacena: Nucl. Phys. B **472**, 591 (1996); G. T. Horowitz, A. Strominger: Phys. Rev. Lett. **77**, 2368 (1996); R. Dijkgraaf, E. P. Verlinde, H. L. Verlinde: Nucl. Phys. B **486**, 77 (1997); D. M. Kaplan, D. A. Lowe, J. M. Maldacena, A. Strominger: Phys. Rev. D **55**, 4898 (1997); J. M. Maldacena: Phys. Lett. B **403**, 20 (1997); J. M. Maldacena, A. Strominger, E. Witten: JHEP **9712**, 002 (1997); M. Bertolini, M. Trigiante: JHEP **0010**, 002 (2000)
9. E. Witten: Nucl. Phys. B **443**, 85 (1995)
10. C. M. Hull, P. K. Townsend: Nucl. Phys. B **438**, 109 (1995)
11. G. Nordström: Proc. Kon. Ned. Akad. Wet. **20**, 1238 (1918); H. Reissner: Ann. Physik **50**, 106 (1916)
12. R. Penrose: Riv. Nuovo Cim. **1**, 252 (1969); Gen. Rel. Grav. **34**, 1141 (2002); R. Penrose: in *General Relativity, an Einstein Centenary Survey,* ed. by S. W. Hawking, W. Israel (Cambridge University Press, Cambridge, 1979)
13. B. Bertotti: Phys. Rev. **116**, 1331 (1959); I. Robinson: Bull. Acad. Pol. Sci. Ser. Sci. Math. Astron. Phys. **7**, 351 (1959)
14. R. Kallosh, T. Ortin: Phys. Rev. D **48**, 742 (1993); E. Bergshoeff, R. Kallosh, T. Ortin: Nucl. Phys. B **478**, 156 (1996)
15. The literature on this topic is quite extended. As a general review, see the lecture notes: K. Stelle: *Lectures on Supergravity p-Branes*, presented at 1996 ICTP Summer School, Trieste, arXiv:hep-th/9701088
16. M. J. Duff, R. R. Khuri, J. X. Lu: *String solitons*, Phys. Rep. **259**, 213 (1995)
17. For recent reviews see: J. H. Schwarz: Nucl. Phys. Proc. Suppl. B **55**, 1 (1997); M. J. Duff: Int. J. Mod. Phys. A **11**, 5623 (1996); A. Sen: Nucl. Phys. Proc. Suppl. **58**, 5 (1997)
18. J. H. Schwarz, A. Sen, Phys. Lett. B **312**, 105 (1993); J. H. Schwarz, A. Sen: Nucl. Phys. B **411**, 35 (1994)
19. M. Gasperini, J. Maharana, G. Veneziano: Phys. Lett. B **272**, 277 (1991); J. Maharana, J. H. Schwarz: Nucl. Phys. B **390**, 3 (1993)
20. J. H. Schwarz: *M theory extensions of T duality*, arXiv:hep-th/9601077; C. Vafa: Nucl. Phys. B **469**, 403 (1996)
21. S. Ferrara, R. Kallosh, A. Strominger: Phys. Rev. D **52**, 5412 (1995)
22. S. Ferrara, R. Kallosh: Phys. Rev. D **54**, 1514 (1996); S. Ferrara, R. Kallosh: Phys. Rev. D **54**, 1525 (1996)
23. A. Strominger: Phys. Lett. B **383**, 39 (1996)
24. J. Polchinski, Y. Cai: Nucl. Phys. B **296**, 91 (1988); C. G. . Callan, C. Lovelace, C. R. Nappi, S. A. Yost: Nucl. Phys. B **308**, 221 (1988); A. Sagnotti: *Open strings and their symmetry groups*, Cargese Summer Inst. (1987) 0521, arXiv:hep-th/0208020; M. Bianchi, A. Sagnotti: Phys. Lett. B **247**, 517 (1990); M. Bianchi, A. Sagnotti: Nucl. Phys. B **361**, 519 (1991); P. Horava: Nucl. Phys. B **327**, 461 (1989); J. Polchinski: Phys. Rev. Lett. **75**, 4724 (1995)
25. S. Ferrara, J. M. Maldacena: Class. Quant. Grav. **15**, 749 (1998); S. Ferrara, M. Gunaydin: Int. J. Mod. Phys. A **13**, 2075 (1998)

26. L. Andrianopoli, R. D'Auria, S. Ferrara: Phys. Lett. B **403**, 12 (1997)
27. R. M. Wald: Phys. Rev. D **48**, 3427 (1993)
28. G. Lopes Cardoso, B. de Wit, T. Mohaupt: Phys. Lett. B **451**, 309 (1999); G. Lopes Cardoso, B. de Wit, T. Mohaupt: Nucl. Phys. B **567**,) 87 (2000)
29. H. Ooguri, A. Strominger, C. Vafa: Phys. Rev. D **70**, 106007 (2004)
30. R. R. Khuri, T. Ortin: Phys. Lett. B **373**, 56 (1996); T. Ortin: Phys. Lett. B **422**, 93 (1998); T. Ortin: *Non-supersymmetric (but) extreme black holes, scalar hair and other open problems*, arXiv:hep-th/9705095
31. S. Ferrara, G. W. Gibbons, R. Kallosh: Nucl. Phys. B **500**, 75 (1997)
32. K. Goldstein, N. Iizuka, R. P. Jena, S. P. Trivedi: Phys. Rev. D **72**, 124021 (2005)
33. R. Kallosh: JHEP **0512**, 022 (2005)
34. P. K. Tripathy, S. P. Trivedi: JHEP **0603**, 022 (2006)
35. A. Giryavets: JHEP **0603**, 020 (2006)
36. R. Kallosh, N. Sivanandam, M. Soroush: JHEP **0603**, 060 (2006)
37. A. Dabholkar, A. Sen, S. Trivedi: *Black hole microstates and attractor without supersymmetry*, arXiv:hep-th/0611143
38. G. G. Gibbons: `http://www.lpthe.jussieu.fr/sugra30/TALKS/Gibbons.pdf`, talk presented at the conference *30 years of Supergravity, journée Joël Scherk*, held in Paris the 18/10/2006
39. M. K. Gaillard, B. Zumino: Nucl. Phys. B **193**, 221 (1981)
40. R. Kallosh: *From BPS to non-BPS black holes canonically*, arXiv:hep-th/0603003
41. For a review, see for instance: S. Bellucci, S. Ferrara, A. Marrani: *Supersymmetric Mechanics P Vol. 2: The Attractor Mechanism and Space Time Singularities*, Lecture Notes in Physics, Vol. 701 (Springer, Berlin/Heidelberg, 2006), Proceedings of the SSM05–Winter School on Modern Trends in supersymmetric Mechanics, INFN-LNF, Italy (2005)
42. G. Lopes Cardoso, B. de Wit, J. Kappeli, T. Mohaupt: JHEP **0412**, 075 (2004)
43. A. Sen: JHEP **0509**, 038 (2005)
44. L. Andrianopoli, R. D'Auria, S. Ferrara, M. A. Lledo: Nucl. Phys. B **640**, 46 (2002)
45. B. Zumino: J.Math.Phys. **3**, 1055 (1962)
46. E. Witten, D. I. Olive: Phys. Lett. B **78**, 97 (1978)
47. For a review on supergravity see for example: L. Castellani, R. D'Auria, P. Fré, *Supergravity and Superstring Theory: A Geometric Perspective* (World Scientific, Singapore, 1990)
48. E. Cremmer: in *Supergravity '81*, ed. by S. Ferrara, J. G. Taylor, p. 313; B. Julia: in *Superspace & Supergravity*, ed. by S. Hawking, M. Rocek (Cambridge, 1981) p. 331
49. A. Salam, E. Sezgin: *Supergravities in Diverse Dimensions*, ed. by A. Salam, E. Sezgin (North-Holland, World Scientific, 1989), Vol. 1
50. L. Andrianopoli, R. D'Auria, S. Ferrara: Int. J. Mod. Phys. A **13**, 431 (1998)
51. L. Andrianopoli, R. D'Auria, S. Ferrara: Int. J. Mod. Phys. A **12**, 3759 (1997)
52. L. Castellani, A. Ceresole, S. Ferrara, R. D'Auria, P. Fré, E. Maina: Nucl. Phys. B **268**, 317 (1986)
53. E. Bergshoeff, I. G. Koh, E. Sezgin: Phys. Lett. B **155**, 71 (1985); M. de Roo, P. Wagemans: Nucl. Phys. B **262**, 644 (1985)
54. S. Ferrara, R. Kallosh: Phys. Rev. D **73**, 125005 (2006)

55. S. Ferrara, A. Strominger: in Proceedings of College Station Workshop *Strings '89'*, ed. by Arnowitt et al. (World Scientific, Singapore, 1989), p. 245; P. Candelas, X. de la Ossa: Nucl. Phys. B **355**, 455 (1991); B. de Wit, A. Van Proeyen: Nucl. Phys. B **245**, 89 (1984); E. Cremmer, C. Kounnas, A. Van Proeyen, J. P. Derendinger, S. Ferrara, B. de Wit, L. Girardello: Nucl. Phys. B **250**, 385 (1985); B. de Wit, P. G. Lauwers, A. Van Proeyen: Nucl. Phys. B **255**, 569 (1985); S. Ferrara, C. Kounnas, D. Lust, F. Zwirner: Nucl. Phys. B **365**, 431 (1991); L. Castellani, R. D'Auria, S. Ferrara: Class. Quant. Grav. **7**, 1767 (1990); L. Castellani, R. D'Auria, S. Ferrara: Phys. Lett. B **241**, 57 (1990)
56. A. Strominger: Commun. Math. Phys. **133**, 163 (1990)
57. L. Andrianopoli, M. Bertolini, A. Ceresole, R. D'Auria, S. Ferrara, P. Fré, T. Magri: J. Geom. Phys. **23**, 111 (1997); L. Andrianopoli, M. Bertolini, A. Ceresole, R. D'Auria, S. Ferrara, P. Fré: Nucl. Phys. B **476**, 397 (1996)
58. E. Cremmer, A. Van Proeyen: Class. Quant. Grav. **2**, 445 (1985)
59. L. Andrianopoli, R. D'Auria, S. Ferrara, P. Fre, M. Trigiante: Nucl. Phys. B **509**, 463 (1998); G. Arcioni, A. Ceresole, F. Cordaro, R. D'Auria, P. Fre, L. Gualtieri, M. Trigiante: Nucl. Phys. B **542**, 273 (1999)
60. K. Behrndt, D. Lust, W. A. Sabra: Nucl. Phys. B **510**, 264 (1998)
61. P. Meessen, T. Ortin: Nucl. Phys. B **749**,) 291 (2006)
62. P. Fré: Nucl. Phys. Proc. Suppl. **57**, 52 (1997)
63. M. Alishahiha, H. Ebrahim: JHEP **0603**, 003 (2006); M. Alishahiha, H. Ebrahim: JHEP **0611**, 017 (2006)
64. P. Kaura, A. Misra: *On the existence of non-supersymmetric black hole attractors for two-parameter Calabi-Yau's and attractor equations*, arXiv:hep-th/0607132
65. S. Bellucci, S. Ferrara, A. Marrani, A. Yeranyan: *Mirror Fermat Calabi-Yau threefolds and Landau-Ginzburg black hole attractors*, arXiv:hep-th/0608091
66. D. Astefanesei, K. Goldstein, S. Mahapatra: *Moduli and (un)attractor black hole thermodynamics*, arXiv:hep-th/0611140
67. G. W. Gibbons, R. Kallosh, B. Kol: Phys. Rev. Lett. **77**, 4992 (1996)
68. S. Ferrara, C. A. Savoy, B. Zumino: Phys. Lett. B **100**, 393 (1981)
69. A. Ceresole, R. D'Auria, S. Ferrara: Nucl. Phys. Proc. Suppl. **46**, 67 (1996)
70. R. Kallosh, B. Kol: Phys. Rev. D **53**, 5344 (1996)
71. S. Ferrara, E. G. Gimon, R. Kallosh: *Magic supergravities, $N = 8$ and black-hole composites*, arXiv:hep-th/0606211
72. M. Gunaydin, O. Pavlyk: JHEP **0508**, 101 (2005)
73. S. Bellucci, S. Ferrara, A. Marrani: Phys. Lett. B **635**, 172 (2006)
74. S. Ferrara, M. Bodner, A. C. Cadavid: Phys. Lett. B **247**, 25 (1990)
75. S. Bellucci, S. Ferrara, M. Gunaydin, A. Marrani: Int. J. Mod. Phys. A **21**, 5043 (2006)
76. E. Cremmer, A. Van Proeyen: Class. Quant. Grav. **2**, 445 (1985)
77. B. de Wit, F. Vanderseypen, A. Van Proeyen: Nucl. Phys. B **400**, 463 (1993)
78. M. R. Douglas: JHEP **0305**, 046 (2003); F. Denef, M. R. Douglas: *Computational complexity of the landscape. I*, arXiv:hep-th/0602072

Expectation Values and Vacuum Currents of Quantum Fields*

G. A. Vilkovisky

Lebedev Physical Institute, Leninsky Prospect 53, Moscow 119991, Russia
vilkov@lebedev.ru

Abstract. Theory of expectation values is presented as an alternative to S-matrix theory for quantum fields. This change of emphasis is conditioned by a transition from the accelerator physics to astrophysics and cosmology. The issues discussed are the time-loop formalism, the Schwinger–Keldysh diagrams, the effective action, the vacuum currents, and the effect of particle creation.

1 Introduction

High-energy physics will probably have to undergo major changes. The accelerators will cease being its experimental base, and it will become a part of astrophysics. Simultaneously, the S-matrix will cease being the central object of high-energy theory because the emphasis on this object is entirely owing to the accelerator setting of the problem. If there is a background radiation that originates from some initial state in the past, then where is the S-matrix here? Astrophysics and cosmology offer the evolution problems rather than the scattering problems. The gravitational collapse is a typical initial-value problem. It is such by its physical setting irrespective of whether the state of the system is classical or quantum. The nature of measurement also changes. No final state is prepared. One measures observables like temperatures or mechanical deflections and subjects these measurements to a statistical treatment to obtain the value of the observable. This means that one measures expectation values in the given initial state. S-matrix theory should give way to expectation-value theory.

There is a proof that accelerator physics is dead: Gabriele Veneziano is leaving CERN for Collège de France. At this historic moment, my mission is to convert him into a new faith. The present preaching consists of four lectures:

* The course of four lectures given at Collège de France in May 2006.

G. A. Vilkovisky: *Expectation Values and Vacuum Currents of Quantum Fields*, Lect. Notes Phys. **737**, 729–784 (2008)
DOI 10.1007/978-3-540-74233-6_23

1. Formal aspects of expectation-value theory.
2. The in-vacuum state and Schwinger–Keldysh diagrams.
3. The effective action.
4. Vacuum currents and the effect of particle creation.

Literature to Lectures 1 and 2 is in [1]–[16]. Additional literature to Lecture 3 is in [17]–[41] and to Lecture 4 in [42]–[56].

2 Formal Aspects of Expectation-Value Theory

2.1 Vocabulary

In these lectures,

$$\hat{\varphi}^i \tag{1}$$

denotes the quantum field. It is an operator function on a given differentiable manifold (referred to below as the base manifold), and i is a point of this manifold. Generally, $\hat{\varphi}^i$ is a collection of fields, and then i is a set containing also the indices labelling these fields. The hat designates an operator. The $\hat{\varphi}^i$ is an operator in a Hilbert space which is not granted. The workers have to build it with their own hands as a representation of the algebra of $\hat{\varphi}$'s. For simplicity, $\hat{\varphi}^i$ will be assumed boson and real (self-adjoint) but otherwise arbitrary.

The starting point is an operator equation for $\hat{\varphi}^i$

$$S_i(\hat{\varphi}) + J_i = 0 \tag{2}$$

which is understood as an expansion. It is meant that there is a c-number function $S_i(\varphi)$ understood as a collection of its Taylor coefficients at some c-number point of configuration space:

$$S_i(\varphi) = \sum_{n=0}^{\infty} \frac{1}{n!} S_{ij_1\cdots j_n}(c)(\varphi - c)^{j_1} \dots (\varphi - c)^{j_n} , \tag{3}$$

and one replaces φ^j in this expansion with an operator. Which c-number field c^j will be used for this expansion does not matter because it will always sum with the operator $(\hat{\varphi}-c)^j$ to make the full quantum field. The expansion point c^j is often called "background field", and there has been much emphasis on it. In fact it is completely immaterial. I shall never make this expansion explicitly, but I shall keep explicit the c-number term of the equation: a source J_i.

Important are only the following three points.

(1) The function $S_i(\varphi)$ is local, i.e., it depends only on φ and its finite-order derivatives at the point i.

(2) The function $S_i(\varphi)$ is a gradient:

$$S_i(\varphi) = \frac{\delta}{\delta\varphi^i} S(\varphi) \,, \tag{4}$$

i.e., there exists an action $S(\varphi)$ generating the operator field equations. For its derivatives the following notation will be used:

$$S_{i_1 \cdots i_n}(\varphi) = \frac{\delta}{\delta\varphi^{i_1}} \cdots \frac{\delta}{\delta\varphi^{i_n}} S(\varphi) \,. \tag{5}$$

Of course, only the total action matters:

$$S_{\text{tot}} = S(\varphi) + \varphi^i J_i \,. \tag{6}$$

(3) There is a special condition on the matrix of second derivatives of $S(\varphi)$. I shall refer to this continuous matrix as S_2:

$$S_{ij}(\varphi) \equiv S_2(\varphi) \,. \tag{7}$$

By locality, S_2 is the kernel of some differential operator on the base manifold for which I shall use the same notation S_2. It is required that S_2 admit a well-posed Cauchy problem in which case it has the unique advanced and retarded inverses (Green's functions) G^+ and G^-:

$$S_{ij} G^{\pm jk} = -\delta_i^k \,, \qquad G^{+jk} = G^{-kj} \,. \tag{8}$$

Because S_2 is symmetric, the advanced inverse is the transpose of retarded.

One may think of S_2 as of a second-order hyperbolic operator which it will in fact be below, but the scheme is more general. It is formalism-insensitive. One's field equations may have the second-order differential form or the first-order differential form – the scheme will work anyway. The importance of the operator S_2 is in the fact that it determines the linear term of the field equations and, therefore, governs the iteration procedures. Commute $\hat{\varphi}^i$ with the field equations. Obtained will be a linear homogeneous equation for the commutator $[\hat{\varphi}^i, \hat{\varphi}^j]$. Consider the respective inhomogeneous equation and its two iterative solutions: one with the advanced inverse for S_2 and the other one with retarded. The equation for the commutator is solved by their difference:

$$[\hat{\varphi}^i, \hat{\varphi}^j] = \mathrm{i}\hbar \left(G^{+ij}(c) - G^{-ij}(c) \right) + O(\hat{\varphi} - c) \,. \tag{9}$$

In this way the algebra of $\hat{\varphi}$'s is built as an operator expansion. This is the quantization postulate.

By the setting of its Cauchy problem, the operator S_2 introduces the concept of causality. If S_2 is a second-order hyperbolic operator, this is the usual relativistic causality. But in any case the base manifold will be foliated with the Cauchy surfaces of the operator S_2. They will be denoted as Σ.

A function of $\hat{\varphi}$ that involves $\hat{\varphi}$ on only one Cauchy surface

$$Q(\hat{\varphi}) = Q(\hat{\varphi}\big|_{\Sigma}) \tag{10}$$

will be called local observable. A state defined as an eigenstate of local observables

$$Q(\hat{\varphi}\big|_{\Sigma})|\ \rangle = q|\ \rangle \tag{11}$$

will be called local state. This latter name may be confusing because the state is, of course, a global concept, and I am using the Heisenberg picture. But the local state is *associated* with a given Σ:

$$|\ \rangle = |\Sigma, q\rangle\ . \tag{12}$$

Of course, for it to be defined, one needs a complete set of commuting local observables. I call the Q's observables, but they may not even be Hermitian. And I shall consider them linear in $\hat{\varphi}$. If they are nonlinear, I shall make a local reparametrization of the field variables so as to make them linear.

In fact, if one has a complete set of commuting local observables, one has already built a Hilbert space. A linear combination

$$|\Sigma\rangle = \int dq\, \Psi(q)|\Sigma, q\rangle \tag{13}$$

is also a local state associated with Σ provided that the function $\Psi(q)$ is external, i.e., independent of the quantum field $\hat{\varphi}^i$.

Our goal is to learn how to calculate expectation values of field observables in a local state, and I shall concentrate on the expectation value

$$\langle\Sigma|\hat{\varphi}^i|\Sigma\rangle\ . \tag{14}$$

However, we shall save the effort if we consider another problem first. Namely, let us recall what would we do in the case of two local states associated with different Cauchy surfaces:

$$|\Sigma_1, q_1\rangle = |1\rangle\ , \qquad |\Sigma_2, q_2\rangle = |2\rangle\ , \tag{15}$$
$$\Sigma_2 > \Sigma_1\ .$$

Here and below, "greater" is a notation for "later".

2.2 The Quantum Boundary-Value Problem

In the problem where given are two local states (15), the field's expectation value is replaced with the scalar product

$$\frac{\langle 2|\hat{\varphi}|1\rangle}{\langle 2|1\rangle} \overset{\text{def}}{=} \langle\varphi\rangle \tag{16}$$

which I shall call mean field although it is not mean in any state.

If our goal was the scalar product (16), we would use the Schwinger principle

$$\delta\langle 2|1\rangle = \mathrm{i}\langle 2|\delta S_{\mathrm{tot}}|1\rangle \text{ or zero} \tag{17}$$

whose meaning is this. Consider a variation in the Taylor coefficients of the field equations, i.e., in the functional form of the total action. The solution for $\hat{\varphi}^i$ will respond and will induce a change in the functions $Q(\hat{\varphi})$ which will induce a change in their eigenstates, and finally there will be a change in the amplitude $\langle 2|1\rangle$ induced by a change in the action. The Taylor coefficients are local. They can be varied in the region between Σ_1 and Σ_2 or outside this region. The Schwinger principle (17) says that, if they are varied outside, the variation of the amplitude is zero. Otherwise, this variation is expressed through the variation of the action by (17).

The Schwinger principle is a consequence of the commutation relations, but it can also be taken for the first principle because one does not need anything else. For many purposes (but not all) it suffices to use a specific case of (17): a freedom of varying the source J. The result of this use is

$$\frac{\delta}{\delta \mathrm{i} J_{j_1}} \cdots \frac{\delta}{\delta \mathrm{i} J_{j_n}} \langle 2|1\rangle = \begin{cases} \langle 2|\overleftarrow{T}\left(\hat{\varphi}^{j_1}\ldots\hat{\varphi}^{j_n}\right)|1\rangle, \text{ if } \Sigma_2 > j_1,\ldots j_n > \Sigma_1\,, \\ 0, \text{ otherwise}\,. \end{cases} \tag{18}$$

Here T orders the operators $\hat{\varphi}^k$, $k \in \Sigma_k$, chronologically, i.e., places them in the order of following of their Σ_k, and the arrow over T points the direction of growth of the time Σ.

Let us come back to the operator field equations. Since all $\hat{\varphi}$'s in these equations are at the same point, one can formally insert in (2) the sign of chronological ordering:

$$\overleftarrow{T} S_i(\hat{\varphi}) + J_i = 0\,. \tag{19}$$

One may worry about additional terms in (19) stemming from the distinction between the chronological and ordinary operator products, and the noncommutativity of $\overleftarrow{T}$ with the derivatives in the Taylor coefficients of the equations. Because the operators in the products are at the same point, these terms are ambiguous expressions whose handling depends on the formalisms and procedures used. There is always a happy end: these terms cancel and help to cancel similar terms appearing in the subsequent calculations. Therefore, it makes sense to use such formalisms and procedures that these terms do not appear at all. This is the approach that I shall follow.

Sandwiching (19) between the states $\langle 2|$ and $|1\rangle$, and using (18), one obtains the following equation for the amplitude:

$$\left(S_i\left(\frac{\delta}{\delta \mathrm{i} J}\right) + J_i\right)\langle 2|1\rangle = 0\,. \tag{20}$$

Multiply it from the left with $\langle 2|1\rangle^{-1}$ and pull the factors $\langle 2|1\rangle$ in the argument of S_i using the fact that this is a unitary transformation:

$$\left(S_i\left(\langle 2|1\rangle^{-1}\frac{\delta}{\delta \mathrm{i}J}\langle 2|1\rangle\right)+J_i\right)1=0\,. \tag{21}$$

In the argument, commute the operators:

$$\left(S_i\left(\frac{\delta\ln\langle 2|1\rangle}{\delta \mathrm{i}J}+\frac{\delta}{\delta \mathrm{i}J}\right)+J_i\right)1=0 \tag{22}$$

and use that by (18)

$$\frac{\delta\ln\langle 2|1\rangle}{\delta \mathrm{i}J_k}=\langle\varphi^k\rangle\,. \tag{23}$$

The result is the following equation for the mean field:

$$\left(S_i\left(\langle\varphi\rangle+\frac{\delta}{\delta \mathrm{i}J}\right)+J_i\right)1=0\,. \tag{24}$$

Equation (24) differs from the classical field equation by the operator addition $\delta/\delta \mathrm{i}J$ to $\langle\varphi\rangle$. When this operator addition acts on 1, its effect is zero, but it will act also on $\langle\varphi\rangle$ because the summands $\langle\varphi\rangle$ and $\delta/\delta \mathrm{i}J$ do not commute. Where in (24) is the Planck constant? It is easy to see by dimension that $\hbar$ is just in front of $\delta/\delta \mathrm{i}J$. Therefore, if one wants to expand the equations in $\hbar$, one should expand them in $\delta/\delta \mathrm{i}J$.

The problem boils down to expanding a function $f(A+B)$ in B when A and B do not commute. It suffices to expand the exponential function since one can write

$$f(A+B)=f\left(\frac{d}{dx}\right)\mathrm{e}^{(A+B)x}\Big|_{x=0} \tag{25}$$

or, equivalently,

$$f(A+B)=\mathrm{e}^{(A+B)d/dx}f(x)\Big|_{x=0}\,. \tag{26}$$

For the exponential function one has the identity

$$\mathrm{e}^{(A+B)x}=\mathrm{e}^{Ax}\left(1+\int_0^x dy\,\mathrm{e}^{-Ay}B\mathrm{e}^{(A+B)y}\right) \tag{27}$$

which makes the expansion possible. This all works well if the series of commutators

$$\mathrm{e}^{-A}B\mathrm{e}^{A}=B+[B,A]+\frac{1}{2!}[[B,A],A]+\frac{1}{3!}[[[B,A],A],A]+\cdots \tag{28}$$

terminates somewhere as in our case. Indeed, if $\langle\varphi\rangle=A$ and $\delta/\delta \mathrm{i}J=B$, then

$$[[B,A],A]=0\,. \tag{29}$$

Under condition (29) one obtains for an arbitrary function:

$$f(A+B) = f(A) + f'(A)B + \frac{1}{2}f''(A)[B,A] + O(B^2)\,. \tag{30}$$

As compared to the ordinary Taylor expansion, there are several additional terms with commutators at each order.

A use of the result above in (24) gives

$$S_i(\langle\varphi\rangle) + \frac{1}{2}S_{ijk}(\langle\varphi\rangle)\frac{\delta\langle\varphi^j\rangle}{\delta \mathrm{i} J_k} + O(\hbar^2) = -J_i\,, \tag{31}$$

$$S_{ij}(\langle\varphi\rangle)\frac{\delta\langle\varphi^j\rangle}{\delta J_k} = -\delta_i^k + O(\hbar)\,. \tag{32}$$

Here the second equation is obtained by differentiating the first one, and it tells us what is $\delta\langle\varphi\rangle/\delta J$. Up to $O(\hbar)$, it is some Green's function of the operator S_2. Denote this Green's function as

$$\frac{\delta\langle\varphi^j\rangle}{\delta J_k} = G^{jk} + O(\hbar)\,. \tag{33}$$

One can work to any order, but I shall stop here. *We obtain closed equations for the mean field:*

$$S_i(\langle\varphi\rangle) + \frac{1}{2\mathrm{i}}S_{ijk}(\langle\varphi\rangle)G^{jk}(\langle\varphi\rangle) + O(\hbar^2) = -J_i\,, \tag{34}$$

$$S_{ij}(\langle\varphi\rangle)G^{jk}(\langle\varphi\rangle) = -\delta_i^k\,. \tag{35}$$

The second term in (34) is the loop

$S_i(\langle\varphi\rangle) +$ 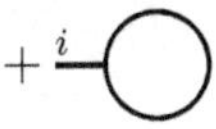 $+\, O(\hbar^2) = -J_i\,,$ (36)

all elements of the loop being functions of $\langle\varphi\rangle$. But two questions remain to be answered:

(i) Which Green's function is G?
(ii) What are the boundary conditions to the mean-field equations?

The answers are again in the Schwinger principle. Equation (18) tells us what are G and $\langle\varphi\rangle$:

$$\frac{1}{\mathrm{i}}G^{jk} = \frac{\langle 2|\overleftarrow{T}\,(\hat\varphi^j\hat\varphi^k)\,|1\rangle}{\langle 2|1\rangle} - \langle\varphi^j\rangle\langle\varphi^k\rangle + O(\hbar)\,, \tag{37}$$

$$\langle\varphi^j\rangle = \frac{\langle 2|\hat\varphi^j|1\rangle}{\langle 2|1\rangle}\,. \tag{38}$$

Multiply these expressions by the coefficients that make the linear Q out of φ:

$$Q(\hat\varphi) = k_j\hat\varphi^j\,, \tag{39}$$

and send j either to Σ_1 or to Σ_2. By the definition of the states $|1\rangle$ and $|2\rangle$, one obtains

$$Q(\langle\varphi\rangle\Big|_{\Sigma_1}) = q_1 \ , \qquad Q(\langle\varphi\rangle\Big|_{\Sigma_2}) = q_2 \ , \tag{40}$$

$$k_j G^{jk}\Big|_{j\in\Sigma_1} = 0 \ , \qquad k_j G^{jk}\Big|_{j\in\Sigma_2} = 0 \ . \tag{41}$$

From (37) it follows also that

$$G^{jk} = G^{kj} \ . \tag{42}$$

The Green's function G is symmetric and completely determined by the boundary conditions (41). This completes the determination of the mean-field equations (34), and for these equations one arrives at a boundary-value problem with the boundary conditions (40). As a result, the quantum boundary-value problem is reduced to a c-number boundary-value problem. I say "c-number" rather than "classical" because there are differences, and one is the presence of terms $O(\hbar)$ in the equations, but, as far as the setting of the problem is concerned, there is no difference. One arrives at the same boundary-value problem for the observable field as in the case of the classical states.

Note that the Green's function G and, thereby, the mean-field equations do not depend on the eigenvalues q. The eigenvalues appear only in the boundary conditions to the equations. However, G depends on the choice of the observables Q themselves and, through them, on the choice of the states $|1\rangle$ and $|2\rangle$. Therefore, the mean-field equations are state-dependent.

Although the Green's function G depends on the choice of the states, it possesses two universal properties. One has already been mentioned: G is always symmetric. The other one is this. Let us make a variation in the operator S_2 and find out how does G respond:

$$S_2 G = -1 \ ,$$

$$S_2 \delta G = -\delta S_2 G \ ,$$

$$\delta G = ?$$

To answer this question, one can use the Schwinger principle again. The result is the following *variational law*:

$$\delta G = G \delta S_2 G \ , \tag{43}$$

and this law is universal. It is the same for all boundary-value problems.

The variational law (43) is remarkable. It is characteristic of finite-dimensional matrices. If a matrix has a unique inverse, then the inverse obeys this law. This law is valid, for example, for the inverse of an elliptic operator, i.e., for the Euclidean Green's function. It is valid also for the advanced and retarded Green's functions:

$$\delta G^+ = G^+ \delta S_2 G^+ \;, \qquad \delta G^- = G^- \delta S_2 G^- \;. \tag{44}$$

But it is not valid generally, and, in the case of S_2, it is exceptional.

The variational law for G has an important implication. Namely, let us differentiate the left-hand side of the mean-field equations

$$\Gamma_i(\varphi) \equiv S_i(\varphi) + \frac{1}{2\mathrm{i}} S_{imn}(\varphi) G^{mn}(\varphi) + O(\hbar^2) \tag{45}$$

to see if the result is symmetric. One obtains

$$\begin{aligned} \frac{\delta \Gamma_i(\varphi)}{\delta \varphi^j} - \frac{\delta \Gamma_j(\varphi)}{\delta \varphi^i} &= \frac{1}{2\mathrm{i}} S_{imn} G^{m\bar{m}} G^{n\bar{n}} S_{\bar{m}\bar{n}j} - (i \leftrightarrow j) + O(\hbar^2) \\ &= 0 + O(\hbar^2) \;. \end{aligned} \tag{46}$$

This means that $\Gamma_i(\varphi)$ is a gradient, i.e., there exists an action generating the mean-field equations:

$$\Gamma_i(\varphi) = \frac{\delta \Gamma(\varphi)}{\delta \varphi^i} \;. \tag{47}$$

There is another way to arrive at the same conclusion. Consider a function of the mean field defined by the Legendre transformation

$$\Gamma(\langle\varphi\rangle) = \frac{1}{\mathrm{i}} \ln\langle 2|1\rangle - \langle\varphi^k\rangle J_k \tag{48}$$

where J is to be expressed through $\langle\varphi\rangle$ by solving equation (23). It is easy to see that this function satisfies the equation

$$\frac{\delta \Gamma(\langle\varphi\rangle)}{\delta\langle\varphi^i\rangle} = -J_i \;, \tag{49}$$

and, therefore, its gradient is the left-hand side of the mean-field equations.

$\Gamma(\varphi)$ is the effective action. Up to $\hbar^2$ it is of the form

$$\Gamma(\varphi) = S(\varphi) + \frac{1}{2\mathrm{i}} \ln\det G(\varphi) + O(\hbar^2) \tag{50}$$

where the second term is the loop without external lines:

$$\Gamma(\varphi) = S(\varphi) + \bigcirc + O(\hbar^2) \;. \tag{51}$$

The effective action exists for any boundary-value problem, but these actions are different for different such problems. Only in the classical approximation, the action and the equations are independent of the boundary conditions.

Let us go over to expectation values.

2.3 The Quantum Initial-Value Problem

In this problem, given is only one local state (which I shall assume normalized). Since the field operators are now sandwiched between the states associated with one and the same Σ:

$$\langle 1|(\cdots)|1\rangle\,, \qquad \langle 1|1\rangle = 1 \tag{52}$$

one cannot apply the Schwinger principle: there is no room for varying the source. One can create this room artificially by inserting a complete set of states associated with some later Σ:

$$\langle 1|1\rangle = \sum_q \langle 1|2q\rangle\langle 2q|1\rangle\,, \tag{53}$$

$$\Sigma_2 > \Sigma_1,$$

but this alone will not help because the source is varied in both amplitudes, and these variations cancel. It will help only if the two amplitudes in (53) are functions of different sources, i.e., if, instead of (53), one introduces a function of two independent sources, J and J^*:

$$Z(J^*, J) = \sum_q \langle 1|2q\rangle_{J^*}\langle 2q|1\rangle_J\,. \tag{54}$$

This amounts to considering two copies of the quantum field: one with the source J and the other one with the source J^*, and using in (54) the amplitudes of both. Then one can vary only one source and, after that, make the sources coincident. Using the Schwinger principle, one obtains

$$\left.\frac{\delta^n Z(J^*, J)}{\delta \mathrm{i} J_{j_1}\cdots\delta\mathrm{i} J_{j_n}}\right|_{J^*=J} = \langle 1|\overleftarrow{T}\left(\hat{\varphi}^{j_1}\ldots\hat{\varphi}^{j_n}\right)|1\rangle\,. \tag{55}$$

In this way the expectation values can be calculated.

The technique of two sources is called time-loop formalism because in expression (54) one goes forward in time, from Σ_1 to some Σ_2, and then back from Σ_2 to Σ_1 but with another copy of the quantum field.

For every partial amplitude in (54) we have (20)

$$\left(S_i\left(\frac{\delta}{\delta \mathrm{i} J}\right) + J_i\right)\langle 2q|1\rangle_J = 0\,. \tag{56}$$

Since the other amplitude in (54) does not depend on J, we can linearly combine (56) to obtain

$$\left(S_i\left(\frac{\delta}{\delta \mathrm{i} J}\right) + J_i\right) Z(J^*, J) = 0\,. \tag{57}$$

Only one source is active in this differential equation. The other one is a parameter. Therefore, we can just repeat the consideration above with

$Z(J^*, J)$ in place of $\langle 2|1\rangle$, and in this way derive the mean-field equations. We obtain the loop expansion of exactly the same form as before:

$$S_i(\langle\varphi\rangle) + \frac{1}{2\mathrm{i}} S_{ijk}(\langle\varphi\rangle) G^{jk}(\langle\varphi\rangle) + O(\hbar^2) = -J_i \; , \tag{58}$$

$$S_{ij}(\langle\varphi\rangle) G^{jk}(\langle\varphi\rangle) = -\delta_i^k \; , \tag{59}$$

and in these loops we must make the sources coincident. There are only two elements in all loops, $\langle\varphi\rangle$ and G. Upon setting $J^* = J$, $\langle\varphi\rangle$ becomes the genuine expectation value

$$\langle\varphi^k\rangle = \left.\frac{\delta \ln Z(J^*, J)}{\delta \mathrm{i} J_k}\right|_{J^*=J} = \langle 1|\hat{\varphi}^k|1\rangle \; , \tag{60}$$

and the matrix G is given by the expression

$$\frac{1}{\mathrm{i}} G^{jk} + O(\hbar) = \left.\frac{\delta^2 \ln Z(J^*, J)}{\delta \mathrm{i} J_j \delta \mathrm{i} J_k}\right|_{J^*=J} = \langle 1|\overleftarrow{T}\left(\hat{\varphi}^j \hat{\varphi}^k\right)|1\rangle - \langle\varphi^j\rangle\langle\varphi^k\rangle \; . \tag{61}$$

I am using for it the same letter G, but it is now a different Green's function of the operator S_2. Equations (58) with this Green's function in all loops are the expectation-value equations.

The solution of the expectation-value equations is specified completely by the initial conditions on Σ_1 following from (60), but it is not easy to write these conditions down in the general terms. Only half of them is obvious: the Q's on Σ_1 are given. To obtain the other half, one would need to find the variables canonically conjugate to Q's and calculate their expectation values on Σ_1.[1] The same concerns the specification of the Green's function G. This issue will be considered in the next lecture where a different approach to it will be used.

Let us consider the state-independent properties of G. First, as seen from (61), G is symmetric for any initial-value problem:

$$G^{jk} = G^{kj} \; . \tag{62}$$

Second, one can apply the Schwinger principle to derive the variational law for G. At this point, the initial-value problem differs significantly from the boundary-value problem. When the operator S_2 is varied in the generating function (54), one can no longer play with only one source because S_2 is the

[1] Let Q's be Hermitian, and let P's have c-number commutators with Q's: $[P, Q] = \mathrm{i}$. Then the expectation values in the state (13) satisfy the initial conditions

$$\langle Q\big|_{\Sigma}\rangle = \int dq \overline{\Psi}(q) q \Psi(q) \; , \qquad \langle P\big|_{\Sigma}\rangle = \mathrm{i} \int dq \overline{\Psi}(q) \frac{\partial}{\partial q} \Psi(q)$$

where the overline means complex conjugation. If both $Q(\hat{\varphi})$ and $P(\hat{\varphi})$ are linear, these are initial conditions directly for $\langle\varphi\rangle$.

same for both copies of the quantum field, and, therefore, both amplitudes in (54) respond. As a consequence, all four matrices of second derivatives are generally involved:

$$\frac{\delta^2 \ln Z}{\delta \mathrm{i} J_j \delta \mathrm{i} J_k}\,, \quad \frac{\delta^2 \ln Z}{\delta \mathrm{i} J^*_j \delta \mathrm{i} J^*_k}\,, \quad \frac{\delta^2 \ln Z}{\delta \mathrm{i} J^*_j \delta \mathrm{i} J_k}\,, \quad \frac{\delta^2 \ln Z}{\delta \mathrm{i} J_j \delta \mathrm{i} J^*_k}\,, \tag{63}$$

i.e., the Green's function G^{jk}, its complex conjugate, and two Wightman functions: $\langle 1|\hat{\varphi}^j\hat{\varphi}^k|1\rangle$ and its transpose. The Wightman functions can be expressed through G^{jk} and the advanced or retarded Green's function:

$$\mathrm{i}\langle 1|\hat{\varphi}^j\hat{\varphi}^k|1\rangle - \mathrm{i}\langle\varphi^j\rangle\langle\varphi^k\rangle = G^{jk} - G^{+jk} + O(\hbar) = G^{kj} - G^{-kj} + O(\hbar)\,. \tag{64}$$

The result of the calculation is the following variational law for G:

$$\delta G = G^- \delta S_2 G + G\delta S_2 G^+ - G^- \delta S_2 G^+\,. \tag{65}$$

It is no more the simple law (43), but it is, nevertheless, universal because G^+ and G^- are state-independent. The variational law (65) is valid for any initial-value problem.

The left-hand side of the expectation-value equations has the form (45) as before but, since the variational law for G is different, the former inference about the symmetry of $\delta\Gamma_i/\delta\varphi^j$ needs to be revised. This inference is no longer valid. The advanced and retarded Green's functions arrange it so that

$$\frac{\delta\Gamma_i(\varphi)}{\delta\varphi^j} = 0 \quad \text{when } i < j \tag{66}$$

and

$$\frac{\delta\Gamma_i(\varphi)}{\delta\varphi^j} \neq 0 \quad \text{when } i > j\,. \tag{67}$$

It follows that there is no action generating the expectation-value equations.

The nonexistence of an action for the initial-value problem is seen also from the consideration of the Legendre transform of the generating function (54). It is now a function of two fields:

$$\Gamma(\varphi^*,\varphi) = \frac{1}{\mathrm{i}}\ln Z(J^*,J) - \varphi J + \varphi^* J^* \tag{68}$$

where

$$\varphi = \frac{\delta \ln Z(J^*,J)}{\delta \mathrm{i} J}\,, \qquad \varphi^* = -\frac{\delta \ln Z(J^*,J)}{\delta \mathrm{i} J^*}\,. \tag{69}$$

The expectation-value equations are obtained as

$$\varphi = \langle 1|\hat{\varphi}|1\rangle\ : \qquad \left.\frac{\delta\Gamma(\varphi^*,\varphi)}{\delta\varphi^i}\right|_{\varphi^*=\varphi} = -J_i\,, \tag{70}$$

and, therefore,

$$\Gamma_i(\varphi) = \left.\frac{\delta\Gamma(\varphi^*,\varphi)}{\delta\varphi^i}\right|_{\varphi^*=\varphi}\,. \tag{71}$$

This is *not* a gradient.

3 The In-Vacuum State and Schwinger–Keldysh Diagrams

3.1 Specification of the State

In order to proceed, I need to specify the state. This will be done in several steps.

Step 1. It will be assumed that S_2 is a second-order hyperbolic operator, and the energy–momentum tensor of the field of small disturbances $\delta\varphi^i$ with the action

$$\frac{1}{2}S_{ij}\delta\varphi^i\delta\varphi^j \tag{72}$$

satisfies the dominant energy condition.

Step 2. The initial-value surface will be shifted to the remote past:

$$\Sigma_1 \to -\infty \,. \tag{73}$$

Consider the operator field equations (2) and (3):

$$J_i + S_i(c) + S_{ij}(c)(\hat{\varphi}-c)^j + \sum_{n=2}^{\infty}\frac{1}{n!}S_{ij_1\cdots j_n}(c)(\hat{\varphi}-c)^{j_1}\dots(\hat{\varphi}-c)^{j_n} = 0 \,. \tag{74}$$

If c^i is some classical solution:

$$S_i(c) = -J_i \,, \tag{75}$$

and $\hat{\phi}^i$ is an operator solution of S_2 against the background c^i:

$$S_{ij}(c)\hat{\phi}^j = 0 \,, \tag{76}$$

then the field

$$\hat{\varphi}^i = c^i + \hat{\phi}^i \,, \qquad i \in \Sigma \to -\infty \tag{77}$$

solves the operator dynamical equations asymptotically in the remote past. It is a property of S_2 that its solution with smooth data having a compact support or decreasing at the spatial infinity decreases also in the time-like directions. Then, as $i \in \Sigma \to -\infty$, the nonlinear terms in (74) decrease even faster and are negligible. Thus, to build a Hilbert space of states, it suffices to build a representation of the algebra of $\hat{\phi}$'s.

Step 3. A Fock space will be built associated with the linear field $\hat{\phi}^i$. This amounts to expanding $\hat{\phi}^i$ in some basis of solutions of $S_2(c)$:

$$S_2(c)\chi_A = 0 \,, \tag{78}$$

$$\hat{\phi}^i = \chi^i_A \hat{a}_{\rm in}{}^A + \overline{\chi}^i_A \hat{a}^{+A}_{\rm in} \tag{79}$$

where the overline means complex conjugation, and the basis functions χ^i_A are normalized with the aid of the inner product:

$$(\chi_A, \chi_B) = 0\,, \qquad (\overline{\chi}_A, \chi_B) = \delta_{AB}\,, \tag{80}$$

$$(\phi_1, \phi_2) \equiv -\mathrm{i} \int\limits_{\Sigma} \phi_1 W_\mu \phi_2 \, d\Sigma^\mu\,. \tag{81}$$

Here W_μ is the Wronskian of S_2. In this way, the concept is introduced of *some* particles detectable in the past. What kind of particles are these, i.e., what kind of detectors detect these particles – depends on the choice of the basis of solutions, but, in any case, the following functions will be chosen for the local observables Q:

$$Q^A(\hat{\varphi}\big|_{\Sigma}) = -\mathrm{i}\delta^{AB} \int\limits_{\Sigma} \chi_B W_\mu (\hat{\varphi} - c)\, d\Sigma^\mu\,, \tag{82}$$

$$\Sigma \to -\infty\,.$$

One needs these observables only on the initial-value surface, and, there, they coincide with the annihilation operators of the introduced particles:

$$Q^A(\hat{\varphi}\big|_{\Sigma\to-\infty}) = \hat{a}_{\mathrm{in}}{}^A\,. \tag{83}$$

The choice of the quantum state will be made in favour of the zero-eigenvalue eigenstate of these observables:

$$\hat{a}_{\mathrm{in}}{}^A|1\rangle = 0\,. \tag{84}$$

This is the vacuum of the introduced particles.

It follows from (77) and (79) that the field's expectation value in the state (84), when taken in the remote past, coincides with the classical solution c^i:

$$\langle 1|\hat{\varphi}^i|1\rangle = c^i\,, \qquad i \in \Sigma \to -\infty\,. \tag{85}$$

The ad hoc classical solution c^i can then be eliminated completely both from the asymptotic form of the quantum field

$$\hat{\varphi}^i = \langle \varphi^i \rangle + \hat{\phi}^i\,, \qquad i \in \Sigma \to -\infty \tag{86}$$

and from the equation defining the Fock modes

$$S_{ij}(\langle\varphi\rangle)\hat{\phi}^j = 0\,, \qquad i \in \Sigma \to -\infty\,. \tag{87}$$

Only the mean field itself figures as a background.

The specification of the state is, however, not completed, because the mean field in the past remains an arbitrary classical solution:

$$S_i(\langle\varphi\rangle) = -J_i\,, \qquad i \in \Sigma \to -\infty \tag{88}$$

and the state itself remains the vacuum of undefined particles. To make the final determination, one more step is needed.

Step 4. The final choice of the state assumes one more limitation on the original action. Namely, it will be assumed that the external source J_i and all the external fields that may be present in the action S are asymptotically static in the past. This means that, asymptotically in the past, there exists a vector field ξ^μ such that it is nowhere tangent to any of the Cauchy surfaces, and the Lie derivative in the direction of ξ^μ of all external fields is zero. Specifically,

$$\mathcal{L}_\xi J_i = 0 \; , \qquad i \in \Sigma \to -\infty \; . \tag{89}$$

If this limitation is fulfilled, then, among the solutions of (88) for the mean field in the past, there is the static one:

$$\mathcal{L}_\xi \langle \varphi^i \rangle = 0 \; , \qquad i \in \Sigma \to -\infty \; . \tag{90}$$

Choose it. Next, use the fact that, with this choice, the operator $S_2(\langle\varphi\rangle)$ commutes with the Lie derivative, and choose for the basis solutions of $S_2(\langle\varphi\rangle)$ the functions that, asymptotically in the past, are eigenfunctions of the Lie derivative:

$$\mathrm{i}\mathcal{L}_\xi \chi^i_A = \varepsilon_A \chi^i_A \; , \quad \varepsilon_A > 0 \; , \quad i \in \Sigma \to -\infty \; . \tag{91}$$

This fixes both the initial conditions for the mean field and the type of particles whose vacuum is the chosen state. These are particles with definite energies.

Since S_2 is a second-order hyperbolic operator, it contains some tensor field, $g^{\mu\nu}$, contracting the second derivatives. The inverse matrix, $g_{\mu\nu}$, can serve and does serve in every respect as a metric on the base manifold. The metric enters the original action S either as a part of the quantum field $\hat{\varphi}^i$ or as an external field. In both cases it is subject to equation (90). When applied to the metric, this is the Killing equation. Thus, we assume the existence, asymptotically in the past, of a time-like Killing vector ξ^μ.

The specification of the quantum initial data is now completed. The notation for the state defined above is

$$|1\rangle = |\text{in vac}\rangle \; , \tag{92}$$

and its full name is relative standard in-vacuum state. It is "relative" because it is relative to the background generated by an asymptotically static source. It is "standard" because it refers to the standard concept of particles. It is "in" because these particles are incoming. And it is "vacuum" because these particles are absent.

The state should not necessarily be chosen as the zero-eigenvalue eigenstate. Since the expectation-value equations do not depend on the eigenvalues, they will have the same form for any eigenstate of the annihilation operators, i.e., for any coherent state

$$\hat{a}_{\text{in}}{}^A |\text{in}\; \alpha\rangle = \alpha^A |\text{in}\; \alpha\rangle \; . \tag{93}$$

Only the initial conditions for the mean field will be different:

$$\langle \alpha \text{ in}|\hat{\varphi}^i|\text{in } \alpha\rangle = c^i + \chi^i_A \alpha^A + \overline{\chi}^i_A \overline{\alpha}^A \,, \qquad i \in \Sigma \to -\infty \,. \tag{94}$$

In addition to the static background c^i generated by a source, the mean field in the past contains now the incoming wave of an arbitrary profile. This is the general setting of the classical evolution problem for an observable field like the electromagnetic or gravitational field. The fact that the nature of the state has changed from classical to quantum did not affect this setting.

It will be useful to keep comparing the initial-value problem with the boundary-value problem. In the latter case, one can define similarly the out-vacuum state and specify the quantum boundary data as

$$|1\rangle = |\text{in vac}\rangle \,, \qquad |2\rangle = |\text{out vac}\rangle \,. \tag{95}$$

3.2 Perturbation Theory

With this specification of the states, let us come back to the mean-field equations. There remains to be obtained the Green's function $G(\varphi)$ that figures in the loops. We need it for an arbitrary background φ, but we have a variational law, (43) or (65), which may be regarded as a differential equation for $G(\varphi)$ with respect to φ. The only thing that is missing and that depends on the choice of states is the initial condition to this equation. It suffices, therefore, to know G for only one background.

Then let us do the simplest: perturbation theory around the trivial background. A second-order hyperbolic operator with the trivial background is the D'Alembert operator with flat metric, $\Box_0$:

$$S_2(\varphi) = \Box_0 + P \,. \tag{96}$$

The remainder is a perturbation P.

In the case of the boundary-value problem, the variational law is (43), and, therefore, the expansion of $G(\varphi)$ is of the form

$$G(\varphi) = G_0 + G_0PG_0 + G_0PG_0PG_0 + \ldots \tag{97}$$

where G_0 is G for the trivial background. This expansion is to be inserted in the loop in the mean-field equations:

$$\frac{1}{2\mathrm{i}} S_{ijk}(\varphi) G^{jk}(\varphi) =$$

(98)

Let for simplicity P be a potential. One obtains the loop expanded in powers of P:

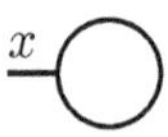

$$= \int dy_1 \ldots dy_n \, F(x|y_1, \ldots y_n) P(y_1) \ldots P(y_n) \,. \tag{99}$$

The coefficients F will be called formfactors. The formfactors are loop diagrams

$F(x|y) =$ 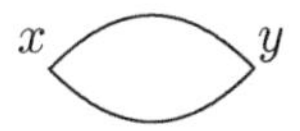

, (100)

$F(x|y_1, y_2) =$ 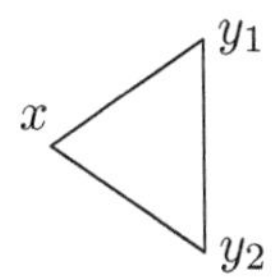

, (101)

....

with the same propagator for all lines: the trivial-background Green's function

$$\text{———} = G_0 \,. \tag{102}$$

What is G_0? With the trivial background and the standard in- and out-vacuum states, it is the Feynman Green's function:

$$G_0 = G_{\text{FEYNMAN}} \,. \tag{103}$$

Let us do the same thing for the initial-value problem. The loop in the expectation-value equations will, in the same way, be expanded in powers of the perturbation, and the expansion will have the same form (99), but the formfactors will be different because the variational law for G is different. It is now (65) rather than (43). Using this law, one obtains for the formfactors three diagrams in place of one:

$F(x|y) =$ 

, (104)

five diagrams in place of one:

$F(x\, y_1, y_2) =$ 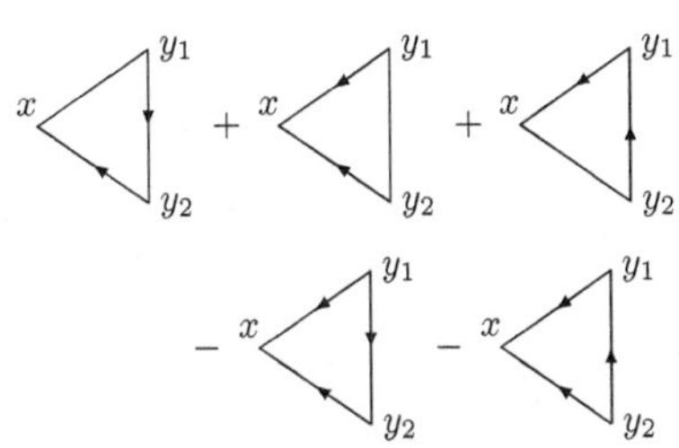

, (105)

and so on. There are two types of propagators in these diagrams: the trivial-background G, and the trivial-background retarded or advanced Green's function. Respectively, there are two types of lines:

$$\text{———} = G_0 \,, \qquad \longleftarrow = G_0^- \text{ or } G_0^+ \,. \tag{106}$$

In the latter case, the arrow points the direction of growth of time. And what is now G_0? In terms of the linear field (76) it is

$$\frac{1}{\mathrm{i}} G_0^{jk} = \langle \text{in vac} | \overleftarrow{T}(\hat{\phi}^j \hat{\phi}^k) | \text{in vac} \rangle \Big|_{\text{trivial background}} \tag{107}$$

and differs from the previous case in that the "⟨out vac|" is replaced by the "⟨in vac|". But, with the trivial background, the vacuum for the linear field is stable. The out-vacuum coincides with the in-vacuum. Therefore,

$$G_0 = G_{\mathrm{FEYNMAN}} \quad \text{(again!)} \,. \tag{108}$$

The diagrams above are called Schwinger–Keldysh diagrams. There is not more than one Feynman propagator in every diagram. The remaining ones are the retarded and advanced Green's functions organized in a special way and with special signs of the diagrams themselves. There is a mystery in this special arrangement. What do these diagrams want to tell us? We must disclose their secret because working with them directly is not what can be recommended.

3.3 Mystery of the Schwinger–Keldysh Diagrams

One thing is obvious right away. In the diagrams above, there is always a chain of retarded Green's functions connecting a given point y with the observation point x. Therefore, the formfactor vanishes if at least one of the y's is in the future of x. This is the *retardation property*

$$F(x|y_1, \dots y_n) = 0 \quad \text{when } y_m > x \,, \ \forall m \,. \tag{109}$$

But this is true of every Schwinger–Keldysh diagram, and why do they appear in the special combinations? What is the role of the Feynman propagator?

Let us make a Fourier transformation of the formfactor with respect to the differences $(x - y_m)$ in the Minkowski coordinates:

$$F(x|y_1, \dots y_n) = \int dk_1 \dots dk_n \, \exp\Big(\mathrm{i} \sum_{m=1}^{n} k_m (x - y_m)\Big) f(k_1, \dots k_n) \,. \tag{110}$$

How come that F possesses the retardation property? It is only that f should admit an analytic continuation to the upper half-plane in the time-like components of k's. Then, for y_m later than x, we shall be able to close the integration contour in the upper half-plane of k_m^0, and the integral will vanish. There should be a function of complex momenta $f(z_1, \dots z_n)$ analytic in the upper half-planes of z_m^0 and such that $f(k_1, \dots k_n)$ is its limiting value on the real axes:

$$f(k_1, \dots k_n) = f(z_1, \dots z_n) \Big|_{z_m^0 = k_m^0 + \mathrm{i}\varepsilon} \,. \tag{111}$$

Let us build this function.

All diagrams in a given-order formfactor are similar. They all are integrals over the momentum circulating in the loop, and the integrands are identical. The difference is only in the integration contours. Thus any diagram in the lowest-order formfactor $f(k)$ is of the form

$$= \int d\boldsymbol{p} \int_{\mathcal{C}} dp^0 \, \frac{\text{polynomial in momenta}}{(-p^{02}+\boldsymbol{p}^2)\,(-(p^0-k^0)^2+(\boldsymbol{p}-\boldsymbol{k})^2)} \, . \tag{112}$$

There are, generally, as many factors in the denominator as there are propagators in the loop, and each factor contains two poles. The contour $\mathcal{C}$ passes round them in accordance with the type of the propagator. One of the three rules applies to each pair of poles:

retardation rule,

advancement rule,

Feynman rule.

Let us now shift the external momentum k^0 to the complex plane. The poles will shift to the complex plane, but we shall also deform smoothly the contour so that it do not cross the poles. In this way one can build a function of complex momenta for each Schwinger–Keldysh diagram. Thus the lowest-order formfactor with complex momentum, $f(z)$, is a sum of three functions:

$$f(z) = \int d\boldsymbol{p} \int_{\mathcal{C}_1} dp^0 \,(\ldots) + \int d\boldsymbol{p} \int_{\mathcal{C}_2} dp^0 \,(\ldots) - \int d\boldsymbol{p} \int_{\mathcal{C}_3} dp^0 \,(\ldots) \, , \tag{113}$$

and the contours $\mathcal{C}_1$, $\mathcal{C}_2$, $\mathcal{C}_3$ for z^0 in the upper half-plane are shown in Fig. 1. By considering the pinch conditions, i.e., the conditions that the poles pinch the integration contour, one can check in each case that these functions can have singularities only on the real axis. Therefore, if we consider them in the upper half-plane, they are analytic, and their limits on the real axis are our original diagrams.

There remains to be understood what are these functions. Since the integrands are identical, the sum of the integrals in (113) is the integral over the sum of the contours

$$f(z) = \int d\boldsymbol{p} \int_{\mathcal{C}_1+\mathcal{C}_2-\mathcal{C}_3} dp^0 \,(\ldots) \, . \tag{114}$$

Sum up the three contours in Fig. 1. The resultant contour is such that every pair of poles is passed round by the Feynman rule. It may be called Feynman contour.

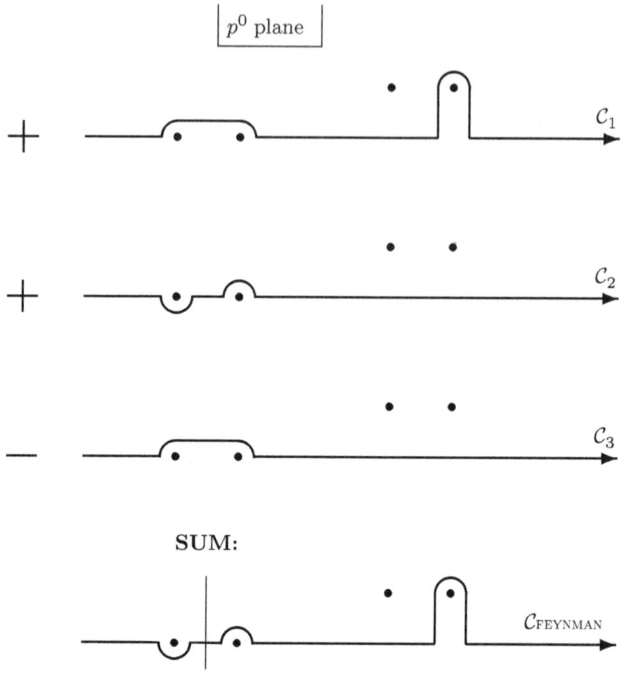

Fig. 1. Integration contours for the three diagrams in the lowest-order formfactor (113). The sum of the contours is the Feynman contour

But the Feynman contour defines also the in–out formfactor (100) in which both propagators are Feynman, except that the in–out formfactor is not the limit of $f(z)$ from the upper half-plane. It is this limit on only half of the real axis, and on the other half it is the limit from the lower half-plane. The in–in and in–out formfactors are different boundary values of the same complex function having a cut on the real axis:

$$\text{in–in}: \qquad f(k) = f(z)\Big|_{z^0 = k^0 + \mathrm{i}\varepsilon}, \tag{115}$$

$$\text{in–out}: \qquad f(k) = f(z)\Big|_{z^0 = (1+\mathrm{i}\varepsilon)k^0}, \tag{116}$$

and the function itself is the integral over the Feynman contour

$$f(z) = \int d\mathbf{p} \int\limits_{C_{\text{FEYNMAN}}} dp^0\,(\ldots)\,. \tag{117}$$

The same is true of all n-th order formfactors, and this is a disclosure of the mystery. In each case, the set of Schwinger–Keldysh diagrams is just a splitting of one Feynman diagram whose purpose is to display the retardation property and in this way to tell us which boundary value is to be taken.

3.4 Reduction to the Euclidean Effective Action

The Feynman contour is famous for the fact that, when the external momenta are on the imaginary axis, the Feynman contour is the imaginary axis itself. With all the momenta imaginary, both the external ones and the one circulating in the loop, this is the Euclidean formfactor. Then we can *start* with the calculation of the Euclidean formfactor and next analytically continue it in momenta from the imaginary axis to the real axis either in the way shown in Fig. 2a or in the way shown in Fig. 2b. In the first case we shall obtain the in-out formfactor, and in the second case the in–in formfactor of Lorentzian theory. It is invaluable that loops can be calculated Euclidean.

Then let us make one more step. A formfactor with the Euclidean momentum can be put in the spectral form

$$f(k) = \int_0^\infty dm^2 \, \frac{\rho(m^2)}{m^2 + k^2} + \text{ a polynomial in } k^2 \,, \tag{118}$$

$$k^2 > 0$$

with some spectral weight $\rho(m^2)$, the resolvent $1/(m^2+k^2)$, and a polynomial accounting for a possible growth of $f(k)$ at $k^2 \to \infty$. There are similar forms for the higher-order formfactors. If the formfactor is in the spectral form, the procedure of analytic continuation boils down merely to replacing the Euclidean resolvent with the retarded or Feynman resolvent:

$$\text{in–in} : \qquad f(k) = \int_0^\infty dm^2 \, \frac{\rho(m^2)}{m^2 - (k^0 + \mathrm{i}\varepsilon)^2 + \boldsymbol{k}^2} + \text{ a polynomial in } k^2 \,, \tag{119}$$

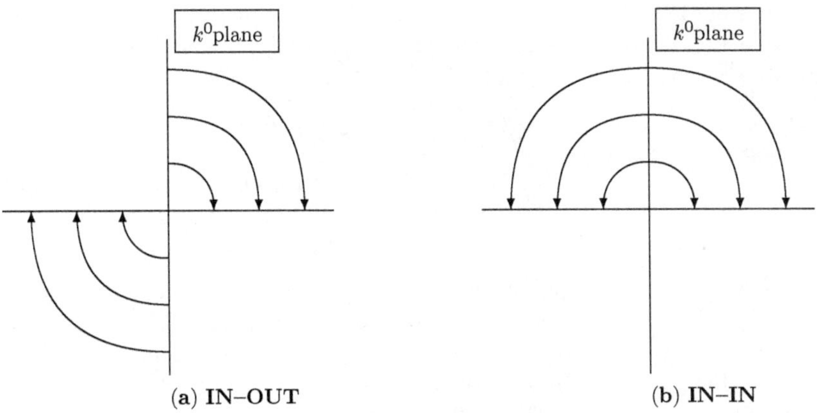

Fig. 2. Analytic continuation of the Euclidean formfactor that gives (**a**) the in–out formfactor and (**b**) the in–in formfactor of Lorentzian theory

$$\text{in–out}: \qquad f(k) = \int_0^\infty dm^2 \frac{\rho(m^2)}{m^2 - k^{02} + \boldsymbol{k}^2 - \mathrm{i}\varepsilon} + \text{ a polynomial in } k^2 \,. \tag{120}$$

Note that the spectral weight is the same in all cases: the one of the Euclidean loop. Thus, the problem boils down to obtaining the spectral weights of the Euclidean formfactors.

Then back from the Fourier-transformed formfactors to the formfactors themselves, and from the formfactors to the mean-field equations. For the loop in these equations expanded in powers of the perturbation, we obtain an expression of the following form:

$$\begin{aligned} &= (c_1 + c_2\Box_0 + \ldots)P(x) \\ &+ \int_0^\infty dm^2\, \rho(m^2) \frac{1}{m^2 - \Box_0} P(x) \\ &+ \int_0^\infty dm_1^2 dm_2^2 dm_3^2\, \rho(m_1^2, m_2^2, m_3^2) \\ &\quad \times \frac{1}{m_1^2 - \Box_0} \left[\left(\frac{1}{m_2^2 - \Box_0} P(x) \right) \left(\frac{1}{m_3^2 - \Box_0} P(x) \right) \right] \\ &+ \ldots \,. \end{aligned} \tag{121}$$

Here the first term is local. It comes from the polynomial in the spectral form. The remaining terms are nonlocal but expressed through the resolvent which is a Green's function of the massive operator $\Box_0 - m^2$. It is initially the Euclidean Green's function since we are calculating the Euclidean loop. For the Lorentzian equations, we arrive at the following rule. To obtain the expectation-value equations in the in-vacuum state, replace all the Euclidean resolvents in (121) with the retarded Green's functions. To obtain the mean-field equations for the in–out problem, replace all the Euclidean resolvents with the Feynman Green's functions:

$$\text{All } \frac{1}{m^2 - \Box_0} \;\longrightarrow\; \begin{cases} \text{Euclidean,} \\ \text{Retarded,} \\ \text{Feynman.} \end{cases} \tag{122}$$

At every level of expectation-value theory, there are proofs that the expectation-value equations possess two basic properties: they are real and causal. Causality is the retardation property discussed above. But it is not enough to have proofs. These properties should be manifestly built into the working formalism. Expression (121) offers such a formalism. Since the retarded resolvent secures the causality and is real, this expression is manifestly real and causal.

But even this is not enough. The theory may possess symmetries, and one may want these symmetries to be manifest. To this end it will be noted that, although expansion (121) is obtained in terms of the trivial-background resolvent $1/(m^2-\Box_0)$, it can be regrouped so as to restore the full-background resolvent

$$\frac{1}{m^2-S_2}=\frac{1}{m^2-\Box_0-P} \tag{123}$$

at each order. It does not matter whether this regrouping will be made in the expectation-value equations or in the Euclidean equations because the retarded and Euclidean Green's functions obey the same variational law (43):

$$\frac{1}{m^2-\Box_0}=\frac{1}{m^2-S_2}-\frac{1}{m^2-S_2}P\frac{1}{m^2-S_2}+\dots\,. \tag{124}$$

This proves that the rule of replacing resolvents applies to the full-background resolvents as well as to the trivial-background ones. The latter fact is important because the Euclidean loops can be calculated covariantly from the outset, and the transition to the expectation-value equations by replacing the full-background resolvents does not break the manifest symmetries. The expectation-value equations are obtained in as good an approximation as the Euclidean equations are.

There remains to be made a final observation. For the Euclidean equations, *there is* an effective action:

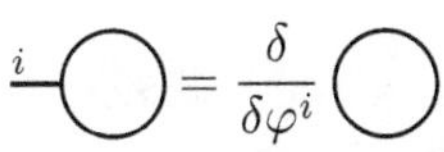

(125)

because the variational law for the Euclidean Green's function is (43). It is invaluable that loops can be calculated without external lines. This reduces the calculations greatly, helps to control symmetries, helps to control renormalizations.

Thus, at the end of the day, we conclude that *there is* an action that generates the expectation-value equations, but it does so indirectly, i.e., *not* through the least-action principle. To make this clear, consider (for the illustrative purposes only) any quadratic action:

$$\Gamma(\varphi)=\frac{1}{2}\int dx\,\varphi f(\Box_0)\varphi\,.$$

Whatever the operator $f(\Box_0)$ is, in the variational derivative it gets symmetrized:

$$\frac{\delta\Gamma(\varphi)}{\delta\varphi}=\frac{1}{2}\left(f(\Box_0)+f^{\mathrm{T}}(\Box_0)\right)\varphi=f^{\mathrm{sym}}(\Box_0)\varphi\,.$$

Assuming that the function $f(\Box_0)$ is in the spectral form

$$f(\Box_0)=\int\limits_0^\infty dm^2\,\rho(m^2)\frac{1}{m^2-\Box_0}\,,$$

one obtains the variational equations with the symmetrized resolvent:

$$\int_0^\infty dm^2\,\rho(m^2)\left(\frac{1}{m^2-\Box_0}\right)^{\text{sym}}\varphi = -J\ .$$

These cannot be the expectation-value equations since they are not causal. But, through the derivation above, we know how to correct this: just to replace the symmetrized resolvent with the retarded resolvent. The corrected equations

$$\int_0^\infty dm^2\,\rho(m^2)\left(\frac{1}{m^2-\Box_0}\right)^{\text{ret}}\varphi = -J\ .$$

do not already follow from any action although indirectly they do. Only if the action $\Gamma(\varphi)$ is local, i.e., the function $f(\Box_0)$ is polynomial, the least-action principle holds directly.

Two precepts should be kept in mind when using the formalism above. First, the replacement rule concerns the resolvents of the formfactors and not the propagators in the loop. The loop should be calculated Euclidean. Hence

First Precept: First do the loop, next replace the resolvents.

Second, the replacement of resolvents is to be made in the equations and not in the action. It does not make sense to make it in the action. Hence

Second Precept: First vary the action, next replace the resolvents.

We thus go over to the calculation of the Euclidean effective action.

4 The Effective Action

4.1 The Operator S_2

The φ^i is a set of fields for which a more explicit notation will now be used:

$$\varphi^i = \varphi^a(x)\ . \tag{126}$$

The operator S_2 acts on a small disturbance of φ^i and is a second-order differential operator

$$S_{ij}\delta\varphi^j = (X^{\mu\nu}_{ab}\partial_\mu\partial_\nu + Y^\mu_{ab}\partial_\mu + Z_{ab})\,\delta\varphi^b(x)\ . \tag{127}$$

The generality of this operator will, however, be restricted by the condition that the coefficient of the senior term factorizes as

$$X^{\mu\nu}_{ab} = \omega_{ab}\,g^{\mu\nu}\ , \qquad \det\omega_{ab}\neq 0\ ,\ \det g^{\mu\nu}\neq 0\ . \tag{128}$$

In this case, the operator (127) is said to be diagonal, or minimal, or nonexotic. Condition (128) is too restrictive and not necessary. It can be replaced by a more general condition

$$\det\left(X^{\mu\nu}_{ab}n_\mu n_\nu\right) = C(g^{\mu\nu}n_\mu n_\nu)^d \quad \forall n_\mu\ , \qquad d = \dim a\ ,\ \ C \neq 0\ ,\ \ \det g^{\mu\nu} \neq 0\ , \tag{129}$$

and even this condition can be generalized. Higher-order and first-order operators can also be considered but, in all of these cases, the Green's functions of S_2 are expressed through the Green's functions of a diagonal second-order operator. The case (128) is basic.

In the case (128), the matrix ω_{ab} can be factored out:

$$S_{ij}\delta\varphi^j = \omega_{ac}H^c_b\delta\varphi^b(x)\ , \tag{130}$$

and a covariant derivative can be introduced:

$$\nabla_\mu\, \delta\varphi^a = (\delta^a_b\partial_\mu + \mathcal{A}_{\mu}{}^a_b)\,\delta\varphi^b \tag{131}$$

so as to absorb the first-order term:

$$H^a_b = \delta^a_b g^{\mu\nu}\nabla_\mu\nabla_\nu + P^a_b\ . \tag{132}$$

This is the final form of S_2. A short notation will be used:

$$H = \Box\hat{1} + \hat{P} \tag{133}$$

where

$$\Box \equiv g^{\mu\nu}\nabla_\mu\nabla_\nu\ , \tag{134}$$

and the hat designates a matrix in a, b:

$$\hat{1} = \delta^a_b\ , \quad \hat{P} = P^a_b\ , \quad \mathrm{tr}\,\hat{P} = P^a_a\ , \quad \text{etc.} \tag{135}$$

The matrix ω_{ab} may be regarded as a local metric in the space of fields. The symmetry of S_2 implies that this matrix is symmetric, covariantly constant, and converts $\hat{P}$ into a symmetric form:

$$\omega_{ab} = \omega_{ba}\ , \qquad \nabla_\mu\omega_{ab} = 0\ , \tag{136}$$

$$P^c_a\omega_{cb} - P^c_b\omega_{ca} = 0\ . \tag{137}$$

The dominant energy condition implies that ω_{ab} is positive definite. The matrix $g^{\mu\nu}$ is the inverse of the metric on the base manifold. Since we are considering Euclidean theory, this metric is positive definite too.

Apart from the algebraic factor ω_{ac} in (130), the operator S_2 contains three background fields:

$$g^{\mu\nu}\ , \quad \nabla_\mu\ , \quad \hat{P}, \tag{138}$$

i.e., the metric, the connection (or covariant derivative), and the matrix potential. And where is the original background φ of $S_2(\varphi)$? When S_2 is calculated from the action S, the metric, connection, and potential are obtained as functions of the original set of fields φ, but from now on it does not matter. The effective action is expressed in a universal manner through the fields (138) only.

The strengths of the fields (138) are respectively the Riemann tensor, the commutator of covariant derivatives, and the potential which is its own strength:

$$R_{\alpha\beta\mu\nu}\,, \quad [\nabla_\mu, \nabla_\nu] = \hat{\mathcal{R}}_{\mu\nu}\,, \quad \hat{P}\,. \tag{139}$$

I shall call these field strengths curvatures and use for them the collective notation

$$\left(R_{\alpha\beta\mu\nu}\,,\ \hat{\mathcal{R}}_{\mu\nu}\,,\ \hat{P} \right) = \Re\,. \tag{140}$$

The following contractions of the curvatures will be called currents:

$$\hat{J}_\mu \equiv \nabla^\nu \hat{\mathcal{R}}_{\mu\nu}\,, \tag{141}$$

$$J_{\mu\nu} \equiv R_{\mu\nu} - \frac{1}{2} g_{\mu\nu} R\,, \quad J \equiv g^{\mu\nu} J_{\mu\nu}\,. \tag{142}$$

The currents are conserved:

$$\nabla^\mu \hat{J}_\mu = 0\,, \qquad \nabla^\mu J_{\mu\nu} = 0\,. \tag{143}$$

If all the curvatures vanish, the background is trivial. The effective action is a functional of the curvatures (140).

4.2 Redundancy of the Curvatures

The effective action is a nonlocal functional of the curvatures, and this fact conditions a certain simplification.

Since the commutator curvature is a commutator, it satisfies the Jacobi identity, and so does the Riemann curvature:

$$\nabla_\gamma \hat{\mathcal{R}}_{\mu\nu} + \nabla_\nu \hat{\mathcal{R}}_{\gamma\mu} + \nabla_\mu \hat{\mathcal{R}}_{\nu\gamma} = 0\,, \tag{144}$$

$$\nabla_\gamma R_{\alpha\beta\mu\nu} + \nabla_\nu R_{\alpha\beta\gamma\mu} + \nabla_\mu R_{\alpha\beta\nu\gamma} = 0\,. \tag{145}$$

Act on these identities with ∇^γ. In the first term, the operator $\Box$ forms, and in the remaining terms commute the covariant derivatives. The commutator brings an extra power of the curvature. The equations obtained

$$\Box \hat{\mathcal{R}}_{\mu\nu} + O(\Re^2) = 2\nabla_{[\nu} \hat{J}_{\mu]}\,, \tag{146}$$

$$\Box R_{\alpha\beta\mu\nu} + O(\Re^2) = 4\nabla_{[\mu} \nabla_{\langle\alpha} \left(J_{\nu]\beta\rangle} - \frac{1}{2} g_{\nu]\beta\rangle} J \right) \tag{147}$$

hold identically and have the form of inhomogeneous wave equations, the role of inhomogeneity being played by the currents. In (146), (147), the brackets of both types [] and $\langle\,\rangle$ denote the antisymmetrization in the respective indices.

The equations (146) and (147) are nonlinear, but they can be solved by iteration. The result is that the commutator and Riemann curvatures get expressed in a nonlocal fashion through their currents and an arbitrary solution of the homogeneous wave equation

$$\Box\hat{\mathcal{R}}^{\text{wave}}_{\mu\nu} = 0 \,, \qquad \Box R^{\text{wave}}_{\alpha\beta\mu\nu} = 0 \,. \tag{148}$$

If the metric is Lorentzian, this solution is fixed by initial data which can be given in the remote past. It follows that the commutator and Riemann curvatures are specified by giving an incoming wave and the current J. This fact underlies the Maxwell and Einstein equations. They fix the currents J. Adding initial conditions to these equations specifies the connection and metric.

In the present case, since the metric is Euclidean, there are no wave solutions:

$$\hat{\mathcal{R}}^{\text{wave}}_{\mu\nu} = 0 \,, \qquad R^{\text{wave}}_{\alpha\beta\mu\nu} = 0 \,, \tag{149}$$

and the Green's function $1/\Box$ is unique. Therefore, the commutator and Riemann curvatures are expressed entirely through their currents:

$$\hat{\mathcal{R}}_{\mu\nu} = \frac{1}{\Box} 2\nabla_{[\nu}\hat{J}_{\mu]} + O(J^2) \,, \tag{150}$$

$$R_{\alpha\beta\mu\nu} = \frac{1}{\Box} 4\nabla_{[\mu}\nabla_{\langle\alpha}\left(J_{\nu]\beta\rangle} - \frac{1}{2} g_{\nu]\beta\rangle} J\right) + O(J^2) \,. \tag{151}$$

Thus, the curvatures are redundant because there are no waves in Euclidean theory. Owing to this fact, the set of field strengths (140) reduces to

$$\left(J_{\mu\nu} \,,\ \hat{J}_{\mu} \,,\ \hat{P}\right) \,, \tag{152}$$

and the effective action is a functional of the reduced set.

4.3 The Axiomatic Effective Action

To what class of functionals does the effective action belong? One can say in advance that this should be a functional analytic in the curvature. Indeed, the first variational derivative of the effective action taken at the trivial background should vanish because, in the absence of an external source, the relative vacuum becomes the absolute vacuum. The trivial background should solve the mean-field equations in the absolute vacuum. Higher-order variational derivatives taken at the trivial background determine the correlation functions in the absolute vacuum. They may not vanish but neither should they blow up.

The analyticity suggests that the effective action can be built as a sum of nonlocal invariants of N-th order in the curvature:

$$\varGamma = \sum_N \varGamma_N \,, \qquad \varGamma_N = O[\Re^N] \,. \tag{153}$$

Nonlocal invariant is, however, an uncertain concept. Even local invariant of N-th order in the curvature is a concept that needs to be refined, but this is easy to do. The most general local monomial that can be built out of the available quantities yields an invariant of the form

$$\int dx\, g^{1/2} \underbrace{(\nabla_1 ... \nabla_1)(\nabla_2 ... \nabla_2)\ldots}_{k} \Re_1 \Re_2 \ldots \Re_N + O[\Re^{N+1}] \,. \tag{154}$$

This monomial is a product of N curvatures and k covariant derivatives, all indices being contracted by the metric. In (154), the labels $1, 2, \ldots$ point out which derivative acts on which curvature, but all the curvatures are at the same point, and the total number of derivatives is finite. Of course, the curvature sits also in the covariant derivatives and in the metric that contracts the indices. Therefore, the N-th order invariant can only be defined up to terms $O[\Re^{N+1}]$. In particular, the covariant derivatives in (154) can be commuted freely because the contribution of a commutator is already $O[\Re^{N+1}]$.

One may now consider a class of nonlocal invariants that can formally be represented as infinite series of local invariants:

$$\mathit{\Gamma}_N = \int dx\, g^{1/2} \sum_{k=0}^{\infty} c_k \underbrace{(\nabla_1 ... \nabla_1)(\nabla_2 ... \nabla_2)\ldots}_{k} \Re_1 \Re_2 \ldots \Re_N + O[\Re^{N+1}] \,. \tag{155}$$

Here c_k are some dimensional constants. It can be seen that this is the needed class.[2] The number of curvatures in (155) is N, but the number of derivatives is unlimited. Only a finite number of derivatives can contract with the curvatures. The remaining ones can only contract among themselves. If two derivatives acting on the same curvature contract, they make a $\Box$ operator acting on this curvature:

$$\nabla_1{}^2 = \Box_1 \,, \quad \nabla_2{}^2 = \Box_2 \,, \quad \ldots \,. \tag{156}$$

If two derivatives acting on different curvatures contract, the contraction can again be written in terms of the $\Box$ operators:

$$\begin{aligned} 2\nabla_1 \nabla_2 &= (\nabla_1 + \nabla_2)^2 - \nabla_1{}^2 - \nabla_2{}^2 \\ &= \Box_{1+2} - \Box_1 - \Box_2, \end{aligned} \tag{157}$$

but there appears a $\Box$ operator acting on the product of two curvatures:

$$\Box_{1+2}\, \Re_1 \Re_2 \Re_3 \ldots = \Box\, (\Re\Re)\, \Re_3 \ldots \,. \tag{158}$$

As a result, (155) takes the form

[2] To see it, consider any diagram with massive propagators and expand it formally in the inverse mass. The method that accomplishes this expansion is known as the Schwinger–DeWitt technique.

$$\mathit{\Gamma}_N = \int dx\, g^{1/2} \left(\sum_{k_1,k_2,\cdots=0}^{\infty} c_k(\Box_1)^{k_1}(\Box_2)^{k_2}(\Box_{1+2})^{k_3}\ldots \right) \times \underbrace{\left(\nabla ... \Re_1 \nabla ... \Re_2 \ldots \nabla ... \Re_N\right)}_{\text{contraction}} + O[\Re^{N+1}]\,. \tag{159}$$

There remains an infinite series in the $\Box$ variables, and these variables themselves are operators acting on the curvatures in a given contraction. The remaining series is some function of the $\Box$ variables:

$$\mathit{\Gamma}_N = \int dx\, g^{1/2} F\left(\Box_1, \Box_2, \Box_{1+2}, \ldots\right) \underbrace{\left(\nabla ... \Re_1 \nabla ... \Re_2 \ldots \nabla ... \Re_N\right)}_{\text{contraction}} + O[\Re^{N+1}]\,. \tag{160}$$

This is the general form of a nonlocal invariant of N-th order in the curvature. The function F is a formfactor.

There is, in addition, the identity

$$\nabla_1 + \nabla_2 + \ldots + \nabla_N = 0 \tag{161}$$

which reduces the number of variables in the function F. The sum in (161) is a derivative acting on the product of all curvatures, i.e., a total derivative. Total derivatives vanish because the curvatures may be considered having compact supports. Thus invariants of first order in the curvature can only be local because any derivative is a total derivative. Therefore, the first-order formfactors are constants:

$$N = 1 \;: \qquad F = \text{const.} \tag{162}$$

At the second order, all formfactors are functions of only one argument because the remaining arguments can be eliminated by integration by parts:

$$N = 2 \;: \qquad F = F(\Box_1)\,, \tag{163}$$

$$\Box_2 = \Box_1\,, \quad \Box_{1+2} = 0\,.$$

At the third order, all formfactors are functions of three individual $\Box$'s because the $\Box$'s acting on pairs can be eliminated:

$$N = 3 \;: \qquad F = F(\Box_1, \Box_2, \Box_3)\,, \tag{164}$$

$$\Box_{1+2} = \Box_3\,, \quad \Box_{1+3} = \Box_2\,, \quad \Box_{2+3} = \Box_1\,.$$

The $\Box$'s acting on pairs appear beginning with the fourth order in the curvature and are parameters of the on-shell scattering amplitudes.

Nonlocal invariants of a given order make a linear space in which all possible contractions of N curvatures and their derivatives make a basis, and the formfactors play the role of coefficients of the linear combining. The basis can be built by listing all independent contractions. The effective action is an expansion in this basis with certain coefficients–formfactors:

$$\Gamma = \Gamma_{\rm I} + \Gamma_{\rm II} + \Gamma_{\rm III} + \dots\ , \tag{165}$$

$$\Gamma_{\rm I} = \int dx\, g^{1/2}\Big[c_1 R + c_2\,{\rm tr}\,\hat{P}\Big]\ , \tag{166}$$

$$\begin{aligned}\Gamma_{\rm II} = \int dx\, g^{1/2}\,{\rm tr}\,\Big[& R_{\mu\nu}\ F_1(\Box)\ R^{\mu\nu} \\ & + R\ F_2(\Box)\ R \\ & + \hat{P}\ F_3(\Box)\ R \\ & + \hat{P}\ F_4(\Box)\ \hat{P} \\ & + \hat{\mathcal{R}}_{\mu\nu}\ F_5(\Box)\ \hat{\mathcal{R}}^{\mu\nu}\Big]\ ,\end{aligned} \tag{167}$$

$$\begin{aligned}\Gamma_{\rm III} = \int dx\, g^{1/2}\,{\rm tr}\ \Big[& F_1(\Box_1,\Box_2,\Box_3)\,\hat{P}_1\hat{P}_2\hat{P}_3 \\ & + F_2(\Box_1,\Box_2,\Box_3)\,\hat{\mathcal{R}}_1{}^{\mu}{}_{\alpha}\hat{\mathcal{R}}_2{}^{\alpha}{}_{\beta}\hat{\mathcal{R}}_3{}^{\beta}{}_{\mu} \\ & + \cdots \\ & + F_{29}(\Box_1,\Box_2,\Box_3)\,\nabla_\lambda\nabla_\sigma R_1^{\alpha\beta}\nabla_\alpha\nabla_\beta R_2^{\mu\nu}\nabla_\mu\nabla_\nu R_3^{\lambda\sigma}\Big]\ .\end{aligned} \tag{168}$$

In the first-order action (166), there are two basis contractions: the Ricci scalar and the trace of the matrix potential, and the formfactors are constants. In the second-order action, there are five independent contractions listed in (167). In the third-order action, there are 29 basis contractions, examples of which are given in (168). Here I shall stop because, for the problems of interest, the third order is sufficient. The reason for that will be explained in the next lecture.

In the expressions above, the basis invariants are written in terms of the curvatures, but they can be rewritten in terms of the conserved currents. Note also that the operator arguments of the third-order formfactors F commute because they act on different objects. Since the arguments commute, the functions F themselves are ordinary functions of three variables.

Thus, even before any calculation, we have an ansatz for the effective action, with unknown formfactors. We need them in the spectral forms

$$F_{\rm k}(\Box) = \int\limits_0^\infty dm^2\,\frac{\rho_{\rm k}(m^2)}{m^2-\Box} + \text{ a polynomial in } \Box\ , \tag{169}$$

$$F_{\rm k}(\Box_1, \Box_2, \Box_3) = \int_0^\infty dm_1^2 dm_2^2 dm_3^2 \, \frac{\rho_{\rm k}(m_1^2, m_2^2, m_3^2)}{(m_1^2 - \Box_1)(m_2^2 - \Box_2)(m_3^2 - \Box_3)} \,, \tag{170}$$

and then we can proceed directly to the expectation-value equations. Unknown are only the spectral weights. These are to be calculated from the loop diagrams, but there is an alternative approach. One can look for the general limitations on the spectral weights stemming from axiomatic theory. These limitations may be sufficient to solve one's expectation-value problem. In this case, the solution will prove to be independent of the details of the quantum-field model and the approximations made in it. Moreover, the effective action above does not refer even to quantum field theory. It is an action for the observable field, and its implications may be valid irrespective of the underlying fundamental theory. Only certain axiomatic properties of the spectral weights may be important. There is an example in which this approach has been implemented [53].

Here, the axiomatic approach will not be considered. Let us see how the effective action is calculated from loops.

4.4 Heat Kernel

Consider any diagram in the effective action

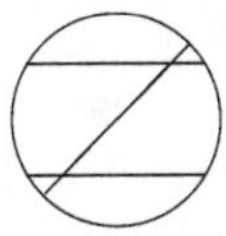

, (171)

and, for every propagator, write

$$\text{———} = -\frac{1}{H} = \int_0^\infty ds \, \mathrm{e}^{sH} \,. \tag{172}$$

The kernel of the exponential operator

$$\mathrm{e}^{sH} \delta(x, y) \equiv \hat{K}(x, y|s) \tag{173}$$

(and the operator itself) is called heat kernel, and the parameter s is often called proper time. Both names are matters of history, and a matter of physics is the fact that H is negative definite. The matrix P in (133) may spoil the negativity but, since it is treated perturbatively, as one of the curvatures, this does not matter.

Upon the insertion of (172), the diagram remains the same as before but with the heat kernels in place of the propagators, and the integrations over the proper times will be left for the last:

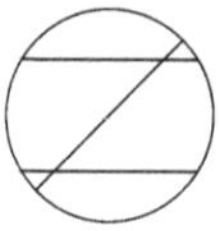

$$= \int_0^\infty ds_1 \ldots \int_0^\infty ds_n$$

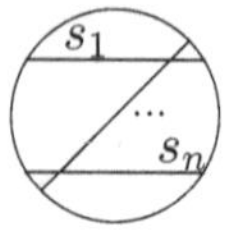

$$. \tag{174}$$

The one-loop effective action is the functional trace of the heat kernel, integrated over s:

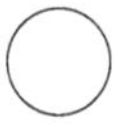

$$= \frac{1}{2} \ln \det \frac{1}{H} = \frac{1}{2} \int_0^\infty \frac{ds}{s} \int dx \, \mathrm{tr}\, \hat{K}(x, x|s) \,. \tag{175}$$

Thus, one is left with diagrams with the heat kernels. It will be seen in a moment why this is better.

The expansion rule for the exponential operator has already been considered in (27). There remains to be presented the lowest-order approximation for the heat kernel:

$$\hat{K}(x, y|s) = \frac{1}{(4\pi s)^{D/2}} \left(\mathrm{e}^{-\sigma(x,y)/2s} \hat{a}(x, y) + O[\Re] \right) , \tag{176}$$

$$D = \text{dimension of the base manifold.} \tag{177}$$

At the lowest order in the curvature, the potential P does not affect this expression, but the metric and connection do. As mentioned above, covariant expansions cannot be rigid. In (176)

$$2\sigma(x, y) = (\text{geodetic distance between } x \text{ and } y)^2 \tag{178}$$

in the metric entering the operator H. The connection entering the operator H defines a parallel transport along a line. Parallel transport is a linear mapping, so there exists a propagator of parallel transport (the matrix that accomplishes this mapping). In (176)

$$\hat{a}(x, y) = \begin{array}{l}\text{propagator of the parallel transport from } y \text{ to } x \\ \text{along the geodesic connecting } y \text{ and } x.\end{array} \tag{179}$$

The geodesic comes from the metric, and the parallel transport from the connection.

The two-point functions (178) and (179) are the main elements of the Schwinger–DeWitt technique mentioned above and the basic building blocks for all Green's functions: of the hyperbolic operator H, and of the elliptic operator H, and the heat kernel. What is special about the heat kernel? Special is the fact that, as seen from expression (176), the heat kernel is finite at the coincident points. Green's functions of the hyperbolic and elliptic operators are singular, and this is normal. Abnormal is the fact that in the loop diagrams they appear at the coincident points. Finiteness of the heat kernel at the coincident points is a bonus owing to which all diagrams with the heat kernels are finite.

The divergences of the loop diagrams reappear in the proper-time integrals in (174). These integrals diverge at the lower limits. At this stage, one more advantage of the heat kernel comes into effect. Namely, the manifold dimension D enters only the overall factor in (176). Apart from this factor, the expansion of the heat kernel in the curvature does not contain D explicitly. Therefore, loops with the heat kernels are calculated once for all dimensions, and then the knowledge of the analytic dependence on D enables one to apply the dimensional regularization to the proper-time integrals. One integrates by parts in s keeping $\Re D < 4$ and next goes over to the limit $D \to 4$. For example,

$$\int_0^\infty \frac{ds}{s^{D/2-1}} f(s) = \frac{1}{2 - D/2} f(0) - \int_0^\infty ds \ln s \frac{df(s)}{ds} + O\left(2 - D/2\right) . \qquad (180)$$

The dimensional regularization annihilates all power divergences. Only the logarithmic divergences survive and take the form of poles in dimension. These poles affect only the polynomial terms in the spectral representations of the formfactors. They appear in the coefficients of the polynomials, thereby making these coefficients indefinite. As a consequence, the local terms of the effective action will have indefinite coefficients. I shall come back to this issue.

After the substitution of the heat kernels for the propagators, the calculation of loops becomes an entertaining geometrical exercise.

4.5 Loops and Geometry

The heat kernel involves σ and $\hat{a}$. The derivative of σ

$$\nabla^\mu \sigma(x, y) \equiv \sigma^\mu(x, y) \qquad (181)$$

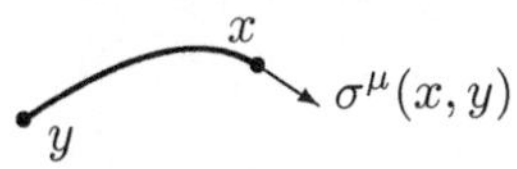

is the vector tangent to the geodesic connecting y and x, directed outwards, and normalized to the geodetic distance between y and x:

$$g_{\mu\nu}\sigma^\mu\sigma^\nu = 2\sigma \ , \qquad \sigma^\mu\Big|_{x=y} = 0 \ , \qquad \det \nabla^\nu \sigma^\mu \Big|_{x=y} \neq 0 \ . \qquad (182)$$

The normalization condition is a closed equation for σ which together with the conditions at the coincident points can serve as the definition of σ. The defining equation for $\hat{a}$ together with the condition at the coincident points is

$$\sigma^\mu \nabla_\mu \hat{a}(x, y) = 0 \ , \qquad \hat{a}\Big|_{x=y} = \hat{1} \ . \qquad (183)$$

The determinant

$$\det\left(\nabla^x_\mu \nabla^y_\nu \sigma(x, y)\right) = g^{1/2}(x) g^{1/2}(y) \varDelta(x, y) \qquad (184)$$

is known as the Van Vleck–Morette determinant. It is responsible, in particular, for a caustic of the geodesics emanating from x or y.

The vector σ^μ can be used to expand any function in a covariant Taylor series. For a scalar, this series is of the form

$$f(y) = \sum_{n=0}^{\infty} \frac{(-1)^n}{n!} \sigma^{\mu_1} \dots \sigma^{\mu_n} \nabla_{\mu_1} \dots \nabla_{\mu_n} f(x) \,. \tag{185}$$

If f is not a scalar, it should at first be parallel transported from y to x:

$$f(y) = \hat{a}(y,x) \sum_{n=0}^{\infty} \frac{(-1)^n}{n!} \sigma^{\mu_1} \dots \sigma^{\mu_n} \nabla_{\mu_1} \dots \nabla_{\mu_n} f(x) \,. \tag{186}$$

The covariant Taylor expansion is a regrouping of the ordinary Taylor expansion. Whatever the connection is, it cancels in this series. The series can formally be written in the exponential form

$$f(y) = \hat{a}(y,x) \exp\left(-\sigma^\mu \nabla_\mu\right) f(x) \tag{187}$$

which will be of use below. Two-point functions expanded in this way get expressed through their covariant derivatives at the coincident points. Thus

$$\varDelta(x,y) = 1 + \frac{1}{6} R_{\mu\nu} \sigma^\mu \sigma^\nu + \dots \,. \tag{188}$$

A loop always involves the ring of $\hat{a}$'s

$$\hat{a}(x,x_1)\hat{a}(x_1,x_2)\dots\hat{a}(x_n,x) \,, \tag{189}$$

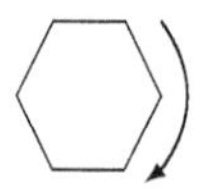

i.e., the parallel transport around a geodetic polygon. The ring of two $\hat{a}$'s is the parallel transport there and back along the same path. Therefore,

$$\hat{a}(x,x_1)\hat{a}(x_1,x) \equiv \hat{1} \,. \tag{190}$$

The ring of three $\hat{a}$'s is the parallel transport around the geodetic triangle. It involves the commutator curvature, and the curvature terms can be calculated:

$$\hat{a}(x,x_1)\hat{a}(x_1,x_2)\hat{a}(x_2,x) = \hat{1} + \frac{1}{2} \hat{\mathcal{R}}_{\alpha\beta} {\sigma_1}^\alpha {\sigma_2}^\beta + \dots \,, \tag{191}$$

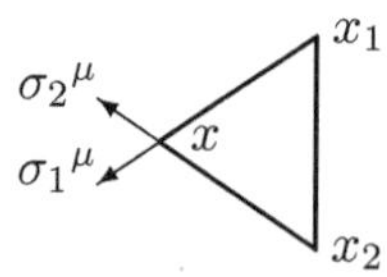

(192)

This is sufficient because any polygon can be broken into triangles:

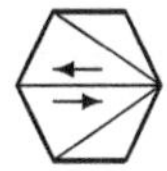

$$. \tag{193}$$

Solution of the geodetic triangle is also involved. In the notation of (192),

$$\left(\sigma^{\mu}(x_1, x_2)\right)^2 = {\sigma_1}^2 + {\sigma_2}^2 - 2\sigma_1\sigma_2 - \frac{1}{3}R_{\mu\alpha\nu\beta}{\sigma_1}^{\mu}{\sigma_1}^{\nu}{\sigma_2}^{\alpha}{\sigma_2}^{\beta} + \dots \,. \tag{194}$$

Here the first two terms make the Pythagorean theorem, the third term accounts for the angle not being the right angle, and the terms with the Riemann curvature can be calculated.

The above is to give a flavour of what loops imply.

4.6 Calculation of Loops

The heat kernel calculates loops with a remarkable elegance. As an example, consider the contribution of the second order in the curvature to the effective action. The respective one-loop diagram contains two curvatures $\Re$ and two heat kernels with the proper times s_1 and s_2:

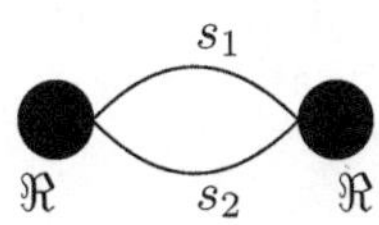

$$+ O[\Re^3]$$

$$= \int dx\, g^{1/2} \int dy\, g^{1/2}\, \Re(x)\hat{K}(x, y|s_1)\hat{K}(x, y|s_2)\Re(y) + O[\Re^3]\,. \tag{195}$$

Suppose that the calculation only needs to be done with accuracy $O[\Re^3]$. Then one can insert in (195) the lowest-order approximation for the heat kernels. In this approximation, the rings of $\hat{a}$'s collapse to $\hat{1}$, and the remaining $\hat{a}$'s always transport the $\Re$'s to the same point arranging their complete contraction. With the $\hat{a}$'s and the numerical coefficients omitted, the diagram (195) is of the form

$$\frac{1}{{s_1}^{D/2}}\frac{1}{{s_2}^{D/2}} \int dx\, g^{1/2} \int dy\, g^{1/2} \times \Re(x) \exp\left(-\frac{\sigma(x, y)}{2s_1}\right) \exp\left(-\frac{\sigma(x, y)}{2s_2}\right) \Re(y)\,. \tag{196}$$

But the exponents here simply add, and the two heat kernels turn into one with a complicated proper-time argument:

$$\frac{1}{(s_1 s_2)^{D/2}} \int dx\, g^{1/2} \int dy\, g^{1/2}\, \Re(x) \exp\left(-\frac{s_1 + s_2}{2s_1 s_2}\sigma(x, y)\right) \Re(y)$$

$$= \frac{1}{(s_1+s_2)^{D/2}} \int dx\, g^{1/2} \int dy\, g^{1/2}\, \Re(x) K\left(x,y\Big|\frac{s_1 s_2}{s_1+s_2}\right) \Re(y)\,. \tag{197}$$

One only needs to rewrite this heat kernel in the operator form:

$$\frac{1}{(s_1+s_2)^{D/2}} \int dx\, g^{1/2}\, \Re \exp\left(\frac{s_1 s_2}{s_1+s_2}\Box\right) \Re(y)\,, \tag{198}$$

and the loop is done. The proper-time integral

$$\int_0^\infty ds_1 \int_0^\infty ds_2 \frac{1}{(s_1+s_2)^{D/2}} \exp\left(\frac{s_1 s_2}{s_1+s_2}\Box\right) = F(\Box) \tag{199}$$

is the formfactor.

What has happened? The propagators in the loop glued together, and the loop turned into a tree:

(200)

This is what means to do the loop. *It means to turn it into a tree.* The role of the propagator in the tree is played by the formfactor $F(\Box)$.

Consider now any multiloop diagram with parallel propagators. It turns into a tree

(201)

in a completely similar way. The inverse proper times add:

$$\frac{1}{s_1} + \frac{1}{s_2} + \ldots = \frac{1}{s_{\text{total}}}$$

(the law of parallel conductors). There is nothing to do.

For more than two curvatures a more powerful method is used. Consider the diagram

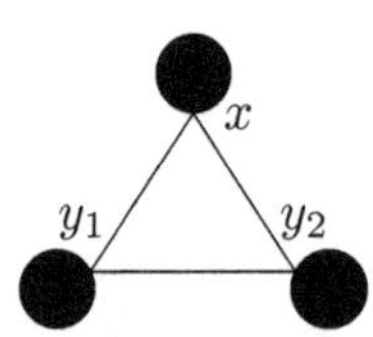

$$+\, O[\Re^4]\,, \tag{202}$$

and suppose again that it is needed only up to the next order in the curvature. Then, with the $\hat{a}$'s and the numerical coefficients omitted, it is of the form

$$\frac{1}{{s_1}^{D/2}} \frac{1}{{s_2}^{D/2}} \frac{1}{{s_3}^{D/2}} \int dx\, g^{1/2} \int dy_1\, g^{1/2} \int dy_2\, g^{1/2}$$

$$\times \exp\left(-\frac{\sigma(x,y_1)}{2s_1}-\frac{\sigma(x,y_2)}{2s_2}-\frac{\sigma(y_1,y_2)}{2s_3}\right)\Re(x)\Re(y_1)\Re(y_2)\,. \quad (203)$$

Choose one of the vertices, say x, to be the observation point of the effective Lagrangian. One of the curvatures, $\Re(x)$, is already there. Shift the remaining curvatures to x using the covariant Taylor series:

$$\Re(y_i) = \exp\left(-\sigma_i{}^{\mu}\nabla_\mu\right)\Re(x)\,, \quad (204)$$

$$\sigma_i{}^{\mu} = \sigma^{\mu}(x,y_i)\,, \quad i=1,2\,. \quad (205)$$

Next, consider the geodetic triangle with the same vertices as in the diagram. For the geodesics connecting x with y_i, write

$$2\sigma(x,y_i) = (\sigma_i)^2\,, \quad (206)$$

and, for the geodesic between the y's, use the Pythagorean theorem:

$$2\sigma(y_1,y_2) = (\sigma_1)^2+(\sigma_2)^2-2\sigma_1\sigma_2+O[\Re]\,. \quad (207)$$

Finally, replace the integration variables:

$$y_1{}^{\mu}\to\sigma_1{}^{\mu}\,, \quad y_2{}^{\mu}\to\sigma_2{}^{\mu}\,. \quad (208)$$

The Jacobian

$$\left|\frac{\partial\sigma^{\mu}(x,y_i)}{\partial y_i{}^{\nu}}\right|^{-1} = \frac{g^{1/2}(x)}{g^{1/2}(y_i)}\Delta^{-1}(x,y_i) = \frac{g^{1/2}(x)}{g^{1/2}(y_i)}\left(1+O[\Re]\right) \quad (209)$$

removes the measure $g^{1/2}$ from the integral in y_i and brings an extra $g^{1/2}$ to the integral in x. Expression (203) takes the form

$$\frac{1}{(s_1s_2s_3)^{D/2}}\int dx\, g^{1/2}\left(g^{1/2}(x)\right)^2\int d\sigma_1 d\sigma_2 \exp\left(-\frac{{\sigma_1}^2}{4s_1}-\frac{{\sigma_2}^2}{4s_2}\right.$$
$$\left.-\frac{{\sigma_1}^2+{\sigma_2}^2-2\sigma_1\sigma_2}{4s_3}-\sigma_1{}^{\mu}\nabla_\mu{}^1-\sigma_2{}^{\mu}\nabla_\mu{}^2\right)\Re(x)\Re_1(x)\Re_2(x)\,. \quad (210)$$

Here the labels $1,2$ on ∇_μ and $\Re$ point out which ∇_μ acts on which $\Re$. The operators ∇_μ figure as parameters in the integral, and, up to the next order in $\Re$, they commute. Since the parameters commute, the integral in $\sigma_1{}^{\mu}$, $\sigma_2{}^{\mu}$ is an ordinary Gaussian integral. Do it. The extra factor $\left(g^{1/2}(x)\right)^2$ cancels, and the result is

$$B(s_1,s_2,s_3)\int dx\, g^{1/2}\exp\left(\sum_{i,k=1}^{2} b_{ik}(s_1,s_2,s_3)\nabla_i\nabla_k\right)\Re(x)\Re_1(x)\Re_2(x) \quad (211)$$

where $B(s_1, s_2, s_3)$ is some function of the proper times, and the exponent is a quadratic form in ∇_1, ∇_2 with s-dependent coefficients. The loop is done. The integral

$$\int_0^\infty ds_1 ds_2 ds_3 \, B(s_1, s_2, s_3) \exp\left(\sum_{i,k=1}^{2} b_{ik}(s_1, s_2, s_3)\nabla_i \nabla_k\right) = F({\nabla_1}^2, {\nabla_2}^2, \nabla_1\nabla_2) \quad (212)$$

is the formfactor. Integration by parts in x brings it to the $\Box$ arguments:

$$F({\nabla_1}^2, {\nabla_2}^2, \nabla_1\nabla_2) \to F({\nabla_1}^2, {\nabla_2}^2, \nabla^2)\,. \quad (213)$$

The effect of the calculation above is again that the loop is turned into a tree:

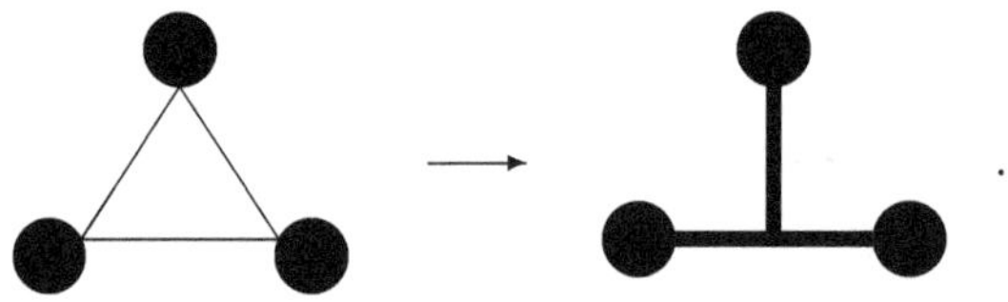

(214)

The vertex of the tree is the formfactor $F({\nabla_1}^2, {\nabla_2}^2, {\nabla_3}^2)$. This method applies to any diagram with the heat kernels. One only needs to do Gaussian integrals, and the result is always the exponential of a quadratic combination of ∇'s. The formfactor is a function of the products $\nabla_i\nabla_k$.

4.7 The One-Loop Formfactors

The result of the proper-time integrations depends essentially on the dimension D. For $D = 4$, the one-loop formfactors in the effective action (165) are as follows.

With one exception, all second-order formfactors are logs:

$$F_1(\Box) = \frac{1}{60}\frac{1}{2(4\pi)^2}\ln(-\Box) + \text{ const.}\,, \quad (215)$$

$$F_2(\Box) = -\frac{1}{180}\frac{1}{2(4\pi)^2}\ln(-\Box) + \text{ const.}\,, \quad (216)$$

$$F_3(\Box) = \frac{1}{18}\frac{1}{2(4\pi)^2}\,, \quad (217)$$

$$F_4(\Box) = \frac{1}{2}\frac{1}{2(4\pi)^2}\ln(-\Box) + \text{ const.}\,, \quad (218)$$

$$F_5(\Box) = \frac{1}{12}\frac{1}{2(4\pi)^2}\ln(-\Box) + \text{ const.} \quad (219)$$

Since

$$-\ln(-\Box) = \int_0^\infty dm^2 \, \frac{1}{m^2 - \Box} + \text{ const.} \,, \tag{220}$$

these expressions have the spectral forms (169) with definite spectral weights and indefinite additive constants (polynomials of the zeroth power). Respectively, the effective action contains a set of local terms with unspecified coefficients:

$$\begin{aligned} \mathit{\Gamma} = \frac{1}{2(4\pi)^2} \int dx \, g^{1/2} \Big(& c_1 R + c_2 \operatorname{tr} \hat{P} + c_3 R_{\mu\nu} R^{\mu\nu} + c_4 R^2 \\ & + c_5 \operatorname{tr}(\hat{P}\hat{P}) + c_6 \operatorname{tr}(\hat{\mathcal{R}}_{\mu\nu} \hat{\mathcal{R}}^{\mu\nu}) + \frac{1}{18} R \operatorname{tr} \hat{P} + \text{ nonlocal terms} \Big) \,. \end{aligned} \tag{221}$$

The nonlocal terms are specified completely.

The third-order formfactors have no polynomial terms and indefinite coefficients. The simplest third-order formfactor is $F_1(\Box_1, \Box_2, \Box_3)$ in (168). It has the spectral form (170), and its spectral weight $\rho_1(m_1^2, m_2^2, m_3^2)$ is obtained as follows. Consider a triangle of three spectral masses

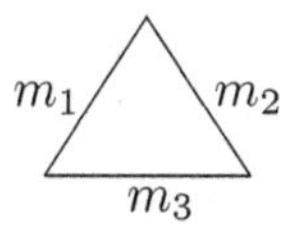

$A =$ area of the triangle.

It can be built only if every mass is smaller than the sum of the two others. The spectral weight ρ_1 is zero if the triangle cannot be built. Otherwise, it is proportional to the inverse area of this triangle:

$$\begin{aligned} \rho_1(m_1^2, m_2^2, m_3^2) = & -\frac{1}{3} \frac{1}{2(4\pi)^2} \frac{1}{4\pi A} \\ & \times \theta(m_1 + m_2 - m_3) \theta(m_1 + m_3 - m_2) \theta(m_2 + m_3 - m_1) \,. \end{aligned} \tag{222}$$

The remaining 28 third-order formfactors are expressed through F_1 and are tabulated [36]. The tables contain various integral representations of the formfactors, and their asymptotics.

The loop of the minimal second-order operator with arbitrary metric, connection, and potential is called standard loop because every calculation with it is done once, and the results can be tabulated. A calculation in any specific model boils down to combining the standard loops and using the tables. A number of recipes for the reduction to minimal operators can be found in [24]. Doing loops becomes a business similar to doing integrals.

The fact that some coefficients in the effective action remain unspecified is none of the tragedy. The effective action is a phenomenological object intended for obtaining the values of observables. The spectral weights are certain phenomenological characteristics of the vacuum like the permittivity of a medium. They are to be calculated from a more fundamental microscopic theory. Some microscopic theory of some level is incapable of specifying some of the coefficients. So what? Classical theory was capable of even less, and, nevertheless, celestial mechanics has been successfully worked up.[3] The only important question is whether the lack of knowledge affects the problems that we want to solve. This will be cleared up in the next lecture.

5 Vacuum Currents and the Effect of Particle Creation

5.1 Vacuum Currents

Consider quantum electrodynamics. In this case, $\varphi^a(x)$ is a set of the vector connection field and the electron–positron field

$$\text{QED:} \qquad \varphi^a = \Big(\mathcal{A}_\mu, \psi\Big)\,. \tag{223}$$

The commutator curvature is, up to a coefficient, the Maxwell tensor, and the operator field equations are of the form

$$\nabla^\nu \mathcal{R}_{\nu\mu}(\hat{\mathcal{A}}) + J_\mu(\hat{\psi}) = -J^{\text{ext}}_\mu \tag{224}$$

where $J_\mu(\hat{\psi})$ is the operator electron–positron current, and J^{ext}_μ is an external source. Averaging these equations over the in-vacuum state, one obtains, according to the general derivation above, the same terms but as functions of the mean field plus a set of loops:

$$\nabla^\nu \mathcal{R}_{\nu\mu}(\langle\mathcal{A}\rangle) + J_\mu(\langle\psi\rangle) + \ldots = -J^{\text{ext}}_\mu\,. \tag{225}$$

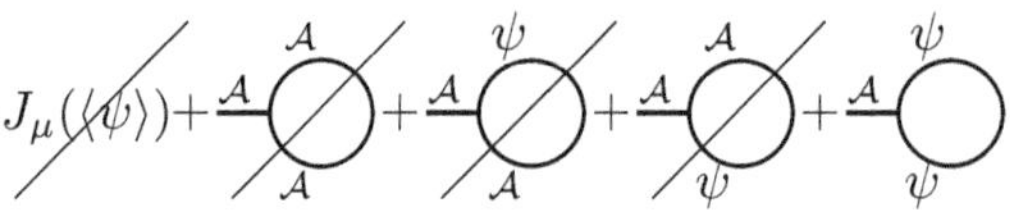

There is another such equation, for ψ, but, since ψ has no external source, its solution is

$$\langle\psi\rangle = 0\,. \tag{226}$$

Then, in (225), $J_\mu(\langle\psi\rangle)$ vanishes, and the loops with the vertices $S_{\mathcal{A}\mathcal{A}\psi}$ vanish. There are no such vertices in QED but, if there were, as in gravidynamics, they would be proportional to $\langle\psi\rangle$ and vanish by (226). The photon loop also vanishes because neither there is a vertex $S_{\mathcal{A}\mathcal{A}\mathcal{A}}$, but this is already a specific property of QED. Only the electron–positron loop survives.

The surviving loop is a function of $\langle\mathcal{A}\rangle$ and, by derivation, is the electron–positron current averaged over the in-vacuum:

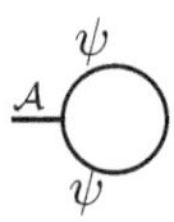

$$= J^{\text{vac}}_\mu(\langle\mathcal{A}\rangle) = \langle \text{in vac}|J_\mu(\hat{\psi})|\text{in vac}\rangle\,. \tag{227}$$

This is the vacuum current. According to (225), the *observable* electromagnetic field satisfies the Maxwell equations with an addition of the vacuum current:

[3] Remarkably, without a knowledge of string theory!

$$\nabla^\nu \mathcal{R}_{\nu\mu}(\mathcal{A}) = -J_\mu^{\text{vac}}(\mathcal{A}) - J_\mu^{\text{ext}} \,. \tag{228}$$

We obtain this current by varying the effective action and next replacing the Euclidean resolvents with the retarded resolvents:

$$J_\mu^{\text{vac}}(\mathcal{A}) = \left.\frac{\delta \Gamma(\mathcal{A})}{\delta \mathcal{A}^\mu}\right|_{\Box \to \Box_{\text{ret}}} , \tag{229}$$

$$\Gamma(\mathcal{A}) = \int dx\, g^{1/2}\Big[\mathcal{R}F(\Box)\mathcal{R} + F(\Box_1,\Box_2,\Box_3)\mathcal{R}_1\mathcal{R}_2\mathcal{R}_3 + \ldots\Big] \,. \tag{230}$$

It is completely similar if $\varphi^a(x)$ is a set of the metric field and any matter fields

$$\text{GRAVITY:} \qquad \varphi^a = \Big(g_{\mu\nu}, \psi\Big) \,. \tag{231}$$

The only difference is that the vertex S_{ggg} is nonvanishing:

$$R_{\mu\nu}(\langle g\rangle) - \frac{1}{2}\langle g_{\mu\nu}\rangle R(\langle g\rangle) +$$
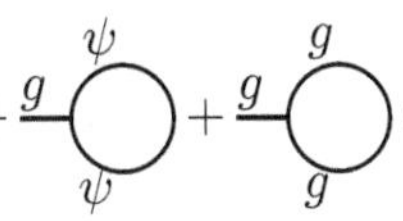

$$= 8\pi T_{\mu\nu}^{\text{ext}} \,, \tag{232}$$

$$\langle\psi\rangle = 0 \,, \tag{233}$$

and it is assumed again that the matter fields have no sources. Again, by derivation, the matter loop is the energy–momentum tensor of the field $\hat{\psi}$ averaged over the in-vacuum, but the vacuum current contains, in addition, the graviton loop:

$$T_{\mu\nu}^{\text{vac}} == \langle \text{in vac}|T_{\mu\nu}(\hat{\psi})|\text{in vac}\rangle + \text{ the graviton loop.} \tag{234}$$

The Einstein equations are replaced by the expectation-value equations in the in-vacuum state:

$$R_{\mu\nu} - \frac{1}{2}g_{\mu\nu}R = 8\pi T_{\mu\nu}^{\text{vac}}(g) + 8\pi T_{\mu\nu}^{\text{ext}} \,. \tag{235}$$

Since the gravitational field couples to everything, the equation (232) should contain loops of all matter fields in Nature. The effective actions for all loops including the graviton loop have the same structure:

$$T_{\mu\nu}^{\text{vac}}(g) = -\left.\frac{2}{g^{1/2}}\frac{\delta \Gamma(g)}{\delta g^{\mu\nu}}\right|_{\Box \to \Box_{\text{ret}}} , \tag{236}$$

$$\Gamma(g) = \int dx\, g^{1/2}\Big[R_{..}F(\Box)R_{..} + F(\Box_1,\Box_2,\Box_3)R_{1..}R_{2..}R_{3..} + \ldots\Big] \,. \tag{237}$$

Only the coefficients of the formfactors are different. To have the correct coefficients, one would need to know the full spectrum of particles. Therefore, in the case of gravity, the axiomatic approach is most suitable.

Now recall that the curvatures are redundant, and the effective action is in fact a functional of the conserved currents (141) and (142). Owing to this fact, the expectation-value equations (228) and (235) close with respect to these currents:

$$\left(\nabla^\nu \mathcal{R}_{\nu\mu}\right) + f(\Box_{\rm ret})\left(\nabla^\nu \mathcal{R}_{\nu\mu}\right) + O\left(\nabla^\nu \mathcal{R}_{\nu\mu}\right)^2 = -J_\mu^{\rm ext}\,, \tag{238}$$

$$\begin{aligned}\left(R_{\mu\nu} - \frac{1}{2}g_{\mu\nu}R\right) + f_1(\Box_{\rm ret})\left(R_{\mu\nu} - \frac{1}{2}g_{\mu\nu}R\right)\\ + f_2(\Box_{\rm ret})(\nabla_\mu\nabla_\nu - g_{\mu\nu}\Box)R + O\left(R_{\mu\nu} - \frac{1}{2}g_{\mu\nu}R\right)^2 = 8\pi T_{\mu\nu}^{\rm ext}\,.\end{aligned} \tag{239}$$

Of course, with respect to the mean fields, these equations are closed from the outset but, at an intermediate stage, they are closed with respect to the Maxwell and Einstein currents. When solved with respect to these currents, they become literally the Maxwell and Einstein equations with some external sources but *not* the original ones. To make this clear, use the fact that the vacuum terms are proportional to the Planck constant and solve the equations by iteration:

$$\nabla^\nu \mathcal{R}_{\nu\mu} == -J_\mu^{\rm ext} + f(\Box_{\rm ret})J_\mu^{\rm ext} + O\left(J_\mu^{\rm ext}\right)^2\,, \tag{240}$$

$$\begin{aligned}R_{\mu\nu} - \frac{1}{2}g_{\mu\nu}R = 8\pi T_{\mu\nu}^{\rm ext} - f_1(\Box_{\rm ret})8\pi T_{\mu\nu}^{\rm ext}\\ + f_2(\Box_{\rm ret})(\nabla_\mu\nabla_\nu - g_{\mu\nu}\Box)8\pi T^{\rm ext} + O\left(T_{\mu\nu}^{\rm ext}\right)^2\,.\end{aligned} \tag{241}$$

These are the Maxwell and Einstein equations with the original sources propagated in a nonlocal and nonlinear manner.

There is an effect in these equations that drives the entire problem.

5.2 Emission of Charges

Consider again QED and suppose that the external source has a compact spatial support. This source is the current of a set of electrically charged particles moving inside a space–time tube, but, since the observable electromagnetic field is the expectation value, only the total current in (228) or (240) is observable:

$$J_\mu^{\rm tot} = J_\mu^{\rm ext} + J_\mu^{\rm vac}(\mathcal{A})\,. \tag{242}$$

And the total current has a noncompact spatial support because the vacuum contribution is nonlocal. One may calculate the flux of charge through the support tube of $J^{\rm ext}$ and even through a wider tube (see Fig. 3), and it will be nonvanishing:

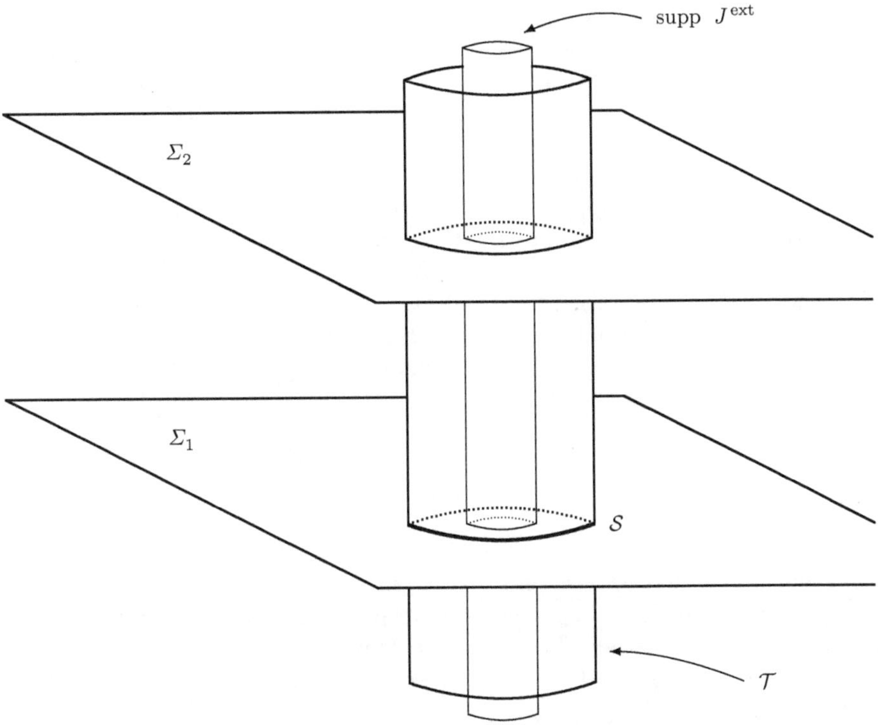

Fig. 3. Support tube of J^{ext} and a wider tube

$$e_{\mathcal{T}}(\Sigma_1) - e_{\mathcal{T}}(\Sigma_2) = \frac{1}{4\pi} \int_{\Sigma_1}^{\Sigma_2} J_\mu^{\text{vac}} \, d\mathcal{T}^\mu \neq 0 \,. \tag{243}$$

Here $e_{\mathcal{T}}(\Sigma)$ is the amount of the electric charge contained inside the tube $\mathcal{T}$ at a given instant Σ. The charge inside the tube is not conserved.

If, when moving away from the support of J^{ext}, the flux (243) falls off rapidly, then its nonvanishing only means that the boundary of the original source gets spread. Because of the creation of virtual pairs, this boundary can never be located precisely. The charges of the external source immersed in the quantum vacuum are always annihilated and created again in a slightly different place. There is no point to worry about. Just step aside a little.

However, one may ask if there is a flux of charge through an infinitely wide tube:

$$e(\Sigma_1) - e(\Sigma_2) = \frac{1}{4\pi} \int_{\Sigma_1}^{\Sigma_2} J_\mu^{\text{vac}} \, d\mathcal{T}^\mu \bigg|_{r\to\infty} \,. \tag{244}$$

In this equation, $e(\Sigma)$ is the total amount of the electric charge in the compact domain of space at a given instant Σ. For (244) to be nonvanishing, J_μ^{vac} should behave as

$$J_\mu^{\text{vac}} = O\left(\frac{1}{r^2}\right), \quad r \to \infty, \tag{245}$$

$$r \propto \sqrt{\text{area of } \mathcal{S}} \tag{246}$$

where $\mathcal{S}$ is the intersection of $\mathcal{T}$ with Σ (Fig. 3). In this case, it would turn out that the charge disappears, i.e., *our source is emitting charge.* But even this may not be a point of concern if the current in (244) oscillates with time, and the oscillations sum to zero for a sufficiently long period between Σ_1 and Σ_2. The expectation values have uncertainties, and these oscillations are a quantum noise. Just do not measure (244) too often.

However, one may ask if the charge emitted for the entire history

$$e(-\infty) - e(+\infty) = \frac{1}{4\pi} \int\limits_{\Sigma \to -\infty}^{\Sigma \to +\infty} J_\mu^{\text{vac}}\, d\mathcal{T}^\mu \Big|_{r \to \infty} \tag{247}$$

is nonvanishing. There will always be oscillations in the current, but they may sum not to zero. Since, as $r \to \infty$, all fields fall off, there are, in this limit, the asymptotic Killing vectors corresponding to all the symmetries of flat and empty space–time. Therefore, one may ask the same questions about the emission of energy and any other charges. Thus the quantity

$$M(-\infty) - M(+\infty) = \int\limits_{\Sigma \to -\infty}^{\Sigma \to +\infty} T_{\mu\nu}^{\text{vac}} \xi^\nu \, d\mathcal{T}^\mu \Big|_{r \to \infty} \tag{248}$$

with ξ^ν the asymptotic time-like Killing vector is the energy emitted by the source for the entire history.

If the total emitted charges are nonvanishing, then this is the real effect, and then the question emerges: What are the carriers of these charges? There should be some real agents carrying them away. But the particles of the original source stay in the tube. Besides them, there is only the electron–positron field, but it is in the in-vacuum state. This means that, at least initially, there are neither electrons nor positrons. There remains to be assumed a miracle: that either the real electrons or the real positrons – depending on the sign of the emitted charge – get created. Then they are created by pairs, and, say, the created positron is emitted while the created electron stays in the compact domain.

This crazy guess can be checked. We have two ways of calculating the vacuum currents: through the effective action and by a direct averaging of the operator currents as in (227) and (234). Specifically, for the in-vacuum of electrons and positrons we have

$$T_{\mu\nu}^{\text{vac}} = \langle \text{in vac} | T_{\mu\nu}(\hat{\psi}) | \text{in vac} \rangle \tag{249}$$

where $T_{\mu\nu}(\hat{\psi})$ is the operator energy–momentum tensor of the electron–positron field $\hat{\psi}$. The equation for $\hat{\psi}$

$$(\not{\partial} + \mu - \mathrm{i}q\langle\not{\mathcal{A}}\rangle)\,\hat{\psi} = 0 \tag{250}$$

contains the electromagnetic field which in (249) figures as an external field but is in fact the mean field solving the expectation-value equations. We know that, in the past, all mean fields are static. In the future, they become static again because, if the total emitted charges are finite, then all the processes should die down. Thus, there are two asymptotically static regions: in the past and in the future. The carriers of the emitted charges should be detectable in the future as particles with definite energies. But then the state in which they are absent is the out-vacuum, whereas their quantum state is the in-vacuum. *It may be the case that the in-vacuum contains the out-particles.* This will be the case if, between the static regions in the past and future, there is a region where $\langle\mathcal{A}\rangle$ is nonstatic because then the basis functions of the Fock modes that are the eigenfunctions of the energy operator in the future and the basis functions that are such in the past are different solutions of the Dirac equation (250).

If we expand $\hat{\psi}$ in the basis solutions of the out-particles, insert this expansion in (249), and then insert (249) in (248), the result will be

$$M(-\infty) - M(+\infty) = \left\langle \text{in vac} \middle| \sum_A \varepsilon_A \, \hat{a}^+_{\text{out}}{}^A \hat{a}_{\text{out}}{}^A \middle| \text{in vac} \right\rangle \tag{251}$$

where ε_A is the energy of the out-mode A, and similarly for the other charges. This result needs no comments. Miracles happen.

5.3 Emission of Charges (Continued)

An important point concerning miracles is that they happen not always. Let us see what is needed for this particular miracle to happen. For that, it is necessary to introduce characteristic parameters of the problem. There are two sets of parameters.

Parameters of the quantum field: q, μ.

Parameters of the external source: e, l, ν.

Here, q and μ are the charge and mass of the vacuum particles (e.g., of the electrons and positrons), e is the charge of the external source, l is the characteristic width of its support tube, and ν is the frequency parameter that characterizes the nonstationarity of the source.

The vacuum current in (240) is of the form

$$J^{\text{vac}} = \int\limits_0^\infty dm^2\, \rho(m^2) \frac{1}{m^2 - \Box_{\text{ret}}} J^{\text{ext}} + O\left(J^{\text{ext}}\right)^2 . \tag{252}$$

Here and above, the notation $\Box_{\rm ret}$ is to record that the resolvent is to be taken retarded. The structure of the nonlinear terms in (252) is similar: There is an overall resolvent acting on a function quadratic in $J^{\rm ext}$ (see (121)). If the vacuum particles are massive, the spectral weight will be proportional to the θ-function:

$$\rho(m^2) \propto \theta(m^2 - 4\mu^2) \tag{253}$$

to tell us that there is a threshold of pair creation. We need to find the behaviour of $J^{\rm vac}$ at a large distance from the support of $J^{\rm ext}$:

$$J^{\rm vac}\Big|_{r\gg l} = ? \tag{254}$$

First we need to calculate the action of the retarded resolvent on a source $J^{\rm ext}$ having a compact spatial support. If $J^{\rm ext}$ is static, the result is

$$\frac{1}{m^2 - \Box_{\rm ret}} J^{\rm ext}\Big|_{r\gg l} = \frac{C}{r}\exp(-mr)\;, \qquad J^{\rm ext} \text{ static.} \tag{255}$$

At a large distance from the source, this is the Yukawa potential. Because the function (255) is static, it does not depend on the spacetime direction in which the limit $r \gg l$ is taken. If $J^{\rm ext}$ is nonstatic, this is no more the case. The limit $r \gg l$ is direction-dependent, and there are directions in which the decrease is slower. Namely, in the directions of the outgoing light rays,

$$\frac{1}{m^2 - \Box_{\rm ret}} J^{\rm ext}\Big|_{r\gg l} = \frac{C}{r}\exp\left(-m\sqrt{rU}\right)\;, \qquad J^{\rm ext} \text{ nonstatic} \tag{256}$$

where U is a function of time[4] whose order of magnitude is

$$U \sim \frac{1}{\nu}\;. \tag{257}$$

Expression (256) is to be inserted in the spectral integral (252), and, since the spectrum is cut off from below, we find that the vacuum current is suppressed by the factor

$$J^{\rm vac} \sim \exp\left(-\frac{\mu\sqrt{r}}{\sqrt{\nu}}\right)\;, \qquad r \gg l\;. \tag{258}$$

This is what constrains miracles. However, we find also that the suppressing factor depends on the frequency of the source and *can be removed by raising the frequency.* The farther from the support of $J^{\rm ext}$, the greater the frequency should be for the current to be noticeable. The pair creation starts as soon as the energy $\hbar\nu$ exceeds the threshold

$$\hbar\nu > 2\mu c^2 \tag{259}$$

[4] Of the retarded time since the surfaces Σ to which the outgoing light rays belong are null.

but, for the source to emit charge, the frequency should be even greater:

$$\hbar\nu > (\mu c^2)\left(\frac{\mu c}{\hbar} l\right) . \tag{260}$$

This is easy to understand. The particles start being created in the support of the source with small momenta and cannot go far away. The extra factor $(\mu c/\hbar)l$ in (260) may be interpreted as the number of created particles for which there is room in the support of the source. If the creation is more violent, the particles get out of the tube. This is the meaning of condition (260). The mechanism of emission and conservation of charge is illustrated in Fig. 4. There are initially the charges of the external source in its support tube. They repel the like particles of the created pairs and, when the number of the latter exceeds $(\mu c/\hbar)l$, push them out of the tube. The unlike particles stay in the tube and diminish its charge.

Since the cause of the vacuum instability is the nonstationarity of the external source, it is interesting to consider the case where the energy $\hbar\nu$ exceeds overwhelmingly all the other energy parameters of the problem. One can then study the strong effect of particle production. It is assumed, in particular, that $\hbar\nu$ exceeds both the rest energy of the vacuum particle and its Coulomb energy in the external field:

$$\hbar\nu \gg \mu c^2 , \tag{261}$$

$$\hbar\nu \gg \frac{qe}{l} . \tag{262}$$

In the limit (261), the flux of charge at a given distance from the source ceases depending on the mass μ, and the vacuum particles can be considered as massless. Condition (262) enables one to get rid of the consideration of the static

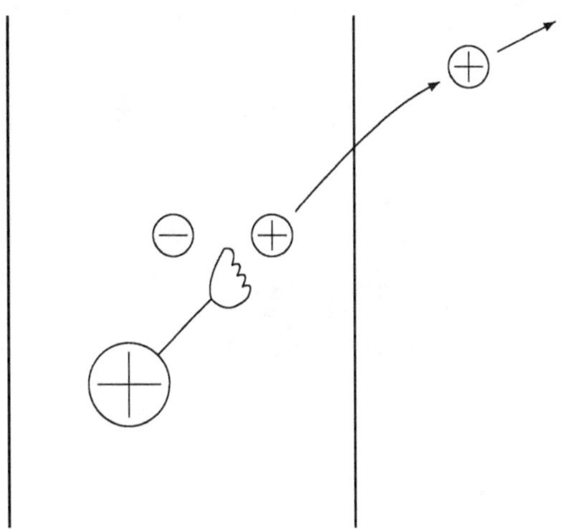

Fig. 4. Mechanism of emission and conservation of charge

vacuum polarization which is irrelevant to the problem. The approximation (261) and (262) is called high-frequency approximation.

The effective action has been calculated above as an expansion in powers of the curvature, but the conditions of validity of this expansion have not been discussed. This lack can now be met. It is the high-frequency approximation in which this expansion is valid. Indeed, consider the series (230). Every next term in this series contains an extra power of $\mathcal{R}$, and, by dimension, its formfactor contains an extra power of $\Box^{-1}$. The commutator curvature is proportional to the charges and to $\hbar^{-1}$:

$$\mathcal{R} \sim \frac{qe}{\hbar l^2} \, . \tag{263}$$

In the limit $r \gg l$ along the outgoing light rays, the operator $\Box$ contains one time derivative:

$$\Box \sim \frac{\nu}{l} \, . \tag{264}$$

As a result, every next term of the series contains, as compared to the previous one, the extra factor

$$\frac{qe}{\hbar \nu l} \ll 1 \, . \tag{265}$$

In addition, the formfactors in (230) can be calculated in the massless limit, as has been done above.

However, the inquest of miracles is not yet completed. Assuming that the vacuum particles are massless or that the high-frequency regime holds, we get rid of the suppressing exponential in (258), but we still need to check the power of decrease of the current. The power should be the one in (245) for the emission of charge to occur. We can readily check this since we know the behaviour of the resolvent. Expression (256) is again to be inserted in the spectral integral (252), but this time assuming that the spectrum begins with zero mass,

$$J^{\mathrm{vac}}\Big|_{r \gg l} = \int\limits_0^\infty dm^2 \, \rho(m^2) \frac{C}{r} \exp\left(-m\sqrt{rU}\right) \, . \tag{266}$$

We see that, for the current to decrease as $O(1/r^2)$, the spectral weight should have a finite and nonvanishing limit at zero mass:

$$\rho(0) = \text{finite} \neq 0 \, . \tag{267}$$

For the respective formfactor, this is a condition on its behaviour at small $\Box$. The behaviour should be

$$F(\Box) = \int\limits_0^\infty dm^2 \, \frac{\rho(m^2)}{m^2 - \Box} \quad \underset{\Box \to 0}{\longrightarrow} \quad -\rho(0) \ln(-\Box) \, . \tag{268}$$

We arrive at the following consistency condition on the vacuum formfactors. In the limit where one (any) of the $\Box$ arguments is small and the others are fixed, the formfactors should not grow faster than $\ln(-\Box)$:

$$F(\Box)\Big|_{\Box\to 0} = \text{const.}\ln(-\Box)\,, \tag{269}$$

$$F(\Box_1,\Box_2,\Box_3)\Big|_{\Box_1\to 0} = f(\Box_2,\Box_3)\ln(-\Box_1)\,, \tag{270}$$

. . . .

If they grow faster, the charges cannot be maintained finite, i.e., an isolated system cannot exist in such a vacuum. If they grow as $\ln(-\Box)$, the theory of isolated systems is consistent, but these systems emit charges. If they grow slower, the charges are conserved.

One can check whether the one-loop formfactors satisfy this consistency condition. The second-order formfactors (215)–(219) do. The third-order formfactors behave generally as [35]

$$F(\Box_1,\Box_2,\Box_3)\Big|_{\Box_1\to 0} = f(\Box_2,\Box_3)\frac{1}{\Box_1} + g(\Box_2,\Box_3)\ln(-\Box_1) + \dots\,. \tag{271}$$

The alarming terms $1/\Box$ appear only in the arguments acting on the gravitational curvatures. Therefore, they can affect only the vacuum energy–momentum tensor, and it has been checked that, in the energy–momentum tensor, these terms coming from different formfactors cancel. In the currents, the one-loop formfactors satisfy strictly the consistency condition. Since, in addition, their asymptotic $\ln(-\Box)$ terms are nonvanishing, the emission of charges in the high-frequency regime is real. The only thing that remains to be checked is that this emission is not a pure quantum noise. It will be checked by a direct calculation.

Now one can answer also the question about the indefinite local terms in the effective action. The coefficients of these terms are the unspecified constants in (215)–(219). In the limit $\Box \to 0$, the values of these constants are immaterial. Only the terms $\ln(-\Box)$, $\Box \to 0$ of the formfactors work, and, therefore, the incompleteness of local quantum field theory does not affect the presently considered problem.

It will be noted that there are now two mechanisms by which an isolated system can emit energy. One is purely classical: a nonstationary source can emit the electromagnetic or gravitational waves. The other is quantum: immersed in the vacuum, a nonstationary source can emit also charged particles. A high-frequency source will generally emit *both.*

5.4 Particle Creation by External Fields

The problem of particle creation by external fields is a part of the expectation-value problem. In the context of the foregoing, it can be set as follows. Consider the quantum field that satisfies a linear second-order equation

$$\left(g^{\mu\nu}\nabla_\mu\nabla_\nu\hat{1} + \hat{P}\right)\phi = 0 \tag{272}$$

containing three external fields: the metric, the connection, and the potential. The external fields are asymptotically static in the past and future but otherwise arbitrary except that their currents

$$J_{\alpha\beta} = R_{\alpha\beta} - \frac{1}{2} g_{\alpha\beta} R \,, \tag{273}$$

$$\hat{J}_{\alpha} = \nabla^{\beta} \hat{\mathcal{R}}_{\alpha\beta} \,, \tag{274}$$

$$\hat{Q} = \hat{P} + \frac{1}{6} R \hat{1} \tag{275}$$

are confined to a space-time tube. The quantum field is in the in-vacuum state. What is the energy of the quanta of the field ϕ created by the external fields for the entire history? In the high-frequency approximation, we have everything to answer this question.

To formulate the answer, I need some preliminary construction. Every current has an associated quantity called its radiation moment. It will now be defined.

Consider a time-like geodesic in the external metric of equation (272). It enters the domain of nonstationarity of external fields with a definite energy and goes out of this domain with a definite energy. Let E be its energy per unit rest mass on going out. I am only interested in the geodesics that escape to $r = \infty$. They have $E > 1$, and, instead of E, I shall use the parameter γ defined as

$$\gamma = \frac{\sqrt{E^2 - 1}}{E} \,, \quad E > 1 \,, \quad 0 < \gamma < 1 \,. \tag{276}$$

At $r = \infty$, the geodesic has a certain spatial direction, or, equivalently, it comes to a certain point of the celestial 2-sphere. I shall denote this sphere as $\mathcal{S}$, its points as θ:

$$\theta = (\theta_1, \theta_2) \,, \qquad \theta \in \mathcal{S} \,, \tag{277}$$

and the integral over the unit 2-sphere as

$$\int d^2\mathcal{S}(\theta) \, (\cdots) \,. \tag{278}$$

A geodesic with given γ and θ will be called γ, θ-geodesic (see Fig. 5).

A γ, θ-geodesic can be emitted from every point of a compact domain. Therefore, the γ, θ-geodesics with *the same values* of γ and θ make a congruence, and it can be proved that this congruence is hypersurface-orthogonal. Let the orthogonal hypersurfaces be

$$T_{\gamma\theta}(x) = \text{const.} \tag{279}$$

Since the parameters γ, θ fix the congruence, they fix also the family of the orthogonal hypersurfaces (279), and the "const." in (279) fixes a member of the family. The function $T_{\gamma\theta}$ is determined up to a transformation $T_{\gamma\theta} \to f\,(T_{\gamma\theta})$. This arbitrariness will be removed by the normalization condition

$$(\nabla T_{\gamma\theta})^2 = -\left(1-\gamma^2\right) \tag{280}$$

and the condition that the vector $\nabla T_{\gamma\theta}$ is past directed. It is a property of the geodetic congruences that the norm in (280) can be chosen constant.

The radiation moment of any scalar current J is the following hypersurface integral:

$$D = \frac{1}{4\pi}\int dx\, g^{1/2}\delta\left(T_{\gamma\theta}(x)-\tau\right) J(x)\,. \tag{281}$$

If the current is not a scalar, it should first be parallel transported from the integration point to $r=\infty$ along the respective γ,θ-geodesic. Thus if the current is a vector, its radiation moment is

$$D^\alpha = \frac{1}{4\pi}\int dx\, g^{1/2}\delta\left(T_{\gamma\theta}(x)-\tau\right) J^\beta(x) a_\beta{}^\alpha(x,\infty) \tag{282}$$

where $a_\beta{}^\alpha(x,\infty)$ is the propagator of parallel transport of vectors to infinity along the γ,θ-geodesic emanating from x. The radiation moment D^α is then a vector at infinity. In the same way, the radiation moment is defined for any current. For the three currents (273)–(275), the radiation moments will be denoted respectively as

$$J_{\alpha\beta},\ \hat{J}_\alpha,\ \hat{Q} \longrightarrow D_{\alpha\beta},\ \hat{D}_\alpha,\ \hat{D}\,. \tag{283}$$

Since the indices of the radiation moments pertain to a point at infinity, their contractions like

$$\hat{D}_\alpha \hat{D}^\alpha = g_{\alpha\beta}\hat{D}^\alpha\hat{D}^\beta\,, \quad \text{etc.,} \tag{284}$$

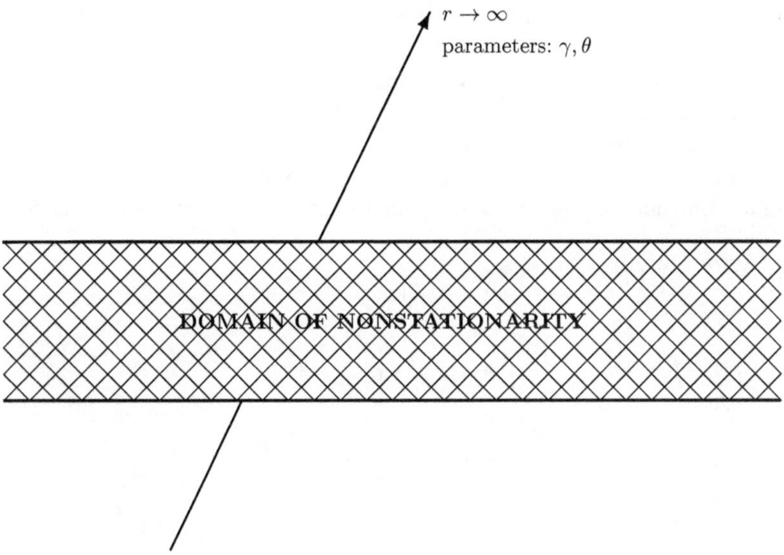

Fig. 5. A γ,θ-geodesic

always assume the flat metric $g_{\alpha\beta}$ at infinity. All radiation moments are functions of four parameters:

$$D = D(\gamma, \theta, \tau)\ . \tag{285}$$

In the limit $\gamma = 1$, the γ, θ-geodesics become null. The orthogonal hypersurfaces (279) also become null, and the geodesics themselves become their generators. For the radiation moments, this is a regular limit. Nothing special happens to them in this limit except that they become very important. The radiation moments at $\gamma = 1$ govern the emission of waves in classical theory. Thus if J_α in (274) is an electric current, then the following expression:

$$\Big(M(-\infty) - M(+\infty)\Big)_{\text{electromagnetic waves}} = \frac{1}{4\pi}\int_{-\infty}^{\infty} d\tau \int d^2\mathcal{S}(\theta)\left[g_{\alpha\beta}\left(\frac{d}{d\tau}D^\alpha\right)\left(\frac{d}{d\tau}D^\beta\right)\right]\Bigg|_{\gamma=1} \tag{286}$$

is the energy of the electromagnetic waves emitted by this current for the entire history. A similar expression with the tensor current (273):

$$\Big(M(-\infty) - M(+\infty)\Big)_{\text{gravitational waves}} = \frac{1}{4\pi}\int_{-\infty}^{\infty} d\tau \int d^2\mathcal{S}(\theta)\,\frac{1}{2}(g_{\alpha\mu}g_{\beta\nu} - \frac{1}{2}g_{\alpha\beta}g_{\mu\nu})\left(\frac{d}{d\tau}D^{\alpha\beta}\right)\left(\frac{d}{d\tau}D^{\mu\nu}\right)\Bigg|_{\gamma=1} \tag{287}$$

is the energy of the gravitational waves emitted by the current $J_{\alpha\beta}$ for the entire history.

The radiation moment is a generating function for the multipole moments. The multipole expansion is the expansion of D at $\gamma = 0$. It makes sense for nonrelativistic systems since γ is proportional to $1/c$.

Expressions (286) and (287) are the solutions of the classical radiation problem. And here is the solution of the quantum radiation problem [50]:

$$\begin{aligned}\Big(M(-\infty) - M(+\infty)\Big)_{\text{created particles}} &= \frac{1}{(4\pi)^2}\int_0^1 d\gamma\,\gamma^2\int_{-\infty}^{\infty} d\tau\int d^2\mathcal{S}(\theta)\,\mathrm{tr}\left[\left(\frac{d^2}{d\tau^2}\hat{D}\right)^2\right.\\ &\quad - \frac{1}{3}\frac{1}{(1-\gamma^2)}g_{\alpha\beta}\left(\frac{d}{d\tau}\hat{D}^\alpha\right)\left(\frac{d}{d\tau}\hat{D}^\beta\right)\\ &\quad \left. + \frac{1}{30}\hat{1}(g_{\alpha\mu}g_{\beta\nu} - \frac{1}{3}g_{\alpha\beta}g_{\mu\nu})\left(\frac{d^2}{d\tau^2}D^{\alpha\beta}\right)\left(\frac{d^2}{d\tau^2}D^{\mu\nu}\right)\right]\ .\end{aligned} \tag{288}$$

This is the energy of the quanta of the field ϕ created by the external fields for the entire history. As compared to the expressions above, there is an extra time derivative in the case of the tensor and scalar moments. It accounts for the dimension of the coupling constant. Also, instead of setting $\gamma = 1$, one needs to integrate over γ. Otherwise, the similarity is striking. The quantum problem of particle creation becomes almost the same thing as the classical problem of emission of waves.

The presence in (288) of an integral over γ is not just a technical detail. The radiation moments have both the longitudinal projections, i.e., the projections on the direction of the geodesic at infinity and the transverse projections. Inspecting the contractions of the moments in (286)–(288), one can see that, at $\gamma = 1$, the longitudinal projections drop out of these contractions. In the integral over γ, also the longitudinal projections survive. Owing to this fact, spherically symmetric sources cannot emit waves but can produce particles from the vacuum.

Now I can explain why, when expanding the effective action, I stopped at the terms cubic in the curvature. In the high-frequency approximation, the expansion (165) needs to be calculated up to the lowest-order terms that give a nonvanishing effect. The terms of first order in the curvature are local and give no effect. The terms of second order in the curvature are nonlocal and contribute to the energy flux at infinity, but it turns out that *their contribution is a pure quantum noise.* The real effect of particle production begins with the third order in the curvature. Expression (288) results from the triangular loop diagrams.

Since varying the action destroys one curvature, a cubic action generates a quadratic current. This gives the radiation energy a chance to be positive definite. Expression (288) is positive definite indeed:

$$\Bigl(M(-\infty) - M(+\infty)\Bigr)_{\text{created particles}} \geq 0 \,. \tag{289}$$

In particular, for the matrix contributions, this follows from relations (136), (137) and the positive definiteness of the matrix ω_{ab}:

$$\operatorname{tr}\left(\frac{d^2}{d\tau^2}\hat{D}\right)^2 \geq 0\,, \quad \operatorname{tr}\left[g_{\alpha\beta}\left(\frac{d}{d\tau}\hat{D}^\alpha\right)\left(\frac{d}{d\tau}\hat{D}^\beta\right)\right] \leq 0\,. \tag{290}$$

The positivity of the gravitational-field contribution can be proved directly.

5.5 The Backreaction Problem

The energy emitted by an isolated system (in all forms) should be bounded both from below and from above: it should be positive and less than the energy stored in the initial state

$$0 \leq \Big(M(-\infty) - M(+\infty) \Big) \leq M(-\infty) \,. \tag{291}$$

In expression (288), the positivity is guaranteed, but the energy conservation is not. The reason is that the setting of the problem with external fields is physically inconsistent. The vacuum current determines the solution of the mean-field equations, and the mean field rather than the external field determines the vacuum current. If the backreaction of the vacuum is neglected, the conservation laws need not be observed.

One case in which the vacuum backreaction may not be neglected is where both mechanisms of the energy emission, classical and quantum, are engaged simultaneously. This concerns particularly the vector connection field. In expression (288), the integral over γ has a pole $(1-\gamma)^{-1}$ in the term with the vector moment. The residue of the integrand in this pole is precisely the quantity (286), i.e., the energy of the outgoing waves of the vector connection field. If it is nonvanishing, e.g., if the external source emits both the electromagnetic waves and the electrically charged particles, the integral in γ diverges. The result is a disaster: The radiation energy appears to be infinite. In fact it should be taken into account that the created charge affects the generation of the electromagnetic waves, and the respective changes in the electromagnetic field affect the creation of charge. In the self-consistent solution, the disaster is removed.

Another example concerns the metric field when it has an event horizon. In this case, the integral in τ diverges at the upper limit. By construction, τ is the time of an external observer. As $\tau \to \infty$, the source moving in the tube hits the event horizon. Its proper time does not turn into infinity. The integrand in (288) is just finite in this limit, and the integral in τ diverges linearly. This is the Hawking constant flux of radiation from the black hole. If its backreaction on the metric is neglected, the total emitted energy is infinite.

But even when the quantity (288) is finite, it depends on the frequency of the source. If the source is external, this frequency is a free parameter. The energy of created quanta grows with frequency, and, typically, the ratio

$$\left. \frac{M(-\infty) - M(+\infty)}{M(-\infty)} \right|_{\nu \to \infty} \sim \ln \nu \tag{292}$$

also grows so that, at a sufficiently high frequency, the energy conservation law will be violated. The backreaction should take into account that, when the source creates real particles, it loses energy and slows down. It then creates less particles, and the process dies away. The conservation laws will then be restored.

The backreaction problem has been solved only in a few cases [51]–[56]. The examples for which it has been solved show that the solution can be unexpected and interesting.

References

1. B. DeWitt: *The Global Approach to Quantum Field Theory*, vols 1,2 (Oxford University Press, Oxford, New York, 2003)
2. B.S. DeWitt: Dynamical theory of groups and fields. In: *Relativity, Groups and Topology. 1963 Les Houches Lectures*, ed. by C. DeWitt, B.S. DeWitt (Gordon and Breach, New York, 1964) pp. 587–820
3. G. Jona-Lasinio: Nuovo Cimento **34**, 1790 (1964)
4. B.S. DeWitt: Phys. Rep. **19**, 295 (1975)
5. E.S. Fradkin, G.A. Vilkovisky: Lett. Nuovo Cimento **19**, 47 (1977)
6. J. Schwinger: Field theory methods in non-field theory contexts. In: *Proc. 1960 Brandeis Summer School* (Brandeis University Press, Brandeis, 1960) pp. 282–285
7. J. Schwinger: J. Math. Phys. **2**, 407 (1961)
8. L.V. Keldysh: Zh. Eksp. Teor. Fiz. **47**, 1515 (1964)
9. Yu.A. Golfand: Yad. Fiz. **8**, 600 (1968)
10. P. Hajicek: Time-loop formalism in quantum field theory. In: *Proc. 2nd Marcel Grossmann Meeting on General Relativity (Trieste, 1979)*, ed. by R. Ruffini (North Holland, Amsterdam, 1982) pp. 483–491
11. E.S. Fradkin, D.M. Gitman: Fortschr. der Phys. **29**, 381 (1981)
12. J.L. Buchbinder, E.S. Fradkin, D.M.Gitman: Fortschr. der Phys. **29**, 187 (1981)
13. R.D. Jordan: Phys. Rev. D **33**, 44 (1986)
14. E. Calzetta, B.L. Hu: Phys. Rev. D **35**, 495 (1987)
15. A.O. Barvinsky, G.A. Vilkovisky: Nucl. Phys. B **282**, 163 (1987)
16. R.C. Hwa, V.L. Teplitz: *Homology and Feynman Integrals* (Benjamin, New York Amsterdam, 1966)
17. G.A. Vilkovisky: Class. Quantum Grav. **9**, 895 (1992)
18. J.S. Schwinger: Phys. Rev. **82**, 664 (1951)
19. J.L. Synge: *Relativity: The General Theory* (North Holland, Amsterdam, 1960)
20. G. 't Hooft, M. Veltman: Ann. Inst. Henri Poincare **XX**, 69 (1974)
21. P.B. Gilkey: J. Diff. Geom. **10**, 601 (1975)
22. L.S. Brown: Phys. Rev. D **15**, 1469 (1977)
23. L.S. Brown, J.P. Cassidy: Phys. Rev. D **15**, 2810 (1977)
24. A.O. Barvinsky, G.A. Vilkovisky: Phys. Rep. **119**, 1 (1985)
25. G.A. Vilkovisky: Heat kernel: rencontre entre physiciens et mathématiciens. In: *R.C.P. 25*, vol. 43 (Publication de l'Institut de Recherche Mathématique Avancée, Strasbourg, 1992) pp. 203–224
26. A.M. Polyakov: Phys. Lett. B **103**, 207 (1981)
27. G.A. Vilkovisky: The Gospel according to DeWitt. In: *Quantum Theory of Gravity*, ed. by S.M. Christensen (Hilger, Bristol 1984) pp 169–209
28. A.A. Ostrovsky, G.A. Vilkovisky: J. Math. Phys. **29**, 702 (1988)
29. I.G. Avramidi: Yad. Fiz. **49**, 1185 (1989)
30. A.O. Barvinsky, G.A. Vilkovisky: Nucl. Phys. B **333**, 471 (1990)
31. A.O. Barvinsky, G.A. Vilkovisky: Nucl. Phys. B **333**, 512 (1990)
32. A.O. Barvinsky, Yu.V. Gusev, G.A. Vilkovisky, V.V. Zhytnikov: J. Math. Phys. **35**, 3525 (1994)
33. A.O. Barvinsky, Yu.V. Gusev, G.A. Vilkovisky, V.V. Zhytnikov: J. Math. Phys. **35**, 3543 (1994)
34. A.O. Barvinsky, Yu.V. Gusev, G.A. Vilkovisky, V.V. Zhytnikov: Nucl. Phys. B **439**, 561 (1995)

35. A.O. Barvinsky, Yu.V. Gusev, V.V. Zhytnikov, G.A. Vilkovisky: Class. Quantum Grav. **12**, 2157 (1995)
36. A.O. Barvinsky, Yu.V. Gusev, G.A. Vilkovisky, V.V. Zhytnikov: Covariant perturbation theory (IV). Third order in the curvature. Report, University of Manitoba, Winnipeg (1993) pp. 1–192
37. A.G. Mirzabekian, G.A. Vilkovisky, V.V. Zhytnikov: Phys. Lett. B **369**, 215 (1996)
38. Y. Nambu: Phys. Rev. **100**, 394 (1955)
39. N. Nakanishi: Prog. Theor. Phys. **24**, 1275 (1960)
40. N. Nakanishi: *Graph Theory and Feynman Integrals* (Gordon and Breach, New York, 1970)
41. J. Schwinger: *Particles, Sources, and Fields*, vol. 2 (Addison-Wesley, Reading, 1973)
42. A.A. Grib, S.G. Mamayev, V.M. Mostepanenko: *Quantum Effects in Intense External Fields* (Atomizdat, Moscow, 1980)
43. N.D. Birrell, P.C.W. Davies: *Quantum Fields in Curved Space* (Cambridge University Press, Cambridge, 1982)
44. N.M.J. Woodhouse: Phys. Rev. Lett. **36**, 999 (1976)
45. A.G. Mirzabekian, G.A. Vilkovisky: Phys. Lett. B **317**, 517 (1993)
46. A.G. Mirzabekian: Zh. Eksp. Teor. Fiz. **106**, 5 (1994) [Engl. trans.: JETP **79**, 1 (1994)]
47. A.G. Mirzabekian, G.A. Vilkovisky: Phys. Rev. Lett. **75**, 3974 (1995)
48. A.G. Mirzabekian, G.A. Vilkovisky: Class. Quantum Grav. **12**, 2173 (1995)
49. A.G. Mirzabekian, G.A. Vilkovisky: Phys. Lett. B **414**, 123 (1997)
50. A.G. Mirzabekian, G.A. Vilkovisky: Ann. Phys. **270**, 391 (1998)
51. G.A. Vilkovisky: Phys. Rev. D **60**, 065012 (1999)
52. G.A. Vilkovisky: Phys. Rev. Lett. **83**, 2297 (1999)
53. R. Pettorino, G.A. Vilkovisky: Ann. Phys. **292**, 107 (2001)
54. G.A. Vilkovisky: Ann. Phys. **321**, 2717 (2006)
55. G.A. Vilkovisky: Phys. Lett. B **634**, 456 (2006)
56. G.A. Vilkovisky: Phys. Lett. B **638**, 523 (2006)

Part VIII

String Cosmology

Dilaton Cosmology and Phenomenology

M. Gasperini

Dipartimento di Fisica , Università di Bari, Via G. Amendola 173, 70126 Bari, Italy, and Istituto Nazionale di Fisica Nucleare, Sezione di Bari, Bari, Italy
gasperini@ba.infn.it

Abstract. This paper is dedicated to Gabriele Veneziano on his 65th birthday. Most of the results reported here are known results, due to Gabriele, or obtained in collaboration with him, or inspired by our joint work on string cosmology. A few new results are also presented concerning the duality invariance of a non-local dilaton coupling to the matter sources, and its possible cosmological applications in the context of the dark energy scenario.

Foreword and Introduction

My collaboration with Gabriele Veneziano has continued, almost uninterruptedly, for more than 15 years (even now we are preparing a joint contribution to the book *Beyond the Big Bang*, which will be published by Springer, as this book). Our first meeting dates back to 1989, when Gabriele came to the University of Turin to give a series of talks and seminars. At that time I was working there as a young researcher at the Department of Theoretical Physics, and I remember that Sergio Fubini (professor at the same Department) introduced me to Gabriele before a seminar. After the seminar we went to my office, and we started talking about cosmology, big bang, inflation, and strings. Gabriele was able to make me feel at ease, in spite of the fact that I was a bit embarrassed, being face to face with such a world-renowned scientist like him: before that meeting, indeed, I knew him only for having seen his name quoted in many books and articles as one of the founders of string theory. I could not imagine that I was about to embark on the most stimulating and important adventure of my scientific life.

After that meeting we started collaborating on string cosmology, and I visited very often the Theory Division (now "TH Unit") at CERN, living in Geneva also for long periods. This has given me the opportunity to appreciate Gabriele not only as a scientist—whose inventiveness, originality, profundity of thought will not be stressed here, because they are well-known to the physicist community—but also for his human qualities. His tireless enthusiasm for

M. Gasperini: *Dilaton Cosmology and Phenomenology*, Lect. Notes Phys. **737**, 787–844 (2008)
DOI 10.1007/978-3-540-74233-6_24

Fig. 1. Gabriele Veneziano (**left**) and the author (**right**), talking about dilatons at CERN (January 1994)

physics, his generosity in sharing knowledge, his intellectual honesty, always make working with him a rewarding and enjoyable experience. I have countless memories of days spent discussing and working out calculations on the blackboard of his office (see Fig. 1), with short "coffee breaks" every now and them, talking about physics even during lunch and dinner. Countless are the things I have learned from him, not only from a scientific but also from a human point of view. I will be grateful to him forever.

Choosing among the lines of research developed in collaboration with Gabriele, I will concentrate the contribution to this book on the possible role played by the dilaton in a cosmological context, with particular attention to the phenomenological aspects of dilaton cosmology. The dilaton is a fundamental scalar field appearing in all models of superstrings, dilaton cosmology is probably the most natural and typical form of "string cosmology," and a direct/indirect confirmation (or disproof) of its predictions could give us important experimental information on string theory in general.

The present contribution contains three lectures. The first lecture (Sect. 1) is devoted to the presentation of a primordial cosmological scenario in which the background evolution is dominated by the dilaton, and the Universe is driven through an accelerated phase representing the "dual" counterpart of the standard, decelerated evolution. The second lecture (Sect. 2) will discuss the possibility that a cosmic background of relic dilaton radiation could have survived until present, and could be detectable by the gravitational antennas that are presently operating (or planned for a near-future operation). Finally, the third lecture (Sect. 3) will suggest a possible "dilatonic" origin of the dark energy fluid dominating the cosmic acceleration recently observed on large scales, stressing the main differences from other, more conventional models of scalar "quintessence."

Notations and Conventions

Unless otherwise stated, the following conventions are used throughout this paper: Greek indices run from 0 to d, Latin indices from 1 to d, where $d = D-1$ is the number of spatial dimensions of the D-dimensional space–time manifold. The metric signature is

$$g_{\mu\nu} = \mathrm{diag}(+, -, -, -, \cdots).$$

The Riemann tensor and its contractions are defined by

$$R_{\mu\nu\alpha}{}^{\beta} = \partial_\mu \Gamma_{\nu\alpha}{}^{\beta} + \Gamma_{\mu\rho}{}^{\beta}\Gamma_{\nu\alpha}{}^{\rho} - (\mu \leftrightarrow \nu),$$
$$R_{\nu\alpha} = R_{\mu\nu\alpha}{}^{\mu}, \qquad R = R_{\mu}{}^{\mu}, \qquad G_{\mu\nu} = R_{\mu\nu} - \frac{1}{2}g_{\mu\nu}R.$$

The conventions for the covariant derivative are

$$\nabla_\mu V_\alpha = \partial_\mu V_\alpha - \Gamma_{\mu\alpha}{}^{\beta}V_\beta, \qquad \nabla_\mu V^\alpha = \partial_\mu V^\alpha + \Gamma_{\mu\beta}{}^{\alpha}V^\beta.$$

Finally, we often use the convenient notation:

$$(\nabla\phi)^2 \equiv \nabla_\mu\phi\nabla^\mu\phi, \qquad \nabla^2\phi \equiv \nabla_\mu\nabla^\mu\phi.$$

1 Dilaton-dominated Inflation: the Pre-big Bang Scenario

If we apply to the "specialized" literature for a description of the birth and of the first moments of our Universe, we may read, in the (probably) most ancient and authoritative book, that

> In the beginning God created the Heaven and the Earth,
> and the Earth was without form, and void;
> and the darkness was upon the face of the deep.
> And the Breath of God
> moved upon the face of the water.
>
> (Genesis, The Holy Bible).

The most impressive aspect of these verses, for a modern cosmologist, is probably the total absence of any reference to the extremely hot, kinetic, explosive state that one could expect at (or immediately after) the "big bang" deflagration. The described state, instead, is somewhat quiet, dark, empty—we can read, indeed, about "void," "darkness," and "the deep" gives us the idea of something enormously desert and empty. In this static configuration there is at most some small fluctuation (the "Breath"), a ripple on the surface of this vacuum.

It is amusing to note that a state of this type (flat, cold, and vacuum, only ruffled by quantum fluctuations), can be obtained as the initial state of our Universe, in a string-cosmology context, under the hypothesis that the Universe evolves in a "self-dual" way with respect to the symmetries of the low-energy string effective action [1, 2].

To introduce this result we start considering the gravi-dilaton sector of the low-energy effective action. To lowest order in the α' (higher-derivative) and g_s^2 (higher-loop) expansion the action is the same for all models of superstrings [3, 4], and is given by

$$S = -\frac{1}{2\lambda_{\mathrm{s}}^{d-1}} \int_{\Omega} \mathrm{d}^{d+1}x\sqrt{-g}\,\mathrm{e}^{-\phi}\left(R + \partial_\mu\phi\,\partial^\mu\phi + V\right). \tag{1}$$

Here ϕ is the dilaton, and $\lambda_{\mathrm{s}} = (2\pi\alpha')^{1/2}$ is the fundamental length parameter of string theory. We have written the action using the so-called String frame (S-frame) metric, i.e., the metric to which a "test" string is minimally coupled and in which its evolution is geodesics. We have also included, for completeness and for further applications, a (possibly non-perturbative) dilaton potential, $V = V(\phi)$.

This action should be completed by the source term $S_m(g, \phi)$, describing the matter fields contributions, and by the Gibbons–Hawking boundary term S_Σ, which is required (as in general relativity) to cancel the variational contributions of the second derivatives of the metric following from the Einstein-Hilbert Lagrangian $\sqrt{-g}\,R$. For the S-frame action (1) the boundary term takes the form [5]

$$S_\Sigma = \frac{1}{2\lambda_{\mathrm{s}}^{d-1}} \int_{\partial\Omega} \sqrt{-g}\;\mathrm{e}^{-\phi} K^\alpha \mathrm{d}\Sigma_\alpha, \tag{2}$$

where $K^\alpha = K n^\alpha$. Here K is the trace of the extrinsic curvature of the d-dimensional hypersurface $\partial\Omega$ bounding the hypervolume Ω over which we are varying the action, and n^α is the unit vector normal to this hypersurface.

The variation of the total action $S + S_m + S_\Sigma$ with respect to $g^{\mu\nu}$ leads then to the equations

$$G_{\mu\nu} + \nabla_\mu\nabla_\nu\phi + \frac{1}{2}g_{\mu\nu}\left(\nabla\phi\right)^2 - g_{\mu\nu}\nabla^2\phi - \frac{1}{2}g_{\mu\nu}V(\phi) = \lambda_{\mathrm{s}}^{d-1}\mathrm{e}^{\phi}\,T_{\mu\nu}, \tag{3}$$

where $G_{\mu\nu}$ is the Einstein tensor and $T_{\mu\nu}$ the gravitational stress tensor of the matter sources, defined as usual by the functional differentiation of S_m as

$$T_{\mu\nu} = \frac{2}{\sqrt{-g}}\frac{\delta S_m}{\delta g_{\mu\nu}}. \tag{4}$$

The variation of the total action with respect to ϕ leads to the dilaton equation of motion,

$$R + 2\nabla^2\phi - (\nabla\phi)^2 + V - \frac{\partial V}{\partial \phi} = \lambda_{\rm s}^{d-1} {\rm e}^{\phi}\sigma, \tag{5}$$

where σ is the (S-frame) density of dilaton charge of the sources, defined by the functional differentiation of S_m with respect to ϕ:

$$\sigma = -\frac{2}{\sqrt{-g}}\frac{\delta S_m}{\delta \phi}. \tag{6}$$

Using the dilaton equation to eliminate the scalar curvature, present in the Einstein tensor, we can eventually rewrite (3) in the convenient (simplified) form

$$R_{\mu\nu} + \nabla_\mu\nabla_\nu\phi - \frac{1}{2}g_{\mu\nu}\frac{\partial V}{\partial \phi} = \lambda_{\rm s}^{d-1}{\rm e}^{\phi}\left(T_{\mu\nu} + \frac{1}{2}g_{\mu\nu}\,\sigma\right). \tag{7}$$

1.1 Scale Factor Duality

We will now consider the particular case in which the space–time manifold described by the S-frame metric is spatially flat, homogeneous (but not necessarily isotropic), and in which the matter fields can be phenomenologically described as perfect fluids, at rest in the comoving frame of the given Robertson–Walker geometry. We can thus set, in the synchronous gauge,

$$\begin{aligned} &g_{\mu\nu} = {\rm diag}(1, -a_i^2\delta_{ij}), \qquad a_i = a_i(t), \qquad \phi = \phi(t), \\ &T_\mu{}^\nu = {\rm diag}(\rho, -p_i\delta_i^j), \qquad \rho = \rho(t), \qquad p_i = p_i(t), \qquad \sigma = \sigma(t). \end{aligned} \tag{8}$$

Separating the time and space components of the gravitational equations we then obtain, from the (00) component of (3),

$$\dot\phi^2 - 2\dot\phi\sum_i H_i + \left(\sum_i H_i\right)^2 - \sum_i H_i^2 - V = 2\lambda_{\rm s}^{d-1}\,{\rm e}^{\phi}\rho \tag{9}$$

(where $H_1 = \dot a_i/a_i$). From the (ii) component of (7) we have

$$\dot H_i - H_i\left(\dot\phi - \sum_k H_k\right) + \frac{1}{2}\frac{\partial V}{\partial\phi} = \lambda_{\rm s}^{d-1}\,{\rm e}^{\phi}\left(p_i - \frac{\sigma}{2}\right). \tag{10}$$

From the dilaton equation (5) we are lead, finally, to

$$2\ddot\phi - \dot\phi^2 + 2\dot\phi\sum_i H_i - \sum_i\left(2\dot H_i + H_i^2\right) - \left(\sum_i H_i\right)^2 + V - \frac{\partial V}{\partial\phi} = \lambda_{\rm s}^{d-1}\,{\rm e}^{\phi}\sigma. \tag{11}$$

We have thus obtained a system of $d+2$ equations for the $2d+3$ unknowns $\{a_i, \phi, \rho, p_i, \sigma\}$: its solution requires the input of $d+1$ "equations of state," $p_i = p_i(\rho)$, $\sigma = \sigma(\rho)$, specifying the properties of the considered matter sources.

Let us now consider the symmetries of this system of equations. There are two symmetries, in particular, that are relevant for the discussion of this section. One of them (also present in the cosmological equations of general relativity) is the invariance under the time-reversal transformation $t \to -t$, which implies

$$H_i \to -H_i, \qquad \dot{H}_i \to \dot{H}_i, \qquad \dot{\phi} \to -\dot{\phi}, \qquad \ddot{\phi} \to \ddot{\phi}. \tag{12}$$

Thanks to this invariance property, if the set of variables $S = \{a_i(t), \phi(t), \rho(t)\}$ represents an exact solution of (9)–(11), then the time-reversed set $\widetilde{S} = \{a_i(-t), \phi(-t), \rho(-t)\}$ also corresponds to an exact solution of the same equations (with different kinematic properties, in general).

The string-cosmology equations, in the particular case $\sigma = 0$ and $V =$ const, are also invariant under other transformations which have no analogue in general relativity, and which include the inversion of an arbitrary number of scale factors of the background geometry (8): the so-called scale-factor duality transformations [1, 6]. For a simple illustration of this property we may conveniently rewrite the equations in terms of the "shifted variables" $\overline{\phi}$, $\overline{\rho}$, $\overline{p}_i$, $\overline{\sigma}$, defined by

$$\overline{\phi} = \phi - \ln \prod_i a_i = \phi - \sum_i \ln a_i, \qquad i = 1, \dots, d,$$
$$\overline{\rho} = \rho \prod_i a_i, \qquad \overline{p}_k = p_k \prod_i a_i, \qquad \overline{\sigma} = \sigma \prod_i a_i. \tag{13}$$

Equations (9)–(11) then become

$$\dot{\overline{\phi}}^2 - \sum_i H_i^2 - V = 2\lambda_{\mathrm{s}}^{d-1}\, \mathrm{e}^{\overline{\phi}}\, \overline{\rho}, \tag{14}$$

$$\dot{H}_i - H_i \dot{\overline{\phi}} + \frac{1}{2}\frac{\partial V}{\partial \phi} = \lambda_{\mathrm{s}}^{d-1}\, \mathrm{e}^{\overline{\phi}} \left(\overline{p}_i - \frac{\overline{\sigma}}{2}\right), \tag{15}$$

$$2\ddot{\overline{\phi}} - \dot{\overline{\phi}}^2 - \sum_i H_i^2 + V - \frac{\partial V}{\partial \phi} = \lambda_{\mathrm{s}}^{d-1}\, \mathrm{e}^{\overline{\phi}}\, \overline{\sigma}. \tag{16}$$

Under the transformation $a \to \widetilde{a} = a^{-1}$, on the other hand, we have

$$H = a^{-1}\frac{\mathrm{d}a}{\mathrm{d}t} \quad \to \quad \widetilde{H} = \widetilde{a}^{-1}\frac{\mathrm{d}\widetilde{a}}{\mathrm{d}t} = a\frac{\mathrm{d}a^{-1}}{\mathrm{d}t} = -H. \tag{17}$$

We can then easily check that (14)–(16), in the particular case $\sigma = 0$ and $\partial V/\partial\phi = 0$, are invariant under the scale-factor duality transformations:

$$a_i \to a_i^{-1}, \qquad \overline{\phi} \to \overline{\phi}, \qquad \overline{\rho} \to \overline{\rho}, \qquad \overline{p}_i \to -\overline{p}_i. \tag{18}$$

This type of transformation is called "dual" as it generalizes to the case of time-dependent backgrounds the T-duality transformation inverting the compactification radius (thus interchanging "winding" and "momentum" modes) in the spectrum of a closed string, quantized in the presence of compact spatial dimensions [7]. For the invariance under the transformations (18), however, there is no need of a compact geometry; what is required, instead, is a non-trivial transformation of the dilaton. Let us suppose, in fact, that we are inverting a number n of scale factors, say $a_1, \dots, a_n$, with $1 \le n \le d$: the condition $\overline{\phi} \to \overline{\phi}$ then implies

$$\phi - \sum_{i=1}^{d} \ln a_i = \widetilde{\phi} - \sum_{i=1}^{d} \ln \widetilde{a}_i = \widetilde{\phi} - \sum_{i=1}^{n} \ln a_i^{-1} - \sum_{i=n+1}^{d} \ln a_i, \qquad (19)$$

from which

$$\phi \to \widetilde{\phi} = \phi - 2 \sum_{i=1}^{n} \ln a_i. \qquad (20)$$

In the presence of sources, their energy density is also non-trivially transformed: the condition $\overline{\rho} \to \overline{\rho}$ implies, in fact,

$$\rho \prod_{i=1}^{d} a_i = \widetilde{\rho} \prod_{i=1}^{n} a_i^{-1} \prod_{i=1+n}^{d} a_i, \qquad (21)$$

from which

$$\rho \to \widetilde{\rho} = \rho \prod_{i=1}^{n} a_i^2. \qquad (22)$$

The transformation of the pressure is similar, but with an additional "reflection" of the equation of state along the spatial directions affected by the duality transformation:

$$p_i \to \widetilde{p}_i = -p_i \prod_{k=1}^{n} a_k^2, \qquad i = 1, \dots, n. \qquad (23)$$

In any case, given a set of variables $S = \{a_i(t), \phi(t), \rho(t), p_i\}$ representing an exact solution of (14)–(16), a new solution can be obtained by inverting an arbitrary number n (between 1 and d) of scale factors, and is represented by

$$\widetilde{S} = \{a_1^{-1}, a_2^{-1}, \dots, a_n^{-1}, a_{n+1}, \dots, a_d, \widetilde{\phi}, \widetilde{\rho}, \widetilde{p}_1, \dots, \widetilde{p}_n, p_{n+1}, \dots, p_d\}, \quad (24)$$

where $\widetilde{\phi}$, $\widetilde{\rho}$, and $\widetilde{p}_i$ are given by (20), (22), and (23), respectively.

The invariance under the transformations (18) is only a particular case of a more general $O(d,d)$ symmetry of the tree-level string cosmology equations [8] (see also the contribution of Meissner [9] to this volume), and can be extended so as to include the NS–NS two-form $B_{\mu\nu}$ in the effective action. Such an extension is also possible in the presence of fluid sources: a homogeneous gas of strings, in particular, provides a realistic example of source which are automatically compatible with the $O(d,d)$ symmetry of the background equations [10].

In addition, the invariance under the transformations (18) can be extended to the case of non-trivial potentials, $\partial V/\partial\phi \neq 0$, and non-zero dilaton couplings to the matter sources, $\sigma \neq 0$. In both cases, however, we need to generalize those parts of the action describing the self-coupling of the dilaton and the coupling of the dilation to the matter fields present in S_m.

In the case of the dilaton potential it is well-known [8, 9, 10] that the invariance under the transformations (18) holds for non-trivial V, provided V depends on ϕ through the variable $\overline{\phi}$. Such a variable, unlike ϕ, *is not* a scalar under general coordinate transformations (as evident from the definition (13)): it is thus impossible, in a generic background, to define a potential which is function of $\overline{\phi}$ and which can be directly inserted as a scalar into the covariant action (1). However, as first pointed out in [11], the action and the corresponding equations of motion can be written in a generalized form which is invariant under general coordinate transformations in any metric background, using for the potential a non-local variable which exactly reduces to $\overline{\phi}$ in the limit of a homogeneous geometry.

Here it will be shown that the invariance under the duality transformations (18) can be restored also in the presence of the dilaton charge σ, provided the dilaton coupling to the matter sources is parametrized by a non-local variable, as in the case of the potential. This result is new, and will be explicitly derived in the following subsection.

1.2 Non-local Dilaton Interactions

The formalism introduced in [11] is based on the non-local variable $\xi(x)$, defined by

$$\xi(x) \equiv \xi[\phi(x)] = -\ln \int \frac{\mathrm{d}^{d+1}y}{\lambda_{\mathrm{s}}^{d}} \left(\sqrt{-g}\,\mathrm{e}^{-\phi}\sqrt{\epsilon(\nabla\phi)^2}\right)_y \delta(\phi_x - \phi_y), \tag{25}$$

where we have explicitly inserted the parameter

$$\epsilon = \mathrm{sign}\{(\nabla\phi)^2\} = \begin{cases} 1, & (\nabla\phi)^2 > 0, \\ -1, & (\nabla\phi)^2 < 0, \end{cases} \tag{26}$$

so as to include in the formalism both time-like and space-like dilaton gradients. Note that we are using the convenient notation in which an index appended to round brackets, $(\dots)_x$, means that all quantities inside the brackets are functions of the appended variable. Similarly, $\phi_x \equiv \phi(x)$. We can immediately check that, for a homogeneous background of the type (8) with spatial sections of finite comoving volume ($\int \mathrm{d}^d y = V_d = \mathrm{const} < \infty$), the variable ξ exactly reduces to the variable $\overline{\phi}$ of (13). In that case, in fact, an explicit integration gives

$$\xi = \phi - \ln \prod_i a_i - \ln\left(\frac{V_d}{\lambda_{\mathrm{s}}^{d}}\right), \tag{27}$$

and the constant volume factor can be simply absorbed by rescaling ϕ, so that $\xi \equiv \overline{\phi}$.

Let us now suppose that the matter couplings and the self-coupling of the dilaton are both parametrized by ξ, according to the effective action

$$\begin{aligned} S = &- \frac{1}{2\lambda_{\mathrm{s}}^{d-1}} \int \mathrm{d}^{d+1}x\, \sqrt{-g}\, \mathrm{e}^{-\phi} \left[R + (\nabla\phi)^2 + V(\mathrm{e}^{-\xi})\right] \\ &+ \int \mathrm{d}^{d+1}x\, \sqrt{-g}\, \mathcal{L}_m(\mathrm{e}^{-\xi}) + S_\Sigma, \end{aligned} \tag{28}$$

which is a (generally covariant) scalar functional of the non-local variable ξ. Note that, without loss of generality, we have written both the potential and the matter Lagrangian $\mathcal{L}_m$ as a function of $\exp(-\xi)$. In higher-dimensional manifolds with compact spatial sections, in fact, the exponential of the shifted dilaton plays the role of a "dimensionally reduced" coupling parameter, and we may thus expect (at least in a perturbative regime) that dilaton interactions appear as a power expansion (or as a a simple function) of such an exponential [11].

The generalized equations of motion can now be obtained by computing the functional derivative of the action (28) with respect to $g^{\mu\nu}$ and ϕ. The derivative with respect to the metric, using the standard definition of gravitational stress tensor, (4), and the properties of the delta distribution, leads to the (integro-differential) equations of motion

$$\begin{aligned} &G_{\mu\nu} + \nabla_\mu\nabla_\nu\phi + \frac{1}{2} g_{\mu\nu} \left(\nabla\phi^2 - 2\nabla^2\phi - V\right) \\ &\quad - \frac{1}{2}\gamma_{\mu\nu}\sqrt{\epsilon(\nabla\phi)^2}\left(\mathrm{e}^{-\phi} I_V - 2\lambda_{\mathrm{s}}^{d-1} I_m\right) = \lambda_{\mathrm{s}}^{d-1}\mathrm{e}^{\phi}\, T_{\mu\nu}, \end{aligned} \tag{29}$$

which generalize (3) (see Appendix A for the details of the derivation). Here

$$\gamma_{\mu\nu} = g_{\mu\nu} - \frac{\nabla_\mu\phi\nabla_\nu\phi}{(\nabla\phi)^2}, \tag{30}$$

$$I_V(x) = \lambda_{\mathrm{s}}^{-d} \int \mathrm{d}^{d+1}y \left(\sqrt{-g}\, V'\right)_y \delta(\phi_y - \phi_x), \tag{31}$$

$$I_m(x) = \lambda_{\mathrm{s}}^{-d} \int \mathrm{d}^{d+1}y \left(\sqrt{-g}\, \mathcal{L}_m'\right)_y \delta(\phi_y - \phi_x), \tag{32}$$

where the prime denotes the derivative with respect to the argument $\exp(-\xi)$, namely:

$$V' = \frac{\partial V}{\partial(\mathrm{e}^{-\xi})} = -\mathrm{e}^{\xi}\frac{\partial V}{\partial\xi}, \qquad \mathcal{L}_m' = \frac{\partial \mathcal{L}_m}{\partial(\mathrm{e}^{-\xi})} = -\mathrm{e}^{\xi}\frac{\partial \mathcal{L}_m}{\partial\xi}. \tag{33}$$

The functional derivative with respect to ϕ leads to the dilaton equation of motion,

$$R+2\nabla^2\phi-(\nabla\phi)^2+V+\epsilon\frac{\gamma_{\mu\nu}\nabla^\mu\nabla^\nu\phi}{\sqrt{\epsilon(\nabla\phi)^2}}\left(e^{-\phi}I_V-2\lambda_s^{d-1}I_m\right)$$
$$+\left(e^{-\xi}-e^{-\phi}J\right)\left(V'-2\lambda_s^{d-1}e^{\phi}\mathcal{L}'_m\right)=0, \quad (34)$$

generalizing (5). Here

$$J(x)=\lambda_s^{-d}\int d^{d+1}y\left(\sqrt{-g}\sqrt{\epsilon(\nabla\phi)^2}\right)_y\delta'(\phi_x-\phi_y), \quad (35)$$

where δ' denotes the derivative of the delta function with respect to its argument (see Appendix A). The combination of (29) and (34) finally leads to the equation

$$R_{\mu\nu}+\nabla_\mu\nabla_\nu\phi$$
$$+\frac{1}{2}\left(e^{-\phi}I_V-2\lambda_s^{d-1}I_m\right)\left(\epsilon\, g_{\mu\nu}\frac{\gamma_{\alpha\beta}\nabla^\alpha\nabla^\beta\phi}{\sqrt{\epsilon(\nabla\phi)^2}}-\gamma_{\mu\nu}\sqrt{\epsilon(\nabla\phi)^2}\right)$$
$$+\frac{1}{2}g_{\mu\nu}\left(e^{-\xi}-e^{-\phi}J\right)\left(V'-2\lambda_s^{d-1}e^{\phi}\mathcal{L}'_m\right)=\lambda_s^{d-1}e^{\phi}\,T_{\mu\nu}, \quad (36)$$

generalizing (7).

We can easily check that these new equations, written for a homogeneous background, are invariant under scale-factor duality transformations even in the presence of non-trivial potentials and dilaton couplings, i.e., for $\partial V/\partial\xi\neq 0$, $\partial\mathcal{L}_m/\partial\xi\neq 0$. Consider, for instance, the background configuration of (8) with time-like dilaton gradients, for which $\epsilon=1$. From (30) we obtain

$$\gamma_0^0=0, \qquad \gamma_i^j=\delta_i^j. \quad (37)$$

The $(0,0)$ component of (29) thus coincides with the $(0,0)$ component of (3), and is given by (9), as before.

For the spatial components we first note that, performing the homogeneous limit in which $\xi\to\overline{\phi}$, we are lead to the identities

$$\sqrt{\epsilon(\nabla\phi)^2}\left(e^{-\phi}I_V-2\lambda_s^{d-1}I_m\right)\equiv e^{-\xi}\left(V'-2\lambda_s^{d-1}e^{\phi}\mathcal{L}'_m\right)$$
$$\longrightarrow-\left(\frac{\partial V}{\partial\overline{\phi}}-2\lambda_s^{d-1}e^{\phi}\frac{\partial\mathcal{L}_m}{\partial\overline{\phi}}\right); \quad (38)$$

$$\frac{\gamma_{\alpha\beta}\nabla^\alpha\nabla^\beta\phi}{\sqrt{\epsilon(\nabla\phi)^2}}\left(e^{-\phi}I_V-2\lambda_s^{d-1}I_m\right)\equiv e^{-\phi}J\left(V'-2\lambda_s^{d-1}e^{\phi}\mathcal{L}'_m\right)$$
$$\longrightarrow-\frac{\sum_i H_i}{\dot{\phi}}\left(\frac{\partial V}{\partial\overline{\phi}}-2\lambda_s^{d-1}e^{\phi}\frac{\partial\mathcal{L}_m}{\partial\overline{\phi}}\right). \quad (39)$$

Using such identities we find that the dependence on V' and $\mathcal{L}'_m$ completely disappears from the spatial components of (36) with $\epsilon=1$, and we obtain the condition

$$R_i{}^j + \nabla_i \nabla^j \phi = \lambda_s^{d-1} e^{\phi} T_i{}^j. \tag{40}$$

Written explicitly, the new spatial equation is given by

$$\dot{H}_i - H_i \left(\dot{\overline{\phi}} - \sum_k H_k \right) = \lambda_s^{d-1} e^{\phi} p_i, \tag{41}$$

and is thus crucially simplified with respect to the corresponding (local) spatial equation (10).

The dilaton equation (34) also simplifies in the homogeneous limit, thanks to the identities (38) and (39) from which we obtain

$$R + 2\nabla^2 \phi - (\nabla \phi)^2 + V - \frac{\partial V}{\partial \overline{\phi}} = -2\lambda_s^{d-1} e^{\phi} \frac{\partial \mathcal{L}_m}{\partial \overline{\phi}}. \tag{42}$$

The explicit form is

$$2\ddot{\phi} - \dot{\phi}^2 + 2\dot{\phi} \sum_i H_i - \sum_i \left(2\dot{H}_i + H_i^2 \right) - \left(\sum_i H_i \right)^2$$
$$+ V(\overline{\phi}) - \frac{\partial V}{\partial \overline{\phi}} = \lambda_s^{d-1} e^{\phi} \sigma(\overline{\phi}), \tag{43}$$

where we have defined, by analogy with (6),

$$\sigma(\overline{\phi}) = -2 \frac{\partial \mathcal{L}_m}{\partial \overline{\phi}}. \tag{44}$$

The new set of equations (9), (41), and (43) is compatible with scale-factor duality for any V and σ, as can be shown by rewriting the equations in terms of the shifted variables of (13). With such variables, (9), (41), and (43) become, respectively,

$$\dot{\overline{\phi}}^2 - \sum_i H_i^2 - V = 2\lambda_s^{d-1} e^{\overline{\phi}} \overline{\rho}, \tag{45}$$

$$\dot{H}_i - H_i \dot{\overline{\phi}} = \lambda_s^{d-1} e^{\overline{\phi}} \overline{p}_i, \tag{46}$$

$$2\ddot{\overline{\phi}} - \dot{\overline{\phi}}^2 - \sum_i H_i^2 + V(\overline{\phi}) - \frac{\partial V}{\partial \overline{\phi}} = \lambda_s^{d-1} e^{\overline{\phi}} \overline{\sigma}(\overline{\phi}). \tag{47}$$

They are manifestly invariant under the generalized transformations

$$a_i \to a_i^{-1}, \qquad \overline{\phi} \to \overline{\phi}, \qquad \overline{\rho} \to \overline{\rho}, \qquad \overline{p}_i \to -\overline{p}_i, \qquad \overline{\sigma} \to \overline{\sigma}, \tag{48}$$

preserving the shifted version of the dilaton-charge density $\overline{\sigma}$.

1.3 The Pre-big Bang Scenario

Let us come back to the cosmological applications of scale-factor duality. Even without using its non-local extensions, the duality symmetry of the equations allows introducing a "dual complement" of the standard cosmological solutions, and suggests new possible scenarios for the primordial evolution of our Universe.

For a simple illustration of this possibility it will be enough to consider a homogeneous, isotropic and spatially flat metric background, sourced by a barotropic perfect fluid with equation of state $p/\rho = \gamma = \text{const}$, with negligible dilaton charge. By imposing $\sigma = 0$, $V = 0$, and assuming $\gamma \neq 0$, one easily finds that (45)–(47) are satisfied by the following particular exact solution:

$$a = \left(\frac{t}{t_0}\right)^{\frac{2\gamma}{1+d\gamma^2}}, \qquad \overline{\rho} = \rho_0 a^{-d\gamma}, \qquad \overline{\phi} = -\frac{2}{1+d\gamma^2} \ln\left(\frac{t}{t_0}\right) + \text{const}, \tag{49}$$

where $t > 0$, and t_0, ρ_0 are positive integration constants. In terms of the non-shifted variables:

$$\begin{aligned} a &= \left(\frac{t}{t_0}\right)^{\frac{2\gamma}{1+d\gamma^2}}, \qquad \rho = \rho_0 a^{-d(1+\gamma)}, \qquad p = \gamma\rho, \\ \phi &= 2\frac{d\gamma - 1}{1+d\gamma^2} \ln\left(\frac{t}{t_0}\right) + \text{const}, \qquad t > 0. \end{aligned} \tag{50}$$

This solution, defined over the real positive semi-axis $t > 0$, describes a Universe evolving from a past curvature singularity at $t \to 0_-$ to an asymptotically flat configuration at $t \to +\infty$. For $\gamma > 0$ we have a phase of decelerated expansion and decreasing curvature,

$$\dot{a} > 0, \qquad \ddot{a} < 0, \qquad \dot{H} < 0, \tag{51}$$

typical of the standard cosmological scenario. Also, for a "realistic" equation of state with $\gamma d \leq 1$, the dilaton turns out to be non-increasing ($\dot{\phi} \leq 0$); in particular, for a radiation fluid with $\gamma = 1/d$, one recovers the radiation-dominated solution at constant dilaton,

$$p = \frac{\rho}{d}, \qquad a = \left(\frac{t}{t_0}\right)^{\frac{2}{1+d}}, \qquad \rho = \rho_0 a^{-(1+d)}, \qquad \phi = \text{const}, \tag{52}$$

which is also an exact solution of the standard Einstein equations.

Thanks to the symmetries of the string cosmology equations we can now obtain new, different solutions (which have no analogue in the context of the Einstein equations) by performing a time reflection $t \to -t$ and, simultaneously, a dual transformation defined by (18). Starting in particular from (50) we are lead to the background

$$a = \left(-\frac{t}{t_0}\right)^{-\frac{2\gamma}{1+d\gamma^2}}, \qquad \rho = \rho_0 a^{-d(1-\gamma)}, \qquad p = -\gamma\rho,$$
$$\phi = -2\frac{1+d\gamma}{1+d\gamma^2}\ln\left(-\frac{t}{t_0}\right) + \text{const}, \qquad t < 0, \qquad (53)$$

which is still a particular exact solution of (45)–(47). It is defined on the negative real semi-axis $t < 0$, and for $\gamma > 0$ it describes a phase of accelerated (i.e., inflationary) expansion and growing curvature:

$$\dot{a} > 0, \qquad \ddot{a} > 0, \qquad \dot{H} > 0. \qquad (54)$$

In this case the Universe evolves from an asimptotically flat initial configuration at $t \to -\infty$ towards a curvature singularity at $t \to 0_-$. The dilaton is always growing ($\dot{\phi} > 0$) for $t \to 0_-$, even if we consider the dual of the radiation-dominated solution (52):

$$p = -\frac{\rho}{d}, \quad a = \left(-\frac{t}{t_0}\right)^{-\frac{2}{1+d}}, \quad \rho = \rho_0 a^{-d+1}, \quad \phi = -\frac{4d}{d+1}\ln\left(-\frac{t}{t_0}\right). \qquad (55)$$

This interesting property of the low-energy string-cosmology equations—i.e., the presence of an inflationary "partner" associated to any standard decelerated solution—is also valid in the absence of sources. Consider, for instance, (45)–(47) with $p = 0 = \rho$, and $V = 0$. In the isotropic limit we find the particular exact solution

$$a = \left(\frac{t}{t_0}\right)^{1/\sqrt{d}}, \qquad \phi = \left(\sqrt{d} - 1\right)\ln\left(\frac{t}{t_0}\right), \qquad t > 0, \qquad (56)$$

describing decelerated expansion and decreasing curvature. By applying the transformations (18) we are lead to the dual solution

$$a = \left(-\frac{t}{t_0}\right)^{-1/\sqrt{d}}, \qquad \phi = -\left(\sqrt{d} + 1\right)\ln\left(-\frac{t}{t_0}\right), \qquad t < 0, \qquad (57)$$

describing accelerated expansion, growing curvature and growing dilaton. Actually, both the vacuum and the fluid-dominated solutions can be obtained as asymptotic limits of the general exact solution of the system of equations (45)–(47) (for barotropic sources, with $V = 0$ and $\sigma = 0$), in the large-curvature and small-curvature limits, respectively [12, 13].

It is well-known that the decelerated configurations, typical of the standard cosmological scenario, cannot be extended back in time without limits: the range of the time coordinate is bounded from below by the presence of the initial singularity (indeed, going back in time, the growth of H is unbounded, and the curvature blows up to infinity in a finite proper-time interval).

The standard scenario, however, is certainly incomplete because it excludes inflation. The inclusion of inflation, on the other hand, modifies the behavior of the curvature scale: during a phase of "slow-roll" inflation [14], for instance, the background geometry can be approximately described by a de Sitter-like metric where $\dot{H} \simeq 0$, and in which the curvature tends to settle at a constant. One might think, therefore, that a complete (and realistic) cosmological scenario could avoid the initial singularity, replacing it with a primordial inflationary phase at constant curvature.

Unfortunately, however, an epoch of accelerated expansion at constant curvature, described by the Einstein equations, and dominated by the potential energy of some "inflaton" scalar field satisfying causality and weak-energy conditions, cannot be "past eternal," as proved in [15]. Thus, the conventional inflationary scenario mitigates the rapid growth of the curvature typical of the standard cosmological evolution, and shifts back in time the position of the initial singularity, without completely removing it, however (namely, without extending in a geodesically complete way the model, back in time, to infinity).

If a constant curvature phase is not appropriate to construct a regular model (fully extended over the whole temporal axis), the alternative we are left is a model in which the curvature, as we go back in time, after reaching a maximum, at some point, starts decreasing; in other words, a model in which the standard evolution is completed and complemented by a primordial phase with a specular behavior of H with respect to the standard one. Remarkably, this is exactly what can be obtained assuming that the cosmological evolution satisfies a principle of "self-duality"—i.e., assuming that the past evolution of our Universe is described by the "dual complement" of the present one [2].

More precisely, if we consider a cosmological model satisfying (at least approximately) the self-dual condition $a(t) = a^{-1}(-t)$, such that the standard decelerated regime at $t > 0$ smoothly evolves, back in time, into the accelerated partner at $t < 0$, we can then obtain a scenario in which the singularity is automatically regularized, and the initial evolution is automatically of the inflationary type. In such a context, the big bang singularity is replaced by an epoch of high (but finite) curvature, characterizing the transition between the standard cosmological phase ($\dot{H} < 0$) and its dual ($\dot{H} > 0$): it comes natural, in such a context, to call *pre-big bang* the initial phase ($t < 0$) at growing curvature and growing dilaton, in contrast to the subsequent *post-big bang* phase ($t > 0$), describing the standard cosmological evolution.

The dilaton, on the other hand, provides an exponential parametrization of the (tree-level) string coupling $g_s = \exp(\phi/2)$, controlling the relative strength of all (gravitational and gauge) interactions [3, 4]. The principle of self-duality thus suggests that the Universe is lead to its present state after a long evolution started from an extremely simple—almost trivial—configuration, characterized by a nearly flat geometry and by a very small coupling parameter,

$$H^2 \to 0, \qquad \phi \to -\infty, \qquad g_s^2 = e^{\phi} \to 0, \tag{58}$$

the so-called string perturbative vacuum (see Fig. 2). In this case, the initial Universe is characterized by a regime of extremely low energies in which the "curvatures" (i.e., the field gradients) are small ($\lambda_s^2 H^2 \ll 1$, $\lambda_s^2 \dot{\phi}^2 \ll 1, \ldots$), the couplings are weak ($g_s^2 \ll 1$), and the background dynamics can be appropriately described by the lowest-order string effective action, at tree level in the α' and quantum loop expansion (also in agreement with the hypothesis of "asymptotic past triviality" [16]). We can talk of "birth of the Universe from the string perturbative vacuum," as also pointed out in a quantum cosmology context (see, e.g., [17, 18]).

This picture is in remarkable contrast with the standard (even inflationary) picture in which the Universe starts evolving from a highly curved geometric state: the more we go back in time, in that context, the more we enter a Planckian and (possibly) trans-Planckian [19] non-perturbative regime of ultra-high energies, requiring the full inclusion of quantum gravity effects, to all orders, for a correct description.

The principle of self-duality, on the contrary, suggests a picture in which the more we go back in time (after crossing the epoch of maximal curvature),

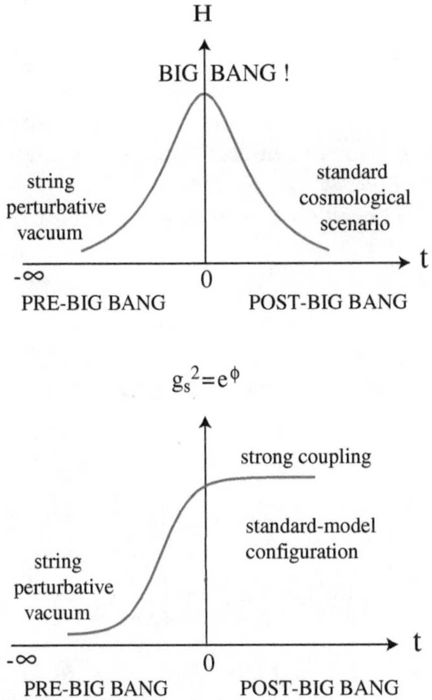

Fig. 2. Qualitative time-evolution of the curvature scale (*upper panel*) and of the string coupling (*lower panel*), for a typical self-dual background which smoothly interpolates between the pre-big bang and the post-big bang phase, starting from the string perturbative vacuum

the more we approach a *flat, cold*, and *vacuum* configuration (strongly reminiscent of the "biblical" scenario quoted at the beginning of Sect. 1), which can be appropriately described by the classical background equations obtained from the action (1). Quantum effects, in the form of higher-curvature and higher-loop contributions, are expected to become important only *toward the end* of the pre-big bang phase, when the background approaches the string scale at $t \to 0_-$. Actually, all studies performed so far have shown that such corrections *must* become dominant, eventually, in order to stop the growth of the curvature [20] and possibly trigger a smooth transition to the post-big bang regime [21].

1.4 A Smooth "Bounce"

The lowest-order string effective action can appropriately describe the phase of primordial background evolution typical of the pre-big bang scenario, but not the transition to the standard decelerated regime occurring at high curvatures and strong coupling, and requiring the introduction of higher-order corrections. Referring the reader to the existing literature for a detailed review of the transition models studied so far (see, for instance, [22]), we shall present here only two simple phenomenological examples, by applying, to this purpose, the formalism introduced in Sect. 1.2 (and Appendix A). In these examples, in fact, the bouncing transition is induced by the presence of a non-local effective potential $V(\overline{\phi})$, expected to simulate the backreaction of the quantum loop corrections in higher-dimensional manifolds with compact spatial sections [11].

The first example is based on a potential which, in the homogeneous limit, takes the form

$$V(\overline{\phi}) = -V_0 e^{4\overline{\phi}}, \qquad V_0 > 0, \tag{59}$$

and which may thus perturbatively interpreted as a four-loop potential. With this potential, the duality-invariant equations (45)–(47), in vacuum ($\rho = p = \sigma = 0$), and in the isotropic limit, are solved by the particular exact solution [18]:

$$\begin{aligned} a(t) &= a_0 \left[\frac{t}{t_0} + \left(1 + \frac{t^2}{t_0^2}\right)^{1/2} \right]^{1/\sqrt{d}}, \\ \overline{\phi} &= -\frac{1}{2} \ln \left[t_0 \sqrt{V_0} \left(1 + \frac{t^2}{t_0^2}\right) \right] + \text{const}, \\ \phi &= \ln \frac{\left[t/t_0 + (1 + t^2/t_0^2)^{1/2} \right]^{\sqrt{d}}}{(1 + t^2/t_0^2)^{1/2}} + \text{const}, \end{aligned} \tag{60}$$

where t_0 and a_0 are positive integration constants. This regular "bouncing" solution is exactly self-dual—as it satisfies $a(t)/a_0 = a_0/a(-t)$ – and is characterized by a bounded, "bell-like" shape of the curvature and of the dilaton kinetic energy (see Fig. 3). The solution smoothly interpolates between the pre- and post-big bang vacuum solutions (57) and (56) (corresponding to the dashed curves of Fig. 3), which are recovered in the asymptotic limits $t \to -\infty$ and $t \to +\infty$, respectively. The bounce of the curvature, and the smooth transition between the two branches of the low-energy solutions, is induced and controlled by the potential (59) which dominates the background evolution in the high-curvature limit $|t| \to 0$, and which becomes rapidly negligible as $t \to \pm\infty$, as illustrated in Fig. 3.

It should be noted that in this solution the dilaton keeps growing, monotonically, even in the limit $t \to +\infty$. In more realistic examples, however, such a growth is expected to be damped by the interaction with the matter/radiation post-big bang sources [23], and/or by the action of a suitable non-perturbative potential appearing in the strong coupling regime.

The second example of bounce is based on a general integration of the duality-invariant equations (45)–(47), in the presence of isotropic fluid sources with $\sigma = 0$ and of a two-loop (non-local) potential which in the homogeneous limit takes the form

$$V(\overline{\phi}) = -V_0 e^{2\overline{\phi}}, \qquad\qquad V_0 > 0. \tag{61}$$

In this case the equations can be integrated exactly not only for barotropic equations of state ($p/\rho = \gamma =$ const), but also for any ratio p/ρ which is an integrable function of an appropriately defined time-like parameter [2].

An interesting example (motivated by the study of the equation of state of a string gas in rolling backgrounds [24]) is the case in which p/ρ smoothly evolves from the value $\gamma = -1/d$ at $t = -\infty$ to the value $\gamma = 1/d$ at $t = +\infty$, thus connecting the radiation equation of state to its dual partner, according to the law:

$$\frac{p}{\rho} = \frac{1}{d}\frac{x}{\sqrt{x_1^2 + x^2}}. \tag{62}$$

Here x_1 is an arbitrary integration constant, and x is a (dimensionless) time-like coordinate defined by

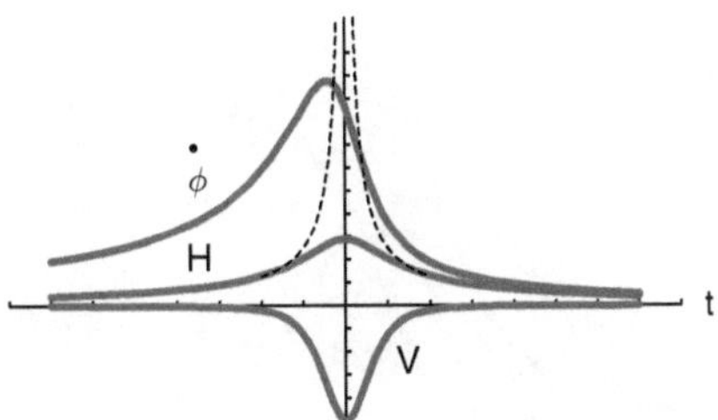

Fig. 3. Plot of the curvature, of the dilaton kinetic energy, and of the potential $V(\overline{\phi})$, for the bouncing solution (60). The dashed curves represent the (singular) vacuum solutions (56), (57), obtained with $V = 0$. All curves are plotted for $t_0 = 1$, $V_0 = 1$, and $d = 3$

$$\frac{\mathrm{d}x}{\mathrm{d}t} = \frac{L}{2}\overline{\rho}, \tag{63}$$

where L is a constant with dimensions of length (we are using units in which $2\lambda_{\mathrm{s}}^{d-1} = 1$, so that $[\rho] = L^{-2}$). Using (61)–(63), and choosing a simplifying set of integration constants (appropriate to the pedagogical purpose of this paper), we can then obtain the following particular exact solution [2]:

$$\begin{aligned}
a &= a_0\left(x+\sqrt{x^2+x_1^2}\right)^{2/(d-1)}, \\
\mathrm{e}^{\phi} &= a_0^d \mathrm{e}^{\phi_0}\left(1+\frac{x}{\sqrt{x^2+x_1^2}}\right)^{2d/(d-1)}, \\
\rho\mathrm{e}^{\phi} &= \frac{d-1}{dL^2}\mathrm{e}^{2\phi_0}\left(x^2+x_1^2\right)^{-(d+1)/(d-1)}, \\
p\mathrm{e}^{\phi} &= \frac{d-1}{d^2L^2}\mathrm{e}^{2\phi_0}x\left(x^2+x_1^2\right)^{-(3d+1)/2(d-1)},
\end{aligned} \tag{64}$$

where a_0 and ϕ_0 are integration constants. The smooth and bouncing behavior of this solution is illustrated in Fig. 4.

The above solution is self-dual, in the sense that $\overline{\phi}(x) = \overline{\phi}(-x)$, $\overline{\rho}(x) = \overline{\rho}(-x)$, and

$$\left[\frac{a(x)}{a_0 x_1^{2/(d-1)}}\right] = \left[\frac{a(-x)}{a_0 x_1^{2/(d-1)}}\right]^{-1} \tag{65}$$

(with an appropriate choice of the integration constant a_0 it is always possible to set to 1 the fixed point of scale-factor inversion). The solution satisfies, asymptotically,

$$\begin{aligned}
x \to -\infty \quad &\Rightarrow \quad a \sim (-x)^{-2/(d-1)} \sim \overline{\rho} \sim \frac{\mathrm{d}x}{\mathrm{d}t}, \\
x \to +\infty \quad &\Rightarrow \quad a \sim x^{2/(d-1)} \sim \frac{1}{\overline{\rho}} \sim \frac{\mathrm{d}t}{\mathrm{d}x}.
\end{aligned} \tag{66}$$

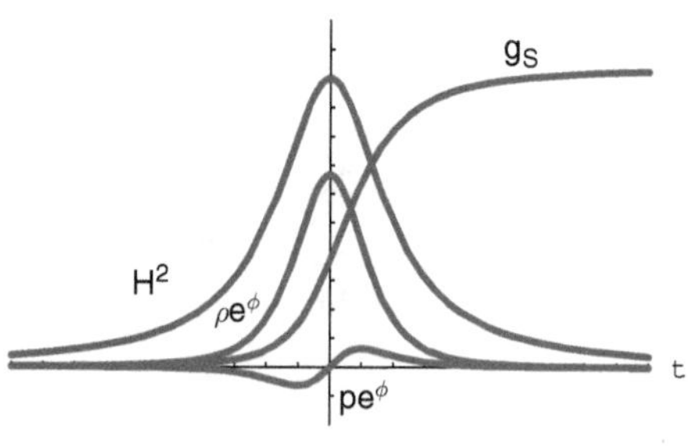

Fig. 4. Plot of the curvature, of the string coupling, of the effective energy density and of the effective pressure for the self-dual solution (64). The curves are plotted for $d = 3$, $L = 1$, $x_1 = 1$, $\phi_0 = 0$, and $a_0 = \exp(-2/3)$

Re-expressing a, ϕ, ρ, and p, in the asymptotic limits $x \to \pm\infty$ in terms of the cosmic time t, we can check that this solution smoothly interpolates between the pre-big bang configuration (55) describing accelerated expansion, growing dilaton, negative pressure, and the final post-big bang configuration (52), describing the radiation-dominated state with frozen dilaton and decelerated expansion. As in the previous case the smoothing out of the tree-level singularity, and the appearance of bouncing transition, is a consequence of the effective potential (61).

1.5 Cosmological Perturbations

The phase of pre-big bang evolution, being accelerated, can amplify the quantum fluctuations of the metric tensor (and of other background fields) just like any other type of inflationary evolution. However, because of the kinematic properties of pre-big bang inflation (associated to the shrinking of the Hubble horizon H^{-1}), the spectral distribution of the metric fluctuations, after their amplification, tends to grow with frequency [25]. This peculiar aspect of the spectrum may be regarded as representing both an advantage and a difficulty of pre-big bang models with respect to other models of inflation.

The advantage is of phenomenological nature, and refers to the transverse and traceless tensor part of the metric fluctuations. Their amplification leads to the formation of a stochastic background of relic gravitational waves whose spectral energy density, Ω_g, grows with frequency

$$\Omega_g(\omega, t) = \left(\frac{H_1}{M_{\rm P}}\right)^2 \Omega_r(t) \left(\frac{\omega}{\omega_1}\right)^\delta, \qquad \delta > 0, \qquad \omega \leq \omega_1. \tag{67}$$

Here $M_{\rm P} = 8\pi G = \lambda_{\rm P}^{-1}$ is the Planck mass, $H_1 \simeq M_{\rm s} = \lambda_{\rm s}^{-1}$ the inflation–radiation transition scale (expected to be controlled by the string mass scale $M_{\rm s}$), $\Omega_r = \rho_r/\rho_c$ is the fraction of critical energy density in radiation, ω_1 is the ultraviolet cutoff (i.e., the maximal amplified frequency) of the spectrum, and δ a model-dependent parameter depending on the background kinematics [25, 26, 27] (see also the contribution of Buonanno and Ungarelli [28] to this volume).

Thanks to the growth of the spectrum, the cosmic graviton background present today as a relic of the inflationary epoch is higher at higher frequencies (in particular, higher than the backgrounds predicted by conventional models of inflation), and thus more easily detectable by current gravitational antennas (see, e.g., [22]). Conversely, however, the spectrum is strongly suppressed in the low-frequency regime: we should thus expect, in particular, a negligible contribution of tensor metric perturbations to the observed CMB anisotropy on large scales (as in the case of the ekpyrotic [29] and "new ekpyrotic [30] scenarios where, however, the gravitational background is expected to be low even in the low-frequency regime [31]). It may be stressed, in this connection, that the possible absence of tensor contributions at large scales emerging from (planned) future measurements of the CMB polarization (such as those of WMAP, PLANCK), in combination with a positive signal possibly detected at high frequency by the next generation of gravitational antennas (such as LIGO/VIRGO, LISA, BBO, and DECIGO), could represent a strong experimental signal in favor of models of pre-big bang inflation (see, e.g., [32]).

The difficulties associated to a growing spectrum refer to the scalar part of the metric perturbations. In fact, a growing scalar spectrum cannot account for the observed peak structure of the temperature anisotropies of the CMB radiation, which requires, instead, a nearly flat (or "scale-invariant") primordial distribution: $\Omega_s(\omega) \sim \omega^{n_s-1}$, with $n_s \approx 1$. There are two possible ways out of this problem.

A first possibility relies on the growth of the dilaton—and thus of the string coupling $g_s^2 = \exp\phi$—during the phase of pre-big bang inflation. Even starting at weak coupling, a pre-big bang background unavoidably evolves toward the strong coupling regime $g_s \sim 1$. If the bounce is not immediate then the Universe, before the transition to the standard regime, enters a strong coupling phase where higher-dimensional extended objects like Dirichlet branes and antibranes [4] (whose tension is proportional to the inverse of the string coupling) become light, and can be copiously produced [33]. The cosmic evolution may become dominated by the presence of these higher-dimensional sources [34] and, in that context, a phase of conventional slow-roll inflation can be triggered by the interaction (and eventual collision) of a brane–antibrane pair [35] (see also the contribution of Tye [36] to this volume). This new inflationary regime may efficiently dilute all pre-existing inhomogeneities and generate a new spectrum of scale-invariant, adiabatic scalar perturbations, as required for a successful explanation of the observed anisotropy. This may resolve the incompatibility between a (growing) spectrum of pre-big bang perturbations and present large-scale observations.

There is, however, a second possibility which avoids introducing additional inflationary epochs besides the initial dilaton-dominated one, and which is based on the so-called "curvaton mechanism" [37]. According to this mechanism the (flat, adiabatic) spectrum of scalar metric perturbations, responsible for the observed anisotropies, is not produced during the primordial evolution: instead, it is the outcome of the post-inflationary decay of a massive scalar field (the curvaton), whose quantum fluctuations are amplified during inflation with a nearly flat spectrum, and are converted into curvature perturbations after its decay. In the context of the pre-big bang scenario the role of the curvaton is possibly played by the Kalb–Ramond axion σ [38], associated—by space–time duality—to the four components of the NS-NS two-form $B_{\mu\nu}$ present in the massless multiplet of the string spectrum.

For a brief discussion of this possibility we should explain, first if all, why axion fluctuations can be amplified by pre-big bang inflation with a flat spectrum [39], unlike metric fluctuations. The reason is that the slope of the spectrum is directly related to kinematic behavior of the effective "pump field" responsible for the amplification, and that metric and axion fluctuations have different pump fields, even in the same given background.

In order to clarify this point let us complete the low-energy action (1) by adding the contribution of the antisymmetric field $B_{\mu\nu}$, considering (for simplicity) a model already dimensionally reduced to four space–time dimensions:

$$S = -\frac{1}{2\lambda_s^2}\int_\Omega \mathrm{d}^4x\sqrt{-g}\,\mathrm{e}^{-\phi}\left(R + \partial_\mu\phi\,\partial^\mu\phi + V - \frac{1}{12}H_{\mu\nu\alpha}H^{\mu\nu\alpha}\right),$$
$$H_{\mu\nu\alpha} = \partial_\mu B_{\nu\alpha} + \partial_\nu B_{\alpha\mu} + \partial_\alpha B_{\mu\nu}. \qquad (68)$$

In the absence of sources the equations of motion for $B_{\mu\nu}$ are automatically satisfied by introducing the "dual" axion field σ, such that

$$H^{\mu\nu\alpha} = \frac{\mathrm{e}^\phi}{\sqrt{-g}}\,\epsilon^{\mu\nu\alpha\beta}\partial_\beta\sigma, \qquad (69)$$

and the last term of the action (68) can be replaced by

$$S = \frac{1}{4\lambda_s^2}\int \mathrm{d}^4x\sqrt{-g}\,\mathrm{e}^\phi(\nabla\sigma)^2. \qquad (70)$$

Perturbing the metric and the axion field,

$$g_{\mu\nu} \to g_{\mu\nu} + h_{\mu\nu}, \qquad \sigma \to \sigma + \delta\sigma, \qquad (71)$$

around a homogeneous, conformally flat metric background, using the conformal time coordinate η (such that $\mathrm{d}t = a\mathrm{d}\eta$), and applying the standard formalism of linear cosmological perturbations (see, e.g., [40]), we obtain for tensor metric and axion fluctuations, respectively, the following quadratic actions:

$$S_h = \frac{1}{2}\int \mathrm{d}^3x\,\mathrm{d}\eta\, z_h^2(\eta)\left(h'^2 + h\nabla^2 h\right),$$
$$z_h = \frac{a}{\sqrt{2}\,\lambda_s}\,\mathrm{e}^{-\phi/2}, \qquad (72)$$
$$S_\sigma = \frac{1}{2}\int \mathrm{d}^3x\,\mathrm{d}\eta\, z_\sigma^2(\eta)\left(\delta\sigma'^2 + \delta\sigma\nabla^2\delta\sigma\right),$$
$$z_\sigma = \frac{a}{\sqrt{2}\,\lambda_s}\,\mathrm{e}^{\phi/2}. \qquad (73)$$

Here h is one of the two physical polarization states of tensor perturbations, the primes denote differentiation with respect to η, and ∇^2 is the flat-space Laplace operator, $\nabla^2 = \delta^{ij}\partial_i\partial_j$. The variation of these actions with respect to h and $\delta\sigma$ leads to the equations of motion, which can be written in terms of the canonical variables $u = (hz_h)$ and $v = (\delta\sigma\, z_\sigma)$ as follows:

$$(hz_h)'' - \left(\nabla^2 + \frac{z_h''}{z_h}\right)(hz_h) = 0, \qquad (74)$$

$$(\delta\sigma z_\sigma)'' - \left(\nabla^2 + \frac{z_\sigma''}{z_\sigma}\right)(\delta\sigma z_\sigma) = 0. \tag{75}$$

The canonical equations are the same for u and v, but the pump fields, z_h and z_σ, are different.

Consider, for instance, the axion equation (75), and recall that during inflation the accelerated evolution of the pump field can be parametrized as a power-law evolution in the negative range of the conformal-time parameter [22, 40], i.e.,

$$z_\sigma(\eta) = \frac{M_{\rm P}}{\sqrt{2}}\left(-\frac{\eta}{\eta_1}\right)^{\alpha_\sigma}, \qquad -\infty \le \eta < 0, \tag{76}$$

where $\eta_1 > 0$ is some appropriate reference timescale. Expanding in Fourier modes, (75) becomes a Bessel equation for the mode v_k,

$$v_k'' + \left[k^2 - \frac{(\nu_\sigma^2 - 1/4)}{\eta^2}\right] v_k = 0, \qquad \nu_\sigma = \frac{1}{2} - \alpha_\sigma, \tag{77}$$

and its general solution can be conveniently written as a combination of first-kind and second-kind Hankel functions [41], of argument $k\eta$ and index ν_σ, as follows:

$$v_k = (-\eta)^{1/2}\left[A_+(k)H_{\nu_\sigma}^{(2)}(k\eta) + A_-(k)H_{\nu_\sigma}^{(1)}(k\eta)\right]. \tag{78}$$

We shall now canonically normalize this general solution by imposing that the initial state of the fluctuations corresponds to a spectrum of quantum vacuum fluctuations [22, 40]. More explicitly, we shall require that the mode v_k, on the initial spatial hypersurface at $\eta \to -\infty$, may represent freely oscillating, positive frequency modes satisfying the canonical normalization

$$v_k v_k'^* - v_k' v_k^* = i, \tag{79}$$

from which

$$v_k \to \frac{{\rm e}^{-ik\eta}}{\sqrt{2k}}, \qquad \eta \to -\infty \tag{80}$$

(modulo an arbitrary phase). Using the large argument limit of the Hankel functions [41],

$$H_\nu^{(2)}(k\eta) = \sqrt{\frac{2}{\pi k\eta}}\,{\rm e}^{-ik\eta - i\epsilon_\nu}, \qquad H_\nu^{(1)}(k\eta) = \sqrt{\frac{2}{\pi k\eta}}\,{\rm e}^{ik\eta + i\epsilon_\nu} \tag{81}$$

($\epsilon_\nu = -\pi/4 - \nu\pi/2$), we obtain $A_+ = \sqrt{\pi/4}$ and $A_- = 0$. The normalized exact solution for the the axion fluctuations $\delta\sigma_k$ can be finally written as

$$\delta\sigma_k = \frac{v_k}{z_\sigma} = \frac{{\rm e}^{i\theta_k}}{M_{\rm P}}\left(\frac{\pi\eta_1}{2}\right)^{1/2}\left(\frac{\eta}{\eta_1}\right)^{\nu_\sigma} H_{\nu_\sigma}^{(2)}(k\eta), \tag{82}$$

where θ_k is an arbitrary phase determined by the choice of the initial conditions.

In order to determine the spectrum of the fluctuations after their inflationary amplification, we must then consider the limit $\eta \to 0_-$, in which $|k\eta| \ll 1$ and the amplitude of the mode k is stretched "outside the horizon." We can use, to this purpose, the small argument limit of the Hankel functions [41], which reads (for $\nu \neq 0$),

$$H_\nu^{(2)}(k\eta) = p_\nu^*(k\eta)^\nu - iq_\nu(k\eta)^{-\nu} + \ldots \tag{83}$$

where q_ν and p_ν are complex (ν-dependent) coefficients (for $\nu = 0$ there are additional logarithmic corrections). We obtain, in this limit,

$$\delta\sigma_k = \frac{v_k}{z_\sigma} \to \frac{e^{i\theta_k}}{M_P}\left(\frac{\pi\eta_1}{2}\right)^{1/2}\left[-iq_{\nu_\sigma}(k\eta_1)^{-\nu_\sigma} + p_{\nu_\sigma}^*(k\eta_1)^{\nu_\sigma}\left(\frac{\eta}{\eta_1}\right)^{2\nu_\sigma}\right]. \tag{84}$$

The cases we are interested here are limited to "conventional" inflationary backgrounds with $\alpha_\sigma \leq 1/2$, i.e., $\nu_\sigma \geq 0$ (see [32] for a detailed discussion of all possibilities). For such backgrounds the time dependence of $\delta\sigma_k$ tends to disappear as $\eta \to 0_-$, the fluctuations become frozen, asymptotically, and their (dimensionless) spectral amplitude $k^3|\delta\sigma_k|^2$, controlling the typical amplitude of the perturbations on a comoving length scale $r = k^{-1}$ [40], has the following k-dependence:

$$k^3 \left|\delta\sigma_k\right|^2 \sim k^{3-2\nu_\sigma} = k^{2+2\alpha_\sigma}. \tag{85}$$

This result also holds in the limiting case $\alpha_\sigma = 1/2$ with the only addition of a mild logarithmic correction [26, 27], i.e., $k^3 \left|\delta\sigma_k\right|^2 \sim k^3 \ln^2(k\eta_1)$.

The above calculations can be exactly repeated, in the same form, for the tensor perturbation variable, starting from (74): the resulting spectrum is formally the same,

$$k^3 \left|h_k\right|^2 \sim k^{3-2\nu_h} = k^{2+2\alpha_h}, \tag{86}$$

with the difference that the spectral slope is now determined by the power α_h, controlling the evolution of the tensor pump field z_h through an equation analogous to (76).

We are now in the position of discussing the possible pre-big bang production of a flat spectrum of axion fluctuations, even if the associated metric fluctuations are amplified (in the same background) with a growing spectrum. Let us consider, to this purpose, an exact anisotropic solution of the string cosmology equations (9)–(11), in vacuum, and without dilaton potential. The solution describes a phase of pre-big bang inflation characterized by the accelerated (isotropic) expansion of three spatial dimensions, with scale factor $a(\eta)$, and by the accelerated contraction of n "internal" spatial dimensions, with scale factors $b_i(\eta)$, $i = 1, \ldots, n$. In conformal time, such a solution can be parametrized for $\eta \to 0_-$ as [13, 22]

$$a = \left(-\frac{\eta}{\eta_1}\right)^{\beta_0/(1-\beta_0)}, \qquad\qquad b_i = \left(-\frac{\eta}{\eta_1}\right)^{\beta_i/(1-\beta_0)},$$

$$\phi_{4+n} = \frac{\sum_i \beta_i + 3\beta_0 - 1}{1-\beta_0} \ln\left(-\frac{\eta}{\eta_1}\right), \tag{87}$$

where the constant coefficients β_0, β_i satisfy the Kasner-like condition

$$\sum_i \beta_i^2 + 3\beta_0^2 = 1, \tag{88}$$

and ϕ_{4+n} is the higher-dimensional dilaton appearing in the full $(4+n)$-dimensional effective action. The four-dimensional dilaton ϕ is related to ϕ_{4+n} by

$$e^{-\phi} = V_n \, e^{-\phi_{4+n}} \equiv e^{-\phi_{4+n}} \prod_i b_i, \tag{89}$$

namely by

$$\phi = \phi_{4+n} - \sum_i \ln b_i = \frac{3\beta_0 - 1}{1-\beta_0} \ln\left(-\frac{\eta}{\eta_1}\right). \tag{90}$$

Let us compute, for this background, the kinematic powers α_h and α_σ controlling the evolution of the pump fields (72) and (73):

$$z_h \sim a e^{-\phi/2} \sim (-\eta)^{\alpha_h}, \qquad \alpha_h = \frac{1}{2}, \tag{91}$$

$$z_\sigma \sim a e^{\phi/2} \sim (-\eta)^{\alpha_\sigma}, \qquad \alpha_\sigma = \frac{5\beta_0 - 1}{2(1-\beta_0)}. \tag{92}$$

It follows, according to (86), that the spectrum of tensor (as well as of scalar) metric perturbations is always characterized by a slope which is cubic (modulo log corrections) [13, 26, 27], and which is also "universal," in the sense that it is insensitive to the background parameters (n, β_0, β_i). For the axion fluctuations, on the contrary, we find from (85) that the spectral slope is strongly dependent on such parameters, and that a scale-invariant spectrum with $2 + 2\alpha_\sigma = 0$ is allowed, in particular, provided $\beta_0 = -1/3$.

We may note, in the special case in which the background is fully isotropic and expanding (i.e., $\beta_0 = \beta_i < 0$), that the Kasner condition (88) implies $\beta_0 = -1/\sqrt{d}$, so that a scale-invariant spectrum corresponds to $d = 9$, i.e., just to the number of spatial dimensions determined by critical superstring theory [3, 4].

In the less special case in which the spatial geometry can be factorized as the product of a three-dimensional and a n-dimensional isotropic subspaces we have, instead, $\beta_i = \beta \neq \beta_0$, and $3\beta_0^2 + n\beta^2 = 1$. The spectral slope, in this case, can be expressed in terms of the parameter

$$r = \frac{1}{2}\left(\frac{\dot{V}_n}{V_n}\right)\left(\frac{\dot{V}_3}{V_3}\right)^{-1} = \frac{n\beta}{6\beta_0}, \tag{93}$$

controlling the relative time evolution of the proper volumes of the internal and external spaces. Eliminating β in terms of β_0 through the Kasner condition, and replacing β_0 with r in (92), one can then parametrize the deviations from a flat axion spectrum as the relative shrinking or expansion of the two subspaces [42].

Given a sufficiently flat spectrum of axion fluctuations, amplified by the phase of pre-big bang inflation, we are then lead to a post-big bang configuration which is initially characterized (at some given time scale η_i) by a primordial sea of "isocurvature" scalar perturbations, dominated on super-horizon scales by the axion fluctuations $\delta\sigma$ (the metric fluctuations are subdominant on such large scales, being strongly suppressed by the steep slope of their spectrum). The axion can play the role of the curvaton provided that the initial configuration, besides containing the initial fluctuations $\delta\sigma_i$, also contains a non-vanishing axion background, $\sigma_i \neq 0$, whose energy density ρ_σ—even if subdominant—is initially determined by an appropriate potential (possibly approximated by $V_\sigma \sim m^2\sigma^2$). In that case the background evolution, after an initial slow-roll regime, leads to a phase where the axion background starts oscillating with proper frequency m, at a curvature scale $H \sim m$, simulating a dust fluid ($\rho_\sigma \sim a^{-3}$) which may become dominant with respect to the radiation fluid, and eventually decay at the typical scale $H \sim \lambda_{\mathrm{P}}^2 m^3$.

In such a type of background the axions fluctuations $\delta\sigma$ become linearly coupled to scalar metric perturbations, and may act as sources for the so-called Bardeen potential Ψ. New metric perturbations can then be generated, starting from $\Psi(\eta_i) = 0$, with the same spectral slope as the axion one, and with a spectral amplitude not smaller, in general, than the axion amplitude. Referring to the literature for a detailed computation [37, 38], we shall recall here that the final spectrum (after the axion decay) of the super-horizon Bardeen potential is related to the initial axion perturbations by

$$
\begin{aligned}
|\Psi_k| &= \lambda_{\mathrm{P}} f(\sigma_i) \, |\delta\sigma_k(\eta_i)| \, , \\
f(\sigma_i) &= c_1 \frac{\Omega_\sigma}{\lambda_{\mathrm{P}}\sigma_i} + c_2 + c_3 \, \lambda_{\mathrm{P}}\sigma_i
\end{aligned}
\tag{94}
$$

(the λ_{P} factors are due to the canonical normalization of the axion field and of its fluctuations). Here σ_i is the initial amplitude of the axion background, $\Omega_\sigma \sim 1$ is the axion fraction of critical density at the axion decay epoch, and c_1, c_2, c_3 are dimensionless numbers of order one (Ω_σ cannot be much smaller than one, to avoid a too strong "non-Gaussianity" of the spectrum" [43]). Thanks to its structure, the "form factor" $f(\sigma_i)$ has a minimum of order one around $\lambda_{\mathrm{P}}\sigma_i \sim 1$. A (nearly) scale-invariant axion spectrum thus reproduces a (nearly) scale-invariant spectrum of scalar metric perturbations.

As discussed in the literature, a curvaton-induced spectrum of scalar metric perturbations provides the right "adiabatic" initial conditions for reproducing the observed temperature anisotropies of the CMB radiation, exactly as in the case of the slow-roll scenario. The only difference is the "indirect" (i.e., post-inflationary) production of the scalar spectrum, triggered by the presence of a non-vanishing axion background. It must be stressed, however, that the direct connection (94) with the axion spectrum of primordial origin gives us the possibility of extracting, from present CMB observations, important constraints on the parameters of pre-big bang models of inflation [38].

In particular, using the experimental normalization of the anisotropy spectrum, and the direct relation between the pre-big bang inflation scale H_1 and the string scale M_s, one can speculate about the possibility of "weighing the string mass with the CMB data" [44]. Another application concerns the slope of the scalar perturbation spectrum which, according to most recent WMAP results [45], is given by

$$n_s \equiv 3 + 2\alpha = 0.951^{+0.015}_{-0.019}. \tag{95}$$

Using (92), and the Kasner condition (88), one obtains

$$\beta_0 \simeq -0.355, \qquad \sum_i \beta_i^2 \simeq 0.62. \tag{96}$$

With $d = 9$ dynamical dimensions this result seems to point out the existence of a small anisotropy between the kinematics of the external and internal spaces during pre-big bang inflation (a fully isotropic expansion would correspond, in fact, to $\beta_0 = -1/\sqrt{9} \simeq -0.33$ and $\sum_i \beta_i^2 = 6/9 \simeq 0.66$). It should be noted, however, that other interpretations of the data are also possible. For instance, the result (95) is also compatible with $\beta_0 = -1/\sqrt{8} \simeq -0.3535$, describing the isotropic expansion of $d = 8$ spatial dimensions! Incidentally, the number (and the kinematics) of the extra spatial dimensions play a crucial role also in the possible production of primordial "seeds" for the large-scale magnetic fields [46].

It should be mentioned, finally, a possible non-Gaussian "contamination" of the statistical properties of the anisotropy spectrum, possibly present in curvaton models with $\Omega_\sigma \ll 1$ [43] (see (94)). A possible detection of non-Gaussianity, in future CMB measurements, could provide support to the curvaton mechanism, and could be used for a direct discrimination between this scenario and other, more standard scenarios based on slow-roll inflation.

2 The Relic Dilaton Background

The accelerated evolution of the Universe, during the phase of pre-big bang inflation, amplifies the quantum fluctuations of all fields present in the string effective action: thus, in particular, it amplifies the dilaton fluctuations, $\delta\phi \equiv \chi$. The formation of a stochastic background of relic gravitational waves, associated to the amplification of the tensor part of metric fluctuations, is thus accompanied by the simultaneous formation of a comic background of relic dilatons [47], whose primordial (high-energy) spectral distribution tends to follow that of tensor metric perturbations [13].

There is, however, a possible important difference in the present intensity of the two cosmic backgrounds, due to the fact that dilatons—unlike gravitons—could become massive in the course of the standard (post-inflationary) evolution. Actually, dilatons *must* become massive if they are non-universally coupled to ordinary matter with gravitational strength (or higher) [48, 49], to avoid the presence of long-range scalar forces which are excluded by the standard gravitational phenomenology (in particular, by the high-precision tests of the equivalence principle). The induced mass may drastically modify the amplitude and the slope of the dilaton spectrum, in the frequency band associated to its non-relativistic sector.

For a simple illustration of the effects of the mass on the spectrum we will consider here the model of vacuum, dilaton-dominated pre-big bang background described by (57), smoothly joined at $\eta = -\eta_1 < 0$ to the standard radiation-dominated background with frozen dilaton, described by (52) (we shall work in $d = 3$ spatial dimensions). Perturbing the background equations [13] one finds, in this case, that the dilaton pump field is the same field $z_h \sim a \exp(-\phi/2)$ governing the amplification of metric fluctuations. Taking into account a possible mass contribution, $m^2 = \partial^2 V/\partial\phi^2$, one then obtains for the Fourier modes χ_k the canonical equation:

$$(\chi z_h)''_k + \left(k^2 + m^2 a^2 - \frac{z''_h}{z_h}\right)(\chi z_h)_k = 0. \tag{97}$$

During the initial pre-big bang regime the potential is negligible ($m^2 = 0$), and the canonically normalized solution for χ_k is that of (82) (with ν_σ replaced by ν_h). In the subsequent radiation-dominated era, ϕ stabilizes to a constant, so that $z_h \sim a \sim \eta$ and the effective potential z''_h/z_h is vanishing. Assuming that the dilaton mass is small enough in string units, and considering the high-frequency sector of the spectrum, associated to the relativistic modes of proper momentum $p = (k/a) \gg m$, we can neglect also the mass term of (97), to obtain the general solution

$$\chi_k = \frac{1}{a\sqrt{2k}}\left[c_+(k)\mathrm{e}^{-ik\eta} + c_-(k)\mathrm{e}^{ik\eta}\right], \qquad \eta \geq -\eta_1. \tag{98}$$

Matching χ and χ' with the pre-big bang solution (82) at η_1, for super-horizon modes with $(k\eta_1) \ll 1$, we are lead to

$$c_\pm(k) = \pm c(k)\mathrm{e}^{\mp ik\eta_1}, \qquad |c(k)| \sim (k\eta_1)^{-\nu_h - 1/2} \tag{99}$$

(modulo numerical factors with modulus of order 1). Thus, at large times $\eta \gg \eta_1$,

$$\chi_k \sim \frac{c(k)}{a\sqrt{k}} \sin k\eta. \tag{100}$$

The spectral energy density for the relativistic sector of the dilaton background, in the radiation era, in then determined by

$$k\frac{\mathrm{d}\rho}{\mathrm{d}k} = \frac{k^3}{2a^2}\left(|X_k'|^2 + k^2|X_k|^2\right)$$
$$\sim \left(\frac{k}{a}\right)^4 |c(k)|^2 \sim \left(\frac{k}{a}\right)^4 \left(\frac{k}{k_1}\right)^{-2\nu_h-1} = p^4\left(\frac{p}{p_1}\right)^{-2\nu_h-1}, \quad (101)$$

where $k_1 \sim \eta^{-1}$ is the high-frequency cutoff scale. In units of critical energy density, $\rho_c = 3M_\mathrm{P}^2 H^2$,

$$\Omega_\chi(p,t) = \frac{p}{\rho_c}\frac{\mathrm{d}\rho_\chi}{\mathrm{d}p} \sim \left(\frac{H_1}{M_\mathrm{P}}\right)^2 \left(\frac{H_1}{H}\right)^2 \left(\frac{a_1}{a}\right)^4 \left(\frac{p}{p_1}\right)^\delta, \qquad m < p < p_1, \quad (102)$$

where we have defined the (model-dependent) slope parameter $\delta = 3-2\nu_h > 0$, and we have introduced the (time-dependent) proper momentum associated to the cutoff scale, $p_1 = k_1/a = H_1 a_1/a$, determined by the background curvature scale H_1 at the end of inflation. In general, $(H_1/H)^2(a_1/a)^4 \equiv \rho_r(t)/\rho_c(t) \equiv \Omega_r(t)$, and we may thus conclude that the relativistic sector of the dilaton spectrum, in the radiation era, is exactly the same as the spectrum of tensor metric perturbations (see (67)), in the same model of background.

However, even if the mass is small, and initially negligible, the proper momentum $p = k/a(t)$ is continuously red shifted with respect to m during the subsequent cosmological evolution, so that all modes tend to become non-relativistic, $p < m$. For non-relativistic modes the solution (98) is no longer valid, and the correct spectrum must refer to the exact solutions of (97) with $m \neq 0$. In the radiation era such a solution can be given in terms of the Weber cylinder functions [50], and one finds that the non-relativistic sector of the spectrum splits into two branches, with different slopes: a first branch of modes becoming non-relativistic at a timescale t_{nr} when they are *already inside* the horizon, with proper momentum p such that $p(t_{nr}) \sim m \gg H(t_{nr})$; and a second branch of modes becoming non-relativistic when they are *still outside* the horizon, with $p(t_{nr}) \sim m \ll H(t_{nr})$. The two branches are separated by the momentum scale p_m of the mode becoming non-relativistic just at the time of horizon crossing, i.e., $p(t_{nr}) = m = H(t_{nr})$, and thus related to the cutoff scale p_1 by

$$\frac{p_m}{p_1} = \frac{m\, a_{nr}}{H_1 a_1} = \frac{m}{H_1}\left(\frac{H_1}{H_{nr}}\right)^{1/2} = \left(\frac{m}{H_1}\right)^{1/2}. \quad (103)$$

Without applying to the explicit form of the massive solutions of (97), a quick estimate of the non-relativistic spectrum can be obtained [51] by noting that, if $p_{nr} > H(t_{nr})$, the number of produced dilatons is the same as in the relativistic case, and the only effect of the non-relativistic transition is a rescaling of the energy density, i.e.,

$$\Omega_\chi^{rel} \to \Omega_\chi^{nr} = \left(\frac{m}{p}\right)\Omega_\chi^{rel}. \quad (104)$$

For this branch of the spectrum we then obtain, from (102),

$$\Omega_\chi(p,t) \sim \left(\frac{m}{H_1}\right)\left(\frac{H_1}{M_{\rm P}}\right)^2\left(\frac{H_1}{H}\right)^2\left(\frac{a_1}{a}\right)^3\left(\frac{p}{p_1}\right)^{\delta-1}, \quad p_m < p < m. \tag{105}$$

In the case $p_{nr} < H_{nr}$, on the contrary, the slope of the spectrum—determined by the background kinematics at the time of horizon exit—has to be the same as that of the relativistic sector, while the time dependence has to be the non-relativistic one ($\rho_\chi \sim a^{-3}$) of (105). Continuity with the branch (105) at $p = p_m$ then gives

$$\Omega_\chi(p,t) \sim \left(\frac{m}{H_1}\right)^{1/2}\left(\frac{H_1}{M_{\rm P}}\right)^2\left(\frac{H_1}{H}\right)^2\left(\frac{a_1}{a}\right)^3\left(\frac{p}{p_1}\right)^{\delta}, \quad p_{\rm eq} < p < p_m. \tag{106}$$

The lower limit $p_{\rm eq} < p$ has been inserted here to recall that we are neglecting the effects of the transition to the matter-dominated phase, i.e., we are considering modes re-entering the horizon during the radiation era, with $p > p_{\rm eq} = H_{\rm eq} \sim 10^{-27}$ eV. We should recall, also, that the spectrum has been computed in a radiation-dominated background, and thus is valid, strictly speaking, only for $t > t_{\rm eq}$.

The three branches (102), (105), and (106) describe the spectrum (between $p_{\rm eq}$ and p_1) of primordial dilatons produced in the simple example of "minimal" pre-big bang model that we have considered. We refer to the literature for a more detailed computation, for a discussion of its transmission to the present epoch t_0, and for the possible modifications induced by generalized background evolutions (see, e.g., [32]). For the pedagogical purpose of this paper, this example provides a sufficiently clear illustration of the effects of the mass on the spectrum: in particular, it clearly illustrates the enhancement produced at lower frequencies because of the reduced spectral slope of the branch (105), which may become even decreasing if $\delta < 1$ (see Fig. 5).

In such a context one is naturally lead to investigate whether this enhanced intensity might favor the detection of a non-relativistic dilaton background, with respect to other, relativistic types of cosmic radiation (such as the relic graviton background).

2.1 Light but Non-relativistic Dilatons

For a phenomenological discussion of this possibility we must start with two important assumptions. The first is that the produced dilaton are light enough to have survived until the present epoch. Supposing that massive dilatons have dominant decay mode into radiation (e.g., two photons), with gravitational coupling strength, i.e., with a decay rate $\Gamma \sim \lambda_{\rm P}^2 m^3$, it follows that the primordial graviton background is still "alive" in the present Universe (characterized by the timescale H_0^{-1}) provided $H_0^{-1} < \Gamma^{-1}$, i.e.,

$$m \lesssim 10^2\,\text{MeV}. \tag{107}$$

The second assumption we need is that the total energy density of the dilaton background, integrated over all modes, turns out to be dominated by its non-relativistic sector. Only in this case we can evade the stringent bound imposed by the nucleosynthesis, which applies to the relativistic part of any cosmic background of primordial origin.

The energy density of a relativistic background, in fact, evolves in time-like the radiation energy density, $\rho^{rel}/\rho_{rad} = \Omega^{rel}/\Omega_{rad} = \text{const}$: the present value of their ratio is thus the same as the value of the ratio at the nucleosynthesis epoch. To avoid disturbing the nuclear processes occurring at that epoch, on the other hand, one must require that $\Omega^{rel}/\Omega_{rad} \lesssim 0.1$ [52]. Using the present value of Ω_{rad}, one is then led to the constraint $\Omega^{rel}(t_0) \lesssim 5 \times 10^{-6}$, which imposes a severe constraint on all relativistic primordial backgrounds. In particular, it imposes an upper limit on the peak value of the graviton background produced in models of pre-big bang inflation, thus determining the minimal level of sensitivity required for its detection [22].

The energy density of a non-relativistic background, on the contrary, evolves like the dark matter density, and grows in time with respect to the radiation background: $\Omega^{nr}/\Omega_{rad} \sim a$. As a consequence, the value of Ω^{nr} can be very large today, even if negligible at the nucleosynthesis epoch. The only constraint we must apply, in this case, is the critical density bound,

$$h^2 \Omega_\chi(t) = h^2 \int^{p_1} \mathrm{d}(\ln p)\, \Omega_\chi(p,t) < 1, \tag{108}$$

to be imposed at any time t, to avoid a Universe overdominated by such a cosmic background of dust matter. Here $h \simeq 0.73$ is the present value of the Hubble parameter H_0 in units of 100 km s^{-1} Mpc.

For the dilaton spectrum of (102)–(106) there are, in particular, two different cases in which the total energy density is dominated by the non-relativistic modes. A first (obvious) possibility is the case in which all modes of the spectrum are presently non-relativistic, namely $p_1(t_0) < m$ (in this case the branch (105) extends from p_m to p_1). This implies, however, that

$$\begin{aligned} m > \frac{H_1 a_1}{a_0} &= H_1 \frac{a_1}{a_{\rm eq}} \frac{a_{\rm eq}}{a_0} = H_1 \left(\frac{H_{\rm eq}}{H_1}\right)^{1/2} \left(\frac{H_0}{H_{\rm eq}}\right)^{2/3} \\ &\simeq \left(\frac{H_1}{M_{\rm P}}\right)^{1/2} 10^{-4}\text{eV}. \end{aligned} \tag{109}$$

For a typical string-inflation scale, $H_1 \sim M_{\rm s}$, we obtain a lower limit on m which is well compatible with the upper limit (107), but which requires mass values too high to be compatible with the sensitivity band of present gravitational antennas (see Sect. 2.2).

The second (more interesting) possibility is the case in which $m < p_1(t_0)$, but the parameter δ is smaller than one, and the slope is flat enough, so that the spectrum is peaked not at p_1 but at $p_m = p_1(m/H_1)^{1/2}$ (see Fig. 5). In that case, the momentum integral (108) is dominated by the peak value $\Omega_\chi(p_m)$, and the critical density bound can be approximated by the condition

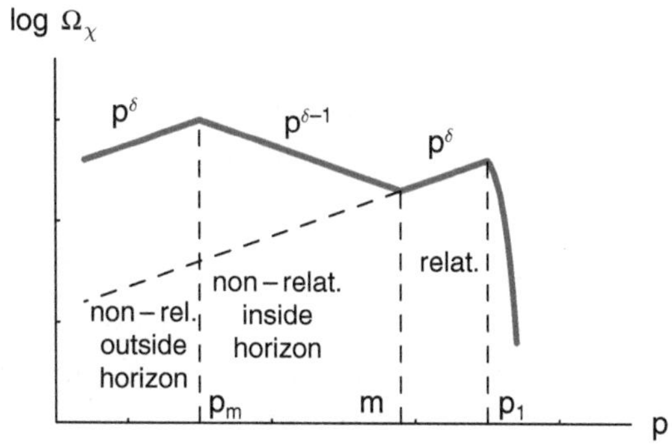

Fig. 5. Example of dilaton spectrum dominated by the non-relativistic sector. The spectrum is peaked at $p = p_m$, and the slope parameter satisfies the condition $\delta < 1$

$\Omega_\chi(p_m, t_0) \lesssim 1$. Using (105), and noting that in the matter-dominated era ($t > t_{\rm eq}$) the value of the non-relativistic spectrum keeps frozen at the equality value $\Omega_\chi(t_{\rm eq})$, we are led to the condition $\Omega_\chi(t_{\rm eq}, p_m) \lesssim 1$, which implies

$$m \lesssim \left(H_{\rm eq} M_{\rm P}^4 H_1^{\delta-4}\right)^{1/(\delta+1)} . \tag{110}$$

For $H_1 \sim M_{\rm s}$, and $\delta \to 0$, this bound can be saturated by masses as small as

$$m \sim H_{\rm eq} \left(\frac{M_{\rm P}}{M_{\rm s}}\right)^4 \sim 10^{-23}\,{\rm eV}. \tag{111}$$

It is quite possible, therefore, to have a dilaton mass small enough to fall within the sensitivity range of present gravitational detectors, even if the energy density of the dilaton background is dominated by non-relativistic modes (thus evading the relativistic upper bound $\Omega^{rel} \lesssim 10^{-6}$), and even if the background intensity is large enough to saturate the critical density bound, $\Omega_\chi \sim 1$.

So small mass values, however, are necessarily associated with long-range dilaton forces: in particular, if the mass satisfies the condition $m < p_1(t_0) \sim (M_{\rm s}/M_{\rm P})^{1/2} 10^{-4}$ eV (as in the example illustrated in Fig. 5), the corresponding force has a range exceeding the centimeter. This might imply macroscopic violations of the equivalence principle (due to the non-universality of the dilaton coupling [48]), and macroscopic deviations from the standard Newtonian form of the low-energy gravitational interactions (which seem to be excluded, however, by present experimental results [53, 54]).

We should recall, in fact, that in the presence of long-range dilaton fields the motion of a macroscopic test body with non-zero dilaton charge is no longer described by a geodesics. There are forces on the test body due to the gradients of the dilaton field, according to the generalized conservation equation

$$\nabla_\nu T_\mu{}^\nu = \frac{\sigma}{2} \nabla_\mu \phi, \tag{112}$$

following from the application of the contracted Bianchi identity to the gravi-dilaton equations (3) and (101). The integration of this conservation equation over a (space-like) $t=$ const hypersurface then gives, in the point particle (or monopole) approximation, the non-geodesic equation of motion [55]

$$\frac{\mathrm{d}u^\mu}{\mathrm{d}\tau}+\Gamma_{\alpha\beta}{}^{\mu}u^\alpha u^\beta=q\nabla^\mu\phi, \tag{113}$$

where q is a dimensionless ratio representing the relative intensity of scalar to tensor forces (i.e., the effective dilaton charge per unit of gravitational mass of the test body).

For the fundamental components of macroscopic matter, such as quark and lepton fields, the value of q (or of the charge density σ) is to be determined from an effective action which includes all relevant dilaton loop corrections [48, 13], and which is of the form

$$S=\frac{1}{2\lambda_\mathrm{s}^2}\int \mathrm{d}^4x\sqrt{-g}\Big[-Z_R(\phi)R-Z_\phi(\phi)(\nabla\phi)^2-V(\phi) \\ +Z_k^i(\phi)(\nabla\psi_i)^2-M_i^2Z_m^i(\phi)\psi_i^2\Big]. \tag{114}$$

Here we have used, for simplicity, a scalar model of matter fields ψ_i, and we have called Z the dilaton "form factors" arising from the loop corrections. The effective dilaton charge, therefore, turns out to be frame-dependent (the charge q appearing in (113), for instance, is referred to the S-frame action and to the S-frame equations (112)). The reason of such a frame dependence is that, in a generic frame, the metric and the dilaton fields are non-trivially mixed through the Z_R and Z_ϕ coupling functions, so that the associated dilaton charge actually controls the matter coupling *not to the pure scalar part*, but *to a mixture of scalar and tensor part* of the gravi-dilaton field.

A frame-independent and unambiguous definition of the dilaton coupling strengths can be given, however, in the canonically rescaled Einstein frame (E-frame), where the full kinetic part of the action (114) (including the matter and gravi-dilaton sector) is diagonalized in terms of the canonically normalized fields $\widehat{g}_{\mu\nu}$, $\widehat{\phi}$ and $\widehat{\psi}_i$ [13]. Assuming that the dilaton is stabilized by its potential, and expanding the Lagrangian term describing the interaction between $\widehat{\phi}$ and $\widehat{\psi}_i$ around the value ϕ_0 which extremizes the potential, we can define, in this rescaled frame, the effective masses $\widehat{m}_i$ and charges $\widehat{q}_i$ for the canonical fields $\widehat{\psi}_i$. In the weak coupling limit in which $Z_R\simeq Z_\phi\simeq\exp(-\phi)$ one then finds, in particular, that the canonical dilaton charge $\widehat{q}_i$ deviates from the standard "gravitational charge" by the dimensionless factor [13]

$$\overline{q}_i\equiv\frac{\widehat{q}_i}{\sqrt{4\pi G}\,\widehat{m}_i}\simeq 1+\left[\frac{\partial}{\partial\phi}\ln\left(\frac{Z_m^i}{Z_k^i}\right)\right]_{\phi=\phi_0}. \tag{115}$$

For a pure Brans–Dicke model of scalar–tensor gravity one has, for instance, $\overline{q}_i = 1$ (because there is no dilaton coupling to the matter fields in the Jordan frame, where $\partial Z^i/\partial\phi = 0$). For a string model, on the contrary, the coupling parameters $\overline{q}_i$ deviate from 1 and are non-universal, in general, since the loop form factors Z^i tend to be different for different fields ψ_i. In particular, in the conventional scenario which assumes that the loop corrections determining the coupling are the same determining also the effective mass of the given particle, one obtains large dilaton charges ($\overline{q}_i \sim 50$) for the confinement-generated components of the hadronic masses [48, 49], and smaller charges ($\overline{q}_i \sim 1$) for the leptonic components. In that case, the total dilaton charge of a macroscopic body tends to be large (in gravitational units) and composition-dependent [55], so that a large dilaton mass ($m \gtrsim 10^{-4}$ eV) is required to avoid conflicting with known gravitational phenomenology.

This conclusion can be avoided if the loop corrections combine to produce a cancellation, in such a way that the value of the coupling parameters $\overline{q}_i$ turns out to be highly suppressed with respect to the natural value of order one (a scenario of this type has been proposed, for instance, in [56]). In that case $\overline{q}_i \ll 1$, and light dilaton masses (as required, for instance, for a resonant interaction with gravitational antennas) may be allowed, without clashing with experimental observations.

In the rest of this section we will focus our attention on this possibility, considering the response of the gravitational detectors to a cosmic background of massive, non-relativistic dilatons, assuming that the background energy density corresponds to large fraction of critical density, and that the dilatons are arbitrarily light and very weakly coupled to ordinary matter.

2.2 Dilaton Signals in Gravitational Antennas

The operation mechanism of all gravitational antennnas is based on the so-called equation of "geodesic deviation" (see, e.g., [57]), which governs the response of the detector to the incident radiation. Such an equation is obtained by computing the relative acceleration between the world lines of two nearby test particles, separated by the infinitesimal space-like vector η^μ, and evolving geodesically in the given gravitational background. The interaction with a dilaton background can be easily included, in this context, by replacing the geodesic paths of the test particles with the world lines described by (113): one is lead, in this way, to a generalized equation of deviation [55],

$$\frac{D^2\eta^\mu}{D\tau^2} + R_{\nu\alpha\beta}{}^\mu \eta^\nu u^\alpha u^\beta = q\,\eta^\nu \nabla_\nu \nabla^\mu \phi, \tag{116}$$

which is at the ground of the response of a detector to a background of gravi-dilaton radiation (the symbol D denotes covariant differentiation along a curve parametrized by the affine time-like variable τ).

This equation implies that a gravitational detector can interact with the scalar radiation in two ways: either

(i) *directly*, through the *non-geodesic* coupling of its scalar charge to the second derivatives of the scalar background [55, 58]; or
(ii) *indirectly*, through the *geodesic* coupling of its gravitational charge to the *scalar part* of the metric fluctuations induced by the dilaton, and contained inside the Riemann tensor [59].

For a precise discussion of the response of the detector we need to compute the "physical strain" $h(t)$ induced by the scalar radiation, which is expressed in terms of the so-called antenna pattern functions $F(\theta, \phi)$, describing the detector sensitivity along the different angular directions. To this purpose, we shall rewrite (116) in the approximation of small displacements ξ^μ around the unperturbed path of the text bodies, by setting $\eta^\mu = L^\mu + \xi^\mu(\tau)$, with $L^\mu =$ const. We then obtain, in the non-relativistic limit,

$$\ddot{\xi}^i = -L^k M_k{}^i, \tag{117}$$

where

$$M_k{}^i = R_{k00}{}^i + q\partial_k\partial^i\phi \tag{118}$$

is the total (scalar–tensor) stress tensor describing the "tidal" forces due to the incident radiation. For the pedagogical purpose of this paper we shall assume that the tensor (i.e., gravity wave) part of the radiation is absent, and that the scalar radiation can be simply described as a linear fluctuation of the Minkowski metric background $\eta_{\mu\nu}$ and of a constant dilaton background ϕ_0: thus, in the longitudinal gauge,

$$\begin{aligned} ds^2 &= (\eta_{\mu\nu} + \delta g_{\mu\nu})\, dx^\mu dx^\nu = (1+2\psi)dt^2 - (1-2\varphi)\delta_{ij}dx^i dx^j, \\ \phi &= \phi_0 + \chi, \end{aligned} \tag{119}$$

so that

$$M_{ij} = \partial_i\partial_j\varphi - \delta_{ij}\ddot{\psi} - q\partial_i\partial_j\chi. \tag{120}$$

To discuss the detection of a stochastic background of massive scalar radiation, it is also convenient to expand the fluctuations in Fourier modes of proper momentum $\boldsymbol{p} = p\widehat{n}$ and frequency $\nu = E(p) = (p^2+m^2)^{1/2}$, where the unit vector $\widehat{n}$ specifies the propagation direction of the given mode on the angular two sphere Ω_2. We obtain

$$\begin{aligned} M_{ij} = \frac{1}{2}\int_{-\infty}^{\infty} \mathrm{d}p \int_{\Omega_2} \mathrm{d}^2\widehat{n}\, (2\pi E)^2 \Big[\delta_{ij}\psi(p,\widehat{n}) - n_i n_j \varphi(p,\widehat{n}) + \frac{m^2}{E^2} n_i n_j \varphi(p,\widehat{n}) \\ + q\frac{p^2}{E^2} n_i n_j \chi(p,\widehat{n}) \Big]\, \mathrm{e}^{2\pi i(p\widehat{n}\cdot\boldsymbol{x} - Et)} + \mathrm{h.c.} \end{aligned} \tag{121}$$

(note that we are using "unconventional" units in which $h = 1$, i.e., $\hbar = 1/2\pi$, for an easier comparison with the experimental variables). We will also assume that the dilaton is the only source of scalar metric perturbations, so that

$\varphi = \psi$ [40]). Introducing the transverse and longitudinal projectors of the scalar stresses, defined, respectively, by

$$T_{ij} = \delta_{ij} - n_i n_j, \qquad L_{ij} = n_i n_j, \tag{122}$$

defining $M_{ij} = -\ddot{F}_{ij}$, and projecting the stress tensor onto the detector tensor D^{ij} (specifying the geometric configuration and the orientation of the arms of the detector), we finally obtain the scalar strain as [58, 60, 61]

$$h(t) \equiv D^{ij} F_{ij} = \frac{1}{2} \int_{-\infty}^{\infty} \mathrm{d}p \int_{\Omega_2} \mathrm{d}^2 \widehat{n} \left[F^{\mathrm{geo}}(\widehat{n}) \psi(p, \widehat{n}) + F^{\mathrm{ng}}(\widehat{n}) \chi(p, \widehat{n}) \right] \mathrm{e}^{2\pi i (p \widehat{n} \cdot \boldsymbol{x} - Et)} + \mathrm{h.c.}. \tag{123}$$

Here

$$F^{\mathrm{geo}} = D^{ij} \left(T_{ij} + \frac{m^2}{E^2} L_{ij} \right), \tag{124}$$

$$F^{\mathrm{ng}} = q \frac{p^2}{E^2} D^{ij} L_{ij}, \tag{125}$$

are the antenna pattern functions corresponding, respectively, to the geodesic (or indirect) and non-geodesic (or direct) interaction of the detector with the scalar radiation background.

It should be noted that the scalar radiation, differently from the case of the tensor component, contributes to the response of the detector also with its longitudinal polarization states. The longitudinal contribution is present also in the ultra-relativistic limit $m \to 0$, $p \to E$, thanks to the non-geodesic coupling (125). In the opposite, non-relativistic limit $p \to 0$, $E \to m$, the geodesic strain tends to become isotropic, $T_{ij} + (m/E)^2 L_{ij} \to \delta_{ij}$, while the non-geodesic one becomes sub-leading.

The results (123) is valid for any type of detector described by the response tensor D^{ij}, and is formally similar to the expression for the strain obtained in the case of tensor gravitational radiation—modulo the presence of different pattern functions, due to the different polarization properties. The scalar strain (123) can thus be processed, following the standard procedure, to correlate the outputs of two detectors and to extract the so-called signal-to-noise ratio (SNR), representing the experimentally relevant variable for the detection of a stochastic background of cosmic radiation [62].

For our scalar massive background, with spectral energy density $\Omega(p)$, we obtain [58, 60, 61], in particular,

$$SNR = \frac{3NH_0^2}{8\pi^3} \left[2T \int_0^{\infty} \frac{\mathrm{d}p}{p^3 (p^2+m^2)^{3/2}} \frac{\gamma^2(p)\, \Omega^2(p)}{P_1(\sqrt{p^2+m^2})\, P_2(\sqrt{p^2+m^2})} \right]^{1/2} \tag{126}$$

(see also [32] for a detailed computation). Here T is the total (experimental) correlation time, N an (irrelevant) normalization factor, P_1 and P_2 the noise power spectra of the two detectors, and $\gamma(p)$ the so-called overlap reduction function, which modulates the correlated signal according to the relative orientation and distance of the detectors, located at the positions $\boldsymbol{x}_1$ and $\boldsymbol{x}_2$:

$$\gamma(p) = \frac{1}{N} \int_{\Omega_2} \mathrm{d}^2\widehat{n}\; F_1(\widehat{n})\; F_2(\widehat{n})\, \mathrm{e}^{2\pi i p \widehat{n}\cdot(\boldsymbol{x}_1 - \boldsymbol{x}_2)}. \tag{127}$$

The overlap is to be calculated with the geodesic pattern function F_i^{geo} of (124) if we are considering the indirect signal due to a spectrum of scalar metric fluctuations, $\Omega_\psi(p)$; it is to be calculated with the non-geodesic pattern function F_i^{ng} of (125) if we are considering, instead, the direct signal due to a spectrum of dilaton fluctuations, $\Omega_\chi(p)$.

We are now in the position of stressing another important difference from the case of pure tensor radiation, due to the presence of the mass in the noise power spectra P_i. For a typical power spectrum, in fact, the minimum level of noise is reached around a rather narrow frequency band ν_0: outside that band the noise rapidly diverges, and the signal (126) tends to zero. As $\nu = (p^2 + m^2)^{1/2}$ we have, in principle, three possibilities.

(1) If $m \gg \nu_0$ then the noise is always outside the sensitivity band $P_i(\nu_0)$, and the signal is always negligible.
(2) If $m \ll \nu_0$ then the sensitivity band may only overlap with the relativistic sector of the spectrum, for $p \sim \nu \sim \nu_0$.
(3) If $m \sim \nu_0$, finally, the whole non-relativistic part of the spectrum $p \lesssim m$ satisfies the condition $P_i(\nu) \sim P_i(m) \sim P_i(\nu_0)$.

It is thus possible to obtain a resonant response to a massive, non-relativistic background of scalar particles, provided the mass lies in the band of maximal sensitivity of the two detectors [58, 60]. Considering the present, Earth-based gravitational antennas, operating between the hertz and the kilohertz range, it follows that the maximal sensitivity is presently in the mass range

$$10^{-15}\,\mathrm{eV} \lesssim m \lesssim 10^{-12}\,\mathrm{eV}. \tag{128}$$

Amusingly enough, it turns out that such small values are not so unrealistic if the dilaton mass is perturbatively generated by the mechanism of radiative corrections. For a scalar particle, gravitationally coupled to fermions of mass M_f with dimensionless strength q, there are, in fact, quantum loop corrections to the mass of order $qM_f(\Lambda/M_\mathrm{P})$, where Λ is the cutoff, which we shall assume typically localized at the tera electron volt scale (see, for instance, [63]). Considering the dilaton coupling to ordinary baryonic matter ($M_f \sim 1$ GeV) the induced mass is then

$$m \sim q \left(\frac{\Lambda}{1\,\mathrm{Tev}}\right) \left(\frac{M_f}{1\,\mathrm{Gev}}\right) \times 10^{-6}\,\mathrm{eV}. \tag{129}$$

Thus, a value of q smaller than (but not very far from) the present upper limits [53] (imposing $q \lesssim 10^{-4}$ in the relevant mass range (128)) is perfectly compatible with the possibility of resonant response of the present detectors.

Quite independently from the possible origin of the dilaton mass, if we assume that the mass is in the resonant range (128), and that the bounds on q are satisfied, we find that a cosmic background of non-relativistic dilatons is possibly detectable by the interferometric antennas of second generation—such as Advanced and Enhanced LIGO—provided the background energy density is sufficiently close to the saturation of the critical density bound [58, 60]. This interesting possibility can be illustrated by considering, for an approximate estimate, the simplified situation of two identical detectors with $P_1 = P_2 = P$, responding non-geodesically with maximal allowed overlap $N\gamma^{ng} \simeq q^2(4\pi/15)$ (the numerical factor is referred to the particular case of interferometric antennas). Let us suppose, also, that the SNR integral (126) is dominated by the peak value Ω_m of the non-relativistic dilaton spectrum, and that such value is reached around $p = m$ (otherwise the response is suppressed by the factor $(p/m)^4$, [60]). Equation (126) gives, in this case,

$$SNR \sim \frac{\sqrt{2T}}{10\pi^2} \frac{q^2 H_0^2 \Omega_m}{m^{5/2} P(m)}, \tag{130}$$

and the condition of detactable background ($SNR \gtrsim 1$) implies

$$m^{5/2} P(m) \lesssim q^2 h^2 \Omega_m \left(\frac{T}{4 \times 10^7 \,\mathrm{s}} \right)^{1/2} \times 10^{-33} \,\mathrm{Hz}^{3/2}. \tag{131}$$

This condition is compared in Fig. 6 with the expected spectral noise of the three LIGO generations (see, e.g., [64]), for $T = 4 \times 10^7$ s. The region of the plane $\{m, P\}$ corresponding to a detectable background is located *above* the bold noise curves (labeled by LIGO I, LIGO II, and LIGO III), and *below* the dashed lines, representing the upper limit (131) for different constant values of the parameter $q^2 h^2 \Omega_m$. This limit may be interpreted either as a constraint on the intensity Ω_m, for backgrounds geodesically coupled ($q^2 = 1$) to the detectors, or as a limit on the non-geodesic coupling strength q^2, for backgrounds of given energy density Ω_m. As shown in the picture, phenomenologically allowed backgrounds are in principle accessible to the sensitivity of next-generation interferometers (see also [32] for a more detailed discussion).

2.3 Enhanced Signals for Flat Non-relativistic Spectra

The result reported in (130) is generally valid for a growing spectrum with a steep enough slope, as typically obtained in "minimal" models of pre-big bang inflation. However, the cross-correlated signal may result strongly enhanced with respect to (130) if the dilaton spectrum is sufficiently flat, and if the considered pair of detectors satisfies the condition $\gamma(p) \to \mathrm{const} \neq 0$ for $p \to 0$.

Let us consider, in fact, the SNR integral (126), which can be written as

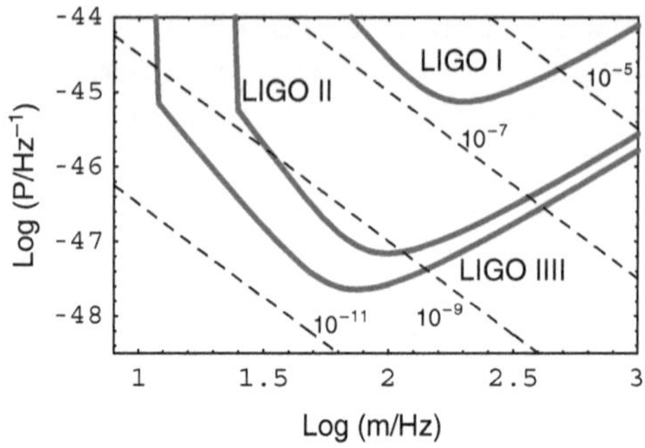

Fig. 6. The noise power spectra of the three LIGO generations (*bold curves*), and the condition of detectable dilaton background (*dashed lines*), plotted at different values of the parameter $q^2h^2\Omega_m$ (ranging from 10^{-5} to 10^{-11})

$$(SNR)^2 \sim T \int_0^{p_1} \mathrm{d}p \frac{\gamma^2(p)\Omega^2(p)}{p^3E^3P_1(E)P_2(E)}, \tag{132}$$

where $E = (p^2 + m^2)^{1/2}$, and where we can assume that $\Omega(p)$ is a power-law function of p, with an ultraviolet cutoff at $p = p_1$. For a massless spectrum ($p = E$), this integral is always convergent (for any slope), even in the infrared limit $p = E \to 0$: in fact, when $p \to 0$, the physical strains are produced outside the sensitivity band of the detectors, and the noises blow up to infinity, $P_i(E) \to P_i(0) \to \infty$. For $m \neq 0$, on the contrary, in the infrared limit $p \to 0$ the noises keep frozen at the frequency scale determined by the mass of the scalar background, $P_i(E) \to P_i(m) = \text{const}$. In this second case, the behavior of the integral dependes on $\gamma(p)$ and $\Omega(p)$.

Suppose now that $\gamma(p) \to \gamma_0 = \text{const}$ for $p \to 0$, and that $\Omega(p) \sim p^\delta$, for $p < m$. For $\delta < 1$ we find that the integral is dominated by the infrared limit, and gives

$$\begin{aligned}(SNR)^2 &\sim \frac{T\gamma_0^2}{m^3P_1(m)P_2(m)} \int_0^m \frac{\mathrm{d}p}{p^3}\Omega^2(p) \\ &= \frac{T\gamma_0^2}{m^3P_1P_2}\left[p^{2(\delta-1)}\right]_0^m .\end{aligned} \tag{133}$$

Thus, the integral is infrared divergent [65] for all spectra (even if blue, $\delta > 0$) with $\delta < 1$!

This divergence is obviously unphysical, and can be removed by noting that the observation time T is finite, and is thus associated to a minimum resolvable frequency interval $\Delta\nu = \Delta E = \Delta(p^2/2m) \gtrsim T^{-1}$, defining the minimum momentum scale

$$p_{\min} = (2m/T)^{1/2} > 0, \tag{134}$$

acting as effective infrared cutoff for the integral (133). This implies a modified dependence of SNR on the correlation time T in the case of flat enough spectra:

$$SNR \sim T^{1/2}\left[p^{\delta-1}\right]^{m}_{p_{\min}} \sim \begin{cases} T^{1/2}, & \delta > 1, \\ T^{1-\delta/2}, & \delta < 1. \end{cases} \tag{135}$$

For $\delta < 1$, in particular, there is a faster growth of SNR with T, which may produce an important enhancement of the sensitivity to a cosmic background of non-relativistic scalar particles, as discussed in [61, 65].

It is important to stress that the case $\gamma(p) \to \gamma_0 = \text{const}$ for $p \to 0$ has not been "invented" ad hoc: it can be implemented, in practice, with detectors already existing and operative (or with detectors planned to be working in the near future, like resonant spheres). A first simple example, studied in [65], refers in fact to spherical, resonant-mass detectors, whose monopole mode is characterized by the "trivial" response tensor $D^{ij} = \delta^{ij}$. In that case the geodesic pattern function (124) is isotropic,

$$F^{\text{geo}} = \frac{2p^2 + 3m^2}{p^2 + m^2}, \tag{136}$$

and the geodesic overlap function (127), for two identical spheres with spatial separation $|\boldsymbol{x}_1 - \boldsymbol{x}_2| = d$, is given by

$$\gamma(p) = \frac{2}{N}\left(\frac{2p^2 + 3m^2}{p^2 + m^2}\right)^2 \frac{\sin(2\pi pd)}{pd}. \tag{137}$$

This function clearly satisfies the requirement $\gamma(p) \to \gamma_0 = \text{const}$ for $p \to 0$.

A second example, studied in [61], refers to the so-called common mode of the interferometric antennas, characterized by the response tensor

$$D^{ij}_{+} = u^i v^j + v^i u^j, \tag{138}$$

where u^i and v^i are the unit vectors specifying the spatial orientation of the axes of the interferometer. Let us consider, for instance, a geometrical configuration where the vectors $\widehat{u}$ and $\widehat{v}$ are coaligned with the x_1 and x_2 axes of a Cartesian frame, respectively, and the direction $\widehat{n}$ ofthe incident radiation is specified (with respect to the axes x_1, x_2 and x_3) by the polar and azimuthal angles φ and θ. The computation of the geodesic pattern function (124) gives, in that case,

$$F^{\text{geo}}_{+} = 2 - \left(\frac{p}{E}\right)^2 \sin^2\theta. \tag{139}$$

The geodesic overlap function (127), for two coplanar interferometers with spatial separation $|\varDelta \boldsymbol{x}| = d$, is [61]

$$\gamma^{\text{geo}}_{+}(p) = \frac{4\pi}{N}\left[\left(4 - 4\frac{p^2}{E^2} + \frac{p^4}{E^4}\right) j_0(\alpha) + \frac{1}{\alpha}\left(4\frac{p^2}{E^2} - 2\frac{p^4}{E^4}\right) j_1(\alpha) + \frac{3}{\alpha^2}\left(\frac{p}{E}\right)^4 j_2(\alpha)\right], \tag{140}$$

where $\alpha = 2\pi pd$, and j_0, j_1, j_2 are spherical Bessel functions. Thus, also in this case, $\gamma \to 16\pi/N = \text{const}$ for $p \to 0$.

3 Late-time Cosmology: Dilaton Dark Energy

In this third lecture we will discuss the possibility that a homogeneous, large-scale dilaton field may be the source of the so-called dark energy which produces the cosmic acceleration first observed at the end of the last century [66], and confirmed by most recent supernovae data [67, 68].

Let us recall, to this purpose, that the initial phase of pre-big bang inflation is characterized by the monotonic growth of the dilaton and of the string coupling g_s (see Sect. 1.3): the subsequent epoch of standard evolution thus opens up in the strong coupling regime, and should be described by an action which includes all relevant loop corrections. Late enough, i.e., at sufficiently low-curvature scales, the higher-derivative corrections can be neglected, and the action can be written in the form of (114). In that context, the loop form factors $Z(\phi)$, and the dilaton potential $V(\phi)$, may play a crucial role in determining the late-time cosmological evolution.

There are, in principle, two possible alternative scenarios.

(i) The dilaton is stabilized by the potential at a constant value $\phi = \phi_0$ which extremizes $V(\phi)$. In this case, the loop corrections induce a constant renormalization of the effective dilaton couplings (as discussed in Sect. 2.1), and the Universe may approach a late-time configuration dominated by the dilaton potential, with $H^2 \sim V(\phi_0)$.

(ii) The dilaton fails to be trapped in a minimum of the potential, and keeps running even during the post-big bang evolution. In this case the late-time cosmological evolution is crucially dependent on the asymptotic behavior of the factors $Z(\phi)$.

These two different possibilities have different impact on the so-called coincidence problem (i.e., on the problem of explaining why the dark matter and dark energy densities are of the same order just at the present epoch), as we shall discuss in the following subsections.

3.1 Frozen Dilaton in the Moderate Coupling Regime

The first type of scenario can be easily implemented [69] using a generic non-perturbative potential which is instantonically suppressed ($V \sim \exp(-1/g_s^2)$) in the weak coupling limit $g_s^2 \to 0$, and which develops a non-trivial structure with a (semi-perturbative) minimum $g_s^2 \sim \alpha_{GUT} \sim (M_{\rm s}/M_{\rm P})^2 \sim 0.1$–$0.01$ in the regime of moderate string coupling. A typical example is the "minimal" potential given, in the E-frame, by [70]

$$\widetilde{V}(\phi) = m_V^2 \left[e^{k_1(\phi-\phi_1)} + \beta e^{-k_2(\phi-\phi_1)} \right] e^{-\epsilon \exp[-\gamma(\phi-\phi_1)]}, \qquad (141)$$

where k_1, k_2, β, ϵ, and γ are dimensionless parameters of order 1 (see Fig. 7). The presence of a local minimum at $\phi_0 \simeq \phi_1$ allows solutions with $\phi = \text{const}$ during the radiation-dominated phase, and (for appropriate values of m_V)

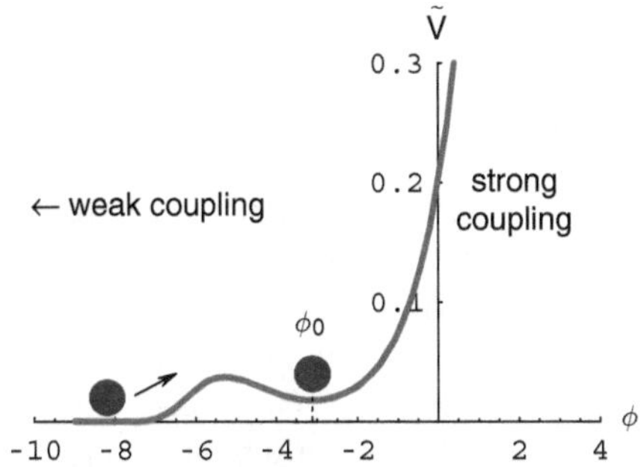

Fig. 7. Plot of the potential (141) for $k_1 = k_2 = \beta = \gamma = 1$, $\epsilon = 0.1$, $\phi_1 = -3$, $m_V = 0.1$, and a local minimum (independent of m_V) at $\phi_0 = -3.112$, corresponding to $g_s^2 = \exp(\phi_0) \simeq 0.045$

may also lead to a late phase of accelerated expansion driven by the potential energy $V(\phi_0)$, *provided* the dilaton is not permanently shifted away from the minimum ϕ_0 by the transition to the matter-dominated epoch [69].

Let us consider, in fact, the equation of motion of a homogeneous dilaton field $\phi(t)$ in the conformally rescaled E-frame (with metric $\widetilde{g}$), where the graviton kinetic energy is canonically normalized, and let us assume that the rescaled matter sources can be described as a perfect fluid of energy density $\widetilde{\rho}$, pressure $\widetilde{p}$, and dilaton charge $\widetilde{\sigma}$. Starting from an action of the type (114) we find that the generalized dilaton equation, for a cosmological background, takes the form

$$A(\phi)\left(\ddot{\phi} + 3\widetilde{H}\dot{\phi}\right) + B(\phi)\dot{\phi}^2 + \frac{\partial \widetilde{V}}{\partial \phi} + \lambda_{\rm P}^2 \left[C(\phi)\left(\widetilde{\rho} - 3\widetilde{p}\right) + \widetilde{\sigma}\right] = 0, \tag{142}$$

where A, B, and C are functions describing the rescaled (E-frame) loop corrections. For a minimally coupled field, for instance, $A = 1$, $B = C = \widetilde{\sigma} = 0$; for the dilaton, at tree level in the string coupling, $A = C = 1$, $B = 0$. In the most general case we find that a stable dilaton configuration with $\dot{\phi} = 0 = \ddot{\phi}$ is possible, in the radiation era ($\widetilde{\rho} = 3\widetilde{p}$), if the scalar charge of the fluid is negligible, $\widetilde{\sigma} = 0$, and the dilaton extremizes the E-frame potential, $\partial \widetilde{V}/\partial \phi = 0$.

When the Universe becomes matter-dominated ($\widetilde{p} = 0$), however, a new acceleration $\ddot{\phi} = -A^{-1}\lambda_{\rm P}^2 C\widetilde{\rho}$ is suddenly generated, which tends to remove the dilaton away from its equilibrium position. Such an acceleration is in competition with the restoring force $\ddot{\phi} = -A^{-1}(\partial \widetilde{V}/\partial \phi)$ (see (142)). The possibility that the dilaton may bounce back to the stable minimum $\phi = \phi_0$, driving the Universe towards a final phase of accelerated, potential-dominated expansion, thus crucially depends on the values of two parameters: the (loop-corrected) strength $\lambda_{\rm P}^2 C(\phi_0)$ of the dilaton coupling to dark matter, and the slope of the dilaton potential (141), determined by the mass scale m_V which also controls the amplitude of the minimum, $V(\phi_0) \sim m_V^2$. Such an amplitude, on the other hand, should correspond to the present Hubble scale ($V(\phi_0) \sim$

H_0^2), in a realistic model able to describe the present phase of accelerated expansion.

It can be shown, with a simple numerical analysis, that the values of the coupling strength allowed by present gravitational phenomenology are compatible with a late-time phase dominated by the potential only for a finite range of values of $V(\phi_0)$, depending on the value of the dilaton coupling at the equality epoch [69]. Using the phenomenological upper limit $|C_{\rm eq}| \simeq 0.1$ one finds that the dilaton, after a smal shift at $t = t_{\rm eq}$, bounces back to the minimum provided $10^{-7} H_{\rm eq} \lesssim m_V \lesssim H_{\rm eq}$ (which includes the realistic case $m_V \sim H_0 \sim 10^{-6} H_{\rm eq}$) [69]. Smaller values of $|C_{\rm eq}|$ correspond to a larger mass interval. We can say, therefore, that the coincidence problem (i.e., why $V(\phi_0) \sim H_0^2$), in this context remains, but is somewhat alleviated because—thanks to the dynamical correlation between the amplitude $V(\phi_0)$ and the matter-dilaton coupling—only a restricted range of values is allowed for $V(\phi_0)$.

3.2 Running Dilaton: Saturation of the Loop Corrections and Asymptotic "Freezing"

The second possibility, which will be discussed here in more detail, in the case in which the dilaton is not stopped by the structures formed by the potential around $g_s^2 = 1$, and keeps rolling towards $+\infty$ along a smoothly decreasing potential. A possible example of non-perturbative potential of this type is given, in the E-frame, by [71]

$$\widetilde{V} = c_1^4 m_V^2 \left(\frac{\mathrm{e}^{\phi}}{b_1 + c_1^2 \mathrm{e}^{\phi}} \right)^2 \left[\mathrm{e}^{-\beta_1 \exp(-\phi)} - \mathrm{e}^{-\beta_2 \exp(-\phi)} \right], \tag{143}$$

where b_1, c_1, β_1, and β_2 are dimensionless parameters, with $0 < \beta_1 < \beta_2$. This potential is instantonically suppressed in the weak coupling limit $\phi \to -\infty$, and is exponentially decaying as

$$\widetilde{V} = m_V^2 \left(\beta_2 - \beta_1\right) \mathrm{e}^{-\phi} + \mathcal{O}\left(\mathrm{e}^{-2\phi}\right) \tag{144}$$

in the limit $\phi \to +\infty$ (see Fig. 8). In this case, as we shall see, we can obtain a scenario of "coupled quintessence" [72] in which the late Universe approaches a (possibly accelerated) state dominated by a mixture of kinetic and potential energy density, and the coincidence problem may find a satisfactory solution thanks to the dilaton–dark matter interactions.

In this case, however, a realistic scenario requires some mechanism of saturation of the loop corrections, so as to keep the present effective values of gravitational and gauge couplings approximately constant and sufficiently "small," even in the large "bare coupling" limit $\phi \to +\infty$. As discussed in [73], such a saturation can be obtained thanks to the large number of fields (e.g., gauge bosons) entering the loop corrections, assuming (as in models of "induced gravity") that the loop form factors of (114) have a finite limit for $\phi \to +\infty$,

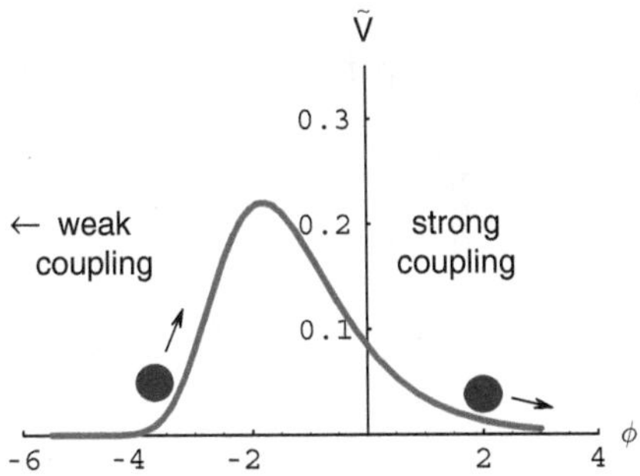

Fig. 8. Plot of the potential (143) for $b_1 = 1$, $c_1 = 10$, $\beta_1 = 0.1$, $\beta_2 = 0.2$, and $m_V = 1$. The dilaton is monotonically growing from the string perturbative vacuum along a "bell-like" non-perturbative potential

and that can be approximated by a Taylor expansion in powers of the inverse bare coupling $g_s^2 = \exp\phi$. Applying these assumptions to the gravi-dilaton form factors, to the potential, and to the dimensionless parameters $q_i(\phi)$ controlling the dilaton charge density of the various matter fields, we can set, for $\phi \to +\infty$,

$$\begin{aligned} Z_R(\phi) &= c_1^2 + b_1 \mathrm{e}^{-\phi} + \mathcal{O}(\mathrm{e}^{-2\phi}), \\ Z_\phi(\phi) &= -c_2^2 + b_2 \mathrm{e}^{-\phi} + \mathcal{O}(\mathrm{e}^{-2\phi}), \\ V(\phi) &= V_0 \mathrm{e}^{-\phi} + \mathcal{O}(\mathrm{e}^{-2\phi}), \\ q_i(\phi) &= q_{0i} + \mathcal{O}(\mathrm{e}^{-2\phi}). \end{aligned} \quad (145)$$

The dimensionless coefficients c_1^2 and c_2^2 of this expansion are typically of order $N \sim 10^{-2}$, because of their quantum-loop origin and of the large number N of gauge bosons in GUT groups like E_8. This is in agreement with the fact that c_1^2 controls (according to the action (114)) the asymptotic value of the ratio between the string and the Planck length scale, $c_1^2 = (\lambda_s/\lambda_P)^2$, which is indeed expected to be a number of the above order. The coefficients $b_1, b_2 \ldots$, on the contrary, are numbers of order 1. Note that the expansion of $V(\phi)$ agrees with the asymptotic form of the potential (144).

We should note, finally, that the asymptotic values of the dilaton charges, q_{0i}, have to be strongly suppressed for the ordinary components of matter (such as baryons) and for electromagnetic radiation: if we want a dilaton field active on a cosmological scale of distances, in fact, we need long-range interactions, and we must avoid unacceptable deviations from the standard gravitational phenomenology by suppressing the dilaton couplings, as discussed in Sect. 2.1. For the (possibly exotic) components of dark matter, however, there is no strict phenomenological bound imposing such suppression: in that case, the asymptotic charge q_0 could be non-vanishing, and of order 1, leading to interesting late-time deviations from the standard cosmological scenario.

For a simpler illustration of this possibility it is convenient to work in the diagonalized E-frame, obtained from the metric g of (114) through the rescaling

$$g_{\mu\nu} = c_1^2\, Z_R^{-1}\, \widetilde{g}_{\mu\nu}. \tag{146}$$

The action (114) becomes, in this new frame

$$S = \frac{1}{2\lambda_{\rm P}^2} \int {\rm d}^4x\sqrt{-\widetilde{g}} \left[-\widetilde{R} + \frac{1}{2}k^2(\phi)\left(\widetilde{\nabla}\phi\right)^2 - \widetilde{V}(\phi)\right] + S_m(\widetilde{g}, \phi, \text{matter}), \tag{147}$$

where

$$k^2(\phi) = 3\left(\frac{\partial \ln Z_R}{\partial \phi}\right)^2 - 2\frac{Z_\phi}{Z_R}, \qquad \widetilde{V}(\phi) = c_1^4 Z_R^{-2}\, V. \tag{148}$$

Assuming that the matter action S_m describes a perfect fluid with a dark matter component $\widetilde{\rho}_m$, a baryon component $\widetilde{\rho}_b$, and a radiation component $\widetilde{\rho}_r = 3\widetilde{p}_r$, the cosmological Einstein equations for the action (147) can then be written (omitting the tilde, and in units $2\lambda_{\rm P}^2 = 1$) as

$$\begin{aligned} 6H^2 &= \rho_r + \rho_b + \rho_m + \rho_\phi, \\ 4\dot{H} + 6H^2 &= -\frac{\rho_r}{3} - p_\phi, \end{aligned} \tag{149}$$

where

$$\rho_\phi = \frac{k^2(\phi)}{2}\dot{\phi}^2 + V, \qquad p_\phi = \frac{k^2(\phi)}{2}\dot{\phi}^2 - V. \tag{150}$$

The associated dilaton equation, assuming a negligible density of dilaton charge for baryons and radiation ($\sigma_r = 0 = \sigma_b$), can be written as [71]

$$k^2(\ddot{\phi} + 3H\dot{\phi}) + kk'\dot{\phi}^2 + V' + \frac{1}{2}\left[\psi'\left(\rho_b + \rho_m\right) + \sigma_m\right] = 0, \tag{151}$$

where we have defined $\psi = -\ln Z_R$, and the prime denotes differentiation with respect to ϕ. The combination of (149)–(151) leads, finally, to the equations of energy–momentum conservation for the various fluid components:

$$\begin{aligned} &\dot{\rho}_r + 4H\rho_r = 0, \\ &\dot{\rho}_b + 3H\rho_b - \frac{\psi'}{2}\dot{\phi}\,\rho_b = 0, \\ &\dot{\rho}_m + 3H\rho_m - \frac{\psi'}{2}\dot{\phi}\,\rho_m - \frac{\sigma_m}{2}\dot{\phi} = 0, \\ &\dot{\rho}_\phi + 3H(\rho_\phi + p_\phi) + \frac{1}{2}\dot{\phi}\left[\psi'\left(\rho_b + \rho_m\right) + \sigma_m\right] = 0 \end{aligned} \tag{152}$$

(the last equation is simply the dilaton equation (151), rewritten in fluido-dynamical form).

Let us now concentrate on the coupled dark matter/dilaton system, and note that there are two types of interactions between these two cosmic sources:

a first one, specific to the particular type of dark matter field, generated by the "intrinsic" dilaton charge σ_m; and a second one, more "universal," generated by the standard dilaton coupling to the trace of the stress tensor, and associated to the ψ' terms of the above equations. Both types of coupling are renormalized by the loop corrections, but with opposite effect according to the asymptotic limits of (145). In fact, the dilaton charge tends to grow, and to reach a constant asymptotic value as $\phi \to +\infty$. The coupling parameter ψ', on the contrary, tends to be exponentially suppressed as

$$\psi' = -(\ln Z_R)' \to \frac{b_1 e^{-\phi}}{c_1^2}, \qquad \phi \to +\infty. \tag{153}$$

As a consequence, after the transition to the matter-dominated phase, the Universe may enter two different types of dynamical regimes [71].

(1) If the dark matter charge σ_m is still negligible at the beginning of the matter-dominated phase (as well as the dilaton potential, expected to become important only near the present epoch), then the Universe enters the so-called *dragging regime*, in which ρ_m is coupled to ϕ through the ψ' terms of (153), and the evolution of the (still subdominant) dilaton kinetic energy ρ_ϕ is "dragged" by ρ_m.

The cosmic evolution, during this regime, can be analytically described (in an approximate way) by noting that the loop factor $k(\phi)$ goes to a constant at late enough timescales,

$$k(\phi) \to k_0 = \sqrt{2}\,\frac{c_2}{c_1}, \qquad \phi \to +\infty, \tag{154}$$

according to (148) and (145). Introducing the canonical variable $\widehat{\phi} = k_0 \phi$ (see the action (147)), and neglecting the subdominant contributions of ρ_r and ρ_b, we can then rewrite the coupled equations (151) and (152), for the dragging regime, as follows:

$$\ddot{\widehat{\phi}} + 3H\dot{\widehat{\phi}} + \frac{\epsilon}{2}\rho_m = 0, \tag{155}$$

$$\dot{\rho}_m + 3H\rho_m - \frac{\epsilon}{2}\rho_m \dot{\widehat{\phi}} = 0, \tag{156}$$

where $\epsilon = \psi'/k_0 \simeq e^{-\phi}/(\sqrt{2}c_1 c_2) \ll 1$ is the effective coupling parameter. Neglecting the time dependence of ϵ with respect to that of H and $\dot{\widehat{\phi}}$ (for small enough time intervals), we find that the system of equations (149) and (155), is satisfied by

$$\dot{\widehat{\phi}} \simeq -2\epsilon H. \tag{157}$$

Thus, from (156),

$$\begin{aligned} &\rho_m \sim a^{-(3+\epsilon^2)} \sim H^2 \sim \dot{\widehat{\phi}}^2 \sim \rho_\phi, \\ &a \sim t^{2/(3+\epsilon^2)}. \end{aligned} \tag{158}$$

During this phase the dark matter and the (kinetic) dilaton dark energy densities are characterized by the same time evolution, which slightly deviates from the standard behavior of a dust-dominated Universe ($\rho \sim a^{-3}$, $a \sim t^{2/3}$). The kinematics, however, remains decelerated (as $\epsilon \ll 1$).

(2) A second, possibly accelerated, *freezing regime* is eventually reached in the limit in which the dilaton potential comes into play, and the coupling induced by the intrinsic charge density σ_m becomes dominant with respect to the exponentially suppressed coupling due to ψ'.

Using again the canonical variable $\widehat{\phi}$, assuming that $\sigma_m = q(\phi)\rho_m$ (for a homogeneous fluid), and considering the asymptotic limits $q(\phi) \to q_0$, $V = V_0 \exp(-\phi)$ of (145), we can rewrite the coupled dilaton–dark matter equations (152), for the freezing regime, as follows:

$$\dot{\rho}_m + 3H\rho_m - \frac{q_0}{2k_0}\rho_m\dot{\widehat{\phi}} = 0,$$
$$\dot{\rho}_\phi + 6H\rho_k + \frac{q_0}{2k_0}\rho_m\dot{\widehat{\phi}} = 0. \tag{159}$$

We have defined the kinetic and potential energy densities, ρ_k and ρ_V, respectively, as

$$\rho_k = \frac{\dot{\widehat{\phi}}^2}{2}, \qquad \rho_V = V(\widehat{\phi}) = V_0 e^{-\widehat{\phi}/k_0}, \qquad \rho_\phi = \rho_k + \rho_V. \tag{160}$$

The system of equations (159) and (149) (with $\rho_r = \rho_b = 0$) can be solved by a late-time configuration in which ρ_m, ρ_ϕ, V and H^2 scale in time in the same way, so that the critical fractions of dark matter density, $\Omega_m = \rho_m/6H^2$, dilaton kinetic energy, $\Omega_k = \rho_k/6H^2$, and potential energy, $\Omega_V = V/6H^2$, are separately frozen at constant values determined by k_0 and q_0 only (i.e., by the parameters c_1, c_2, and q_0 of the asymptotic expansion (145)). A simple analysis gives [71]

$$\Omega_k = \frac{3k_0^2}{(q_0+2)^2}, \qquad \Omega_V = \frac{3k_0^2 + q_0(q_0+2)}{(q_0+2)^2},$$
$$\Omega_\phi = \Omega_k + \Omega_V, \qquad \Omega_m = 1 - \Omega_\phi, \tag{161}$$

where k_0 is given by (154) (see also [32] for a detailed computation).

In this asymptotic state the Universe is thus dominated by a fixed mixture of dark matter and dilaton (kinetic plus potential) energy density. The dilaton fluid has equation of state

$$w = \frac{p_\phi}{\rho_\phi} = \frac{\Omega_k - \Omega_V}{\Omega_k + \Omega_V} = -\frac{q_0(q_0+2)}{6k_0^2 + q_0(q_0+2)}, \tag{162}$$

and can play the role of the dark energy fluid responsible for the observed cosmic acceleration, provided $q_0 > 1$.

In fact, by rewriting the Einstein equations (149) for $\dot{H}$ in the form

$$1+\frac{2\dot{H}}{3H^2}=\Omega_V-\Omega_k, \tag{163}$$

we obtain

$$\frac{\ddot{a}}{aH^2}=1+\frac{\dot{H}}{H^2}=\frac{3}{2}\left(\Omega_V-\Omega_k\right)-\frac{1}{2}=\frac{q_0-1}{q_0+2}. \tag{164}$$

The expansion is accelerated ($\ddot{a}>0$) for $q_0>1$ or $q_0<-2$. The second case (corresponding to an acceleration of superinflationary type, with $\dot{H}>0$) is to be excluded, however, in our context, as it would imply $\Omega_m<0$ according to (161). Thus, acceleration is only possible for $q_0>1$. The explicit form of this asymptotic solution can be finally obtained through the integration of (164), which gives

$$a\sim t^{(q_0+2)/3}, \qquad H\sim a^{-3/(q_0+2)}, \tag{165}$$

from which

$$\rho_m\sim H^2\sim\frac{\dot{\widehat{\phi}}^2}{2}\sim V_0\,\mathrm{e}^{-\widehat{\phi}/k_0}\sim a^{-6/(q_0+2)}. \tag{166}$$

To illustrate the smooth background evolution from the initial radiation phase to the intermediate dragging phase, and to the final freezing regime, we shall conclude this subsection by presenting the results of an exact numerical integration of the string cosmology equations (149)–(152). For our illustrative purpose, we will assume that Z_R and Z_ϕ are given by the expansion (145) truncated to first order in $\exp(-\phi)$, with $b_1=b_2=1$, $c_1^2=100$ and $c_2^2=30$. We will adopt the model of dilaton coupling already used in [71], parametrized by the time-dependent charge

$$q(\phi)=q_0\,\frac{\mathrm{e}^{q_0\phi}}{c^2+\mathrm{e}^{q_0\phi}}, \tag{167}$$

with $c^2=150$ and $q_0=2.5$. We will also use the E-frame potential (144), with $\beta_1=0.1$, $\beta_2=0.2$, and $c_1^2 m_V=10^{-3}H_{\mathrm{eq}}$. The last choice, which implies $m_V\sim H_0$, is crucial to obtain a realistic scenario in which the asymptotic accelerated regime starts at a phenomenologically acceptable epoch (see [71, 32] for a discussion of the mass scale of the non-perturbative dilaton potential, and of the degree of fine-tuning possibly required for realistic cosmological applications). Finally, we will integrate our equations imposing the initial conditions $\rho_\phi(t_i)=\rho_r(t_i)$, $\rho_m(t_i)=10^{-20}\rho_r(t_i)$, $\rho_b(t_i)=7\times10^{-21}\rho_r(t_i)$, $\phi(t_i)=-2$, at the initial scale $H(t_i)=10^{40}H_{\mathrm{eq}}$.

The obtained scaling evolution $\rho=\rho(a)$ is illustrated in Fig. 9 for the various cosmic components. We can note that, at large enough times, baryons (full line) and radiation (dotted line) are fully decoupled from the dilaton, and obey the standard scaling behavior ($\rho_r\sim a^{-4}$, $\rho_b\sim a^{-3}$). The late-time dark matter evolution, on the contrary, is closely tied to the dilaton evolution, and the ratio of their energy densities becomes asymptotically frozen at a constant. With the particular numerical values used in this example we obtain an asymptotic configuration characterized by $\Omega_\phi\simeq0.73$ and $\Omega_m\simeq0.27$, with a dark energy equation of state $w\simeq-0.76$.

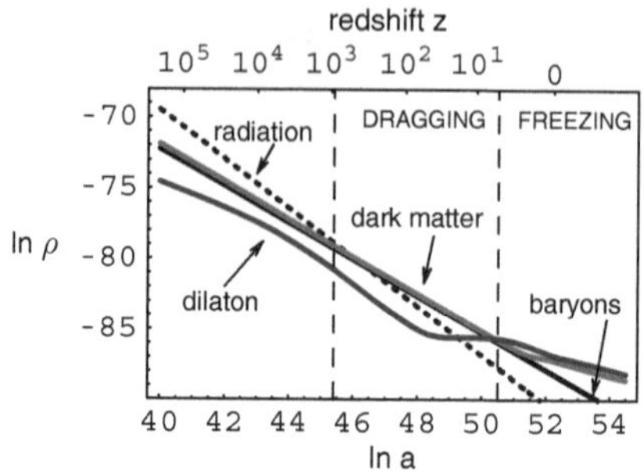

Fig. 9. Late-time evolution (on a logarithmic scale) of the various components of the cosmic energy density. The plots are the result of a numerical integration of (149)–(152)

3.3 Non-local Coupling and Pressure Back Reaction

Another interesting asymptotic configuration can be obtained in the case in which the dilaton is non-locally coupled to the dark matter components, as discussed in Sect. 1.2. In that case, the fractions Ω_m and Ω_ϕ may also become frozen at constant asymptotic values, but the background evolution turns out to be decelerated for any values of q_0, as the dark matter develops an effective pressure which tends to compensate the accelerating action of the dilaton potential. This effect is new, and will be illustrated in some details in this subsection.

We start assuming that, in the matter part of the action (158), the dilaton is non-locally coupled to the sources through the variable $\xi(\phi)$, as in the action (168). However, differently from the action (28), we will assume (for simplicity) that the dilaton potential is local, $V = V(\phi)$: the gravi-dilaton part of the action is thus identical to that of (114), and our model is described by

$$S = -\frac{1}{2\lambda_s^2}\int d^4x\sqrt{-g}\left[Z_R(\phi)R + Z_\phi(\phi)(\nabla\phi)^2 + V(\phi)\right] + \int d^4x\sqrt{-g}\,\mathcal{L}_m(e^{-\xi}). \tag{168}$$

Let us vary this action with respect to g and ϕ, and evaluate the resulting (general covariant) field equations in the limit of a homogeneous, isotropic, spatially flat background, using the results of Sect. 1.2. We obtain a set of equations similar to (45)–(47) for what concerns the dilaton charge density $\sigma(\overline{\phi})$, but different for the potential (which now is local), and for the presence of the loop corrections Z_R, Z_ϕ.

Let us finally transform the equations in the E-frame (using the rescaling (146)), and consider the asymptotic limit in which ρ_r, ρ_b are negligible, and the dark matter is coupled to the dilaton only through its intrinsic dilaton charge (namely, the limit in which $\psi' \simeq 0$). The resulting equations (omitting the tilde, in units $2\lambda_P^2 = 1$) can be written as

$$6H^2 = \rho_m + \rho_\phi, \tag{169}$$

$$4\dot{H} + 6H^2 = -p_\phi - \frac{\sigma_m}{2}, \tag{170}$$

$$\dot{\rho}_m + 3H\left(\rho_m + \frac{\sigma_m}{2}\right) - \frac{\sigma_m}{2}\dot{\phi} = 0, \tag{171}$$

$$\dot{\rho}_\phi + 3H(\rho_\phi + p_\phi) + \frac{\sigma_m}{2}\dot{\phi} = 0, \tag{172}$$

with ρ_ϕ and p_ϕ defined by (150), as before. A comparison with the asymptotic limit of (149) and (152), shows that the genuinely new effect of the non-local interactions is the appearance of an effective pressure term $\sigma_m/2$ for the dark matter component. Indeed, the new terms present in (170) and (171), can also be obtained from the standard Einstein equations through the shift $p_m \to p_m + \sigma_m/2$.

We are now in the position of asking whether or not this modification (of non-local origin) may change the results of the previous subsection, in particular those concerning the asymptotic freezing configuration. We shall consider, to this purpose, the limit in which $\sigma_m \to q_0\rho_m$ and $V = V_0 \exp(-\phi)$, using the canonical variable $\widehat{\phi}$ as in the previous computations.

Let us look for solutions of (169)–(172) by requiring for ρ_m, ρ_k, and ρ_V the same scaling behavior, and thus imposing, as a first condition, that

$$\frac{\dot{\rho}_m}{\rho_m} = \frac{\dot{\rho}_\phi}{\rho_\phi}. \tag{173}$$

Using (171) and (172) for ρ_m and ρ_ϕ, and the Einstein equation (169), we obtain

$$\frac{\dot{\widehat{\phi}}}{H} = \frac{6k}{q_0}\left[\overline{\Omega}_V\left(1 + \frac{q_0}{2}\right) - \overline{\Omega}_k\left(1 - \frac{q_0}{2}\right)\right]. \tag{174}$$

We are denoting with a bar the fractions of critical energies for the new configuration associated to the non-local equations, to distinguish it from the "local" freezing solution of (161). We also impose, as a second condition, that

$$\frac{\dot{\rho}_m}{\rho_m} = \frac{\dot{\rho}_V}{\rho_V}. \tag{175}$$

The definition (160) of ρ_V, together with (171), gives then

$$\frac{\dot{\widehat{\phi}}}{H} = 3k_0, \tag{176}$$

which, combined with (174), leads to

$$\overline{\Omega}_V = \overline{\Omega}_k \frac{2-q_0}{2+q_0} + \frac{q_0}{q_0+2}. \tag{177}$$

From the definition of $\overline{\Omega}_k$ and (176), on the other hand, we have

$$\overline{\Omega}_k = \frac{\dot{\hat{\phi}}}{12\,H^2} = \frac{3}{4}\,k_0^2. \tag{178}$$

The insertion of this result into (177) finally gives

$$\overline{\Omega}_V = \frac{3k_0^2(2-q_0)+4q_0}{4(2+q_0)}. \tag{179}$$

The combination of $\overline{\Omega}_k$ and $\overline{\Omega}_V$ provides now the values of $\overline{\Omega}_\phi$, $\overline{\Omega}_m$, and the equation of state $\overline{w}$, according to the definitions (161) and (162). As we are interested in the kinematical properties of the solution we shall compute, in particular, the acceleration parameter $\ddot{a}/(aH^2)$: dividing by $6H^2$ the modified equation (170) we obtain

$$\frac{\dot{H}}{H^2} = \frac{3}{2}\left(\overline{\Omega}_V - \overline{\Omega}_k\right) - \frac{3}{4} q_0\,\overline{\Omega}_m - \frac{3}{2}, \tag{180}$$

from which

$$\frac{\ddot{a}}{aH^2} \equiv 1 + \frac{\dot{H}}{H^2} = -\frac{1}{2}, \tag{181}$$

quite independently of the values of k_0 and q_0! The integration of $\dot{H}$ finally provides $a \sim t^{2/3}$ and $H^2 \sim \rho \sim a^{-3}$, as in the standard phase of dark matter-dominated evolution.

The considered model of non-local coupling is thus associated to an asymptotic freezing phase which is decelerated, and in which the dilaton energy density has the same dynamical behavior of a dust fluid, $\rho_\phi \sim \rho_m \sim a^{-3}$, in spite of a pressure which is non-vanishing, in general:

$$\overline{w} = \frac{q_q}{2}\,\frac{3k_0^2-2}{3k_0^2+q_0}. \tag{182}$$

This result can be understood by noting (180) and (181) together imply

$$\overline{\Omega}_k - \overline{\Omega}_V + \frac{q_0}{2}\,\overline{\Omega}_m \equiv \frac{1}{6H^2}\left(p_\phi + \frac{\sigma_m}{2}\right) = 0, \tag{183}$$

namely a zero total pressure for the coupled dilaton–dark matter fluid (see (170). The dark matter pressure associated to the non-local effects thus generates a backreaction which exactly compensates—at least in this model—the dilaton pressure, leading the system to restore, asymptotically, the standard dust matter configuration.

3.4 Main Differences from Uncoupled Models

Let us come back to the class of models in which the dilaton is locally coupled to the dark matter components, as discussed in Sect. 3.2. If we identify the accelerated freezing phase with our present cosmological phase, and thus the energy density of the dilaton field with the "dark energy" density responsible for the present cosmic acceleration, we are lead to a dilaton model of dark energy which is substantially different from the conventional models of quintessence [74] based on a rolling scalar field, uncoupled to dark matter.

A first, important (conceptual) difference concerns the mentioned problem of the cosmic coincidence. In the considered class of dilaton models this problem, if not solved, is at least relaxed: in fact, the dark energy and dark matter densities are of the same order not only today but also in the future (forever), and also in the past for a significantly amount of time, depending on the beginning of the freezing epoch (see below).

A second, more phenomenological difference concerns the scaling behavior of the baryonic and dark matter components of the dust fluid during the freezing epoch. Because of the coupling to the dilaton, the dilution in time of the dark matter density ρ_m is slower than the standard baryon dilution, $\rho_b \sim a^{-3}$: in particular, the ratio ρ_b/ρ_m decreases in time as

$$\frac{\rho_b}{\rho_m} \sim a^{-3q_0/(2+q_0)} \tag{184}$$

(see (166)). This could explain why the present fraction of baryons is small ($\sim 10^{-2}$) in critical units—provided the accelerated epoch has an early enough beginning. Direct/indirect measurements of the past value of the ratio ρ_b/ρ_m, compared with its present value, could provide unambiguous tests of this class of models.

Finally, concerning the beginning of the accelerated epoch, it is important to stress that in dilaton models the acceleration can start much earlier than in models of uncoupled quintessence [75, 76].

For a simple illustration of this point we may consider a model in which, during the accelerated regime, there are two types of sources with different dynamical behavior: (i) an *uncoupled* component ρ_u, with pressure $p_u = 0$ and scaling behavior $\rho_u \sim a^{-3}$ (represented by baryons and, possibly, by a fraction of non-baryonic dark matter uncoupled to the dilaton); (ii) a *coupled* component ρ_c, with pressure $p_c = w\rho_c$, and a slower scaling behavior $\rho_c \sim a^{-3(1+w)}$ (represented by the dilaton and by the fraction of dark matter coupled to the dilaton). Thus, even if today ρ_c dominates, and drives an accelerated evolution, at ealy enough times the Universe was dominated by ρ_u, and decelerated. From the Einstein equations

$$\begin{aligned} 6H^2 &= \rho_u + \rho_c, \\ 4\dot{H} + 6H^2 &= -p_c = -w\rho_c, \end{aligned} \tag{185}$$

we obtain that the acceleration switches off at the scale a_{acc} such that

$$\left(\frac{\ddot{a}}{aH^2}\right)_{\rm acc} = 1 + \left(\frac{\dot{H}}{H^2}\right)_{\rm acc} = -\frac{1}{2}\left[\Omega_u - (1+3w)(\Omega_u - 1)\right]_{\rm acc} = 0, \tag{186}$$

where $\Omega_u = \rho_u/6H^2$, $\Omega_c = 1 - \Omega_u$. In terms of the present values Ω_u^0, Ω_c^0 of these fractions the above condition becomes

$$\Omega_u^0 \left(\frac{a_{\rm acc}}{a_0}\right)^{-3} = (1+3w)\left(\Omega_u^0 - 1\right)\left(\frac{a_{\rm acc}}{a_0}\right)^{-3(1+w)}, \tag{187}$$

and fixes the beginning of the acceleration at the redshift scale z_{acc} such that

$$z_{\rm acc} \equiv \frac{a_0}{a_{\rm acc}} - 1 = \left[(1+3w)\left(\frac{\Omega_u^0 - 1}{\Omega_u^0}\right)\right]^{-1/3w} - 1. \tag{188}$$

Consider now a model of uncoupled quintessence, in which the uncoupled component corresponds to the totality of the dark matter fluid (plus

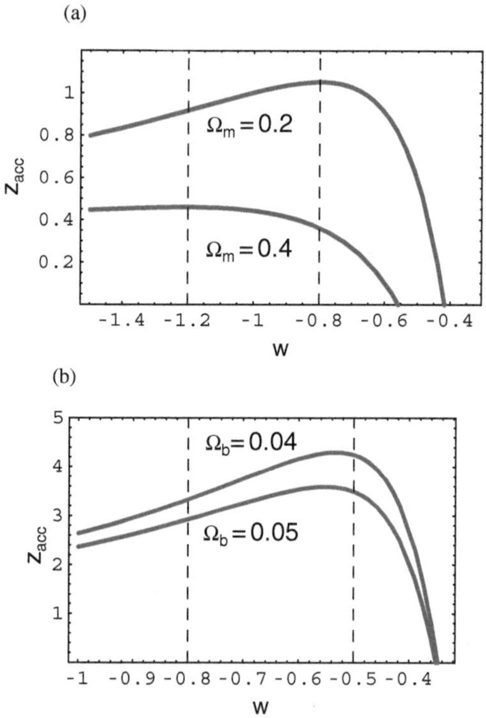

Fig. 10. Beginning of the accelerated epoch for dark-energy models with uncoupled (**top**) and fully coupled (**bottom**) dark matter, according to the observations of type Ia supernovae in the SNLS data set. The plotted curves are obtained from (188), for constant values of the present fraction of the uncoupled dust matter Ω_u^0

subdominant contributions), i.e., $\Omega_u^0 = \Omega_m^0$. Using the recent analysis of the SNLS Collaboration [68], based on present observations of supernovae Ia and large-scale structure, one finds $0.2 \leq \Omega_m^0 \leq 0.4$, and $-1.2 \leq w \leq 0.8$. One then obtains, from (188), $0.4 \lesssim z_{acc} \lesssim 1$ (see Fig. 10, top panel).

If we consider instead a model of dilaton dark energy, then the uncoupled component may range from the baryon component Ω_b to some fraction of the dark matter component Ω_m. In the "maximally coupled" version of the model, in which $\Omega_u^0 = \Omega_b^0$, we can re-apply the supernovae results of SNLS with $\Omega_b^0 \simeq 0.04 - -0.05$, to obtain $w \simeq -0.65 \pm 0.15$. Equation (188) then implies [76] $z_{acc} \simeq 3 - -4$ (see Fig. 253, bottom panel).

Thus, dilaton models of dark energy are compatible with a beginning of the cosmic accelerations at epochs much earlier than those suggested by other models, according to the most recent supernovae data. The extension back in time of the accelerated regime might have a significant impact on the dilution of baryons, according to (184). Finally, strongly coupled models tend to be compatible with a "less negative" parameter w (see Fig. 10), thus alleviating the need for "phantom" dark energy [77] with "supernegative" ($w < -1$) equation of state.

Acknowledgements

I am very grateful to all friends and collaborators contributing to the results reported in this paper. First of all, I would like to thank Gabriele Veneziano for many years of collaboration, friendship, and support. In addition, I am grateful to Valerio Bozza, Massimo Giovannini, and Jnan Maharana for their collaboration on the results presented in the first lecture; to Nicola Bonasia, Eugenio Coccia and Carlo Ungarelli for their collaboration on the results presented in the second lecture; to Luca Amendola, Federico Piazza, and Carlo Ungarelli for their collaboration on the results presented in the third lecture.

Appendix A

In this appendix we present a detailed derivation of the equations of motion (29) and (34), starting from the action (28) which includes non-local dilaton interactions.

The functional derivation of the action with respect to the metric $g^{\mu\nu}(x)$ contains, besides the standard contributions leading to (3), the new non-local contributions $V_{\mu\nu}(x)$ and $L_{\mu\nu}(x)$, and can be written as follows:

$$\frac{\delta S}{\delta g^{\mu\nu}(x)} = -\frac{\left(\sqrt{-g}\,\mathrm{e}^{-\phi}\right)_x}{2\lambda_{\mathrm{s}}^{d-1}}\left[G_{\mu\nu} + \nabla_\mu\nabla_\nu\phi + \frac{1}{2}g_{\mu\nu}\left(\nabla\phi^2 - 2\nabla^2\phi - V\right)\right]_x + \frac{1}{2}\sqrt{-g}\,T_{\mu\nu}(x) + V_{\mu\nu}(x) + L_{\mu\nu}(x), \tag{A.1}$$

where

$$V_{\mu\nu}(x) = -\frac{1}{2\lambda_{\mathrm{s}}^{d-1}} \int \mathrm{d}^{d+1}x' \left(\sqrt{-g}\,\mathrm{e}^{-\phi}V'\right)_{x'} \frac{\delta}{\delta g^{\mu\nu}(x)} \mathrm{e}^{-\xi(x')}, \quad \text{(A.2)}$$

$$L_{\mu\nu}(x) = \int \mathrm{d}^{d+1}x' \left(\sqrt{-g}\mathcal{L}'_m\right)_{x'} \frac{\delta}{\delta g^{\mu\nu}(x)} \mathrm{e}^{-\xi(x')} \quad \text{(A.3)}$$

(V' and $\mathcal{L}'_m$ are defined by (33)). We need now to compute the functional derivation of $\exp(-\xi)$. Using the definition (25) we obtain

$$\begin{aligned}\frac{\delta}{\delta g^{\mu\nu}(x)} \mathrm{e}^{-\xi(x')} &= -\frac{1}{\lambda_{\mathrm{s}}^{d}} \int \mathrm{d}^{d+1}y\, \delta^{d+1}(x-y)\delta(\phi_{x'}-\phi_y)\,\mathrm{e}^{-\phi_y} \\ &\quad \left[-\frac{1}{2}\left(\sqrt{-g}\,g_{\mu\nu}\sqrt{\epsilon(\nabla\phi)^2}\right) + \frac{1}{2}\sqrt{-g}\sqrt{\epsilon(\nabla\phi)^2}\,\frac{\partial_\mu\phi\partial_\nu\phi}{(\nabla\phi)^2}\right]_y \\ &= -\frac{1}{2\lambda_{\mathrm{s}}^{d}}\left(\gamma_{\mu\nu}\sqrt{-g}\,\sqrt{\epsilon(\nabla\phi)^2}\,\mathrm{e}^{-\phi}\right)_x \delta(\phi_{x'}-\phi_x), \end{aligned} \quad \text{(A.4)}$$

where $\gamma_{\mu\nu}$ is defined in (30). Thus,

$$V_{\mu\nu} = \frac{1}{2\lambda_{\mathrm{s}}^{d-1}} \sqrt{-g}\,\mathrm{e}^{-2\phi}\frac{1}{2}\gamma_{\mu\nu}\sqrt{\epsilon(\nabla\phi)^2}\, I_V, \quad \text{(A.5)}$$

$$L_{\mu\nu} = -\sqrt{-g}\,\mathrm{e}^{-\phi}\frac{1}{2}\gamma_{\mu\nu}\sqrt{\epsilon(\nabla\phi)^2}\, I_m, \quad \text{(A.6)}$$

where I_V and I_m are defined in (31) and (32). Inserting these results into (A.1), multiplying by $(-2\lambda_{\mathrm{s}}^{d-1})\exp(-\phi)/\sqrt{-g}$, and imposing the condition of zero functional derivative, one is finally lead to (29).

Let us now consider the functional derivative with respect to $\phi(x)$. Separating the local and non-local terms, as before, we obtain

$$\begin{aligned}\frac{\delta S}{\delta\phi(x)} &= \frac{\left(\sqrt{-g}\,\mathrm{e}^{-\phi}\right)_x}{2\lambda_{\mathrm{s}}^{d-1}}\left[R + 2\nabla^2\phi - (\nabla\phi)^2 + V\right]_x \\ &\quad + A(x) + B(x), \end{aligned} \quad \text{(A.7)}$$

where

$$A(x) = -\frac{1}{2\lambda_{\mathrm{s}}^{d-1}} \int \mathrm{d}^{d+1}x' \left(\sqrt{-g}\,\mathrm{e}^{-\phi}V'\right)_{x'} \frac{\delta}{\delta\phi(x)} \mathrm{e}^{-\xi(x')}, \quad \text{(A.8)}$$

$$B(x) = \int \mathrm{d}^{d+1}x' \left(\sqrt{-g}\mathcal{L}'_m\right)_{x'} \frac{\delta}{\delta\phi(x)} \mathrm{e}^{-\xi(x')}. \quad \text{(A.9)}$$

The functional derivative of the variable (25) leads to

$$\begin{aligned}&\frac{\delta}{\delta\phi(x)} \mathrm{e}^{-\xi(x')} \\ &= \frac{1}{\lambda_{\mathrm{s}}^{d}} \int \mathrm{d}^{d+1}y \Big\{ -\left(\sqrt{-g}\,\mathrm{e}^{-\phi}\sqrt{\epsilon(\nabla\phi)^2}\right)_y \delta(\phi_{x'}-\phi_y)\delta^{d+1}(x-y)\end{aligned}$$

$$+\left(\sqrt{-g}\,\mathrm{e}^{-\phi}\sqrt{\epsilon(\nabla\phi)^2}\right)_y \delta'(\phi_{x'}-\phi_y)\left[\delta^{d+1}(x-x')-\delta^{d+1}(x-y)\right]$$

$$\left.-\partial_\mu\left[\frac{\sqrt{-g}\,\mathrm{e}^{-\phi}\epsilon\,\partial^\mu\phi}{\sqrt{\epsilon(\nabla\phi)^2}}\delta(\phi_{x'}-\phi_y)\right]_y \delta^{d+1}(x-y)\right\}, \tag{A.10}$$

where $\partial_\mu = \partial/\partial y^\mu$, and δ' denotes the derivative of the delta function with respect to its argument.

The first term of this integral exactly cancels the term containing $\partial_\mu \mathrm{e}^{-\phi}$ in the last part of the integral; also, the third term exactly cancels the term containing $\partial_\mu[\delta(\phi_{x'}-\phi_y)]$ in the last part of the integral. Thus, we are left with

$$\frac{\delta}{\delta\phi(x)}\mathrm{e}^{-\xi(x')} = \frac{\delta^{d+1}(x-x')}{\lambda_\mathrm{s}^d}\int \mathrm{d}^{d+1}y\left(\sqrt{-g}\mathrm{e}^{-\phi}\sqrt{\epsilon(\nabla\phi)^2}\right)_y \delta'(\phi_{x'}-\phi_y)$$
$$-\,\epsilon\,\frac{\mathrm{e}^{-\phi}}{\lambda_\mathrm{s}^d}\delta(\phi_{x'}-\phi_x)\partial_\mu\left(\frac{\sqrt{-g}\,\partial^\mu\phi}{\sqrt{\epsilon(\nabla\phi)^2}}\right)_x . \tag{A.11}$$

The second term on the right-hand side of the above equation can be conveniently rewritten as

$$-\epsilon\frac{\mathrm{e}^{-\phi}}{\lambda_\mathrm{s}^d}\delta(\phi_{x'}-\phi_x)\sqrt{-g}\,\nabla_\mu\left(\frac{\partial^\mu\phi}{\sqrt{\epsilon(\nabla\phi)^2}}\right)_x$$
$$= -\epsilon\frac{\mathrm{e}^{-\phi}}{\lambda_\mathrm{s}^d}\delta(\phi_{x'}-\phi_x)\frac{\sqrt{-g}}{\sqrt{\epsilon(\nabla\phi)^2}}\,\gamma_{\mu\nu}\nabla^\mu\nabla^\nu\phi. \tag{A.12}$$

For the first term containing δ' we can exploit the properties of the delta function, and the identities

$$\mathrm{d}y_0 = \frac{1}{\dot\phi_y}\mathrm{d}\phi_y, \qquad \frac{\mathrm{d}}{\mathrm{d}\phi_y} = \frac{1}{\dot\phi_y}\frac{\mathrm{d}}{\mathrm{d}y_0}, \tag{A.13}$$

to obtain

$$\lambda_\mathrm{s}^{-d}\int \mathrm{d}^d y\,\frac{\mathrm{d}\phi_y}{\dot\phi_y}\left(\sqrt{-g}\,\mathrm{e}^{-\phi}\sqrt{\epsilon(\nabla\phi)^2}\right)_y \delta'(\phi_x-\phi_y)$$
$$= \lambda_\mathrm{s}^{-d}\int \mathrm{d}^d y\,\frac{\mathrm{d}\phi_y}{\dot\phi_y}\frac{\mathrm{d}}{\mathrm{d}y_0}\left(\frac{\sqrt{-g}\,\mathrm{e}^{-\phi}\sqrt{\epsilon(\nabla\phi)^2}}{\dot\phi}\right)_y \delta(\phi_x-\phi_y)$$
$$= -e^{-\xi(x)} + \lambda_\mathrm{s}^{-d}\mathrm{e}^{-\phi(x)}\int \mathrm{d}^d y\,\frac{\mathrm{d}\phi_y}{\dot\phi_y}\left(\sqrt{-g}\,\sqrt{\epsilon(\nabla\phi)^2}\right)_y \delta'(\phi_x-\phi_y)$$
$$= -e^{-\xi(x)} + e^{-\phi(x)}J(x), \tag{A.14}$$

where J is the integral defined in (35). Inserting the results (A.12) and (A.14) into (A.11), using the definitions of A and B, and integrating over x', we finally obtain

$$A(x) + B(x) = \left(\frac{\sqrt{-g}\,\mathrm{e}^{-\phi}}{2\lambda_{\mathrm{s}}^{d-1}} V' - \sqrt{-g}\,\mathcal{L}'_m \right)_x \left(e^{-\xi} - e^{-\phi} J\right)_x$$
$$+ \epsilon \frac{\sqrt{-g}\,\mathrm{e}^{-\phi}}{\sqrt{\epsilon(\nabla\phi)^2}} \gamma_{\mu\nu} \nabla^\mu \nabla^\nu \phi \left(\frac{\mathrm{e}^{-\phi}}{2\lambda_{\mathrm{s}}^{d-1}} I_V - I_m \right)_x . \quad \text{(A.15)}$$

Summing this contribution to the other terms of (A.7), multiplying by $(2\lambda_{\mathrm{s}}^{d-1} \exp\phi/\sqrt{-g})$, and imposing the vanishing of the functional derivative, we are lead to the equation of motion (34) for the dilaton.

References

1. G. Veneziano: Phys. Lett. B **265**, 287 (1991)
2. M. Gasperini, G. Veneziano: Astropart. Phys. **1**, 317 (1993)
3. M. B. Green, J. Schwartz, E. Witten: *Superstring Theory* (Cambridge University Press, Cambridge, 1987)
4. J. Polchinski: *String Theory* (Cambridge University Press, Cambridge, 1998)
5. M. Gasperini: in *Gravitational Waves*, ed. by I. Ciufolini et al. (IOP Publishing, Bristol, 2001), p. 280
6. A. A. Tseytlin: Mod. Phys. Lett. A **6** 1721 (1991)
7. K. Kikkawa, M. Y. Yamasaki: Phys. Lett. B **149** 357 (1984)
8. K. A. Meissner, G. Veneziano: Mod. Phys. Lett. A **6**, 3397 (1991); Phys. Lett. B **267**, 33 (1991)
9. K. Meissner: *Dualities in string cosmology*, this volume
10. M. Gasperini, G. Veneziano: Phys. Lett. B **277**, 256 (1992)
11. M. Gasperini, M. Giovannini, G. Veneziano: Phys. Lett. B **569**, 113 (2003); Nucl. Phys. B **694**, 206 (2004)
12. M. Gasperini, G. Veneziano: Mod. Phys. Lett. A **8**, 3701 (1993)
13. M. Gasperini, G. Veneziano: Phys. Rev. D **50**, 2519 (1994)
14. A. D. Linde: Phys. Lett. B **129**, 177 (1983)
15. A. Vilenkin: Phys. Rev. D **46**, 2355 (1992); A. Borde, A. Vilenkin: Phys. Rev. Lett. **72**, 3305 (1994)
16. A. Buonanno, T. Damour, G. Veneziano: Nucl. Phys. B **543**, 275 (1999)
17. M. Gasperini, G. Veneziano: Gen. Rel. Grav. **28**, 1301 (1996)
18. M. Gasperini, J. Maharana, G. Veneziano: Nucl. Phys. B **472**, 349 (1996)
19. R. H. Brandenberger, J. Martin: Phys. Rev. D **71**, 023504 (2005)
20. M. Gasperini, M. Maggiore, G. Veneziano: Nucl. Phys. B **494**, 315 (1997)
21. R. Brustein, R. Madden: Phys. Lett. B **410**, 110 (1997); Phys. Rev. D **57**, 712 (1998); C. Cartier, E. J. Copeland, R. Madden, JHEP **0001**, 035 (2000)
22. M. Gasperini, G. Veneziano: Phys. Rep. **373** 1 (2003)
23. M. Gasperini: Mod. Phys. Lett. A **14**, 1059 (1999)
24. M. Gasperini, M. Giovannini, K. A. Meissner, G. Veneziano: Nucl. Phys. (Proc. Suppl.) B **49**, 70 (1996)
25. M. Gasperini, M. Giovannini: Phys. Lett. B **282**, 36 (1992); Phys. Rev. D **47**, 1519 (1993)
26. R. Brustein, M. Gasperini, M. Giovannini, V. Mukhanov, G. Veneziano: Phys. Rev. D **51**, 6744 (1995)

27. R. Brustein, M. Gasperini, M. Giovannini, G. Veneziano: Phys. Lett. B **361**, 45 (1995)
28. A. Buonanno and C. Ungarelli: *Primordial gravitational radiation in string cosmology*, this volume
29. J. Khoury, B. A. Ovrut, P. J. Steinhardt, N. Turok: Phys. Rev. D **64**, 123533 (2001)
30. E. I. Buchbinder, J. Khoury, B. A. Ovrut: *New ekpyrotic cosmology*, hep-th/0702154
31. L. A. Boyle, P. J. Steinhardt, N. Turok: Phys. Rev. D **69**, 127302 (2002)
32. M. Gasperini: *Elements of String Cosmology* (Cambridge University Press, Cambridge, 2007)
33. M. Maggiore, A. Riotto: Nucl. Phys. B **548**, 427 (1999)
34. S. Alexander, R. Brandenberger, D. Easson: Phys. Rev. D **62**, 103509 (2000)
35. C. Burgess et al: JHEP **0107**, 047 (2001); S. Kachru et al.: JCAP **0310**, 013 (2003)
36. H. Tye: *Brane inflation: string theory viewed from the cosmos*, this volume
37. K. Enqvist, M. Sloth: Nucl. Phys. B **626**, 395 (2002); D. H. Lyth, D. Wands: Phys. Lett. B **524**, 5 (2002); T. Moroi, T. Takahashi: Phys. Lett. B **522**, 215 (2001)
38. V. Bozza, M. Gasperini, M. Giovannini, G. Veneziano: Phys. Lett. B **543**, 14 (2002); Phys. Rev. D **67**, 063514 (2003)
39. E. J. Copeland, R. Easther, D. Wands: Phys. Rev. D **56**, 874 (1997); E. J. Copeland, J. E. Lidsey, D. Wands: Nucl. Phys. B **506**, 407 (1997)
40. V. F. Mukhanov, H. A. Feldman, R. H. Brandenberger: Phys. Rep. **215**, 203 (1992)
41. M. Abramowitz, I. A. Stegun: *Handbook of Mathematical Functions* (Dover, New York, 1972)
42. A. Melchiorri, F. Vernizzi, R. Durrer, G. Veneziano: Phys. Rev. Lett. **83**, 4464 (1999)
43. D. Lyth, C. Ungarelli, D. Wands: Phys. Rev. D **67**, 023503 (2003)
44. M. Gasperini: in Proc. of the *Fifth Paris Cosmology Colloquium*, ed. by H. J. De Vega and N. Sanchez (Publication Observatoire de Paris, Paris, 1999), p. 317
45. D. N. Spergel et al.: astro-ph/0603449
46. M. Gasperini, M. Giovannini, G. Veneziano: Phys. Rev. D **52**, 6651 (1995); M. Gasperini, M. Giovannini, G. Veneziano: Phys. Rev. Lett. **75**, 3796 (1995); D. Lemoine, M. Lemoine: Phys. Rev. D **52**, 1995 (1995); M. Gasperini, S. Nicotri: Phys. Lett. B **633**, 155 (2006)
47. M. Gasperini: Phys. Lett. B **327**, 314 (1994)
48. T. Taylor, G. Veneziano: Phys. Lett. B **213**, 459 (1988)
49. J. Ellis et al: Phys. Lett. B **228**, 264 (1989)
50. R. Durrer, M. Gasperini, M. Sakellariadou, G. Veneziano: Phys. Rev D **59**, 43511 (1999)
51. M. Gasperini, G. Veneziano: Phys. Rev. D **59**, 43503 (1999)
52. R. Brustein, M. Gasperini, G. Veneziano: Phys. Rev. D **55**, 3882 (1997)
53. E. Fischbach, C. Talmadge: Nature **356**, 207 (1992)
54. C. D. Hoyle et al: Phys. Rev. D **70**, 042004 (2004)
55. M. Gasperini: Phys. Lett. B **470**, 67 (1999)
56. T. Damour, A. M. Polyakov: Nucl. Phys. B **423**, 352 (1994); Gen. Rel. Grav. **26**, 1171 (1994)

57. C. Misner, K. Thorne, J. A. Wheeler: *Gravitation* (Freeman, San Francisco 1973)
58. M. Gasperini: Phys. Lett. B **477**, 242 (2000)
59. M. Bianchi et al.: Phys. Rev. D **57**, 4525 (1998); M. Maggiore, A. Nicolis: Phys. Rev. D **62**, 024004 (2000)
60. M. Gasperini, C. Ungarelli: Phys. Rev. D **64**, 064009 (2001)
61. N. Bonasia, M. Gasperini: Phys. Rev. D **71**, 104020 (2005)
62. B. Allen, J. D. Romano: Phys. Rev. D **59**, 102001 (1999)
63. M. Doran, J. Jackel: Phys. Rev. D **66**, 043519 (2002)
64. B. J. Owen, B. S. Sathyaprakash: Phys. Rev. D **60**, 022002 (1999)
65. E. Coccia, M. Gasperini, C. Ungarelli: Phys. Rev. D **65**, 067101 (2002)
66. S. Perlmutter et al.: Nature **391**, 51 (1998); A. G. Riess et al.: Astron. J. **116**, 1009 (1998)
67. A. G. Riess et al.: Astrophys. J. **607**, 665 (2004)
68. P. Astier et al.: Astron. Astrophys. **447**, 31 (2006)
69. M. Gasperini: Phys. Rev. D **64**, 043510 (2001)
70. N. Kaloper, K. A. Olive: Astropart. Phys. **1**, 185 (1993)
71. M. Gasperini, F. Piazza, G. Veneziano: Phys. Rev. D **65**, 023508 (2002)
72. L. Amendola: Phys. Rev. D **62**, 043511 (2000); L. Amendola, D. Tocchini-Valentini: Phys. Rev. D **64**, 04359 (2001); Phys. Rev. D **66** 043528 (2002)
73. G. Veneziano: JHEP **0206**, 051 (2002)
74. B. Ratra, P. J. E. Peebles: Phys. Rev. D **37** 3406 (1988); C. Wetterich: Nucl. Phys. B **302**, 668 (1988); M. S. Turner, C. White: Phys. Rev. D **56**, 4439 (1997); R. R. Caldwell, R. Dave, P. J. Steinhardt: Phys. Rev. Lett. **80**, 1582 (1998); I. Zlatev, L. Wang, P. J. Steinhardt: Phys. Rev. Lett. **82**, 896 (1999); Phys. Rev. D **59**, 123504 (1999)
75. L. Amendola, M. Gasperini, D. Tocchini-Valentini, C. Ungarelli: Phys. Rev. D **67**, 043512 (2003); L. Amendola, M. Gasperini, F. Piazza: JCAP **09** , 014 (2004)
76. L. Amendola, M. Gasperini, F. Piazza: Phys. Rev. D **74**, 127302 (2006)
77. R. R. Caldwell, M. Kamionkowski, N. N. Weinberg: Phys. Rev. Lett. **91**, 07130 (2003)

Relic Gravitons and String Pre-big-bang Cosmology

A. Buonanno[1] and C. Ungarelli[2]

[1] Physics Department, University of Maryland, College Park, MD 20742, USA
buonanno@umd.edu
[2] Physics Department "Enrico Fermi", University of Pisa, Largo Pontecorvo 3, 56127 Pisa, Italy and Geosciences and Earth Resources Institute, CNR, via Moruzzi 1, 56124 Pisa, Italy
carlo.ungarelli@igg.cnr.it

Abstract. In this paper, after discussing the mechanism of graviton production during an early phase of accelerated expansion, we will review the main features of the spectrum of primordial gravitational radiation for the class of string-inspired models called *pre-big-bang* models. Furthermore, we will also outline the implications on pre-big-bang models of current and future searches of gravitational waves with ground-based detectors.

Foreword

This contribution reviews one of the many research topics originally pioneered by Gabriele Veneziano in *String theory and fundamental interactions*. We are thankful to Gabriele for having taught us not only physics, but also how to be good physicists, choosing and tackling problems with deepness and seriousness. He will continue to be for us a precious source of inspiration.

1 Introduction

In recent years, a number of detectors have been designed and built to search for gravitational waves (GWs). Ground-based interferometers aimed at detecting GWs in the frequency range between 10 Hz and 1 kHz, such as LIGO [1], VIRGO [2], GEO600 [3], and TAMA [4] are now operating at design sensitivity (or close to it in the case of VIRGO). The design of a space-based three-arm interferometer, the Laser Interferometer Space Antenna (LISA) [5], will explore the frequency window between 0.1 and 10 mHz. A second generation of space-based detector probing primordial GWs [6, 7] is under planning. Following earlier theoretical works [8], prototypes for detecting high-frequency

A. Buonanno and C. Ungarelli: *Relic Gravitons and String Pre-big-bang Cosmology*, Lect. Notes Phys. **737**, 845–861 (2008)
DOI 10.1007/978-3-540-74233-6_25

GWs, in the millihertz band, have been developed [9]. Finally, the large number of millisecond pulsar detectable with the square kilometer array (SKA) [10] would provide an ensemble of clocks that can be used as multiple arms of a GW detector in the frequency range around 10^{-9} Hz.

One of the possible targets of such search is a stochastic gravitational-wave background (SGWB). Depending on its origin, the stochastic background can be broadly divided into two classes (for a review see, e.g., [11, 12]): the astrophysically generated background due to the incoherent superposition of gravitational radiation emitted by large populations of astrophysical sources that cannot be resolved individually, and the primordial GW background generated by processes taking place in the early stages of the Universe. A primordial component of such background is especially interesting, since it would carry unique information about the state of the primordial Universe. Here we focus our attention on a particular type of primordial stochastic background, namely the relic radiation produced by the parametric amplification of metric tensor perturbations during an early stage of accelerated expansion (inflationary stage) [13]. Leaving the detailed analysis of the production mechanism to the following section, the energy and spectral content of such radiation is encoded in the spectrum, defined as follows:

$$\Omega_{\rm GW} = \frac{1}{\rho_c}\, f \tilde{\rho}_{\rm GW}(f)\,, \tag{1}$$

where f is the frequency, ρ_c is the critical energy density of the Universe ($\rho_c = 3H_0^2/8\pi G$) and $\tilde{\rho}_{\rm GW}$ is the GWs energy density per unit frequency, i.e.,

$$\rho_{\rm GW} = \int_0^\infty df\, \tilde{\rho}_{\rm GW}(f)\,. \tag{2}$$

For a spectrum produced during an early stage of *slow-roll* inflation, the spectrum decreases as f^{-2} in the frequency window $10^{-18} - 10^{-16}$ Hz, and then slowly decreases up to a frequency corresponding to modes whose physical frequency becomes less than the maximum causal distance during the reheating phase (which is of order of a few gigahertz). For this class of models, the spectral content of the SGWB is fixed in terms of the shape parameters of the inflaton potential. Its magnitude depends on both the value of the Hubble parameter during inflation and a number of features characterizing the Universe evolution after the inflationary era—for example, tensor anisotropic stress due to free-streaming relativistic particles, equations of state [14, 15], and so on. An upper bound on the spectrum can be obtained from the measurement of the quadrupole anisotropy of the cosmic microwave background (CMB). Through the Sachs–Wolfe effect, a SGWB at large scales (i.e., at wavelengths comparable to the present value of the Hubble radius) would induce stochastic anisotropies in the CMB temperature. This yields an upper limit of $h_0^2\, \Omega_{\rm GW} \sim 5 \times 10^{-15}$ at $f \sim 10^{-16}$ Hz [16]. Since for a generic slow-roll inflationary model the spectrum is (weakly) decreasing with frequency (for a recent review see, e.g., [15, 17]), this implies an upper bound $h_0^2\,\Omega_{\rm GW} \sim 5\times10^{-16}$ at frequencies around $f \sim 100$ Hz , where ground-based detectors such as LIGO reach the best sensitivity. For a flat spectrum, the recent LIGO results [18] sets an upper limit $h_0^2\,\Omega_{\rm GW} < 6.5\times10^{-5}$. For frequency-independent spectra, the expected upper limit for the current LIGO configuration is $h_0^2\,\Omega_{\rm GW} < 5\times10^{-6}$, while the advanced LIGO project design sensitivity is $h_0^2\,\Omega_{\rm GW} \sim 8\times10^{-9}$ (see, e.g., [19]).

The spectrum predicted by the class of single-field inflationary models is then too low to be observed by ground-based detectors. It is therefore evident that a background satisfying the bound imposed by the observed amount of CMB anisotropies at large scales could be detected at frequencies relevant for ground-based GW detectors provided that its spectrum grows significantly with frequency.

Pre-big-bang (PBB) models, originally proposed by Veneziano [20], and then Gasperini and Veneziano [21] (for a detailed review, see [22]), represent an interesting class of inflationary models alternative to the standard slow-roll ones. In particular, the presence in the inflationary phase of fields like the dilaton or moduli, can have important consequences on the spectral properties of the SGWB, thus affecting the possibility of detection by earth-based interferometers. As first shown in [23], at low frequencies, say $f \gtrsim 10^{-16}$ Hz, the SGWB spectrum grows as $\Omega_{\rm GW} \sim f^3$. Hence, the COBE bound is easily evaded and the spectrum can peak at frequencies around $10 - 10^3$ Hz, still satisfying the bound from big-bang nucleosynthesis (BBN) [24] and CMB [25].

The aim of this paper is both to review the general mechanism of cosmological graviton production, describing its key features, and discuss the prospect of detection within the class of PBB cosmological models. The paper is organized as follows. In Sect. 2 we review the mechanism of parametric amplification of metric perturbations, and discuss examples in De Sitter and slow-roll inflation. In Sect. 3 we compute the SGWB in non-minimal PBB models and discuss the main modifications in non-minimal models. In Sect. 4 we review the implications of current and future results of ground-based detector (in particular the LIGOs) on PBB models. Finally, in Sect. 5 we draw some conclusions.

2 Graviton Production in Cosmology

One of the most relevant aspects of inflationary models is that they provide a natural mechanism for generating perturbations in all matter fields. Such primordial perturbations can then be considered as seeds for the observed CMB anisotropies and large-scale structures, and can also yield to a SGWB. Those observable consequences are all related to the well-known phenomenon of amplification of quantum–vacuum fluctuations in cosmological backgrounds [13]. In this section, starting with the simple toy model of a one-dimensional harmonic oscillator [26], we shall compute the SGWB in De Sitter inflation and slow-roll inflation.

Let us consider a one-dimensional harmonic oscillator moving in an expanding background described by a scale factor $a(t)$. The Lagrangian is

$$L = \frac{a^2 m}{2} \left(\dot{x}^2 - \omega^2 x^2\right) , \tag{1}$$

the canonical momentum and the corresponding Hamiltonian—computed as the Legendre transformation of the Lagrangian (3)—read

$$p = a^2 m \dot{x} , \tag{2}$$

$$H = \frac{1}{2} \left(\frac{p^2}{a^2 m} + a^2 m \omega^2 x^2\right) . \tag{3}$$

The corresponding equations of motion are

$$\ddot{x} + 2\frac{\dot{a}}{a}\dot{x} + \omega^2 x = 0 , \tag{4}$$

$$\ddot{y} + \left(\omega^2 - \frac{\ddot{a}}{a}\right) y = 0 , \tag{5}$$

where we denote with a dot the derivative with respect to the cosmic time t and $y = a\,x$ is the proper physical amplitude of the harmonic oscillator. Without specifying the details of the cosmological evolution, the properties of the solutions of (6) and (7) can be derived by analyzing their behavior in two different regimes:

(a) When $\omega^2 \gg \ddot{a}/a$, the comoving amplitude and momentum are adiabatically damped

$$x \sim \frac{1}{a} e^{i\omega t} , \qquad p \sim a\omega e^{i\omega t} . \tag{6}$$

Hence, in this regime the proper physical amplitude and momentum are approximately constant (as well as the Hamiltonian (5));

(b) For $\omega^2 \ll \ddot{a}/a$ the comoving amplitude and momentum are *frozen*

$$x \sim B + C \int_0^t dt' \frac{1}{a^2(t')} , \qquad p \sim C . \tag{7}$$

Notice that in this freeze-out regime

$$\frac{d}{dt} \left(\frac{\lambda_{\text{phys}}}{H^{-1}}\right) > 0 , \tag{8}$$

where $\lambda_{\text{phys}} = 2\pi a/\omega$ is the proper physical wavelength of the oscillator and $H = \dot{a}/a$ is the expansion rate. The condition (10) implies that the background expansion is accelerating (as it occurs during an inflationary phase) and the proper wavelength characteristic of the oscillator is expanding faster than the maximum causal distance H^{-1}. Furthermore, during such freeze-out regime the value of the energy is asymptotically dominated by the term proportional to x^2 and is due to the *stretching* of the oscillator produced by the rapidly accelerated expansion. Let us now consider a cosmological evolution characterized by the following three different phases:

$$\omega^2 > \ddot{a}/a \qquad t < t_{\rm ex}\,, t > t_{\rm re}\,, \tag{9}$$

$$\omega^2 < \ddot{a}/a \qquad t_{\rm ex} < t < t_{\rm re}\,. \tag{10}$$

By smoothly joining the solution of the equations of motion (6) and (7) in the three different phases, it is straightforward to show that the final energy $E_{\rm fin}$ of the harmonic oscillator (which is asymptotically constant during the initial and final phases) is enhanced during the intermediate, accelerating phase by a factor proportional to $[a(t_{\rm re})/a(t_{\rm ex})]^2$

$$E_f \sim \left[\frac{a(t_{\rm re})}{a(t_{\rm ex})}\right]^2 E_{in}. \tag{11}$$

Note that for a classical oscillator initially at rest ($x_{\rm in} = p_{\rm in} = 0$) the initial energy is zero and no amplification takes place. Within the same cosmological evolution described by (11) and (12), let us consider instead a one-dimensional quantum mechanical oscillator initially in the ground state. The initial wave function is

$$\psi_{\rm in}(x) = \left(\frac{\alpha_{\rm in}}{\pi}\right)^{1/4} e^{-\alpha_{\rm in} x^2/2}\,, \tag{12}$$

where $\alpha_{\rm in} = a^2(t_{\rm ex}) m\omega^2/\hbar$. In the final stage of the cosmological evolution, the harmonic oscillator will be in a high occupation number state.[1] This can be shown by computing the expectation value of the Hamiltonian (defined in the final stage) with respect to the initial vacuum state defined by the wave function (14). In the final stage of the cosmological evolution, the Hamiltonian operator can be approximated as

$$\hat{H}_f = -\frac{\hbar^2}{2m_{\rm ex}}\frac{d^2}{dx^2} + \frac{m_{\rm ex}\omega^2 x^2}{2}\,, \tag{13}$$

where $m_{\rm ex} = a^2(t_{\rm ex})\, m$. Expressing the wavefunction (14) in terms of the eigenfunctions of (15), one obtains (see, e.g., [11])

$$\psi_{\rm in}(x) = \sum_{n=0}^{\infty} \beta_{2n} \psi_{\rm fin}^{(n)}(x) \tag{14}$$

[1] More precisely in a *squeezed* state.

$$\beta_{2n} = (\alpha_{\text{in}}\alpha_{\text{fin}})^{1/4} \sqrt{\frac{2(2n)!}{\alpha_{\text{in}} + \alpha_{\text{fin}}} \frac{1}{n!}} \left[\frac{\omega_{\text{fin}} - \omega_{\text{in}}}{2(\omega_{\text{fin}} + \omega_{\text{in}})} \right]^n , \tag{15}$$

$$\psi_{\text{fin}}^{(n)}(x) = N_n \, H_n(\sqrt{\alpha_{\text{fin}}} x) \, e^{-\alpha_{\text{fin}} x^2/2} , \tag{16}$$

where $\alpha_{\text{fin}} = a(t_{\text{re}})^2 m\omega^2/\hbar$, $\omega_{\text{in,fin}} = \omega/a(t_{\text{ex,re}})$, H_n are Hermite polynomials, and N_n is a normalization constant. Using (18) it is straightforward to show that the expectation value of the Hamiltonian (15) on the initial state described by the wavefunction (14) is given by

$$E_{\text{fin}} = \hbar\omega_{\text{fin}} \left[\frac{1}{2} + \left(\frac{\omega_{\text{fin}} - \omega_{\text{in}}}{2\sqrt{\omega_{\text{in}}\omega_{\text{fin}}}} \right)^2 \right] . \tag{17}$$

Hence, for a sufficiently long intermediate phase (for which $\omega_{\text{fin}} \ll \omega_{\text{in}}$) the harmonic oscillator final state is a semiclassical state characterized by a large number of created quanta

$$N_f \sim \frac{1}{4} \left(\frac{\omega_{\text{in}}}{\omega_{\text{fin}}} \right) . \tag{18}$$

This simple example shows how a period of accelerated expansion (i.e., an inflationary phase) can generate large scale inhomogeneities and anisotropies. Barring some technical issues (the conditions that guarantee the existence of a well-defined vacuum state at the onset of the inflationary phase), every quantum field in the vacuum state can be described as a collection of harmonic oscillators. The occurrence of a phase of accelerated expansion produces an amplification of the vacuum energy of those oscillators, *stretching* their corresponding wavelengths to scales larger than the horizon ($k < aH$). Such modes then eventually *re-enter* the horizon ($k > aH$) later, during the radiation/matter-dominated phases of the Universe evolution. Since those matter fields are gravitationally coupled (at least minimally) to the gravitational field, such enhancement of vacuum fluctuations is transferred to the background metric, therefore yielding to perturbations. Within the class of homogeneous and isotropic cosmological models, the perturbations can be classified in terms of their properties under coordinate transformations corresponding to the space–time isometries. (The latter are represented by the group $SO(3)$ of three-dimensional rotations, for a review, see [28].) In particular, scalar perturbations (described by fields invariant under rotations) are coupled to ordinary matter and radiation fields during the radiation/matter dominated phases, thus they are the seeds for large-scale structures and CMB anisotropies. Tensorial perturbations which are described by a field that transforms as a rank 2 tensor under rotations, produce a characteristic spectrum of stochastic gravitational radiation, whose energy and spectral content both depend on the background evolution and carries a unique imprint of the inflationary phase.

2.1 De Sitter Inflation

For the sake of simplicity, let us consider a simplified *two-stage* cosmological model where the epoch of accelerated expansion is described by a De Sitter phase smoothly connected to a standard radiation dominated phase. In conformal coordinates, the space–time metric is given by

$$ds^2 = a(\eta)^2(d\eta^2 - d\boldsymbol{x}^2)\,, \tag{19}$$

where η is the conformal time. The background evolution is specified by the following expressions for the scale factor:

$$a(\eta) = -\frac{1}{H_{\rm ds}\eta}\,, \qquad \eta < \eta_1 < 0 \tag{20}$$

$$a(\eta) = \frac{1}{H_{\rm ds}\eta_1^2}\,(\eta - 2\eta_1) \tag{21}$$

For each comoving wave number k, we add transverse and traceless fluctuations of the metric described by the following tensor:

$$h^{(A)}_{ab}(\boldsymbol{k},\eta) = e^{A}_{ab}(\boldsymbol{k})\tilde{h}^{(A)}_{k}(\eta)\mathrm{e}^{i\boldsymbol{k}\cdot\boldsymbol{x}}\,, \tag{22}$$

where $a, b = 1, 2, 3$, $A = +, \times$ labels the polarization state described by the tensor $e^{(A)}_{ab}$ and $\boldsymbol{k}$ is the comoving wave vector. The amplitude $h^{(A)}_k$ satisfies the following equation:

$$\left(\frac{d^2}{d\eta^2} + 2\mathcal{H}\frac{d}{d\eta} + |\boldsymbol{k}|^2\right) h^{(A)}_k = 0\,, \tag{23}$$

where $\mathcal{H} = (1/a)da/d\eta$. The general solution of (25) can be written in terms of elementary functions (in this case half-integer Hankel functions). In particular, imposing that for $\eta \to -\infty$ the solution of (25) corresponds to a vacuum state, one obtains [27]

$$h^{(A)}_k = \frac{a(\eta_1)}{a(\eta)}\left[1 + H_{\rm ds}\omega^{-1}\right] e^{-ik(\eta-\eta_1)}\,, \quad \eta < \eta_1\,, \tag{24}$$

$$h^{(A)}_k = \frac{a(\eta_1)}{a(\eta)}\left[\alpha_k\, e^{-ik(\eta-\eta_1)} + \beta_k\, e^{ik(\eta-\eta_1)}\right]\,, \quad \eta > \eta_1, \tag{25}$$

where $\omega = ck/a$ and α_k, β_k are the so-called Bogoliubov coefficients relative to the transition from a De Sitter to the radiation-dominated regime. In particular, for $\eta \to +\infty$ $|\beta_k|^2$ represents the number of gravitons created per unit cell of the phase space. The Bogoliubov coefficients can be computed by imposing the continuity of the amplitude and its time derivative on the space-like surface $\eta = \eta_1$ [27]:

$$\alpha_k = 1 + i\frac{\sqrt{H_0\, H_{\rm ds}}}{\omega} - \frac{H_0\, H_{\rm ds}}{2\omega^2}\,, \tag{26}$$

$$\beta_k = \frac{H_0\, H_{\rm ds}}{2\omega^2}\,. \tag{27}$$

The graviton energy density per unit cell of phase is therefore

$$d\rho_{\rm GW} = 2\hbar\omega \left(\frac{\omega^2 d\omega}{2\pi^2}\right) |\beta_k|^2 = \frac{\hbar\, H_0^2\, H_{\rm ds}^2}{4\pi^2}\frac{df}{f}\,, \tag{28}$$

where $f = \omega/2\pi$ is the physical frequency and H_0 is the present value of the Hubble constant. Using the definition (1) the spectrum turns out to be scale-invariant and its value reads

$$\Omega_{\rm GW} = \frac{16}{9}\left(\frac{M_{\rm infl}}{M_{\rm pl}}\right)^4\,, \tag{29}$$

where $M_{\rm pl} = G^{-1/2} = 1.22 \times 10^{19}\,{\rm GeV}$ is the Planck mass and $M_{\rm infl}$ is the inflationary scale defined by $H_{\rm ds}^2 = 8\pi M_{\rm infl}^4/3M_{\rm pl}^2$. This result cannot be directly compared with experimental sensitivities, since the presence of a matter dominated and a dark energy phase is not properly taken into account. However, for modes that at the time of radiation–matter equality have physical wavelengths smaller than the horizon (corresponding to frequencies today $f > (H_0/2\pi)(1+z_{\rm eq})^{1/2}$, $z_{\rm eq}$ being the redshift of matter–radiation equality), the frequency dependence is not affected by the presence of matter and dark energy-dominated eras. The corresponding spectrum is reduced by a factor $1/(1+z_{\rm eq})$ with respect to (31) and is given by

$$\Omega_{\rm GW} = \frac{16}{9}\left(\frac{M_{\rm infl}}{M_{\rm pl}}\right)^4 \left(\frac{\Omega_r}{1-\Omega_{\rm de}}\right)\,, \tag{30}$$

where Ω_r are the current fractions of radiation and dark energy densities in units of the critical energy density, respectively. Current WMAP data place an upper limit on the inflation scale around $M_{\rm inf} \sim 2 \times 10^{16}\,{\rm GeV}$ [29]. Since the total energy in radiation is approximately $h_0^2\Omega_r \sim 4.15 \times 10^{-5}$, assuming $\Omega_{\rm de} = 0.7$, one finds for the spectrum (32)

$$h_0^2\, \Omega_{\rm GW} < 1.7 \times 10^{-15}\,. \tag{31}$$

2.2 Slow-roll Inflation

In the previous section we have focused our attention to a simplified inflationary model where the phase of accelerated expansion is driven by a constant energy density. However, a more general class of inflationary models is characterized by a scalar field Φ slowly rolling in a potential $V(\Phi)$. For such models, the expansion rate is not constant during the accelerating period and this feature produces a small tilt in the spectrum. In particular, for frequencies $f > (H_0/2\pi)(1+z_{\rm eq})^{1/2}$ the spectrum has a power-law frequency dependence

$$\Omega_{\rm GW} \sim f^{n_T} . \tag{32}$$

For single-field models characterized by a slowly varying potential the spectral slope n_T is given by [30]

$$n_T = -\frac{M_{\rm pl}^2}{8\pi}\left(\frac{V_*'}{V_*}\right)^2 , \tag{33}$$

where V_* is the value of the inflationary potential when the scale associated to the present size of the horizon (corresponding to a frequency $f_0 = (1/2\pi)H_0$) crossed the horizon during the inflationary phase and V_*' is the first derivative of the inflaton potential at that point.Taking into account the frequency dependence of the spectral slope (35), for frequencies $f \gg (H_0/2\pi)(1+z_{\rm eq})^{1/2}$ the spectrum reads [30]

$$\Omega_{\rm GW} = \frac{5}{2}\left(\frac{M_*}{M_{\rm pl}}\right)^4\left(\frac{\Omega_r}{1-\Omega_{\rm de}}\right)\left(\frac{f}{f_0}\right)^{n_{\rm GW}} , \tag{34}$$

where $M_* = V_*^{1/4}$ and

$$n_{\rm GW} = n_T\left\{1 - \frac{1}{2}\left[(n_S - 1) - n_T\right]\log\frac{f}{f_0}\right\} , \tag{35}$$

where n_S is the spectral index for scalar perturbations. A detailed analysis [16] using solutions of inflationary flow equations shows that for single-field slow-roll inflationary models the maximum of the spectrum (36) compatible with WMAP data is $h_0^2\Omega_{\rm GW} \sim 5 \times 10^{-16}$ for frequencies $f \sim 100\,{\rm Hz}$ (see also [14, 17]).

3 Gravitational-wave Background in Pre-big-bang Inflation

In slow-roll inflation, the horizon and flatness problems are solved by postulating the presence of an epoch during which the energy–momentum tensor is dominated by the potential energy of a scalar field. This potential energy drives the phase of accelerated expansion, during which the field slowly rolls towards the minimum of the potential. In the 1990s, several attempts of building such cosmological setup in string theory encountered a number of problems [31][2]. Due to the presence of other fields, superstring theory at low energy

[2] For more recent successful attempts to build slow-roll inflationary models within string theory see, e.g., [32].

does not give Einstein general relativity—e.g., heterotic string theory in four dimensions is described by the action

$$\Gamma_{\text{eff}} = \frac{1}{2\lambda_s^2} \int d^4x \sqrt{|g|}\, e^{-\varphi} \left[\mathcal{R} + g^{\mu\nu}\, \partial_\mu \varphi\, \partial_\nu \varphi - \frac{1}{12}(dB)^2 - V(\varphi) \right], \quad (1)$$

where φ is the dilaton field, related to the string coupling by $g^2 = e^{\varphi}$; $dB = \partial_\mu B_{\nu\rho} + \partial_\nu B_{\rho\mu} + \partial_\rho B_{\mu\nu}$, where $B_{\mu\nu}$ is the two-form gauge field or antisymmetric field; $V(\phi)$ is a non-perturbative potential; and where λ_s is the string scale. In writing (38) we disregard for simplicity the internal dimensions, whose dynamics can be described in terms of moduli fields [22]. Henceforth, we limit the discussion to the homogeneous and isotropic case with $B = 0$ and $V = 0$ $[ds^2 = -dt^2 + a^2(t)\, d\boldsymbol{x}^2, \varphi = \varphi(t)]$.

In 1991 Veneziano [20] discovered that the solution of the low-energy string-effective action (38) satisfies the scale factor duality symmetry: $a(t) \to 1/a(t), \varphi(t) \to \varphi(t) - 6 \log a(t)$,[3] with $a(t) \sim t^{1/\sqrt{3}}$ and $\varphi(t) \sim -\log t$. Noticing this property, Veneziano [20] conceived the idea of implementing the inflationary phase at times before the *would-be* big-bang singularity, proposing the *pre-big-bang* scenario. Indeed, it can be easily shown that for $t < 0$, $\dot{a} > 0, \ddot{a} > 0$, thus the Universe undergoes a (super) inflationary phase. Two different but physically equivalent descriptions of the PBB phase exist: either the string-frame picture described by (38), where the Universe undergoes an accelerated expansion $(H > 0, \dot{H} > 0, \dot{\varphi} > 0)$, or the Einstein-frame picture, where the action (38) has the standard Hilbert–Einstein form and the evolution of the Universe is described by an accelerated contraction, or gravitational collapse $(H < 0, \dot{H} < 0, \dot{\varphi} > 0)$.

This new kind of inflation, which can be shown to solve the homogeneity and flatness conundra, is driven by the kinetic energy of the dilaton field and forces both the string coupling ($\dot{g} > 0$) and the spacetime curvature to grow toward the future (i.e., toward *the stringy* phase). As a consequence, at least in the homogeneous case, the inflationary stage lasts for ever ($t \to -\infty$) and the initial state of the Universe is nearly flat, cold, and decoupled: $g \ll 1$, $\mathcal{R}\lambda_s^2 \ll 1$.

The scale factor duality symmetry has constituted the basis of a class of inflationary models subsequently investigated by Gasperini and Veneziano [21], and it is also at the basis of the so-called ekpyrotic cosmological scenario [33, 34].

As far as metric perturbations are concerned, the most striking feature of the PBB spectra is a strong tilt toward high frequencies [36]. In [23], Veneziano and collaborators estimated the SGWB, obtaining

$$\Omega_{\text{GW}} \sim \frac{1}{z_{\text{eq}}} g_s^2 \left(\frac{f}{f_s}\right)^3 \left\{ \left[1 + \frac{1}{2} \log \frac{f_s}{f}\right] + \frac{1}{z_s^3} \left(\frac{g_1}{g_s}\right)^2 \right\}, \quad (f < f_s,) \quad (2)$$

[3] Here for convenience we fix the origin of time at $t = 0$.

$$\Omega_{\rm GW} \sim \frac{g_1^2}{z_{\rm eq}} \left[\left(\frac{f}{f_1} \right)^{6-2\beta} + \left(\frac{f}{f_1} \right)^{2\beta} \right], \qquad (f_s < f < f_1), \tag{3}$$

where g_s is the value of the string coupling at the onset of the high-curvature stringy phase, f_s is the frequency corresponding to the lowest scale exiting during the dilaton-driven phase, $f_1 \sim 10^{11}$ Hz is the ultraviolet cutoff, $g_1 = M_s/M_{\rm pl}$, z_s is the redshift of the stringy phase, $\beta = -\log(g_s/g_1)/\log z_s$, and $z_{\rm eq}$ is the redshift of matter–radiation equality.

3.1 Pre-big-bang Minimal Model

Here, we derive more in detail the SGWB following [35]. In the string frame, where strings follow geodesic trajectories, it is straightforward to show that the canonical variable $\Psi_{\mu\nu}$ associated to tensor perturbations is related to the metric by

$$g_{\mu\nu} = a^2(\eta_{\mu\nu} + h_{\mu\nu}) = a^2 \left(\eta_{\mu\nu} + \frac{g}{a} \Psi_{\mu\nu} \right) . \tag{4}$$

The Fourier modes of the two physical traceless and transverse polarization states satisfy the following wave equation:

$$\Psi_k'' + (k^2 - V)\Psi_k = 0 , \tag{5}$$

where prime denotes differentiation with respect to the conformal time and $V = (g/a)''/(g/a)$. In the following, we shall restrict our attention to a class of minimal PBB models characterized by an initial accelerated, dilaton-driven phase followed by a stringy phase (during which H and $d\varphi/dt$ are assumed to be approximately constant [37]) eventually evolving towards a standard radiation-dominated phase. During the dilaton-dominated regime ($-\infty < \eta < \eta_s < 0$) the scale factor and the dilaton field read

$$a(\eta) = -\frac{1}{H_s \eta_s} \left(\frac{\eta - (1-\alpha)\eta_s}{\alpha \eta_s} \right)^{-\alpha} , \tag{6}$$

$$\varphi(\eta) = \varphi_s - \gamma \log \frac{\eta - (1-\alpha)\eta_s}{\alpha \eta_s} , \tag{7}$$

where $\alpha = 1/(1+\sqrt{3})$, $\gamma = \sqrt{3}$. During the stringy phase ($\eta_s < \eta < \eta_1$) one expects that higher-order terms saturate the growth of the curvature [37]. Hence during this phase the scale factor and the dilaton can be parametrized as follows:

$$a(\eta) = -\frac{1}{H_s \eta} , \tag{8}$$

$$\varphi(\eta) = \varphi_s - 2\beta \ln \frac{\eta}{\eta_s} . \tag{9}$$

Finally, assuming that a non-perturbative dilaton potential sets in stabilizing the dilaton, the radiation phase ($\eta_1 < \eta < \eta_r$) is described by

$$a(\eta) = \frac{1}{H_s \eta_1^2} (\eta - 2\eta_1) \,. \tag{10}$$

For those three different phases the potential V reads

$$V(\eta) = \frac{1}{4}\left(4\nu^4 - 1\right)\left[\eta - (1-\alpha)\eta_s\right]^{-2}, \quad -\infty < \eta < \eta_s, \tag{11}$$

$$V(\eta) = \frac{1}{4}\left(4\mu^4 - 1\right)\eta^{-2}, \quad \eta_s < \eta < \eta_1, \tag{12}$$

$$V(\eta) = 0, \quad \eta_1 < \eta < \eta_r \,. \tag{13}$$

where, $2\nu = |2\alpha - \gamma + 1|$, $2\mu = |2\beta - 3|$. The exact solutions of (42) in the three phases are

$$\Psi_k = \sqrt{|\eta - (1-\alpha\eta_s)|}\, H_\nu^{(2)}(k|\eta - \alpha\eta_s|)\,, \quad -\infty < \eta < \eta_s, \tag{14}$$

$$\Psi_k = \sqrt{|\eta|}\left[A_+ H_\mu^{(2)}(k|\eta|) + A_- H_\mu^{(1)}(k|\eta|)\right], \quad \eta_s < \eta < \eta_1, \tag{15}$$

$$\Psi_k = i\sqrt{\frac{2}{\pi k}}\left[B_+ e^{-ik\eta} - B_- e^{ik\eta}\right], \quad \eta_1 < \eta < \eta_r, \tag{16}$$

where $H_{\mu,\nu}^{(1,2)}$ are Hankel's functions of the first and second kind. The Bogoliubov coefficients $A_\pm\,, B_\pm$ can be computed by requiring the continuity of the Fourier modes and its first derivative on the space-like surfaces $\eta = \eta_s$ and $\eta = \eta_1$. The result for the spectrum is

$$\Omega_{\rm GW}(f) = a(\mu)\,\frac{(2\pi f_s)^4}{H_0^2 M_{\rm pl}^2}\left(\frac{f_1}{f_s}\right)^{2\mu+1}\left(\frac{f}{f_s}\right)^{5-2\mu}\left|H_\nu^{(2)\prime}\left(\frac{\alpha f}{f_s}\right) J_\mu\left(\frac{f}{f_s}\right)\right.$$
$$\left. - H_\nu^{(2)}\left(\frac{\alpha f}{f_s}\right) J'_\mu\left(\frac{f}{f_s}\right) + \frac{(1-\alpha)}{2\alpha}\frac{f_s}{f}\, H_\nu^{(2)}\left(\frac{\alpha f}{f_s}\right) J_\mu\left(\frac{f}{f_s}\right)\right|^2 \tag{17}$$

where

$$a(\mu) = \frac{\alpha}{48}\, 2^{2\mu}\,(2\mu - 1)^2\,\Gamma^2(\mu)\,.$$

For the class of cosmological models under consideration $\nu = 0$; hence, using the identity $H_0^{(2)\prime}(z) = -H_1^{(2)}(z)$, the spectrum is given by

$$\Omega_{\rm GW}(f) = a(\mu)\,\frac{(2\pi f_s)^4}{H_0^2 M_{\rm pl}^2}\left(\frac{f_1}{f_s}\right)^{2\mu+1}\left(\frac{f}{f_s}\right)^{5-2\mu}\left|H_0^{(2)}\left(\frac{\alpha f}{f_s}\right) J'_\mu\left(\frac{f}{f_s}\right)\right.$$
$$\left. + H_1^{(2)}\left(\frac{\alpha f}{f_s}\right) J_\mu\left(\frac{f}{f_s}\right) - \frac{(1-\alpha)}{2\alpha}\frac{f_s}{f}\, H_0^{(2)}\left(\frac{\alpha f}{f_s}\right) J_\mu\left(\frac{f}{f_s}\right)\right|^2 . \tag{18}$$

Assuming that the curvature scale at the onset of the string scale is $H_s \sim 1/\lambda_s \sim g_{\rm gut} M_{\rm pl} \sim 0.015\, M_{\rm Pl}$ and that the cosmic time value at which the stringy phase ends is $t_1 \sim \lambda_s$ the peak frequency is $f_1 \sim 4.3 \times 10^{10}$ Hz. Hence the spectrum depends on two arbitrary parameters, f_s and β. (Note that (3) can be recovered with the following mapping: $z_s = f_1/f_s$ and $g_s/g_1 = (f_s/f_1)^\beta$, with β given by $2\mu = |2\beta - 3|$.)

From (55), one finds that the maximum value of the spectrum compatible with the BBN and CMB bounds is

$$h_0^2\, \Omega_{\rm GW}^{\rm max} \sim 3.0 \times 10^{-7}\,. \tag{19}$$

Such value is quite interesting, since it is about one order of magnitude below the sensitivity of first-generation LIGO interferometers, and well above the sensitivity of second-generation interferometers, such as advanced LIGO.

3.2 Pre-big-bang Non-minimal Models

The SGWB in the minimal PBB model was originally evaluated neglecting the higher-curvature corrections in the equation of tensorial fluctuations during the stringy phase. Gasperini [38] evaluated the higher-order equation for tensorial fluctuations and showed that these corrections modify the amplitude of the perturbation *only* by a factor of order 1.

In [39, 40] the authors examined the effect of radiation produced during reheating processes occurring below the string scale. Such processes may be needed in the PBB model to dilute relic particles produced during (or at the very end of) the PBB phase. The abundance of those particles could spoil the BBN predictions [41]. Depending on when and for how long the entropy is produced, it can change the shape and reduce the amplitude of the SGWB. If we assume that the reheating process occurs at the end of the stringy phase (i.e., all the entropy is produced at the end of the stringy phase), then the effect of the process is a simple scaling of the original spectrum by the factor $(1 - \delta s)^{4/3}$, where δs is the fraction of the present thermal entropy density that the reheating process produced.

Finally, as first noticed in [39], it is well possible that many more cosmological phases are present between the pre- and the post-big-bang cosmological phases (see, e.g., [39, 42]). If this is the case, the GW spectra during the high-curvature and/or strong coupling region will be characterized by several branches with increasing and decreasing slopes. Due to the dependence of the spectra on a larger number of parameters, it would be more difficult to constrain these non-minimal scenarios using GW detectors.

4 Accessibility of LIGO to Pre-big-bang Models

We shall now discuss the implications of recent and future analysis of LIGO on PBB models, notably on its parameter space.

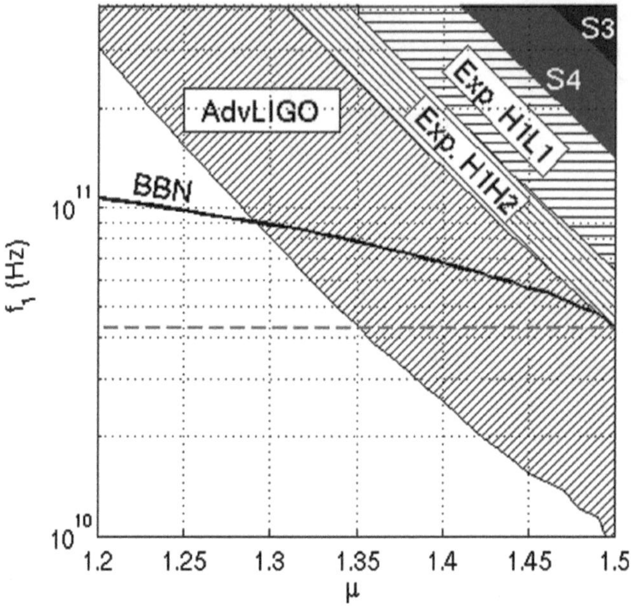

Fig. 1. The $f_1 - \mu$ plane with $f_s = 30$ Hz [18]. The shaded regions are excluded by the LIGO S3 upper limit (darker) and by the LIGO S4 limit. The hatched regions are accessible to future LIGO runs, assuming an observation time of 1 year: the predicted sensitivity for the H1L1 pair, assuming first design configuration (−); expected LIGO sensitivity for the H1H2 pair, assuming first design configuration (\); expected Advanced LIGO sensitivity for the H1H2 pair, assuming interferometer configuration optimized for the binary neutron star inspiral search (/). The solid black curve is the exclusion curve consistent with the nucleosynthesis limit (the excluded region is above the curve). The horizontal dashed line denotes the value of $f_1 = 4.3 \times 10^{10}$ Hz (courtesy of LIGO)

As already pointed out in the previous section, in minimal PBB models the spectrum (55) is characterized by the following parameters: (i) the dimensionless quantity $\mu = |2\beta - 3|$ which is positive definite and constrained to be $\mu \leq 3/2$, since for $\mu > 3/2$ the spectrum would be incompatible with the existing experimental bounds and (ii) the frequency parameter f_s which is defined so that $0 < f_s < f_1$. Since the spectrum (55) behaves as $(f/f_s)^3$ for $f < f_s$, a comparison with LIGO data is insensitive to values of f_s above the interferometer band. In particular, one finds that LIGO data can scan values $f_s < 30\,\text{Hz}$ [18]. Furthermore, in the high-frequency limit, the value of the spectrum is independent on f_s. The other parameter is the frequency f_1 (which defines the peak frequency of the spectrum)[4]

[4] In the more common version of the minimal PBB model [22, 23, 42], the frequency f_1 is obtained by imposing that the energy density becomes critical at the beginning of the radiation phase and that the photons we observe

$$f_1 \sim 4.3 \times 10^{10}\,\mathrm{Hz} \left(\frac{H_s}{0.15 M_{\rm pl}}\right) \left(\frac{t_1}{\lambda_s}\right)^{1/2} , \tag{1}$$

where H_s is the curvature scale at the onset of the intermediate stringy phase and t_1 is the value of the cosmic time at which such phase ends. Even assuming $H_s \sim \lambda_s^{-1}, t_1 \sim \lambda_s$ since the spectrum (55) has a strong dependence on f_1 ($\Omega_{\rm GW} \propto f_1^4$), an order of magnitude combined variation in t_1 , H_s can yield quite a large variation in the spectrum.

Based on a previous analysis carried out in [19], during the last scientific run, the LIGO scientific collaboration has scanned the three-dimensional parameter space μ , f_s , f_1 using 192 s—long intervals with 1/32 Hz resolution—assessing the accessibility of LIGO to each of the PBB parameters describing the spectrum (18). Furthermore, the design sensitivity of the initial and advanced LIGO configuration was taken into account. A summary of the results is shown in Fig. 1, where the $f_1 - \mu$ plane is considered (fixing $f_s = 30\,\mathrm{Hz}$). The results pertaining the third (S3) and fourth (S4) LIGO scientific runs provide a first, albeit quite restricted, scanning of the parameter space. The indirect BBN bound is still quite a strong constraint, but future and longer runs of the LIGO interferometers are expected to enlarge the available part of the parameter space, eventually overcoming the BBN bound.

5 Conclusions

The most interesting and robust feature of the relic background of gravitational radiation predicted by the PBB model is the positive spectral slope ($n_T = 3$) at low frequency, i.e., for modes that exit the horizon during the dilaton-driven (super) inflationary phase and re-enter during radiation-dominated era. Such attribute is a consequence of the Universe's equation of state during the (super) inflationary PBB phase and is shared by other non-conventional cosmological models. For example in quintessential inflationary model [43], where the standard radiation-dominated era is preceded by a phase characterized by a stiffer equation of state, the SGWB increases linearly with frequency. A blue primordial spectrum of GWs could be produced in a class of models with superluminal ($w = p/\rho < 1$) equation of state [44], as discussed in [45], where the inflaton field is characterized by a non-local Lagrangian. Furthermore, in other cosmological setups based upon superstring theory, as the the cyclic/ekpyrotic models [46], the GW spectrum increases as function of frequency, although its amplitude normalization makes it unobservable by ground- and space-based detectors.

---

today originated from the amplified vacuum fluctuations during the dilaton-driven inflationary phase. Within these assumptions (57) can be re-written as $f_1 \simeq g_1^{1/2}\,(H_s/(0.15\,M_{\rm Pl}))^{1/2}\,(H_0\,M_{\rm Pl})^{1/2}\,\Omega_\gamma^{1/4}$, where $\Omega_\gamma = 4\times 10^{-5}\,h_0^{-2}$ and g_1 is the string coupling at the end of the stringy phase.

The blue spectrum predicted by PBB models yields to a lack of tensorial contributions to the CMB temperature and polarization anisotropies. Therefore, the detection of a tensorial component of primordial origin in the CMB fluctuations would rule out the current version of the PBB model.

Finally, it is worth to notice that during the initial dilaton-driven inflationary phase characteristic of PBB models the initial (classical) tensor inhomogeneities are not de-amplified [47], as it occurs in slow-roll inflation. This result, though paradoxical, does not imply that the initial value of tensor inhomogeneities must be fine tuned to an unnaturally small value. Indeed, it can be shown [47] that the energy density of such tensor classical fluctuations is indeed de-amplified. However, in order to solve the homogeneity problem in those superstring-inspired models, more tighter constraints [47] than in slow-roll inflation models ought to be imposed.

All that said, future ground- and space-based detectors will be in the unique and privileged position of either detect the PBB SGWB or put relevant bounds on the parameter space of the string–cosmology scenario originally proposed by Gabriele Veneziano.

References

1. `http://www.ligo.org`
2. `http://www.virgo.pi.infn.it`
3. `http://www.geo600.uni-hannover.de`
4. `http://www.tama.mtk.nao.ac.jp`
5. `http://lisa.jpl.nasa.gov`
6. `http://science.hq.nasa.gov/universe/science/bang.html`
7. N. Seto, S. Kawamura, T. Nakamura: Phys. Rev. Lett. **87**, 221103 (2001)
8. E. Iacopini, E. Picasso, F. Pegoraro, L. A. Radicati: Phys. Lett. A **73**, 140 (1979); C.M. Caves: Phys. Lett. B **80**, 323 (1979)
9. A.M. Cruise: Class. Quant. Grav. **17**, 2525 (2000); A.M. Cruise, R. Ingley: Class. Quant. Grav. **22**, S497 (2004)
10. `http://www.skatelescope.org`
11. B. Allen, "The stochastic gravity-wave background: sources and detection," [gr-qc/9604033]
12. M. Maggiore Phys. Rep. **331**, 283 (2000)
13. L.P. Grishchuk: Sov. Phys. JEPT **40**, 409 (1975); A.A. Starobinski: JEPT Lett. **30**, 682 (1979); V.A. Rubakov, M. Sazhin, A. Veryaskin: Phys. Lett. B **115**, 189 (1982); R. Fabbri, M. Pollock: Phys. Lett. B **125**, 445 (1983); L.F. Abbott, D.D. Harari: Nucl. Phys. B **264**, 487 (1986); B. Allen: Phys. Rev. D **37**, 2078 (1988); V. Sahni: Phys. Rev. D **42**, 453 (1990)
14. L. A. Boyle, P. J. Steinhardt, N. Turok: Phys. Rev. Lett. **96**, 311101 (2006)
15. L. A. Boyle, P. J. Steinhardt: "Probing the early universe with inflationary gravitational waves," [astro-ph/0512014]
16. B. C. Friedman, A. Cooray, A. Melchiorri: "WMAP-normalized inflationary model predictions and the search for primordial gravitational waves with direct detection experiments," [astro-ph/0610220].

17. T.L. Smith, M. Kamionkowski, A. Cooray: Phys. Rev. D **73** 023504 (2006)
18. B. Abbott et al.: "Searching for a stochastic background of gravitational waves with LIGO," [astro-ph/0608606]
19. V. Mandic, A. Buonanno: Phys. Rev. D **73**, 063008 (2006)
20. G. Veneziano: Phys. Lett. B **265**, 287 (1991)
21. M. Gasperini, G. Veneziano: Astropart. Phys. **1**, 317 (1993); Mod. Phys. Lett. A **8**, 3701 (1993); Phys. Rev. D **50**, 251 (1994)
22. M. Gasperini, G. Veneziano: Phys. Rep. **373**, 1 (2003)
23. R. Brustein, M. Gasperini, M. Giovannini, G. Veneziano, Phys. Lett. B **361**, 45 (1995)
24. C.J. Copi, D.N. Schramm, M.S. Turner: Phys. Rev. D **55**, 3389 (1997).
25. T. Smith, E. Pierpaoli, M. Kamionkowski: Phys. Rev. Lett. **97**, 021301 (2006)
26. G. Veneziano: "String cosmology: concepts and consequences," [hep-th/9512091]
27. B. Allen: Phys. Rev. D **37**, 2078 (1988)
28. H. Kodama, M. Sasaki: Suppl. Prog. Theor. Phys. **78**, 1 (1984); V. F. Mukhanov, H. A. Feldman, R. H. Brandenberger: Phys. Rep. **215**, 203 (1992)
29. W. H. Kinney, E. W. Kolb, A. Melchiorri, A. Riotto: Phys. Rev. D **74**, 023502 (2006)
30. M. Turner: Phys. Rev. D **55**, 435 (1997)
31. B.A. Campbell, A.D. Linde, K.A. Olive: Nucl. Phys. B **335**, 146 (1991); R. Brustein, P.J. Steinhardt: Phys. Lett. B **302**, 196 (1993)
32. S. Kachru, R. Kallosh, A. Linde, J. Maldacena, L. McAllister, S.P. Trivedi: JHEP **0408**, 030 (2004)
33. J. Khoury, B.A. Ovrut, P.J. Steinhardt, N. Turok: Phys. Rev. D **64**, 123522 (2001)
34. J. Khoury, B.A. Ovrut, N. Seiberg, P.J. Steinhardt and N. Turok: Phys. Rev. D **65**, 086007 (2002)
35. A. Buonanno, M. Maggiore, C. Ungarelli: Phys. Rev. D **55**, 3330 (1997)
36. R. Brustein, M. Gasperini, M. Giovannini, V. F. Mukhanov, G. Veneziano, Phys. Rev. D **51**, 6744 (1995)
37. M. Gasperini, M. Maggiore, G. Veneziano: Nucl. Phys. B **494**, 315 (1996).
38. M. Gasperini: Phys. Rev. D **56**, 4815 (1997).
39. M. Gasperini, "Relic gravitons from the pre-big bang: what we know and what we do not know," [hep-th/9607146]
40. R. Brustein, M. Gasperini, G. Veneziano: Phys. Rev. D **55**, 3882 (1997)
41. A. Buonanno, M. Lemoine, K.A. Olive: Phys. Rev. D **62**, 083513 (2000)
42. A. Buonanno, K. Meissner, C. Ungarelli, G. Veneziano: JHEP **001**, 004 (1998)
43. P.J.E. Peebles, A. Vilenkin: Phys. Rev. D **59**, 063505 (1999); M. Giovannini: Phys. Rev. D **60**, 123511 (1999)
44. L. Grishchuk: "Relic gravitational waves and cosmology," [astro-ph/0504018]
45. M. Baldi, F. Finelli, S. Matarrese: Phys. Rev. D **72**, 083504 (2005)
46. L. A. Boyle, P. Steinhardt, N. Turok: Phys. Rev. D **69**, 127302 (2004)
47. A. Buonanno, T. Damour: Phys. Rev. D **50**, 3713 (2001)

Magnetic Fields, Strings and Cosmology

M. Giovannini

Centro "Enrico Fermi", Via Panisperna 89/A, 00184 Rome, Italy,
and Department of Physics, Theory Division, CERN, 1211 Geneva 23, Switzerland
massimo.giovannini@cern.ch

Abstract. The main motivations and challenges related with the physics of large-scale magnetic fields are briefly reviewed. The interplay between large-scale magnetic fields and scalar CMB anisotropies is addressed with specific attention on recent progresses.

1 Half a Century of Large-Scale Magnetic Fields

1.1 A Premise

The content of the present contribution is devoted to large-scale magnetic fields whose origin, evolution and implications constitute today a rather intriguing triple point in the phase diagram of physical theories. Indeed, sticking to the existing literature (and refraining from dramatic statements on the historical evolution of theoretical physics), it appears that the subject of large-scale magnetization thrives and prospers at the crossroad of astrophysics, cosmology and theoretical high-energy physics.

Following the kind invitation of Jnan Maharana and Maurizio Gasperini, I am delighted to contribute to this set of lectures whose guideline is dictated by the inspiring efforts of Gabriele Veneziano in understanding the fundamental forces of Nature. My voice joins the choir of gratitude proceeding from the whole physics community for the novel and intriguing results obtained by Gabriele through the various stages of his manifold activity. I finally ought to convey my personal thankfulness for the teachings, advices and generous clues received during the last 15 years.

1.2 Length Scales

The typical magnetic field strengths, in the Universe, range from few μG (micro-Gauss in the case of galaxies and clusters) to few Gauss (in the case of planets, like the earth or Jupiter) and up to 10^{12}G in neutron stars. Magnetic

M. Giovannini: *Magnetic Fields, Strings and Cosmology*, Lect. Notes Phys. **737**, 863–939 (2008)
DOI 10.1007/978-3-540-74233-6_26

fields are not only observed in planets and stars but also in the interstellar medium, in the intergalactic medium and, last but not least, in the intracluster medium.

Magnetic fields whose correlation length is larger than the astronomical unit ($1\,\mathrm{AU} = 1.49 \times 10^{13}$cm) will be named *large-scale magnetic fields*. In fact, magnetic fields with approximate correlation scale comparable with the earth–sun distance are not observed (on the contrary, both the magnetic field of the sun and the one of the earth have a clearly distinguishable localized structure). Moreover, in magnetohydrodynamics (MHD), the magnetic diffusivity scale (i.e. the scale below which magnetic fields are diffused because of the finite value of the conductivity) turns out to be, amusingly enough, of the order of the AU.

1.3 The Early History

In the 1940s large-scale magnetic field had no empirical evidence. For instance, there was no evidence of magnetic fields associated with the galaxy as a whole with a rough correlation scale of 30 kpc.[1] More specifically, the theoretical situation can be summarized as follows. The seminal contributions of Alfvén [1] convinced the community that magnetic fields can have a very large lifetime in a highly conducting plasma. Later on, in the 1970s, Alfvén will be awarded by the Nobel prize "for fundamental work and discoveries in magnetohydrodynamics with fruitful applications in different parts of plasma physics".

Using the discoveries of Alfvén, Fermi [2] postulated, in 1949, the existence of a large-scale magnetic field permeating the galaxy with approximate intensity of micro-Gauss and, hence, in equilibrium with the cosmic rays. [2]

Alfvén [3] did not react positively to the proposal of Fermi, insisting, in a somehow opposite perspective, that cosmic rays are in equilibrium with stars and disregarding completely the possibility of a galactic magnetic field. Today we do know that this may be the case for low-energy cosmic rays but certainly not for the most energetic ones around, and beyond, the knee in the cosmic ray spectrum.

At the historical level it is amusing to notice that the mentioned controversy can be fully understood from the issue 75 of *Physical Review* where it is

[1] Recall that $1\,\mathrm{kpc} = 3.085 \times 10^{21}$cm. Moreover, $1\mathrm{Mpc} = 10^3\,\mathrm{kpc}$. The present size of the Hubble radius is $H_0^{-1} = 1.2 \times 10^{28}\mathrm{cm} \equiv 4.1 \times 10^3\,\mathrm{Mpc}$ for $h = 0.73$.

[2] In this contribution magnetic fields will be expressed in Gauss. In the SI units $1\,\mathrm{T} = 10^4$G. For practical reasons, in cosmic ray physics and in cosmology it is also useful to express the magnetic field in GeV^2 (in units $\hbar = c = 1$). Recalling that the Bohr magneton is about $5.7 \times 10^{-11}\mathrm{MeV/T}$ the conversion factor will then be $1\,\mathrm{G} = 1.95 \times 10^{-20}\mathrm{GeV}^2$. The use of Gauss (G) instead of Tesla (T) is justified by the existing astrophysical literature where magnetic fields are typically expressed in Gauss.

possible to consult the article of Fermi [2], the article of Alfvén [3] and even a paper by Richtmyer and Teller [4] supporting the views and doubts of Alfvén.

In 1949 Hiltner [5] and, independently, Hall [6] observed polarization of starlight which was later on interpreted by Davis and Greenstein [7] as an effect of galactic magnetic field aligning the dust grains.

According to the presented chain of events it is legitimate to conclude that

- the discoveries of Alfvén were essential in the Fermi proposal who was pondering on the origin of cosmic rays in 1938 before leaving Italy[3] because of the infamous fascist legislation and
- the idea that cosmic rays are in equilibrium with the galactic magnetic fields (and hence that the galaxy possesses a magnetic field) was essential in the correct interpretation of the first, fragile, optical evidence of galactic magnetization.

The origin of the galactic magnetization, according to [2], had to be somehow primordial. It should be noticed, for sake of completeness, that the observations of Hiltner [5] and Hall [6] took place from November 1948 to January 1949. The paper of Fermi [2] was submitted in January 1949, but it contains no reference to the work of Hiltner and Hall. This indicates the Fermi was probably not aware of these optical measurements.

The idea that large-scale magnetization should somehow be the remnant of the initial conditions of the gravitational collapse of the protogalaxy idea was further pursued by Fermi in collaboration with Chandrasekar [8, 9] who tried, rather ambitiously, to connect the magnetic field of the galaxy to its angular momentum.

1.4 The Middle Ages

In the 1950s various observations on polarization of Crab nebula suggested that the Milky Way (MW) is not the only magnetized structure in the sky. The effective new twist in the observations of large-scale magnetic fields was the development (through the 1950s and 1960s) of radio-astronomical techniques. From these measurements, the first unambiguous evidence of radio-polarization from the Milky Way was obtained (see [10] and references therein for an account of these developments).

It was also soon realized that the radio-Zeeman effect (counterpart of the optical Zeeman splitting employed to determine the magnetic field of the sun) could offer accurate determination of (locally very strong) magnetic fields in the galaxy. The observation of Lyne and Smith [11] that pulsars could be used to determine the column density of electrons along the line of sight opened the possibility of using not only synchrotron emission as a diagnostic of the presence of a large-scale magnetic field, but also Faraday rotation. For

[3] The author is indebted to Prof. G. Cocconi who was so kind to share his personal recollections of the scientific discussions with E. Fermi.

a masterly written introduction to pulsar physics the reader may consult the book of Lyne and Smith [12].

In the 1970s all the basic experimental tools for the analysis of galactic and extragalactic magnetic fields were ready. Around this epoch also extensive reviews on the experimental endeavors started appearing and a very nice account could be found, for instance, in the review of Heiles [13].

It became gradually evident in the early 1980s that measurements of large-scale magnetic fields in the MW and in the external galaxies are two complementary aspects of the same problem. While MW studies can provide valuable information concerning the *local* structure of the galactic magnetic field, the observation of external galaxies provides the only viable tool for the reconstruction of the *global* features of the galactic magnetic fields.

Since the early 1970s, some relevant attention has been paid not only to the magnetic fields of the galaxies but also to the magnetic fields of the *clusters*. A cluster is a gravitationally bound system of galaxies. The *local group* (i.e. *our* cluster containing the MW, Andromeda together with other fifty galaxies) is an *irregular* cluster in the sense that it contains fewer galaxies than typical clusters in the Universe. Other clusters (like Coma, Virgo) are more typical and are then called *regular* or Abell clusters. As an order of magnitude estimate, Abell clusters can contain 10^3 galaxies.

1.5 New Twists

In the 1990s magnetic fields have been measured in single Abell clusters but around the turn of the century these estimates became more reliable, thanks to improved experimental techniques. In order to estimate magnetic fields in clusters, an independent knowledge of the electron density along the line of sight is needed. Recently, Faraday rotation measurements obtained by radio telescopes (like VLA[4]) have been combined with independent measurements of the electron density in the intracluster medium. This was made possible by the maps of the x-ray sky obtained with satellites measurements (in particular ROSAT[5]). This improvement in the experimental capabilities seems to have partially settled the issue confirming the measurements of the early 1990s and implying that also clusters are endowed with a magnetic field of micro-Gauss strength which is *not associated with individual galaxies* [14, 15].

While entering the new millennium the capabilities of the observers are really confronted with a new challenge: the possibility that also superclusters are endowed with their own magnetic field. Superclusters are (loosely) gravitationally bound systems of clusters. An example is the local supercluster formed by the local group and by the VIRGO cluster. Recently a large

[4] The Very Large Array Telescope consists of 27 parabolic antennas spread over a surface of 20 km^2 in Socorro (New Mexico).

[5] The Roengten SATellite (flying from June 1991 to February 1999) provided maps of the x-ray sky in the range 0.1–2.5 keV. A catalog of x-ray bright Abell clusters was compiled.

new sample of Faraday rotation measures of polarized extragalactic sources has been compared with galaxy counts in Hercules and Perseus-Pisces (two nearby superclusters) [16]. First attempts to detect magnetic fields associated with superclusters have been reported [17]. A cautious and conservative approach suggests that these fragile evidences must be corroborated with more conclusive observations (especially in light of the, sometimes dubious, independent determination of the electron density[6]). However, it is not excluded that as the 1990s gave us a firmer evidence of cluster magnetism, the new millennium may give us more solid understanding of supercluster magnetism. In the present historical introduction various experimental techniques have been swiftly mentioned. A more extensive introductory description of these techniques can be found in [18].

1.6 Hopes for the Future

The hope for the near future is connected with the possibility of a next generation radio-telescope. Along this line the SKA (square kilometer array) has been proposed [15] (see also [19]). While the technical features of the instrument cannot be thoroughly discussed in the present contribution, it suffices to notice that the collecting area of the instrument, as the name suggest, will be of $10^6\,\mathrm{m}^2$. The specifications for the SKA require an angular resolution of 0.1 arcsec at 1.4 GHz, a frequency capability of 0.1–25 GHz and a field of view of at least $1\,\mathrm{deg}^2$ at 1.4 GHz [19]. The number of independent beams is expected to be larger than 4 and the number of instantaneous pencil beams will be roughly 100 with a maximum primary beam separation of about 100 deg at low frequencies (becoming 1 deg at high frequencies, i.e. of the order of 1 GHz). These specifications will probably allow full-sky surveys of Faraday rotation.

The frequency range of SKA is rather suggestive if we compare it with the one of the Planck experiment [20]. Planck will operate in nine frequency channels from 30 to, approximately, 900 GHz. While the three low-frequency channels (from 30 to 70 GHz) are not sensitive to polarization, the six high-frequency channels (between 100 and 857 GHz) will be definitely sensitive to CMB polarization. Now, it should be appreciated that the Faraday rotation signal *decreases* with the frequency ν as ν^{-2}. Therefore, for lower frequencies the Faraday rotation signal will be larger than in the six high-frequency channels. Consequently, it is legitimate to hope for a fruitful interplay between the next generation of SKA-like radio-telescopes and CMB satellites. Indeed, as suggested above, the upper branch of the frequency capability of SKA almost

[6] In [21] it was cleverly argued that information on the plasma densities from direct observations can be gleaned from detailed multifrequency observations of few giant radio-galaxies (GRG) having dimensions up to 4 Mpc. The estimates based on this observation suggest column densities of electrons between 10^{-6} and $10^{-5}\ \mathrm{cm}^{-3}$.

overlaps with the lower frequency of Planck so that possible effects of large-scale magnetic fields on CMB polarization could be, with some luck, addressed with the combined action of both instruments. In fact, the same mechanism leading to the Faraday rotation in the radio leads to a Faraday rotation of the CMB *provided* the CMB is linearly polarized. These considerations suggest, as emphasized in a recent topical review, that CMB anisotropies are germane to several aspects of large-scale magnetization [18]. The considerations reported so far suggest that during the next decade the destiny of radio-astronomy and CMB physics will probably be linked together and not only for reasons of convenience.

1.7 Few Burning Questions

In this general and panoramic view of the history of the subject we started from the relatively old controversy opposing E. Fermi to H. Alfvén with the still uncertain but foreseeable future developments. While the nature of the future developments is inextricably connected with the advent of new instrumental capabilities, it is legitimate to remark that, in more than 50 years, magnetic fields have been detected over scales that are progressively larger. From the historical development of the subject a series of questions arises naturally:

- What is the origin of large-scale magnetic fields?
- Are magnetic fields primordial as assumed by Fermi more than 50 years ago?
- Even assuming that large-scale magnetic fields are primordial, is there a theory for their generation?
- Is there a way to understand if large-scale magnetic fields are really primordial?

In what follows we will not give definite answers to these important questions, but we shall be content of outlining possible avenues of new developments.

The plan of the present lecture will be the following. In Sect. 2 the main theoretical problems connected with the origin of large-scale magnetic fields will be discussed. In Sect. 3 the attention will be focused on the problem of large-scale magnetic field generation in the framework of string cosmological model, a subject where the pre-big-bang model, in its various incarnations, plays a crucial role. But, finally, large-scale magnetic fields are really primordial? Were they really present prior to matter–radiation equality? A modest approach to these important questions suggests to study the physics of magnetized CMB anisotropies which will be introduced, in its essential lines, in Sect. 4. The concluding remarks are collected in Sect.5.

2 Magnetogenesis

While in the previous section the approach has been purely historical, the experimental analysis of large-scale magnetic fields prompts a collection of interesting theoretical problems. They can be summarized by the following chain of evidences (see also [18]):

- In spiral galaxies magnetic fields follow the orientation of the spiral arms, where matter is clustered because of differential rotation. While there may be an asymmetry in the intensities of the magnetic field in the northern and southern emisphere (like it happens in the case of the Milky Way), the typical strength is in the range of the micro-Gauss.
- Locally magnetic fields may even be in the milli-Gauss range and, in this case, they may be detected through Zeeman splitting techniques.
- In spiral galaxies the magnetic field is predominantly toroidal with a poloidal component present around the nucleus of the galaxy and extending for, roughly, 100 pc.
- The correlation scale of the magnetic field in spirals is of the order of 30 kpc.
- In elliptical galaxies magnetic fields have been measured at the micro-Gauss level, but the correlation scale is shorter than in the case of spirals: this is due to the different evolutionary history of elliptical galaxies and to their lack of differential rotation.
- Abell clusters of galaxies exhibit magnetic fields present in the so-called intracluster medium: these fields, always at the micro-Gauss level, are not associated with individual galaxies;
- Superclusters *might* also be magnetized even if, at the moment, conclusions are premature, as partially explained in Sect. 1 (see also [17] and [18]).

The statements collected above rest on various detection techniques ranging from Faraday rotation, to synchrotron emission, to Zeeman splitting of clouds of molecules with an unpaired electron spin. The experimental evidence swiftly summarized above seems to suggest that different and distant objects have magnetic fields of comparable strength. The second suggestion seems also to be that the strength of the magnetic fields is, in the first (simplistic) approximation, independent on the physical scale.

These empirical coincidences remind a bit of one of the motivations of the standard hot big-bang model, namely, the observation that the light elements are equally abundant in rather different parts of our Universe. The approximate equality of the abundances implies that, unlike the heavier elements, the light elements have primordial origin. The four light isotopes D, ^{3}He, ^{4}He and ^{7}Li are mainly produced at a specific stage of the hot big bang model named nucleosynthesis occurring below the typical temperature of 0.8 MeV when neutrinos decouple from the plasma and the neutron abundance evolves via free neutron decay [23]. The abundances calculated in the simplest big-bang nucleosythesis model agree fairly well with the astronomical observations.

In similar terms it is plausible to argue that large-scale magnetic fields have comparable strengths at large scales because the initial conditions for their evolutions were the same, for instance at the time of the gravitational collapse of the protogalaxy. The way the initial conditions for the evolution of large-scale magnetic fields are set is generically named *magnetogenesis* [18].

There is another comparison which might be useful. Back in the 1970s the so-called Harrison–Zeldovich spectrum was postulated. Later, with the developments of inflationary cosmology the origin of a flat spectrum of curvature and density profiles has been justified on the basis of a period of quasi-de Sitter expansion named *inflation*. It is plausible that in some inflationary models not only the fluctuations of the geometry are amplified but also the fluctuations of the gauge fields. This happens if, for instance, gauge couplings are effectively dynamical. As the Harrison–Zeldovich spectrum can be used as initial condition for the subsequent Newtonian evolution, the primordial spectrum of the gauge fields can be used as initial condition for the subsequent MHD evolution which may lead, eventually, to the observed large-scale magnetic fields. The plan of the present section is the following. In Sect. 2.1 some general ideas of plasma physics will be summarized with particular attention to those tools that will be more relevant for the purposes of this lecture. In Sect. 2.2 the concept of dynamo amplification will be introduced in a simplified perspective. In Sect. 2.3 it will be argued that the dynamo amplification, in one of its potential incarnations, necessitates some *initial conditions* or as we say in the jargon, some *seed field*. In Sect. 2.4 a panoramic view of astrophysical seeds will be presented with the aim of stressing the common aspects of, sometimes diverse, physical mechanisms. Sects. 2.5 and 2.6 the two basic approaches to cosmological magnetogenesis will be illustrated. In the first case (see Sect. 2.5) magnetic fields are produced inside the Hubble radius at a given stage in the life of the Universe. In the second case (see Sect. 2.6) vacuum fluctuations of the hypercharge field are amplified during an inflationary stage of expansion. Section 2.7 deals with the major problem of inflationary magnetogenesis, namely, conformal (Weyl) invariance whose breaking will be one of the themes of string cosmological mechanisms for the generation of large-scale magnetic fields.

2.1 Magnetized Plasmas

Large-scale magnetic fields evolve in a plasma, i.e. a system often illustrated as the *fourth state of matter*. As we can walk in the phase diagram of a given chemical element by going from the solid to the liquid and to the gaseous state with a series of diverse phase transitions, a plasma can be obtained by ionizing a gas. A typical example of weakly coupled plasma is therefore an ionized gas. Examples of strongly coupled plasmas can be found also in solid-state physics. An essential physical scale that has to be introduced in the description of plasma properties is the so-called Debye length that will be discussed in the following paragraph.

Different descriptions of a plasma exist and they range from effective fluid models of charged particles [24, 25, 26, 27] to kinetic approaches like the ones pioneered by Vlasov [28] and Landau [29]. From a physical point of view, a plasma is a system of charged particles which is globally neutral for typical lengthscales larger than the Debye length $\lambda_{\rm D}$:

$$\lambda_{\rm D} = \sqrt{\frac{T_0}{8\pi n_0 e^2}}, \tag{1}$$

where T_0 is the kinetic temperature and n_0 the mean charge density of the electron–ion system, i.e. $n_{\rm e} \simeq n_{\rm i} = n_0$. For a test particle the Coulomb potential will then have the usual Coulomb form, but it will be suppressed, at large distances by a Yukawa term, i.e. $e^{-r/\lambda_{\rm D}}$. In the interstellar medium there are three kinds of regions which are conventionally defined:

- H_2 regions, where the hydrogen is predominantly in molecular form (also denoted by HII);
- H^0 regions (where hydrogen is in atomic form);
- and H^+ regions, where hydrogen is ionized (also denoted by HI).

In the H^+ regions the typical temperature T_0 is of the order of 10–20 eV while for n_0 let us take, for instance, $n_0 \sim 3 \times 10^{-2}{\rm cm}^{-3}$. Then $\lambda_{\rm D} \sim 30\,{\rm km}$.

For $r \gg \lambda_{\rm D}$ the Coulomb potential is screened by the global effect of the other particles in the plasma. Suppose now that particles exchange momentum through two-body interactions. Their cross section will be of the order of $\alpha_{\rm em}^2/T_0^2$ and the mean free path will be $\ell_{\rm mfp} \sim T_0^2/(\alpha_{\rm em}^2 n_0)$, i.e. recalling (1) $\lambda_{\rm D} \ll \ell_{\rm mfp}$. This means that the plasma is a weakly collisional system which is, in general, not in local thermodynamical equilibrium and this is the reason why we introduced T_0 as the kinetic (rather than thermodynamic) temperature.

The last observation can be made even more explicit by defining another important scale, namely, the plasma frequency which, in the system under discussion, is given by

$$\omega_{\rm pe} = \sqrt{\frac{4\pi n_0 e^2}{m_{\rm e}}} \simeq 2\left(\frac{n_0}{10^3\ {\rm cm}^{-3}}\right)^{1/2}\ {\rm MHz}, \tag{2}$$

where $m_{\rm e}$ is the electron mass. Notice that, in the interstellar medium (i.e. for $n_0 \simeq 10^{-2}\ {\rm cm}^{-3}$) (2) gives a plasma frequency in the giga hertz range. This observation is important, for instance, in the treatment of Faraday rotation since the plasma frequency is typically much larger than the Larmor frequency, i.e.

$$\omega_{\rm Be} = \frac{eB_0}{m_{\rm e}} \simeq 18.08\left(\frac{B_0}{10^{-3}\ {\rm G}}\right)\ {\rm kHz}, \tag{3}$$

implying, for $B_0 \simeq \mu{\rm G}$, $\omega_{\rm Be} \simeq 20\,{\rm Hz}$. The same hierarchy holds also when the (free) electron density is much larger than in the interstellar medium, and, for instance, at the last scattering between electrons and photons for a redshift $z_{\rm dec} \simeq 1100$ (see Sect. 4).

The plasma frequency is the oscillation frequency of the electrons when they are displaced from their equilibrium configuration in a background of approximately fixed ions. Recalling that $v_{\rm ther} \simeq \sqrt{T_0/m_{\rm e}}$ is the thermal velocity of the charge carriers, the collision frequency $\omega_{\rm c} \simeq v_{\rm ther}/\ell_{\rm mfp}$ is always much smaller than $\omega_{\rm pe} \simeq v_{\rm ther}/\lambda_{\rm D}$. Thus, in the idealized system described so far, the following hierarchy of scales holds

$$\lambda_{\rm D} \ll \ell_{\rm mfp}, \qquad \omega_{\rm c} \ll \omega_{\rm pe}, \tag{4}$$

which means that before doing one collision the system undergoes many oscillations, or, in other words, that the mean free path is not the shortest scale in the problem. Usually one defines also the *plasma parameter* $\mathcal{N} = n_0^{-1}\lambda_{\rm D}^{-3}$, i.e. the number of particles in the Debye sphere. In the approximation of weakly coupled plasma, $\mathcal{N} \ll 1$ which also imply that the mean kinetic energy of the particles is larger than the mean inter-particle potential.

The spectrum of plasma excitations is a rather vast subject and it will not strictly necessary for the following considerations (for further details see [24, 25, 26]). It is sufficient to remark that we can envisage, broadly speaking, two regimes that are physically different:

- typical length-scales much *larger* than $\lambda_{\rm D}$ and typical frequencies much *smaller* than $\omega_{\rm pe}$;
- typical length-scales smaller (or comparable) with $\lambda_{\rm D}$ and typical frequencies much *larger* than $\omega_{\rm pe}$.

In the first situation reported above it can be shown that a single-fluid description suffices. The single-fluid description is justified, in particular, for the analysis of the dynamo instability which occurs for dynamical times of the order of the age of the galaxy and length-scales larger than the kilo parsec. In the opposite regime, i.e. $\omega \geq \omega_{\rm pe}$ and $L \geq \lambda_{\rm D}$ the single-fluid approach breaks down and a multi-fluid description is mandatory. This is, for instance, the branch of the spectrum of plasma excitation where the displacement current (and the related electromagnetic propagation) cannot be neglected. A more reliable description is provided, in this regime, by the Vlasov–Landau (i.e. kinetic) approach [28, 29] (see also [25]).

Consider, therefore, a two-fluid system of electrons and protons. This system will be described by the continuity equations of the density of particles, i.e.

$$\frac{\partial n_{\rm e}}{\partial t} + \boldsymbol{\nabla}\cdot(n_{\rm e}\boldsymbol{v}_{\rm e}) = 0, \qquad \frac{\partial n_{\rm p}}{\partial t} + \boldsymbol{\nabla}\cdot(n_{\rm p}\boldsymbol{v}_{\rm p}) = 0, \tag{5}$$

and by the momentum conservation equations

$$m_{\rm e}n_{\rm e}\left[\frac{\partial}{\partial t} + \boldsymbol{v}_{\rm e}\cdot\boldsymbol{\nabla}\right]\boldsymbol{v}_{\rm e} = -en_{\rm e}\left[\boldsymbol{E} + \boldsymbol{v}_{\rm e}\times\boldsymbol{B}\right] - \boldsymbol{\nabla}p_{\rm e} - \mathcal{C}_{\rm ep}, \tag{6}$$

$$m_{\rm p}n_{\rm p}\left[\frac{\partial}{\partial t} + \boldsymbol{v}_{\rm p}\cdot\boldsymbol{\nabla}\right]\boldsymbol{v}_{\rm p} = en_{\rm p}\left[\boldsymbol{E} + \boldsymbol{v}_{\rm p}\times\boldsymbol{B}\right] - \boldsymbol{\nabla}p_{\rm p} - \mathcal{C}_{\rm pe}. \tag{7}$$

Equations (5), (6) and (7) must be supplemented by Maxwell equations reading, in this case

$$\nabla \cdot \boldsymbol{E} = 4\pi e(n_{\mathrm{p}} - n_{\mathrm{e}}), \tag{8}$$
$$\nabla \cdot \boldsymbol{B} = 0, \tag{9}$$
$$\nabla \times \boldsymbol{E} + \frac{\partial \boldsymbol{B}}{\partial t} = 0, \tag{10}$$
$$\nabla \times \boldsymbol{B} = \frac{\partial \boldsymbol{E}}{\partial t} + 4\pi e(n_{\mathrm{p}}\boldsymbol{v}_{\mathrm{p}} - n_{\mathrm{e}}\boldsymbol{v}_{\mathrm{e}}). \tag{11}$$

The two-fluid system of equations is rather useful to discuss various phenomena like the propagation of electromagnetic excitations at finite charge density both in the presence and in the absence of a background magnetic field [24, 25, 26]. The previous observation implies that a two-fluid treatment is mandatory for the description of Faraday rotation of the cosmic microwave background (CMB) polarization. This subject will not be specifically discussed in the present lecture (see, for further details, [30] and references therein).

Instead of treating the two fluids as separated, the plasma may be considered as a single fluid defined by an appropriate set of *global* variables:

$$\boldsymbol{J} = e(n_{\mathrm{p}}\boldsymbol{v}_{\mathrm{p}} - n_{\mathrm{e}}\boldsymbol{v}_{\mathrm{e}}), \tag{12}$$
$$\rho_{\mathrm{q}} = e(n_{\mathrm{p}} - n_{\mathrm{e}}), \tag{13}$$
$$\rho_{\mathrm{m}} = (m_{\mathrm{e}}n_{\mathrm{e}} + m_{\mathrm{p}}n_{\mathrm{p}}), \tag{14}$$
$$\boldsymbol{v} = \frac{m_{\mathrm{e}}\, n_{\mathrm{e}}\boldsymbol{v}_{\mathrm{e}} + n_{\mathrm{p}}m_{\mathrm{p}}v_{\mathrm{p}}}{m_{\mathrm{e}}n_{\mathrm{e}} + m_{\mathrm{p}}n_{\mathrm{p}}}, \tag{15}$$

where $\boldsymbol{J}$ is the global current and ρ_{q} is the global charge density, ρ_{m} is the total mass density and $\boldsymbol{v}$ is the so-called bulk velocity of the plasma. From the definition of the bulk velocity it is clear that $\boldsymbol{v}$ is the center-of-mass velocity of the electron–ion system. The interesting case is the one where the plasma is globally neutral, i.e. $n_{\mathrm{e}} \simeq n_{\mathrm{p}} = n_0$, implying, from Maxwell and continuity equations the following equations

$$\nabla \cdot \boldsymbol{E} = 0, \qquad \nabla \cdot \boldsymbol{J} = 0, \qquad \nabla \cdot \boldsymbol{B} = 0. \tag{16}$$

The equations reported in (16) are the first characterization of MHD equations, i.e. a system where the total current as well as the electric and magnetic fields are all solenoidal. The remaining equations allow to obtain the relevant set of conditions describing the long-wavelength modes of the magnetic field, i.e.

$$\nabla \times \boldsymbol{B} = 4\pi \boldsymbol{J}, \tag{17}$$
$$\nabla \times \boldsymbol{E} = -\frac{\partial \boldsymbol{B}}{\partial t}. \tag{18}$$

In (17), the contribution of the displacement current has been neglected for consistency with the solenoidal nature of the total current (16). Two other relevant equations can be obtained by summing and subtracting the momentum conservation equations, i.e. (6) and (7). The result of this procedure is

$$\rho_{\mathrm{m}}\left[\frac{\partial \boldsymbol{v}}{\partial t} + \boldsymbol{v}\cdot\boldsymbol{\nabla}\boldsymbol{v}\right] = \boldsymbol{J}\times\boldsymbol{B} - \boldsymbol{\nabla}P \tag{19}$$

$$\boldsymbol{E} + \boldsymbol{v}\times\boldsymbol{B} = \frac{\boldsymbol{J}}{\sigma} + \frac{1}{e n_{\mathrm{q}}}(\boldsymbol{J}\times\boldsymbol{B} - \boldsymbol{\nabla}p_{\mathrm{e}}), \tag{20}$$

where $n_{\mathrm{q}} \simeq n_0 \simeq n_{\mathrm{e}}$ and $P = p_{\mathrm{e}} + p_{\mathrm{p}}$. Equation (19) is derived from the sum of (6) and (7) and in (19) $\boldsymbol{J}\times\boldsymbol{B}$ is the Lorentz force term which is quadratic in the magnetic field. In fact using (17)

$$\boldsymbol{J}\times\boldsymbol{B} = \frac{1}{4\pi}(\boldsymbol{\nabla}\times\boldsymbol{B})\times\boldsymbol{B}. \tag{21}$$

Note that to derive (20) the limit $m_{\mathrm{e}}/m_{\mathrm{p}} \to 0$ must be taken, at some point. There are some caveats related to this procedure since viscous and collisional effects may be relevant [25]. Equation (20) is sometimes called one-fluid generalized Ohm law. In (20) the term $\boldsymbol{J}\times\boldsymbol{B}$ is nothing but the *Hall current* and $\boldsymbol{\nabla}p_{\mathrm{e}}$ is often called thermoelectric term. Finally, the term $\boldsymbol{J}/\sigma$ is the resistivity term and σ is the conductivity of the one-fluid description. In (20) the pressure has been taken to be isotropic. Neglecting the Hall and thermoelectric terms (that may play, however, a role in the Biermann battery mechanism for magnetic field generation), the Ohm law takes the form

$$\boldsymbol{J} = \sigma(\boldsymbol{E} + \boldsymbol{v}\times\boldsymbol{B}). \tag{22}$$

Using (22) together with (17) it is easy to show that the Ohmic electric field is given by

$$\boldsymbol{E} = \frac{\boldsymbol{\nabla}\times\boldsymbol{B}}{4\pi\sigma} - \boldsymbol{v}\times\boldsymbol{B}. \tag{23}$$

Substituting then (23) into (18) and exploiting known vector identities, we can get the canonical form of the magnetic diffusivity equation

$$\frac{\partial \boldsymbol{B}}{\partial t} = \boldsymbol{\nabla}\times(\boldsymbol{v}\times\boldsymbol{B}) + \frac{1}{4\pi\sigma}\nabla^2\boldsymbol{B}, \tag{24}$$

which is the equation to be used to discuss the general features of the dynamo instability.

MHD can be studied into two different (but complementary) limits

- the ideal (or superconducting) limit where the conductivity is set to infinity (i.e. the $\sigma \to \infty$ limit) and
- the real (or resistive) limit where the conductivity is finite.

The plasma description following from MHD can also be phrased in terms of the conservation of two interesting quantities, i.e. the magnetic flux and the magnetic helicity [27, 31]:

$$\frac{d}{dt}\left(\int_{\Sigma} \boldsymbol{B} \cdot d\boldsymbol{\Sigma}\right) = -\frac{1}{4\pi\sigma} \int \boldsymbol{\nabla} \times \boldsymbol{\nabla} \times \boldsymbol{B} \cdot d\boldsymbol{\Sigma}, \tag{25}$$

$$\frac{d}{dt}\left(\int_{V} d^3x \boldsymbol{A} \cdot \boldsymbol{B}\right) = -\frac{1}{4\pi\sigma} \int_{V} d^3x \boldsymbol{B} \cdot \boldsymbol{\nabla} \times \boldsymbol{B}. \tag{26}$$

In (25), Σ is an arbitrary closed surface that moves with the plasma. In the ideal MHD limit the magnetic flux is exactly conserved and the flux is sometimes said to be frozen into the plasma element. In the same limit also the magnetic helicity is conserved. In the resistive limit the magnetic flux and helicity are dissipated with a rate proportional to $1/\sigma$ which is small provided the conductivity is sufficiently high. The term appearing at the right-hand side off (26) is called magnetic gyrotropy.

The conservation of the magnetic helicity is a statement on the conservation of the *topological* properties of the magnetic flux lines. If the magnetic field is completely stochastic, the magnetic flux lines will be closed loops evolving independently in the plasma and the helicity will vanish. There could be, however, more complicated topological situations where a single magnetic loop is twisted (like some kind of Möbius stripe) or the case where the magnetic loops are connected like the rings of a chain. In both cases the magnetic helicity will not be zero since it measures, essentially, the number of links and twists in the magnetic flux lines. The conservation of the magnetic flux and of the magnetic helicity is a consequence of the fact that, in ideal MHD, the Ohmic electric field is always orthogonal both to the bulk velocity field and to the magnetic field. In the resistive MHD approximation this is no longer true [27].

2.2 Dynamos

The dynamo theory has been developed starting from the early 1950s through the 1980s and various extensive presentations exist in the literature [32, 33, 34]. Generally speaking, a *dynamo* is a process where the kinetic energy of the plasma is transferred to magnetic energy. There are different sorts of dynamos. Some of the dynamos that are currently addressed in the existing literature are large-scale dynamos, small-scale dynamos, nonlinear dynamos, α-dynamos etc.

It would be difficult, in the present lecture, even to review such a vast literature and, therefore, it is more appropriate to refer to some review articles where the modern developments in dynamo theory and in mean field electrodynamics are reported [35, 36]. As a qualitative example of the dynamo action it is practical do discuss the magnetic diffusivity equation obtained, from general considerations, in (24).

Equation (24) simply stipulates that the first-time derivative of the magnetic fields intensity results from the balance of two (physically different) contributions. The first term at the right-hand side of (24) is the the *dynamo* term and it contains the bulk velocity of the plasma $\boldsymbol{v}$. If this term dominates the magnetic field may be amplified, thanks to the differential rotation of the plasma. The dynamo term provides then the coupling allowing the transfer of the *kinetic* energy into *magnetic* energy. The second term at the right-hand side of (24) is the *magnetic diffusivity* whose effect is to damp the magnetic field intensity. Defining then as L the typical scale of spatial variation of the magnetic field intensity, the typical time-scale of resistive phenomena turns out to be

$$t_\sigma \simeq 4\pi\sigma L^2. \tag{27}$$

In a nonrelativistic plasma the conductivity σ goes typically as $T^{3/2}$ [24, 25]. In the case of planets, like the earth, one can wonder why a sizable magnetic field can still be present. One of the theories is that the dynamo term regenerates continuously the magnetic field which is dissipated by the diffusivity term [32]. In the case of the galactic disk the value of the conductivity[7] is given by $\sigma \simeq 7 \times 10^{-7}$Hz. Thus, for $L \simeq$ kpc $t_\sigma \simeq 10^9 (L/\text{kpc})^2$sec.

Equation (27) can also give the typical resistive length-scale once the time-scale of the system is specified. Suppose that the time-scale of the system is given by $t_U \sim H_0^{-1} \sim 10^{18}$sec where H_0 is the present order of magnitude of the Hubble parameter. Then

$$L_\sigma = \sqrt{\frac{t_U}{\sigma}}, \tag{28}$$

leading to $L_\sigma \sim$ AU. The scale (28) gives then the upper limit on the diffusion scale for a magnetic field whose lifetime is comparable with the age of the Universe at the present epoch. Magnetic fields with typical correlation scale larger than L_σ are not affected by resistivity. On the other hand, magnetic fields with typical correlation scale $L < L_\sigma$ are diffused. The value $L_\sigma \sim$ AU is consistent with the phenomenological evidence that there are no magnetic fields coherent over scales smaller than 10^{-5} pc.

The dynamo term may be responsible for the origin of the magnetic field of the galaxy. The galaxy has a typical rotation period of 3×10^8 years and comparing this figure with the typical age of the galaxy, $\mathcal{O}(10^{10}\text{years})$, it can be appreciated that the galaxy performed about 30 rotations since the time of the protogalactic collapse.

The effectiveness of the dynamo action depends on the physical properties of the bulk velocity field. In particular, a necessary requirement to have a potentially successful dynamo action is that the velocity field is non-mirror-symmetric or that, in other words, $\langle \boldsymbol{v} \cdot \boldsymbol{\nabla} \times \boldsymbol{v} \rangle \neq 0$. Let us see how this statement can be made reasonable in the framework of (24). From (24) the usual structure of the dynamo term may be derived by carefully averaging over the velocity field according to the procedure of [37, 38]. By assuming that the motion of the fluid is random and with zero mean velocity the average is taken over the ensemble of the possible velocity fields. In more physical terms

[7] It is common use in the astrophysical applications to work directly with $\eta = (4\pi\sigma)^{-1}$. In the case of the galactic disks $\eta = 10^{26}\text{cm}^2$ Hz.

this averaging procedure of (24) is equivalent to average over scales and times exceeding the characteristic correlation scale and time τ_0 of the velocity field. This procedure assumes that the correlation scale of the magnetic field is much bigger than the correlation scale of the velocity field which is required to be divergence-less ($\boldsymbol{\nabla} \cdot \boldsymbol{v} = 0$). In this approximation the magnetic diffusivity equation can be written as

$$\frac{\partial \boldsymbol{B}}{\partial t} = \alpha(\boldsymbol{\nabla} \times \boldsymbol{B}) + \frac{1}{4\pi\sigma}\nabla^2 \boldsymbol{B}, \tag{29}$$

where

$$\alpha = -\frac{\tau_0}{3}\langle \boldsymbol{v} \cdot \boldsymbol{\nabla} \times \boldsymbol{v}\rangle, \tag{30}$$

is the so-called α-term in the absence of vorticity. In (29) and (30) $\boldsymbol{B}$ is the magnetic field averaged over times longer that τ_0 which is the typical correlation time of the velocity field.

The fact that the velocity field must be globally non-mirror-symmetric [33] suggests, already at this qualitative level, the deep connection between dynamo action and fully developed turbulence. In fact, if the system would be, globally, invariant under parity transformations, then the α term would simply be vanishing. This observation may also be related to the turbulent features of cosmic systems. In cosmic turbulence the systems are usually rotating and, moreover, they possess a gradient in the matter density (think, for instance, to the case of the galaxy). It is then plausible that parity is broken at the level of the galaxy since terms like $\boldsymbol{\nabla}\rho_{\rm m} \cdot \boldsymbol{\nabla} \times \boldsymbol{v}$ are not vanishing [33].

The dynamo term, as it appears in (29), has a simple electrodynamical meaning, namely, it can be interpreted as a mean Ohmic current directed along the magnetic field

$$\boldsymbol{J} = -\alpha \boldsymbol{B}. \tag{31}$$

Equation stipulates that an ensemble of screw-like vortices with zero mean helicity is able to generate loops in the magnetic flux tubes in a plane orthogonal to the one of the original field. As a simple (and known) application of (29), it is appropriate to consider the case where the magnetic field profile is given by a sort of Chern–Simons wave

$$B_x(z,t) = f(t)\sin kz, \qquad B_y = f(t)\cos kz, \qquad B_z(k,t) = 0. \tag{32}$$

For this profile the magnetic gyrotropy is nonvanishing, i.e. $\boldsymbol{B}\cdot\boldsymbol{\nabla}\times\boldsymbol{B} = kf^2(t)$. From (29), using (32) $f(t)$ obeys the following equation

$$\frac{df}{dt} = \left(k\alpha - \frac{k^2}{4\pi\sigma}\right) f \tag{33}$$

admits exponentially growing solutions for sufficiently large scales, i.e. $k < 4\pi|\alpha|\sigma$. Notice that in this naive example the α term is assumed to be constant. However, as the amplification proceeds, α may develop a dependence upon $|\boldsymbol{B}|^2$, i.e. $\alpha \to \alpha_0(1-\xi|\boldsymbol{B}|^2)\alpha_0[1-\xi f^2(t)]$. In the case of (33) this modification will introduce nonlinear terms whose effect will be to stop the growth of the magnetic field. This regime is often called *saturation of the dynamo* and the nonlinear equations appearing in this context are sometimes called Landau equations [33] in analogy with the Landau equations appearing in hydrodynamical turbulence.

In spite of the fact that in the previous example the velocity field has been averaged, its evolution obeys the Navier–Stokes equation which we have already written but without the diffusion term

$$\rho_{\rm m}\left[\frac{\partial \boldsymbol{v}}{\partial t} + (\boldsymbol{v}\cdot\boldsymbol{\nabla})\boldsymbol{v} - \nu\nabla^2\boldsymbol{v}\right] = -\boldsymbol{\nabla}P + \boldsymbol{J}\times\boldsymbol{B}, \tag{34}$$

where ν is the thermal viscosity coefficient. There are idealized cases where the Lorentz force term can be neglected. This is the so-called force-free approximation. Defining the kinetic helicity as $\boldsymbol{\Omega} = \boldsymbol{\nabla}\times\boldsymbol{v}$, the magnetic diffusivity and Navier–Stokes equations can be written in a rather simple and symmetric form

$$\begin{aligned} \frac{\partial \boldsymbol{B}}{\partial t} &= \boldsymbol{\nabla}\times(\boldsymbol{v}\times\boldsymbol{B}) + \frac{1}{4\pi\sigma}\nabla^2\boldsymbol{B}, \\ \frac{\partial \boldsymbol{\Omega}}{\partial t} &= \boldsymbol{\nabla}\times(\boldsymbol{v}\times\boldsymbol{\Omega}) + \nu\nabla^2\boldsymbol{\Omega}. \end{aligned} \tag{35}$$

In MHD various dimensionless ratios can be defined. The most frequently used are the magnetic Reynolds number, the kinetic Reynolds number and the Prandtl number:

$$\mathrm{R_m} = vL_B\sigma, \tag{36}$$

$$\mathrm{R} = \frac{vL_v}{\nu}, \tag{37}$$

$$\mathrm{Pr} = \frac{\mathrm{R_m}}{\mathrm{R}} = \nu\sigma\left(\frac{L_B}{L_v}\right), \tag{38}$$

where L_B and L_v are the typical scales of variation of the magnetic and velocity fields. If $\mathrm{R_m} \gg 1$ the system is said to be *magnetically* turbulent. If $\mathrm{R} \gg 1$ the system is said to be *kinetically* turbulent. In realistic situations the plasma is both kinetically and magnetically turbulent and, therefore, the ratio of the two Reynolds numbers will tell which is the dominant source of turbulence. There have been, in recent years, various studies on the development of magnetized turbulence (see, for instance, [27]) whose features differ slightly from the ones of hydrodynamic turbulence. While the details of this discussion will be left aside, it is relevant to mention that, in the early Universe, turbulence may develop. In this situation a typical phenomenon, called inverse cascade, can take place. A direct cascade is a process where energy is transferred from large to small scales. Even more interesting, for the purposes of the present lecture, is the opposite process, namely the inverse cascade where the energy transfer goes from small to large length-scales. One can also generalize the the concept of energy cascade to the cascade of any conserved quantity in the plasma, like, for instance, the helicity. Thus, in general terms, the transfer process of a conserved quantity is a cascade.

The concept of cascade (either direct or inverse) is related with the concept of turbulence, i.e. the class of phenomena taking place in fluids and plasmas at high Reynolds numbers. It is very difficult to reach, with terrestrial plasmas, the physical situation where the magnetic and the kinetic Reynolds numbers are both large but in such a way that their ratio is also large, i.e.

$$\mathrm{R_m} \gg 1, \qquad \mathrm{R} \gg 1, \qquad \mathrm{Pr} = \frac{\mathrm{R_m}}{\mathrm{R}} \gg 1. \tag{39}$$

The physical regime expressed through (39) rather common in the early Universe. Thus, MHD turbulence is probably one of the key aspects of magnetized plasma dynamics at very high temperatures and densities. Consider, for instance, the plasma at the electroweak epoch when the temperature was of the order of 100 GeV. One can compute the Reynolds numbers and the Prandtl number from their definitions given in (36)–(38). In particular,

$$\mathrm{R_m} \sim 10^{17}, \qquad \mathrm{R} = 10^{11}, \qquad \mathrm{Pr} \simeq 10^{6}, \tag{40}$$

which can be obtained from (36)–(38) using as fiducial parameters $v \simeq 0.1$, $\sigma T/\alpha$, $\nu \simeq (\alpha T)^{-1}$ and $L \simeq 0.01\ H_{\mathrm{ew}}^{-1} \simeq 0.03$ cm for $T \simeq 100$ GeV.

If an inverse energy cascade takes place, many (energetic) magnetic domains coalesce giving rise to a magnetic domain of larger size but of smaller energy. This phenomenon can be viewed, in more quantitative terms, as an effective increase of the correlation scale of the magnetic field. This consideration plays a crucial role for the viability of mechanisms where the magnetic field is produced in the early Universe inside the Hubble radius (see Sect. 2.5).

2.3 Initial Conditions for Dynamos

According to the qualitative description of the dynamo instability presented in the previous subsection, the origin of large-scale magnetic fields in spiral galaxies can be reduced to the three keywords: *seeding*, *amplification* and *ordering*. The first stage, i.e. the seeding, is the most controversial one and will be briefly reviewed in the following sections of the present review. In more quantitative terms the amplification and the ordering may be summarized as follows:

- During the 30 rotations performed by the galaxy since the protogalactic collapse, the magnetic field should be amplified by about 30 e-folds;
- If the large-scale magnetic field of the galaxy is, today, $\mathcal{O}(\mu\mathrm{G})$ the magnetic field at the onset of galactic rotation might have been even 30 e-folds smaller, i.e. $\mathcal{O}(10^{-19}\mathrm{G})$ over a typical scale of 30–100 kpc.

- Assuming perfect flux freezing during the gravitational collapse of the protogalaxy (i.e. $\sigma \to \infty$), the magnetic field at the onset of gravitational collapse should be $\mathcal{O}(10^{-23})$ G over a typical scale of 1 Mpc.

This picture is oversimplified and each of the three steps mentioned above can be questioned. In what follows the main sources of debate, emerged in the last 10 years, will be briefly discussed.

There is a simple way to relate the value of the magnetic fields right after gravitational collapse to the value of the magnetic field right before gravitational collapse. Since the gravitational collapse occurs at high conductivity, the magnetic flux and the magnetic helicity are both conserved (see, in particular, (25)). Right before the formation of the galaxy a patch of matter of roughly 1 Mpc collapses by gravitational instability. Right *before* the collapse the mean energy density of the patch, stored in matter, is of the order of the critical density of the Universe. Right *after* collapse the mean matter density of the protogalaxy is, approximately, six orders of magnitude larger than the critical density.

Since the physical size of the patch decreases from 1 Mpc to 30 kpc, the magnetic field increases, because of flux conservation, of a factor $(\rho_{\mathrm{a}}/\rho_{\mathrm{b}})^{2/3} \sim 10^4$ where ρ_{a} and ρ_{b} are, respectively the energy densities right after and right before gravitational collapse. The correct initial condition in order to turn on the dynamo instability would be $|\boldsymbol{B}| \sim 10^{-23}$ Gauss over a scale of 1 Mpc, right before gravitational collapse.

The estimates presented in the last paragraph are based on the (rather questionable) assumption that the amplification occurs over 30 e-folds while the magnetic flux is completely frozen in. In the real situation, the achievable amplification is much smaller. Typically a good seed would not be 10^{-19} G after collapse (as we assumed for the simplicity of the discussion) but rather [35]

$$|\boldsymbol{B}| \geq 10^{-13}\mathrm{G}. \tag{41}$$

The galactic rotation period is of the order of 3×10^8 years. This scale should be compared with the typical age of the galaxy. All along this rather large dynamical time-scale the effort has been directed, from the 1950s, to the justification that a substantial portion of the kinetic energy of the system (provided by the differential rotation) may be converted into magnetic energy amplifying, in this way, the seed field up to the observed value of the magnetic field, for instance in galaxies and in clusters. In recent years a lot of progress has been made both in the context of the small- and of large-scale dynamos [36, 39] (see also [40, 41, 42]). This progress was also driven by the higher resolution of the numerical simulations and by the improvement in the understanding of the largest magnetized system that is rather close to us, i.e. the sun [36]. More complete accounts of this progress can be found in the second article of [39] and, more comprehensively, in [36]. Apart from the aspects involving solar physics and numerical analysis, better physical understanding of the role of the magnetic helicity in the dynamo action has been reached. This point is crucially connected with the two conservation laws arising in MHD, i.e. the magnetic flux and magnetic helicity conservations whose relevance has been already emphasized, respectively, in (25) and (26). Even if the rich interplay between small-and large-scale dynamos is rather important, let us focus on the problem of large-scale dynamo action that is, at least superficially, more central for the considerations developed in the present lecture.

Already at a qualitative level it is clear that there is a clash between the absence of mirror-symmetry of the plasma, the quasi-exponential amplification of the seed and the conservation of magnetic flux and helicity in the high (or more precisely infinite) conductivity limit. The easiest clash to understand, intuitively, is the flux conservation versus the exponential amplification: both flux freezing and exponential amplification have to take place in the *same* superconductive (i.e. $\sigma^{-1} \to 0$) limit. The clash between helicity conservation and dynamo action can also be understood in general terms: the dynamo action implies a topology change of the configuration since the magnetic flux lines cross each other constantly [39].

One of the recent progress in this framework is a more consistent formulation of the large-scale dynamo problem [39, 39]: large-scale dynamos produce small-scale helical fields that quench (i.e. prematurely saturate) the α effect. In other words, the conservation of the magnetic helicity can be seen, according to the recent view, as a fundamental constraint on the dynamo action. In connection with the last point, it should be mentioned that, in the past, a rather different argument was suggested [43]: it was argued that the dynamo action leads to the amplification not only of the large-scale field but also of the random field component. The random field would then suppress strongly the dynamo action. According to the considerations based on the conservation of the magnetic helicity, this argument seems to be incorrect since the increase of the random component would also entail and increase of the rate of the topology change, i.e. a magnetic helicity nonconservation.

The possible applications of dynamo mechanism to clusters are still under debate and it seems more problematic. The typical scale of the gravitational collapse of a cluster is larger (roughly by one order of magnitude) than the scale of gravitational collapse of the protogalaxy. Furthermore, the mean mass density within the Abell radius ($\simeq 1.5\,h^{-1}$ Mpc) is roughly 10^3 larger than the critical density. Consequently, clusters rotate much less than galaxies. Recall that clusters are formed from peaks in the density field. The present overdensity of clusters is of the order of 10^3. Thus, in order to get the intracluster magnetic field, one could think that magnetic flux is exactly conserved and, then, from an intergalactic magnetic field $|\boldsymbol{B}| > 10^{-9}$ G an intracluster magnetic field $|\boldsymbol{B}| > 10^{-7}$ G can be generated. This simple estimate shows why it is rather important to improve the accuracy of magnetic field measurements in the intracluster medium: The change of a single order of magnitude in the estimated magnetic field may imply rather different conclusions for its origin.

2.4 Astrophysical Mechanisms

Many (if not all) the astrophysical mechanisms proposed so far are related to what is called, in the jargon, a *battery*. In short, the idea is the following. The explicit form of the generalized Ohmic electric field in the presence of thermoelectric corrections can be written as in (20) where we set $n_{\rm q} = n_{\rm e}$ to stick to the usual conventions[8]

$$\boldsymbol{E} = -\boldsymbol{v} \times \boldsymbol{B} + \frac{\boldsymbol{\nabla} \times \boldsymbol{B}}{4\pi\sigma} - \frac{\boldsymbol{\nabla} P_{\rm e}}{e n_{\rm e}}. \tag{42}$$

By comparing (23) with (42), it is clear that the additional term at the right-hand side receives contribution from a temperature gradient. In fact, restoring for a moment the Boltzmann constant k_B we have that since $P_{\rm e} = k_B\, n_{\rm e}\, T_{\rm e}$, the additional term depends upon the gradients of the temperature, hence the name thermoelectric. It is interesting to see under which conditions the curl of the electric field receives contribution from the thermoelectric effect. Taking the curl of both sides of (42), we obtain

$$\boldsymbol{\nabla} \times \boldsymbol{E} = \frac{1}{4\pi\sigma} \nabla^2 \boldsymbol{B} + \boldsymbol{\nabla}(\boldsymbol{v} \times \boldsymbol{B}) - \frac{\boldsymbol{\nabla} n_{\rm e} \times \boldsymbol{\nabla} P_{\rm e}}{e n_{\rm e}^2} = -\frac{\partial \boldsymbol{B}}{\partial t}, \tag{43}$$

where the second equality is a consequence of Maxwell's equations. From (43) it is clear that the evolution of the magnetic field inherits a source term iff the gradients in the pressure and electron density are not parallel. If $\boldsymbol{\nabla} P_{\rm e} \parallel \boldsymbol{\nabla} n_{\rm e}$, a fully valid solution of (43) is $\boldsymbol{B} = 0$. In the opposite case a seed magnetic field is naturally provided by the thermoelectric term. The usual (and rather general) observation that one can make in connection with the geometrical properties of the thermoelectric term is that cosmic ionization fronts may play an important role. For instance, when quasars emit ultraviolet photons, cosmic ionization fronts are produced. Then the intergalactic medium may be ionized. It should also be recalled, however, that the temperature gradients are usually normal to the ionization front. In spite of this, it is also plausible to think that density gradients can arise in arbitrary directions due to the stochastic nature of density fluctuations.

In one way or in another, astrophysical mechanisms for the generation of magnetic fields use an incarnation of the thermoelectric effect [44] (see also [45, 46]). In the 1960s and 1970s, for instance, it was rather popular to

[8] For simplicity, we shall neglect the Hall contribution arising in the generalized Ohm law. The Hall contribution would produce in (42) a term $\boldsymbol{J} \times \boldsymbol{B}/n_{\rm e}e$ that is of higher order in the magnetic field and that is proportional to the Lorentz force. The Hall term will play no role in the subsequent considerations. However, it should be borne in mind that the Hall contribution may be rather interesting in connection with the presence of strong magnetic fields like the ones of neutron stars (i.e. 10^{13} G). This occurrence is even more interesting since in the outer regions of neutron stars strong density gradients are expected.

think that the correct "geometrical" properties of the thermoelectric term may be provided by a large-scale vorticity. As it will also be discussed later, this assumption seems to be, at least naively, in contradiction with the formulation of inflationary models whose prediction would actually be that the large-scale vector modes are completely washed out by the expansion of the Universe. Indeed, all along the 1980s and 1990s the idea of primordial vorticity received just a minor attention.

The attention then focused on the possibility that objects of rather small size may provide intense seeds. After all we do know that these objects may exist. For instance the Crab nebula has a typical size of a roughly 1 pc and a magnetic field that is a fraction of the multi Gauss. These seeds will then combine and diffuse leading, ultimately, to a weaker seed but with large correlation scale. This aspect may be, physically, a bit controversial since we do observe magnetic fields in galaxies and clusters that are ordered over very large length-scales. It would then seem necessary that the seed fields produced in a small object (or in several small objects) undergo some type of dynamical self-organization whose final effect is a seed coherent over length-scales 4 or 5 orders of magnitude larger than the correlation scale of the original battery.

An interesting idea could be that qualitatively different batteries lead to some type of conspiracy that may produce a strong large-scale seed. In [44] it has been suggested that Population III stars may become magnetized, thanks to a battery operating at stellar scale. Then if these stars would explode as supernovae (or if they would eject a magnetized stellar wind), the pregalactic environment may be magnetized and the remnants of the process incorporated in the galactic disk. In a complementary perspective, a similar chain of events may take place over a different physical scale. A battery could arise in fact in active galactic nuclei at high redshift. Then the magnetic field could be ejected leading to intense fields in the lobes of "young" radio-galaxies. These fields will be somehow inherited by the "older" disk galaxies and the final seed field may be, according to [44], as large as 10^{-9} G at the pregalactic stage.

In summary, we can therefore say that

- both the primordial and the astrophysical hypothesis for the origin of the seeds demand an efficient (large-scale) dynamo action;
- due to the constraints arising from the conservation of magnetic helicity and magnetic flux the values of the required seed fields may turn out to be larger than previously thought at least in the case when the amplification is only driven by a large-scale dynamo action;[9]
- magnetic flux conservation during gravitational collapse of the protogalaxy may increase, by compressional amplification, the initial seed of even 4 orders of magnitude;

[9] The situation may change if the magnetic fields originate from the combined action of small- and large-scale dynamos like in the case of the two-step process described in [44].

- compressional amplification and large-scale dynamo are much less effective in clusters: therefore, the magnetic field of clusters is probably connected to the specific way the dynamo saturates, and, in this sense, harder to predict from a specific value of the initial seed.

2.5 Magnetogenesis: Inside the Hubble Radius

One of the weaknesses of the astrophysical hypothesis is connected with the smallness of the correlation scale of the obtained magnetic fields. This type of impasse led the community to consider the option that the initial conditions for the MHD evolution are dictated not by astrophysics but rather by cosmology. The first ones to think about cosmology as a possible source of large-scale magnetization were Zeldovich [47, 48] and Harrison [49, 50, 51].

The emphasis of these two authors was clearly different. While Zeldovich thought about a magnetic field which is *uniform* (i.e. homogeneous and oriented, for instance, along a specific Cartesian direction), Harrison somehow anticipated the more modern view by considering the possibility of an *inhomogeneous* magnetic field. In the scenario of Zeldovich the uniform magnetic field would induce a slight anisotropy in the expansion rate along which the magnetic field is aligned. So, for instance, by considering a constant (and uniform) magnetic field pointing along the $\hat{x}$ Cartesian axis, the induced geometry compatible with such a configuration will fall into the Bianchi I class

$$ds^2 = dt^2 - a^2(t)dx^2 - b^2(t)[dy^2 + dz^2]. \tag{44}$$

By solving Einstein equations in this background geometry, it turns out that, during a radiation-dominated epoch, the expansion rates along the $\hat{x}$ and the $\hat{y} - \hat{z}$ plane change and their difference is proportional to the magnetic energy density [47, 48]. This observation is not only relevant for magnetogenesis but also for cosmic microwave background anisotropies since the difference in the expansion rate turns out to be proportional to the temperature anisotropy. While we will get back to this point later, in Sect. 4, as far as magnetization is concerned we can just remark that the idea of Zeldovich was that a uniform magnetic field would modify the initial condition of the standard hot big bang model where the Universe would start its evolution already in a radiation-dominated phase.

The model of Harrison [49, 50, 51] is, in a sense, more *dynamical.* Following earlier work of Biermann [52], Harrison thought that inhomogeneous MHD equations could be used to generate large-scale magnetic fields *provided* the velocity field was turbulent enough. The Biermann battery was simply a battery (as the ones described above in this session) but operating prior to decoupling of matter and radiation. The idea of Harrison was instead that vorticity was already present so that the effective MHD equations will take the form

$$\frac{\partial}{\partial\tau}(a^2\boldsymbol{\Omega} + \frac{e}{m_{\rm p}}\boldsymbol{B}) = \frac{e}{4\pi\sigma m_{\rm p}}\nabla^2\boldsymbol{B}, \tag{45}$$

where, as previously defined, $\boldsymbol{\Omega} = \nabla \times \boldsymbol{v}$ and $m_{\rm p}$ is the ion mass. Equation (45) is written in a conformally flat Friedmann–Robertson–Walker (FRN) metric of the form

$$ds^2 = G_{\mu\nu} dx^{\mu} dx^{\nu} = a^2(\tau)[d\tau^2 - d\boldsymbol{x}^2], \tag{46}$$

where τ is the conformal time coordinate and where, in the conformally flat case, $G_{\mu\nu} = a^2(\tau)\eta_{\mu\nu}$, $\eta_{\mu\nu}$ being the four-dimensional Minkowski metric. If we now postulate that some vorticity was present prior to decoupling, then (45) can be solved and the magnetic field can be related to the initial vorticity as

$$\boldsymbol{B} \sim -\frac{m_{\rm p}}{e}\boldsymbol{\omega}_{\rm i}\left(\frac{a_i}{a}\right)^2 . \tag{47}$$

If the estimate of the vorticity is made prior to equality (as originally suggested by Harrison [49]) or after decoupling as also suggested, a bit later, in [53], the result can change even by two orders of magnitude. Prior to equality $|\boldsymbol{\Omega}(t) \simeq 0.1/t$ and, therefore, $|\boldsymbol{B}_{\rm eq}| \sim 10^{-21}$G. If a similar estimate is made after decoupling, the typical value of the generated magnetic field is of the order of 10^{-18} G. So, in this context, the problem of the origin of magnetic fields is circumvented by postulating an appropriate form of vorticity whose origin must be explained.

The Harrison mechanism is just one of the first examples of magnetic field generation *inside the Hubble radius*. In cosmology we define the Hubble radius as the inverse of the Hubble parameter, i.e. $r_H = H^{-1}(t)$. The first possibility we can think of implies that magnetic fields are produced, at a given epoch in the life of the Universe, inside the Hubble radius, for instance by a phase transition or by any other phenomenon able to generate a charge separation and, ultimately, an electric current. In this context, the correlation scale of the field is much smaller than the typical scale of the gravitational collapse of the protogalaxy which is of the order of mega parsecs. In fact, if the Universe is decelerating and if the correlation scale evolves as the scale factor, the Hubble radius grows much faster than the correlation scale. Of course, one might invoke the possibility that the correlation scale of the magnetic field evolves more rapidly than the scale factor. A well-founded physical rationale for this occurrence is what is normally called inverse cascade, i.e. the possibility that magnetic (as well as kinetic) energy density is transferred from small to large scales. This implies, in real space, that (highly energetic) small-scale magnetic domains may coalesce to form magnetic domains of smaller energy but over larger scales. In the best of all possible situations, i.e. when inverse cascade is very effective, it seems rather hard to justify a growth of the correlation scale that would eventually end up into a mega parsec scale at the onset of gravitational collapse.

In Fig. 1 we report a schematic illustration of the evolution of the Hubble radius R_H and of the correlation scale of the magnetic field as a function of the scale factor. In Fig. 1 the horizontal dashed line simply marks the end of the radiation–dominated phase and the onset of the matter-dominated phase:

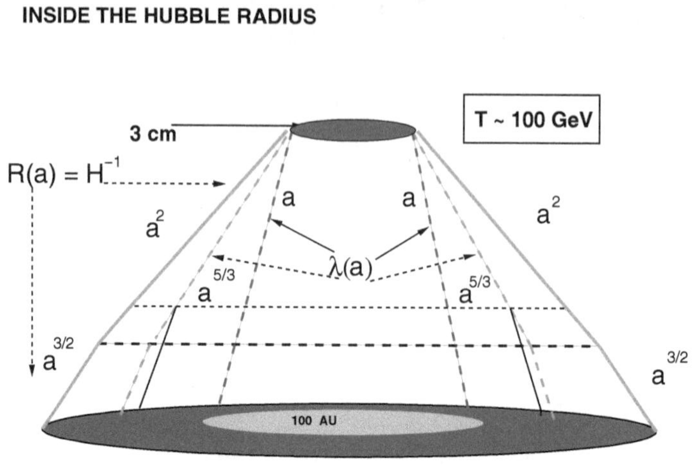

Fig. 1. Evolution of the correlation scale for magnetic fields produced inside the Hubble radius. The horizontal thick dashed line marks the end of the radiation-dominated phase and the onset of the matter-dominated phase. The horizontal thin dashed line marks the moment of e^+–e^- annihilation (see also footnote 2). The full (vertical) lines represent the evolution of the Hubble radius during the different stages of the life of the Universe. The dashed (vertical) lines illustrate the evolution of the correlation scale of the magnetic fields. In the absence of inverse cascade the evolution of the correlation scale is given by the (inner) vertical dashed lines. If inverse cascade takes place, the evolution of the correlation scale is faster than the first power of the scale factor (for instance $a^{5/3}$) but always slower than the Hubble radius

while above the dashed line the Hubble radius evolves as a^2 (where a is the scale factor), below the dashed line the Hubble radius evolves as $a^{3/2}$.

We consider, for simplicity, a magnetic field whose typical correlation scale is as large as the Hubble radius at the electroweak epoch when the temperature of the plasma was of the order of 100 GeV. This is roughly the regime contemplated by the considerations presented around (40). If the correlation scale evolves as the scale factor, the Hubble radius at the electroweak epoch (roughly 3 cm) projects today over a scale of the order of the astronomical unit. If inverse cascades are invoked, the correlation scale may grow, depending on the specific features of the cascade, up to 100 AU or even up to 100 pc. In both cases the final scale is too small if compared with the typical scale of the gravitational collapse of the protogalaxy. In Fig. 1 a particular model for the evolution of the correlation scale $\lambda(a)$ has been reported.[10]

[10] Notice, as it will be discussed later, that the inverse cascade lasts, in principle, only down to the time of $e^+ - e^-$ annihilation (see also thin dashed horizontal

2.6 Inflationary Magnetogenesis

If magnetogenesis takes place inside the Hubble radius, the main problem is therefore the correlation scale of the obtained seed field. The cure for this problem is to look for a mechanism producing magnetic fields that are coherent over large scales (i.e. mega parsec and, in principle, even larger). This possibility may arise in the context of inflationary models. Inflationary models may be conventional (i.e. based on a quasi-de Sitter stage of expansion) or unconventional (i.e. not based on a quasi-de Sitter stage of expansion). Unconventional inflationary models are, for instance, pre-big-bang models that will be discussed in more depth in Sect. 3.

The rationale for the previous statement is that, in inflationary models, the zero-point (vacuum) fluctuations of fields of various spin are amplified, typically fluctuations of spin 0 and spin 2 fields. The spin 1 fields enjoy however of a property, called Weyl invariance, that seems to forbid the amplification of these fields. While Weyl invariance and its possible breaking will be the specific subject of the following subsection, it is useful for the moment to look at the kinematical properties by assuming that, indeed, also spin 1 field can be amplified.

Since during inflation the Hubble radius is roughly constant (see Fig. 2), the correlation scale evolves much faster than the Hubble radius itself and, therefore, large-scale magnetic domains can naturally be obtained. Notice that, in Fig. 2 the (vertical) dashed lines illustrate the evolution of the Hubble radius (that is roughly constant during inflation) while the full line denotes the evolution of the correlation scale. Furthermore, the horizontal (dashed) lines mark, from top to bottom, the end of the inflationary phase and the onset of the matter-dominated phase. This phenomenon can be understood as the gauge counterpart of the superadiabatic amplification of the scalar and tensor modes of the geometry. The main problem, in such a framework, is to get large amplitudes for scale of the order of mega parsec at the onset of gravitational collapse. Models where the gauge couplings are effectively dynamical (breaking, consequently, the Weyl invariance of the evolution equations of Abelian gauge modes) may provide rather intense magnetic fields.

The two extreme possibilities mentioned above may be sometimes combined. For instance, it can happen that magnetic fields are produced by superadiabatic amplification of vacuum fluctuations during an inflationary stage of expansion. After exiting the horizon, the gauge modes will reenter at different moments all along the radiation- and matter-dominated epochs. The spectrum of the primordial gauge fields after reentry will not only be determined by the amplification mechanism but also on the plasma effects. As soon as the magnetic inhomogeneities reenter, some other physical process, taking place inside the Hubble radius, may be triggered by the presence of large-scale magnetic

line in Fig. 1) since for temperatures smaller than $T_{e^+-e^-}$ the Reynolds number drops below 1. This is the result of the sudden drop in the number of charged particles that leads to a rather long mean free path for the photons.

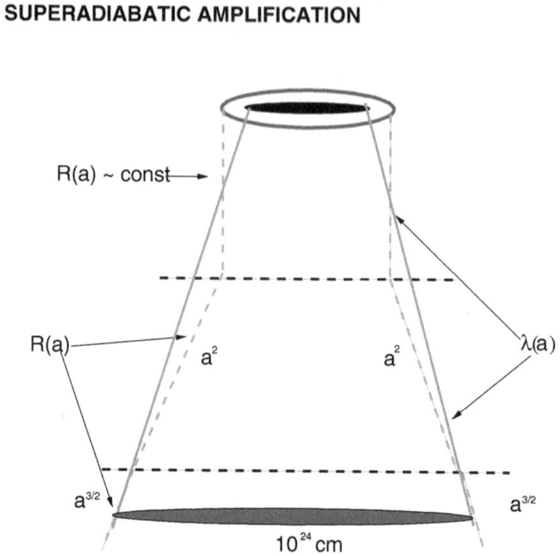

Fig. 2. Evolution of the correlation scale if magnetic fields would be produced by superadiabatic amplification during a conventional inflationary phase. The dashed vertical lines denote, in the present figure, the evolution of the Hubble radius, while the full line denotes the evolution of the correlation scale (typically selected to smaller than the Hubble radius during inflation)

fields. An example, in this context, is the production of topologically non-trivial configurations of the hypercharge field (hypermagnetic knots) from a stochastic background of hypercharge fields with vanishing helicity [54, 55, 56] (see also [57, 58, 59, 60, 61]).

2.7 Breaking of Conformal Invariance

Consider the action for an Abelian gauge field in four-dimensional curved space–time

$$S_{\rm em} = -\frac{1}{4}\int d^4x \sqrt{-G} F_{\mu\nu} F^{\mu\nu}. \tag{48}$$

Suppose, also, that the geometry is characterized by a conformally flat line element of Friedmann–Robertson–Walker type as the one introduced in (46). The equations of motion derived from (48) can be written as

$$\partial_{\mu}\left(\sqrt{-G} F^{\mu\nu}\right) = 0. \tag{49}$$

Using (46) and recalling that $\sqrt{-G} = a^4(\tau)$, we will have

$$\sqrt{-G} F^{\mu\nu} = a^4(\tau) \frac{\eta^{\mu\alpha}}{a^2(\tau)} \frac{\eta^{\nu\beta}}{a^2(\tau)} F_{\alpha\beta} = F^{\mu\nu} \tag{50}$$

where the second equality follows from the explicit form of the metric. Equation (50) shows that the evolution equations of Abelian gauge fields are the same in flat space–time and in a conformally flat FRW space–time. This property is correctly called Weyl invariance or, more ambiguously, conformal invariance. Weyl invariance is realized also in the case of chiral (massless) fermions always in the case of conformally flat space–times.

One of the reasons of the success of inflationary models in making predictions is deeply related with the lack of conformal invariance of the evolution equations of the fluctuations of the geometry. In particular it can be shown that the tensor modes of the geometry (spin 2) as well as the scalar modes (spin 0) obey evolution equations that are *not* conformally invariant. This means that these modes of the geometry can be amplified and eventually affect, for instance, the temperature autocorrelations as well as the polarization power spectra in the microwave sky.

To amplify large-scale magnetic fields, therefore, we would like to break conformal invariance. Before considering this possibility, let us discuss an even more conservative approach consisting in studying the evolution of Abelian gauge fields coupled to another field whose evolution is *not* Weyl invariant. An elegant way to achieve this goal is to couple the action of the hypercharge field to the one of a complex scalar field (the Higgs field). The Abelian–Higgs model, therefore, leads to the following action

$$S = \int d^4x\sqrt{-G}\left[G^{\mu\nu}(\mathcal{D}_\mu)^*\phi\mathcal{D}_\nu\phi - m^2\phi^*\phi - \frac{1}{4}\mathcal{F}_{\mu\nu}\mathcal{F}^{\mu\nu}\right], \qquad (51)$$

where $\mathcal{D}_\mu = \partial_\mu - ie\mathcal{A}_\mu$ and $\mathcal{F}_{\mu\nu} = \partial_{[\mu}\mathcal{A}_{\nu]}$. Substituting (46) into (51) and assuming that the complex scalar field (as well as the gauge fields) are not a source of the background geometry, the canonical action for the normal modes of the system can be written as

$$S = \int d^3x d\tau\left[\eta^{\mu\nu}(D_\mu\Phi)^* D_\nu\Phi + \left(\frac{a''}{a} - m^2a^2\right)\Phi^*\Phi - \frac{1}{4}F_{\alpha\beta}F^{\alpha\beta}\right], \qquad (52)$$

where $\Phi = a\phi$, $D_\mu = \partial_\mu - ieA_\mu$ and $F_{\mu\nu} = \partial_{[\mu}A_{\nu]}$. From (52) it is clear that also when the Higgs field is massless the coupling to the geometry breaks explicitly Weyl invariance. Therefore, current density and charge density fluctuations will be induced. Then, by employing a similar Vlasov–Landau description the resulting magnetic field will be of the order of $B_{\rm dec} \sim 10^{-40}T_{\rm dec}^2$ [62] which is, by far, too small to seed any observable field even assuming, optimistically, perfect flux freezing and maximal efficiency for the dynamo action. The results of [62] disproved earlier claims (see [63] for a critical review), neglecting the role of the conductivity in the evolution of large-scale magnetic fields after inflation.

The first attempts to analyze the Abelian–Higgs model in de Sitter space have been made by Turner and Widrow [66] who just listed such a possibility as an open question. These two authors also analyzed different scenarios where conformal invariance for spin 1 fields could be broken in four space–time dimensions. Their first suggestion was that conformal invariance may be broken, at an effective level, through the coupling of photons to the geometry [67]. Typically, the breaking of conformal invariance occurs through products of gauge-field strengths and curvature tensors, i.e.

$$\frac{1}{m^2}F_{\mu\nu}F_{\alpha\beta}R^{\mu\nu\alpha\beta}, \qquad \frac{1}{m^2}R_{\mu\nu}F^{\mu\beta}F^{\nu\alpha}g_{\alpha\beta}, \qquad \frac{1}{m^2}F_{\alpha\beta}F^{\alpha\beta}R, \tag{53}$$

where m is the appropriate mass scale, $R_{\mu\nu\alpha\beta}$ and $R_{\mu\nu}$ are the Riemann and Ricci tensors and R is the Ricci scalar. If the evolution of gauge fields is studied during phase of de Sitter (or quasi-de Sittter) expansion, then the amplification of the vacuum fluctuations induced by the couplings listed in (53) is minute. The price in order to get large amplification should be, according to [66], an explicit breaking of gauge invariance by direct coupling of the vector potential to the Ricci tensor or to the Ricci scalar, i.e.

$$RA_{\mu}A^{\mu}, \qquad R_{\mu\nu}A^{\mu}A^{\nu}. \tag{54}$$

In [66] two other different models were proposed (but not scrutinized in detail), namely, scalar electrodynamics and the axionic coupling to the Abelian field strength.

Dolgov [68] considered the possible breaking of conformal invariance due to the trace anomaly. The idea is that the conformal invariance of gauge fields is broken by the triangle diagram where two photons in the external lines couple to the graviton through a loop of fermions. The local contribution to the effective action leads to the vertex $(\sqrt{-g})^{1+\epsilon}F_{\alpha\beta}F^{\alpha\beta}$, where ϵ is a numerical coefficient depending upon the number of scalars and fermions present in the theory. The evolution equation for the gauge fields, can be written, in Fourier space, as

$$\mathcal{A}_k'' + \frac{\epsilon}{8}\mathcal{H}\mathcal{A}_k' + k^2\mathcal{A}_k = 0, \tag{55}$$

and it can be shown that only if $\epsilon > 0$ the gauge fields are amplified. Furthermore, only $\epsilon \sim 8$ substantial amplification of gauge fields is possible.

In a series of papers [69, 70, 71] the possible effect of the axionic coupling to the amplification of gauge fields has been investigated. The idea here is that conformal invariance is broken through the explicit coupling of a pseudoscalar field to the gauge field (see Sect. 5), i.e.

$$\sqrt{-g}c_{\psi\gamma}\alpha_{\rm em}\frac{\psi}{8\pi M}F_{\alpha\beta}\tilde{F}^{\alpha\beta}, \tag{56}$$

where $\tilde{F}^{\alpha\beta}$ is the dual field strength and $c_{\psi\gamma}$ is a numerical factor of order 1. Consider now the case of a standard pseudoscalar potential, for instance $m^2\psi^2$, evolving in a de Sitter (or quasi-de Sitter space–time). It can be shown, rather generically, that the vertex given in (56) leads to negligible amplification at large length-scale(s). The coupled system of evolution equations to be solved in order to get the amplified field is

$$\boldsymbol{B}'' - \nabla^2 \boldsymbol{B} - \frac{\alpha_{\rm em}}{2\pi M}\psi' \boldsymbol{\nabla} \times \boldsymbol{B} = 0, \tag{57}$$

$$\psi'' + 2\mathcal{H}\psi' + m^2 a^2 \psi = 0, \tag{58}$$

where $\boldsymbol{B} = a^2 \boldsymbol{B}$. From (57), there is a maximally amplified physical frequency

$$\omega_{\rm max} \simeq \frac{\alpha_{\rm em}}{2\pi M}\dot{\psi}_{\rm max} \simeq \frac{\alpha_{\rm em}}{2\pi} m \tag{59}$$

where the second equality follows from $\psi \sim a^{-3/2} M \cos mt$ (i.e. $\dot{\psi}_{\rm max} \sim mM$). The amplification for $\omega \sim \omega_{\rm max}$ is of the order of $\exp\left[m\alpha_{\rm em}/(2\pi H)\right]$ where H is the Hubble parameter during the de Sitter phase of expansion. From the above expressions one can argue that the modes which are substantially amplifed are the ones for which $\omega_{\rm max} \gg H$. The modes interesting for the large-scale magnetic fields are the ones which are in the opposite range, i.e. $\omega_{\rm max} \ll H$. Clearly, by lowering the curvature scale of the problem, the produced seeds may be larger and the conclusions much less pessimistic [71].

Another interesting idea pointed out by Ratra [72] is that the electromagnetic field may be directly coupled to the inflaton field. In this case the coupling is specified through a parameter α, i.e. $e^{\alpha\varphi} F_{\alpha\beta} F^{\alpha\beta}$ where φ is the inflaton field in Planck units. In order to get sizable large-scale magnetic fields the effective gauge coupling must be larger than one during inflation (recall that φ is large, in Planck units, at the onset of inflation).

In [73] it has been suggested that the evolution of the Abelian gauge coupling during inflation induces the growth of the two-point function of magnetic inhomogeneities. This model is different from the one previously discussed [72]. Here the dynamics of the gauge coupling is not related to the dynamics of the inflaton which is not coupled to the Abelian field strength. In particular, r_B(Mpc) can be as large as 10^{-12}. In [73] the MHD equations have been generalized to the case of evolving gauge coupling. Recently, a scenario similar to [73] has been discussed in [74].

In the perspective of generating large-scale magnetic fields Gasperini [75] suggested to consider the possible mixing between the photon and the graviphoton field appearing in supergravity theories (see also, in a related context [76]). The graviphoton is the massive vector component of the gravitational supermultiplet and its interaction with the photon is specified by an interaction term of the type $\lambda F_{\mu\nu} G^{\mu\nu}$, where $G_{\mu\nu}$ is the field strength of the massive vector. Large-scale magnetic fields with $r_B(\text{Mpc}) \geq 10^{-34}$ can be obtained if $\lambda \sim \mathcal{O}(1)$ and for a mass of the vector $m \sim 10^2$TeV.

Bertolami and Mota [77] argue that if Lorentz invariance is spontaneously broken, then photons acquire naturally a coupling to the geometry which is not gauge-invariant and which is similar to the coupling considered in [66].

3 Why String Cosmology?

The moment has come to review my personal interaction with Gabriele Veneziano on the study of large-scale magnetic fields. While we had other 15 joined papers with Gabriele (together with different combinations of authors), two of them [80, 81] (both in collaboration with Maurizio Gasperini) are directly related to large-scale magnetic fields. Both papers reported in [80, 81] appeared in 1995 while I was completing my PhD at the Theory Division of CERN.

My scientific exchange with Gabriele Veneziano started at least 4 years earlier and the first person mentioning Gabriele to me was Sergio Fubini. At that time Sergio was Professor of Theoretical Physics at the University of Turin and I had the great opportunity of discussing physics with him at least twice a month. Sergio was rather intrigued by the possibility of getting precise measurements on macroscopic quantum phenomena like superfluidity, superconductivity, and quantization of the resistivity in the (quantum) Hall effect. I started working, under the supervision of Maurizio Gasperini, on the spectral properties of relic gravitons and we bumped into the concept of squeezed state [82], a generalization of the concept of coherent state (see, for instance, [83, 84, 85]). Sergio got very interested and, I think, he was independently thinking about possible applications of squeezed states to superconductivity, a topic that became later on the subject of a paper [86]. Sergio even suggested a review by Rodney Loudon [87], an author that I knew already because of his inspiring book on quantum optics [88]. Reference [87] together with a physics report of Schumaker [89] was very useful for my understanding of the subject. Nowadays a very complete and thorough presentation of the intriguing problems arising in quantum optics can be found in the book of Mandel and Wolf [90].

It is amusing to notice the following parallelism between quantum optics and the quantum treatment of gravitational fluctuations. While quantum optics deals with the coherence properties of systems of many photons, we deal, in cosmology, with the coherence properties of many gravitons (or phonons) excited during the time evolution of the background fields. The background fields act, effectively, as a "pump field." This terminology, now generally accepted, is exactly borrowed by quantum optics where the pump field is a laser. In the 1960s and 1970s the main problem of optics can be summarized by the following question: Why is *classical* optics so precise? Put into different words, it is known that the interference of the amplitudes of the radiation field (the so-called Young interferometry) can be successfully treated at a classical level. Quantum effects, in optics, arise not from the first-order interference effects (Young interferometry) but from the second-order interference effects, i.e. the so-called Hanbury–Brown–Twiss interferometry [90], where the quantum nature of the radiation field is manifest since it leads, in the jargon introduced by Mandel [90], to light which is either bunched or antibunched. A similar problem also arises in the treatment of cosmological perturbations when we ask the question of the classical limit of a quantum mechanically generated fluctuation (for instance relic gravitons).

The interaction with Sergio led, few years later, to a talk that I presented at the physics department of the University of Torino. The title was *Correlation properties of many photons systems.* I mentioned my interaction with Sergio Fubini since it was Sergio who suggested that, eventually, I should talk to Gabriele about squeezed states.

During the first few months of 1991, Gabriele submitted a seminal paper on the cosmological implications of the low-energy string effective action [91]. This paper, together with another one written in collaboration with Maurizio Gasperini [92], represents the first formulation of pre–big-bang models. A relatively recent introduction to pre–big-bang models can be found in [93].

In [80, 81] it was argued that the string cosmological scenario provided by pre-big-bang models [91, 92] would be ideal for the generation of large-scale magnetic fields. The rationale for this statement relies on two different observations:

- in the low-energy string effective action gauge fields are coupled to the dilaton whose expectation value, at the string energy scale, gives the unified value of the gauge and gravitational coupling;
- from the mathematical analysis of the problem it is clear that to achieve a sizable amplification of large-scale magnetic fields it is necessary to have a pretty long phase where the gauge coupling is sharply growing in time [80].

Let us therefore elaborate on the two mentioned points. In the string frame the low-energy string effective action can be schematically written as [94, 95, 96]

$$S_{\rm eff} = -\int d^4x\sqrt{-G}\left[\frac{e^{-\varphi}}{2\lambda_{\rm s}^2}\left(R + G^{\alpha\beta}\partial_\alpha\varphi\partial_\beta\varphi - \frac{1}{12}H_{\mu\nu\alpha}H^{\mu\nu\alpha}\right)\right.$$
$$\left.+\frac{e^{-\varphi}}{4}F_{\alpha\beta}F^{\alpha\beta} + e^{-\varphi}\overline{\psi}\left(\frac{i}{2}\gamma^{\rm a}D_{\rm a}\psi + {\rm h.c.}\right) + \mathcal{R}^2 + \;.......\right] + \mathcal{O}(g^2) + \;.... \tag{1}$$

In (60) the ellipses stand, respectively, for an expansion in powers of $(\lambda_{\rm s}/L)^2$ and for an expansion in powers of the gauge coupling constant $g^2 = e^{\varphi}$. This action is written in the so-called string frame metric where the dilaton field φ is coupled to the Einstein–Hilbert term.

Concerning the action (60) few general comments are in order:

- the relation between the Planck and string scales depends on time and, in particular, $\ell_{\rm P}^2 = e^{\varphi}\lambda_{\rm s}^2$; the present ratio between the Planck and string scales gives the value, i.e. $g(\tau_0) = e^{\varphi_0/2} = \ell_P(\tau_0)/\lambda_{\rm s}$;

- in four space–time dimensions the antisymmetric tensor field $H^{\mu\nu\alpha}$ can be written in terms of a pseudoscalar field, i.e.

$$H^{\mu\nu\alpha} = e^{\varphi} \frac{\epsilon^{\mu\nu\alpha\rho}}{\sqrt{-G}} \partial_{\rho}\sigma; \tag{2}$$

In critical superstring theory the dilaton field must have a potential that vanishes in the weak coupling limit (i.e. $\varphi \to -\infty$). Moreover, from the direct tests of Newton law at short distances it should also happen that the mass of the dilaton is such that $m_{\varphi} > 10^{-4}$. This requirement may be relaxed by envisaging nonperturbative mechanisms where the dilaton is effectively decoupled from the matter fields and where a massless dilaton leads to observable violations of the equivalence principle.

From the structure of the action (60), Abelian gauge fields are amplified if the gauge coupling is dynamical. Consider, in fact, the equations of motion for the hypercharge field strength

$$\partial_{\mu}\left(e^{-\varphi}\sqrt{-G}F^{\mu\nu}\right) = 0, \tag{3}$$

where $F_{\mu\nu} = \partial_{[\mu}A_{\nu]}$. In the Coulomb gauge where $A_0 = 0$ and $\nabla \cdot \boldsymbol{A} = 0$ the equation for the rescaled vector potential $\mathcal{A}_{\mu} = e^{\varphi/2}A_{\mu}$ becomes, for each independent polarization and in Fourier space,

$$\mathcal{A}_k'' + \left[k^2 - g\left(\frac{1}{g}\right)''\right]\mathcal{A}_k = 0, \tag{4}$$

where, as usual, the prime denotes a derivation with respect to the conformal time coordinate. In (63) k denotes the comoving wave-number. From the structure of (63) there exist two different regimes. For $k^2 \gg |g(g^{-1})''|$ the solution off (63) is oscillatory. In the opposite limit, i.e. $k^2 \ll |g(g^{-1})''|$, the general solution can be written as

$$\mathcal{A}_k(\tau) = \frac{C_1(k)}{g(\tau)} + \frac{C_2(k)}{g(\tau)} \int^{\tau} g^2(\tau')d\tau', \tag{5}$$

where $C_1(k)$ and $C_2(k)$ are two arbitrary constants. These two constants can be fixed by imposing quantum mechanical initial conditions for $\tau \to -\infty$. Thus, depending on the evolution of $g(\tau)$ the Fourier amplitude $\mathcal{A}_k$ can be amplified.

It can be shown [80, 81] that the amplified magnetic energy density depends on the ratio between the value of the gauge coupling at the reentry and at the exit of the typical scale of the gravitational collapse, i.e.

$$r(k) = \frac{1}{\rho_{\gamma}} \frac{d\rho_{\rm B}}{d\ln k} \simeq \frac{k^4}{a^4 \rho_{\gamma}} \left(\frac{g_{\rm re}}{g_{ex}}\right)^2. \tag{6}$$

The parameter $r(k)$ measures the relative weight of the magnetic energy density in units of the radiation background. To turn on the galactic dynamo in its simplest realization, one should require that $r(k_{\rm G}) \geq 10^{-34}$ for a typical comoving wave-number corresponding to the typical scale of the gravitational collapse of the protogalaxy. As explained before, this requirement seems to be too optimistic in light of the most recent understanding of the dynamo theory. The limit $r(k_{\rm G}) \geq 10^{-24}$ seems more reasonable.

The fact that the gauge coupling must be sharply growing in order to produce large-scale magnetic fields fits extremely well with the pre–big-bang dynamics where, indeed, the gauge coupling is expected to grow. The second requirement to obtain a phenomenologically viable mechanism for the amplification of large-scale gauge fields turned out to be the existence of a pretty long stringy phase.

The "stringy" phase is simply the epoch where quadratic curvature corrections start being important and lead to an effective dynamics where the dilaton field is linearly growing in the cosmic time coordinate (see [93] and references therein). Towards the end of the stringy phase the dilaton freezes to its (constant) value and the Universe gets dominated by radiation. One possibility for achieving the transition to radiation is represented by the back-reaction effects of the produced particles [102]. In particular, the short-wavelength modes play, in this context a crucial role. It is interesting that while the magnetic energy spectrum produced during the stringy phase is quasi-flat and the value of $r(k_{\rm G})$ can be as large as 10^{-8} implying a protogalactic magnetic field of the order of 10^{-10} G. Under these conditions the dynamo mechanism would even be superfluous since the compressional amplification alone can amplify the seed field to its observed value.

The results reported above may be "tested" in a framework where the pre-big-bang dynamics is solvable. Consider, in particular, the situation where the evolution of the dilaton field as well as the one of the geometry is treated in the presence of a nonlocal dilaton potential [97, 98, 99, 100, 101].

In the Einstein frame description, the asymptotics of the (four-dimensional) pre-big-bang dynamics can be written as [102]

$$
\begin{aligned}
&a(\tau) \simeq a_{-}\sqrt{-\frac{\tau}{2\tau_0}}, \qquad a_{-} = e^{-\varphi_0/2}\sqrt{\frac{2(\sqrt{3}+1)}{\sqrt{3}}},\\
&\varphi_{-} = \varphi_0 - \ln 2 - \sqrt{3}\ln\left(\frac{\sqrt{3}+1}{\sqrt{3}}\right) - \sqrt{3}\ln\left(-\frac{\tau}{2\tau_0}\right),\\
&\mathcal{H}_{-} = \frac{1}{2\tau}, \qquad \varphi'_{-} = -\frac{\sqrt{3}}{\tau},
\end{aligned}
\tag{7}
$$

for $\tau \to -\infty$, and

$$
a(\tau) \simeq a_{+}\sqrt{\frac{\tau}{2\tau_0}}, \qquad a_{+} = e^{\varphi_0/2}\sqrt{\frac{2(\sqrt{3}-1)}{\sqrt{3}}}
$$

$$\varphi_+ = \varphi_0 - \ln 2 - \sqrt{3}\ln\left(\frac{\sqrt{3}-1}{\sqrt{3}}\right) + \sqrt{3}\ln\left(\frac{\tau}{2\eta_0}\right),$$
$$\mathcal{H}_+ = \frac{1}{2\tau}, \qquad \varphi_+' = \frac{\sqrt{3}}{\tau}, \tag{8}$$

for $\tau \to +\infty$. In (66) and (67), $\mathcal{H} = a'/a$ and, as usual, the prime denotes a derivation with respect to τ. The branch of the solution denoted by minus describes, in the Einstein frame, an accelerated contraction, since the first derivative of the scale factor is negative, while the second is positive. The branch of the solution denoted with plus describes, in the Einstein frame, a decelerated expansion, since the first derivative of the scale factor is positive while the derivative is negative. In both branches the dilaton grows and its derivative is always positive-definite (i.e. $\varphi_\pm' > 0$) as required by the present approach to bouncing solutions. The numerical solution corresponding to the asymptotics given in (66) and (67) is reported in Fig. 3.

In the Schrödinger description the vacuum state evolves, unitarily, to a multimode squeezed state, in full analogy with what happens in the case of relic gravitons [82, 103, 104]. In the following the same process will be discussed within the Heisenberg representation. The two physical polarizations of the photon can then be quantized according to the standard rules of quantization in the radiation gauge in curved space–times:

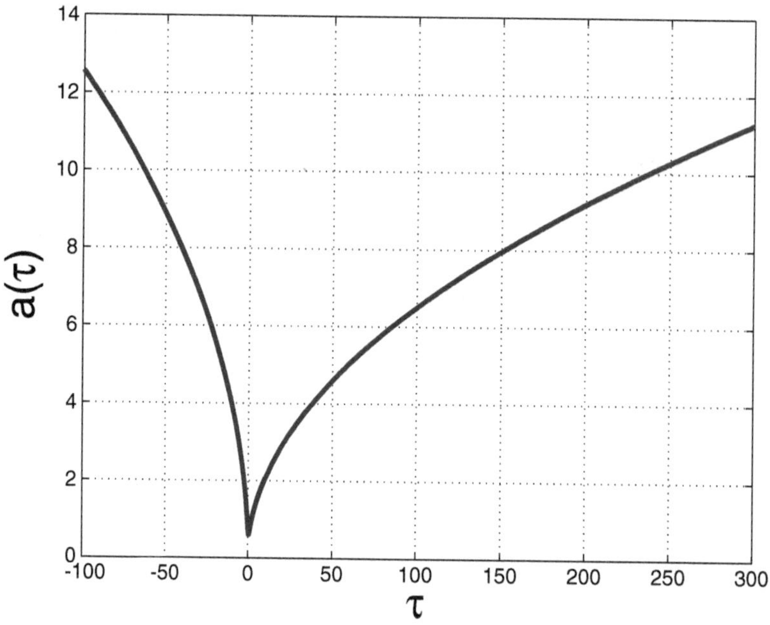

Fig. 3. The evolution of the scale factor in conformal time for a bouncing model regularized via nonlocal dilaton potential in the Einstein frame

$$\hat{\mathcal{A}}_i(\boldsymbol{x},\tau) = \sum_{\alpha} \int \frac{d^3k}{(2\pi)^{3/2}} \left[\hat{a}_{k,\alpha} e_i^{\alpha} \mathcal{A}_k(\tau) e^{-i\boldsymbol{k}\cdot\boldsymbol{x}} + \hat{a}^{\dagger}_{k,\alpha} e_i^{\alpha} \mathcal{A}_k(\tau)^{\star} e^{i\boldsymbol{k}\cdot\boldsymbol{x}} \right], \tag{9}$$

and

$$\hat{\pi}_i(\boldsymbol{x},\tau) = \sum_{\alpha} \int \frac{d^3k}{(2\pi)^{3/2}} \left[\hat{a}_{k,\alpha} e_i^{\alpha} \Pi_k(\tau) e^{-i\boldsymbol{k}\cdot\boldsymbol{x}} + \hat{a}^{\dagger}_{k,\alpha} e_i^{\alpha} \Pi_k(\tau)^{\star} e^{i\boldsymbol{k}\cdot\boldsymbol{x}} \right], \tag{10}$$

where $e_i^{\alpha}(k)$ describe the polarizations of the photon and

$$\Pi_k(\tau) = \mathcal{A}'_k(\tau), \qquad [\hat{a}_{k,\alpha}, \hat{a}^{\dagger}_{p,\beta}] = \delta_{\alpha\beta}\delta^{(3)}(\boldsymbol{k}-\boldsymbol{p}). \tag{11}$$

The evolution equation for the mode functions will then be, in Fourier space,

$$\mathcal{A}''_k + \left[k^2 - g(g^{-1})'' \right] \mathcal{A}_k = 0, \tag{12}$$

i.e. exactly the same equation obtained in (63). The pump field can also be expressed as

$$g(g^{-1})'' = \left(\frac{\varphi'^2}{4} - \frac{\varphi''}{2} \right). \tag{13}$$

The maximally amplified modes are then the ones for which

$$k^2_{\rm max} \simeq |g(g^{-1})''|. \tag{14}$$

The Fourier modes appearing in (71) have to be normalized while they are inside the horizon for large and negative τ. In this limit the initial conditions provided by quantum mechanics are

$$\mathcal{A}_k(\tau) = \frac{1}{\sqrt{2k}} e^{-ik\tau}, \qquad \Pi_k(\tau) = -i\sqrt{\frac{k}{2}} e^{-ik\tau}. \tag{15}$$

In the limit $\tau \to +\infty$ the positive and negative frequency modes will be mixed, so that the solution will be represented in the plane-wave orthonormal basis as

$$\mathcal{A}_k(\tau) = \frac{1}{\sqrt{2k}} \left[c_+(k) e^{-ik\tau} + c_-(k) e^{ik\tau} \right],$$
$$\mathcal{A}'_k(\tau) = -i\sqrt{\frac{k}{2}} \left[c_+(k) e^{-ik\tau} - c_-(k) e^{ik\tau} \right]. \tag{16}$$

where $c_{\pm}(k)$ are the (constant) mixing coefficients. The following two relations fully determine the square modulus of each of the two mixing coefficients in terms of the complex wave-functions obeying (71):

$$|c_+(k)|^2 - |c_-(k)|^2 = i(\mathcal{A}^{\star}_k \Pi_k - \mathcal{A}_k \Pi^{\star}_k), \tag{17}$$

$$|c_+(k)|^2 + |c_-(k)|^2 = \frac{1}{k^2}\left(|\Pi_k|^2 + k^2|\mathcal{A}_k|^2\right). \quad (18)$$

After having numerically computed the time evolution of the properly normalized mode functions, (76) and (77) can be used to infer the value of the relevant mixing coefficient (i.e. $c_-(k)$). Equation (76) is, in fact, the Wronskian of the solutions. If the second-order differential equation is written in the form (71), the Wronskian is always conserved throughout the time evolution of the system. Since, from (74), the Wronskian is equal to 1 initially, it will be equal to 1 all along the time evolution. Thus, from (76) $|c_+(k)|^2 = |c_-(k)|^2 + 1$. The fact that the Wronskian must always be equal to 1 is the measure of the precision of the algorithm.

In Figs. 4 and 5 the numerical calculation of the spectrum is illustrated for different values of k. In Fig. 5 the mixing coefficients are reported for modes $k \ll k_{\rm max}$. In Fig. 4 the mixing coefficients are reported for modes around $k_{\rm max}$. Clearly, from Fig. 5 a smaller k leads to a larger mixing coefficient which means that the spectrum is rather blue. Furthermore, by comparing the amplification of different modes, it is easy to infer that the scaling law is $|c_+(k)|^2 + |c_-(k)|^2 \propto (k/k_{\rm max})^{-n_g}$, with $n_g \sim 3.46$, which is in excellent agreement with the analytical determination of the mixing coefficients leading to $n_g = 2\sqrt{3} \sim 3.46$ [see below (88)].

The second-piece information that can be drawn from Fig. 4 concerns $k_{\rm max}$, whose specific value

$$k_{\rm max} \simeq \frac{\sqrt{5} - 0.5}{\tau_0}. \quad (19)$$

can be determined numerically for different values of τ_0.

For the value of $k_{\rm max}$ reported in (78), the obtained mixing coefficient is 1, i.e. $|c_-(k_{\rm max})| \simeq 1$. According to Fig. 4 as we move from $k_{\rm max}$ to larger

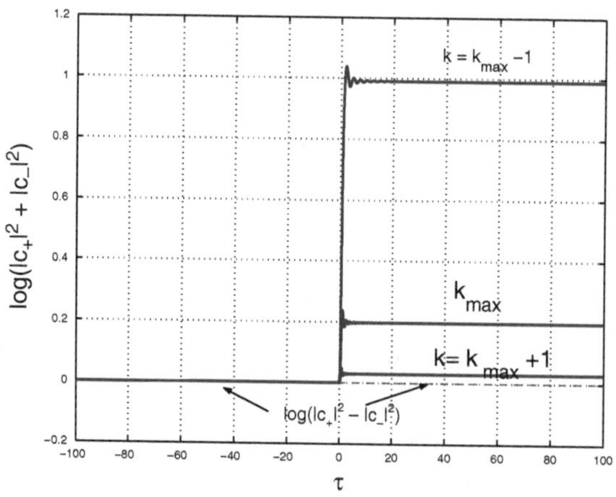

Fig. 4. The evolution of the mixing coefficients for $k \simeq k_{\rm kmax}$ in units of τ_0

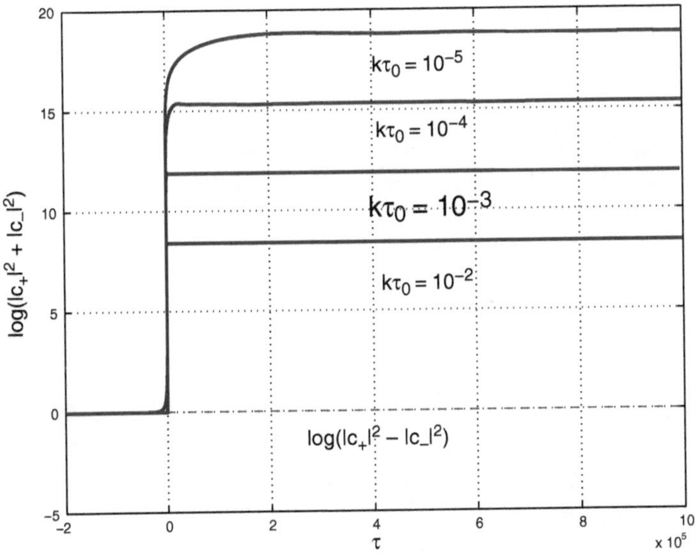

Fig. 5. The numerical estimate of the mixing coefficients in the case $k\tau_0 \ll 1$

k, $(|c_+(k)|^2 + |c_-(k)|^2) \simeq (|c_+(k)|^2 - |c_-(k)|^2)$, implying that $|c_-(k)| \sim 0$. Moreover, from the left plot of Fig. 5 it can be appreciated that

$$|c_-(k_{\max})|^2 = 1, \qquad \log\left(|c_+(k_{\max})|^2 + |c_-(k_{\max})|^2\right) = \log 3 \simeq 0.477. \tag{20}$$

Thus the absolute normalization and slope of the relevant mixing coefficient can be numerically determined to be

$$|c_-(k)|^2 = \left(\frac{k}{k_{\max}}\right)^{-2\sqrt{3}}. \tag{21}$$

It can be concluded that (80) is rather accurate as far as both the slope and the absolute normalization are concerned. The numerical estimates presented so far can also be corroborated by the usual analytical treatment based on the matching of the solutions for the mode functions before and after the bounce. The evolution of the modes described by (71) can be approximately determined from the exact asymptotic solutions given in (66) and (67), and implying that $\varphi'_\pm \simeq \pm\sqrt{3}/\tau$. Thus the solutions of (71) can be obtained in the two asymptotic regimes, i.e. for $\tau \leq -\tau_1$

$$\mathcal{A}_{k,-}(\tau) = \frac{\sqrt{-\pi\tau}}{2} e^{i\frac{\pi}{2}(\nu+1/2)} H_\nu^{(1)}(-k\tau), \tag{22}$$

and for $\tau \geq \tau_1$

$$\mathcal{A}_{k,+}(\eta) = \frac{\sqrt{\pi\tau}}{2} e^{i\frac{\pi}{2}(\mu+1/2)} \left[c_- H_\mu^{(1)}(k\tau) + c_+ e^{-i\pi(\mu+1/2)} H_\mu^{(2)}(k\tau)\right], \quad \tau \geq -\tau_1, \tag{23}$$

where $H_\alpha^{(1,2)}(z)$ are Hankel functions of first and second kind whose related indices are

$$\nu = \frac{\sqrt{3}-1}{2}, \qquad \mu = \frac{\sqrt{3}+1}{2}. \tag{24}$$

The time-scale τ_1 defines the width of the bounce and, typically, $\tau_1 \sim \tau_0$.

The phases appearing in (81) and (82) are carefully chosen so that

$$\lim_{\tau\to-\infty} \mathcal{A}_k = \frac{1}{\sqrt{2k}} e^{-ik\tau}. \tag{25}$$

Using then the appropriate matching conditions

$$\mathcal{A}_{k,-}(-\tau_1) = \mathcal{A}_{k,+}(\tau_1),$$
$$\mathcal{A}'_{k,-}(-\tau_1) = \mathcal{A}'_{k,+}(\tau_1), \tag{26}$$

and defining $x_1 = k\tau_1$, the obtained mixing coefficients are

$$c_+(k) = i\frac{\pi}{4} x_1 e^{i\pi(\nu+\mu+1)/2}\left[-\frac{\nu+\mu+1}{x_1} H_\mu^{(1)}(x_1) H_\nu^{(1)}(x_1)\right.$$
$$\left. + H_\mu^{(1)}(x_1) H_{\nu+1}^{(1)}(x_1) + H_{\mu+1}^{(1)}(x_1) H_\nu^{(1)}(x_1)\right], \tag{27}$$

$$c_-(k) = i\frac{\pi}{4} x_1 e^{i\pi(\nu-\mu)/2}\left[-\frac{\nu+\mu+1}{x_1} H_\mu^{(2)}(x_1) H_\nu^{(1)}(x_1)\right.$$
$$\left. + H_\mu^{(2)}(x_1) H_{\nu+1}^{(1)}(x_1) + H_{\mu+1}^{(2)}(x_1) H_\nu^{(1)}(x_1)\right], \tag{28}$$

satisfying the exact Wronskian normalization condition $|c_+(k)|^2 - |c_-(k)|^2 = 1$. In the small argument limit, i.e. $k\tau_1 \sim k\tau_0 \ll 1$, the leading term in (87) leads to

$$c_-(k) \simeq \frac{i\, 2^{\mu+\nu}}{4\pi} e^{i\pi(\nu-\mu)/2} x_1^{-\mu-\nu} (\nu+\mu-1)\Gamma(\mu)\Gamma(\nu) \tag{29}$$

If we now insert the values given in (83), it turns out that $c_-(k) \simeq 0.41\ |k\tau_1|^{-\sqrt{3}}$. The spectral slope agrees with the numerical estimate, as already stressed. The absolute normalization cannot be determined from (88), where the small argument limit has already been taken. In order to determine the absolute normalization, the specific value of $k_{\rm max}\tau_1$ has to be inserted in (87). The result of this procedure, taking $\tau_1 \sim \tau_0$ is $|c_-(k_{\rm max})|^2 = 0.14$, which is roughly a factor of 10 smaller than the interpolating formula given in (80).

The observation that a dynamical gauge coupling implies a viable mechanism for the production of large-scale magnetic fields can be interesting in general terms and, more specifically, in the context of the pre-big bang models. In fact, in pre-big bang models, not only the fluctuations of the hypercharge field are amplified. In the minimal case we will have to deal with the fluctuations of the tensor [105, 138] and scalar [106] modes of the geometry and with the fluctuations of the antisymmetric tensor field [107, 108].

The amplified tensor modes of the geometry lead to a stochastic background of gravitational waves (GW) with violet spectrum in both the GW amplitude and energy density. In Fig. 6 the GW signal is parametrized in terms of the logarithm of $\Omega_{\mathrm{GW}} = \rho_{\mathrm{GW}}/\rho_{\mathrm{c}}$, i.e. the fraction of critical energy density present (today) in GW. On the horizontal axis of Fig. 6 the logarithm of the present (physical) frequency ν is reported. In conventional inflationary models, for $\nu \geq 10^{-16}$ Hz, Ω_{GW} is constant (or slightly decreasing) as a function of the present frequency. In the case of string cosmological models, $\Omega_{\mathrm{GW}} \propto \nu^3 \ln \nu$, which also implies a steeply increasing power spectrum. This possibility spurred various experimental groups to analyze possible directs limits on the scenario arising from specific instruments such as resonant mass detectors [109] and microwave cavities [110, 111]. These attempts are justified since the signal of pre-big bang models may be rather strong at high frequencies and, anyway, much stronger than the conventional inflationary prediction

The sensitivity of a pair of VIRGO detectors to string cosmological gravitons has been specifically analyzed [112] with the conclusion that a VIRGO pair, in its upgraded stage, can certainly probe wide regions of the parameter space of these models. If we maximize the overlap between the two detectors [112] or if we reduce (selectively) the pendulum and pendulum's internal modes contribution to the thermal noise of the instruments, the visible region (after 1 year of observation and with SNR = 1) of the parameter space will get even larger. Unfortunately, as in the case of the advanced LIGO detectors, the

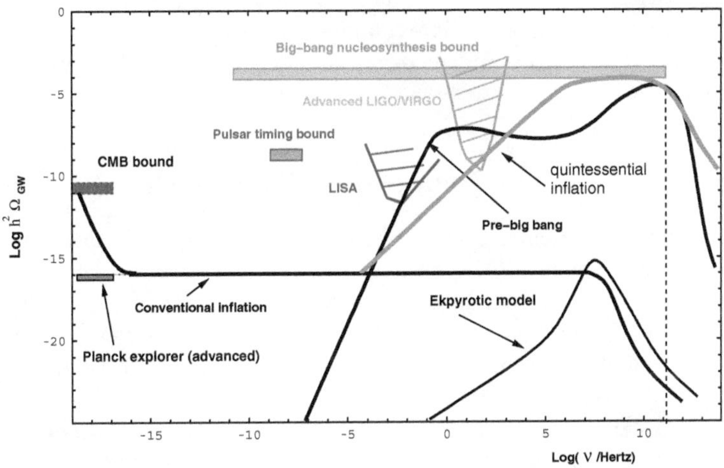

Fig. 6. The spectrum of relic gravitons from various cosmological models presented in terms of $h^2\Omega_{\mathrm{GW}}$

sensitivity to a flat $\Omega_{\rm GW}$ will be irrelevant for ordinary inflationary models also with the advanced VIRGO detector. It is worth mentioning that growing energy spectra of relic gravitons can also arise in the context of quintessential inflationary models [113, 114]. In this case $\Omega_{\rm GW} \propto \nu \ln^2 \nu$ (see [114] for a full discussion).

The spectra of gravitational waves have features that are, in some sense, complementary to the ones of the large-scale magnetic fields. The parameter space leading to a possible signal of relic (pre-big bang) gravitons with wide-band interferometers has only a small overlap with the region of the parameter space leading to sizable large-scale magnetic fields. This conclusion can be evaded if the coupling of the dilaton to the hypercharge field is, in the action, of the type $e^{-\beta\varphi} F_{\mu\nu} F^{\mu\nu}$ [115] where the parameter β has values 1 and 1/2, respectively, for heterotic and type I superstrings. In particular, in the case $\beta = 1/2$, it is possible to find regions where both large-scale magnetic fields and relic gravitons are copiously produced.

Let us finally discuss the scalar fluctuations of the geometry. The spectrum of the scalar modes is determined by the spectrum of the Kalb–Ramond axion(s). If the axions would be neglected, the spectrum of the curvature fluctuations would be sharply increasing, or as we say in the jargon, the spectrum would be violet in full analogy with the spectrum of the tensor modes of the geometry. This result [106] has been recently analyzed in light of a recent controversy (see [97, 98] and references therein).

If the Kalb–Ramond axions are consistently included in the calculation, it is found that the large-scale spectrum of curvature perturbations becomes flat [108] and essentially inherits the spectrum of the Kalb–Ramond axions. If the axions decay (after a phase of coherent oscillations), the curvature perturbations will be adiabatic as in the case of conventional inflationary models but with some important quantitative differences [108] since, in this case, the CMB normalization is explained in terms of the present value of the string curvature scale and in terms of the primordial slope of the axion spectrum.

4 Primordial or Not Primordial, This Is the Question...

While diverse theoretical models for the origin of large-scale magnetism can certainly be questioned on the basis of purely theoretical considerations, direct observations can tell us something more specific concerning the epoch of formation of large-scale magnetic fields. It would be potentially useful to give some elements of response to the following burning question: Are really magnetic fields primordial?

The plan of the present section is the following. In Sect. 4.1 different meanings of the term *primordial* will be discussed. It will be argued that CMB physics can be used to constrain large-scale magnetic fields possibly present prior to matter–radiation equality. In Sect. 4.2 the scalar CMB anisotropies will be specifically discussed by deriving the appropriate set of evolution equations accounting for the presence of a fully inhomogeneous magnetic field. In Sect. 4.3 the evolution of the different species composing the pre-decoupling plasma will be solved, in the tight-coupling approximation and in the presence of a fully inhomogeneous magnetic field. Finally Sect. 4.4 contains various numerical results and a strategy for parameter extraction.

4.1 Pre-equality Magnetic Fields

The term primordial seems to have slightly different meanings depending on the perspective of the various communities converging on the study of large-scale magnetic fields. Radio-astronomers have the hope that by scrutinizing the structure of magnetic fields in distant galaxies it would be possible, in the future, to understand if the observed magnetic fields are the consequence of a strong dynamo action or if their existence precedes the formation of galaxies.

If the magnetic field does not flip its sign from one spiral arm to the other, then a strong dynamo action can be suspected [116]. In the opposite case the magnetic field of galaxies should be *primordial*, i.e. present already at the onset of gravitational collapse. In this context, primordial simply means protogalactic. An excellent review on the evidence of magnetism in nearby galaxies can be found in [117]. In Fig. 7 a schematic view of the Milky Way is presented. The magnetic field follows the spiral arm. There have been claims, in the literature, of three to five field reversals. The arrows in Fig. 7 indicate one of the possible field reversals. One reversal is certain beyond any doubt. Another indication that would support the primordial nature of the magnetic field of galaxies would be, for instance, the evidence that not only spirals but also elliptical galaxies are magnetized (even if the magnetic field seems to have correlation scale shorter than in the case of spirals). Since elliptical galaxies have a much less efficient rotation, it seems difficult to postulate a strong dynamo action. We will not pursue here the path of specific astrophysical signatures of a truly pregalactic magnetic field and we refer the interested reader to [116, 117].

As a side remark, it should also be mentioned that magnetic fields may play a role in the analysis of rotation curves of spiral galaxies. This aspect has been investigated in great depth by Battaner, Florido and collaborators also in connection with possible effects of large-scale magnetic fields on structure formation [119,120,121,122] (see also [123] and references therein).

The large-scale magnetic fields produced via the parametric amplification of quantum fluctuations discussed earlier in the present lecture may also be defined primordial but, in this case, the term primordial has a much broader signification embracing the whole epoch that precedes the equality between matter and radiation taking place, approximately, at a redshift $z_{eq} = 3230$ for $h^2 \Omega_{\rm m0} = 0.134$ and $h^2 \Omega_{\rm r0} = 4.15 \times 10^{-5}$. Consequently, large-scale magnetic fields may affect, potentially, CMB anisotropies [18]. Through the years, various studies have been devoted to the effect of large-scale magnetic fields on the vector and tensor CMB anisotropies [124, 125] (see also [126] and references therein for some recent review articles).

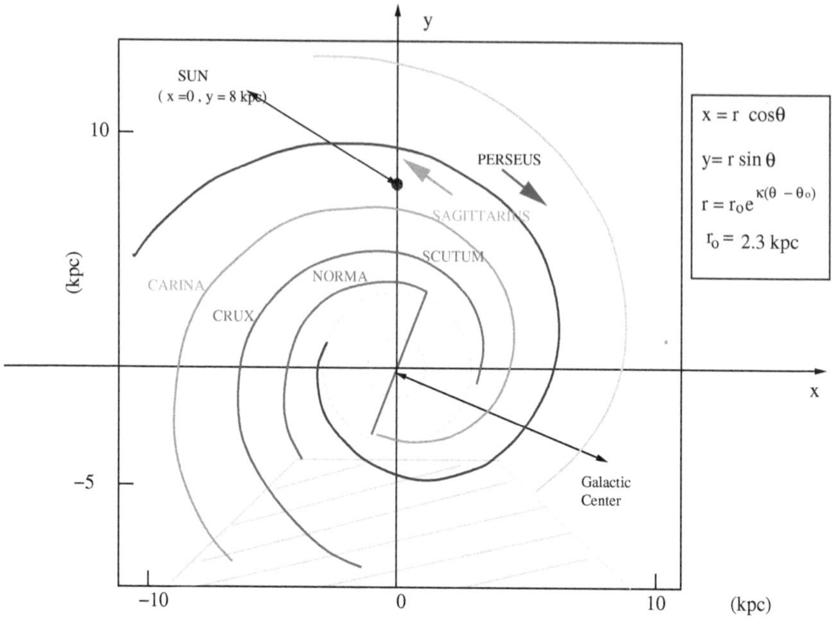

Fig. 7. The schematic map of the MW is illustrated. Following [118] the origin of the two-dimensional coordinate system are in the galactic center. The two large arrows indicate one of the possible (three or five) field reversals observed so far

The implications of fully inhomogeneous magnetic fields on the scalar modes of the geometry remain comparatively less explored. By fully inhomogeneous we mean stochastically distributed fields that do not break the spatial isotropy of the background [22, 23].

CMB anisotropies are customarily described in terms of a set of carefully chosen initial conditions for the evolution of the brightness perturbations of the radiation field. One set of initial conditions corresponds to a purely adiabatic mode. There are, however, more complicated situations where, on top of the adiabatic mode there is also one (or more) nonadiabatic mode(s). A *mode*, in the present terminology, simply means a consistent solution of the governing equations of the metric and plasma fluctuations, i.e. a consistent solution of the perturbed Einstein equations and of the lower multipoles of the Boltzmann hierarchy.

The simplest set of initial conditions for CMB anisotropies implies, in a ΛCDM framework, that a nearly scale-invariant spectrum of adiabatic fluctuations is present after matter–radiation equality (but before decoupling) for typical wavelengths larger than the Hubble radius at the corresponding epoch [127].

It became relevant, through the years, to relax the assumption of exact adiabaticity and to scrutinize the implications of a more general mixture of adiabatic and nonadiabatic initial conditions (see [128, 129, 130] and references therein). In what follows it will be argued, along a similar perspective, that large-scale magnetic fields slightly modify the adiabatic paradigm so that their typical strengths may be constrained. To achieve such a goal, the first step is to solve the evolution equations of magnetized cosmological perturbations well before matter–radiation equality. The second step is to follow the solution through equality (and up to decoupling). On a more technical ground, the second step amounts to the calculation of the so-called transfer matrix [131] whose specific form is one of the subjects of the present analysis.

4.2 Basic Equations

Consider then the system of cosmological perturbations of a flat Friedmann–Robertson–Walker Universe, characterized by a conformal time-scale factor $a(\tau)$ (see (46)), and consisting of a mixture of photons, baryons, CDM particles and massless neutrinos. In the following the basic set of equations used in order to describe the magnetized curvature perturbations will be introduced and discussed. The perspective adopted here is closely related to the recent results obtained in [132, 133] (see also [134, 135] for interesting developments).

In the conformally Newtonian gauge [136, 137, 138, 139, 140], the *scalar* fluctuations of the metric tensor $G_{\mu\nu} = a^2(\tau)\eta_{\mu\nu}$ are parametrized in terms of the two longitudinal fluctuations, i.e.

$$\delta G_{00} = 2a^2(\tau)\phi(\tau, \boldsymbol{x}), \qquad \delta G_{ij} = 2a^2(\tau)\psi(\tau, \boldsymbol{x})\delta_{ij}, \tag{1}$$

where δ_{ij} is the Kroeneker δ. While the spatial curvature will be assumed to vanish, it is straightforward to extend the present considerations to the case when the spatial curvature is not negligible.

In spite of the fact that the present discussion will be conducted within the conformally Newtonian gauge, it can be shown that gauge-invariant descriptions of the problem are possible [133]. Moreover, specific nonadiabatic modes (like the ones related to the neutrino system) may be more usefully described in different gauges (like the synchronous gauge). The rationale for the last statement is that the neutrino isocurvature modes may be singular in the conformally Newtonian gauge. These issues will not be addressed here but have been discussed in the existing literature (see, for instance, [139, 140] and references therein). Furthermore, for the benefit of the interested reader it is appropriate to mention that the relevant theoretical tools used in the present and in the following paragraphs follows the conventions of a recent review [140].

Hamiltonian and Momentum Constraints

The Hamiltonian and momentum constraints, stemming from the (00) and (0i) components of the perturbed Einstein equations are

$$\nabla^2\psi - 3\mathcal{H}(\mathcal{H}\phi + \psi') = 4\pi G a^2[\delta\rho_{\rm t} + \delta\rho_{\rm B}], \tag{2}$$

$$\nabla^2(\mathcal{H}\phi + \psi') = -4\pi G a^2\left[(p_{\rm t} + \rho_{\rm t})\theta_{\rm t} + \frac{\boldsymbol{\nabla}\cdot(\boldsymbol{E}\times\boldsymbol{B})}{4\pi a^4}\right], \tag{3}$$

where $\mathcal{H} = a'/a$ and the prime denotes a derivation with respect to the conformal time coordinate τ. In writing (90) and (91) the following set of conventions has been adopted

$$\delta\rho_{\rm t}(\tau,\boldsymbol{x}) = \delta\rho_\gamma(\tau,\boldsymbol{x}) + \delta\rho_\nu(\tau,\boldsymbol{x}) + \delta\rho_{\rm c}(\tau,\boldsymbol{x}) + \delta\rho_{\rm b}(\tau,\boldsymbol{x}), \tag{4}$$

$$\delta\rho_{\rm B}(\tau,\boldsymbol{x}) = \frac{B^2(\boldsymbol{x})}{8\pi a^4(\tau)}, \tag{5}$$

$$\begin{aligned}(p_{\rm t} + \rho_{\rm t})\theta_{\rm t}(\tau,\boldsymbol{x}) &= (p_\gamma + \rho_\gamma)\theta_\gamma(\tau,\boldsymbol{x}) + (p_\nu + \rho_\nu)\theta_\nu(\tau,\boldsymbol{x})\\ &\quad + (p_{\rm c} + \rho_{\rm c})\theta_{\rm c}(\tau,\boldsymbol{x}) + (p_{\rm b} + \rho_{\rm b})\theta_{\rm b}(\tau,\boldsymbol{x}).\end{aligned} \tag{6}$$

Concerning (92), (93) and (94) the following comments are in order:

- In (92) the total density fluctuation of the plasma, i.e. $\delta\rho_{\rm t}(\tau,\boldsymbol{x})$ receives contributions from all the species of the plasma.
- In (93) the fluctuation of the magnetic energy density $\delta\rho_{\rm B}(\tau,\boldsymbol{x})$ is quadratic in the magnetic field intensity.
- In (94) $\theta_{\rm t}(\tau,\boldsymbol{x}) = \partial_i v^i_{\rm t}$ is the divergence of the total peculiar velocity while $\theta_\gamma(\tau,\boldsymbol{x})$, $\theta_\nu(\tau,\boldsymbol{x})$, $\theta_{\rm c}(\tau,\boldsymbol{x})$ and $\theta_{\rm b}(\tau,\boldsymbol{x})$ are the divergences of the peculiar velocities of each individual species, i.e. photons, neutrinos, CDM particles and baryons.

The second term appearing at the right-hand side of (91) is the divergence of the Poynting vector. In MHD the Ohmic electric field is subleading and, in particular, from the MHD expression of the Ohm law we will have

$$\boldsymbol{E}\times\boldsymbol{B} \simeq \frac{(\boldsymbol{\nabla}\times\boldsymbol{B})\times\boldsymbol{B}}{4\pi\sigma}. \tag{7}$$

Since the Universe, prior to decoupling, is a very good conductor, the ideal MHD limit can be safely adopted in the first approximation (see also [130]); thus for $\sigma \to \infty$ (i.e. infinite conductivity limit) the contribution of the Poynting vector vanishes. In any case, even if σ would be finite but large, the second term at the right-hand side of (91) would be suppressed in comparison with the contribution of the divergence of the total velocity field.

The total (unperturbed) energy density and pressure of the mixture, i.e.

$$\begin{aligned}\rho_{\rm t} &= \rho_\gamma + \rho_\nu + \rho_{\rm c} + \rho_{\rm b} + \rho_\Lambda,\\ p_{\rm t} &= p_\gamma + p_\nu + p_{\rm c} + p_{\rm b} + p_\Lambda,\end{aligned} \tag{8}$$

determine the evolution of the background geometry according to Friedmann equations:

$$\mathcal{H}^2 = \frac{8\pi G}{3} a^2 \rho_{\rm t}, \tag{9}$$

$$\mathcal{H}^2 - \mathcal{H}' = 4\pi G a^2 (\rho_{\rm t} + p_{\rm t}), \tag{10}$$

$$\rho_{\rm t}' + 3\mathcal{H}(\rho_{\rm t} + p_{\rm t}) = 0. \tag{11}$$

Notice that in (96) the contribution of the cosmological constant has been included. If the dark energy is parametrized in terms of a cosmological constant (i.e. $p_\Lambda = -\rho_\Lambda$), then, $\delta\rho_\Lambda' = 0$. Furthermore, the contribution of ρ_Λ to the background evolution is negligible prior to decoupling. Slightly different situations (not contemplated by the present analysis) may arise if the dark energy is parametrized in terms of one or more scalar degrees of freedom with suitable potentials.

Dynamical Equation and Anisotropic Stress(es)

The spatial components of the perturbed Einstein equations imply instead

$$\left[\psi'' + \mathcal{H}(\phi' + 2\psi') + (\mathcal{H}^2 + 2\mathcal{H}')\phi + \frac{1}{2}\nabla^2(\phi - \psi)\right]\delta_i^j$$
$$-\frac{1}{2}\partial_i\partial^j(\phi - \psi) = 4\pi G a^2 \left[(\delta p_{\rm t} + \delta p_{\rm B})\delta_i^j - \Pi_i^j - \tilde{\Pi}_i^j\right]. \tag{12}$$

Equation (100) contains, as source terms, not only the total fluctuation of the pressure of the plasma, i.e. $\delta p_{\rm t}$, but also

$$\delta p_{\rm B}(\tau, \boldsymbol{x}) = \frac{B^2(\boldsymbol{x})}{24\pi a^4(\tau)} = \frac{\delta\rho_{\rm B}(\tau, \boldsymbol{x})}{3}. \tag{13}$$

$$\tilde{\Pi}_i^j(\tau, \boldsymbol{x}) = \frac{1}{4\pi a^4}\left(B_i B^j - \frac{1}{3}B^2\delta_i^j\right). \tag{14}$$

Moreover, in (100), $\Pi_i^j(\tau, \boldsymbol{x})$ is the anisotropic stress of the fluid. As it will be mentioned in a moment (and later on heavily used) the main source of anisotropic stress of the fluid is provided by neutrinos which free-stream from temperature smaller than mega electronvolts. Notice that both the anisotropic stress of the fluid, i.e. $\Pi_i^j(\tau, \boldsymbol{x})$, and the magnetic anisotropic stress, i.e. $\tilde{\Pi}_i^j(\tau, \boldsymbol{x})$, are, by definition, traceless.

Using this last observation, (100) can be separated into two independent equations. Taking the trace of (100) we do get

$$\psi'' + \mathcal{H}(\phi' + 2\psi') + (2\mathcal{H}' + \mathcal{H}^2)\phi + \frac{1}{3}\nabla^2(\phi - \psi) = 4\pi G a^2(\delta p_{\rm t} + \delta p_{\rm B}). \tag{15}$$

By taking the difference between (100) and (103), the following (traceless) relation can be obtained:

$$\partial_i\partial^j(\phi-\psi)-\frac{1}{3}\delta_i^j\nabla^2(\phi-\psi)=8\pi Ga^2(\Pi_i^j+\tilde{\Pi}_i^j). \qquad (16)$$

By applying the differential operator $\partial_j\partial^i$ to both sides of (104), we do obtain the following interesting relation:

$$\nabla^4(\phi-\psi)=12\pi Ga^2[(p_\nu+\rho_\nu)\nabla^2\sigma_\nu+(p_\gamma+\rho_\gamma)\nabla^2\sigma_{\rm B}], \qquad (17)$$

where the parametrization

$$\partial_j\partial^i\Pi_i^j=(p_\nu+\rho_\nu)\nabla^2\sigma_\nu, \qquad \partial_j\partial^i\tilde{\Pi}_i^j=(p_\gamma+\rho_\gamma)\nabla^2\sigma_{\rm B}, \qquad (18)$$

has been adopted. In (105) $\sigma_\nu(\tau,\boldsymbol{x})$ is related with the quadrupole moment of the (perturbed) neutrino phase–space distribution. In (105) $\sigma_{\rm B}(\tau,\boldsymbol{x})$ parametrizes the (normalized) magnetic anisotropic stress. It is relevant to remark at this point that in the MHD approximation adopted here the two main sources of scalar anisotropy associated with magnetic fields can be parametrized in terms of $\sigma_{\rm B}(\tau,\boldsymbol{x})$ and in terms of the dimensionless ratio

$$\Omega_{\rm B}(\tau,\boldsymbol{x})=\frac{\delta\rho_{\rm B}(\tau,\boldsymbol{x})}{\rho_\gamma(\tau)}. \qquad (19)$$

Since both $\Omega_{\rm B}(\tau,\boldsymbol{x})$ and $\sigma_{\rm B}(\tau,\boldsymbol{x})$ are quadratic in the magnetic field intensity, a non-Gaussian contribution may be expected. $\Omega_{\rm B}(\tau,\boldsymbol{x})$ is the magnetic energy density referred to the photon energy density, and it is constant to a very good approximation if magnetic flux is frozen into the plasma element.

There is, in principle, a third contribution to the scalar problem coming from magnetic fields. Such a contribution arises in the evolution equation of the photon–baryon peculiar velocity and amounts to the divergence of the Lorentz force. While the mentioned equation will be derived later in this section, it is relevant to point out here that the MHD Lorentz force can be expressed solely in terms of $\sigma_{\rm B}(\tau,\boldsymbol{x})$ and $\Omega_{\rm B}(\tau,\boldsymbol{x})$. In fact a well-known vector identity stipulates that

$$\partial_i B_j\partial^j B^i=\boldsymbol{\nabla}\cdot[(\boldsymbol{\nabla}\times\boldsymbol{B})\times\boldsymbol{B}]+\frac{1}{2}\nabla^2 B^2. \qquad (20)$$

From the definition of $\sigma_{\rm B}$ in terms of $\tilde{\Pi}_i^j$, i.e. (106), it is easy to show that

$$\nabla^2\sigma_{\rm B}=\frac{3}{16\pi a^4\rho_\gamma}\partial_i B_j\partial^j B^i-\frac{1}{2}\nabla^2\Omega_{\rm B}. \qquad (21)$$

Using then (108) into (109) and recalling that

$$4\pi\boldsymbol{\nabla}\cdot[\boldsymbol{J}\times\boldsymbol{B}]=\boldsymbol{\nabla}\cdot[(\boldsymbol{\nabla}\times\boldsymbol{B})\times\boldsymbol{B}], \qquad (22)$$

we obtain

$$\nabla^2\sigma_{\rm B}=\frac{3}{16\pi a^4\rho_\gamma}\boldsymbol{\nabla}\cdot[(\boldsymbol{\nabla}\times\boldsymbol{B})\times\boldsymbol{B}]+\frac{\nabla^2\Omega_{\rm B}}{4}. \qquad (23)$$

Curvature Perturbations

Two important quantities must now be introduced. The first one, conventionally denoted by ζ, is the density contrast on uniform curvature hypersurfaces,[11] i.e.

$$\zeta = -\psi - \mathcal{H}\frac{(\delta\rho_{\rm t} + \delta\rho_{\rm B})}{\rho_{\rm t}'}. \tag{24}$$

The definition (112) is invariant under infinitesimal coordinate transformations. In fact, while $\delta\rho_{\rm B}$ is automatically gauge-invariant (since the magnetic field vanishes at the level of the background), ψ and $\delta\rho_{\rm t}$ transform as [140]

$$\begin{aligned} \psi &\to \tilde{\psi} = \psi + \mathcal{H}\epsilon, \\ \delta\rho_{\rm t} &\to \tilde{\delta\rho_{\rm t}} - \rho_{\rm t}'\epsilon, \end{aligned} \tag{25}$$

for

$$\begin{aligned} \tau &\to \tilde{\tau} = \tau + \epsilon^0 \\ x^i &\to \tilde{x}^i = x^i + \partial^i\epsilon. \end{aligned} \tag{26}$$

Recalling (99), (112) can also be written as

$$\zeta = -\psi + \frac{\delta\rho_{\rm t} + \delta\rho_{\rm B}}{3(\rho_{\rm t} + p_{\rm t})}. \tag{27}$$

The second variable we want to introduce, conventionally denoted by $\mathcal{R}$ is the curvature perturbation on comoving orthogonal hypersurfaces,[12] i.e.

$$\mathcal{R} = -\psi - \frac{\mathcal{H}(\mathcal{H}\phi + \psi')}{\mathcal{H}^2 - \mathcal{H}'}. \tag{28}$$

Inserting (115) and (116) into (90), the Hamiltonian constraint takes then the form

$$\zeta = \mathcal{R} + \frac{\nabla^2\psi}{12\pi G a^2(p_{\rm t} + \rho_{\rm t})}. \tag{29}$$

Equation (117) is rather interesting in its own right and it tells that, in the long wavelength limit,

$$\zeta \simeq \mathcal{R} + \mathcal{O}(k^2\tau^2). \tag{30}$$

When the relevant wavelengths are larger than the Hubble radius (i.e. $k\tau \ll 1$), the density contrast on uniform curvature hypersurfaces and the

[11] Since, as it will be discussed, ζ is gauge-invariant, we can also interpret it as the curvature fluctuation on uniform density hypersurfaces, i.e. the fluctuation of the scalar curvature on the hypersurface where the total density is uniform.

[12] It is clear, from the definition (116) that the second term at the right-hand side is proportional, by the momentum constraint (91), to the total peculiar velocity of the plasma which is vanishing on comoving (orthogonal) hypersurfaces.

curvature fluctuations on comoving orthogonal hypersurfaces coincide. Since the ordinary Sachs–Wolfe contribution to the gauge-invariant temperature fluctuation is dominated by wavelengths that are larger than the Hubble radius after matter–radiation equality (but before radiation decoupling), the calculation of ζ (or $\mathcal{R}$), in the long-wavelength limit, will essentially give us the Sachs–Wolfe (SW) plateau.

A remark on the definition given in (112) is in order. The variable ζ must contain the *total* fluctuation of the energy density. This is crucial since the Hamiltonian constraint is sensitive to the total fluctuation of the energy density. If the magnetic energy density $\delta\rho_{\mathrm{B}}$ is correctly included in the definition of ζ, then the Hamiltonian constraint (117) maintains its canonical form.

Equations (117) and (118) can be used to derive the appropriate transfer matrices, allowing, in turn, the estimate of the Sachs–Wolfe plateau. For this purpose it is important to deduce the evolution equation for ζ. The evolution of ζ can be obtained from the evolution equation of the total density fluctuation which reads, in the conformally Newtonian gauge,

$$\delta\rho_{\mathrm{t}}' - 3\psi'(p_{\mathrm{t}} + \rho_{\mathrm{t}}) + (p_{\mathrm{t}} + \rho_{\mathrm{t}})\theta_{\mathrm{t}} + 3\mathcal{H}(\delta p_{\mathrm{t}} + \delta\rho_{\mathrm{t}}) + 3\mathcal{H}\delta p_{nad} = \frac{\boldsymbol{E}\cdot\boldsymbol{J}}{a^4}. \tag{31}$$

The technique is now rather simple. We can extract $\delta\rho_{\mathrm{t}}$ from (27)

$$\delta\rho_{\mathrm{t}} = 3(\rho_{\mathrm{t}} + p_{\mathrm{t}})(\zeta + \psi) - \delta\rho_{\mathrm{B}}. \tag{32}$$

Inserting (120) into (119) we get to the wanted evolution equation for ζ. Before doing that it is practical to discuss the case when the relativistic fluid receives contributions from different species that are simultaneously present. In the realistic case, considering that the cosmological constant does not fluctuate, we will have four different species.

For deriving the evolution equation of ζ, it is practical (and, to some extent, conventional) to separate the pressure fluctuation into an adiabatic component supplemented by a nonadiabatic contribution:

$$\delta p_{\mathrm{t}} = \left(\frac{\delta p_{\mathrm{t}}}{\delta\rho_{\mathrm{t}}}\right)_{\varsigma}\delta\rho_{\mathrm{t}} + \left(\frac{\delta p_{\mathrm{t}}}{\delta\varsigma}\right)_{\rho_{\mathrm{t}}}\delta\varsigma. \tag{33}$$

In a relativistic description of gravitational fluctuations, the pressure fluctuates both because the energy density fluctuates (first term at the right-hand side of Eq. (121)) and because the specific entropy of the plasma, i.e. ς, fluctuates (first term at the right-hand side of (121)). The subscripts appearing in the two terms at the right-hand side of (121) simply mean that the two different variations must be taken, respectively, at constant ς (i.e. $\delta\varsigma = 0$) and at constant ρ_{t} (i.e. $\delta\rho_{\mathrm{t}} = 0$).

Here is an example of the usefulness of this decomposition. Consider, for instance, a mixture of CDM particles and radiation. In this case the coefficient of the first term at the right-hand side of (121) can be written as

$$\left(\frac{\delta p_{\rm t}}{\delta \rho_{\rm t}}\right)_{\varsigma} = \frac{1}{3}\left(\frac{\delta \rho_{\rm r}}{\delta \rho_{\rm c} + \delta \rho_{\rm r}}\right)_{\varsigma}, \tag{34}$$

where we simply used the fact that $\delta p_{\rm r} = \delta \rho_{\rm r}/3$ and that $\delta \rho_{\rm t} = \delta \rho_{\rm r} + \delta \rho_{\rm c}$. Now, the quantity appearing in (122) must be evaluated at constant ς, i.e. for $\delta\varsigma = 0$. The specific entropy, in the CDM radiation system, is given by $\varsigma = T^3/n_{\rm c}$ where T is the temperature and $n_{\rm c}$ is the CDM concentration. The relative fluctuations of the specific entropy can then be defined and they are

$$\mathcal{S} = \frac{\delta\varsigma}{\varsigma} = \frac{3}{4}\frac{\delta \rho_{\rm r}}{\rho_{\rm r}} - \frac{\delta \rho_{\rm c}}{\rho_{\rm c}}, \tag{35}$$

where it has been used that $\rho_{\rm r} \simeq T^4$ and that $\rho_{\rm c} \simeq m\, n_{\rm c}$ (m is here the typical mass of the CDM particle). Requiring now that $\mathcal{S} = 0$ we do get $\delta \rho_{\rm c} = (3/4)(\rho_{\rm c}/\rho_{\rm r})\delta \rho_{\rm r}$. Thus, inserting $\delta \rho_{\rm c}$ into (122), the following relation can be easily obtained:

$$\left(\frac{\delta p_{\rm t}}{\delta \rho_{\rm t}}\right)_{\varsigma} = \frac{4\rho_{\rm r}}{3(3\rho_{\rm c} + 4\rho_{\rm r})} \equiv \frac{p_{\rm t}'}{\rho_{\rm t}'} = c_{\rm s}^2. \tag{36}$$

The second and third equalities in (124) follow from the definition of the total sound speed for the CDM-radiation system. This occurrence is general and it is not a peculiarity of the CDM-radiation system so that we can write, for an arbitrary mixture of relativistic fluids:

$$\left(\frac{\delta p_{\rm t}}{\delta \rho_{\rm t}}\right)_{\varsigma} = \frac{p_{\rm t}'}{\rho_{\rm t}'} = c_{\rm s}^2. \tag{37}$$

The definition of relative entropy fluctuation proposed in (123) is invariant under infinitesimal gauge transformations [140] and it can be generalized by introducing two interesting variables, namely,

$$\zeta_{\rm r} = -\psi - \mathcal{H}\frac{\delta \rho_{\rm r}}{\rho_{\rm r}'} \textit{and} \qquad \zeta_{\rm c} = -\psi - \mathcal{H}\frac{\delta \rho_{\rm c}}{\rho_{\rm c}'}. \tag{38}$$

Using the continuity equations for the CDM and for radiation, i.e. $\rho_{\rm r}' = -4\mathcal{H}\rho_{\rm r}$ and $\rho_{\rm c}' = -3\mathcal{H}\rho_{\rm c}$, (126) can be also written as

$$\zeta_{\rm r} = -\psi + \frac{\delta_{\rm r}}{4}, \qquad \zeta_{\rm c} = -\psi + \frac{\delta_{\rm c}}{3}, \tag{39}$$

where $\delta_{\rm r} = \delta \rho_{\rm r}/\rho_{\rm r}$ and $\delta_{\rm c} = \delta \rho_{\rm c}/\rho_{\rm c}$. Thus, using (127), the relative fluctuation in the specific entropy introduced in (123) can also be written as

$$\mathcal{S} = -3(\zeta_{\rm c} - \zeta_{\rm r}). \tag{40}$$

It is a simple exercise to verify that (123) and (128) have indeed the same physical content.

Up to now the coefficient of *the first term* at the right-hand side of (121) has been computed. Let us now discuss *the second term* appearing at the right-hand side of (121). Conventionally, the whole second term is often denoted by $\delta p_{\rm nad}$, i.e. nonadiabatic pressure variation. From (123) defining the relative fluctuation in the specific entropy, i.e. $\mathcal{S} = \delta\varsigma/\varsigma$, the following equation can be written:

$$\delta p_{\rm nad} = \left(\frac{\delta p_{\rm t}}{\delta\varsigma}\right)_{\rho_{\rm t}}, \delta\varsigma \equiv \left(\frac{\delta p_{\rm t}}{\mathcal{S}}\right)_{\rho_{\rm t}} \mathcal{S}. \tag{41}$$

Now, $\mathcal{S}$ must be evaluated, inside the round bracket, for $\delta\rho_{\rm t} = 0$. The result will be

$$\left(\frac{\delta p_{\rm t}}{\mathcal{S}}\right)_{\rho_{\rm t}} = \frac{4}{3}\frac{\rho_{\rm c}\,\rho_{\rm r}}{3\rho_{\rm c} + 4\rho_{\rm r}}. \tag{42}$$

Recalling the definition of sound speed and using (130) into (129), we do get

$$\delta p_{\rm nad} = c_{\rm s}^2 \rho_{\rm c} \mathcal{S}. \tag{43}$$

If the mixture of fluids is more complicated, the discussion presented so far can be easily generalized. If more than two fluids are present, we can still separate, formally, the pressure fluctuation as

$$\delta p_{\rm t} = c_{\rm s}^2 \delta\rho_{\rm t} + \delta p_{\rm nad}. \tag{44}$$

However, if more than two fluids are present, the nonadiabatic pressure density fluctuation has a more complicated form that reduces to the one previously computed in the case of two fluids:

$$\delta p_{\rm nad} = \frac{1}{6\mathcal{H}\rho_{\rm t}'} \sum_{\rm ij} \rho_{\rm i}' \rho_{\rm j}' (c_{\rm s\,i}^2 - c_{\rm s\,j}^2) \mathcal{S}_{\rm i\,j},$$
$$\mathcal{S}_{\rm ij} = -3(\zeta_{\rm i} - \zeta_{\rm j}), \qquad c_{\rm s\,i}^2 = \frac{p_{\rm i}'}{\rho_{\rm i}'}, \tag{45}$$

where $\mathcal{S}_{\rm i\,j}$ are the relative fluctuations in the entropy density that can be computed in terms of the density contrasts of the individual fluids. The indices i and j run over all the components of the plasma. Assuming a plasma formed by photons, neutrinos, baryons and CDM particles, we will have that various entropy fluctuations are possible. For instance

$$\mathcal{S}_{\gamma{\rm c}} = -3(\zeta_\gamma - \zeta_{\rm c}), \qquad \mathcal{S}_{\gamma\nu} = -3(\zeta_\gamma - \zeta_\nu), \quad ... \tag{46}$$

where the ellipses stand for all the other possible combinations. From the definition of relative entropy fluctuations it appears that $\mathcal{S}_{\gamma\nu} = -\mathcal{S}_{\nu\gamma}$. Finally, with obvious notations, while $c_{\rm s}^2$ denotes the *total* sound speed, $c_{\rm s\,i}^2$ and $c_{\rm s\,i}^2$ denote the sound speeds of a generic pair of fluids contributing $\mathcal{S}_{\rm ij}$ to $\delta p_{\rm nad}$, i.e.

$$c_{\rm s}^2 = \frac{p_{\rm t}'}{\rho_{\rm t}'}, \qquad c_{\rm s\,i}^2 = \frac{p_{\rm i}'}{\rho_{\rm i}'}, \qquad c_{\rm s\,j}^2 = \frac{p_{\rm j}'}{\rho_{\rm j}'}. \tag{47}$$

In light of (134), also the physical interpretation of (132) becomes more clear. The contribution of $\delta p_{\rm nad}$ arises because of the inherent multiplicity of fluid present in the plasma. Thanks to (132) using (120) in (119), we can obtain the evolution equation for ζ which becomes

$$\zeta' = -\frac{\mathcal{H}}{p_{\rm t}+\rho_{\rm t}}\delta p_{\rm nad} + \frac{\mathcal{H}}{p_{\rm t}+\rho_{\rm t}}\left(c_{\rm s}^2 - \frac{1}{3}\right)\delta\rho_{\rm B} - \frac{\theta_{\rm t}}{3}. \tag{48}$$

The evolution equation for $\mathcal{R}$ can also be directly obtained by taking the first-time derivative of (117), i.e.

$$\zeta' = \mathcal{R}' + \frac{\nabla^2\psi'}{12\pi G a^2(p_{\rm t}+\rho_{\rm t})} + \frac{\mathcal{H}(3c_{\rm s}^2+1)\nabla^2\psi}{12\pi G a^2(p_{\rm t}+\rho_{\rm t})}. \tag{49}$$

By now inserting (137) into (136) and by using the momentum constraint of (91) to eliminate $\theta_{\rm t}$ we do get the following expression:

$$\begin{aligned}\mathcal{R}' = &-\frac{\mathcal{H}}{p_{\rm t}+\rho_{\rm t}}\delta p_{nad} + \frac{\mathcal{H}}{p_{\rm t}+\rho_{\rm t}}\left(c_{\rm s}^2 - \frac{1}{3}\right)\delta\rho_{\rm B}\\ &-\frac{\mathcal{H}c_{\rm s}^2\nabla^2\psi}{4\pi G a^2(p_{\rm t}+\rho_{\rm t})} + \frac{\mathcal{H}\nabla^2(\phi-\psi)}{12\pi G a^2(p_{\rm t}+\rho_{\rm t})}.\end{aligned} \tag{50}$$

It could be finally remarked that (138) can be directly derived from (103). For this purpose, The definition (116) can be derived once with respect to τ. The obtained result, once inserted back into (103) reproduces (138).

4.3 Evolution of Different Species

Up to now the global variables defining the evolution of the system have been discussed in a unified perspective. The evolution of the global variables is determined by the evolution of the density contrasts and peculiar velocities of the different species. Consequently, in the following paragraphs, the evolution of the different species will be addressed.

Photons and Baryon

The evolution equations of the lowest multipoles of the photon–baryon system amount, in principle, to the following two sets of equations:

$$\delta_{\rm b}' = 3\psi' - \theta_{\rm b}, \tag{51}$$

$$\theta_{\rm b}' + \mathcal{H}\theta_{\rm b} = -\nabla^2\phi + \frac{\boldsymbol{\nabla}\cdot[\boldsymbol{J}\times\boldsymbol{B}]}{a^4\rho_{\rm b}} + \frac{4}{3}\frac{\rho_\gamma}{\rho_{\rm b}}a n_{\rm e} x_{\rm e}\sigma_{\rm T}(\theta_\gamma - \theta_{\rm b}), \tag{52}$$

and

$$\delta'_\gamma = 4\psi' - \frac{4}{3}\theta_\gamma, \tag{53}$$

$$\theta'_\gamma + \frac{\nabla^2\delta_\gamma}{4} + \nabla^2\phi = a n_e x_e \sigma_T(\theta_b - \theta_\gamma). \tag{54}$$

Equation (140) contains, as a source term, the divergence of the Lorentz force that can be expressed in terms of $\sigma_B(\tau, \boldsymbol{x})$ and $\Omega_B(\tau, \boldsymbol{x})$, as already pointed out in (111).

At early times photons and baryons are tightly coupled by Thompson scattering, as it is clear from (140) and (142) where σ_T denotes the Thompson cross section and $n_e\ x_e$ the concentration of ionized electrons. To cast light on the physical nature of the tight-coupling approximation, let us subtract (142) and (140). The result will be

$$(\theta_\gamma - \theta_b)' + a n_e\ x_e \left[1 + \frac{4}{3}\frac{\rho_\gamma}{\rho_b}\right](\theta_\gamma - \theta_b) = -\frac{\nabla^2\delta_\gamma}{4} + \mathcal{H}\theta_b - \frac{\boldsymbol{\nabla}\cdot[\boldsymbol{J}\times\boldsymbol{B}]}{a^4\rho_b}. \tag{55}$$

From (143) it is clear that any deviation of $(\theta_\gamma - \theta_b)$ swiftly decays away. In fact, from (143), the characteristic time for the synchronization of the baryon and photon velocities is of the order of $(x_e n_e \sigma_T)^{-1}$ which is small compared with the expansion time. In the limit $\sigma_T \to \infty$ the tight coupling is exact and the photon–baryon velocity field is a unique physical entity which will be denoted by $\theta_{\gamma b}$. From the structure of (143), the contribution of the magnetic fields in the MHD limit only enters through the Lorentz force, while the damping term is always provided by Thompson scattering.

To derive the evolution equations for the photon–baryon system in the tight-coupling approximation, we can add (140) and (142) taking into account that $\theta_b \simeq \theta_\gamma = \theta_{\gamma b}$. Of course, also the evolution equations of the density contrasts will depend upon $\theta_{\gamma b}$. Consequently, the full set of tightly coupled evolution equations for the photon–baryon fluid can be written as

$$\delta'_\gamma = 4\psi' - \frac{4}{3}\theta_{\gamma b}, \tag{56}$$

$$\delta'_b = 3\psi' - \theta_{\gamma b}, \tag{57}$$

$$\theta'_{\gamma b} + \frac{\mathcal{H}R_b}{(1+R_b)}\theta_{\gamma b} + \frac{\nabla^2\delta_\gamma}{4(1+R_b)} + \nabla^2\phi = \frac{3}{4}\frac{\boldsymbol{\nabla}\cdot[\boldsymbol{J}\times\boldsymbol{B}]}{a^4\rho_\gamma(1+R_b)}, \tag{58}$$

where

$$R_b(\tau) = \frac{3}{4}\frac{\rho_b(\tau)}{\rho_\gamma(\tau)} = \left(\frac{698}{z+1}\right)\left(\frac{h^2\Omega_b}{0.023}\right). \tag{59}$$

The set of equations (144), (145) and (146) have to be used in order to obtain the correct initial conditions to be imposed on the evolution for the integration of the brightness perturbations.

If we assume, effectively, that $\sigma_T \to \infty$ we are working to lowest order in the tight-coupling approximation. This means that the CMB is effectively isotropic in the baryon rest frame. To discuss CMB polarization in the presence of magnetic fields, one has to go to higher order in the tight-coupling expansion. However, as far as the problem of initial conditions is concerned, the lowest order treatment suffices, as it will be apparent from the subsequent discussion.

Neutrinos

After neutrino decoupling the (perturbed) neutrino phase–space distribution evolves according to the collisionless Boltzmann equation. This occurrence implies that to have a closed system of equations describing the initial conditions it is mandatory to *improve* the fluid description by adding to the evolution of the monopole (i.e. the neutrino density contrast) and of the dipole (i.e. the neutrino peculiar velocity), and also of the quadrupole, i.e. the quantity denoted by σ_ν and appearing in the expression of the anisotropic stress of the fluid (see (105) and (106)).

The derivation of the various multipoles of the perturbed neutrino phase–space distribution is a straightforward (even if a bit lengthy) calculation and it has been performed, for the set of conventions employed in the present lecture, in [140]. The result is, in Fourier space,

$$\delta_\nu' = 4\psi' - \frac{4}{3}\theta_\nu, \tag{60}$$

$$\theta_\nu' = \frac{k^2}{4}\,\delta_\nu + k^2\phi - k^2\sigma_\nu, \tag{61}$$

$$\sigma_\nu' = \frac{4}{15}\theta_\nu - \frac{3}{10}k\mathcal{F}_{\nu\,3}. \tag{62}$$

In (150) $\mathcal{F}_{\nu\,3}$ is the octupole of the (perturbed) neutrino phase–space distribution. The precise relation of the multipole moments of $\mathcal{F}_\nu$ with the density contrast and the other plasma quantities is as follows:

$$\delta_\nu = \mathcal{F}_{\nu\,0}, \qquad \theta_\nu = \frac{3}{4}k\mathcal{F}_{\nu\,1}, \qquad \sigma_\nu = \frac{\mathcal{F}_{\nu\,2}}{2}. \tag{63}$$

For multipoles larger than the quadrupole, i.e. $\ell > 2$, the Boltzmann hierarchy reads

$$\mathcal{F}_{\nu\ell}' = \frac{k}{2\ell+1}[\ell\mathcal{F}_{\nu(\ell-1)} - (\ell+1)\mathcal{F}_{\nu(\ell+1)}]. \tag{64}$$

In principle, to give initial conditions, we should specify, at a given time after neutrino decoupling, the values of *all* the multipoles of the neutrino phase–space distribution. In practice, if the initial conditions are set deep in the radiation epoch, the relevant variables only extend, for the purpose of the initial conditions, up to the octupole. Specific examples will be given in a moment.

CDM Component

The CDM component is, in some sense, the easier. In the standard case the evolution equations do not contain neither the magnetic field contribution nor the anisotropic stress. The evolution of the density contrast and of the peculiar velocity are simply given, in Fourier space, by the following pair of equations:

$$\delta_{\rm c}' = 3\psi' - \theta_{\rm c}, \tag{65}$$

$$\theta_{\rm c}' + \mathcal{H}\theta_{\rm c} = k^2\phi. \tag{66}$$

Magnetized Adiabatic and Nonadiabatic Modes

The evolution equations of the fluid and metric variables will now be solved deep in the radiation-dominated epoch and for wavelengths much larger than the Hubble radius, i.e. $|k\,\tau| \ll 1$. In the present lecture only the magnetized adiabatic mode will be discussed. However, the treatment can be usefully extended to the other nonadiabatic modes. For this purpose we refer the interested reader to [132] (see also [139]). Moreover, since this lecture has been conducted within the conformally Newtonian gauge, there is no reason to change. However, it should be noticed that fully gauge-invariant approaches are possible [133]. To give the flavor of the possible simplifications obtainable in a gauge-invariant framework, we can just use gauge-invariant concepts to classify more precisely the adiabatic and nonadiabatic modes. For this purpose, in agreement with (126), let us define the gauge-invariant density contrasts on uniform curvature hypersurfaces for the different species of the pre-decoupling plasma:

$$\zeta_{\gamma} = -\psi + \frac{\delta_{\gamma}}{4}, \qquad \zeta_{\nu} = -\psi + \frac{\delta_{\nu}}{4}, \tag{67}$$

$$\zeta_{\rm c} = -\psi + \frac{\delta_{\rm c}}{3}, \qquad \zeta_{\rm b} = -\psi + \frac{\delta_{\rm b}}{3}. \tag{68}$$

In terms of the variables of (155) and (158) the evolution equations for the density contrasts, i.e. (144), (148), (154) and (154), acquire a rather symmetric form:

$$\zeta_{\gamma}' = -\frac{\theta_{\gamma{\rm b}}}{3}, \qquad \zeta_{\nu}' = -\frac{\theta_{\nu}}{3}, \tag{69}$$

$$\zeta_{\rm c}' = -\frac{\theta_{\rm c}}{3}, \qquad \zeta_{\rm b}' = -\frac{\theta_{\gamma{\rm b}}}{3}. \tag{70}$$

From (157) and (158) we can easily deduce a rather important property of fluid mixtures: in the long-wavelength limit the relative fluctuations in the specific entropy are conserved. Consider, for instance, the CDM-radiation mode. In this case the nonvanishing entropy fluctuations are

$$\mathcal{S}_{\gamma c} = -3(\zeta_\gamma - \zeta_c), \qquad \mathcal{S}_{\nu c} = -3(\zeta_\nu - \zeta_c). \tag{71}$$

Using (157) and (158) the evolution equations for $\mathcal{S}_{\gamma c}$ and $\mathcal{S}_{\nu c}$ can be readily obtained and they are

$$\mathcal{S}_{\gamma c}' = -(\theta_{\gamma b} - \theta_c), \qquad \mathcal{S}_{\nu c}' = -(\theta_\nu - \theta_c). \tag{72}$$

Outside the horizon the divergence of the peculiar velocities is $\mathcal{O}(|k\tau|^2)$, so the fluctuations in the specific entropy are approximately constant in this limit. This conclusion implies that if the fluctuations in the specific entropy are zero, they will still vanish at later times. Such a conclusion can be evaded if the fluids of the mixture have a relevant energy–momentum exchange or if bulk viscous stresses are present [143, 144].

A mode is therefore said to be adiabatic iff $\zeta_\gamma = \zeta_\nu = \zeta_c = \zeta_b$. Denoting by ζ_i and ζ_j two generic gauge-invariant density contrasts of the fluids of the mixture, we say that the initial conditions are nonadiabatic if, at least, we can find a pair of fluids for which $\zeta_i \neq \zeta_j$.

As an example, let us work out the specific form of the magnetized adiabatic mode. Let us consider the situation where the Universe is dominated by radiation after weak interactions have fallen out of thermal equilibrium but before matter–radiation equality. This is the period of time where the initial conditions of CMB anisotropies are usually set both in the presence and in the absence of a magnetized contribution. Since the scale factor goes, in conformal time, as $a(\tau) \simeq \tau$ and $\mathcal{H} \simeq \tau^{-1}$, (90) can be solved for $|k\tau| \ll 1$. The density contrasts can then be determined, in Fourier space, to lowest order in $k\tau$ as

$$\begin{aligned} \delta_\gamma &= \delta_\nu = -2\phi_i - R_\gamma \Omega_B, \\ \delta_b &= \delta_c = -\frac{3}{2}\phi_i - \frac{3}{4}R_\gamma \Omega_B, \end{aligned} \tag{73}$$

where the fractional contribution of photons to the radiation plasma, i.e. R_γ has been introduced and it is related to R_ν, i.e. the fractional contribution of massless neutrinos, as

$$\begin{aligned} R_\gamma &= 1 - R_\nu, \qquad R_\nu = \frac{r}{1+r}, \\ r &= \frac{7}{8}N_\nu\left(\frac{4}{11}\right)^{4/3} \equiv 0.681\left(\frac{N_\nu}{3}\right). \end{aligned} \tag{74}$$

In (161) $\phi_i(k)$ denotes the initial value of the metric fluctuation in Fourier space. It is useful to remark that we have treated neutrinos as part of the radiation background. If neutrinos have a mass in the meV range, they are nonrelativistic today, but they will be counted as radiation prior to matter–radiation equality. Concerning (161) the last remark is that, of course, we just kept the lowest order in $|k\tau| < 1$. It is possible, however, to write the solution to arbitrary order in $|k\tau|$ as explicitly shown in [139].

Let us then write (105) in Fourier space and let us take into account that the background is dominated by radiation. The neutrino quadrupole is then determined to be

$$\sigma_\nu = -\frac{R_\gamma}{R_\nu}\sigma_{\mathrm{B}} + \frac{k^2\tau^2}{6R_\nu}(\psi_{\mathrm{i}} - \phi_{\mathrm{i}}), \tag{75}$$

where $\psi_{\mathrm{i}}(k)$ is the initial (Fourier space) value of the metric fluctuation defined in (89).

Let us then look for the evolution of the divergences of the peculiar velocities of the different species. Let us therefore write (146), (149) and (153) in Fourier space. By direct integration, the following result can be obtained:

$$\theta_{\gamma\mathrm{b}} = \frac{k^2\tau}{4}[2\phi_{\mathrm{i}} + R_\nu\Omega_{\mathrm{B}} - 4\sigma_{\mathrm{B}}], \tag{76}$$

$$\theta_\nu = \frac{k^2\tau}{2}\left[\phi_{\mathrm{i}} - \frac{R_\gamma\Omega_{\mathrm{B}}}{2}\right] + k^2\tau\frac{R_\gamma}{R_\nu}\sigma_{\mathrm{B}}, \tag{77}$$

$$\theta_{\mathrm{c}} = \frac{k^2\tau}{2}\phi_{\mathrm{i}}. \tag{78}$$

As a consistency check of the solution, (164), (165) and (166) can be inserted into (91). Let us therefore write (91) in Fourier space

$$k^2\mathcal{H}\phi_{\mathrm{i}} = 4\pi G a^2\left[\frac{4}{3}\rho_\gamma(1+\rho_{\mathrm{b}})\theta_{\gamma\mathrm{b}} + \frac{4}{3}\rho_\nu\theta_\nu + \rho_{\mathrm{c}}\theta_{\mathrm{c}}\right], \tag{79}$$

where we used that $\psi_{\mathrm{i}}' = 0$ and we also used the tight-coupling approximation since $\theta_\gamma = \theta_{\mathrm{b}} = \theta_{\gamma\mathrm{b}}$. Notice that in (91) the term arising from the Poynting vector has been neglected. This approximation is rather sound within the present MHD treatment. In (167) $R_{\mathrm{b}} \ll 1$ (see (147) for the definition of R_{b}) since we are well before matter–radiation equality. The same observation can be made for the CDM contribution which is negligible in comparison with the radiative contribution provided by photons and neutrinos. Taking into account these two observations, we can rewrite (167) as

$$k^2\mathcal{H}\phi_{\mathrm{i}} = 2\mathcal{H}^2(R_\gamma\theta_{\gamma\mathrm{b}} + R_\nu\theta_\nu), \tag{80}$$

where (97) and (98) have been used. Inserting then (164) and (165) into (168), it can be readily obtained that the left-hand side exactly equals the right-hand side, so that the momentum constraint is enforced.

The final equation to be solved is the one describing the evolution of the anisotropic stress, i.e. (150). Inserting (150) and (165) into (62), we do get an interesting constraint on the initial conditions on the two longitudinal fluctuations of the geometry introduced in (89), namely:

$$\psi_{\mathrm{i}} = \phi_{\mathrm{i}}\left(1 + \frac{2}{5}R_\nu\right) + \frac{R_\gamma}{5}(4\sigma_{\mathrm{B}} - R_\nu\Omega_{\mathrm{B}}). \tag{81}$$

Concerning the magnetized adiabatic mode, the following comments are in order:

- The peculiar velocities are always suppressed, with respect to the other terms of the solution, by a factor $|k\tau|$ which is smaller than 1 when the wavelength is larger than the Hubble radius.
- In the limit $\sigma_{\rm B} \to 0$ and $\Omega_{\rm B} \to 0$ the magnetized adiabatic mode presented here reproduces the well-known standard results (see for instance [138]).
- The difference between the two longitudinal fluctuations of the metric is due to both the presence of magnetic and fluid anisotropic stresses.
- The longitudinal fluctuations of the geometry are both constant outside the horizon and prior to matter–radiation equality; this result still holds in the presence of a magnetized contribution as it is clearly demonstrated by the analytic solution presented here.

The last interesting exercise we can do with the obtained solution is to compute the important variables $\mathcal{R}$ and ζ introduced, respectively, in (165) and (166). Since both ψ and ϕ are constants for $|k\tau| < 1$ and for $\tau < \tau_{\rm eq}$, also $\mathcal{R}$ will be constant. In particular, by inserting (169) into (116), the following expression can be obtained:

$$\mathcal{R}_{\rm i} = -\frac{3}{2}\left(1 + \frac{4}{15}R_{\nu}\right)\phi_{\rm i} - \frac{R_{\gamma}}{5}(4\sigma_{\rm B} - R_{\nu}\Omega_{\rm B}), \tag{82}$$

where $\mathcal{R}_{\rm i}(k)$ denotes the initial value, in Fourier space, of the curvature perturbations. In numerical studies it is sometimes useful to relate the initial values of ϕ and ψ, i.e. $\phi_{\rm i}$ and $\psi_{\rm i}$ to $\mathcal{R}_{\rm i}$. This relation is expressed by the following pair of formulae that can be derived by inverting (170) and by using (169):

$$\begin{aligned}\phi_{\rm i} &= -\frac{10}{15 + 4R_{\nu}}\mathcal{R}_{\rm i} - \frac{2R_{\gamma}(4\sigma_{\rm B} - R_{\gamma}\Omega_{\rm B})}{15 + 4R_{\nu}},\\ \psi_{\rm i} &= -2\frac{5 + 2R_{\nu}}{15 + 4R_{\nu}}\mathcal{R}_{\rm i} - \frac{2}{5}\frac{R_{\gamma}(5 + 2R_{\nu})}{15 + 4R_{\nu}}(4\sigma_{\rm B} - R_{\gamma}\Omega_{\rm B}).\end{aligned} \tag{83}$$

From the Hamiltonian constraint written in the form (117) it is easy to deduce in the limit $|k\tau| \ll 1$ that $\zeta_{\rm i}(k) = \mathcal{R}_{\rm i}(k)$. The same result can be obtained through a different, but also instructive, path. Consider the definition of ζ given either in (112) or in (115). The variable ζ can be expressed in terms of the partial density contrasts defined in (155) and (156). More precisely, from the definitions of the two sets of variables it is easy to show that

$$\zeta = \frac{\rho_{\nu}'\zeta_{\nu} + \rho_{\gamma}'\zeta_{\gamma} + \rho_{\rm c}'\zeta_{\rm c} + \rho_{\rm b}'\zeta_{\rm b}}{\rho_{\rm t}'} + \zeta_{\rm B}, \qquad \zeta_{\rm B} = \frac{\delta\rho_{\rm B}}{3(p_{\rm t} + \rho_{\rm t})}. \tag{84}$$

Thus, to obtain ζ it suffices to find ζ_{γ}, ζ_{ν}, $\zeta_{\rm b}$ and $\zeta_{\rm c}$ evaluated at the initial time and on the adiabatic solution. Using (161) and (169) into (155) and (156) we obtain, as expected,

$$\zeta_\gamma = \zeta_\nu = \zeta_\mathrm{c} = \zeta_\mathrm{b} = -\left(\psi_\mathrm{i} + \frac{\phi_\mathrm{i}}{2}\right) + \frac{R_\gamma}{4}\Omega_\mathrm{B}. \tag{85}$$

This result was expected, since, as previously stressed, for the adiabatic mode all the partial density contrasts must be equal. Inserting now (173) into (172) and recalling that the CDM and baryon contributions vanish deep in the radiation epoch, we do get

$$\zeta = -\left(\psi_\mathrm{i} + \frac{\phi_\mathrm{i}}{2}\right) = \mathcal{R}_\mathrm{i}, \tag{86}$$

where the last equality follows from the definition of (116) evaluated deep in the radiation epoch and for the adiabatic solution derived above.

Up to now, as explained, attention has been given to the magnetized adiabatic mode. There are, however, also other nonadiabatic modes that can enter the game. We will not go, in this lecture, through the derivation of the various nonadiabatic modes. It is, however, useful to give at least the result in the case of the magnetized CDM-radiation mode. In such a case the full set of equations admitting the adiabatic solution can be solved, in the limit $\tau < \tau_1$ and $k\tau < 1$, by

$$\begin{aligned}
&\phi = \phi_1\left(\frac{\tau}{\tau_1}\right), \qquad \psi = \psi_1\left(\frac{\tau}{\tau_1}\right),\\
&\delta_\gamma = \delta_\nu = 4\psi_1\left(\frac{\tau}{\tau_1}\right) - R_\gamma\Omega_\mathrm{B},\\
&\delta_\mathrm{c} = -\left[\mathcal{S}_* + \frac{3}{4}R_\gamma\Omega_\mathrm{B}\right] + 3\psi_1\left(\frac{\tau}{\tau_1}\right),\\
&\delta_\mathrm{b} = 3\psi_1\left(\frac{\tau}{\tau_1}\right) - \frac{3}{4}R_\gamma\Omega_\mathrm{B},\\
&\theta_\mathrm{c} = \frac{k^2\tau_1}{3}\phi_1\left(\frac{\tau}{\tau_1}\right)^2,\\
&\theta_{\gamma\mathrm{b}} = \frac{k^2\tau_1}{2}(\phi_1 + \psi_1)\left(\frac{\tau}{\tau_1}\right)^2 + \frac{k^2\tau}{4}[R_\nu\Omega_\mathrm{B} - 4\sigma_\mathrm{B}],\\
&\theta_\nu = \frac{k^2\tau_1}{2}(\phi_1 + \psi_1)\left(\frac{\tau}{\tau_1}\right)^2 + \frac{k\tau}{4}\left(4\frac{R_\gamma}{R_\nu}\sigma_\mathrm{B} - \Omega_\mathrm{B}\right),\\
&\mathcal{F}_{\nu 3} = \frac{8}{9}k\tau\left[4\frac{R_\gamma}{R_\nu}\sigma_\mathrm{B} - \Omega_\mathrm{B}\right],\\
&\sigma_\nu = -\frac{R_\gamma}{R_\nu}\sigma_\mathrm{B} + \frac{k^2\tau_1^2}{6R_\nu}(\psi_1 - \phi_1)\left(\frac{\tau}{\tau_1}\right)^3,
\end{aligned} \tag{87}$$

where

$$\psi_1 = \frac{15 + 4R_\nu}{8(15 + 2R_\nu)}\left[\mathcal{S}_* + \frac{3}{4}R_\gamma \Omega_{\rm B}\right],$$
$$\phi_1 = \frac{15 - 4R_\nu}{8(15 + 2R_\nu)}\left[\mathcal{S}_* + \frac{3}{4}R_\gamma \Omega_{\rm B}\right]. \quad (88)$$

In (175) the following notation for the nonvanishing entropy fluctuations has been employed:

$$\mathcal{S}_{c\gamma} = \mathcal{S}_{c\nu} = \mathcal{S}_*. \quad (89)$$

In deriving (175) it is practical to use a form of the scale factor (obtained by solving (97), (98) and (99) for a mixture of matter and radiation) which explicitly interpolates between a radiation-dominated regime and a matter-dominated regime:

$$a(\tau) = a_{\rm eq}\left[\left(\frac{\tau}{\tau_1}\right)^2 + 2\left(\frac{\tau}{\tau_1}\right)\right], \qquad 1 + z_{\rm eq} = \frac{1}{a_{\rm eq}} = \frac{h^2 \Omega_{\rm m0}}{h^2 \Omega_{\rm r0}}, \quad (90)$$

where $\Omega_{\rm m0}$ and $\Omega_{\rm r0}$ are evaluated at the present time and the scale factor is normalized in such a way that $a_0 = 1$. In (178) $\tau_1 = (2/H_0)\sqrt{a_{\rm eq}/\Omega_{\rm m0}}$. In terms of τ_1 the equality time is

$$\tau_{\rm eq} = (\sqrt{2} - 1)\tau_1 = 119.07\left(\frac{h^2\Omega_{\rm m0}}{0.134}\right)^{-1} \text{Mpc}, \quad (91)$$

i.e. $2\tau_{\rm eq} \simeq \tau_1$. In this framework the total optical depth from the present to the critical recombination epoch, i.e. $800 < z < 1200$ can be approximated analytically, as discussed in [145]. By defining the redshift of decoupling as the one where the total optical depth is of order 1, i.e. $\kappa(z_{\rm dec}, 0) \simeq 1$, we will have approximately

$$z_{\rm dec} \simeq 1139\left(\frac{\Omega_{\rm b}}{0.0431}\right)^{-\alpha_1}, \qquad \alpha_1 = \frac{0.0268}{0.6462 + 0.1125\ln(\Omega_{\rm b}/0.0431)}, \quad (92)$$

where $h = 0.73$. From (180) and (178) it follows that for $1100 \leq z_{\rm dec} \leq 1139$, $275\ \text{Mpc} \leq \tau_{\rm dec} \leq 285\ \text{Mpc}$.

Equations (179) and (180) will turn out to be relevant for the effective numerical integration of the brightness perturbations which will be discussed later on. For numerical purposes the late-time cosmological parameters will be fixed, for a spatially flat Universe, as [13]

[13] The values of the cosmological parameters introduced in (181) are compatible with the ones estimated from WMAP-3 [127, 146, 147] in combination with the "Gold" sample of SNIa [148] consisting of 157 supernovae (the furthest being at redshift z = 1.75). We are aware of the fact that WMAP-3 data alone seem to favor a slightly smaller value of $\omega_{\rm m}$ (i.e. 0.126). Moreover, WMAP-3 data may also have slightly different implications if combined with supernovae of the SNLS project [149]. The values given in (181) will just be used for a realistic numerical illustration of the methods developed in the present investigation.

$$\omega_\gamma = 2.47 \times 10^{-5}, \qquad \omega_{\rm b} = 0.023, \qquad \omega_{\rm c} = 0.111, \qquad \omega_{\rm m} = \omega_{\rm b} + \omega_{\rm c}, \quad (93)$$

where $\omega_X = h^2 \Omega_X$ and $\Omega_\Lambda = 1 - \Omega_{\rm m}$; the present value of the Hubble parameter H_0 will be fixed, for numerical estimates, to 73 in units of km/(sec Mpc).

Transfer Matrix and Sachs–Wolfe Plateau

Before presenting some numerical approaches suitable for the analysis of magnetized CMB anisotropies, it is useful to discuss a class of analytical estimates that allow the calculation of the so-called Sachs–Wolfe plateau. The idea, in short, is very simple. We have the evolution equation for ζ given in (136). This evolution equation can be integrated across the matter–radiation transition using the interpolating form of the scale factor proposed in (178).

Consider, first, the case of the magnetized adiabatic mode where $\delta p_{\rm nad} = 0$. Deep in the radiation-dominated epoch, for $\tau \ll \tau_{\rm eq}$, $c_{\rm s}^2 \to 1/3$ and, from (136), $\zeta' = 0$, so that

$$\zeta = \zeta_{\rm i} \simeq \mathcal{R}_{\rm i}, \qquad \zeta_{\rm i} = -\frac{3}{2}\phi_{\rm i}\left(1 + \frac{4}{15}R_\nu\right) - \frac{R_\gamma}{5}(4\sigma_{\rm B} - R_\nu \Omega_{\rm B}). \qquad (94)$$

When the Universe becomes matter-dominated, after $\tau_{\rm eq}$, $c_{\rm s}^2 \to 0$ and the second term at the right-hand side of (136) does contribute significantly at decoupling (recall that for $h^2\Omega_{\rm matter} = 0.134$, $\tau_{\rm dec} = 2.36\,\tau_{\rm eq}$). Consequently, from (136), recalling that $c_{\rm s}^2 = 4a_{\rm eq}/[3(3a + 4a_{\rm eq})]$, we obtain

$$\zeta_{\rm f} = \zeta_{\rm i} - \frac{3\,a\,R_\gamma\,\Omega_{\rm B}}{4(3a + 4a_{\rm eq})}, \qquad \Omega_{\rm B\,f} = \Omega_{\rm B\,i}. \qquad (95)$$

The inclusion of one or more nonadiabatic modes changes the form of (136) and, consequently, the related solution (183). For instance, in the case of the CDM-radiation nonadiabatic mode the relevant terms arising in the sum (133) are $\mathcal{S}_{{\rm c}\gamma} = \mathcal{S}_{{\rm c}\nu} = \mathcal{S}_{\rm i}$ where $\mathcal{S}_i$ is the (constant) fluctuation in the relative entropy density initially present (i.e. for $\tau \ll \tau_{\rm eq}$). If this is the case, $\delta p_{\rm nad} = c_{\rm s}^2 \rho_{\rm c} \mathcal{S}_i$ and (136) can be easily solved. The transfer matrix for magnetized CMB anisotropies can then be written as

$$\begin{pmatrix} \zeta_{\rm f} \\ \mathcal{S}_{\rm f} \\ \Omega_{\rm B\,f} \end{pmatrix} = \begin{pmatrix} \mathcal{M}_{\zeta\zeta} & \mathcal{M}_{\zeta\mathcal{S}} & \mathcal{M}_{\zeta{\rm B}} \\ 0 & \mathcal{M}_{\mathcal{S}\mathcal{S}} & \mathcal{M}_{\mathcal{S}{\rm B}} \\ 0 & 0 & \mathcal{M}_{\rm BB} \end{pmatrix} \begin{pmatrix} \zeta_{\rm i} \\ \mathcal{S}_{\rm i} \\ \Omega_{\rm B\,i} \end{pmatrix}. \qquad (96)$$

In the case of a mixture of (magnetized) adiabatic and CDM-radiation modes, we find, for $a > a_{\rm eq}$

$$\mathcal{M}_{\zeta\zeta} \to 1, \qquad \mathcal{M}_{\zeta\mathcal{S}} \to -\frac{1}{3}, \qquad \mathcal{M}_{\zeta{\rm B}} \to -\frac{R_\gamma}{4},$$
$$\mathcal{M}_{\mathcal{S}\mathcal{S}} \to 1, \qquad \mathcal{M}_{\mathcal{S}{\rm B}} \to 0, \qquad (97)$$

and $\mathcal{M}_{\rm BB} \to 1$. Equations (184) and (185) may be used, for instance, to obtain the magnetized curvature and entropy fluctuations at photon decoupling in terms of the same quantities evaluated for $\tau \ll \tau_{\rm eq}$. A full numerical analysis of the problem confirms the analytical results summarized by (184) and (185). The most general initial condition for CMB anisotropies will then be a combination of (correlated) fluctuations receiving contribution from $\delta p_{\rm nad}$ and from the fully inhomogeneous magnetic field. To illustrate this point, the form of the Sachs–Wolfe plateau in the sudden decoupling limit will now be discussed.

To compute the SW contribution, we need to solve the evolution equation of the monopole of the temperature fluctuations in the tight-coupling limit, i.e. from (145) and (146),

$$\delta_\gamma'' + \frac{\mathcal{H}R_{\rm b}}{1+R_{\rm b}}\delta_\gamma' + \frac{k^2}{3}\frac{\delta_\gamma}{1+R_{\rm b}} = 4\psi'' + \frac{4\mathcal{H}R_{\rm b}}{1+R_{\rm b}}\psi' - \frac{4}{3}k^2\phi - \frac{k^2}{3(1+R_{\rm b})}(\Omega_{\rm B} - 4\sigma_{\rm B}). \tag{98}$$

In the sudden decoupling approximation the visibility function, i.e. $\mathcal{K}(\tau) = \kappa'(\tau)e^{-\kappa(\tau)}$ and the optical depth, i.e. $\epsilon^{-\kappa(\tau)}$ are approximated, respectively, by $\delta(\tau - \tau_{\rm dec})$ and by $\theta(\tau - \tau_{\rm dec})$ (see [150, 151] for an estimate of the width of the last scattering surface). The power spectra of ζ, $\mathcal{S}$ and $\Omega_{\rm B}$ are given, respectively, by

$$\mathcal{P}_\zeta(k) = \mathcal{A}_\zeta\left(\frac{k}{k_{\rm p}}\right)^{n_r-1}, \qquad \mathcal{P}_{\mathcal{S}}(k) = \mathcal{A}_{\mathcal{S}}\left(\frac{k}{k_{\rm p}}\right)^{n_s-1}, \tag{99}$$

$$\mathcal{P}_\Omega(k) = \mathcal{F}(\varepsilon)\overline{\Omega}^2_{{\rm B}\,L}\left(\frac{k}{k_L}\right)^{2\varepsilon}, \tag{100}$$

where $\mathcal{A}_\zeta$, $\mathcal{A}_{\mathcal{S}}$ and $\overline{\Omega}_{{\rm B}\,L}$ are constants and

$$\mathcal{F}(\varepsilon) = \frac{4(6-\varepsilon)(2\pi)^{2\varepsilon}}{\varepsilon(3-2\varepsilon)\Gamma^2(\varepsilon/2)},$$

$$\overline{\Omega}_{{\rm B}\,L} = \frac{\rho_{{\rm B}\,L}}{\overline{\rho}_\gamma}, \qquad \rho_{{\rm B}\,L} = \frac{B_L^2}{8\pi}, \qquad \overline{\rho}_\gamma = a^4(\tau)\rho_\gamma(\tau). \tag{101}$$

To deduce (187), (188) and (189) the magnetic field has been regularized, according to a common practice [22, 124, 126], over a typical comoving scale $L = 2\pi/k_L$ with a Gaussian window function and it has been assumed that the magnetic field intensity is stochastically distributed as

$$\langle B_i(\boldsymbol{k},\tau)B^j(\boldsymbol{p},\tau)\rangle = \frac{2\pi^2}{k^3}\,P_i^j(k)\,P_{\rm B}(k,\tau)\,\delta^{(3)}(\boldsymbol{k}+\boldsymbol{p}), \tag{102}$$

where

$$P_i^j(k) = \left(\delta_i^j - \frac{k_i k^j}{k^2}\right), \qquad P_{\rm B}(k,\tau) = A_{\rm B}\left(\frac{k}{k_{\rm p}}\right)^{\varepsilon}. \tag{103}$$

As a consequence of (190) the magnetic field does not break the spatial isotropy of the background geometry. The quantity $k_{\rm p}$ appearing in (187) and (191) is conventional pivot scale that is 0.05 Mpc (see [128, 129, 130] for a discussion of other possible choices). Equations (188) and (189) hold for $0 < \varepsilon < 1$. In this limit the $\mathcal{P}_{\Omega}(k)$ (see (188)) is nearly scale-invariant (but slightly bluc). This means that the effect of the magnetic and thermal diffusivity scales (related, respectively, to the finite value of the conductivity and of the thermal diffusivity coefficient) do not affect the spectrum [22]. In the opposite limit, i.e. $\varepsilon \gg 1$, the value of the mode-coupling integral appearing in the two-point function of the magnetic energy density (and of the magnetic anisotropic stress) is dominated by ultraviolet effects related to the mentioned diffusivity scales [22]. Using then (187),(188) and (189), the C_ℓ can be computed for the region of the SW plateau (i.e. for multipoles $\ell < 30$):

$$\begin{aligned} C_\ell = \biggl[&\frac{\mathcal{A}_\zeta}{25}\,\mathcal{Z}_1(n_r,\ell) + \frac{9}{100}\,R_\gamma^2\overline{\Omega}_{{\rm B}\,L}^2\mathcal{Z}_2(\epsilon,\ell) - \frac{4}{25}\sqrt{\mathcal{A}_\zeta\mathcal{A}_{\mathcal{S}}}\mathcal{Z}_1(n_{rs},\ell)\cos\gamma_{rs} \\ &+\frac{4}{25}\mathcal{A}_{\mathcal{S}}\,\mathcal{Z}_1(n_s,\ell) - \frac{3}{25}\sqrt{\mathcal{A}_\zeta}\,R_\gamma\,\overline{\Omega}_{{\rm B}\,L}\,\mathcal{Z}_3(n_r,\varepsilon,\ell)\cos\gamma_{br} \\ &+\frac{6}{25}\sqrt{\mathcal{A}_{\mathcal{S}}}\,R_\gamma\overline{\Omega}_{{\rm B}\,L}\,\mathcal{Z}_3(n_s,\varepsilon,\ell)\cos\gamma_{bs}\biggr], \end{aligned} \qquad (104)$$

where the functions $\mathcal{Z}_1$, $\mathcal{Z}_2$ and $\mathcal{Z}_3$

$$\mathcal{Z}_1(n,\ell) = \frac{\pi^2}{4}\left(\frac{k_0}{k_{\rm p}}\right)^{n-1} 2^n \frac{\Gamma(3-n)\Gamma\left(\ell+\frac{n-1}{2}\right)}{\Gamma^2\left(2-\frac{n}{2}\right)\Gamma\left(\ell+\frac{5}{2}-\frac{n}{2}\right)}, \qquad (105)$$

$$\mathcal{Z}_2(\varepsilon,\ell) = \frac{\pi^2}{2}2^{2\varepsilon}\mathcal{F}(\varepsilon)\left(\frac{k_0}{k_L}\right)^{2\varepsilon}\frac{\Gamma(2-2\varepsilon)\Gamma(\ell+\varepsilon)}{\Gamma^2\left(\frac{3}{2}-\varepsilon\right)\Gamma(\ell+2-\varepsilon)}, \qquad (106)$$

$$\begin{aligned} \mathcal{Z}_3(n,\varepsilon,\ell) = &\frac{\pi^2}{4}2^{\varepsilon}2^{\frac{n+1}{2}}\sqrt{\mathcal{F}(\varepsilon)}\left(\frac{k_0}{k_L}\right)^{\varepsilon}\left(\frac{k_0}{k_{\rm p}}\right)^{\frac{n+1}{2}} \\ &\times\frac{\Gamma\left(\frac{5}{2}-\varepsilon-\frac{n}{2}\right)\Gamma\left(\ell+\frac{\varepsilon}{2}+\frac{n}{4}-\frac{1}{4}\right)}{\Gamma^2\left(\frac{7}{4}-\frac{\varepsilon}{2}-\frac{n}{4}\right)\Gamma\left(\frac{9}{4}+\ell-\frac{\varepsilon}{2}-\frac{n}{4}\right)}, \end{aligned} \qquad (107)$$

are defined in terms of the magnetic tilt ε and of a generic spectral index n which may correspond, depending on the specific contribution, either to n_r (adiabatic spectral index), or to n_s(nonadiabatic spectral index) or even to $n_{rs} = (n_r + n_s)/2$ (spectral index of the cross-correlation). In (192) γ_{rs}, γ_{br}

and γ_{sb} are the correlation angles. In the absence of magnetic and nonadiabatic contributions and for (192) and (193) imply that for $n_r = 1$ (Harrison–Zeldovich spectrum) $\ell(\ell+1)C_\ell/2\pi = \mathcal{A}_\zeta/25$ and WMAP data [127] would imply that $\mathcal{A}_\zeta = 2.65 \times 10^{-9}$. Consider then the physical situation where on top of the adiabatic mode there is a magnetic contribution. If there is no correlation between the magnetized contribution and the adiabatic contribution, i.e. $\gamma_{br} = \pi/2$, the SW plateau will be enhanced in comparison with the case when magnetic fields are absent. The same situation arises when the two components are anticorrelated (i.e. $\cos\gamma_{br} < 0$). However, if the fluctuations are positively correlated (i.e. $\cos\gamma_{br} > 0$), the cross-correlation adds negatively to the sum of the two autocorrelations of ζ and $\Omega_{\rm B}$ so that the total result may be an overall reduction of the power with respect to the case $\gamma_{br} = \pi/2$. In (193),(194) and (195) $k_0 = \tau_0^{-1}$ where τ_0 is the present observation time.

4.4 Numerical Analysis

The main idea of the numerical analysis is rather simple. Its implementation, however, may be rather complicated. In order to capture the simplicity out of the possible complications, we will proceed as follows. We will first discuss a rather naive approach to the integration of CMB anisotropies. Then, building up on this example, the results obtainable in the case of magnetized scalar modes will be illustrated.

Simplest Toy Model

Let us therefore apply the Occam razor and let us consider the simplest situation we can imagine, that is to say the case where

- magnetic fields are absent;
- neutrinos are absent;
- photons and baryons are described within the tight-coupling approximation to lowest order (i.e. $\sigma_{\rm T} \to \infty$);
- initial conditions are set either from the adiabatic mode or from the CDM-radiation mode.

This is clearly the simplest situation we can envisage. Since neutrinos are absent, there is no source of anisotropic stress and the two longitudinal fluctuations of the metric are equal, i.e. $\phi = \psi$. Consequently, the system of equations to be solved becomes

$$\mathcal{R}' = \frac{k^2 c_s^2 \mathcal{H}}{\mathcal{H}^2 - \mathcal{H}'}\psi - \frac{\mathcal{H}}{p_{\rm t} + \rho_{\rm t}}\delta p_{\rm nad}, \tag{108}$$

$$\psi' = -\left(2\mathcal{H} - \frac{\mathcal{H}'}{\mathcal{H}}\right)\psi - \left(\mathcal{H} - \frac{\mathcal{H}'}{\mathcal{H}}\right)\mathcal{R}, \tag{109}$$

$$\delta_{\gamma}' = 4\psi' - \frac{4}{3}\theta_{\gamma \mathrm{b}}, \tag{110}$$

$$\theta_{\gamma \mathrm{b}}' = -\frac{\mathcal{H} R_{\mathrm{b}}}{R_{\mathrm{b}}+1}\theta_{\gamma \mathrm{b}} + \frac{k^2}{4(1+R_{\mathrm{b}})}\delta_{\gamma} + k^2 \psi, \tag{111}$$

$$\delta_{\mathrm{c}}' = 3\psi' - \theta_{\mathrm{c}}, \tag{112}$$

$$\theta_{\mathrm{c}}' = -\mathcal{H}\theta_{\mathrm{c}} + k^2 \psi. \tag{113}$$

We can now use the explicit form of the scale factor discussed in (178) which implies:

$$\begin{aligned} \mathcal{H} &= \frac{1}{\tau_1}\frac{2(x+1)}{x(x+2)}, \\ \mathcal{H}' &= -\frac{2}{\tau_1^2}\frac{x^2+2x+4}{x^2(x+2)^2}, \\ \mathcal{H}^2 - \mathcal{H}' &= \frac{1}{\tau_1^2}\frac{2(3x^2+6x+4)}{x^2(x+2)^2}, \end{aligned} \tag{114}$$

where $x = \tau/\tau_1$. With these specifications the evolution equations given in (196)–(201) become

$$\frac{d\mathcal{R}}{dx} = \frac{4}{3}\frac{x(x+1)(x+2)}{(3x^2+6x+4)^2}\kappa^2\psi, \tag{115}$$

$$\frac{d\psi}{dx} = -\frac{3x^2+6x+4}{x(x+1)(x+2)}\mathcal{R} - \frac{5x^2+10x+6}{x(x+1)(x+2)}\psi, \tag{116}$$

$$\frac{d\delta_{\gamma}}{dx} = -\frac{4(3x^2+6x+4)}{x(x+1)(x+2)}\mathcal{R} - \frac{4(5x^2+10x+6)}{x(x+1)(x+2)}\psi - \frac{4}{3}\tilde{\theta}_{\gamma \mathrm{b}}, \tag{117}$$

$$\frac{d\tilde{\theta}_{\gamma \mathrm{b}}}{dx} = -\frac{2R_{\mathrm{b}}}{R_{\mathrm{b}}+1}\frac{(x+1)}{x(x+2)} + \frac{\kappa^2}{4(1+R_{\mathrm{b}})}\delta_{\gamma} + \kappa^2\psi, \tag{118}$$

$$\frac{d\delta_{\mathrm{c}}}{dx} = -\frac{3(3x^2+6x+4)}{x(x+1)(x+2)}\mathcal{R} - \frac{3(5x^2+10x+6)}{x(x+1)(x+2)}\psi - \tilde{\theta}_{\mathrm{c}}, \tag{119}$$

$$\frac{d\tilde{\theta}_{\mathrm{c}}}{dx} = -\frac{2(x+1)}{x(x+2)}\tilde{\theta}_{\mathrm{c}} + \kappa^2\psi. \tag{120}$$

In (203)–(208) the following rescalings have been used:

$$\kappa = k\tau_1, \qquad \tilde{\theta}_{\gamma\mathrm{b}} = \tau_1 \theta_{\gamma\mathrm{b}}, \qquad \tilde{\theta}_{\mathrm{c}} = \tau_1 \theta_{\mathrm{c}}. \tag{121}$$

The system of equations (203)–(208) can be readily integrated by giving initial conditions for at $x_{\mathrm{i}} \ll 1$. In the case of the adiabatic mode (which is the one contemplated by (203)–(208) since we set $\delta p_{nad} = 0$) the initial conditions are as follows

$$\begin{aligned} &\mathcal{R}(x_{\mathrm{i}}) = \mathcal{R}_*, &&\psi(x_{\mathrm{i}}) = -\frac{2}{3}\mathcal{R}_*, \\ &\delta_\gamma(x_{\mathrm{i}}) = -2\psi_*, &&\tilde{\theta}_{\gamma\mathrm{b}}(x_{\mathrm{i}}) = 0, \\ &\delta_{\mathrm{c}}(x_{\mathrm{i}}) = -\frac{3}{2}\psi_*, &&\tilde{\theta}_{\mathrm{c}}(x_{\mathrm{i}}) = 0. \end{aligned} \tag{122}$$

It can be shown by direct numerical integration that the system (203)–(208) gives a reasonable semiquantitative description of the acoustic oscillations. To simplify initial conditions even further, we can indeed assume a flat Harrison–Zeldovich spectrum and set $\mathcal{R}_* = 1$.

The same philosophy used to get to this simplified form can be used to integrate the full system. In this case, however, we would miss the important contribution of polarization since, to zeroth order in the tight-coupling expansion, the CMB is not polarized.

Integration of Brightness Perturbations

To discuss the polarization, we have to go (at least) to first order in the tight-coupling expansion [152, 153, 154]. For this purpose, it is appropriate to introduce the evolution equations of the brightness perturbations of the I, Q and U Stokes parameters characterizing the radiation field. Since the Stokes parameters Q and U are not invariant under rotations about the axis of propagation the degree of polarization $P = (Q^2 + U^2)^{1/2}$ is customarily introduced [155, 156]. The relevant brightness perturbations will then be denoted as Δ_{I}, Δ_{P}. This description reproduces to zeroth order in the tight coupling expansion, the fluid equations that have been presented before to set initial conditions prior to equality. For instance, the photon density contrast and the divergence of the photon peculiar velocity are related, respectively, to the monopole and to the dipole of the brightness perturbation of the intensity field, i.e. $\delta_\gamma = 4\Delta_{\mathrm{I}0}$ and $\theta_\gamma = 3k\Delta_{\mathrm{I}1}$. The evolution equations of the brightness perturbations can then be written, within the conventions set by (89)

$$\Delta_{\mathrm{I}}' + (ik\mu + \kappa')\Delta_{\mathrm{I}} + ik\mu\phi = \psi' + \kappa'\left[\Delta_{\mathrm{I}0} + \mu v_{\mathrm{b}} - \frac{1}{2}P_2(\mu)S_{\mathrm{P}}\right], \tag{123}$$

$$\Delta_{\mathrm{P}}' + (ik\mu + \kappa')\Delta_{\mathrm{P}} = \frac{\kappa'}{2}[1 - P_2(\mu)]S_{\mathrm{P}}, \tag{124}$$

$$v_{\rm b}' + \mathcal{H} v_{\rm b} + ik\phi + \frac{ik}{4R_{\rm b}}[\Omega_{\rm B} - 4\sigma_{\rm B}] + \frac{\kappa'}{R_{\rm b}}(v_{\rm b} + 3i\Delta_{\rm I1}) = 0. \quad (125)$$

Equation (213) is nothing but the second relation obtained in (140) having introduced the quantity $ikv_{\rm b} = \theta_{\rm b}$. The source terms appearing in (211) and (212) include a dependence on $P_2(\mu) = (3\mu^2 - 1)/2$ ($P_\ell(\mu)$ denotes, in this framework, the ℓ-th Legendre polynomial); $\mu = \hat{k} \cdot \hat{n}$ is simply the projection of the Fourier wave-number on the direction of the photon momentum. In (211) and (212) the source term $S_{\rm P}$ is defined as

$$S_{\rm P}(k,\tau) = \Delta_{\rm I2}(k,\tau) + \Delta_{\rm P0}(k,\tau) + \Delta_{\rm P2}(k,\tau). \quad (126)$$

The evolution equations in the tight-coupling approximation will now be integrated numerically. More details on the tight coupling expansion in the presence of a magnetized contribution can be found in [132].

The normalization of the numerical calculation is enforced by evaluating, analytically, the Sachs–Wolfe plateau and by deducing, for a given set of spectral indices of curvature and entropy perturbations, the amplitude of the power spectra at the pivot scale. Here is an example of this strategy. The Sachs–Wolfe plateau can be estimated analytically from the evolution equation of $\mathcal{R}$ (or ζ) by using the technique of the transfer matrix appropriately generalized to the case where on top of the adiabatic and nonadiabatic contributions the magnetic fields are consistently taken into account. The main result is expressed by (192).

If the SW plateau is determined by an adiabatic component supplemented by a (subleading) nonadiabatic contribution both correlated with the magnetic field intensity, the obtainable bound may not be so constraining (even well above the nano-Gauss range) due to the proliferation of parameters. A possible strategy is therefore to fix the parameters of the adiabatic mode to the values determined by WMAP-3 and then explore the effect of a magnetized contribution which is not correlated with the adiabatic mode. This implies in (192) that $\mathcal{A}_{\mathcal{S}} = 0$ and $\gamma_{\rm br} = \pi/2$. Under this assumption, in Figs. 8 and 9 the bounds on B_L are illustrated. The nature of the constraint depends, in this case, both on the amplitude of the protogalactic field (at the present epoch and smoothed over a typical comoving scale $L = 2\pi/k_L$) and on its spectral slope, i.e. ε. In the case $\varepsilon < 0.5$ the magnetic energy spectrum is nearly scale-invariant. In this case, diffusivity effects are negligible (see, for instance, [18, 126]). As already discussed, if $\varepsilon \gg 1$, the diffusivity effects (both thermal and magnetic) dominate the mode-coupling integral that lead to the magnetic energy spectrum [18, 126].

In Fig. 8 the magnetic field intensity should be below the different curves if the adiabatic contribution dominates the SW plateau. Different choices of the pivot scale $k_{\rm p}$ and of the smoothing scale k_L are also illustrated. In Fig. 8 the scalar spectral index is fixed to $n_r = 0.951$ [144]. In Fig. 9 the two curves corresponding, respectively, to $n_r = 0.8$ and $n_r = 1$ are reported.

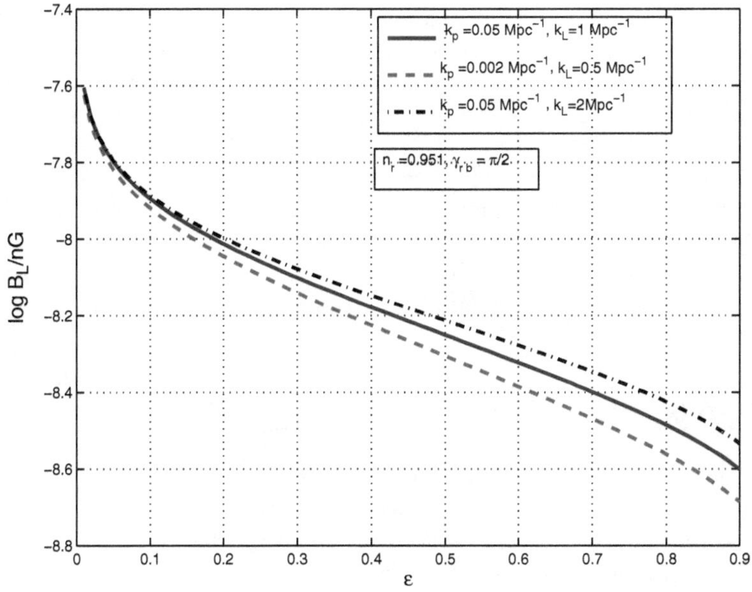

Fig. 8. Bounds on the protogalactic field intensity as a function of the magnetic spectral index ε for different values of the parameters defining the adiabatic contribution to the SW plateau

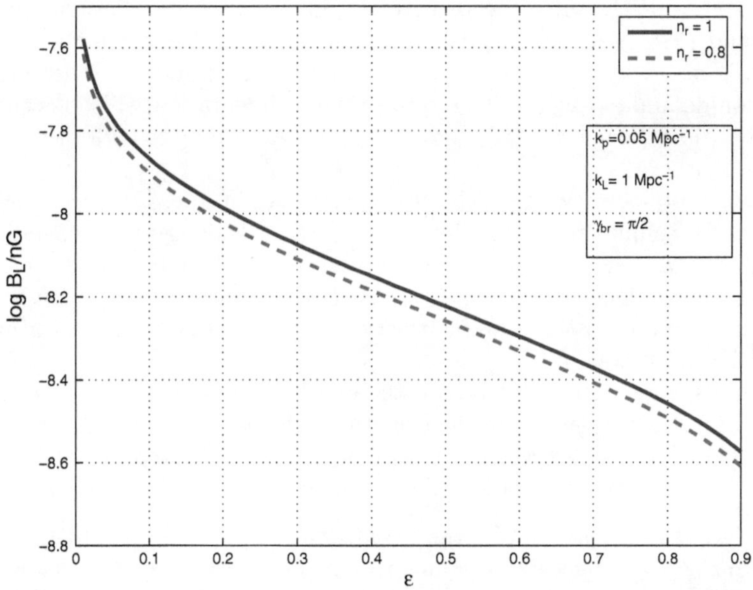

Fig. 9. Same plot as in Fig. 8 but with emphasis on the variation of $n_{\rm r}$

If $\varepsilon < 0.2$, the bounds are comparatively less restrictive than in the case $\varepsilon \simeq 0.9$. The cause of this occurrence is that we are here just looking at the largest wavelengths of the problem. As it will become clear in a moment, intermediate scales will be more sensitive to the presence of fully inhomogeneous magnetic fields.

According to Figs. 8 and 9 for a given value of the magnetic spectral index and of the scalar spectral index, the amplitude of the magnetic field has to be sufficiently small not to affect the dominant adiabatic nature of the SW plateau. Therefore Figs. 8 and 9 (as well as other similar plots) can be used to normalize the numerical calculations for the power spectra of the brightness perturbations, i.e.

$$\frac{k^3}{2\pi^2}|\Delta_{\rm I}(k,\tau)|^2, \qquad \frac{k^3}{2\pi^2}|\Delta_{\rm P}(k,\tau)|^2, \qquad \frac{k^3}{2\pi^2}|\Delta_{\rm I}(k,\tau)\Delta_{\rm P}(k,\tau)|. \qquad (127)$$

Let us then assume, for consistency with the cases reported in Figs. 8 and 9, that we are dealing with the situation where the magnetic field is not correlated with the adiabatic mode. It is then possible to choose a definite value of the magnetic spectral index (for instance $\epsilon = 0.1$) and a definite value of the adiabatic spectral index, i.e. n_r (for instance $n_r = 0.951$, in agreement with [144]). By using the SW plateau the normalization can be chosen in such a way the adiabatic mode dominates over the magnetic contribution. In the mentioned case, Fig. 8 implies $B_L < 1.14 \times 10^{-8}$ G for a pivot scale $k_p = 0.002\,\mathrm{Mpc}^{-1}$. Since the relative weight of the power spectra given in (187) and (188) is fixed, it is now possible to set initial conditions for the adiabatic mode according to (161)–(163), (164)–(166) and (167) deep in the radiation-dominated phase. The initial time of integration will be chosen as $\tau_{\rm i} = 10^{-6}\tau_1$ in the notations discussed in (178). According to (179), this choice implies that $\tau_{\rm i} \ll \tau_{\rm eq}$.

The power spectra of the brightness perturbations, i.e. (215), can be then computed by numerical integration. Clearly, the calculation will depend upon the values of $\omega_{\rm m}$, $\omega_{\rm b}$, $\omega_{\rm c}$ and R_ν. We will simply fix these parameters to their fiducial values reported in (181) (see also (147)) and we will take $N_\nu = 3$ in (162) determining in this way the fractional contribution of the neutrinos to the radiation plasma.

The first interesting exercise, for the present purposes, is reported in Fig. 10 where the power spectra of the brightness perturbations are illustrated for a wave-number $k = 0.1\ \mathrm{Mpc}^{-1}$. Concerning the results reported in Fig. 10 different comments are in order:

- For $\varepsilon = 0.1$ and $n_r = 0.951$, the SW plateau imposes $B_L < 1.14 \times 10^{-8}$ G; from Fig. 10 it follows that a magnetic field of only 30 nG (i.e. marginally incompatible with the SW bound) has a large effect on the brightness perturbations as it can be argued by comparing, in Fig. 10, the dashed curves (corresponding to 30 nG) to the full curves which illustrate the case of vanishing magnetic fields.

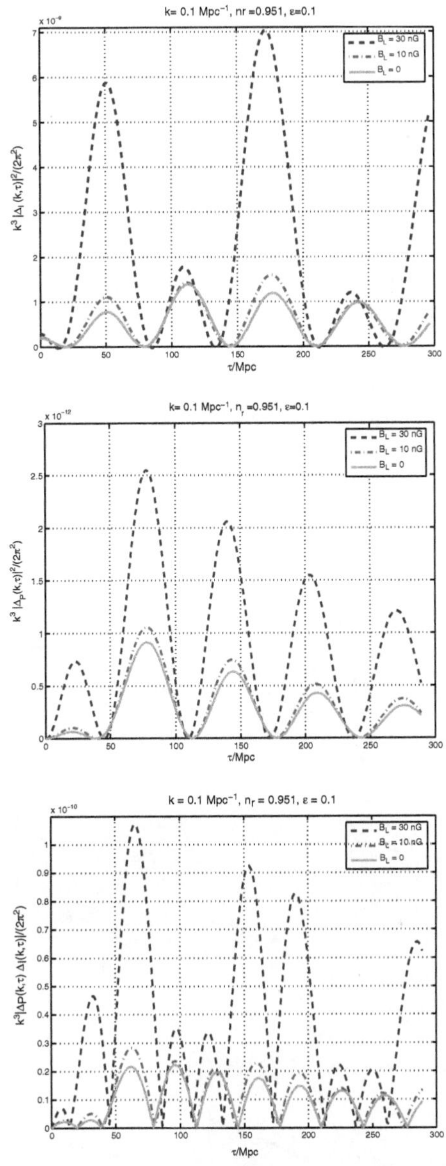

Fig. 10. The power spectra of the brightness perturbations for a typical wavenumber $k = 0.1\mathrm{M}pc^{-1}$. The values of the parameters are specified in the legends. The pivot scale is $k_{\mathrm{p}} = 0.002\ \mathrm{M}pc^{-1}$ and the smoothing scale is $k_L = \mathrm{Mpc}^{-1}$ (see Figs. 8 and 9)

- The situation where $B_L > \mathrm{n}G$ cannot be simply summarized by saying that the amplitudes of the power spectra get larger since there is a combined effect which both increases the amplitudes and shifts slightly the phases of the oscillations.
- From the qualitative point of view, it is still true that the intensity oscillates as a cosine, the polarization as a sine.
- The phases of the cross-correlations are, comparatively, the most affected by the presence of the magnetic field.

The features arising in Fig. 10 can be easily illustrated for other values of ϵ and for different choices of the pivot or smoothing scales. The general lesson that can be drawn is that the constraint derived only by looking at the SW plateau are only a necessary condition on the strength of the magnetic field. They are, however, not sufficient to exclude observable effects at smaller scales. This aspect is illustrated in the plot at the left in Fig. 11 which captures a detail of the cross-correlation. The case when $B_L = 0$ can be still distinguished from the case $B_L = 0.5$ nG. Therefore, recalling that for the same choice of

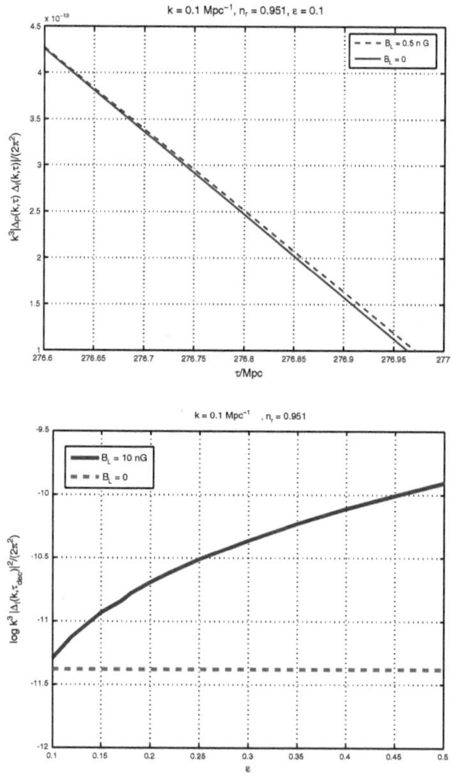

Fig. 11. A detail of the cross-correlation (**top**). The autocorrelation of the intensity at τ_{dec} as a function of ε, i.e. the magnetic spectral index (**bottom**)

parameters the SW plateau implied that $B_L < 11.4$ nG, it is apparent that the intermediate scales lead to more stringent conditions even for nearly scale-invariant spectra of magnetic energy density. For the range of parameters of Fig. 11 we will have that $B_L < 0.5$ nG which is more stringent than the condition deduced from the SW platea by, roughly, one order of magnitude.

If ε increases to higher values (but always with $\varepsilon < 0.5$) by keeping fixed B_L (i.e. the strength of the magnetic field smoothed over a typical lengthscale $L = 2\pi/k_L$), the amplitude of the brightness perturbations gets larger in comparison with the case when the magnetic field is absent. This aspect is illustrated in the bottom plot of Fig. 11 where the logarithm (to base 10) of the intensity autocorrelation is evaluated at a fixed wave-number (and at $\tau_{\rm dec}$) as a function of ε. The full line (corresponding to a $B_L = 10$ nG) is progressively divergent from the dashed line (corresponding to $B_L = 0$) as ε increases.

In Fig. 12 the power spectra of the brightness perturbations are reported at $\tau_{\rm dec}$ and as a function of k. In the two plots at the top the autocorrelation of the intensity is reported for different values of B_L (left plot) and for different values of ε at fixed B_L (right plot). In the two plots at the bottom the polarization power spectra are reported always at $\tau_{\rm dec}$ and for different values of B_L at fixed ε. The position of the first peak of the autocorrelation of the intensity is, approximately, $k_{\rm d} \simeq 0.017$ Mpc^{-1}. The position of the first peak of the cross-correlation is, approximately, 3/4 of $k_{\rm d}$. From this consideration, again, we can obtain that $B_L < 0.3$ nG which is more constraining than the SW condition.

Up to now the adiabatic mode has been considered in detail. We could easily add, however, nonadiabatic modes that are be partially correlated with the adiabatic mode. It is rather plausible, in this situation, that by adding new parameters, also the allowed value of the magnetic field may increase. Similar results can be achieved by deviating from the assumption that the magnetic field and the curvature perturbations are uncorrelated. This aspect can be understood already from the analytical form of the SW plateau (192). If there is no correlation between the magnetized contribution and the adiabatic contribution, i.e. $\gamma_{br} = \pi/2$, the SW plateau will be enhanced in comparison with the case when magnetic fields are absent. The same situation arises when the two components are anticorrelated (i.e. $\cos\gamma_{br} < 0$). However, if the fluctuations are positively correlated (i.e. $\cos\gamma_{br} > 0$), the cross-correlation adds negatively to the sum of the two autocorrelations of $\mathcal{R}$ and $\Omega_{\rm B}$ so that the total result may be an overall reduction of the power with respect to the case $\gamma_{br} = \pi/2$.

From Fig. 12 various features can be appreciated. The presence of magnetic fields, as already pointed out, does not affect only the amplitude but also the phases of oscillations of the various brightness perturbations. Moreover, an increase in the spectral index ε also implies a quantitative difference in the intensity autocorrelation.

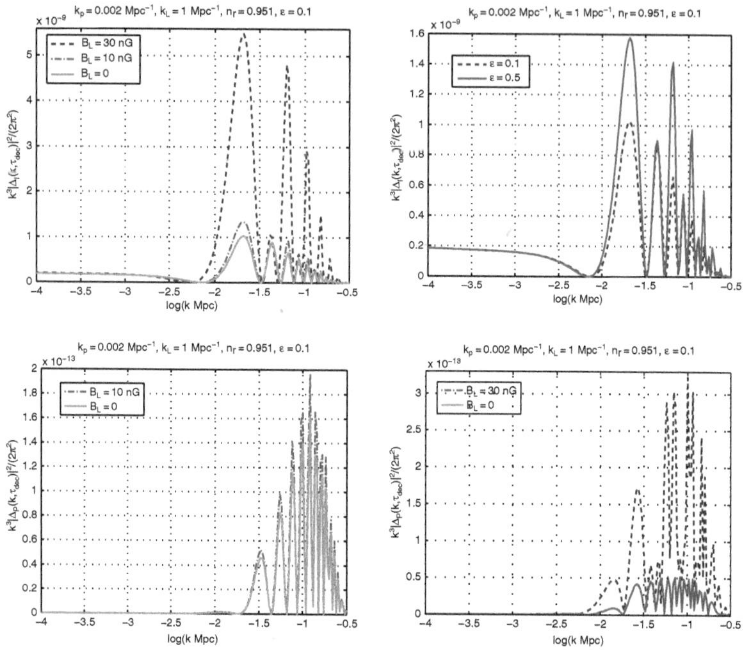

Fig. 12. The power spectra of the brightness perturbations at τ_{dec} for the parameters reported in the legends

5 Concluding Remarks

There is little doubts that large-scale magnetic exist in nature. These fields have been observed in a number of different astrophysical systems. The main question concerns therefore their origin. String cosmological models of pre-big-bang type still represent a viable and well-motivated theoretical option.

Simple logic dictates that if the origin of the large-scale magnetic fields is primordial (as opposed to astrophysical) it is plausible to expect the presence of magnetic fields in the primeval plasma also *before* the decoupling of radiation from matter. CMB anisotropies are germane to several aspect of large-scale magnetization. CMB physics may be the tool that will finally enable us either to confirm or to rule out the primordial nature of galactic and clusters magnetic field seeds. In the next 5 to 10 years the forthcoming CMB precision polarization experiments will be sensitive in, various frequency channels between 30 GHz and roughly 900 GHz. The observations will be conducted both via satellites (like the Planck satellite) and via ground based detectors (like in the case of the QUIET arrays). In a complementary view, the SKA telescope will provide full-sky surveys of Faraday rotation that may even get close to 20 GHz.

In an optimistic perspective the forthcoming experimental data together with the steady progress in the understanding of the dynamo theory will hopefully explain the rationale for the ubiquitous nature of large-scale magnetization. In a pessimistic perspective, the primordial nature of magnetic seeds will neither be confirmed nor be ruled out. It is wise to adopt a model-independent approach by sharpening those theoretical tools that may allow, in the near future, a direct observational test of the effects of large-scale magnetic fields on CMB anisotropies. Some efforts along this perspective have been reported in the present lecture. In particular, the following results have been achieved:

- Scalar CMB anisotropies have been described in the presence of a fully inhomogeneous magnetic field.
- The employed formalism allows the extension of the usual CMB initial conditions to the case when large-scale magnetic fields are present in the game.
- By going to higher order in the tight-coupling expansion the evolution of the brightness perturbations has been computed numerically.
- It has been shown that the magnetic fields may affect not only the amplitude but also the relative phases of the Doppler oscillations.
- From the analysis of the cross-correlation power spectra it is possible to distinguish, numerically, the effects of a magnetic field as small as 0.5 nG.

It is interesting to notice that a magnetic field in the range 10^{-10}–10^{-11} G is still viable according to the present considerations. It is, therefore, not excluded that large-scale magnetic fields may come from a primordial field of the order of 0.1–0.01 nG present prior to gravitational collapse of the protogalaxy. Such a field, depending upon the details of the gravitational collapse, may be amplified to the observable level by compressional amplification. The present problems in achieving a large dynamo amplification may therefore be less relevant than for the case when the seed field is in the range $10^{-9} - -10^{-18}$nG. To confirm this type of scenario, it will be absolutely essential to introduce the magnetic field background into the current strategies of parameter extraction.

The considerations reported in the present lecture provide already the framework for such an introduction. In particular, along a minimalist perspective, the inclusion of the magnetic field background boils down to add two new extra parameters: the spectral slope and amplitude of the magnetic field (conventionally smoothed over a typical comoving scale of mega parsec size). The magnetic field contribution will then slightly modify the adiabatic paradigm by introducing, already at the level of initial conditions, a subleading non-Gaussian (and quasi-adiabatic) correction.

References

1. H. Alfvén: Arkiv. Mat. F. Astr., o. Fys. **29 B**, 2 (1943)
2. E. Fermi: Phys. Rev. **75**, 1169 (1949)
3. H. Alfvén: Phys. Rev. **75**, 1732 (1949)

4. R. D. Richtmyer, E. Teller: Phys. Rev. **75**, 1729 (1949)
5. W. A. Hiltner: Science **109**, 165 (1949)
6. J. S. Hall: Science **109**, 166 (1949)
7. L. J. Davis J. L. Greenstein: Astrophys. J. **114**, 206 (1951)
8. E. Fermi, S. Chandrasekar: Astrophys. J. **118**, 113 (1953)
9. E. Fermi, S. Chandrasekar: Astrophys. J. **118**, 116 (1953)
10. R. Wielebinski, J. Shakeshaft: Nature**195**, 982 (1962)
11. A. G. Lyne, F. G. Smith: Nature **218**, 124 (1968)
12. A. G. Lyne, F. G. Smith: *Pulsar Astronomy* (Cambridge University Press, Cambridge, 1998)
13. C. Heiles: Annu. Rev. Astron. Astrophys. **14**, 1 (1976)
14. F. Govoni, L. Feretti : Int. J. Mod. Phys. D **13**, 1549 (2004)
15. B. M. Gaensler, R. Beck, L. Feretti: New Astron. Rev. **48**, 1003 (2004)
16. Y. Xu, P. P. Kronberg, S. Habib, Q. W. Dufton: Astrophys. J. **637**, 19 (2006)
17. P. P. Kronberg : Astron. Nachr. **327**, 517 (2006)
18. M. Giovannini: Int. J. Mod. Phys. D **13**, 391 (2004)
19. http://www.skatelescope.org
20. http://www.rssd.esa.int
21. P. P. Kronberg: Rep. Prog. Phys. **57**, 325 (1994)
22. M. Giovannini: Class. Quant. Grav. **23**, R1 (2006)
23. J. Bernstein, L. S. Brown, G. Feinberg: Rev. Mod. Phys. **61**, 25 (1989)
24. T. J. M. Boyd, J. J. Serson: *The Physics of Plasmas* (Cambridge University Press, Cambridge, 2003)
25. N. A. Krall, A. W. Trivelpiece: *Principles of Plasma Physics* (San Francisco Press, San Francisco, 1986)
26. F. Chen: *Introduction to Plasma Physics* (Plenum Press, New York, 1974)
27. D. Biskamp: *Non-linear Magnetohydrodynamics* (Cambridge University Press, Cambridge, 1994)
28. A. Vlasov: Zh. Éksp. Teor. Fiz. **8**, 291 (1938); J. Phys. **9**, 25 (1945)
29. L. D. Landau: J. Phys. U.S.S.R. **10**, 25 (1945)
30. M. Giovannini: Phys. Rev. D **71**, 021301 (2005)
31. M. Giovannini: Phys. Rev. D **58**, 124027 (1998)
32. E. N. Parker: *Cosmical Magnetic Fields* (Clarendon Press, Oxford, 1979)
33. Ya. B. Zeldovich, A. A. Ruzmaikin, D. D. Sokoloff: *Magnetic Fields in Astrophysics* (Gordon Breach Science, New York, 1983)
34. A. A. Ruzmaikin, A. M. Shukurov, D. D. Sokoloff : *Magnetic Fields of Galaxies* (Kluwer Academic Publisher, Dordrecht, 1988)
35. R. Kulsrud: Annu. Rev. Astron. Astrophys. **37**, 37 (1999)
36. A. Brandenburg, K. Subramanian: Phys. Rept **417**, 1 (2005)
37. S. I. Vainshtein, Ya. B. Zeldovich: Usp. Fiz. Nauk. **106**, 431 (1972)
38. W. H. Matthaeus, M. L. Goldstein, S. R. Lantz: Phys. Fluids **29**, 1504 (1986)
39. A. Lazarian, E. Vishniac, J. Cho: Astrophys. J. **603**, 180 (2004); Lect. Notes Phys. **614**, 376 (2003)
40. A. Brandenburg, A. Bigazzi, K. Subramanian: Mon. Not. Roy. Astron. Soc. **325**, 685 (2001)
41. K. Subramanian, A. Brandenburg: Phys. Rev. Lett. **93**, 205001 (2004)
42. A. Brandenburg, K. Subramanian: Astron. Astrophys. **439**, 835 (2005)
43. R. Kulsrud, S. W. Anderson: Astrophys. J. **396**, 606 (1992)
44. M. J. Rees: Lect. Notes Phys. **664**, 1 (2005)

45. K. Subramanian, D. Narashima, S. Chitre: Mon. Not. Roy. Astron. Soc. **271**, L15 (1994)
46. N. Y. Gnedin, A. Ferrara, E. G. Zweibel: Astrophys. J. **539**, 505 (2000)
47. Ya. Zeldovich, I. Novikov: *The Structure Evolution of the Universe* (Chicago University Press, Chicago, 1971), Vol. 2
48. Ya. Zeldovich: Sov. Phys. JETP **21**, 656 (1965)
49. E. Harrison: Phys. Rev. Lett. **18**, 1011 (1967)
50. E. Harrison: Phys. Rev. **167**, 1170 (1968)
51. E. Harrison: Mon. Not. R. Astr. Soc. **147**, 279 (1970)
52. L. Biermann: Z. Naturf. **5A**, 65 (1950)
53. I. Mishustin, A. Ruzmaikin: Sov. Phys. JETP **34**, 223 (1972)
54. M. Giovannini: Phys. Rev. D **61**, 063004 (2000)
55. M. Giovannini: Phys. Rev. D **61**, 063502 (2000)
56. G. Piccinelli, A. Ayala: Lect. Notes Phys. **646**, 293 (2004)
57. D. Boyanovsky, H. J. de Vega, M. Simionato: Phys. Rev. D **67**, 123505 (2003)
58. D. Boyanovsky, M. Simionato, H. J. de Vega: Phys. Rev. D **67**, 023502 (2003)
59. M. Giovannini, M. E. Shaposhnikov: Phys. Rev. D **57**, 2186 (1998)
60. K. Bamba: arXiv:hep-ph/0611152
61. A. Sanchez, A. Ayala, G. Piccinelli: arXiv:hep-th/0611337
62. M. Giovannini, M. E. Shaposhnikov: Phys. Rev. D **62**, 103512 (2000)
63. M. Giovannini, M. Shaposhnikov: *Proc. CAPP2000* (July 2000, Verbier Switzerland), eprint Archive [hep-ph/0011105]
64. E. Calzetta, A. Kus, F. Mazzitelli: Phys. Rev. D, **57**, 7139 (1998)
65. A. Kus, E. Calzetta, F. Mazzitelli, C. Wagner: Phys. Lett. B **472**, 287 (2000)
66. M. S. Turner, L. M. Widrow: Phys. Rev. D **37**, 2734 (1988)
67. I. Drummond, S. Hathrell: Phys. Rev. D **22**, 343 (1980)
68. A. Dolgov: Phys. Rev. D **48**, 2499 (1993)
69. S. Carroll, G. Field, R. Jackiw: Phys. Rev. D **41**, 1231 (1990)
70. W. D. Garretson, G. Field, S. Carroll: Phys. Rev. D **46**, 5346 (1992)
71. G. Field, S. Carroll: Phys. Rev. D **62**, 103008 (2000)
72. B. Ratra: Astrophys. J. Lett. **391**, L1 (1992)
73. M. Giovannini: Phys. Rev. D **64**, 061301 (2001)
74. K. Bamba, J. Yokoyama: e-print Archive [astro-ph/0310824]
75. M. Gasperini: Phys. Rev. D **63**, 047301 (2001)
76. L. Okun: Sov. Phys. JETP **56**, 502 (1982)
77. O. Bertolami, D. Mota: Phys. Lett. B **455**, 96 (1999)
78. M. Giovannini: Phys. Rev. D **62**, 123505 (2000)
79. L. H. Ford: Phys. Rev. D **31**, 704 (1985)
80. M. Gasperini, M. Giovannini, G. Veneziano: Phys. Rev. Lett. **75**, 3796 (1995)
81. M. Gasperini, M. Giovannini, G. Veneziano: Phys. Rev. D **52**, 6651 (1995)
82. M. Gasperini, M. Giovannini: Phys. Rev. D **47**, 1519 (1993)
83. D. Stoler: Phys. Rev. D **1**, 3217 (1970); D. Stoler: Phys. Rev. D **4**, 2309 (1971)
84. A. O. Barut, L. Girardello: Commun. Math. Phys. **21**, 41 (1971)
85. H. Yuen: Phys. Rev. A **13**, 2226 (1976)
86. S. Fubini, A. Molinari: Nucl. Phys. Proc. Suppl. **33C**, 60 (1993)
87. R. Loudon: J. Mod. Opt. **34**, 709 (1987)
88. R. Loudon: *The Quantum Theory of Light* (Clarendon Press, Oxford, 1983)
89. B. L. Schumaker: Phys. Rep. **135**, 318 (1986)
90. L. Mandel, E. Wolf: *Optical Coherence and Quantum Optics* (Cambridge University Press, Cambridge, 1995)

91. G. Veneziano: Phys. Lett. B **265**, 287 (1991)
92. M. Gasperini, G. Veneziano: Astropart. Phys. **1**, 317 (1993)
93. M. Gasperini, G. Veneziano: Phys. Rep. **373**, 1 (2003)
94. C. Lovelace: Phys. Lett. B **135**, 75 (1984)
95. E. Fradkin, A. Tseytlin: Nucl. Phys. B **261**, 1 (1985)
96. C. Callan et al.: Nucl. Phys. B **262**, 593 (1985)
97. M. Gasperini, M. Giovannini, G. Veneziano: Phys. Lett. B **569**, 113 (2003)
98. M. Gasperini, M. Giovannini, G. Veneziano: Nucl. Phys. B **694**, 206 (2004)
99. K. A. Meissner, G. Veneziano: Mod. Phys. Lett. A **6**, 3397 (1991)
100. K. A. Meissner G. Veneziano: Phys. Lett. B **267**, 33 (1991)
101. M. Gasperini, J. Maharana, G. Veneziano: Nucl. Phys. B **472**, 349 (1996)
102. M. Giovannini: Class. Quant. Grav. **21**, 4209 (2004)
103. M. Gasperini, M. Giovannini: Phys. Lett. B **301**, 334 (1993)
104. M. Giovannini: Phys. Rev. D **61**, 087306 (2000)
105. R. Brustein, M. Gasperini, M. Giovannini, G. Veneziano: Phys. Lett. B **361**, 45 (1995)
106. R. Brustein, M. Gasperini, M. Giovannini, V. F. Mukhanov, G. Veneziano, Phys. Rev. D **51**, 6744 (1995)
107. K. Enqvist M. S. Sloth: Nucl. Phys. B **626**, 395 (2002); M. S. Sloth: Nucl. Phys. B **656**, 239 (2003)
108. V. Bozza, M. Gasperini, M. Giovannini, G. Veneziano: Phys. Rev. D **67** (2003) 063514; V. Bozza, M. Gasperini, M. Giovannini, G. Veneziano: Phys. Lett. B **543**, 14 (2002)
109. P. Astone et al.: Astron. Astrophys. **351**, 811 (1999)
110. Ph. Bernard, G. Gemme, R. Parodi, E. Picasso: Rev. Sci. Instrum. **72**, 2428 (2001)
111. A. M. Cruise: Class. Quantum Grav. **17**, 2525 (2000); A. M. Cruise: Mon. Not. R. Astron. Soc **204**, 485 (1983)
112. D. Babusci and M. Giovannini: Int. J. Mod. Phys. D **10** 477 (2001); D. Babusci and M. Giovannini: Class. Quant. Grav. **17**, 2621 (2000)
113. P. J. E. Peebles, A. Vilenkin: Phys. Rev. D **59**, 063505 (1999)
114. M. Giovannini: Class. Quant. Grav. **16**, 2905 (1999); M. Giovannini: Phys. Rev. D **60**, 123511 (1999); D. Babusci and M. Giovannini: Phys. Rev. D **60**, 083511 (1999); M. Giovannini: Phys. Rev. D **58**, 083504 (1998)
115. M. Gasperini, S. Nicotri: Phys. Lett. B **633**, 155 (2006)
116. R. Beck: Astron. Nachr. **327**, 512 (2006)
117. R. Beck, A. Brenburg, D. Moss, A. Skhurov, D. Sokoloff: Annu. Rev. Astron. Astrophys. **34**, 155 (1996)
118. J. Vallée: Astrophys. J. **566**, 261 (2002)
119. E. Battaner, E. Florido: Mon. Not. R. Astron. Soc **277**, 1129 (1995)
120. E. Battaner, E. Florido, J. Jimenez-Vincente: Astron. Astrophys. **326**, 13 (1997)
121. E. Florido, E. Battaner: Astron. Astrophys. **327**, 1 (1997)
122. E. Florido et al.: arXiv:astro-ph/0609384
123. E. Battaner, E. Florido: Fund. Cosmic Phys. **21**, 1 (2000)
124. J. Barrow, K. Subramanian: Phys. Rev. Lett. **81**, 3575 (1998); J. Barrow, K. Subramanian: Phys. Rev. D **58**, 83502 (1998); C. Tsagas, R. Maartens: Phys. Rev. D **61**, 083519 (2000); A. Mack, T. Kahniashvili, A. Kosowsky: Phys. Rev. D **65**, 123004 (2002); A. Lewis: Phys. Rev. D **70**, 043518 (2004); T. Kahniashvili, B. Ratra: Phys. Rev. D **71**, 103006 (2005)

125. G. Chen et al.: Astrophys. J. **611**, 655 (2004); P. D. Naselsky et al.: Astrophys. J. **615**, 45 (2004); L. Y. Chiang, P. Naselsky: Int. J. Mod. Phys. D **14**, 1251 (2005); L. Y. Chiang, P. D. Naselsky, O. V. Verkhodanov, M. J. Way: Astrophys. J. **590**, L65 (2003); D. G. Yamazaki et al.: Astrophys. J. **625**, L1 (2005)
126. K. Subramanian: Astron. Nachr. **327**, 399 (2006)
127. H. V. Peiris et al. [WMAP Collaboration]: Astrophys. J. Suppl. **148**, 213 (2003)
128. K. Enqvist, H. Kurki-Suonio, J. Valiviita: Phys. Rev. D **62**, 103003 (2000)
129. H. Kurki-Suonio, V. Muhonen, J. Valiviita: Phys. Rev. D **71**, 063005 (2005)
130. K. Moodley, M. Bucher, J. Dunkley, P. G. Ferreira, C. Skordis: Phys. Rev. D **70**, 103520 (2004)
131. M. Giovannini: Phys. Rev. D **73**, 101302 (2006)
132. M. Giovannini: Phys. Rev. D **74**, 063002 (2006)
133. M. Giovannini: Class. Quant. Grav. **23**, 4991 (2006)
134. J. D. Barrow, R. Maartens, C. G. Tsagas: arXiv:astro-ph/0611537
135. T. Kahniashvili, B. Ratra: arXiv:astro-ph/0611247.
136. E. Harrison: Rev. Mod. Phys. **39**, 862 (1967)
137. J. M. Bardeen: Phys. Rev. D **22**, 1882 (1980)
138. C.-P. Ma E. Bertschinger: Astrophys. J. **455**, 7 (1995)
139. M. Giovannini: Phys. Rev. D **70**, 123507 (2004)
140. M. Giovannini: Int. J. Mod. Phys. D **14**, 363 (2005)
141. J. Bardeen, P. Steinhardt, M. Turner: Phys. Rev. D **28**, 679 (1983)
142. R. Brandenberger, R. Kahn, W. Press: Phys. Rev. D **28**, 1809 (1983)
143. M. Giovannini: Phys. Lett. B **622**, 349 (2005)
144. M. Giovannini: Class. Quant. Grav. **22**, 5243 (2005)
145. D. Spergel et al. [WMAP Collaboration]: arXiv:astro-ph/0603449
146. W. Hu N. Sugiyama: Astrophys. J. **444**, 489 (1995); ibid. **471**, 30 (1996)
147. L. Page et al. [WMAP collaboration]: arXiv:astro-ph/0603450
148. A. G. Riess et al.: Astrophys. J. **607**, 665 (2005)
149. P. Astier et al.: astro-ph/0510447
150. P. Naselsky, I. Novikov: Astrophys. J. **413**, 14 (1993)
151. H. Jorgensen, E. Kotok, P. Naselsky, I. Novikov: Astron. Astrophys. **294**, 639 (1995)
152. P. J. E. Peebles, J. T. Yu: Astrophys. J. **162**, 815 (1970)
153. A. G. Doroshkevich, Ya. B. Zeldovich, R. A. Sunyaev: Sov. Astron. **22**, 523 (1978)
154. M. Zaldarriaga, D. D. Harari: Phys. Rev. D **52** (1995) 3276.
155. S. Chandrasekar: *Radiative Transfer* (Dover, New York, 1966)

Cosmological Singularities and a Conjectured Gravity/Coset Correspondence

T. Damour

Institut des Hautes Etudes Scientifiques, 35 route de Chartres,
F-91440 Bures-sur-Yvette, France
damour@ihes.fr

Abstract. We review the recently discovered connection between the Belinsky–Khalatnikov–Lifshitz-like "chaotic" structure of generic cosmological singularities in 11-dimensional supergravity and the "last" hyperbolic Kac–Moody algebra E_{10}. This intriguing connection suggests the existence of a hidden "correspondence" between supergravity (or even M-theory) and null geodesic motion on the infinite-dimensional coset space $E_{10}/K(E_{10})$. If true, this gravity/coset correspondence would offer a new view of the (quantum) fate of space (and matter) at cosmological singularities.

1 Introduction

It is a pleasure to participate in the celebration of the seminal accomplishments of Gabriele Veneziano. I will try to do so by reviewing a line of research which is intimately connected with several of Gabriele's important contributions, being concerned with the cardinal problem of String Cosmology: the fate of the Einstein-like space–time description at big crunch/big bang cosmological singularities. Actually, the work described below started as a by-product of the string cosmology program initiated by Gasperini and Veneziano [1]. While collaborating with Gabriele on the possible birth of "pre–big bang bubbles" from the gravitational collapse instability of a *generic* string vacuum made of a stochastic bath of incoming gravitational and dilatonic waves [2], an issue raised itself: What is the structure of a *generic* spacelike (i.e. big crunch or big bang) singularity within the effective field theory approximation of (super-) string theory (when keeping all fields, and not only the metric and the dilaton)? The answer turned out to be surprisingly complex, and rich of hidden structures. It was first found [3, 4] that the general solution, near a space-like singularity, of the massless bosonic sector of all superstring models ($D = 10$, IIA, IIB, I, HE, HO), as well as that of M theory ($D = 11$ supergravity), exhibits a never-ending oscillatory behaviour of the Belinsky–Khalatnikov–Lifshitz (BKL) type [5]. However, it was later realized that behind this seeming entirely *chaotic* behaviour there was a *hidden symmetry structure* [6, 7, 8].

T. Damour: *Cosmological Singularities and a Conjectured Gravity/Coset Correspondence*, Lect. Notes Phys. **737**, 941–948 (2008)
DOI 10.1007/978-3-540-74233-6_26

This led to the conjecture of the existence of a hidden equivalence (i.e. a *correspondence*) between two seemingly very different dynamical systems: on the one hand, 11-dimensional supergravity (or even, hopefully, "M-theory"), and, on the other hand, a *one-dimensional* $E_{10}/K(E_{10})$ nonlinear σ model, i.e. the geodesic motion of a massless particle on the infinite-dimensional coset space[1] $E_{10}/K(E_{10})$ [8]. The intuitive hope behind this conjecture is that the BKL-type *near spacelike singularity limit* might act as a tool for revealing a hidden structure, in analogy to the much better established AdS/CFT correspondence [9], where the consideration of the *near horizon limit* of certain black D-branes has revealed a hidden equivalence between 10-dimensional string theory in AdS space–time on one side, and a lower-dimensional CFT on the other side. If the (much less firmly established) "gravity/coset correspondence" were confirmed, it might provide both the basis of a new definition of M-theory, and a description of the "de-emergence" of space near a cosmological singularity (see [10] and below).

2 Cosmological Billiards

Let us start by summarizing the BKL-type analysis of the "near spacelike singularity limit", that is, of the asymptotic behaviour of the metric $g_{\mu\nu}(t, \mathbf{x})$, together with the other fields (such as the 3-form $A_{\mu\nu\lambda}(t, \mathbf{x})$ in supergravity), near a singular hypersurface. The basic idea is that, near a spacelike singularity, the time derivatives are expected to dominate over spatial derivatives. More precisely, BKL found that spatial derivatives introduce terms in the equations of motion for the metric which are similar to the "walls" of a billiard table [5]. To see this, it is convenient [11] to decompose the D-dimensional metric $g_{\mu\nu}$ into nondynamical (lapse N, and shift N^i, here set to zero) and dynamical ($e^{-2\beta^a}$, θ^a_i) components. They are defined so that the line element reads

$$ds^2 = -N^2 dt^2 + \sum_{a=1}^{d} e^{-2\beta^a} \theta^a_i \theta^a_j dx^i dx^j . \tag{1}$$

Here $d \equiv D-1$ denotes the spatial dimension ($d = 10$ for SUGRA_{11}, and $d = 9$ for string theory), $e^{-2\beta^a}$ represent (in an Iwasawa decomposition) the "diagonal" components of the spatial metric g_{ij}, while the "off diagonal" components are represented by the θ^a_i, defined to be upper triangular matrices with 1's on the diagonal (so that, in particular, $\det\theta = 1$).

The Hamiltonian constraint, at a given spatial point, reads (with $\tilde{N} \equiv N/\sqrt{\det g_{ij}}$ denoting the "rescaled lapse")

[1] Here $K(E_{10})$ denotes the (formal) "maximal compact subgroup" of the hyperbolic Kac–Moody group E_{10}.

$$\mathcal{H}(\beta^a, \pi_a, P, Q)$$
$$= \tilde{N}\left[\frac{1}{2}\, G^{ab}\pi_a\pi_b + \sum_A c_A(Q, P, \partial\beta, \partial^2\beta, \partial Q) \exp\big(-2w_A(\beta)\big)\right]. \quad (2)$$

Here π_a (with $a = 1, ..., d$) denote the canonical momenta conjugate to the "logarithmic scale factors" β^a, while Q denote the remaining configuration variables (θ^a_i, 3-form components $A_{ijk}(t, \mathbf{x})$ in supergravity), and P their canonically conjugate momenta (P^i_a, π^{ijk}). The symbol ∂ denotes *spatial* derivatives. The (inverse) metric G^{ab} in (2) is the DeWitt "superspace" metric induced on the β's by the Einstein–Hilbert action. It endows the D-dimensional[2] β space with a Lorentzian structure $G_{ab}\,\dot{\beta}^a\dot{\beta}^b$.

One of the crucial features of (2) is the appearance of Toda-like exponential potential terms $\propto \exp(-2w_A(\beta))$, where the $w_A(\beta)$ are *linear forms* in the logarithmic scale factors: $w_A(\beta) \equiv w_{Aa}\,\beta^a$. The range of labels A and the specific "wall forms" $w_A(\beta)$ that appear depend on the considered model. For instance, in SUGRA$_{11}$ there appear: "symmetry wall forms" $w^S_{ab}(\beta) \equiv \beta^b - \beta^a$ (with $a < b$), "gravitational wall forms" $w^g_{abc}(\beta) \equiv 2\beta^a + \sum_{e \neq a,b,c} \beta^e$ ($a \neq b$, $b \neq c$, $c \neq a$), "electric 3-form wall forms", $e_{abc}(\beta) \equiv \beta^a + \beta^b + \beta^c$ ($a \neq b$, $b \neq c$, $c \neq a$), and "magnetic 3-form wall forms", $m_{a_1 .... a_6} \equiv \beta^{a_1} + \beta^{a_2} + ... + \beta^{a_6}$ (with indices all different).

One then finds that the near-spacelike-singularity limit amounts to considering the *large β limit* in (2). In this limit a crucial role is played by the linear forms $w_A(\beta)$ appearing in the "exponential walls". Actually, these walls enter in successive "layers". A first layer consists of a subset of all the walls called the *dominant walls* $w_i(\beta)$. The effect of these dynamically dominant walls is to confine the motion in β-space to a *fundamental billiard chamber* defined by the inequalities $w_i(\beta) \geqslant 0$. In the case of SUGRA$_{11}$, one finds that there are 10 dominant walls: 9 of them are the symmetry walls $w^S_{12}(\beta), w^S_{23}(\beta), ..., w^S_{910}(\beta)$, and the 10th is an electric 3-form wall $e_{123}(\beta) = \beta^1 + \beta^2 + \beta^3$. As noticed in [6] a remarkable fact is that the fundamental cosmological billiard chamber of SUGRA$_{11}$ (as well as type II string theories) is the *Weyl chamber* of the hyperbolic Kac–Moody algebra E_{10}. More precisely, the 10 dynamically dominant wall forms $\{w^S_{12}(\beta), w^S_{23}(\beta), ..., w^S_{910}(\beta), e_{123}(\beta)\}$ can be identified with the 10 *simple roots* $\{\alpha_1(h), \alpha_2(h), ..., \alpha_{10}(h)\}$ of E_{10}. Here h parametrizes a generic element of a Cartan subalgebra (CSA) of E_{10}. [Let us also note that for heterotic and type I string theories the cosmological billiard is the Weyl chamber of another rank 10 hyperbolic Kac–Moody algebra, namely BE_{10}]. In the Dynkin diagram of E_{10}, Fig. 1, the 9 "horizontal" nodes correspond to the 9 symmetry walls, while the characteristic "exceptional" node sticking

[2] 10 dimensional for SUGRA$_{11}$; but the various superstring theories also lead to a 10-dimensional Lorentz space because one must add the (positive) kinetic term of the dilaton $\varphi \equiv \beta^{10}$ to the nine-dimensional DeWitt metric corresponding to the nine spatial dimensions.

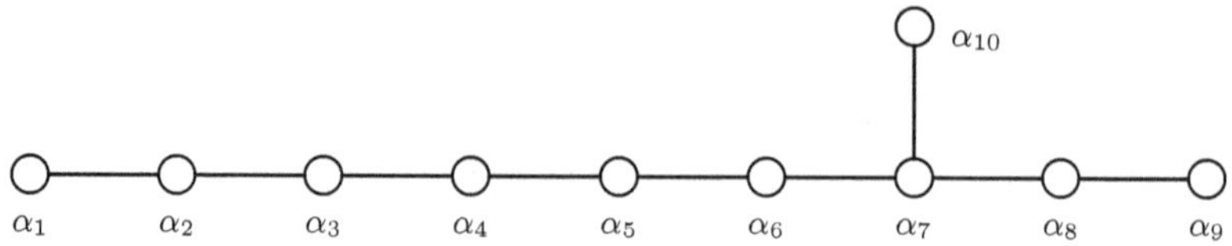

Fig. 1. *Dynkin diagram of* E_{10}

out "vertically" corresponds to the electric 3-form wall $e_{123} = \beta^1 + \beta^2 + \beta^3$. (The fact that this node stems from the *3rd* horizontal node is then seen to be directly related to the presence of the *3-form* $A_{\mu\nu\lambda}$, with electric kinetic energy $\propto g^{i\ell} g^{jm} g^{kn} \dot{A}_{ijk} \dot{A}_{\ell mn}$.)

The appearance of E_{10} in the BKL behaviour of SUGRA$_{11}$ revived an old suggestion of Julia [12] about the possible role of E_{10} in a *one-dimensional reduction* of SUGRA$_{11}$. A posteriori, one can view the BKL behaviour as a kind of spontaneous reduction to one dimension (time) of a multidimensional theory. Note, however, that we are always discussing generic *inhomogeneous* 11-dimensional solutions, but that we examine them in the near-spacelike-singularity limit where the spatial derivatives are subdominant: $\partial_x \ll \partial_t$. Note also that the discrete $E_{10}(\mathbb{Z})$ was proposed as a U-duality group of the full (T^{10}) spatial toroidal compactification of M-theory by Hull and Townsend [13].

3 Gravity/Coset Correspondence

References [8, 14] went beyond the leading-order BKL analysis just recalled by including the first three "layers" of spatial-gradient-related subdominant walls $\propto \exp(-2w_A(\beta))$ in (2). The relative importance of these subdominant walls, which modify the leading billiard dynamics defined by the 10 dominant walls $w_i(\beta)$, can be ordered by means of an expansion which counts how many dominant wall forms $w_i(\beta)$ are contained in the exponents of the subdominant wall forms $w_A(\beta)$, associated to *higher spatial gradients*. By mapping the dominant gravity wall forms $w_i(\beta)$ onto the corresponding E_{10} simple roots $\alpha_i(h), i = 1, ..., 10$, the just described BKL-type *gradient expansion* becomes mapped onto a Lie algebraic *height expansion* in the roots of E_{10}. It was remarkably found that, up to height 30 (i.e. up to small corrections to the billiard dynamics associated to the product of 30 leading walls $e^{-2w_i(\beta)}$), the SUGRA$_{11}$ dynamics for $g_{\mu\nu}(t, \mathbf{x})$, $A_{\mu\nu\lambda}(t, \mathbf{x})$ considered at some given spatial point $\mathbf{x}_0$, could be identified to the geodesic dynamics of a *massless particle* moving on the (infinite-dimensional) coset space $E_{10}/K(E_{10})$. Note the "holographic" nature of this correspondence between an 11-dimensional dynamics on one side, and a 1-dimensional one on the other side.

A point on the coset space $E_{10}(\mathbb{R})/K(E_{10}(\mathbb{R}))$ is coordinatized by a time-dependent (but spatially independent) element of the $E_{10}(\mathbb{R})$ group of the

(Iwasawa) form: $g(t) = \exp h(t) \exp \nu(t)$. Here, $h(t) = \beta^a_{\rm coset}(t) H_a$ belongs to the 10-dimensional CSA of E_{10}, while $\nu(t) = \sum_{\alpha>0} \nu^\alpha(t) E_\alpha$ belongs to a Borel subalgebra of E_{10} and has an infinite number of components labelled by a *positive root* α of E_{10}. The (null) geodesic action over the coset space $E_{10}/K(E_{10})$ takes the simple form

$$S_{E_{10}/K(E_{10})} = \int \frac{dt}{n(t)} (v^{\rm sym}|v^{\rm sym}), \tag{1}$$

where $v^{\rm sym} \equiv \frac{1}{2}(v + v^T)$ is the "symmetric"[3] part of the "velocity" $v \equiv (dg/dt)g^{-1}$ of a group element $g(t)$ running over $E_{10}(\mathbb{R})$.

The correspondence between the gravity (2) and coset (3) dynamics is best exhibited by decomposing (the Lie algebra of) E_{10} with respect to (the Lie algebra of) the $GL(10)$ subgroup defined by the horizontal line in the Dynkin diagram of E_{10}. This allows one to grade the various components of $g(t)$ by their $GL(10)$ level ℓ. One finds that, at the $\ell = 0$ level, $g(t)$ is parametrized by the Cartan coordinates $\beta^a_{\rm coset}(t)$ together with a unimodular upper triangular zehnbein $\theta^a_{{\rm coset}\, i}(t)$. At level $\ell = 1$, one finds a 3-form $A^{\rm coset}_{ijk}(t)$; at level $\ell = 2$, a 6-form $A^{\rm coset}_{i_1 i_2 \ldots i_6}(t)$, and at level $\ell = 3$ a 9-index object $A^{\rm coset}_{i_1|i_2\ldots i_9}(t)$ with Young-tableau symmetry $\{8,1\}$. The coset action (3) then defines a coupled set of equations of motion for $\beta^a_{\rm coset}(t)$, $\theta^a_{{\rm coset}\, i}(t)$, $A^{\rm coset}_{ijk}(t)$, $A^{\rm coset}_{i_1\ldots i_6}(t)$, $A^{\rm coset}_{i_1|i_2\ldots i_9}(t)$. By explicit calculations, it was found that these coupled equations of motion could be identified (modulo terms corresponding to potential walls of height at least 30) to the SUGRA$_{11}$ equations of motion, considered at some given spatial point $\mathbf{x}_0$.

The *dictionary* between the two dynamics says essentially that (i) $\beta^a_{\rm gravity}(t, \mathbf{x}_0) \leftrightarrow \beta^a_{\rm coset}(t)$, $\theta^a_i(t, \mathbf{x}_0) \leftrightarrow \theta^a_{{\rm coset}\, i}(t)$, (ii) $\partial_t A^{\rm coset}_{ijk}(t)$ corresponds to the electric components of the 11-dimensional field strength $F_{\rm gravity} = d\, A_{\rm gravity}$ in a certain frame e^i, (iii) the conjugate momentum of $A^{\rm coset}_{i_1\ldots i_6}(t)$ corresponds to the *dual* (using $\varepsilon^{i_1 i_2 \ldots i_{10}}$) of the "magnetic" frame components of the 4-form $F_{\rm gravity} = d\, A_{\rm gravity}$, and (iv) the conjugate momentum of $A_{i_1|i_2\ldots i_9}(t)$ corresponds to the ε^{10} dual (on jk) of the structure constants C^i_{jk} of the coframe e^i $(d\, e^i = \frac{1}{2} C^i_{jk}\, e^j \wedge e^k)$.

The fact that at levels $\ell = 2$ and $\ell = 3$ the dictionary between supergravity and coset variables maps the *first spatial gradients* of the SUGRA variables $A_{ijk}(t, \mathbf{x})$ and $g_{ij}(t, \mathbf{x})$ onto (time derivatives of) coset variables suggested the conjecture [8] of a hidden *equivalence* between the two models, i.e. the existence of a dynamics-preserving map between the infinite tower of (spatially independent) coset variables $(\beta^a_{\rm coset}, \nu^\alpha)$, together with their conjugate momenta $(\pi^{\rm coset}_a, p_\alpha)$, and the infinite sequence of spatial Taylor coefficients $(\beta(\mathbf{x}_0),\ \pi(\mathbf{x}_0),\ Q(\mathbf{x}_0),\ P(\mathbf{x}_0),\ \partial Q(\mathbf{x}_0),\ \partial^2 \beta(\mathbf{x}_0),\ \partial^2 Q(\mathbf{x}_0), \ldots, \partial^n Q(\mathbf{x}_0), \ldots)$

[3] Here the transpose operation T denotes the negative of the Chevalley involution ω defining the real form $E_{10(10)}$ of E_{10}. It is such that the elements k of the Lie subalgebra of $K(E_{10})$ are "T-antisymmetric": $k^T = -k$, which is equivalent to them being fixed under $\omega : \omega(k) = +\omega(k)$.

formally describing the dynamics of the gravity variables ($\beta(\mathbf{x}), \pi(\mathbf{x}), Q(\mathbf{x}), P(\mathbf{x})$) around some given spatial point $\mathbf{x}_0$.[4]

It has been possible to extend the correspondence between the two models to the inclusion of fermionic terms on both sides [15, 16, 17]. Moreover, [18] found evidence for a nice compatibility between some high-level contributions (height −115!) in the coset action, corresponding to *imaginary* roots,[5] and M-theory *one-loop corrections* to SUGRA_{11}, notably the terms quartic in the curvature tensor. (See also [19] for a study of the compatibility of an underlying Kac–Moody symmetry with quantum corrections in various models.)

4 A New View of the (quantum) Fate of Space at a Cosmological Singularity

Let us now, following [10], sketch the physical picture suggested by the gravity/coset correspondence. That is, let us take seriously the idea that, upon approaching a spacelike singularity, the description in terms of a spatial continuum, and space–time based (quantum) field theory breaks down, and should be replaced by a purely abstract Lie algebraic description. More precisely, we suggest that the information previously encoded in the spatial variation of the geometry and of the matter fields gets transferred to an infinite tower of spatially independent (but time-dependent) Lie algebraic variables. In other words, we are led to the conclusion that space actually "disappears" (or "de-emerges") as the singularity is approached.[6] In particular (and this would be bad news for Gabriele's pre–big bang scenario), we suggest no (quantum) "bounce" from an incoming collapsing universe to some outgoing expanding universe. Rather it is suggested that "life continues" for an infinite "affine time" at a singularity, with the double understanding, however, that (i) life continues only in a totally new form (as in a kind of "transmigration") and (ii) an infinite affine time interval (measured, say, in the coordinate t of (3) with a coset lapse function $n(t) = 1$) corresponds to a sub-Planckian interval of geometrical proper time.[7]

Let us also comment on some expected aspects of the "duality" between the two models. It seems probable (from the AdS/CFT paradigm) that, even

[4] One, however, expects the map between the two models to become spatially non-local for heights ≥ 30.

[5] i.e. such that $(\alpha, \alpha) < 0$, by contrast to the "real" roots, $(\alpha, \alpha) = +2$, which enter the checks mentioned above.

[6] We have in mind here a "big crunch", i.e. we conventionally consider that we are tending *towards* the singularity. *Mutatis mutandis*, we would say that space "appears" or "emerges" at a big bang.

[7] Indeed, it is found that the coset time t (with $n(t) = 1$) corresponds to a "Zeno-like" gravity coordinate time (with rescaled lapse $\tilde{N} = N/\sqrt{g} = 1$) which tends to $+\infty$ as the proper time tends to zero.

if the equivalence between the "gravity" and the "coset" descriptions is formally exact, each model has a natural domain of applicability in which the corresponding description is sufficiently "weakly coupled" to be trustable as is, even in the leading approximation. For the gravity description this domain is clearly that of curvatures smaller than the Planck scale. One then expects that the natural domain of validity of the dual coset model would correspond (in gravity variables) to that of curvatures larger than the Planck scale. In addition, it is possible that the coset description should primarily be considered as a quantum model, as now sketched.

The coset action (3) describes the classical motion of a massless particle on the symmetric space $E_{10}(\mathbb{R})/K(E_{10}(\mathbb{R}))$. Quantum mechanically, one should consider a quantum massless particle, i.e., if we neglect polarization effects[8] a Klein–Gordon equation,

$$\Box \Psi(\beta^a, \nu^\alpha) = 0\,, \tag{1}$$

where $\Box$ denotes the (formal) Laplace–Beltrami operator on the infinite-dimensional Lorentz-signature curved coset manifold $E_{10}(\mathbb{R})/K(E_{10}(\mathbb{R}))$. Equation (4) would apply to the case considered here of uncompactified M-theory. In the case where all spatial dimensions are toroidally compactified, it has been suggested [20, 21] that Ψ satisfy (4) together with a condition of periodicity over the discrete group $E_{10}(\mathbb{Z})$. In other words, Ψ would be a "modular wave form" on $E_{10}(\mathbb{Z})\backslash E_{10}(\mathbb{R})/K(E_{10}(\mathbb{R}))$.

Let us emphasize (still following [10]) that all reference to space and time has disappeared in (1). The disappearance of time is common between (4) and the usual Wheeler-DeWitt equation in which the "wave function(al) of the universe" $\Psi[g_{ij}(\mathbf{x})]$ no longer depends on any *extrinsic* time parameter. (As usual, one needs to choose among all the dynamical variables a specific "clock field" to be used as an *intrinsic* time variable parametrizing the dynamics of the remaining variables.) The interesting new feature of (4) (when compared to a Wheeler–DeWitt type equation) is the disappearance of any notion of geometry $g_{ij}(\mathbf{x})$ and its replacement by the infinite tower of Lie algebraic variables (β^a, ν^α).[9] This quantum de-emergence of space, and the emergence of an infinite-dimensional symmetry group E_{10}[10] which *deeply intertwines space-time with matter degrees of freedom*, might be radical enough to get us closer to an understanding of the fate of space–time and matter at cosmological singularities.

---

[8] Actually, [15, 16, 17] indicate the need to consider a *spinning* massless particle, i.e. some kind of Dirac equation on $E_{10}/K(E_{10})$.

[9] Note that this is conceptually very different from the E_{11}-based proposal of [22].

[10] Let us note that E_{10} enjoys a similarly distinguished status among the (infinite-dimensional) *hyperbolic* Kac–Moody Lie groups as E_8 does in the Cartan–Killing classification of the *finite-dimensional* simple Lie groups [23].

Acknowledgments

It is a pleasure to dedicate this review to Gabriele Veneziano, a dear friend and a great physicist from whom I have learned a lot. I am also very grateful to my collaborators Marc Henneaux and Hermann Nicolai for the (continuing) E_{10} adventure. I also wish to thank Maurizio Gasperini and Jnan Maherana for their patience.

References

1. M. Gasperini, G. Veneziano: Phys. Rep. **373**, 1 (2003)
2. A. Buonanno, T. Damour, G. Veneziano: Nucl. Phys. B **543**, 275 (1999)
3. T. Damour, M. Henneaux: Phys. Rev. Lett. **85**, 920 (2000)
4. T. Damour, M. Henneaux: Phys. Lett. B **488**, 108 (2000) [Erratum-ibid. B **491**, 377 (2000)]
5. V. A. Belinsky, I. M. Khalatnikov, E. M. Lifshitz: Adv. Phys. **19**, 525 (1970)
6. T. Damour, M. Henneaux: Phys. Rev. Lett. **86**, 4749 (2001)
7. T. Damour, M. Henneaux, B. Julia, H. Nicolai: Phys. Lett. B **509**, 323 (2001)
8. T. Damour, M. Henneaux, H. Nicolai: Phys. Rev. Lett. **89**, 221601 (2002)
9. O. Aharony, S. S. Gubser, J. M. Maldacena, H. Ooguri, Y. Oz: Phys. Rep. **323**, 183 (2000)
10. T. Damour, H. Nicolai: "Symmetries, Singularities and the De-emergence of Space", essay submitted to the Gravity Research Foundation (March 2007)
11. T. Damour, M. Henneaux, H. Nicolai: Class. Quant. Grav. **20**, R145 (2003)
12. B. Julia: in *Lectures in Applied Mathematics*, Vol. 21 (1985), AMS-SIAM, p. 335; preprint LPTENS 80/16
13. C. M. Hull, P. K. Townsend: Nucl. Phys. B **438**, 109 (1995)
14. T. Damour, H. Nicolai: arXiv:hep-th/0410245
15. T. Damour, A. Kleinschmidt, H. Nicolai: Phys. Lett. B **634**, 319 (2006)
16. S. de Buyl, M. Henneaux, L. Paulot: JHEP **0602**, 056 (2006)
17. T. Damour, A. Kleinschmidt, H. Nicolai: JHEP **0608**, 046 (2006)
18. T. Damour, H. Nicolai: Class. Quantum. Grav. **22**, 2849 (2005)
19. T. Damour, A. Hanany, M. Henneaux, A. Kleinschmidt, H. Nicolai: Gen. Rel. Grav. **38**, 1507 (2006)
20. O. J. Ganor: arXiv:hep-th/9903110
21. J. Brown, O. J. Ganor, C. Helfgott: JHEP **0408**, 063 (2004)
22. P. C. West: Class Quantum Grav. **18**, 4443 (2001)
23. V. G. Kac: *Infinite Dimensional Lie Algebras*, 3rd edition (Cambridge University Press, Cambridge, 1990)

Brane Inflation: String Theory Viewed from the Cosmos*

S.-H. H. Tye

Newman Laboratory for Elementary Particle Physics, Cornell University, Ithaca, NY 14853, USA
tye@lepp.cornell.edu

Abstract. Brane inflation is a specific realization of the inflationary universe scenario in the early universe within the brane world framework in string theory. The naturalness and robustness of this realistic scenario is explained. Its predictions on the cosmological observables in the cosmic microwave background radiation, especially possible distinct stringy features, such as large non-Gaussianity or large tensor mode that deviates from that predicted in the slow-roll scenario, are discussed. Stringy Kaluza–Klein (KK) modes as hidden dark matter is also a possibility. Another generic consequence of brane inflation is the production of cosmic strings towards the end of inflation. These cosmic strings are nothing but superstrings stretched to cosmological sizes. The properties of these cosmic superstrings and their subsequent cosmological evolution into a scaling network open up their possible detections in the near future, via cosmological, astronomical and/or gravitational wave measurements. At the moment, cosmological data are already imposing strong constraints on the details of the scenario. Finding distinctive stringy signatures in cosmological observations will go a long way in revealing the specific brane inflationary scenario and validating string theory as well as the brane world picture. Precision measurements may even reveal the structures of the flux compactification. Irrespective of the final outcome, we see that string theory is confronting data and making predictions.

1 Introduction

It is believed by many that superstring theory is the fundamental theory of all matter and forces, including a consistent quantum gravity sector. In fact, it is the only known theory that incorporates general relativity in a quantum mechanically consistent way around the near Minkowski spacetime that describes our universe today. The theory is also extraordinarily intricate, revealing numerous deep and rich mathematical and physical structures. However, the string scale is believed to be so high that it is almost hopeless to find stringy signatures at any high-energy experiments in the conceivable future. Since

* In celebration of the 65th birthday of Gabriele Veneziano, teacher and friend.

S.-H. Henry Tye: *Brane Inflation: String Theory Viewed from the Cosmos*, Lect. Notes Phys. **737**, 949–974 (2008)
DOI 10.1007/978-3-540-74233-6_26

such a high-energy scale was probably once reached in the early universe, it is natural to look for stringy signatures in cosmology. Looking towards the sky for information and tests on fundamental physics has a long tradition. This follows the route taken by, for example, the discovery of Newton's gravitational force law and Einstein's theory of general relativity.

The inflationary universe was proposed to solve a number of fine-tuning problems such as the flatness problem, the horizon problem and the defect problem [1]. Besides providing an origin for the hot big bang (the ultimate free lunch), its prediction of an almost scale-invariant density perturbation power spectrum (which is responsible for structure formation in our universe) has received strong observational support from the temperature fluctuation and polarization in the cosmic microwave background radiation (CMBR), e.g., COBE [2] and WMAP [3]. However, the origin of the key ingredients of the inflationary scenario, namely, the scalar field known as the inflaton and its potential, remains undetermined. In this sense, the inflationary universe scenario is considered by many to be a paradigm or framework, not quite a theory. As the cosmological data keep improving in a very impressive fashion, it becomes urgent to find a specific model that has a solid theoretical foundation.

If string theory is the theory of everything, we should be able to find a natural inflationary scenario there. This will allow us to identify the inflaton and its properties, while at the same time cosmological measurements will help us to determine the precise stringy description of our universe. With some luck, we may even find distinct stringy signatures in this framework in the cosmological data to confirm our faith in the theory. Since the inflationary scale turns out to be comparable to the string scale, such an investigation is clearly very worthwhile. If the scenario is natural, one should be able to explain why many e-folds of expansion are generic (without fine-tuning). A good test requires the scenario/model to be over-constrained, i.e., the number of measurements should eventually exceed the number of parameters in the model. We shall explain how (and in what sense) brane inflation, a specific realization of the inflationary universe scenario in the early universe within the brane world framework in string theory, satisfies these two criteria; that is, it is both natural and testable.

Since the discovery of D-branes in string theory [4], a natural realization of nature in string theory is the brane world. In the brane world, all standard model particles are open string modes. Since each end of an open string must end on a brane, the standard model (SM) particles (being light) are stuck on a stack of Dp-brane, where 3 of the p dimensions span our universe of standard model particles, while the remaining $p-3$ dimensions are wrapping some cycles in the bulk (the remaining $9-p$ spatial dimensions) where closed string modes such as the graviton live (Fig. 1a). Suppose our today's universe is described by such a brane world solution in string theory. A simple, realistic and well-motivated inflationary model is the brane inflation, where the inflaton is simply the position a Dp-brane moving in the bulk [5]. In the simple $D3$-$\bar{D}3$-brane inflation [6], inflation takes place while the $D3$-brane is moving

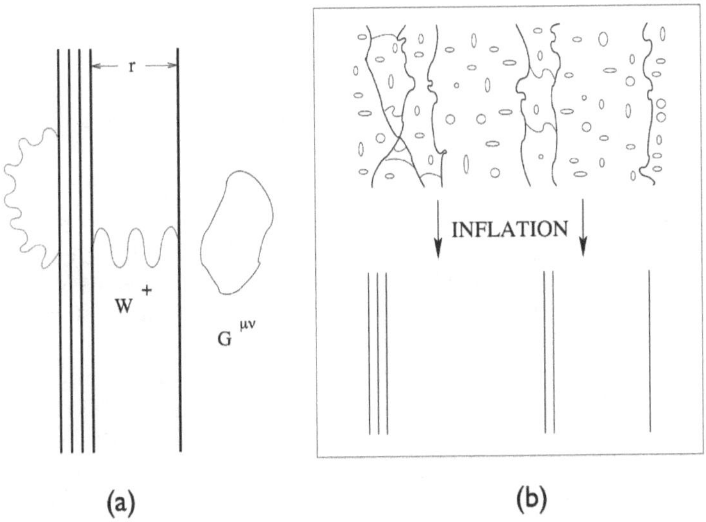

Fig. 1. (**a**) The brane world scenario. Here, as light open string modes with each end of an open string ending on a brane, the standard model particles are stuck to the branes, while closed string modes such as a graviton are free to roam the bulk. (**b**) During brane inflation, a tiny region of the branes (i.e., our universes) grows by an exponentially large factor. Fluctuations such as defects, radiation or matter will be inflated away. Also, the differences in spacing between branes as well as the curvature decreases rapidly

towards the $\bar{D}3$-brane (i.e., anti-$D3$-brane, which has the same tension but opposite RR charge as a $D3$-brane) inside the six-dimensional bulk (due to the attractive force between them), and inflation ends when they collide and annihilate each other. Fluctuations that are present before inflation, such as defects, radiation or matter, will be inflated away (see Fig. 1b). Here, the relative $D3$-$\bar{D}3$-brane position ϕ is the inflaton and the inflaton potential $V(\phi)$ comes from their tensions and interactions. The annihilation releases the brane tension energy that heats up the universe to start the hot big bang epoch. Typically, strings of all sizes and types may be produced during the collision. Large fundamental strings and/or $D1$-branes (or D-strings) become cosmic superstrings.

In a more realistic brane world scenario, all moduli of the six extra spatial dimensions are dynamically stabilized via flux compactification [7, 8], and the presence of RR fluxes introduces intrinsic torsion and warped geometry, so there are regions in the bulk with warped throats. They are six-dimensional versions of the Randall–Sundrum (RS) warped geometry. There are numerous such solutions in string theory, some with a small positive vacuum energy (cosmological constant). This is known as the string landscape. Presumably the standard model particles are open string modes; they can live either on $D7$-branes wrapping a 4-cycle in the bulk or (anti-)$D3$-branes at the bottom of a warped throat (Fig. 2). In the early universe, there is an extra pair of $D3$-$\bar{D}3$-branes. Due to the attractive forces present, the $\bar{D}3$-brane is expected to sit at the bottom of a throat. Here again, inflation takes place as the $D3$-brane moves down the throat towards the $\bar{D}3$-brane, and inflation ends when they collide and annihilate each other, allowing the universe to settle down to the string vacuum state that describes our universe today. This is the KKLMMT scenario [9]. Although the original toy model version encounters some fine-tuning problems, the scenario becomes substantially better as we make it more realistic: It is surprisingly robust, that is, many e-folds of inflation are a generic feature. This is very encouraging. Briefly speaking, there are two key stringy ingredients that come into play:

- Because of the warped geometry, a consequence of flux compactification, a mass M in the bulk becomes $h_A M$ at the bottom of a warped throat, where $h_A \ll 1$ is the warped factor (Fig. 2). This warped geometry tends to flatten, by orders of magnitude, the inflaton potential $V(\phi)$, so the attractive $D3$-$\bar{D}3$-brane potential is rendered exponentially weak in the warped throat. The potential takes the form

$$V(\phi) = V_K + V_A + V_{D\bar{D}} = \frac{1}{2}\beta H^2\phi^2 + 2T_3 h_A^4 (1 - \frac{1}{N_A}\frac{\phi_A^4}{\phi^4}) + ... \qquad (1)$$

where the first term $V_K(\phi) = m^2\phi^2/2 + ...$ receives contributions from the Kähler potential and various interactions in the superpotential [9] as well as

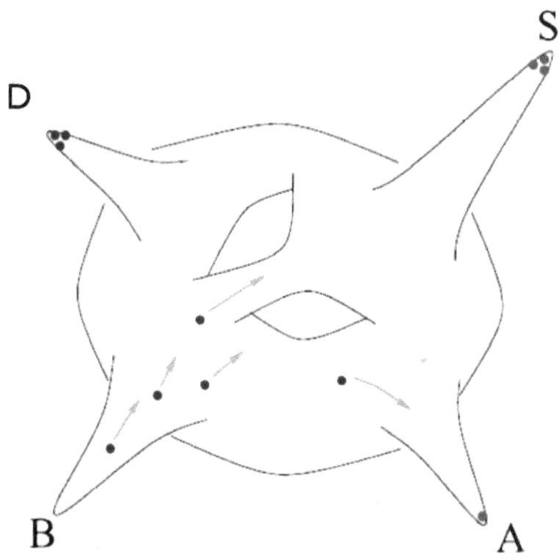

Fig. 2. A pictorial sketch of the compactified bulk. Besides some warped throats, there are $D7$-branes wrapping a 4-cycle. The $D3$-$\bar{D}3$-brane inflationary scenario in a generic flux compactified six-dimensional bulk. The blue dots stand for mobile $D3$-branes, while the red dots are $\bar{D}3$-branes sitting at the bottoms of throats. After inflation and the annihilation of the last $D3$-brane with the $\bar{D}3$-brane in A-throat, the remaining $\bar{D}3$-branes in S-throat may be the standard model branes

possible D-terms [10]. H is the (initial) Hubble parameter so this interaction term behaves like a conformal coupling. Here, β, and more generally V_K, probes the structure of the flux compactification [11, 12]. The warp factor depends on the details of the throat. Crudely, $h(\phi) \sim \phi/\phi_{edge}$, where $\phi = \phi_{edge}$ when the $D3$-brane is at the edge of the throat, so $h(\phi_{edge}) \simeq 1$. At the bottom of the throat, where $\phi = \phi_A$, $h_A = h(\phi_A) = \phi_A/\phi_{edge}$. T_3 is the $D3$-brane tension and the effective tension is warped to a very small value $T_3 h_A^4$ (as we shall see, $h_A \sim 10^{-2}$). The attractive gravitational (plus RR) potential is further warped to a very small value : $N_A \gg 1$ is the $D3$ charge of the throat. If the last 55 e-folds of inflation takes place inside the throat, then $\phi_{edge} \geq \phi \geq \phi_A$ during this period of inflation. Note that β is expected to be of order unity, $\beta \sim 1$. Despite the warped geometry effect, the above potential yields enough inflation only if β is small enough, $\beta \lesssim 1/5$ [13]. We see in Fig. 3 that the data can easily over-constrain the model. However, this is not the end of the story.

• Because the inflaton is an open string mode, its kinetic term appears inside the Dirac–Born–Infeld (DBI) action. For slow-roll, this term reduces to the usual kinetic term. However, when the inflaton is moving relativistically, the full effect of the DBI action must be taken into account [14]. The DBI action in brane inflation leads to the "Lorentz factor"

$$\gamma(\phi) = \frac{1}{\sqrt{1 - \dot{\phi}^2/T(\phi)}}, \tag{2}$$

where $T(\phi) = T_3 h(\phi)^4$ is the warped $D3$-brane tension and the limiting speed, $c(\phi) = \sqrt{T(\phi)}$, is decreasing rapidly as the $D3$-brane moves down the throat $c \sim \phi^2 \to \phi_A^2$. This means the speed $\dot{\phi}$ of ϕ is limited by the rapidly decreasing limiting speed irrespective of the steepness of the inflaton potential. In the warped throat, even for a steep potential, the inflaton motion must slow down considerably towards the bottom of the throat as it is becoming ultra-relativistic, so it takes a while before it reaches the bottom of the throat.

As a result, the warped geometry of the throat combined with the DBI action generically allows for many e-folds of inflation. Robustness of the overall scenario suggests that we are in the right direction. A few comments are in order here :

(i) Since the inflaton is an open string mode that stretches between the branes, it no longer exists as a physical degree of freedom after the $D3$-$\bar{D}3$-brane annihilation.

(ii) The above scenario does not guarantee enough inflation; however, it does yield enough inflation for a large region in the parameter space. Once CMBR and other cosmological data are introduced, constraints on the parameters will sharpen the predictions. At the moment, data are already putting strong constraints on the parameter space. Future data will constrain the parameters further and tell us about the structure of the bulk as well as the throat.

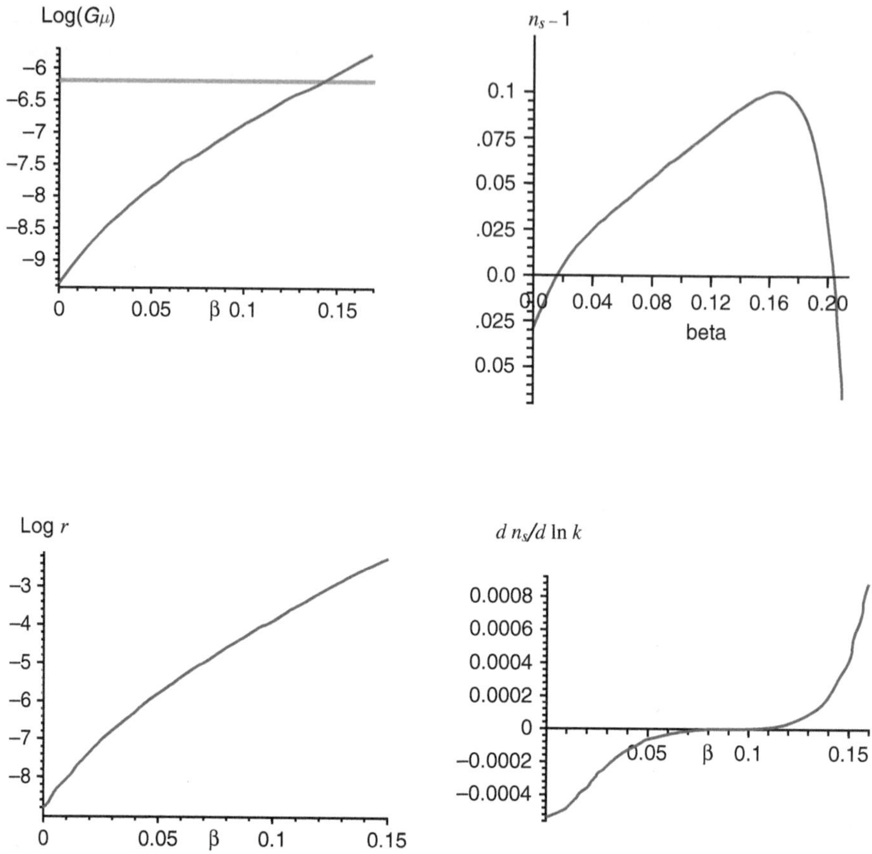

Fig. 3. The predictions of the slow-roll brane inflationary scenario [9, 13]: the cosmic string tension μ, the power spectrum index n_s, the ratio r of the tensor to the scalar density perturbations and the running of n_s

(iii) The presence of a $D3$-$\bar{D}3$-brane pair explicitly breaks supersymmetry. Although this breaking is large, it is very soft, as we shall see. Furthermore, the warping exponentially suppresses the breaking terms. So it is justified to study the scenario within the supergravity approximation when the string scale is much smaller than the Planck scale.
(iv) The interplay between cosmology and gauge/gravity duality should receive more attention, since cosmological data may provide valuable information about strongly coupled gauge theory (via structures of throats and cosmic string properties).
(v) There are many variations of the above scenario. For large m [15], or for a modified warped throat [16], enough inflation can be obtained without the $\bar{D}3$-brane. Multi-throat and/or multi-brane scenarios are also very easy to envision [17, 18]. It is beyond the scope of this review to discuss the large set of

multi-brane inflationary models under the name "assisted inflation". Clearly, they should be fully explored.
(vi) The six-dimensional (or seven-dimensional in M theory) compactification typically introduces many light closed string modes known as moduli. The resulting effective potential involving these bulk modes is in general complicated enough so, with some fine-tuning, one can find a flat enough direction to carry out inflation. It is entirely possible that nature takes this path and moduli inflation should be and has been extensively studied. However, the moduli inflationary scenario does not seem to have distinct stringy signatures, or as compelling and predictive as brane inflation.

The rest of this chapter discusses the various aspects of the above scenario:
• Inflation. For small m or β, the model reduces to the slow-roll scenario (Fig. 3). In this case, WMAP and other cosmological data impose the constraint $\beta < 0.05$ [19]. That is, $0.05 \lesssim \beta \lesssim 0.2$ is ruled out. For large inflaton mass m, the DBI action comes into play and new stringy features such as non-Gaussianity will appear [15]. Furthermore, the three-point correlation function (or bispectrum) has a distinct distribution that is clearly different from what may appear in a slow-roll scenario [20, 21]. For intermediate values of m, the tensor mode perturbation may be large [22]. It can also be distinguished from that coming from the slow-roll scenario. This is encouraging since, unlike the scalar mode perturbation, the metric perturbation directly probes the very early universe.
• Heating at the end of inflation. The $D3$-$\bar{D}3$-brane annihilation produces only closed strings, with the graviton as the lightest mode. The transfer of energy from closed string modes to the standard model particles which are open string modes seems problematic, since gravitational radiation can make up at most a few percent of the density of the standard model particles during big bang nucleosynthesis. Naively, this problem seems most severe if inflation takes place in one throat (the A-throat), while the standard model branes are in another throat (the S-throat). It is satisfying that an analysis of what happens indicates that heating will work out nicely. In fact, the situation improves dramatically when one considers a realistic (i.e., flux compactification) scenario instead of a toy model version based on the Randall–Sundrum scenario. It also offers some possibilities of specific features (such as KK modes as hidden dark matter [23]) that may be tested.
• Production and properties of cosmic strings.
• Evolution of the cosmic string network and its possible detection. Here, we discuss our present knowledge of the scaling cosmic string network and some of its observational consequences.

The history of cosmic strings is a long one [24, 25]. First proposed by Kibble and others, it was applied to generate density perturbations that seeded the structure formation. This requires a tension of $G\mu \sim 10^{-6}$. This was ruled out by the CMBR data. The possibility of superstrings as cosmic strings was first studied by Witten [26]. However, in the heteroric string framework,

$G\mu \sim 10^{-3}$, which is far too big to be compatible with observations. In any case, either these cosmic strings would have been inflated away, or they are unstable to breakage. In brane inflation in Type IIB theory, we see that they are produced after inflation [27, 28], with much lower tensions due to the warped geometry [9, 13]. They are stable under a variety of situations [29, 30], so they can survive to form a scaling cosmic string network. Cosmic superstrings will also have non-trivial tension spectrum and junctions can appear [29]. Of course, the presence of a cosmic string network is not guaranteed. However, if they are around, the chances of detecting them are very promising. Irrespective of the final outcome, we see that string theory is confronting data and making predictions.

2 Brane Inflation

It is possible (in fact one may argue likely) that the inflaton potential has relatively flat directions outside the throat, allowing substantial inflation. Unfortunately, the precise potential is rather dependent on the detailed structures of the compactification and remains to be explored more carefully. To avoid this issue, we shall assume here that the $D3$-brane starts close to or inside the throat. If we have enough e-folds in the throat, then the physics outside the throat need not concern us. As explained earlier, this is an easy condition to satisfy.

First, let us consider the potential $V(y)$ per unit volume between a parallel Dp-$\bar{D}p$-brane pair separated by a distance y, where the Dp-branes are BPS with respect to each other. We shall consider $p < 7$, where T_p is the Dp-brane tension. We may view $V(y)$ as coming from the closed string exchanges between the branes (Fig. 4a). In the closed string channel, at large y, when the massive mode exchanges are Yukawa-suppressed,

$$V(y) \simeq -\frac{\kappa^2 T_p^2}{\pi^{(9-p)/2}} \Gamma((7-p)/2) \frac{1}{y^{7-p}}, \tag{1}$$

where $\kappa^2 = 8\pi G_{10}$ and $T_p = (2\pi\alpha')^{-(p+1)/2}$ is the Dp-brane tension. Here $\alpha' = m_s^{-2}$ is the Regge slope and m_s is the string scale. For $p < 7$, $V(y)$ vanishes as $y \to \infty$. This is simply the attractive gravitational (NS-NS) plus massless RR interaction between the branes. At short distances, the exchange of the massive closed string modes are not Yukawa-suppressed and the evaluation of $V(y)$ is somewhat subtle. Because of the exponentially growing degeneracy (as a function of mass) in the closed string spectrum, a naive summation yields an oscillating divergent result. Looking at Fig. 4a, we see that we may evaluate $V(y)$ as a one-loop radiative correction in the open string channel by including the whole tower of open string modes. The particular way of grouping the contributions should be dictated by the soft supersymmetry breaking [31, 32]. When the two branes are parallel, there is no potential between them

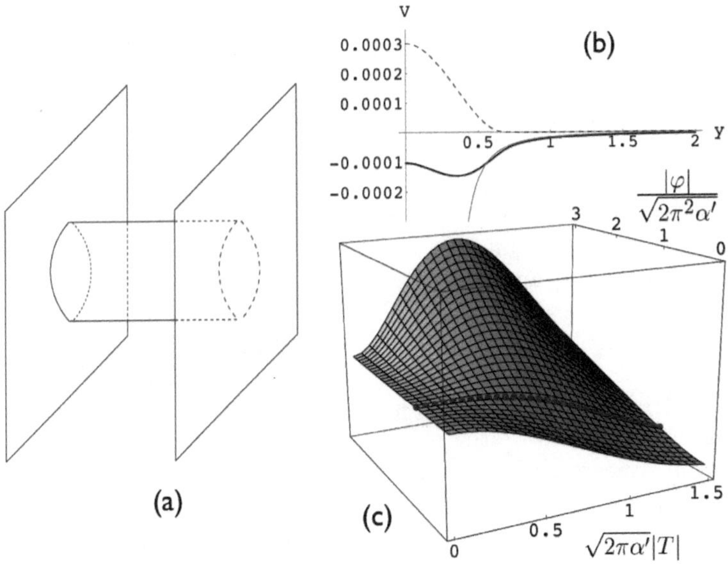

Fig. 4. (a) The exchange of closed strings between two branes. In the dual channel, this describes the one-loop radiative effect of the open strings stretching between two branes. **(b)** The potential $V(y)$ between the $D3$-brane and the $\bar{D}3$-brane due to the diagram (a), as a function of the separation y for the brane pair, where $\alpha' = 1$ [32]. The dashed curve is the imaginary part of $V(y)$. The thick line is the real part of $V(y)$. The Coulombic potential (the thin red curve) is shown for comparison. **(c)** The potential $V(\phi, T)$ as a function of the inflaton $y \sim \phi$ and the tachyon expectation value T [33]. Brane inflation is a hybrid inflationary scenario

because of supersymmetry. Each mass level contains a set of supermultiplets. The contribution to the potential $V(y)$ from the open string bosons is exactly cancelled by the contribution from the open string fermions, mass level by mass level. Now we consider the $\bar{D}p$-brane as a Dp-brane rotated by π. Supersymmetry broken by this rotation is large, in the sense that level crossings take place. However, the supersymmetry breaking is very soft, that is, the open string spectrum follows the spectral flow. For each broken supermultiplet,

$$\sum_i (-1)^F m_i^{2n} = 0, \qquad n = 1, 2, 3, \tag{2}$$

where i runs over the spectrum in each large but "softly broken" supermultiplet (and F is the fermion number). Keeping this grouping in the sum over the open string spectrum yields a finite $V(y)$ (Fig. 4b). This very soft SUSY breaking also justifies the continuous use of the supergravity formulation.

In the open string one-loop channel, a tachyon appears at short distances,

$$\alpha' m_{tachyon}^2 = \frac{y^2}{4\pi^2\alpha'} - \frac{1}{2}, \tag{3}$$

which contributes an imaginary part to $V(y)$. We see that the Coulombic form is a very good approximation before the tachyon appears, by which time inflation is over anyway. With $\phi = \sqrt{T_3}y$, the tachyon appears when $\phi = \phi_E$, and the annihilation process begins. The potential $V(\phi, T)$ in Fig. 4c is evaluated using boundary superstring field theory method [33]. So we have $\phi_i > \phi_{55} > \phi_E > \phi_A$, where ϕ_i is the initial $D3$-brane position when inflation starts and ϕ_{55} is the value of ϕ at 55 e-folds before inflation ends. So the scenario is a hybrid inflation. In the more realistic KKLMMT scenario, $V(\phi)$ becomes $V_{D\bar{D}}(\phi)$ given in (1).

Warped throats such as the Klebanov–Strassler (KS) warped deformed conifold [34] are generic in any flux compactification that stabilizes the moduli. The DBI action for the inflaton field follows simply because the inflaton is an open string mode. By now it is clear that enough inflation is generic in this scenario, thanks to (i) the warped geometry of the throat in a realistic string compactification, which tends to flatten (by orders of magnitude) the attractive Coulombic potential between the $D3$-brane and the $\bar{D}3$-brane [9]. The warped geometry also reduces the vacuum energy that breaks supersymmetry, so the supergravity approximation is expected to be valid. (ii) The warped geometry of the throat combined with the DBI action, which forces the inflaton to move slowly as it falls towards the bottom of the throat, as pointed out by Silverstein and Tong [14]. In fact, one may get enough e-folds just from around the bottom of the throat [35].

Inside the throat, the metric takes the form

$$ds^2 = h^2(r)(-dt^2 + a(t)^2 dx^2) + h^{-2}(r)(dr^2 + r^2 ds_5^2), \tag{4}$$

and the potential takes the simple approximate form (1),

$$V(\phi) = V_K(\phi) + V_0 + V_{D\bar{D}}(\phi) \simeq \frac{m^2}{2}\phi^2 + V_0\left(1 - \frac{vV_0}{4\pi^2}\frac{1}{\phi^4}\right), \tag{5}$$

where the constant term $V_0 = 2T_3 h_A^4 = 2T_3 h(\phi_A)^4$ is the effective vacuum energy. The factor v depends on the properties of the warped throat, with $v = 27/16$ for the KS throat. With some warping (say, $h_A \simeq 1/5$ to 10^{-3}), the attractive Coulombic potential $V_C(\phi)$ can be very weak (i.e., flat). The quadratic term $V_K(\phi)$ receives contributions from a number of sources and is rather model-dependent. However, m^2 is expected to be comparable to $H_0^2 = V_0/3M_p^2$, where M_p is the reduced Planck mass ($G^{-1} = 8\pi M_p^2$). This sets the canonical value for the inflaton mass $m_0 = H_0$ (which turns out to be around $10^{-7}M_p$).

The scale of the throat R is given by

$$R^4 = \frac{27\pi g_s N_A \alpha'^2}{4}. \tag{6}$$

For a generic value of m, usual slow-roll inflation will not yield enough e-folds of inflation. Reference [13] shows that $m \lesssim m_0/3$ will be needed. Naïvely, a substantially larger m will be disastrous, since the inflaton will roll fast, resulting in very few e-folds in this case. However, for a fast-roll inflaton, string theory dictates that we must include higher powers of the time derivative of ϕ, in the form of the DBI action

$$S = -\int d^4x\, a^3(t) \left[T\sqrt{1-\dot{\phi}^2/T} + V(\phi) - T \right], \tag{7}$$

where $T(\phi) = T_3 h(\phi)^4$ is the warped $D3$-brane tension at ϕ. For the usual slow-roll, $T\sqrt{1-\dot{\phi}^2/T} - T \simeq \dot{\phi}^2/2$, reproducing the standard kinetic term. It is quite amazing that the DBI action now allows enough e-folds even when the inflaton potential is steep [14, 15]. As the $D3$-brane approaches $\bar{D}3$-brane, ϕ and $T(\phi)$ decrease, and $h(\phi) \to h(\phi_A)$. The key is that $\dot{\phi}$ is bounded by the limiting speed, and this bound gets tighter as $T(\phi)$ decreases. This happens even if the potential is steep, for example, when $m > H_0$. So the inflaton rolls slowly either because the potential is relatively flat (so $\gamma \simeq 1$ in the usual slow-roll case) or because the warped tension $T(\phi)$ is small (so $1 \ll \gamma < \infty$). As a result, it can take many e-folds for ϕ to reach the bottom of the throat. When $\gamma \gg 1$, the kinetic energy is enhanced by a Lorentz factor of γ. Note that the inflaton is actually moving slowly down the throat even in the ultra-relativistic limit. However, the characteristics of this scenario are very different from the usual slow-roll limit, where $\gamma \simeq 1$. To draw a distinction, we call this the ultra-relativistic regime.

In general, there are three parameters, namely, m, λ and ϕ_A (note that V_0 is a function of λ and ϕ_A), plus the constraint that the $D3$-brane should be inside the throat. We find that the power spectrum can be red-tilted in all three scenarios.

(1) $\beta \ll 1$, $\gamma \simeq 1$, the slow-roll case, when $m^2 \simeq 0$. Here, there are essentially two parameters : m and V_0. After fitting the COBE density perturbation data [2], the predictions are reduced to a one-parameter, namely β, analysis [13]. For small β, $n_s \sim 0.98 + \beta$, $\log r \sim -8.8 + 60\beta$, $\log G\mu \sim -9.4 + 30\beta$. The cosmological data restrict the relevant range to $0 \leq \beta < 0.05$ [19].

(2) $\beta \sim 1$, $\gamma \simeq 1$ at $N_e \sim 55$, but increases to a large value towards the end of inflation; this corresponds to some intermediate values of m^2. In this case, the tensor mode can be large, i.e., as large as saturating the present observational bound $r < 0.3$ [3]. Here, the DBI introduces a deviation from the slow-roll relation between R and the tensor power spectrum index n_t [22],

$$n_t = -\frac{r}{8}\left(\frac{\gamma}{1-\epsilon-\kappa}\right), \tag{8}$$

where ϵ is the usual slow-roll parameter divided by γ and κ measures the running of γ. For large ϕ, the parameterization of the potential should probably include a ϕ^4 term.

(3) $\beta \gg 1$, γ is large throughout. In this ultra-relativistic case, m is large so V_K dominates (i.e., V_0 can be ignored), and the model is again reduced to the

above three parameters before imposing the COBE normalization. In this scenario, ensuring that all 55 e-folds of inflation take place, while the $D3$-brane is inside the throat becomes a strong constraint; that is, the "initial" position ϕ_i (at 55 e-folds before the end of inflation) should satisfy $\phi_i \leq \phi_e$ where ϕ_e is the value at the edge of the throat, i.e., $h(\phi_e) \simeq 1$. To implement this condition, we need to introduce the $D3$-brane tension T_3, or the string scale α'. Since V_0 can be ignored in this case, one may obtain all the inflationary properties without the $\bar{D}3$-brane. $|f_{NL}| \simeq 0.32\gamma^2 \lesssim 300$ yields $\gamma \lesssim 31$ [36]. However, one should check if reheating or preheating can be successfully realized in such a scenario. The structure of the non-Gaussianity from this UV DBI model is different from that due to slow-roll. The three-point correlation function $A(k_1, k_2)/k_1 k_2 k_3$ (where $k_1 + k_2 + k_3 = 0$) [20] is shown in Fig. 5.

Note that n_s is quite sensitive to the warped factor. This point is clearly illustrated by the two different predictions of n_s using two different approximations to the KS warp factor: an AdS cut-off (very slightly blue tilt) [22] and a mass gap cut-off (red tilt) [35]. For large R, we have to consider a highly orbifolded version of the throat in order to fit it inside the bulk.

(4) For tachyonic inflaton mass ($m^2 < 0$), the scenario becomes the multi-throat brane inflation scenario proposed by Chen [17, 38]. The Coulombic term $V_{D\bar{D}}$ is negligible and inflation takes place as the $D3$-brane moves out of a throat (see Fig. 2). For small tachyonic mass, this is simply a slow-roll

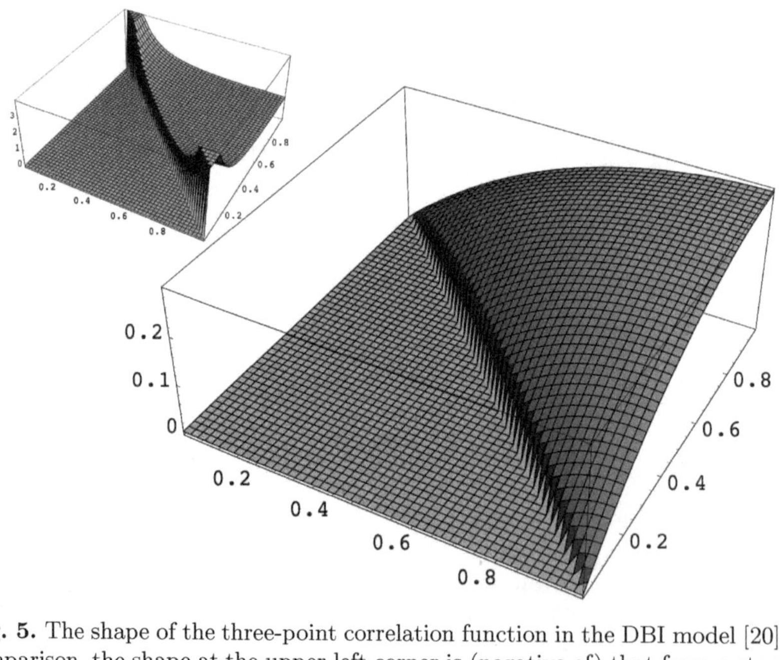

Fig. 5. The shape of the three-point correlation function in the DBI model [20]. For comparison, the shape at the upper left corner is (negative of) that from a standard slow-roll model

model. This IR DBI inflation can happen when inflaton mass takes a generic value, $m \approx H$ ($\beta \approx 1$). The distance the inflaton travels through during inflation, $\Delta\phi \approx HR^2\sqrt{T_3}$, is always sub-Planckian. This model may be realized in a multi-throat compactification starting with a number of antibranes settled down at the ends of various throats. These antibranes are classically stable but can annihilate against the fluxes quantum mechanically [37]. The end products of such a phase transition are many $D3$-branes in, say, the B-throat, which is sufficiently long (typically more than twice longer than the A-throat). The IR model predicts large non-Gaussianity with the same shape as in the UV model. The difference is the running, $f_{NL} \approx 0.036\beta^2 N_e^2$, that is, f_{NL} decreases with k, while f_{NL} increases with k in the UV model. The power spectrum index undergoes an interesting phase transition at a critical e-fold from red ($n_s - 1 \approx -4/N_e$) at small scales to blue ($n_s - 1 \sim 4/N_e$) at large scales [38, 39]. This transition is due to the Hagedorn phase when the red-shifted string scale drops below the Hubble constant. If such a transition falls into the observable range of CMBR, it predicts a large running of n_s around the transition point, i.e., a large negative $dn_s/d\ln k$. Outside of this transition region, $dn_s/d\ln k$ is unobservably small.

If the brane inflationary scenario is correct, it will provide a great probe to both the origin of our early universe and the particular compactification in string theory, i.e., where we are in the cosmic landscape. For example, the inflaton is actually a six-component field. So far, we have only considered the radial mode. When a 4-cycle is close to the A-throat, the symmetry of the throat ($S^3 \times S^2$ for the KS case) would be broken by the 4-cycle's position, shape and orientation, generating a richer inflaton potential [12]. This may also tell us whether eternal inflation is happening or not. Since ϕ is bounded by the size of the bulk, eternal inflation is far from a given in brane inflation [40].

3 Graceful Exit

The crucial step that links the inflationary epoch to the hot big bang epoch is the heating at the end of inflation. This is known as the graceful exit, namely, how the inflationary energy can be efficiently transferred to heat up the standard model particles, and be compatible with the well-understood late-time cosmological evolution? This is the heating problem (also called reheating or preheating problem). To see why this is quite a non-trivial issue, we first look at the end process of brane inflation.

In the above brane inflationary scenario, inflation ends when the $D3$-brane annihilates with the $\bar{D}3$-brane. Significant insights have been gained into such a process [41]. Tachyonic modes appear when the brane–antibrane distance approaches the string scale and the annihilation process may be described by tachyon rolling [42, 43]. (The decay width is signified by the imaginary part of the potential $V(\phi)$.) No matter whether there are adjacent extra branes surviving such an annihilation (e.g., a $D3$-brane colliding with a stack of $\bar{D}3$-branes), the initial end product is expected to be dominated by non-relativistic heavy closed strings [44, 45, 46]. These will then go to lighter closed strings, light KK modes, gravitons and open strings. We know from observations that, during big bang nucleosynthesis (BBN), the density of gravitons can be no more than a few percent of the total energy density of the universe. The rest is contributed by the standard model particles (mostly photons, neutrinos and electrons), which are open strings attached to a stack of SM (anti-)branes. We also know that the density of any non-relativistic relics can be no more than about 10 times that of the baryons. Therefore, the question becomes how the brane annihilation products, originally dominated by the closed string degrees of freedom, can eventually become the required light open string degrees of freedom living on the SM branes, with a negligible graviton density and a non-lethal amount of stable relics. This question is particularly sharp in the multi-throat scenario, where the inflationary branes annihilate in one throat (A-throat), while the SM branes are sitting in another throat (S-throat). Let us discuss this case and then comment on the other cases.

A number of studies have been done to address this heating problem [47, 48, 49, 50]. An important observation is that, because the KK mode wave function is peaked at the bottom of the throat, its interaction with particles located there is much enhanced compared to that with the graviton, whose wavefunction spreads throughout the bulk. This is essentially along the line of Randall–Sundrum warped geometry. Because of this, the graviton emission branching ratio during the brane decay and KK evolution is suppressed by powers of warp factors [51]. In a realistic compactification, throats are typically separated in the bulk, which tends to generate resonance effects in the tunnelling from one throat to another. We expect the compactification volume to be dominated by the bulk, with typical size $L \gg R$, another important ingredient in the success of the graceful exit. Again, the realistic scenario of heating improves in a number of ways over the RS scenario. The discussions below follows [23] and relies on Fig. 2.

First, we note that that the cross sections for KK self-interaction and KK interactions with SM particles in a throat with size R and a warp factor h goes like

$$\sigma \sim \left(\frac{L}{R}\right)^6 \frac{1}{M_P^2 h^2}. \tag{1}$$

This is much bigger than that for the graviton, where the corresponding $\sigma \sim M_P^{-2}$. Note that the factor $(L/R)^6$ comes from the six-dimensional bulk. Next, it is important to follow the thermal history of the KK modes as the universe expands. Because of the above warped enhanced KK self-interactions, it is easy to see that the KK modes become non-relativistic before they decouple. So, instead of a tower of non-relativistic KK modes, only the lightest few stable KK modes remain. As a result, their relic density is very much suppressed. The qualitative picture of heating goes as follows.

Massive closed string modes produced during the $D3$-$\bar{D}3$-brane annihilation rapidly decay to light KK modes and gravitons. Among the light KK modes in a throat are ones with conserved angular momenta, so they are quite stable against further decay, with typical mass of order h_A/R. Due to the self-interaction, the relic density in the non-relativistic KK modes is very much suppressed. Due to the red-shift and the low tunnelling rate, the universe enters a matter-dominated phase with these KK modes, which then tunnel to the S-throat and other throats, if present. To ameliorate the hierarchy problem, we expect the S-throat to have a much smaller warped factor $h_S \ll h_A$. Generically, we expect the tunnelling rate from A-throat to S-throat to be enhanced by the bulk resonance effect (for $R/L \lesssim h_A$) [52],

$$\Gamma_{A \to S} \sim h_A^9/R \gg h_A^{17}/R, \tag{2}$$

where the second rate is that for the case when there is no bulk resonance effect. Once the KK modes reach the S-throat, they rapidly decay to open string modes and heat up the universe, starting the hot big bang epoch. For a successful scenario, (i) the matter-dominated duration should be long enough to red-shift away the gravitational radiation away, but not so long as over-cool the universe. This condition requires $h_A \sim 10^{-1}$ to 10^{-3}. It is very encouraging that these values are precisely those required to fit the CMBR data. (ii) The decay of KK modes in the S-throat should go to open string modes instead of to gravitational radiation. This is guaranteed because the coupling of KK modes to gravitons is dictated by the Newton's constant $G_4 = 8\pi/M_P^2$, while their couplings to open strings modes, i.e., SM particles, are enhanced by the localization of both the KK modes and the SM branes in the throat, as shown in (11).

It is interesting to point out some novel features in this heating scenario:

- There is a matter-dominated epoch between the end of inflation and the beginning of the hot big bang era. The cosmic scale factor can grow by a large factor (10^5 or more) during this epoch. As a result, both the gravitational radiation and the gravitino density will be substantially suppressed. It will be interesting to study other cosmological consequences of such an epoch.
- There is a dynamical process that selects a long throat to be heated. This is because the dense spectrum in a long throat makes the level matching of the energy eigenstates, a necessary condition for tunnelling between throats, easier to satisfy. This may provide a dynamical explanation of the selection of the RS type (i.e., with very large warping that solves the hierarchy problem) warp space as our standard model throat in the early universe.
- Although KK modes as dark matter have been considered in the literature, we see the possibility of KK modes as hidden dark matter. These are almost stable KK modes in another throat (say, the B-throat in Fig. 2), which interacts only via gravitons with SM particles. This hidden dark matter has many unusual properties compared to the usual dark matter candidates, e.g., it may tunnel to the S-throat and generate a cosmic ray that violates the GZK bound.

4 Production and Properties of Cosmic Superstrings

Although the production of domain walls and monopoles at the grand unified (GUT) scale will over-close the universe by many orders of magnitude, cosmic strings do not suffer from the same problem. This is a consequence of the intercommutation properties of strings, which leads to a scaling cosmic string network that tracks the radiation (matter) during the radiation-(matter)-dominated era. A key property of cosmic string is its tension μ. In fact, cosmic strings around the GUT scale, i.e., $G\mu \sim 10^{-6}$, was originally proposed as an alternative to inflation in generating density perturbation for structure formation [25]. However, the properties of CMBR data, in particular the acoustic peaks, ruled out this possibility. It is these same data that strongly support inflation. In fact, all defects present before inflation would have been inflated away. So we need to consider only defects that are produced after inflation.

The topological properties of defect formation in tachyon condensation are well understood in superstring theory [53]. The spontaneous symmetry breaking will support defects with even codimension (i.e., $2k$), as classified by K theory. In particular, $D3$-$\bar{D}3$-brane annihilation yields $D1$-branes and fundamental $F1$-strings, when the large massive ones appear as cosmic strings in our universe [27, 28, 29]. Qualitatively, it is easy to see how this takes place. There is a $U(1)$ gauge theory associated with each brane, and the tachyon couples to one combination $U(1)_-$. This is simply the Abelian Higgs model in the field theory approximation. Tachyon rolling results in spontaneous symmetry breaking and the resulting vortices are $D1$-strings. So they are cosmologically produced via the Kibble mechanism. The other $U(1)_+$ becomes confining, and the resulting flux tubes become the fundamental closed strings [54]. So cosmic strings are generically produced towards the end of brane inflation. It is quite amazing that string theory dictates that the dangerous domain walls and monopole-like defects are not produced. In the Type IIB theory that we are studying, there is simply no $D0$- or $D2$-branes.

We find that the cosmic string tension μ roughly satisfies $10^{-13} < G\mu < 10^{-6}$. Fundamental string (F-string) tension in 10 dimensions defines the string scale α' via $T_{F1} = 1/2\pi\alpha'$. In Type IIB theory, there are branes including $D1$-branes, or D-strings, with tension $T_{D1} = 1/2\pi\alpha' g_s$, where g_s is the string coupling. In the light of all the progress coming from dualities in string theory, we now know that the D-strings and the F-strings should be considered on the same footing and a general string state in Type IIB is the bound state of these two types of strings. In 10 flat dimensions, supersymmetry dictates that the tension of the bound state of p F-strings and q D-strings is given by [55],

$$T_{p,q} = T_{F1}\sqrt{p^2 + \frac{q^2}{g_s^2}}\,. \tag{1}$$

This tension spectrum (for coprime (p,q)) allows junctions to be formed [29]. Since the $D3$-$\bar{D}3$-brane annihilation most likely takes place at the bottom of a throat, that will be where the cosmic superstrings are. To be specific, we consider the KS throat [34] whose properties are relatively well understood. On the gravity side, this is a warped deformed conifold. Inside the throat, the geometry is a shrinking S^2 fibred over an S^3. The tensions of the bound state of p F-strings and that of q D-strings were individually computed for the KS throat [56]. The tension formula for the (p,q) bound states is given by [57]

$$T_{p,q} \simeq \frac{h_A^2}{2\pi\alpha'}\sqrt{\frac{q^2}{g_s^2}+(\frac{bM}{\pi})^2\sin^2(\frac{\pi p}{M})}, \tag{2}$$

where $b = 0.93$ is a number numerically close to one and M is the number of fractional D3-branes, that is, the units of 3-form RR flux F_3 through the S^3. For $M \to \infty$ and $b = h_A = 1$, it reduces to (13). Very interestingly, the F-strings are charged in $\mathbb{Z}_M$ and are non-BPS. The D-string on the other hand is charged in $\mathbb{Z}$ and is BPS with respect to each other. Because p is Z_M-charged with non-zero binding energy, binding can take place even if (p,q) are not coprime. Since it is a convex function, i.e., $T_{p+p'} < T_p + T_{p'}$, the p-string will not decay into strings with smaller p. The interpretation of these strings in the gauge theory dual is known. The F-string is dual to a confining string between a quark and an anti-quark, while the D-string is dual to an axionic string. M fundamental strings can terminate to a point-like baryon (with mass $\sim Mh_A/\sqrt{\alpha'}$), irrespective of the number of D-strings around.

Besides the above Kibble and confining mechanisms, there are other possible ways to produce cosmic strings which may evolve to a cosmic string network :

• Consider another throat with warped factor h_C. If the temperature at the beginning of the hot big bang is T_i, then strings in C-throat will be excited if $T_i > h_C m_s$.

• D-strings can be stable inside $D3$-branes [30]. Such D-strings can be pair-produced inside the horizon at the end of inflation when a small stack of $D3$-branes collide with a larger stack of $\bar{D}3$-branes.

• One may also consider the situation when a single brane move towards the bottom of the A-throat. Assuming that heating is not a problem for such a scenario, stable D-strings can be pair-produced if $T_i > m_s h_A$.

Isolated loops would just decay via gravitational radiation. However, if the density of loops is high enough so that they overlap and tangle with each other, then their reconnections will yield long strings and lead to a scaling cosmic string network. For the C-throat, this probably requires $T_i \gg h_C m_s$. This is more likely for small $G\mu$, since the decay rate is proportional to $G\mu \sim G m_s^2 h_C^2$, so light tension sting loops will be quite long lived. In addition to cosmic strings in A-throat, the universe may have cosmic strings with much smaller tensions if throats with large warping exist in the bulk. These cosmic strings interact very weakly with cosmic strings in A-throat.

5 Evolution and Detection of Cosmic Superstrings

The cosmological evolution of cosmic superstrings is a very challenging problem. For slow-moving cosmic strings that stretch across the horizon, the energy density naively scales like a^{-2}. For cosmic string loops, the naive energy density is similar to that for monopoles, scaling like a^{-3}. So, naively, the cosmic string density is a problem. However, their interactions substantially suppress the density. The intercommutation of intersecting cosmic strings and the decay of the resulting cosmic string loops (to gravitational waves) reduce the density so that it decrease like radiation (matter) during the radiation-(matter)-dominated era [25]. Furthermore, the resulting scaling cosmic string network energy density is insensitive to the initial density, i.e., the network rapidly approaches the scaling solution. As a consequence, the physics is essentially dictated by the single parameter $G\mu$ in the Nambu–Goto or the Abelian Higgs model, and by the tension spectrum for a more complicated model.

Although the cosmic string network reaches a scaling solution, the fraction of energy density in cosmic string loops has been an outstanding question [25]. Early simulations did not reach fine enough resolution to determine the role played by string loops [58]. The basic assumption is that once a loop is produced by the intersection of long strings (including self intersection), they decay quickly via gravitational radiation. More recent analysis seems to change the story.

Let us first consider the Nambu–Goto case. The fraction of energy density in the string network is given by

$$\Omega_s = \Omega_\infty + \Omega_{loops} \sim \Gamma G\mu + \chi\sqrt{\alpha G\mu}, \tag{1}$$

where the first term is the contribution of long strings, with $\Gamma \sim 10^2$ for Nambu–Goto strings. The second term is the contribution of string loops within the horizon. Very crudely speaking, $\chi \sim 10^3$. The value of α, the ratio of characteristic loop size to the horizon size, is poorly understood. It has been estimated to be as small as $\alpha < 10^{-12}$, or $\alpha \sim G\mu$ or even $(G\mu)^{5/2}$. Recently, both numerical simulations [59, 60] and analytic studies [61] have indicated that there are more energies in the string loops than previously thought. That is, α may be as big as 0.25, although $\alpha \sim 10^{-4}$ seems to be more likely. For small $G\mu$, the increase in the energy density in the string network can be very substantial.

As mentioned earlier, cosmic superstrings will have different properties than vortices in the Abelian Higgs model. Although a simulation is not available, one can analyse the evolution of the string network by solving a set of coupled equations. As shown in Fig. 6b, recent analysis on the tension spectrum (i) strongly suggests that cosmic superstrings also evolve dynamically to a scaling solution (with a stable relative distribution of strings with different quantum numbers) [62, 63], very much like usual cosmic strings (either coming from the Abelian Higgs model or from Nambu-Goto type) [25]. This is due to the rapid decrease in the density of strings with large tensions,

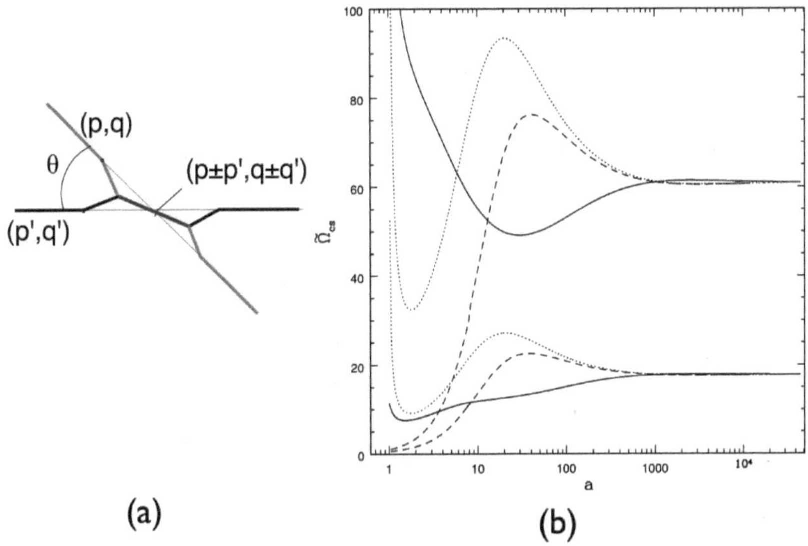

Fig. 6. (**a**) The (p,q) string binding generates junctions [29]. (**b**) (p,q) string network evolution as a function of the cosmic scale factor. The top three lines stand for total density, while the bottom three lines stand for the corresponding $(p,q)=(1,0)$ string density. We see that, irrespective of the initial densities, both the total density and the $(1,0)$ density approach rapidly the scaling solutions [63]

which goes roughly like $\mu(p,q)^{-N}$, where $N \sim 8$. We shall consider a scenario where the cosmic strings are stable enough to allow such a scaling solution. The inter-commutation probability of vortices is known to be around unity, $P \simeq 1$, while that of superstrings is rather complicated, but $P \sim g_s^2$ [62], where the string coupling $g_s \sim 1/10$. Also, the tension spectrum tells us that cosmic superstrings will come in a variety of tensions and charges. A simple analysis indicates that a number of species of cosmic strings will be around in the string network [63], so

$$\Omega_s \to \frac{n}{P}\Omega_s,$$

where n is the effective number of types, $n \sim 5$. For very small P, it is argued that $1/P \to 1/P^{2/3}$ [60]. It is not clear how the presence of baryons in the tension spectrum (ii) will impact on the evolution of the string network. It is clear that further studies, the properties of cosmic string spectrum (including baryons), their productions, stabilities and interactions, and the cosmic evolution of the network as well as their possible detections will be most interesting to watch. It is reasonable to be optimistic about the detectability of cosmic superstrings, but this is far from guaranteed.

Originally proposed as an alternative to inflation, the detection of cosmic strings has been extensively studied [25]. Since the cosmic superstrings interact with the SM particles only via gravity, all detection involves the gravitational interactions of cosmic strings. Recent understanding on the importance of string loops will certainly enhance the detectability of cosmic strings. Since the particular brane inflationary scenario is not yet known, the cosmic string tensions are only loosely constrained. We shall be open-minded in comparing with observation. Many ways to detect cosmic strings have been suggested. Here let us discuss some of them :

• Gravitational lensing is probably most direct. Cosmic string introduces a deficit angle, so a galaxy behind a long cosmic string will appear as a double (undistorted) image. The image separation is roughly $5 \times 10^6 G\mu$ arcsec. For $G\mu \ll 10^{-7}$, this approach becomes very challenging. Finding a lensing by a junction will be quite definitive [29, 64].

• Micro-lensing. This was first studied in [65]. For small string tension, string loops are expected to be dominant. They can lens stars by watching the brightness of a star doubling for a short period of time. Since there are more string loops for smaller tension, non-observation may put a lower bound on the string tension [66].

• In brane inflation, the density perturbation (and CMBR anisotropy) comes from two sources: the usual quantum fluctuation (scalar and tensor modes) during inflation and the fluctuations (scalar and vector modes) induced by the cosmic string network. The density perturbation coming from the cosmic string network is active and incoherent, so there is no acoustic peaks that are prominent in the density perturbation coming from inflation. The COBE data roughly yield $G\mu \simeq 10^{-6}$ if the scaling solution of the cosmic string network is the sole source of the density perturbation. Using WMAP data, one finds that the contribution from cosmic strings is bounded by about 10%, which translates to about $G\mu \sim 7 \times 10^{-7}$. So the cosmic string production towards the end of brane inflation is perfectly compatible with the present CMBR data [3], while future data may be able to test this scenario [67, 68].

• Since the density perturbation coming from cosmic string is continuously being produced, its magnitude in CMBR anisotropy at large l will not be attenuated as much as that coming from inflation. For $G\mu \sim 7 \times 10^{-7}$, the contribution from cosmic strings may become comparable to (bigger than) that from inflation at $l > 2000$ ($l > 3000$). This may be measurable if $G\mu$ is not too small. Polarization in CMB will also be measured. In particular, the B (i.e., curl) mode due to the tensor mode perturbation will be tested, reaching $\Delta T \simeq 0.5\mu K$. Here the gravitational wave anisotropy density is much higher than that in a pure inflationary scenario, so passage through space will presumably yield a B-mode polarization clearly larger than that coming from a purely inflationary scenario [68]. Figure 7 illustrates this possibility.

• As a cosmic string moves with velocity **v** across the sky, a shift in the CMB temperature may be observed, $\Delta T/T \simeq 8\pi G\mu v\gamma$ [69]. A careful analysis of the CMBR data may probe $G\mu \simeq 10^{-10}$. It is important to see what bound on $G\mu$ the data can eventually reach. Detection may be possible for as small as $G\mu \simeq 10^{-13}$.

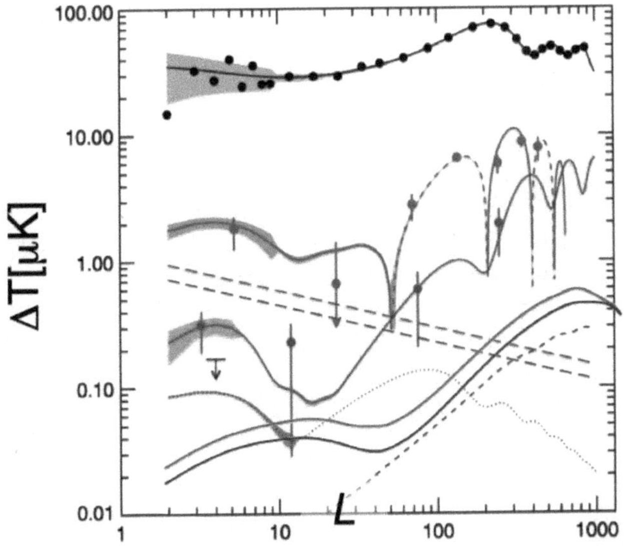

Fig. 7. The CMBR power spectrum from WMAP [3]. They are (from *top*) the temperature TT correlation (*black*), the temperature-electric-mode polarization TE correlation (*red*), the EE correlation (*green*), possible B-mode polarization BB correlation (*blue*) and possible BB correlation (*red/blue*) from cosmic strings [68]. The dashed lines are likely background/foreground that should be subtracted

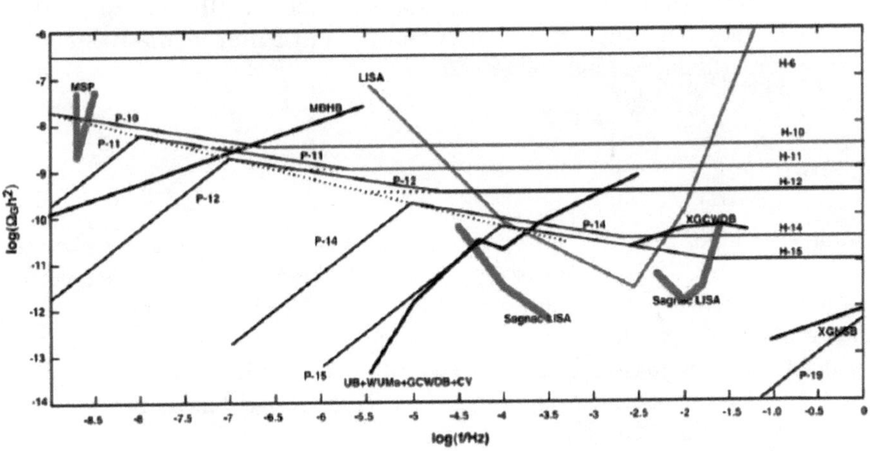

$P-n$-$H-n$ is signal for cosmic strings with tension $G\mu \sim 10^{-n}$. Various other signals are also shown: MBHB is for massive black hole binary system etc.. The sensitivity of LISA is shown in red. So is that for milli-second pulsar timing (MSP) at low frequency. We see that LISA can reach $G\mu \sim 10^{-15}$.

Fig. 8. The detectability of cosmic strings by LISA via gravitational radiation, both background and bursts for Nambu–Goto strings [72]

•The cosmic string network also generates gravitational waves that may be observable. This has been studied extensively in the literature. The stochastic gravitational wave spectrum has an almost flat region that extends from $f \sim 10^{-8}$ Hz to $f \sim 10^{10}$ Hz. Within this frequency range, both ADVANCED LIGO/VIRGO (sensitive at $f \sim 10^2$ Hz) and LISA (sensitive at $f \sim 10^{-3}$ Hz) may have a chance. Following [70], we obtain $\Omega_{gw}h^2 \simeq 0.04G\mu$ coming from long strings. Since LIGO II/VIRGO can reach $\Omega_{gw}h^2 \simeq 10^{-10}$ at $f \simeq 100$ Hz, it can reach $G\mu \geq 2 \times 10^{-9}$. Such stochastic gravitational wave also influences the very precise pulsar timing measurements. Although present pulsar timing measurement is compatible with $G\mu < 10^{-6}$, a modest improvement on the accuracy may detect a network of cuspy cosmic string loops down to $G\mu \simeq 10^{-11}$.

Cusps and kinks are quite common in oscillating cosmic strings. Strongly focused beams of relatively high-frequency gravitational waves are emitted by these cusps and kinks. The sharp bursts of gravitational waves have very distinctive waveform: $t^{1/3}$ (cusps) and $t^{2/3}$ (kinks) [71]. ADVANCED LIGO/VIRGO may detect them for values down to $G\mu \geq 10^{-13}$ and LISA to 10^{-15} [71, 72, 73], so this may be the most sensitive test of cosmic strings. At the moment, theoretical uncertainties (such as string tension, tension spectrum, interactions and cosmic string loops) must be better understood. Figure 8 takes into account the recent analysis where the string loops are important.

• Cusps also introduces temperature shifts in the CMBR that should be searched. They may appear as a sharp down and then up temperature shift that is quite distinctive [66, 74].

6 Remarks

Brane inflation is a natural realization of inflation in the brane world scenario in string theory. If the string scale is close to the GUT scale, as expected, cosmology offers a powerful approach to study and test string theory. We see that brane inflation offers a variety of possible distinct stringy signatures to be detected. Existing data are perfectly compatible with brane inflation. It is exciting that near-future experiments/observations will likely provide non-trivial tests of the scenario.

Many interesting problems remain. Here is a partial list. On the theoretical side:

• Search for other inflationary scenarios in string theory.

• Search for other distinct stringy signatures that can be detected.

• We have seen that the structure of the bulk as well as the properties of the warped deformed throat impacts on the CMBR predictions, e.g., the power spectral index. Flux compactifications must be studied in much greater detail than currently known.

• The gauge/gravity duality has played an important role in studying the properties of throats and the cosmic string tension spectrum. One may actually apply cosmology to study strongly coupled gauge theory via gauge/gravity duality.
• Non-Gaussianity in CMBR and its more detailed properties.
• Understand better the properties of cosmic strings, such as the tension spectrum and their interactions, their production and stability, and the cosmological evolution of the string network that may include baryons and/or light domain walls bounded by the cosmic strings.
• Gott finds that closed time-like curves appear when two cosmic strings move ultra-relativistically towards each other [75]. He proposed to use this as a time machine. It is argued that energetics would prevent the appearance of such closed time-like curves in our universe under any realistic situation [76]. This important issue certainly deserves further analysis.
On the observational side:
• Searching for cosmic string signatures, large tensor mode and/or non-Gaussianity that differs from that predicted in slow-roll inflation in CMBR will be important.
• Astronomical searches for lensing, micro-lensing, temperature shifts due to moving strings and string cusps can be both challenging and exciting. Some of these searches need not be dedicated searches, i.e., they can be part of other programs.
• Gravitational wave detection of the stochastic background gravitational radiation due to cosmic strings as well as bursts coming from string cusps will be valuable.

One should consider the discovery of cosmic strings as another verification of the inflationary paradigm. This will shed light on the specific brane inflationary scenario that took place, providing a valuable probe to the brane world picture before inflation. That is, information on the early universe before inflation may not be totally lost. To my knowledge, this is the best observational window into supertstring theory. Irrespective of the final outcome, whether brane inflation or some other stringy scenario is eventually proved correct or not, we see that string theory is confronting data and making a number of distinctive predictions that can be tested in the near future. This is exciting.

Acknowledgment

I thank Rachel Bean, Xingang Chen, David Chernoff, Gia Dvali, Hassan Firouzjahi, Girma Hailu, Nick Jones, Louis Leblond, Levon Pogosian, Sash Sarangi, Sarah Shandera, Gary Shiu, Ben Shlaer, Horace Stoica, Ira Wasserman, Mark Wyman and Jiajun Xu for collaborations and valuable discussions. Discussions with Cliff Burgess, Jim Cline, Shamit Kachru, Renata Kallosh, Igor Klebanov, Andre Linde, Liam McAllister, Juan Maldacena,

Irit Maor, Ken Olum, Joe Polchinski, Fernando Quevedo, Eva Silverstein, Bret Underwood and Alex Vilenkin are gratefully acknowlegded. This work is supported by the National Science Foundation under grant PHY-0355005.

References

1. A. H. Guth: Phys. Rev. D **23**, 347 (1981); A. D. Linde: Phys. Lett. B **108**, 389 (1982); A. Albrecht, P. J. Steinhardt: Phys. Rev. Lett. **48**, 1220 (1982)
2. G. F. Smoot et al.: Astrophys. J. **396**, L1 (1992); C. L. Bennett et al.: Astrophys. J. **464**, L1 (1996)
3. D. N. Spergel et al.: astro-ph/0603449
4. J. Polchinski: Phys. Rev. Lett. **75**, 4727 (1995)
5. G. R. Dvali, S.-H. H. Tye: Phys. Lett. B **450**, 72 (1999)
6. C. P. Burgess, M. Majumdar, D. Nolte, F. Quevedo, G. Rajesh, R. J. Zhang: JHEP **0107**, 047 (2001); G. R. Dvali, Q. Shafi, S. Solganik: hep-th/0105203; S. Buchan, B. Shlaer, H. Stoica, S.-H. H. Tye: JCAP **0402**, 013 (2004)
7. S. B. Giddings, S. Kachru, J. Polchinski: Phys. Rev. D **66**, 106006 (2002)
8. S. Kachru, R. Kallosh, A. Linde, S. P. Trivedi: Phys. Rev. D **68**, 046005 (2003)
9. S. Kachru, R. Kallosh, A. Linde, J. Maldacena, L. McAllister, S. P. Trivedi: JCAP **0310**, 013 (2003)
10. C. P. Burgess, R. Kallosh, F. Quevedo: JHEP **0310**, 056 (2003)
11. M. Berg, M. Haack, B. Kors: Phys. Rev. D **71**, 026005 (2005); hep-th/0409282; JHEP **0511**, 030 (2005)
12. D. Baumann, A. Dymarsky, I. R. Klebanov, J. Maldacena, L. McAllister, A. Murugan: hep-th/0607050
13. H. Firouzjahi, S.-H. H. Tye: JCAP **0503**, 009 (2005)
14. E. Silverstein, D. Tong: Phys. Rev. D **70**, 103505 (2004)
15. M. Alishahiha, E. Silverstein, D. Tong: Phys. Rev. D **70**, 123505 (2004)
16. A. Dymarsky, I. R. Klebanov, N. Seiberg: JHEP **0601**, 155 (2006)
17. X. Chen: Phys. Rev. D **71**, 063506 (2005)
18. S. Dimopoulos, S. Kachru, J. McGreevy, J. G. Wacker: hep-th/0507205
19. U. Seljak, A. Slosar: Phys. Rev. D **74**, 063523 (2006)
20. X. Chen, M. X. Huang, S. Kachru, G. Shiu: hep-th/0605045
21. J. M. Maldacena: JHEP **0305**, 013 (2003)
22. S. E. Shandera, S.-H. H. Tye: JCAP **0605**, 007 (2006)
23. X. Chen, S.-H. H. Tye: JCAP **0606**, 011 (2006)
24. E. W. Kolb, M. S. Turner: *The Early Universe* (Addison-Wesley Publ. Co., Redwood City, 1990)
25. A. Villenkin, E. P. S. Shellard: *Cosmic Strings and Other Topological Defects* (Cambridge University Press, Cambridge, 2000)
26. E. Witten: Phys. Lett. B **153**, 243 (1985)
27. N. Jones, H. Stoica, S.-H. H. Tye: JHEP **0207**, 051 (2002); S. Sarangi, S.-H. H. Tye: Phys. Lett. B **536**, 185 (2002); N. T. Jones, H. Stoica, S.-H. H. Tye: Phys. Lett. B **563**, 6 (2003)
28. G. Dvali, A. Vilenkin: JCAP **0403**, 010 (2004)
29. E. J. Copeland, R. C. Myers, J. Polchinski: JHEP **0406**, 013 (2004)
30. L. Leblond, S.-H. H. Tye: JHEP **03**, (2004) 055
31. J. Garcia-Bellido, R. Rabadan, F. Zamora: JHEP **0201**, 036 (2002)

32. S. Sarangi, S.-H. H. Tye: Phys. Lett. B **573**, 181 (2003)
33. N. T. Jones, S.-H. H. Tye: JHEP **0301**, 012 (2003)
34. I. R. Klebanov, M. J. Strassler: JHEP **0008**, 052 (2000)
35. S. Kecskemeti, J. Maiden, G. Shiu, B. Underwood: JHEP **0609**, 076 (2006)
36. P. Creminelli, A. Nicolis, L. Senatore, M. Tegmark, M. Zaldarriaga: JCAP **0605**, 004 (2006)
37. S. Kachru, J. Pearson, H. L. Verlinde: JHEP **0206**, 021 (2002)
38. X. Chen: JHEP **0508**, 045 (2005)
39. X. Chen: Phys. Rev. D **72**, 123518 (2005)
40. X. Chen, S. Sarangi, S.-H. H. Tye, J. Xu: hep-th/0608082
41. A. Sen: JHEP **0204**, 048 (2002); JHEP **0207**, 065 (2002)
42. G. Shiu, S.-H. H. Tye, I. Wasserman: Phys. Rev. D **67**, 083517 (2003)
43. J. M. Cline, H. Firouzjahi, P. Martineau: JHEP **0211**, 041 (2002)
44. N. Lambert, H. Liu, J. Maldacena: hep-th/0303139
45. X. Chen: Phys. Rev. D **70**, 086001 (2004)
46. L. Leblond: JHEP **0601**, 033 (2006)
47. N. Barnaby, C. P. Burgess, J. M. Cline: JCAP **0504**, 007 (2005)
48. L. Kofman, P. Yi: Phys. Rev. D **72**, 106001 (2005)
49. D. Chialva, G. Shiu, B. Underwood: JHEP **0601**, 014 (2006)
50. A. R. Frey, A. Mazumdar, R. Myers: Phys. Rev. D **73**, 026003 (2006)
51. S. Dimopoulos, S. Kachru, N. Kaloper, A. E. Lawrence, E. Silverstein: Phys. Rev. D **64**, 121702 (2001)
52. H. Firouzjahi, S.-H. H. Tye: JHEP **0601**, 136 (2006)
53. A. Sen: JHEP **9808**, 010 (1998); JHEP **9809**, 023 (1998); E. Witten: JHEP **9812**, 019 (1998); P. Horava: Adv. Theor. Math. Phys. **2**, 1373 (1999)
54. O. Bergman, K. Hori, P. Yi: Nucl. Phys. B **580**, 289 (2000)
55. J. H. Schwarz: Phys. Lett. B **360**, 13 (1995) [Erratum-ibid. B **364**, 252 (1995)]
56. S. S. Gubser, C. Herzog, I. R. Klebanov: JHEP **09**, 036 (2004)
57. H. Firouzjahi, L. Leblond, S.-H. H. Tye: JHEP **0605**, 047 (2006)
58. A. Albrecht, N. Turok: Phys. Rev. Lett. **54**, 1868 (1985); D. P. Bennett, F. R. Bouchet: Phys. Rev. Lett. **60**, 257 (1988); B. Allen, E. P. S. Shellard: Phys. Rev. Lett. **64**, 119 (1990)
59. V. Vanchurin, K. Olum, A. Vilenkin: Phys. Rev. D **72**, 063514 (2005); gr-qc/0511159; C. Ringeval, M. Sakellariadou, F. Bouchet: astro-ph/0511646; C. J. A. Martins, E. P. S. Shellard: Phys. Rev. D **73**, 043515 (2006)
60. A. Avgoustidis, E. P. S. Shellard: Phys. Rev. D **73**, 041301 (2006)
61. J. Polchinski, J. V. Rocha: Phys. Rev. D **74**, 083504 (2006)
62. M. G. Jackson, N. T. Jones, J. Polchinski: JHEP **0510**, 013 (2005)
63. S.-H. H. Tye, I. Wasserman, M. Wyman: Phys. Rev. D **71**, 103508 (2005) [Erratum-ibid. D **71**, 129906 (2005)]
64. B. Shlaer, M. Wyman: Phys. Rev. D **72**, 123504 (2005)
65. C. Hogan, R. Narayan: MNRAS **211**, 575 (1984)
66. D. Chernoff, S.-H. H. Tye: to appear.
67. M. Landriau, E. P. S. Shellard: Phys. Rev. D **69** 023003 (2004); L. Pogosian, M. C. Wyman, I. Wasserman: astro-ph/0403268; astro-ph/0604141; E. Jeong, G. F. Smoot: Astrophys. J. **624**, 21 (2005)
68. L. Pogosian, S.-H. H. Tye, I. Wasserman, M. Wyman: Phys. Rev. D **68**, 023506 (2003)
69. N. Kaiser, A. Stebbin: Nature **310**, 391 (1984); J. R. Gott: Ap. J. **288**, 422 (1985)

70. R. R. Caldwell, B. Allen: Phys. Rev. D **45**, 3447 (1992)
71. T. Damour, A. Vilenkin: Phys. Rev. D **64**, 064008 (2001); Phys. Rev. D **71**, 063510 (2005)
72. C. J. Hogan: Phys. Rev. D **74**, 043526 (2006)
73. X. Siemens, J. Creighton, I. Maor, S. R. Majumder, K. Cannon, J. Read: Phys. Rev. D **73**, 105001 (2006)
74. A. A. de Laix, T. Vachaspati: Phys. Rev. D **54**, 4780 (1996); A. Stebbin: Astrophys. J. **327**, 584 (1988); F. Bernardeau, J.-P. Uzan: Phys. Rev. **D63**, 023004 (2001); **D63**, 023005 (2001)
75. J. R. I. Gott: Phys. Rev. Lett. **66**, 1126 (1991).
76. B. Shlaer, S.-H. H. Tye: Phys. Rev. D **72**, 043532 (2005)

SPRINGER NATURE

GPSR Compliance

The European Union's (EU) General Product Safety Regulation (GPSR) is a set of rules that requires consumer products to be safe and our obligations to ensure this.

If you have any concerns about our products, you can contact us on ProductSafety@springernature.com

In case Publisher is established outside the EU, the EU authorized representative is:

Springer Nature Customer Service Center GmbH
Europaplatz 3
69115 Heidelberg, Germany

FSC
www.fsc.org
MIX
Papier | Fördert
gute Waldnutzung
FSC® C083411

Kolibri 360 GmbH
Ferdinand-Jühlke-Straße 7
99095 Erfurt, Deutschland
produktsicherheit@kolibri360.de